# OPERATIONS ENGINEERING AND MANAGEMENT

# OPERATIONS ENGINEERING AND MANAGEMENT

*Concepts, Analytics, and Principles for Improvement*

**SEYED M.R. IRAVANI**
*Northwestern University*

New York   Chicago   San Francisco
Athens   London   Madrid
Mexico City   Milan   New Delhi
Singapore   Sydney   Toronto

Library of Congress Cataloging-in-Publication Data

Names: Iravani, Seyed, author.
Title: Operations engineering and management : concepts, analytics, and
    principles for improvement / Seyed M. R. Iravani.
Description: New York : McGraw Hill Education, [2021] | Includes
    bibliographical references and index. | Summary: "A new textbook for
    upper level undergraduate to graduate level industrial and manufacturing
    engineering students in courses of Operations Engineering or Management
    applying engineering problem solving data analytics to business
    operations"—Provided by publisher.
Identifiers: LCCN 2020017970 | ISBN 9781260461831 (hardcover ; acid-free
    paper) | ISBN 9781260461848 (ebook)
Subjects: LCSH: Operations research.
Classification: LCC T57.6 .I73 2021 | DDC 658.5—dc23
LC record available at https://lccn.loc.gov/20200179700

McGraw Hill books are available at special quantity discounts to use as premiums and sales promotions, or for use in corporate training programs. To contact a representative please visit the Contact Us page at www.mhprofessional.com.

**Operations Engineering and Management: Concepts, Analytics, and Principles for Improvement**

1 2 3 4 5 6 7 8 9   LWI   25 24 23 22 21 20

ISBN    978-1-260-46183-1
MHID    1-260-46183-1

The pages within this book were printed on acid-free paper.

**Sponsoring Editor**
Robert Argentieri

**Editorial Supervisor**
Donna M. Martone

**Acquisitions Coordinator**
Elizabeth Houde

**Project Managers**
Jyoti Shaw and Ishan Chaudhary, MPS Limited

**Copy Editor**
Kirti Dogra, MPS Limited

**Proofreader**
Bindu Singh, MPS Limited

**Indexer**
Seyed M.R. Iravani

**Production Supervisor**
Lynn M. Messina

**Composition**
MPS Limited

**Art Director, Cover**
Jeff Weeks

*To Sonya and Mina*

## About the Author

**Seyed M.R. Iravani** is a professor of Industrial Engineering and Management Sciences at the McCormick School of Engineering and professor of Operations Management (Courtesy) at the Kellogg School of Management at Northwestern University. His research focuses on improving operations in manufacturing, supply chains, service, and healthcare. His articles have appeared in *Management Science, Operations Research, Manufacturing and Service Operations Management, Marketing Science, Production and Operations Management, Queueing Systems,* and *IIE Transactions,* among others. At Northwestern University, he teaches manufacturing and service operations management courses at the undergraduate and graduate levels to engineering students as well as the core operations management course to MBA students. He received his PhD in Industrial Engineering at the University of Toronto. Before joining Northwestern University in 1998, he was a postdoctoral fellow in the Department of Industrial and Operations Engineering at the University of Michigan.

# Contents

# Preface

Operations of a firm is a complex system consisting of a collection of resources—people, facility, equipment, material, information—used to produce, transfer, and deliver products (goods or services) to consumers. Like any complex system, to improve operations one first needs to understand the basic elements of the operations and discover how those elements interact with each other, and how their interactions impact performance. Managing and controlling operations has become more complex as firms are getting larger, global competition is becoming more intense, technology is getting more advanced, and customers are becoming more selective. In their book[1] *Competing on Analytics: The New Science of Winning*, Davenport and Harris argue that, in this competitive business environment, analytics can give firms a competitive advantage. They describe *analytics* as extensive use of data, statistics, and quantitative analysis, descriptive and prescriptive models, and fact-based management to drive decisions and actions.

## Engineering and Management Views of Operations

One discipline that heavily relies on analytics to design and control complex systems and solve problems is "Engineering." Engineering is defined as the application of scientific and mathematical methods to the design, manufacture, and operations of efficient and economical structures, machines, processes, and systems. Engineers are trained to acquire two important skills:

- *Engineering Thinking:* The ability to define the real problem clearly, identify the criterion by which the solution is measured, incorporate constraints into the solution, and develop solutions that take into account the uncertainty in the environment.
- *Analytical Skills:* The ability to use scientific tools and methodologies such as modeling and mathematical analysis during the engineering thinking process to obtain reliable and robust solutions to real-world problems.

The main focus of operations management (OM) courses in management schools has been on presenting a broad and strategic view of operations, with less emphasis on analytics and more emphasis on principles of managing operations and understanding the intuition behind them—and this is indeed what MBAs need to know. The operations courses in engineering schools, while presenting some operations management principles, put more focus on developing engineering thinking and analytics to support making tactical and control-level decisions. Teaching operations management–related courses at both engineering and management schools at Northwestern University, I always tried to take more managerial view of operations when teaching engineering students and to promote engineering thinking and analytics (in an easy-to-understand and simple-to-use format) when teaching management students. This book grew out of the material I developed through my teaching

---

[1]Davenport, T.H. and J.G. Harris, *Competing on Analytics: The New Science of Winning*.

of those courses. The goal of the book is to provide readers with a balanced view of operations, from both management and engineering perspectives. I believe that managers with engineering view of operations, and engineers with management view of operations would be more successful in management and control of manufacturing and service operations systems.

In fact, the data also support this. For example in 2008, Spencer Stuart—an executive search consulting firm—performed a study on the age, education, and work experience of chief executive officers (CEOs) of S&P 500 companies. Their study revealed that 67 percent of all CEOs have earned some type of advanced degree (MBA, master's, law degree, doctorate, etc.). The most common advanced degree among the CEOs was MBA with 39 percent. However, the interesting finding (which was also shown in their previous study in 2006) was that the most common undergraduate degree among the CEOs was not Business Administration or Economics or Accounting; it was Engineering (22 percent), followed by Economics (16 percent), Business Administration (13 percent), Accounting (9 percent), and Liberal Art (6 percent). This is an indication that an engineering approach (i.e., engineering thinking and analytical skills) contributes to developing successful managers.

Academic and industry leaders have also recognized that in today's complex and competitive business environment, skilled engineers who understand the essential principles of management have a tremendous competitive advantage. Many top-tier academic institutions—including Duke, Northwestern, MIT, Dartmouth, John Hopkins, Purdue, Tufts, USC, and Cornell—have started to offer Master of Engineering Management (or MEM) degree to fill this gap. Many firms are sending their managers to these full-time and part-time programs. The curriculum of an MEM degree is designed to integrate analytical skills of an engineer with business skills of a manager. The goal is to train people who can develop innovative solutions to complex business problems.

## Role of Intuition

*The Merriam-Webster Dictionary* defines intuition as "the power or faculty of attaining to direct knowledge or cognition without evident rational thought and inference." Intuition is not contrary to analytics. It is a form of reasoning that draws on previously acquired experiences and observations to recognize the problem and come up with a solution. Intuition can also be a helpful sanity check for a solution derived from analytics.

While using intuition without analytics is not an effective approach in making decisions, using analytics without intuition is also risky. In today's competitive environment, operations management decisions cannot be made solely based on business experience, intuition, or some thoughtful guess. Thus, when it comes to solving operations management problems, analytics and intuition go hand-in-hand. When either one is absent or weak, the chance of success is diminished.

## The Plan of This Book

Operations Management (OM) is a technical subject relying heavily on analytics. There is a famous quote that says: "Give a man a fish; you have fed him for today. Teach a man to fish; and you have fed him for a lifetime." Presenting OM principles without explaining how they were developed and without explaining the intuition behind each principle is like giving a man a fish and not teaching him how to fish. The main goal of this book is to present concepts and principles of operations management, with emphasis on *both analytics and intuition,* and with a sharp focus on *improving operations.* I use the title *Operations Engineering and Management* for this book to highlight the focus on both the engineering approach to operations (e.g., analytics and engineering thinking) and the management principles of operations.

To facilitate solid understanding of the subjects, the book divides the science of operations management into three elements: (i) *Concepts*, (ii) *Analytics*, and (iii) *Principles*. To acquire knowledge in any domain, one must first learn the basic concepts in that domain. This is also true in operations management. For example, without a solid understanding of the concepts of Effective and Theoretical Capacities of a process, one cannot make good capacity improvement decisions. Analytics are, of course, essential in solving operations management problems. Finally, solving operations management problems using analytics often reveals some management principles that help the operations engineers and managers improve their processes. One goal of this book is to present concepts, analytics, and management principles in an *easy-to-understand* and *easy-to-implement* manner.

The *Process-Flow* view of operations has been successfully used in teaching operations management courses in management schools. The two books that have presented this view in an effective manner are *Managing Business Process Flows* by Anupindi, Chopra, Deshmukh, Van Mieghem, and Zemel, and *Matching Supply with Demand* by Cachon and Terwiesch. These books are developed for students in management schools. This book takes the same approach with more rigor and more detailed treatment of each topic. Other features of this book are as follows.

- *Learning "Why" and "How," Not Just "What":* A solid understanding of a concept goes beyond just knowing *what* the concept is. One also needs to know: *why* operations engineers and managers should know the concept, and *how* it helps them make a better decision. This book uses examples to illustrate the *need* for each concept and how they are used in making decisions. For instance, before introducing the two concepts of Effective and Theoretical Capacities of a process, the book uses several examples that lead readers to realize the need for two different concepts to evaluate the capacity of a process. The book then introduces the two concepts and shows how they can help identify opportunities for capacity improvement.

- *Presenting Analytics along with Analytical (Engineering) Thinking:* Analytics, as we discussed, can play an important role in making good management decisions. Using analytical methods requires a basic knowledge of mathematics. However, in the core of analytics is logic and analytical thinking, that is, engineering thinking. Therefore, to understand the analytics of a class of operations management problems, one should learn both the *logic* and the *mathematics* behind it. This book achieves this by first describing the underlying logic behind the trade-offs in an operations management problem through simple numerical examples. It then develops analytics (i.e., mathematical model) based on the described logic.

- *Presenting Modeling and Solution in a Step-by-Step Approach:* The modeling and solution methodologies corresponding to operations management decisions are often complex. This book presents them in stand-alone, step-by-step algorithms that can be easily followed by both engineering and management students. To illustrate the algorithm, the book uses an example and shows how each step of the algorithm is implemented in the example. Also, the algorithms are written such that they can be easily modeled in an Excel spreadsheet.

- *Not Just Presenting Operations Management Principles, Also Developing Them:* One of the main goals of this book is to present the principles that are essential for improving process performance. Instead of just presenting those principles, this book takes readers through a thought process that leads to those principles. For example, there is a principle that states that the throughput of processes with limited buffer size can increase if the variability in the process decreases. The book derives this principle through an example of a production line with limited buffer space, and explains how the limited buffer results in blocking and starvation, which in turn leads to lower utilization of resources, and thus lower throughput. This not only facilitates learning of the principle, but also helps readers recognize the contexts in which the principle is applicable.

- *Building Intuition:* As mentioned earlier, intuition draws on previously acquired experiences and observations to identify problems and find solutions. Operations managers develop intuition through years of dealing with different management situations and learning from their right and wrong decisions. This book helps readers build intuition by presenting those management situations and discuss the thought process that leads to *right* answers as well as those leading to *wrong* answers.

- *Stimulate Thinking:* Thinking is driven by questions not by answers. Questions stimulate thinking while answers often result in stopping the thought process. A common theme in the narrative of this book is that it relies on questioning the readers first, before illustrating the thought process that results in answers. This method is used in all stages of developing concepts, analytics, principles, and building intuition.

- *Focusing on Improving Operations:* The main goal of the book is to provide readers with a concrete set of tools and strategies to improve performance of operations systems. Thus, one complete chapter is devoted to each performance measure of operations, that is, throughput, flow time, inventory, and quality. Each chapter provides a list of actions that operations engineers and managers can take to improve each measure.

- *"Want To Know More?"* For interested readers some chapters have online supplements that extend their content either by presenting more applications of the topics, or by developing more advanced analytics. For instance, Chapter 10 presents lot-size reorder-point policy for systems in which there is a fixed cost for each backlogged customer. The corresponding online supplement of this chapter extends this model to systems in which the backlog costs are charged per unit time that customers are waiting to receive their orders. Another example is the line balancing problem in Chapter 13. This chapter presents a simple heuristic that is commonly used to design paced assembly lines. The solution of the heuristic is not guaranteed to give the best design. The online supplement of this chapter, therefore, presents two mathematical models that yield the optimal design for two different line design problems. Chapters and their supplements are linked through sections called "Want to Know more?." In those sections readers can find about the content of chapter supplement and how it is related to the subject that they have just finished reading in the chapter.

## Who Should Read This Book?

The book is designed to serve the following purposes:

- A textbook for operations management–related courses in industrial engineering and manufacturing engineering programs. The book is suited for advanced undergraduate (e.g., senior-level) students and as a first course in operations engineering and management for graduate students.

- A textbook or a reference book for dual-degree programs in engineering and management schools, for example, Master of Engineering Management (MEM).

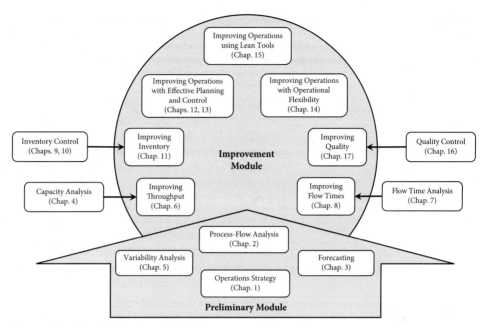

**Figure 1** Overview of the chapters of the book.

The book is also suited for operations management courses in more technical MBA programs.

- This book can also serve as a reference for technical professionals who work in the areas of manufacturing, service operations, or supply chain management.

- Each individual chapter or a collection of chapters of the book can be used as a part of different courses. For example, Chap. 14, which is about Operational Flexibility, can be used as a part of any operations management or supply chain management–related course in both management and engineering schools. See overview of the book below for more details.

## Overview of the Book

Figure 1 presents an overview of the book. As the figure shows, the core of the book is on *Performance Improvement* of operations in manufacturing and service systems.

### Preliminary Modules

The chapters in this module introduce the preliminary concepts and tools that provide readers with a better understanding of the two sides of any operations: (i) demand, and (2) supply. Chapter 1 (Operations Strategy) describes how firms gain competitive advantage by better aligning their process (i.e., supply) with the market (i.e., demand). Chapter 2 (Process-Flow Analysis) presents the process-flow perspective of operations and introduces process-flow measures—throughput, flow time, inventory, and quality—that have significant impact on financial performance of a firm and thus on its success in the market. Chapter 3 (Forecasting) discusses the demand side of the operations and provides several analytics to forecast the demand of a product. Finally, Chap. 5 (Variability Analysis) describes how variability in demand or in supply affects management and control of operations.

### Improvement Modules

The focus of Improvement Module is on providing a wide range of strategic and tactical decision principles that can improve performance of an operations. This module can be divided into two groups: Analysis and Improvement.

- *Analysis:* The Analysis part includes chapters that provide basic concepts and analytics required for improvement of four process-flow measures of a process—throughput, flow time, inventory, and quality. Chapter 4 (Capacity Analysis) is required for improving throughput of a process; Chap. 7 (Flow Time Analysis) is required for improving flow times; Chaps. 9 and 10 (Inventory Control) are required for improving inventory; and Chap. 16 (Quality Control) is required for improving quality.

- *Improvement*: Specific to each of the four main process-flow measures, there is a corresponding chapter that provides several strategies to improve each measure, namely Chap. 6 (Improving Throughput), Chap. 8 (Improving Flow Time), Chap. 11 (Improving Inventory), and Chap. 17 (Improving Quality). In addition, the Improvement Module includes several chapters that provide effective practices that can improve several performance measures of an operations. Improving operations with Operational Flexibility in Chap. 14 describes the design and control principles for using flexible resources and postponement strategy to improve performance. Lean Operations in Chap. 15 presents lean operations tools and shows how they improve all aspects of operations. Finally, Chaps. 12 and 13 focus on developing efficient production planning and scheduling policies to improve performance.

## Acknowledgment

I owe a debt of gratitude to all my teachers, from my first-grade teachers who taught me how to read and write, to my PhD advisors who taught me how to do research. Special thanks to all my professors in Iran University of Science and Technology, especially the late professor Aryanezhad who gave me my first class to teach. It was then that I experienced the joy of teaching for the first time and decided to become a teacher. I would also like to thank my PhD advisors, Morton J.M. Posner and John A. Buzacott, for their support during my PhD years and for giving me the freedom to peruse various projects in both theoretical and application domains.

During my early years at Northwestern University, I had the pleasure of working with my mentor, Wally Hopp, from whom I learned a lot. I would like to thank him for giving me the opportunity to teach Operations courses in our dual-degree MBA program. The variability chapter of this book is greatly influenced by the elegant treatment of variability in Wally's Factory Physics book.

I would also like to acknowledge all students and faculty who provided me with valuable feedback to improve the book. Special thanks to Barry Nelson for his constructive feedback on many chapters of the book, and Daniel Apley and Bruce Ankenman for their reviews of the quality and forecasting chapters. I would also like to thank my colleagues in Kellogg School of Management at Northwestern University, Sunil Chopra, Robert Bray, Martin Lariviere, Jan Van Mieghem, and Achal Bassamboo, for their encouraging and constructive comments. All simulations in this book were performed using Simul8 software and I would like to thank Simul8 for providing me with their software.

Several people at McGraw Hill spent time and effort to bring this textbook through the publication process. I would like to gratefully acknowledge the outstanding members of McGraw Hill publishing team: Lynn Messina (Senior Production Supervisor),

Donna Martone (Editing Manager), Elizabeth Houde (Editorial Coordinator), and Jyoti Shaw (Project Manager) and Ishan Chaudhary (Project Manager). Special thanks to Robert Argentieri (Editorial Director—Engineering) for his patience and dedication to creating a high-quality product and for addressing all my comments and suggestions.

Last, but certainly not least, I thank my wife, Sonya, for tolerating me when I holed up in my home office to work on this book.

## Your feedback is appreciated

The First Edition of a book always benefits from the comments and suggestions of readers. I would greatly appreciate hearing from you. Please send me an email at s-iravani@north western.edu and let me know if there is something missing in the book, which part of the book you like or you do not like, the errors that you detect, and also about the strengths and weaknesses of the book.

# OPERATIONS ENGINEERING AND MANAGEMENT

# Operations Strategy

## 1.0 Introduction

The main three functions of a firm are Marketing, Operations, and Finance. *Marketing* establishes the demand for products (goods or services) and communicates the value of the product to consumers. *Finance* provides the capital, and *Operations* actually makes the product. Marketing and Finance functions of a firm are there to support Operations.

But what exactly does Operations of a firm do? Operations is a complex system consisting of a collection of resources—people, facility, equipment, material, information, technology—used to produce and deliver products (goods or services) to customers. Let's consider operations in a car assembly plant as an example. Raw material and parts, such as tires, seats, electrical and electronic components, chassis, body parts, engine, transmission, etc., are delivered to the assembly plant from different suppliers or from other plants of the car manufacturer every day (and sometimes every hour). These parts are stored in buffers in different parts of the plant or are directly sent to the assembly line for assembly. There are a large number of resources—assembly workers, welders, robots—that perform the task of assembly using different types of equipment and tools. The cars are assembled and sent to finished goods inventory. The entire process from receiving raw material and parts to shipping assembled cars to dealers constitutes the operations in the assembly plant.

What do operations engineers and managers do? Well, they need to make decision and find solutions to a variety of problems involving workforce, equipment, material, facilities, suppliers, customers, etc. These include both strategic and tactical decisions. *Strategic* decisions are long-term decisions with broad scope that can affect the entire firm. Examples of strategic decisions are: deciding to introduce new products, build a new factory, or to outsource some of the activities. *Tactical* decisions, on the other hand, are day-to-day short-term decisions with limited scope; for example, deciding how many parts to order from a supplier each time, how to schedule overtime hours next week, how many items to produce next week, where to locate the quality control station, etc. Underlying all these decisions is the goal of doing more with less, building quality into products and processes, creating a high-performance operations, and continual improvement of operations.

## 1.1 Operations Strategy Decisions

Each firm has a *Mission Statement*. The mission statement is a clear statement of the purpose that serves as a guide for overall strategy of the firm and decisions that shape those strategies. Some examples of a company's mission statement are:

**Apple:**   "Apple designs Macs, the best personal computers in the world, along with OS X, iLife, iWork and professional software. Apple leads the digital music revolution with its iPods and iTunes online store. Apple has reinvented the mobile phone with its revolutionary iPhone and App store, and is defining the future of mobile media and computing devices with iPad."

**Walmart:**   "We save people money so they can live better."

**Google:**   "Google's mission is to organize the world's information and make it universally accessible and useful."

The mission statement is the foundation of firm's Business Strategy. The *Business Strategy* of a firm is its long-term plan that identifies firm's targeted customers and sets the performance objective of the firm. The *Operations Strategy* is a set of decisions that shape the capabilities of operations and their contribution to the business strategy of a firm.

Obviously, the main goal of a firm is to make money by selling its products. In order for customers to buy a firm's product, the product must have a particular feature (or features) that makes it a better option than other existing products. But this is only one side of the story; the other side is whether the firm has the capabilities to make such a product with low cost. Operations strategy decisions consider these two important factors: they are called Value Proposition and Operations Capabilities of a firm.

---

**CONCEPT**

**Value Proposition and Operations Capabilities**

- *Value Proposition* of a product offered by a firm is the particular value (i.e., feature) of the *product* that is different from the existing products, and thus makes the product competitive.

- *Operations Capabilities and Competencies* correspond to the features of the *operations* that enable it to make and deliver the product.

---

Obviously to be successful, a firm must have the capability to deliver the value proposition it would like to offer to its customers. Operations strategy decisions are high-level decisions that are made to align a firm's operations capabilities with its value proposition.

These decisions are important decisions that affect a firm's long-term performance. Examples of these *strategic* decisions, as mentioned earlier, are increasing capacity by building a new plant, introducing a new product, or decision about outsourcing or in-house production of a part. In contrast, *tactical* decisions are concerned with detailed decisions with a shorter time scale such as increasing capacity of a workstation in a plant by assigning 2 hours overtime next week, or changing the delivery schedule of a customer because of the rush delivery of another customer.

### 1.1.1 Value Propositions

Value proposition is what makes a product sell in the market.[1] From an operations strategy point of view, the major value proposition offered by firms is one of the following: (i) Innovativeness, (ii) Cost, (iii) Flexibility, (iv) Quality, (v) Time, and (vi) Sustainability.

---

**CONCEPT**

**Value Proposition—Features and Innovativeness**

In general, *Features and Innovativeness* refer to any unique or innovative feature or features of a product (goods or service) that makes the product to be perceived by customers as more desirable than competitors' product.

---

There are many firms that have Innovativeness as their value proposition. Apple is an example of such a firm that achieved success with its innovative features in mobile phones (e.g., iPhone), portable music players (e.g., iPod and iPod Touch), and personal computers (e.g., iPad) products. Zipcar brought the innovative idea of car sharing to the United States that allowed customers to rent a car for a number of hours instead of the minimum of 1 day required by major car rental companies. While innovativeness can create initial success for firms, it is often difficult to maintain innovativeness as the main value proposition. The reason is that other firms start copying these innovative ideas (especially in the service industry) and then other factors (e.g., lower price) become the value proposition.

---

**CONCEPT**

**Value Proposition—Cost**

*Cost* of a product (goods or service) refers to the cost of acquiring and ownership of the product. This includes the purchase price and the lifetime cost of owning, using, and maintaining and disposal of the product.

---

Note that depending on the product, customers put different emphasis on the cost of purchasing and the cost of ownership of the product. For food items, customers focus on the cost of purchasing—price—than cost of ownership. On the other hand, for customers who buy cars, the cost of ownership—maintenance cost, insurance, gas-mileage—is as important as the price of the car.

Walmart offers cost as its main value proposition to its customers. The company opened its first Walmart store in 1962 and started to grow. In late 1980s, Walmart was behind

---

[1] Of course marketing and advertising also increase sales; however, what they mainly do is to highlight the value proposition offered by the product.

K-mart—K-mart had twice the number of stores than Walmart. By offering lower prices, Walmart was able to capture the market share and became the largest retail store in the world. Offering lower cost than the competition is not always possible. The ability to make products with lower cost depends on the capabilities of the firm, as we will discuss in Sec. 1.1.4.

Another value proposition that helps firms distinguish their products from the competition is Flexibility. Flexibility is a broad concept that covers different aspects of satisfying customer demand.

---

**Value Proposition—Flexibility**

*Flexibility* refers to being able to quickly respond to the changes in customer demand. Some types of flexibilities are:

- *Product Flexibility:* The ability to develop and make novel products (goods or services) or to modify the existing one.
- *Mix Flexibility:* The ability to offer a variety of products or change variety.
- *Volume Flexibility:* The ability to produce whatever volume the customer needs.
- *Delivery Flexibility:* The ability to change planned delivery times.

Mix flexibility, for example, has been one of the key success of many companies. One of the factors in the success of Dell Computers was that it gave customers the option to custom-build their computers by choosing among several options of memory, hard-drives, monitors, etc. Another example of mix and volume flexibility is semiconductor manufacturing. To survive, semiconductor manufacturers must be able to adjust the volume and mix of their products in a short time period. This is because most of the electronic products have short life cycles and unpredictable demands. The suppliers of components to these manufacturers should also have mix and volume flexibility as well as delivery flexibility, so that they can provide the manufacturers with the required components before the deadline.

---

**Value Proposition—Quality**

*Quality* refers to the ability of a product (goods or service) to deliver features and characteristics that satisfy customers' needs and expectations.

Quality is also a broad concept which relates to many features of the product such as performance, reliability and durability, its appeal and serviceability, among other features. Quality of a car, for example, is evaluated through several dimensions including its gas mileage (i.e., performance), its required maintenance (i.e., reliability and serviceability), its useful lifetime (i.e., durability), and its interior and outside appearance (i.e., appeal). Rolex, Ritz-Carlton Hotels, and Singapore Airlines are examples of service firms that focus on providing high-quality products and service as their main value propositions.

---

**CONCEPT**

**Value Proposition—Time**

*Time*, as a value proposition, refers to the ability to rapidly produce and deliver the product (goods or service) to customers. Time corresponds to two basic features: Speed and Dependability.

- *Speed* refers to the time when a customer requests a product or a service to the time the customer receives it.
- *Dependability* refers to the ability to deliver the right product in the right quantity and at the time it was promised.

---

Time has become one of the most important value propositions in today's competitive market. Customers demand quick response and short waiting time. Firms in all industries are therefore competing to develop, produce, and deliver products as quickly as possible. FedEx, Mc Donald's, Domino's Pizza, and the emergency departments of hospitals offer speed of delivery as their value propositions. Most companies charge a higher cost for fast delivery. For example, Amazon charges a higher cost for rush deliveries, and dry cleaners charge a higher cost for 1-hour dry clean.

---

**CONCEPT**

**Value Proposition—Sustainability**

*Sustainability* or *Green Operations* is based on a simple principle: Everything that we need for our survival and well-being depends, either directly or indirectly, on our natural environment. Sustainability creates and maintains the conditions under which humans and nature can exist in productive harmony, and permits fulfilling the social, economic, and other requirements of present and future generations.[2] This includes, but not limited to, pollution prevention, energy conservation, water conservation, and solid waste reduction. Sustainability requires continuous search for newer and better ways to make the business as a whole more environmentally friendly.

---

Sustainability has recently become an important value proposition that firms use to distinguish their products and services from their competitors. For example, Eco Gym is a health club that advertises itself as the firm that follows global initiative of energy sustainability and Earth friendliness. Their treadmills, for example, generate electricity when used by club members. The electricity is used to power other equipment and utilities in the club.

Sustainability has also become one of the important factors affecting operations strategy decisions. ISO 14001, for example, is a management system that helps businesses reduce their environmental impact. To get ISO 14001 certified, firms need to show that their entire business operations is committed to reducing harmful effects on the environment, and provide evidence of continual improvement in their Environmental Management Systems. ISO 14001 is expected to become a prerequisite for doing business worldwide.

### 1.1.2 Strategic Position

While offering several value propositions, such as offering high-quality innovative products with low cost, is desirable, this rarely occurs in practice. The reason is that there are *trade-offs* among these values. For example, producing high-quality products requires high-quality material and highly experienced workers which result in higher production cost,

---

[2]Definition from United States Environmental Protection Agency (EPA).

preventing firms from offering a low-cost product. Thus, firms should not aim to provide all value propositions to their customers.

An interesting analogy that has been used to illustrate this important fact is the difference between an Olympic gold medal sprinter and marathon runner. The sprinter is trained for speed and short distance running, while the marathon runner is trained for slow but steady and long distance running. Both are best in their fields, but none of them compete in both events. So, firms should prioritize the value propositions they would like to offer and must remember that offering one value may conflict with another.

This, however, does not mean that firms should pick one value and completely ignore the others. They should first identify Order Qualifiers and Order Winners.

---

**CONCEPT**

### Trade-Offs—Order Winners and Order Qualifiers

*Trade-off* is the need to focus more on one value over others based on the fact that superior performance in some values may conflict with superior performance in others.

- *Order Qualifiers* are the characteristics of a product (goods or service) that qualify it to be considered for purchase by a customer. Order qualifiers are the minimum industry-standard characteristics that customers expect in a product.

- *Order Winner* is the characteristic of a product (goods or service) that wins order in marketplace. It is the final factor in customers' purchasing decisions. Order winner is the one that differentiates a firm from its competitors.

Order qualifiers and order winners change over time. For example, in 1970s the order winner for a car buyer in U.S. market was price. The Japanese auto makers entered the U.S. market and, by offering a competing price, they changed the order winner from price to quality. The price became an order qualifier. Then during 1980s, American auto manufacturers raised their quality and became competitive with Japanese cars. Then quality became an order qualifier. Today, the order winners mostly depend on the model. Within a model, additional features and innovative designs are among the order winners.

After choosing their order qualifiers and order winners, a firm should make sure that it offers order qualifiers, and chooses order winners that will identify firm's *Strategic Position* in the market. To illustrate this, let's look at two value propositions of quality and price. In any industry, if we map all the firms with respect to their price and quality in two-dimensional space—price versus quality—we get something like the graph in Fig. 1.1. This graph is called

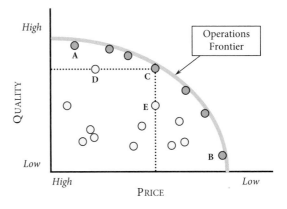

**Figure 1.1    Strategic position and operations frontier.**

*Competitive Product Space* of quality and price. We will find some firms such as Firm B in this space offering low-quality products but with very low prices. In contrast, we also see firms such as Firm A that offers high-quality products and charge higher prices. Other firms in the industry are somewhere between these two firms. By choosing to produce a product at a certain level of quality and a certain price, a firm chooses its strategic position in space of quality and price.

What does the boundary line in the figure represent? and why there are no firms above that line? The line is called *Operations Frontier*. It is the minimal curve containing all current positions of firms in the industry. Firms on this curve have the highest operational performance in that industry with respect to the value propositions of quality and price. Consider Firm C, Firm D, and Firm E in the graph, and let's compare Firms C and D first. Firm C offers products with the same quality as those of Firm D, but with lower prices. How? Firm C most probably has a more cost-effective operations than that of Firm D. Now compare Firm C and Firm E. Firms C and E offer the same price for their products but Firm C is capable of producing higher quality products. This, again, is because Firm C must have a better operations that produces higher quality products than those produced by Firm E. Furthermore, note that there is no firm in the graph that can produce a higher quality product with the same price as Firm C, or a cheaper product with the same quality as Firm C. Thus, with respect to quality and price, Firm C is said to be on the operations frontier.

The reason there are no firms above the operations frontier is that no firm can achieve better combination of quality and price than those firms on the operations frontier. With the existing technology and cutting edge management practices, firms cannot achieve a better trade-off between quality and price beyond operations frontier. Once in a while, however, a firm develops a novel practice or a breakthrough technology that pushes the operations frontier higher. For example, the lean manufacturing practices of Toyota in 1970s and 1980s allowed Toyota to make higher quality cars with very competitive price. That pushed the operations frontier in auto industry up and made Toyota very successful in capturing the global market.

The graph in Fig. 1.1 is an example of operation frontiers for quality and price. Similar graphs can be developed for other value propositions such as time versus cost, or variety versus time, etc. Below we summarize the concepts of strategic position and operations frontier.

CONCEPT

**Strategic Position and Operations Frontiers**

*Strategic Position* of a firm is the position that the firm wants to be in the competitive product space. *Operations Frontier* in any competitive product space is the minimal curve that contains all current industry positions. The firms on the operations frontier are the firms that have superior operational performance for the level of trade-off they have chosen.

### 1.1.3 Focused Factory and Plant-Within-a-Plant

As mentioned previously, while providing all value propositions of high quality, fast delivery, low cost, high variety is desirable, it is not possible or economical. Wickham Skinner, who introduced the concept of focused factory, writes: "A factory cannot perform well on every yardstick," and emphasizes on the following:

A factory that focuses on a narrow product mix for a particular market niche will outper-form the conventional plant, which attempts a broader mission. Because its equipment, supporting systems, and procedure can concentrate on a limited task for one set of customers, its costs and especially its overhead are likely to be lower than those of the conventional plant. But, more important, such a plant can become a competitive weapon because its entire apparatus is focused to accomplish the particular manufacturing task demanded by the company's overall strategy and marketing objectives.

Skinner describes the case of a company that was producing two different products in the same factory: fuel gauges and automatic-pilot instruments. Fuel gauges were standard products produced in high volume, and automatic-pilot instruments were mainly customized low-volume products. The company was losing money on fuel gauges, until it decided to separate production process of the two products and build a wall to separate them. The firm allocated separate equipment, labor, quality control, and management crews to each product. Thus, the company actually established two plants within its plant. The plant for fuel gauge focused on producing high-volume, low-cost, standard products, and the plant for automatic-pilot instruments focused on producing low-volume, customized products. After four months of creating plants within the plant, the fuel gauge production became profitable and the profit of the automatic-pilot instrument also increased.

This further emphasizes on the lesson that a firm cannot compete in all dimensions such as low cost, high quality, speed, and high variety (product mix). A firm should focus on a limited set of value propositions to serve a limited market segment. The firm's process should also be focused on the attributes of the product (e.g., standard/customized, low cost/low quality, or high cost/high quality) it offers. In cases where firms offer products with different attributes, the concept of plants-within-a-plant (PWP) suggests that the entire process (the plant) be segregated into smaller focused processes (mini-plants), each serving a specific family of products with similar attributes.

### 1.1.4 Operations Capabilities and Competencies

In the previous section, we discussed the set of value propositions that a firm can offer to its customers. Those values are product attributes by which the firm would compete in the market and determine the firm's strategic position in the market. The value proposition and strategic position are all about the market. While choosing the right value or set of values to offer is important, more important is whether a firm can offer those values, and offer it with profit. Consider the value proportion "Speed" to offer direct sales through company's website. Before making such a decision, the firm should determine whether its current operations (i.e., information technology, distribution, and transportation resources) are capable of delivering such value proposition.

Being able to offer a value or values is not about market anymore, it is about the operations. As mentioned earlier, the goal of operations strategy is to align operations capabilities with market requirement (i.e., value proposition), as shown in Fig. 1.2.[3]

Offering high-volume products with low cost, for example, requires a different type of operations than the one that offers low volume and high variety. But, what are the characteristics of an operations that enable it to produce high volume of products, or high variety of products in a cost-effective manner?

The ability of an operations to provide a particular capability, as shown in Fig. 1.2, depends on the capabilities of its *resources* and its *processes*. Resources are divided into Tangible (e.g., facilities, equipment, staff, material) and Intangible (e.g., expertise, know-how, culture). Processes, on the other hand, are divided into Primary Processes (e.g., production

---

[3]The figure is based on the framework in Slack and Lewis (2003).

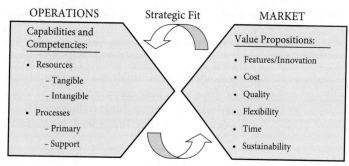

Figure 1.2    **Market and operations factors affecting operations strategy.**

of goods and services) and Support Processes (e.g., HR, accounting, planning, and control). Researchers have found that while the type of resources needed in an operations greatly depends on the *type* of the product, the type of processes needed generally depends on the *volume* and the *variety* of the product. For example, compare the operations in a hospital with that in a tool-and-die shop. The product in the hospital is a wide range of services offered to patients, while the product in a tool-and-die shop is a wide range of dies designed for particular customers. The equipment (e.g., x-ray machines, operating rooms) and staff (e.g., doctors, nurses) in a hospital are completely different from the equipment (e.g., drills, welding machines) and staff (e.g., machine operators) in a tool-and-die shop. However, the underlying (primary) process of the two operations is the same. In both systems, the entire operations is divided into different departments, each performing a specialized task using a group of special equipment and staff. Examples of these departments in a hospital are radiology and urology, and examples of departments in a tool-and-die shop would be the machine tool and welding departments. Similar to a hospital in which patients move between the departments, products move between the departments in a tool-and-die shop. What operations in hospitals and in tool-and-die shops have in common is that the underlying process is the same. It is called a Job Shop process. The job shop process (which is discussed in detail in the following section) has the capability to provide a wide range of products in cost-effective ways.

How many other processes are there and what can those processes offer? To answer this question, in the following sections, we classify operations processes from two different perspectives: (i) classifying based on the level of product customization, and (ii) classifying based on material flow.

### 1.1.5 Operations Processes: Classifying by Level of Product Customization

*Product Customization* is the process whereby the company makes the product based on customer's individual requirements and preferences. Based on the degree and point of customization, processes can be divided into four groups: (i) Engineer-to-order, (ii) Make-to-order, (iii) Make-to-stock, and (iv) Assemble-to-order.

**Engineer-to-Order (ETO)**
ETO processes are processes in which product design starts only after an order is received from a customer. Orders are for highly customized product (or products) that is suited for the individual customer's needs. The product is designed in close collaboration with the customer to ensure that the final design satisfies all their needs. The product may need non-standard parts or material that must be identified and acquired only for that product. Since

the product is often one-of-a-kind and complex, customer remains involved throughout the manufacturing phases of the process to ensure customer's specifications are met. Examples of ETO products are space rovers, special-purpose equipment, or customized cabinets and bookshelves that fit a customer's space. Because ETO products are one-of-a-kind products, they often require highly engineered and unique parts that are difficult or expensive to acquire. Thus, one challenge in managing ETO processes is the difficulty in acquiring those components before and during production. Another challenge is the significant uncertainty in operations times, since the product is new and has not been produced before.

### Make-to-Order (MTO)

In a MTO production process, production of a product starts after an order for the product is received from a customer. The difference between MTO and ETO processes is that in ETO processes the design of a product starts after a customer order arrives. However, in MTO processes, predesigned standard component and material are used in all orders, and thus the process can start production when customer order (for a certain configuration of standard components and material) is received. Upon receiving an order, the labor and equipment are used to complete the order. In general, MTO production processes produce each customer order in small quantities. Example of MTO systems are custom-made furniture or jewelry, ships, specialized equipment manufacturers, custom tailoring, etc.

The advantage of MTO systems is that they can provide a wide range of products tailored to customer orders. They also have low inventory, since they do not produce an item until they receive an order. In other words, MTO systems are capable of offering Product-mix Flexibility as a value proposition. The main performance measure of MTO system is the time it takes to make and deliver the product. The big management challenge is therefore to shorten this time to meet the product delivery date promised to the customers.

### Make-to-Stock (MTS)

In a MTS production process, production of a product starts based on the future forecast of orders. As opposed to MTO processes where customers decide about the features of their products (e.g., the color, the fabric, and style of a vanity table), in MTS processes, the firm offers a limited number of standard products and keeps them in its inventory in anticipation of customer orders. When customers place orders, their orders are immediately satisfied from firm's inventory. If the item is not available in inventory, the customer may go to another firm (i.e., the order is lost), or the customer waits for the product (i.e., the order is backlogged). Examples of products produced in MTS production processes are appliances, TVs, cola drinks, shoes, and clothing. The main advantage of MTS process is that it can fulfill customers' orders immediately from inventory; that is, MTS processes are capable of offering the value proposition of Time (i.e., Speed). The operational challenge in MTS processes is to have an accurate forecast of future demand and to lower inventory while minimizing the number of lost and backlogged orders.

### Assemble-to-Order (ATO)

MTO processes provide a wide range of products (i.e., product-mix flexibility), but may lack speed in producing and delivering them. MTS processes, on the other hand, satisfy a customer order immediately, but customer choices are limited to the fixed products designed by the firm. ATO processes are a hybrid process of MTS and MTO systems, and are designed to achieve both variety and speed. ATO processes build product's parts and subassemblies (e.g., different computer hard drives, different computer memory, monitors of different sizes) in advance of an order arrival. The subassemblies are designed such that different combinations of them create products with different features (e.g., different combinations of hard drives

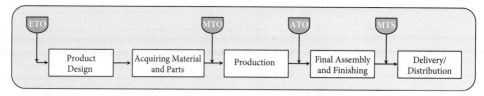

**Figure 1.3    Point of customization in operations processes in ETO, MTO, ATO, and MTS processes.**

and memories create different personal computers with different capabilities). Customers design their products by choosing from different features and place their orders. When an order is received, the firm can assemble the subassemblies in a very short time and deliver the product to the customer.

Examples of firms that use ATO processes are Dell Computers, Subway sandwich shops, machine tool shops, etc. ATO processes, if well-designed, are capable of providing both Product-Mix Flexibility and Speed value propositions. Note, however, that the range of products produced in ATO production processes is not as large as MTO processes and their production and delivery times are not as small as MTS processes. ATO processes bear the same operational challenges as MTO and MTS, namely creating high customer service (low wait times, low lost or backlogged orders) and lowering inventory.

As discussed, the above processes offer different degree of customization (i.e., product variety or product-mix flexibility). What makes a clear distinction between these processes is *when* and *where* in the process a customer's order specifications impact the process. Figure 1.3 gives an overview of point of customization in the these four processes. As the figure shows, customer order initiates product design activity in ETO processes, while it initiates production in MTO processes using predesigned parts and components. In ATO systems, a customer order triggers final assembly and finishing activities using preassembled modules, and in MTS processes a customer order initiates the distribution and delivery of the products that have already been produced and kept in stock.

### 1.1.6  Operations Processes: Classifying by Material Flow

Another way to classify the operations processes is based on the structure of the process and the way units flow within the process. From this perspective, the operation processes can be divided into five main processes: (i) project process, (ii) job shop process, (iii) batch process, (iv) flow-line process, and (v) continuous-flow process.

#### Project Process

The project process is suitable for producing unique and highly customized products (or services) in a very low quantity, sometimes just one unit. This process usually works in ETO and MTO modes. Examples of a project process are construction projects, software development, medical procedures, and production process of large ships and aircrafts. In a project process, the process of making a product is broken into many smaller tasks, each done by different group of resources. The major operational problem with project processes is the coordination among resources and tasks so that the project is completed on time and within the specified budget. We discuss the coordination techniques for project processes in Chap. 13.

Another feature of project processes, mainly observed in manufacturing and constructions setting, is that the physical layout of the process is a *Fixed-Position Layout*. That is, the product is stationary and does not move within the process; it is assembled in a fixed location. Resources (material, parts, and labor) are brought to the product, and the product

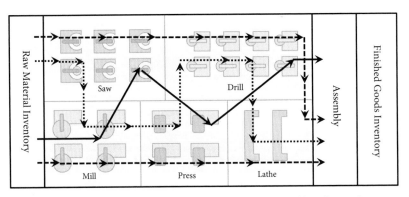

**Figure 1.4**  In process or functional layout, similar resources located in same positions (i.e., work centers or departments) and different products move through different work centers.

stays in the same position until it is completed. Project processes are difficult to automate, and highly skilled labor is used along with general-purpose equipment.

**Job Shop Process**
Similar to the project processes, job shop processes make a wide variety of customized and unique products in very small quantities. The difference, however, is that products are of smaller sizes and thus products move between different shops (or departments) for processing. Each shop consists of highly skilled workers that perform similar functions (e.g., drilling or painting) using general-purpose (i.e., flexible) equipment. Job shops have *Functional Layout*, which is also called *Process Layout*, where resources are physically grouped by their functions. Each group can be physically located in separate sections of a building, or in different buildings within a factory. Figure 1.4 shows an example of a functional layout.

In job shops, the product design is not standardized and job shop managers often need to work closely with customers to determine the exact characteristics of the product. Products are MTO and each product may follow a different routing (e.g., a different sequence of machines) within the process. Job shop processes are also used in ETO mode. Examples of products that are made by job shop processes are custom-made furniture, special-purpose machine tools (e.g., robots), and dies used by manufacturers in stamping operations. As mentioned, hospital operations are also job shop processes. Similar to project processes, the degree of automation is very low in job shops and the process relies on highly skilled labor and flexible equipment. The operational challenges are production scheduling and inventory management. Often large inventory of different jobs wait in work centers for labor or equipment for processing. This problem becomes more significant as job shop processes become highly utilized.

**Batch Process**
Batch processes are similar to job shop processes, in that products are moved between work centers with functional layout. In batch processes, items are moved through the process in batches. A setup time or a changeover time is often required in each work center before the process of the next batch starts. The main differences between batch and job shop processes are the following: First, batch processes produce in higher volume than job shops but in lower variety. Second, products in batch processes are more standardized than those in job shops. In general, batch processes work in MTS mode, while job shops

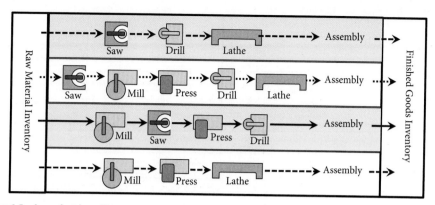

Figure 1.5    In product-based layout, resources are arranged sequentially according to the activities required to make the product.

work in ETO or MTO mode. Examples of batch processes are bakery, printing shops, musical instrument, and boat production systems. The main operational challenges, similar to job shop processes, are production and inventory scheduling, especially in highly utilized processes.

### Flow-Line Process

Flow-line processes are capable of producing a large volume of products at a faster speed, but with much less variety (i.e., less product-mix flexibility). These processes are also called flow shops, flow lines, production lines, or assembly lines. This process was first used by Henry Ford in 1913. After implementing a flow-line process, Henry Ford was able to reduce the time for producing a car (Model T) from 13 hours to less than 6 hours. Today, using flow-line processes, cars exit assembly lines at the rate of one every minute (or less). Flow-line processes work in MTS mode. Examples of flow-line processes are assembly plants for automobiles, appliances, electronics, etc. Flow-line processes have *Product Layout* in which resources are arranged sequentially according to the sequence of operations required to make the product (see Fig. 1.5). Items are moved between consecutive stages of production using a conveyor system.

Flow-line processes are suited for mass production of a single product, or products that are similar in design and requirement of material and equipment. For example, auto assembly lines can produce the same model car but with different options (e.g., different engines, different transmission, different color seats, etc.). This is because assembling these options requires same equipment and same labor skills. In contrast to job shops and batch processes that have very little or no automation, the degree of automation in flow-line processes can be high. Some lines use robots in addition to workers to perform some of their assembly operations. The main operational challenge in managing flow-line processes is to minimize the line stoppage. This is because flow-line processes often require a very high investment in building the line. For example, the capital investment at a car assembly plant is in order of billions. To justify the large capital investment, the line must be highly utilized to produce a high volume of products. The second challenge is that flow-line processes are not flexible to produce products with different designs. Even when the product design attributes are minimally changed (e.g., a new car model year is introduced), a long time (e.g., several weeks in car assembly plants) is required to set up the line to produce the new product efficiently and at a constant rate. A more modular product design, however, can enable a flow-line process to produce different products on the same line (e.g., mixed-model car assembly lines), see Chap. 14.

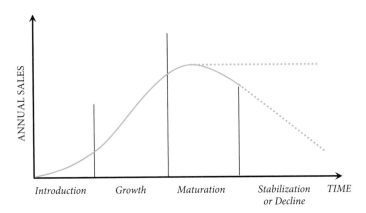

Figure 1.6   The product life cycle curve.

**Continuous-Flow Process**

Continuous-flow processes are very similar to flow-line processes, in that they are designed to produce standard products in very high volumes. Items move through consecutive stages of operations at a constant rate. Both processes build items in MTS fashion. The difference between the two, however, is the following. In flow-line processes, items are moved one-by-one in a discrete manner, for example, car chassis are discrete work items that move through assembly line. In continuous-flow processes, however, items cannot be broken into separate discrete units. The products are often liquids or semi-solids that can flow continuously through the process.

Examples of continuous-flow processes are oil refineries, paper and steel mills, paint factories, and beer factories. Continuous-flow processes are even less flexible than flow-line processes, in that they produce lower variety, the process is highly automated, and thus the capital investment is even higher. The process uses highly specialized and automated equipment often controlled by computers. The role of labor is mainly to load, unload, and monitor the process. Since it is very difficult to differentiate the products produced by continuous-flow processes from those of other firms (e.g., gasoline made by two different firms), low-cost production is the main factor in firm's profitability. Thus, operational challenge is to operate in high utilization—most work 24 hours a day, 7 days a week—and minimize inventory and supply chain costs to reduce the total manufacturing and delivery costs.

### 1.1.7 Product-Process Matrix

Classifying operations processes by the degree of product customization and by product flow provides useful insights into the relationship between product volume, product variety, and point of customization. This helps managers make a better process selection decision that enables them to offer the desired value proposition.

One other important factor that should be considered in operations process selection is product life cycle. Every new product goes through four stages during its life cycle. A typical product life cycle is shown in Fig. 1.6.

As the figure shows, a product's life cycle includes four major stages: (i) introduction stage, (ii) growth stage, (iii) maturation stage, and (iv) stabilization or decline stage. For some products, the period from introduction to stabilization/decline stages can be several weeks and for some products several years.

During the introduction stage, the product is still in the process of undergoing minor design changes based on feedback from market. The manufacturing process of the product is also adjusted to reduce production cost. During this stage, sales revenue is low and firms often lose money on the product. It is essential that more demand is created through marketing and advertising, since competition also starts to grow during this stage.

During the growth stage, the demand for the product increases rapidly. The design of the product is stabilized, and the unit cost of production is reduced due to standardization of the production process and economies of scale. All these result in rapid increase in sales. On the other hand, competition pressure is high, and hence the company must establish the product in the market as firmly as possible.

During the maturation stage, demand reaches its maximum. While the problems with the product design and quality as well as with the production process should have been resolved during the introduction and growth stages, the firm must still consider additional improvement. The competition is tough during this stage and thus firm should focus on increasing market share by improving brand loyalty and providing competitive pricing.

The final stage of the product life cycle depends on the type of the product. The sales for some products (e.g., automobiles, appliances, orange juice) stabilize with potential for some growth. For other products (e.g., DVD players, landline phones) sales decline because the product becomes obsolete, or the market for the product is saturated because there are many competitors and competing products. In these cases, firms, while reducing their marketing and advertising for the product, must divert their resources to other emerging products.

## CONCEPT

**Product Life Cycle**

*Product life cycle* shows the characteristics of a product during its life cycle. It describes how sales of a product increase and then decline (or stabilize) during its four stages of production: introduction, growth, maturation, and decline/stabilization.

Hayes and Wheelwright (1979) studied several industries and summarized the relationship between product life cycle and operations processes to what is now known as Product-Process Matrix—shown in Fig. 1.7.

The product-process matrix emphasizes on two important points. First, the product-process matrix illustrates the alignment of operations processes with the characteristics of the product. It provides firms with the strategic product and process selection. The top of product-process matrix corresponds to product characteristics moving from the left (low-volume, one-of-a-kind) to the right (high-volume, standardized) products. On the other hand, the left side of the matrix corresponds to the processes, from project processes to continuous-flow processes.

Processes often start from project/job shop type processes (e.g., start-up firms) and move to line- and continuous-flow processes (e.g., more established firms) as the product volume increases. The most suitable alignment between the product and the process, therefore, occurs in the diagonal of the product-process matrix. The upper left corner of the matrix corresponds to the project processes which are suited for very low volume, but very high variety (low standardization). As one moves from the upper left to the lower right corner of the matrix, the emphasis of the process moves from low volume/high variety, to high volume/low variety. The lower right corner of the matrix corresponds to the continuous

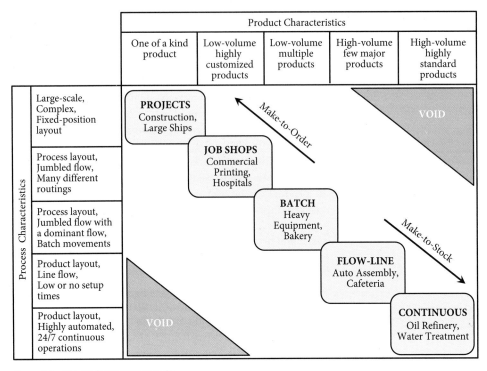

Figure 1.7    The product-process matrix.

process, which is appropriate for very high volume but very low variety. Also, moving along the diagonal, the processes move from ETO, to MTO, ATO, and MTS. Firms that operate off-diagonal should clearly understand the reasons for and implications of doing so.

The second important point emphasized by the product-process matrix is that different processes are used to produce products during different stages of their life cycles. In general, a product is initially produced in a job shop process, because of very low demand and uncertainty about the success of the product in the market. As the demand for the product increases and product enters its maturation, the process moves from job shop process (i.e., the upper-left corner of the product-process matrix) to flow-line and continuous processes (i.e., lower-right corner of the product-process matrix). For example, automobiles or electronics were initially produced in job shops and batch processes in their introduction stages. As the demand for those products increased (i.e., maturation stage), their production processes switched to flow-line processes that are more efficient and economical in producing high volumes. Thus, firms should make sure that their products match with their processes. If the product and process characteristics do not match, the firm will find itself off the diagonal of the product-process matrix, which makes firm unable to remain competitive.

Before we end this section, we would like to emphasize that production processes of firms may not match exactly with one of the processes we discussed in the product-process matrix; they are sometimes a combination of these processes. One example is *Group Technology* (GT) developed in the former Soviet Union in the 1940s and 1950s. It was developed to reduce the difficulties of managing large job shops. In GT, products are divided into several groups, called a *Family of Products*. Each family of products has similar design and requires a similar manufacturing sequence. Each family of products is then assigned to individual cells. In each cell, equipment is arranged according to manufacturing sequence, often in a U-shaped line called *Cellular Layout*. GT processes tend to achieve the efficiency of flow-line processes (through line configuration of each cell) and the product-mix flexibility of job shops and batch processes. GT processes are shown to improve the performance of systems with a large product-mix, where the large number of products can be divided into smaller number of family of products.

Another example is when firms utilize the above standard processes in different stages of operations. For example, production of frozen meal such as frozen pizza in high volume is a combination of batch processes and continuous-flow processes. Pizza doughs are made in batches, and are loaded into a conveyor for the remaining process of making pizzas. This includes flattening the dough, adding sauce, adding toppings, precooking, freezing, and packaging, all done using fully automatized continuous-flow processes.

## 1.2 Automation Technology

Automation technology has given firms competitive advantage and increased their market shares. It has been a major factor in increasing efficiency and productivity and reducing costs of operations processes in both manufacturing and service systems. Automation technology is nothing but using computers and information technology in design, manufacturing, warehousing, and distribution of products. In this section, we discuss some of the commonly used automation technologies that have been developed in the past decades.

### 1.2.1 Computer Numerical Control (CNC)

One of the first significant technological advances in manufacturing was the automation of machine tools, called *Computer Numerical Control* or CNC. In CNC machines, a computer controls the movement of the tool or the workpiece, speed, feed, and depth of cuts according to a computer program. The operator develops the computer program, sets up the machine, and attends the machine while tasks are performed automatically by the machine. Besides speed and consistency in dimension (and quality) of parts, CNC machines have another important feature—setup or changeover reduction between tasks. The ability of CNC machines to automatically change tools in a very short time and the use of high-pressure coolant to flush away chips (and cool the cutting area) significantly reduces the changeover time from one machining task to the next.

Computer-controlled machine tools performing machining operations such as cutting, milling, drilling, and sanding led to the development of robots—another programmable machine—that were designed to perform other tasks. The world's first industrial robot was installed at the GM Ternstedt plant in Trenton, NJ. It performed spot welding and extracted die castings. Robots began to replace human workers in factories and automation began to change the way manufacturing was done. Robots were able to perform tasks with consistent quality, more efficiently, and faster than humans. Today robots perform a large variety of tasks such as spray painting, inspection, welding, and working with hazardous material (e.g., nuclear power plants). ATM machines in banks and tickets kiosks in airports are some examples of programmable machines that perform activities in service systems.

### 1.2.2 Automated Guided Vehicle (AGV)

The use of automated machines is not limited to performing manufacturing tasks such as machining and welding. Automation also found its way in to material handling systems in factories and warehouses. In the past, conveyors were the main method for automating the movement of products from one part of factory to another. In 1953, the first *Automated*

*Guided Vehicle* (AGV) was built and used in a grocery warehouse. It was nothing but a modified towing tractor pulling a trailer following an overhead wire in the warehouse. Today AGVs are more advanced; they can move items of different sizes, following optical paths and going anywhere within a factory or warehouse (or even in hospitals) avoiding obstacles (e.g., pallets of items) in their ways.

### 1.2.3 Automated Storage and Retrieval System (AS/RS)

Another automation technology that had a significant impact on warehouse operations was *Automated Storage and Retrieval Systems* (AS/RS). This recent technology uses AGV technology to move items in or out of warehouse and uses computer-controlled storage bins and racks to store the items. AS/RS systems generally consist of computer-controlled machines that move up and down on parallel storage aisles, storing arriving products delivered by AGVs or retrieving products and materials and putting them on AGVs to exit the warehouse. The computer keeps track of how many items are in the storage and where exactly they are located in the warehouse. Some advantages of AS/RS systems are efficient and accurate inventory control and tracking, lower labor cost, increased workforce safety, and improved warehouse space utilization (vertically and horizontally).

### 1.2.4 Flexible Manufacturing Systems (FMS)

*Flexible Manufacturing System* (FMS) consists of a set of numerically controlled machine tools or robots that are used to manufacture a wide range of products. The operations performed by the machines are mainly machine tool operations such as cutting, forming, drilling, and also assembly operations. Parts move from one machine to another using automated material handling systems. A central computer controls the movements of parts among machines, selects the tool required to process each part, and controls the machine operations on each part. The advantages of FMS include the ability to provide a wide variety of products with low work-in-process (WIP) inventory, low production lead time, and low labor cost. These benefits come from the fact that FMS has very low machine setup and tool changeover times (done automatically and very fast) and specialized automated part transfer systems with limited number of fixtures, each holding one part.[4]

### 1.2.5 Computer-Aided Design and Manufacturing (CAD/CAM/CIMS)

Impact of computer on operations goes beyond just manufacturing products; it also advanced the product design process. *Computer-Aided Design* or CAD is a computer interactive graphical system that helps designers in creation, modification, and analysis of a design. CAD software has increased the productivity of the product designer and improved the quality of the designed product. CAD software creates electronic date files corresponding to the design that can later be used by computers to control the manufacturing process. This is called *Computer-Aided Manufacturing* or CAM. The collection of all computer-controlled hardware and software in product design, manufacturing, warehousing, and distribution is referred to as *Computer-Integrated Manufacturing Systems* or CIMS.

## 1.3 Information Technology

While automation technologies have made the flow of *products* more efficient, advances in *Information Technology* (IT) allowed firms to make the flow of *information* within different levels of supply chains more efficient. Here we refer to some of those technologies.

---

[4]The number of parts in FMS is limited to the number of fixtures or special pallets in the system.

### 1.3.1 Bar Code and RFID

One of the first successful use of IT was bar codes. Bar codes were originally developed to help retail stores speed up their checkout processes and to keep more accurate inventory records. A bar code contains information about the product, the manufacturer who produced the product, and its country of origin. Scanner receives the information by reading the bars by the intensity of light reflected back from the bars and spaces between the bars. Walmart was among the first companies that used bar code technology very efficiently. Bar codes enabled Walmart to track sales on specific items at any specific time of the day. This helps Walmart to identify their popular items and to set the optimal prices, and hence increase its sales.

The advances in wireless communications led to *Radio Frequency Identification* (RFID) technology. This technology employs small electronic chips that are attached to items. These chips are capable of transmitting a variety of data about each item's location and movement. For example, RFID attached to an automobile in a factory shows the car's progress through the assembly line. RFID systems are now tracking consumer products in the supply chains worldwide. Firms are now able to track the location of each product they make from the time it is made until a customer purchases the item. For example, Walmart transmit the sales data at its retail locations to its replenishment warehouses using its own satellites. Such information helps operations engineers and managers to make more effective inventory and distribution decisions resulting in a more efficient product flow within their supply chains.

The impact of new IT has gone beyond the operations functions of firms, and now integrates all other functions of firms such as accounting, customer relationship, sales, human resource, etc. *Enterprise Resource Planning* or ERP is a powerful IT that is used for planning and control of resources throughout the firm. It provides analysis and reporting of all activities of a firm including sales, inventory, customer, accounting, manufacturing, and human resources.

### 1.3.2 E-Commerce

E-commerce refers to a type of business model (or a part of a business model) in which a firm conducts its business using electronic network, mainly Internet. In e-commerce, activities such as purchases and sales of products (goods or services) are done via Internet. E-commerce was started in 1960 by using electronic data interchange (EDI) and became popular with the increase in Internet access. Firms such as Amazon started selling products online in late 1990s, with Jeff Bezos selling books and shipping them from his garage. Nowadays, any product or services can be bought online, for example, shoes, books, clothes, groceries, ticket for a flight, ordering a taxi.

Due to success of e-commerce, most firms have adopted the model as a segment of their business. For example, while retailers such as Target, Walmart, and Home Depot have physical stores to sell their products, they also established Internet channels for their direct sales to customers.

To be successful in using e-commerce business model, firms need to have an efficient IT system to coordinate physical flow of material and goods as well as financial flow and information flow. The three basic IT systems that provide the infrastructure for e-commerce are Enterprise Resource Planning (ERP), Supplier Relationship Management (SRM), and Customer Relationship Management (CRM).

**Enterprise Resource Planning**
Enterprise Resource Planning or ERP is a powerful source of IT that is used for planning and control of resources throughout the firm. It provides analysis and reporting of

all activities of a firm including sales, inventory, customer, accounting, manufacturing, and human resources. ERP also includes decision support systems that help managers store data efficiently and make more informed operations decisions. ERP will be discussed in detail in Chap. 13.

### Supplier Relationship Management

While ERP's main focus is on management and organization of operations within the firm, Supplier Relationship Management (SRM) system's focus is on managing and organizing the inputs to the operations—what the firm buys, and from whom it buys it. SRM is a data management and decision support system that helps a firm understand the level of risk as well as profitability of working with each supplier. SRM, therefore, maps suppliers against their exposure to risk and their profitability. These will help the firm evaluate its suppliers with respect to their quality, price, on-time delivery, and their ability to deliver emergency orders. By recording the past performance of its suppliers, a firm can make better tactical decisions such as consolidation of purchasing across different division, as well as strategic decision of whether to partner with a supplier, or decide whether to replace a supplier with another.

### Customer Relationship Management

While the SRM system is concerned with the supply (input) side of the operations, Customer Relationship Management (CRM) systems deals with the customer (output) of the operations. CRM is a data management and decision support system for managing a firm's interactions with its current customers as well as future potential customers. By gathering and analyzing information about customers' purchasing behavior with respect to frequency of purchase, response to promotions and sales, and the location of their purchases, firms can make more informed decisions on prices, promotion, opening or closing a sales channel, offering a new product or stop offering a product, suggesting substitute products for an order, and identifying profitable customer segment. These decisions have direct impact on firm's profit.

The above three information systems—ERP, SRM, and CRM—integrate, manage, and coordinate activities in the supply chain of e-commerce systems and are essential in the success of e-commerce business.

## 1.4 Summary

Each firm has a mission statement that serves as the foundation of a firm's business strategy. Business strategy of a firm is its long-term plan that identifies firm's targeted customers and sets the performance objective of the firm. Operations strategy is a set of decisions that shape the capabilities of operations and their contribution to business strategy of a firm. To stay competitive and make profit, firms must offer a product (goods or service) that provides values that are different or are superior to the competitor's products. Specifically, value proposition of a product offered by a firm is the particular value (i.e., feature) of the product that is different from the existing products, and thus makes the product competitive. Main value propositions are features and innovativeness, cost, quality, flexibility, time, and sustainability.

It is important to understand that superior performance in some values may conflict with superior performance in others. Therefore, firms need to make trade-off between these values. Knowing what customers consider as order qualifiers and order winners helps managers make a more informed trade-offs between values offered by their products and identify the strategic position of their firm in the competitive product space.

While choosing the right value or set of values to offer is important, more important is whether a firm can offer those values, and offer it with profit. Thus, before choosing any value proposition to offer, firms should determine whether their current operations are capable of delivering such value proposition. The ability of an operations to provide particular capabilities depends on the capabilities of its processes. Some processes are effective in offering different levels of product customization (e.g., ETO, MTO, MTS, and ATO), and some processes are effective in providing different levels of speed and product variety (e.g., project process, job shop process, batch process, flow-line process, and continuous-flow process).

Another important factor that firms must consider when making operations strategy decision is the level of automation. Automation technology has given firms competitive advantage and has been a major factor in increasing efficiency and productivity and reducing costs of operations processes in both manufacturing and service systems. With the rapid growth of super computers and artificial intelligence (AI), firms have been facing the challenge of how to effectively integrate automation technology and AI in their product design, manufacturing, warehousing, and distribution of goods and services.

## Discussion Questions

1. What are the main functions of a manufacturing or service firm, and what are the responsibilities of each function?

2. Give an example of an operations of a firm and identify the inputs, outputs, and resources used to produce products (goods or services).

3. Using an example, describe the differences in strategic decisions and tactical decisions that operations engineers and managers make.

4. What are mission statement, business strategy, and operations strategies of firms and how are they related?

5. What are the six major value propositions that a firm can offer? Provide examples of firms that offer each value proposition.

6. Describe four different types of flexibility that a firm can offer as its value proposition and provide an example for each type.

7. What are order winners and order qualifiers and how do they relate to the trade-offs a firm should make when it identifies its value proposition.

8. Use an example to describe strategic position, competitive product space, and operations frontier.

9. What is plant-within-a-plant and how does that relate to value propositions offered by a firm?

10. What are operations capabilities and competencies and what do they depend on?

11. Using examples describe the four main types of processes that provide different levels of product customization.

12. What are the five main types of process, if the processes are classified based on how material flows within the process? How do these processes differ with respect to the characteristics of their products and their main operational challenges?

13. What is product life cycle? Using an example describe the four stages of the product life cycle.

14. What is product-process matrix and how does it help managers make decisions regarding their products and processes?

15. Describe the following automation technology and their applications in manufacturing or service operations:

    – CNC      – AGV      – CAD
    – CAM      – AS/RS     – FMS
    – CIMS

16. What is RFID and how does it improve the performance of operations systems and supply chains?

17. What is e-commerce and what are the three basic information technology systems that are essential in e-commerce? Explain the functions of each system.

# CHAPTER 2

# Process-Flow Analysis of Operations

## 2.0 Introduction

A *process* is a collection of activities that transform inputs to outputs. Operations in manufacturing and service firms is a process that transforms inputs such as raw material, components, customers, customer orders, electricity, and money into outputs such as finished products or served customers. The outputs of operations processes have values to either customers (outside of the firm) or other processes (inside the firm).

Consider a furniture manufacturer as an example. The manufacturer's operations process has many inputs such as woods, nails, screws, leather, fabrics, among others. Using their labor and equipment, different departments make different parts such as legs or back supports that are used in the assembly of chairs and tables. These parts have values for the assembly department of the firm. The assembly department uses these parts as well as other inputs and assembles the final product. The output of the assembly process—the finished product—has value to customers. Another example of a process is a supermarket that has many items that it orders from suppliers as well as customers who buy those items as its input and has satisfied customers who bought the product as its output.

One aspect of operations processes is that the main output of the process—products (goods or services)—flows through different activities within the process and is gradually transformed into the final product before it exits the process. *Process-Flow Analysis* is about the analysis of how inputs flow within the process before they exit the process as outputs. Its goal is to improve the process performance by finding new ways to perform activities or by changing the way inputs flow within the process.

In this chapter, we show how process-flow analysis can be used to analyze operations of a firm. We also introduce process-flow measures that affect revenue or expenses of a firm. The process-flow measures we consider are Throughput, Flow Time, Inventory, Inventory Turn, Utilization, Quality, and Productivity. We start by introducing process-flow analysis framework.

## 2.1 Process-Flow Analysis Framework

We use a simple example to illustrate the steps in performing a process-flow analysis of an operations. Consider a manager of a retail store who wants to make a decision about the number of cashiers needed in its checkout counter in each hour of the store's operations. The main goal is to make sure that there are enough registers open so that 90 percent of the customers wait less than a minute in the line for cashiers. Fig. 2.1 shows a schematic of checkout operations with three cash registers opened.

As the figure shows, customers who already picked up their items from store go to the three open cash registers to pay for their items. The first cashier is assigned to customers who have 10 items or less. The other two cashiers are for customers with more than 10 items. As shown in the figure, three customers are being served by the cashiers (one by each cashier) and other customers are waiting in line. To analyze the operations of a checkout counter in a retail store, the process-flow analysis takes several steps that are discussed in the following sections.

### 2.1.1 Step 1: Determine the Process Boundaries

The first and the most important step is to identify the boundaries of the operations we would like to analyze. The operations in a retail store include many activities such as ordering product's, storing product's, filling up the shelves with products, serving customers who return products, bringing shopping carts from outside (parking) to the store, among others. Checking out customers at the cash registers is only one of the activities that the manager in our example is focusing on. For this particular operation, the process boundaries start from when the customers join the checkout line to when they leave the store after paying for their items.

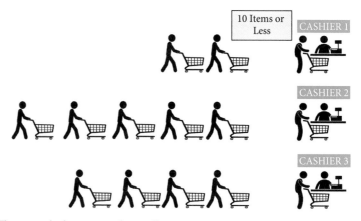

Figure 2.1    Three open checkout counters in a retail store.

The main factor to determine process boundaries is the objective of the analysis. For example, suppose the goal is to increase the profit of selling a particular type of toothpaste. In this case, the process boundaries start from the supplier of the toothpaste and ends when a customer picks up the toothpaste to buy—the boundaries end where the boundaries of the checkout counter starts.

### 2.1.2 Step 2: Determine the Flow Unit

*Flow units* are units that enter the process as inputs, flow through the process while being transformed, and exit the process as outputs. As we mentioned above, production systems have many different types of inputs such as raw material, components, people, money, among others. Which one of these inputs does process-flow analysis consider as a flow unit (or flow units)? The answer depends on the operations decisions involved as well as the desired goal.

In our customer checkout example, the decision is about the number of cashiers required in each hour, and the goal is to have enough cashiers so that 90 percent of the customers wait in line for less than a minute. Both the decision and the goal directly depend on the number of customers joining the line (i.e., entering the process) and the time it takes to serve them (i.e., exiting the process). Therefore, the flow unit in checkout process is customers. In our toothpaste example, the flow unit is, of course, a unit of toothpaste that enters the retailer's supply chain and exits when purchased by a customer.

Most manufacturing systems produce more than one type of product (e.g., TV producers make TVs of different sizes) and most service systems offer more than one type of services (e.g., hospitals provide different services such as radiology, surgery, blood test). In such cases, the process-flow analysis determines more than one type of flow unit. In our customer checkout example, there are, in fact, two types of flow units: (i) Type-1 are customers who bought 10 items or less, and (ii) Type-2 are customers who bought more than 10 items.

We call each type of the flow unit a *Product* of the process and we call a process with more than one type of flow units a *Multi-Product Process*.

### 2.1.3 Step 3: Determine the Resources

Each flow unit that enters a manufacturing or service system will be transformed before it exits the system. The transformation is done through a single or a sequence of activities performed by one or several resources. In our customer checkout example, there is a single activity "checking out" that is performed by the cashiers—the resources. In manufacturing systems, products (i.e., flow units) go through different stages in which activities such as cutting drilling, assembling, welding, and painting are performed on the products. Resources are workers or equipment that perform those activities. In accounting firms, for example, filing a tax is an activity that is performed by the accountants—the resources.

Sometimes more than one resource is required to perform an activity. For example, drilling a hole in a part requires two resources: a drilling machine and the drill operator. In contrast, there are cases where one resource can perform more than one activity. In our customer checkout example, each cashier performs three activities of scanning the item, charging the customer, and bagging customers' items. In our process-flow analysis, we consider these activities as one activity of "Checking Out" a customer. In some stores, however, the bagging is done by a different worker (i.e., by another resource).

### 2.1.4 Step 4: Determine the Routings and Buffers

Flow unit (or flow units) often go through a sequence of activities to become the final product (or products) of the process. The order of the activities in these sequences is important, since there is often a precedence relationship within these activities. For example, when making potato chips, the activity of sealing the potato chips bag must be done after the activity of filling the bag with potato chips. The sequence of the activities that each type of flow unit must go through until it exits the process is called *Routing*. For example, the routing of customers (i.e., flow units) in retail stores starts with customers first performing the activity of "shopping" (i.e., picking up items from shelves and putting them in their shopping carts), and then "checking out." Another example is the auto-assembly process in which painting is performed on body parts after those parts are made by stamping presses.

One common feature of the operations system is that in each routing, after completion of one activity, the flow units must often wait before the next activity can start. In process-flow analysis terminology, flow units that are finished with one activity wait in the *Buffer* for their next activity in the routing. In manufacturing systems, buffers are physical locations in which raw material, work-in-process, or finished products are stored. In service systems such as doctor's office, the buffer is the waiting room in which the patients wait before their appointments with the doctor. In our customer checkout example, the buffer is the area where customers form a line awaiting their turns to check out. Buffers can also be virtual buffers with no physical location. For example, customers' orders placed online for pizza at Domino's Pizza wait in Domino's computer system (i.e., in a virtual buffer) until the pizza is delivered. The number of flow units waiting in the buffer is called inventory (see Sec. 2.3).

In this book, we call a process a *Single-Stage Process* if flow units go through only one buffer and one activity before they depart the system. On the other hand, we call a process a *Multi-Stage Process* if flow units go through a routing with several activities and buffers. Of course, whether a process would be a single-stage or a multi-stage process depends on how process boundaries and activities are defined. For example, if we define our process boundaries to only include the checkout activity, then we have a single-stage process in which flow units (i.e., customers) wait in a buffer (the line for cashiers) and are processed for one activity (i.e., checking out). However, if we are interested in finding how long a customer stays in the retail store, we must define the process boundaries to start from when the customers enter the store until they leave the store. In this two-stage process each customer will go through a routing with two activities: (i) shopping, and (ii) checking out.

### 2.1.5 Step 5: Draw Process-Flow Chart

A *Process-Flow Chart* visualizes the main elements of a process including flow units and their corresponding routings, buffers, and activities as well as process boundaries. In its most basic form, process-flow chart uses three symbols to map the activities and routings of flow units in a process. These symbols are shown in Fig. 2.2.

Fig. 2.3 shows the process-flow chart developed for our customer checkout example. As the figure shows, there are two types of flow units: Type-1 and Type-2, each going through a different routing. Type-1 flow units are customers with 10 items or less, who go to Cashier 1 for "Checking Out" activity. Type-2 flow units are customers with more than 10 items who are going to Cashier 2 or 3 for "Checking Out" activity. The buffers in front of each checking out activity imply that flow units (i.e., customers) may wait before their activities start. In process-flow charts, all activities have a buffer, unless all flow units never wait for the activity.

### 2.1.6 Step 6: Determine the Control Policy

The resources in all operations follow some policies that govern the way they process flow units. These policies are called *Control Policies*. Control policies determine rules such as which flow units in a buffer should be processed next, where a flow unit must be routed

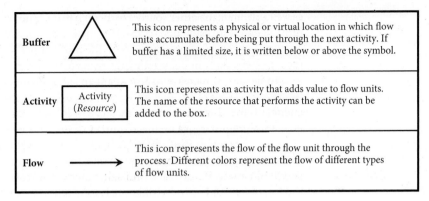

Figure 2.2    **Three basic symbols used in process-flow charts.**

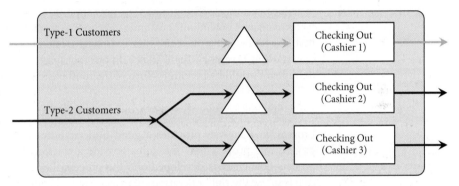

Figure 2.3    **Process-flow chart of the checkout counter operations in the retail store example.**

after being processed by a resource, when the processing of a flow unit should start, and how many flow units should be processed each time.

Consider our customer checkout process. The process has the following control policies: (i) customers (i.e., flow units) with less than 10 items are routed to a different cashier (i.e., different resource), (ii) each cashier chooses the customer from the head of the line (i.e., the buffer) to serve—a first-come-first-served (FCFS) policy. In our toothpaste example, the routing of the toothpaste in the retailer's supply chain starts from the supplier of the toothpaste, to the retailer's Distribution Center (DC), and then to the retail store. In this setting, control policy determines how many times in a year the DC should purchase toothpaste from the supplier and how many toothpastes should be purchased each time; when and how many toothpastes should be shipped from DC to the retail store; and when should store's staff check the shelves and put more toothpastes on the shelves (if needed).

Without knowing these control policies, one cannot determine the sources of inefficiencies in a process and recommend strategies for improvement.

### 2.1.7  Step 7: Determine the Process-Flow Measure

The main goal of process-flow analysis is to find ways to improve operations and thus improve the profitability of a firm. To see how process-flow analysis can improve a firm's profit, we first need to know how the profit of a firm is calculated. Firms compute their profit (or loss) during a specific period by subtracting firm's expenses from its revenue during that period. These are presented in a financial statement called "Income Statement" or "Statement of Income." The income statement lists the firm's revenues from sales of goods and services as well as from firm's other sources of income that are not central part of firm's business (e.g., firm's financial investments). It also lists the firm's expenses, including the cost of goods (or services) sold and operating expenses. The profit or net income of the firm during a period is computed simply by subtracting the firm's total expenses from its total revenue.

To be able to increase its profit, a firm should either increase its revenue, or decrease its expenses. It turns out that, as we will discuss in the following sections, the measures of operations that directly or indirectly relate to either revenue or expenses correspond to how flow units are processed by the resources of the firm and how the units flow through the operations process of the firm. We call these measures *Process-Flow Measures*. They include measures such as Throughput, Flow Time, Inventory, Inventory Turn, Utilization, Quality, and Productivity. We describe these measures in the rest of this chapter and show how they can be computed. In process-flow analysis of any operations, a critical goal is to determine which one of the process-flow measure (or measures) must be improved. For example, the goal might be to improve throughput of a process to respond to the increase in demand, or to reduce customer waiting time (i.e., flow time) to improve customer satisfaction, which is the manager's goal in our customer checkout example. We will show later in this chapter that one process-flow measure that is directly related to the profitability of items in retail operations is Inventory Turn. Hence, in our toothpaste example, the process-flow measure to improve would be Inventory Turn of the toothpaste (see Sec. 2.7).

### 2.1.8  Step 8: Use Operations Principles and Analytics to Improve Process Performance

Process-flow analysis provides a good understanding of the elements of the operations process and process-flow measures provide a clear goal for process improvement. After process-flow chart is developed and the process-flow measures are determined, the next step is to use the right operations management and control principles as well as analytics to improve the desired process-flow measure.

Operations principles and analytics improve process performance by: (i) revising the control policies or developing new policies, or (ii) redesigning some part or the entire process flow. In our customer checkout example, the waiting time of the customers might be improved as follows: (i) Allowing the cashier assigned to customers with 10 items or less to also serve other customers, if there is no customer in that line. When a cashier is idle, it is easy to simply ask customers in other lines to switch to the cashier's line to relieve any congestion. (ii) Redesigning the process flow by having only one line (instead of two separate lines) for the two cashiers that serve customers with more than 10 items. It has been shown that, in most cases, having a single line for multiple servers reduces customer waiting time compared to having multiple dedicated lines for each server (see vertical pooling in Chap. 8).

In our toothpaste example, operations principle and analytics can help us reduce the cost of purchasing and holding inventory. There are many analytics that have been developed to find the optimal ordering policies. These analytics—that we will discuss in Chaps. 9, 10, and 11—determine how many times in a year the retailer should purchase toothpaste from the supplier and how many toothpastes should the retailer buy each time and store in its DC. They also determine when and how many toothpastes the retail store should order from the DC.

The main goal of this book is to provide principles and analytics that improve different process-flow measures in manufacturing and service operations systems. In other words, this book focuses mainly on Step 8 of process-flow analysis.

Now that we have illustrated the main steps of process-flow analysis of an operations, the rest of this chapter is devoted to describing process-flow measures. The remaining chapters of the book then focus on management principles and analytics that help improve these process-flow measures.

## 2.2 Throughput

Throughput is an important process-flow measure that is directly related to revenue. Throughput of a process is a measure that captures the performance of a process, focusing on the rate at which the process is producing good (non-defective) outputs. The metric related to throughput of a process is average throughput, also called throughput rate.

### 2.2.1 Throughput Rate

To better understand the concept of the Throughput of a process, we first need to highlight the difference between the throughput and the output of a process. We utilize the following example to illustrate these concepts.

 **Example 2.1**

> VH Plastics produces and supplies plastic products for computer accessories such as monitor stands, keyboard trays, among others. The monitor stand production process consists of two main departments, that is, monitor stand production (MSP) and quality control (QC) departments. The MSP department makes the top and the bottom parts of a stand using injection molding. The top and the bottom parts are assembled together with a mechanism that allows the adjustment of the height of the stand. Finished stands are then sent to the QC department for quality check. To evaluate the performance of the MSP department, vice president of production has ordered a study of the number of stands that the MSP department is producing per day. The summary of a 10-day study shows that the MSP department has produced a total of 1440 monitor stands in the last 10 days. Among the 1440 stands, 72 stands were rejected and discarded by the QC department due to major quality problems that cannot be fixed. What do you think would be a good measure to evaluate the performance of the MSP department with respect to its output?

The MSP department had an output of 1440 monitor stands in 10 days. Thus, on average, the department produced 144 ($= 1440/10$) stands per day. However, among those 1440 stands, 72 stands were defective. Hence, the total number of good (non-defective) stands produced in 10 days was $1440 - 72 = 1368$ stands. This means that the MSP department produced $1368/10 = 136.8$ non-defective stands per day. Which number is a better measure of the daily performance of the MSP department? 144 or 136.8?

While both numbers carry some information about the daily production rate of the MSP department, operations engineers and managers prefer 136.8 (over 144) since it presents the number of monitor stands that they can actually sell. This is called "Throughput Rate" of the MSP department, while 144 stands per day represents the "Output Rate" of the department.

**CONCEPT**

**Throughput Rate and Output Rate**

Consider a process that processes $N$ units from time 0 to time $t$, that is, a period of length $t$. Among those $N$ units, $N_g$ units are good (non-defective) units, and the remaining $N - N_g$ units are defective units that are discarded.

- *Output Rate* of the process during the time period of 0 to $t$, which we denote by $R_{\text{out}}^{\text{avg}}(0, t)$ is the total *average* number of flow units that exit the process per unit time:

$$R_{\text{out}}^{\text{avg}}(0, t) = \frac{N}{t}$$

- *Throughput Rate* of the process during the time period of 0 to $t$, which we denote by $\text{TH}^{\text{avg}}(0, t)$ is the total *average* number of *good (non-defective)* flow units that exit the process per unit time:

$$\text{TH}^{\text{avg}}(0, t) = \frac{N_g}{t}$$

When the length of the time period is very large, the output and the throughput during the $t$ time units are, respectively, called *long-run output rate* and *long-run throughput rate*, or simply *output rate* and *throughput rate* of the process. We let

$$R_{\text{out}} = \text{Long-run output rate}$$

$$\text{TH} = \text{Long-run throughput rate}$$

When the process does not produce defective units that are discarded, the throughput rate of the process is equal to its output rate. In most textbooks, the word "Throughput" is used to refer to "Throughput Rate." In this book also, we use "Throughput" for simplicity.

Note that in Example 2.1, $N = 1440$, $N_g = 1368$, and $t = 10$ days. Thus, the average throughput of VH plastics during the 10-day period is $\text{TH}^{\text{avg}}(0, 10) = 1368/10 = 136.8$ stands per day.

The term "long-run" refers to the fact that throughput should be computed for a long period of time to get an accurate estimate of the throughput of a process. For example, in the case of VH Plastics, the output rate and throughput should be estimated for samples of more than 10 days.

Throughput is an operational performance measure that has a direct impact on firm's profit. When there is market for selling what is produced, then increasing throughput results in increase in sales, which, in turn, results in increase in profit. Thus, improving throughput of a process is one of the main goals of operations engineers and managers. In Chap. 6, we present different ways to improve throughput of a process.

### 2.2.2  Due Date Quoting Using Throughput Data

In addition to finding ways to improve throughput, analyzing the throughput of a process also allows us to estimate the due date for delivering a product to a customer. We explain this in the case of LuxBoat company below.

**Due Date Quoting at LuxBoat**

LuxBoat produces luxury fiberglass boats. Joseph, the production manager of LuxBoat, receives a call from Bob, the vice president of marketing, about a large order from a sailing club. Specifically, the club would like to order 10 boats. However, before placing an order, the club manager wants to know when his order will be completed and shipped. Furthermore, he wants all of the 10 boats to be shipped together when they are all finished. Bob is asking Joseph to give him an estimate for the shipping date of this order.

Joseph realizes that this is not an easy task. Looking at LuxBoat's current order list, he finds that there are 10 orders (each requesting one boat) for which the production has already started. These 10 boats are in several stages of production. Furthermore, there are five orders (one boat per order) in the order list that have not even started. So, Joseph realizes that, if the sailing club places an order now, the club's order will be ready to ship when 25 (= 10 + 5 + 10) boats are produced. This is because orders are processed first-come-first-served.

To get an estimate of how long it takes to finish making 25 boats, Joseph calls his plant manager, Tom, and asks him for his estimate of how long it takes to make one boat. Tom informs Joseph that the time from when the plant starts working on an order for a boat until the time that the boat exits the production line is about 4 days. Joseph does a simple calculation and computes the average time to produce 25 boats to be around 25 × 4 = 100 days. However, 100 days seems too long to him. Joseph immediately realizes his mistake. The shipping date of 100 days is only true if the plant does not start the production of the next boat until it finishes making the current boat. However, this is not the case in his plant. Each boat goes through several stages of production, and thus at any time there are several orders (boats) in different stages of the plant.

Joseph's first reaction is to ask Tom for detailed data of: (i) where (i.e., in which stage of production) each of the 10 boats (for which production has started) are, (ii) how long it takes to process a boat in each stage, (iii) how long it takes to finish each of those 10 boats, (iv) when the plant starts working on the next five orders, and (v) how long will it take to finish those five orders. But, he realizes that even if he gets all these data, the process of estimating the shipping time of the club's order would not be easy.

While drinking his tea and thinking, Joseph recognizes that he does not need to know how long it takes to produce a boat (at each stage of production or from beginning to the end). All he needs to know is the throughput of the plant. If the throughput is, for example, 50 boats per month, then the plant can finish the production of 25 boats in half a month. Joseph then calls Tom and asks him about the throughput of his plant. Tom sends him data given in Table 2.1, which shows the exact date and time of production completion of 30 boats

**Table 2.1    Production Completion Dates in September and October at LuxBoat**

| Boat no. | Production completion | | Boat no. | Production completion | |
|----------|------|------|----------|------|------|
|          | Date | Time |          | Date | Time |
| B01 | Sept. 1 | 10:00 am | B16 | Sept. 25 | 12:00 pm |
| B02 | Sept. 2 | 6:30 pm | B17 | Sept. 27 | 1:30 am |
| B03 | Sept. 3 | 6:00 am | B18 | Sept. 28 | 7:30 am |
| B04 | Sept. 4 | 10:00 pm | B19 | Sept. 29 | 3:30 pm |
| B05 | Sept. 6 | 12:30 pm | B20 | Oct. 1 | 2:00 am |
| B06 | Sept. 7 | 6:00 pm | B21 | Oct. 2 | 12:00 pm |
| B07 | Sept. 9 | 7:00 am | B22 | Oct. 4 | 3:00 pm |
| B08 | Sept. 10 | 11:00 pm | B23 | Oct. 6 | 3:00 pm |
| B09 | Sept. 13 | 12:00 am | B24 | Oct. 8 | 8:30 am |
| B10 | Sept. 14 | 8:00 pm | B25 | Oct. 10 | 12:00 am |
| B11 | Sept. 16 | 5:30 am | B26 | Oct. 11 | 12:00 pm |
| B12 | Sept. 18 | 1:30 am | B27 | Oct. 12 | 7:00 pm |
| B13 | Sept. 19 | 3:00 pm | B28 | Oct. 14 | 7:00 am |
| B14 | Sept. 21 | 2:00 pm | B29 | Oct. 16 | 12:00 am |
| B15 | Sept. 23 | 3:00 pm | B30 | Oct. 17 | 10:00 am |

in September and October until today, October 17. Joseph looks at his watch, which shows 10:05 am, and realizes that, according to the data that Tom sent, the plant has just finished a boat 5 minutes ago at 10:00 am.

Using the data in the table, Joseph determines that the plant has produced 29 boats from the morning (i.e., 10:00 am) of September 1 to the morning (i.e., 10:00 am) of October 17, a period of 47 days. So, he computes the throughput of the plant as

$$\text{TH}^{\text{avg}}(\text{Sept. 1, Oct. 17}) = \frac{29 \text{ boats}}{47 \text{ days}} = 0.617 \text{ boats per day}$$

So, if the plant works at the same pace as in the last 47 days, Joseph thinks the number of days required to finish making 25 boats would be around $25/0.617 = 40.5$ days.

Encouraged by his estimate for the shipping day of the club's order, Joseph calls the vice president of marketing and tells him that he can ship club's order in about 40.5 days. "Great!" Bob replies, "but Joe, how sure are you about this date? I don't want to promise them a shipping date that we cannot deliver." Joseph explains that 40.5 days is the *average time* to complete their order. The actual time may be smaller or larger than 40.5 days. Bob responds with some sarcasm, "I cannot tell them that we *may* be able to do it *before* or *after* 40.5 days! This does not mean much." He continues, "I know that it is difficult to get the exact shipping date, but can you at least give me a date that we can be 90 percent sure we can ship their order on or before that date?" Joseph has no answer to Bob's question. He tells Bob that he needs some time to think, and that he will get back to him with some date.

Joseph recognizes that knowing the average throughput 0.617 boats a day is not enough to get a shipping date with 90 percent certainty. He needs more data about the throughput of his plant. After some thinking, Joseph realizes that he can extract more information from the throughput data that Tom sent. Specifically, he can compute the time between two outputs of the plant. For example, in September, the first finished boat exited the production line on September 1 at 10:00 am. The second boat exited the line at 6:30 pm on September 2. Considering the fact that the plant works for three shifts (i.e., 24 hours a day), the time between the first and the second departure from the plant is 32.5 hours. This is called "inter-throughput time."

---

**CONCEPT**

### Inter-throughput Time

*Inter-throughput time* of a process is the time between two consecutive throughputs of the process.

---

Joseph does some algebra and computes the inter-throughput times of the plant in September and October using the data that Tom had sent him. Table 2.2 summarizes his calculations.

Contemplating about the inter-throughput times in the table, Joseph sees why due date quoting is such a hassle. The inter-throughput times vary a lot. They range from 29.5 hours to 51 hours, with an average of 38.9 hours. It was easier to quote an accurate shipping date, if inter-throughput times did not vary that much. For example, if the inter-throughput times were exactly 38.9 hours with zero variability, then the time it takes to finish the production of 25 items would be $38.9 \times 25 = 972.5$ hours. Considering that the plant works 24 hours a day, then the time to produce 25 boats would have been $972.5/24 = 40.5$ days. This is exactly what Joseph had found as an "average" time to ship the order when he was using the average throughput rate of 0.617 boats per day.

**Table 2.2    Inter-Throughput Times in September and October**

| Boat no. | Inter-throughput time (hours) | Boat no. | Inter-throughput time (hours) |
|---|---|---|---|
| B01 | — | B16 | 45 |
| B02 | 32.5 | B17 | 37.5 |
| B03 | 35.5 | B18 | 30 |
| B04 | 40 | B19 | 32 |
| B05 | 38.5 | B20 | 34.5 |
| B06 | 29.5 | B21 | 34 |
| B07 | 37 | B22 | 51 |
| B08 | 40 | B23 | 48 |
| B09 | 49 | B24 | 41.5 |
| B10 | 44 | B25 | 39.5 |
| B11 | 33.5 | B26 | 36 |
| B12 | 44 | B27 | 31 |
| B13 | 37.5 | B28 | 36 |
| B14 | 47 | B29 | 41 |
| B15 | 49 | B30 | 34 |

Joseph thinks that he is back to square one, which makes him frustrated. While taking a 10-minute break, he remembers some similar problems in his statistics course in college. He also remembers his statistics professor's advice to him, when he had trouble doing his assignments: "always translate the question into mathematical language, and I promise it will lead you to the solution." Following the advice, Joseph formally defines the inter-throughput time for the $i$th throughput as follows:

$$X_i = \text{Inter-throughput time between the } (i-1)\text{th and } i\text{th throughputs}$$

Joseph then realizes that the time from now until 25 boats exit the production line (i.e., the time to finish the club's order) can be written as follows:

$$\text{Time to finish club's order} = X_1 + X_2 + \cdots + X_{25}$$

where $X_1$ is the inter-throughput time of the first boat that will exit the production line, that is, the time between the last throughput of the plant (at 10:00 am today) until the first throughput after now. Similarly, $X_2$ is the time between the first throughput of the plant after now until the time of the second throughput, and so on.

Joseph also defines $T_{\text{due}}$ to be the *Due Date* for the club's order. Specifically, $T_{\text{due}}$ is the time it takes (from now) until the club's order is ready for shipment. Joseph feels that he can now write his due date quoting problem in a mathematical form. Specifically, he needs to find a due date that satisfies the following:

$$P(\text{Time to finish club's order} \leq \text{Quoted due date}) = 90\%$$
$$P(X_1 + X_2 + \cdots + X_{25} \leq T_{\text{due}}) = 90\% \qquad (2.1)$$

After writing the mathematical form of his problem, Joseph gets an idea about how to approach the problem. He recognizes these types of equations and remembers something about Central Limit Theorem. After searching Internet for Central Limit Theorem, he

H5 = CORREL(B2:AC2,B3:AC3)

| | A | B | C | D | E | F | G | H | I | J | K | L | M | N | O |
|---|---|---|---|---|---|---|---|---|---|---|---|---|---|---|---|
| 1 | | | | | | | | | | | | | | | |
| 2 | 33 | 36 | 40 | 39 | 29.5 | 37 | 40 | 49 | 44 | 34 | 44 | 38 | 47 | 49 | 45 |
| 3 | | 33 | 36 | 40 | 38.5 | 30 | 37 | 40 | 49 | 44 | 34 | 44 | 38 | 47 | 49 |
| 4 | | | | | | | | | | | | | | | |
| 5 | | | | | Correlation = | | | 0.382 | | | | | | | |
| 6 | | | | | | | | | | | | | | | |

| | P | Q | R | S | T | U | V | W | X | Y | Z | AA | AB | AC | AD |
|---|---|---|---|---|---|---|---|---|---|---|---|---|---|---|---|
| 1 | | | | | | | | | | | | | | | |
| 2 | 37.5 | 30 | 32 | 35 | 34 | 51 | 48 | 41.5 | 40 | 36 | 31 | 36 | 41 | 34 | |
| 3 | 45 | 38 | 30 | 32 | 35 | 34 | 51 | 48 | 42 | 39.5 | 36 | 31 | 36 | 41 | 34 |
| 4 | | | | | | | | | | | | | | | |
| 5 | | | | | | | | | | | | | | | |
| 6 | | | | | | | | | | | | | | | |

**Figure 2.4 Finding lag-1 autocorrelation of inter-throughput times using Excel.**

learns that the theorem basically states that the distribution of the summation of a *large number* of *independent* and *identically distributed* variables will be approximately Normal, regardless of the underlying distribution. Joseph becomes more optimistic that he is getting somewhere. First, his inter-throughput times are random variables. Second, Eq. (2.1) is about summation of a large number of variables (i.e., 25 variables), and third, all of his 25 inter-throughput times have the same distribution, since they are generated from the same process—the plant's production line. However, he is not sure if the inter-throughput times are independent of each other.

Before doing any more math, Joseph thinks why inter-throughput time $X_2$ may or may not depend on $X_1$. Specifically, if it takes longer time to have the next throughput exit the production line (i.e., if $X_1$ takes longer than average), would the throughput after the next one be also expected to take a longer time (i.e., would $X_2$ also be longer)? Joseph believes that this is indeed the case. He remembers last month's equipment failure in the assembly department when one of the assembly robots failed due to mechanical problems. This happened on September 19. That was the reason why inter-throughput time for Boat no. B14 took longer than usual, that is, took around 47 hours. The problem with slow production continued for 2 more days and also prolonged the inter-throughput time of Boat no. B15 to around 49 hours. Thus, Joseph comes to the conclusion that there is a relationship between the inter-throughput of two consecutive throughputs of the production line. But he is wondering how he can check his intuition.

One way to check whether there is a dependence between two consecutive inter-throughput times, Joseph realizes, is to find the correlation between them. By remotely logging into his computer at home, Joseph starts searching through the files related to statistics course he took in college. After a while, he finds exactly what he was looking for: "lag-1 autocorrelation," also called "first-order correlation." Going through his notes, he finds out that lag-1 autocorrelation is nothing but the correlation between all pairs of consecutive observations of a sample. So all he needs to do is to find lag-1 autocorrelation for his sample of 29 inter-throughput times.

Joseph is good with Excel spreadsheet. So, he types the 29 consecutive inter-throughput times of Table 2.2 in one row (i.e., Row 2) of the spreadsheet starting from Column A to Column AC (see Fig. 2.4). He then copies the 29 numbers in Row 2 and pastes them in Row 3, but with one lag, starting from Column B. Now Joseph has two series of identical numbers in the second and third rows of the spreadsheet with one lag from their starting numbers.

Excluding 33 from the beginning of the first series and 34 from the end of the second series, all Joseph needs to do is to find the correlation between the two series, which in fact becomes the lag-1 autocorrelation of the inter-throughput times. In cell "H5" he types "=CORREL(B2:AC2,B3:AC3)," which gives him the correlation of 0.382.

Joseph finds that his intuition was not that off. Lag-1 autocorrelation of 0.382, while not a strong correlation, still implies that there is some dependence between two consecutive inter-throughput times. Also, as Joseph thought, the correlation is positive, which means that in general, when the inter-throughput of a boat exiting the production line is larger (than average), the inter-throughput of the next boat departing the production line will probably be also larger (i.e., above average). Joseph decides to take the dependence among consecutive inter-throughput times into account when computing the due date for the club. This means that Joseph needs to find a version of the Central Limit Theorem or something similar that identifies the distribution of sum of a large number of random variables when the variables are not independent. After spending some time going through his notes and checking Internet, Joseph does not find anything useful. He sends an email to his statistics professor in his old college and asks him about his due date quoting problem. In less than half an hour, Joseph gets a reply email from his professor with an attachment that makes Joseph very happy. It basically solves Joseph's problem. The attachment goes something like the following:

## ANALYTICS

### Due Date Quoting

Consider a process with *stationary* throughput. Suppose we observe $n$ consecutive inter-throughput times of the process, and we compute the mean, the variance, and lag-1 autocorrelation of the inter-throughput times as follows:

$$\text{Mean of inter-throughput times} = \overline{X}$$
$$\text{Variance of inter-throughput times} = S^2$$
$$\text{Lag-1 autocorrelation} = \rho_1$$

Assuming that the process has just had a throughput, the time until the process has $b$ more throughputs is estimated to have a Normal distribution with mean ($\mu_b$) and variance ($\sigma_b^2$), where

$$\mu_b = b\overline{X} \qquad \sigma_b^2 = \left(\frac{1 + \rho_1}{1 - \rho_1}\right)bS^2$$

Therefore, with probability $\alpha$ the process will generate $b$ throughputs before due date $T_{\text{due}}$, where

$$T_{\text{due}} = \mu_b + z_\alpha \sigma_b$$

where $z_\alpha$ is the value of $z$ in Standard Normal table where $\Phi(z) = \alpha$, or using Excel:

$$T_{\text{due}} = \text{NORMINV}(\alpha, \mu_b, \sigma_b)$$

So, to use the above results, Joseph first needs to make sure that the throughput of his process—the throughput of the plan—is stationary. But is the throughput of his plant a stationary throughput?

Going back to the document that his professor sent, he finds a simple explanation about stationary throughput. Joseph learns that when the throughput of a process is stationary, the throughput rate of the process does not have an upward or downward trend. In the context of his plant, this implies that, while the plant may have some slow days and some fast days, the weekly or monthly throughput rates should not be increasing or decreasing in time. Furthermore, the throughput of the plant should not exhibit a seasonal trend. For example, the throughput of his plant should not be higher or lower in particular days of a week or particular shifts of a day. Joseph is sure that there is no significant difference in the speed and efficiency of the plant in the three shifts of a day, or in the 7 days of the week. So, he is confident that the throughput of his plant is a stationary throughput.

Having lag-1 autocorrelation of $\rho_1 = 0.382$ for his sample of $n = 29$ consecutive inter-throughput times, Joseph finds the mean and the variance of the inter-throughput times in Table 2.2 using the spreadsheet model in Fig. 2.4 as follows:

$$\overline{X} = \text{AVERAGE(A2:AC2)} = 38.9 \qquad S^2 = \text{VAR(A2:AC2)} = 37.614$$

To find due date to produce 25 boats, Joseph first needs to compute the mean ($\mu_b$) and variance ($\sigma_b^2$) of the Normal distribution corresponding to the sum of $b = 25$ future inter-throughput times as follows:

$$\mu_b = b\overline{X} = 25(38.9) = 972.5 \text{ hours}$$
$$\sigma_b^2 = \left(\frac{1 + \rho_1}{1 - \rho_1}\right)bS^2 = \left(\frac{1 + 0.382}{1 - 0.382}\right)25(37.614) = 2102.85$$

Thus, standard deviation of the Normal distribution is $\sigma_b = \sqrt{\sigma_b^2} = \sqrt{2102.85} = 45.86$ hours. Joseph can now obtain due date $T_{\text{due}}$ that ensures with probability $\alpha = 90\%$ that $b = 25$ boats are finished before the date. He simply uses the formula that his professor sent him and finds due date $T_{\text{due}}$ to be

$$T_{\text{due}} = \mu_b + z_\alpha \sigma_b = 972.5 + (1.28)45.86 = 1031.2$$

since $z_\alpha = 1.28$, because $\Phi(z = 1.28) = 0.9$ in Standard Normal table (see the Appendix). Considering that the plant works three shifts a day (i.e., 24 hours a day), the $T_{\text{due}} = 1031.2/24 = 42.97$ days—almost 43 days.

Compared to his first due date estimate of 40.5 days, Joseph is more confident about the due date of 43 days. First, his new estimate takes into account the variability of and dependence among inter-throughput times. Second, the additional time $2.5(= 43 - 40.5)$ above the average 40.5 days, which is called "safety time," ensures that LuxBoat finishes club's order before the quoted due date with probability of 90 percent.

---

**CONCEPT**

**Safety Time**

*Safety Time* is the additional time above the average time to complete the order of a customer to guarantee achieving the desired service level.

---

Joseph picks up his phone and calls Bob and gives him the new shipping date of 43 days. "While this new shipping date is 2.5 days larger than my first estimate," Joseph proudly adds, "the chance of us shipping the club's order before the due date is 90 percent." Bob answers, "Great! I will call the club manager right now and tell him about our estimate of delivering his order. But, let's have lunch next week and talk about this. I really would like to know how you got the due date. May be we can talk to our IT people, and see if they can add your due date quoting procedure into our system. We can certainly use this in marketing when customers ask for a firm delivery date." Joseph replies, "Sure, this is easy to do. If we set a service level, say 90 percent, we can add the procedure that I developed to our IT system and generate due dates for future customers, to guarantee on-time delivery at least 90 percent of times. We can even make the system flexible to compute due date for any service level larger or smaller than 90 percent."

## 2.3 Inventory

Inventory is another process-flow measure that has a direct and significant impact on profitability of a process. Both purchasing inventory and storing it require money. Obviously, lower inventory results in lower operating costs. Furthermore, lower inventory can also result in higher quality products and faster response times. We will discuss this in more detail in Chaps. 9 and 15. Hence, one of the important tasks of operations engineers and managers is to find ways to reduce inventory of a process.

In this section, we first start by defining inventory of a process, and we use an example to further illustrate how inventory of a process can be measured.

---

**CONCEPT**

**Inventory of a Process**

Inventory of a process at time $t$ is the total number of flow units within process boundaries at time $t$. The inventory includes both the flow units in the buffer as well as the flow units being processed by the resources. We let

$$I(t) = \text{Inventory of a process at time } t$$

---

### Projects at Solar for Life

Solar for Life (SFL) is a company that installs solar panels in residential and commercial structures. The company has one team that installs solar panels. When the team finishes working on a project, it starts working on the next project in the FCFS order. Specifically, it starts the project that has the contract signing date before others. Recently, the company has received several complaints from its clients about the long time it took SFL to finish their projects. To improve its performance, SFL is thinking of hiring a new team of workers and engineers. However, before making any hiring decision, they would like to get a sense of the total number of unfinished projects that the company is dealing with each month. This includes the project that has started but not yet finished as well as the projects that are contracted but have not yet started.

The company has some data about 14 projects that were completed in a 29-day period in last June. The period starts from June 1 when SFL finished all of its projects with no projects to work on, and ends on June 29 when SFL finished all 14 projects it received during that period. The data is presented in Table 2.3. The contract for a project is often negotiated throughout a day and is signed at the end of the day. Thus, the contract signing date in Table 2.3 refers to the end of the day. Project completion dates also correspond to the end of the day. What was the number of projects that the company had been dealing with in June?

Considering a project as the flow unit and SFL operations as the process, a flow unit enters the process when the contract for the project is signed. Also, a flow unit departs the

**Table 2.3    Projects' Contract and Completion Dates**

|  | Date | |
| --- | --- | --- |
| Project no. | Contract | Completion |
| P001 | June 1 | June 6 |
| P002 | June 3 | June 8 |
| P003 | June 4 | June 9 |
| P004 | June 5 | June 11 |
| P005 | June 7 | June 13 |
| P006 | June 7 | June 17 |
| P007 | June 11 | June 18 |
| P008 | June 12 | June 20 |
| P009 | June 14 | June 21 |
| P010 | June 15 | June 22 |
| P011 | June 15 | June 23 |
| P012 | June 17 | June 25 |
| P013 | June 26 | June 28 |
| P014 | June 27 | June 29 |

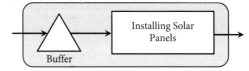

**Figure 2.5    Process-flow chart of Solar for Life.**

process when the flow unit (the project) is completed by SFL, see Fig. 2.5. Thus, managers of SFL would like to know about the total inventory of projects in June.

To compute the inventory of a process, we first need to understand the relationship between inventory of a process and its arrivals and departures.

ANALYTICS

**Computing Inventory of a Process**

Consider a process that is empty at time 0. Also, consider *Cumulative Arrival* of the process at time $t$ as the total number of arrivals to the process up to time $t$ and *Cumulative Departure* of the process at time $t$ as the total number of departures from the process up to time $t$, where

$$AR(t) = \text{Cumulative arrival until time } t$$
$$DP(t) = \text{Cumulative departure until time } t$$

Then, $I(t)$, the inventory of a process at time $t$, can be obtained as follows:

$$I(t) = AR(t) - DP(t)$$

For example, based on Table 2.3, the total number of contracts signed until the end of June 16—the cumulative arrival until the end of June 16—is $AR(\text{June } 16) = 11$ projects. On

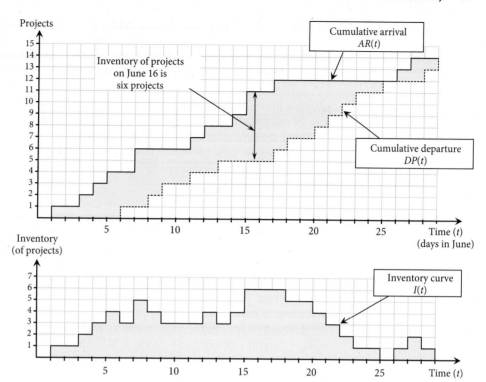

Figure 2.6    *Top:* Cumulative arrival and departure. *Bottom:* Inventory curve from June 1 to June 29 at SFL.

the other hand, the cumulative departure until June 16 is $DP$(June 16) $= 5$ projects. Hence, the inventory of the projects at SFL on the day of June 16 is

$$I(\text{June 16}) = AR(\text{June 16}) - DP(\text{June 16})$$
$$= 11 - 5 = 6 \text{ projects}$$

This is also shown in the top graph in Fig. 2.6. Specifically, the top graph in Fig. 2.6 depicts the cumulative arrival and cumulative departure from June 1 to June 29 at SFL. At any time $t$, the vertical distance between the cumulative arrival and cumulative departure in the top graph is the inventory of the process at time $t$. As the graph shows, on June 16, the vertical distance between the cumulative arrival and cumulative departure is 6, which is the number of projects at SFL at that day.

One can draw the inventory of a process during a period in a separate graph as shown at the bottom of Fig. 2.6. This is called *Inventory Curve* or *Inventory Buildup Diagram*. The graph shows the inventory of projects from June 1 to June 29. As the inventory curve shows, SFL has a maximum of six projects on June 16, 17, and 18; and has no projects on June 1 and June 26.

Inventory curves provide detail information about the inventory level of a process during a period. For example, as the inventory curve shows, on June 28 and from June 4 to June 23—a total of 21 days—the number of projects at SFL were more than one project. Considering that SFL has only one team that works on projects, this implies that for a period of 21 days, there was at least one project waiting to be started. On June 16, 17, and 18 the number of projects at SFL that has not been started is five. On the other hand, the inventory curve also shows that there are 8 days in June when the number of projects at SFL was either one or zero. This implies that if SFL had hired a second team, the team would have been idle for 8 days in June. To decide whether to hire a new team or not, SFL should consider both the information about the number of days that the inventory is large (i.e., several projects are waiting) and the days that the inventory is small (i.e., SFL teams would be idle). This is exactly the type of information inventory curve provides for SFL.

A single metric that provides an overall view of the inventory of a process during a period is the average inventory. The average inventory of the projects at SFL from June 1 to June 29 can be computed using inventory curve as follows. Note from inventory curve that on June 1 the inventory at SFL was zero. The inventory was 1 on June 2 and 3, and so on. The inventory on June 28 and 29 were 2 and 1, respectively. Thus, the average number of projects—the average inventory—in the 29-day period in June can be computed as

$$\text{Average inventory} = \frac{0 + 1 + 1 + 2 + 3 + \cdots + 1 + 0 + 1 + 2 + 1}{29 \text{ days}}$$

$$= \frac{86}{29} = 2.96 \text{ projects} \tag{2.2}$$

So, on average, SFL was dealing with about three projects (ongoing or waiting to be started) in the first 29 days of June.

Inventory curve and average inventory, respectively, provide *detailed* and *overall* information about the number of projects at SFL. The inventory information along with other information such as the cost of having a second team and the penalty for missing due dates can help managers of SFL to make a more informed decision about whether or not to hire a new team.

As Eq. (2.2) shows, we obtained the average inventory during the 29-day period by adding the inventories of projects in all those 29 days and then dividing it by 29. We call the sum of all inventories in the 29-day period "Total Inventory" in that period.

---

**ANALYTICS**

**Total and Average Inventory of a Process**

- *Total Inventory* of a process from time 0 to time $t$ is the summation of inventories in that time interval. We use $I^{\text{tot}}(0, t)$ to denote total inventory from time 0 to time $t$. Note that total inventory can be obtained by using cumulative arrival and cumulative departure or by using inventory curve as follows:

$$I^{\text{tot}}(0, t) = \text{Area between } AR(t) \text{ and } DP(t) \text{ from time 0 to } t$$

$$= \int_0^t [AR(s) - DP(s)] \, ds$$

$$= \text{Area under inventory curve } I(t) \text{ from time 0 to } t = \int_0^t I(s) ds$$

- *Average Inventory* of a process from time 0 to time $t$, which we denote by $I^{\text{avg}}(0, t)$, can be obtained as follows:

$$I^{\text{avg}}(0, t) = \frac{I^{\text{tot}}(0, t)}{t}$$

$$= \frac{\text{Area under inventory curve } I(t) \text{ from time 0 to } t}{t}$$

When the average inventory is computed for a large time period (i.e., when $t$ is very large), the average inventory during the period is called *Long-Run Average Inventory,* and we use $I$ to denote it:

$$I = \text{Long-run average inventory}$$

Recall that the total and average inventory of projects in SFL from June 1 to June 29 are

$$I^{\text{tot}}(\text{June 1, June 29}) = 86 \qquad I^{\text{avg}}(\text{June 1, June 29}) = 2.96$$

Since business processes work every day for months and years, the long-run average inventory is a good performance measure that corresponds to the average inventory of those processes. For the case of SFL, for example, the average number of projects in 1 month—June—may not be a good proxy for the overall average number of projects. The average number of projects throughout a year, as we discussed, would be a better representation of the average number of projects at SFL.

## 2.4 Continuous Approximation of Inventory Curves

As Fig. 2.6 shows, cumulative arrival and cumulative departure as well as inventory curve of SFL are stepwise functions. This is because, as in most cases in practice, the flow units are discrete objects (e.g., parts in a work station, patients in a hospital, projects at SFL). The variability in arrival and departure makes these stepwise functions more complex. It is often easier to approximate these stepwise functions with some continuous linear functions to make the computation easier. One approach known as *Continuous Approximation* or *Fluid Approximation* estimates the cumulative arrival and departure functions with simple linear functions using only arrival (i.e., input) and departure (i.e., output) *rates.*

**ANALYTICS**

**Continuous (Linear) Approximation of Inventory Curve**

Define

$$R_{\text{in}}(t_i, t_j) = \text{Constant } Input\ Rate \text{ to a process during time interval } t_i \text{ to } t_j$$
$$R_{\text{out}}(t_i, t_j) = \text{Constant } Output\ Rate \text{ from a process during time interval } t_i \text{ to } t_j$$

Then, we have

$$R_{\text{inv}}(t_i, t_j) = \text{Inventory Buildup Rate during time interval } t_i \text{ to } t_j$$
$$= R_{\text{in}}(t_i, t_j) - R_{\text{out}}(t_i, t_j)$$

Considering $I(t_j)$ as the inventory at time $t_j$, then inventory at time $t_j$ will be

$$I(t_j) = I(t_i) + (t_j - t_i)R_{\text{inv}}(t_i, t_j) \tag{2.3}$$

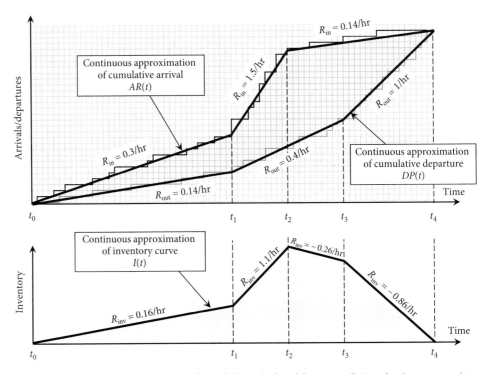

Figure 2.7    *Top:* Continuous approximation of cumulative arrivals and departures. *Bottom:* Continuous approximation of inventory curve.

Fig. 2.7 shows an example in which the stepwise cumulative arrivals and cumulative departure curves are approximated in four intervals $(t_0 = 0, t_1)$, $(t_1, t_2)$, $(t_2, t_3)$, and $(t_3, t_4)$. Consider time period $(t_1, t_2)$ as an example. During that time period, the input rate is $R_{in}(t_1, t_2) = 1.5$ per hour. During the same interval, the output rate is $R_{out}(t_1, t_2) = 0.4$ per hour. Thus, between time $t_1$ and $t_2$ the inventory in the process will build up at the rate

$$R_{inv}(t_1, t_2) = R_{in}(t_1, t_2) - R_{out}(t_1, t_2) = 1.5 - 0.4 = 1.1 \text{ per hour}$$

As shown in the inventory curve in Fig. 2.7, the inventory *increases* at the rate of 1.1 per hour during time interval $(t_1, t_2)$.

Now consider time interval $(t_2, t_3)$ during which the input rate is $R_{in}(t_2, t_3) = 0.14$ per hour and the output rate is $R_{out}(t_2, t_3) = 0.4$ per hour. Therefore,

$$R_{inv}(t_2, t_3) = R_{in}(t_2, t_3) - R_{out}(t_2, t_3) = 0.14 - 0.4 = -0.26 \text{ per hour}$$

Since the inventory buildup rate is negative (there are more outputs than inputs), as shown in Fig. 2.7, the inventory *decreases* at the rate of 0.26 per hour during time interval $(t_2, t_3)$.

The continuous approximation of inventory curves has large applications in design and control of inventory systems, as will be discussed in Chaps. 9, 10, and 11. The following example presents such applications.

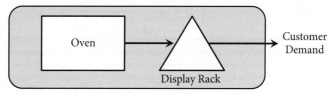

**Figure 2.8    Process-flow chart of Absolute Calzone in Example 2.2.**

### Example 2.2

Absolute Calzone is a bakery in a shopping mall that is well-known for its fresh calzones. The store uses its single oven to make calzones and sells them during the lunch hours of 11:00 am to 2:00 pm. While the number of customers arriving at the store is variable, the average arrival rates have been estimated as follows. It is found that customers come to the store to buy calzones at the rate of one customer per minute from 11:00 am to 12:00 pm; at the rate of two customers per minute between 12:00 pm and 1:00 pm; and about 0.5 customers per minute (i.e., one customer every 2 minutes) from 1:00 pm to 2:00 pm. There is no demand for calzones after 2:00 pm. The store uses its oven to make different products. The oven is used to make calzones starting at 11:00 am at the rate of two calzones per minute until it had produced enough calzones to serve lunch hour demand (i.e., a produce-to-stock process). Produced calzones are put on a display rack until they are sold to customers.

- To free up the oven for other bakery products, how long the oven should be used to make calzones to satisfy all lunch-hour demands with no leftover calzones at 2:00 pm?
- What are the average and maximum number of calzones that are stored on the rack?

To answer the above questions, we need to develop inventory curve for the calzones stored in the display rack. Fig. 2.8 shows the process-flow chart for the calzones. As the figure shows, calzones (i.e., the flow units) are produced by the oven (i.e., the resource) and are stored in the rack (i.e., the buffer).

The total demand for calzones during the lunch hours of 11:00 am to 2:00 pm can be obtained as follows. Customers arrive at the rate of one per minute from 11:00 am to 12:00 pm. Thus, the total demand during that time period is $1 \times 60 = 60$ calzones (assume each customer buys one calzone). The arrival rate from 12:00 pm to 1:00 pm is two customers per minute, thus the total demand during that hour is $2 \times 60 = 120$ calzones. Finally, customers arrive at rate 0.5 customers per minute from 1:00 pm to 2:00 pm, so the demand during the last hour is $0.5 \times 60 = 30$ calzones. Therefore, the total demand during the 3 hours of lunch is $60 + 120 + 30 = 210$ calzones.

Since the store would like to produce all the calzones it needs to satisfy the demand during lunch hours (with no leftover calzones), and since the oven has the capacity of making two calzones per minute, then the total number of minutes that the oven must be used to produce calzones would be $210/2 = 105$ minutes, that is, 1 hour and 45 minutes. Therefore, the store should start making calzones at 11:00 am until 12:45 pm. The store can then use the oven to produce other products.

To answer the second question regarding the average and maximum number of calzones on display rack, note that the arrivals (inputs) to the display rack come from the oven that

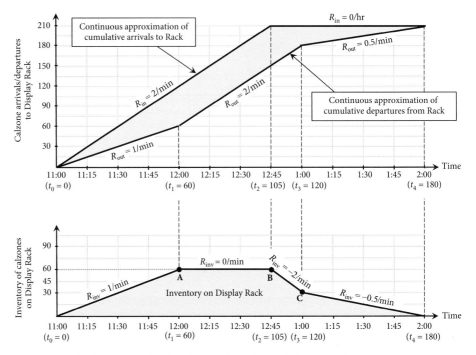

**Figure 2.9** *Top:* **Continuous approximation of cumulative arrivals and departures.** *Bottom:* **Continuous approximation of inventory curve for calzones in Example 2.2.**

produces at the rate of two calzones per minute. Hence, the input rates to the display rack are

$$R_{in}(\text{11:00 am, 12:45 pm}) = 2 \text{ per minute}$$
$$R_{in}(\text{12:45 pm, 2:00 pm}) = 0 \text{ per minute}$$

On the other hand, the departures (outputs) from the display rack are indeed the arrival of customers who purchase calzones. Hence,

$$R_{out}(\text{11:00 am, 12:00 pm}) = 1 \text{ per minute}$$
$$R_{out}(\text{12:00 pm, 1:00 pm}) = 2 \text{ per minute}$$
$$R_{out}(\text{1:00 pm, 2:00 pm}) = 0.5 \text{ per minute}$$

Consider time period (12:45 pm to 1:00 pm) as an example. During that period the input rate is 0 calzones per minute (since the oven stopped producing calzones) and the output rate (i.e., demand rate) is two calzones per minute. Therefore, the inventory buildup rate is $R_{inv}(\text{12:45 pm, 1:00 pm}) = 0 - 2 = -2$ calzones per minute. This, and other inventory buildup rates, are shown in Fig. 2.9.

Having the inventory curve in Fig. 2.9, we can find estimates for the number of calzones on display rack at any time between 11:00 am and 2:00 pm. Using Eq. (2.3), we find the number of calzones on display rack at 12:00 pm, that is, point A in the figure. Note that setting time 11:00 am as $t_0 = 0$, then 12:00 pm is 60 minutes later corresponding to time $t_1 = 60$ (minutes). Also, the inventory at 11:00 am is $I(t_0 = 0) = 0$. Hence,

$$I(t_1 = 60) = I(t_0) + (t_1 - t_0)R_{inv}(t_0, t_1)$$
$$= 0 + (60 - 0)(1) = 60 \qquad \text{Point A on Inventory Curve}$$

Number of calzones on the rack at time 12:45 pm (i.e., $t_2 = 105$) is

$$I(t_2 = 105) = I(t_1) + (t_2 - t_1)R_{inv}(t_1, t_2)$$
$$= 60 + (105 - 60)(0) = 60 \qquad \text{Point B on Inventory Curve}$$

The inventory at 1:00 pm (i.e., time $t_3 = 120$) can be found as follows:

$$I(t_3 = 120) = I(t_2) + (t_3 - t_2)R_{inv}(t_2, t_3)$$
$$= 60 + (120 - 105)(-2) = 30 \qquad \text{Point C on Inventory Curve}$$

Thus, the store will have a maximum of 60 calzones on its display rack from 12:00 pm to 12:45 pm. We can also compute the average number of calzones on the rack during lunch hours from 11:00 am (time $t_0 = 0$) to 2:00 pm (time $t_4 = 180$) as follows:

$$I^{avg}(t_0 = 0, t_4 = 180) = \frac{\text{Area under inventory curve } I(t) \text{ from time } t_0 = 0 \text{ to } t_4 = 180}{180}$$
$$= \frac{6075}{180} = 33.75 \text{ calzones}$$

Continuous approximation of inventory curves is particularly useful in the analysis of processes that are non-stationary and go through the same cycle in each period. Absolute Calzone in Example 2.2 is a good representation of such cases. The store goes through the same cycle of producing and selling calzones from 11:00 am to 2:00 pm each day. The inventory at the beginning and at the end of each cycle is often zero.

## 2.5 Flow Time

Another important process-flow measure that has a significant impact on firm's profit is Flow Time.

---

**CONCEPT**

**Flow Time**

*Flow Time* of a flow unit in a process (also called *Cycle Time* in manufacturing systems) is the time that the flow unit spends within process boundaries. This includes the total time that the flow unit waits in the buffer (or buffers) and the time that the flow unit is being processed by resource (or resources) of the process.

---

Flow time is one of the process-flow measures that customers directly experience, especially in service systems where customers are sensitive to waiting. For example, suppose you arrive at a hotel at 8:30 pm after a long flight and you go directly to check-in desk. You observe a large number of people waiting in line for checking in. You wait in the line for 15 minutes until you get to the desk. The hotel staff starts your check-in. Your check-in takes 3 minutes. You get your key and you leave the check-in desk to go to your room. From process-flow analysis perspective, you (the flow unit) entered the process (the hotel's check-in) at 8:30 pm and left the process at 8:48 pm. Thus, your flow time at the check-in is 18 minutes. This includes your 15 minutes of waiting in the line for check-in (i.e., in buffer) and the 3 minutes of your check-in time.

How does the flow time affect firms' profits? Shorter flow times can increase firms' profits in different ways. For example, wouldn't you be happier if your flow time in check-in was shorter? You bet. Would the hotel managers be happier if your flow time was shorter? Absolutely. The more pleasant experience you have with your stay at a hotel, the more chance that you choose the hotel in the future, and the higher chance of recommending the hotel to your friends. From the hotel manager's perspective, this means not losing your future business and possibly getting more demand (e.g., your friends) in the future. More demand means more revenue and thus more profit for the hotel.

The benefit of shorter flow times is not limited to the service industry. Suppose you are looking to buy a sofa for your living room. You have checked several stores and you narrowed down your choices to two sofas: a modern style sofa and a traditional sofa. You are uncertain between the two. Both will look good in your living room and have almost the same price. You go back to the store that sells both sofas to buy one of them. The salesperson tells you that the manufacturer who makes the modern style sofa will deliver the sofa 8 weeks after you place your order. Note that the 8 weeks start from the time your order is sent to the manufacturer (i.e., your flow unit enters the process) until the sofa is manufactured and sent to you (i.e., the flow unit departs the process). So, 8 weeks is, in fact, the flow time of your modern style sofa in the manufacturer's process. The salesperson also tells you that this time—the flow time—for the traditional sofa is around 5 weeks. Which sofa would you order?

When other characteristics of products (e.g., quality, price) are the same, most people prefer the product with the shorter flow times. Thus, shorter flow times often increase manufacturers' demands and thus result in more profits. Shorter flow times also reduce cost as shorter flow times often result in lower inventory. We discuss this in more detail in Sec. 2.6.

### Flow Times of Projects at SFL

Recall SFL, the company that installs solar panels for residential and commercial structures. The company was receiving complaints from its clients about the long time it took from when they signed a contract with SFL until their projects were finished. Recall that SFL knows that in June the average number of projects that were signed but not finished was around three projects—2.96 projects to be exact. While this number is informative about *how many* projects are waiting to be finished, it does not tell SFL *how long* those projects waited until they were finished. Thus, managers of SFL would like to know how long it takes from when they sign a contract for a project until the project is completed.

From process-flow analysis perspective, SFL managers would like to know the flow time of projects in their system. We use the data in Table 2.3 for the 29-day period in June to get the flow times of projects during that period. For example, the contract for Project P002 was signed at the end of June 3 and was completed later at the end of June 8. Thus, the flow time of Project P002 is 5 days. The flow times of all 14 projects that SFL completed in June are shown in Table 2.4.

As the table shows, time from signing a contract for a project until the project is completed—the flow time of the project—ranges from 2 days up to 10 days. Except for two cases (i.e., P013 and P014), all of the flow times are 5 days or more.

To get a general sense of the flow times for the 14 projects in June, let's compute the average flow time of the 14 projects. We can do this by summing all 14 flow times and dividing it by 14. Note that the sum of the flow times of all 14 projects is given as 86 in the last row of Table 2.4. Therefore,

$$\text{Average flow time of 14 projects} = \frac{86}{14} = 6.14 \text{ days}$$

**Table 2.4    Flow Times (in Days) of SFL's Projects in June**

| Project no. | Date Contract | Date Completion | Flow time (days) |
|---|---|---|---|
| P001 | June 1 | June 6 | 5 |
| P002 | June 3 | June 8 | 5 |
| P003 | June 4 | June 9 | 5 |
| P004 | June 5 | June 11 | 6 |
| P005 | June 7 | June 13 | 6 |
| P006 | June 7 | June 17 | 10 |
| P007 | June 11 | June 18 | 7 |
| P008 | June 12 | June 20 | 8 |
| P009 | June 14 | June 21 | 7 |
| P010 | June 15 | June 22 | 7 |
| P011 | June 15 | June 23 | 8 |
| P012 | June 17 | June 25 | 8 |
| P013 | June 26 | June 28 | 2 |
| P014 | June 27 | June 29 | 2 |
| | | Total: | 86 |

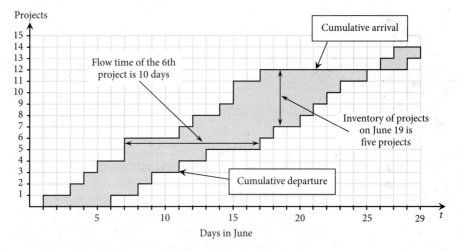

**Figure 2.10    Flow time, inventory, and cumulative arrival and departure of projects.**

In other words, the average time from when a contract is signed until the project is completed by SFL in the month of June was 6.14 days.

You might have noticed by now that the sum of the flow times of all 14 projects in the 29-day period in June, that is, 86, is exactly the same as the sum of all inventory of projects (i.e., total inventory) during the same period. In other words, sum of all flow times in the 29-day period is also equal to the area under inventory curve during that period. A coincidence? or this is always the case?

Recall that the area under the inventory curve is equal to the area between the cumulative arrival and cumulative departure. Let's have a closer look at this area in Fig. 2.10. As the figure shows, while the *vertical distance* between cumulative arrival and cumulative departure at time $t$ is the inventory of the process at that time, the *horizontal distance* between the cumulative arrival and cumulative departure is the flow time of a flow unit. So, if we add all inventories (i.e., all vertical distances) or if we add all flow times (i.e., all horizontal distances) we get to the same number (i.e., 86), which is the area between cumulative arrival and departure curves. This is also the area under inventory curve.

**Average Flow Times**

Consider a process that is empty at time 0. Flow units arrive and are processed by the resources of the process until time $t$, when the process becomes empty after serving $N$ arriving flow units. Let $T^{\text{tot}}(0, t)$ be the total flow times of all $N$ flow units. Then

$$T^{\text{tot}}(0, t) = I^{\text{tot}}(0, t)$$

The *Average Flow Time* of the process during the time period of 0 to $t$, which we denote by $T^{\text{avg}}(0, t)$, can be obtained as follows:

$$T^{\text{avg}}(0, t) = \frac{T^{\text{tot}}(0, t)}{N}$$
$$= \frac{\text{Area under in-process inventory curve } I(t) \text{ from time 0 to } t}{N}$$

When the average flow time is computed over a long time period, the average flow time is called *long-run average flow time,* and we use $T$ to denote it:

$$T = \text{long-run average flow time}$$

The case of SFL shows the importance of inventory curves which not only contain information about the *inventory* of a process, but also contain information about the *flow times* in the process. We can find the "total" and "average" inventory as well as the "total" and "average" flow times using the area under the inventory curve. This is because of the simple fact that the flow units in inventory are also *waiting*. In other words, at the same time that a flow unit is counted as inventory, it is also accumulating waiting time.

Although we showed the above property for a case such as SFL in which flow units are processed in the order they arrive, the result goes beyond this FCFS discipline. For any given discipline, regardless of how the next flow unit in the buffer is chosen for processing (e.g., last-come-first-served or chosen randomly), the area under the inventory curve is equal to the total inventory as well as total flow times of all flow units.

 **Example 2.3**

Consider Absolute Calzone bakery in Example 2.2. The manager of the store is worried that if the calzones stay on the rack for a long time, they get cold and may not taste as good as freshly baked calzones. How long, on average, a calzone stays on the rack before it is sold to a customer?

We use the inventory curve for calzones in Fig. 2.9 to compute the average flow time of calzones. Note that during lunch hours of 11:00 am ($t_0 = 0$) to 2:00 pm ($t_4 = 180$) a total of 210 calzones are made and sold. Therefore, the average time a calzone stays on the rack before it is sold is

$$T^{\text{avg}}(t_0 = 0, t_4 = 180) = \frac{\text{Area under inventory curve } I(t) \text{ from time } t_0 = 0 \text{ to } t_4 = 180}{210}$$

$$= \frac{6075}{210} = 28.93 \text{ minutes}$$

Thus, on average, a calzone stays on the display rack for about 29 minutes before it is sold to a customer.

## 2.6 Little's Law: Relating the Averages

So far we have introduced three process-flow measures—throughput, inventory, and flow time—and discussed how improving these measures can improve financial performance (e.g., profit) of a firm. But, is it possible, for example, to improve throughput and, at the same time, decrease inventory? or decrease flow times? To answer this question, first we need to understand the relationship between these three performance measures.

Consider SFL as an example. For SFL we found the average inventory in the 29-day period in June to be $I^{\text{avg}}(1, 29) = 2.96$ projects. We also computed the average flow time of the 14 projects completed in those 29 days to be $T^{\text{avg}}(1, 29) = 6.14$ days. The throughput of SFL during the 29-day period can be obtained as $\text{TH}^{\text{avg}}(1, 29) = 14/29 = 0.482$ projects per day. Is there any relationship among 2.96 projects (average inventory), 6.14 days (average flow time), and 0.48 projects per day (average throughput)? Well, it seems so! $2.96 = 6.14 \times 0.482$; in other words,

$$I^{\text{avg}}(1, 29) = T^{\text{avg}}(1, 29) \times \text{TH}^{\text{avg}}(1, 29)$$

which implies that in a period of 29 days in which 14 projects were completed, the average inventory of SFL was equal to the average flow time at SFL multiplied by the throughput of SFL in those 29 days. A coincidence? or a fact? It is indeed a fact. In 1961, John D.C. Little wrote a paper and provided a mathematical proof for this relationship. After that, this relationship is known as "Little's Law."

**ANALYTICS**

**Little's Law**

Consider a process that is empty at time 0. Flow units arrive and are processed by the resources of the process until time $t$, when the process becomes empty after serving $N$ arriving flow units. The average inventory during time interval 0 to $t$ is

$$I^{\text{avg}}(0, t) = \frac{I^{\text{tot}}(0, t)}{t} = \frac{I^{\text{tot}}(0, t)}{t} \times \frac{N}{N} = \frac{I^{\text{tot}}(0, t)}{N} \times \frac{N}{t}$$

Since $I^{\text{tot}}(0, t) = T^{\text{tot}}(0, t)$, we have

$$I^{\text{avg}}(0, t) = \frac{T^{\text{tot}}(0, t)}{N} \times \frac{N}{t}$$

and because fraction $N/t$ is in fact $R_{\text{out}}^{\text{avg}}(0, t)$, the average output rate of the process during time period 0 to $t$, we have

$$I^{\text{avg}}(0, t) = T^{\text{avg}}(0, t) \times R_{\text{out}}^{\text{avg}}(0, t)$$

which is Little's law.

Recall that throughput rate, or simply throughput, is the average number of good (non-defective) flow units that depart the system per unit time. Thus, when the process does not produce defective flow units, the throughput rate of the process is equal to its output rate. Little's law for those processes becomes

$$I^{\mathrm{avg}}(0, t) = T^{\mathrm{avg}}(0, t) \times \mathrm{TH}^{\mathrm{avg}}(0, t)$$

which links together the average inventory, average flow time, and average throughput of a process during time interval of 0 to $t$.

In the remainder of this chapter and this book, unless it is specifically mentioned, we assume that the throughput of the process is equal to its output rate—all flow units produced by the process is non-defective. We, therefore, use Little's law for throughput, inventory, and flow times.

Note that Little's law is simple but very general. It holds for all processes regardless of the probability distribution of inter-arrival or processing times. It does not depend on how a flow unit is chosen for processing from among those waiting in the buffer (e.g., first-come-first-served, last-come-first-served, service in random order, or priority). It is independent of the number of resources of a process or how those resources are assigned to flow units. All it requires is that the process being empty at time 0 and at time $t$. For such a system, Little's law holds for the time period of 0 to $t$.

Consider the case of Absolute Calzone in Example 2.2 in which for lunch hours between 11:00 am ($t_0 = 0$) and 2:00 pm ($t_4 = 180$), we found the average inventory $I^{\mathrm{avg}}(0, 180) = 33.75$ calzones, and average flow time $T^{\mathrm{avg}}(0, 180) = 28.93$ minutes. Since there were $N = 210$ calzones made (i.e., arrived) and sold (i.e., departed) during the 180 minutes of lunch hours, we have $\mathrm{TH}^{\mathrm{avg}}(0, 180) = 210/180 = 1.167$ per minute. Because the inventory at the beginning and at the end of the lunch hours is zero, Little's law, as expected, holds in this case:

$$I^{\mathrm{avg}}(0, 180) = T^{\mathrm{avg}}(0, 180) \times \mathrm{TH}^{\mathrm{avg}}(0, 180) = 28.93 \times 1.167 = 33.75$$

Little's law is also shown to hold when the number of flow units at the beginning and at the end of a time period are not zero, but are equal. But does it hold for long-run averages, that is, when time period 0 to $t$ is very long ($t \to \infty$)?

To have Little's law hold for long-run averages, the number of arrivals during the long period should be equal to the number of departures during the long period. This guarantees that the number of flow units at time $t$ be the same as that at time 0, which is the condition for Little's law to hold. Having the same number of arrivals and departures during the period of 0 to $t$ when $t \to \infty$ corresponds to processes in which their input rates are equal to their output rates. These processes are called *Stable processes*.

---

**CONCEPT**

**Stable Process**

A *Stable Process* is a process in which the input rate of the process is equal to its output rate. This implies that all arriving flow units are eventually processed and leave the process.

In a non-stable process where the input rate is larger than the output rate, the number of flow units in the process (i.e., inventory of the process) increases as time $t$ becomes larger. Mathematically speaking, as $t \to \infty$, the inventory of the process also grows to infinity.

The good news is that almost all business processes are stable processes. All customers who enter a bank eventually leave at the end of the day. All jobs released into a production process will eventually be completed and leave the system. For these stables processes, Little's law also holds for the long-run averages.

---

**ANALYTICS**

**Little's Law for Long-Run Averages**

In a stable process, Little's law holds for long-run average inventory, flow time, and throughput. Specifically, in a stable process:

$$I = T \times \text{TH}$$

---

The intuition behind Little's law is simple. Suppose you enter your local state office to renew your driving license. Assume that your state office is a stable process, and its input rate of 20 people per hour is equal to its output rate, that is, throughput TH = 20. Suppose after you enter the office, it takes an average of $T = 2$ hours for you to wait in line, renew your license, and leave the office. The average number of people in the office when you leave the office—the average inventory $I$—would be equal to the average number of people who arrived after you during the 2 hours that you were in the office. Since people arrive at the rate of TH = 20 people per hour in each of the $T = 2$ hours you were in the office, then $I = T \times \text{TH} = 2 \times 20 = 40$ people, which is Little's law.

**Example 2.4**

> Arya Cylinder Inc. has a plant that produces cylinder blocks for diesel and gasoline engines among others. The machining department of the plant, which works 8 hours per day, has been machining an average of 40 cylinder blocks per day at a steady rate. The average number of cylinder blocks in the machining department is around 81. To compute the average time a cylinder spends in the machining department, the firm needs to perform a time study of the arrival and departure times for a large number of cylinders at that department. This is a time-consuming task. Is there any other way to compute the average time that a cylinder block spends in the machining department without doing a time study?

Yes, we can use Little's law. Considering a cylinder block as a flow unit in the machining department, as the problem states, the throughput of the machining department is TH = 40 cylinder blocks per day or TH = 5 cylinder blocks per hour. The average number of cylinder blocks in the department is $I = 81$. Thus, using Little's law we can compute the average flow time of a block in the department as follows:

$$I = T \times \text{TH} \implies T = \frac{I}{\text{TH}} = \frac{81}{5} = 16.2 \text{ hours}$$

Therefore, it takes an average of 16.2 hours from the time a cylinder block enters the machining department until its machining is completed, whereupon it leaves the department.

**Example 2.5**

The number of technicians in the machining department in Example 2.4 is 48. However, the department has a high employee turnover. Specifically, the average time that a newly hired technician works at the machining department before he quits his job or he is transferred to other departments of the plant is around 18 months. To keep the number of technicians at the required level of six, how many technicians should be hired every year?

While the process in this example is the same process as in Example 2.4—the machining department—we consider a different flow unit in this example. Specifically, we consider a technician as a flow unit that enters the machining department and leaves the department after an average of 18 months. Thus, the average flow time of a technician is $T = 18$ months. On the other hand, the number of technicians in the machining department is $I = 6$. Thus, the average number of technicians that leave the machining department—the throughput of the department—can be computed using Little's law:

$$I = T \times \text{TH} \implies \text{TH} = \frac{I}{T} = \frac{6}{18} = 0.333 \text{ technicians per month}$$

which is $0.333 \times 12 = 4$ technicians per year. In other words, to maintain the number of technicians in the machining department at six, the department should hire around four technicians per year, since an average of four technicians are leaving the department per year.

**Example 2.6**

When Arya Cylinder delivers cylinder blocks to a customer (mainly an auto manufacturer), it also submits the bill to the customer. It often takes a long time for the customer to process the bill and pay Arya Cylinder for the purchase. Arya Cylinder takes the record of unpaid bills in its "Accounts Receivable." Arya Cylinder had an annual sales of around $150 million per year in the last 10 years. Its accounts receivable in the last 10 years has been around $20 million. How long Arya Cylinder must wait, on average, to collect its sales dollars from a customer?

If we consider "Accounts Receivable" as a process, then the flow unit would be a dollar bill. Since Arya Cylinder's annual sales is $150 million, the input rate of the accounts receivable is $150 million per year. On the other hand, assuming all customers will eventually pay their bills (i.e., a stable process), the throughput of the accounts receivable would also be TH = $150 million per year. The average dollar amount at accounts receivable is $I = $20 million. Hence, using Little's law, the average time $1 spends in accounts receivable until it exits the account (i.e., paid by the customer) can be found as follows:

$$T = \frac{I}{\text{TH}} = \frac{\$20 \text{ million}}{\$150 \text{ million}} = 0.1333 \text{ years}$$

Hence, after Arya Cylinder makes a sales to a customer, it takes on average 0.1333 years or $0.1333 \times 365 = 47.5$ days until it receives its sales dollar from the customer.

### 2.6.1 Managerial Implications of Little's Law

As we discussed, average flow time, average inventory, and throughput of a process are key process-flow measures that have direct impact on a firm's profit. Some managers would like to increase throughput to get more sales, others would like to reduce the average flow time to increase customer satisfaction, and some would like to reduce average inventory in their processes to reduce costs. Little's law, as John C. Little puts it, "locks these three measures together." The implication is that a change in one measure can have an impact on the other two. For example, if the goal of the managers is to increase the throughput of their process by 20 percent and if the average flow time of the process is not reduced, the managers should expect a 20 percent increase in average inventory of the process as a result of throughput improvement. Or, if the throughput cannot be improved, but managers' goal is to reduce the average flow time in their process by 20 percent, this is possible only if the average inventory in the process is reduced by 20 percent.

Little's law is also useful for measuring average flow times of a process. This is important because it is often harder to measure flow times than to measure inventory or throughput. Measuring flow times requires keeping record of when each flow unit enters and exits the process. It is, however, easier to keep track of inventory and throughput of a process. For example, most manufacturing systems keep track of the number of items that they produce in a day (i.e., throughput) and their inventory. Even if this information is not available, it is easier to take a sample of inventory in random points in a day than following several flow units throughout the process to compute their flow times. When the average inventory and the throughput of a process are measured, the average flow time can be easily computed using Little's law.

### 2.6.2 Little's Law for In-Buffer and In-Process Flow Units

We defined inventory of a process as the number of flow units within process boundaries. From a process-flow analysis perspective, these flow units are either in the buffer (or buffers) of the process or they are being processed by the resources of the process. In most circumstances, it is essential to separate the inventory in buffer from those being processed. When a resource is processing a flow unit, it is adding value to the unit, which eventually generates revenue. However, no processing is performed on the flow units in the buffer; thus, those flow units only incur inventory holding cost for the system. One goal of Lean Operations, for example, is to reduce the inventory in buffers to zero (see chapter 15 for Lean Operations).

Similarly, the flow time of a flow unit within process boundaries can be divided into the time that the flow unit spends waiting in the buffers of the process, and the time the flow unit spends being processed by the resources of the process. Long waiting times in lines (i.e., in the buffers) of service systems is known to have more negative impact on customer satisfaction than long customer processing times. For example, we do not like it when our waiting time in our doctor's office is long. However, we do not mind if our processing time (i.e., our time with the doctor) is long. Thus, in some cases, it is important to distinguish between the flow time in the buffer and in the process.

---

**CONCEPT**

**Average In-Buffer and In-Process Inventory and Flow Times**

- *In-Buffer Inventory* of a process is the total number of flow units in the buffer (or buffers) of the process. *In-Process Inventory*, on the other hand, is the total number of flow units that are being processed by the resources of the process. We use $I_b$ and $I_p$ to, respectively, denote the long-run average in-buffer and in-process inventory. Thus, we have

$$I = I_b + I_p$$

- *In-Buffer Flow Time* of a flow unit is the time that the flow unit spends in the buffer (or buffers) of a process. *In-Process Flow Time* of a flow unit is the time that the flow unit spends being processed by the resources of the process. We denote $T_b$ and $T_p$, respectively, to denote the long-run average in-buffer and in-process flow times. Thus, we have

$$T = T_b + T_p$$

The application of Little's law goes beyond the average inventory and the average flow time and extends to in-buffer and in-process inventory and flow times as well.

---

**ANALYTICS**

**Little's Law for In-buffer and In-Process Flow Units**

In a stable process, Little's law also holds for long-run average inventory, flow time, and throughput for both in-buffer and in-process flow units. Specifically,

$$I_b = T_b \times \text{TH} \qquad I_p = T_p \times \text{TH}$$

We use the following example to illustrate this.

 **Example 2.7**

> Consider the machining department of Arya Cylinder's plant in Example 2.4. Recall that the department has a throughput of $\text{TH} = 5$ cylinder blocks per hour, an average inventory of $I = 81$ blocks, and an average flow time of $T = 16.2$ hours. When the technicians at the machining department start machining a block, it takes an average of 2.2 hours for them to finish the block. What is the average number of cylinder blocks waiting in the buffer of the machining department?

The average number of cylinder blocks in the buffer of the machining department, $I_b$, can be obtained as $I_b = I - I_p$. Since the average inventory in the machining department is $I = 81$, then $I_b = 81 - I_p$. Hence, to find $I_b$, we only need to compute the average in-process inventory $I_p$.

Since the average time that a cylinder block is being machined—the average in-process flow time—is $T_p = 2.2$ hours, using Little's law, we can find the average in-process inventory of cylinder blocks as follows:

$$I_p = T_p \times \text{TH} = 2.2 \times 5 = 11 \text{ cylinder blocks}$$

Thus, $I_b = 81 - I_p = 81 - 11 = 70$ cylinder blocks.

### 2.6.3 Little's Law for Processes with Multiple Products

Little's law also holds in processes with more than one type of flow units, that is, processes with multiple products. Each type may have a different throughput and may require a different processing time. The following example shows how Little's law works in such processes too.

**Example 2.8**

Consider the machining department of Arya Cylinder in Example 2.4 that produces diesel and gasoline cylinder blocks. From the analysis in Example 2.7, the production manager knows the in-buffer and in-process inventory and flow times for a cylinder block in his machining department to be $I_b = 70$ blocks, $I_p = 11$ blocks, $T_b = 14$ hours, and $T_p = 2.2$ hours. However, these are overall flow times and inventories for cylinder blocks, regardless of their types—diesel or gasoline. The production manager would like to do a more detailed study and find flow times and inventories of each cylinder block type separately. The information that is available about each cylinder block type is limited to the following. On average, 80 percent of blocks machined in the machining department are gasoline cylinder blocks and the remaining 20 percent are diesel blocks. The average time to machine a gasoline cylinder block—the in-process flow time for gasoline blocks—is 2 hours. This number is 3 hours for diesel blocks. The average in-buffer inventory of gasoline blocks is around 30. Can the manager compute the average flow time for each type of cylinder block using this information?

The machining department is a process with two types of products: gasoline cylinder blocks, which we call *type-1*, and diesel cylinder blocks, which we call *type-2*. We use superscripts (1) and (2) to refer to the process-flow measures of type-1 and type-2 blocks, respectively.

Let's start with finding the average flow time of diesel (i.e., type-2) blocks. The average time that a type-2 block spends in the machining department—the average flow time of type-2 blocks—is $T^{(2)}$. On the other hand, we know that $T^{(2)} = T_b^{(2)} + T_p^{(2)}$. Since we know that the average processing time of type-2 blocks is $T_p^{(2)} = 3$ hours, to compute $T^{(2)}$, we only need to find $T_b^{(2)}$, the average in-buffer flow time for type-2 blocks.

We know that the overall average in-buffer inventory of blocks is $I_b = 70$. This includes both type-1 and type-2 blocks. We also know that among these blocks, $I_b^{(1)} = 30$ blocks are type-1 blocks. Thus, $70 - 30 = 40$ would be the average in-buffer inventory of type-2 blocks, that is, $I_b^{(2)} = 40$ type-2 blocks. Furthermore, since 20 percent of the blocks processed in the machining department are type-2 blocks, this implies that the throughput of type-2 blocks is $TH^{(2)} = 20\% \times 5 = 1$ type-2 blocks per hour.

So far, for type-2 blocks we found $I_b^{(2)} = 40$ and $TH^{(2)} = 1$. Can we use Little's law for in-buffer performance measures of only type-2 blocks? In other words, is it true that

$$I_b^{(2)} \overset{?}{=} T_b^{(2)} \times TH^{(2)}$$

The answer is yes. Below we present the analytics of Little's law for processes with multiple products.

### Little's Law for Processes with Multiple Products

Consider the process in the following figure that processes $K$ different flow units of type $1, 2, \ldots, K$.

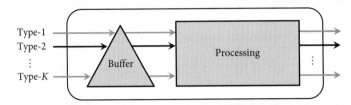

Suppose that fraction $p_k$ of the throughput of the process (i.e., TH) is of type $k$. Let $I^{(k)}$, $T^{(k)}$, and $\mathrm{TH}^{(k)}$ denote, respectively, the long-run average inventory, flow time, and the throughput of the process for type-$k$ flow units. Then, for type-$k$ flow units ($k = 1, 2, \ldots, K$), we have

$$\mathrm{TH}^{(k)} = p_k \times \mathrm{TH}$$

- *Little's Law for Type-k:* Little's law holds for type-$k$ flow units as follows:

$$I^{(k)} = T^{(k)} \times \mathrm{TH}^{(k)}$$

- *Little's Law for In-Buffer and In-Process Flow Units of Type-k:* Little's law also holds for in-buffer and in-process flow units of each type. Specifically, for all $k = 1, 2, \ldots, K$, we have

$$I_b^{(k)} = T_b^{(k)} \times \mathrm{TH}^{(k)} \qquad I_p^{(k)} = T_p^{(k)} \times \mathrm{TH}^{(k)}$$

- *Little's Law for Overall System:* Finally, considering $I$ as the long-run average inventory of all types of flow units, and TH as the total throughput, we have

$$I = I^{(1)} + I^{(2)} + \cdots + I^{(K)} \qquad \mathrm{TH} = \mathrm{TH}^{(1)} + \mathrm{TH}^{(2)} + \cdots + \mathrm{TH}^{(K)}$$

  Then Little's law also holds for the overall system as follows:

$$I = T \times \mathrm{TH}$$

  where

$$T = p_1 T^{(1)} + p_2 T^{(2)} + \cdots + p_K T^{(K)}$$

  Note that overall average flow time of the process, $T$, is the weighted average of the flow times of all $K$ types of flow units.

- *Little's Law for In-buffer and In-Process Flow Units in the Overall System:* Similarly, for the overall in-buffer and in-process flow units, we have

$$I_b = T_b \times \mathrm{TH} \qquad I_p = T_p \times \mathrm{TH}$$

  where

$$I_b = I_b^{(1)} + I_b^{(2)} + \cdots + I_b^{(K)} \qquad I_p = I_p^{(1)} + I_p^{(2)} + \cdots + I_p^{(K)}$$

  and

$$T_b = p_1 T_b^{(1)} + p_2 T_b^{(2)} + \cdots + p_K T_b^{(K)} \qquad T_p = p_1 T_p^{(1)} + p_2 T_p^{(2)} + \cdots + p_K T_p^{(K)}$$

Table 2.5    Inventory, Flow Times, and Throughput in Example 2.8

|  | Inventory | | | Flow time | | | Throughput |
|---|---|---|---|---|---|---|---|
|  | $I_b$ | $I_p$ | $I$ | $T_b$ | $T_p$ | $T$ |  |
| Gasoline cylinder | 30 | 8 | 38 | 7.5 | 2 | 9.5 | 4 |
| Diesel cylinder | 40 | 3 | 43 | 40 | 3 | 43 | 1 |
| Overall | 70 | 11 | 81 | 14 | 2.2 | 16.2 | 5 |

Therefore, we can obtain the average in-buffer flow time of type-2 blocks in the machining department in Example 2.8 as follows:

$$I_b^{(2)} = T_b^{(2)} \times \text{TH}^{(2)} \implies T_b^{(2)} = \frac{I_b^{(2)}}{\text{TH}^{(2)}} = \frac{40}{1} = 40 \text{ hours}$$

By knowing the average in-buffer flow time of type-2 blocks, we can compute the average flow time of type-2 blocks in the department as follows:

$$T^{(2)} = T_b^{(2)} + T_p^{(2)} = 40 + 3 = 43 \text{ hours}$$

On the other hand, we know that the overall average flow time for a block is $T = 16.2$ hours, which is the weighted average of the average flow times of each type. Specifically,

$$T = 16.2 = p_1 T^{(1)} + p_2 T^{(2)} = 0.8\, T^{(1)} + 0.2(43) \implies T^{(1)} = 9.5 \text{ hours}$$

Applying Little's law for each type of cylinder we can find the average in-buffer and in-process inventory and flow times for both cylinder types as well as for the overall system as shown in Table 2.5.

It is worth to reemphasize, as also shown in Table 2.5, that

- The overall average inventory is the sum of the inventories of the two types of cylinders, while the overall average flow times are not the sum of the flow times of both types. It is the weighted average of the flow times of both types of cylinders.
- Since 20 percent of blocks produced by the machining department (i.e., the throughput) are diesel cylinder blocks, one might expect that 20 percent of the overall in-buffer inventory of blocks, that is, $I_b = 20\% \times 70 = 14$ blocks, to be diesel cylinder blocks. But, as the table shows, the in-buffer inventory of diesel cylinder blocks is $I_b^{(2)} = 40$. This is also the case for average in-process inventory.

### 2.6.4 Little's Law for Processes with Multiple Stages

In most processes with multiple products, each type of product goes through a different sequence of processing (i.e., different routing) within process boundaries. In this section, we show how Little's law applies to such processes.

Example 2.9

Recall that Arya Cylinder's plant produces several types of cylinder blocks, two of which are gasoline and diesel cylinder blocks. Also, recall that the throughput of the plant for gasoline and diesel blocks are 4 and 1 per day, respectively. After finding the average

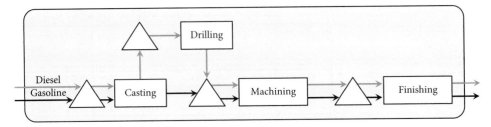

Figure 2.11    Process-flow chart of Arya Cylinder plant in Example 2.9.

flow times of diesel and gasoline cylinder blocks in the machining department of their plant, the plant's production managers would like to also find the average flow times and inventories of each type of cylinder in other departments of their plant. They have constructed the process-flow chart of the operations in their plant as shown in Fig. 2.11.

As the figure shows, the plant consists of four main departments: casting, drilling, machining, and finishing. Also note that diesel cylinder blocks follow a different routing than that of gasoline cylinder blocks. After casting, diesel blocks are sent to the drilling department, while the gasoline blocks are sent to the machining department. In addition to the average flow times and average inventories of the machining department (obtained in Example 2.5), the production manager also has the following information about the casting, drilling, and finishing departments. The average inventory of gasoline and diesel blocks in the casting department are 40 and 12 blocks, respectively. The average inventory in the drilling department is around 20 diesel blocks. Finally, the average inventory of gasoline and diesel blocks in the finishing department are 12 and 16 blocks, respectively. How long, on average, a cylinder block spends in Arya Cylinder's plant until its production is finished? How does this average time differ for gasoline and diesel cylinders?

Having average inventories and throughput of each department for each type of cylinder, we can use Little's law to find the average flow times of each type of cylinder in each department. For example, consider the finishing department. It has an average inventory of 12 gasoline cylinder blocks. Since the plant's throughput for gasoline is 4 and since all gasoline blocks go through the finishing department, the throughput of the finishing department is 4 gasoline blocks per hour. We can use Little's law and find the average flow time of gasoline blocks in the finishing department as $T^{(1)} = I^{(1)}/\text{TH}^{(1)} = 12/4 = 3$ hours. Table 2.6 shows the average flow time and average inventories of both types of blocks in each department of Arya Cylinder's plant.

Table 2.6    Inventory, Flow Times, and Throughput in Arya Cylinder's Plant

| Departments | Gasoline cylinder | | | Diesel cylinder | | |
|---|---|---|---|---|---|---|
| | $I^{(1)}$ | $T^{(1)}$ | $\text{TH}^{(1)}$ | $I^{(2)}$ | $T^{(2)}$ | $\text{TH}^{(2)}$ |
| Casting | 40 | 10 | 4 | 12 | 12 | 1 |
| Drilling | — | — | — | 20 | 20 | 1 |
| Machining | 38 | 9.5 | 4 | 43 | 43 | 1 |
| Finishing | 12 | 3 | 4 | 16 | 16 | 1 |

Since gasoline cylinder blocks—type-1 blocks—must go through the casting, machining, and finishing departments, their average flow time in the plant is

$$T^{(1)}_{\text{plant}} = T^{(1)}_{\text{casting}} + T^{(1)}_{\text{machining}} + T^{(1)}_{\text{finishing}} = 10 + 9.5 + 3 = 22.5 \text{ hours}$$

Another approach to find the average flow time for gasoline cylinders in the plant is to use Little's law for the entire plant. This approach does not require computing the average flow times of all departments. Specifically, we know that the total average inventory of gasoline cylinder blocks in the plant is

$$I^{(1)}_{\text{plant}} = I^{(1)}_{\text{casting}} + I^{(1)}_{\text{machining}} + I^{(1)}_{\text{finishing}} = 40 + 38 + 12 = 90 \text{ gasoline blocks}$$

On the other hand, we know that the throughput of the plant is $\text{TH}^{(1)}_{\text{plant}} = 4$ gasoline blocks per hour. Thus, using Little's law for gasoline blocks in the entire plant we get

$$T^{(1)}_{\text{plant}} = \frac{I^{(1)}_{\text{plant}}}{\text{TH}^{(1)}_{\text{plant}}} = \frac{90}{4} = 22.5 \text{ hours}$$

Using the same approach, we can compute the average flow time of diesel blocks by first finding the average inventory of diesel blocks in the plan,

$$I^{(2)}_{\text{plant}} = I^{(2)}_{\text{casting}} + I^{(2)}_{\text{drilling}} + I^{(2)}_{\text{machining}} + I^{(2)}_{\text{finishing}} = 12 + 20 + 43 + 16 = 91 \text{ diesel blocks}$$

Since for diesel blocks the throughput of the plant is $\text{TH}^{(2)}_{\text{plant}} = 1$ block per hour, using Little's law we get

$$T^{(2)}_{\text{plant}} = \frac{I^{(2)}_{\text{plant}}}{\text{TH}^{(2)}_{\text{plant}}} = \frac{91}{1} = 91 \text{ hours}$$

Thus, on average, diesel blocks spend $91 - 22.5 = 68.5$ more hours in the plant than gasoline blocks.

To get an overall average of how long a block stays in the plant, we can use the flow times of both types of blocks as follows:

$$T_{\text{plant}} = p_1 T^{(1)}_{\text{plant}} + p_2 T^{(2)}_{\text{plant}} = 0.8(22.5) + 0.2(91) = 36.2 \text{ hours}$$

Another approach, again, is to use the average inventory of both blocks in the plant to get the average flow time of a block. If we ignore the type of the block, the total average inventory in the plant is

$$I_{\text{plant}} = I^{(1)}_{\text{plant}} + I^{(2)}_{\text{plant}} = 90 + 91 = 181 \text{ blocks}$$

Since the throughput of the plant is

$$\text{TH}_{\text{plant}} = \text{TH}^{(1)}_{\text{plant}} + \text{TH}^{(2)}_{\text{plant}} = 4 + 1 = 5 \text{ blocks per hour}$$

Using Little's law, we have

$$T_{\text{plant}} = \frac{I_{\text{plant}}}{TH_{\text{plant}}} = \frac{181}{5} = 36.2 \text{ hours}$$

## 2.7 Inventory Turn

Inventory turn is another operational performance measure of a process that corresponds to the efficiency of the process regarding how fast it can turn its inventory into sales. This measure is directly related to the average flow time. In fact, it is the reciprocal of the average flow time. Inventory turn is a particularly important performance measure in Retail settings.

To understand the idea behind inventory turn, assume that you work for a large firm that imports and distributes around 150 different types of exotic food items such as specialty cheese, fresh meat, and seafood from European countries. The food items are transferred from the port of origin by ships in large refrigerated containers to a port of entry in the United States. These containers are then moved by trucks to your Distribution Center (DC) in Arizona. After sorting and loading to trucks, the items are shipped to DCs of supermarkets in different states.

One item that you import from France is Abbaye De Belloc cheese which has become a popular item. Your only customer for this item is a major high-end supermarket chain in California, which requires 2000 pounds of the cheese on 10th, 20th, and 30th of each month. You receive one shipping container containing 2000 pounds on the 10th of every month. When the container is unloaded from the ship in the port of entry, you pay the supplier in France $20 per pound, which adds up to a total of $20 \times 2000 = \$40{,}000$. It takes 10 days from the time the container is unloaded until you deliver the cheese to your main customer in California. Specifically, it takes 2 days to bring the container from the port to your DC in Arizona. In your DC, the container is unloaded, the cheese is sorted, and loaded to another truck for California. Due to the high volume of work at your DC, this takes an average of 7 days. It then takes 1 day for you to deliver the cheese to your customer's DC in California—a total of $2 + 7 + 1 = 10$ days. When you deliver the cheese, you receive a payment of $30 \times 2000 = \$60{,}000$, that is, $30 per pound.

If we consider the container with 2000 pounds of cheese as one flow unit, and your firm as the process, the flow unit enters your process (at the port) and becomes your inventory when you pay the supplier. The container stays in your process for 10 days and exits the process when you deliver it to the customer (in California) and receive your payment. Thus, the average flow time of your process is $T = 10$ days. In other words, it takes your firm 10 days to turn its input of one container of Abbaye De Belloc cheese (which stays in your system as inventory) into sold output. Your firm's inventory turn is

$$\text{Inventory Turn} = \frac{1}{T} = \frac{1}{10} = 0.1 \text{ turns per day}$$
$$= 0.1 \times 30 = 3 \text{ turns per month}$$

In other words, your firm turns its inventory of Abbaye De Belloc cheese three times a month.

Now suppose you improved your operations at your DC, and reduced the time the shipment of Abbaye De Belloc cheese spends in your DC from 7 days to 2 days. Hence, you only need a total of $2 + 2 + 1 = 5$ days to pick up the container from the port, sort it in your DC, and deliver it to your customer in California. You call your supplier in France and ask him to send you the shipment on 5th, 15th, and 25th of each month. Thus, under the improved

DC operations, a container spends only $T = 5$ days with your firm. Consequently, it now takes your firm 5 days to turn its input of one container of Abbaye De Belloc cheese into sold output. Your firm's new inventory turn is, therefore,

$$\text{Inventory Turn} = \frac{1}{T} = \frac{1}{5} = 0.2 \text{ turns per day}$$
$$= 0.2 \times 30 = 6 \text{ turns per month}$$

In other words, with the improved DC operations, your firm "*can*" turn its inventory of Abbaye De Belloc cheese six times a month. But does that really matter? The company is still paying $40,000 for each shipment and sells it for $60,000, and does it three times a month.

Well, let's check the operational performance of your firm. The throughput in both cases are the same—three containers a month. However, your firm's inventory of Abbaye De Belloc cheese has reduced after the improvement in your DC operations. Before the improvement, using Little's law, the average inventory of the cheese was $I = T \times \text{TH} = 10/30$ (month) $\times 3$ (containers per month) $= 1$ container. After the improvement, the average flow time is reduced to $T = 5$ days, and thus the average inventory is reduced to half, that is, $I = (5/30) \times 3 = 0.5$ containers. So, for the same throughput of 3 containers per month, the higher inventory turn of 6 resulted in lower average inventory of 0.5 containers. Hence, your firm is saving money since it reduced the inventory holding cost of Abbaye De Belloc cheese to half. So, yes. It does matter. Higher inventory turns result in lower inventory costs.

Now suppose that you would like to know how improvement in DC affected the inventory turn of other items that your firm sells. One way is to collect data about the throughput and the average flow time of all 150 items that your firm imports and distributes. This will give you detail information about how long it takes your firm to turn the dollar it invests on each item into sales revenue. You can then identify items that have slow turns and those with high turns. You can compare this with those before improvement in DC.

To get an *overall* view of your operations after the improvement in DC, you may like to combine the 150 inventory turns and come up with one aggregate measure for inventory turn. This means that you need to find a way to aggregate the flow times of all 150 units into one, and then use the reciprocal of the aggregated flow units as the aggregated inventory turn of the 150 items to simply represent the inventory turn of your firm. Accountants and financial analysts have found one easy way to do this. They define inventory turn of a firm as follows:

$$\text{Inventory Turn} = \frac{\text{Cost of Goods Sold}}{\text{Inventory in Dollars}}$$

If you notice, the above is nothing but using Little's law to compute inventory turn. Note that, since $T = I/\text{TH}$, then

$$\text{Inventory Turn} = \frac{1}{T} = \frac{\text{TH}}{I}$$

Considering "one cost dollar" as the flow unit, we can measure the total inventory of a firm by adding the cost of all units in inventory. On the other hand, the total number of cost dollars leaving the firm in a period (i.e., the cost of goods sold) would be the throughput of the firm in that period. Therefore, the cost of goods sold divided by the average inventory of a firm (measured in dollars) is nothing but $\text{TH}/I$, which is the inventory turn.

So, all you need to compute the aggregated inventory turn of your firm is the cost of goods sold (in a month or in a year) and the average inventory of your firm (in a month or in a year) in dollar amount. You call your accounting department and they give you those numbers for a period of 3 months before and after the improvement in DC. Specifically, the cost of goods sold and the total inventory in a quarter before the DC improvement were $24.68 million and $7.84 million, respectively. These numbers are $23.11 million and $5.36 million, respectively, in a quarter after DC improvement. Thus, the inventory turn before the improvement was $24.68/7.84 = 3.15$ turns in a quarter. After the improvement, the inventory turn is increased to $23.11/5.36 = 4.31$ turns in a quarter. So, in an aggregated level, after the improvement in DC, your firm is able to turn $1 investment into sales dollars much faster (i.e., 3.15 times a quarter compared to 4.31 times in a quarter).

Firms publish their financial statements at least once a year to inform public and investors about their performance. Using these financial statements (and using cost dollars as flow units), we can easily compute firms' inventory turn. Cost of goods sold, also called cost of sales, is reported in a firm's "Statement of Income" and inventory is reported in a firm's "Balance Sheet." For example, Walmart, the largest and the most successful firm in retail industry reported in its consolidated statement of income a cost of sales of $360.98 billion for year 2016. In their consolidated balance sheets, Walmart reported the value of their inventory (as of January 31, 2016) to be $44.47 billion.[1] Thus, we can compute Walmart's inventory turn in 2016 as follows:

$$\text{Inventory Turn} = \frac{\text{Cost of Goods Sold}}{\text{Inventory in Dollars}} = \frac{\$360.98}{\$44.47} = 8.12 \text{ turns per year}$$

Target, another firm in retail industry, reported a cost of sales of $54.21 billion with an inventory of $8.60 billion for year 2016. Target's inventory turn in 2016 is, therefore,

$$\text{Inventory Turn} = \frac{\text{Cost of Goods Sold}}{\text{Inventory in Dollars}} = \frac{\$54.21}{\$8.6} = 6.30 \text{ turns per year}$$

These numbers show that Walmart is capable of turning its $1 investment in its items (i.e., inventory) into revenue much faster than Target. This is due to Walmart's efficient logistics and supply chain management, among other factors. In fact, Walmart has the highest inventory turn of any (discount) retailer.

What is a good inventory turn for a firm? Well, it depends on the type of industry and the type of product. For example, the inventory turn of consumer electronics is around 4, while it is more than 20 for bakery products. Inventory turn also seems to change from year to year. For example, Walmart had an inventory turn of 6.94 in 2001. Fifteen years later, as we computed, it increased to 8.12. This is mainly due to advances in information technology and logistics among others.

<hr>

**CONCEPT**

### Inventory Turn

*Inventory Turn*, also called *Inventory Turnover Ratio*, is the reciprocal of the average flow time of a process. Considering Little's law, this is equal to the ratio of throughput to average inventory:

$$\text{Inventory Turn} = \frac{1}{T} = \frac{\text{TH}}{I}$$

<hr>

[1] Note that $44.47 billion is not the average inventory during year 2016. It is the inventory on January 31, 2016. However, it is used as proxy for average inventory in computing inventory turn.

Inventory turn of a process in a given period refers to the number of times that the process *can* sell and replace its inventory during that period. The higher the inventory turn, the faster the process can turn dollars invested (in inventory) into sales dollars.

For a process that has inventory of a large number of different flow units, the overall inventory turn for the process can be computed using cost of goods sold (also known as cost of sales) and the value of the inventory (in dollars) as follows:

$$\text{Inventory Turn} = \frac{\text{Cost of Goods Sold}}{\text{Inventory in Dollars}}$$

### Weeks of Supply

*Weeks of Supply* or *Days of Supply* is another way of measuring inventory. This measure is a critical measure in retail industry. Week of supply tells us how long the current on hand inventory will last based on current sales. For example, 2 weeks of supply at a DC implies that, at the current sales (i.e., throughput), the center has inventory enough to satisfy 2 weeks of its demand. Considering cost of goods sold and average inventory in a year, weeks of supply is computed as follows:

$$\text{Weeks of Supply} = \frac{\text{Inventory in Dollars}}{\text{Cost of Goods Sold}} \times 52 \text{ weeks}$$

Comparing the above formula with the formula for inventory turn, you find that weeks of supply is nothing but the average flow time (expressed in weeks). Thus, weeks of supply is indeed the average flow time, interpreted differently.

## 2.8 Utilization

In the previous sections, we defined throughput, flow time, and inventory as the three main process-flow measures—note that inventory turn is nothing but the inverse of average flow time. These performance measures are all about the arrivals to, departures from, and movement of *flow units* within the process. Specifically, inventory is the number of flow units in the different stages of a process. Flow time is the time a flow unit spends in the process, and throughput is the rate at which flow units exit the process. While lower inventory and flow times and higher throughput are desirable, they are only one side of process performance. The other side is about *resources* of the process, and how effectively they can deliver the desirable operational performance. Anyone can increase throughput by investing lots of money to acquire more resources. The key, however, is to achieve better performance with lower investment. The concept of utilization tends to capture this. We elaborate on this using an example.

Consider a new product with demand of 8000 units per month. The demand is forecasted to be steady in the next 5 to 10 years. Your company is planning to build a new plant dedicated to the production of this new product. It is considering the following two alternatives: (i) building a large plant with capacity of 16,000 units per month that costs $10 million, or (ii) building a smaller plant with capacity of 10,000 units per month that costs $6 million. It is estimated that both plants will have the same average inventory to generate the throughput of 8000 units per month. Which plant do you recommend?

Note that both plants will have the same throughput of 8000 units per month, and are predicted to have the same average inventory. Using Little's law, they both will have the same average flow times. Thus, in terms of our three process-flow measures, the two plants look the same. However, as you noticed, there is a big difference between the two plants. While both plants have enough capacity to satisfy the monthly demand of 8000 units, in the long run, the larger plant will be used only 50 percent (=8000/16,000) of the time. The remaining 50 percent of the time the plant will be idle. The smaller plant, on the other hand, will be used 80 percent (=8000/10,000) of the time. Is it worth spending $10 million in building a plant that is idle half of the time?

Most probably not. In most cases, investing in resources that are not used that often does not make business and operational sense. One performance measure that captures what percentage of time a resource is used is "utilization."

**CONCEPT**

**Utilization**

*Utilization* of a resource (or a process) is the percentage of time the resource (or the process) is not idle due to lack of flow units (e.g., lack of demand). This includes the fraction of time that the resource is processing a flow unit, and the fraction of time that the units are waiting, but the resource cannot process them because it is broken or being set up for processing. Utilization of a resource can also be viewed as the percentage of time the resource is being used to satisfy demand.

$$u = \text{Utilization} = \frac{\text{Demand}}{\text{Capacity}}$$

The numerator corresponds to what the resource *is doing* (i.e., processing units to satisfy demand), and the denominator corresponds to what the resource *can do*. We use $u$ to denote utilization.

Utilization cannot increase beyond 100 percent—a resource cannot be, for example, 110 percent busy. Thus, when the demand is larger than capacity, utilization is considered to be 100 percent.

Hence, the utilization of the larger plant is $u = 8000/16,000 = 50$ percent, and utilization of the smaller plant is $u = 8000/10,000 = 80$ percent. From operational perspective, the capacity of the smaller plant is better utilized. But, what does a financial analyst say about this?

Accountants and financial analysts have a similar view. One of the financial measure of how profitable a company is relative to its total assets is Return On Assets. Return On Assets or ROA is an indicator of how efficient management is in using its assets to generate profit. Specifically,

$$\text{ROA} = \frac{\text{Net Income}}{\text{Total Assets}}$$

As the formula implies, ROA shows how much profit a company generated for each $1 in its total assets. Total assets, among other things, include fixed assets. Fixed assets are also referred to as PPE, which is short for property, plant, and equipment. Thus, all things being equal, a company that has a higher utilization of its resources (e.g., plants and equipment) will have lower fixed assets and thus a higher ROA.

What is the appropriate utilization level for a resource or a process? The answer depends on the variability in the demand and in process times of resources. The higher the variability in the process, the lower the utilization should be in order for the process to achieve acceptable operating costs and customer service. This is discussed in detail in Chaps. 5 and 7.

## 2.9 Quality

Suppose you are planning to buy a 42-inch TV of a particular brand. If someone tells you that he has that TV and it is of low quality, would you still buy the TV? If there are better quality TVs in the market with prices close to this one, you most probably would not buy the TV (I know I wouldn't). Firms are also aware of this.

Quality of goods and services have become one of the main factors in the success or failure of firms. Why? First of all, higher quality results in lower costs. Specifically, higher quality products or higher quality processes result in lower rework, lower scraps, lower warranty, and recall costs. These costs can be very significant. For example, in 2006, Dell computers found a quality issue with the battery packs of its laptops. The batteries were overheating and sometimes causing fire. The company issued a recall and had to replace 4.1 million battery packs. This was estimated to cost Dell $200 to $400 million. Toyota also had quality issues with their cars that resulted in a significant loss. In 2009 to 2010 Toyota announced a series of recalls for more than 8 million of its sold vehicles for a potential problem involving sticky accelerator and floor mats that could trap the gas pedal. *The Wall Street Journal* reported on March 9, 2010 that the financial loss of Toyota due to the recall could exceed $5 billion over the next fiscal year.

Second, higher quality not only reduces costs, but also results in higher profit. Researchers have shown that the profitability of firms that have initiated quality improvement efforts and won quality awards increases more (approximately 25% to 50%) compared to other firms in the same industry (Boyer and Verma, 2010).

The third reason why companies should care about quality is customer satisfaction. Customer satisfaction has a direct impact on long-term profitability of a firm. For example, the American Customer Satisfaction (ACS) Index (www.theacsi.org), which was developed at the University of Michigan, measures customer satisfaction on a scale of 0 to 100 based on interviews with customers. The index is generated based on factors such as customer loyalty and customer complaints. Data has shown that firms with a consistently higher ACS Index than their competitors have a higher increase in their revenues.

How can operations support high quality? Lean operations and Six Sigma quality—two key operations engineering and management tools—have been developed through the past decades to improve product quality. These tools are designed to reduce scraps, rework, recalls, and the overall cost of quality. They use methods to reveal the root cause of quality problems, which can be bad product design or problems with the process that produces the product. While principles of lean operations and Six Sigma quality have been widely used to improve the quality of conformance, recently it has started to expand to the process of product development and design. This is not surprising because from process-flow analysis perspectives, product development is nothing but a process in which flow units are ideas and information.

We conclude this section by reemphasizing that quality is an important process-flow measure that can have a significant impact on firms' profits. In Chaps. 15 and 16 we will discuss how operations engineers and managers can use principles of lean operations and Six Sigma to improve the quality of a process and its outputs. The objective of this section was to only introduce quality as one of the process-flow measures.

## 2.10 Productivity

Recall that operations is the process of transferring inputs into outputs. Productivity tends to measure how efficiently inputs are converted to outputs.

---

**CONCEPT**

**Productivity**

*Productivity* is a broad performance measure that evaluates how efficiently resources of a process are used to convert inputs to outputs. It is simply defined as the ratio of outputs to inputs.

$$\text{Productivity} = \frac{\text{Outputs}}{\text{Inputs}}$$

In general, productivity measures are divided into three groups:

- *Total Productivity* is the ratio of *all* outputs (i.e., goods and services produced) to *all* inputs used to produce those outputs (e.g., all equipment, labor, material, etc.). Usually both inputs and outputs are measured in monetary values.
- *Partial Productivity* is the ratio of output (or outputs) to only *one* input.
- *Multifactor Productivity* is the ratio of output (or outputs) to *some* (but not all) inputs.

In all three cases, both inputs and outputs must be computed over the same time period.

---

In a hospital, for example, a partial productivity measure can be the ratio of patients served in a month to the number of hospital beds or to the number of doctor-hours in a month. In a manufacturing system, the number of products produced per labor-hour (i.e., labor productivity) or the number of products produced per machine-hour (i.e., machine productivity) or the number of products produced per kilowatt-hour (i.e., energy productivity) are examples of partial productivity. Examples of multi-factor productivity are dollar value of outputs to labor plus material cost, or the ratio of the dollar value of outputs to the sum of capital and energy costs.

One must be careful when interpreting productivity measures. The productivity measures alone do not provide useful information unless they are compared to similar productivity measures. For example, consider a call center agent who works in the customer service center of a financial institution and handles customers' calls. The agent has a labor productivity of 10 calls per hour. Is that a high labor productivity? or is it low?

Well, the answer is not clear unless we compare the productivity of this agent with some benchmark. One option is to compare the agent's productivity to that of another agent who performs the same task, or to the standard labor productivity in the same industry (if one exists). If the productivity of other agents in the service center or if the industry average labor productivity is 12 calls per hour, then labor productivity of 10 means something, that is, our agent has a low productivity.

The productivity of 10 calls per hour is also informative if it is compared with the productivity of the same agent over time. Suppose, 2 months ago, the productivity of the agent was 9 calls per hour. The agent was sent to a training program to learn how to respond to calls more efficiently. Now, the productivity of 10 calls per hour implies that the training program has been successful in increasing the productivity of this agent. However, the productivity of the agent is still low compared to other agents and industry average.

## 2.11 Summary

One effective approach to analyze and improve performance of operations systems is process-flow analysis. From a process-flow perspective, in operations processes, the main output of the process—products (goods or services)—flow through different activities within the process and is gradually transformed into the final product before it exits the process. Process-flow analysis improves operations performance by finding new ways to perform activities or by changing the way inputs flow within the process.

Process-flow analysis focuses on identifying and improving key operational performance measures such as throughput, flow time, inventory, inventory turn, utilization, quality, and productivity. Improving these operational measures improves the financial performance measures of firms.

Throughput is the rate by which flow units flow throughout and exit the process. Throughput of a firm is therefore goods or services that the firm produces per unit time (e.g., a day). Thus improving throughput—if there is a market to sell it—will increase a firm's profit. Inventory is the number of flow units (e.g., parts, subassemblies, customers) within process boundaries. Reducing inventory decreases the money tied-up with the flow units waiting in different stages of the process. Flow time is the time that the flow unit spends within process boundaries. This includes the total time that the flow unit waits in the buffer (or buffers) and the time that the flow unit is being processed by resource (or resources) of the process. Inventory turn of a process in a given period refers to the number of times that the process can sell and replace its inventory during that period. The higher the inventory turn, the smaller the flow time, and the faster the process can turn dollars invested (in inventory) into sales dollars. Improving quality increases demand for the product and thus increases firm's sales and revenue. Productivity evaluates how efficiently resources of a process are used to convert inputs to outputs. It is simply defined as the ratio of outputs to inputs. Improving productivity of a process results in more effective use of the resources and leads to lower costs per throughput.

Little's law is a very useful tool in managing and controlling operational performance measures. Little's law connects three main process-flow measures of a process, namely throughput, inventory, and flow time. Specifically, in stable systems, the average inventory of a process is equal to the products of the throughput and the average flow time of the process. The managerial implication of Little's law is that one cannot improve one measure without affecting the other two. Little's law is very broad and holds in small processes such as the buffer of a work station in an assembly line, to larger processes such as the entire assembly line or the entire supply chain.

## Discussion Questions

1. Describe flow time, inventory, and throughput of a process. How do they affect a firm's financial performance?
2. What is a stable process and how Little's law connects the operational performance of a process?
3. Explain two managerial implications of Little's law?
4. What is inventory turn and how can it be estimated using a firm's financial statements?
5. What is utilization and how does it affect a firm's financial performance?
6. What is productivity? Describe three different groups of productivity measures and provide an example for each.

## Problems

1. Custom Fire (CF) produces a wide range of high-end custom-designed fireplaces for large houses and mansions in make-to-order fashion. Customers choose a product from firm's website and place their orders through the site. All products go through several stages of production, including fabrication, assembly, packaging, etc. The average time it takes for an order to go through these several stages and be completed is about 3 days with standard deviation of 0.5 days. The completed orders are shipped the same day. Table 2.7 shows the time that CF finished their last 30 orders. The completion time of the last fireplace produced before the first one in the table was at 8:00 am on Monday.

Table 2.7    Production Completion Times of Last 30 Fireplaces at Custom Fire

| No. | Production completion Day | Time | No. | Production completion Day | Time | No. | Production completion Day | Time |
|-----|------|------|-----|------|------|-----|------|------|
| 1 | Monday | 10:20 am | 11 | Tuesday | 4:25 am | 21 | Tuesday | 10:19 pm |
| 2 | Monday | 12:10 pm | 12 | Tuesday | 6:39 am | 22 | Wednesday | 12:14 am |
| 3 | Monday | 1:28 pm | 13 | Tuesday | 8:49 am | 23 | Wednesday | 2:20 am |
| 4 | Monday | 3:35 pm | 14 | Tuesday | 10:11 am | 24 | Wednesday | 4:25 am |
| 5 | Monday | 5:24 pm | 15 | Tuesday | 11:55 am | 25 | Wednesday | 6:34 am |
| 6 | Monday | 6:59 pm | 16 | Tuesday | 1:47 pm | 26 | Wednesday | 9:06 am |
| 7 | Monday | 8:28 pm | 17 | Tuesday | 3:20 pm | 27 | Wednesday | 11:01 am |
| 8 | Monday | 10:45 pm | 18 | Tuesday | 5:27 pm | 28 | Wednesday | 1:07 pm |
| 9 | Tuesday | 12:50 am | 19 | Tuesday | 6:57 pm | 29 | Wednesday | 3:33 pm |
| 10 | Tuesday | 2:54 am | 20 | Tuesday | 8:13 pm | 30 | Wednesday | 5:22 pm |

The firm receives different number of orders in different days and it works on them in a first-come-first-served basis. The data shows that the number of outstanding orders in the last 6 months was never above 72.

a. Managers of CF would like to add a delivery guarantee date to their website promising customers that their orders will be delivered within 6 days after they place their orders (assume each customer orders one fireplace). If the firm adds that to its website, what would be the maximum fraction of customers who will receive their orders beyond the promised date of 6 days?

b. After weeks of discussion, CF managers came to the conclusion that announcing the same due date of 6 days to all customers without considering the status of their factory (i.e., the number of outstanding orders in the factory) is not a good strategy. Also, they have started working with several construction companies that often order more than one fireplace for their big projects, which should be delivered to them in one shipment when they are all finished. The new strategy is to give individual delivery date to each customer depending on the status of the factory and the number of products in the customer order. CF managers all agree that the promised delivery date should be such that the firm satisfies its promises 95 percent of the time. Under this new policy, what due date the firm should promise to a customer who places an order for three fireplaces, while the firm already has 22 outstanding orders consisting of 35 fireplaces in their factory?

2. Fast Postal Service (FPS) is a mail delivery firm that has its main mail processing center in south of Chicago. The center sorts the arriving mail to be shipped to different destinations. The center has two overlapping working shifts: the first shift starts from 7:00 am to 3:00 pm and the second shift starts from 11:00 am to 7:00 pm. Each day, 20 workers work in the first shift and 40 workers work in the second shift. Each worker can process 600 mails in an hour.

Mails arrive at the processing center from 7:00 am to 7:00 pm in a constant but different rates. Specifically, from 7:00 am to 10:00 am mails are delivered to the processing center at the rate of 24,000 pieces an hour. This rate is 34,000 per hour from 10:00 am to 3:00 pm, and is 11,500 per hour from 3:00 pm to 7:00 pm.

    a. What time of the day (in each day) the center has the maximum number of unprocessed mail, and what is that maximum number?

    b. What is the average number of unprocessed mail in the processing center?

    c. How long, on average, a piece of mail waits in the center until it is processed?

3.  Organilk has a large milk-processing plant producing organic dairy products such as cheese, creams, and different types of pasteurized milk. The plant works from 7:00 am to 4:00 pm. There are 72 trucks that bring milk from farms to Organilk. They arrive at the plant starting from 7:00 am to 1:00 pm at almost a constant rate of 12 trucks per hour. Each truck brings a full tank of milk which is around 5000 gallons (a total of $72 \times 5000 = 360,000$ gallons). Milk is then unloaded into the outdoor tanks to be used by the processing plant. Outdoor tanks have a total capacity of storing 50,000 gallons of milk. The unloading process is very fast, if the space is available in outdoor tanks. If outdoor tanks are full, trucks cannot be unloaded and they must wait until space becomes available in outdoor tanks. After unloading a truck, the truck leaves the plant. The plant has the capacity of processing 40,000 gallons of milk per hour, which are taken directly from outdoor tanks. The plant starts processing at 7:00 am when the first truck arrives and works until it finishes processing all 360,000 gallons of milk delivered on the same day. Use the continuous approximation of inventory curve and find the answers to the following questions:

    a. When does the queue of trucks begin to build?

    b. What is the maximum number of trucks waiting in the queue to unload their milk?

    c. What is the average gallons of milk waiting in the tanks or in the trucks?

    d. How many of the 72 trucks end up waiting and what is the average waiting time for those trucks?

4.  Chrome Wheels (CW) Inc. makes gear alloy wheels for mid- and large-sized cars and trucks. CW uses a base gear that is machined and reshaped based on customers' orders from CW online catalogs. Orders for custom-made GAWs can be divided into three groups: (i) standard, (ii) ex-chrome, and (iii) ultra-chrome. The company receives an average of 20, 4, and 8 orders per day for standard, ex-chrome, and ultra-chrome. Since ultra is CW's high-end product which also has a higher profit margin, CW gives the highest priority to producing ultra-chrome orders, and starts producing other types of orders only when there is no outstanding ultra-chrome orders. Ex-chrome orders have second priority and standard orders have the lowest priority. CW starts producing standard orders only when there are no ultra- and ex-chrome orders outstanding. When an order is finished, it is shipped to the customer on the same day. Production data shows that the average time it takes CW to finish production of an order after it receives the order is about 6 days. Data also shows that there are an average of 10 ultra-chrome and 8 ex-chrome orders outstanding.

    a. What is the total average number of outstanding orders at CW's production system?

    b. What is the average waiting times for customers of CW who order standard wheels?

    c. What is the average waiting time for ex-chrome and ultra-chrome customers?

5.  Global Trans Co, a logistics company, has 30 ships that deliver goods from a port in France to other parts of Europe and vice versa. The firm also owns one loading and unloading (L&U) station in a port in France. When a ship returns to the port in France, it waits behind other ships (if there are any) in L&U station for its turn to unload its shipment and load new shipment. After the ship is loaded, it leaves the port to deliver its shipment and returns to the port for its next shipment. Data shows that there are an average of six ships in the port

in France, and a ship spends an average of 2 days in the port before it is loaded and ready to leave.

a. If the firms earns about $50,000 for each ship delivering its shipment to Europe, what is the monthly earning of the firm (assume 30 days in a month)?

b. What is the average travel time of firm's ships? Specifically, how long it takes, on average, for a ship to return to France, after it leaves the port in France?

6. J.C. Nickel is a retail store that has a checkout counter with six cashiers and a self-checkout counter with two machines. Shoppers form a single line to go to one of the six cashiers, and another single line to use the two self-checkout machines. A study of the checkout process has shown that, on average, 120 customers per hour need to check out their items, some of whom use the self-checkout. The study also provided the following information: (i) the average time a cashier spends with a shopper to scan, receive payment, and bag her items is about 3 minutes, and on average, there are about one of the six cashiers idle. (ii) There are an average of 10 shoppers waiting in the line to go to cashiers. (iii) Self-checkout also has an average of one machine idle. (iv) The average time from when a shopper joins the line for self-checkout until she bags her items and leaves the self-checkout counter is about 5 minutes.

a. How long should shoppers who do not use the self-checkout expect to wait in the cashier's line until they get to one of the cashiers?

b. What fraction of customers is using the self-checkout counter?

c. What is the average time a shopper uses a self-checkout machine?

d. What is the average time a shopper spends checking out her items at J.C. Nickel?

e. Store manager is worried about the long waiting time for those shoppers who do not use self-checkout and is thinking of ways to reduce that time. John, the assistant manager, claims that if they hire two more cashiers, the line for cashiers will be reduced to half and the waiting time in the line will be shortened to less than 2 minutes. What do you think about his claim? Assume the number of customers using self-checkout and cashiers do not change.

7. Car Doctor is a car service shop that provides only two types of services—engine services and wheel services. The shop has two main departments: (i) engine department that performs services related to engine (e.g., oil change, fuel and electrical system repair), and (ii) wheel department that performs services on wheels (e.g., tire rotations, brake services). Cars are served in both departments on the first-come-first-served basis. The customers of Car Doctor can be classified into three groups: (i) Group 1 are those who only require engine services, (ii) Group 2 are those who only require wheel services, and (iii) Group 3 are those who require both engine and wheel services. Cars in Group 3 first get their engine serviced and are then sent to wheel department for wheel services. On average, 20 cars per hour arrive at the shop, and 20 percent of those are Group 1 customers who require only engine services. The shop's records show that there are an average of 24 and 2 cars, respectively, in the engine and wheel departments (being repaired or waiting to be repaired) and cars stay in the engine department for about 2 hours (before they leave the shop or are sent to the wheel department).

a. What is the average times Group 1 customers wait for their cars to be fixed? What is this time for Group 2 and Group 3 customers?

b. If the average profits that Car Doctor makes by serving a Group 1, a Group 2, and a Group 3 customers are $60, $40, and $90, respectively, then what is the total average profit that Car Doctor makes in an hour?

8. Davidson Realty is a firm that owns several parking structures in a large metropolitan area. One of their parking structure in the business area of downtown has the capacity of accommodating 400 cars. Cars arrive at the rate of 60 per hour and enter the parking if the parking

is not full. If the parking is full, cars go to another parking structure in downtown. Data shows that the parking is full about 10 percent of the time. The time cars park in the structure varies between 30 minutes and 12 hours, with an average of 7 hours. Define "Parking Utilization" as the percentage of spaces in the parking that are occupied by cars.

    a. What is parking utilization for this parking structure?

    b. The firm charges $2 per hour (25 cents every 15 minutes) that a car parks in the structure. What is the average hourly revenue that the firm makes in an hour?

    c. What would be the firm's revenue if it charges a fixed cost of $30 per car, regardless of how long cars park in the structure. The firm believes that this new pricing will have two major impacts: (i) it will reduce parking utilization to 85 percent, and (ii) it will increase the average time cars park in the structure from 7 to 10 hours. However, because customers are people who work in downtown, they will not park their cars for more than 24 hours. What would be the daily revenue of the firm if it adopts this new pricing strategy?

9. The number one challenge in management of call centers is employee turnover. Limited opportunity for career advancement and employee burnout are among many reasons that call center agents quit their jobs after a few months of work. JCD Insurance experiences this first hand in its call center. JCD call center hires agents and trains them to answer customer calls about different types of insurance. After training, these agents work at the *entry level* with an annual salary of $45,000. Data has shown that 60 percent of these hires quit their jobs after an average of 4 months. The remaining stay with the company, and after about 1 year, they are promoted to the *senior level* after going through a short training. The annual salary of a senior-level agent is about $52,000. Data also shows that senior-level agents stay with JCD for about 2 years before they find better jobs and leave the firm. The average number of senior-level employees has been about 20 and the firm would like to maintain it at that level.

    a. What is the average number of entry-level agents that are working for JCD?

    b. To maintain the number of senior-level agents at 20, how many entry-level agents must JCD hire each year?

    c. How many agents, on average, are leaving the firm each year?

    d. If training of a new hire costs JCD $3500 and training for a senior-level position costs JCD $4000, what is JCD's total annual personnel cost (including salary and training costs)?

10. A car rental company located at a major airport has a fleet of 230 cars, including 80 compact cars, 120 sedans, and 30 SUVs. The firm charges rental fee of $35, $45, and $60 per day for a compact, a sedan, and a SUV, respectively. When a car is returned to the company, it goes through a service (e.g., cleaning and fixing possible issues) which costs company an average of $10 per car. A study has shown that when a compact car is returned by a customer, the car stays in company's lot at the airport for an average of 2 days before it is rented to another customer. This number is 3 days and 4 days for sedans and SUVs, respectively. The study also shows that the company has an average of 28 compacts, 40 sedans, and 8 SUVs available in their lot for customers to rent.

    a. On average, how long customers keep (rent) compact cars? What about sedans and SUVs?

    b. Defining fleet utilization as the percentage of cars that are rented (i.e., not sitting idle at the airport), what is the fleet utilization for compact, sedan, and SUV vehicles?

    c. How many cars, on average, the company rents each day?

    d. If the company's fixed operating costs (i.e., employees' salary, facility and equipment cost, etc.) is $3000 per day, what is the company's daily profit?

11. Home Depot and Lowes are two American home improvement firms with a large number of retail stores that sell tools, home appliances, construction products and services. If you

search for their financial performance, you find the following figures (in thousands) in their 2018 income statements and balance sheets.

|  | Lowes | Home Depot |
|---|---|---|
| Total revenue | 68,619,000 | 100,904,000 |
| Cost of revenue | 45,210,000 | 66,548,000 |
| Gross profit | 23,409,000 | 34,356,000 |
| Inventory | 11,393,000 | 12,748,000 |

One of the products that both companies sell is a table saw which is sold for $499 in both stores. Suppose that both companies buy the table saw from the same supplier at the same price. If it costs both companies $75 to keep one table saw in inventory for 1 year, which company makes more profit selling a table saw? How much more? Assume that other operating costs are the same for each table saw for both companies.

# CHAPTER 3

# Forecasting

## 3.0 Introduction

Operations Engineering and Management corresponds to decisions of matching supply with demand to achieve a certain performance (e.g., minimize cost or maximize revenue, or both). This is often an easy task when supply and demand are known with 100 percent certainty. For example, suppose that an appliance store sells exactly 10 units of a particular refrigerator each week. Also suppose that the refrigerator manufacturer delivers refrigerators to the store every Monday morning. How many refrigerators should the store order for the next week? Because, there is no uncertainty in demand (i.e., demand is exactly 10 units

| Week | Demand | Week | Demand | Week | Demand | Week | Demand | Week | Demand |
|------|--------|------|--------|------|--------|------|--------|------|--------|
| 1 | 11 | 21 | 10 | 41 | 5 | 61 | 9 | 81 | 10 |
| 2 | 10 | 22 | 7 | 42 | 14 | 62 | 11 | 82 | 8 |
| 3 | 11 | 23 | 13 | 43 | 10 | 63 | 9 | 83 | 8 |
| 4 | 9 | 24 | 8 | 44 | 14 | 64 | 11 | 84 | 10 |
| 5 | 10 | 25 | 7 | 45 | 10 | 65 | 7 | 85 | 15 |
| 6 | 10 | 26 | 8 | 46 | 11 | 66 | 14 | 86 | 9 |
| 7 | 11 | 27 | 7 | 47 | 6 | 67 | 6 | 87 | 14 |
| 8 | 12 | 28 | 11 | 48 | 11 | 68 | 8 | 88 | 8 |
| 9 | 4 | 29 | 10 | 49 | 12 | 69 | 5 | 89 | 12 |
| 10 | 6 | 30 | 8 | 50 | 9 | 70 | 10 | 90 | 7 |
| 11 | 7 | 31 | 4 | 51 | 13 | 71 | 3 | 91 | 14 |
| 12 | 16 | 32 | 10 | 52 | 9 | 72 | 3 | 92 | 8 |
| 13 | 8 | 33 | 12 | 53 | 9 | 73 | 7 | 93 | 8 |
| 14 | 7 | 34 | 12 | 54 | 19 | 74 | 9 | 94 | 8 |
| 15 | 8 | 35 | 9 | 55 | 7 | 75 | 12 | 95 | 6 |
| 16 | 10 | 36 | 6 | 56 | 7 | 76 | 9 | 96 | 13 |
| 17 | 13 | 37 | 8 | 57 | 6 | 77 | 8 | 97 | 17 |
| 18 | 5 | 38 | 15 | 58 | 8 | 78 | 9 | 98 | 13 |
| 19 | 6 | 39 | 11 | 59 | 6 | 79 | 9 | 99 | 7 |
| 20 | 6 | 40 | 8 | 60 | 5 | 80 | 10 | 100 | 6 |

Figure 3.1    Demand for refrigerators in the last 100 weeks.

per week) and there is no uncertainty in supply (i.e., the supplier always delivers what is ordered on Monday morning), the decision is easy: The store should ask the supplier to deliver exactly 10 refrigerators next week. However, such cases do not occur in real world, because there is always uncertainty in demand or in supply or in both.

For example, suppose the numbers in Fig. 3.1 show the weekly demand for refrigerators in the last 100 weeks. How many refrigerators should the store order now for the next week?

Well, the answer depends on what the demand is going to be next week. As the data shows, the demand in each week of the last 100 weeks was somewhere between 3 and 19 units. Furthermore, looking at demand numbers, it is not easy to predict what the demand would be next week. Hence, to make a more informed ordering decision, we must first capture the underlying demand behavior and use it to develop an estimate (i.e., a forecast) for next week's demand. Without *modeling* the demand data into some particular form, we cannot develop analytics to *forecast* next week's demand and therefore cannot develop cost-effective ordering polices for products. These two are the focus of this chapter. Specifically, in this chapter we first illustrate how demand data are modeled so that they can be used in making informed operations management decisions. We then present principles and analytics for forecasting demand in future periods.

## 3.1 Modeling Demand

There are three commonly used demand models in Operations Engineering and Management: (i) Empirical Distribution Model, (ii) Poisson Distribution Model, and (iii) Normal Distribution Model.

### 3.1.1 Empirical Distribution Model

One approach is to develop the frequency histogram for the demand data, as shown in Fig. 3.2. For our refrigerator example, *Frequency histogram* simply represents the number of times during the last 100 weeks that each particular demand occurred.

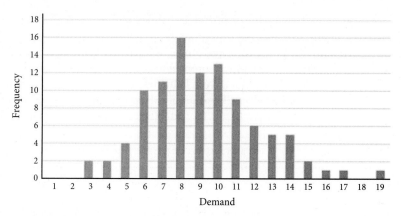

**Figure 3.2    Frequency histogram of weekly demand for refrigerators.**

The demand frequency histogram provides several useful information. For example, it shows which demand values occurred more often than others. As the figure shows, the demand of 8 refrigerators occurred in 16 out of the last 100 weeks, more than any other demands. The demand with the second highest frequency is 10, which was observed in 13 out of the last 100 weeks.

### Empirical Probability Distribution

The frequency histogram can also be used to estimate the probability of having any value of demand in a week. For each given value of demand, this is done by dividing the frequency of observing such demand in the last 100 weeks by 100. For example, the probability that the demand for refrigerators in a week is 8 can be computed as $16/100 = 0.16$. This probability is $13/100 = 0.13$ for demand of 10 units. Hence, if we define $D$ as the random variable representing demand for refrigerators in a week, then we have $P(D = 8) = 0.16$ and $P(D = 10) = 0.13$. These probabilities are called *Empirical Probabilities*. The empirical probabilities for all values of demand (i.e., $D = x$ for all values of $x$) are computed in Fig. 3.3.

| Demand value | Empirical probability | Cumulative probability |
|:---:|:---:|:---:|
| $x$ | $P(D = x)$ | $P(D \leq x)$ |
| 0 | 0 | 0 |
| 1 | 0 | 0 |
| 2 | 0 | 0 |
| 3 | 0.02 | 0.02 |
| 4 | 0.02 | 0.04 |
| 5 | 0.04 | 0.08 |
| 6 | 0.1 | 0.18 |
| 7 | 0.11 | 0.29 |
| 8 | 0.16 | 0.45 |
| 9 | 0.12 | 0.57 |
| 10 | 0.13 | 0.7 |
| 11 | 0.09 | 0.79 |
| 12 | 0.06 | 0.85 |
| 13 | 0.05 | 0.9 |
| 14 | 0.05 | 0.95 |
| 15 | 0.02 | 0.97 |
| 16 | 0.01 | 0.98 |
| 17 | 0.01 | 0.99 |
| 18 | 0 | 0.99 |
| 19 | 0.01 | 1 |

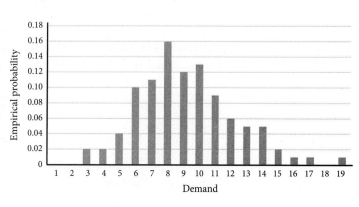

**Figure 3.3    *Left:* Empirical probability and cumulative probability distribution for refrigerator demand. *Right:* Empirical probability histogram.**

The empirical probability distribution can also be represented in a graph similar to the frequency histogram by simply replacing the frequency of occurrence in Y-axis in Fig. 3.2 by the corresponding empirical probabilities, as shown in Fig. 3.3.

Empirical probability distribution can also be used to compute the probability of demand in a week being less than or equal to a certain number. This is called *Cumulative Probability*. For example, the probability of having a demand of 8 for refrigerators in a week is $P(D = 8) = 0.16$, while the cumulative probability of demand in a week being 8 or less is

$$P(D \leq 8) = P(D = 0) + P(D = 1) + P(D = 2) + \cdots + P(D = 7) + P(D = 8)$$
$$= 0 + 0 + 0 + \cdots + 0.11 + 0.16 = 0.45$$

In other words, in 45 percent of the past 100 weeks the demand for refrigerators has been 8 or less. The cumulative probability distribution for all values of demand is shown in Fig. 3.3.

Empirical probability distributions of demand are useful when the number of possible weekly demand is not large. As the number of possible values of demand in a week increases, it becomes more and more difficult to use the empirical distribution. For example, suppose that the weekly demand for refrigerators ranges from 100 to 1100. In this case, the histogram will include 1000 values for demand (1000 columns in the histogram). But that is not the only problem. It is also more difficult to develop analytical models when the demand is modeled by empirical distributions. Due to these difficulties, empirical distributions are often approximated by well-known distributions that have more tractable structures and are easier to work with. Two of the commonly used distributions are Poisson distribution and Normal distribution.

### 3.1.2 Poisson Distribution Model

One distribution that is often used for modeling demand is Poisson distribution. Poisson random variable represents the number of events in a given period (e.g., the number of sales in a week, or the number of defective items produced in one shift). The distribution is characterized by only one parameter, that is, $\lambda > 0$, and has the probability mass function as follows:

$$P(x) = \frac{e^{-\lambda}\lambda^x}{x!}; \qquad \text{for } x = 0, 1, 2, 3 \ldots$$

Poisson distribution has mean and standard deviation $\lambda$ and $\sqrt{\lambda}$, respectively. In other words, when the demand in a period is modeled by Poisson distribution, the average demand in the period is $\lambda$ and the standard deviation of the demand in the period is $\sqrt{\lambda}$.

To use Poisson distribution to model the demand for refrigerators in a week, we need to find an estimate for Poisson parameter $\lambda$. Recall that $\lambda$ is the average demand in a period. As

$$\overline{D} = \text{Average demand in a period}$$

then $\lambda = \overline{D}$. Hence, an estimate for $\lambda$ would be the average of weekly demand in the last 100 weeks. Taking the average of 100 demands in the table in Fig. 3.1, we find $\overline{D} = 9.25$. Then, the Poisson distribution that can be used to model weekly demand for refrigerators has parameter $\lambda = 9.25$, that is,

$$P(D = x) = \frac{e^{-9.25}(9.25)^x}{x!}; \qquad \text{for } x = 0, 1, 2, 3 \ldots$$

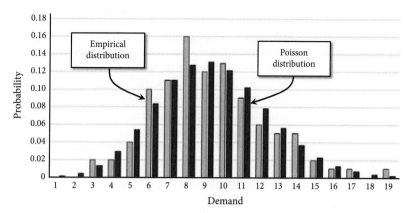

**Figure 3.4    Empirical probability distribution for refrigerator demand and its Poisson distribution approximation.**

Based on the above model, the probability that the demand for refrigerators in a week be 10 can be computed as follows:

$$P(D=10) = \frac{e^{-9.25}(9.25)^{10}}{10!} = 0.121$$

To simplify the computation of probabilities for Poisson distribution, statistics textbooks provide tables in which these probabilities are computed for different values of $\lambda$ and $x$. Excel also has function "POISSON (x, mean, False)" that computes Poisson distribution probabilities. For example, to compute the probability of demand in a week being $x = 10$, in an Excel cell we simply type

$$= POISSON(10, 9.25, False)$$

and Excel returns the value of 0.121.

Notice that according to empirical distribution the probability of demand being 10 is 0.13, which is about 0.009 larger than what the Poisson distribution model yields—not a big difference. Figure 3.4 compares the probabilities of weekly demand estimated by empirical and Poisson distribution side-by-side. As the figure shows, Poisson distribution is a good approximation for the weekly demand for refrigerators.

For Poisson distribution, the probability of demand being 10 or less (i.e., the cumulative probability) can be computed as follows:

$$\begin{aligned} P(D \leq 10) &= \sum_{x=0}^{10} P(D=x) = \frac{e^{-9.25}(9.25)^0}{0!} + \frac{e^{-9.25}(9.25)^1}{1!} + \cdots + \frac{e^{-9.25}(9.25)^9}{9!} \\ &\quad + \frac{e^{-9.25}(9.25)^{10}}{10!} \\ &= 0.676 \end{aligned}$$

The cumulative probability for demand of 10 units, using our empirical distribution shown in the table in Fig. 3.3, is 0.7, which is close to what is estimated by the Poisson distribution.

Excel also provides a function for computing cumulative probabilities of Poisson distribution. It is of the form "POISSON (x, mean, True)." We can compute the cumulative probability for demand value $x = 10$ by simply typing the following in a cell in Excel:

$$= POISSON(10, 9.25, True)$$

and Excel returns the value of 0.676.

Poisson distribution fits demand in many situations in practice. It has been shown that if there are a large number of customers making decisions of when to buy a product randomly and independent of each other, then the demand for the product in a period follows a Poisson distribution. This is the case for demand for refrigerators. The retail store most probably serves a large number of customer population. Customers make their decision of when to buy a refrigerator independent of other customers. The timing of the decision is random depending on when their refrigerators break, or when they renovate their kitchens, or when they get enough savings to pay for a new one. So, it is not surprising that, as shown in Fig. 3.4, Poisson distribution approximates the empirical distribution reasonably well. In fact, the question is whether the Poisson distribution is a better model for the demand than the empirical distribution. It is possible that by taking more samples of weekly demands we find that the empirical distribution resembles Poisson distribution.

### 3.1.3 Normal Distribution Model

Another important and commonly used distribution to model demand is Normal distribution. As opposed to Poisson distribution, which is a discrete probability distribution (i.e., demand can only take integer values), Normal distribution is a continuous distribution in which demand can take any values—any real number. While Poisson distribution has only one parameter (i.e., its mean $\lambda$), Normal distribution is characterized by two parameters: its mean $\mu$ and its standard deviation $\sigma$, and it has the following probability density function:

$$f(x) = \frac{1}{\sigma\sqrt{2\pi}}\, e^{-\frac{1}{2}[(x-\mu)/\sigma]^2} : \quad -\infty < x < \infty$$

Suppose that demand in a period has an average of $\overline{D}$ and standard deviation $\sigma_D$, then the demand can be approximated by a Normal distribution with the following probability density function:

$$f(x) = \frac{1}{\sigma_D\sqrt{2\pi}}\, e^{-\frac{1}{2}[(x-\overline{D})/\sigma_D]^2} : \quad -\infty < x < \infty$$

Therefore, if we decide to model the weekly demand for refrigerators with a Normal distribution, we first need to compute the average demand (i.e., $\overline{D}$) and the standard deviation of the demand (i.e., $\sigma_D$) in a week using the last 100 weekly demands in Fig. 3.1. Denoting demands in Week 1 to Week $N$ by $D_1, D_2, \ldots, D_N$, we can use the following formulas to compute the mean and standard deviation of weekly demand:

$$\overline{D} = \frac{D_1 + D_2 + \cdots + D_N}{N}, \qquad \sigma_D = \sqrt{\frac{1}{N-1}\sum_{i=1}^{N}(D_i - \overline{D})^2}$$

For our $N = 100$ demands, using the above formulas, we get $\overline{D} = 9.25$ and $\sigma_D = 3.05$. Hence, the Normal distribution model for weekly demand of refrigerators is

$$f(x) = \frac{1}{3.05\sqrt{2\pi}}\, e^{-\frac{1}{2}[(x-9.25)/3.05]^2} : \quad -\infty < x < \infty$$

Figure 3.5 shows our Normal distribution model of demand for refrigerators as well as the empirical probability distribution.

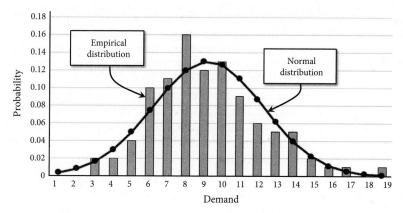

The cumulative distribution function of Normal distribution is

$$F(x) = P(D \le x) = \int_{-\infty}^{x} f(y)dy = \int_{-\infty}^{x} \frac{1}{3.05\sqrt{2\pi}}\, e^{-\frac{1}{2\pi}[(y-9.25)/3.05]^2}\, dy$$

As it is clear from the above equation, computing the cumulative probability for Normal distribution is not simple. However, we can "standardize" normal random variable and use the available look-up tables for standard Normal distribution to find the cumulative distribution for any value of demand. The idea behind standardization of Normal random variable is as follows. If random variable $D$ is Normally distributed with mean $\overline{D}$ and standard deviation $\sigma_D$, then the standardize random variable $Z$

$$Z = \frac{D - \overline{D}}{\sigma_D}$$

is Normally distributed with mean 0 and standard deviation 1. Standard Normal distribution, therefore, has density function $\phi(z)$ and cumulative distribution function $\Phi(z)$ as follows:

$$\phi(z) = \frac{1}{\sqrt{2\pi}}\, e^{-\frac{1}{2}(z)^2}$$

$$\Phi(z) = P(Z \le z) = \int_{-\infty}^{z} \phi(u)du$$

The relationship between the cumulative Normal distribution with mean $\overline{D}$ and standard deviation $\sigma_D$ and cumulative distribution of standard Normal is as follows:

$$F(x) = P(D \le x) = \Phi(z) \quad \text{where} \quad z = \frac{x - \overline{D}}{\sigma_D}$$

The above relationship allows us to compute probability of having certain demand values as well as the cumulative probabilities for any demand values. Suppose we would like to use Normal distribution to approximate the probability that the demand for refrigerators in a week is 10. Since Normal is a continuous probability distribution, but demand is a discrete random variable, the probability that discrete demand is 10 can be approximated by the probability that continuous Normal random variable $D$ is between 9.5 and 10.5, that is,

$$P(\text{Demand} = 10) = P(9.5 \leq D \leq 10.5)$$

By subtracting $\overline{D}$ from both sides of the above inequalities and dividing both sides by $\sigma_D$, we have

$$
\begin{aligned}
P(9.5 \leq D \leq 10.5) &= P\left(\frac{9.5 - 9.25}{3.05} \leq \frac{D - \overline{D}}{\sigma_D} \leq \frac{10.5 - 9.25}{3.05}\right) \\
&= P(0.082 \leq Z \leq 0.401) \\
&= P(Z \leq 0.401) - P(Z \leq 0.082) \\
&= \Phi(0.401) - \Phi(0.082) \\
&= 0.6554 - 0.5320 = 0.1234
\end{aligned}
$$

From Standard Normal table in The Appendix, we find that $\Phi(0.401) = 0.6554$ and $\Phi(0.082) = 0.5320$. Thus, using Normal distribution, we can approximate the probability of demand for refrigerators in a week being 10 to be 0.123, which is close to what we found when we used Poisson distribution (i.e., 0.121) and when we used empirical distribution (i.e., 0.13).

Now suppose we would like to find the cumulative probability that the demand for refrigerators in a week is 12 or less. Again, since we are using Normal distribution (i.e., a continuous probability distribution) to approximate the demand distribution (i.e., a discrete distribution) we have

$$P(\text{Demand} \leq 12) = P(D \leq 12.5)$$

By subtracting $\overline{D}$ from both sides of the inequality and dividing both sides by $\sigma_D$, we have

$$
\begin{aligned}
P(D \leq 12.5) &= P\left(\frac{D - \overline{D}}{\sigma_D} \leq \frac{12.5 - 9.25}{3.05}\right) \\
&= P(Z \leq 1.066) \\
&= \Phi(1.066)
\end{aligned}
$$

From Standard Normal table in Appendix N, we find that $\Phi(1.066) = 0.857$.

Similar to the case for Poisson distribution, Excel also has a function to compute the cumulative probabilities for Normal distribution. The function has the form "NORMDIST ($x$, mean, standard_dev, True)." Therefore, to find the cumulative probability for demand of $x = 12.5$, we must type the following in a cell in Excel:

$$= \text{NORMDIST}(12.5, 9.25, 3.05, \text{True})$$

and the Excel returns 0.857.

Another feature of Normal distribution is that it is also a good approximation model for Poisson distribution. The reason is that, as parameter $\lambda$ of Poisson distribution increases, the shape of Poisson distribution looks more like a Normal distribution. Considering the mean (i.e., $\lambda$) and standard deviation (i.e., $\sqrt{\lambda}$) of Poisson process, then a Normal distribution with mean $\mu = \lambda$ and standard deviation $\sigma = \sqrt{\lambda}$ is shown to be a good approximation of a Poisson distribution when $\lambda$ is larger than 20. This is another reason that Normal distribution is widely used to model demand in many operations management analytics. We also use Normal distribution to model demand in most of the inventory management models in this book.

One last point of caution is that, since Normal random variable can also take negative numbers, it must be used carefully when modeling demand. Because demand cannot be negative, Normal distribution can only be used to model demand if the probability of a negative value of demand (i.e., $P(D \leq 0)$) is very small (i.e., less than 0.01). This occurs when mean of demand is at least 2.4 times larger than the standard deviation of demand.

Empirical, Poisson, and Normal distributions try to capture the underlying demand behavior using past observations of demand. For example, based on the demand in the last 100 weeks, we can conclude that Normal distribution is a reasonably good model for weekly demand for refrigerators.[1] We also estimated that the average demand and standard deviation of demand in the last 100 weeks were 9.25 and 3.05, respectively. But, what would be the average and standard deviation of demand in the future weeks? Can we just simply use 9.25 and 3.05 as average and standard deviation of demand in future weeks? Or, is there a better way to find more accurate values for these parameters?

These questions bring us to the important task of forecasting future values of demand. While modeling demand provides us with the distributional form of demand per unit time, forecasting provides us with a more accurate estimate of the *parameters* of those distributions (e.g., accurate estimates of $\lambda$ for Poisson distribution model and $\mu$ and $\sigma$ for Normal distribution model). Hence, the rest of this chapter is devoted to forecasting future values of demand.

## 3.2 Forecasting

Many firms have invested millions of dollars on systems that help them better forecast and plan for the future. Forecasting is indeed one of the most critical functions of a firm that is essential for many business decisions. Consider forecasting demand as an example. If a forecast for demand of a product in the next month is not known, a firm cannot develop a cost-effective plan for the next month's production, inventory, and labor requirements. Forecasts must be as accurate as possible. Poor forecasting can have significant consequences on a firm's profit. If, for example, the forecast underestimates the actual demand, the firm would face lost sales and overtime costs as well as unsatisfied customers. On the other hand, if the forecast overestimates the demand, the firm would end up with unused capacity and excess inventory. Hence, it is important that firms develop accurate forecasts for their future demand.

One successful example of using forecasts in making operational decision is Taco Bell. In 1980s Taco Bell had several issues with their labor management. Specifically, because 52 percent of sales occur during the 3-hour lunch period, scheduling labor during a day was difficult. To find a better labor scheduling for each hour, Taco Bell installed a computerized system to gather data and forecast demand in every 15-minute interval of each day. The forecast was then used to develop a cost-effective labor allocation in each hour. In its first 4 years, the system saved about $40 million in labor cost.[2] Interestingly, the best forecasting model that fitted Taco Bell's demand was to average the demand of the same 15 minutes of the same day of the last 6 weeks to get an estimate of demand for the same 15 minutes of the same day of the next week. This is called a 6-week moving average method and will be discussed later in this chapter.

---

[1] There are many statistical tests such as Goodness-of-Fit test that can be used to check whether a distribution fits data.
[2] From Winston, W.L. and S.C. Albright, *Practical Management Science*. South-Western, Cengage Learning, 2009.

Forecasting is the process of estimating the future value of some variable. The focus of this chapter is to present different methods and analytics to forecast demand. While the focus is on forecasting demand, the methodology can also be used to forecast other variables such as price, capacity, or supply. Before we introduce forecasting methods, we first need to present some fundamental principles of forecasting.

## 3.3 Principles of Forecasting

Forecasting is difficult. The main reason is that there are many factors that affect the demand for a product. Consider the demand for a large screen TV, as an example. Its demand depends on several factors such as the quality of the TV, its features, its selling price, the competition price, state of economy, and other factors that are unknown. Forecasting models are designed to capture the effects of some of these factors and develop a reasonably accurate prediction of future demand. These models are different in their complexity, the type of data they need, and in the way they develop forecasts. However, they all share the following principles. These principles give a better understanding of the benefits and limitation of forecasting methods and prevent misinterpretation of the forecasts developed by these methods.

*1. Forecasts are usually wrong, but they are useful:* It is important to point out that even the forecasts made by the most sophisticated forecasting methods are wrong more than being right. As mentioned before, there are many known and unknown factors that affect the demand of a product. Hence, it is almost impossible to predict the exact demand. Managers should, therefore, not consider a forecast as a known and 100 percent accurate information. The goal of forecasting methods is to find a close estimate of the demand, which is then used in making planning decisions. But, the planning process should take into account the possibility of errors in the forecast.

*2. A good forecast also includes a measure of error:* Because forecasts are usually wrong, a good forecast should also provide some measure for the accuracy of the forecast. The measure should correspond to the magnitude of the error that the forecaster should expect. Examples are sum of squared errors, or variance of errors, see Sec. 3.8.3.

*3. A good forecast is more than just a single number:* A forecast for demand in the next period is a single number that provides an estimate for what the demand in the next period is expected to be. This single number is called *Point Forecast* for demand. However, because there is variability in demand, we know that the probability that next week's demand will be exactly what we forecasted is very low. So, it is useful to also forecast a range that contains next week's demand with high probability (e.g., 95 percent). This is called *Prediction Interval Forecast*. A good forecast should include a point forecast for the *average demand* in the next period, as well as a prediction interval forecast that contains the *actual demand* with high probability (see Sec. 3.8.5).

*4. Forecast for a group of products is more accurate than the forecast for each individual product:* For example, forecast for demand of blue-color two-door car of a particular model with leather seats in a year is less accurate than the forecast for the total demand of all car models of different colors with different options. The reason is that the demand for a specific item is often affected by more factors. The demand for the blue color and leather seat car in a year, for example, is affected by the popularity of the blue color and leather seat in that year. However, the impacts of these factors—color and seat options—disappear when the demand for all cars are forecasted. When total number of cars are considered, the impact of the popularity of a certain color is canceled out by the unpopularity of other colors. Hence, the data for a group of products is often more stable, even when each individual product has very variable and unstable demand.

**5. Forecasts for shorter time horizons are more accurate:** This is quite clear. It is easier to forecast next month's demand for a DVD player than the next year's demand for the DVD player. The reason is that, as we go further into the future, there is a higher chance that some factors affecting the demand change, but this is less likely in shorter time horizons. For example, in longer time horizons, there is a higher possibility that new DVD players with better features are introduced to the market.

**6. Forecasts are not substitute for derived values:** There are cases in practice that the demand for an item can be calculated using some available information. In such cases, forecasting methods should not be used. One common case is predicting demand for a part used in assembly of a product. As an example, suppose that the demand for cars is forecasted and a production schedule is set to assemble 500 cars per day. Considering that each car requires four tires, the demand for tires would be $4 \times 500 = 2000$ tires a day. In this case, there is no need for forecasting demand for tires.

## 3.4 Forecasting Methods

The decision of which forecasting method to use depends on the availability of data. The availability of the data, as shown in Fig. 3.6, divides the forecasting methods into the two groups of *Qualitative* (or subjective) and *Quantitative* (or objective) methods. Quantitative methods use past demand data to create forecasts, while qualitative methods do not use any data to do so.

As Fig. 3.6 also shows, there are two types of quantitative forecasting methods: (i) Causal Forecasting Methods, and (ii) Time Series Forecasting Methods.

*Causal Forecasting Methods* use the past relationship between demand and other variables to forecast future demand. For example, the demand for cold drink in each week may depend on the price of the drink, the competition price, and the amount of advertising for the drink (measured in dollars) in the local media. Causal methods use past data to capture the relationship between the demand in the past weeks and their corresponding prices and advertising dollars, to forecast the demand in the next week. Because causal methods use the association between demand and other variables (e.g., price) to develop forecast,

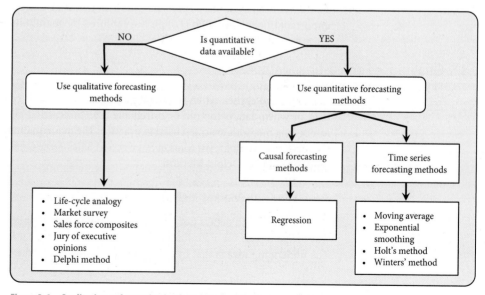

Figure 3.6   Qualitative and quantitative forecasting methods.

they are also called "Associative Methods." The main causal method that is vastly used in forecasting is Regression.

*Time Series Forecasting Methods,* however, use a different approach. The idea behind time series forecasting methods is that demand for any product has a particular pattern. Time series forecasting methods, therefore, try to capture the patterns of demand from past demand data and use that to forecast demand for future. In the case of the cold drink, time series forecasting methods use only weekly demand in the past several weeks to develop a forecast for the demand in the next week. Since, time series forecasting methods develop forecast by extrapolating the past patterns of demand into the future, they are also called "Extrapolative Methods." The main time series forecasting methods that are used, as shown in Fig. 3.6, are Moving Average, Exponential Smoothing, Holt's Method, and Winters' Method.

While quantitative methods rely on data to develop a forecast, qualitative methods are primarily based on judgment opinion of experts in the industry. Qualitative methods are used when there are no historical demand data available, or when the available data is not reliable. One example is when a firm introduces a new product to the market. In that case, there is no historical demand data available for the product. Another example is when the forecast is made far into the future and hence the current data might not be reliable. Consider the producer of the cold drink that wants to forecast demand for its drink in the next 10 years. The forecast will be used to make a decision whether to build another plant to increase capacity. In this case, the demand data in the last years may not be reliable to forecast demand 10 years into the future. The reason is that, during the next 10 years, factors that affect demand can change. This may include changes in customers' drinking habits, health concerns about having less sugar, new competition entering the market, etc.

One important point to emphasize is that a better forecast can be obtained if quantitative and qualitative methods are combined. The reason is that there are always some factors that are not or cannot be captured by quantitative methods. However, experts and people who have experience with the product and the market often have insights on how those factors impact forecast. Hence, they can further improve the forecast suggested by the quantitative methods. One example is the case of a manufacturer of food-service equipment.[3] Using a quantitative method, the firm was able to reduce its forecast errors by 60 percent. When a qualitative method was used together with the quantitative method, the error was further reduced by about 40 percent. This is an evidence that qualitative method can capture important information that is often not captured by quantitative methods.

## 3.5 Qualitative Forecasting Methods

As mentioned earlier, when no quantitative data is available (e.g., demand for a new product), or when data is too old or unreliable (i.e., forecasting far into the future), qualitative forecasting methods are often used in practice. The main qualitative forecasting methods are: (i) Life-Cycle Analogy, (ii) Market Survey, (iii) Sales-force Composite, (iv) Jury of Executive Opinion, and (v) Delphi Method.

### 3.5.1 Life-Cycle Analogy Method

When the demand data is not available for a new product or service, one can use the historical demand data for a similar existing product to provide a forecast for the new product. The underlying idea is that by analyzing the demand data during different stages of the life

---

[3] Wilson, J. H. and D. Alison-Koerber, "Combining Subjective and Objective Forecasts Improve Results." *The Journal of Business Forecasting Methods & Systems.* Vol 11, No.3 (1992), p. 15.

cycle of the existing product (e.g., introduction, growth, maturity, and decline), one can get an estimate for the demand of the new product. For example, it is possible to get a reasonably accurate estimate for demand of 3D televisions in different stages of their life cycle, by analyzing the demand of LED televisions during the same stages of their life cycles. Life-cycle analogy method is used to get forecasts for long-range sales for a new product in order to determine capacity and plan for new production facilities. Their accuracy has been shown to be good for medium (i.e., weeks to months) and long-range (i.e., months to years) forecasts.

### 3.5.2 Market Survey Method

The demand for a product obviously depends on customers' decisions to buy the product. Therefore, to forecast demand, it would be very useful to survey market and find customers' opinions of the existing products, their buying habits, their likes and dislikes of the competitors' products, and their opinion about new product ideas. Market survey is done by marketing departments using questionnaires, interviews, and customer focus groups.

Market survey is often used by firms to forecast their total sales for a family of products or an individual product. While the cost of performing a market survey is high, it provides good short- and medium-term forecasts. To have a more accurate forecast, market survey should be carefully designed. Questions should be carefully chosen to capture the desired information and eliminate any biases. Customers and focused groups should also be chosen carefully so that they are representative of the customer base. If market surveys are poorly designed, they may result in wrong conclusions.

### 3.5.3 Sales-Force Composite Method

Besides customers who buy the product, sales-force who are in close contact with customers are another good source of information about the demand of a product. In fact, they are usually the closest contact that firms have with their customers. To develop a forecast for a product using sales-force composite, members of the sales-force are asked to submit their forecasts of the products that they expect to sell in the next period (e.g., next quarter or next year). They may be asked to provide a single forecast, or several forecasts to represent their optimistic, pessimistic, or most likely forecasts for the product. A person—often the sales manager—then gathers these forecasts, aggregates them, and develops an overall forecast for the demand of the product.

When aggregating sales-force forecasts, some important points must be considered. First, due to Recency Effect[4] the most recent demand often has more impact on sales-force's forecast. For example, it is more likely that sales-force provide a high forecast (a low forecast) when their most recent sales was high (was low). This introduces a bias in their forecasts of future demand. Second, if the sales-force are paid bonuses based on whether they achieve their sales quota, they will have incentive to provide a lower forecast for demand, which increases their chances of getting the bonus.

### 3.5.4 Jury of Executive Opinion Method

To develop a forecast, Jury of Executive Opinion method relies on opinions of experts who have expertise in firm's products, competitors' products, market condition, and general business environment. These include top managers and senior executives in the related industry,

---

[4]Recency effect is the principle that the most recent experiences will most likely be remembered best.

marketing consulting firms, and academics. Also, to consider different aspects of the business that may affect forecast, it is useful to select experts from different functional areas of the firm (e.g., finance, marketing, operations).

To develop a forecast, this method combines the opinion of the experts in different ways. One approach is for the person in charge of the forecast to interview the executives directly and develop a forecast according to the results of the interviews. Another approach is to have all executives meet as a group, discuss their various perspectives of the market, and come to a forecast that all or most agree.

The accuracy of the forecast developed by the Jury of Executive Opinion method depends on the knowledge of the executives about the product, the overall market, and the industry. Also, one issue with the Jury of Executive Opinion method is the possibility that one person's opinion dominates the discussion. This can occur when the person has more power than others in the meeting (e.g., the person has a higher executive position in the firm or in the industry), or if the person has a strong and dominating personality.

### 3.5.5 Delphi Method

The Delphi method was developed by the RAND corporation in the 1950s. Similar to the Jury of Executive Opinion method, this method relies on knowledge and insights of a group of experts to develop a forecast. The Delphi method has the advantage of anonymity among the participants. The experts in the Delphi method never actually meet, they may not even be in the same country, they do not know the identity of other members, and never discuss their views with each other. The idea behind the Delphi method is that experts would have different ideas that might or might not agree with others' opinions. The basic premise is that, while experts may disagree on some things, but what they do agree on will have a higher probability of happening. Hence, the job of the person in charge of the Delphi process is to determine what the experts do agree on and turn that into a forecast.

Here is how the Delphi method works. The person in charge of the forecast asks each member of the panel to submit their individual forecast, mainly by asking them to complete a questionnaire. The results of the questionnaire are then compiled and a summary of the forecasts of all members is shared with all experts on the panel. The panelists are then allowed to change their initial forecasts based on the forecasts and the opinions of other panelists written in the summary they received. They are asked to send their revised forecasts with their opinions back to the person in charge. This process continues in the same manner and the forecast is further refined in each iteration until (ideally) the experts reach a consensus.

The Delphi method has several advantages. First, because experts do not need to physically be in one location, it has a large pool of experts to choose from, even those who are in other countries—without much cost or effort. Second, because experts do not physically meet and are anonymous, panelists get equal chance of expressing their opinions. This eliminates the risk of one member's opinion dominating the result. Third, because panelists have longer time to think about their response and must submit a written response, their response would be more carefully prepared and better thought of (compared to a verbal response). Fourth, under the Delphi method more ideas about the forecast are generated and discussed, since each panelist introduces their ideas in their subsequent forecasts. Some of these ideas, however, could be overlooked in face-to-face meeting because of interruptions and group dynamics.

The Delphi method has some disadvantages too. The method is highly sensitive to the way the questionnaires are designed and the way the responses are compiled and summarized. Also, because some of the ideas are eliminated in each iteration, experts may not get a chance to defend or justify those ideas. This does not occur in face-to-face meeting, since experts get a chance to clarify and defend their ideas. Finally, the Delphi method is a time-consuming method for the participants. Having time lags between successive questionnaires may result in participants lose their focus. Also, there is a possibility that the panel does not reach a consensus. However, when it is designed and implemented properly, the Delphi method has been shown to be a great method for long-term demand forecasting, capacity and facility planning, and predicting technological advances.

Having discussed the qualitative forecasting methods, the rest of this chapter presents different quantitative methods that use time series data.

## 3.6 Types of Time Series

*Time series* is a sequence of observations about a quantity taken at successive and equally spaced points in time. The number of calls a call center receives every 15 minutes, the number of customers arriving to a fast-food restaurant every hour, the number of TVs sold in an electronic retail store every week, and the monthly demand for a car model are all examples of time series. As mentioned before, time series forecasting methods use the pattern in past values of the time series to develop forecasts for the future values of the series. One essential first step to use time series forecasting methods is to identify the type of time series. Time series are classified in four main types according to how the values of the series change as a function of time. As Fig. 3.7 shows, there are four main types of time series: (i) stationary time series, (ii) time series with trend, (iii) seasonal times series, and (iv) cyclic time series.

### 3.6.1 Stationary Time Series

Stationary time series is a class of time series in which the values of the series fluctuate around a constant mean (see Fig. 3.7). This type of time series represents demand for products that

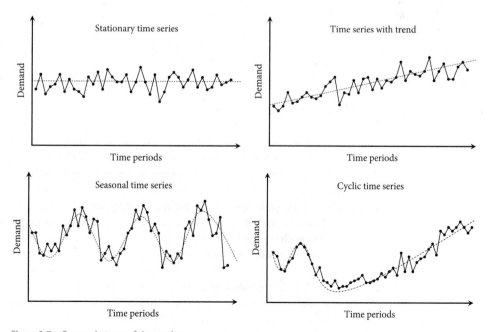

**Figure 3.7    Four main types of time series.**

are in the mature stage of their life cycle; hence, they have a stable demand with no significant increase or decrease over time. Examples of products that exhibit stationary pattern in their demand are milk, cereal, bread, toothpaste, and soap. The (weekly or monthly) demand for these products is relatively constant, unless there is a significant increase or decrease in customer population.

### 3.6.2 Time Series with Trend

While in stationary time series the average demand does not change over time, time series with trend exhibit increase or decrease in average demand over time. Specifically, in time series with trend, the demand fluctuates around a mean that is gradually increasing or decreasing over time (see Fig. 3.7). This gradual shift is often the result of population growth or age, gradual cultural shifts, or change in customer preferences. In general, the demand for a product has an increasing trend in introduction and growth phases of its life cycle and has a decreasing trend in the decline phase of its life cycle. For example, when wireless mouse for personal computers was introduced to the market, its demand had an increasing trend, while the demand for wired mouse had a decreasing trend.

### 3.6.3 Seasonal Time Series

In seasonal time series, as shown in Fig. 3.7, the pattern of the data has an increasing trend followed by a decreasing trend (or vice versa). This increase and decrease pattern—a cycle—regularly repeats itself and each cycle has a constant length. Seasonal time series represents the demand for products that are affected by seasonal factors such as the quarter or month of the year, day of the week, or hours of the day. For example, the sales of ice cream or cold drinks have a quarterly seasonal pattern. Each year their sales increase and peak in summer, then start decreasing, and hit their lows in winter. Restaurants have a daily seasonal pattern: their demands change from high demand on Fridays and Saturdays to low demand on Mondays and medium demand on other days. Call centers of a credit card company, for example, have an hourly seasonal demand pattern. In each day, the demand changes from medium demand (from 5:00 am to 8:00 am), to high demand (from 8:00 am to 3:00 pm), to medium demand (from 3:00 pm to 10:00 pm), and to very low demand (from 10:00 pm to 5:00 am). Note that in all these examples, the seasonal pattern repeats itself—every year for ice cream, every week for the restaurants, and every day for call centers.

### 3.6.4 Cyclic Time Series

Similar to seasonal time series, cyclic time series also has a cycle consisting of high and low demands. However, there are two main differences between seasonal and cyclic time series: (i) The length of the cycles is known and is the same in seasonal time series (e.g., a year, a week, or a day); however, in cyclic time series, the length of a cycle is not clearly known and each cycle may have a different length. (ii) The length of cycles in a cyclic time series is often longer than that in a seasonal time series. Examples of cyclic time series are the demand for products that are significantly affected by factors such as fashion trends, politics, war, economic conditions, or sociocultural influences. These are factors that are extremely difficult, if not impossible, to forecast when they would occur or how long they would take. In seasonal time series, on the other hand, it is known when the demand would be high (e.g., Fridays and Saturdays in a restaurant) and when it would happen (e.g., each week). Due to the complexity and that it requires forecasting long into the future, qualitative forecasting methods are often used to generate forecasts for cyclic time series.

What all types of time series—stationary, trend, seasonal, and cyclic—have in common is the component of *randomness*. Randomness in a time series is generated by many unknown factors that cannot be predicted. This randomness is the underlying reason for having variability in demand and is what makes forecasting demand difficult. It makes demand fluctuate around its pattern and gives it the zig-zag shape, as shown in all four patterns in Fig. 3.7.

## 3.7 Time Series Forecasting Framework

Time series forecasting methods use the demand information in past periods to develop forecasts for future periods. In all forecasting methods in this chapter we assume that the current period is Period $t$ and data about the demand in last $t$ periods, that is, Periods $1, 2, 3, \ldots, t$ are known and denoted by $D_1, D_2, D_3, \ldots, D_t$, respectively, see Fig. 3.8.

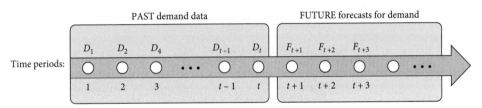

**Figure 3.8    Available data and future forecasts in forecasting methods.**

These methods use this past demand information to develop *One-Period-Ahead Forecast* for Period $t + 1$, denoted by $F_{t+1}$ and *Multiple-Period-Ahead Forecasts* for Periods $t + 2$, $t + 3$, etc., denoted by $F_{t+2}, F_{t+3}$, etc.

Figure 3.9 shows five steps that are required for generating, monitoring, and improving forecasts made using time series forecasting methods. The first step is to determine the type of the time series. This is important, since each type of time series requires a different method for forecasting. Graphing the time series data along with the understanding of the product and the market can help identify the type of a time series.

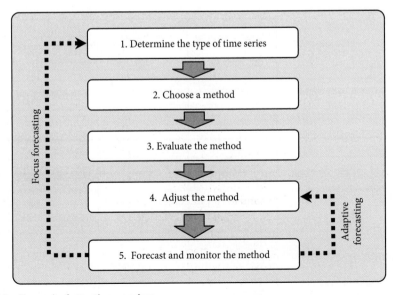

**Figure 3.9    Time series forecasting procedure.**

After the type of time series is determined, the next step is to choose a method—among those applicable for the time series—to develop a forecast. In the third step, an evaluation criterion must be chosen and used to evaluate the forecasting method. The evaluation criterion is usually the errors made by the forecasting method. The larger the error, the less accurate the forecast. An example of an evaluation criterion is the sum of squared errors of past forecasts made by the method. To improve the performance, in Step 4, the parameters of the forecasting method are adjusted to result in better forecasts based on the evaluation criterion chosen (e.g., resulting in lower sum of squared errors). Finally, after the method is adjusted, the forecast method can be used to generate forecast.

This, however, is not the end of the process. The errors in the forecasts must be monitored constantly and actions must be taken to improve forecasting methods. This is done through Adaptive Forecasting and Focus Forecasting that are described in Sec. 3.12.

In the following section, we focus on developing forecasts for stationary time series and we explain the above framework in more detail.

## 3.8 Forecasting Stationary Time Series

To show how time series forecasting methods can be used to develop forecasts, we use the following example.

 **Example 3.1**

Oatplex is a breakfast cereal with steady sales in the last 3 years in a food department store. Table 3.1 shows the demand for Oatplex in each week of the last 100 weeks. To manage and control the inventory of Oatplex, manager of the store must forecast the demand for the cereal in the next 2 weeks. What do you think the demand in the next 2 weeks would be?

Table 3.1    Demand for Oatplex Cereal Boxes in the Last 100 Weeks in Example 3.1

| Week | Demand | Week | Demand | Week | Demand | Week | Demand | Week | Demand |
|------|--------|------|--------|------|--------|------|--------|------|--------|
| 1 | 150 | 21 | 212 | 41 | 212 | 61 | 277 | 81 | 193 |
| 2 | 199 | 22 | 288 | 42 | 221 | 62 | 233 | 82 | 255 |
| 3 | 187 | 23 | 257 | 43 | 185 | 63 | 221 | 83 | 191 |
| 4 | 145 | 24 | 253 | 44 | 231 | 64 | 217 | 84 | 220 |
| 5 | 155 | 25 | 248 | 45 | 242 | 65 | 260 | 85 | 191 |
| 6 | 140 | 26 | 242 | 46 | 240 | 66 | 285 | 86 | 187 |
| 7 | 168 | 27 | 237 | 47 | 246 | 67 | 220 | 87 | 244 |
| 8 | 160 | 28 | 220 | 48 | 316 | 68 | 258 | 88 | 182 |
| 9 | 226 | 29 | 177 | 49 | 252 | 69 | 190 | 89 | 137 |
| 10 | 261 | 30 | 185 | 50 | 265 | 70 | 204 | 90 | 121 |
| 11 | 198 | 31 | 244 | 51 | 228 | 71 | 185 | 91 | 125 |
| 12 | 215 | 32 | 224 | 52 | 215 | 72 | 225 | 92 | 227 |
| 13 | 254 | 33 | 151 | 53 | 267 | 73 | 215 | 93 | 212 |
| 14 | 185 | 34 | 173 | 54 | 198 | 74 | 225 | 94 | 204 |
| 15 | 298 | 35 | 171 | 55 | 239 | 75 | 147 | 95 | 194 |
| 16 | 237 | 36 | 198 | 56 | 230 | 76 | 203 | 96 | 204 |
| 17 | 250 | 37 | 173 | 57 | 193 | 77 | 210 | 97 | 211 |
| 18 | 222 | 38 | 194 | 58 | 226 | 78 | 227 | 98 | 101 |
| 19 | 245 | 39 | 180 | 59 | 191 | 79 | 225 | 99 | 210 |
| 20 | 299 | 40 | 234 | 60 | 198 | 80 | 171 | 100 | 217 |

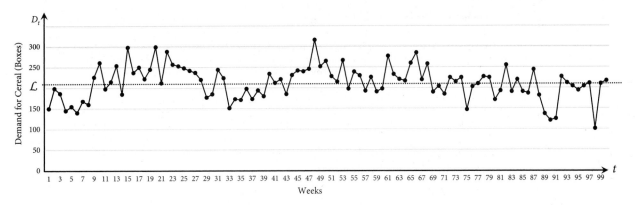

**Figure 3.10     Demand for Oatplex cereal boxes in the last 100 weeks.**

As mentioned before, time series forecasting methods identify the pattern in past values of the time series and use it to develop forecasts for the future values of the series. Hence, the first step, as shown in Fig. 3.9, is to identify the type of the time series.

### 3.8.1  Step 1—Determine the Type of Time Series

One approach that makes it easier to recognize the pattern in a time series is to graph the past data. Figure 3.10 graphs the demand data in Table 3.1 for Oatplex. As the figure shows, the demand in the last 100 weeks does not show any upward or downward trend or any seasonal pattern. The demand exhibits some temporary decrease or increase for a short number of periods, but overall, it fluctuates around a certain value, shown by $\mathcal{L}$ in Fig. 3.10. Hence, it is reasonable to assume that the weekly demand for Oatplex is a stationary time series. Another reason that makes stationary time series a good fit for Oatplex demand is that, as stated in the example, sales of Oatplex have been steady in the last 3 years. Oatplex seems to be in the mature phase of its life cycle. Choosing stationary time series as the demand pattern, we go to Step 2.

### 3.8.2  Step 2—Choose a Method

Since the underlying assumption of stationary time series is that the demand fluctuates around a mean that is constant, time series forecasting methods decompose stationary time series into two components:

- *Level:* Level corresponds to the underlying constant component (i.e., mean) of the data around which the data fluctuates.

- *Noise:* Noise or random variation corresponds to deviation of time series data in each period from its true level (i.e., true mean). Noise corresponds to the variability in demand. It obscures the underlying pattern—the constant level—of the stationary time series.

The noise in time series is the result of many unknown factors and cannot be predicted. For example, the demand for cereal in a week, among others, depends on how many customers run out of cereal in that week or the week before and decide to go to the store to buy cereal. One model that captures the pattern in stationary time series is

$$D_t = \mathcal{L} + \epsilon_t \qquad (3.1)$$

where

$$D_t = \text{Demand in Period } t$$

$$\mathcal{L} = \text{Average demand in a period, that is, the level}$$

$$\epsilon_t = \text{Random variation or noise in Period } t$$

While in some periods the noise results in demand being larger than its level—positive noise—in other periods the noise may result in demand being lower than its level—negative noise. However, regardless of the magnitude of the noise, a good forecasting model must have the following two features:

(i)   The positive and negative noise should cancel each other out and the noise must have a mean of zero. If the noise does not have a mean of zero, then it is not noise; it is the part of the underlying process and the method must be revised to capture this feature.

(ii)  The number of periods in which the noise is positive should be about the same as the number of periods in which the noise is negative.

Based on the above two points, it is reasonable to assume that noise in our model follows a *symmetric* distribution with mean $\mu_\epsilon = 0$ and a constant variance $\sigma_\epsilon$. It has been shown that considering Normal distribution for noise is a reasonable and robust assumption, even for cases where the noise does not exactly follow Normal distribution.

Returning to Example 3.1, the actual demand in (future) Weeks 101 and 102 would be the constant level $\mathcal{L}$ (which we do not know) plus noise components $\epsilon_{101}$ and $\epsilon_{102}$ (which we also do not know and can be positive or negative). So, how should managers in Example 3.1 develop forecasts for Weeks 101 and 102?

There are several methods that they can use. All of those methods work by averaging the past values of demand and hence are called "Averaging Methods."

### Averaging Methods

A good forecast finds an estimate for $\mathcal{L}$ that is not affected by the unpredictable noise that can be negative or positive—after all, noises have a mean of zero. *Averaging Methods*, also called *Smoothing Methods*, use the average value of past data to estimate $\mathcal{L}$. Therefore, averaging methods are not affected by the noise. This is because, when combined into an average, positive and negative noises offset each other. Hence, forecast based on average of past data shows less variability than the variability generated by noises in the actual data. In each Period $t$, an averaging method finds estimate $L_t$ to approximate $\mathcal{L}$. The estimate $L_t$ might be different in each Period $t$, since as $t$ increases, new demand is observed and a better estimate can be developed. The estimate $L_t$ is called

$$L_t = \textit{Average} \text{ or } \textit{smoothed} \text{ estimate of demand at the end of Period } t$$

After demand $D_t$ is observed in Period $t$, averaging methods compute $L_t$ and use it as the forecast for the demand in future periods. Considering

$$F_{t+1} = \text{Forecast for Period } t+1 \text{ (i.e., an estimate for } \mathcal{L}\text{)}$$

then averaging methods provide the one-period-ahead forecast of

$$F_{t+1} = L_t$$

Based on the assumption that the demand is stationary, and since we cannot predict the noise in future Periods $t+2, t+3, \ldots$, then the best multiple-period-ahead forecasts for these future periods are

$$F_{t+\tau} = L_t \quad \text{for} \quad \tau = 2, 3, \ldots$$

For averaging methods, the question is how to average past demand data $D_1, D_2, \ldots, D_t$ to compute the smoothed estimate $L_t$. Different methods use different approaches. In the following sections we discuss some of these methods, including: (i) simple average method, (ii) moving average method, (iii) weighted moving average method, and (iv) exponential smoothing method.

### Simple Average Method

One averaging method is to simply use the average of all past observations $D_1, D_2, \ldots, D_t$ to get the estimate $L_t$ and use it as the forecast for future demands. In the case of Oatplex cereal in Example 3.1, this method results in the following forecast:

$$L_{100} = \frac{150 + 199 + 187 + \cdots + 210 + 217}{100} = 212.74$$

which is the average of all past 100 observations of demand. The demand forecast for Weeks 101 and 102 are, therefore,

$$F_{101} = F_{102} = L_{100} = 212.74$$

We can update our forecast for Week 102 at the end of Week 101 when we observe the actual demand $D_{101}$.

### Moving Average Method

The simple average method assumes that, for forecasting demand in Week 101, the old data (e.g., demand for Oatplex in Week 1) is as important as the most recent data (e.g., demand for Oatplex in Week 100). While the main assumption of the stationary time series is that the base demand (i.e., level $\mathcal{L}$) is constant, in practice, however, the level is not constant all the time and may change gradually in a small amount. This change is not large and is not in one direction to create an upward or downward trend, and is not cyclic to create a seasonal pattern. Hence, the demand is still stationary, but the most recent data contains more information about the most recent changes in base demand ($\mathcal{L}$) than an old data.

An averaging method, called "moving average method," takes this into account and takes the average of the most recent (i.e., last $N$) observations to compute the smoothed forecast $L_t$ at Period $t$.

---

**ANALYTICS**

**Moving Average Method**

Consider a stationary time series of demands $D_1, D_2, \ldots, D_t$ in past $t$ periods. If the current period is $t$, the *moving average method with span N* computes the smoothed estimate $L_t$ after observing demand in Period $t$ as follows:

$$L_t = \frac{D_t + D_{t-1} + D_{t-2} + \cdots + D_{t-N+1}}{N} \tag{3.2}$$

- *One-Period-Ahead Forecast:* The forecast for the next period, that is, Period $t+1$, is

$$F_{t+1} = L_t$$

- *Multiple-Period-Ahead Forecast:* Since the time series is stationary, the forecast for all future Periods $t + 2$, $t + 3$, etc., is

$$F_{t+\tau} = L_t : \quad \text{for all} \quad \tau = 2, 3, \ldots$$

For example, if we use a moving average method with span $N = 3$, using data in Table 3.1, the smoothed estimate $L_{100}$ would be

$$L_{100} = \frac{D_{100} + D_{99} + D_{98}}{3} = \frac{217 + 210 + 101}{3} = 176$$

Therefore, $F_{101} = L_{100} = 176$. Again, since the demand for Oatplex is stationary, the forecast for Week 102 is also $F_{102} = L_{100} = 176$.

### Impact of Forecasting Span $N$

One important question is how the value of span $N$ of the moving average method affects the forecast generated by the method. To explore this, we add and subtract term $D_{t-N}/N$ from the right-hand side of Eq. (3.2) as follows:

$$
\begin{aligned}
L_t &= \frac{D_t + D_{t-1} + D_{t-2} + \cdots + D_{t-N+1}}{N} + \frac{D_{t-N} - D_{t-N}}{N} \\
&= \frac{D_{t-1} + D_{t-2} + \cdots + D_{t-N+1} + D_{t-N}}{N} + \frac{D_t - D_{t-N}}{N} \\
&= L_{t-1} + \frac{D_t - D_{t-N}}{N}
\end{aligned}
$$

Since $F_{t+1} = L_t$ and $F_t = L_{t-1}$, the above becomes

$$F_{t+1} = F_t + \frac{D_t - D_{t-N}}{N}$$

In other words, in moving average method with span $N$, the forecast for Period $t + 1$ is obtained by adding $(D_t - D_{t-N})/N$ to the forecast for Period $t$. As span $N$ gets larger, the moving average method becomes less responsive to the changes in demand. Specifically, for large $N$, the forecast for Period $t + 1$ (i.e., $F_{t+1}$) would still be close to the forecast for Period $t$ (i.e., $F_t$) even if the demand realized in Period $t$ (i.e., $D_t$) is significantly large or significantly small. This is because large (or small) demand is divided by a large $N$ in fraction $(D_t - D_{t-N})/N$. Figure 3.11 shows the forecast made for Oatplex cereal in Example 3.1 by two moving average methods with $N = 3$ and $N = 10$.

As the figure shows, the forecast with span $N = 3$ is more responsive to changes in the demand of Oatplex and reacts more quickly to random fluctuations in recent demands. Hence, choosing the right span for the moving average method is a trade-off between how important the most recent random fluctuations of demand is and how quickly we want the forecast to respond to changes in demand. For processes that are relatively stable, a large $N$ is more appropriate.

### Weighted Moving Average Method

The moving average method uses the average of $N$ most recent data to find smoothed estimate $L_t$ at Period $t$. The method assumes that, in predicting next week's demand, the demand in $N$ weeks ago ($D_{t-N}$) is as important as the last week's demand ($D_t$). Hence, it gives all $N$ demand data the same weight (i.e., $1/N$) when taking the average. Similar to the moving

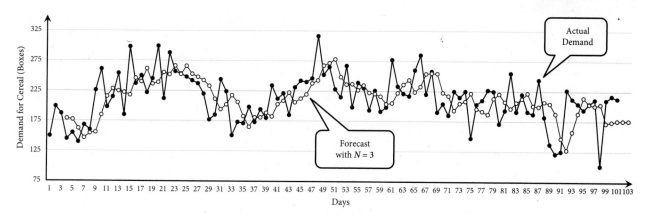

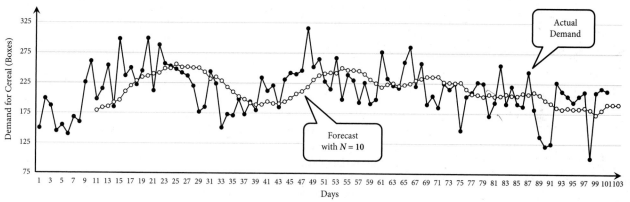

**Figure 3.11**    Actual demand and forecast made by moving average methods with span $N = 3$ (top) and $N = 10$ (bottom).

average method, the weighted moving average method considers the last $N$ observations; however, it gives higher weights to most recent data.

---

**ANALYTICS**

**Weighted Moving Average Method**

Consider a stationary time series of demands $D_1, D_2, \ldots, D_t$ for the past $t$ periods. If the current period is $t$, the *weighted moving average method* with *span N* and *weights* $w_1, w_2, \ldots, w_N$ obtains smoothed estimate $L_t$ after observing demand $D_t$ as follows:

$$L_t = w_1 D_t + w_2 D_{t-1} + w_3 D_{t-2} + \cdots + w_N D_{t-N+1} \qquad (3.3)$$

Note that $w_1 \geq w_2 \geq \cdots \geq w_N$, and $\sum_{i=1}^{N} w_i = 1$.

- *One-Period Ahead Forecast:* The forecast for the next period, that is, Period $t + 1$, is

$$F_{t+1} = L_t$$

- *Multiple-Period Ahead Forecast:* The forecast for all future periods, that is, Periods $t + 2, t + 3$, etc., is

$$F_{t+\tau} = L_t : \quad \text{for all } \tau = 2, 3, \ldots$$

For Oatplex cereal in Example 3.1, a weighted moving average with span $N = 3$ and weights $w_1 = 0.5$, $w_2 = 0.4$, and $w_3 = 0.1$ provides smoothed estimate in Week 100 as follows:

$$L_{100} = w_1 D_{100} + w_2 D_{99} + w_3 D_{98} = (0.5)217 + (0.4)210 + (0.1)101 = 202.6$$

and the demand forecasts for Weeks 101 and 102 would be $F_{101} = F_{102} = L_{100} = 202.6$.

### Simple Exponential Smoothing Method

Simple exponential smoothing method is another averaging method that recognizes the fact that the most recent observations are more important in forecasting future demand. Like simple average method, this method takes the average of *all* past data; and similar to the weighted moving average method, it takes the weighted average of those data and assigns more weights to more recent data.

Let's assume that we want to take a weighted average of the 100 observations for demand of Oatplex in Example 3.1. To do that, we need to determine 100 different weights $w_1, w_2, \ldots, w_{100}$, where $\sum_1^{100} w_i = 1$ and $w_1 > w_2 > \cdots > w_{100}$. Having these weights, the smoothed estimate in Week 100 would be

$$L_{100} = w_1 D_{100} + w_2 D_{99} + \cdots + w_{100} D_1$$

However, finding 100 different numbers between 0 and 1—which add up to 1 and are in decreasing order—is not easy. The simple exponential smoothing method uses a simple way to do this. By using only one number $0 < \alpha \leq 1$, which is the weight for the last observation, it generates the other weights as follows:

$$w_1 = \alpha, \quad w_2 = \alpha(1 - \alpha), \quad w_3 = \alpha(1 - \alpha)^2, \quad \ldots, \quad w_{100} = \alpha(1 - \alpha)^{99}$$

Notice that, since $0 < \alpha \leq 1$, weight $w_i = \alpha(1 - \alpha)^{i-1}$ decreases as $i$ increases. Also, simple algebra can show that when $0 < \alpha \leq 1$ we have $\sum_{i=1}^{\infty} w_i = \sum_{i=1}^{\infty} \alpha(1 - \alpha)^{i-1} = 1$. For $\alpha = 0.4$, for example, simple exponential smoothing method provides the smoothed estimated $L_{100}$ as follows:

$$\begin{aligned} L_{100} &= \alpha D_{100} + \alpha(1 - \alpha)D_{99} + \alpha(1 - \alpha)^2 D_{98} + \cdots + \alpha(1 - \alpha)^{99} D_1 \\ &= 0.4(217) + 0.4(1 - 0.4)(210) + 0.4(1 - 0.4)^2(101) + \cdots + 0.4(1 - 0.4)^{99}(150) \\ &= 195.72 \end{aligned}$$

Thus, the forecast for week 101 is $F_{101} = L_{100} = 195.72$.

In the first glance, it seems that simple exponential smoothing method requires all past demand observations to compute smoothed estimate $L_t$. However, a simple algebra shows that, besides the value of $\alpha$, simple exponential smoothing method needs only two quantities to make a forecast for the next period. Consider current Period $t$ and suppose we want to compute the smoothed estimate $L_t$ at Period $t$. According to simple exponential smoothing method:

$$\begin{aligned} L_t &= \alpha D_t + \alpha(1 - \alpha)D_{t-1} + \alpha(1 - \alpha)^2 D_{t-2} + \cdots + \alpha(1 - \alpha)^{99} D_1 \\ &= \alpha D_t + (1 - \alpha)\left[\alpha D_{t-1} + \alpha(1 - \alpha)^1 D_{t-2} + \cdots + \alpha(1 - \alpha)^{98} D_1\right] \\ &= \alpha D_t + (1 - \alpha)L_{t-1} \end{aligned}$$

Since $F_{t+1} = L_t$ and $F_t = L_{t-1}$, the above equation becomes

$$F_{t+1} = \alpha D_t + (1 - \alpha)F_t \tag{3.4}$$

In other words,

$$\text{Next period's forecast} = \alpha(\text{current period's demand})$$
$$+ (1 - \alpha)(\text{forecast made for current period's demand})$$

Computing the value of $F_1$ using Eq. (3.4) requires $F_0$—the forecast for period zero—which does not exist. One common approach used is to set $F_1 = D_1$ (i.e., assuming a 100 percent accurate forecast for Period 1).

ANALYTICS

### Simple Exponential Smoothing Method

Consider a stationary time series of demands $D_1, D_2, \ldots, D_t$ for past $t$ periods. *Simple exponential smoothing method with smoothing constant $0 < \alpha \leq 1$* obtains a forecast for future periods as follows:

- *Initializing Step:* The method assumes that the forecast for Period 1 was accurate. Hence, it sets

$$F_1 = D_1$$

  and goes to the next step.

- *Forecasting Step:* Having the actual demand $D_t$ in Period $t$ and also the forecast $F_t$ made by the method for Period $t$, simple exponential smoothing method finds a smoothed estimate of the level of the stationary series as follows:

$$L_t = \alpha D_t + (1 - \alpha)F_t \tag{3.5}$$

  - *One-Period-Ahead Forecast:* The forecast for the next period, that is, Period $t + 1$, is

$$F_{t+1} = L_t \tag{3.6}$$

  - *Multiple-Period-Ahead Forecast:* The forecast for all future periods, that is, Periods $t + 2, t + 3$, etc., is

$$F_{t+\tau} = L_t : \quad \text{for all} \quad \tau = 2, 3, \ldots$$

Let's define the forecast error in Period $t$ as follows:

$$e_t = F_t - D_t \tag{3.7}$$

then rearranging Eq. (3.4) provides some interesting insight into how simple exponential smoothing method works:

$$\begin{aligned} F_{t+1} &= \alpha D_t + (1 - \alpha)F_t \\ &= F_t - \alpha(F_t - D_t) \\ &= F_t - \alpha\, e_t \end{aligned}$$

which implies that the forecast for Period $t + 1$ is equal to the forecast for Period $t$ minus a fraction (i.e., fraction $\alpha$) of the forecast error in Period $t$. Hence, if for example, the forecast for Period $t$ was higher than the actual demand—the error $e_t$ is positive—then simple exponential smoothing method adjusts its forecast for the next period by reducing the forecast for current period by fraction $\alpha$ of the error. On the other hand, if the forecast was lower than the actual demand—the error $e_t$ is negative—the method adjusts its forecast for the next period by increasing the forecast for current period by fraction $\alpha$ of the error.

Let's use simple exponential smoothing method with $\alpha = 0.25$ to find $F_{101}$, the forecast for Oatplex cereal boxes in Week 101 using the data in Table 3.1. To compute $F_{101}$ we need to compute $F_{100}$, for which we need to compute $F_{99}$, for which we need to compute $F_{98}$, and so on, leading to the need to compute $F_1$. As suggested in the initializing step of the analytics, we set $F_1 = D_1 = 150$. Hence, using Eq. (3.4) we can compute $F_2$ as follows:

$$F_2 = \alpha D_1 + (1 - \alpha)F_1$$
$$= 0.25(150) + (1 - 0.25)150 = 150$$

Therefore, using Eq. (3.4), and since demand in Table 3.1 for Period 2 is $D_2 = 199$, we can compute $F_3$ as follows:

$$F_3 = \alpha D_2 + (1 - \alpha)F_2$$
$$= 0.25(199) + (1 - 0.25)150 = 162.25$$

By continuing this way we can compute $F_4$, $F_5$, ..., $F_{100}$. Figure 3.12 shows an Excel spreadsheet model developed to obtain these forecasts. Note that $\alpha$ in cell D2 is the input value. Except for cell D5 that contains the initial forecast for Period 1, that is, $F_1 = 150$, all other cells D6 to D104 use Eq. (3.4) to compute forecasts for Periods 2 to 100.

Using $F_{100} = 183.05$, we can now compute the forecast for Week 101 as follows:

$$F_{101} = \alpha D_{100} + (1 - \alpha)F_{100}$$
$$= 0.25(217) + (1 - 0.25)183.05 = 191.54$$

So simple exponential smoothing method with smoothing constant $\alpha = 0.25$ predicts that the demand in Week 101 is around 191 boxes.

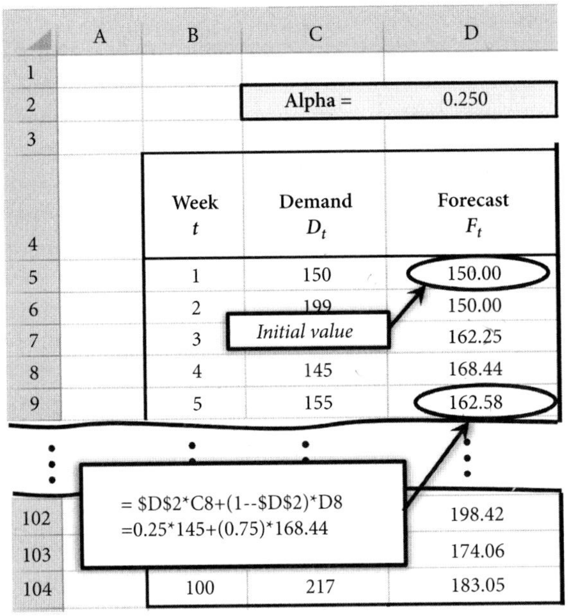

**Figure 3.12**   Demand forecast for Oatplex cereal in Example 3.1, using simple exponential smoothing method with $\alpha = 0.25$.

***Impact of Smoothing Constant*** $\alpha$

Recall that the value of span $N$ in the moving average method controls the responsiveness of the forecast method to changes in most recent demands. Smoothing constant $\alpha$ plays the same role in simple exponential smoothing method. Larger values of $\alpha$ make the method more responsive to changes in demand. This is clear from Eq. (3.4), where $F_{t+1} = \alpha D_t + (1 - \alpha)F_t$. As $\alpha$ gets larger, $(1 - \alpha)$ gets smaller, and the new forecast would be more affected by the most recent value of demand (i.e., $D_t$) than by previous smoothed forecast (i.e., $F_t$). Hence, with larger $\alpha$, simple exponential smoothing method becomes more responsive to most recent changes in demand.

Two questions come to mind about forecast $F_{101} = 191.54$ we made for Oatplex: (i) How good the forecast $F_{101} = 191.54$ suggested by simple exponential smoothing method with $\alpha = 0.25$ is? and (ii) Is $\alpha = 0.25$ the best constant to choose? In other words, are there any values of $\alpha$ that yield a better forecast than that with $\alpha = 0.25$?

The answer to both questions depends on what we mean by a "good" forecast. In a more general sense, the question is: how forecasts generated by a forecasting method are evaluated to be a good forecast?

### 3.8.3 Step 3—Evaluate the Method

To evaluate the performance of a forecasting method, the first question is: What is the criterion by which the method is evaluated? The answer, of course, is the "accuracy" of the forecasts generated by the method. The more accurate a forecast is, the more helpful it would be for planning operations. Hence, forecasting errors are the main criteria by which a forecasting method should be evaluated.

Recall that we defined forecast error made by a forecasting method at Period $t$ as follows:

$$e_t = F_t - D_t$$

Suppose we turn the clock back for 100 weeks to the end of Week 1, when we observe the demand for Oatplex is 150. Suppose we start using the simple exponential smoothing method to forecast demand as we observe demand in each period. The resulting forecasts would be those shown in Fig. 3.12.

Having these forecasts, we can compute what the errors would have been, if we had used simple exponential smoothing method with $\alpha = 0.25$. For example, the forecast error in Week 2 is $e_2 = F_2 - D_2 = 150 - 199 = -49$. Errors for all periods are shown in Fig. 3.13.

The figure includes 100 different forecast errors made by the method. Some of the errors are small and some of them are large. Obviously, it is difficult to decide about the accuracy of the forecasting method when we have 100 different values of errors. One common approach is to aggregate these 100 errors into a simple number—a metric that can be used to evaluate the accuracy of the forecasting method. Four most commonly used metrics are: (i) BIAS, (ii) Mean Absolute Deviation (MAD), (iii) Mean Squared Error (MSE), and (iv) Mean Absolute Percentage Error (MAPE).

***Bias (BIAS)***

To define bias, we first need to define cumulative error. *Cumulative error* (CE) is the sum of all observed errors:

$$\text{CE} = \sum_{i=1}^{t} e_i \tag{3.8}$$

Bias is then the average of cumulative error per period:

$$\text{BIAS} = \frac{\text{CE}}{t} \tag{3.9}$$

| | A | B | C | D | E | F | G | H |
|---|---|---|---|---|---|---|---|---|
| 1 | | | | | | | | |
| 2 | | | Alpha = | 0.250 | | | | |
| 3 | | | | | | | | |
| 4 | | Week $t$ | Demand $D_t$ | Forecast $F_t$ | Error $e_t$ | Absolute error $|e_t|$ | Squared error $(e_t)^2$ | Absolute percentage error $|e_t|/D_t$ |
| 5 | | 1 | 150 | 150.00 | 0.00 | 0.00 | 0.00 | 0.000 |
| 6 | | 2 | 199 | 150.00 | −49.00 | 49.00 | 2401.00 | 0.246 |
| 7 | | 3 | 187 | 162.25 | −24.75 | 24.75 | 612.56 | 0.132 |
| 8 | | 4 | 145 | 168.44 | 23.44 | 23.44 | 549.32 | 0.162 |
| 9 | | 5 | 155 | 162.58 | 7.58 | 7.58 | 57.43 | 0.049 |
| | | ⋮ | ⋮ | ⋮ | ⋮ | ⋮ | ⋮ | ⋮ |
| 102 | | 98 | 101 | 198.42 | 97.42 | 97.42 | 9490.10 | 0.965 |
| 103 | | 99 | 210 | 174.06 | −35.94 | 35.94 | 1291.48 | 0.171 |
| 104 | | 100 | 217 | 183.05 | −33.95 | 33.95 | 1152.80 | 0.156 |
| 105 | | | | Total = | −166.14 | 2843.40 | 133015.31 | 14.328 |
| 106 | | | | Mean = | −1.66 | 28.43 | 1330.15 | 0.14 |

**Figure 3.13** Forecasting errors for simple exponential smoothing method with $\alpha = 0.25$.

which is the sum of all $t$ observed forecast errors in the last $t$ periods divided by $t$. What is a good BIAS for a forecasting method? If a forecasting method is consistently underestimating the actual demand, then we will have more negative errors than positive errors, and hence the sum of the errors—the numerator of Eq. (3.9)—becomes negative. On the other hand, if the forecasting method is consistently overestimating the actual demand, then we will have more positive errors than negative ones, and the numerator of Eq. (3.9) becomes positive. With a good forecasting method, positive errors should cancel out negative errors, resulting in a CE and thus BIAS close to zero. For our Oatplex cereal in Example 3.1, we have BIAS $= -166.14/100 = -1.66$.

BIAS is actually mean of the forecast errors (see the mean of all errors in Fig. 3.13). Also, note that forecast errors $e_t$ are estimates of the noise $\epsilon_t$ in our stationary time series model, see Eq. (3.1). If the demand for Oatplex is really stationary, and if the simple exponential smoothing method with $\alpha = 0.25$ is a good method to forecast Oatplex demand, then the forecast errors should have a mean (i.e., bias) close to zero.

Figure 3.14 shows the forecast errors of the simple exponential smoothing method in forecasting demand for Oatplex in the last 100 weeks. As the figure shows, the errors are fluctuating randomly above and below zero—with a mean of $-1.60$, which is close to zero. This further confirms that our assumption of Oatplex demand being a stationary demand is valid.

### Mean Absolute Deviation (MAD)

While BIAS is a good metric to evaluate whether a forecasting method tends to always overestimate or underestimate the actual demand, it does not capture the accuracy of a forecast. For example, consider two forecasting methods used to forecast demand. Both methods have a BIAS very close to zero. However, the positive and negative errors in one forecast range between $-20$ and $+20$, while in the other one range from $-100$ to $+100$. Which forecasting method is more accurate? Although the errors in both forecasting methods have the mean close to zero, the latter results in larger forecast errors and thus is less accurate.

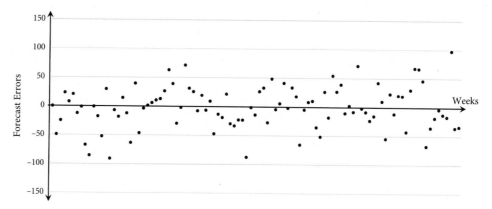

**Figure 3.14** Forecasting errors for exponential smoothing method with $\alpha = 0.25$.

To aggregate the errors into one metric such that positive and negative errors do not offset each other, we can use absolute values of the errors. Mean Absolute Deviation (MAD) measures the magnitude of errors by their absolute values. Specifically, MAD computes the mean of the absolute values of errors as follows:

$$\text{MAD} = \frac{\sum_{i=1}^{t} |e_i|}{t} \tag{3.10}$$

For the Oatplex cereal in Example 3.1, MAD, as shown in Fig. 3.13, is MAD $= 2843.4/100 = 28.43$.

### Mean Squared Error (MSE)

Another metric used to measure the accuracy of a forecasting method is Mean Squared Error (MSE). To obtain MSE, we first need to compute *Sum of Squared Errors* or SSE as follows:

$$\text{SSE} = \sum_{i=1}^{t} e_i^2 \tag{3.11}$$

Hence, *Mean Squared Error* (MSE) is defined as follows:

$$\text{MSE} = \frac{\text{SSE}}{t} \tag{3.12}$$

Similar to MAD, MSE prevents negative errors to cancel out positive errors by aggregating the squared values of the errors. However, there is a difference between how individual errors impact the aggregated metrics MAD and MSE.

While MAD is influenced equally by small or large errors, MSE is influenced much more by large forecast errors than by small errors. This is because errors in MSE are squared. Hence, MSE is an appropriate metric for situations where larger forecast errors result in significantly higher costs.

In Example 3.1, MSE for the simple exponential smoothing method with $\alpha = 0.25$, as shown in Fig. 3.13, is MSE $= 133015/100 = 1330.15$.

### Mean Absolute Percentage Error (MAPE)

Both MAD and MSE are very useful criteria to compare the accuracy of two forecasting methods. For example, consider two different forecasting methods used to forecast monthly demand for a product. Method 1 and Method 2 result in $\text{MAD}_1 = 120$ and $\text{MAD}_2 = 90$, respectively. Which method provides more accurate forecast for demand? Since Method 2 has a lower MAD than that of Method 1, we can conclude that Method 2 is a more accurate forecasting method. But, how accurate is forecasting Method 2? Is MAD of 90 small enough to consider Method 2 an accurate method? or is it too large and Method 2 is not an accurate method—even though it is more accurate than Method 1?

The answer depends on the magnitude of the demand. If the average monthly demand for the product is about 10,000, then MAD = 90 is very small, since it is, on average, about $90/10{,}000 = 0.9$ percent of the average demand. In other words, forecasting Method 2 is considered a very accurate forecasting method, since on average its forecast is less than 1 percent above or below the actual demand. However, if the monthly demand has an average of 200, then forecasting Method 2 is not an accurate method; it results in forecasts that are, on average, about $90/200 = 45$ percent above or below the actual demand. Hence, the key to evaluating the accuracy of a forecasting method is to measure its forecast errors relative to the actual value.

One metric that takes the relative value of forecast errors to the actual values is called *Absolute Percentage Error* or APE, which is

$$\text{APE}_i = \frac{|e_i|}{D_i}$$

*Mean Absolute Percentage Error* or MAPE is the average of all APE of the past $t$ observations. Specifically,

$$\text{MAPE} = \frac{\sum_{i=1}^{t} \text{APE}_i}{t} \times 100\% \tag{3.13}$$

MAPE is a commonly used measure to evaluate the accuracy of forecasting methods, since it has a simple interpretation. A MAPE = 8, for example, implies that, on average, the forecast is off by 8 percent. For our Oatplex cereal in Example 3.1, we have (see Fig. 3.13)

$$\text{MAPE} = \frac{\sum_{i=1}^{t} \text{APE}_i}{t} \times 100\% = \frac{14.328}{100} \times 100\% = 14.3\%$$

Hence, we should expect that the forecast generated by the simple exponential smoothing method with $\alpha = 0.25$ to be, on average, about 14 percent below or above the actual demand.

### 3.8.4 Step 4—Adjust the Method

Let's return to Example 3.1 and assume that the manager of the store decides to use simple exponential smoothing method to forecast the weekly demand for Oatplex. A forecast that significantly overestimates demand results in the store placing large orders for the cereal and ends up with large inventory. On the other hand, a forecast that significantly underestimates demand results in not ordering enough cereal and having a large stockout. Thus, the manager of the store would choose MSE that penalizes large errors (positive or negative) as an evaluation criterion for the accuracy of the forecasting method.

In the previous section, we used simple exponential smoothing method with smoothing constant $\alpha = 0.25$ to forecast demand for Oatplex. Choosing MSE as the evaluation criterion, the simple exponential smoothing method with $\alpha = 0.25$ resulted in MSE = 1330.15. But is $\alpha = 0.25$ the best smoothing constant for forecasting weekly demand for Oatplex in order to have a low MSE? If not, what value of $\alpha$ results in the lowest MSE? This is the goal of this section. Specifically, in this section we illustrate how one can adjust the value of $\alpha$ such that it results in the lowest MSE.

To find the best value for smoothing constant $\alpha$, we need to compute the value of MSE for all possible values of $\alpha$, and then find the value of $\alpha$ that yields the lowest MSE. This can be done using the "Solver" option of Excel (see Appendix K for how Solver can be used to solve mathematical programming problems). The problem we are trying to solve is the following optimization problem in which $\alpha$ is the decision variable:

$$\text{Min } Z = \text{MSE}$$

subject to:

$$0 \leq \alpha \leq 1$$

Figure 3.15 shows how Excel spreadsheet and Excel Solver can be set up to solve the above optimization problem. Note that cell G106, which is the mean squared error MSE, is assigned to "Set Objective" of the Solver. The objective, as shown in the figure, is to minimize this value by changing variable cells D2, which is the value of $\alpha$. In the "Subject to the Constraints" part of the Solver we have two constraints. First is "$D$2 <= 1" corresponding to constraint $\alpha \leq 1$ and the second is "$D$2 >= 0," corresponding to $\alpha \geq 0$.

Note that in "Selecting a Solving Method" part of Solver, we choose "GRG Nonlinear," since our objective function of the problem—MSE—is nonlinear (i.e., is squared) function of the errors, which are themselves a function of $\alpha$. After solving the problem, and choosing the option "Keep the Solver Solution," we get the answer as shown in Fig. 3.16. In the figure, the best value for smoothing constant $\alpha$ is $\alpha = 0.317$. This results in MSE = 1316.97, which is smaller than MSE = 1330.15 for our previous choice of $\alpha = 0.25$.

Having the optimized value of $\alpha$ and the data in Fig. 3.16, we can develop a forecast for Weeks 101 and 102 for Oatplex in Example 3.1.

### 3.8.5 Step 5—Develop a Forecast

Having $\alpha = 0.317$ and $F_{100} = 182.19$ (see Fig. 3.16), we can compute the forecast for week 101 as follows:

$$F_{101} = \alpha D_{100} + (1 - \alpha)F_{100} = 0.317(217) + (1 - 0.317)182.19 = 193.2$$

| | A | B | Week $t$ | Demand $D_t$ | Forecast $F_t$ | Error $e_t$ | Absolute Error $|e_t|$ | Squared Error $(e_t)^2$ |
|---|---|---|---|---|---|---|---|---|
| 1 | | | | | | | | |
| 2 | | | Alpha = | 0.250 | | | | |
| 3 | | | | | | | | |
| 4 | | | 1 | 150 | 150.00 | 0.00 | 0.00 | 0.00 |
| 5 | | | 1 | 150 | 150.00 | 0.00 | 0.00 | 0.00 |
| 6 | | | 2 | 199 | 150.00 | -49.00 | 49.00 | 2401.00 |
| 7 | | | 3 | 187 | 162.25 | -24.75 | 24.75 | 612.56 |
| 8 | | | 4 | 145 | 168.44 | 23.44 | 23.44 | 549.32 |
| 9 | | | 5 | 155 | 162.58 | 7.58 | 7.58 | 57.43 |
| 102 | | 98 | 101 | 198.42 | 97.42 | 97.42 | 9490.10 | |
| 103 | | 99 | 210 | 174.06 | -35.94 | 35.94 | 1291.48 | |
| 104 | | 100 | 217 | 183.05 | -33.95 | 33.95 | 1152.80 | |
| 105 | | | | Total = | -166.14 | 2843.40 | 133015.31 | |
| 106 | | | | Mean = | -1.66 | 28.43 | 1330.15 | |

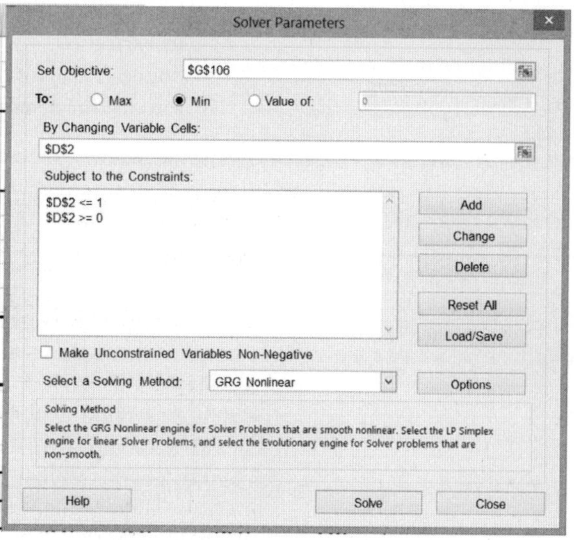

**Figure 3.15** The spreadsheet model and Solver window for adjusting simple exponential smoothing method for Oatplex cereal in Example 3.1.

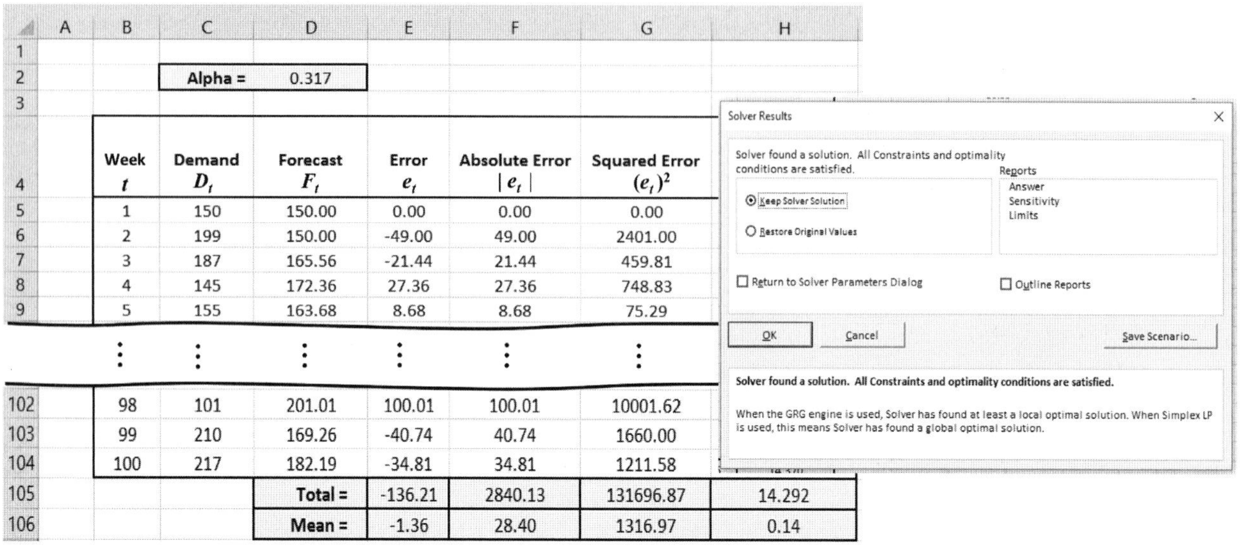

**Figure 3.16**    Solver solution for the optimal value of $\alpha$ that minimizes MSE for forecasting Oatplex cereal in Example 3.1.

Hence, we expect that the demand for Oatplex in Week 101 to be about 193.2 boxes. But the question is: how confident are we that the demand in Week 101 would be exactly 193.2? This is a very important question that distinguishes between the following two types of forecasts:

- *Point Forecast:* Point forecast is our best estimate for the future demand in Period $t$ and is specified by only one number, that is, $F_t$: the *average demand* in Period $t$. For example, the statement that "the forecasted demand for Oatplex in Week 101 is $F_{101} = 193.2$," is a point forecast or point prediction, referring to the *average demand* for Oatplex in Week 101, since it is obtained using "averaging methods" that take the average of the available demand data to generate a forecast.

- *Prediction Interval Forecast:* Prediction interval forecast is an interval of numbers for which we are very confident (e.g., 95-percent confident) that the actual demand will be contained in that interval. A two-sided prediction interval forecast for demand in Period $t + 1$ is specified by three numbers: a lower bound ($\underline{F}_{t+1}$) and an upper bound ($\overline{F}_{t+1}$) for future demand, and a probability. A 95-percent prediction interval forecast for demand in Period $t + 1$ implies

$$P(\underline{F}_{t+1} \leq D_{t+1} \leq \overline{F}_{t+1}) = 0.95$$

For example, a 95-percent prediction interval forecast $[\underline{F}_{101}, \overline{F}_{101}] = [121, 265]$ for Oatplex demand in Week 101 means that with probability 0.95 the demand for Oatplex in Week 101 would be between 121 and 265. In other words, we have $D_{101} \in [121, 265]$ with probability 0.95.

Prediction interval forecasts contain more information about the future demand beyond its average. It also provides an indication of the magnitude of noise (i.e., variability). For example, consider two different prediction interval forecasts for two different Products A and B: (i) For Product A, the point forecast is 200, and the 95-percent prediction interval forecast is $[190, 210]$. (ii) For Product B, the point forecast is also 200, but the 95-percent forecast interval is $[150, 250]$. While forecasts predict that one would expect the future demand for both products to be about 200, the actual demand for Product B—with probability 95 percent—can be anywhere between 150 and 250, a wider range than that for Product A. The wider prediction interval forecast of Product B implies that the noise in demand for Product B is higher than that for Product A. Hence, for a product with higher variability in demand, one should expect a larger prediction interval forecast. The wider the prediction interval, the less reliable the point forecast.

The following analytics shows how a prediction interval forecast can be developed for a stationary demand when forecasts are made using simple exponential smoothing.

---

**ANALYTICS**

**Two-Sided Prediction Interval Forecast for Stationary Time Series**

Consider a stationary time series of demands $D_1, D_2, \ldots, D_t$ of past $t$ periods. Suppose *simple exponential smoothing* method with constant $\alpha$ was used to develop forecast $F_1, F_2, \ldots, F_t$ for those $t$ periods. Hence, the errors generated by the forecast method are $e_1, e_2, \ldots, e_t$, where $e_i = F_i - D_i$. Finally, suppose that the method will be used to generate forecast for future Periods $t + \tau$, as $F_{t+\tau}$, for $\tau = 1, 2, 3, \ldots$.

A $100(1 - x)$-percent prediction interval forecast $[\underline{F}_{t+\tau}, \overline{F}_{t+\tau}]$ for the actual demand in Period $t + \tau$ is

$$D_{t+\tau} \in [F_{t+\tau} \pm \sigma_e \, z_{(x/2)} \sqrt{1 + \mathbb{D}_\tau}] \quad \text{for} \quad \tau = 1, 2, 3 \ldots \tag{3.14}$$

where $z_{(x/2)}$ is the value of standard Normal distribution for which its cumulative probability distribution satisfies $\Phi(z) = x/2$, and $\sigma_e$ is *Standard Error*, which is an estimate of standard deviation of the noise obtained using observed errors $e_1, e_2, \ldots, e_t$. For simple exponential smoothing method, the standard error is

$$\sigma_e = \sqrt{\frac{\text{SSE} - t \, (\text{BIAS})^2}{t - 1}} \tag{3.15}$$

where SSE is sum of squared errors as in Eq. (3.11) and BIAS is the bias of the observed errors as in Eq. (3.9).

The value of $\mathbb{D}_\tau$ depends on whether the prediction interval is for the next period (i.e., one-period-ahead forecast), or is it for the periods beyond the immediate next period (i.e., multiple-period-ahead forecast):

- *One-Period-Ahead Forecast:* For one-period-ahead forecast, where $\tau = 1$, we have

$$\mathbb{D}_1 = 0$$

- *Multiple-Period-Ahead Forecast:* When $\tau \geq 2$, we have

$$\mathbb{D}_\tau = (\tau - 1)\alpha^2 \quad \text{for} \quad \tau = 2, 3, \ldots \tag{3.16}$$

---

To see the intuition behind the prediction interval forecast, let's have a closer look at Eq. (3.14). Note that the prediction interval in that equation consists of the following terms:

- *Point forecast $F_{t+\tau}$:* Point forecast $F_{t+\tau}$ is our forecast for the average demand in future Period $t + \tau$. The point forecast $F_{t+\tau}$ is the center (i.e., in the middle) of the two-sided prediction interval. The reason is as follows. If our forecasting model fits the data reasonably well, the errors will be positive and negative and are randomly spread out around the mean of zero. In other words, the number of times that the method overestimates the demand and the amount it overestimates the demand should be—on average—almost the same as those when the method underestimates the demand. Thus, we should expect our point forecast to have the same probability of overestimating or underestimating the actual demand and with the same amount. Therefore, the lower and upper bounds of prediction intervals should be, respectively, lower and higher than the point forecast by the same amount. This is why forecast $F_{t+\tau}$ is in the center of the two-sided prediction interval.

- *Accuracy of the foresting method:* The less accurate a forecasting method is (i.e., the larger the errors), the wider the prediction interval forecasts would be. One measure that relates to the size of errors is the standard deviation of errors generated by the method (i.e., standard error). As prediction interval in Eq. (3.14) shows, the larger the standard error $\sigma_e$, the wider the prediction interval.

- *Measure of confidence:* Recall that if a time series is indeed a stationary time series, then the noise $\epsilon_t$ would have an average $\mu_\epsilon = 0$ and standard deviation $\sigma_\epsilon$, which we estimate by $\sigma_e$. Furthermore, the distribution of the noise is well approximated by a Normal distribution. The term $z_{(x/2)}$, which is the value of $z$ for which $\Phi(z) = x/2$, corresponds to the confidence level. For a $100(1-x)$-percent prediction interval forecast as confidence level $100(1-x)$ increases, the absolute value of $z_{(x/2)}$ increases, resulting in a wider prediction interval.

- *Forecasting horizon:* One of the forecasting principles is that forecasts for longer time horizons are less accurate. Hence, the larger the time horizon $\tau$, the less accurate the forecast $F_{t+\tau}$ should be. This is captured by the term $\mathbb{D}_\tau$, which is called *Distance Value*. In time series forecasting method, distance value is a measure of how far beyond the next period the forecast is made. As Eq. (3.16) shows, the distance value $\mathbb{D}_\tau$ is increasing in $\tau$, resulting in a wider prediction interval in periods beyond the next period.

Let's use the above analytics to find the 95-percent prediction interval forecast for Oatplex demand in Week 101 based on our new estimate of $F_{101} = 193.2$. Consider Fig. 3.16, which shows the errors for the optimal value of $\alpha = 0.317$ as follows:

$$\text{SSE} = 131696.87, \quad \text{BIAS} = -1.36$$

So, using Eq. (3.15), we can compute standard error as follows:

$$\sigma_e = \sqrt{\frac{\text{SSE} - t\,(\text{BIAS})^2}{t-1}} = \sqrt{\frac{131696.87 - 100\,(-1.36)^2}{100-1}} = 36.44$$

On the other hand, for the 95-percent prediction interval forecast, we have $100(1-x) = 95$; therefore, $x = 0.05$ and $x/2 = 0.025$. Using standard Normal table, we find that $\Phi(z = -1.96) = 0.025$. Hence, $z_{(0.025)} = -1.96$. Also, forecast for Week 101 is a one-period-ahead forecast and thus $\mathbb{D}_1 = 0$. Therefore, the 95-percent prediction interval forecast for the demand in Week 101 would be

$$D_{101} \in [F_{101} \pm \sigma_e\, z_{(0.025)} \sqrt{1+0}] = [193.2 \pm 36.44(-1.96)] = [121.77,\ 264.62]$$

In other words, with probability 0.95, the demand for Oatplex in Week 101 would be between 121 and 265 boxes.

A prediction interval forecast for Oatplex demand 2 weeks from now—Week 102—can also be found using the above analytics. First, since demand for Oatplex is assumed to be stationary, the point forecast for the average demand in Week 102 would be the same as that for Week 101, that is, $F_{102} = 193.2$. Since forecast for Week 102 is a multiple-period-ahead forecast with $\tau = 2$, using Eq. (3.16), we have

$$\mathbb{D}_2 = (\tau - 1)\alpha^2 = (2 - 1)(0.317)^2 = 0.1005$$

Hence, the prediction interval forecast for Week 102 would be

$$D_{102} \in [F_{102} \pm \sigma_e\, z_{(0.025)} \sqrt{1 + \mathbb{D}_2}] = [193.2 \pm 36.44(-1.96)(\sqrt{1 + 0.1005})]$$
$$= [118.27\, ,\ 268.12]$$

So, with 95-percent probability, the demand in Period $t = 102$ will be between 118 and 269 boxes. Note that, as expected, the prediction interval forecast for Week 102 is a wider interval than that for Week 101. This is consistent with the forecasting principle that the forecast for longer time horizons is less accurate. Because the forecast for Week 102 is less accurate, it has a wider prediction interval to include the actual demand with probability 95 percent.

## 3.9 Forecasting Time Series with Trend

In time series with trend, the demand exhibits an increasing or decreasing trend over time. The averaging methods introduced for stationary time series do not work when there is a trend. In this section, we explain how new averaging methods can be developed to obtain forecasts for time series with trend. We start with the following example.

 **Example 3.2**

> A manufacturer has a separate production line used for making specialized wheelchairs for people with disabilities. The firm produces to order after receiving orders from different organizations. This is because each order is different depending on the individual's need. The demand in the last 12 months have been 57, 254, 335, 307, 347, 450, 434, 479, 621, 597, 641, and 793. To plan next month's production, the firm needs to know how many orders it should expect to receive next month?

### 3.9.1 Step 1—Determine the Type of Time Series

Step 1 of the procedure is to identify the type of the time series. There is an obvious increase in the demand during the last 12 months, and hence we can conclude that monthly demand for wheelchairs constitutes a time series with an increasing trend. This is also shown in Fig. 3.17 (see the "Actual Demand" in the figure).

### 3.9.2 Step 2—Choose a Method

Recall that we modeled stationary time series as $D_t = \mathcal{L} + \epsilon_t$, where $\mathcal{L}$ was the base or level of the series and $\epsilon_t$ was the random noise in Period $t$. The idea behind this model is that the demand in each period does not change significantly—it is around $\mathcal{L}$—and the main change is due to random noise. To develop a more accurate model of demand that also includes trend, the stationary demand model can be changed into the following:

$$D_t = \mathcal{L} + \mathcal{G}t + \epsilon_t \tag{3.17}$$

The above model fits situations where there is a *Linear Trend* in demand such that the demand increases at a fixed *Growth Rate* $\mathcal{G}$, which can be positive or negative. Based on this model, the demand in the next period should be approximately $\mathcal{G}$ units higher (if $\mathcal{G}$ is positive) or lower (if $\mathcal{G}$ is negative) than the demand in the current period. Similar to the

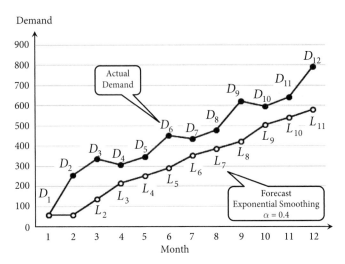

**Figure 3.17**    Actual demand and forecasts made by simple exponential smoothing method with $\alpha = 0.4$.

stationary time series, if the demand indeed has a linear trend, then noise $\epsilon_t$ should have a mean of zero with a constant variance, and should be well-approximated by a Normal distribution.

The trend in a time series may or may not be linear. This section, however, focuses only on linear trends, since they are fairly common patterns in data. Nonlinear trends require more advanced analytics that are beyond the scope of this text. In the following sections, we show how the simple exponential smoothing method can be extended to forecast time series with trend.

### Double Exponential Smoothing (Holt's) Method

The double exponential smoothing method, also called Holt's method, tries to find estimates for the growth rate $\mathcal{G}$ in our trend model in Eq. (3.17). It updates the estimate each time a new demand is realized. Double exponential smoothing method is also called *Trend-Adjusted Exponential Smoothing Method*, since it uses the simple exponential smoothing method for stationary demand and adjusts its forecast to account for the trend. To illustrate the idea behind this method, let's use the simple exponential smoothing method with $\alpha = 0.4$ and initial forecast $F_1 = D_1 = 57$ to develop forecasts for the demand for wheelchairs in Example 3.2. This is shown in Fig. 3.17.

Note that in simple exponential smoothing method, the smoothed estimated $L_t$ is used as the forecast for Period $t + 1$, that is, $F_{t+1} = L_t$. The figure therefore shows the smoothed estimate $L_{t-1}$ as the forecast for month $t$. From Fig. 3.17 we make the following observations:

- Forecasts for month $t$ are smaller than the actual demand in month $t$ (i.e., $L_{t-1} < D_t$) for all months $t = 2, 3, \ldots, 12$. This is because the actual demand $D_t$ has an increasing trend (corresponding to growth rate $\mathcal{G}$).[5] Since simple exponential smoothing method does not include trend when generating forecasts, if one wants to use simple exponential smoothing method to develop forecasts for a demand with increasing trend, one must add an estimate of the growth rate to the forecast obtained by simple exponential smoothing method. Hence, considering

$$G_t = \text{Smoothed estimate of growth rate } \mathcal{G} \text{ at Period } t$$

---

[5]If the demand had a decreasing trend (i.e., a negative growth), then forecasts would have been larger than the actual demand.

we can get a forecast for demand with trend in Period $t+1$ by simply adding the estimate $G_t$ for the growth to the smoothed forecast $L_t$ obtained using simple exponential smoothing method:

$$F_{t+1} = L_t + G_t$$

But how do we get the estimate $G_t$ for the growth? This brings us to the next observation.

- While simple exponential smoothing forecasts are smaller than the actual demand, as the figure shows, they indeed have an increasing trend (i.e., $L_t$ is increasing in $t$). This is because smoothed estimate of demand is the weighted average of the past demands, with the largest weights (i.e., $\alpha$) on the last one. Hence, if there is an increasing trend in demand $D_t$, there would be an increasing trend in $L_t$. Furthermore, as the figure also shows, the increasing trend in $L_t$ is almost the same as the increasing trend in actual demand $D_t$. Hence, difference $(L_t - L_{t-1})$, $(L_{t-1} - L_{t-2}), \ldots, (L_2 - L_1)$ can provide approximations for the growth rate $G$ in Periods $t, t-1, \ldots, 2$, respectively. To get a good estimate and reduce the impact of noise, we can use the averaging method again. Specifically, we can take the average of all these differences through another simple exponential smoothing method with smoothing constant $\beta$ to get smoothed estimate $G_t$ as follows:

$$G_t = \beta(L_t - L_{t-1}) + (1 - \beta)G_{t-1} \qquad (3.18)$$

The above two observations are the main idea behind double exponential smoothing. The forecast for Period $t+1$ is $F_{t+1} = L_t + G_t$. At Period $t$, the method uses the simple exponential smoothing method with constant $\alpha$ to compute $L_t$ assuming there is no trend. Specifically, having the actual demand at Period $t$ (i.e., $D_t$) and the forecast for Period $t$ (i.e., $F_t = L_{t-1} + G_{t-1}$), the new estimate for $L_t$ is computed as follows:

$$L_t = \alpha D_t + (1 - \alpha)(L_{t-1} + G_{t-1})$$

It then approximates the trend using another simple exponential smoothing method with constant $\beta$ using Eq. (3.18). Finally, by adding $G_t$ to $L_t$, the forecast $F_{t+1}$ for Period $t+1$ is computed.

Similar to simple exponential smoothing method, double exponential smoothing method also requires some initial values to start the method. Specifically, it requires the initial estimates $L_0$ and $G_0$. A common approach is to set $L_0 = D_1$ and $G_0 = (D_t - D_1)/(t-1)$. Note that $D_t - D_1$ is the growth of the series in $(t-1)$ periods; hence, $(D_t - D_1)/(t-1)$ is an estimate for the average growth rate of the series from period to period. The analytics below summarizes the double exponential smoothing method.

**ANALYTICS**

**Double Exponential Smoothing (Holt's) Method**

Consider a time series with trend with demands $D_1, D_2, \ldots, D_t$ in past $t$ periods. The *double exponential smoothing* method, also called *Holt's* method, with smoothing constants $0 < \alpha \leq 1$ and $0 < \beta \leq 1$ obtains a forecast for future Period $t+1$ as follows:

- *Initializing Step:* The initial values for level and growth rate can be set as follows:

$$L_0 = D_1, \qquad G_0 = \frac{D_t - D_1}{t - 1}$$

and therefore the forecast for Period $t = 1$ would be $F_1 = L_0 + G_0$.

- *Forecasting Step:* Using the demand $D_t$ in Period $t$, and having estimates $L_{t-1}$ and $G_{t-1}$, the method updates these estimates as follows:

$$L_t = \alpha D_t + (1 - \alpha)(L_{t-1} + G_{t-1}) \tag{3.19}$$
$$G_t = \beta(L_t - L_{t-1}) + (1 - \beta)G_{t-1} \tag{3.20}$$

- *One-Period-Ahead Forecast:* Using the updated estimates for level and growth rate, the forecast for Period $t + 1$ is, therefore,

$$F_{t+1} = L_t + G_t \tag{3.21}$$

- *Multiple-Period-Ahead Forecast:* Since at Period $t$ the only available estimate for growth rate $G$ is $G_t$, then the forecast for all future Periods $t + 2, t + 3$, etc., would be

$$F_{t+\tau} = L_t + \tau\, G_t: \quad \text{for all} \quad \tau = 2, 3, \ldots \tag{3.22}$$

Let's use Holt's method with $\alpha = 0.5$ and $\beta = 0.3$ to find a forecast for the demand for wheelchairs (Example 3.2) in month $t = 13$. Following the Initializing Step, we get the initial values needed to start the method as follows:

$$L_0 = D_1 = 57, \quad G_0 = \frac{D_{12} - D_1}{12 - 1} = \frac{793 - 57}{11} = 66.91$$

The forecast for month $t = 1$ is then $F_1 = L_0 + G_0 = 57 + 66.91 = 123.91$.

The forecast for month $t = 2$ using Eq. (3.21) is $F_2 = L_1 + G_1$. Using Eqs. (3.19) and (3.20) to obtain values of $L_1$ and $G_1$, respectively, we have

$$\begin{aligned}
L_1 &= \alpha D_1 + (1 - \alpha)(L_0 + G_0) \\
&= 0.5(57) + (1 - 0.5)(57 + 66.91) = 90.45 \\
G_1 &= \beta(L_1 - L_0) + (1 - \beta)G_0 \\
&= 0.3(90.45 - 57) + (1 - 0.3)66.91 = 56.87
\end{aligned}$$

Hence, the forecast for month $t = 2$ is $F_2 = L_1 + G_1 = 90.46 + 56.87 = 147.33$. Figure 3.18 shows the forecasts for all 12 periods, resulting in $L_{12} = 746.17$ and $G_{12} = 59.53$. Hence, the forecast for month 13 is $F_{13} = L_{12} + G_{12} = 746.17 + 59.53 = 805.7$.

Comparing the forecasts of the Holt's method (the graph in Fig. 3.18) with those of simple exponential smoothing method (in Fig. 3.17), we see that the new forecasts are now closer to the actual demand values. These forecasts are made based on our arbitrary values of $\alpha = 0.5$ and $\beta = 0.3$. In the next steps, we will look for the best values to improve the performance of our method.

### 3.9.3 Step 3—Evaluate the Method

Let's choose MSE as our accuracy criterion, and evaluate our method based on that. As Fig. 3.18 shows, the MSE for the Holt's method with $\alpha = 0.5$ and $\beta = 0.3$ is MSE $= 4582.40$.

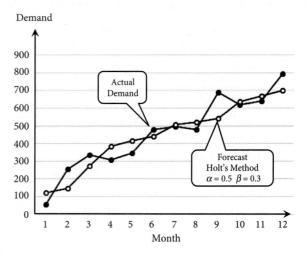

| | A | B | C | D | E | F | G | H |
|---|---|---|---|---|---|---|---|---|
| 1 | | | | | | | | |
| 2 | | | alpha = | 0.5000 | | | | |
| 3 | | | beta = | 0.3000 | | | | |
| 4 | | Month $t$ | Demand $D_t$ | $L_t$ | $G_t$ | $F_t$ | $e_t$ | $(e_t)^2$ |
| 5 | | 1 | 57 | 90.45 | 56.87 | 123.91 | 66.91 | 4476.83 |
| 6 | | 2 | 254 | 200.66 | 72.87 | 147.33 | -106.67 | 11379.07 |
| 7 | | 3 | 335 | 304.27 | 82.09 | 273.54 | -61.46 | 3777.67 |
| 8 | | 4 | 307 | 346.68 | 70.19 | 386.36 | 79.36 | 6298.28 |
| 9 | | 5 | 347 | 381.93 | 59.71 | 416.87 | 69.87 | 4881.77 |
| 10 | | 6 | 450 | 445.82 | 60.96 | 441.64 | -8.36 | 69.84 |
| 11 | | 7 | 434 | 470.39 | 50.04 | 506.78 | 72.78 | 5297.43 |
| 12 | | 8 | 479 | 499.72 | 43.83 | 520.44 | 41.44 | 1716.95 |
| 13 | | 9 | 621 | 582.27 | 55.45 | 543.55 | -77.45 | 5998.97 |
| 14 | | 10 | 597 | 617.36 | 49.34 | 637.72 | 40.72 | 1658.15 |
| 15 | | 11 | 641 | 653.85 | 45.48 | 666.70 | 25.70 | 660.44 |
| 16 | | 12 | 793 | 746.17 | 59.53 | 699.33 | -93.67 | 8773.41 |
| 17 | | | | | | Total = | 49.1676 | 54988.80 |
| 18 | | | | | | Mean = | 4.0973 | 4582.40 |

**Figure 3.18    Forecasts for demand in Example 3.2 using Holt's method with $\alpha = 0.5$ and $\beta = 0.3$.**

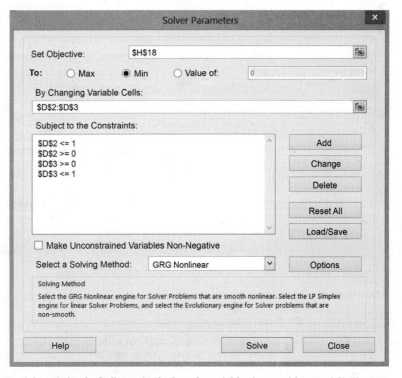

**Figure 3.19    Solver window for finding optimal values of $\alpha$ and $\beta$ for the spreadsheet model in Fig. 3.18.**

### 3.9.4  Step 4—Adjust the Method

Similar to the case of stationary time series we can use Excel Solver to find the best initial values of $\alpha$ and $\beta$ to minimize the MSE. This is shown in Fig. 3.19.

The result of Solver is shown in Fig. 3.20. As the figure shows, the best values for smoothing constants are $\alpha = 0.397$ and $\beta = 0.034$. These values result in MSE = 4006.66, which is

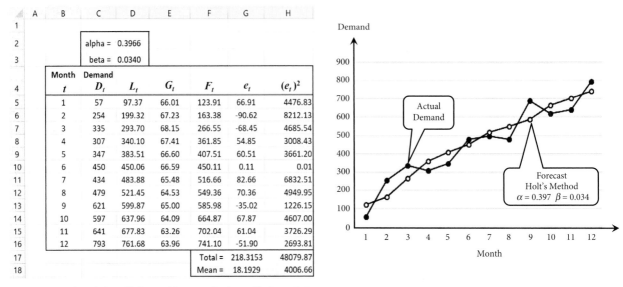

**Figure 3.20** The solution of Solver and forecasts for demand in Example 3.2.

much smaller than what we had before adjusting the parameters of the method (i.e., much smaller than MSE = 4582.40).

Figure 3.20 also shows a graph of the forecasts against the actual values of demand. Compared with the graph in Fig. 3.18, Holt's method with the optimized parameters generates forecasts that are very close to the actual demand.

### 3.9.5 Step 5—Develop a Forecast

Having the new parameters we can now develop point and prediction interval forecasts for month 13 in Example 3.2. From Fig. 3.20, for the optimized parameters we have $L_{12} = 761.68$, $G_{12} = 63.96$. Hence, the firm should expect to receive an average of $F_{13} = L_{12} + G_{12} = 761.68 + 63.96 = 825.64$ orders for wheelchairs next month.

**ANALYTICS**

**Two-Sided Prediction Interval Forecast for Time Series with Trend**

Consider a time series with trend with demands $D_1, D_2, \ldots, D_t$ in the past $t$ periods. Suppose that the *double exponential smoothing method* with constants $\alpha$ and $\beta$ was used to develop forecast $F_1, F_2, \ldots, F_t$. Hence, the errors generated by the forecast method for the past $t$ periods are $e_1, e_2, \ldots, e_t$, where $e_i = F_i - D_i$. Finally, suppose that the method is used to generate forecast for future Period $t + \tau$, as $F_{t+\tau}$, for $\tau = 1, 2, 3, \ldots$.

A $100(1 - x)$-percent prediction interval forecast $[\underline{F}_{t+\tau}, \overline{F}_{t+\tau}]$ for the actual demand in Period $t + \tau$ is

$$D_{t+\tau} \in [F_{t+\tau} \pm \sigma_e \, z_{(x/2)} \sqrt{1 + \mathbb{D}_\tau}] \quad \text{for} \quad \tau = 1, 2, 3 \ldots \tag{3.23}$$

where $z_{(x/2)}$ is the value of standard Normal distribution with its cumulative probability distribution $\Phi(z) = x/2$ and $\sigma_e$ is the standard error for double exponential smoothing method as follows:

$$\sigma_e = \sqrt{\frac{\text{SSE} - t \, (\text{BIAS})^2}{t - 2}} \tag{3.24}$$

- *One-Period-Ahead Forecast:* For one-period-ahead forecast, where $\tau = 1$, we have

$$\mathbb{D}_1 = 0$$

- *Multiple-Period-Ahead Forecast:* When $\tau \geq 2$, we have

$$\mathbb{D}_\tau = \sum_{j=1}^{\tau-1} \alpha^2 (1 + j\beta)^2 \tag{3.25}$$

Figure 3.20 provides $\text{BIAS} = 18.19$ and $\text{SSE} = 48079.87$. Hence, the standard error is

$$\sigma_e = \sqrt{\frac{\text{SSE} - t\,(\text{BIAS})^2}{t - 2}} = \sqrt{\frac{48079.87 - 12\,(18.19)^2}{12 - 2}} = 66.41$$

For the 95-percent prediction interval, we have $z_{(0.025)} = -1.69$ and $\mathbb{D} = 0$; therefore,

$$[F_{13} \pm \sigma_e\, z_{(0.025)}] = [825.64 \pm 66.41(-1.96)] = [695.48, 955.80]$$

In other words, with probability 0.95, the number of orders received next month is expected to be between 695 and 956 boxes. But what is the forecast in 2 months from now, that is, month $t = 14$?

Using Eq. (3.22), the average demand in month $t = 14$ would be

$$F_{14} = L_{12} + 2\,G_{12} = 761.68 + 2(63.96) = 889.60$$

Using Eq. (3.25), for $\tau = 2$ we have

$$\mathbb{D}_2 = \sum_{j=1}^{2-1} \alpha^2 (1 + j\beta)^2 = \alpha^2 (1 + \beta)^2$$

Hence,

$$\mathbb{D}_2 = (0.397)^2 (1 + 0.034)^2 = 0.168$$

Therefore, the 95-percent prediction interval forecast for month $t = 14$ would be

$$[F_{14} \pm \sigma_e\, z_{(x/2)} \sqrt{1 + \mathbb{D}_2}\,] = [889.60 \pm 66.41(-1.96)\sqrt{1 + 0.168}\,] = [748.93, 1030.27]$$

In other words, with probability 0.95, the number of orders received in month $t = 14$ is expected to be between 748 and 1031 boxes.

## 3.10 Forecasting Seasonal Demand

In seasonal time series, data has an increasing pattern followed by a decreasing pattern (or vice versa). This increase and decrease repeats regularly in cycles, and all cycles have the same length. Examples of seasonal demand are demand for snowblowers which is high in winter followed by low demand in summer, or hourly demand in fast-food restaurants which is high during lunch hours followed by low demand until dinner time when the demand increases again. The repeating cycle for the snowblower is a year and for the fast-food restaurant is a day.

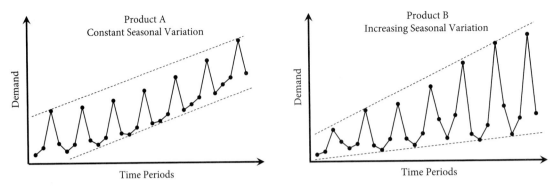

**Figure 3.21**   Seasonal demand patterns with constant seasonal variation (*left*), and increasing seasonal variation (*right*).

The difficulty in forecasting seasonal time series is that in addition to the base demand and trend, they also have seasonal variations. Hence, a good forecasting method should take all these three components of time series into account.

### 3.10.1 Step 1—Determine the Type of Time Series

Consider the demand for two Products A and B in Fig. 3.21. Both products show a seasonal pattern with the same length (of four periods) and an increasing trend. However, there is a fundamental difference between the demand patterns.

- *Constant Seasonal Variation:* As shown in the series for Product A, the seasonal variation—the difference between peak and low demands in a season—does not change significantly and remains relatively constant in different seasons.

- *Increasing (or Decreasing) Seasonal Variation:* For Product B, however, seasonal variation is not constant. It gets larger from one season to the next. For cases where the demand has a decreasing trend, the seasonal variation becomes smaller from a season to the next.

Therefore, the first step to generate forecast for a seasonal time series is to identify whether the seasonal variation is constant (as for Product A) or changes with trend (as for Product B). Each of these cases requires a different method for forecasting future demand.

### 3.10.2 Step 2—Choose a Method

Seasonal time series with constant seasonal variation are modeled using "Additive Models," and series with increasing or decreasing seasonal variation are modeled using "Multiplicative Models." We start by illustrating the multiplicative model using the following example.

 **Example 3.3**

A firm that produces electrical snowblowers (among other products) is considering hiring more part-time workers for its assembly operations in next fall and winter. This is because the demand for snowblowers is usually higher in fall and winter. Also, due to a comprehensive marketing and advertising campaign, the demand for snowblowers has also been increasing in recent years. Table 3.2 shows the demand for snowblowers in each quarter of the last 3 years. What would you predict the demand for snowblowers be next fall and winter (in Year 4)? The firm tries to avoid large errors in forecasts, since large errors have a significantly higher impact on labor hiring and layoffs costs than smaller errors.

To develop a method to forecast the demand for snowblowers, we first need to identity if the seasonal demand for snowblowers has a constant seasonal variation or increasing (or decreasing) seasonal variation. One approach is to simply graph the demand and visually check the seasonal pattern in demand. This is shown in Fig. 3.22.

As the figure shows, the demand has an increasing trend, and the seasonal variation in demand becomes larger in later seasons. This is similar to the demand for Product B in Fig. 3.21.

To develop a method to forecast seasonal time series (multiplicative or additive), two main characteristics of the series must be clearly defined:

- *Seasons:* In seasonal time series, a season is defined as the cycle that repeats itself. In the demand for snowblowers in Fig. 3.22 a season refers to a year, since the demand has the same pattern in all 3 years. Note that this is different from the common use of the word "season" which refers to fall, winter, spring, and summer.

- *Periods:* Each season consists of the same number of periods for which we have demand data. In Fig. 3.22 each season (i.e., each year) has four periods (summer, fall, winter, and spring). The number of periods in a season is called the "*Length of the Season*" and is shown by $N$. As shown in Fig. 3.22, the seasons for snowblower demand has length $N = 4$.

Note that a season in a time series can be a year, a week, or a day. For example, the hourly demand in fast-food restaurants repeats itself every day. Hence, the season can be defined as a day. The length of a season is therefore the number of (working) hours during the day. If a restaurant works from 10:00 am to 10:00 pm, then each season will have $N = 12$ periods. We

**Table 3.2  Demand for Snowblowers in the Last 3 Years in Example 3.3**

| Year 1 | | Year 2 | | Year 3 | |
|---|---|---|---|---|---|
| Quarter | Demand | Quarter | Demand | Quarter | Demand |
| Summer | 21 | Summer | 44 | Summer | 30 |
| Fall | 41 | Fall | 80 | Fall | 69 |
| Winter | 91 | Winter | 151 | Winter | 212 |
| Spring | 48 | Spring | 54 | Spring | 75 |

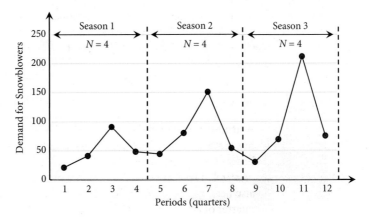

**Figure 3.22  Demand for snowblowers in Example 3.3.**

can, if needed, define the periods to be every 15 minutes. This results in each season having $N = 12 \times 4 = 48$ periods.

## Multiplicative Model

The main idea behind the seasonal time series forecasting methods is to separate the variation due to seasonality, from those due to base demand and trend. This is often done by estimating what is called as "Seasonal Index" or "Seasonal Factor" for each period of a season. These indices capture the seasonal impact of each period on average demand.

The multiplicative model also uses seasonal factor to model seasonality. To illustrate this, consider the demand in four periods of Season 1 in Table 3.2, which are 21, 41, 91, and 48. Obviously, the demand in winter—the third period—is higher than the demand in other periods. This is because of the seasonality in demand where more people buy snowblowers, since it snows more in winter. But how much the demand in winter is higher than the average demand in a period in Season 1? Well, the average demand in a period in Season 1 is

$$\text{Average demand in a period in Season 1} = \frac{21 + 41 + 91 + 48}{4} = 50.25$$

Since the demand in winter in Season 1 is 91, then the demand in winter is $91/50.25 = 1.81$ times larger than the average demand in a period in that season. On the other hand, the demand in summer is $21/50.25 = 0.42$ times larger than the average demand in Season 1.

---

**CONCEPT**

**Multiplicative Seasonal Factor**

Multiplicative seasonal factor $\mathcal{C}_t$ of Period $t$ in a season is a number that captures *how many times* the demand in Period $t$ of a season is larger (or smaller) than the average demand in that season. Specifically,

$$\text{Multiplicative } \mathcal{C}_t = \frac{\text{Demand in Period } t \text{ of the season}}{\text{Average demand in a period of the season}}$$

Table 3.3 shows the multiplicative seasonal factors for all periods of the three seasons. As the table shows, the demand in the first period of Season 1 (i.e., summer) is 42 percent

Table 3.3   Demand for Snowblowers in the Last 3 Years

| Period $t$ | Demand $D_t$ | Multiplicative seasonal factor |
|---|---|---|
| 1 | 21 | 0.42 |
| 2 | 41 | 0.82 |
| 3 | 91 | 1.81 |
| 4 | 48 | 0.96 |
| 5 | 44 | 0.53 |
| 6 | 80 | 0.97 |
| 7 | 151 | 1.84 |
| 8 | 54 | 0.66 |
| 9 | 30 | 0.31 |
| 10 | 69 | 0.72 |
| 11 | 212 | 2.20 |
| 12 | 75 | 0.78 |

of the average demand of a period in Season 1. This number is 53 percent and 31 percent in Seasons 2 and 3, respectively.

The multiplicative model for forecasting seasonal time series assumes that the seasonal demand in Period $t$ follows the following model:

$$D_t = (\mathcal{L} + \mathcal{G}\,t)\,.\,\mathcal{C}_t\,.\,\epsilon_t \tag{3.26}$$

Similar to time series with trend, in the above model $(\mathcal{L} + \mathcal{G}\,t)$ captures the level $(\mathcal{L})$ and the trend $(\mathcal{G})$ component of demand in Period $t$. The model is called *multiplicative* since it incorporates the seasonal variation of the demand by multiplying average demand $(\mathcal{L} + \mathcal{G}\,t)$ by seasonal factor $\mathcal{C}_t$. The term $\epsilon_t$ is to represent noise as in previous models.

Similar to the previous models, the above multiplicative model assumes that the trend has a linear growth. Also, the model assumes that multiplicative seasonal factor $\mathcal{C}_t$ for Period $t$ is stationary and does not have an increasing or decreasing pattern. In other words, it assumes that seasonal factors for the first period of all seasons are almost the same. It also assumes the same for seasonal factors for the second, third, ..., and $N$th periods of all seasons.

Having a closer look at the multiplicative model in Eq. (3.26), we can see why the model captures increase (or decrease) in seasonal variation, even though the multiplicative seasonal factors are the same for the same periods of each season. As an example, suppose that the $(\mathcal{L} + \mathcal{G}\,t)$ in the first period of the first and the second seasons of a series are 100 and 200 (i.e., the series has an increasing trend). Also suppose the first period of the seasons has multiplicative seasonal factor 1.2, which is the highest among all periods, and the second period has a seasonal factor of 0.4 which is the lowest among all periods. According to multiplicative model, the demand for the first and the second periods in Season 1 are $100 \times 1.2 = 120$ and $100 \times 0.4 = 40$, respectively, and the demand for the first and the second periods of Season 2 are $200 \times 1.2 = 240$ and $200 \times 0.4 = 80$, respectively. Notice that even though both seasons have the same seasonal factors, the difference between the highest and lowest demand (i.e., seasonal variation) in Season 2 is $240 - 80 = 160$, which is much larger than that in Season 1, which is $120 - 40 = 80$. Hence, because of the increasing trend, the seasonal variation increases in later seasons.

The model in Eq. (3.26) has two main applications in forecasting: (i) obtaining deseasonalized data, and (ii) forecasting seasonal time series.

### Multiplicative Model—Deseasonalizing Data

Seasonal factors can be used to deseasonalize data in order to remove the seasonal component. If we divide both sides of the model in Eq. (3.26) by multiplicative seasonal factor $\mathcal{C}_t$, we get the *Deseasonal Data* which is

$$\frac{D_t}{\mathcal{C}_t} = (\mathcal{L} + \mathcal{G}\,t)\,.\,\epsilon_t'$$

where $\epsilon_t' = \epsilon_t/\mathcal{C}_t$ is the noise. As the above equation shows, the deseasonal demand—the right-hand side of the above equation—captures the base and the trend components of the demand. This deseasonalized data, also called *Seasonally Adjusted* data, takes seasonal pattern out of the data and shows a clearer picture of other patterns or lack of patterns in the data. For example, the seasonally adjusted unemployment removes the seasonality in the unemployment data and reveals whether the unemployment data has any upward or downward trend (besides seasonality).

Let's convert the seasonal demand for snowblowers in Table 3.2 to deseasonalized demand to explore if there is any trend in the demand. To do that, we need to divide demand

$D_t$ in Period $t$ by its corresponding multiplicative seasonal factor $\mathcal{C}_t$. We start with the first period of the first season—summer. Recall that in Table 3.3, we found three different estimates 0.42, 0.53, and 0.31 for the seasonal factor for summer in the last 3 years. Under the assumption that the actual seasonal factors at Period $t$ does not change from season to season, we can get a good estimate for the seasonal factor for summer by simply taking the average of the three factors:

$$\text{Multiplicative seasonal factor for summer} = \frac{0.42 + 0.53 + 0.31}{3} = 0.42$$

Similarly, we can get the seasonal factors for fall, winter, and spring to be 0.83, 1.95, and 0.80, respectively. Having these multiplicative seasonal factors, we can now deseasonalize demand for snowblowers by dividing the demand in each period by its corresponding seasonal factor. This is shown in Fig. 3.23. The figure also shows the graph of the deseasonalized demand.

By removing the seasonality in demand, the deseasonalized demand in Fig. 3.23 clearly shows that the demand for snowblowers indeed has an increasing trend.

### Multiplicative Model—Forecasting Demand

The second important application of the multiplicative seasonal factors is that it enables us to use our forecasting methods for time series with trend to find a forecast for seasonal time series. This involves two steps:

- *Step 1:* Deseasonalizing past demands, and using methods for times series with trend to find a forecast for deseasonalized demand for future periods.

- *Step 2:* Using multiplicative seasonal factors to incorporate seasonality into the deseasonalized forecast for future periods (obtained in Step 1). Considering $DF_t$ as the forecast for deseasonalized demand for Period $t$, and $F_t$ as the forecast for Period $t$, then

$$F_t = DF_t \times \mathcal{C}_t$$

One commonly used method that follows the above steps to generate forecast for seasonal time series is *Winters' Method,* which is discussed next.

The multiplicative Winters' method is appropriate when a time series has a linear trend with increasing or decreasing seasonal variations for which the level, growth rate, and the seasonal pattern may be changing rather than being fixed.

| Periods | Seasonal Demand | Multiplicatve Seasonal Factors | Deseasonalized Demand |
|---|---|---|---|
| 1 | 21 | 0.46 | 45.80 |
| 2 | 41 | 0.83 | 49.42 |
| 3 | 91 | 1.98 | 45.90 |
| 4 | 48 | 0.77 | 62.10 |
| 5 | 44 | 0.46 | 95.96 |
| 6 | 80 | 0.83 | 96.42 |
| 7 | 151 | 1.98 | 76.17 |
| 8 | 54 | 0.77 | 69.86 |
| 9 | 30 | 0.46 | 65.43 |
| 10 | 69 | 0.83 | 83.16 |
| 11 | 212 | 1.98 | 106.93 |
| 12 | 75 | 0.77 | 97.03 |

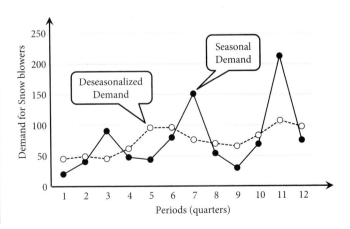

**Figure 3.23    Deseasonalizing the demand for snowblowers in Example 3.3.**

Winters' method starts by estimating multiplicative seasonal factors and uses them to deseasonalize demand. It then uses Holt's method to develop a forecast for the deseasonalized demand for the next period.[6] Finally, it uses the seasonal factors to convert the deseasonalized forecast to the forecast for the seasonal demand.

To be more specific, at the end of each Period $t$—after observing demand $D_t$—Winters' method follows the following steps:

- *Deseasonalizing demand:* It deseasonalizes the demand $D_t$ using the latest estimate for seasonal factor in Period $t$, which is $C_{t-N}$. Note that since the length of the season is $N$, the last estimate for seasonal factor for Period $t$ was made $N$ period ago. The deseasonalized demand would therefore be $D_t/C_{t-N}$.

- *Developing forecast using Holt's method:* Winters' method then uses Holt's method to make a forecast for deseasonalized demand for Period $t+1$ using double exponential smoothing equations as follows:

$$L_t = \alpha(D_t/C_{t-N}) + (1-\alpha)(L_{t-1} + G_{t-1})$$
$$G_t = \beta(L_t - L_{t-1}) + (1-\beta)G_{t-1}$$

and

$$DF_{t+1} = L_t + G_t$$

Using the latest estimate of multiplicative seasonal factor for Period $t+1$, which is $C_{t+1-N}$, Winters' method generates the forecast for Period $t+1$ as follows:

$$F_{t+1} = DF_{t+1} \cdot C_{t+1-N} = (L_t + G_t)\,C_{t+1-N}$$

At the end of Period $t$, if the forecast for the next $N-1$ periods (i.e., Periods $t+2, t+3, \ldots, t+N$) is also needed, it would be

$$F_{t+\tau} = (L_t + \tau\, G_t)\,C_{t+\tau-N}: \quad \text{for } \tau = 2, 3, \ldots, N$$

- *Updating seasonal factors:* Finally, Winters' method uses the observed demand $D_t$ to get $C_t$, the new estimate for the multiplicative seasonal factor in Period $t$. This is done through a simple exponential smoothing method with parameter $\gamma$. Specifically, the new simple exponential smoothing method uses the newly observed multiplicative seasonal factor in Period $t$—approximated by $D_t/L_t$—and the last estimate for the seasonal factor, that is, $C_{t-N}$, to get a new estimate for the factor in Period $t$ as follows:

$$C_t = \gamma(D_t/L_t) + (1-\gamma)C_{t-N}$$

This process is repeated and new estimates for $L_{t+1}$, $G_{t+1}$, and $C_{t+1}$ are obtained after observing the actual demand $D_{t+1}$ in Period $t+1$.

### Multiplicative Model—Initializing Winters' Method

Similar to the simple and double exponential smoothing methods, Winters' method also requires initial values. For Winters' method, the required initial values are: (i) initial estimate for the base component of the demand, that is, $\mathcal{L}$, (ii) initial estimate for the growth rate, that is, $\mathcal{G}$, and (iii) initial estimates for multiplicative seasonal factors, $\mathcal{C}_1, \mathcal{C}_2, \ldots, \mathcal{C}_N$.

- *Initial Estimate for Level $\mathcal{L}$:* To get an initial estimate for the level of the seasonal series, data of at least one season is needed. Having demands $D_1, D_2, \ldots, D_N$ in Season 1, Winters' method uses the average of these demands to get a smoothed

---

[6]That is the reason the method is also called Holt-Winters method.

estimate for level $\mathcal{L}$ of the demand at the end of Season 1 (i.e., end of Period $N$) as $L_N$:

$$L_N = \frac{D_1 + D_2 + \cdots + D_N}{N}$$

In Example 3.3, with the length of seasons being $N = 4$, the initial value for the level would be

$$L_4 = \frac{D_1 + D_2 + D_3 + D_4}{4} = \frac{21 + 41 + 91 + 48}{4} = 50.25$$

- *Initial Estimate for Growth Rate $\mathcal{G}$:* To find an initial estimate for the growth rate, Winters' method requires demand data for at least two seasons. Suppose demand data is available for $K$ seasons, where $K \geq 2$. One approach to get an estimate for the growth rate is to find how much the average demand in a season increases (or decreases) from first season to the last (i.e., the $K$th) season. Suppose $V_1$ and $V_K$ are the average demand in a period of Seasons 1 and $K$, respectively, then $V_K - V_1$ is an estimate for the growth in average demand after $K - 1$ seasons. Hence, $(V_K - V_1)/(K - 1)$ is an estimate for the growth from one season to the next. However, note that $\mathcal{G}$ is the growth rate from period to period (not from season to season). Since each season has $N$ periods, then an estimate for growth rate in a period can be found by dividing the growth rate in a season by $N$. This gives us an initial estimate for growth rate of the seasonal series. Winters' method uses this estimate as the estimate at the end of Season 1 (i.e., the end of Period $N$) as $G_N$:

$$G_N = \frac{V_K - V_1}{N(K - 1)}$$

In Example 3.3, we have data for three seasons; hence, $K = 3$. The average demand in Season 1 and Season 3 is

$$V_1 = \frac{D_1 + D_2 + D_3 + D_4}{4} = \frac{21 + 41 + 91 + 48}{4} = 50.25$$
$$V_3 = \frac{D_9 + D_{10} + D_{11} + D_{12}}{4} = \frac{30 + 69 + 212 + 75}{4} = 96.5$$

Hence, the initial estimate for growth rate $\mathcal{G}$ at Period $N$ is

$$G_N = \frac{V_K - V_1}{N(K - 1)} = \frac{96.5 - 50.25}{4(3 - 1)} = 5.78$$

- *Initial Estimate for Seasonal Factors:* To start, Winters' method needs the initial estimates for multiplicative seasonal factors for all periods of Season 1. A simple estimate can be obtained using the demand data in Season 1. By simply dividing the demand in each period of Season 1 by the average demand in Season 1, we can get an initial estimate for the seasonal factors as follows:

$$C_1 = \frac{D_1}{V_1}, \quad C_2 = \frac{D_2}{V_1}, \quad \cdots, \quad C_N = \frac{D_N}{V_1}$$

In Example 3.3, the initial estimates for seasonal factors will be

$$C_1 = \frac{21}{50.25} = 0.418, \quad C_2 = \frac{41}{50.25} = 0.816, \quad C_3 = \frac{91}{50.25} = 1.811, \quad C_4 = \frac{48}{50.25} = 0.955$$

We summarize the Winters' multiplicative method for forecasting seasonal time series in the following analytics.

### Winters' Multiplicative Method

Consider a seasonal time series with seasons of length $N$, where $N \geq 3$. Suppose that the demand data for $K$ $(K \geq 2)$ seasons are available. Specifically, we know the actual demands $D_1, D_2, \ldots, D_{KN-1}, D_{KN}$. Also, suppose that the demand has a linear (increasing or decreasing) trend and seasonal variation changes (increases or decreases) from one season to the next. The Winters' multiplicative method, also called *Triple Exponential Smoothing method* with parameters $0 < \alpha \leq 1$, $0 < \beta \leq 1$, and $0 < \gamma \leq 1$ finds the forecast for future periods as follows:

- *Initializing Step:* Winters' method requires initial estimates $L_N$, $G_N$ and multiplicative seasonal factors $C_1, C_2, \ldots, C_N$ to start forecasting. The following can be used for these initial estimates:

$$L_N = \frac{D_1 + D_2 + \cdots + D_N}{N}$$

$$G_N = \frac{V_K - V_1}{N(K-1)}$$

$$C_i = \frac{D_i}{V_1} \qquad \text{for } i = 1, 2, \ldots, N$$

  where $V_1$ and $V_K$ are the average demand in a period in Season 1 and Season $K$, respectively,

$$V_1 = \frac{D_1 + D_2 + \cdots + D_N}{N}, \qquad V_K = \frac{D_{(K-1)N+1} + D_{(K-1)N+2} + \cdots + D_{KN}}{N}$$

  Hence, the forecast for Period $N + 1$ is

$$F_{N+1} = (L_N + G_N) \, C_1$$

- *Forecasting Step:* After observing demand $D_t$ in Period $t$, and having estimates $L_{t-1}$, $G_{t-1}$, and $C_{t-N}$, the method computes

$$L_t = \alpha(D_t/C_{t-N}) + (1-\alpha)(L_{t-1} + G_{t-1}) \tag{3.27}$$

$$G_t = \beta(L_t - L_{t-1}) + (1-\beta)G_{t-1} \tag{3.28}$$

$$C_t = \gamma(D_t/L_t) + (1-\gamma)C_{t-N} \tag{3.29}$$

  - *One-Period-Ahead Forecast:* Using $C_{t+1-N}$ the forecast for Period $t + 1$ is, therefore,

$$F_{t+1} = (L_t + G_t) \, C_{t+1-N} \tag{3.30}$$

  - *Multiple-Period-Ahead Forecast:* Since at Period $t$ the latest estimates are $L_t$ and $G_t$, then the forecast for future Periods $t + 2, t + 3, \ldots, t + N$ would be

$$F_{t+\tau} = (L_t + \tau \, G_t) \, C_{t+\tau-N} : \quad \text{for } \tau = 2, 3, \ldots, N$$

Let's use the above analytics with $\alpha = 0.4$, $\beta = 0.1$, and $\gamma = 0.3$ to develop a forecast for demand in the four periods of the next year. Note that we have data for $K = 3$ seasons, where the length of the seasons is $N = 4$. We have already found the initial values of $L_4 = 50.25$,

| | Period | Demand | | | | | | |
|---|---|---|---|---|---|---|---|---|
| | $t$ | $D_t$ | $L_t$ | $G_t$ | $C_t$ | $F_t$ | $e_t$ | $(e_t)^2$ |
| | 1 | 21 | | | 0.418 | | | |
| | 2 | 41 | | | 0.816 | | | |
| | 3 | 91 | | | 1.811 | | | |
| | 4 | 48 | 50.25 | 5.78 | 0.955 | | | |
| | 5 | 44 | 75.73 | 7.75 | 0.467 | 23.42 | -20.58 | 423.70 |
| | 6 | 80 | 89.31 | 8.33 | 0.840 | 68.12 | -11.88 | 141.21 |
| | 7 | 151 | 91.94 | 7.76 | 1.760 | 176.83 | 25.83 | 667.10 |
| | 8 | 54 | 82.43 | 6.04 | 0.865 | 95.24 | 41.24 | 1700.61 |
| | 9 | 30 | 78.79 | 5.07 | 0.441 | 41.30 | 11.30 | 127.72 |
| | 10 | 69 | 83.18 | 5.00 | 0.837 | 70.43 | 1.43 | 2.04 |
| | 11 | 212 | 101.08 | 6.29 | 1.861 | 155.22 | -56.78 | 3223.61 |
| | 12 | 75 | 99.10 | 5.46 | 0.833 | 92.89 | 17.89 | 320.12 |

alpha = 0.4, beta = 0.1, Gamma = 0.3

Total = 8.44, 6606.11
Mean = 1.06, 825.76

**Figure 3.24**    Winters' multiplicative method to forecast demand for snowblowers in Example 3.3.

$G_4 = 5.78$, and $C_1 = 0.418$, $C_2 = 0.816$, $C_3 = 1.811$, and $C_4 = 0.955$. Using these estimates, we can develop a forecast for Period $N + 1 = 4 + 1 = 5$ as follows:

$$F_5 = (L_4 + G_4)\, C_1 = (50.25 + 5.78)0.418 = 23.42$$

Now suppose we observe demand for Period $t = 5$, which is $D_5 = 44$. We can then compute $L_5$, $G_5$, and $C_5$ using Eqs. (3.27), (3.28), and (3.29) as follows:

$$\begin{aligned}
L_5 &= \alpha(D_5/C_{5-4}) + (1 - \alpha)(L_4 + G_4) \\
&= 0.4(44/0.418) + (1 - 0.4)(50.25 + 5.78) = 75.73 \\
G_t &= \beta(L_5 - L_4) + (1 - \beta)G_4 \\
&= 0.1(75.73 - 50.25) + (1 - 0.1)5.78 = 7.75 \\
C_5 &= \gamma(D_5/L_5) + (1 - \gamma)C_{5-4} \\
&= 0.3(44/75.73) + (1 - 0.3)0.418 = 0.467
\end{aligned}$$

The forecast for Period $t = 6$ using Eq. (3.30) is, therefore,

$$F_6 = (L_5 + G_5)\, C_{6-4} = (75.73 + 7.75)0.816 = 68.12$$

Similarly, after observing demand in Period $t = 6$, which is $D_6 = 80$, we can use Eqs. (3.27), (3.28), and (3.29) to compute $L_6$, $G_6$, and $C_6$. These, along with the computations for all 12 periods, are shown in Fig. 3.24.

### 3.10.3  Step 3—Evaluate the Method

Recall that the firm in Example 3.3 tries to avoid large errors in demand forecasts, since they have a significantly high impact on labor hiring and layoffs. Hence, MSE that penalizes larger errors much more than small errors would be a good evaluation criterion for forecasting

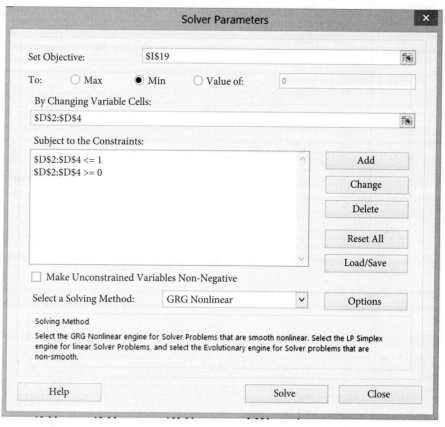

**Figure 3.25**  Solver window for optimizing values $\alpha$, $\beta$, and $\gamma$ to minimize MSE for Winters' multiplicative method in Example 3.3.

method used to predict the demand for snowblowers. For this criterion, as Fig. 3.24 shows, Winters' method with constants $\alpha = 0.4$, $\beta = 0.1$, and $\gamma = 0.3$ results in an MSE = 825.76.

### 3.10.4  Step 4—Adjust the Method

Similar to the case of double exponential smoothing (i.e., Holt's) method, we can use Excel Solver to find the best values for smoothing constants $\alpha$, $\beta$, and $\gamma$ that minimize the MSE. The setup for Solver is shown in Fig. 3.25.

The solution obtained by Solver is shown in Fig. 3.26. As the figure shows, the best values for smoothing constants are $\alpha = 0.038$, $\beta = 0.721$, and $\gamma = 0.237$. This results in MSE = 449.85, which is about half of MSE = 825.76 we found in Fig. 3.24, before adjusting the method.

### 3.10.5  Step 5—Develop a Forecast

Having the new smoothing constants, we can now develop a point and a prediction interval forecast for the demand for snowblowers next year. For the adjusted values of $\alpha = 0.038$, $\beta = 0.721$, and $\gamma = 0.237$, as shown in Fig. 3.26, we have $L_{12} = 106.88$, $G_{12} = 5.96$. Also, the figure gives seasonal factors for the last year (i.e., Season 3) as $C_9 = 0.461$, $C_{10} = 0.866$,

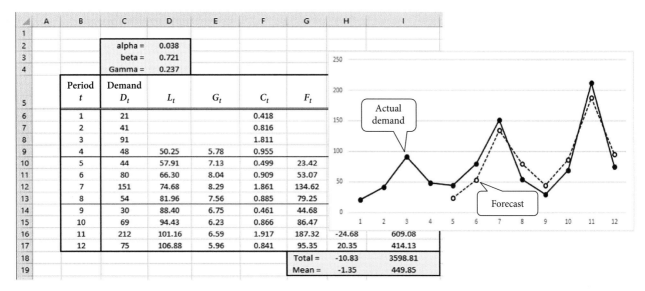

**Figure 3.26** Forecasts for snowblowers obtained using Winters' multiplicative method with optimal parameters $\alpha = 0.038$, $\beta = 0.721$, and $\gamma = 0.237$.

$C_{11} = 1.917$, and $C_{12} = 0.841$. Hence, the point forecasts are:

$$F_{13} = (L_{12} + (1)G_{12})C_9 = (106.88 + (1)5.96)0.461 = 52.02$$
$$F_{14} = (L_{12} + (2)G_{12})C_{10} = (106.88 + (2)5.96)0.886 = 105.26$$
$$F_{15} = (L_{12} + (3)G_{12})C_{11} = (106.88 + (3)5.96)1.917 = 239.16$$
$$F_{16} = (L_{12} + (4)G_{12})C_{12} = (106.88 + (4)5.96)0.841 = 109.93$$

Therefore, the firm should expect an average demand of about 52, 105, 239, and 110 for snowblowers in summer, fall, winter, and spring of the next year, respectively.

The following analytics shows how prediction interval forecasts can be established for demand using Winters' multiplicative method.

---

**ANALYTICS**

**Two-Sided Prediction Interval Forecast for Winters' Multiplicative Method**

Suppose that Winters' multiplicative method with smoothing constants $\alpha$, $\beta$, and $\gamma$ was applied on a seasonal time series in which the length of a season is $N$. The method generates forecasts $F_1, F_2, \ldots, F_t$ for demand in the last $t$ periods, resulting in errors $e_1, e_2, \ldots, e_t$. Finally, suppose that point forecasts $F_{t+1}, F_{t+2}, \ldots, F_{t+N}$ are computed for $N$ future periods of the next season.

A $100(1 - x)$-percent prediction interval forecast $[\underline{F}_{t+\tau}, \overline{F}_{t+\tau}]$ for the actual demand in future Period $t + \tau$ is

$$D_{t+\tau} \in \left[ F_{t+\tau} \pm \sigma_r \, z_{(x/2)} \; C_{t+\tau-N} \sqrt{(L_t + \tau G_t)^2 + \mathbb{D}_\tau} \right] \quad \text{for} \quad \tau = 1, 2, \ldots, N \tag{3.31}$$

where $z_{(x/2)}$ is the value of Standard Normal distribution with its cumulative probability distribution $\Phi(z) = x/2$ and $\sigma_r$ is an estimate of standard deviation of the *relative* errors computed for the last $t$ periods as follows:

$$\sigma_r = \sqrt{\frac{\sum_{i=1}^{t}(e_i/F_i)^2}{t - 3}} \tag{3.32}$$

The value of $\mathbb{D}_\tau$ depends on the forecast horizon as follows:

- *One-Period-Ahead Prediction Interval Forecast:* When $\tau = 1$, we have

$$\mathbb{D}_1 = 0$$

- *Multiple-Period-Ahead Prediction Interval Forecast:* When $\tau = 2, 3, \ldots, N$, we have

$$\mathbb{D}_\tau = \sum_{j=1}^{\tau-1} \alpha^2 (1 + [\tau - j]\beta)^2 (L_t + jG_t)^2 \qquad \text{for} \quad \tau = 2, 3, \ldots N \tag{3.33}$$

Let's use the above analytics to compute 95-percent prediction interval forecast for snowblower demand in summer and fall of the next year. To do that, we first need to compute the standard deviation of the relative error, that is, $\sigma_r$. Note that, as Fig. 3.26 shows, we have eight forecasts for the eight periods in Seasons 2 and 3. Hence,

$$
\begin{aligned}
\sigma_r &= \sqrt{\frac{\sum_{i=1}^{8}(e_i/F_i)^2}{8-3}} \\
&= \sqrt{\frac{(-20.58/23.42)^2 + (-26.93/53.07)^2 + \cdots + (20.35/95.35)^2}{5}} \\
&= 0.272
\end{aligned}
$$

Also, note that the point forecast for Period $t = 13$ (i.e., next summer) was $F_{13} = 52.03$, and we computed seasonal factor $C_{t+\tau-N} = C_{12+1-4} = C_9 = 0.461$ (see Fig. 3.26). Having $\mathbb{D}_1 = 0$, we use Eq. (3.31) to compute the one-period-ahead prediction interval forecast as follows:

$$
\begin{aligned}
D_{13} &\in \left[ F_{13} \pm \sigma_r z_{(0.025)} C_9 \sqrt{(L_{12} + G_{12})^2 + \mathbb{D}_1} \right] \\
&\in \left[ 52.03 \pm 0.272(-1.96)(0.461)\sqrt{(106.88 + 5.96)^2 + 0} \right] \\
&\in [24.30, \ 79.76]
\end{aligned}
$$

In other words, with probability 0.95, the demand for snowblowers in next summer would be between 24 and 80.

To compute a prediction interval forecast for next fall (i.e., $\tau = 2$), we need to compute $\mathbb{D}_2$ using Eq. (3.33) as follows:

$$
\begin{aligned}
\mathbb{D}_2 &= \sum_{j=1}^{2-1} \alpha^2 (1 + [2-j]\beta)^2 (L_{12} + jG_{12})^2 \\
&= \alpha^2 (1 + \beta)^2 (L_{12} + G_{12})^2 \\
&= (0.038)^2 (1 + 0.721)^2 (106.88 + 5.96)^2 \\
&= 54.46
\end{aligned}
$$

Since we already calculated the forecast for next fall to be $F_{14} = 102.93$, and since we have estimated $C_{t+\tau-N} = C_{12+2-4} = C_{10} = 0.866$, we can compute the prediction interval forecast for fall as follows:

$$D_{14} \in [F_{14} \pm \sigma_r z_{(0.025)} C_{10} \sqrt{(L_{12} + 2G_{12})^2 + \mathbb{D}_2}]$$
$$\in [102.93 \pm 0.272(-1.96)(0.866) \sqrt{(106.88 + 2(5.96))^2 + 54.46}]$$
$$\in [47.98, \ 157.88]$$

Thus, with probability 0.95, we should expect that the demand for the snowblower in fall to be somewhere between 47 and 158 units.

### Want to Learn More?

**Additive Seasonal Forecasting Models**

When a seasonal time series has a *constant* seasonal variation, as shown in Fig. 3.21-left for Product A, a different model is used for forecasting. The model is called *Additive Model*, since it incorporates the seasonal variation into the forecast by simply adding a seasonal factor to the demand. So, if you plot the past demand for a seasonal product and you observe a pattern similar to that in Fig. 3.21-left, what do you need to do? You need to check the supplement of this chapter that describes how you can develop forecasts for your seasonal product.

## 3.11 Forecasting Cyclic Time Series

Similar to seasonal time series, cyclic time series also have a cycle consisting of high and low demands. However, the length of cycles in cyclic time series is not clearly known. Furthermore, each cycle may have a different length. Also, the length of the cycles is often longer than those in seasonal time series. Therefore, developing forecasts for cyclic time series is difficult and requires advanced mathematical and statistical techniques.

One commonly used approach for forecasting demand that has cyclic pattern is to find another variable that is related to the demand. These are variables whose changes lead to the changes in demand, and hence they are called *Leading Indicators*. Changes in the values of leading indicators often precede changes in the economy, which affects changes in demand. For example, the number of permits issued in a month for building houses is often a leading indicator of demand in later months for products and services related to construction of new houses as well as for the demand of durable goods such as washers, dryers, and refrigerators. Examples of other leading indicators are stock market prices, and economic indices such as consumer sentiment.

Analytical models developed for forecasting demand with cyclic behavior try to capture the correlation between the value of the leading indicators and the value of demand. If there is a strong correlation between the leading indicators and demand, then the model provides a more accurate forecast. One class of models that is developed to forecast demand using its relationship with other variables is called causal forecasting model.

### Want to Learn More?

**Causal Forecasting Models**

While time series forecasting models use only data about past demands of a product to forecast future demand for the product, causal forecasting models use the past relationship between demand and other variables to forecast future demand. For example, the demand for children diapers is related to the population of children of 2 years old or younger, and the price of the diapers. Suppose you are trying to forecast the demand for a cold drink that your firm produces. Do you think that in warmer days people buy more cold drinks? Do you think if you offer price discount through advertising coupons, you will sell more drinks? If you believe so—I certainly do—you should check the supplement of this chapter to learn how one can use data about next week's forecasted temperature and next week's advertising expenditure to forecast the demand for a cold drink next week.

## 3.12 Monitoring and Controlling Forecasts

The fact that a forecasting method has been performing very accurately so far does not necessarily mean that it will also perform well in the future. In fact, the ability of a given forecasting method to accurately forecast demand diminishes over time. This is because customers' tastes change over time, or the product may become popular resulting in an increasing trend, or a competitor may enter the market resulting in a decreasing trend. Hence, after a future demand is realized, a good forecasting process traces the performance of the forecasting method by monitoring the errors made by the method, especially the most recent errors. This, as shown in Fig. 3.9, is done through Adaptive Forecasting and Focus Forecasting.

- *Adaptive Forecasting:* This corresponds to continuously monitoring the errors made by the forecast and adjusting parameters of the method as it is done in Step 4.
- *Focus Forecasting:* This corresponds to periodic review of the errors and the entire procedure. For example, every month it must be checked whether the type of the time series has changed or not (i.e., Step 1). Also, it must be examined whether a different method of forecasting gives a better forecast.

The key task in both adaptive and focus forecasting is *monitoring* the performance of the forecasting method by checking the errors generated by the method. There are three approaches that are used to monitor the performance of forecasting methods: (i) Tracking Signals, (ii) Control Charts, and (iii) Visual Inspection.

### 3.12.1 Monitoring Bias Using Tracking Signal

Recall from Sec. 3.8.3 that cumulative error (CE) is the sum of all observed errors in the past $t$ periods, and BIAS is the average cumulative error, that is, $CE/t$. As mentioned, if a forecast method is consistently underestimating the actual demand, then we will have more negative errors than positive errors. In this case, the CE becomes negative, resulting in a negative bias. On the other hand, if the forecast method is consistently overestimating the actual demand, then the CE becomes positive, resulting in a positive bias. Within a good forecasting method positive errors should cancel out negative errors, resulting in a CE and thus a bias of zero.

However, due to the noise in the demand, it is very unlikely that a forecasting method has a CE of zero. Hence, it is acceptable for a good forecasting method to have a non-zero CE, as long as it is not very large. A large CE implies that the forecasting method is *out of control*. But how large the CE should be to conclude that a forecasting method is out of control?

To answer this question, we need to define Tracking Signal. *Tracking Signal* in Period $t$ for a forecasting method resulting in errors $e_1, e_2, \ldots, e_t$ is defined as follows:

$$\text{Tracking Signal in Period } t = \frac{e_1 + e_2 + \cdots + e_t}{(|e_1| + |e_2| + \cdots + |e_t|)/t} = \frac{CE_t}{MAD_t}$$

where $CE_t$ and $MAD_t$ are CE and MAD up to Period $t$. Hence, tracking signal is nothing but CE measured in MAD units. A tracking signal of 2 in Period 10, for example, means that CE of all 10 periods is positive and is twice as large as MAD of the 10 periods. A tracking signal of $-2$, on the other hand, means that CE is negative and it is twice as large as MAD. The question is whether the tracking signal in a period is too large (positive or negative) that the forecasting method should be considered out of control.

The notion of "too large" is decided by setting an upper limit for positive values of tracking signal and a lower limit for negative values of tracking signal. When tracking signal in a period goes beyond these limits, we can conclude that the forecasting method is out of control. But what are these upper and lower limits? Well, there is no single answer, because the limits are usually based on judgment and experiment as well as on the impact of large errors on costs. If limits are chosen too low, the tracking signal responds to every small forecast error. On the other hand, if the limits are chosen too high, the tracking signal does not detect bad forecasts, unless they are very bad. In practice, firms use limits that range from $\pm 3$ to $\pm 8$. Some inventory experts recommend to use a limit of $\pm 4$ for high-volume stock items and limits $\pm 8$ for low-volume items.[7] It is important to recognize that having tracking signals within its control limits conjectures—but not guarantees—that the forecast is performing properly.

The only numbers needed to compute tracking signal are the errors made by the forecasting method. Hence, tracking signal can be used in all of the forecasting methods discussed earlier including time series forecasting methods for stationary, trend, and seasonal demands. Tracking signal is recomputed each Period $t$ after demand $D_t$ in that period is realized and its corresponding error is computed. The new error is used to compute new $CE_t$ and new $MAD_t$ and new tracking signal. The tracking signal is then compared with the control limits.

To show how tracking signal works, we use the following example.

### Example 3.4

John works for a producer of dairy products and is in charge of forecasting demand for several items including 2-gallon bottles of lactose-free milk. He has been using simple exponential smoothing to forecast demand in the last 12 weeks, in which the actual demand in Week 1 to Week 12 have been 387, 407, 411, 404, 420, 431, 435, 401, 444, 450, 460, and 541. Simple exponential smoothing method resulted in forecasts 420, 403, 420, 392, 385, 402, 405, 412, 421, 422, 450, and 480 for those weeks. In some weeks, the method resulted in small errors and in some weeks the method resulted in large errors. For example in Week 2, the difference between the forecast and demand was only 4, where in the last week, the difference was about 60. John knows that the first rule of forecasting is that forecasts are always wrong, and therefore there will always be some random small or large errors in the forecasts. For his case, however, he is wondering if the errors his method produces are too large, and whether he needs to use another forecasting method or to readjust his method. This is important to him, because lactose-free milk is one of their profitable products. What do you suggest? Do you think John's forecasting method is producing reasonable errors, or is it getting out of control and is it time for John to revise his method?

We use tracking signal to check if John's simple exponential smoothing method produces reasonable errors. Table 3.4 shows the computation of tracking signal for all 12 weeks.

---

[7] Plossi, G.W. and O.W. Wight, *Production and Inventory Control*. Prentice Hall, Englewood Cliffs, NJ, 1967.

**Table 3.4    Computing Tracking Signals for Example 3.4**

| Week $t$ | Demand $D_t$ | Forecast $F_t$ | Error $e_t$ | Absolute error, $|e_t|$ | Cumulative error, $CE_t$ | Mean absolute deviation, $MAD_t$ | Tracking signal |
|---|---|---|---|---|---|---|---|
| 1 | 387 | 420 | 33 | 33 | 33 | 33 | 1.00 |
| 2 | 407 | 403 | −4 | 4 | 29 | 18.50 | 1.57 |
| 3 | 411 | 420 | 9 | 9 | 38 | 15.33 | 2.48 |
| 4 | 404 | 392 | −12 | 12 | 26 | 14.50 | 1.79 |
| 5 | 420 | 385 | −35 | 35 | −9 | 18.60 | −0.48 |
| 6 | 431 | 402 | −29 | 29 | −38 | 20.33 | −1.87 |
| 7 | 435 | 405 | −30 | 30 | −68 | 21.71 | −3.13 |
| 8 | 401 | 412 | 11 | 11 | −57 | 20.38 | −2.80 |
| 9 | 444 | 421 | −23 | 23 | −80 | 20.67 | −3.87 |
| 10 | 450 | 422 | −28 | 28 | −108 | 21.40 | −5.05 |
| 11 | 460 | 450 | −10 | 10 | −118 | 20.36 | −5.79 |
| 12 | 541 | 480 | −61 | 61 | −179 | 23.75 | −7.54 |

Let's compute the tracking signal for Week 4 as an example. The cumulative error, $CE_4$, is

$$CE_4 = e_1 + e_2 + e_3 + e_4 = 33 + (-4) + 9 + (-12) = 26$$

On the other hand,

$$MAD_4 = \frac{|e_1| + |e_2| + |e_3| + |e_4|}{4} = \frac{33 + 4 + 9 + 12}{4} = 14.5$$

Hence,

$$\text{Tracking signal for Week 4} = \frac{CE_4}{MAD_4} = \frac{26}{14.5} = 1.79$$

Since lactose-free milk is one of John's profitable products, we use a tighter bound $\pm 4$ on tracking signal to check if his forecasting method is out of control. Figure 3.27 depicts these bounds and tracking signals.

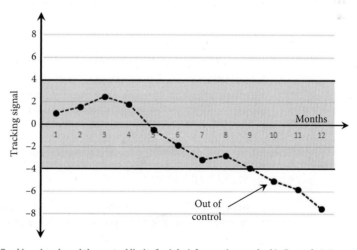

**Figure 3.27    Tracking signals and the control limits for John's forecasting method In Example 3.4.**

As the figure shows, John's simple exponential smoothing method is out of control, since tracking signals in Weeks 10, 11, and 12 are outside the control limits. Since tracking signals are all negative in the last 7 weeks, it indicates that the forecasting method is underestimating the demand. This might be because the demand for lactose-free milk is increasing with a positive trend, and hence, simple exponential smoothing method—which is for stationary demand—is lagging behind the demand.

### 3.12.2 Monitoring Individual Forecasts Using Control Charts

By monitoring whether cumulative error (CE) is too large or too small, tracking signal shows if a forecast is out of control by consistently overestimating or underestimating the demand. However, CE can be low, while it is a summation of very large positive and very large negative errors. In other words, individual errors can be very high, while CEs and the resulting tracking signals are very small.

Recall that the errors of a good forecasting method should be due to random noise and thus should exhibit a random pattern. Therefore, the key to check whether a forecasting method performs well is to check whether its forecast errors are random. This is exactly what a control chart does. The idea behind a control chart is as follows: If errors are random and the method captures the demand pattern well, then with a high probability errors should be within an acceptable range that random errors belong to. A very large error outside the acceptable range may point to the fact that the error is not due to random noise, but may be due to a significant change in the demand (e.g., appearance of a trend). Hence, one should reevaluate and adjust the method if needed. But, similar to the tracking signal, the question is: how to set the acceptable range for each individual error?

The answer turned out to be simple for time series forecasting methods. Recall that in time series forecasting methods errors are Normally distributed with mean zero and standard deviation:

$$\sigma_e = \sqrt{\frac{\text{SSE} - t\,(\text{BIAS})^2}{t - p}}$$

For Normal distribution, we know that about 99.7 percent of the values (i.e., forecast errors) are expected to fall within $\pm 3$ standard deviation of the distribution (i.e., $\pm 3\,\sigma_e$). Therefore, if errors are due to random noise—if the forecast is *in control*—99.7 percent of the errors must fall within the lower control limit (LCL) and upper control limit (UCL),

$$\text{LCL} = -3\,\sigma_e \qquad \text{UCL} = 3\,\sigma_e$$

If a forecast error falls outside the above control limits, then with a high probability this forecast error might not be due to random noise and must be considered as an evidence that the forecasting method might be out of control and thus may need to be reevaluated. We summarize this procedure in the following analytics.

---

**ANALYTICS**

**Control Chart for Time Series Forecasting Methods**

A time series forecasting method was used to forecast demands in Periods $1, 2, \ldots, t$, and resulted in errors $e_1, e_2, \ldots, e_t$. Control limits to check whether the forecasting method has been out of control in the past $t$ periods can be established as follows:

- *Step 1:* Compute the standard error $\sigma_e$ as

$$\sigma_e = \sqrt{\frac{\text{SSE} - t\,(\text{BIAS})^2}{t - p}} \tag{3.34}$$

where $p = 1$ for simple moving average and simple exponential smoothing; $p = 2$ for double exponential smoothing (Holt's) method; and $p = 3$ for triple exponential smoothing (Winters') method.

- *Step 2:* The lower control limit (LCL) and the upper control limit (UCL) for the errors $e_1, e_2, \ldots, e_t$ are:

$$\text{LCL} = -3\,\sigma_e \qquad \text{UCL} = 3\,\sigma_e$$

- *Step 3:* If any of the errors $e_1, e_2, \ldots, e_t$ are outside the control limits (i.e., if $e_i \leq \text{LCL}$ or $e_i \geq \text{UCL}$), then there is a high probability that the forecasting method is out of control.

Let's use the above analytics to check whether the individual forecasts generated by the forecasting method that John used in Example 3.4 is out of control. Since the forecasts are generated by a simple exponential smoothing method, we use $p = 1$ in Eq. (3.34) to compute $\sigma_e$. To do that, we also need BIAS and SSE. Hence,

$$\text{BIAS} = [33 + (-4) + 9 + \cdots + (-10) + (-61)]/12 = -14.92$$
$$\text{SSE} = (33)^2 + (-4)^2 + (9)^2 + \cdots + (-10)^2 + (-61)^2 = 9551$$

Therefore,

$$\sigma_e = \sqrt{\frac{\text{SSE} - t\,(\text{BIAS})^2}{t - p}} = \sqrt{\frac{9551 - 12(-14.92)^2}{12 - 1}} = 25.01$$

Thus, the lower and the upper control limits for the errors are as follows:

$$\text{LCL} = -3\,\sigma_e = -3(25.01) = -75.03 \qquad \text{UCL} = 3\,\sigma_e = 3(25.01) = 75.03$$

Figure 3.28 shows a control chart that includes the errors of the simple exponential smoothing method that John used and the corresponding control limits.

As the figure shows, the errors generated by the forecasting method are all within the control limits. Hence, the individual errors made by the simple exponential smoothing method are due to random noise. But, if the forecasting method is in control, then why

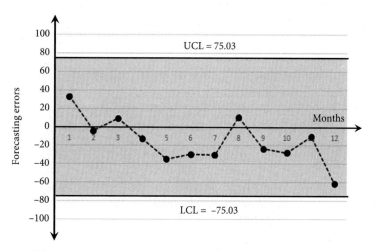

**Figure 3.28    Control chart for monitoring individual errors in Example 3.4.**

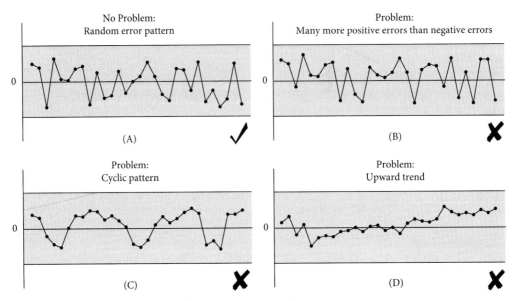

Figure 3.29   Different examples of possible patterns for forecast errors in control charts.

tracking signal analysis suggests that the forecasting method is out of control. We know that the method is underestimating demand, so why control chart did not catch this issue?

Well, the control chart in Fig. 3.28 also points to the issue of underestimating demand, but not through its control limits. It actually shows it through visual presentation of the errors, which is discussed in the following section.

### 3.12.3  Monitoring Error Pattern by Visual Inspection

Another approach, which complements tracking signal and control chart approaches, is the visual inspection of errors. Recall that in good forecasting methods errors are due to random noise and hence they should appear completely random. One way to check if the errors are random is to plot them and visually inspect them.[8] If the errors are random, they should be randomly spread around zero and do not show any detectable pattern. If the errors exhibit a detectable pattern, it implies that the errors are "predictable," and therefore are not random. Figure 3.29 shows some examples of the most common patterns observed in practice.

Pattern A in Fig. 3.29 shows that the errors are randomly spread around zero; hence, there is no sign of issues with the forecasting method. Pattern B, although appears random, indicates bias. Specifically, there are more positive errors than negative errors, indicating that the forecasting method tends to overestimate the demand more frequently. Pattern C shows a periodic upward and downward trend in errors, implying that in some periods the method overestimates the demand while in subsequent periods it underestimates the demand in a cyclic fashion. This implies that the demand has a seasonal or cyclic pattern that is not captured by the forecasting method (e.g., a seasonal demand is forecasted by simple exponential smoothing method which is suited for stationary demand). Finally, Pattern D indicates that the forecasting method is overestimating the demand and the errors are getting larger. This is a sign that the demand has a decreasing trend, but the forecasting method cannot capture it. Therefore, the forecasting method should be revised to include trends in its forecast.

---

[8]There are also statistical tests that can check the randomness of the errors.

Let's return to the control chart for the lactose-free milk in Fig. 3.28. The error patterns in the figure have a decreasing trend, similar to Pattern D in Fig. 3.29 but in the opposite direction. Therefore, while all the errors in Fig. 3.28 are within the control limits, the decreasing pattern of errors is an indication that the simple exponential smoothing method is constantly underestimating the demand for lactose-free milk and should be revised.

One important point is that forecasting methods that generated errors in Patterns B, C, and D in Fig. 3.29 could have been considered "in control" by both tracking signal and control charts, since they are all within their control limits. This emphasizes on the importance of visual inspection. Without a visual inspection, one may not be able to identify the problem with forecasting methods and hence lose the opportunity to improve the forecasts.

## 3.13 Forecasting Methods and Demand Distribution

As mentioned earlier, one of the frequent applications of forecasting methods is to forecast demand. Without an estimate for future demand, operations engineers and managers cannot develop cost-effective plans for production, inventory, and resources. To develop such plans, most analytical models need probability distribution of the demand. On the other hand, forecasting methods provide point and prediction interval forecasts, but not demand distribution. This raises the question of whether forecasting methods can also be used to get the demand distribution in next periods and how?

To answer this question, as discussed in Sec. 3.1, the most common distributions that can capture demand behavior are Poisson and Normal distribution. Also recall that Normal distribution can be used as an approximation of Poisson distribution (when the average demand is high). That is the reason why Normal distribution is widely used to model demand, and is the reason we also use Normal distribution in most of the analytics in this book.

Normal distribution is fully specified by its mean $\mu$ and its standard deviation $\sigma$ (or variance $\sigma^2$). Below we explain how the forecasting methods discussed earlier can be used to estimate $\mu$ and $\sigma$.

**Estimating the Mean of Demand Distribution**
Recall the point forecasts obtained by our forecasting methods are the forecasts for the average demand in future periods. Hence, forecasts $F_{t+1}$, for example, would be the mean of demand distribution in Period $t + 1$.

**Estimating the Variance of Demand Distribution**
But what would be the variance of the demand in Period $t + 1$? One approach is to use the actual demand data in the last $t$ periods and compute the variance of the $t$ observed demands. The problem, however, is that the underlying demand distribution is often not constant and changes through time. That is the reason forecasting methods (e.g., moving average and simple exponential smoothing) put more emphasis on the most recent demand. In these cases, what would be a good estimate for the variance of demand? The answer is the variance of error forecasts.

Considering $\sigma_e^2 = \mathrm{VAR}(e_t)$ and

$$e_t = F_t - D_t \quad \Longrightarrow \quad \mathrm{VAR}(e_t) = \mathrm{VAR}(F_t) + \mathrm{VAR}(D_t)$$
$$\sigma_e^2 \geq \sigma \qquad (\text{since } \mathrm{VAR}(F_t) \geq 0)$$

which implies that the variance of error forecasts is greater than the variance of the actual demand. Why? Because the forecasts developed by forecasting methods use a limited sample size (i.e., a sample of most recent demands) and hence introduces a sampling error to the forecast. When the sample size becomes larger and approaches infinity, then the variance of errors and the variance of actual demand become equal.

The question is then, for smaller sample sizes, why should we use the variance of error forecasts as an estimate for variance of demand? Why not try to estimate the variance of demand? Well, the answer is simple. If we use mean of the actual demand (i.e., $\mu$) to make our planning decisions for the next period, then it would be more appropriate to use variance of the actual demand ($\sigma^2$)—if we have a good estimate for that. However, we do not know the actual demand in future Period $t + 1$, and we are using our forecast $F_{t+1}$ as its estimate. In this case, we should make our decision based on the variations (i.e., errors) in the forecast. This is why if $F_{t+1}$ is used as an estimate for mean of the demand; it is more accurate to use the variance of error forecasts (i.e., $\sigma_e^2$) as the estimate for the variance in the demand.

The standard deviations of forecast errors are different for different forecasting methods and for one-period, or multiple-period-ahead forecasts. They are all given in the analytics for computing the prediction interval forecasts for each method.

### Example 3.5

The supplier of the Oatplex cereal in Example 3.1 has a distribution center very close to the store. Thus, the store manager can place an order by the end of a day and receive orders the morning of the next day. The store places an order every week on Sunday afternoon before they close the store to satisfy the demand of the next week. Store manager would like to avoid stockout of Oatplex, so that the store does not face lost sales of an Oatplex during the week. Specifically, the store manager would like to order enough amount of Oatplex such that the chance of facing a stockout next week be less than 5 percent. How many boxes of Oatplex should the manager order for next week, if the store already has 23 boxes in inventory on Sunday night?

Consider $I$ to be the store inventory at the beginning of Week 101 after the store receives its order placed the night before. Also, consider the actual demand in the next week to be $D_{101}$, which is a random variable. To have the possibility of stockout being less than 5 percent, we must have

$$P(D_{101} > I) = 0.05 \implies P(D_{101} \leq I) = 0.95$$

Hence, assuming that the demand in Week 101 can be approximated by a Normal distribution, we need to estimate its mean and standard deviation. Recall from Example 3.1 in Sec. 3.8 that the point forecast for the next week, Week 101, is $F_{101} = 193$. Hence, an estimate for the mean of the demand distribution in the next week is 193.

Also, recall that the forecasts were made by a simple exponential smoothing method with $\alpha = 0.317$. When computing prediction interval in Sec. 3.8.5, we computed the standard deviation of forecast error to be $\sigma_e = 36.44$. Hence, to find the required inventory $I$, we must have

$$P\left(\frac{D_{101} - F_{101}}{\sigma_e} \leq \frac{I - 193}{36.44}\right) = 0.95$$

$$P\left(z \leq \frac{I - 193}{36.44}\right) = 0.95$$

Using Standard Normal table we find $\Phi(1.65) = 0.95$, hence

$$I = 193 + 1.65(36.44) = 253.1$$

Therefore, if the store has $I = 254$ boxes of Oatplex at the beginning of the next week, the probability of not facing a stockout would be at least 95 percent. Since the store already has

23 boxes in inventory, then it should order $I - 23 = 254 - 23 = 231$ boxes of Oatplex for the next week.

## 3.14 Summary

Effective matching of supply and demand is not possible until accurate information about the demand or supply is available. However, due to changes in the marketplace, future demand for a product is not known with certainty. Thus, forecasting future demand has become a critical part of all business processes. While the focus of this chapter is on developing methods for modeling and forecasting demand, the methods can also be used to forecast other variables such as supply level and prices.

Using past data, the demand process can be modeled with empirical distribution, Poisson distribution, or Normal distribution. Each distribution has parameters that must be estimated to develop forecasts for future demand. The main focus of this chapter is to present models for forecasting the parameters of these distributions. Forecasts are often wrong, but they are useful. A good forecast also includes a measure of error and is more than just a single number; it includes a point forecast and a prediction interval forecast.

Based on the availability of data, forecasting methods can be divided into the two groups of qualitative (or subjective) and quantitative (or objective) methods. Quantitative methods use past demand data to create forecasts, while qualitative methods do not use any data to do so. There are two main types of quantitative forecasting methods: causal forecasting methods and time series forecasting methods. Causal forecasting methods use the past relationship between demand and other variables to forecast future demand (see the online supplement of this chapter). The idea behind time series forecasting methods is that demand for any product has a particular pattern. Time series forecasting methods, therefore, try to capture the patterns of demand from past demand data and use that to forecast demand for future. There are four main patterns for demand, resulting in four different time series: (i) stationary time series, (ii) time series with trend, (iii) seasonal times series, and (iv) cyclic time series. This chapter provides several methods that can be used to forecast demand in each of the four types of time series.

To develop an accurate forecast for demand, this chapter presents a five-step approach that is applicable to both causal and time series forecasting methods. The first step is to determine the type of data and study which type of time series is suitable as forecasting model. The second step is to choose a method—among those applicable for the time series—to develop a forecast. In the third step, an appropriate evaluation criterion must be chosen, so that it can be used to evaluate the accuracy of forecasting method. The evaluation criterion is usually the errors made by the forecasting method, for example, MAD, SSE, and MAPE. The larger the error, the less accurate the forecast. In Step 4, the parameters of the forecasting method are adjusted to yield more accurate forecasts. Finally, after the method is adjusted, the forecast method can be used to generate forecasts. The errors in the forecasts must be monitored each time a new demand is observed. Tracking signals, control charts, and visual inspection can be used to monitor the forecasts and the forecasting method should be revised if the forecasts are out of control.

## Discussion Questions

1. Describe the main six principles of forecasting.
2. What is the difference between qualitative and quantitative forecasting methods? Name five different qualitative forecasting methods and briefly describe each method.

3. Describe Delphi's forecasting method and discuss its similarities and differences with Jury of Executive Opinion method.

4. What is a time series? Describe the differences between stationary, trend, seasonal, and cyclic time series.

5. What is the difference between time series forecasting methods and causal forecasting methods?

6. Describe the five steps of the time series forecasting procedure and discuss the role of adaptive and focus forecasting.

7. Name four different measures that are used to evaluate the accuracy of a forecasting method and explain what each measure represents.

8. What is the difference between point forecast and prediction interval forecast?

9. Both tracking signals and control charts are used to monitor forecast accuracy. What are the similarities and differences of these two methods?

## Problems

1. An electronics store sells only one brand of smart speaker. The following table shows the demand for the speaker in the last 60 days:

| | | | | | | | | | |
|---|---|---|---|---|---|---|---|---|---|
| 8 | 5 | 12 | 10 | 3 | 9 | 9 | 10 | 9 | 9 |
| 12 | 1 | 10 | 12 | 10 | 9 | 11 | 8 | 11 | 5 |
| 11 | 8 | 8 | 7 | 4 | 7 | 9 | 7 | 8 | 9 |
| 4 | 8 | 9 | 7 | 12 | 13 | 12 | 5 | 13 | 6 |
| 8 | 7 | 10 | 10 | 8 | 8 | 14 | 8 | 10 | 6 |
| 7 | 9 | 10 | 5 | 10 | 8 | 9 | 8 | 4 | 8 |

   a. At the end of today, the store has only nine speakers in its inventory and is supposed to receive a new shipment of speakers in 2 days. Using empirical distribution find an estimate for the probability that the store faces stockout of the smart speaker tomorrow?

   b. Use a Poisson distribution to answer Part (a).

   c. Use a Normal distribution to answer Part (a).

2. A major airline uses an airport in a large city as a hub for its connecting flights. Each day the airline faces a number of claims from its passengers who lost their baggage. The number of lost baggage claims in the last 14 days have been 67, 25, 57, 58, 35, 29, 77, 44, 38, 36, 23, 69, 53, and 45

   a. Use moving average method with $N = 3$ to forecast the expected number of claims for tomorrow.

   b. Use exponential smoothing method with parameter $\alpha = 0.7$ to forecast the number of lost baggage claims that the airline should expect in the next 2 days.

   c. If the criterion for accuracy is the mean squared error (MSE), which of the forecasting methods in parts (a) and (b) are preferred?

   d. If you become in charge of providing forecasts for the number of claims in the next 2 days using exponential smoothing method, what would be your point and 90-percent prediction interval forecasts for the number of claims in each of the next 2 days? Use MSE as your evaluation criterion.

3. An opera center has a concert hall that has 200 seats. The center has a performance every Saturday night. The number of tickets sold on Saturdays of the last 5 months were: 381, 306, 216, 283, 207, 270, 360, 319, 313, 373, 354, 303, 393, 228, 252, 211, 326, 291, 257, and 205. A ticket is sold for an average price of $80.

    a. Use exponential smoothing method and find an estimate for the average sales of the next week. Use MSE as your measure of accuracy.

    b. Develop point and 90-percent prediction interval forecasts for the ticket sales next Saturday.

    c. What is the chance that the concert hall's sales for next Saturday exceeds $20,000?

4. A firm sells home security systems online. Customers place their orders through firm's website and receive the system in 3 days. These systems can be easily installed by home owners. Total number of systems sold in the last 12 weeks were 810, 991, 1067, 942, 977, 729, 1052, 941, 804, 847, 754, and 915.

    a. Use exponential smoothing method to develop point and 95-percent prediction interval forecasts for the sales of home security systems in the next month. Use MSE as your evaluation criterion.

    b. The firm has a total of 800 systems in its inventory next month. What is the chance that the firm will have shortage of systems to satisfy next month's demand?

5. Harington Inn is a hotel in the business district of a large city with 300 rooms. Almost all of the hotel's customers are people who come for business with one of several companies in town. Data shows that the hotel has its highest occupancy rate on Mondays. The occupancy rate in the last 12 Mondays have been 212, 262, 270, 285, 260, 252, 204, 283, 250, 211, 254, and 290.

    a. Use exponential smoothing method and find an estimate for the number of rooms that will be occupied in the next two Mondays. Use MSE as the evaluation metric for accuracy of your estimate.

The hotel manager receives a request from a university that wants to have a small 1-day conference on Tuesday after the next Tuesday. It is expected that conference attendees will make room reservation for the night before (i.e., for Monday night). Also, university has requested that the hotel keep 60 rooms aside for conference attendees who will make their reservation during the next 2 weeks before the conference.

    b. If the manager holds 60 rooms for conference attendees, what would be the probability that the hotel becomes full on the Monday night of the conference and has to reject its regular customers?

6. A pharmaceutical company has recently introduced a new pain killer into the market. The pain killer is sold in boxes that include 24 pills. The sales of the pain killer in the last 16 weeks have been 11797, 6724, 15346, 12825, 15939, 9997, 13671, 21994, 17774, 14587, 11190, 18083, 25320, 15129, 17092, and 27652 boxes.

    a. Use Holt's method with smoothing constants $\alpha = 0.7$ and $\beta = 0.1$ and develop a sales forecast for the next 2 weeks.

    b. Adjust the method and find the best values for $\alpha$ and $\beta$ that result in the least MSE.

    c. Use the method in part (b) to develop point and 90-percent prediction interval forecasts for the next 2 weeks' sales.

7. Feeding Neighbors (FN) is a local nonprofit organization that receives donations to purchase food and distribute it in soup kitchens to homeless and needy people. The amount of donation (in thousands of dollars) in the last 12 months have been $479, $500, $469, $470, $334, $333, $373, $356, $329, $242, $335, and $213. Managers of FN are worried about the decline in donations and the fact that they may not be able to provide enough food to serve people.

    a. What amount of donation should FN expect to receive next month? Use MSE as your accuracy measure.

    b. Provide a 90-percent prediction interval forecast for your answer to part (a).

c. FN needs at least $200,000 in a month to provide minimum service to its community. What is the likelihood that FN will not be able to provide its minimum service next month?

8. Sharp Tools Inc. (ST) has a large stamping press used in the production of heavy duty industrial equipment. The press is old and often incurs repair and maintenance costs. A company specialized in maintenance of machine tools is offering ST a maintenance contract with the annual fixed cost of $10,000. The maintenance costs in the last 12 years have been $11026, $9529, $7949, $11265, $5712, $10161, $8614, $9286, $15723, $11884, $7868, and $15992.

   a. Provide a forecast for the next year's maintenance cost. Use MSE as the evaluation criterion for the accuracy of your forecast.

   b. Should the firm purchase the contract for the next year?

9. Ships arrive at a port of entry in a European country in random fashion. Arriving ships are unloaded and their goods are transferred to the storage areas in the port. The number of ships that arrived at the port in the last 12 months were 199, 300, 433, 244, 384, 407, 406, 574, 289, 504, 624, and 747 ships. Recently, the port authorities have been receiving complaints from ships about their long waiting times for unloading. Thus, they are planning to increase the unloading capacity of the port to 650 ships in a month. If the capacity is increased, what would be the probability that the number of ships arriving next month would still be more than the increased capacity of 650? Use MSE as the criterion for your prediction.

10. A restaurant that serves pastry and dinner opens for business from 3:00 pm to 11:00 pm every day. To plan their working shifts and to determine the number of required workforce in a day, the restaurant needs to forecast the demand for each day of the week. The following table shows the number of customers—demand—during four time intervals on Saturdays of the last 4 weeks. Saturday is the busiest days for the restaurant.

| | Demand on Saturdays of | | |
| --- | --- | --- | --- |
| | Week 1 | Week 2 | Week 3 |
| 3:00 pm–5:00 pm | 15 | 22 | 13 |
| 5:00 pm–7:00 pm | 29 | 39 | 29 |
| 7:00 pm–9:00 pm | 64 | 74 | 89 |
| 9:00 pm–11:00 pm | 34 | 28 | 32 |

Develop point and 90-percent prediction interval forecasts for the four time periods of next Saturday. Use MSE as your forecast accuracy metric.

11. Hinter Ski is one of few ski resorts that is open 365 days a year. The number of people who have visited Hinter Ski in the last 3 years is shown in the following table

| | Year 1 | Year 2 | Year 3 |
| --- | --- | --- | --- |
| Summer | 2080 | 3150 | 1820 |
| Fall | 4120 | 5600 | 4220 |
| Winter | 9150 | 10500 | 11000 |
| Spring | 4800 | 3980 | 4530 |

   a. Develop a forecasting model to estimate the number of visitors that Hinter Ski should expect to see in each season of the next year. Use MSE as your measure for forecast accuracy.

   b. What is the probability that the number of visitors in the next summer will exceed 3000?

12. As your first project in your new job at a major clothing company, you are asked to evaluate whether the firm needs to change its forecasting method used to forecast its weekly sales of blue jeans. You are presented with the following table that includes the actual demand for jeans in the last 12 weeks as well the forecasts made by simple exponential smoothing method used by your firm.

| Week | Demand | Forecast |
| --- | --- | --- |
| 1 | 503 | 498 |
| 2 | 529 | 564 |
| 3 | 534 | 520 |
| 4 | 525 | 549 |
| 5 | 546 | 521 |
| 6 | 560 | 563 |
| 7 | 566 | 580 |
| 8 | 521 | 577 |
| 9 | 577 | 482 |
| 10 | 679 | 591 |
| 11 | 690 | 630 |
| 12 | 710 | 672 |

a. Use tracking signals to investigate whether the company should have revised its forecasting model.

b. Develop a control chart for the forecasts and determine whether the forecasting model should have been revised.

# CHAPTER 4

# Capacity Concepts and Measures

## 4.0 Introduction

One of the process-flow measures that has a direct impact on sales, and therefore on profit, is Throughput. When there is a market for products or services of a firm, increasing the throughput of the firm results in higher profits.

Throughput of a firm, on the other hand, is closely related to a firm's capacity. Hence, capacity decisions such as acquiring new equipment, hiring new workers, or outsourcing some activities of the firm are critical decisions that operations engineers and managers face in manufacturing and service operations systems. What makes capacity management critical is that both excess and shortage of capacity are costly. Excess capacity is an investment with no return, and shortage of capacity simply means loss of profit due to loss of sales. Furthermore, capacity decisions are often costly and are difficult to reverse. For example, adding a new airplane to the fleet of an airline or opening a new factory is an expensive investment. On the other hand, when these investments are made, they are costly to reverse—salvage the airplane or close the factory.

One key factor in making a good capacity decision is to know what the capacity of the current system is, and how it is affected by different capacity expansion decisions. In this chapter, we first present the fundamental concepts and measures of capacity. We start by introducing different capacity measures of a process, and then we discuss how interruptions such as setups, resource breakdowns, and quality issues such as defective flow

**141**

units and rework impact the capacity of the process. Finally, we illustrate how capacity of more complex processes—processes with multiple stages and multiple products—can be computed.

## 4.1  What Is Capacity of a Process?

Finding the capacity of a process is often not easy, not only because tasks performed by a process may be complex, but also because the concept of "capacity" itself is very broad. We use the following example to show different aspects of the capacity of a process.

Suppose that your department that issues credit cards has one copy machine. The machine is used by all employees to make copies of documents in credit card application files. You are thinking of buying another copy machine, since you often see two or three people waiting to make copies. But, before you make a decision whether to purchase a new copy machine, you would like to first know what the capacity of the current copy machine is. So, you send an email to your assistant, Jane, and ask her to find out what the capacity of the copy machine is. Jane replies that she could not find the instruction manual of the copy machine; hence, she needs some time to gather some data to estimate the capacity of the machine.

Three weeks later, Jane comes to your office and gives you a one-page report that describes how she estimated the capacity of the copier. She explains that she gathered data about the number of application files that the copier made during a 10-day period. She has done that by posting a form close to the copy machine and asking everyone to write the number of application files that they copy before using the copy machine. She mentions that the number of applications copied in an hour in those 10 days ranges from 45 to 80 applications per hour. In other words, the maximum number of copies that the copier made in an hour in the 10-day period was 80. To get a better estimate, she repeated this experiment for another 5 days, and the number of application files copied in an hour in the 5-days period was still less than the maximum 80 applications. She therefore concludes that the copier has the maximum capacity of copying 80 applications per hour. She also mentions that, obviously, having more data about the number of applications copied in an hour results in a better estimate of the capacity of the copier, but she believes it would still be somewhere around 80 files per hour.

You thank Jane for her report, and she leaves your office. However, you do not believe that 80 files per hour is the true representation of the capacity of the copy machine. First, you disagree with taking the maximum of the 15 days as the capacity. You know that the number of pages in an application file that must be copied are different in different files, ranging from 2 pages to around 10 pages per file. So the maximum of 80 files copied in 1 hour of the 15-day period might be because most applications copied in that hour required less copies per application. And, the minimum of 45 applications per hour might be because most of the applications copied in that hour required more copies per application. Thus, it would be better to use the *average* (not the maximum) of the 15-day period as a proxy for the capacity.

Second, the maximum (or even the average) of the number of applications copied per hour does not represent the capacity of the copier; it represents the "throughput" of the copier. For example, consider the hour in which the maximum of 80 files were copied. The question is whether in that hour the copy machine was ever idle. If in that hour, the copy machine had, for example, a total of 5-minute idle time (since there was no demand for copy) then 80 is not the maximum number of application files that the copier can make in an hour. In other words, 80 is the number of application files that the copier *did* copy in that particular hour, not the number of applications that the copier *could* copy in that hour. Consequently, the maximum and the average of the hours in the 15-day period experiment represent the maximum and the average "throughput" of the copier during that period. Only if the copier was fully utilized without any idle time in every hour of the 15-day period, the average of all the hours in the 15-day period represents the true capacity of the copier.

Thus, to find the capacity of the copier, one should take data regarding the number of applications copied per hour, only on those hours that the machine is fully utilized. Then, the average of those numbers would be a good estimate of the capacity of the copier. You realize, however, that this would not be easy. It is rare, you believe, to have an hour in which the copy machine is never idle. So, taking a large sample of those hours may not be practical. You soon realize that there is an easier way to do this. If you can find out how long it takes, on an average, to copy the required pages in one application file, you can easily compute how many applications, on an average, the copier can copy in an hour if it is fully utilized. Suppose it takes an average of 1 minute to copy the required documents in a file. Then, if the copy machine is fully utilized for an hour without any idling, it has the capacity to copy 60 files in an hour and $60 \times 8 = 480$ files in an 8-hour working day.

To get an estimate for the average time it takes to copy a file, you decide to go and make copies of some application files yourself. You randomly choose 10 files of different sizes and plan to find how long it takes to copy all of them. Dividing the total time by 10, you get the average time to copy one file.

You take 10 randomly chosen application files and your cell phone (which has a stopwatch) and go to the copy machine. When you get there, you find the copy machine idle. You start your stopwatch and start copying. Before you start copying the second file, you get an urgent call from your secretary reminding you about your meeting in 30 minutes. You tell her that you will be back in your office by then. That takes 30 seconds. In the middle of copying the third file, the copier stops and you have a paper jam that takes you 3 minutes and 40 seconds to fix. After clearing the paper jam, the copy machine starts copying the remaining pages of the third file.

You are almost finished with copying the ninth file, when you notice that the copy machine is about to go out of paper. You cannot fill the copy machine with papers during copying, so you need to wait until either the machine stops when it runs out of paper, or it finishes copying the ninth file. Like every thing else that went wrong today, the copy machine stops and a light flashes asking you to fill in papers. You go to the cabinet where the papers are stored, but you find that there are no papers there. You then call your secretary and ask her to fill the cabinet with papers. This takes 10 minutes. After papers arrive, you put some papers in the machine. That takes you another 2 minutes. You resume copying the remaining documents in the ninth application file. You copy the last application file thinking what else could go wrong now! The last file is copied without any problem and you stop your stopwatch. You summarize your time study in Table 4.1.

Now, while rushing to your next meeting, you are thinking with all those disruptions, how should you calculate the average time to make copy of a file? After coming back from your meeting, you do some thinking and decide to separate the total time you spent copying 10 files into three categories:

- Time related to interruptions that had nothing to do with the processing time of the copy machine. For example, talking to your secretary for 30 seconds. During that 30 seconds the copy machine was available for copying file number 2, but you were not. So, you discard this from your calculation. Also, waiting for 10 minutes (i.e., 600 seconds) for your secretary to bring papers so you can put them into the copier. You exclude that too.

- Time of the main task of copying the documents. This includes the task of putting the documents in the machine and making the copies. These times are shown in Table 4.1 as "Copying Time (Main Task)."

Table 4.1   Time Study of Copying 10 Files

| File number | Copying time (main task) (*seconds*) | Interruption time (supporting task) (*seconds*) | Total time (*seconds*) | Source of interruption |
|---|---|---|---|---|
| 1 | 60 | 0 | 60 | |
| 2 | 40 | ~~30~~ | 40 | *Talking to secretary* |
| 3 | 75 | 220 | 295 | *Fixing paper jam* |
| 4 | 54 | 0 | 54 | |
| 5 | 58 | 0 | 58 | |
| 6 | 62 | 0 | 62 | |
| 7 | 75 | 0 | 75 | |
| 8 | 48 | 0 | 48 | |
| 9 | 62 | ~~600~~ + 120 | 182 | *Putting paper in the machine* |
| 10 | 66 | 0 | 66 | |
| Total: | 600 | 340 | 940 | |
| Average: | 60 | 34 | 94 | |

- Time related to the supporting tasks that are needed to perform the main task of copying. This includes filling the copy machine with paper (120 seconds) and fixing paper jams (220 seconds). Without these supporting tasks, you were not able to copy 10 files. These are shown in Table 4.1 under column "Interruption Times (Supporting Tasks)."

Considering the data in Table 4.1, how do you calculate the average time to copy a file? Is the average time to copy a file the average time of the main task of copying, that is, 600/10 = 60 seconds? or is it 940/10 = 94 seconds (which includes the main and the supporting tasks)? Well, operations engineers and managers are interested to know both averages, as we will describe later in this chapter.

## 4.2 Theoretical and Effective Process Times

An activity performed on a flow unit by a resource includes two groups of tasks: Main Tasks and Supporting Tasks.

*Main Tasks:*   Performing the main tasks is the primary objective of a resource. For example, the main objective of the copy machine is to copy documents. The main task of a drill is to make a hole in a part. These main tasks are also called *Value-Added Tasks*, since they are the tasks that directly add economic value to the flow unit. We call the time required to perform the main tasks *Theoretical Process Time*.

*Supporting Tasks:*   Performing supporting tasks is not the main objective of a resource; they are tasks that are essential to complete the main tasks. Fixing the paper jam (i.e., repair) and filling the machine with new papers (i.e., setups) are examples of supporting tasks. In the case of a drilling activity, sharpening the tool or fixing the drill when it breaks down are supporting tasks. The supporting tasks are also called *Non-Value-Added Tasks*, since they do not directly add value to the flow unit. Examples of non-value-added tasks are setting up a resource, maintenance and repair of a resource, doing rework to fix quality problems, etc. We call the time needed to perform the non-value-added tasks *Supporting Process Times*.

In your case, the theoretical process times of copying activities are the 10 numbers in the second column of Table 4.1. The average of these theoretical process times is called "theoretical mean process time."

### Theoretical Process Time of an Activity

- *Theoretical Process Time* of an *Activity* is the time of the main task of the activity for one flow unit without taking into account interruptions such as setups, rework, failures, etc.
- *Theoretical Mean process Time* of an *Activity* is the total long-run average of theoretical process times of the activity. We let

$$T_0(activity\ j) = \text{Theoretical mean process time of Activity } j$$

Hence, the theoretical mean process time of activity "Copying" is $T_0(copying) = 60$ seconds per file. This implies that in the best-case scenario when all supporting tasks such as feeding papers and fixing paper jams, etc., are eliminated, the average time to copy a file is 60 seconds. But this does not seem realistic, does it? The copy machine is old, it will have paper jams once in a while, and it cannot copy if it is not filled with papers. Therefore, a better estimate would be the one that considers the total of theoretical process time (copying) and supporting process times (filling papers and fixing paper jams). This leads us to the concept of "Effective Process Times."

### Effective Process Time of an Activity

- *Effective Process Time* of an *activity* is the sum of *theoretical* process time (for the main task) and *supporting* process time (for the supporting tasks, e.g., rework, setup, repair) of the activity.
- *Effective Mean Process Time* of an *activity* is the total long-run average of the activity to produce one *good* (non-defective) flow unit. We let

$$T_{\text{eff}}(activity\ j) = \text{Effective mean process time of activity } j$$

One important distinction of "Effective Mean Process Time" is that it is the average time to make one non-defective unit. If a process, for example, produces 10 units and 2 of those units are defective, then the effective mean process time is the sum of 10 effective process times divided by 8. This is critical, since it allows effective mean process time take the process quality issues (i.e., producing unusable products) into account.

Let's return to the copy machine example. The effective mean process time of activity "copying" is the average of the 10 numbers in the last column of Table 4.1, that is, $T_{\text{eff}}(coping) = 940/10 = 94$ seconds. Ok, it seems that you got what you wanted—It takes the copy machine *an average* of 94 seconds to copy a file. This, you believe, is a more realistic estimate of the time needed to make copies. You, of course, understand that your sample of 10 files is small, and all the problems that occurred during your experience of copying 10 files (paper jams, filling the paper tray, etc.) might not happen that frequently. Thus, to get a more accurate estimate, you ask one of your staff to do a similar study for several days and for larger number of files and send you a report.

Ten days later, you receive the report. The report includes Fig. 4.1 which shows the list of copy machine activities and their corresponding tasks. As the figure shows, in addition to copying, the copy machine performs another activity—scanning the credit report of each

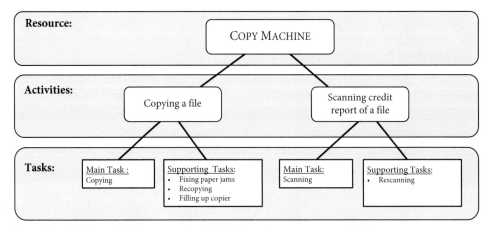

**Figure 4.1    Activities and their corresponding tasks of the copier to process one application.**

application. Since scanning does not require papers, the only supporting task is "rescanning," that sometimes is required due to occasional issues with the quality of scanned pages. The supporting task "recopying" corresponds to the case where some or all the documents of an application are copied again due to damages caused by paper jams.

The task times in the reports, which is the average of 100 samples, are as follows:

- *Copying:* The average time of the main task of copying all required documents in a file (i.e., theoretical mean process time) is 40 seconds, and the average of the total time including supporting process times (i.e., effective mean process time) is 48 seconds.

- *Scanning:* The average time of the main task of scanning the credit report in a file (i.e., theoretical mean process time) is 10 seconds, and the average of the total time (i.e., effective mean process time) is 12 seconds.

Thus, you have

$$T_{\text{eff}}(copying) = 48 \text{ seconds per file}; \qquad T_{\text{eff}}(scanning) = 12 \text{ seconds per file}$$

Having these more accurate estimates of activity times, it seems easy to compute the capacity of your copy machine. Your copier performs two activities on each application: copying and scanning. Copying takes an average of $T_{\text{eff}}(coping) = 48$ seconds, and scanning takes an average of $T_{\text{eff}}(scanning) = 12$ seconds. Thus, it takes a total of

$$T_{\text{eff}}(\text{copier}) = T_{\text{eff}}(copying) + T_{\text{eff}}(scanning) = 48 + 12 = 60 \text{ seconds} = 1 \text{ minute}$$

for the copier to process a file.[1] This is called effective mean process time of the copier.

---

**CONCEPT**

**Theoretical and Effective Mean Process Time of a Resource**

- Theoretical Mean Process Time of a *Resource* is the sum of the theoretical mean process times of all activities performed on a flow unit by the resource within process boundaries.

---

[1] We use *italic* fonts to present activities, for example, *copying* and *scanning*, and use regular fonts to present resources, for example, copier.

● Effective Mean Process Time of a *Resource* is the sum of the effective mean process times of all activities that the resource performs on one flow unit within process boundaries.

Now that you have the average time it takes your copy machine to copy and scan one application file, it is easy to compute the capacity of your copy machine.

## 4.3  Theoretical and Effective Capacities

The capacity of your copy machine in terms of the number of applications that the copy machine can process (i.e., copy and scan) in an hour can be computed as follows. Since it takes the copy machine 1 minute (= 1/60 hours) to process one application, the capacity of the copy machine is one file per minute (= 60 files per hour). This is called Effective Capacity of the copier (i.e., your resource), since it also takes into account the supporting tasks. Theoretical capacity of a resource is also defined in the same manner.

---

**CONCEPT**

**Theoretical and Effective Capacities of a Resource**

● *Theoretical Capacity of a resource* is the long-run average number of outputs that the resource can produce per unit time if the resource is fully utilized and does not idle due to unavailability of input units, or unavailability of other resources. Also, the resource never breaks down, does not require setup, does not produce defective outputs that are scraped or require rework, etc. We use $C_0(\text{resource } i)$ to show the theoretical capacity of resource $i$, and therefore we have

$$C_0(\text{resource } i) = \frac{1}{T_0(\text{resource } i)}$$

● *Effective Capacity of a resource* is the long-run average number of *good* (non-defective) outputs that the resource can produce per unit time if the resource is fully utilized and does not idle due to unavailability of input units or unavailability of other resources. We use $C_{\text{eff}}(\text{resource } i)$ to show the effective capacity of resource $i$, and we have

$$C_{\text{eff}}(\text{resource } i) = \frac{1}{T_{\text{eff}}(\text{resource } i)}$$

---

It is important to remember that, as stated above, the effective capacity of your copier, $C_{\text{eff}}(\text{copier}) = 60$ files per hour, does not take into account the time that the resource is idle due to unavailability of other resources (e.g., you talking to your secretary) or unavailability of flow unit (e.g., the copy machine is idle since there is no demand for copying).

To find whether you need to buy a new copier or not, you ask your staff about the average hourly demand for the copier. They tell you that the average demand for the copier is around 45 application files per hour. Obviously, your copier's capacity of 60 files per hour is 33.33 percent higher than the demand of 45 files per hour, so there is no need to buy a new copier.

Now consider another scenario in which the average demand per hour is 80 files per hour, which is higher than the effective capacity of your copier. While you are now thinking of buying another copier, your assistant tells you that there is no need for a new copier.

The assistant argues that if we take better care of the copy machine and train our staff how to properly use the machine, the number of paper jams and reworks will be reduced, which will increase the average number of files that the machine can copy per hour. You think about this and you realize that your assistant is actually suggesting ways to reduce supporting process times (i.e., non-value-added tasks). Do you think the assistant is right?

Well, in an ideal case that the time of all supporting tasks are reduced to zero, the average time to copy a file is the theoretical mean process time. If that is the case, the effective capacity of your copier becomes equal to its theoretical capacity. Since you have $T_0(coping) = 40$ seconds, and $T_0(scanning) = 10$ seconds, and

$$T_0(\text{copier}) = T_0(coping) + T_0(scanning) = 40 + 10 = 50 \text{ seconds}$$

the theoretical mean process time of your copy machine is $T_0(\text{copier}) = 50$ seconds $= 50/60 = 0.833$ minutes per file. This implies that in the best-case scenario when all supporting tasks such as rework and paper jams, etc., are eliminated, the copier can process $1/0.833 = 1.2$ files per minute, or $1.2 \times (60 \text{ minutes}) = 72$ files per hour.

Clearly, even in the best-case scenario in which the copy machine performs only the main tasks, the capacity of the copier is $C_0(\text{copier}) = 72$ files per hour, which is still less than the demand of 80 files per hour. Thus, training staff or better maintenance of the copier (which should definitely be implemented) will not solve the capacity issue of your copier. You still need a new copier.

Now consider a third scenario in which the demand is 65 files per hour. In this case the effective capacity $C_{eff}(\text{copier}) = 60$ is less than the demand of 65 files per hour, but theoretical capacity of $C_0(\text{copier}) = 72$ is greater than the average hourly demand. Hence, there is a possibility that the effective capacity of the copier can be increased enough to match the demand, for example, by implementing your assistant's suggestions.

The above example demonstrates why operations engineers and managers—in addition to the effective capacity of a process (which is the actual capacity of a process)—are also interested in theoretical capacity of a process (which is not the actual capacity). Theoretical capacity of a process presents a good benchmark for the best-case scenario. The gap between effective and theoretical capacities of a process is an indication of the maximum potential increase in resource capacity that can occur if all non-value-added tasks, such as setups, rework, failure, repairs, etc., are eliminated.

---

**CONCEPT**

### Capacity Improvement Potential

Capacity Improvement Potential (CIP) of resource $i$ is the gap between the theoretical and effective capacities of resource $i$:

$$\text{CIP}(\text{resource } i) = C_0(\text{resource } i) - C_{eff}(\text{resource } i)$$

Hence, CIP is the *absolute* increase in effective capacity if all the supporting tasks performed by the resource can be eliminated.

For example, the CIP for the copier is

$$\text{CIP}(\text{copier}) = C_0(\text{copier}) - C_{eff}(\text{copier}) = 72 - 60 = 12 \text{ files per hour}$$

which implies that if all non-value-added tasks of the copier can be eliminated, the maximum capacity improvement that one can expect is 12 files per hour.

While the gap between the theoretical and effective capacities of a resource is representative of the maximum opportunity for capacity improvement, it might not present the entire picture. Suppose that the gap between the effective and theoretical capacities of a stamping press is 30 flow units an hour. Does this present a large opportunity for capacity improvement? Well it might, if the effective capacity of the press is 60 flow units an hour. However, 30 flow units per hour increase in the capacity of the press may not be considered a large opportunity for improvement if the effective capacity of the press is 900 flow units per hour. In other words, increasing capacity by 30 units from 900 to 930 might not be considered as significant as increasing capacity from 60 to 90. So, in addition to absolute gap between the theoretical and effective capacities of a resource, it is also important to get a sense of how large this gap is compared to the current capacity (i.e., effective capacity) of the resource. The Percent Capacity Improvement Potential does exactly that.

<br>

**CONCEPT**

**Percent Capacity Improvement Potential**

Percent Capacity Improvement Potential (%CIP) of resource $i$ is the *percent* increase in effective capacity of resource $i$ if all the supporting tasks performed by the resource can be eliminated. Specifically,

$$\%\text{CIP}(\text{resource } i) = \frac{\text{Gap between theoretical and effective capacities of resource } i}{\text{Effective capacity of resource } i}$$

$$= \frac{\text{CIP}(\text{resource } i)}{C_{\text{eff}}(\text{resource } i)}$$

In case of the copier, for example, the %CIP is

$$\%\text{CIP}(\text{copier}) = \frac{72 - 60}{60} = 0.2 = 20\%$$

This implies that if all non-value-added tasks (e.g., setups, rework, failures, and repairs) can be eliminated, the effective capacity of the copier can be increased by 20 percent and the copier can reach its theoretical capacity of 72 files per hour.

It should be emphasized here that CIP or %CIP presents benchmarks for the *potential* in capacity improvement by reducing non-value-added tasks. It might not always be possible to reduce all non-value-added tasks of a resource to zero to achieve %CIP improvement. Nevertheless, the main goal of Lean Operations, as we will discuss in Chap. 15, is to make the process ideal with no waste—zero setups, zero failures, zero defectives. Thus, capacity improvement potential is considered a useful goal for lean operations practitioners.

**Example 4.1**

Motopower is a firm that produces front wheel drive gas-powered lawn mowers and snow plowers. As a production manager of the company, you would like to evaluate the opportunities for capacity improvement in the painting station of your lawn mower production line. The painting station currently has one spray painting machine that paints springs used in the suspension of lawn mowers. The machine is operated by one worker. Springs

Table 4.2    Time Study of Processing 21 Springs in the Painting Station in Example 4.1

| Part number | Theoretical process time (*minutes*) | Supporting process time (*minutes*) | Effective process time (*minutes*) | Note |
|---|---|---|---|---|
| M001 | 2 | 0 | 2 | |
| M002 | 2 | 0 | 2 | |
| M003 | 1.7 | 0 | 1.7 | |
| M004 | 2.1 | 5 | 7.1 | *Machine setup:* The paint was refilled |
| M005 | 1.8 | 0 | 1.8 | |
| M006 | 2.1 | 1.2 | 3.3 | *Quality issue:* M006 required rework |
| M007 | 2.4 | 0 | 2.4 | |
| M008 | 2 | 0 | 2 | |
| M009 | 1.8 | 0 | 1.8 | |
| M010 | 1.7 | 13 | 14.7 | *Machine failure:* M010 waiting for repair |
| M011 | 2.3 | 0 | 2.3 | |
| M012 | 2 | 0 | 2 | |
| M013 | 2 | 0 | 2 | |
| M014 | 2.4 | 0 | 2.4 | |
| M015 | 1.7 | 0 | 1.7 | |
| M016 | 2 | 0 | 2 | *Quality issue:* M016 was scrapped due to paint quality |
| M017 | 2.1 | 0 | 2.1 | |
| M018 | 1.8 | 0 | 1.8 | |
| M019 | 2.1 | ~~3~~ | 2.1 | *Worker unavailability:* Worker took 3-minute break |
| M020 | 2 | 0 | 2 | |

are painted one-by-one and are sent to the next station in batches of size 10. You would like to get an idea of the maximum possible improvement in the capacity of the painting machine, if you can eliminate supporting tasks such as setups, quality problems, failures, etc. A time study has been performed to evaluate the current capacity of the painting machine. The summary of the time study is presented in Table 4.2. The "Theoretical Process Time" in the table refers to the painting time of one spring by the machine. The quality issues with springs are detected at the quality control (QC) station downstream of the painting station.

To evaluate the potential for capacity improvement, the effective and the theoretical capacities of the painting station must be computed.

Since spray painting machine performs a single activity of painting springs, the theoretical (and effective) process time of the spray painting machine (i.e., resource) is equal to the theoretical (and effective) process time of its single activity—painting. Table 4.2 provides the theoretical process time of painting activity of the machine for 20 springs. Hence, we can get an estimate for theoretical mean process time of the machine by simply taking the average of the 20 theoretical process times in the table.

$$T_0(\text{machine}) = T_0(painting) = \frac{2 + 2 + 1.7 + \cdots + 2.1 + 2}{20} = \frac{40}{20} = 2 \text{ minutes per spring}$$

or $2/60 = 0.0333$ hours per spring. Thus, theoretical capacity of the machine is

$$C_0(\text{machine}) = \frac{1}{T_0(\text{machine})} = \frac{1}{0.0333} = 30 \text{ springs per hour}$$

To find the effective mean process time of the painting activity, we need to find the average of 20 total process times (which includes both theoretical and supporting process times) in the table.

$$T_{\text{eff}}(\text{machine}) = T_{\text{eff}}(painting) = \frac{2 + 2 + 1.7 + 7.1 + \cdots + 2.1 + 2}{19}$$

$$= \frac{59.2}{19} = 3.116 \text{ minutes per spring}$$

Note that while 59.2 minutes is the sum of total process times of 20 springs, we divide it by 19 (not 20) to get the effective mean process time. Why? The answer is in the definition of the effective mean process time. It is the long-run average time of the activity to produce one "good" (non-defective) flow unit. During the total time of 59.2 minutes, only 19 non-defective springs were produced (i.e., M016 item was scrapped). Thus, it takes an average of $59.2/19 = 3.116$ minutes to make one non-defective spring.

Also, note that we did not include the 3-minute machine idle time due to worker unavailability in our calculations. This is because the machine idle time during those 3 minutes has nothing to do with the machine capacity (it only impacts machine's through-put). Finally, note that the above estimates of the effective and theoretical capacities were made using a small sample of only 20 springs. Larger samples should be used to find a more accurate estimate of these quantities.

Considering both theoretical and effective capacities of the painting machine, we find the Capacity Improvement Potential for the painting machine as follows:

$$\text{CIP}(\text{machine}) = C_0(\text{machine}) - C_{\text{eff}}(\text{machine}) = 30 - 19.26 = 10.74$$

In other words, the maximum improvement in capacity that one can expect by eliminating all non-value-added tasks (i.e., supporting tasks) is 10.74 spring per hour. This represents a potential to increase the effective capacity of the painting machine by

$$\%\text{CIP}(\text{machine}) = \frac{C_0(\text{machine}) - C_{\text{eff}}(\text{machine})}{C_{\text{eff}}(\text{machine})} = \frac{30 - 19.26}{19.26} = 55.8\%$$

The %CIP of 55.8 percent indicates a significant opportunity for capacity improvement, and should motivate operations engineers and managers to seriously consider reducing the non-value-added activities through lean operations principles.

## 4.4 Six Points to Remember about Effective Capacity

Finding an accurate and useful estimate of the effective capacity of a resource requires a good understanding of what the effective capacity is. The following six points are useful hints to remember:

*1. Determine the flow unit before you start computing capacity.* As we mentioned in Chap. 2, to analyze a system from a process-flow perspective, one should specify the flow unit of the process. The flow units are mainly the output of a resource. The effective capacity of a resource is measured by the "number" of good outputs processed by the resource per unit time, so it is very important to first decide what the unit of output is. In the painting station in Example 4.1, one can consider one spring as one flow unit, or a batch of size 10 of those springs as one flow unit of output. Although, this does not make much difference in finding the effective capacity of a simple process, choosing the right unit for the output is very important when finding effective capacity of more complex processes with multiple products and multiple stages of operations. We will discuss this in Sec. 4.10.

*2. Effective capacity is not throughput.* The effective capacity of a resource is the long-run average number of non-defective outputs per unit time that the process *can process,* as opposed to what it *has been processing.* For instance, in Time Study in Table 4.2, we found that the machine has produced 19 non-defective springs in 62.2 ($= 59.2 + 3$) minutes, that is, $19/62.2 = 0.305$ springs per minute or $0.305 \times 60 = 18.3$ springs per hour. This, as we have discussed, is not the capacity of the painting machine. It is the *throughput* of the painting machine during the 62.2 minutes of the study (in which the machine was idle for 3 minutes). The capacity of the machine is higher than its average throughput.

*3. Effective capacity is a "long-run average" measure.* The effective capacity of a process is the maximum long-run average number of non-defective outputs per unit time. Specifically,

(i)   It is the *average* capacity. It ignores the variations in capacity due to variability in process times, repair times, setup times, etc. The reason is that making capacity decisions based on the numbers that represent the good days of operations (e.g., days with no machine failures), or the bad days of operations (e.g., days that the machine was mainly down due to failure or repair) are too optimistic or too pessimistic, respectively. The average of those days is a better representative of the actual capacity of the process.

(ii)  It is the *long-run* average, that is, average of a large sample. For example, the data presented in Table 4.2 was only for a 64.5 ($= 42.3 + 21 + 1.2$) minute study. More accurate estimates of effective capacity can be obtained by studying the painting operations for a longer time.

*4. Effective capacity is independent of the input.* The effective capacity of a resource is calculated assuming that the process "never idles due to lack of inputs." As an example, consider a copier that can copy 60 application files per hour. When there is no application to copy during an hour, it does not mean the capacity of the copier is zero in that hour. It only means that the throughput of the copier is zero in that hour due to lack of input.

*5. Effective capacity of a resource is independent of other resources.* The effective capacity of a resource is computed assuming that the resource is not idle due to the unavailability of other resources. Again, consider the copier that can copy 60 files per hour. If there is no staff available to copy the files, it does not mean that the capacity of the copier is reduced to zero. It means that the throughput of the copier is reduced to zero due to unavailability of other resources.

*6. Effective capacity is independent of the control policy.* Effective capacity is computed independent of the policies used to control resources and movements of flow units (e.g., material, parts). For example, an inefficient *inventory control* policy that does not order enough raw material (resulting in resource idle time) should not affect the computation of effective capacity—it affects the throughput. Also, inefficient *production control* policies that fail to coordinate the production of all subassemblies (resulting in resources idle time waiting for subassemblies) should not affect the effective capacity—again, it affects the throughput.

## 4.5 Capacity of a Single Resource

There are often data about the theoretical process times and supporting task times available in different databases of a firm. For example, the maintenance departments keep detailed data of machine failures and their repair times. Thus, they can provide information such as average uptime and downtimes of a machine. Production managers have data about the number of items produced in a shift, and quality control departments have data about number of defective items produced by a resource. While these data are very useful, they are gathered independent of each other, and are not in the format of Table 4.2 in Example 4.1. Since we have already discussed how to use data such as those in Table 4.2 to compute theoretical and effective capacities of a resource, in this section, we discuss how to use independent data sets to compute those capacities.

As discussed, the main reason for the gap between theoretical (i.e., ideal) capacity and effective (i.e., actual) capacity is the time spent on supporting tasks. The supporting tasks that are common in most manufacturing and service operations systems correspond to (i) setups, (ii) quality issues resulting in defective items and items that require rework, and (iii) resource interruption such as machine failure and repair. We further discuss these below.

### Setups

A large number of processes require some setup and adjustment before they start performing the main operations on a flow unit. Filling papers in a copy machine before making copies, changing the die of a stamping press before starting the production of door panels of a different car model, and seating customers and making final safety checks before a ride in an amusement park are different examples of setup operations. The setup is often performed before or after a certain number of flow units—called a batch of flow units—are processed. Setup operations do not add value to the flow unit because they do not process the unit. Setups only prepare resources (or flow units) for performing the main tasks.

### Rework

There are not many processes in manufacturing and service systems that do not generate defective flow units. Defective flow units are units that have quality problems since they do not meet their design specifications. For example, potato chips bagging equipment often produce potato chips bags that are lighter than the minimum required weight, credit card companies make errors about customers' transactions, and online retailers sometimes ship wrong items to their customers.

Defective flow units with quality problems can generally be classified into two groups: scrap units and rework units. *Scrap Units* are flow units with quality problems that are not possible or not economical to fix, and therefore these units are often scrapped. Scrap units are part of what is known as Yield Loss (we will discuss yield loss in detail in Sec. 4.11). *Rework Units*, on the other hand, are flow units with quality problems that can be fixed through rework done by one or more resources. For example, if two parts are not assembled properly, they are sent back to the assembly worker to fix the assembly issue.

### Resource Failure

Resource breakdown is one of the major causes of interruptions in processes. Interruptions such as failure and repair time reduce resource availability during working hours. Resource availability is defined below.

---

**CONCEPT**

**Availability of a Resource**

*Availability* of resource $i$ is the percentage of time that resource $i$ is up and capable of processing flow units. The availability of resource $i$ can be computed as follows:

$$A_i = \frac{\text{Average uptime of resource } i}{\text{Average uptime of resource } i + \text{Average downtime of resource } i}$$

For example, if a machine works for an average of 180 hours before it breaks down, and if it takes an average of 20 hours to repair the machine, the machine has availability

$$A_{mach} = \frac{\text{Average uptime of machine}}{\text{Average uptime of machine} + \text{Average downtime of machine}} = \frac{180}{180 + 20} = 90\%$$

We use the following example to show how to compute the effective capacity of a resource with setups, breakdowns, and rework. In Sec. 4.11 we show how to compute the capacity of processes that also have yield loss.

 **Example 4.2**

A cutting station has a single punch press that is used to cut metal sheets into exterior panels used in the production of small dehumidifiers. The press works three shifts—24 hours a day. It takes an average of 20 seconds to cut a panel in one strike. After producing a batch of 100 panels, the press is stopped so the operator can remove the small pieces of metal generated during the operations and clean the press bench to prevent possible damage to the press die. This takes an average of 2 minutes. Each panel is shaped through one strike of the press. About 5 percent of panels produced by the press have quality issues that require rework. Specifically, after the first strike, if some of the holes on the panel are not completely cut, the press does another strike. This rework takes an average of 10 seconds. The press is old and has frequent breakdowns. The maintenance department data shows that after the press is repaired, it works an average of 190 hours until it breaks down and requires repair again. Data also shows that the repair time has an average of 10 hours. What is the effective capacity of the press in a day?

To find the effective capacity of the punch press in Example 4.2, we need the following analytics to compute the effective mean process time of activities performed by the press.

**Effective Process Time of an Activity with Setup, Rework, and Resource Failures**

Consider activity $j$ performed by resource $i$ which has the following features:

- *Main task:* Activity $j$ has a main task that takes an average of $T_0(activity\,j)$ units of time *per* flow unit.
- *Setups:* Activity $j$ requires setups, which takes an average of $T_s$ units of time, and processes $N_s$ flow units between two setups.
- *Rework:* Fraction $\alpha_r$ of the flow units processed by activity $j$ requires rework that takes an average of $T_r$ units of time, and is performed by resource $i$. Performing rework does not require setups.
- *Failure:* Resource $i$ has an average time to failure of $m_f$ and an average repair time of $m_r$, and thus availability $A_i = m_f/(m_f + m_r)$.

The effective mean process time of activity $j$ is (see online Appendix A)

$$T_{eff}(activity\,j) = \frac{[T_s + N_s T_0(activity\,j)] + N_s \alpha_r T_r}{A_i N_s} \tag{4.1}$$

Theoretical mean process time of activity $j$, on the other hand, is simply $T_0(activity\,j)$.

In Example 4.2, the press (the resource) is performing activity "cutting" the metal sheets. The press has an average time to failure of $m_f = 190$ hours with an average repair time of $m_r = 10$ hours. Thus, its availability is

$$A_{press} = \frac{190}{190 + 10} = 0.95 = 95\%$$

About $\alpha_r = 5\%$ of the panels produced by the press require rework that takes an average of $T_r = 10$ seconds. Finally, the cutting activity requires a setup after processing $N_s = 100$ panels, and the setup takes an average of $T_s = 2$ minutes $= 120$ seconds. The theoretical mean process time of cutting one panel is $T_0(cutting) = 20$ seconds. Putting these numbers into Eq. (4.1), we obtain the effective mean process time of activity "cutting" as follows:

$$\begin{aligned}
T_{eff}(cutting) &= \frac{[T_s + N_s T_0(cutting)] + N_s \alpha_r T_r}{A_i N_s} \\
&= \frac{[120 + 100 \times 20] + 100(0.05) \times 10}{0.95(100)} = 22.8 \text{ seconds}
\end{aligned}$$

Since the punch press only performs the cutting activity, its effective mean process time is $T_{eff}(press) = T_{eff}(cutting) = 22.8$ seconds. Thus, the effective capacity of the punch press is

$$C_{eff}(press) = \frac{1}{T_{eff}(press)} = \frac{1}{22.8} = 0.0438 \text{ panels per second}$$

Since we are interested in computing the effective capacity per day, and since the punch press works 24 hours a day, the effective capacity of the punch press is $0.0438 \times 3600 \times 24 = 3782.5$ panels per day.

Note that Eq. (4.1) is flexible and works for cases that do not have all of the supporting tasks. For example, if there is no setup, we set $N_s = 1$ (batch of size one) and $T_s = 0$; if there is no failure, we set $A_i = 1$; and if there is no rework, we set $\alpha_r = 0$ and $T_r = 0$.

## 4.6 Capacity of a Resource Pool

Processes often use several resources that perform the *same* tasks. For example, there are often more than one cashier in supermarkets, more than one teller in banks, and workstations often have more than one worker. The set of resources that perform the same activities is called "resources pool."

We use the following example to show how the effective capacity of a pool of resources can be computed.

 **Example 4.3**

Consider the punch press workstation in Example 4.2. To increase the capacity of the workstation, another punch press was added to the station. The new punch press also works three shifts, but produces less defective panels. Specifically, only 2 percent of its remaining panels require rework. How many panels the pool of the two presses can produce in a day?

The new press has the same features of the current one, except for percent rework of $\alpha_r = 2\%$. Using Eq. (4.1) for the new press we find

$$T_{\text{eff}}(cutting) = \frac{[120 + 100 \times 20] + 100(0.02) \times 10}{0.95(100)} = 22.5 \text{ seconds}$$

Since the new punch press only performs the cutting activity, its effective mean process time is $T_{\text{eff}}(\text{press}) = T_{\text{eff}}(cutting) = 22.5$ seconds. The effective capacity of the new punch press is, therefore,

$$C_{\text{eff}}(\text{press}) = \frac{1}{T_{\text{eff}}(\text{press})} = \frac{1}{22.5} = 0.044 \text{ panel per seconds}$$

or $0.044 \times 3600 \times 24 = 3801.6$ panels per day. Consequently, the pool of two punch presses in the station has the effective capacity $C_{\text{eff}}(\text{station}) = 3782.5 + 3801.6 = 7584.1$ panels per day.

## 4.7 Capacity of a Process with a Single Product

After discussing ways to compute the effective capacity of a *resource* or a *resource pool*, we now illustrate how to compute the effective capacity of a *process*. Depending on the scope and complexity of a process, the steps to compute its capacity are different. One main factor that relates to the complexity of a process is the number of different types of flow units that are processed within process boundaries. For example, the cutting station in Example 4.2 is a process with a single flow unit—the exterior panel of small dehumidifiers. On the other hand, a financial institution that offers different types of loans (i.e., home loan, or loan for small businesses) is a process with multiple types of flow units.

We use the term *Single-Product* or *Multi-Product* to refer to processes with single type or multiple types of flow units, respectively. In a manufacturing process, the term "product" refers to the final output of the process, while in a service process it refers to the type of service offered by a service firm. For example, in automobile manufacturing processes, product type mainly refers to the brand and model of a car, while in airlines it refers to the type of seats or tickets (i.e., coach or first class), or in an insurance company it refers to the type of insurance offered (i.e., home or life insurance).

It is worth mentioning that when analyzing capacity, determining the number of product types of a process depends on the level of details in the analysis. For example, when analyzing the capacity of the emergency department (ED) of a hospital, one might consider the ED as a process with five different types of flow units, since upon arrival (and after registration) patients are categorized by a triage nurse into five priority groups based on the severity of their conditions. Each priority group may receive a different treatment and follow a different route within the ED. However, when analyzing the capacity of the hospital (which includes the ED and many other departments), one may consider the ED of the hospital as a process with two types of patients—those who are discharged from the ED, and those who are admitted to the hospital. Hence, when computing capacity of a process, it is important to first determine the level of the capacity analysis to decide whether to model the process as a single-product or a multi-product process.

In this section, we study single-product processes. We use Example 4.4 to describe a step-by-step procedure to compute the capacity of a single-product process. In Sec. 4.10 we extend this procedure to processes with multiple products.

 **Example 4.4**

> A drilling station has a numerically controlled (NC) drill with one operator. Drilling station works two 8-hour shifts a day, and 25 days a month. Each shift is run by a different worker. Each shift has a 1-hour break (for lunch or dinner). Activities performed in the station include drilling a hole of 1-inch diameter on steel disks and sanding the disk to smoothen the surface of the cut. Both the drill and the worker are required for drilling operations, but sanding is done by the worker. The drill is idle when the worker sands a disk. The time to drill a disk does not vary much and has an average of 2 minutes. The sanding time has an average of 15 seconds. This average is consistent between the two workers in the two shifts and among different hours of those shifts. What is the capacity of the drilling station in a month?

The following analytics shows how the effective capacity of a process with a single product can be obtained.

## ANALYTICS

### Finding Capacity of a Process with a Single Product

**Step 1:** (*Process Boundaries, Product, and Time Unit*) Draw process-flow chart and determine process boundaries. Also, determine the product of the process and the time unit by which you want to measure capacity.

**Step 2:** (*Resources Characteristics*) Construct the resource table. Specifically,
- Identify the resources of the process.
- For each resource, determine the list of activities it performs within process boundaries.
- For each activity, determine its flow unit, and its main and supporting tasks.

**Step 3:** (*Effective Mean Process Time of Resources*) Compute the effective mean process time of each resource as follows. For each resource:
- Compute the effective mean process time of each activity performed by the resource using Eq. (4.1).
- Sum the effective mean process times of all the activities required for the product (specified in Step 2) to get the effective mean process time of the resource.

**Step 4:** (*Effective Capacities of Resources and Resource Pools*) For each resource,
- Determine the working schedule during the time unit specified in Step 1.
- Using the working schedule, compute the effective capacity of each *resource* in terms of the product and time unit determined in Step 1.
- If there is a pool of identical resources performing the same activity, compute the effective capacity of the *resource pool* by summing their effective capacities.

**Step 5:** (*Effective Capacity of the Process*)
- Determine the *bottleneck resource*—the resource or resource pool with the lowest effective capacity among all resources.
- The effective capacity of the *process* is equal to the effective capacity of its bottleneck resource.

We follow the above five steps to compute the effective capacity of the drilling station in Example 4.4.

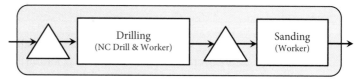

**Figure 4.2    Process-flow chart of the drilling station in Example 4.4.**

### 4.7.1  Step 1: Process Boundaries, Product, and Time Unit

In this step, we determine process boundaries, the product of the process, and the time unit by which we want to measure the capacity. A process-flow chart often helps to better understand the boundaries of the process and the flow dynamics within the boundaries. Figure 4.2 shows the process-flow chart for the drilling station. Obviously, process boundaries are the boundaries of the drilling station. Process starts from when a disk enters the station until it exits the station (after drilling and sanding).

Choosing the product of the process and time unit is often easy in single-product processes. The drilling station in Example 4.4 produces only one type of product—a 1-inch finished (i.e., drilled and sanded) disk. Thus, we use "finished disk" as the product. The time unit can be an hour, a day, or a month. It is easier to choose the time unit to be the unit by which we would like to measure the capacity of a process. Since we would like to know how many *finished disks* the drilling station can produce per *month*, we choose 1 month as *time unit*. Hence,

$$\text{Product} = \textit{finished disk} \qquad \text{Time Unit} = \textit{month}$$

### 4.7.2  Step 2: Resources Characteristics

In this step, we need to identify the resources of the drilling station, the flow units that they process, and the list of all the activities they perform on their flow units within process boundaries. Also, for each of those activities, we must determine the list of main and supporting tasks.

The drilling station has two main resources: the NC drill and the worker. Note that one can also consider sand paper or other tools (e.g., wrenches used to fix a disk on the drill) as other resources. This raises the question of which resource should be considered when measuring the capacity. The answer is not to include resources that are consumable, are abandon, and are not critical in the process, for example, sand papers and screw drivers, etc. If it is not clear whether one resource is a critical resource or not, one can always include it in capacity analysis.

Worker performs two activities: drilling and sanding. Both activities are performed on a disk—the flow unit. The activity of "Drilling a Disk" includes the main task of drilling, and no supporting tasks (i.e., no setups, rework, etc.). The activity of "Sanding a Disk" includes the main task of sanding with no supporting tasks. The NC drill performs only one activity of "Drilling a Disk." This activity has the main task of drilling and no supporting tasks. We summarize the information about the resources in Table 4.3.

### 4.7.3  Step 3: Effective Process Times of Resources

In this step, we first need to compute the effective mean process times of each activity performed by each resource. By adding effective mean process times of all the activities that a resource performs to make one product (a finished disk), we obtain the effective mean process time of the resource.

**Table 4.3    Resources Table for the Drilling Station in Example 4.4**

| Resource | Activity | Flow unit | Main task | Supporting tasks |
|---|---|---|---|---|
| NC drill | Drilling a disk | Disk | Drilling | None |
| Worker | Drilling a disk | Disk | Drilling | None |
|  | Sanding a disk | Disk | Sanding | None |

### Effective Mean Process Time of the Worker

The worker's main activities are drilling a disk and sanding a disk. The activity of "Drilling a Disk," or "drilling" for brevity, includes only the main task of drilling that takes an average of 2 minutes, that is, $T_0(drilling) = 2$ minutes per disk. Since the worker does not do any supporting tasks, the effective mean process time of drilling activity is also 2 minutes, that is, $T_{\text{eff}}(drilling) = 2$ minutes per disk.

The worker also performs the activity "Sanding a Disk," or "sanding" for brevity. This activity has the main task of sanding, which takes $T_0(sanding) = 0.25$ minutes per disk. Similarly, sanding does not have any supporting tasks, and thus the effective mean process time of sanding activity is $T_{\text{eff}}(sanding) = 0.25$ minutes per disk.

Since the worker performs both drilling and sanding activities to make a finished disk, the effective mean process time of the worker (i.e., the resource) to process one product (i.e., a disk) is the sum of the effective mean process times of drilling and sanding activities:

$$T_{\text{eff}}(\text{worker}) = T_{\text{eff}}(drilling) + T_{\text{eff}}(sanding)$$
$$= 2 + 0.25 = 2.25 \text{ minutes per disk}$$

### Effective Mean Process Time of the NC Drill

The NC drill performs only one activity "Drilling a Disk" or "drilling." This activity consists of only the main task of drilling with theoretical mean process time of $T_0(drilling) = 2$ minutes per disk. Since there are no supporting tasks, $T_{\text{eff}}(drilling) = 2$ minutes per disk. Furthermore, since the NC drill performs only one activity, the effective mean process time of the NC drill (i.e., the resource) to process one product (i.e., a disk) is

$$T_{\text{eff}}(\text{drill}) = T_{\text{eff}}(drilling) = 2 \text{ minutes per disk}$$

Table 4.4 summarizes the effective mean process times of the resources.

**Table 4.4    Effective Mean Process Times of Activities at the Drilling Station in Example 4.4**

| Resource ($i$) | Activity ($j$) | $T_0(j)$ (*minutes per activity*) | Supporting task | $T_{\text{eff}}(j)$ (*minutes per activity*) |
|---|---|---|---|---|
| NC drill | Drilling | 2 | None | 2 |
| Worker | Drilling | 2 | None | 2 |
|  | Sanding | 0.25 | None | 0.25 |

### 4.7.4 Step 4: Effective Capacities of Resources and Resource Pools

In this step, we obtain the effective capacity of each resource in terms of the product (finished disk) and time unit (month) specified in Step 1. To do this, we first need to specify the working schedule of each resource in a month. Both resources of the drilling station—the worker and the NC drill—work two 8-hour shifts a day, and 25 days a month. The worker, however, has a 1-hour lunch break in each shift. So, if we compute the effective capacity of the worker and the drill in a minute or in an hour, we can then—using their working schedule in a month—find their effective capacities in a month.

**Effective Capacity of the Worker**

Let's first obtain the effective capacity of the worker in an hour. If the worker spends $T_{\text{eff}}(\text{worker}) = 2.25$ minutes $= 0.0375$ hours to process one disk, the number of disks that the worker can process in an hour—the effective capacity of the worker in an hour—would be

$$C_{\text{eff}}(\text{worker}) = \frac{1}{0.0375} = 26.67 \text{ disks per hour}$$

Having the effective capacity of the worker in an hour, we can simply calculate the effective capacity of the worker in a month. As mentioned, the worker works two 8-hour shifts a day (with 1-hour break in each shift), and 25 days a month.[2]

$$C_{\text{eff}}(\text{worker}) = 26.67 \times 2 \times 7 \times 25 = 9334.5 \text{ disks per month}$$

**Effective Capacity of the NC Drill**

Similarly, if it takes the NC drill $T_{\text{eff}}(\text{drill}) = 2$ minutes $= 0.0333$ hours to process one disk, the effective capacity of the NC drill in an hour is

$$C_{\text{eff}}(\text{drill}) = \frac{1}{0.0333} = 30 \text{ disks per hour}$$

Having the effective capacity of the NC drill in an hour, we can compute its effective capacity in a month. Note that the drill is available 8 hours in each shift (no lunch break for the drill); therefore,

$$C_{\text{eff}}(\text{drill}) = 30 \times 2 \times 8 \times 25 = 12{,}000 \text{ disks per month}$$

Now that we have obtained the capacities of all the resources of the drilling station, we can compute its capacity.

### 4.7.5 Step 5: Effective Capacity of the Process

Finding the effective capacity of a process is very simple when the effective capacities of its resources are computed in terms of the same product (e.g., finished disks) and the same time unit (e.g., 1 month). In this case, the effective capacity of the process is simply the effective capacity of its bottleneck.

**CONCEPT**

**Bottleneck Resource**

Bottleneck Resource of a process is the resource of the process with the smallest capacity. The theoretical bottleneck resource is the resource with the smallest theoretical capacity, and the effective bottleneck resource is the resource with the smallest effective capacity. A process may have more than one bottleneck resource.

Therefore, $C_{\text{eff}}(\text{station})$, the effective capacity of the drilling station is the effective capacity of its bottleneck. Since the worker has a lower effective capacity than the NC drill,

---

[2]Note that although two different workers work in two different shifts, from capacity analysis perspective we can consider them as one resource, unless there is a difference in their characteristics (e.g., average process times).

that is, $C_{\text{eff}}(\text{worker}) < C_{\text{eff}}(\text{drill})$, then the worker is the bottleneck resource with effective capacity $C_{\text{eff}}(\text{worker}) = 9334.5$. Therefore,

$$C_{\text{eff}}(\text{station}) = C_{\text{eff}}(\text{worker}) = 9334.5 \text{ disks per month}$$

The reason behind the above argument is simple. Every disk must be processed by both the worker and the drill. If the worker cannot process more than 9334.5 disks during 1 month of work at the drilling station, regardless of the higher capacity of the drill during that month, the drilling station—the process—cannot produce more than 9334.5 disks per month. In other words, the capacity of the process is constrained by the capacity of its bottleneck resource.

The same five steps can be used to find the *theoretical capacity* of the drilling station by simply using theoretical process times instead of effective process times.

## 4.8 Capacity of a Multi-Stage Process with Single Product

The drilling station in Example 4.4 has two resources where both process the same flow unit—a disk. There are more complex processes with more than two resources, all performing different activities to make a single product. In this section, we show that our five-step approach can also be used in such more complex multi-stage processes. We use the Flowcet case to illustrate this.

### Order Processing in Flowcet Distribution Center

Flowcet is a small manufacturer of faucets and showerheads with several distribution centers (DC) in the East Coast. One of their DCs is in Jackson, Michigan, and receives orders from three large retailers within a 50-mile radius. Orders are faxed or emailed to Jackson's DC. The DC works 8 hours a day (from 8:00 am to 5:00 pm with lunch break from 12:00 to 1:00), 5 days a week.

*Order Processing:* The operations at Jackson's DC are as follows. Faxes and emails from retailers are received by Jeannette, who enters the order to the retailer's account and then uploads the quantity and the due dates of the order in DC's computer system. The order then, through DC's computer system, is routed to pick up, package, inspection, and boxing sections of the DC. Each order is for different quantities of several items. The average time that takes Jeannette to process an order is around 20 minutes.

*Pick Up Section:* In the pick up section, for each order, Bruce first prints the "Item List" generated by the computer. The item list includes the list of all the items and their quantities in each order. Bruce then picks up all the items in the item list from different shelves of the DC, puts them in a pallet, and moves them to the inspection section. The time that takes Bruce to do this varies between 40 and 180 minutes depending on the size of the order. The average time is around 60 minutes.

*Package Section:* In the package section, folded packages (i.e., boxes) of products are separately stored for each retailer. Although all retailers order same products, they use slightly different packages. This is because each package has the retailer's barcodes (used in the retailer's inventory system) and, sometimes, slightly different presentation of the product specification. This is common in retail business, since it makes it difficult for customers to compare the price of the same product offered by different retailers. For each order, Karen first prints the "Package List." Package list of an order has the list of all the packages needed for the products in that order. Karen picks up the right packages, unfolds the packages and makes boxes, and then takes them to the inspection section. Depending on the size of the order, it takes Karen between 20 and 200 minutes. But overall, it takes Karen an average of 45 minutes per order.

*Inspection Section:* Harry works at the inspection section. For each order, Harry first makes sure that the item list and package list match with the items and boxes that are sent by the pick up and package sections. It takes Harry an average of 15 minutes to check that items match with the item list. However, data has shown that in 10 percent of orders, items do not match with their item lists. When this occurs, Harry calls Bruce and explains the inconsistencies, and discusses how to fix the problem. This takes an average of 5 minutes. It takes Bruce, on the other hand, an average of 15 minutes to resolve the issue by bringing the right items and taking back the wrong ones. Harry also checks the boxes to make sure they are consistent with the package list. This takes an average of 10 minutes. Harry has found that for 5 percent of orders, boxes do not match with the package lists. In those cases, Harry calls Karen and discusses the issue. The call takes an average of 5 minutes, and it takes Karen an average of 7 minutes to bring the right packages and take back the wrong ones. When the package list and item list of an order are checked and issues (if any) are resolved, Harry sends the order to Jamil at the boxing section.

*Boxing Section:* At the boxing section, Jamil puts each item in its corresponding box, staples and tapes the boxes, and stacks the boxes on pallets. The time to finish boxing an order also depends on the size of the order. But, overall, it has an average of 50 minutes per order. When boxing of the order is finished, it is sent to the shipping department. The shipping department then prints the delivery address for each order, loads the order on their trucks, and delivers the order. Flowcet has outsourced its shipping operations to a trucking company that guarantees the delivery of the order within the next day of receiving the order from the DC.

Managers of Flowcet have been receiving complaints from retailers of Jackson's DC regarding the late deliveries of their orders. Therefore, they are thinking of expanding the capacity of Jackson's DC. To do that, they first need to know the current capacity of the DC in terms of the number of orders that they can process per week. Flowcet is not worried about the capacity of the shipping department, since it can revise its contract with the trucking company (if needed) after it determines the capacity of Jackson's DC.

### Finding Capacity of the Distribution Center

To compute the effective capacity of the Flowcet DC, we use our five-step approach.

**Step 1:** (*Process Boundaries, Product, and Time Unit*) Figure 4.3 exhibits the process-flow chart of Jackson's DC. The process starts when an order arrives at the DC and ends when the order is boxed at the boxing section and sent to the shipping department. The final product of the DC is a completed order that is shipped to a retailer. Thus, we choose a

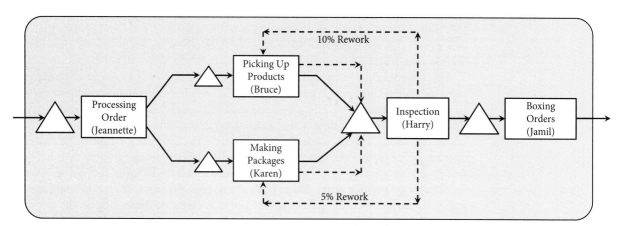

**Figure 4.3    Process-flow chart of Jackson distribution center.**

*completed order* as our product. Note that one order—one product—consists of several items ordered by one retailer. With respect to time unit, since Flowcet's managers are interested in the number of orders per week, we choose *week* as our time unit.

$$\text{Product} = completed\ order \qquad \text{Time Unit} = week$$

Note that the product we chose is different from the flow unit of some of the resources. For example, the actual output of Karen is empty packages (boxes) of an order, and the actual output of Bruce is items of the order, which both are different from the product we chose to measure the capacity (i.e., a completed order). This does not make any difference in our five-step approach. In fact, this is what makes finding the capacity of a process easy. To be able to compare the capacities of resources and identify the bottleneck, the capacity of all resources must be computed in terms of the same unit, that is, the number of orders that they can process per unit time.

**Step 2:** (***Resource Characteristics***) The DC has five resources that are involved with processing orders. The list of resources and their main and supporting tasks are presented in Table 4.5.

**Table 4.5    Activities at Jackson DC; All Resources have $A_i = 100\%$ (*Time is in Minutes*)**

| Resource (i) | Activity (j) | Main task $T_0(j)$ | Setups $N_s$ | $T_s$ | Rework $\alpha_r$ | $T_r$ | $T_{eff}(j)$ |
|---|---|---|---|---|---|---|---|
| Jeannette | Processing order | 20 | 1 | 0 | 0% | 0 | 20 |
| Bruce | Picking up items | 60 | 1 | 0 | 10% | 20 | 62 |
| Karen | Making packages | 45 | 1 | 0 | 5% | 12 | 45.6 |
| Harry | Inspecting item list | 15 | 1 | 0 | 10% | 5 | 15.5 |
|  | Inspecting package list | 10 | 1 | 0 | 5% | 5 | 10.25 |
| Jamil | Boxing | 50 | 1 | 0 | 0% | 0 | 50 |

**Step 3:** (***Effective Mean Process Times of Resources***) Effective mean process times of resources for their corresponding flow units are also presented in Table 4.5. These process times were computed using Eq. (4.1). Consider Bruce's activity of "Picking Up Items" in Table 4.5 as an example. This activity is done one-by-one for each order (i.e., $N_s = 1$) and there is no setup time required between preparing two orders (i.e., $T_s = 0$). The activity of Picking Up Items takes $T_0(pick\ up) = 60$ minutes. However, 10 percent of the orders required fixing (i.e., $\alpha_r = 0.1$). This takes an average of $T_r = 20$ minutes to resolve the issue (i.e., 5 minutes discussion with Henry and 15 minutes of fixing the order). There is no interruptions during Bruce's work (i.e., $A_{Bruce} = 100\%$). Using these numbers in Eq. (4.1) we find the effective mean process time of "Picking Up Items" activity performed by Bruce as $T_{eff}(pick\ up) = 62$ minutes per order.

**Step 4:** (***Effective Capacities of Resources and Resource Pools***) Effective capacities of resources are shown in Table 4.6. For example, effective mean process time of Henry to process one order (i.e., one final product) is the summation of his two activities of inspecting item list and inspecting package list:

$$T_{eff}(\text{Henry}) = T_{eff}(item\ list) + T_{eff}(package\ list)$$
$$= 15.5 + 10.25 = 25.75 \text{ minutes per order} = 0.429 \text{ hours per order}$$

Thus, Henry's effective capacity per hour is

$$C_{eff}(\text{Henry}) = \frac{1}{T_{eff}(\text{Henry})} = \frac{1}{0.429} = 2.33 \text{ orders per hour}$$

Considering working schedule of 40 hours a week—8 hours per day and 5 days a week—Henry's effective capacity in a week is

$$C_{\text{eff}}(\text{Henry}) = 2.33 \times 40 = 93.2 \text{ orders per week}$$

Table 4.6    Effective Capacities of Resources in Jackson DC

| Resources | Effective process time (*minute per ord.*) | Effective capacity (*ord. per hour*) | Work schedule (*hours per week*) | Effective capacity (*ord. per week*) |
|---|---|---|---|---|
| Jeannette | 20 | 3 | 40 | 120 |
| Bruce | 62 | 0.968 | 40 | 38.72 |
| Karen | 45.6 | 1.316 | 40 | 52.64 |
| Henry | 25.75 | 2.330 | 40 | 93.20 |
| Jamil | 50 | 1.200 | 40 | 48 |

**Step 5:** (*Effective Capacity of the Process*) Finally, the effective capacity of the DC in Jackson is equal to the effective capacity of its bottleneck resource. As Table 4.6 shows, Bruce is the bottleneck resource with effective capacity of 38.72—the lowest among all resources. Therefore,

$$C_{\text{eff}}(\text{DC}) = 38.72 \text{ orders per week}$$

Hence, the effective capacity of the Jackson DC is 38.72 orders per week.

## 4.9 Capacity of a Process with Multiple Products and a Single Resource

So far, we presented methodologies for computing capacity of simple processes that produce only one type of product and the product follows the same routing within process boundaries. In practice, however, there are a large number of firms that offer more than one type of products or services. In those systems, each product requires different activities and thus follows a different routing. For example, a kitchen appliances manufacturer produces multiple products ranging from small bread toaster to large gas or electrical ovens and stoves. The manufacturing process for a toaster is completely different from that of a gas oven, so it follows a different route than the oven does. This is also true for service firms. Patients that enter a hospital, depending on their medical problems, follow a different routing and visit different departments of the hospital.

When processes produce more than one product, computing their capacities requires different methodologies, even if the process has only one resource. The reason is that when a resource processes only one type of flow unit, the capacity of the resource can be simply stated as the maximum number of flow units that the resource can produce per unit time. However, when the resource processes more than one type of flow units, stating the capacity of the resource is more complicated. We utilize the following example to illustrate this and show how to compute the capacity of a single resource processing multiple products.

 **Example 4.5**

Suppose a time study in a small bank reveals that customers who use the only ATM (automated teller machine) of the bank can be divided into two types: Type-1 customers who use the ATM for only one transaction—cash withdrawal. This type constitutes 80

percent of the customers of the ATM machine. Type-2 customers use the ATM for more than one transaction (e.g., depositing checks, bill payment, cash withdrawal, etc.). This type constitutes 20 percent of the ATM customers. Based on the time study, the effective mean process time of the ATM for type-1 and type-2 have found to be 0.5 and 2 minutes, respectively. What is the effective capacity of this ATM?

Considering that the effective mean process time of type-1 customers is 0.5 minutes, the ATM—the single-resource—is capable of serving $60/0.5 = 120$ customers in an hour, if it only serves type-1 customers. Similarly, if the ATM serves only type-2 customers, its effective capacity is $60/2 = 30$ customers per hour. However, the key question is how many customers the ATM machine can serve in an hour if it serves *both* types of customers?

One method that can be used to compute the effective capacity of processes with multiple products is Product Aggregation method. This method is based on aggregating the multiple products of a process into one product called "Aggregate Product." For each resource, the effective mean process time of the aggregate product is the weighted average of the effective mean process times of the products based on the product mix.

---

CONCEPT

### Product Mix

For a process that produces $K$ different types of flow units, *Product Mix* is a set of $K$ numbers $(p_1, p_2, \ldots, p_K)$, where $p_k$ is the fraction of the total number of flow units produced by the process that are of type $k$. Note that $0 \leq p_k \leq 1$, and $p_1 + p_2 + \cdots + p_K = 1$.

---

To show how Product Aggregation method uses Product Mix to compute the effective capacity of processes with multiple products, let's return to our ATM example. Note that the product mix for the ATM is $(p_1, p_2) = (0.8, 0.2)$. Now consider the total average time that takes the ATM to serve 100 customers. Since an average of 80 of those customers are type-1 customers (with effective mean process time of 0.5 minutes), and the remaining customers are type-2 customers (with effective mean process time of 2 minutes), the total average time to serve those 100 customers is $80 \times 0.5 + 20 \times 2 = 80$ minutes. Since 80 minutes correspond to serving 100 customers, the average time to process one customer is $80/100 = 0.8$ minutes.

Note that the average process time of 0.8 minutes is the weighted average of the effective mean process times of two types of customers, based on the product mix, that is, $0.8 = p_1(2) + p_2(0.5) = 0.2 \times 2 + 0.8 \times 0.5$. So, if one considers another ATM machine serving only one type of customer, which we call "aggregate customer," with effective mean process time of 0.8 minutes, then this ATM also finishes the processing time of 100 customers, on an average, in 80 minutes. This implies that the ATM with two types of customers and product mix $(0.8, 0.2)$ and the ATM with one type of "aggregate customer" with effective mean process time of 0.8 minutes have the same effective capacity of serving a total of 100 customers in 80 minutes, that is, the effective capacity of $100/80 = 1.25$ customers per minute, or $1.25 \times 60 = 75$ customers per hour.

As this example shows, one can use the concept of "aggregate flow unit" or "aggregate product" to compute the effective capacity of a resource with multiple products. To simplify notation, in this and in the remaining chapters of this book we often omit "activity" from $T_{\text{eff}}(\text{activity } j)$ and we omit "resource" from $T_{\text{eff}}(\text{resource } i)$, and instead, we use $T_{\text{eff}}(j)$ and $T_{\text{eff}}(i)$, respectively. We do the same for effective and theoretical capacities of a resource. But we still use $i$ to refer to resources and $j$ to refer to activities. We also use $k$ to refer to the type of the flow unit.

**Aggregating Multiple Flow Units**

Consider resource $i$ that processes $K$ different types of flow units with product mix $(p_1, p_2, \ldots, p_K)$. If effective mean process time of resource $i$ to process one flow unit of type $k$ is $T_{\text{eff}}^{(k)}(i)$, then the effective mean process time of resource $i$ to process one *aggregate flow unit* (i.e., an aggregate product) is

$$T_{\text{eff}}^{(agg)}(i) = p_1 T_{\text{eff}}^{(1)}(i) + p_2 T_{\text{eff}}^{(2)}(i) + \cdots + p_K T_{\text{eff}}^{(K)}(i)$$

Consequently, the effective capacity of resource $i$ in terms of the number of aggregate flow units per unit time is

$$C_{\text{eff}}^{(agg)}(i) = \frac{1}{T_{\text{eff}}^{(agg)}(i)}$$

For example, the effective mean process time of the ATM for an "aggregate customer" is

$$
\begin{aligned}
T_{\text{eff}}^{(agg)}(\text{ATM}) &= p_1 T_{\text{eff}}^{(1)}(\text{ATM}) + p_2 T_{\text{eff}}^{(2)}(\text{ATM}) = 0.8 \times 0.5 + 0.2 \times 2 \\
&= 0.8 \text{ minutes per aggregate customer} \\
&= 0.8/60 = 0.0133 \text{ hours per aggregate customer}
\end{aligned}
$$

Therefore, the capacity of the ATM is

$$C_{\text{eff}}^{(agg)}(\text{ATM}) = \frac{1}{T_{\text{eff}}^{(agg)}(\text{ATM})} = \frac{1}{0.0133} = 75 \text{ aggregate customers per hour}$$

It is easy to see that a change in the product mix results in a change in the effective capacity of a process. For example, if the product mix in the ATM example changes to $(0.7, 0.3)$, then the effective mean process time for the aggregate product will be $0.7(0.5) + 0.3(2) = 0.95$ minutes $= 0.0158$ hours, resulting in effective capacity of $1/0.0158 = 63.16$ aggregate products (i.e., total number of customers) per hour. This points to the following important principle.

**Impact of Product Mix on Capacity**

Changing product mix of a process can change both theoretical and effective capacities of the process.

The effective capacity of $C_{\text{eff}}^{(agg)}(\text{ATM}) = 75$ aggregate customers represents the *total* number of customers (of types 1 and 2) that the ATM can serve in an hour under product mix $(0.8, 0.2)$. The question is: how does this total capacity translate to the effective capacity of the ATM to process each type of customers?

The answer is simple. Since 80 percent of the 75 customers are type-1 customers, an average of $0.8 \times 75 = 60$ of the 75 customers are type-1 customers and $0.2 \times 75 = 15$ of the 75 customers are type-2 customers. Hence, under product mix $(0.8, 0.2)$, we can conclude that the ATM has the effective capacity of serving 60 and 15 customers of type 1 and type 2 in an hour, respectively.

As this example shows, having the effective capacity for an aggregate product, one can use the product mix to disaggregate the aggregate product and obtain the effective capacity for each product type.

**Disaggregating the Aggregate Flow Unit**

Consider resource $i$ that processes $K$ different types of flow units with product mix $(p_1, p_2, \ldots, p_K)$. If effective capacity of resource $i$ to process the aggregate flow unit is $C_{\text{eff}}^{(agg)}(i)$, then $C_{\text{eff}}^{(k)}(i)$, the effective capacity of resource $i$ for type-$k$ flow units $(k = 1, 2, \ldots, K)$ is as follows:

$$C_{\text{eff}}^{(1)}(i) = p_1 C_{\text{eff}}^{(agg)}(i) \qquad C_{\text{eff}}^{(2)}(i) = p_2 C_{\text{eff}}^{(agg)}(i) \qquad \cdots \qquad C_{\text{eff}}^{(K)}(i) = p_K C_{\text{eff}}^{(agg)}(i)$$

The ATM station is an example of a simple process with multiple products, where both products require processing by a single resource. Most operations, however, are more complex and have multiple resources processing multiple products. We discuss this in the following section.

## 4.10 Capacity of a Process with Multiple Products and Multiple Resources

Most manufacturing and service processes have the following characteristics: (i) they have more than one resource, (ii) each product type may require processing by different resources, and (iii) each type of product may go through different paths (i.e., routing) in the process. Although these processes are complex, the Product Aggregation method can still be used—with some small changes—to compute the effective capacity of such processes. We use the case of Café Minoo in Example 4.6 to illustrate this.

 **Example 4.6**

Café Minoo is located in a very busy subway station and offers fast, but high-quality, drinks and sandwiches for takeout. Around 60 percent of the customers order drink, and around 10 percent of the customers order sandwich. The remaining 30 percent order sandwich and drink. Upon arrival, a customer first places the order with the cashier and pays for it. The cashier, Henry, asks the name of the customer and enters the order into the computer. The order then appears on the monitors in the drink and sandwich stations. Taking an order takes an average of 1 minute. Sam is in charge of making drinks in the drink station. The average time to prepare a drink for an order is about 1.5 minutes. Jill and Mat work in the sandwich station. The average time to prepare an order for a sandwich by Jill or Mat is around 6 minutes per order. When the order of a customer (drink or sandwich) is ready, the customer's name is called and the customer picks up the order. The Café opens at 6:00 am and closes at 8:00 pm, Monday through Friday, and is closed on weekends. Since the Café's demand is higher during rush hours, the owner of the Café Minoo would like to know what is the maximum number of customers that the Café can serve in an hour with its current employees.

The following analytics shows how product aggregation method can be used to determine the effective capacity of Café Minoo.

### Finding Capacity of Multi-Product Processes with Multiple Resources

**Step 1:** (*Process Boundaries, Products, Product Mix, and Time Unit*) Draw process-flow chart and determine process boundaries. Also, determine products and their product mix. Finally, choose the time unit by which you want to measure the capacity of producing each product type.

**Step 2:** (*Resources Characteristics*) Construct the resource table. Specifically,
- Identify the resources of the process.
- For each resource, determine the list of activities that it performs on each type of product.
- For each activity, determine the main and supporting tasks.

**Step 3:** (*Effective Mean Process Times of Resources for Each Type of Product*) Compute the effective mean process time of a resource to process each type of product as follows. For each resource,
- Compute the effective mean process time of all of its activities using Eq. (4.1).
- Sum the effective mean process times of all activities to get the effective mean process time of the resource to process each product type.

**Step 4:** (**Aggregation**): (*Effective Mean Process Times of Resources for the Aggregate Product*) For each resource, compute its effective mean process time for the aggregate product. Specifically, use the product mix to get the weighted average of the effective mean process time of the product types.

**Step 5:** (*Effective Capacities of Resources and Resource Pools*) For each resource,
- Determine the working schedule during the time unit specified in Step 1.
- Using the working schedule, compute the effective capacity of each *resource* and each *resource pool* in terms of the *aggregate* product and the time unit (chosen in Step 1).

**Step 6:** (*Effective Capacity of the Process for the Aggregate Product*) The effective capacity of the *process* (in terms of aggregate products per unit time) is the effective capacity of its bottleneck, that is, the effective capacity of the resource with the minimum effective capacity among resources and resource pools.

**Step 7:** (**Disaggregation**): (*Effective Capacity of the Process for Each Type of Product*) Determine the effective capacity of the process for each type of product by allocating the capacity of the process (for the aggregate product) to each type of product according to the product mix.

As the process-flow chart in Fig. 4.4 shows, service process in Café Minoo is a process with multiple resources in which all flow units (i.e., customers) do not follow the same path.[3] While all customers visit the cashier to place their orders, some customers do not visit the drink station and some other customers do not visit the sandwich station.

We follow our analytics to compute the effective capacity of Café Minoo.

**Step 1, *Process Boundaries, Products, Product Mix, and Time Unit:*** One simple way to determine the product types in a process is to divide the flow units into those that follow the same routing. In case of Café Minoo, the products (i.e., served customers) can be divided

---

[3] The process-flow chart assumes that customers who buy drink and sandwich first get their drinks and then their sandwiches. This does not have any effect on computing the effective capacity of resources of the process.

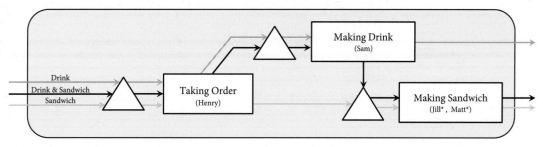

**Figure 4.4    Process-flow chart of Café Minoo in Example 4.6.**

into three groups:

Type-1 Product $=$ *Customers who only buy drink*

Type-2 Product $=$ *Customers who only buy sandwich*

Type-3 Product $=$ *Customers who buy drink and sandwich*

The product mix for the three types of products, as mentioned in the example, is $(p_1, p_2, p_3) = (0.6, 0.1, 0.3)$. Also, since we are interested to find the hourly capacity of Café Minoo, we choose our time unit as 1 hour or 60 minutes.

**Table 4.7    Activities of Resources in Café Minoo (*Time is in Minutes*)**

| Resource (i) | Activity (j) | Main task $T_0(j)$ | Supporting task | $T_{\text{eff}}(i)$ |
|---|---|---|---|---|
| **Cashier** Henry | | | | |
| | Taking order from type-1 | 1 | None | 1 |
| | Taking order from type-2 | 1 | None | 1 |
| | Taking order from type-3 | 1 | None | 1 |
| **Drink station** Sam | | | | |
| | Making drink for type-1 | 1.5 | None | 1.5 |
| | Making drink for type-2 | — | None | 0 |
| | Making drink for type-3 | 1.5 | None | 1.5 |
| **Sandwich station** Jill* | | | | |
| | Making sandwich for type-1 | — | None | 0 |
| | Making sandwich for type-2 | 6 | None | 6 |
| | Making sandwich for type-3 | 6 | None | 6 |
| Mat* | | | | |
| | Making sandwich for type-1 | — | None | 0 |
| | Making sandwich for type-2 | 6 | None | 6 |
| | Making sandwich for type-3 | 6 | None | 6 |

**Step 2, *Resources Characteristics:*** Café Minoo has four resources: (i) Henry, (ii) Sam, (iii) Jill, and (iv) Mat, where Mat and Jill constitute a pool of resource, both working on the same task of making sandwiches. While Henry—the cashier—processes all three types of customers (because all customers need to first place their orders with him), Sam who makes drinks only serves customers of types 1 and 3, and Jill and Mat serve the customers of types 2 and 3. Table 4.7 shows the list of the activities (with their main and supporting tasks) that each resource performs on each type of product (i.e., customers). We use "*" for Matt and Jill to show that they construct a pool of two resources.

**Step 3, *Effective Process Times of Resources for Each Type of Product:*** As Table 4.7 shows, activities that resources perform on products have no supporting tasks. Therefore,

**Table 4.8**    Computing the Effective Capacity of Café Minoo (*Time is in Minutes*)

| Resources | Effective mean process time | | | Effective mean process time of the aggregate product | Effective capacity (per hour) | Number of resources in the resources pool | Effective capacity of the resource pool |
|---|---|---|---|---|---|---|---|
| | Type 1 | Type 2 | Type 3 | | | | |
| Henry | 1 | 1 | 1 | 1 | 60 | 1 | 60 |
| Sam | 1.5 | 0 | 1.5 | 1.35 | 44.44 | 1 | 44.44 |
| Pool of Jill/Matt | 0 | 6 | 6 | 2.4 | 25 | 2 | 50 |

the effective mean process time of the activities are the same as the time to perform the main task. The last column of the table is the effective mean process time of each resource to process each product.

**Step 4 (Aggregation),** *Effective Mean Process Times of Resources for the Aggregate Product:* Having effective mean process times of resources to process each type of product, we can compute the effective mean process time of the aggregate product using product mix $(p_1, p_2, p_3) = (0.6, 0.1, 0.3)$. For instance, for Sam, who only serves type-1 and type-3 products we have

$$\begin{aligned} T_{\text{eff}}^{(agg)}(\text{Sam}) &= p_1 T_{\text{eff}}^{(1)}(\text{Sam}) + p_2 T_{\text{eff}}^{(2)}(\text{Sam}) + p_3 T_{\text{eff}}^{(3)}(\text{Sam}) \\ &= 0.6(1.5) + 0.1(0) + 0.3(1.5) \\ &= 1.35 \text{ minutes} \end{aligned}$$

The effective mean process times for the aggregate product for all resources are presented in Table 4.8.

**Step 5,** *Effective Capacities of Resources and Resource Pools:* Using the effective mean process time of each resource to process the aggregate product, we can compute the effective capacity of each resource in an hour. For example, Sam has effective mean process time $T_{\text{eff}}^{(agg)}(\text{sam}) = 1.35$ minutes. Thus, he has the effective capacities of processing $C_{\text{eff}}^{(agg)}(\text{sam}) = 60/T_{\text{eff}}^{(agg)}(\text{sam}) = 60/1.35 = 44.44$ aggregate products per hour. This is shown in Table 4.8. The effective capacities of other resources are also shown in the table. Note that the effective capacity of the pool of Jill and Matt is the sum of their individual effective capacities, that is, $25 + 25 = 50$.

**Step 6,** *Effective Capacity of the Process for the Aggregate Product:* The effective capacity of Café Minoo is equal to the effective capacity of its bottleneck resource—Sam. Sam has the effective capacity of processing 44.44 aggregate products per hour, which is the minimum among all resources of the process.

$$C_{\text{eff}}^{(agg)}(\text{cafe}) = 44.44 \text{ aggregate products per hour}$$

**Step 7 (Disaggregation),** *Effective Capacity of the Process for Each Type of Product:* Having the capacity of Café Minoo for the aggregate product, and product mix $(p_1, p_2, p_3) = (0.6, 0.1, 0.3)$, we can compute the effective capacity of Café Minoo for each type of product as follows:

$$\begin{aligned} C_{\text{eff}}^{(1)}(\text{cafe}) &= p_1 C_{\text{eff}}^{(agg)}(\text{cafe}) = 0.6(44.44) = 26.67 \text{ Type-1 customers per hour} \\ C_{\text{eff}}^{(2)}(\text{cafe}) &= p_2 C_{\text{eff}}^{(agg)}(\text{cafe}) = 0.1(44.44) = 4.44 \text{ Type-2 customers per hour} \\ C_{\text{eff}}^{(3)}(\text{cafe}) &= p_3 C_{\text{eff}}^{(agg)}(\text{cafe}) = 0.3(44.44) = 13.33 \text{ Type-3 customers per hour} \end{aligned}$$

Hence, Café Minoo has the effective capacity of serving a maximum of 26.67 type-1 customers (who only buy drink), 4.44 type-2 customers (who only buy sandwich), and 13.33 type-3 customers (who buy drink and sandwich).

## 4.11 Capacity of Processes with Yield Loss

As mentioned in Sec. 4.5, scrap units are defective flow units that have quality problems that cannot be fixed with rework and are therefore scrapped. Scrap units produced by a process are of great concern to operations engineers and managers. First, all the material and resource time spent on scrapped units are lost, since the unit cannot be used. Second, recall that the effective capacity of a process is the maximum number of "non-defective" units produced by the process per unit time. Hence, scrap (i.e., defective ) units reduce the effective capacity of the process and result in Yield Loss.

---

**CONCEPT**

**Yield and Yield Loss**

*Yield* of a resource (or a process) is the fraction of good (non-defective) flow units that the resource (the process) produces per unit time.

$$\text{Yield} = \frac{\text{Number of good (non-defective) units produced}}{\text{Total number of unit produced}}$$

*Yield Loss* of a resource (or a process), on the other hand, is the fraction of scrap flow units generated by the resource (by the process).

$$\text{Yield Loss} = 1 - \text{Yield}$$

---

A worker with a yield loss of 5 percent is a worker that produces 5 percent defective items that cannot be fixed and are therefore scrapped. This worker has a yield of 95 percent.

When resources of a process have yield loss, the effective capacity of the process—the maximum number of non-defective products produced per unit time—may be less than the effective capacity of its bottleneck. We use Example 4.7 to illustrate this and describe how the capacity of a process with yield loss can be computed.

 **Example 4.7**

Cool Pens is a firm that produces color pencils, where 70 percent of its products are pencils with erasers and 30 percent are pencils without erasers. Figure 4.5 shows the process-flow chart of the firm's plant. In the first stage of the process—wood casting—wooden slats are cut by a cutting machine that makes nine semicircular grooves on the slats. Then, nine cut graphites are laid in the grooves and two slats are glued together. The slats go to the next stage—pencil shaping. In this stage, the slats are cut into nine pencil blocks. The pencil blocks are sanded, colored, and dried. Some of the pencils are sent to the packaging department to be sold as pencils without erasers and the remaining are sent to eraser punching where erasers are glued and punched to the end of the pencils.

The effective process time of a slat (a batch of size 9) in wood casting is 5 seconds. However, 5 percent of the slats produced in wood casting are defective and are scrapped. The effective process time for a slat in the pencil shaping stage is 10 seconds, and effective process time for batch of size 9 at the eraser punching stage is 3 seconds. About 3 percent of the pencils made in this stage are scrapped because some pencils crack during the punching process. What is the maximum number of pencils of each type that Cool Pens can make in its plant in an hour?

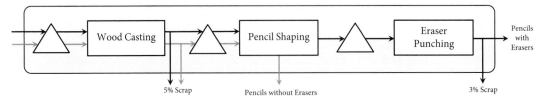

Figure 4.5    Process-flow chart of Cool Pens in Example 4.7.

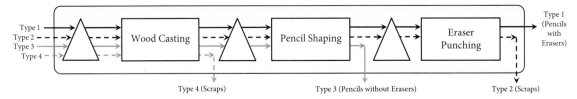

Figure 4.6    Process-flow chart of Cool Pens in Example 4.7 with four product types.

Note that the Cool Pens production process is a multi-stage process with two products and yield loss. There are two types of defective pencils, one is produced at the wood casting stage and the other at the eraser punching stage. One approach to compute the effective capacity of the plant is to consider those defective pencils as product types that are not sold to customers and are scrapped. Thus, the plant will have four types of products:

$$\text{Type-1 Product} = \textit{Good pencils with erasers}$$
$$\text{Type-2 Product} = \textit{Defective pencils with erasers at eraser punching}$$
$$\text{Type-3 Product} = \textit{Good pencils without erasers}$$
$$\text{Type-4 Product} = \textit{Defective pencil slats at wood casting}$$

Figure 4.6 shows how each product type flows within the plant.

Following our analytics for finding capacity of processes with multiple stages and multiple products, the first step is to define our flow unit and time unit. We consider our flow unit to be a batch of size 9 pencils of each type, and our time unit to be 1 hour. We also need to determine the product mix. If there was no yield loss, the product mix was $(0.7, 0.3)$, representing the fraction of pencils produced with and without erasers, respectively. However, because we now have four product types, we need to revise the product mix to incorporate yield loss—fraction of defective products.

Recall that product mix $p_i$ for product $i$ is the fraction of the total output of a process that are of type $i$. Because we have four products, we have product mix $(p_1, p_2, p_3, p_4)$ that must satisfy the following constraints:

- Of course, sum should add up to one:

$$p_1 + p_2 + p_3 + p_4 = 1$$

- *Proportion of good products:* We know that 70 percent of the good pencils that will be sold (i.e., type-1 and type-3 products) must be pencils with erasers, thus,

$$p_1 = 0.7(p_1 + p_3)$$

- *Yield loss at wood casting:* As Fig. 4.6 shows, all four products are processed at the wood casting stage. With yield loss of 5 percent, 5 percent of all products processed

Table 4.9    Computing the Effective Capacity of Cool Pens in Example 4.7

| Resources | Effective mean process time | | | | Effective mean process time of the aggregate product (*in seconds*) | Effective capacity (*batches per hour*) | Effective capacity (*pencils per hour*) |
|---|---|---|---|---|---|---|---|
| | Type 1 | Type 2 | Type 3 | Type 4 | | | |
| Wood casting | 5 | 5 | 5 | 5 | 5 | 720 | 6480 |
| Pencil shaping | 10 | 10 | 10 | 0 | 9.5 | 378.95 | 3410.5 |
| Eraser punching | 3 | 3 | 0 | 0 | 2.01 | 1791.04 | 16119.4 |

in this stage will be defective (will be product type 4). Hence,

$$p_4 = 0.05(p_1 + p_2 + p_3 + p_4)$$

- *Yield loss at eraser punching:* As Fig. 4.6 also shows, two products (type-1 and type-2 products) are processed at the eraser punch stage. With yield loss of 3 percent, 3 percent of those products will be defective (will be type-2 products). Therefore,

$$p_2 = 0.03(p_1 + p_2)$$

Putting all these together, we have

$$
\begin{aligned}
p_1 + \quad p_2 + \quad p_3 + \quad p_4 &= 1 \\
0.3p_1 \quad\quad - \quad 0.7p_3 \quad\quad &= 0 \\
0.05p_1 + 0.05p_2 + 0.05p_3 - 0.95p_4 &= 0 \\
0.03p_1 - 0.97p_2 \quad\quad\quad &= 0
\end{aligned}
$$

Solving the above four equations with four unknown $p_1, p_2, p_3$, and $p_4$, we find the product mix for the four products as follows:

$$(p_1, p_2, p_3, p_4) = (0.651, 0.020, 0.279, 0.050)$$

With the above product mix, we can follow the analytics for multi-stage, multi-product processes and compute the effective capacity of the Cool Pens' plant. Note that all product types have the same effective process time at each stage. Table 4.9 shows the effective mean process times as well as the effective capacity of each resource (each stage).

As Table 4.9 shows, the bottleneck of the process is the pencil shaping stage with the capacity of making 3410.5 pencils per hour. Thus, the capacity of Cool Pens' plant is

$$C_{\text{eff}}^{(agg)}(\text{plant}) = 3410.5 \text{ aggregate good and defective pencils per hour}$$

Recall, however, that *effective capacity* is the maximum number of *good* (non-defective) units that a process can produce. Hence, to get the effective capacity of the plant, we need to find its capacity to produce good products. Using product mix $(p_1, p_2, p_3, p_4) = (0.651, 0.020, 0.279, 0.050)$, we have

$$
\begin{aligned}
C_{\text{eff}}^{(1)}(\text{plant}) &= p_1\, C_{\text{eff}}^{(agg)}(\text{plant}) = 0.651(3410.5) = 2220.2 \quad \text{Type-1 products per hour} \\
C_{\text{eff}}^{(2)}(\text{plant}) &= p_2\, C_{\text{eff}}^{(agg)}(\text{plant}) = 0.020(3410.5) = 68.2 \quad\;\; \text{Type-2 products per hour} \\
C_{\text{eff}}^{(3)}(\text{plant}) &= p_3\, C_{\text{eff}}^{(agg)}(\text{plant}) = 0.279(3410.5) = 951.5 \quad\;\; \text{Type-3 products per hour} \\
C_{\text{eff}}^{(4)}(\text{plant}) &= p_4\, C_{\text{eff}}^{(agg)}(\text{plant}) = 0.050(3410.5) = 170.5 \quad\;\; \text{Type-4 products per hour}
\end{aligned}
$$

Therefore, the plant has the effective capacity of producing 2220.2 good pencils with erasers and 951.5 good pencils without erasers per hour—a total of 3171.7 pencils per hour. On the other hand, the plant produces $68.2 + 170.5 = 238.7$ scrapped pencils in an hour. Thus, the plant has yield of

$$\text{Plant Yield} = \frac{\text{effective capacity of producing good units}}{\text{total effective capacity of the plant}} = \frac{3171.7}{238.7 + 3171.7} = 0.93$$

Consequently, the plant's yield loss is $1 - 0.93 = 0.07 = 7$ percent.

One final note to make is that the effective capacity of the process (the plant) is 3171.7 per hour, which is not equal to the effective capacity of its bottleneck—pencil shaping—which has effective capacity of 3410.5 per hour. The reason is that the process loses some of the pencils produced at the bottleneck (i.e., about 68.2 pencils per hour) in the eraser punching stage. Hence, as this example shows, in processes with yield loss, the effective capacity of the process may be lower than the effective capacity of its bottleneck.

## 4.12 Summary

Throughput of a process has a direct impact on sales, and therefore on profit. Throughput of a firm, on the other hand, is closely related to a firm's capacity. Excess capacity is an investment with no return, and shortage of capacity simply means loss of profit due to loss of sales. Therefore, setting the right capacity for a process is of critical importance. One factor in making good capacity decision is to know what the capacity of the current system is. This chapter focuses on fundamental concepts and measures of capacity in processes with multiple products and multiple stages.

One benchmark measure of capacity is theoretical capacity. Theoretical capacity of a resource is the long-run average number of outputs that the resource can produce per unit time in ideal situations where the resource is fully utilized and does not idle because of any reasons (e.g., lack of material, breakdowns, setups). Effective capacity of a resource, on the other hand, is the long-run average number of good (non-defective) outputs that the resource can produce per unit time if the resource is fully utilized and, as in real world, is subject to interruptions such as breakdowns and setups. The gap between theoretical capacity (the ideal case) and effective capacity (the actual case) represents the potential for the maximum increase in the capacity of a resource.

The capacity of a process can be computed by finding the capacity of each of its resources. The bottleneck resource of a process is the resource of the process with the smallest capacity. When a process produces different products, each having a different processing time with the resources of the process, product aggregation method can be used to compute the capacity of the process. The insights provided by the method is that the change in the product mix can change the capacity of the process. The product aggregation method for processes with multiple products can also be used to find the capacity of processes with yield loss. The fundamental methods and concepts developed in this chapter lead us to identify strategies to improve the capacity of a process discussed in Chap. 6.

## Discussion Questions

1. What is the difference between theoretical process time and effective process time of a resource?

2. What is the difference between theoretical capacity and effective capacity of a resource?

3. Describe percent capacity improvement potential and explain how it can help managers make capacity improvement decisions.

4. Briefly explain how product aggregation method determines the capacity of a process with multiple products.

5. What is yield loss of a process? How product aggregation method can be used to compute the capacity of a process with yield loss?

## Problems

1. Consider a stamping press that makes large metal rings (i.e., washers) for nuts and bolts. The coil of steel sheet is fed to the process automatically and each strike of the process, which takes about 3 seconds, makes 4 washers. Each coil is enough to make 8000 washers. Thus, after making 8000 washers, the process stops and a new coil is fed to the process. This takes about 5 minutes. The press is subject to occasional breakdowns and requires some minor repair. Specifically, the data shows that after working for an average of 20 hours, the press breaks and must be repaired. The repair takes an average of 2 hours. The press is used 8 hours in each day to produce washers.
   a. Compute theoretical and effective capacity of the press based on the number of washers the press can produce in a day.
   b. What is the capacity improvement potential for this press?

   The company is considering purchasing a new press that can work with larger coils producing 5 washers in each strike, which also takes 3 seconds. Similar to the current press, the new press requires a new coil after making 8000 washers. Feeding the new coil into the press takes an average of 5 minutes. The new press is not expected to fail if the recommended preventive maintenance operations are performed. Specifically, the press must be stopped after every 2 hours of operations and some adjustments and lubrication must be performed, which takes about 10 minutes.
   c. Putting the purchasing and replacement costs aside, should the company replace the current press with the new press?

2. The shipping department of an appliances manufacturer in Chicago works three shifts, 24 hours a day. The department has two loading stations to load products (e.g., refrigerators, washers, dryers) to trucks. When a truck arrives, it takes about 20 minutes for the truck to park at the loading station and is set up for loading. The products are then loaded to the truck, which takes 1 minute per product. Trucks have maximum capacity of holding 20 products. After being loaded, trucks leave the shipping department to deliver the products to retailers in and outside Chicago. It takes an average of 8 hours for a truck to deliver the products and return to the shipping department for loading. The company has a fleet of 30 trucks. However, the trucks are rather old, and have occasional breakdowns. Data shows that a truck spends about 230 hours on the road before it requires some substantial repair or maintenance, which takes between 3 hours and 2 days, with an average of 20 hours.
   a. Draw the process-flow chart of the operations of the shipping department.
   b. What is the capacity of the shipping department? Specifically, how many products the department can ship in a day?
   c. The manufacturer would like to increase its shipping capacity to 2000 products per day. To do that, it is considering to add more loading stations and to purchase new trucks that have the capacity of carrying 30 products. The new trucks do not break. The setup time to load these trucks is still 20 minutes. What is the minimum number of loading stations and trucks that the manufacturer needs to achieve the shipment capacity of 2000 products per day?

Table 4.10    Activities for Assembling Humid-X10 in Problem 3

| Activity | Description | Process time (minutes) | Predecessor activities |
|----------|-------------|------------------------|------------------------|
| A | Set up the base and power knob | 5 | — |
| B | Install the heating device | 2 | A |
| C | Connect and test wirings | 5 | B |
| D | Assemble thermostat | 6 | B |
| E | Install demineralization cartridge | 6 | A |
| F | Test thermostat and power knob | 3 | C,D |
| G | Assemble the water tank | 3 | E |
| H | Final assembly and inspection | 4 | F,G |

3. FreshAir is a producer of a wide range of commercial and industrial humidifiers. One of their product that is designed for small rooms is called Humid-X10. The demand for the product has increased and is estimated to be 420 units per week. The main activities required for assembling Humid-X10 and their mean effective process times are presented in Table 4.10. Since the assembly operations are mainly manual operations performed by workers using simple tools, workers are able to perform any of the activities in Table 4.10.

In the table, the predecessors of Activity F, for example, are Activities C and D, indicating that Activity F cannot be performed until Activities C and D have already been completed. Currently, Activities A and B are done by Worker 1, Activities C and D by Worker 2, Activities E and F by Worker 3, and Activities G and H by Worker 4. FreshAir works in two shifts (i.e., 12 hours a day) and 7 days a week.

    a. With the current four workers, can FreshAir satisfy the demand of 420 per week?

    b. Can FreshAir satisfy its weekly demand with only three workers doing the assembly?

4. Consider J.C. Nickel store in Problem 6 of Chap. 2 that has a checkout counter with six cashiers and a self-checkout counter with two machines. Recall that, on an average, 120 customers per hour need to check out their items. Since more customers are using the self-checkout machines, the store is planning to close one of its six cashier stations and instead add a third self-checkout machine. With this change, it is expected that about 30 out of 120 customers will use the self-checkout machines. Also, after training cashiers and redesigning the bagging process, it is expected that the average time a cashier spends to scan, to receive payment, and bag a shopper's items will be reduced to 2 minutes. The average time a customer uses a self-checkout machine is expected to remain at 3 minutes. Self-checkout machines are subject to random occasional problems. Data shows that a self-checkout machine stops working, on an average, every 20 minutes. It takes an average of 2 minutes for some store staff to go to the machine and fix the error. If the store closes one of its cashier stations and adds a self-checkout machine, what would be the maximum number of customers that check-out process (cashiers and self-checkout together) can serve in an hour?

5. Great Italiano is a small Italian restaurant that has 30 tables, each table seating between three and six customers. When a party of customers arrive—which is often less than six people—a hostess greets them and seats them, if a table is available. This takes about 1.5 minutes. The customers wait for an average of 3 minutes on their table, until one of the three servers shows up to take their orders. It takes an average of 3 minutes to take customers' orders. It takes an average of 10 minutes for a chef in the kitchen to prepare an order for a party. The kitchen has four chefs, each works on a different order. The server then takes the food to the customers, which takes about 3 minutes. Data shows that depending on the size of the party, it takes between 20 and 60 minutes—with an average of 40 minutes—for a party to finish eating and

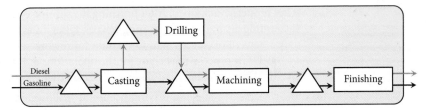

**Figure 4.7    Process-flow chart of cylinder blocks in Problem 6.**

ask for the check. The cashier brings the check, processes the payment, and provides receipt. This takes about 2 minutes. The customers then leave the restaurant, and the server cleans the table for the next customers. This takes about 2 minutes. If the restaurant has one hostess and one cashier, how many parties the restaurant can serve in an hour?

6.  A manufacturer produces cylinder blocks for diesel and gasoline engines in its Detroit plant. About 80 percent of its demand is for gasoline engine blocks and 20 percent is for diesel engine blocks. Blocks are produced in four departments in the plant as shown in Fig. 4.7. Gasoline blocks are produced in batches of size 4 and diesel blocks are produced in batches of size 2. Before the casting department starts to produce any batch of gasoline blocks, a setup time of 30 minutes is needed to start the production. The setup time for producing any batch of diesel blocks in the casting department is 15 minutes. After setting up, it takes an average of 45 minutes to cast a gasoline block and 60 minutes to cast a diesel block. About 10 percent of casted blocks require some rework, which is also done in the casting department. Performing rework does not need setups and takes an average of 10 minutes. The drilling department performs specific drilling operations only on diesel blocks. The average drilling time is about 15 minutes for each block. The machining department has a single CNC machine that processes both types of blocks. The average machining times for gasoline and diesel blocks are 90 minutes and 75 minutes, respectively. The CNC machine, however, is subject to random failures, with an average time to failure of 8 hours and average repair time of 30 minutes. The processing time of a block in the finishing department is 2 hours for gasoline and diesel blocks. No setups are required in the drilling, machining, and finishing departments.

The plant works for two 8-hour shifts in each day with 1-hour break during each shift.

   a.  What is the maximum number of cylinder blocks that the manufacturer can produce each day? How many of those cylinder blocks are diesel blocks?

   b.  A consulting firm claims that by implementing lean operations principles, they can increase the capacity of the Detroit plant by eliminating setups and rework in the casting department and eliminating the CNC breakdowns at the machining department. What do you think about this claim?

7.  Happy Wash (HW) is a car wash that advertises on its detailed cleaning performed by professional staff. HW offers three services: (i) wash, (ii) wash & vac, and (iii) full service. Data shows that 50 percent of customers choose wash service, 30 percent choose wash & vac service, and 20 percent choose full service.

Wash service includes washing and drying. Washing is done by an automatic car wash equipment and drying is done by workers using towels. Wash & vac service includes wash service (i.e., washing and drying) and then vacuuming inside the car. Full service includes wash & vac service and shine service, which is performed after wash & vac. Shine service includes cleaning and waxing interior of the car and its wheels.

The time it takes to wash a car is 2 minutes; it takes an average of 3 minutes to dry the car; vacuuming the interior takes an average of 4 minutes; and shining interior and wheels takes an average of 5 minutes. When shinning the interior of a car, workers often find that the car is not properly vacuumed. After finishing the shining operations, those cars are sent back for vacuuming again. This vacuuming rework takes an average of 2 minutes. Data shows that only 5 percent of cars are sent to vacuum for rework.

    a. Draw process-flow chart for the operations at Happy Wash.

    b. What is the maximum number of cars that Happy Wash can serve in an hour?

8. The process of renewing driving license at Willowglen County DMV office is as follows. Upon arrival, applicants wait in the line for a clerk to review their documents and check if they have violations and restrictions on their driving license. If that is the case, the applicant leaves the DMV office to resolve those issues. It takes a clerk about 2.5 minutes to review each applicant's documents. Those applicants who have proper documents with no violations, that constitutes 85% of the applicants, are then sent to the cashier to pay the renewal cost. It takes the cashier an average of 1 minute to process the payment. Applicants then go for vision screening test, which takes an average of 2 minutes. About 10 percent of the applicants who take the vision screening test fail their tests. Those applicants are required to leave and bring an eye report from an eye care professional and come back and take the test with glasses or contact lenses. Those applicants come back after about 3 days and take the vision screening test again, and this time they pass the test. The last stage after vision screening is taking photos and printing the license. This takes an average of 3 minutes. The DMV office has two clerks who review the documents, two cashiers who process payments, two agents who perform the vision screening test, and four agents that use four cameras and printers to issue the license.

    a. Draw process-flow chart of Willowglen DMV office for renewing driving license.

    b. What is the maximum number of driving licenses that the DMV office can renew in an hour?

    c. If 45 people come to DMV office in each hour, what fraction of time each resource (i.e., clerks, cashiers, vision screening agents, cameras, and printers) are working?

9. Heavenly Treat is a hair and body treatment salon that provides hair services such as haircuts, perms, colors, shampoo, curling; and facial treatment such as removing facial hair, fixing eyebrows, waxing, and facial massage. Customers can be divided into three groups: (i) Group-1 customers want facials and hair styling, (ii) Group-2 customers want hair styling, and (iii) Group-3 customers are walk-in customers who want a simple haircut. Groups 1, 2, and 3 constitute 20 percent, 70 percent, and 10 percent of the customers, respectively.

    All arriving customers must first go to the receptionist to register. It takes an average of 1 minute for the receptionist to register each customer. The registration time, however, is 2 minutes for Group-1 customers who require both facial and hair styling. After registration, Group-1 customers are sent to a skin care therapist for facial services. Facials take an average of 30 minutes. After receiving their facials, Group-1 customers go to the shampoo station to have their hair washed. Washing and drying hair takes about 5 minutes. They are then sent to a senior hair stylist to have their hair done. It takes a senior hair stylist an average of 40 minutes to finish serving a customer. The customer then goes to the cashier to pay for the services, which takes about 3 minutes.

    After the reception, a Group-2 customer goes to the shampoo station, a senior hair stylist, and the cashier. The service times of Group-2 customers at this stages are the same as those for Group-1 customers. Group-3 customers—after reception and the shampoo station—are sent to a junior hair stylist to get their haircuts. The average time for a haircut is about 20 minutes. There are one receptionist, one hair washer, five senior hair stylists, one junior hair stylist, and one cashier.

    a. Draw process-flow chart of the operations in Heavenly Treat.

    b. Where is the bottleneck of the operations and what is the maximum number of customers that the salon can serve in an hour?

10. Security check in a small airport works as follows. Passengers wait in a line for about 3 minutes to show their boarding pass and ID to one of the two security officers. It takes the officers about 15 seconds to check a passenger's ID. The passengers then wait in another line for about 1 minute to put their luggage on one of four conveyors for checking with one of the four x-ray machines. The passenger then goes to one of the two metal detector machines. It takes an x-ray machine 40 seconds to check a passenger's luggage. After x-ray machines, about 10 percent of the passengers are asked to open their luggage for further checking. This is done by an agent and it takes the agent about 60 seconds to check the luggage. The time it takes a metal detector machine to check a passenger is 20 seconds. The passenger then takes the luggage and goes to the gate.

    a. Draw process-flow chart of the airport security.
    b. What is the maximum number of passengers that the airport security can check every hour?

11. Cert Mobile is a subsidiary of a major mobile phone manufacturer that refurbishes used phones and sells them as certified used phones with 2-year warranty. The operations at Cert Mobile are performed in its main repair center in Chicago that includes the following four departments: (i) inspection, (ii) repair, (iii) testing, and (iv) refurbish and packaging. Used phones are first inspected in 1 of the 10 inspection stations in the inspection department. It takes 30 minutes for a station to check whether a phone needs major repair or minor repair and maintenance. If the phone requires minor repair, the phone is sent to 1 of the 20 repair stations in the repair department of Cert Mobile. It takes an average of 60 minutes for a repair station to repair a phone. The repaired phone is then sent to the testing department for final check. The testing department has eight testing stations. The time it takes for a testing station to test a phone is about 30 minutes. The testing department data shows that about 5 percent of the repaired phones have issues and must be sent back to the repair department for rework. It takes about 20 minutes for one of the repair stations in the repair department to fix the issue. After the issue is fixed, the phone is sent directly to one of the eight stations in the refurbish and packaging department.

    If the inspection department finds that a phone requires major repair, the phone is shipped to the manufacturer's facility in Detroit to be fixed. Data shows that about 15 percent of the phones are sent to the manufacturer. Also, 10 percent of the phones that are sent to the manufacturer are found to be non-repairable (i.e., too costly to repair) and are discarded. The manufacturer has the capacity of processing (i.e., repairing) 20 phones in an hour, including those that are discarded. After the manufacturer fixes a phone, the phone is shipped back to the refurbish and packaging department of Cert Mobile.

    It takes an average of 30 minutes for a station in the refurbish and packaging department to process a phone. The department has eight stations. At the end of the working day, the packaged phones are sent to the Cert Mobile's distribution center to be distributed to its retailers. The repair center works two shifts—a total of 14 hours a day.

    a. Draw process-flow chart for the operations in Cert Mobile repair center.
    b. What is the maximum number of phones that Cert Mobile can ship to its distribution center in a day?
    c. Cert Mobile is planning to establish additional stations in its four departments to be able to ship 280 phones to its distribution center. What is the minimum number of additional stations needed in each department to achieve this?

# CHAPTER 5

# Variability

## 5.0 Introduction

Variability is perhaps the most difficult issue to deal with in managing and controlling business processes. If it is not properly managed, variability can have significant impact on all process-flow measures. Without variability, businesses can make more profit and customers would be more satisfied. Unfortunately, variability is present in all processes to some extent. Hence, it is essential for the operations engineers and managers to understand variability and its sources and to find ways to either reduce variability, or better yet eliminate it. If variability cannot be eliminated or reduced, then they need to find strategies to diminish the negative impact of variability on operational performance measures.

This chapter focuses on introducing the concept of variability, measuring variability, identifying the common sources of variability, and suggesting ways to eliminate or reduce variability. We start by discussing what variability is.

## 5.1 What Is Variability?

Variability is defined as the quality of nonuniformity of a class of entities.[1] The concept of variability is closely related to "Variation." To illustrate this, consider the following three

---

[1] Hopp, W.J. and M.L. Spearman, *Factory Physics*. Third Edition, Waveland Press, Inc., 2008.

firms that produce the same product and have the same capacity of producing 700 units per month. The total demand for the product of the three firms has the same average of 650, but each firm has a different demand pattern as follows:

**Firm A:**    The demand for the product in each of the next 7 months is known with 100 percent certainty to be 650 units, that is,

Firm A's demand    :    650, 650, 650, 650, 650, 650, 650

with a total of 4550 units.

**Firm B:**    Firm B's demand in each of the next 7 months are known with 100 percent certainty, but it varies from month to month. Specifically, it is known for sure that the demand in the next 7 months are

Firm B's demand    :    600, 1000, 850, 400, 300, 500, 900

a total of 4550 units.

**Firm C:**    While the *average* monthly demand is known to be 650 units—the same as that for Firms A and B—Firm C's monthly demand for the product is not known with 100 percent certainty. Its monthly demand can be any number between 300 and 1000.

Let's compare the demand in each of the above three cases. While there is no *variation* in the monthly demand of Firm A, the demands of Firms B and C have variations. However, there is an important difference between the variation in Firm B's and Firm C's demands. Firm B's demand has *known variation*, that is, variation in Firm B's demand in the next 7 months is known with certainty. However, Firm C's demand has *random variation*, that is, the demand in each of the 7 months is random and can be any number between 300 and 1000 units.

Both types of known and random variations degrade system's performance. However, dealing with random variation is more difficult and requires different types of analytics to manage. We refer to both known and random variations in this book as "variability."

---

**CONCEPT**

**Variability**

Variability of an entity refers to the variation in the characteristics of the entity. The variations are either known in advance or are random and unpredictable.

---

While the above example discusses variability in demand, variability can be present in many places in a process. For example, there is variability in the process times (e.g., the time it takes to weld a piece varies each time); there is variability in the delivery times (e.g., how long it takes to receive your order varies each time); there is variability in the quality of a product (e.g., the weight of 1.5-ounce potato chips bag varies from bag to bag). These variations, as we mentioned, make it very difficult for the operations engineers and managers to plan and control their processes and to provide high-quality products and services to the customers.

## 5.2 Impact of Variability on Process Performance

To show how variability affects process performance, we return to our example of the three firms.

**Firm A's Operations:** The production planning and control is easy for Firm A. With monthly demand of exactly 650 units and monthly capacity of exactly 700 units, the firm is capable of satisfying the monthly demand with the current capacity. There is no need to carry inventory in a month to satisfy the demand in the future months. There is also no possibility of having shortage to satisfy the demand. An ideal operations!

**Firm B's Operations:** Similar to Firm A, Firm B has the same average monthly demand of 650 units and knows the exact demand in each of the next 7 months. Firm B, however, faces more challenging operations. This is due to variability in demand. Recall that Firm B has a monthly capacity of 700 units. But, the demand in the second of the 7 months is 1000, which is 300 units above the firm's capacity. How can the firm satisfy this demand? Well, it needs to produce more than the demand of 600 units in the first month and then carry the additional inventory to satisfy the demand in the second month. Since the total demand in the first 2 months is 1600, which is 200 units above the firm's 2-month capacity, the firm also needs to acquire an additional 200 units of capacity in the first month, or in the second month, or in both the months. For example, the firm may need to work overtime or to hire additional part-time workers to increase its capacity. If the firm cannot find ways to increase its capacity (at least temporarily), it fails to satisfy some of the demand in the second month. Hence, compared to Firm A, Firm B will have additional inventory and capacity costs, and possibly a delay in satisfying the demand. These are all because Firm B faces variability in its demand—even though the variation in the demand is known with certainty.

**Firm C's Operations:** Because of the variability in its demand, Firm C will face similar dilemma as Firm B, resulting in an increase in inventory and capacity costs. In addition to these issues caused by *variation* in the demand, Firm C also needs to deal with the issues caused by *randomness* in the demand. As opposed to Firm B that knows the demand in the next 7 months when it plans its monthly production, Firm C does not even know the exact demand in the future months when it makes its production plan for the current month. Thus, to safeguard against shortages of products in the next month, Firm C may need to carry some additional inventory in case the demand in the next month is large. Thus, compared to Firm B, Firm C will end up carrying more inventory, incurring more capacity cost, and having a higher chance of unsatisfied or delayed demand.

The above examples lead to the following important principle:

## PRINCIPLE

### Impact of Variability on Process Performance

Variability anywhere in a process—regardless of being due to known or random variation—degrades process-flow measures such as inventory, flow times, or throughput. The randomness in variation further exacerbates the impact and adds additional complexity in managing and controlling the process.

Before we conclude this section, we should emphasize that there are situations where adding variability can result in an increase in a firm's profit. For example, while offering a variety of products increases the variability in production times—since products are different—it can boost demand and increase a firm's market share, resulting in higher profit. Nevertheless, increase in variability of production times would still degrade process-flow measures and create more issues in managing and controlling the process. The challenge is to find effective strategies to improve process-flow measures while still offering a wide range of products and services. We will discuss some of these strategies in Chap. 14.

## 5.3 Measuring Variability

Several metrics are used to measure variations in an entity. Consider the 7-month demands for Firm B in our example, that is, 600, 1000, 850, 400, 300, 500, and 900. How should we measure variability in this demand? One measure, which is called *Range* (shown by $R$), computes the difference between the maximum and minimum values of an entity as a measure of variation. In our example, the range for 7-month demands of Firm B is $R_B = 1000 - 300 = 700$. The larger the range, the larger the variability. But there is an issue that makes Range not a good candidate for measuring variability in the operations systems.

The issue is that Range does not use all the data and only uses the maximum and minimum values to measure variability. For example, consider Firm D with the following 7-month demands: 1000, 300, 1000, 1000, 1000, 1000, and 1000. The maximum and minimum values for Firm D's demand are also 300 and 1000, respectively, resulting in the same $R_D = 1000 - 300 = 700$. But can we conclude that variability in demand of Firm B and Firm D is the same? Of course not. Firm D has a much smaller variability in its demand than Firm B. Specifically, in 6 out of 7 months, Firm D's demand is 1000 (i.e., no variation). But Firm B's demand varies every month. Range does not take into account how all seven demands vary and only considers the maximum and minimum demands which are the same for both the firms.

A measure of variation that takes into account all the demand data to measure the variability is *Variance* (shown by $\sigma^2$). The square root of variance is called *Standard Deviation* (shown by $\sigma$). Instead of measuring how spread the minimum and maximum values are from each other—as Range does—variance measures how much demand in each month deviates (i.e., varies) from the mean of the demand. Consider Firm B's demand. The mean—the average—of the 7-month demands is $(600 + 1000 + \cdots + 900)/7 = 650$. Thus, the first month deviation from the mean is $600 - 650 = -50$, the second month deviation from the mean is $1000 - 650 = 350$, and so on. The variance is then the average of all squared deviations,[2]

$$\sigma_B^2 = \frac{(600 - 650)^2 + (1000 - 650)^2 + \cdots + (900 - 650)^2}{6} = 72{,}500$$

and thus the standard deviation of Firm B's demand is $\sigma_B = \sqrt{72,500} = 269.26$. The higher the standard deviation, the higher the variation in the demand.

While the standard deviation is a better representation of variations of an entity than range, it has its own issue. Consider the following two series of numbers representing 7 months of demand for two different firms

| Firm E's demand | : | 70, 80, 90, 100, 110, 120, 130 |
| Firm F's demand | : | 970, 980, 990, 1000, 1010, 1020, 1030 |

Two questions: (i) Do both firms have the same standard deviation in their demands? (ii) Do both firms have the same variability in their demands?

If you calculate the standard deviation for both demands, you find that they have the same standard deviation of 20. The reason is that both demands have the same deviations from their means. Note that the mean monthly demand for Firms E and F are 100 and 1000, respectively. For Firm E, the first month's deviation from its mean of 100 is $70 - 100 = -30$. Firm F's first month deviation from its mean of 1000 is also $970 - 1000 = -30$. This is the same for all the other 6 months.

---

[2] Variance uses the squared deviation to prevent negative and positive deviations from canceling each other out. One can use the average of the absolute deviations. The resulting measure is called Mean Absolute Deviation (MAD).

But, does having the same standard deviation imply same variability in demand? No. Note that for Firm E, the first month's demand is about $30/100 = 30$ percent lower than its mean monthly demand. Thus, the monthly demand varies between 30 percent lower than mean (in the first month) and 30 percent higher than mean (in the seventh month). For Firm F, however, the demand varies between 3 percent lower than its mean and 3 percent higher than its mean, which is a significantly lower variation compared to the mean. In fact, this small 3-percent variation does not affect system performance as much compared to the 30-percent variation and makes it easier to manage and control the process.

Therefore, a better way to measure the variability is a relative measure that describes the magnitude of the variation compared to the mean. This is called *Coefficient of Variation* and is shown by CV as follows:

$$\text{Coefficient of Variation} = CV = \frac{\text{Standard Deviation}}{\text{Mean}}$$

Overall, a random variable (e.g., demand for a product or process time of an activity) is considered to have *low variability* if its coefficient of variation is less than 0.75; it is considered to have a *moderate variability* if its coefficient of variation is between 0.75 and 1.33; and it is considered to have a *high variability* if it has a coefficient of variation above 1.33.

Coefficient of variation is shown to have a critical role in the analysis and management of manufacturing and service operations systems. It appears alongside of mean demand and effective capacity in most of the analytics used to manage these systems and it reveals very critical insights that are useful in understanding the dynamics of the operations.

## 5.4 Variability in Demand

One source of variability in almost all the operations systems is the demand. Retailers do not know exactly how many customers will come to each of their stores each day, auto manufacturers do not know exactly what the demand will be for each of their car models next year, and airlines do not know how many passengers will buy tickets for each of their flights next week.

Since we know that variability degrades system performance, it is important that we also know what causes variability in demand and to explore strategies that can help reduce variability. These are the goals of this and the following sections. We first identify the sources of variability in demand. As mentioned in Chap. 3, there are four typical demand patterns: (i) stationary demands, (ii) demands with trend, (iii) seasonal demand, and (iv) cyclic demand. The underlying reasons for variability in cyclic demand are factors such as fashion trends, politics, war, economic conditions, or sociocultural influences, which are very difficult to predict. Hence, here we focus on the other three demand patterns (see Fig. 5.1).

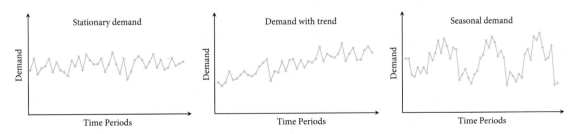

Figure 5.1    Three demand patterns: stationary demand, demand with trend, and seasonal demand.

*Sources of Variability in Stationary Demand*

Recall from Chap. 3 that the stationary demands are those for which the mean demand is steady and does not change significantly over time. Figure 5.1-left shows a typical stationary demand pattern. Examples of stationary demand are the weekly demands for products such as bread, milk, detergent, etc. in a supermarket. The main source of variability in the stationary demand is customers making independent decisions, each having a different probability of needing the product each week. Consider the demand for a box of breakfast cereal at your supermarket. There are probably a limited number of customers who go to your supermarket to buy that cereal. The customers' consumptions of cereal and their purchasing habits are independent and different from each other, resulting in choosing different times to purchase cereal. Therefore, the demand for cereal becomes random and varies from week to week. We call this inherent variability in demand *Natural Variability* in demand.

*Sources of Variability in Demand with Trend*

Compared with the stationary demand in Fig. 5.1-left, we realize that there are two main sources of variability in the demand with trend in Fig. 5.1-middle: (i) natural variability in demand, and (ii) variability due to increase in mean demand. Consider our cereal example and suppose that the supermarket now also sells a new cereal brand. The producer of the new cereal runs an advertising campaign on TV to introduce its new cereal. If a successful product, the weekly demand for the new cereal would have an increasing pattern as in Fig. 5.1-middle. There is still natural variability in the weekly demand—random variation in the number of customers who buy the new cereal each week. However, there is also variability in the average demand in the consecutive weeks; the average weekly demand is increasing in time. As opposed to stationary demand in which the average weekly demand of, say first 5 weeks, is about the same as the average demand of last 5 weeks, when demand has an increasing trend, the average weekly demand in the last 5 weeks is higher than the average demand in the first 5 weeks.

Note that a demand can also have a decreasing trend—the mean demand decreases with time. This corresponds to the demand at the end of product life cycle or when the product is not successful in the market. Regardless of being increasing or decreasing, the changes in mean demand, as mentioned above, is the second source of variability in demand with trend.

*Sources of Variability in Demand with Seasonality*

Consider the seasonal demand in Fig. 5.1-right. There are similarities and differences between this and the demand with trend in Fig. 5.1-middle. Both demand patterns have natural variability. Also, in both the patterns the mean demand changes over time. However, while the average demand is increasing in demand with trend, in seasonal demand the average demand is increasing and then decreasing in a cyclic manner. This cyclic behavior—the seasonality—is another source of variability which only occurs in seasonal demand.

Consider the demand for ice cream, cold drinks, or bag of ice in a supermarket. As the weather becomes warmer (in spring and summer), the average monthly demand for these products increases. Then, as the cold seasons (fall and winter) approach, their average monthly demand decreases. This pattern repeats itself each year in a cyclic manner.

A seasonal demand can also have an increasing (or decreasing) trend. Consider a new flavor of ice cream that is introduced to the market and is becoming very popular. While the average monthly demand for this ice cream is higher in summer than in winter (i.e., seasonality), the average monthly demand in summer would be higher than that in previous summer—an increasing trend.

## 5.5 Measuring Variability in Demand

Variability in demand can be measured in two different ways: (i) measuring variability in the number of demands per unit time, and (ii) measuring variability in the demand interarrival times. To illustrate this, three different demand scenarios for 25 orders received by a pizza shop are considered in Table 5.1.

The data in the table corresponds to the arrival times of each order between 10:00 am and 4:00 pm, except for the first order that arrived before 10:00 am.

### 5.5.1 Variability in the Number of Demands per Unit Time

Let's find the variability in the number of demands per unit time. To do this, we first need to choose our unit of time. The unit can be a minute, a 15-minute period, a 30-minute period, or an hour. The choice of the unit depends on the goal of the analysis and how detailed the analysis is. Let's choose an hour as our time period in this example.

Table 5.1    Order Arrival Times to Pizza Shop under Scenarios A, B, and C

| Order number | Scenario A | Scenario B | Scenario C |
|:---:|:---:|:---:|:---:|
| 1 | 9:45 | 9:56 | 9:50 |
| 2 | 10:45 | 10:50 | 10:15 |
| 3 | 10:55 | 11:05 | 10:44 |
| 4 | 11:05 | 11:19 | 11:05 |
| 5 | 11:22 | 11:22 | 11:20 |
| 6 | 11:24 | 11:40 | 11:28 |
| 7 | 11:43 | 11:42 | 11:42 |
| 8 | 11:54 | 11:48 | 11:53 |
| 9 | 12:05 | 12:09 | 12:05 |
| 10 | 12:10 | 12:25 | 12:12 |
| 11 | 12:17 | 12:40 | 12:18 |
| 12 | 12:19 | 12:45 | 12:27 |
| 13 | 12:21 | 12:56 | 12:34 |
| 14 | 12:35 | 1:09 | 12:43 |
| 15 | 12:44 | 1:38 | 12:50 |
| 16 | 12:47 | 1:48 | 12:57 |
| 17 | 12:59 | 1:55 | 12:59 |
| 18 | 1:12 | 2:06 | 1:13 |
| 19 | 1:32 | 2:13 | 1:20 |
| 20 | 1:42 | 2:30 | 1:52 |
| 21 | 2:16 | 2:52 | 2:12 |
| 22 | 2:22 | 2:55 | 2:45 |
| 23 | 3:07 | 3:10 | 3:05 |
| 24 | 3:20 | 3:15 | 3:40 |
| 25 | 3:28 | 3:41 | 3:48 |

Define random variable

$$D_t = \text{Number of orders received in time period } t$$

and assign time Period $t = 1$ as the hour between 10:00 am and 11:00 am, Period $t = 2$ as the hour between 11:00 am and 12:00 pm, and so on. By counting the number of orders received within each hour in Table 5.1, we find

$$\text{Scenario A:} \quad D_1 = 2, \quad D_2 = 5, \quad D_3 = 9, \quad D_4 = 3, \quad D_5 = 2, \quad D_6 = 3$$
$$\text{Scenario B:} \quad D_1 = 1, \quad D_2 = 6, \quad D_3 = 5, \quad D_4 = 4, \quad D_5 = 5, \quad D_6 = 3$$
$$\text{Scenario C:} \quad D_1 = 2, \quad D_2 = 5, \quad D_3 = 9, \quad D_4 = 3, \quad D_5 = 2, \quad D_6 = 3$$

Note that the first order in the three scenarios occurred before 10:00 am, so we did not count that. We can find mean $(\overline{D})$ and variance $(\sigma_D^2)$ of the number of orders placed in an hour in Scenario A as follows[3]:

$$\overline{D} = \frac{2 + 5 + 9 + 3 + 2 + 3}{6} = 4$$

$$\sigma_D^2 = \frac{(2-4)^2 + (5-4)^2 + (9-4)^2 + (3-4)^2 + (2-4)^2 + (3-4)^2}{5} = 7.2$$

Hence, the demand in an hour has standard deviation $\sigma_D = \sqrt{7.2} = 2.7$. So, we can measure the variability in the demand in Scenario A by the coefficient of variation of the number of demand per unit time as $\text{CV}_A = 2.7/4 = 0.68$. Similarly, for Scenarios B and C we find

$$\text{Scenario B}: \overline{D} = 4, \quad \sigma_B = 1.8, \quad \text{CV}_B = 1.8/4 = 0.45$$
$$\text{Scenario C}: \overline{D} = 4, \quad \sigma_C = 2.7, \quad \text{CV}_C = 2.7/4 = 0.68$$

The above numbers show that while all three scenarios have the same average demand during an hour, Scenarios A and C have the same variability (coefficient of variation) of 0.68 in the number of arrivals in an hour, while Scenario B has a lower variability of 0.45.

### 5.5.2 Variability in the Demand Interarrival Times

Our above analysis shows that Scenarios A and C have the same demand variability. However, looking at the order arrival times, we observe that the arrival times for the 25 orders are different under these two scenarios. Then, how can these scenarios have the same variability?

To further analyze these two scenarios, let's get a visual picture of the arrivals in all three scenarios, as in Fig. 5.2.

As Fig. 5.2 depicts, while the number of order arrivals during an hour is the same in all six time periods in Scenarios A and C, the arrivals have different patterns. Consider time period between 10:00 am and 11:00 am (Period $t = 1$) as an example. In that period, both Scenarios A and C have two order arrivals. The two order arrivals are spread out during the period in Scenario C, but they are clamped at the end of the period in Scenario A (both occur in the last 15 minutes of the period). So, to have a more detailed analysis of the demand arrivals, we need to take into account how demands are spread out within a time period. There are two ways to capture this. One approach is to choose a smaller time unit, for example, measuring variability in the number of orders in 15-minute periods. Another approach

---

[3]We use sample variance, in which the denominator in the formula is $n - 1 = 6 - 1 = 5$. This is because for small sample sizes, sample variance is an unbiased estimator of the variance of the population.

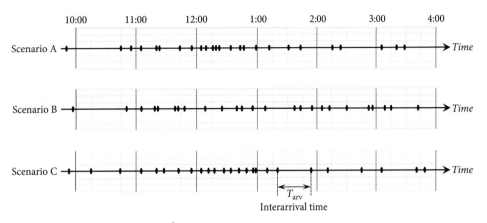

**Figure 5.2**    Demand (Order) arrival epochs to pizza shop for Scenarios A, B, and C.

is to measure variability in the time between two consecutive order arrivals. Specifically, if we define the interarrival time of an order as follows:

$$T_{arv} = \text{Time between the arrival of an order and the arrival of the next order}$$

then we can compute the interarrival times of all orders as shown in Table 5.2. The table clearly shows that Scenarios A and C, while having the same variation in the number of orders arriving in an hour, have different variations in their order interarrival times.

Table 5.2 also shows mean, standard deviation (STD), and coefficient of variation (CV) of interarrival times for the three scenarios. As these numbers show, although Scenarios A and C have the same variability in the number of demand per unit time—in an hour—Scenario C has lower variability in its interarrival times than Scenario A.

### 5.5.3  Which Variability Measure to Choose?

We discussed two different metrics to measure the variability in the demand of a process—one is the coefficient of variation of the number of demands per unit time, and the other is the coefficient of variation of demand interarrival times. Which one is better for measuring variability in demand?

Well, each metric measures the variability in demand from a different perspective, and thus is used to model variability in demand in different operations processes. In processes where the number of arrivals in each time period has more significant impact on process performance, the variability in demand per unit time becomes more critical. As an example, consider a supermarket that orders fresh bread from a local bakery every day and sells them in its store. The supermarket sends an order to the bakery every day and the bakery delivers the order early morning of the next day when the supermarket opens. To prevent stockout of bread and to decide how many breads to order each day, the manager of the supermarket would like to know the variability in demand for bread. Which variability measure is critical to the manager: Variability in the number of customers who buy bread each day (i.e., demand per day) or the variability in interarrival times of the customers who want to buy bread each day?

Obviously, the number of demands in a day is more important to the manager than the interarrival times of demand for bread. For example, if the manager orders 15 loaves of bread for tomorrow, and tomorrow's demand is for 18 loaves, the store will have a shortage of 3 loaves, resulting in loss of profit. The variability in demand interarrival times does not matter in this case, since the store would have shortage of 3 loaves, regardless of whether all

Table 5.2    Interarrival Times (in Minutes) of Orders to Pizza Shop in Scenarios A, B, and C

| Order Number | Scenario A | Scenario B | Scenario C |
|---|---|---|---|
| 1 | — | — | — |
| 2 | 47 | 54 | 25 |
| 3 | 10 | 15 | 29 |
| 4 | 10 | 14 | 21 |
| 5 | 17 | 3 | 15 |
| 6 | 2 | 18 | 8 |
| 7 | 19 | 2 | 14 |
| 8 | 11 | 6 | 11 |
| 9 | 11 | 21 | 12 |
| 10 | 5 | 16 | 7 |
| 11 | 7 | 15 | 6 |
| 12 | 2 | 5 | 9 |
| 13 | 2 | 11 | 7 |
| 14 | 14 | 13 | 9 |
| 15 | 9 | 29 | 7 |
| 16 | 3 | 10 | 7 |
| 17 | 12 | 7 | 2 |
| 18 | 13 | 11 | 14 |
| 19 | 20 | 7 | 7 |
| 20 | 10 | 17 | 32 |
| 21 | 34 | 22 | 20 |
| 22 | 6 | 3 | 33 |
| 23 | 45 | 15 | 20 |
| 24 | 13 | 5 | 35 |
| 25 | 8 | 26 | 8 |
| Mean | 13.75 | 14.38 | 14.92 |
| STD | 12.16 | 11.14 | 9.69 |
| Variability (CV) | 0.88 | 0.78 | 0.65 |

18 demands arrive in the hour after the store opens, or they are spread out evenly during the 14 hours that supermarket is open during the day.

Now consider our pizza shop example. Which one of the two variability measures is more critical for the shop manager: Variability in the number of orders placed in an hour or variability in the interarrival time of orders? In this case, the latter is the key measure. Consider an hour in which 10 orders arrive. Does it matter to the manager of pizza shop if all these 10 orders arrive at the first 10 minutes of the hour (e.g., with 1-minute interarrival times between orders), or if they arrive throughout the hour (e.g., with 6-minute interarrival times)? In both cases, the pizza shop faces the demand of 10 orders in the hour.

To investigate this, suppose it takes the pizza shop 5 minutes to make a pizza for an order (suppose pizzas are made one-by-one). In the latter case with interarrival times of 6 minutes, each order is finished 1 minute before the next order arrives. In other words, no order would wait in the buffer of the process. In the former case with interarrival times of 1 minute, however, except for the first order, all other orders must wait in the buffer for some time before the shop can start working on them. For example, the last order which arrives at the 10th minute of the hour, must wait for the completion of 9 pizzas, which occurs at time $9 \times 5 = 45$th minute of the hour. Hence, the waiting time of the 10th order in the buffer is

$45 - 10 = 35$ minutes. This simple example shows why variability in interarrival times of the demand has more significant impact on pizza shop operations than variability in the number of demands per unit time (i.e., an hour).

The example of the supermarket and pizza shop are typical examples of cases in which variability in demand per unit time and variability in demand interarrival times, respectively, are the key demand variability measures. In processes such as the supermarket in which the main goal is to minimize *inventory* and corresponding costs, the demand for products is often defined as demand per unit time (e.g., daily demand, weekly demand). Consequently, the variability in the number of demands per unit time is more critical than variability in interarrival times. In fact, standard deviation and coefficient of variation (variability) of demand per unit time appear in all analytical models of inventory management and control systems in Chaps. 9, 10, and 11.

On the other hand, in processes such as the pizza shop in which the main goal is to minimize *flow time* (e.g., order waiting time) and its corresponding costs, the variability in interarrival times is more critical. This is because, to compute flow times in such processes, one needs to have information about when each demand arrives in the process. Demand interarrival times provide such information. That is the reason why standard deviation and coefficient of variation (variability) of interarrival times appear in all of the analytical models used in managing and controlling of flow times in Chaps. 7 and 8.

Before we conclude this section, we must emphasize that, obviously, there is a relationship between variability in demand interarrival times, and variability in demand per unit time. The demands with lower variability in their interarrival times would have lower variability in their demands per unit time. For example, if the interarrival times have a mean of 5 minutes with a very small variability (e.g., interarrival times vary between 4.95 and 5.05 minutes), one would expect to see about 12 demands per hour with very low variability.

There is no exact formula to capture the relationship between these two demand variability measures, but there is an approximate relationship that becomes more accurate as the number of demand during a time period increases. Specifically, if the demand in each period has the same distribution and is independent of each other, and if the demand is stationary over time with mean $\overline{D} \geq 10$ per unit time, then we have

$$CV_{arv} \simeq CV \times \sqrt{\overline{D}}$$

where $CV_{arv}$ is the coefficient of variation (i.e., variability) of interarrival times and CV is the coefficient of variation (i.e., variability) of demand in a period.

## 5.6 Reducing Variability in Demand

In Sec. 5.2, we explained that variability in demand degrades system performance, and in Sec. 5.4, we identified the sources of variability in demand. In this section, we introduce strategies to deal with variability in demand.

We divide these strategies into two groups: (i) strategies that are developed to eliminate or reduce variability in demand, and (ii) strategies that are developed to reduce the negative impact of variability on process performance by synchronizing capacity with demand.

### 5.6.1 Strategies to Reduce Variability in Demand

There are several strategies that operations engineers and managers can take to eliminate or reduce variability in demand for goods and services, some of which we present here.

**Scheduling Demand**

Demand scheduling refers to practices that are developed to directly manage the time a demand arrives. Two examples of demand scheduling are Appointment and Reservation Systems and Job Release schedule.

- *Appointment and Reservation Systems:* Many service industry such as health care, hospitality (e.g., hotels and restaurants), and airlines use appointment and reservation systems, for example, making an appointment to see a doctor, or making a reservation for a room in a hotel or for a seat in a flight. Appointment and reservation systems have the following two goals: (i) spreading demand arrivals into certain time intervals (i.e., reducing variability in interarrival times), and (ii) making demand arrivals more predictable. For example, by allocating 20 minutes for a doctor to see a patient, doctor offices arrange appointments such that one patient arrives (approximately) every 20 minutes. In some cases, for example, dentist offices, each patient may need different service times. Some patients need 20 minutes for a simple procedure and some need 45 minutes for more complex procedures. In these cases, patient arrivals are scheduled to match their service times. Note that, although there are still variations in patient interarrival times, the variation is predictable (assuming patients do not come too early or too late). In other words, the random variation in demand arrivals is eliminated or reduced.

- *Job Release Scheduling Systems:* In manufacturing systems, an analogous strategy to Appointment and Reservation systems is Job Release Scheduling systems. Consider a produce-to-order system with variability in orders it receives from its customers (i.e., variability in demand). To start processing an order, raw material and parts (i.e., jobs) are released into the process. These jobs go through several stages of production and assembly to make the final product. Job Release Schedule refers to the timing of releasing jobs corresponding to orders into the system. One strategy is to release them immediately after an order is placed by the customer. If there is high variability in order arrivals (i.e., demand), this creates large inventories in the process and long flow times for each order (see Chap. 7). Another option is to release material for each order in evenly spaced time intervals (e.g., every minute or every hour). This reduces the variability in job interarrival times and thus the variability in the number of jobs released into the process per unit time, and consequently reduces inventory and flow time in the system.

Several points are worth mentioning here. The impact of demand scheduling practices such as appointment systems in reducing variability depends on how much customers stick to their appointment times. Late arrivals or no shows result in some variability in arrivals. One practice to prevent this is to send customers reminders for their appointments. Most doctor offices and hospitals send automatic reminders to their patients the day before the appointment day.

Second, appointment and reservation systems raise customer expectations. We do not mind to go to a hair salon at 11:00 am (without appointment) and wait for 30 minutes to get a haircut. But if we already have an appointment at 11:00 am, we would be very unhappy if we end up waiting for 30 (or even 15) minutes to get a haircut. Thus, if a process adopts an appointment system, it must adhere to the appointment schedule as much as possible.

Third, job release scheduling systems are designed to deal with natural variability in demand. However, while they do eliminate or reduce natural variability in *demand arrivals to a process*, it does not affect the natural variability in demand itself. For example, by releasing a job corresponding to an order every hour to a factory, a firm can eliminate the natural variability in job (i.e., order) arrival to its factory. However, this does not eliminate the natural variability in customer orders that the firm receives. In other words, there might not be many orders under process in the factory, but there might be lots of orders waiting (in firm's computer systems) to be released into the factory. What is the implication? Well, the short flow times of orders in the factory do not represent the flow time of a customer order. The flow time of a customer order consists of two parts: (i) the time the order is waiting until it is released to the factory, and (ii) the time the order spends in the factory until it is shipped to the customer. By eliminating variability in order release to factory through job release schedules, the latter is reduced. To reduce the former, other strategies (e.g., increasing factory capacity) can be considered.

### Providing Incentives

While demand scheduling strategies such as appointment systems tend to reduce natural variability in demand, incentive-based strategies try to reduce variability due to seasonality. By providing monetary or nonmonetary incentives, firms try to convince customers to use their services or products during the periods with low demand instead of periods with high demand. For example, airlines and hotels offer cheaper price in low seasons, restaurants offer discount in happy hours during which the demand is low. These monetary incentives are designed to shift some of the demand from periods with high demand to periods with low demand and hence reduce the variability in demand in a period—that is, reduce seasonality in demand.

One example of nonmonetary incentive is Fastpass in Disneyland. The incentive provided to the customers is the lower waiting time. Customers get a Fastpass ticket on which a time window for customer "return time" is identified. When customers return during that time, they will have priority to enter the attraction over those customers without Fastpass. Thus, the Fastpass system can shift the demand during busy hours of an attraction to hours with lower demand.

Other examples of nonmonetary incentive are online and self-service facilities that offer lower or no waiting times. By encouraging customers to use these services, firms can eliminate some of the demand in their peak-demand hours, and thus reduce the variability due to seasonality. Examples are ATM machines, online banking, and online government services such as driving license and immigration services.

One source of variability in demand is that customers often switch their purchases from one firm to another, resulting in unpredictable customer demand. One way to reduce this phenomenon is through *loyalty program* under which customers receive some discount in their next purchase if their past accumulated purchases reach a certain level. One example is frequent-flyer programs that provide a free ticket after a customer accumulated certain points (i.e., miles). Another example is loyalty programs in restaurants that offer free food after customers' past purchases accumulate to a certain level. Note that while providing incentive shifts the demand from one period to another, loyalty programs tend to generate more demand and reduce variability by keeping a steady set of customers.

### Reducing Batch Size

Batch arrivals inflate the variability in the arrival process. For example, consider a workstation in which one job is released to the station exactly every minute. Hence, the job interarrival times are 1, 1, 1, 1, 1, . . ., which have zero variability. Suppose that the worker at the station has effective process time of exactly 0.9 minutes (no variability). Hence, the job arrival rate is one job per minute and the worker's utilization is 0.9—the worker is working 90 percent of time. In this station, no jobs wait in the buffer (i.e., zero in-buffer inventory and flow times), since each job enters the station 0.1 minutes after the last job exits the station.

Now suppose that instead of releasing a job every minute, a batch of four jobs are released at exactly every 4 minutes. Similar to the above, the job arrival rate has an average of one per minute, and the worker utilization is 0.9. However, the variability in job interarrival times is not zero. The interarrival times between the last job of a batch and the first job of the next batch is 4 minutes, while the interarrival times between the first and the second, the second and the third, and the third and the last job of a batch is zero. Hence, the interarrival times would be 4, 0, 0, 0, 4, 0, 0, 0, 4, 0 , . . ., that have variability with known variations.

What does this variability do to the station's process-flow measures? Well, upon each batch arrival the station will have an inventory of 4 jobs. The inventory is reduced by one every 0.9 minutes and reaches zero after $4 \times 0.9 = 3.6$ minutes. As opposed to the case with single arrivals, under the batch arrivals all jobs in a batch must wait in the buffer for some time—except for the first job of the batch. Therefore, the station will have a higher average inventory and flow times compared to those in the case with single arrivals. Reducing the batch size from 4 to 2 or to 1 improves these process-flow measures.

As this example shows, batch size reduction reduces the variability in arrivals and thus improves process performance. Reducing batch sizes has other important benefits that we will discuss in Chap. 15.

### Aggregating Demand

One commonly used strategy to reduce variability in demand is "Demand Aggregation." The premise is that the variability in the *sum* of several demands is lower than the variability of each individual demand. To illustrate this, we need the following analytics.

---

**ANALYTICS**

**Variability of Aggregated Demand**

Suppose demand for Item $i$ per unit time (e.g., per day, per week, or per month) is random variable $X_i$ that follows distribution $f_i(x)$ with mean $\mu_i$, standard deviation $\sigma_i$, and coefficient of variation $CV_i = \sigma_i/\mu_i$. Also, there is a correlation $\rho_{ij}$ between the demands for Item $i$ and Item $j$, where $i, j = 1, 2, \ldots, n$. Then, the total demand of all $n$ items per unit time is random variable $Y$, where

$$Y = X_1 + X_2 + \cdots + X_n.$$

The total demand per unit time $Y$, has mean $\mu_y$, standard deviation $\sigma_y$, and coefficient of variations $CV_y$ as follows:

$$\mu_y = \sum_{i=1}^{n} \mu_i \tag{5.1}$$

$$\sigma_y = \sqrt{\sum_{i=1}^{n} \sigma_i^2 + 2 \sum_{j>i} \rho_{ij}\sigma_i\sigma_j} \tag{5.2}$$

and variability of the aggregated demand is $CV_y = \sigma_y/\mu_y$.

To see how aggregating demand reduces variability in demand per unit time, let's assume that all $n$ items have demand per unit time with the same mean $\mu$, standard deviation

$\sigma$, and variability $CV = \sigma/\mu$. Then, according to the above analytics, we have $\mu_y = n\mu$, and

$$\sigma_y = \sqrt{n + 2\sum_{j>i} \rho_{ij}} \times \sigma \qquad (5.3)$$

and hence,

$$CV_y = \frac{\sigma_y}{\mu_y} = \frac{\left(\sqrt{n + 2\sum_{j>i} \rho_{ij}}\right)\sigma}{n\mu} = \sqrt{\frac{1}{n} + \frac{2}{n^2}\sum_{j>i} \rho_{ij}} \times CV$$

As the above equation shows, the variability of the aggregated demand (i.e., $CV_y$) is always less than the variability in demand of each item (i.e., $CV$). But why? This is because there is a good possibility that when one item has peak demand in a period, another item (or items) has low demand in that period. The low demand of an item and the high demand of another item cancel each other, resulting in less variability in the total (aggregated) demand. This possibility is higher if the demand for items are negatively correlated. Only when the demand for all $n$ items are perfectly positively correlated, that is, $\rho_{ij} = 1$ for all $i$ and $j$, that the aggregated demand in a period will have the same coefficient of variation (variability) as that of the demand for each item. Note that the case of a perfect positive correlation between the demand for two items is very rare in practice. Thus, one should expect that, in real world, aggregated demand will always have lower variability.

When demands for all $n$ items in a period are independent of each other (i.e., $\rho_{ij} = 0$ for all items $i$ and $j$), the standard deviation of the aggregated demand is simply $\sigma_y = \sqrt{n} \times \sigma$, and variability of the aggregated demand is $CV_y = (1/\sqrt{n})\, CV$.

Reducing variability in demand by aggregating demand has been widely used in different industries. When Walmart closes two of its stores (that are in close proximity) and opens a superstore, the demand of the superstore will be the aggregated demand, and will have lower variability than the demand in each of the two stores. Car manufacturers aggregate demand in different ways. When a manufacturer reduces the number of colors offered from 20 to only 8 colors, it is indeed aggregating the demand for 20 items to demands for 8 items. The benefits of demand aggregation is discussed in more detail in Chap. 11.

The analytics in this section shows how the variability in the number of demand per unit time can be computed if several demands are aggregated into one. But how is the variability in demand interarrival times computed for the aggregated demand? For that, see Sect. 5.8.3.

### 5.6.2 Strategies to Reduce the Impact of Demand Variability by Synchronizing Capacity

Demand scheduling and incentive strategies aim to directly reduce variability in demand. Synchronizing capacity, however, refers to strategies that tend to reduce the impact of demand variability on operational performance—they do not affect demand variability. The main idea is to change capacity to match the variation in demand. Two main approaches to synchronize capacity are: (i) using flexible resources, and (ii) changing capacity.

**Using Flexible Resources**

Flexible resources such as cross-trained workers can be used to match capacity with demand, especially in dealing with natural variability in demand. Almost all checkout counters in fast-food restaurants use this strategy. When demand increases in a short period of time resulting in longer lines for checkout, store calls some of its staff performing other tasks to open a new checkout counter. When the line becomes short, the counter is closed and the staff returns to his or her other tasks.

**Changing Capacity**

One major difference between natural variability and variability due to seasonality is the following. While variation due to natural variability is not predictable, the variation due to seasonality (i.e., variation in mean demand) is predictable to some extent. For example, a fast-food restaurant does not know exactly when the next customer enters its store, but it has some idea about the changes in the average demand during low-demand and high-demand periods. Hence, while having two flexible workers might be enough to handle natural variability in demand during low-demand hours, this flexibility alone is not enough to handle demand during peak-demand hours. The store needs to increase its capacity during those hours. How do managers provide the additional capacity to deal with peak-demand periods? The answer depends on whether one can predict when peak-demand periods occur. By having a forecast of demand during the low- and peak-demand periods, managers can schedule the capacity (e.g., the number of checkout staff) needed during those intervals. Call centers, for example, use their forecast of the call arrival rate in each 15- or 30-minute period during each day of the week and determine the number of agents needed to keep the waiting times below a certain level. By designing shift schedules, managers of call centers provide the additional capacity needed to deal with the high demand. Another approach is to use part-time employees during the peak-demand periods. In manufacturing systems, overtime and subcontracting components are other common strategies used.

Now that we have identified sources of variability in demand and discussed strategies to either reduce its variability or to mitigate its impact on process performance, in the following section we focus on sources of variability in process times.

## 5.7 Variability in Process Times

There is always variability in process times in all operations systems. There is variability in how long it takes to cut or to drill a part, to assemble, to pack, or to ship a product. There is always variability in how long it takes to file a tax report, or to make a hamburger for a customer in a restaurant. Unless a task is performed by a robot that is never interrupted (e.g., due to failure or setups or defective material) and unless the robot performs the same task on all jobs, there would always be variability in process times.

### 5.7.1 Sources of Variability in Effective Process Times

Recall from Chap. 4 that an activity consists of main tasks and supporting tasks. Main tasks are the primary objective of performing an activity and supporting tasks are essential to perform the main task. Also, recall that the time it takes to perform the main task is called Theoretical Process Time and the time it takes to do a supporting task is called Supporting Process Time. Effective process time of a task is the sum of theoretical and supporting process times.

As an example, consider a sanding station that performs the activity of sanding the surface of 40-inch wooden square pieces used in making dining tables. In the sanding station, the sanding activity is performed by a worker using an oscillating edge sanding machine. The activity of sanding consists of the main task of sanding, and supporting tasks of repairing the sanding machine (if it fails) and cleaning the machine's work table (when needed). Table 5.3 presents a time study of sanding 20 pieces. Note that it includes theoretical process times, supporting process times, and total process times (i.e., effective process time) of the sanding activity for the 20 pieces.

The supporting tasks are done either in the middle of the main task, or before the main task starts. For example, before the sanding of piece number 4 started, it was interrupted because the worker spent 2 minutes to clean the work table of the sanding machine. Then

Table 5.3    **Study of Process Times for Sanding 20 Pieces**

| Piece number | Theoretical process time (*minutes*) | Supporting process time (*minutes*) | Effective process time (*minutes*) | Supporting tasks |
|---|---|---|---|---|
| 1 | 1 | 0 | 1 | |
| 2 | 1.2 | 0 | 1.2 | |
| 3 | 1.4 | 0 | 1.4 | |
| 4 | 1 | 2 | 3 | *Cleaning work table* |
| 5 | 1 | 0 | 1 | |
| 6 | 1.2 | 20 | 21.2 | *Repairing sanding machine* |
| 7 | 1.1 | 0 | 1.1 | |
| 8 | 1.2 | 0 | 1.2 | |
| 9 | 1.4 | 0 | 1.4 | |
| 10 | 1 | 1.8 | 2.8 | *Cleaning work table* |
| 11 | 1.1 | 0 | 1.1 | |
| 12 | 1 | 0 | 1 | |
| 13 | 1.2 | 0 | 1.2 | |
| 14 | 1 | 0 | 1 | |
| 15 | 1.2 | 0 | 1.2 | |
| 16 | 1.2 | 0 | 1.2 | |
| 17 | 1.1 | 0 | 1.1 | |
| 18 | 1 | 0 | 1 | |
| 19 | 1.3 | 3 | 4.3 | *Cleaning work table* |
| 20 | 1.1 | 0 | 1.1 | |
| Average | 1.13 | | 2.47 | |
| STD | 0.13 | | 4.49 | |
| CV | 0.11 | | 1.81 | |

sanding of the piece started and took 1 minute. Hence, the effective process time of the sanding activity for piece number 4 is $1 + 2 = 3$ minutes (also shown in the table), which is the sum of theoretical and supporting process times.

The situation with piece number 6 was a bit different. Its sanding (i.e., the main task) was started without interruption. However, the sanding machine failed in the middle of sanding and it took the worker 20 minutes to repair the machine. The worker then resumed sanding piece number 6. The total theoretical process time (before and after repair) was 1.2 minutes. So, the effective process time for piece number 6 was $1.2 + 20 = 21.2$ minutes.

Recall that we defined the average of the total process time of an activity, as Effective Mean Process Time, which is $T_{\text{eff}}(sanding) = 2.47$, as shown in Table 5.3. But, what is the variability in the process time of the sanding activity?

**CONCEPT**

**Effective Variability of an Activity**

Effective variability of activity $j$ corresponds to the variations in the total time it takes to perform that activity. Considering $T_{\text{eff}}(j)$ as effective mean process time, and $\sigma_{\text{eff}}(j)$ as the standard deviation of the effective process time of the activity, the effective variability of the activity is

$$CV_{\text{eff}}(j) = \frac{\sigma_{\text{eff}}(j)}{T_{\text{eff}}(j)}$$

As the table shows, the effective process times of sanding of the 20 pieces has a standard deviation (denoted by STD in Table 5.3) $\sigma_{\text{eff}}(sanding) = 4.49$; hence, the effective variability of the sanding activity is

$$\text{CV}_{\text{eff}}(sanding) = \frac{\sigma_{\text{eff}}(sanding)}{T_{\text{eff}}(sanding)} = \frac{4.49}{2.47} = 1.81$$

which implies a high variability (i.e., $\text{CV} = 1.81 > 1.33$). What are the sources of this high variability in the effective process time?

Well, since effective process time includes time to perform the main task (i.e., theoretical process time) and time to perform the supporting tasks (supporting process times), we need to explore the sources of variability in these tasks. We start with theoretical process time.

### Sources of Variability in the Main Task

As Table 5.3 shows, there is some variability in theoretical process times corresponding to the variability in performing the main task of sanding. In general, the sources of variability in main tasks are:

- *Human Performance:* One source of the variation in theoretical process time is the fact that it is done by a worker. As opposed to robots, humans cannot finish performing a task repeatedly with the same quality and in exactly the same amount of time. While a robot finishes sanding each of the 20 pieces in exactly 1 minute, it takes the worker between 1 and 1.4 minutes to sand a piece.
- *Job Condition:* Another factor that often results in variation in theoretical process time is the condition of the job. For example, there might be variation in theoretical process time of sanding because some pieces are smoother (need less sanding) and some are rougher (need more sanding).
- *Task Complexity:* Complexity of a task is another factor that contributes to the variations in theoretical process times. Routine and standard tasks have less variations but complex and not well-defined tasks have higher variations. For example, there is low variability in sticking a shipping address on a package, but there is a high variability in the task of diagnosing why a machine failed.

We call the variability in the theoretical process time Natural Variability.

---

**CONCEPT**

### Natural Variability of an Activity

Natural variability of an activity corresponds to the variations in the theoretical process time of the activity, which is the time to perform the main task. Considering $T_0(j)$ as the theoretical mean process time of activity $j$ and $\sigma_0(j)$ as the standard deviation of theoretical process time of activity $j$, the natural variability of the activity is

$$\text{CV}_0(j) = \frac{\sigma_0(j)}{T_0(j)}$$

The sanding activity has theoretical mean process time $T_0(sanding) = 1.13$ with standard deviation $\sigma_0(sanding) = 0.13$; thus, the natural variability of the sanding activity is

$$\text{CV}_0(sanding) = \frac{\sigma_0(sanding)}{T_0(sanding)} = \frac{0.13}{1.13} = 0.11$$

which implies a low variability (i.e., $\text{CV}_0 = 0.11 < 0.75$).

Which variability is more important, natural variability or effective variability? Well, while effective variability presents the variability in performing an activity, knowing natural variability can help managers in planning variability reduction effort. For example, suppose that effective variability of an activity is $\text{CV}_{\text{eff}} = 2$. This implies that the activity has high variability. If you also know that the natural variability of the activity is $\text{CV}_0 = 1.8$, then you know that the majority of the variability in the activity is in its main task. Supporting tasks only add 0.2 to the coefficient of variation. However, if the natural variability of the activity is $\text{CV}_0 = 0.4$, then the variability reduction effort should mainly be directed to eliminating or reducing variability in the supporting tasks.

In conclusion, it is worth mentioning that another source of natural variability—which is the direct consequence of management decision—is product mix. Producing more than one type of products increases variability of the main task. Consider our sanding example and suppose the worker now sands two different pieces: small and large pieces. Sanding a large piece takes an average of 2 minutes and sanding a small piece takes an average of 1.1 minutes. In this case, the natural variability is now due to: (i) (natural) variation in the process time of each type, and (ii) variation in the process times when the worker switches from one piece to another.

### Sources of Variability in Supporting Tasks

In addition to variation in the main task (i.e., natural variability), supporting tasks also contribute to effective variability of an activity. Supporting tasks bring two additional sources of variability (in addition to natural variability) to effective process times:

(i) There is a random variation in the time the supporting task is needed. As Table 5.3 depicts, after cleaning the work table for piece number 4, six pieces were sanded until it was time to clean the table again (i.e., before sanding piece number 10). However, the next time the cleaning was needed was after sanding 9 pieces (i.e., before sanding piece number 19). Thus, it is not known for sure how many pieces will be sanded until the worker needs to clean the work table. This is also true for supporting task of repairing the machine. After repairing the machine and finishing piece number 6, it is not known for certain when exactly the machine will fail again and needs repair.

(ii) There is also random variation in the length of the supporting task. Consider the supporting task of cleaning the work table. As Table 5.3 depicts, this task takes 2, 1.8, and 3 minutes for pieces number 4, 10, and 19, respectively. It is not known, with certainty, how long the next cleaning task will take. Same is true for supporting task of repairing the machine, which took 20 minutes when it interrupted sanding piece number 6. There is often high variability in repair tasks and thus it is not clear how long the next repair task would take.

The above two points explain, in general, how supporting tasks contribute to the effective variability of an activity. However, to better understand the relationship between the parameters of a supporting task of an activity (e.g., the length of a repair operation or the length of a setup operation) and the effective variability of the activity, we need to develop some analytics. These analytics will also help us develop strategies to reduce variability.

Note that supporting tasks are nothing but interruptions in preforming the main task. We can divide these interruptions into two groups. In both groups the interruption is inevitable, but the difference is the time they occur.

- *Time-Based Interruptions:* Time-based interruptions are those that can occur according to a time schedule. The time schedule can be known in advance, or can be random. For example, most machine failures are time-based interruptions with random time schedule. We know that the machine is going to fail some time in the future, but we do not know exactly when. In contrast, preventive maintenance that stops a machine every 120 minutes and performs some adjustment for 10 minutes is an example of a time-based interruption with known schedule—we know exactly when the next interruption will occur, that is, in 120 minutes. The time-based interruptions can interrupt a task at the beginning, in the middle, or at the end of it. Some other examples of time-based interruptions are power outages and running out of material (e.g., in the middle of assembling parts, the worker runs out of one of the parts).

- *Task-Based Interruptions:* Task-based interruptions will also occur some time in the future, but they occur according to a task schedule, that is, occur after finishing a number of tasks. The task schedule can be known in advance or can be random. Consider, again, a preventive maintenance operation consisting of lubricating a machine after finishing every 20 tasks. Hence, the number of tasks between two maintenance is exactly 20. This is an example of a task-based interruption with known interruption schedule. Now consider another preventive maintenance under which the worker is told to lubricate the machine—before staring the next task—if the worker hears friction noises or detects excessive heat. This is an example of a random interruption schedule. It is not known in advance when exactly the worker will hear friction noise. Thus, the number of tasks between two maintenance is random. But what is not random is the fact that the maintenance will be performed at the end of a task—not during a task. Other examples of task-based interruptions are setups, worker breaks (e.g., lunch breaks or breaks for meetings), rework, etc., which all occur mostly after finishing a job, not during processing of a job.

We first present the analytics to find the effective variability of an activity which is subject to time-based interruptions. Since machine failures and repairs are good examples of time-based interruptions, we present our analytics in the context of a machine failure.

---

**ANALYTICS**

**Effective Variability for an Activity with Time-Based Interruptions**

Consider an activity performed by a machine. The machine is subject to failure. Specifically, after a machine failure is repaired, the time until next failure is random and is exponentially distributed with mean $m_f$. (Note that exponential distribution has coefficient of variation 1, representing a moderate variability.) When the machine fails, the time it takes to repair the machine has an average of $m_r$, standard deviation $\sigma_r$, and coefficient of variation $CV_r$. Hence, machine availability, as we discussed in Chap. 4, is

$$A = \frac{m_f}{m_f + m_r}$$

The machine performs an activity with theoretical mean process time $T_0$, standard deviation $\sigma_0$, and coefficient of variation $CV_0 = \sigma_0 / T_0$. The activity has no supporting task, expect for repairing the machine when it breaks. The effective mean process time, variance, and squared coefficient of variation of the effective process time of the activity can be computed as follows:

$$T_{eff} = \frac{T_0}{A} \tag{5.4}$$

$$\sigma_{eff}^2 = \left(\frac{\sigma_0}{A}\right)^2 + \frac{(m_r^2 + \sigma_r^2)(1-A)T_0}{Am_r} \tag{5.5}$$

$$CV_{eff}^2 = CV_0^2 + (1 + CV_r^2)A(1-A)\frac{m_r}{T_0} \tag{5.6}$$

### Example 5.1

Metal-Work is a firm that produces a wide range of metal components used by auto manufacturers, with specialty in friction welding. A friction welding machine in one of their bottleneck stations is operated by a worker and is used to make driver and trailer axles. It takes an average of 6 minutes with standard deviation 2 minutes for the worker to process a job. The machine, however, is very old and requires repair frequently. Specifically, the time to failure for the machine has a moderate variability (coefficient of variation 1) with an average time to failure of 36 hours. The time until the machine is repaired and put back to production has an average of 4 hours with standard deviation of 1 hour. Maintenance department has come up with a preventive maintenance (PM) plan that is expected to prevent current frequent random failures. The plan recommends that the operator should stop the machine every 90 minutes and perform some maintenance operations (e.g., inspection and adjustment, if needed). The maintenance operation is expected to take about 10 minutes with standard deviation 3 minutes. How does this PM plan affect the welding machine's performance?

The machine's performance is characterized by the machine's capacity and its effective variability. We first compute the machine's current capacity. We recognize that the machine is subject to time-based interruption with random interruption schedule following exponential distribution. So, we can use the above analytics to compute the effective mean process time (and thus effective capacity) as well as the effective variability of the machine.

Since machine works an average of $m_f = 36$ hours until it breaks and the average repair time is $m_r = 4$ hours, then it has availability

$$A = \frac{m_f}{m_f + m_r} = \frac{36}{36 + 4} = 0.9$$

Considering theoretical mean process time $T_0 = 6$ minutes, the machine has an effective mean process time and effective capacity

$$T_{eff} = \frac{T_0}{A} = \frac{6}{0.9} = 6.667 \text{ minutes} \implies C_{eff} = \frac{60}{6.667} = 9 \text{ jobs per hour}$$

What about the effective capacity under the PM plan? The PM plan is also a time-based interruption that interrupts the operations every 90 minutes to perform a 10-minute adjustment—a time-based interruption with known interruption schedule. The machine availability under the PM plan is $A = 9/(9+1) = 0.9$—the same as the current availability. Thus, under the PM plan the machine will also have the same effective capacity of 9 jobs per hour. Therefore, the PM plan does not affect the machine's effective capacity. But what about its effective variability?

Under current situation we know that the repair time has an average of $T_r = 4$ hour $= 60 \times 4 = 240$ minutes, and standard deviation $\sigma_r = 1$ hour $= 60$ minutes, and thus coefficient of variation $CV_r = 60/(240) = 0.25$. On the other hand, the machine has natural variability $CV_0 = \sigma_0/T_0 = 2/6 = 0.33$. Using Eq. (5.6), we can compute the effective variability of the machine as follows:

$$CV_{eff}^2 = CV_0^2 + (1 + CV_r^2)A(1 - A)\frac{m_r}{T_0}$$

$$= (0.33)^2 + (1 + 0.25^2)0.9(1 - 0.9)\frac{240}{6} = 3.93 \implies CV_{eff} = \sqrt{3.94} = 1.98$$

which is a high variability.

We now compute the effective variability of the machine under the PM plan. But, there are bad news and good news. The bad news is that our analytics is for the time-based interruption in which the interruption schedule (i.e., time to machine failure) is a random variable following exponential distribution. However, the interruption schedule under the PM plan is not random—it is known for certainty. What is the good news? The good news is that we can still use the analytics to have an upper bound on the effective variability as follows. We assume that the time to the next PM is exponentially distributed with an average of 90 minutes. We know that the PM operations has an average of $m_r = 10$ minutes and standard deviation of $\sigma_r = 3$ minutes. Hence, $CV_r = 3/10 = 0.33$. Using Eq. (5.6), we can compute the effective variability of the machine under the PM plan as follows:

$$CV_{eff}^2 = CV_0^2 + (1 + CV_r^2)A(1 - A)\frac{m_r}{T_0}$$

$$= (0.33)^2 + (1 + 0.33^2)0.9(1 - 0.9)\frac{10}{6} = 0.275 \implies CV_{eff} = \sqrt{0.275} = 0.52$$

which is a low variability.

Therefore, if our time-to-PM plan has moderate variability of one (i.e., exponentially distributed), the machine would have had an effective variability of 0.52. Since there is no variability in the time-to-PM plan—it is exactly 90 minutes—the effective variability of the machine would be even lower than 0.52. Hence, the PM plan would reduce the effective variability of the friction welding machine significantly, at least by a factor of about 4.

What is the benefit of this variability reduction? As we will discuss in Chap. 7, such a dramatic variability reduction—even though the capacity has not changed—can significantly improve process-flow measures of the machine such as inventory, flow time, and throughput. For example, if one job is released every 7.5 minutes to the welding machine, under current situation there would be a total average inventory of 15 jobs in the station and each job would wait an average of 1.75 hours in the buffer for its process to start. Under the PM plan, however, the total average inventory would be around two jobs and jobs would wait in the buffer for about 7.3 minutes for their process to start. This is a significant improvement in both inventory and flow time—all due to variability reduction.

Example 5.1 reveals an interesting and useful phenomenon about the time-based interruptions. Note that in both cases of current and after the PM plan, the machine has the same theoretical and effective capacities and the same availability. The only difference is that in the current situation interruptions are less frequent than those under the PM plan (i.e., an interruption occurs every 4 hours rather than every 90 minutes), but the length of interruption (i.e., repair times) is larger than that under the PM plan (4 hours rather than 10 minutes). This points to the fact that, from variability perspective, more frequent but shorter interruptions are preferred over less frequent but longer interruptions (given the same resource availability).

With this insight, we proceed with presenting our analytics for task-based interruptions. We also use an example of a machine with setups to derive the analytics regarding the effective variability for task-based interruptions. However, the analytics is general and works for other cases of task-based interruptions such as rework.

**ANALYTICS**

**Effective Variability for an Activity with Task-Based Interruptions**

Consider an activity performed by a machine to process a job. The machine requires a setup after processing an *average* of $N_s$ jobs. Setup time takes an average $T_s$ and standard deviation $\sigma_s$. Also, suppose that the probability of needing a setup after processing a job is the same. With an average number of jobs between setups being $N_s$, this implies that the probability of doing a setup after finishing a job is $1/N_s$.

The activity performed by the machine has theoretical mean process time $T_0$ with theoretical standard deviation $\sigma_0$ and coefficient of variation $CV_0 = \sigma_0/T_0$. The activity has no supporting task, except for the setup.

With this task-based interruption, the activity's effective mean process time, variance, and coefficient of variation (i.e., effective variability) of the machine can be computed as follows:

$$T_{\text{eff}} = T_0 + \frac{T_s}{N_s} \tag{5.7}$$

$$\sigma_{\text{eff}}^2 = \sigma_0^2 + \frac{\sigma_s^2}{N_s} + \left(\frac{N_s - 1}{N_s^2}\right) T_s^2 \tag{5.8}$$

$$CV_{\text{eff}} = \frac{\sigma_{\text{eff}}}{T_{\text{eff}}} \tag{5.9}$$

We use the following example to illustrate the above analytics.

 **Example 5.2**

LB is the manufacturer of small kitchen and home appliances. They use a metal sheet bending machine to make cases for their wide range of small- and medium-sized electrical ovens that are produced to order. If there is a large difference between the size of the current and the next case, the worker needs to perform a setup to make the machine ready for the next case. Past data shows that the worker performs a setup after processing an average of six cases, with the probability of needing a setup after finishing a case to be almost the same. The setup takes an average of 6 minutes with standard deviation of 3 minutes. The time to make a case of any size has an average of 2 minutes with standard deviation 1 minute. The firm is considering the following two approaches in their lean operations initiatives to reduce setups:

- *Setup Time Reduction:* Reduce the setup time by using additional offline tools and fixtures. This will reduce the average setup time from 6 minutes to 3 minutes with standard deviation of 1 minute. However, the setup will be performed more often after processing an average of three jobs.

- *Setup Elimination:* Redesign their cases such that there would be no need for setups. This, however, increases the (theoretical) mean process time at the machine from 2 minutes to 3 minutes and the standard deviation from 1 minute to 1.5 minutes.

How do the above two approaches affect the performance of the bending machine?

The setup operations in the bending machine is a task-based interruption in which the interruption schedule is random. In the current situation, the machine requires a setup after processing an average of $N_s = 6$ jobs. The setup time has mean and standard deviation of $T_s = 6$ and $\sigma_s = 3$ minutes, respectively. The time to process a job has mean and standard deviation $T_0 = 2$ and $\sigma_0 = 1$ minutes, respectively. Hence, the effective mean process time of the machine is

$$T_{\text{eff}} = T_0 + \frac{T_s}{N_s} = 2 + 6/6 = 3 \text{ minutes}$$

and the effective capacity is, therefore, $C_{\text{eff}}(\text{machine}) = 60/3 = 20$ jobs per hour.

Under setup time reduction, we have $N_s = 3$ and $T_s = 3$ minutes. Thus, the machine will have effective mean process time of $T_0 + T_s/N_s = 2 + 3/3 = 3$ minutes, resulting in the same effective capacity of 20 jobs per hour.

If the setup operations is eliminated, we have $T_s = 0$. However, the theoretical mean process time of the machine is increased to $T_0 = 3$ minutes. Since there are no setups, the effective mean process time of the machine would be equal to the theoretical mean process time. This results in the same effective capacity of 20 jobs an hour.

So, the two approaches do not affect the effective capacity of the machine. But, they certainly impact the effective variability. For the current situation, the variance of the effective process times can be computed using Eq. (5.8) as follows:

$$\sigma_{\text{eff}}^2 = \sigma_0^2 + \frac{\sigma_s^2}{N_s} + \left(\frac{N_s - 1}{N_s^2}\right) T_s^2 = 1^2 + \frac{3^2}{6} + \left(\frac{6-1}{6^2}\right) 6^2 = 7.5$$

The effective variability of the machine is then $CV_{\text{eff}} = \sigma_{\text{eff}}/T_{\text{eff}} = \sqrt{7.5}/3 = 0.91$.

Now, if the mean setup time is reduced to $T_s = 3$ minutes, with standard deviation $\sigma_s = 1.5$ minutes, and the setup operations needed after an average of $N_s = 3$ jobs, then the variance of the effective process time of the machine would be

$$\sigma_{\text{eff}}^2 = \sigma_0^2 + \frac{\sigma_s^2}{N_s} + \left(\frac{N_s - 1}{N_s^2}\right) T_s^2 = 1^2 + \frac{1.5^2}{3} + \left(\frac{3-1}{3^2}\right) 3^2 = 3.75$$

The effective variability of the machine is then $CV_{\text{eff}} = \sigma_{\text{eff}}/T_{\text{eff}} = \sqrt{3.75}/3 = 0.64$, which is lower than that in the current situation.

If setup operations is completely eliminated, the effective variability of the machine is equal to its natural variability, which is $CV_{\text{eff}} = \sigma_0/T_0 = 1.5/3 = 0.5$, which is lower than that under setup time reduction. Thus, the option of eliminating setups is preferred, since it reduces the effective variability of the machine.

Before we conclude our discussion, we must emphasize on the following:

- In the above example, setup time reduction or setup elimination did not affect the effective mean process time and thus did not change the effective capacity. In most cases in practice, however, setup time reduction or elimination do reduce the effective mean process times (by reducing or eliminating supporting task times) and thus result in increase in effective capacity. That is one of the reasons that setup time reduction is one of the important tools of lean operations (see Chap. 15 for more details).

- Comparing the machine's current situation with the option of reducing the setup in the above example, we reach to the same insights as in time-based interruptions: If theoretical and effective capacities remain unchanged, more frequent interruption but with smaller length of interruption is preferred over less frequent interruption with longer interruption time. Specifically, in the current situation the average time between two interruptions (i.e., time to process an average of six jobs) is longer than that under setup time reduction (i.e., time to process an average of three jobs). However, the interruption time in the current situation (the average setup time of 6 minute) is longer than that under setup time reduction option (the average setup time of 3 minutes). Consequently, setup time reduction resulted in lower effective variability.

We conclude this section by presenting the following principle:

---

**PRINCIPLE**

**Frequent versus Infrequent Interruptions**

More frequent but short interruptions result in lower effective variability than less frequent but long interruptions, given that capacity (theoretical and effective) is unchanged.

---

### Variability Due to Rework

Reworks are also task-based interruptions, that is, workers start rework tasks after they finish their main tasks. Our analytics for task-based interruptions can therefore be used to compute the effective variability in processes with rework. For example, consider a worker that has a rework rate of $\alpha_r = 10\%$. Specifically, when a job is finished, there is probability of 0.1 that the finished job requires rework. In this case, the worker performs the rework.

Note that the above rework situation fits our analytics for setups. The probability of $\alpha_r$ is analogous to the probability of needing a setup. The average number of jobs between two reworks is therefore $1/\alpha_r = 1/0.1 = 10$, which is analogous to $N_s$—the average number of jobs between setups. The time to do a rework is analogous to the time to perform a setup. Hence, $T_s$ and $\sigma_s$ would be the mean and standard deviation of the rework task.

The above relationship implies that our insights regarding setups extend to the cases with rework. Specifically, reducing the time to perform a rework, and better yet eliminating the need for rework, can reduce the effective variability of a resource, resulting in better process-flow measures. This is in addition to the fact that reducing or eliminating rework also increases effective capacity of the resource—a win-win situation.

### 5.7.2 Reducing Variability in Process Times

To reduce variability in process times the effort should focus on reducing variability in both the main tasks and supporting tasks. Some strategies to reduce variability in process times are as follows:

- *Automation Technology:* Automatic machines (e.g., welding robots used in auto assembly) have close to zero variability in performing the main tasks. Hence, one way to reduce variability in process times is to use automation technology. Two points, however, are worth mentioning here. First, the automated machines may still have some variability due to supporting tasks such as setups and breakdowns. Second, the decision to acquire automation technology is a strategic decision that should be made by also considering other factors than just reducing variability.

- *Worker Training:* Experienced workers perform main (or supporting) tasks faster and more consistently (i.e., with less variability) than less experienced workers. Thus, one way to decrease variability is to have well-designed training programs for workers to perform tasks faster and with less variability. The worker turnover should also be minimized, since newly hired workers will have large variability in their process time until they gain more experience.

- *Standardize Work:* Some of the variability in process times of long tasks comes from the worker's decision of how (e.g., in what sequence) to perform those tasks. One can standardize work by breaking tasks into smaller subtasks, define the sequence, and train the worker to perform each task in one standard way. This will reduce the worker's discretion and hence the variability in the process time. This can be used to reduce variability in both the main and supporting tasks. As an example, consider an assembly task with mean process time of 4 minutes and moderate variability $CV = 1$. Now suppose we create a standard worksheet that breaks the assembly task into four subtasks, each with mean process time of $T_s = 1$ minute and standard deviation $\sigma_s = 0.8$. Thus, each subtask has variability of $CV_s = 0.8$. How much does following the worksheet affect the variability of the assembly task? Well, the assembly task still has mean process time of $T_s + T_s + T_s + T_s = 4$ minutes. But the assembly tasks will now have standard deviation $\sqrt{4}(\sigma_s) = 2(0.8) = 1.6$. Thus, the variability of the assembly task would be reduced to $CV = 1.6/4 = 0.4$—much less than variability $CV = 1$ before standardizing the work. Even if the four subtasks have the same moderate variability as the main assembly task (i.e., coefficient of variation 1), following standard worksheet would still reduce the variability of the assembly task from 1 to 0.5.

- *Reduce interruptions:* Recall that interruptions such as setups, yield loss, rework, and machine failures increase effective process times and hence reduce the effective capacity of a process. As shown here, these interruptions also inflate the variability of process times. Therefore, to reduce the variability in process times, we need to reduce or eliminate these supporting tasks. Section 6.5.1 of Chap. 6 presents practices that help eliminate or reduce these supporting tasks.

## 5.8 Variability in Flows

So far we discussed variability in arrivals (i.e., demand) and variability in process times (i.e., supply). In a single-stage process in which flow units exit the system after being processed, these are the two main sources of variability. In multi-stage processes in which a flow unit, after departure from one stage, enters the next stage, the variability in flow between stages becomes very important.

Overall, there are three different scenarios in which the flow variability needs further investigation:

- *Output Flow:* The departure of one stage of the process may be the arrival to (demand for) the other stages of the process. Hence, it becomes essential to study the factors that affect variability in the output of a process.

- *Splitting Flows:* A flow may split to several flows and each split flow establishes a different routing. For example, the flow of order arrivals can split to several flows, each corresponding to a different product going through different stages of a process. Hence, we need some analytics to compute the variability in each split flow.

- *Composing Flows:* Finally, there may be cases in which several flows merge with each other and compose a single flow. For example, the flow of patients needing x-ray coming from different departments of a hospital merge together when they all go to the radiology department. Therefore, we also need analytics to compute how variability in several flows constitutes the variability in the composed flow.

We describe each of the above in detail in the following sections.

### 5.8.1  Variability in the Output of a Single-Stage Process

There are several situations where operations engineers and managers are interested to know the variability in the output flow of a process. A common case, as mentioned previously, is in multi-stage processes when the output of an upstream stage becomes the input of a downstream stage. Even in processes in which the output is the final product shipped to the customers, the variability of the output process is important in setting delivery dates for the customers. The higher the variability of the output flow, the longer delivery dates must be promised to the customer.

Below we present an example to illustrate the analytics for computing the variability in the output of a single-stage process.

Example 5.3

> Figure 5.3 shows a two-stage process of packaging and shipping of online sales of a small firm. At the first station (i.e., packaging) two workers—each works independently on different orders—put the items of each order in a box with packaging materials and the order packing list and send the box to the next station. At the next station (i.e., shipping), a worker checks the items in the box with the packing list, closes the box using packing tapes, prints the shipping address, and sticks it to the box. The box is then put on a conveyor that takes the box to the trucking department for shipping. The effective process time of each worker at the first station has mean of 2 minutes and standard deviation 2 minutes per order. The effective process time at the second station, on the other hand, has a mean of 1 minute with standard deviation 1.5 minutes. There is only one worker working in the shipping station.
>
> The current demand for online orders is such that orders arrive at the first station with mean interarrival times of 1.1 minutes and coefficient of variation 1. Data has shown that the variability of the order arrivals (i.e., coefficient of variation of interarrival times) at the shipping station is about 1, and the variability of order arrivals at the trucking department is about 1.43. The data also shows that the average number of orders at the packaging and shipping stations are about 20 and 17, respectively.
>
> However, it is expected that, due to the firm's marketing campaign and due to the changes in competition prices, the variability in the orders received by the firm doubles, while the average number of orders remain the same. This means that the packaging station will receive a flow of orders with average order interarrival times of 1.1 minutes, but with coefficient of variation 2. Considering the relatively high variability of orders that are currently sent to the trucking department, managers are wondering if and how much the change in online demand would affect the variability of orders sent to the trucking department?

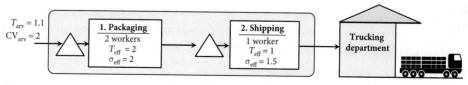

**Figure 5.3**    Two-stage process of packing and shipping after an increase in demand variability.

Since the arrival process to the trucking department is indeed the output of the shipping station, to answer the above questions, we need to present the following analytics. It shows how the variability in the output of a process relates to the variability in its arrivals as well as variability in its process times.

**ANALYTICS**

### Output of Single-Stage Process with Infinite Buffer Size

Consider a single-stage process with ample (infinite) buffer size in which flow units are processed by $s$ resources. All $s$ resources are identical and have effective mean process time $T_{eff}$ and effective variability $CV_{eff}$, see the figure below.

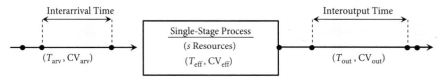

The arrival to the process has interarrival times with mean $T_{arv}$ and variability $CV_{arv}$. If utilization

$$u = \frac{T_{eff}}{s\,T_{arv}} < 1$$

then the process is stable and the time between two consecutive outputs of the process—the interoutput time—has mean $T_{out}$ and variability $CV_{out}$, where

$$T_{out} = T_{arv}$$

$$CV_{out}^2 = 1 + (1 - u^2)(CV_{arv}^2 - 1) + \frac{u^2}{\sqrt{s}}(CV_{eff}^2 - 1) \tag{5.10}$$

If the process has only one resource, that is, $s = 1$, Eq. (5.10) simplifies to

$$CV_{out}^2 = (1 - u^2)CV_{arv}^2 + u^2 CV_{eff}^2 \tag{5.11}$$

From the above analytics we observe that when the process is stable, the mean interoutput time is equal to the mean interinput (interarrival) time. The mean interoutput time does not depend on the number of servers or the capacity of the servers, or the variability in the arrival or process times. This is expected, since in stable processes input rate is equal to output rate (i.e., $R_{in} = R_{out}$) and thus the average time between two consecutive outputs would be equal to the average time between two consecutive inputs.

To explain the intuition behind the analytics, consider Eq. (5.11) for the case of single resource in two extreme cases of $u \simeq 1$ and $u \simeq 0$. When the resource is almost always busy processing flow units (i.e., utilization $u \simeq 1$), the time between two consecutive outputs is almost always the effective process time of the resource. This is what Eq. (5.11) gives: when we set $u = 1$, we get $CV_{out} = CV_{eff}$. On the other hand, if the resources is almost always idle (i.e., utilization $u \simeq 0$), it will process a job after a very long time and an arriving unit never waits (i.e., the in-buffer inventory is always zero). In this case, the time between two consecutive outputs is the time between two consecutive arrivals lagged by the time it takes to process a unit—effective process time. However, since the effective process times are very small compared to interarrival times (since $u \simeq 0$), the impact of the variation in lag is insignificant and thus the time between two outputs becomes almost the same as the time between two arrivals. Again, we get the same result $CV_{out} = CV_{arv}$ from Eq. (5.11), if we put $u = 0$.

For cases in between the two extreme cases, Eq. (5.11) provides the variability in the output as the weighted average of the variability in arrivals and variability in process times, weighted on the resource utilization—squared utilization to be exact. Hence, the variability in the output of the process becomes closer to the variability of effective process times for processes with higher utilization.

Equations (5.10) and (5.11) provide a useful principle with respect to variability reduction.

---

**PRINCIPLE**

**Reducing Variability in the Output of a Single-Stage Process**

Variability in the output of a process depends on both the variability in the arrivals to the process as well as the variability in the effective process times. Hence,

- To reduce variability in the output of a process, one can reduce the variability in the arrivals and/or variability in effective process time of the process.

- In processes with high utilization, reducing variability in effective process times has more impact on the variability reduction of the output than reducing variability in arrivals.

---

We now return to Example 5.3. Under the new demand, the packaging station with $s = 2$ workers will have

$$T_{arv}(\text{pckg}) = 1.1, \quad CV_{arv}(\text{pckg}) = 2, \quad T_{eff}(\text{pckg}) = 2, \quad CV_{eff}(\text{pckg}) = 1$$

Hence, the packaging station will have utilization

$$u = \frac{T_{eff}}{s\, T_{arv}} = \frac{2}{2(1.1)} = 0.909 < 1$$

So, the packaging station is stable and we can use Eq. (5.10) to compute its mean and variability of interoutput times as follows:

$$T_{out} = T_{arv} = 1.1 \text{ minutes}$$
$$CV_{out}^2 = 1 + (1 - u^2)(CV_{arv}^2 - 1) + \frac{u^2}{\sqrt{s}}(CV_{eff}^2 - 1)$$
$$= 1 + (1 - (0.909)^2)(2^2 - 1) + \frac{0909^2}{\sqrt{2}}(1^2 - 1)$$
$$= 1.52$$

Hence, under the new demand, the packaging station will have mean interoutput times of 1.1 minutes, with coefficient of variation $CV_{out} = \sqrt{1.52} = 1.23$. Since the output of the packaging station is the arrivals to the shipping station, then for the shipping station we have

$$T_{arv}(\text{ship}) = 1.1, \quad CV_{arv}(\text{ship}) = 1.23, \quad T_{eff}(\text{ship}) = 1, \quad CV_{eff}(\text{ship}) = 1.5$$

The utilization at the shipping station, which has only $s = 1$ worker, is $u = T_{\text{eff}}/T_{\text{arv}} = 1/1.1 = 0.909 < 1$; therefore, the shipping station is also stable, and

$$
\begin{aligned}
\text{CV}_{\text{out}}^2 &= (1 - u^2)\text{CV}_{\text{arv}}^2 + u^2\text{CV}_{\text{eff}}^2 \\
&= (1 - 0.909^2)1.23^2 + (0.909)^2 1.5^2 \\
&= 2.12
\end{aligned}
$$

Consequently, the variability of order interarrival times at the trucking department—which is the variability of the output of the shipping station—is $\text{CV}_{\text{arv}}(\text{truk}) = \sqrt{2.12} = 1.46$.

In conclusion, if the variability of the demand for the online sales of the firm doubles, the variability of order arrivals at the packaging station increases from 1 to 1.52 and the variability in order arrivals at the trucking department increases by only 0.03 from 1.43 to 1.46. Hence, the change in the variability in demand will not significantly impact the operations in the trucking department. However, it will have a significant impact on process-flow measures of the packaging and shipping stations. Using analytics that we will present in Chap. 7, we can estimate the average number of orders under the new demand in the packaging and shipping stations to increase from 10 to 24 and from 15 to 18, respectively—240 percent and 20 percent increase, respectively. This again points to the fact that an increase in variability can significantly degrade operational performance measures of a process.

### 5.8.2 Random Splitting of a Flow

In some operations systems, a flow—arriving or output flow—often splits to several flows each constituting a different routing. Consider arrivals of customers to an amusement park. Upon arrival, customers split, with each group of customers going to a different attraction. In a factory producing multiple products, the output of each station splits and each product follows a different route. To analyze these operations systems, we need to compute how variability in a flow changes when the flow splits into several flows. We use the following example to illustrate this.

 **Example 5.4**

> Department of motor vehicles (DMV) of a small town is facing long waiting times for its customers. The data has shown that, on an average, 90 customers arrive at DMV in an hour. The interarrival times have variability (i.e., coefficient of variation) of 1.6. To reduce the waiting times, DMV is considering to provide some of its services online. It is expected that, after the online services are offered, approximately 25 percent of the customers will use the online services and will not come to DMV. How does providing the online services affect the customer arrivals at DMV?

To find the mean and variability of the number of customers arriving at the DMV office, we need the following analytics.

**ANALYTICS**

**Random Splitting of a Flow**

Consider Flow $f$, a flow of events (arrivals or outputs) with rate $R$ in which inter-event times have a mean of $T$ and coefficient of variation CV. Also, suppose that Flow $f$ is *randomly* split into $K$ flows $f_1, f_2, \ldots, f_K$, according to fractions $p_1, p_2, \ldots p_K$, as shown in the following figure, where $0 < p_k \leq 1$ for all $k = 1, 2, \ldots, K$:

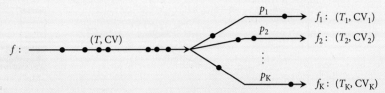

and $p_1 + p_2 + \cdots + p_K = 1$. Then Flow $f_k$ has flow rate $R_k$, inter-event times with mean $T_k$, and squared coefficient of variation of $CV_k^2$ as follows:

$$R_k = p_k R, \quad T_k = \frac{1}{R_K} = \frac{T}{p_k}, \quad CV_k^2 = p_k CV^2 + (1 - p_k) \tag{5.12}$$

Figure 5.4 shows how the arrival process to DMV in Example 5.4 would be split after online services are provided.

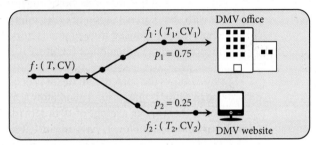

**Figure 5.4    Splitting of the arrival to DMV and its website in Example 5.4.**

Since the total arrival rate to DMV is $R = 90$ customers per hour, and since fraction $p_1 = 0.75$ will be coming to DMV office after online services start, the average number of customers coming to DMV would be $R_1 = p_1 R = 0.75 \times 90 = 67.5$ customers. The average interarrival times at DMV would, therefore, be $T_1 = 1/R_1 = 1/67.5 = 0.015$ hours $\times$ $60 = 0.9$ minutes. The squared coefficient of variation of customer interarrival times at DMV would be

$$CV_1^2 = p_1 CV^2 + (1 - p_1) = 0.75(1.6^2) + (1 - 0.75) = 2.17$$

Hence, after online services start, DMV office will have an average of 67.5 customers arriving per hour, with variability of interarrival times of $\sqrt{2.17} = 1.47$.

On the other hand, DMV website will have about $R_2 = p_2 R = 0.25 \times 90 = 22.5$ customers an hour with

$$CV_2^2 = p_2 CV^2 + (1 - p_2) = 0.25(1.6^2) + (1 - 0.25) = 1.39$$

Thus, DMV website will have the variability of interarrival times of $\sqrt{1.39} = 1.18$.

### 5.8.3 Composing Flows

Finally, many operations systems face situations in which several flows with different variabilities merge together and compose one flow that enters the process or enters one of the stages of a process. One example is when several demands are aggregated into one (see Sec. 5.6.1). Other examples include different parts from different stations coming to a painting station in a factory, or different patients coming from different departments of a hospital to the radiology department. To be able to analyze such situations, we need to know the variability of the composed flow.

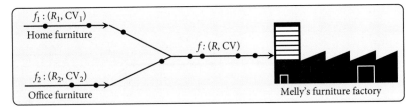

Figure 5.5    Composing demand for home and office furniture in Example 5.5.

 Example 5.5

Melly's Furniture is a firm that produces home furniture. The firm has extended its production capacity and decided to also produce office furniture. The demand for home furniture has the average demand of 10 orders per day with coefficient of variation of 1 for interarrival times of orders. It is expected that the orders for office furniture would be independent of the orders for home furniture and will have an average of five orders per day with standard deviation of order interarrival time of 0.5 days. Both office and home furniture are produced in the same factory. What would be the new demand for Melly's Furniture factory when office furniture is added to the firm's product mix?

Figure 5.5 shows the current and the new demand for Melly's factory. Flow $f_1$ represents the order arrivals for home furniture and Flow $f_2$ represents the order arrivals for the office furniture. Since there are an average of 10 orders (demand) per day for home furniture, the order interarrival times in Flow $f_1$ have mean $1/10 = 0.1$ days and coefficient of variation 1. The demand for office furniture (Flow $f_2$), on the other hand, will have an average order interarrival times of $1/5 = 0.2$ days and coefficient of variation $0.5/0.2 = 2.5$. The total demand for Melly's factory, as shown in the figure, is the merging of the two demands, and the question is about finding the mean and variability of the total demand.

We need the following analytics to answer this question.

## ANALYTICS

### Composing Several Independent Flows

Consider $K$ flows of events (arrivals or outputs) $f_1, f_2, \ldots, f_K$ which are independent of each other. Flow $k$ has rate $R_k$, and inter-event times with mean $T_k$ and coefficient of variation $CV_k$, $(k = 1, 2, \ldots, K)$. If these $K$ flows are composed into one, which we call Flow $f$, as shown in the following graph:

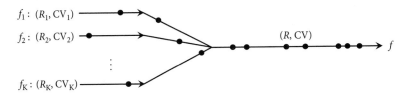

then the composed Flow $f$ has rate $R$, and mean inter-event times $T$, where

$$R = R_1 + R_2 + \cdots + R_K, \quad T = \frac{1}{R}$$

The squared coefficient of variation of inter-event time of the composed Flow $f$ is $CV^2$, where

$$CV^2 = \frac{R_1}{R}CV_1^2 + \frac{R_2}{R}CV_2^2 + \cdots + \frac{R_K}{R}CV_K^2 \tag{5.13}$$

For the flow of orders for the home furniture (Flow $f_1$), we have $R_1 = 10$ per day, and $CV_1 = 1$. The flow of orders for office furniture (Flow $f_2$) has $R_2 = 5$ per day, with $CV_2 = 2.5$. Hence, the factory will face a total demand with an average rate of $R = R_1 + R_2 = 10 + 5 = 15$ orders per day. The squared coefficient of variation of order interarrival times would be

$$CV^2 = \frac{R_1}{R}CV_1^2 + \frac{R_2}{R}CV_2^2 = \frac{10}{15}(1^2) + \frac{5}{15}(2.5^2) = 2.75$$

Hence, the variability of the demand that the factory faces is $CV = \sqrt{2.75} = 1.66$.

## 5.9 Summary

Variability is one of the main factors that makes managing and controlling operations very difficult. Variability refers to the variation in the characteristics of an entity. The variations are either known in advance or are random and unpredictable. Variability anywhere in a process—regardless of being due to known or random variation—degrades process-flow measures such as inventory, flow times, or throughput. One useful measure of variability is coefficient of variation (CV), which measures how large standard deviation of a process is compared to its mean.

To improve performance, variability in both demand and process times must be decreased or eliminated. Reducing variability in demand can be done through several strategies such as using appointment and reservation systems, scheduling job release, providing incentives for customers to change their arrival times, reducing batch size (if demand arrivals are in batches), and aggregating demand. Another strategy is to reduce the impact of demand variability by synchronizing capacity using flexible resources and changing capacity levels according to demand arrivals.

Variability in process times is often due to human performance and task complexity as well as interruptions such as setup operations, rework, and resource failures. The strategies to reduce variability in process times include automation, worker training, work standardization, reducing rework, setup time reduction, and preventive maintenance.

Variability in the output of a process also degrades performance, since it is often the input to another stage of the process, or it is directly connected to the customers. The variability in the output of a process depends on both the variability in the demand arrivals to the process as well as the variability in process times. Hence, variability in the output of a process can be reduced by reducing the variability in the demand and/or variability in process time of the process. In processes with high utilization, reducing variability in effective process times has more impact on the variability reduction of the output than reducing variability in arrivals.

## Discussion Questions

1.  What is variability and how is it measured in operations systems?

2.  Describe two different ways to measure variability in demand, and explain when each measure is used.

3.  Describe four strategies that can be used to reduce the variability in demand and provide an example for each strategy.

4.  How can operations engineers and managers utilize the capacity of a process to reduce the negative impacts of variability in demand? Provide two examples commonly used in practice.

5.  What is the difference between natural variability and effective variability of an activity? What are the sources of variability in effective process time of an activity?

6. Describe four different approaches to reduce variability in process time of an activity.

7. What are the three factors that affect the variability in output of a process?

## Problems

1. The following table presents the call answering times of three agents A, B, and C in a call center that provides technical support for a software company. Specifically, these times correspond to the time (in seconds) an agent talks to a caller and answers the caller's questions and resolves callers' issues with the software.

   Customer survey ranked the three agents the same in customer satisfaction. Assume that there is no significant difference in the questions asked from the agents and that the 12 samples in the table are good representatives of call answering times of the agents.

   | Call no. | Agent A | Agent B | Agent C |
   |----------|---------|---------|---------|
   | 1 | 151 | 177 | 162 |
   | 2 | 240 | 163 | 131 |
   | 3 | 213 | 239 | 151 |
   | 4 | 190 | 140 | 148 |
   | 5 | 189 | 162 | 184 |
   | 6 | 274 | 206 | 211 |
   | 7 | 174 | 253 | 160 |
   | 8 | 140 | 186 | 167 |
   | 9 | 317 | 146 | 192 |
   | 10 | 238 | 212 | 169 |
   | 11 | 219 | 223 | 152 |
   | 12 | 237 | 171 | 146 |

   a. How do you compare these three agents in terms of efficiency and consistency in their call answering times?

   b. Call center manager wants to send one of the agents for training to improve call answering time. Which agent do you recommend for training?

2. Consider a painting station in which a spray painting machine is used to paint steel panels that have industrial applications. Panels arrive at the rate of 12 per hour with interarrival times with coefficient of variation of 1. The painting time of a panel has an average of 4 minutes with standard deviation of 4 minutes.

   a. What fraction of an hour the painting spray is idle waiting for panels to arrive at the painting station?

   b. Answer Part (a), if the painting machine requires an adjustment after working for an average of 90 minutes with standard deviation of 90 minutes. The adjustment takes an average of 10 minutes with moderate variability, that is, coefficient of variation 1.

   c. Answer Part (a), if after completing a panel, there is an 8 percent chance that the spray machine needs an adjustment that takes 5 minutes with standard deviation of 12 minutes.

   d. Which one of the cases in Parts (b) or (c) results in longer waiting times for panels in the painting station?

3. John is working in the claim processing center of a health insurance company. If a claim has all required information (i.e., a complete claim), it takes John an average of 3 minutes with standard deviation of 2 minutes to process the claim. However, for claims with incomplete information, John needs to call the customer to complete the claim. This takes an average

of 5 minutes with standard deviation of 4 minutes. After completing the information in the claim, the claim is processed like other complete claims, which takes about 3 minutes with standard deviation of 2 minutes. Data shows that about 20 percent of the claims are incomplete.

Managers are considering to adopt a new data management system that eliminates the need for calling customers. This is expected to improve performance. However, the claim processing time under this new system would be larger than the current system. Specifically, it takes about 4 minutes with standard deviation of 2 minutes to process a claim under the new system. Focusing only on process-flow measures, do you support the idea of adopting the new system? Provide analytical support for your answer.

4. It takes an average of 2 minutes for a laser cutting machine to cut a piece. Jobs are released to this machine from upstream stations in a random fashion. To prevent quality problems, the machine is stopped after cutting an average of 12 jobs and some cleaning and adjustment are performed, which take exactly 6 minutes (with no variability). The work-in-process (WIP) inventory of jobs in the cutting station is large. One suggestion to reduce the WIP is to stop the machine earlier after processing about six jobs and do cleaning which takes exactly 3 minutes (with no variability). The shop floor manager rejects this suggestion, since it does not change the number of jobs that the machine can cut in an hour. Do you agree with the shop floor manager? Provide analytical support for your answer.

5. GND is a firm that offers a wide range of dietary supplements in its three stores in a small city. GND is planning to stop offering one of the following two products in its stores and only offer it online: vitamin E cream and fiber supplement pills. The demand for the two products are variable. Table 5.4 shows the daily demand for the two products in the three stores in the last 20 days. The data is a good representative of the daily demand in all 12 months of a year.

Table 5.4   Demand for Fiber and Vitamin E at GND Stores in Problem 5

| Day | Fiber supplement pills | | | Vitamin E cream | | |
|---|---|---|---|---|---|---|
| | Store 1 | Store 2 | Store 3 | Store 1 | Store 2 | Store 3 |
| 1 | 79 | 119 | 207 | 148 | 203 | 138 |
| 2 | 87 | 59 | 123 | 118 | 166 | 124 |
| 3 | 36 | 56 | 84 | 128 | 137 | 70 |
| 4 | 130 | 214 | 64 | 118 | 110 | 141 |
| 5 | 151 | 132 | 172 | 129 | 128 | 140 |
| 6 | 179 | 80 | 179 | 52 | 141 | 177 |
| 7 | 87 | 137 | 119 | 128 | 175 | 161 |
| 8 | 79 | 168 | 97 | 127 | 98 | 183 |
| 9 | 163 | 51 | 157 | 61 | 193 | 177 |
| 10 | 46 | 125 | 132 | 113 | 99 | 79 |
| 11 | 128 | 121 | 107 | 101 | 140 | 110 |
| 12 | 53 | 194 | 187 | 65 | 112 | 108 |
| 13 | 156 | 149 | 138 | 115 | 197 | 118 |
| 14 | 22 | 94 | 201 | 105 | 176 | 135 |
| 15 | 133 | 171 | 164 | 149 | 166 | 61 |
| 16 | 21 | 87 | 118 | 77 | 107 | 90 |
| 17 | 153 | 133 | 177 | 59 | 186 | 139 |
| 18 | 94 | 48 | 81 | 103 | 130 | 114 |
| 19 | 141 | 200 | 148 | 136 | 151 | 120 |
| 20 | 24 | 66 | 41 | 133 | 113 | 120 |

By offering a product online, instead of holding inventory in its three stores, GND will keep the inventory of the product in its distribution center from which the product will be shipped to the customers. GND believes that all customers who buy the products from the stores will still buy it, if it is offered online. If other costs are the same for both products, which product do you recommend to be offered online?

6. Consider the process in Fig. 5.6 in which all units require three activities A, B, and C, performed in stations A, B, and C, respectively.

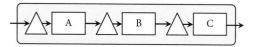

**Figure 5.6**    Process-flow chart of Problem 6.

Jobs arrive at Station A at the rate of 12 jobs per hour with standard deviation of 5 minutes for interarrival times. Process time of Activity A has mean and standard deviation of 4.5 and 8 minutes, respectively. These numbers are 4 and 7 minutes for Station B, and are 4.8 and 8 minutes for Station C. Activities A and B are performed by Workers A and B, respectively. Activity C is performed by an expensive equipment operated by Worker C. Activity C is the bottleneck activity and thus Station C has a large inventory and long flow times. Because it is too costly to increase the capacity of Station C by acquiring another machine, the following alternatives are suggested to decrease inventory and flow time at the bottleneck Station C:

- Reducing variability in Station A by reducing its standard deviation from 8 to 5 minutes.
- Reducing variability in Station B by reducing its standard deviation from 7 to 4 minutes.
- Training Worker A to do Activity B and training Worker B to do Activity A. Then, redesigning the process as a two-station process as follows. In the first station with two workers—Workers A and B—each worker performs Activity B immediately after finishing Activity A. The unit is then sent to the second station, that is, Station C.

Which one of the above three alternatives is expected to reduce inventory and flow times in the bottleneck Station C more than the other two?

7. Consider the process in Fig. 5.7 that receives jobs at the rate of 12 jobs per hour according to a Poisson process. About 60 percent of arriving units require processing at Station A and then Station C. The remaining units require processing at Station B and then Station C. The process time of Activity A has mean and standard deviation of 6.5 and 13 minutes, respectively. These numbers are 4 and 8 minutes for Activity B, and are 4.8 and 8 minutes for Activity C. Activity C is the bottleneck activity.

Consider the following three alternatives:

- Reducing the variability in Station A by reducing its standard deviation from 13 to 10 minutes. This will reduce the variability (i.e., coefficient of variation) at Station A from 2 to 1.54.

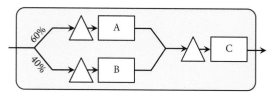

**Figure 5.7**    Process-flow chart of Problem 7.

- Reducing the variability in Station B by reducing its standard deviation from 8 to 4 minutes. This will reduce the variability at Station B from 2 to 1.
- Training Worker A to do Activity B and training Worker B to do Activity A. Then, redesigning the process as a two-station process as follows. In the first station with two workers—Workers A and B—each worker performs Activity B immediately after finishing Activity A. The unit is then sent to the second station, that is, Station C.

Which one of the above three alternatives is expected to reduce inventory and flow times at the bottleneck Station C more than the other two?

# Throughput Improvement

## 6.0 Introduction

One of the process-flow measures that has a direct impact on profit is Throughput. In fact, throughput of a production process (goods or services) is what is sold to the customers. If there is demand for a product of a process, increasing the throughput of the process increases sales and thus increases profit. In Chap. 4 we showed how the effective capacity of a process can be obtained using the information about the activities performed by each resource of the process. The goal of this chapter is to study the relationship between capacity and the throughput of a process, and to understand how this relationship is affected by the demand and variability in the process. This will help identify strategies that can improve throughput. We start by discussing the relationship between Throughput, Capacity, and Demand.

## 6.1 Throughput, Capacity, and Demand

Understanding the underlying dynamics between demand, effective capacity, and throughput provides us with insights into the ways one can improve the throughput of a process. We start with the case of Campus Crust Pizza.

### Campus Crust Pizza
Joseph owns a small pizza shop—Campus Crust Pizza—close to the campus of a state university. The shop is open from 10:00 am to 10:00 pm, and sells pizzas only for takeout. Joseph

offers only one size—8-inch pizza. The pizza is larger than a slice, but smaller and cheaper than the standard 10-inch pizzas offered by other pizza shops. It is the right size for one meal, therefore, all of his customers order one. Pizzas are made-to-order, and customers can choose among a variety of toppings when they place their orders. To speed up the process of making pizzas, Joseph does not make the dough or sauce in his shop. He buys large quantity of premade dough (from a local bakery) and pizza sauce and toppings (from a large distributor) at a discounted price.

Joseph knows that his shop has an effective capacity of 24 pizzas an hour—the effective capacity of the pool of two Conveyor Pizza Ovens, which are the effective bottleneck resources. Furthermore, his past sales data shows that the demand for pizza varies among different hours, but has an overall average of 18 customers per hour. What is the throughput of the Campus Crust Pizza?

Since every customer orders one pizza, we can consider a customer as a flow unit. The effective capacity of 24 customers per hour implies that when pizza shop is fully utilized (i.e., works constantly during an hour), it can produce 24 pizzas to serve 24 customers. On the other hand, only 18 customers arrive per hour and all get their pizzas and leave the shop. While the shop has the capacity of serving 24 customers, the throughput of the pizza shop cannot be more than the input rate of 18 customers per hour. Thus,

$$\text{Throughput} = \text{Input Rate} = 18 \text{ customers per hour} \tag{6.1}$$

Now consider a different case in which the demand for pizza increases from 18 to 30 customers per hour, larger than the effective capacity of the process. In this case, while the demand (the input rate) is 30 customers per hour, the shop does not have the capacity of serving more than 24 customers an hour. The pizza shop must work all the time and therefore the throughput of the pizza shop would be equal to its effective capacity of 24 customers per hour. In other words,

$$\text{Throughput} = \text{Effective Capacity} = 24 \text{ customers per hour} \tag{6.2}$$

From the above two cases, and from the relationship among Input Rate, Capacity, and Throughput in Eqs. (6.1) and (6.2), we can conclude that

$$\text{Throughput} = \min\{\text{Input Rate}, \text{Effective Capacity}\} \tag{6.3}$$

For example, in the former case, when the demand for pizza was 18 per hour, we have TH $= \min\{18, 24\} = 18$ customers per hour as in Eq. (6.1). In the latter case, when the demand was 30 per hour, we have TH $= \min\{30, 24\} = 24$ customers per hour as in Eq. (6.2).

Equation (6.3) implies that the throughput of a process is constrained by either the Input Rate (i.e., demand) of the process or by the Effective Capacity of the process.

---

**CONCEPT**

### Demand-Constrained and Capacity-Constrained Processes

- A *Demand-Constrained* process is a process in which the demand for the process is smaller than the effective capacity of the process. In these processes, the throughput of the process is constrained by the demand (input rate) of the process.

- A *Capacity-Constrained* process is a process in which the effective capacity of the process is smaller than the demand for the process. In these processes, the throughput of the process is constrained by the capacity of the effective bottleneck of the process.

To improve the throughput of a process, it is critical to know if the process is demand-constrained or capacity-constrained. This is because strategies to improve throughput are different for demand- and capacity-constrained processes. We will discuss these strategies in Sec. 6.3. But, first we need to describe factors that limit the throughput of a process and result in throughput loss.

## 6.2 Throughput Loss

While Eq. (6.3) illustrates the fundamental relationship between throughput, effective capacity, and input rate in a simple single-stage process like the Campus Crust Pizza, in general, computing the throughput of a process is difficult and may require complex mathematical models, computer simulation, or extensive data gathering. One reason is that the "Input Rate" in Eq. (6.3) is sometimes not easy to compute. Consider a long line for coffee in a coffee shop. Often times customers who arrive at the coffee shop (i.e., demand) decide not to join the long line. In this case, the coffee shop loses inputs, and hence throughput, due to its long line. Other times customers join the line, but after waiting for some time, they become impatient and leave the system before being served. In this case, the process again loses inputs that results in loss of throughput.

Therefore, to improve the throughput of a process, in addition to the fundamental relation in Eq. (6.3), it is essential that one also identifies the underlying reasons behind throughput loss of a process. In the following sections, we provide examples of several common reasons for loss of throughput in manufacturing and service operations systems.

### 6.2.1 Throughput Loss Due to Lack of Supply

Some processes lose throughput not because they do not have enough capacity or there is no flow unit (i.e., no demand) available to process, but because there is no supply of materials or parts to finish the processing of a flow unit. For instance, consider the Campus Crust Pizza with the demand of 18 pizzas an hour and the effective capacity of 24 pizzas an hour. Recall that the pizza shop orders pizza doughs from a local bakery. Specifically, to have fresh pizza doughs, Joseph orders doughs twice a day. He receives 120 pizza doughs from the bakery at 8:00 am and another 120 doughs at 2:00 pm. Recently, Joseph has noticed that the bakery has delays in delivering doughs. As a result, Joseph has occasionally been running out of doughs and had to reject customers, while waiting for delivery from the bakery. Joseph has realized that he has been losing 8 percent of his customers due to lack of pizza doughs.

Joseph's pizza shop is now a demand-constrained process in which some of the input rate of 18 customers per hour is lost because the process runs out of supply (of pizza doughs). Specifically, $8\% \times 18 = 1.44$ customers per hour or $1.44 \times 12 = 17.28$ customers per day are lost due to lack of supply. This reduces the throughput of Campus Crust pizza from 18 to 16.56 customers per hour, and consequently reduces Joseph's sales.

The throughput loss due to lack of supply is not limited to demand-constrained processes. Capacity-constrained processes can also lose throughput because of the lack of supply. In other words, the bottleneck of a capacity-constrained process may remain idle because the material or parts required for processing a flow unit are not available. This reduces the throughput of the bottleneck, which in turn reduces the throughput of the capacity-constrained process.

### 6.2.2 Throughput Loss Due to Balking and Abandonment

A 24-hour convenience store has recently opened near Campus Crust Pizza that also sells packaged sandwiches. While Joseph thinks that cold packaged sandwiches are not substitute for Joseph's hot and made-to-order pizzas, nevertheless he would like to investigate this to make sure. In a 2-week study of arrivals and queue formation at Campus Crust Pizza, he finds the average number of customers who come to his shop is around 219 customers per day. However, among these 219 who intend to go to his shop, an average of 19 customers do not enter, since they observe a long waiting line for ordering pizzas. They, instead, go to the convenience store. Furthermore, among the remaining $219 - 19 = 200$ customers who have already joined the line, 10 customers abandoned the line after a long waiting and went to the convenience store.

Joseph's observations of customers not entering his shop, or leaving his shop before ordering pizza, respectively, refer to the concepts of Balking and Abandonment.

---

**CONCEPT**

**Balking and Abandonment**

- *Balking* refers to the event that upon arrival, a customer (a flow unit) does not join the queue (i.e., the buffer) because the queue is too long.

- *Abandonment (or Reneging)* refers to the event that customers who have already entered the process and are waiting in the queue lose patience and leave the process.

---

To compute the throughput of Campus Crust Pizza per hour, recall that an average of 219 customers arrive at Campus Crust Pizza during a 12-hour period (i.e., a working day). Thus, the average hourly arrival rate to the pizza shop is $219/12 = 18.25$ customers per hour, which is close to what Joseph had estimated. However, an average of only $219 - 19 - 10 = 190$ customer stay in the line and buy pizza. So, the throughput of the pizza shop is 190 customers per day, or $190/12 = 15.83$ customers per hour.

Campus Crust Pizza, in this case, illustrates situations in which throughput of a process is reduced due to long waiting time in the input buffer of the process. This is a critical issue in most of the service operations systems in which flow units are people, and people do not like long waiting times. In Chap. 8 we will discuss how the waiting times in a process can be reduced, or how the waiting time can be made less annoying in order to reduce the throughput loss due to balking and abandonment.

### 6.2.3 Throughput Loss Due to Limited Space at Input Buffer

Even in the absence of balking and abandonment, some processes may still lose throughput. One such situation is when the arriving flow units that intent to enter the process cannot enter because of the lack of space in the input buffer. We discuss this in the case of Michigan Yacht Club.

**Throughout Loss in Michigan Yacht Club**

Michigan Yacht Club has a powerboat launch ramp in a Park Municipal Marina for people who trailer their powerboats to the park. The club has a waiting area for four boat trailers (in addition to the one being launched). There are also several other ramps in the park operated by others; however, since Michigan Club is in a better location and provides a better service, all arriving trailers first go to Michigan's Club ramp. Boat trailers arrive at the club at an average rate of 10 trailers an hour. On the other hand, the effective mean process time of launching a boat has an average of 5 minutes, which implies that the club has an effective capacity of launching 12 boats an hour. Therefore, the throughput of the club should be an average of 10 trailers an hour, that is, $TH = \min\{10, 12\} = 10$. However, this is not the case. Based on the daily revenue, Jack, the manager of the club, realized that the number of boats served in an hour—the throughput—is much less than 10 boats per hour.

At first, Jack thought that the limited space at the waiting area was the problem. Jack has occasionally observed that if the club's waiting area is full, arriving customers do not wait and go to the other ramps in the park. But he does not expect that he would lose many customers because club's waiting area becomes full. After all, a boat arrives, on an average, every 6 minutes, while the club can serve that boat, on an average, in only 5 minutes. In other words, the service of the arriving trailer is expected to be completed 1 minute before the next boat arrives. Puzzled by these thoughts, Jack decides to perform a study and find how often he loses customers because his waiting area is full. His 2-week study shows that his waiting area is full 20 percent of the time, whereupon arriving customers go to other ramps.

While surprised, Jack does a simple calculation and finds that, an average of $20\% \times 10 = 2$ of its 10 arriving customers face the full buffer and only an average of $10 - 2 = 8$ customers enter the club in an hour. Since, club has the effective capacity of serving 12 $(= 60/5)$ customers per hour, the throughput of the club is therefore

$$TH = \min\{\text{Input Rate}, \text{Effective Capacity}\} = \min\{8, 12\} = 8 \text{ boats an hour}$$

Jack always thought that having a waiting area to hold 4 trailers is more than enough, since he has the capacity of serving 12 trailers an hour, while the demand is for only 10 trailers per hour. He is not happy about losing an average of two customers per hour (i.e., 20 percent of his business). He thinks: Why is this number too high, and how can he reduce the loss in his revenue (i.e., reduce throughput loss)?

Jack, therefore, decides to observe its operations one more time with a goal of investigating why and when its buffer gets full. So, he watches how the queue in its buffer forms and when it reaches the limit of four trailers. He observes that the queue in his waiting area is less than four trailers most of the time. However, the queue reaches four and the club starts losing customers in the following cases:

- Sometimes the waiting area becomes full when the time to launch a boat takes much longer than the average of 5 minutes. For example, in one case the owner's boat got stranded on the ramp, which took around 15 minutes to fix. In contrast, Jack also observed many cases when the launching time was very short, that is, less than 2 minutes. Using all of his data, Jack finds that the time to launch a boat as he expected has an average of 5 minutes, but with standard deviation of 10 minutes.

- Sometimes the waiting area gets full when in a short period of time many cars arrive at his club. When this occurs, regardless of whether the launching times are short or long, the queue of trailers increases rapidly and reaches four and the club starts losing customers. For example, in one case, six cars arrived in a 5-minute period and Jack lost four of those customers. Jack also observed periods of time with a few arrivals. For example, there were cases when no trailers arrived for more than 20 minutes. His numbers show that the time between two arrivals has an average of 6 minutes with standard deviation of 6 minutes.

Jack now understands that the variability in the arrival and its boat launching process is causing him to lose customers. After all, if there was no variability in the interrraival times or launching times, each customer was served 1 minute before the next customer arrives, and thus there was not even one trailer in the waiting area. Having standard deviation of 10 and 6 minutes, variability $CV_{eff} = 2$ and $CV_{arv} = 1$ for launching and interarrival times, respectively, results in the waiting area being full 20 percent of the time.

Happy that he discovered the underlying reasons for losing throughput, Jack is now thinking of ways to reduce its throughput loss by reducing the chance of having a full input buffer. As his first attempt, he is thinking of the following three plans, which he believes can increase his throughput.

*Alternative 1: Increasing Buffer Size*    Jack now thinks that his waiting area is too small for his operations. He believes that if he, for example, could double the size of the waiting area, he will be able to cut his throughput loss to half (from two customers per hour to one customer per hour).

*Alternative 2: Increasing Capacity*    Another alternative Jack is thinking of is to increase the capacity of its launching operations. Since he only owns the space for one ramp, he cannot add a second ramp. Hence, the only option to increase his capacity is to reconstruct the layout of his ramp to make it easier and faster for customers to park their trailers in the launching position. Jack thinks that, under the new layout that he has in mind, his effective capacity will be increased by 25 percent, since the average launching time will be reduced from 5 to 4 minutes per trailer and standard deviation from 10 to 8 minutes. Since he is losing 20 percent of his customers, he believes that a 25-percent increase in the capacity should be more than enough to eliminate throughput loss.

*Alternative 3: Reducing Variability*    Jack understands that he cannot reduce the variability in his operations to zero. Arriving boats have different sizes and trailers have different releasing mechanisms, so there will always be variability in the time to launch a boat. However, Jack thinks that he can reduce the variability of the launching time from $CV_{eff} = 2$ to $CV_{eff} = 1$. The simplest way is to let his trained staff drive the owners' cars and help owners launch their boats. This will reduce the number of accidents that require long time to fix. Since Jack will still launch boats of different sizes, and since these accidents may still occur, this is not expected to significantly affect the average launching time of 5 minutes. Jack is not optimistic if this alternative can make a significant increase in his throughput, since it does not increase the effective capacity or size of the waiting area.

To see the impact of the three alternatives on Michigan Club's throughput, we develop a computer simulation model to estimate the throughput of the club under different alternatives.[1] The results are shown in Fig. 6.1.

As the figure shows, Jack's intuition about the impact of the buffer size, effective capacity, and variability on throughput of the club was right and wrong! First, as the figure shows, increasing the size of the input buffer increases the throughput of the club. This can be seen under all three alternatives. As the input buffer size gets larger, the throughput of the club is getting closer to 10 units per hour, which corresponds to the case that all arrivals enter the club. This confirms Jack's intuition about the positive impact of the input buffer size on increasing the throughput. However, the increase in throughput shows a diminishing return not a linear relationship as Jack thought. Specifically, the improvement in throughput by increasing the buffer size by one unit is larger for smaller buffer sizes. For example, when we increase the input buffer by one unit from size 4 to 5, we observe a higher increase in throughput compared to when we increase it by one unit from 10 to 11. Thus, doubling the size of the buffer, as Jack thought, will not cut the throughput loss by half. There is a nonlinear and diminishing return relationship between increase in buffer size and increase in throughput.

---

[1] In this simulation all times are assumed to follow a gamma distribution, which allows us to model coefficient of variations of less than, equal to, or greater than one.

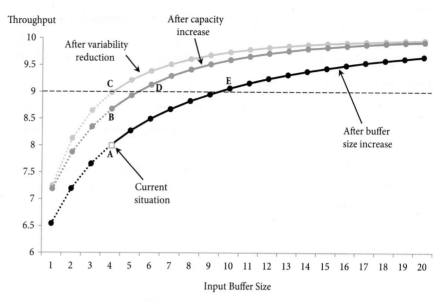

**Figure 6.1**  **Throughput improvement alternatives for Michigan Club.**

Jack was also right that increasing the effective capacity of its launching operation increases the throughput of the club. As Fig. 6.1 shows, for the case with a 25-percent higher effective capacity, the throughput of the club is higher than that in the current system. However, Jack was wrong thinking that the 25-percent higher effective capacity will eliminate the 20-percent throughput loss. As the figure shows, increasing the effective capacity by 25 percent (i.e., Point B in the figure) results in throughput of 8.7 boats per hour—a 0.7 increase in throughput—which reduces the throughput loss from 2 to 1.3, but does not eliminate it.

Finally, Jack was right that the only way to eliminate the throughput loss is to completely eliminate variability in the process. He was also right that reducing variability can improve its throughput. However, he was wrong about the significance of the impact of variability reduction from $CV_{eff} = 2$ to $CV_{eff} = 1$. As the figure shows, reducing the variability in launching process results in a throughput of nine boats an hour (Point C in the figure). In fact, the throughput after variability reduction (Point C) is higher than that under 25-percent capacity increase (Point B). The only way to have a throughput of nine boats an hour under 25-percent capacity increase is to also increase the size of the waiting area to six (Point D).

One implication of the above observation is that processes with lower variability or higher effective capacity can achieve the same throughput with lower input buffer size. Again, consider the throughput of nine trailers per hour in Fig. 6.1, as an example. The system with higher capacity (Alternative 2) and the system with lower variability (Alternative 3) can achieve this throughput with input buffer size of 6 (Point D) and 4 (Point C), respectively, while this can be achieved in Alternative 1 (with higher variability and lower capacity) with input buffer of size 10 (Point E).

Another important observation in the Michigan Club case is that variability reduction can (in some cases) have a significant impact on throughput of a process with finite input buffer. Variability reduction is often less costly to implement and may result in higher throughput improvement than more expensive alternatives of capacity expansion or adding more buffer space. Thus, capacity and buffer size expansion should be considered as last resorts after exhausting possibilities of variability reduction. We further emphasize on the importance of variability reduction in Chaps. 8 and 15.

Our observations lead us to the following principles:

**Throughput and Input Buffer Size**

In processes in which the input buffer has a limited space:

- Throughput increases as the size of the input buffer increases. The increase in throughput has the diminishing return property.
- Throughput increases as the effective capacity of the process increases, or the variability (in interarrival or processing times) decreases.
- Systems with lower variability or higher effective capacity, in general, can achieve a desired throughput with smaller input buffer sizes.

Before we conclude this section, we must note that the above principles that we established by studying Michigan Yacht Club—a demand-constrained process—also hold for capacity-constrained processes. For example, if Michigan Club had an arrival rate of 15 trailers per hour, which is higher than its effective capacity of 12 trailers per hour, the club will still lose throughput, and increasing capacity or buffer size, or reducing variability, will still improve throughput. However, the magnitude of improvement will be different when the club is a demand-constrained process.

### 6.2.4  Throughput Loss Due to Limited Intermediate and Output Buffer Space

While the limited space in the *input buffer* of a process results in losing throughput because of losing input flow rate, the limited space in *intermediate* and *output* buffers of a process results in losing throughput because of idling of the resource due to blocking and starvation. We illustrate this using the case of Museart.

**Finishing Department at Museart**

Museart is a company that produces replica of large famous sculptures for outdoors and indoors. The final stage of the production of these sculptures is performed in the finishing department. Knowing that the finishing department is the bottleneck department, John, the production manager of Museart, has taken couple of actions to make sure that the finishing department will never stay idle awaiting arrival of sculptures from other departments. First, he increased the size of the input buffer of the finishing department to 40, so that it can hold a large number of sculptures. Second, he developed a production schedule that guarantees that the input buffer of the finishing department is never empty (i.e., the finishing department always has work to do).

The finishing department has three stations: quality control station, touch-up station, and packaging station. The process-flow chart of the finishing department is shown in Fig. 6.2.

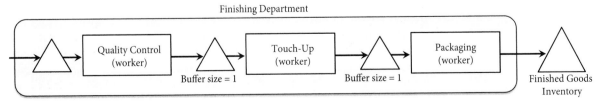

Figure 6.2    Process-flow chart of the finishing department at Museart.

*Quality Control Station:* In the quality control station, using an overhead crane, one worker transfers a sculpture waiting in the input buffer of the station to the work table, which has the space for only one sculpture. The worker then checks the quality of the sculpture (e.g., looking for cracks, scratches, or paint defects). Using an erasable red marker, the worker draws a circle around the problem areas and takes the piece to the touch-up station using the overhead crane. The total effective mean process time of checking and transferring the sculpture is 5 minutes with standard deviation of 5 minutes. While there is some space between the quality control station and touch-up station, because of the large size of the sculptures, this space (i.e., the input buffer of the touch-up station) can only hold one sculpture. When the input buffer of the touch-up station is full, to prevent damage during transfer, the worker is ordered not to transfer the sculpture from his work table to the touch-up station. When this occurs, the worker has to wait until a space becomes available in the touch-up station, so he can remove the sculpture from his working table and start working on a new sculpture.

*Touch-up Station:* The touch-up station has an input buffer which can store one sculpture. The worker in the touch-up station is a highly skilled worker who fixes the quality problems of the sculptures and transfers the sculpture to the packaging station using the overhead crane. The input buffer of the packaging station also has a limited space for only one sculpture. Similar to the quality control station, the worker does not transfer a finished sculpture (and remains idle) if the input buffer of the packaging station is full. The total effective mean process time of fixing the quality problem and transferring a piece to the packaging station is 6 minutes with standard deviation of 6 minutes.

*Packaging Station:* The packaging station has an input buffer that can store only one sculpture. In the packaging station, the piece is very carefully packaged using bubble wraps, custom pads, and ethafoam sheets to prevent damage during shipment. The effective process time for packaging is also variable with an average of 5.5 minutes and standard deviation of 5.5 minutes. Upon completion, a lift truck transfers the packaged sculptures to the finished goods inventory outside the finishing department.

## Throughput of the Finishing Department

John knows that the effective capacity of the finishing department is 10 sculptures per hour—the capacity of the touch-up station, which is the bottleneck. He also knows that there are always a large number of sculptures in the input buffer of the finishing department, so the finishing department should never run out of work. He concludes that the throughput of the finishing department should, therefore, be 10 sculptures per hour. The data, however, shows that the finishing department has not been able to process more than eight jobs an hour. This does not make sense to John. To investigate this inconsistency, John orders a comprehensive time study of the finishing department. The summary of the study is shown in Table 6.1. This study confirms the previous data that the throughput of the finishing department is less than eight sculptures an hour, more specifically, around seven sculptures per hour. This is not good news for John. He now needs to find the underlying reasons of why the finishing department cannot achieve the throughput of 10 sculptures per hour?

Looking carefully at Table 6.1, John makes several useful observations about the performance of the finishing department. First, the table confirms that John's actions have been successful in making sure that the finishing department is never idle waiting for jobs from other departments. The likelihood of having an empty input buffer at the finishing department (i.e., having an empty buffer at the quality control station) is 0.16 percent, which is almost zero.

Second, throughput, that is, the number of completed sculptures per hour in the three stations of the finishing department (6.99, 7.04, and 7.03) is almost the same. This is due to the fact that, in the long run, all sculptures that enter a station will eventually exit the

Table 6.1    Time Study of the Finishing Department at Museart (*Flow Unit = One Sculpture*)

|  | Quality control station | Touch-up station | Packaging station |
|---|---|---|---|
| **Input buffer** | | | |
| Input buffer size | 40 | 1 | 1 |
| Avg. sculptures in input buffer | 38.62 | 0.6 | 0.32 |
| Percent input buffer is empty | 0.16% | 13.11% | 37.51% |
| **Resource (worker)** | | | |
| Effective capacity per hour | 12 | 10 | 10.91 |
| Percent working | 58.50% | 70.40% | 64.44% |
| Percent idle (starved) | 0.08% | 11.25% | 35.56% |
| Percent idle (blocked) | 41.42% | 18.35% | 0% |
| Sculptures completed per hour | 6.99 | 7.04 | 7.03 |

station, and enter the next station (or exit the department). Therefore, the throughput of the finishing department is around seven sculptures per hour.

Third, focusing on the touch-up station—the bottleneck of the department—John tries to understand why the bottleneck cannot process 10 jobs per hour. John realizes that the touch-up station is only 70.40 percent utilized. Thus, even though the station has an effective capacity of 10 sculptures an hour, it only works 70.4 percent of time and therefore processes $70.4\% \times 10 = 7.04$ sculptures per hour.

Finally, the table also shows that the idling at the touch-up station is caused by either starvation or blocking of the station.

## CONCEPT

### Starvation and Blocking

- *Starvation* of a resource refers to the idling of the resource due to lack of input flow units. This idling continues until a new flow unit enters the input buffer of the resource, whereupon the resource starts processing the unit.

- *Blocking* of a resource refers to the idling of the resource due to lack of space in the buffer of the next stage of the process. Specifically, it refers to situations in which the resource has just finished processing a flow unit, but the resource cannot start the processing of the next unit because there is no space in the next stage to put the completed unit. Therefore, the resource remains idle until a space becomes available in the next stage. When this occurs, the completed unit is moved to the next stage, and processing of a new unit starts.

The touch-up station is starved 11.25 percent of the time, during which the station is idle awaiting arrival of the next job. On the other hand, the worker in the touch-up station is blocked 18.35 percent of the time. This is because, when the worker in the touch-up station completes processing of a job, if the buffer of the packaging station is full, the worker stays idle until a space becomes available at that buffer, which occurs 18.35 percent of the time.

Finding the reasons behind the low throughput of the bottleneck station—starvation and blocking—let's use a computer simulation to investigate the impacts of buffer size,

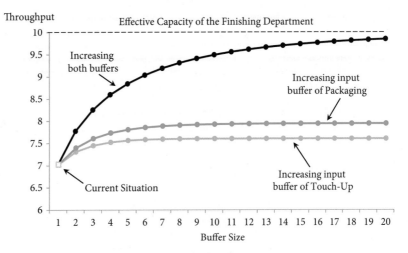

Figure 6.3   Impact of buffer size on throughput of the finishing department.

capacities of bottleneck and non-bottleneck stations, and their variabilities on the throughput of the finishing department.

**Impact of Buffer Size on Throughput**

In this section, we investigate the impact of the sizes of the buffers before and after the bottleneck station on the throughput of the finishing department. Using a computer simulation of the finishing department, we compute the throughput of the finishing department in the following three cases: (i) increasing the buffer of the touch-up station, (ii) increasing the buffer of the packaging station, and (iii) simultaneously increasing the buffers of the touch-up and the packaging stations. The results are shown in Fig. 6.3. From the figure, we observe the following:

- Throughput of the finishing department increases as the size of the input buffer at the touch-up or the packaging or both stations increases. The increase in throughput is higher when both buffers increase. For example, as the input buffer of the touch-up station is increased from 1 to a large number (e.g., 20 in the figure), the throughput of the department increases from 7 to around 7.6 sculptures per hour. This is due to reducing the starvation of the touch-up station. The reason is that a larger input buffer at the touch-up station can hold more units and thus reduces the chance of starvation of the station (the bottleneck). Now, if only the input buffer of the packaging station is increased to a very large number (e.g., 20), the throughput of the department increases to about 7.9 sculptures per hour. The reason is that a larger input buffer at the packaging station will have more space available for the worker at the touch-up station (the bottleneck) to put the finished sculptures—reducing the worker's idle time due to blocking. Finally, when the size of both buffers becomes very large, the throughput of the finishing department approaches its effective capacity of 10 sculptures per hour—the effective capacity of the touch-up station. This is because, in this case, both starvation and blocking of the touch-up station are reduced. Note that the throughput becomes exactly 10 sculptures per hour, only when all buffer sizes become infinitely large.

- Similar to the case of increasing *input buffer* at Michigan Yacht Club, the increase in throughput shows a diminishing return with respect to the increase in *intermediate buffers*.

The above observations lead to the following principle:

---

**Impact of Intermediate and Output Buffers on Throughput**

In processes in which intermediate and output buffers have limited space, the throughput of the process increases as the size of the buffers increases. The increase in throughput has the diminishing return property.

---

**Impact of Capacity on Throughput**

To investigate the impact of capacity on throughput of the finishing department, we consider three cases of (i) increasing the effective capacity of the touch-up station (the bottleneck station), (ii) increasing the effective capacity of the quality control station, that is, a non-bottleneck station that feeds the bottleneck station and can starve it, and (iii) increasing the effective capacity of the packaging station, that is, a non-bottleneck station that receives flow units from the bottleneck station and can block it.

- *Increasing Capacity of the Touch-Up Station:* To study the impact of an increase in the effective capacity of the touch-up station, we reduce its effective mean process time by 0.5 minutes from 6 minutes to 5.5 minutes per sculpture. This increases the effective capacity of the touch-up station from 10 to 10.91 sculptures per hour. This also increases the effective capacity of the finishing department to 10.91 sculptures per hour. Using a computer simulation, we find that the increase in the effective capacity of the touch-up station increases the throughput of the finishing department from 7.04 to 7.29 sculptures per hour.

- *Increasing Capacity of the Quality Control Station:* To study the impact of an increase in the effective capacity of the station that feeds the bottleneck station, we reduce the effective mean process time of the quality control station, similarly, by 0.5 minutes from 5 to 4.5 minutes per sculpture. This increases the effective capacity of the quality control station from 12 to 13.3 sculptures per hour. This, however, does not increase the *effective capacity* of the finishing department, since the quality control station is not the bottleneck. Therefore, the question is whether this will increase the *throughput* of the finishing department. The answer is "yes." Using our computer simulation we find that it increases the throughput of the finishing department from 7.04 to 7.18. Why? The reason is that increasing the effective capacity of the quality control station that feeds the touch-up station (i.e., the bottleneck station) reduces the idling of the touch-up station due to starvation. This is because the quality control station can now feed the touch-up station at a faster rate. Note that this increase in throughput is not the result of an increase in the effective capacity of the finishing department, since the effective capacity of the finishing department remains at 10 sculptures per hour—the effective capacity of its bottleneck. It is the result of reducing the throughput loss by reducing the starvation of the bottleneck station.

- *Increasing Capacity of the Packaging Station:* Similarly, we increase the effective capacity of the packaging station by reducing its effective mean process time by 0.5 minutes from 5.5 to 5 minutes per sculpture. This increases the effective capacity of the packaging station from 10.9 to 12 sculptures per hour. This does not increase the effective capacity of the finishing department; but, will it increase the throughput of the finishing department? Yes. Increasing the effective capacity of the packaging station (i.e., the non-bottleneck station) that receives units from the touch-up station (i.e., the bottleneck station) also increases the throughput of the finishing department from 7.04 to 7.23 sculptures per hour, without increasing the effective capacity of the finishing department. As opposed to the previous case, this is done by reducing the blocking of the touch-up station. Specifically, when the capacity of the packaging station increases, the station can process units faster, which results in reduction in the likelihood of having a full input buffer. This reduces the blocking of the touch-up station, resulting in an increase in its utilization. Higher utilization of the bottleneck station results in higher throughput of the finishing department.

Finally, note that increasing the capacity of the packaging station results in higher throughput (i.e., 7.23 sculptures per hour) than that when the capacity of the quality control station is increased. In general, increasing the capacity of the non-bottleneck station with the smallest capacity among all non-bottleneck stations results in higher increase in throughput than increasing the capacity of any other non-bottleneck station. This is because the non-bottleneck station with the smallest effective capacity is more likely to cause blocking and starvation than other non-bottleneck stations.[2]

Our observations from the study of Museart's finishing department lead us to the following principle:

---

**PRINCIPLE**

**Impact of Capacity on the Throughput of Processes with Limited Buffers**

In processes with limited intermediate and output buffers,

- The throughput of the process increases as the effective capacity of the bottleneck resource increases. The throughput also increases as the effective capacity of non-bottleneck resources increases; however, the increase in throughput is not as significant as that when the capacity of bottleneck resource is increased.

- Among non-bottleneck resources, increasing the effective capacity of the non-bottleneck resource with the smallest capacity, in general, results in higher increase in throughput than increasing the effective capacity of any other non-bottleneck resource.

---

**Impact of Variability on Throughput**

To investigate the impact of variability on the throughput of a process with limited intermediate buffers, we use our computer simulation program and compute the throughput of the finishing department under three cases of: (i) reducing the variability of effective process time ($CV_{eff}$) of the quality control station from 1 to 0.5, by reducing its standard deviation to 2.5 minutes, (ii) reducing the variability of effective process time ($CV_{eff}$) of the touch-up station from 1 to 0.5, by reducing its standard deviation to 3 minutes, and (iii) reducing the variability of effective process time ($CV_{eff}$) of the packaging station from 1 to 0.5, by reducing its standard deviation to 2.75 minutes. The results of the simulation are presented in Table 6.2.

As the table shows, reducing variability in any station results in an increase in throughput of the finishing department, since it results in reduction in bottleneck's starvation or blocking. The variability reduction at the bottleneck station—the touch-up station—results

---

[2]Hopp, W.J. and M. L. Spearman, *Factory Physics*. Third Edition, Waveland Press, Inc, 2011.

Table 6.2    Impact of Reducing Variability on Throughput of the Finishing Department

|  | Current situation | Reducing variability in | | |
|---|---|---|---|---|
|  |  | Quality control | Touch-up | Packaging |
| **Touch-up station** |  |  |  |  |
| Percent working | 70.40% | 73.66% | 75.84% | 74.76% |
| Percent idle (starved) | 11.25% | 5.81% | 9.76% | 12.25% |
| Percent idle (blocked) | 18.35% | 20.53% | 14.40% | 12.99% |
| Throughput of the finishing department | 7.04 | 7.35 | 7.57 | 7.43 |

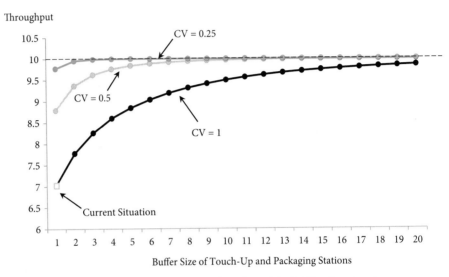

Figure 6.4    Impact of variability and buffer size on the throughput of the finishing department.

in reduction of both blocking (from 18.35 to 14.4 percent) and starvation (from 11.25 to 9.76 percent) and thus increases the bottleneck utilization by 5.44% (from 70.40 to 75.84 percent). This, in turn, reduces the throughput loss (due to blocking and starvation) and therefore increases the throughput of the finishing department from 7.04 to 7.57 sculptures per hour.

The table also shows that reducing variability at the bottleneck station yields a higher throughput than reducing variability in a non-bottleneck station. While this is true in the case of the finishing department of Museart, and is often generally true when there is a large gap between the effective capacity of the bottleneck station and that of non-bottleneck stations, it may not always be true in other settings. The impact of variability reduction on throughput of processes with finite buffers is complex and depends on many factors such as flow routings, the number of resources in each station, and the level of variability itself. Nevertheless, in general, variability reduction at the bottleneck resource should always be seriously considered as an alternative for improvement since, at the minimum, it reduces inventory and flow time in that station (see Chap. 7).

Finally, we investigate the impact of both variability reduction and buffer size increase on the throughput of the finishing department as shown in Fig. 6.4. In the figure, we simultaneously decrease the variability of all three stations from $CV_{eff} = 1$ to $CV_{eff} = 0.5$ and to $CV_{eff} = 0.25$. Furthermore, for each case, we increase the buffer sizes of both the touch-up and the packaging stations.

As the figure shows, regardless of the size of the buffers, decreasing the variability in the finishing department increases its throughput. Furthermore, when the variability of all stations is reduced to $CV_{eff} = 0.25$, the throughput of the finishing department gets very close to its effective capacity of 10 sculptures per hour, even with small buffer sizes of 1 or 2. We also observe that, when the variability is reduced, the throughput of the finishing department approaches its effective capacity faster as the buffer sizes increase.

The above observations lead us to the following principle:

---

**PRINCIPLE**

**Impact of Variability on the Throughput of Processes with Limited Buffers**

In processes with limited intermediate buffers, the throughput of the process increases as the variability in the system decreases. The throughput of processes with lower variability approach their effective capacities faster as their buffer size increases.

---

**Throughput Improvement at Museart**

John, the production manager of Museart, is now asked to find ways to increase the throughput of Museart's finishing department by at least 25 percent from current throughput of 7.04 to at least 8.8 sculptures per hour. He is also told that he can use part-time workers from other departments at Museart to increase the capacity of his department by 15 percent from 10 to 11.5 sculptures per hour. However, John is told that he can only ask for additional labor if there are no other alternatives to increase throughput.

John does not like this. How could they ask for a 25-percent increase in throughput, he thinks, but are willing to provide labor that can increase the capacity of the finishing department by only 15 percent. He, however, has no choice but to, at least, explore possible alternatives for getting throughput of 8.8 sculptures per hour. After considering different actions, he narrows them down to the following three alternatives:

*Alternative 1: Increasing Buffer Size*    John is considering making the buffer of both the touch-up and the packaging stations larger. John believes that this will reduce his throughput loss due to the idling at the touch-up station (i.e., starvation and blocking), and thus the throughput of the station will get closer to its effective capacity of 10 sculptures per hour. This will have no impact on the effective capacity or variabilities of the stations.

*Alternative 2: Increasing Capacity*    Another option that John is considering is to increase the capacity of the finishing department by 15 percent to 11.5 sculptures per hour without increasing the buffer sizes. This alternative is attractive since it is difficult to increase the buffer size of each station due to the limited space of the finishing department. John knows that increasing the capacity does not have a significant impact on the variabilities of activities in each station. To increase the capacity of the finishing department to 11.5 sculptures per hour, John must increase the effective capacities of both the touch-up and the packaging stations, respectively, from 10 and 10.91, to 11.5 sculptures per hour.[3]

---

[3]Note that when the capacity of the touch-up station—the bottleneck—increases to 10.91, then both the packaging and the touch-up stations become bottlenecks. So, to increase the effective capacity of the department to 11.5 sculptures per hour, the effective capacity of both the touch-up and the packaging stations must be increased to 11.5.

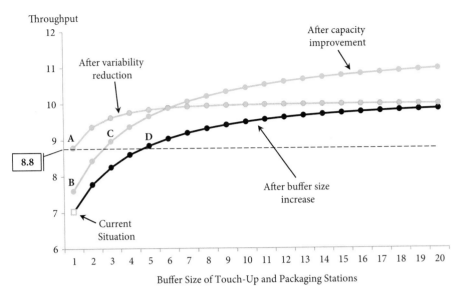

Figure 6.5   Three alternatives to improve the throughput of the finishing department.

***Alternative 3: Reducing Variability***   The third option is to decrease the variability of the stations in the finishing department under the current capacity. John is considering this option to see if it is possible to increase throughput without increasing the capacity (since he may not get the additional labor) or without increasing buffer sizes (since it might be too costly to do so). John thinks that he can reduce the variability of all three stations at the finishing department from CV = 1 to CV = 0.5.

Again, we use our computer simulation program to evaluate the impact of John's three alternatives on the throughput of the finishing department. For Alternative 1, we simultaneously increase the size of buffers at the touch-up and the packaging stations and obtain the throughput under each scenario using our simulation model. The increase in throughput is shown in Fig. 6.5 as "after buffer size increase." The increase in throughput of the finishing department under Alternative 2 is shown in the figure as "after capacity improvement." Finally, the throughput under Alternative 3 is shown as "after variability reduction." From Fig. 6.5, we can make the following observations:

- To implement Alternative 1, both buffers of the touch-up and the packaging stations should be expanded to include a minimum of five sculptures each. This corresponds to Point D in the figure, for which the throughput of the finishing department is estimated to be 8.84 sculptures per hour.

- Alternative 2 that increases the effective capacity of the finishing department by 15 percent cannot achieve the throughput of 8.8 sculptures per hour. This corresponds to Point B in the figure. The throughput of the finishing department in this case is 7.6 sculptures per hour. To achieve the throughput of 8.8 under the improved capacity, the size of both buffers at the touch-up and the packaging stations must be increased to at least 3. This corresponds to Point C in the figure.

- Alternative 3 that reduces the variability in the three stations achieves the desired throughput of 8.8 sculptures per hour without adding capacity or increasing buffer sizes anywhere in the finishing department. This corresponds to Point A in the figure.

Of course, there are other combinations of the above three policies that one can consider, but those are not discussed here. The important point we would like to make is that similar to the case of Michigan Yacht Club, variability reduction can have a significant impact on throughput of processes with intermediate finite buffers. The good news is that variability reduction is usually a less costly alternative than increasing capacity or increasing buffer sizes.

### 6.2.5 Throughput Loss Due to Control Policies

In the previous sections, we showed how constraints such as limited physical space in buffers, or customer impatience, may result in loss of throughput. In this section we show that even when these constraints do not exist, some processes may still lose throughput due to control policies they use to govern activities in the process. We use the case of H-Foods to illustrate this.

**Fresh Bread at H-Food**
H-Foods is a supermarket that sells organic food. Its bakery department makes a walnut-oat bread using organic ingredients every morning and sells them during the day. The store's policy is to sell fresh bread every day, so it does not sell the leftover bread that are not sold during the day. The unsold bread is donated to a soup kitchen that serves needy people. The demand for the bread on each day is independent of the demands on other days and is found to be well-approximated by a Normal distribution with an average of 50 and standard deviation of 15 loaves. The store makes 50 loaves of bread every day, equal to the expected demand on each day.

Due to variability in demand, some days H-Foods sells all 50 loaves that it makes, and some other days it ends up with some unsold bread. Data of the last 100 days show that under daily production quantity of 50 loaves, the average daily sales (i.e., throughput) has been about 44 loaves, and the average inventory of unsold bread at the end of the day has been about 6 loaves. Note that the average demand in the market is for 50 loaves on each day, and H-Foods makes 50 loaves of bread each day. However, the throughput of the store is only TH = 44 loaves of bread—a throughput loss of 50 − 44 = 6 loaves. What is the underlying reason for the 6 units of throughput loss?

Well, one reason for the throughput loss is the variability in demand. If the demand was exactly 50 units every day (with no variability), then the store would sell all of its 50 loaves (i.e., TH = 50) and there would be no leftover inventory and no throughput loss. In the presence of variability, however, the reason for having 6 units of throughput loss is not all because of variability, it is also because of the production control policy of the store. If the store produces 60 units each day, for instance, its sales (i.e., throughput) is increased to about 48 units[4] and thus the throughput loss is reduced to 50 − 48 = 2 units. Figure 6.6 shows the throughput and the inventory of H-Foods under different production control policies.

From Fig. 6.6, we make the following observations:

- Similar to balking, abandonment, and limited buffer spaces, process control policies used to manage operations in a process can also result in throughput loss. As the figure shows, by increasing production quantity, the throughput loss is reduced. For example, by making 80 loaves of bread each day, H-Foods can reduce its throughput loss to almost zero.

- Reducing throughput loss comes at the cost of increasing inventory. As the figure also shows, making 80 loaves of bread each day results in the average inventory of

---

[4]The problem that H-Foods faces is called newsvendor problem. We will present an analytics in Chap. 10 that finds the sales (throughput) and leftover inventory for this class of problems.

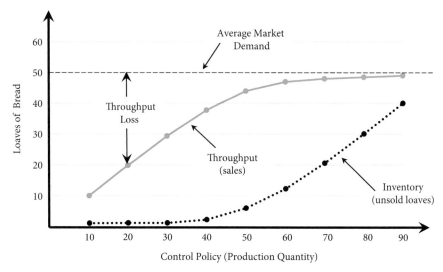

Figure 6.6   **Throughput and inventory of H-Foods under different production control policies.**

about 30 loaves of unsold bread each day. Hence, higher production quantities that result in higher throughput—and thus less throughput loss—also result in larger inventory in the process.

Control policies are rules that authorize resources to take actions regarding processing flow units in a process. In the case of H-Foods, the control policy authorizes the production of 50 loaves of bread every day. Control policies can increase throughput loss and therefore reduce the throughput of a process, often by limiting the inventory in the process. The challenge is, therefore, to find the best process control policy that balances the cost of carrying inventory and that of throughput loss. This is one of the main goals of operations engineers and managers, and is the focus of the most chapters of this book.

We conclude this section with the following throughput management principle:

---

**PRINCIPLE**

**Impact of Process Control Policies on Throughput**

Process control policies that limit inventory in the process by not authorizing resources to process more flow units, or by not allowing flow units to enter the process, result in throughput loss.

Now that we have determined major factors that can result in throughput loss, in the following section we introduce actions that can be taken to reduce throughput loss and therefore improve throughput.

## 6.3 Improving Throughput of a Process

Increasing throughput of a process has a direct impact on increasing sales and thus on a firm's profit. In the previous sections, we learned the fundamental relationship between throughput of a process and its demand, capacity, buffer size, and variability. In this section, we provide a framework for improving the throughput of a process as shown in Fig. 6.7.

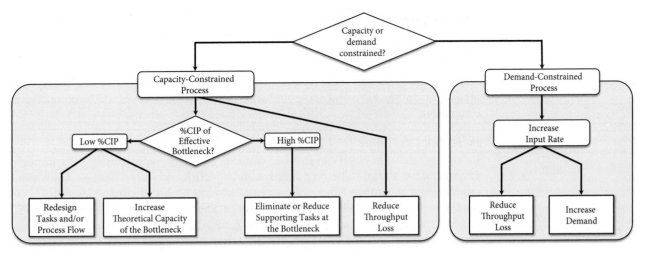

**Figure 6.7    Throughput improvement strategies for demand-constrained and capacity-constrained processes.**

The framework in the figure is based on whether the process is a demand-constrained process or a capacity-constrained process.

Recall that when the demand of a process is less than its effective capacity, the process is demand-constrained. In these processes, increasing the effective capacity of the process does not increase throughput. The focus should be on increasing the input rate of the process by increasing demand and also on decreasing throughput loss, as shown in Fig. 6.7.

In a capacity-constrained process, on the other hand, the effective capacity of the process is less than the demand, so all the efforts to improve throughput should focus on increasing the effective capacity of the process as well as on reducing throughput loss. As Fig. 6.7 depicts, increasing the effective capacity of a process requires increasing the effective capacity of its bottleneck resource. In the following sections, we present strategies that can improve throughput in demand- and capacity-constrained processes.

## 6.4 Improving Throughput of Demand-Constrained Processes

To increase the throughput of demand-constrained processes, we must increase the input rate of the process. The increase in input rate can be achieved (i) by increasing the demand of the process, or (ii) by reducing the throughput loss due to losing arriving demand. We discuss these two strategies in this section.

### 6.4.1 Increasing Demand of the Process

Demand for a product or a service can be increased by providing lower prices and better quality products or by an effective advertising campaign.

**Decreasing Price**

Offering lower prices for a wide range of items was one of the main reasons of Walmart's success and its large market share in retail industry. Besides reducing the original price, the demand for a product can be increased through other pricing mechanisms such as quantity discount (e.g., buy one, get one free), or discount for members (e.g., mileage accumulation for air travel).

### Improving Quality

Quality of a product has become almost as important as its price when a customer chooses a product. Using resources such as Internet, consumers have the advantage of reading about the quality of a product before they make a decision to buy it. For example, online retailers such as Amazon provide these reviews to its customers. Therefore, improving quality of a product can increase demand of the product, especially when prices of similar products are comparable.

### Increasing Sales Effort

Increasing sales effort also can increase demand. For example, increasing sales through advertising has shown to be very effective. In late 1970s and early 1980s, Reebok's sport products sold better than Nike's, who were selling a line of shoes for marathon runners. In the 1980s, they started their "Just Do It" campaign, and produced a series of smart, humorous, and cool advertising campaigns. These all resulted in an increase in their market share from 18 to 43 percent, and their sales increased from $800 million a year in 1988 to around $9.2 billion in 1998.[5]

## 6.4.2 Reducing Throughput Loss

Even when demand for a product is increased, the process can still lose input rate, and thus lose throughput, because of the lack of supply, balking, or reneging.

### Improving the Supply Mechanism

As we illustrated in the case of Campus Crust Pizza, lack of supply (i.e., pizza doughs) can reduce throughput of a process, even when the process has enough capacity and demand. Thus, an effective inventory control of supplies is essential in maintaining a high throughput. In Chaps. 10 and 11, we will discuss efficient and cost-effective inventory control mechanisms that reduce the possibility of supply stockouts and thus reduce the loss of throughput due to lack of supply.

### Reducing Balking

In general, balking occurs when an arriving customer decides not to enter the process because (i) she observes a long queue (e.g., a long queue outside a restaurant), or (ii) she does not see the queue but is informed about a long waiting time (e.g., told by the hostess that the waiting time to get a table is long), or (iii) she just gets the impression that the waiting time would be long. To reduce balking, one must either reduce the long queue or the long waiting time, or make the impression that waiting time would not be long. The former can be done by increasing the effective capacity of the process, or using a better queue control policy, or reducing variability in the interarrival or process times. We will discuss all these in detail in Chap. 8, when we discuss flow time improvement.

### Reducing Abandonment

To reduce abandonment, one must implement policies that either (i) shorten the waiting time in the queue or (ii) make the waiting time in the queue less annoying. Similar to balking, the former can be done by increasing the effective capacity or reducing variability. The latter can be done using the waiting time management principles that change customers' perception of waiting times, so that the customers do not feel they have been waiting for a long time—for example, installing TVs in waiting areas to make customers busy so that they are less annoyed by long waitings. We will also discuss these principles in detail in Chap. 8.

---

[5]Source: Internet, Top ten most successful advertising campaigns.

**Revising the Control Policy**

As in the case of H-Foods in Sec. 6.2.5, processes can lose throughput because of their control policies that do not authorize processing, when the flow unit, resource, and supply are available. For finding profit-maximizing process control policies (that often reduce throughput loss) in production or service systems, see Chaps. 10, 11, 12 and 13.

## 6.5 Improving Throughput of Capacity-Constrained Processes

Strategies used in demand-constrained processes are aimed at increasing input rate of the process by increasing demand or reducing throughput loss caused by factors that affect input rate. In capacity-constrained processes, however, throughput improvement strategies are focused on improving the effective capacity of the bottleneck of the process or by reducing throughput loss due to factors that result in idling of the bottleneck.

Recall from Chap. 4 that %CIP is a good indicator of potential for capacity improvement of a resource. A high %CIP indicates that a large portion of the capacity of the resource is spent (i.e., wasted) performing supporting tasks such as setups, rework, repairs, etc. Thus, in these cases, the focus of the capacity improvement should be on eliminating or reducing supporting tasks of activities performed by the bottleneck resource (see Fig. 6.7). On the other hand, a low %CIP implies that the resource spends a small portion of its time performing supporting tasks. Thus, eliminating or reducing supporting tasks, while always a good strategy, does not significantly increase the capacity of the resource. In this case, the focus should be on increasing the theoretical capacity of the bottleneck resource. This may be done by directly reducing the process time of the main task performed by the bottleneck resource (i.e., reducing the theoretical process time), by revising the tasks performed by the bottleneck resource or, in some cases, by redesigning the process flow (i.e., routings) within process boundaries. We now discuss these strategies in detail.

### 6.5.1 Eliminating or Reducing Supporting Tasks

For resources with high %CIP, reducing or eliminating supporting tasks such as setups, rework, repair, etc. can often significantly increase the capacity of the resource.

**Eliminate or Reduce Setup Times**

There have been a considerable amount of effort on finding ways to reduce setup times in manufacturing systems. These efforts resulted in a series of guidelines known as *Single-Minute Exchange of Dies* or SMED. These guidelines allowed manufacturers to significantly reduce and often eliminate setups. We discuss this in detail in Chap. 15 in which we discuss Lean Operations. Another approach to reduce the impact of setup time is to increase the number of flow units processed between setups. This also increases the effective capacity of a process. The only drawback of this approach is that it may increase the inventory because of having larger batch sizes. So, one should carefully balance the benefit of increasing the effective capacity through increasing the batch size, and the cost of having more inventory. We will further discuss this in Chap. 8.

**Eliminating or Reducing Yield Loss and Rework**

Quality issues in processing a flow unit result in need for rework, or even scraping the unit. The first step toward eliminating or reducing a quality issue is to find its root cause. After the cause is identified, it should be eliminated or its impact should be reduced through better training and/or utilizing appropriate tools. While quality issues such as rework and scraps are often visible in manufacturing settings, they are hidden in service settings, or even built right into the process as a required activity under titles such as "review," "revision," "editing," or "correction." To reduce rework and scrap, the goal should be "to do it right one time only, the first time." We discuss strategies for improving quality in Chap. 15.

**Increasing Availability of Bottleneck Resource**

Availability of a resource can be increased by increasing the average uptime (i.e., average time to failure) or decreasing the average downtime of the resource or both. The average time to failure of an equipment, for example, can be improved by implementing preventive maintenance (PM) policies, which are regular inspections and maintenance (e.g., cleaning, lubricating, adjustments) that keep equipment operational and in good condition. These scheduled maintenance can be performed outside of working hours so as not to interrupt the effective capacity of the bottleneck resource. The average downtime of the equipment, on the other hand, consists of the average time until the repair of the equipment starts plus the average repair time. The average time until the repair starts can be reduced by having enough repair crew available or by giving high priority to repairing the bottleneck equipment. The repair time of an equipment can be reduced by having standby (spare) parts or modules of the equipment ready to be used in case of a failure, and also by better training repair crew to handle repair of the bottleneck equipment in faster and more effective ways. A common practice to increase availability of equipment is to make workers responsible for routine maintenance of their own equipment and to create a culture that develops workers' pride in keeping their machines in top condition. We discuss this in more detail when we discuss Total Productive Maintenance (TPM) in Chap. 15.

### 6.5.2 Increasing Theoretical Capacity of Bottleneck

When %CIP of the bottleneck resource is low, then there is not much room to increase the effective capacity of the bottleneck resource by eliminating non-value-added (supporting) tasks. In this case, the effective capacity of the resource may be improved using different strategies that increases the theoretical capacity of a resource. Some of these strategies are discussed in this section.

**Reducing Theoretical Process Time**

Recall that the theoretical capacity of a resource is the long-run average number of flow units that the resource can process per unit time, if processing of a unit does not require any supporting task. It is the inverse of the sum of the theoretical process times of all the main activities performed by the resource. Hence, one way to increase the theoretical capacity of the bottleneck resource is to find ways to reduce the processing times of its main tasks.

- *Work and Time Study:* Increasing effectiveness of workers to perform tasks has been the focus of industrial engineers for many years. One of the main reasons for development of time and work study was to design more efficient operations by making them faster. The general approach to achieve this is to break the main tasks into smaller steps, eliminate the unnecessary steps, and find ways to make the necessary steps shorter.

- *Simultaneous Processing:* Finding ways to simultaneously work on several flow units decreases the theoretical process time of the resource, and thus increases its capacity. For example, baking four pizzas in the oven takes less time (per pizza) than baking pizzas one-by-one, one after another.

- *Worker Training:* Training workers with the appropriate skills needed to perform the main task increases workers' speed and decreases the variability of workers' process times. Faster processing of flow units increases the capacity of the worker, and lower variability can reduce throughput loss.

- *New Technology:* Equipment manufacturers invest a lot of money to design and manufacture faster and more flexible equipment. Faster computers, faster printing devices, and faster and more powerful machine tools are introduced by their manufacturers every day. Thus, one should always look for the latest technology for service or manufacturing processes. Of course, a feasibility study should always be performed to consider the costs and benefits of using the new technology.

- *Worker Incentives:* Worker incentives, if properly designed, can motivate workers to work faster. Firms often set a quota for workers' production, beyond which workers receive a bonus for each unit of production. Caution, however, should be taken not to sacrifice quality to achieve higher throughput. Working faster may come at the price of making lower quality products (i.e., higher scraps and rework).

### Increasing the Number of Resources

Increasing the number of resources (labor or equipment) and creating a pool of resources at the bottleneck increase the bottleneck's capacity. The only drawback is the cost of acquiring additional resources. Therefore, it is important that the cost of additional resources be evaluated and compared with the benefit of the increase in throughput. In general, investment on additional resources should be the last resort after all other strategies are considered. Obviously when additional resources are cheap and easy to acquire, this option becomes more attractive. Also, one should consider increasing the capacity carefully and step-by-step, since the benefit of increasing the capacity has a diminishing return. For example, doubling the capacity of the bottleneck may not double the capacity of the process. One reason is that increasing the capacity of the bottleneck beyond a certain point makes another station a bottleneck.

### Increasing Working Hours

Another way to increase the theoretical capacity of a resource is to increase its scheduled working hours. Many firms assign overtime to their bottleneck stations to increase their throughput. However, overtime is costly. In most cases, the cost of an hour of overtime is at least 1.5 times higher than the cost of a regular working hour. Furthermore, there are often rules and regulations (e.g., union regulations) that limit the number and frequency of overtime that can be requested from workers.

### Outsourcing Activities or Units

Outsourcing the bottleneck activity or flow unit refers to buying similar capacity or flow units from other firms to compensate for the low capacity of the bottleneck. Sometimes firms decide to completely outsource a part of their process. For example, Taco Bell, a fast-food restaurant, has outsourced much of its food preparation to other companies, who process food more efficiently and in larger volumes than Taco Bell restaurants. While outsourcing can be a good alternative for increasing the capacity, it has its own drawbacks. It requires a high level of planning and coordination and involves risks. For example, outsourcing critical activities may require sharing intellectual property and know-hows, which in the long run, may backfire and result in the firm losing its technological advantage. Even, when intellectual property is not an issue, quality can be. For example, in December 2006, more than 60 customers of Taco Bell became sick because of the *E. coli* in Taco Bell restaurants. The cause was found to be the food processed and delivered to Taco Bell by other firms.[6]

---

[6]Boyer and Verma (2010).

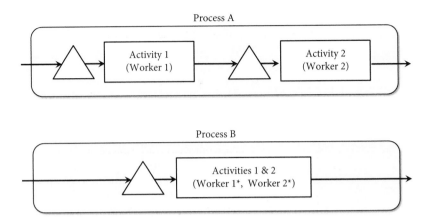

Figure 6.8    Horizontal pooling; combining Activities 1 and 2 and making a resource pool.

### 6.5.3 Redesigning Activities and Process Flow

In addition to increasing bottleneck capacity, the theoretical capacity of a process can also be increased by redesigning activities performed by the resources or changing the flow of units (i.e., routing) within the process. Below we present some of these strategies.

**Making Activities Self-Service**

Making activities self-service can significantly increase the capacity of a process. It eliminates or reduces activities that are performed by resources of a process and assigns them to the customers. For example, checkout cashiers are the bottleneck at grocery stores, and there are often long queues in front of them. Stores have increased the capacity of their checkout stations using self-checkout stations in which customers scan, weight, and pay for their groceries using self-checkout machines. Other examples of self-service activities are airlines online check-ins and check-in kiosks at the airport that allow passengers to check-in and print their boarding passes.

**Assigning Some Bottleneck Activities to Non-bottleneck Resources**

As discussed above, self-service activities can increase the capacity of the bottleneck resource by moving some of its activities to the customers. If it is not possible to move bottleneck activities to the customer (e.g., in manufacturing settings), the capacity of the bottleneck resource can still be increased by assigning some of its activities to non-bottleneck resources. Note that non-bottleneck resources have unused capacity which can be utilized for this purpose. This may require some investment or training of the non-bottleneck resources, which should be considered when making such decisions.

**Pooling Bottleneck Resource with Non-bottleneck Resources**

Another way to increase the capacity is to combine the activity performed by the bottleneck resource with that performed by a non-bottleneck resource, and then assign the two activities to the pool of bottleneck and non-bottleneck resources. Consider the simple Process A in Fig. 6.8 which consists of Activities 1 and 2, performed by Worker 1 and Worker 2, respectively. The theoretical process times of workers are

$$T_0(\text{worker } 1) = T_0(activity\ 1) = 10 \text{ minutes per flow unit}$$
$$T_0(\text{worker } 2) = T_0(activity\ 2) = 5 \text{ minutes per flow unit}$$

All units processed by Worker 1 are sent to Worker 2 for processing. Obviously, the theoretical bottleneck is Worker 1; thus, Process A will have theoretical capacity of $C_0(\text{process A}) = 60/10 = 6$ flow units per hour. Now suppose that we combine Activities 1 and 2 into one activity, which we call Activity 1&2, and assign both workers to this larger activity. Specifically, we ask both Worker 1 and Worker 2 to, independently and in parallel, perform both Activities 1 and 2 on their flow units, before they start working on the next flow unit. This new process is shown as Process B in Fig. 6.8. For this process we have

$$T_0(\text{worker 1}) = T_0(\text{activity 1}) + T_0(\text{activity 2})$$
$$= 10 + 5 = 15 \text{ minutes per flow unit}$$

Therefore, the theoretical capacity of Worker 1 is $C_0(\text{worker 1}) = 4$ flow units per hour. Similarly, since Worker 2 performs the same activities, the theoretical capacity is $C_0(\text{worker 2}) = 4$ flow units per hour. Thus, the theoretical capacity of Process B is

$$C_0(\text{process B}) = C_0(\text{worker 1}) + C_0(\text{worker 2})$$
$$= 4 + 4 = 8 \text{ flow units per hour}$$

By combining the two activities—one having theoretical mean process time twice larger than the other—and assigning the task to the pool of two workers, we were able to increase the theoretical capacity of the process by one-third from 6 to 8 flow units per hour. This is referred to as Horizontal Pooling.

## CONCEPT

### Horizontal Pooling of Resources

Consider two Resources $R_1$ and $R_2$, and two Activities $A_1$ and $A_2$. Resource $R_2$ performs Activity $A_2$ on a flow unit after Activity $A_1$ is completed by Resource $R_1$ on that flow unit. All units require processing by both resources. Under horizontal pooling of these two resources, each one of the resources performs both activities on a flow unit independent of the other resource. Specifically, Resource $R_1$ performs activities $A_1$ and $A_2$ on a flow unit and Resource $R_2$ also performs activities $A_1$ and $A_2$, but on another flow unit.

As the example in Fig. 6.8 shows, by implementing horizontal pooling, one can increase the capacity of the process. How much the capacity of a process increases depends on the processing times of the activities performed by the resources. The following analytics obtains the percent increase in capacity after horizontal pooling of two resources.

## ANALYTICS

### Percent Increase in Capacity with Horizontal Pooling

Consider two resources $R_b$ and $R_o$ in Process I below, who perform Activities $A_b$ and $A_o$, respectively. Activity $A_b$ has mean (theoretical or effective) process time $T_b$, and thus Resource $R_b$ has capacity $C_b = 1/T_b$. Activity $R_o$ has mean process time $T_o$ and hence Resource $A_o$ has capacity $C_o = 1/T_o$.

Suppose $T_b > T_o$, which implies that Activity $A_b$ is the bottleneck activity. Therefore, the capacity of Process I is

$$\text{Capacity of Process I} = \min\{1/T_o, 1/T_b\} = 1/T_b$$

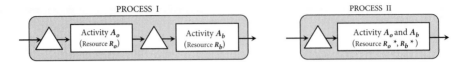

Now suppose that we implement horizontal pooling of the two resources as shown above in Process II. In Process II, both resources perform the two activities, and thus each resource has mean (effective or theoretical) process time $T_b + T_0$ and capacity $1/(T_b + T_o)$. Therefore, the capacity of Process II will be the sum of the capacities of the two resources:

$$\text{Capacity of Process II} = \frac{2}{T_b + T_o} = \frac{2}{T_b(1 + T_o/T_b)}$$

The percent increase in capacity of Process I due to horizontal pooling (resulting in Process II) is, therefore,

$$\begin{aligned}
\text{Percent Increase in Capacity} &= \frac{\text{Capacity of Process II} - \text{Capacity of Process I}}{\text{Capacity of Process I}} \\
&= \frac{\text{Capacity of Process II}}{\text{Capacity of Process I}} - 1 = \frac{2/[T_b(1 + T_o/T_b)]}{1/T_b} - 1 \\
&= \frac{2}{1 + T_o/T_b} - 1
\end{aligned} \tag{6.4}$$

Note that Worker 1 in Fig. 6.8 has process time $T_1 = 10$ minutes, and Worker 2 has process time $T_2 = 5$ minutes. Hence, Worker 1 is the bottleneck. Using Eq. (6.4), the percent increase in capacity due to horizontal pooling of the two workers would be

$$\text{Percent Increase in Capacity} = \frac{2}{1 + T_o/T_b} - 1 = \frac{2}{1 + 5/10} - 1 = 0.333 = 33.33\%$$

While horizontal pooling corresponds to pooling resources that work on flow units in series, vertical pooling corresponds to pooling of resources that work in parallel. Consider Process A in Fig. 6.9 in which fraction $p_1 = 0.6$ of the flow units is sent to Worker 1, and fraction $p_2 = 0.4$ is sent to Worker 2. The theoretical mean process times of workers are

$$T_0(\text{worker } 1) = T_0(\text{activity } 1) = 10 \text{ minutes per flow unit}$$
$$T_0(\text{worker } 2) = T_0(\text{activity } 2) = 5 \text{ minutes per flow unit}$$

Note that Process A is a process with two types of products—one being processed by Worker 1 and the other one being processed by Worker 2—with product mix $(0.6, 0.4)$. Thus, we use product aggregation method to compute the capacity of Process A. Following this method, the theoretical mean process time of Worker 1 for the aggregate product is $T_0^{(agg)}(\text{worker } 1) = 0.6 \times 10 + 0.4 \times 0 = 6$ minutes. Hence, the theoretical capacity of Worker 1 is $C_0^{(agg)}(\text{worker } 1) = 60/6 = 10$ per hour. Similarly, the theoretical mean process time of Worker 2 for the aggregate product is $T_0^{(agg)}(\text{worker } 2) = 0.6 \times 0 + 0.4 \times 5 = 2$ minutes, and the worker has theoretical capacity of $C_0^{(agg)}(\text{worker } 2) = 60/2 = 30$ aggregate

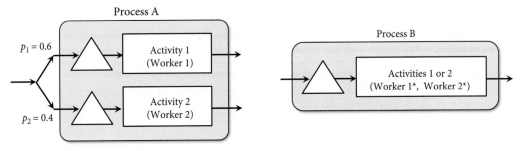

Figure 6.9    Vertical pooling: aggregating Activities 1 and 2 and making a resource pool.

products per hour. Hence, the theoretical capacity of Process A is

$$C_0^{(agg)} \text{ (process A)} = \min\{C_0^{(agg)} \text{ (worker 1)}, C_0^{(agg)} \text{ (worker 2)}\}$$
$$= \min\{10, 30\} = 10 \text{ aggregate products per hour}$$

Now consider Process B in Fig. 6.9 in which Workers 1 and 2 are pooled. Specifically, both workers can process type-1 and type-2 products. In this case, the theoretical mean process time of Worker 1 for the aggregate product is $T_0^{(agg)}$ (worker 1) $= 0.6 \times 10 + 0.4 \times 5 = 8$ minutes. The theoretical capacity of Worker 1 is therefore $C_0^{(agg)}$ (worker 1) $= 60/8 = 7.5$ aggregate products per hour. Since Worker 2 also processes the same two products with the same product mix, the worker will have the same theoretical capacity of 7.5 aggregate products per hour. Consequently, the theoretical capacity of Process B is

$$C_0^{(agg)} \text{ (process B)} = C_0^{(agg)} \text{ (worker 1)} + C_0^{(agg)} \text{ (worker 2)}$$
$$= 7.5 + 7.5 = 15 \text{ flow units per hour}$$

As this example shows, by pooling two workers to do Activities 1 or 2, when needed, we were able to increase the capacity of Process A by 50 percent from 10 to 15 units per hour. We refer to this as Vertical Pooling.

---

**CONCEPT**

**Vertical Pooling of Resources**

Consider two Resources $R_1$ and $R_2$ who perform two different Activities $A_1$ and $A_2$, respectively. Each resource performs its activity independent of the other resource, and each flow unit requires only one of the two activities. Specifically, fraction $p_1$ of flow units, which we call type-1 flow units, require only Activity $A_1$, and fraction $p_2$ of flow units, which we call type-2 flow units, require only Activity $A_2$. Under vertical pooling of resources, both resources can process two types of flow units, if needed. Hence, a flow unit of type 1 can be processed by any of the two resources (whoever is available). After being processed, the flow unit leaves the process.

Similar to horizontal pooling, as the example in Fig. 6.9 shows, vertical pooling can also result in increasing the capacity of a process. The percent increase in capacity depends on the processing times of the resources as well as on the product mix, as illustrated in the following analytics.

**Percent Increase in Capacity with Vertical Pooling**

Consider Process I in which two Activities $A_o$ and $A_b$ are performed, respectively, by Resources $R_o$ and $R_b$ in parallel and independent of each other. Fraction $p_o$ of flow units requires only Activity $A_o$, and fraction $p_b$ of flow units requires only Activity $A_b$. We call flow units that require Activity $A_o$ type-$o$ flow units and those requiring Activity $A_b$, type-$b$ flow units. After an activity is performed on a flow unit, the unit leaves the process.

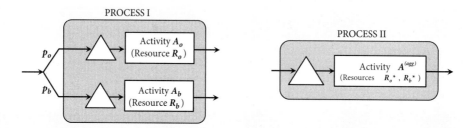

Also suppose the mean (effective or theoretical) process time of Activities $A_o$ and $A_b$ are $T_o$ and $T_b$, respectively. Using the product aggregation method, Resource $R_o$'s process time for the aggregate product will be $T_o^{(agg)} = p_o T_o$, and Resource $R_b$'s process time for the aggregate product will be $T_b^{(agg)} = p_b T_b$. Thus, Resource $R_o$ has the capacity of $C_o^{(agg)} = 1/(p_o T_o)$ for the aggregate product, and Resource $R_b$ has the capacity of $C_b^{(agg)} = 1/(p_b T_b)$ for the aggregate product.

Suppose that $p_o T_o < p_b T_b$, which implies that Resource $R_b$ is the bottleneck. Hence, the capacity of Process I will be

$$\text{Capacity of Process I} = \min\{C_o^{(agg)}, C_b^{(agg)}\} = C_b^{(agg)} = 1/p_b T_b$$

Now consider Process II above in which we implement vertical pooling. Specifically, we pool the two resources such that each resource can perform any one of the activities, depending on which type of flow unit is available. In this case, both resources have the same mean process time to process the aggregate product as follows:

$$T^{(agg)} = p_o T_o + p_b T_b \tag{6.5}$$

Thus, the capacity of each resource is $1/T^{(agg)} = 1/(p_o T_o + p_b T_b)$ aggregate products per hour. Consequently,

$$\text{Capacity of Process II} = \frac{2}{p_o T_o + p_b T_b} = \frac{2}{p_b T_b (1 + p_o T_o / p_b T_b)}$$

The percent increase in capacity of Process I due to vertical pooling (resulting in Process II) is, therefore,

$$\begin{aligned}
\text{Percent Increase in Capacity} &= \frac{\text{Capacity of Process II}}{\text{Capacity of Process I}} - 1 \\
&= \frac{2/[p_b T_b (1 + p_o T_o / p_b T_b)]}{1/p_b T_b} - 1 \\
&= \frac{2}{1 + p_o T_o / p_b T_b} - 1
\end{aligned} \tag{6.6}$$

Recall that Worker 1 in Fig. 6.9 processes fraction $p_1 = 0.6$ of the units with process time $T_1 = 10$ minutes. Worker 2, on the other hand, processes fraction $p_2 = 0.4$ of the units with process time $T_2 = 5$ minutes. Since Worker 1 is the bottleneck, we set $p_b = p_1 = 0.6$ and $T_b = T_1 = 10$ minutes, and since Worker 2 is not the bottleneck, we set $p_o = p_2 = 0.4$ and $T_o = T_2 = 5$ minutes. Using Eq. (6.6), Percent Increase in Capacity due to vertical pooling of the workers will be

$$\text{Percent Increase in Capacity} = \frac{2}{1 + p_o T_o / p_b T_b} - 1$$
$$= \frac{2}{1 + 0.4(5)/0.6(10)} - 1 = 50\%$$

Ratio $T_o/T_b$ in Eq. (6.4) is equal to ratio $C_b/C_o$, that is, the ratio of capacities of bottleneck and non-bottleneck resources before horizontal pooling. Similarly, ratio $p_o T_o / p_b T_b$ in Eq. (6.6) is equal to $C_b^{(agg)}/C_o^{(agg)}$, which is the ratio of capacities of bottleneck and non-bottleneck resources (for the aggregate product) before vertical pooling. Thus, we can summarize Eqs. (6.4) and (6.6) as follows. When a bottleneck resource is pooled with a non-bottleneck resource (vertically or horizontally) the percent increase in capacity would be

$$\text{Percent Increase in Capacity} = \frac{2}{1 + \dfrac{\text{capacity of the } \textit{bottleneck} \text{ resource}}{\text{capacity of the } \textit{non-bottleneck} \text{ resource}}} - 1$$

If we define

$$u_i^* = \frac{\text{Capacity of the } \textit{bottleneck} \text{ resource}}{\text{Capacity of resource } i} \tag{6.7}$$

then $u_i^*$ is, in fact, the utilization of resource $i$, if the process works at its maximum capacity— the capacity of the bottleneck resource.[7] So, pooling non-bottleneck resource $i$ with the bottleneck resource will result in percent increase in capacity:

$$\text{Percent Increase in Capacity} = \frac{2}{1 + u_i^*} - 1 \tag{6.8}$$

**Which Resources to Pool?**
Often times there are more than one non-bottleneck resource that can be pooled with the bottleneck resource to increase process capacity. So, the question is which resource to pool? The answer depends on the cost and effectiveness of the pooling (vertical or horizontal).

- *Effectiveness of Pooling:* From Eq. (6.8) it becomes clear that pooling—vertical or horizontal—becomes more effective in increasing the capacity of a process when the gap in the capacities of bottleneck and non-bottleneck resources that are pooled is large, that is, when $u_i^*$ is small.
- *Cost of Pooling:* Using a non-bottleneck resource to perform a new activity (i.e., bottleneck activity) often incurs costs such as worker training costs or cost of retooling a machine. Thus, when considering options for pooling, one should consider the cost of pooling and its effectiveness in increasing the capacity.

---

[7]Note that bottleneck resource will have utilization of 100 percent and non-bottleneck resources will have utilization of less than 100 percent.

We summarize the above observation in the following principle:

**Impact of Resource Capacity on the Benefit of Pooling**

As the gap between the effective capacities of the bottleneck and non-bottleneck resources increases, pooling—vertical or horizontal—results in higher percent increase in capacity. Thus, processes with sharper bottlenecks benefit more from pooling resources than those with more balanced (i.e., closer) effective mean process times.

### Changing Product Mix

As we discussed in Chap. 4, in multi-product processes, changing product mix can change the capacity of a process. Consider a bottleneck resource that processes two types of products with product mix $(p_1, p_2) = (0.8, 0.2)$. Also, suppose effective mean process times for products types 1 and 2 are 8 and 4 minutes, respectively. The effective mean process time for the aggregate product will be $0.8 \times 10 + 0.2 \times 5 = 9$ minutes per aggregate product. Thus, the effective capacity of the bottleneck resource will be $60/9 = 6.67$ aggregate products per hour. This implies that the maximum throughput of the bottleneck resource is a total of 6.67 products per hour, that is, 5.33 products of type 1, and 1.34 products of type 2.

Now, suppose the product mix is changed to $(p_1, p_2) = (0.2, 0.8)$. Under this new product mix, the effective mean process time of the aggregate product is $0.2 \times 10 + 0.8 \times 5 = 6$ minutes per aggregate product, and the effective capacity of the bottleneck resource will be $60/6 = 10$ aggregate products per hour, that is, two products of type 1, and eight products of type 2. This is a 50-percent increase in the effective capacity of the bottleneck resource, due to change in product mix.

We need to emphasize, however, that when choosing the right product mix, one should consider the profit margin of each product in addition to how it affects the throughput of a process. For example, if profit margin of products 1 and 2 are \$10 and \$4, respectively, the total profit under the product mix $(p_1, p_2) = (0.8, 0.2)$ is $\$10 \times 5.33 + \$4 \times 1.34 = \$58.66$ per hour, while the profit of the process with product mix $(p_1, p_2) = (0.2, 0.8)$ is $\$10 \times 2 + \$4 \times 8 = \$52$ per hour. In other words, the process with lower throughput results in higher profit than the process with higher throughput. We discuss product mix decisions in detail in Chap. 12.

### Redesigning the Process Flow

Another approach to increase the capacity of a process is to redesign the routings of the flow units in the process. As an example, consider Process A in Fig. 6.10 in which three workers perform three different activities of stamping, assembly, and inspection. Worker who does the inspection detects the quality issues of the final product, which results in scrapping of the product. Data of past inspections has shown that 10 percent of the products are scrapped due to quality problems resulting from the stamping activity, while the assembly activity results in almost no quality issues. For this process, theoretical (or effective) mean process times of stamping, assembly, and inspection are 15, 30, and 5 minutes, respectively. Thus, theoretical capacities of workers at stamping, assembly, and inspection are 4, 2, and 12 products per hour, respectively. The bottleneck is Worker 2 with theoretical capacity of 2 products per hour, $0.2 (= 10\% \times 2)$ of which are discarded due to quality issues with stamping. Hence, the theoretical capacity of Process A is $2 - 0.2 = 1.8$ products per hour.[8]

---

[8] Recall that the capacity of a process is defined as the maximum number of good (non-defective) products that the process can produce per unit time.

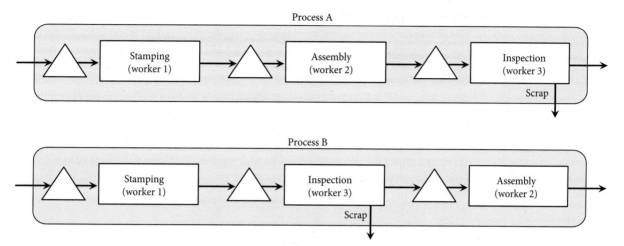

**Figure 6.10**   Redesigning Process A to Process B by moving the inspection before assembly.

Now, suppose that we redesign the process flow as shown in Process B in Fig. 6.10. Specifically, we move the inspection activity immediately after the stamping and before the assembly activity. In Process B, all quality issues are detected after stamping, and defective units are scrapped.

As Fig. 6.10 shows, Process B can be considered as a multi-product process. Specifically, we can divide products into two types: (i) non-defective products and (ii) defective products, and we consider the product mix of $(p_1, p_2) = (0.9, 0.1)$.

**Table 6.3**   Process Times of Activities for Process B in Fig. 6.10

|  |  | Theoretical mean process time | | |
|---|---|---|---|---|
| Resources | Activity | Type 1 (*minutes*) | Type 2 (*minutes*) | Aggregate (*minutes*) |
| Worker 1 | Stamping | 15 | 15 | 15 |
| Worker 3 | Inspection | 5 | 5 | 5 |
| Worker 2 | Assembly | 30 | 0 | 27 |

Having the product mix $(p_1, p_2) = (0.9, 0.1)$, we can compute the theoretical process time of each resource for an aggregate product, as shown in Table 6.3. For example, the theoretical mean process time of Worker 2 who performs assembly for the aggregate product is

$$T_0^{(agg)}(\text{Worker 2}) = 0.9 \times 30 + 0.1 \times 0$$
$$= 27 \text{ minutes per aggregate product}$$

Therefore, the theoretical capacity of Worker 2 is $60/27 = 2.222$ aggregate products per hour. As Table 6.3 depicts, the theoretical mean process times for Workers 1 and 3 who perform stamping and inspection, respectively, are 15 and 5 minutes per aggregate product, which implies theoretical capacities of 4 and 12 aggregate products per hour. Hence, the theoretical bottleneck of the process is Worker 2 in assembly with theoretical capacity of 2.222 aggregate products per hour. The capacity of Process B is therefore equal to the capacity of its bottleneck, that is, 2.222 aggregate products per hour, which translates into $2.222 \times 0.9 = 2$ type-1 (non-defective) and $2.222 \times 0.1 = 0.222$ type-2 (defective) products per hour. Thus, Process B will have theoretical capacity of 2 products per hour, which is $2 - 1.8 = 0.2$ products higher than the capacity of Process A. The question is, how can Process B have a higher theoretical capacity, without any change in the capacity of its bottleneck (i.e., assembly station)? Worker 2 in the assembly station is doing the same task of assembling products. How did his theoretical capacity increase from 2 to 2.222 products per hour?

To answer this question, consider the product mix (0.9, 0.1). Under this product mix, Process B has the theoretical capacity of processing a total of 2.222 aggregate products per hour, where $0.9 \times 2.222 = 2$ products are of type 1 and $0.1 \times 2.222 = 0.222$ products are of type 2. However, among the 2.222, only 2 products are sent to Worker 2 (the bottleneck worker). This is equal to Worker 2's theoretical capacity in Process A. So, Worker 2's theoretical capacity is still 2 products (of type 1 or type 2) per hour. However, Worker 2 only receives 2 type-1 (i.e., non-defective) products per hour. The remaining 0.222 products are type-2 (i.e., defective) products that are processed (i.e., stamped, inspected, and discarded) without taking bottleneck time. In Process A, type-2 products were also processed by the assembly (bottleneck) worker, which was a waste of the worker's capacity, since these products were discarded later in the inspection station. That is the main reason why Process B has a theoretical capacity of 2 products per hour, which is 0.2 products higher than that of Process A. This is an example of cases in which redesigning process flow within the process can increase the capacity of the process.

### 6.5.4 Reducing Throughput Loss

Capacity-constrained processes can lose throughput when their bottleneck resource becomes idle. As we described earlier, several reasons can cause idleness of the bottleneck, including lack of supply, blocking, and starvation. To decrease throughput loss, one should reduce or eliminate these causes of idleness.

**Reducing Blocking and Starvation**
Blocking and Starvation can be reduced in the following ways:

- If the output buffer of the bottleneck resource (i.e., the input buffer of the stage after the bottleneck) is limited, and if there is variability in the process, then the bottleneck resource may have periods of idling due to blocking. Thus, increasing the size of the output buffer of the bottleneck resource reduces blocking. Similarly, increasing the size of the input buffer of the bottleneck resource can decrease starvation. It should be noted that increasing the buffer size of other non-bottleneck stations can also decrease the blocking and starvation of the bottleneck resource, as we described in Sec. 6.2.4. But, the increase may not be as significant.

- Decreasing variability in the process (i.e., variability in arrivals or processing times of bottleneck and non-bottleneck resources) can also decrease blocking and starvation. To learn about different ways to decrease variability, see Chap. 5.

**Revising the Supply Mechanism**
Recall that processes can lose throughput because they do not have supply (e.g., material, parts) to complete the process of a flow unit. For capacity-constrained processes, the strategies for revising the supply mechanism to prevent throughput loss due to lack of supply are similar to those for demand-constrained processes. We will discuss these strategies in Chaps. 9, 10, and 11.

**Revising the Control Policy**
Similar to demand-constrained processes, capacity-constrained processes can also lose throughput due to the control policy they use to manage activities in their processes. See section "Revising the Control Policy" in Sec. 6.2.4.

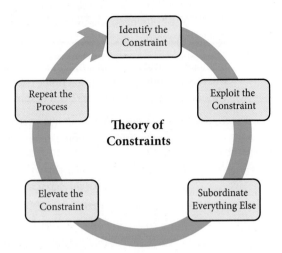

Figure 6.11    The five focusing steps of the Theory of Constraints.

### 6.5.5 Theory of Constraints

The Theory of Constraints was introduced recently as a systematic approach to manage capacity and increase throughput of a process.[9] According to the Theory of Constraints, the throughput of a process is determined by three types of constraints: (i) Market constraints, (ii) Internal resource constraints, and (iii) Policy constraints. Processes with market constraints are processes in which the demand for the process is less than process capacity. This is analogous to our demand-constrained processes. Similarly, processes with internal resource constraints are analogous to our capacity-constrained processes. Finally, processes with policy constraints are processes in which process control policy (e.g., the production control policy of H-Foods in Sec. 6.2.5) limits the throughput of the process.

The foundation of the Theory of Constraints, similar to what we did in this chapter, is to improve system performance (e.g., throughput) by focusing on these constraints. It includes the following steps, also shown in Fig. 6.11. Note that the Theory of Constraints is a general problem-solving approach with applications in many different contexts. Figure 6.11 only illustrates its five focusing steps in removing constraints in the context of improving throughput of a process.

Step 1    *Identify the Constraint:* The first step is to identify the constraint that limits the throughput of the process. In other words, identify whether the process is demand-constrained, capacity-constrained, or whether (in either case) the throughput is limited due to the control policy that governs the process.

Step 2    *Exploit the Constraint:* Focus on the constraint and, using existing resources, make quick improvements without committing to expensive changes. For example, if the process is capacity-constrained, try to reassign bottleneck activities to non-bottleneck resources, or combine these activities with non-bottleneck activities and make resource pools before deciding to utilize new technology or hire new worker—expensive options—to increase the capacity of the bottleneck resource.

Step 3    *Subordinate Everything Else:* Effective utilization of the constraint has the highest priority. With a plan to exploit the constraint, review all other activities in the process to make sure that those activities support the needs of the constraint. For example, schedule the non-bottleneck resources to maximize the effectiveness of

[9]Goldratt, Eliyahu M., *The Goal.* 2nd Edition, North River Press, Great Barrington, MA, 1992.

the bottleneck resource, or make enough inventory available so that the bottleneck does not suffer from lack of supply.

Evaluate the outcomes of implementing Steps 1 to 3 and check whether the constraint is still the limiting constraint that holds back throughput. If this is the case, then go to Step 4. Otherwise, the constraint has been eliminated and go to Step 5.

**Step 4** *Elevate the Constraint:* Take whatever action needed to eliminate the constraint. This may include capital investment. For example, in demand-constrained processes, this may include a big advertising campaign to boost demand. In capacity-constrained processes, this may include buying new equipment or hiring new workers to increase the capacity of the bottleneck.

**Step 5** *Repeat:* When a constraint is relaxed, the next constraint should immediately be identified. Thus, go to Step 1 and repeat the steps. While repeating these steps, monitor how changes in new constraints affect the already improved constraints.

Theory of Constraints emphasizes on the following important points:

- *Look for Inexpensive Fixes First:* To improve the constraints, one should first look for quick and inexpensive solutions using the existing resources before making any substantial investment to remove the constraint.

- *Bottleneck May Shift:* After implementing throughput improvement actions, bottleneck can shift from one resource to another, or the process can shift from being demand-constrained to capacity-constrained (or vice versa). Thus, one should take this into account when taking actions to exploit or elevate the constraint.

- *Never Stop:* Improving throughput of a process should be a continuous and ongoing improvement cycle. When one constraint is removed, one should always identify and remove the next constraint.

Now that we presented the list of strategies that can be used to improve the throughput of a process, in the following section we show how some of these strategies can be used to increase the throughput of a firm that offers loans for home buyers.

## 6.6 Improving Throughput at Easy Mortgage

Easy Mortgage is a financial institution that offers, among other financial products, mortgage (or loan) to home buyers. The home loan division of the company handles home loan applications from home buyers. Applicants submit their applications online through Easy Mortgage's website. Currently, the loan review team includes six employees: John, Jim, Tom, Frank, Stacy, and Ravi. Ravi is the head of the home loan division. The team (including Ravi) works 8 hours a day (from 8:00 am to 5:00 pm with lunch break from 12:00 to 1:00), and 5 days a week. Figure 6.12 exhibits the process-flow chart of the home loan division. The process starts when an application arrives at the division and ends when an approve/disapprove decision is made.

**Initial Review**

The home loan division processes applications according to first-come-first-served order. The review process starts by an initial review of applications, performed by John. He makes sure that all applications have proper documents and uploads the application file to the company's information system called MorSys. This takes John an average of 20 minutes. After uploading, the application becomes accessible to all members of the review team. To complete the review and make an approve/disapprove decision, four more documents are

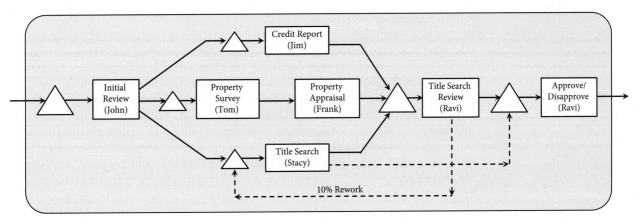

**Figure 6.12    Process-flow chart of the home loan division at Easy Mortgage.**

needed: (i) applicant's credit report, (ii) property survey, (iii) property appraisal, and (iv) title search. These documents are prepared, independently, by Jim, Tom, Frank, and Stacy.

### Credit Report

Jim is in charge of writing the credit report. Jim first sends requests through MorSys to credit score firms and asks for applicant's credit score and credit report. Easy Mortgage have contracts with major credit bureau firms (that keep credit scores) and receives response to its request in minutes. The process of sending requests and uploading the responses takes Jim an average of 10 minutes. Using credit scores and credit report of the applicant as well as other documents in the application, Jim writes his credit report and uploads that to the application file on MoreSys. Writing report and uploading takes Jim an average of 70 minutes.

### Property Survey and Appraisal

Tom and Frank are in charge of property survey and property appraisal. The property survey helps Easy Mortgage determine whether there is any violations that may affect legal title. Property appraisal, on the other hand, is used in assessing the value of the house. This helps Easy Mortgage to determine the amount and the rate of the loan. Until last year, Tom was in charge of property appraisal and Frank was writing property survey reports. According to job rotation policy at Easy Mortgage, and since both Tom and Frank have expertise in both the areas, this year Tom works on property survey, and Frank is in charge of property appraisal. The average time that takes Tom to prepare the property survey and upload it to MorSys is 90 minutes. Frank then prepares property appraisal report based on the initial appraisal and other documents (submitted by the applicant) and uploads the report to the application file. This takes him an average of 50 minutes.

### Title Search

The last document—title search—is prepared by Stacy, who has a real estate law degree. This takes Stacy an average of 100 minutes. Specifically, Stacy first types the address of the property in MorSys, which is linked to different sources of public records. The system automatically downloads all public records corresponding to the property. This takes an average of 15 minutes. Analyzing this data, Stacy determines, in her report, whether there are any claims on the property. She then uploads her report to the application file on MorSys. Data analysis, writing report, and uploading it take Stacy an average of 85 minutes.

### Approve/Disapprove Decision

When all four documents are added to the application file, MorSys routes the application to Ravi. Ravi makes the final decision of approve/disapprove of an application. Easy Mortgage has had several cases of approving home loans that later were found to have some legal issues and claims on the property. This, in some cases, resulted in significant losses. Thus, Ravi first performs an initial screening of the title search report to see if, and to what extent, the property has legal issues and claims. This takes an average of 10 minutes. The data has shown that Ravi requires some clarification on 10 percent of title search reports provided by Stacy. When this occurs, Ravi calls Stacy and clarifies the issue on the phone. The average time that takes Ravi and Stacy to resolve the issue on the phone is about 10 minutes. After checking the title search report, Ravi performs a detailed review of the application and makes a final decision of whether to approve or disapprove an application. In case of approval, Ravi also makes decisions about the terms of the mortgage contract. The time of this detailed review is highly variable. Ravi first checks the credit score of the application. If it is below an acceptable point (decided by the upper management), Ravi rejects the application without further review. Ravi changes the status of the application on MorSys to "Rejection due to Low Credit." This takes an average of 10 minutes. The system then automatically sends the standard rejection letter to the applicant. Data has shown that about 15 percent of applications that Easy Mortgage receives are rejected due to low credit score. If the applicant's credit score is higher than the acceptable threshold, Ravi checks all other reports and makes his final approval/disapproval decision. This process takes an average of 40 minutes. Thus, overall, the average time for Ravi to process an application is $85\% \times (10 + 40) + 15\% \times 10 = 44$ minutes per application.

### New Advertising Campaign

Charles Jackson, the vice president of Easy Mortgage, has just contacted Ravi about the firm's plan for an extensive TV advertising campaign. The campaign will announce the new low interest rate loan offered by Easy Mortgage. The campaign is expected to increase the average number of applications for home loan division by 50 percent from its current 20 applications per week to 30 applications per week. Charles emphasized that the upper management wants to make sure that the home loan division is capable of handling the new demand. The bad news, said Charles, is that they are very reluctant to authorize new hirings at any division, unless it is absolutely needed to respond to the new demand. Hence, Charles asked Ravi to send him a report in which he describes whether his division can handle the new demand, and if not, what are the steps he will take to achieve this. Charles believes that it is easier to make a case for new hires at the home loan division, if the division shows that they have already taken some steps that have improved their ability to handle more applications per week.

After his conversation with Charles, Ravi begins to think about how to write the report that Charles wants. To do that, Ravi realizes that he first needs to find whether his division is capable of handling the new demand of 30 applications per week; and if not, how he can increase the capacity of his division.

### Finding Capacity of the Division

To compute the effective capacity of the division, we first need to define the product and the time unit. The product of the division is an *application* (approved or disapproved) and we choose *week* as our time unit, so we can compare that with the new weekly demand of 30 applications.

The division has five resources that process applications. The list of resources and the parameters of their main and supporting tasks are presented in Table 6.4. Effective process

Table 6.4    Resources and Activities at the Home Loan Division (*Time is in minutes; $A_i = 100\%$*)

| Resources (i) | Activity (j) | Main task $T_0(j)$ | Setups $N_s$ | Setups $T_s$ | Rework $\alpha_r$ | Rework $T_r$ | $T_{\text{eff}}(j)$ |
|---|---|---|---|---|---|---|---|
| John | Initial Review | 20 | 1 | 0 | 0% | 0 | 20 |
| Jim | Credit Report | 80 | 1 | 0 | 0% | 0 | 80 |
| Tom | Survey Report | 90 | 1 | 0 | 0% | 0 | 90 |
| Frank | Appraisal Report | 50 | 1 | 0 | 0% | 0 | 50 |
| Stacy | Title Search | 100 | 1 | 0 | 10% | 10 | 101 |
| Ravi | Title Search Review | 10 | 1 | 0 | 10% | 10 | 11 |
|  | Approve/Disapprove | 44 | 1 | 0 | 0% | 0 | 44 |

times of resources are also presented in Table 6.4. The process time of each resource is computed by substituting its parameters into Eq. (4.1).

Using the effective mean process time of each resource and considering the working schedule of 40 hours per week, the effective capacity of each resource is computed in Table 6.5. As the table shows the effective capacity of the division is 23.76, the effective capacity of its bottleneck—Stacy. Resource utilization $u_i^*$ is computed using Eq. (6.7) that corresponds to resource utilization if the process works at its maximum capacity of 23.76.

Table 6.5    Capacities of Resources of the Home Loan Division

| Resources (i) | Effective mean process time (*minute per appl.*) | Effective capacity (*appl. per hour*) | Work schedule (*hours per week*) | Effective capacity (*appl. per week*) | Resource utilization $u_i^*$ |
|---|---|---|---|---|---|
| John | 20 | 3 | 40 | 120 | 19.8% |
| Jim | 80 | 0.750 | 40 | 30 | 79.2% |
| Tom | 90 | 0.667 | 40 | 26.68 | 89.1% |
| Frank | 50 | 1.200 | 40 | 48 | 49.5% |
| Stacy | 101 | 0.594 | 40 | 23.76 | 100% |
| Ravi | 55 | 1.091 | 40 | 43.64 | 54.4% |
| **Home Loan Division Effective Capacity:** | | | | 23.76 | |

Easy Mortgage is expecting that after the advertising campaign, its demand will increase from the current 20 applications per week to 30 applications per week. The current effective capacity of the home loan division is 23.76 applications per week, which is higher than its current weekly demand of 20. Thus, the home loan division is currently a demand-constrained process.

However, after the advertising campaign, the division will be a capacity-constrained process, since its new demand of 30 will be higher than its effective capacity of 23.76 applications per week. So, to be able to handle the new demand, Ravi must find ways to increase the number of applications that his division can process in a week. According to Fig. 6.7, Ravi needs to check the list of ways that one can increase the throughput of a capacity-constrained process. For his capacity-constrained process, as the figure shows, Ravi needs to either increase the effective capacity of its bottleneck or to decrease the throughput loss in his process, or both. Considering that Stacy (who performs the title search activity) is the bottleneck resource of the home loan division, we go over our list of throughput improvement actions.

**Improving Throughput at Home Loan Division**

The strategies for improving throughput by reducing throughput loss in capacity-constrained processes are the following:

- *Reduce Blocking and Starvation:* Blocking and starvation occur in systems with variability and limited buffer, and their impact is higher in processes with smaller buffer sizes and higher variability. While there is certainly variability in the application arrivals and their processing times, there are no limited buffers in the process at the home loan division. This is because MorSys, the information system of Easy Mortgage, has no limit on the number of applications that it can store at each stage of the process. Thus, Stacy—the bottleneck resource—and also other employees of the home loan division are not idle due to blocking and starvation that are caused by having limited buffers.

- *Revise the Supply Mechanism:* There are no indications that the division is losing throughput due to lack of supply. Processing applications at each stage of the process does not require supply of any materials or parts, etc. Resources (e.g., employees) at each stage, therefore, do not idle due to lack of supply.

- *Revise the Control Policy:* The control policy at each stage of the process is a simple first-come-first-served (FCFS) policy. Furthermore, at each stage, the process of the next application can start immediately after the process of the current application ends. Therefore, the control policy at each stage of the process does not result in idling of resources, when an application is waiting in that stage. Consequently, there is no throughput loss because of the control policies.

Now that we checked the strategies for reducing throughput loss, we now go over our list of throughput improvement strategies that focus on improving the bottleneck operations. Theoretical and effective mean process times of Stacy—the bottleneck—are 100 and 101 minutes, respectively. Her theoretical and effective capacities are therefore 24 and 23.76 applications per week. Hence, the percent capacity improvement potential for Stacy is

$$\%\text{CIP} = \frac{24 - 23.76}{23.76} = 1\%$$

which is very small. Thus, reducing Stacy's supporting task—rework—will increase the effective capacity of Stacy by only 1 percent, which is not a significant increase in the capacity of the division. Therefore, while eliminating rework should always be considered, the major effort should be on our other two strategies, namely, (i) increasing theoretical capacity of the bottleneck, and (ii) redesigning activities and process flow to improve theoretical capacity of Stacy and the process. Without improving the theoretical capacity of Stacy (which will also improve her effective capacity), the home loan division will not be able to process 30 applications per hour under the future demand.

**Increasing Theoretical Capacity of Stacy**

Some of the strategies that we mentioned for increasing the theoretical capacity can be used here to increase the theoretical capacity of Stacy. For example, a work study that breaks Stacy's activities into smaller steps, eliminating the unnecessary steps and shortening some of the steps, can reduce her theoretical mean process time and increase her theoretical capacity. Stacy's activity consists of two major tasks of downloading public records from MorSys (which takes an average of 15 minutes), and analyzing and writing a report (which takes an average of 85 minutes). A work study might be able to identify ways to reduce either of these tasks. Another option is to outsource the title search. Specifically, ask applicants to provide a title search from a reliable outside source which is approved by Easy Mortgage.

While this eliminates the title search activity, it is a strategic decision that should be made by upper management at Easy Mortgage. Some other capacity improvement strategies are not feasible. For example, asking Stacy to work simultaneously on two title searches (i.e., simultaneous processing) is not feasible.

### Redesigning Activities and Process Flow at the Home Loan Division

We now examine actions that can increase the capacity by changing activities performed by Stacy or changing the process flow of applications. Specifically, we focus on the following three actions: (i) reassigning some of the bottleneck activities to non-bottleneck resources, (ii) horizontal pooling, and (iii) redesigning process flow.

### (i) Reassigning Stacy's Tasks (Bottleneck Resource) to John (Non-bottleneck Resource)

One option for increasing Stacy's capacity—the bottleneck resource—is to reassign some of her tasks to other non-bottleneck resources. Recall that Stacy's activities consist of two tasks: (i) typing the address of the property in MorSys, and downloading all public records corresponding to the property (15 minutes), and (ii) analyzing the data and writing a report (85 minutes). While the latter requires specific knowledge and expertise, the former is easy and can be done (with a small training) by anyone at the home loan division.

One obvious candidate is John, whose main task is to check the completeness of applications and to upload them on MorSys, which takes him 20 minutes. In fact, as Table 6.5 shows, John has a very low utilization of $u_i^* = 19.8\%$—the lowest utilization among everyone. After finishing each application, John can easily type the property address in MorSys, get the pubic record, and upload those records to MorSys. This adds 15 minutes to John's effective mean process time, while reducing Stacy's effective mean process time from 101 minutes to 86 minutes. Consequently, John's effective capacity decreases from 120 to 68.56 applications per week, while Stacy's effective capacity increases from 23.76 to 27.92 applications per week.

This, as Table 6.6 shows, will increase the effective capacity of the home loan division from 23.76 to 26.68, the effective capacity of Tom—the new bottleneck. The improved capacity of 26.68 is still less than the weekly demand of 30. We, therefore, still need to look for other actions to increase the capacity of the division.

Table 6.6    Effective Capacities after Reassigning Stacy's Uploading Task to John

| Resources (i) | Effective mean process time (minute per appl.) | Effective capacity (appl. per hour) | Work schedule (hours per week) | Effective capacity (appl. per week) | Resource utilization $u_i^*$ |
|---|---|---|---|---|---|
| John | 35 | 1.714 | 40 | 68.56 | 38.9% |
| Jim | 80 | 0.750 | 40 | 30 | 88.9% |
| Tom | 90 | 0.667 | 40 | 26.68 | 100% |
| Frank | 50 | 1.200 | 40 | 48 | 55.6% |
| Stacy | 86 | 0.698 | 40 | 27.92 | 95.6% |
| Ravi | 55 | 1.091 | 40 | 43.64 | 61.1% |
| | Home Loan Division Effective Capacity: | | | 26.68 | |

### (ii) Horizontal Pooling

We now focus on Tom, the new bottleneck. Tom prepares the property survey and uploads it to MorSys. This takes him an average of 90 minutes. Some time after Tom uploads the property survey, Frank starts working on property appraisal report, which is used in assessing

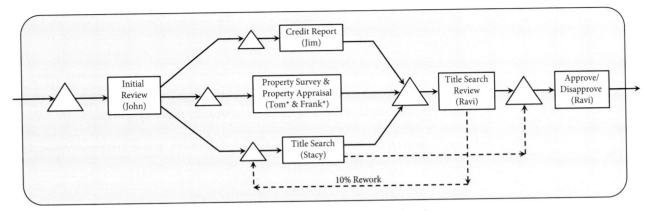

**Figure 6.13    Process-flow chart of the home loan division after horizontal pooling of Tom and Frank.**

**Table 6.7    Effective Capacities after Reassigning Stacy's Task and Horizontal Pooling**

| Resources (i) | Effective mean process time (minute per appl.) | Effective capacity (appl. per hour) | Work schedule (hours per week) | Effective capacity (appl. per week) | Resource utilization $u_i^*$ |
|---|---|---|---|---|---|
| John | 35 | 1.714 | 40 | 68.56 | 40.7% |
| Jim | 80 | 0.750 | 40 | 30 | 93.1% |
| Pool of Tom and Frank | 140 | 0.856 | 40 | 34.24 | 81.5% |
| Stacy | 86 | 0.698 | 40 | 27.92 | 100% |
| Ravi | 55 | 1.091 | 40 | 43.64 | 64% |
| | | **Home Loan Division Effective Capacity:** | | 27.92 | |

the value of the house. This helps Easy Mortgage to determine the amount and the rate of the loan. Frank prepares property appraisal report based on the initial appraisal and other documents (submitted by the applicant) and uploads his report to the application file. This takes an average of 50 minutes.

One way to increase the capacity is to do a horizontal pooling of Frank (the non-bottleneck) and Tom (the bottleneck). First, because both Tom and Frank can do both property survey and property appraisal, this will not result in additional (training) costs. Second, as Table 6.6 shows, Frank has utilization $u_i^* = 55.6\%$, which is low. Hence, according to Eq. (6.8), we should expect a large percent increase in the capacity of the pool of Frank and Tom—about $[2/(1 + 0.556)] - 1 = 28\%$. Figure 6.13 shows the new process-flow chart after Tom and Frank are pooled horizontally.

Specifically, we ask Tom to write both property survey and property appraisal reports on an application. We also ask Frank to do the same (but on different applications, of course). This combines two activities of property survey and property appraisal into one activity with a total effective mean process time of $90 + 50 = 140$ minutes per application, and assigns it to a pool of two workers (Tom and Frank) to perform the activity. Therefore, the effective capacity of Tom (or Frank) is $0.428 \ (= 60/140)$ applications per hour, and thus the pool of Tom and Frank has the effective capacity of $2(0.428) = 0.856$ per hour, or $0.856 \times 40 = 34.24$ applications per week. Table 6.7 depicts the effective capacity of the home loan division after reassigning Stacy's task and horizontal pooling of Tom and Frank.

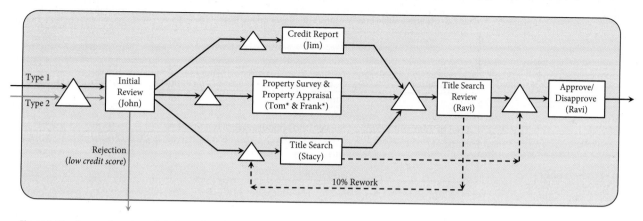

**Figure 6.14    Process-flow chart of the home loan division after redesigning the flow.**

As the table depicts, the effective capacity of the home loan division is increased from 26.68 to 27.92 applications per week. Tom is not the effective bottleneck of the division anymore and Stacy becomes the bottleneck again. Because the weekly capacity of 27.92 is still lower than future weekly demand of 30 applications per week, we still need to look for more ways to increase the capacity of the division.

Since we have already reassigned some of Stacy's tasks to others, increasing the capacity of Stacy should focus on alternatives such as making her activity time shorter (which might be difficult) or asking her to work overtime (which is costly). On the other hand, horizontal pooling of Stacy with another person might not be possible, since others do not have her expertise. Title search is an important document that prevents future legal battles and costs for Easy Mortgage. So, we need to find other ways, besides increasing Stacy's capacity, to increase the capacity of the home loan division. This may not seem possible. How can one increase the capacity of a process without directly increasing the capacity of its bottleneck? Well, one way to do that is to redesign the flow process to send less work to the bottleneck station.

### (iii) Redesigning Flow Process in the Home Loan Division

Recall that, on an average, 15 percent of applications have a credit score lower than a minimum required score and are rejected by Ravi without further review of their other documents. As explained, for these applications the review of the credit score and the rejection process takes Ravi an average of 10 minutes. Thus, it seems logical to identify these applications as early as possible, and reject them before sending them to Stacy or other staff of the home loan division for further processing. The best stage to identify and reject these applications is during the initial review performed by John. To make this possible, John needs to perform two additional tasks: (i) sending a request to credit report firms and upload their response (10 minutes), and (ii) comparing the score of the applicant with minimum required score and reject the applications with lower than minimum score (10 minutes). The good news is that both of these activities are straightforward and easy activities that do not require expertise and are done through MorSys. Performing these activities increases John's effective mean process time by 20 minutes from 35 minutes to 55 minutes per application, while decreasing Jim's effective mean process time by 10 minutes from 80 to 70 minutes. It also eliminates Ravi's task of rejecting applications due to low credit score. Figure 6.14 shows the new process flow.

Under the redesigned flow, the home loan division is a process in which flow units follow two different routings. As we explained in Chap. 4, we can compute the capacity of the

Table 6.8    Effective Mean Process Times after Redesigning the Process Flow

| | | Effective mean process time | | |
| | | Type 1 (*minutes*) | Type 2 (*minutes*) | Aggregate (*minutes*) |
| Resource | Activity | | | |
|---|---|---|---|---|
| John | Initial review | 55 | 55 | 55 |
| Jim | Credit report | 70 | 0 | 59.5 |
| Pool of Tom and Frank | Survey and appraisal reports | 140 | 0 | 119 |
| Stacy | Title search | 86 | 0 | 73.1 |
| Ravi | Title search review | 11 | 0 | 9.35 |
| | Approve/Disapprove | 50 | 0 | 42.5 |

Table 6.9    Effective Capacities of Resources after Redesigning the Process Flow

| Resources (*i*) | Effective mean process time (*minute per appl.*) | Effective capacity (*appl. per hour*) | Work schedule (*hours per week*) | Effective capacity (*appl. per week*) | Resource utilization $u_i^*$ |
|---|---|---|---|---|---|
| John | 55 | 1.901 | 40 | 43.64 | 75.3% |
| Jim | 59.5 | 1.008 | 40 | 40.32 | 81.4% |
| Pool of Tom and Frank | 119 | 1.008 | 40 | 40.32 | 81.4% |
| Stacy | 73.1 | 0.821 | 40 | 32.84 | 100% |
| Ravi | 51.85 | 1.157 | 40 | 46.28 | 71% |
| | | **Home Loan Division Effective Capacity:** | | 32.84 | |

process with different routing by modeling them as a process with multiple products. In this case, type-1 products are applications with minimum or higher than minimum required credit score, and type-2 products are applications with lower than minimum required credit score. The product mix is $(0.85, 0.15)$. To compute the effective capacity of the process, we use Product Aggregation method in Chap. 4 for processes with multiple products. Table 6.8 depicts the effective mean process time of each resource for each type of product.

Note in the table that type-2 products are only processed by John, that is, the effective mean process time of type-2 products is 55 minutes for John, and is zero for others. Having the effective mean process time of the aggregate product, we compute the effective capacity of the home loan division (in terms of the aggregate product) under the redesigned process flow, as shown in Table 6.9.

From Table 6.9, we find that the effective capacity of the division after revising the process flow is 32.84 aggregate products per week. Recall that the capacity for the aggregate product represents the capacity of the division in terms of the *total* average number of products (applications) that it can process in a week. This is a significant increase from 27.92 applications per week. The question is: how can the process have such a significant increase without any change in the capacity of its bottleneck (i.e., Stacy)? Stacy is still working the same number of hours performing the same task of title search. How did her effective capacity increase from 27.92 to 32.84 applications per hour?

We have answered this question in the section "Redesigning the Process Flow" and discussed why process-flow changes that reduce the load of the bottleneck station can increase the capacity of a process. For the case of the home loan division, under the product mix $(0.85, 0.15)$ the division has the effective capacity of processing a total of 32.84 applications per week, where $0.85 \times 32.84 = 27.92$ applications are of type 1 and $0.15 \times 32.84 = 4.92$ applications are of type 2. However, among the 32.84, only 27.92 applications are sent to

Stacy. This is equal to Stacy's effective capacity 27.92 before changing the process flow. Thus, the Stacy's effective capacity is still 27.92 (type-1 or type-2) applications. While before the change Stacy processed both type-1 and type-2 applications, after the change she only receives type-1 applications. The remaining 4.92 applications are type-2 applications that are handled by John and leave the system without getting to Stacy. This increases the effective capacity of the process to a total of 32.84 (type-1 and type-2) applications per week.

**Conclusion**

By implementing the three actions of (i) reassigning Stacy's task to John, (ii) horizontal pooling of Tom and Frank, and (iii) redesigning the process flow, Ravi will be able to increase the effective capacity of his division from 23.76 applications per week to 32.84 applications per week. This will be higher than the new weekly demand of 30 applications after the advertising campaign. The good news is that this improvement can be achieved without any significant cost such as cost of hiring a new staff or cost of overtime. The only action needed is to train John to perform his new tasks.

## 6.7 Summary

Throughput is one of the operational performance measure that has a direct impact on a firm's revenue, because the throughput of a firm (goods or services) is what is sold to its customers. The throughput of a firm is constrained either by the demand for the firm's products or by the production capacity of the firm.

Strategies to increase throughput in demand-constrained processes can be divided into two categories: one is to increase the demand of the process and the other is to decrease throughput loss. A firm can increase its demand by decreasing its price, improving the quality of its products, and by increasing its sales effort (e.g., marketing and advertising). The main sources of throughput loss in demand-constrained processes that must be eliminated or reduced are: inefficient supply mechanism that results in lack of material, balking and reneging of customers due to long waiting time to get their products, and control policies that result in inefficient use of resources.

In capacity-constrained processes, the main factor that limits the throughput is the capacity. Thus, efforts to increase throughput focus on increasing the effective capacity of the process. Strategies to increase the effective capacity include eliminating supporting tasks such as setups, rework, yield loss, or increasing number of resources or their working hours, or outsourcing. Other strategies to increase effective capacity correspond to redesigning activities and flow in the process. Examples include making activities self-service, assigning some bottleneck activities to non-bottleneck resources, changing product mix, and redesigning flow such that the workload of the bottleneck resource is reduced.

Finite buffers within a process result in throughput loss in both demand- and capacity-constrained processes. Buffers with limited space to hold flow units lead to blocking and starvation, both of which reduce resource utilization and thus reduce throughput. Strategies to reduce blocking and starvation include reducing variability in the process, increasing buffer sizes, and increasing resource capacity.

## Discussion Questions

1. What is the relationship between throughput, input rate, and capacity of a process?

2. What is the difference between capacity-constrained and demand-constrained processes? Why is it important for managers to know if their process is a demand- or capacity-constrained process?

3.  Describe five factors that can affect throughput loss of a process.

4.  Briefly explain the following terms and discuss how they affect the throughput of a process:
    - Balking
    - Abandonment
    - Starvation
    - Blocking

5.  Describe two main strategies to increase throughput of demand-constrained processes.

6.  Discuss three strategies that can increase demand of a process.

7.  What are four ways to increase the capacity of a bottleneck station?

8.  Present six different approaches that can increase the capacity of a process by redesigning its activities or the flow within the process.

9.  What is the difference between horizontal pooling and vertical pooling of resources? When does pooling—vertical or horizontal—result in higher improvement in capacity of a process?

10. What is Theory of Constraints (TOC)? What are its steps and what are the three main ideas behind TOC?

## Problems

1.  Consider five activities A, B, C, D, and E, each performed by Workers A, B, C, D, and E, respectively, as shown in Fig. 6.15. Worker D performs a quality control (QC) activity and discards about 5 percent of units that do not pass quality control test. The process time of each activity (in minutes) is also shown in the figure. Process times have no variability. The buffers between each two activities have the maximum limit of holding 10 units. The following questions refer to the process in Fig. 6.15 and are independent of each other.
    a.  If Activities B and C are performed simultaneously in parallel (instead of serial), how much does that increase the capacity of the process?
    b.  If Activity D is performed before Activity C but after Activity B, how much does that affect the capacity of the process?
    c.  If the size of the buffer between Activities B and C is increased to 15 units, how does that affect the capacity and throughput of the process?
    d.  Suppose that the process time of Activity C is either 12 or 20 minutes with equal probability—the average is still 16 minutes. How does this affect the capacity and the throughput of the process?
    e.  How much does the capacity of the process increase if Activities A and B are pooled? How much does the capacity of the process increase if Activities B and C are pooled?

2.  Southshore Medical Group has an office building for immediate care that accepts emergency walk-in patients without appointment from 7:00 am to 12:00 pm. All arriving patients first go to the receptionist to register. The registration takes 0.5 minutes. Patients are then given a form to fill out. The form has questions about patients' medical history. It takes patients an average of 3 minutes to fill out the form and return it to the receptionist. The receptionist then

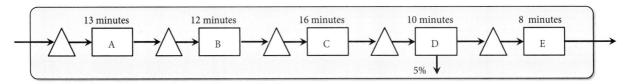

**Figure 6.15**    Process-flow chart of activities in Problem 1.

enters the data in the form into the computer, which takes about 2 minutes. Patients then wait in the waiting room until they are called. After some time, a nurse calls a patient and takes the patient to the visit room. The nurse takes patient's blood pressure and temperature and enters those into the computer. This takes about 4 minutes. The patient is then seen by a doctor. The time a doctor spends with a patient varies, but has an average of 20 minutes. After the doctor leaves, the patients get their visit summary from the nurse. It takes the nurse about 2 minutes to print the summary and deliver it to the patient. The patient then leaves the office. There are one receptionist, six doctors, and three nurses.

    a. What is the maximum number of patients that the Medical Group can serve in a day?

    b. Can the Medical Group reduce its labor cost (i.e., having less number of employees) without affecting its daily capacity of serving patients?

3. A pharmacy operates with two technicians and two pharmacists. The activities to fill a prescription are as follows. First, John—one of the technicians—receives the prescription from a customer and enters it into the computer. This takes about 60 seconds. John then bills the insurance company for the prescription. This takes an average of 30 seconds. Jill—the second technician—prints the prescription labels for the pharmacists. This takes about 120 seconds. A pharmacist starts filling the medication, which takes about 150 seconds, and then attaches the label and does the final check, which takes about 30 seconds. When the prescription is ready, Jill calls the customer and charges the customer for the prescription. This takes about 45 seconds. Jill also asks the customer if there are any questions for the pharmacist. About 80 percent of customers say no, and the remaining 20 percent consult with one of the pharmacist for about 60 seconds. The customer then leaves the pharmacy.

    a. What is the maximum number of prescriptions that the pharmacy can fill in an hour, and who is the bottleneck?

    b. The pharmacy would like to increase its capacity to 28 customers per hour. Recommend strategies for achieving this capacity without increasing the pharmacy's operations costs.

4. Consider the Heavenly Treat (HT) hair salon in Problem 9 of Chap. 4. Suppose that the cashier is paid $20 per hour, receptionist is paid $20 per hour, skin care professional is paid $30 per hour, hair washer is paid $18 per hour, senior hair stylists are paid $35 per hour, and junior hair stylist is paid $25 per hour. The net profit—excluding the above labor cost—of the services that HT offers are: $45 for facial services, $50 for hair styling, and $30 for simple haircuts.

    To improve the operations, HT is considering the following two actions: (i) lay off the cashier and ask the receptionist to do the cashier's job, and (ii) lay off the junior hair stylist and instead hire a senior hair stylist, and ask all senior hair stylists to also do simple haircut if needed. Does implementing these strategies improve HT's profit? Assume that there is always demand for the services that HT offers.

5. Consider the Willowglen County DMV office in Problem 8 in Chap. 4. The demand at the DMV office is expected to reach 70 customers per hour. Suggest strategies that can increase the capacity of the DMV office to 70 customers per hour. Suppose that all the agents in the office, if needed, can be easily trained to do other activities in the office.

# CHAPTER 7

# Fundamentals of Flow Time Analysis

## 7.0 Introduction

It is 4:20 pm, and you have a flight to catch at 7:00 pm to go to visit your sister in New York, who has just given birth to a baby boy. You call for a taxi, and you get the usual message: "All of our agents are currently busy helping other customers. Please continue to hold and your call will be answered in the order it was received." You wait for 1 minute until you get connected to an agent. You place an order for a taxi to pick you up at 5:00 pm. It takes you half an hour to pack and get ready. It is now 5:00 pm. While getting out of your apartment, you see a package behind your door. It is the sports jacket that you ordered online 2 weeks ago and you have been waiting for it all week. You put the package in your apartment, lock the door, and go to the elevator. You push the elevator button and after waiting for an unusually long time, one of the two elevators arrives. You get to the elevator and go to the lobby and see the taxi waiting.

You tell the driver to take you to the airport. It is 5:15 pm and you are stuck in traffic waiting for a green light. The traffic moves slowly and finally you arrive at the airport at 6:15 pm. You rush to check in your luggage and get your boarding pass, but you see a long line of passengers waiting in the line for the boarding pass. You wait in the line for 7 minutes, check in your luggage, and get your boarding pass. Now the big challenge comes: waiting in a long line for security check. The time is 6:35 pm. You pass the security check and you get to the gate for your flight, where you find another line for boarding the plane. You wait in the line, and get to the plane and find your seat. It is 6:55 pm, and you are happy that you finally made it.

Doors are closed and the plane starts moving at exactly 7:00 pm. However, the plane stops 5 minutes later and the pilot announces that because there are only three lanes used for takeoff and because a large number of flights are waiting for takeoff, your plane needs to wait for 15 minutes for its turn to takeoff. Your flight finally takes off and, 2 hours later, you are above New York airport. Happy that the flight is about to be over, you hear the pilot's announcement that due to a large number of flights arriving at the airport, your plane needs to hover above the airport for another 10 minutes for its turn to land. The plane finally lands, but you know that it is not yet over. You need to wait to get your luggage back, wait for taxi to come, and wait in traffic before you finally arrive at your sister's and see your newborn nephew.

Waiting in lines is a part of every day life. Do you remember any day that you did not wait in a line? We wait in our cars in traffic jams, we wait in line in retail stores for check out, we wait on hold to talk to an operator when we call a firm, we wait in line in banks to do our transactions, and we wait several days to receive the item we ordered online. We wait in line in post office, cafeterias, emergency departments of hospitals, etc. It has been estimated that Americans spend 37,000,000,000 hours per year waiting in lines,[1] and the average American will spend 5 years of their lifetime waiting in lines.[2]

In Chap. 2 we introduced waiting time of a flow unit in a process—the flow time—as an important measure and discussed how it can affect financial performance measures of firms. The undesirable impact of long flow times on a firm's performance is more directly observable in service operations systems where customers are physically present in lines waiting to receive service. Waiting is annoying, so the longer the waiting time, the more dissatisfied a customer will be.

Waiting is not exclusive to people. Airplanes wait for landing or takeoff, items in supermarkets are waiting for customers to buy them, files in computer systems wait for processing by the central processors, jobs in manufacturing systems are waiting (as inventory) to be processed by a worker or by a machine, and broken machines in factories are waiting for repairperson (or spare parts) to arrive. Long waiting times (i.e., long flow times) usually imply large waiting lines (large inventory). Both customers and managers of manufacturing and service operations systems do not like large flow times and large waiting lines. The question are: Why do these waiting lines exist? How can they be reduced or eliminated?

In this chapter, we study the underlying reasons that make a flow unit wait in a buffer of a process until its processing starts. The chapter also presents analytics that are very useful in estimating average flow times as well as other process-flow measures of a process. Finally, we conclude the study of flow times in Chap. 8 where we present different strategies for improving flow times in a process.

## 7.1 Why Do Waiting Lines Exist?

To be able to find estimates for waiting times in production and service operations systems and also to gain insights into how to improve them, we need to understand why waiting lines exist in the first place, and what makes the wait in some lines shorter than others? To answer these questions we study a series of simple examples that lead us to the underlying principles that govern the waiting time dynamics in all manufacturing and service operations systems.

Consider a packaging station at the end of a mobile phone assembly line in which a worker puts mobile phones he receives from the assembly lines in a package.

---

[1] Hillier, S.F. and M.S. Hillier, *Introduction to Management Science*. Third Edition, McGraw-Hill Irwin, 2008.
[2] U.S. News & World Report, January 30, 1989, p. 81.

**Case 1:** *(Stable Process with No Variability)* Suppose that a phone exits the assembly line and arrives at the packaging station *exactly* every minute, and also assume that the worker can finish packaging a phone in *exactly* 45 seconds. Would there be a queue of phones (i.e., in-buffer inventory) in the packaging station?

The arrival rate of phones to (i.e., demand for) the packaging station is 60 phones per hour, while the effective capacity of the station is 80 ($= 3600/45$) phones per hour, which is higher than the arrival rate. The worker has utilization of $u = 60/80 = 75\%$. In this system, 45 seconds after the arrival of a phone, its packaging is completed and the worker would be idle for 15 seconds until the next phone arrives. Thus, there would be no queue of phones (i.e., no in-buffer inventory) at the packaging station at any time.

**Case 2:** *(Unstable Process with No Variability)* Suppose that a phone exits the assembly line and arrives at the packaging station *exactly* every 40 seconds. The packaging time is still *exactly* 45 seconds. Would there be a queue of phones in the packaging station?

In this case, the arrival rate (i.e., demand) is 90 ($= 3600/40$) phones per hours, which is larger than the effective capacity of 80 phones per hour. So, before the worker finishes the packaging of a phone, another phone arrives and waits in the queue. Therefore, there would be a queue of phones waiting in the packaging station. Furthermore, since the worker has the capacity of packaging only 80 out of 90 arriving phones per hour, the queue length increases at the rate of $90 - 80 = 10$ phones per hour. And, in the long run, the queue grows to infinity if the process continues in this manner. As we discussed in Chap. 2, the packaging station is not a stable process, since its input rate is larger than its output rate.

**Case 3:** *(Stable Process with Variability)* Now suppose that the time between two consecutive arrivals at the packaging station is either 30 seconds or 90 seconds with equal probability of 0.5. This implies an *average* interarrival time of $0.5 \times 30 + 0.5 \times 90 = 60$ seconds. Suppose also that it takes *exactly* 45 seconds to package a phone. Would there be a queue of phones in the packaging station in this case?

Figure 7.1 shows the number of phones in the queue of the packaging station (i.e., in-buffer inventory) during a 100-hours (360,000-second) period. The graph was generated using a computer simulation of the station. As the figure shows, during the 100-hour operation, the queue of the packaging station can go up to five phones. This occurs during the periods in which a large number of arriving phones have interarrival times of 30 seconds. Note that

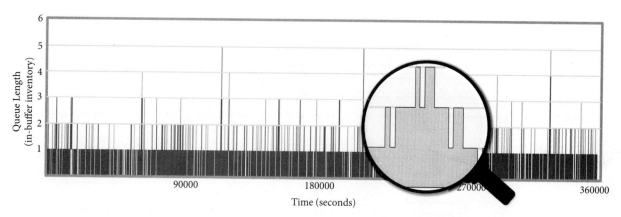

**Figure 7.1**    Number of phones in the buffer of the packaging station in Case 3 during 100 hours of operations.

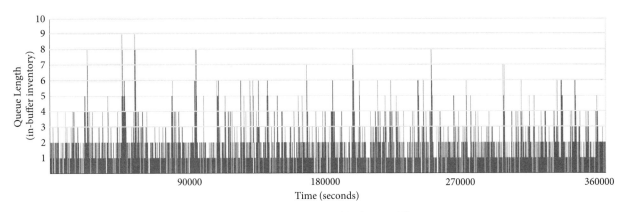

Figure 7.2 **Number of phones in the buffer of the packaging station in Case 4 with higher variability.**

the interarrival times of 30 seconds corresponds to arrival rate of $3600/30 = 120$ phones per hour. While the *average* arrival rate is still 60 phones per hour, when more interarrival times are 30 seconds, the input rate of the packaging station becomes temporarily larger than its average of 60. When this temporary arrival rate becomes larger than 80—the capacity of the packaging station—the queue of phones starts to build in the buffer. The temporary increase in the arrival rate is the result of the variability in the arrival process and is the main underlying reason for having a queue in the packaging station. Of course, there are also periods of time in which more jobs have interarrival times of 90 seconds, resulting in a temporary arrival rate of less than 60 phones per hour. During those periods, the queue length is reduced. This corresponds to time intervals in Fig. 7.1 in which queue length is smaller (e.g., no queue).

Thus, while both Case 1 and Case 3 have the same effective capacity of 80 phones per hour, and receive the same *average* of 60 phones per hour from the assembly line, the station in Case 1 has no queue of phones in its buffer, but there is a queue in Case 3 due to variability in the arrivals.

**Case 4:** *(Stable Process with Higher Variability)* Consider Case 3, but suppose that the time between two consecutive arrivals at the packaging station is either 15 seconds or 105 seconds with equal probability of 0.5. Note that, as in Case 3, the average interarrival time is still $0.5 \times 15 + 0.5 \times 105 = 60$ seconds, and therefore the average input rate is still 60 per hour. How would this affect the queue (in-buffer) inventory of the packaging station?

The main difference between Case 3 and Case 4 is that there is a higher variability in interarrival times of the packaging station in Case 4. In Case 4, the interarrival times can be as small as 15 seconds, while this number is 30 seconds in Case 3. On the other hand, they can be as large as 105 seconds in Case 4, while in Case 3 they can be as large as 90 seconds.[3] How does this increase in variability affect the queue length? Well, Fig. 7.2 shows the queue length in Case 4 during 100 hours of operations of the packaging station.

As Fig. 7.2 shows, the queue length in the buffer of the packaging station in Case 4 with higher variability can be as high as nine phones. Also, comparing Figs. 7.1 and 7.2, we observe that the queue length in Case 4 is longer than that in Case 3 most of the time during the 100 hours of operations. This also implies that, in Case 4, phones wait longer time in the buffer of the packaging station—phones in Case 4 have higher average in-buffer flow times.

---

[3]Standard deviation of interarrival times for Case 3 is 30 seconds, resulting in $CV = 30/60 = 0.5$, while it is 45 seconds for Case 4, corresponding to $CV = 45/60 = 0.75$.

From Cases 3 and 4 we observe that when there is variability in the arrival process, resulting in input rate becoming *temporarily* larger than the effective capacity, then phones begin to accumulate in the buffer and a queue is constructed. It is important to note that the variability in process times can also result in the effective capacity being temporarily smaller than the input rate. In other words, variability either in arrival or in effective capacity of a process can result in queues (i.e., in-buffer inventory) if it results in the effective capacity of the process becoming temporarily smaller than its input rate.

The observations we made from Cases 1 to 4 lead us to the following principle:

---

**PRINCIPLE**

**Impact of Variability on Flow Time and Inventory**

- In-buffer Inventory (i.e., queue) exists when the input rate to a process temporarily or permanently exceeds the effective capacity of the process. If the input rate of a process is always larger than the effective capacity of the process, the in-buffer inventory grows without a limit.

- For a given input rate and effective capacity, as variability in interarrival or process times increases, the in-buffer inventory (i.e., queue) and in-buffer flow time increase.

---

## 7.2 Stability and Safety Capacity

To introduce the concept of safety capacity, we start with the following case in our packaging station.

**Case 5:** *(Fully Utilized Process with No Variability)* Consider the packaging station in Case 1 with no variability in which phone interarrival times are *exactly* 60 seconds. Now suppose that the line produces a different type of phone that requires more accessories in its package. This increases the packaging time in the station from 45 seconds to *exactly* 60 seconds. How does that affect the queue (i.e., the in-buffer inventory) of the packaging station?

In this new setting, the effective capacity of the station is 60 phones per hour, which is now equal to the (demand) input rate of 60 phones per hour. In this case, 60 seconds after the arrival of a phone, its packaging is completed. At the time that a phone leaves the packaging station, the next phones arrive. Hence, similar to Case 1, starting from an empty station, there will not be any queue of phones in the buffer of the packaging station.

In Case 1, the packaging station has the effective capacity of 80 phones per hour and arrival rate (i.e., demand) of 60 phones per hour. This corresponds to a utilization of 75 percent. However, in Case 5 the station has the effective capacity of 60 phones per hour to handle the demand of 60 phones per hour—a utilization of 100 percent. Case 5 represents an ideal process: a process in which the resource is 100 percent utilized with in-buffer inventory and in-buffer flow time of zero. But does such an ideal case occur in the real world?

Well, to have such a process, there must be no variability in arrival and in process times. For example, the setting in Case 5 requires that a phone is delivered to the packaging station exactly every 60 seconds. But this is very unlikely, since it is most likely that there is variability in the process times of stations in the assembly line, resulting in variability in the throughput of the assembly line. Even if the assembly line uses fully automated machines (e.g., robots) in all of its stations, factors such as breakdowns or quality problems (e.g., scraps or rework) will still result in variability in the throughput of the line. Another source of variability might be

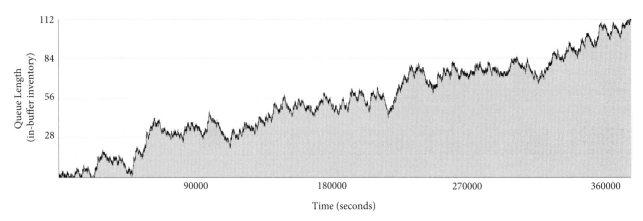

**Figure 7.3    Number of phones in the buffer of packaging station in Case 6 with variability and 100% utilization.**

in the time it takes to transfer phones from assembly line to the packaging station. Even if we assume the unlikely situation that there is no variability in the input rate to the packing station, there is still variability in the time to package a phone. This is specially true if, as in our case, packaging is done by a worker (not by an automated machine). The bottom line of all this is that there are not many processes in the real world with no variability in arrival or in process times. Case 6 presents the packaging station in Case 5 with variability in its arrival process.

**Case 6:** *(Fully Utilized Process with Variability)* Consider Case 5, but assume that there is variability in the arrival process. Specifically, the time between two consecutive arrivals at the packaging station is either 30 seconds or 90 seconds with equal probability of 0.5. Thus, the phone interarrival times have an *average* of 60 seconds. Time to package a phone is still *exactly* 60 seconds, as in Case 5. How does the variability in arrivals affect the queue (i.e., in-buffer inventory) at the packaging station?

Similar to Case 5, the input rate and the effective capacity in Case 6 are both 60 phones per hour. In other words, as in Case 5, the worker in Case 6 is also 100 percent utilized. The only difference is that there is variability in the arrival process in Case 6.

Based on our previous observations, since there is variability in the arrival process in Case 6, we can predict that, as opposed to Case 5, there will be a queue of phones in the buffer of the station. Figure 7.3 shows the inventory of phones in the buffer of the packing station for Case 6 during 100 hours of operations when the station starts with no phones at time zero.

Comparing Fig. 7.3 (for Case 6) with Figs. 7.2 (for Case 4) and 7.1 (for Case 3), for example, we observe the following:

- The queue in Case 6 is much longer than that in Cases 3 and 4. This is expected because for the same input rate of 60 phones per hour, stations in Cases 3 and 4 have a higher effective capacity of 80 phones per hour, compared to the effective capacity of 60 phones per hour in Case 6.

- But what is troubling is that in Case 6, the queue exhibits an *increasing trend* with time. Specifically, while the station started with no phone in its buffer at time zero, after 50 hours of operations the station has an in-buffer inventory of around 56, and after 100 hours of operations the in-buffer inventory reaches 112. This behavior was not observed in Cases 3 and 4. In those cases, while the queue of phones varied between zero and some maximum number, the queue length was stable and did not exhibit an upward trend.

**Table 7.1**    **Input to and Output from the Piggy Bank in 2 Days**

| Possible scenarios | Coin toss | | Probability of occurrence | Input ($) on | | Output ($) on | | Money left at the end of | |
|---|---|---|---|---|---|---|---|---|---|
| | Day 1 | Day 2 | | Day 1 | Day 2 | Day 1 | Day 2 | Day 1 | Day 2 |
| 1 | Head | Head | 0.25 | 10 | 10 | 0 | 0 | 10 | 20 |
| 2 | Head | Tail | 0.25 | 10 | 10 | 0 | 20 | 10 | 0 |
| 3 | Tail | Head | 0.25 | 10 | 10 | 10 | 0 | 0 | 10 |
| 4 | Tail | Tail | 0.25 | 10 | 10 | 10 | 10 | 0 | 0 |

The second observation implies that in processes with 100 percent utilization—input rate being equal to effective capacity—if there exists variability in the process, then the in-buffer inventory increases as the process operates more.[4]

One might think that since the packaging station in Case 6 has the effective capacity of packaging 60 phones per hour, it should be capable of handling the demand of 60 phones per hour. One can argue that while due to variability there would be periods of time (about 50 percent of time) that the input rate is larger than the process capacity—causing the accumulation of phones—there will also be periods of time that the input rate is smaller than the capacity—during which the accumulated phones are packaged and the queue is reduced. Thus, in the long run, the station should be able to reduce the accumulated queue and prevent the queue from increasing in time. But that is not happening in Case 6. Why? The answer becomes clear after we discuss the piggy bank dilemma.

**The Piggy Bank Dilemma**

Suppose you decide to save money for down payment to buy a car in the next 5 years. So you put $10 in a piggy bank every morning before you leave home. However, you change your mind every day when you come back home and take the $10 out of the piggy bank. How much money will you have in 5 years? Ok, I know; obviously this is not going to work. But what do you think of the next idea? Putting $10 in the piggy bank every morning, but when you come home in the evening, you flip a coin. If it comes head (H), you take nothing from the piggy bank. However, if it comes tail (T), then you take $20, if there is $20 in the bank. If there is only $10 in the bank, you take the $10, and if there is nothing there, you take nothing. You repeat this every day. How much money will you have in the piggy bank in 5 years?

Well, you are putting in $10 each day, and taking an average of $20 \times 0.5 + \$0 \times 0.5 = $10 at the end of each day (if there is money in the bank). So, on an average, there should be no money left at the end of 5 years, right? Wrong! Surprisingly, similar to Fig. 7.3, the average money in the piggy bank grows as days pass.

To illustrate this, let's compute the *average* money in the bank after only 2 days. This is shown in Table 7.1.

As the table shows, there are four possible scenarios for the outcomes of a coin toss in 2 consecutive days. Consider Scenario 2 as an example. In the morning of the Day 1, you put $10 in the piggy bank. When you come back, you flip a coin and you get a head. Thus, you do not take any money from the bank (i.e., the output is zero). Thus, at the end of Day 1, the money left in the bank is $10. On Day 2, you put another $10 in the bank, so there is a total of $20 in the bank. When you come home, you flip a coin and you get a tail. Thus, you take $20 from the bank and the money left in the bank at the end of Day 2 is zero.

Now consider Scenario 3, in which you get a tail on Day 1. You want to take $20 from the bank, but there is only $10 in the bank—the money you put in the morning. You take

---

[4]The in-buffer inventory, as shown in Fig. 7.3 may decrease in short time periods, but in the long term it has an increasing trend.

the $10 and the money in the bank at the end of Day 1 is zero. On Day 2 you put another $10 in the bank, and when you come back your coin gives a head, which means you are not taking any money from the bank. Thus, at the end of Day 2, there is $10 left in the bank.

In summary, under Scenarios 1, 2, 3, and 4, the money left in the bank at the end of Day 2 are $20, $0, $10, and $0, respectively. Since each scenario occurs with probability 0.25, the *average* amount of money in the piggy bank at the end of 2 days is

$$Average \text{ money after 2 days} = 0.25(\$20) + 0.25(\$0) + 0.25(\$10) + 0.25(\$0) = \$7.5$$

So, although the average input rate to the piggy bank (i.e., $10) is equal to the average potential output rate of the piggy bank (i.e., 0.5 ($0)+0.5 ($20) = $10) per day, the average money in the bank after 2 days is not zero. It is $7.5.

What happens if we flip the coin for more days? Does the average money go down to zero? No, the average money gets even larger. For example, if we repeat our experiment for a 3-day period, we find that the average money left in the piggy bank increases to $10. In fact, repeating the experiment for more days, we find that the average money in the piggy bank increases. Why? The answer becomes clear when we look at Scenarios 3 and 4 in Table 7.1. In Scenario 3, when you get a tail and you are supposed to take $20 from the bank, there is only $10 available to take. So, you lose the opportunity to take another $10 from the bank. This is even worse in Scenario 4, in which you get two tails. In that scenario, each day you can only take $10 from the bank—a total of $20—while you are supposed to take $20 each day—a total of $40. Thus, while the chance of getting head and tail is the same, there is less opportunity to take money from the bank than adding money, so the average money in the bank increases gradually. This is because the number of lost opportunities to take money out of the bank becomes larger, for example, we have larger number of times that we get tails but there is only $10 or no money in the bank.

The situation of the packaging station in Case 6 is similar to our piggy bank example.[5] Recall that in Case 6, the interarrival times have an average of $90 \times 0.5 + 30 \times 0.5 = 60$ seconds, resulting in arrival rate of 60 phones per hour. The worker has the capacity of packaging 60 phones per hour. Due to the variability in arrivals, the average number of phones in the packaging station increases with time.

The observation from Case 6 leads us to the following principle:

---

**PRINCIPLE**

**Impact of Variability on Flow Time and Inventory in Fully Utilized Processes**

In a process in which the input rate is equal to the effective capacity of the process—processes with 100 percent utilization—when there is variability in the arrival process or in processing times, the in-buffer inventory and thus the flow time increase without limit.

What is the implication of the above principle in practice? It implies that operations engineers and managers should not release jobs into a process with the rate equal to the effective capacity of the process. Just because a resource (e.g., a worker or a machine) has the effective capacity of 100 jobs per hour, for example, one should not assign 100 jobs per hour to it. By doing so, as days of operations pass, the in-buffer inventory will gradually

---

[5] In Case 6 the variability is in arrivals and in the piggy bank the variability is in the output process. This does not matter. You can create a table similar to Table 7.1 for the case where you flip a coin to put zero or $20 in the bank and you always take $10 at the end of the day. You will find that the average money in the bank increases as you repeat the experiment for more days.

increase without a limit. In the real world, however, the queue of processes does not grow very large for two reasons.

The first reason is because of Balking and Abandonment, see Sec. 6.2.2 of Chap. 6. For example, when queue of a process becomes very large, customers who see the long queue will not join the queue (i.e., balking), and often times customers who join the queue decide to leave the queue after waiting for some time (i.e., abandonment). This prevents the queue from growing very large.

The second reason is that managers often use temporary measures to increase the capacity of their processes when they observe long waiting times. In their book, Hopp and Spearman (2011) refer to a phenomenon called "Overtime Vicious Cycle" that illustrates this practice. They mention that plants often go through a series of cycles as follows: After the effective capacity of a plant is computed, a master production schedule is developed to release jobs into the plant according to the effective production capacity. As plants continue working for days and weeks, inventories in the plant (i.e., in effective bottleneck resource) increase. Since the bottleneck is 100 percent utilized, its throughput, and thus the plant throughput remains constant. Based on Little's law, an increase in inventory results in an increase in flow times. This results in missing customer due dates and customers begin to complain. To reduce inventory and flow times, management takes a "one-time" action by allowing overtime (i.e., temporarily increasing the effective capacity). This will reduce inventory and flow time and thus customer service improves. Overtime days end, everyone breathes a sigh of relief, and things go back to normal, which means releasing jobs equal to the effective capacity of the plant; and the cycle repeats.

Balking and abandonment as described in Chap. 6 result in loss of throughput, and temporary measures of increasing and decreasing capacity (e.g., overtime, subcontracting, hiring part-time workers) increase costs, both of which result in lower profit for a firm. Hence, to avoid profit loss, the effective capacity of a process should always be set greater than the input rate of the process—processes should not be 100 percent utilized—unless there is no variability anywhere in the process (which is very unlikely in the real world). The additional capacity which is set beyond the input rate is called safety capacity.

---

**CONCEPT**

**Safety Capacity of a Resource**

If we define $R_{in}$(resource $i$) as the long-run input rate that resource $i$ receives per unit time, and also denote $C_{sft}$(resource $i$) as the *Safety Capacity* of resource $i$, also called *Capacity Cushion*, then

$$C_{sft}(\text{resource } i) = C_{eff}(\text{resource } i) - R_{in}(\text{resource } i)$$

---

### Safety Capacity—Creating Stability

Thus, when there is variability in a process, all resources of a process must have safety capacities to prevent the inventory and flow time of the process increase in time without limit. The main reason for having safety capacity is to eliminate or diminish the negative impact of variability on process performance.

Now, let's return to the packaging station in Case 6 in which the station has no safety capacity and decrease the packaging time by only 1 second from 60 seconds to 59 seconds. This corresponds to increasing the effective capacity of the station by only 1.7 percent from 60 phones per hour to around 61 phones per hour. Considering the input rate of 60 phones per hour, the safety capacity of the station would be $61 - 60 = 1$ phone per hour. Can this small increase in capacity stop the increasing trend of the in-buffer inventory in Case 6?

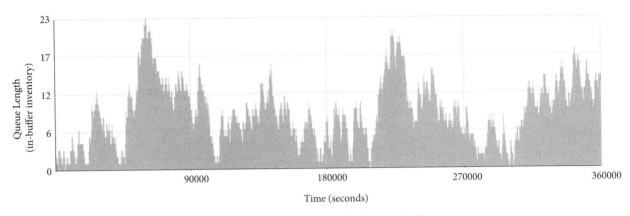

Figure 7.4   Number of phones in the buffer of the packaging station in Case 6 with safety capacity of one.

Figure 7.4 shows the queue—the in-buffer inventory—of the packaging station during 100 hours of operations if we have a safety capacity of only one phone per hour in Case 6.

Comparing Fig. 7.4 with Fig. 7.3, we observe that having even a small safety capacity stops the increasing trend in the in-buffer inventory and prevents it from growing without limit. The maximum in-buffer inventory during the 100 hours operations in the system with safety capacity of one (Fig. 7.4) is 23 phones and the in-buffer inventory seems to have an average of 10 phones.

**Safety Capacity—Improving Performance**

While one reason for having safety capacity is to prevent inventory from growing without limit, safety capacity also serves another purpose. By increasing the safety capacity of a process, one can further improve performance measures such as inventory and flow time of the process. Consider, for example, the packaging station in Case 3 that has the same input rate of 60 phones per hour as in Case 6, but the packaging time is 45 seconds—corresponding to the effective capacity of 80 phones per hour. Hence, the safety capacity in Case 3 is $80 - 60 = 20$ phones per hour. Comparing the in-buffer inventory of Case 3 (Fig. 7.1) with safety capacity of 20 with that with safety capacity of 1 (Fig. 7.4), we can clearly see that the station with higher safety capacity has a significantly lower inventory. This observation points to the following principle, after which we conclude this section.

**PRINCIPLE**

**Impact of Safety Capacity on Flow Time and Inventory**

Resources with variability (in their arrivals or process times) must have safety capacity to prevent their inventory and flow time grow without limits. As the safety capacity of a resource increases, the inventory and flow time of the resource improve (i.e., decrease).

## 7.3 Queueing Theory and Flow Time Measures

As we will discuss later in this chapter, one can use several strategies to improve flow time measures (e.g., hiring more workers). However, before spending any money on any improvement strategy, the big challenge is to evaluate how much will implementing the suggested strategies improve those measures.

One class of models that is used to estimate flow times of a process under different improvement strategies is simulation models. *Simulation models* use computer to generate all the events that occur during operations, mimic the flow dynamics within the process, and keep track of the flow times of all flow units for a large number of simulation hours. This is how we generated the graphs in the previous sections. However, developing simulation models requires expertise in probability and statistics and computer coding or in special-purpose simulation languages. We do not discuss simulation models in this book.

The other option is to use a large number of very useful *mathematical models* that have been developed in the last 60 years to estimate flow times for a wide range of processes. These models constitute a field in applied mathematics called "queueing theory." Queueing theory has many applications in telecommunication, computer science, transportation, manufacturing, and service operations, among others. Before we describe queueing models, we first need to introduce the foundational elements of a queueing system.

In Queueing Theory terminology, flow units are referred to as "customers," resources are called "servers," processing times are referred to as "service times," in-buffer inventory is referred to as "queue," and flow times are referred to as "waiting times."

## 7.4 Elements of Queueing Systems

Using limited information about the process, queueing theory provides simple mathematical expressions that can be used to estimate several process-flow measures such as average flow times, average inventory, and throughput. The required information includes information about the basic elements of a process: (i) arrival to the process, (ii) process times, (iii) number of resources, (iv) maximum buffer size, (v) the size of the population from which the (demand) arrivals are generated, and (vi) the order by which flow units are chosen from the buffer for processing (i.e., the control policy). We utilize the following example to further describe these.

 **Example 7.1**

> Recall the Michigan Yacht Club in Sec. 6.2.3 of Chap. 6 Jack, the owner of the club, was able to convince its adjacent property owner to rent a small part of his property to Jack, so Jack could expand his business by adding one additional powerboat launch ramp to its current one. The club still has a waiting area for four boat trailers (in addition to those being launched). Jack was able to also improve his launching process, and as expected, the average launching time was reduced to about 4 minutes. New data also shows that boat trailers arrive at the club at an average rate of 12 trailers an hour. However, as before, when the club's waiting area is full, arriving customers do not wait and go to the other ramps in the park. Before making his final decision of renting the property, Jack would like to know how much adding a new launching ramp will improve his revenue. He is currently charging $45 for launching a boat.

Jack would like to have an estimate of the impact of implementing a change (i.e., adding capacity) on his process performance. Specifically, he would like to know how much adding a new ramp reduces his customers' waiting times (i.e., flow times) and increases his throughput. To be able to use queueing theory to answer these questions, we first need to identify the elements of the underlying queueing systems of Jack's operations. We start with arrivals.

Table 7.2    Interarrival Times of 49 Customers at Yacht Club in Example 7.1

| Interarrival times (*time is in seconds*) | | | | | | | |
|---|---|---|---|---|---|---|---|
| 163 | 852 | 89 | 309 | 199 | 183 | 506 | 143 |
| 41 | 140 | 1001 | 527 | 344 | 475 | 274 | 1107 |
| 68 | 195 | 4 | 723 | 255 | 181 | 55 | 73* |
| 244 | 98 | 78 | 751 | 149 | 87 | 1033 | 440 |
| 137 | 558 | 174 | 276 | 69* | 80 | 252 | 348 |
| 135 | 214 | 107 | 129 | 27 | 138 | 82 | 927 |

## 7.4.1 Arrivals

Arrival to a process is one of the main elements that affect the flow times of a process. The higher the number of arrivals to a process, the more congested the process will be. This, based on Little's law, results in longer flow times. Arrivals to a process usually corresponds to the demand for the process. As described in Chaps. 3 and 5, the arrival to a process is modeled by either the probability distribution of the number of arrivals during a period (e.g., the number of customers arriving to Michigan Yacht Club in an hour), or by the probability distribution of the interarrival times (e.g., the distribution of time between two consecutive arrivals to Michigan Yacht Club). While both distributions are useful in modeling arrivals, most analytical models of flow time use the latter. In fact, most queueing models only require the mean and variability (measured by the coefficient of variation) of interarrival times to a process and not its distribution.

Let's return to Example 7.1, and suppose that the Michigan Yacht Club would like to model its demand arrivals. The first step is to get data about time between two consecutive arrivals. Note that the number of arrivals at the club is different from the number of arrivals that enter the queue (i.e., the input) of the club. Recall that those arrivals that find no space in the waiting area go to other facilities to launch their boats. The question is: Which type of data should the club collect: the interarrival times, or interinput times? The answer is the interarrival times, since it contains more information about the actual arrivals. Interinput times miss information about those arrivals that did not enter the club due to lack of space.

Table 7.2 represents the interarrival times of 49 customers to the club. The sign "*" corresponds to those arrivals that did not enter the club, because the line was full. Most of the analytics that estimate the flow time of a process use only the mean and coefficient of variation of interarrival times. If we define

$$T_{\text{arv}} = \text{Mean of interarrival times}$$
$$\sigma_{\text{arv}} = \text{Standard deviation of interarrival times}$$
$$\text{CV}_{\text{arv}} = \text{Coefficient of variation of interarrival times}$$

then for the interarrival time data in Table 7.2, we have

$$T_{\text{arv}} = 300.8 \text{ seconds}, \qquad \sigma_{\text{arv}} = 292.4 \text{ seconds}, \qquad \text{CV}_{\text{arv}} = \frac{\sigma_{\text{arv}}}{T_{\text{arv}}} = \frac{292.4}{300.8} = 0.972$$

Hence, the interarrival times have a mean of 300.8 seconds, which is about 5 minutes, and variability (i.e., coefficient of variation) 0.972. We will show later in this chapter how these two quantities can be used to estimate process-flow measures of a process.

There is a class of analytical models called *Markovian models* that provide more detailed and more accurate estimates of flow times for a wide range of processes. These models, however, can be used only if the interarrival times (and service times) follow exponential distribution. If a continuous random variable $T$ (e.g., interarrival time) follows an *exponential distribution* with parameter $\lambda$, then its density function is

$$f(t) = \lambda e^{-\lambda t} \qquad \text{for} \quad t \geq 0$$

The cumulative probability distribution of exponential random variable is

$$P(T \leq x) = 1 - e^{-\lambda x}$$

Exponential distribution has some interesting features:

- *Mean and standard deviation of exponential distribution are equal.* The mean, standard deviation, and coefficient of variation of an exponential distribution with parameter $\lambda$ are as follows:

$$T_{\text{arv}} = \frac{1}{\lambda} \qquad \sigma_{\text{arv}} = \frac{1}{\lambda} \qquad \text{CV}_{\text{arv}} = 1$$

Note that the mean and standard deviation of exponential distribution are the same, and hence, exponential distribution has coefficient of variation of one—a moderate variability. Also, note that if the mean of interarrival times is $1/\lambda$ unit time, the number of arrivals per unit time is $\lambda$.

- *Exponential distribution is memoryless.* Suppose that interarrival times follow an exponential distribution with an average of 5 minutes. This means that $1/\lambda = 5$ minutes and thus $\lambda = 0.2$ arrivals per minute. The probability that the next demand arrives in the next 5 minutes is

$$P(T \leq 5) = 1 - e^{-\lambda(5)} = 1 - e^{-0.2(5)} = 0.63$$

The memoryless property states that the probability of the next demand arriving in the next 5 minutes (i.e., 0.63) is independent of whether the last demand arrived 1 minute ago or 10 minutes ago or 100 minutes ago. This memoryless property points to the fact that the timing of the next demand arrival is *completely random* and is independent of the timing of the previous demand arrivals.

- *Exponential distribution is related to Poisson distribution.* Recall from Sec. 3.1.2 of Chap. 3 that the number of arrivals (i.e., demand) per unit time can be modeled by Poisson distribution when demand arrivals are completely random, that is, when a large number of customers make decisions of when to arrive randomly and independent of each other. So, one expects that there should be a relationship between exponential distribution (that has completely random interarrival times) and Poisson distribution (that has completely random number of arrivals per unit time). There is, indeed, a tight relationship between these two distributions. If the number of arrivals to a process per unit time follows a Poisson distribution with parameter $\lambda$, then the interarrival times follow an exponential distribution with parameter $\lambda$, and vice versa. This is one reason we used $\lambda$ as the parameter of both Poisson and exponential distributions. This relationship is useful when we analyze arrivals to a process. For example, if the interarrival times follow an exponential distribution with an average of 5 minutes (i.e., $\lambda = 0.2$ per minute $\times 60 = 12$ per hour), then the probability of having $D = 10$ customers in the next hour can be found using a Poisson distribution with parameter $\lambda = 12$ per hour as follows:

$$P(D = 10) = \frac{e^{-\lambda}(\lambda)^x}{x!} = \frac{e^{-12}(12)^{10}}{10!} = 0.105$$

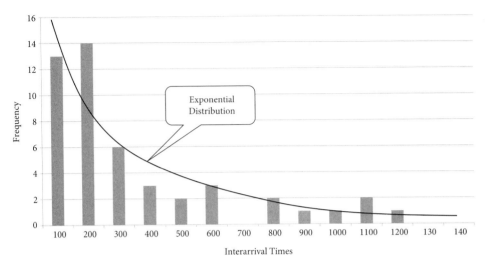

Figure 7.5    Histogram of interarrival times at Yacht Club and its exponential distribution approximation.

Exponential distribution has a decreasing density function (see Fig. 7.5), implying that smaller value of random variable $T$ has a higher probability of occurrence than larger values of $T$.

So, to take advantage of the variety of results that Markovian models offer, it is always a good idea to first check whether the interarrival times of a process follow an exponential distribution, and if they don't, check to see if interarrival times can be approximated by an exponential distribution. To check whether the interarrival times at Michigan Yacht Club follows exponential distribution, we can draw the histogram of the interarrival times of the data in Table 7.2. This is shown in Fig. 7.5. The figure also shows an exponential distribution.

As Fig. 7.5 shows, we can assume that the exponential distribution can be a reasonably good approximation for the interarrival times at Michigan Yacht club.[6] Two more evidences that support this assumption are the following. First the coefficient of variation of interarrival times at the club (i.e., 0.972) is very close to 1, the coefficient of variation of an exponential distribution. Second, in each day, the arrivals at the Yacht Club is the result of a large number of people making decisions of whether to go for a boat ride, independent of each other, resulting in complete random arrivals. As we mentioned, exponential distribution models completely random arrivals.

In conclusion, we model the *interarrival times* at the Yacht Club as an exponential distribution with mean $1/\lambda = 5$ minutes (which we found from our data). This is equivalent to say that we model the *number of arrivals per hour* to the club with a Poisson distribution with mean $\lambda = 12$ per hour.

### 7.4.2 Process Times

Another essential piece of information required in queueing models is information about how long it takes to process a flow unit. In Chap. 4 we introduce this as effective process times. The longer the effective mean process times, the lower the capacity, resulting in more congestion (i.e., longer flow time) in the process.

---

[6]Note that statistical tests such as goodness-of-fit test must be used to further support that exponential distribution is a good fit for interarrival times. These tests can be found in statistics textbooks and are out of scope of this book.

**Table 7.3**    Effective Process Times of 48 Customers in Yacht Club in Example 7.1

| Times to Launch a Boat (*time is in seconds*) | | | | | | | |
|------|------|------|------|------|------|------|------|
| 252  | 89   | 63   | 414  | 185  | 491  | 891  | 184  |
| 972  | 23   | 401  | 228  | 61   | 320  | 21   | 71   |
| 390  | 123  | 277  | 22   | 29   | 168  | 83   | 31   |
| 782  | 25   | 311  | 24   | 27   | 35   | 318  | 173  |
| 114  | 121  | 128  | 93   | 199  | 297  | 216  | 216  |
| 207  | 148  | 355  | 520  | 81   | 1012 | 128  | 223  |

While the probability distribution of effective process times carries all the information about the effective process time, similar to the arrivals, most analytical models for estimating flow times require only the mean and coefficient of variation of effective process times. Hence, to get an estimate for the mean and the coefficient of variation of effective process times at Michigan Yacht Club in Example 7.1, we need to gather information about how long it takes for the club to launch a boat. Table 7.3 shows a sample of 48 launching times at the club. The times are effective process times that include both the main task and supporting tasks of launching a boat.

Computing the mean and standard deviation of the 48 effective process times in Table 7.3, we find

$$T_{\text{eff}} = 240.5 \text{ seconds}, \qquad \sigma_{\text{eff}} = 244.1 \text{ seconds}, \qquad \text{CV}_{\text{eff}} = \frac{\sigma_{\text{eff}}}{T_{\text{eff}}} = \frac{244.1}{240.5} = 1.01$$

Therefore, the effective mean process time of launching a boat at Michigan Yacht Club is about $240.5/60 = 4$ minutes, with coefficient of variation 1.01, which is very close to one. This raises the question of whether the effective process times follow an exponential distribution. If they do, then we will be able to use a wider range of Markovian models to analyze the flow times in Michigan Yacht Club. If we draw the histogram of the launching times, we will see that an exponential distribution can provide a good approximation for the distribution of processing times at the club.

Thus, we model the effective process times at the Michigan Yacht Club in Example 7.1 by an exponential distribution. To distinguish the exponential distribution of process times from that of the interarrival times, we use $\mu$ as the parameter of the exponential distribution of effective process time. Therefore, the effective process time follows an exponential distribution with mean $1/\mu = 4$ minutes, which represents the effective capacity of $\mu = 15$ boats per hour.

### 7.4.3  Number of Resources

The third factor that influences flow times in a process is the number of resources. In the language of queueing theory, resources that perform the task of processing flow units are called *Servers*. A server can be a bank teller, a drill in a drilling station, a lane for takeoff in an airport, or a central processor in a computer.

Increasing the number of servers (resources) in a queueing system, increases the capacity of the system and thus reduces the queue length. This in turn reduces the flow time in the system. To estimate flow times in a process, in addition to mean and standard deviation of interarrival and effective process times, queueing models need the number of servers that process flow units.

The number of servers in the Michigan Yacht Club in Example 7.1 is currently 1. If an additional launching ramp is added, the number of servers will increase to 2. Number of servers in queueing systems can vary between 1 and a very large number. For example, when you call your credit card company, your call is directed to a call center where a group of agents (i.e., servers) are available to answer the calls. Your call often waits on hold (i.e., in the queue) and finally gets connected to an agent. The number of agents in these call centers can be as high as 300 to 400 agents.

There are also queueing systems in which the number of servers is assumed to be infinite. These are queueing systems in which, for every arriving customer, there is always a server available. Thus, customers never wait in the queue. An example of such a system is TV channels. When you push the channel number on your TV, you are immediately connected to the channel—your service starts immediately. You do not need to wait for minutes for your turn to get connected to a TV channel. These queueing models are called "infinite-server queues."

### 7.4.4 Maximum System Size

There are situations in practice when the number of customers in the system (i.e., total in-buffer plus in-process inventory) does not or cannot exceed a certain level. When the number of customers in the system reaches that level, new arrivals cannot enter the system and thus are lost. The Michigan Yacht Club is an example of such a system. Recall that the club's waiting area can accommodate at most four boat trailers, and the club currently has one launching ramp—space for maximum of five boats. When the club is full (i.e., there are five boats in the system), arriving customers do not wait and go to the other ramps in the park. Thus, the club's maximum system size is 5.

### 7.4.5 Arrival Population Size

Another important factor is the size of the population of potential customers who may decide to use the process. The larger the number of potential customers, the higher the arrival rate. Consider, for example, two fitness clubs that operate on a membership basis: One has 200 members and the other has 100 members. When both clubs are empty, the club with the larger number of members can expect more arrivals than the smaller club. In both clubs, as the number of customers in the club increases, the number of potential customers (i.e., members) who may come to the club decreases, which in turn decreases the arrival rate to the club. For example, if 100 members of the smaller club are currently using the club, then the arrival rate to the club would be zero, since all the customers (members) are already in the system.

As the size of potential customers increases, the arrival rate increases, which results in higher congestion and thus larger flow times. Except for a small number of processes (that often involve membership), most processes have a large population of potential customers. For example, the population of customers who use a supermarket or a fast-food restaurant is large enough such that the number of customers inside these processes does not have a significant impact on the number of cars arriving to these systems. In the language of queueing theory, such processes are considered to have a potential customer population of size infinite. The Michigan Yacht Club also is a queueing system with a large customer population that can be considered infinite.

### 7.4.6 Queue Discipline—The Control Policy

Queue discipline refers to the rule by which customers are chosen from the queue to receive service when a server becomes available. The most common discipline is first-come-first-served (FCFS). FCFS is used in most queues in which flow units are people, since it preserves fairness in offering service. For example, the queue discipline in Michigan Yacht Club is FCFS.

Another queue discipline is last-come-first-served (LCFS). This discipline is not appropriate for processes in which flow units are people. However, there are many applications of such discipline in manufacturing systems. For example, consider a worker who finishes sanding disks and puts them in his output buffer, which also serves as input buffer for the next worker in line. When the next worker takes a disk from the buffer, he always takes the one from the top, which is the disk that arrived to the buffer last—LCFS discipline.

Priority (PRIO) is another service discipline under which some group of customers are given priority over others. One example is the Emergency Departments of hospitals that give priority to patients with more critical conditions. Another example is giving priority to first-class and business-class passengers during boarding of airplanes.

Finally, service in random order (SIRO) is a discipline that chooses next customer from the queue to serve in no particular order such as FCFS, LCFS, or PRIO—the customers are chosen randomly.

### 7.4.7 Kendall's Notation for Queueing Systems

In 1953 Kendall suggested a coding system to present the six basic elements of queueing systems. The code consists of six symbols separated by slashes as follows:

$$\mathcal{A}1/\mathcal{A}2/\mathcal{A}3/\mathcal{A}4/\mathcal{A}5/\mathcal{A}6$$

$\mathcal{A}1$:  The first part of Kendall's notion, $\mathcal{A}1$, corresponds to the arrival process. Kendall's notation uses letter $M$ in location $\mathcal{A}1$ if the interarrival times follow exponential distribution. Letter $D$ is used if interarrival times are deterministic (i.e., interarrival times have no variability). Letter G is used if the interarrival times follow a general distribution (i.e., a distribution other than deterministic and exponential).

$\mathcal{A}2$:  The second part corresponds to the process times. Similar to arrivals, $M$ is used if the process times follow exponential distribution, $D$ is used for deterministic process times, and $G$ is used for all other distributions.

$\mathcal{A}3$:  The third part corresponds to the number of servers. The number of servers of a queueing system can be between one server and infinite servers, that is, $\mathcal{A}3 \in \{1, 2, 3, \ldots, \infty\}$.

$\mathcal{A}4$:  The fourth part corresponds to the maximum number of customers that can be held in the system, that is, maximum system size. The maximum system size can be between one and infinite, that is, $\mathcal{A}4 \in \{1, 2, 3, \ldots, \infty\}$.

$\mathcal{A}5$:  The fifth part corresponds to the size of the population of potential customers, which can be from one to infinite, that is, $\mathcal{A}5 \in \{1, 2, 3, \ldots, \infty\}$.

$\mathcal{A}6$:  The sixth part corresponds to service discipline $\mathcal{A}6 \in \{\text{FCFS, LCFS, SIRO, PRIO}\}$.

So an $M/D/3/10/\infty/\text{FCFS}$ represents a queueing system in which customers' interarrival times are exponentially distributed (i.e., arrival is Poisson process), service times are deterministic (i.e., always takes the same time with no variability), with three servers. The queueing system can accommodate a maximum of 10 customers (i.e., 3 being served and a maximum of 7 in the queue). The population of potential customers is very large, that is, infinite, and arriving customers are served according to first-come-first-served discipline.

To shorten the notation, system size and population size are omitted from the notation, if they are infinite, and service discipline is omitted from the notation if it is FCFS. Therefore, $M/D/3/10/\infty/\text{FCFS}$ can be simply presented as $M/D/3/10$. Similarly, $M/G/3$ represents a queueing system in which interarrival times are exponentially distributed, service time is generally distributed, and the system has three servers. System size and population size are infinite and service discipline is FCFS. An $M/G/\infty$ is similar to $M/G/3$, except that it has infinite servers. Also, $M/G/3/15$ is similar to $M/G/3$, except that it cannot hold more than 15 customers in the system. Arrivals who find 15 customers in the $M/G/3/15$ system cannot enter the system and are lost.

Based on Kendall's notation, the current operations in Michigan Yacht Club is an $M/M/1/5$ queueing system, and after adding the new launching ramp, it becomes an $M/M/2/6$ queueing system. Since both interarrival and processing times are exponentially distributed, as we said before, queueing systems at Michigan Yacht Club belong to the class of Markovian Queues, which we will discuss in detail in Sec. 7.6. But first, in the following section, we present the details of the models that provide estimates for flow times in single-stage processes with generally distributed interarrival and process times, that is, $G/G/s$ queues.

## 7.5 Single-Stage Processes with Infinite Buffer

A single-stage process is a process in which all tasks required to process a flow unit are performed by one resource or one pool of resources. For example, the Michigan Yacht Club in Example 7.1 is a single-stage process in which all tasks needed to launch a boat are done by launching equipment.

The infinite buffer size corresponds to having ample space to keep arriving flow unit in the buffer. When a process has ample buffer space, all arriving flow units can enter the process. For example the buffer of the security at airports have almost infinite size, and all passengers who arrive join the line. The Michigan Yacht Club, however, is a process with finite buffer of size four. As we mentioned, if there are four boat trailers waiting in line—five boat trailers in the club—the arriving trailers cannot enter the club and hence go to other clubs.

Recall from Chap. 2, that $I_b$ and $T_b$ are, respectively, the average in-buffer inventory and the average in-buffer flow times. Throughout Chap. 2, these values were either given or computed using the sample data about the actual cumulative arrivals and departures of the process. In this section we provide queueing models that can be used to estimate these values in situations where the sample data about actual arrivals and actual departures are not available. For example, before adding an additional launch ramp, Jack at the Michigan Yacht Club cannot gather data about the actual departures, and hence cannot estimate what the waiting times would be after he adds a new ramp.

Before we show how one can estimate flow times of a process using queueing models, we first need to make an analogy between process-flow chart of a single-stage process and its corresponding queueing model. This is shown in Fig. 7.6 for a single-stage process with three resources and infinite buffer size. Flow units that arrive and enter the buffer of the process are analogous to the customers who arrive and enter the queue of the queueing model. The flow unit in the buffer, therefore, corresponds to the customers waiting in line of the queueing

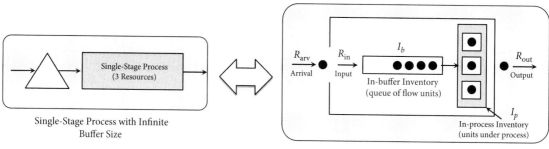

**Figure 7.6    A single-stage process with infinite buffer size and its corresponding queueing model.**

system. The pool of three resources corresponds to the three servers in the queueing system. Thus, the average in-buffer inventory ($I_b$) and the average in-buffer flow time ($T_b$) are indeed the average queue length and the average waiting time in the queue of the queueing model, respectively. The average in-process inventory ($I_p$) is the average number of customers being served, and the average in-process flow time ($T_p$) is the average service time in the queueing model.

With this introduction, we now show how one can estimate the process-flow measures of a single-stage process with $s$ identical servers and infinite buffer size using a $G/G/s$ queueing model. The queueing analytics in this chapter is for processes with no rework and no yield loss. Queueing models with rework and yield loss are complex and are out of the scope of this book.

---

**ANALYTICS**

**Estimating Performance Measures of a Single-Stage Process with Infinite Buffer Size—The $G/G/s$ Model**

Consider a process with no rework and no yield loss and a pool of $s$ resources that process flow units independent of each other. The resources are identical in that they all have the same effective mean process time $T_{eff}(\text{resource } i) = T_{eff}$ with the same standard deviation $\sigma_{eff}(\text{resource } i) = \sigma_{eff}$ for all $i = 1, 2, \ldots, s$. Hence, the process has the effective capacity

$$C_{eff}(\text{process}) = \frac{s}{T_{eff}}$$

Arriving units require processing by only one of the $s$ resources, whichever is available, and they are processed FCFS. Suppose interarrival times of the flow units to the buffer of the process have mean of $T_{arv}$ and standard deviation $\sigma_{arv}$. Since the process has infinite buffer size, all the arriving flow units enter the process, therefore,

$$R_{in} = R_{arv} = \frac{1}{T_{arv}}$$

- **Utilization:** Utilization of the process, $u$, is

$$u = \frac{R_{in}(\text{process})}{C_{eff}(\text{process})} = \frac{1/T_{arv}}{s/T_{eff}} = \frac{T_{eff}}{s\,T_{arv}} \tag{7.1}$$

- **Throughput:** If $u < 1$, the process is stable, and its throughput will be equal to its input rate,

$$\text{TH} = R_{in}(\text{process}) = R_{arv}(\text{process})$$

If $u \geq 1$ and there is variability in the arrivals or processing times, then the process is not stable, and the throughput of the process is equal to its effective capacity.

- **Average Flow Times:** In a stable process, the *average in-buffer flow time* can be estimated as follows:

$$T_b = \left(\frac{CV_{arv}^2 + CV_{eff}^2}{2}\right)\left(\frac{u^{\sqrt{2(s+1)}-1}}{s(1-u)}\right)T_{eff} \tag{7.2}$$

where $CV_{arv}$ and $CV_{eff}$ are the coefficient of variation of interarrival times and effective process times, respectively,

$$CV_{arv} = \sigma_{arv}/T_{arv} \qquad CV_{eff} = \sigma_{eff}/T_{eff}$$

The *average flow time* is, therefore,

$$T = T_b + T_{eff}$$

- **Average Inventory:** Using Little's law, the *average in-buffer* and *in-process inventories* are as follows:

$$I_b = \text{TH} \times T_b \qquad I_p = \text{TH} \times T_p = \text{TH} \times \text{T}_{\text{eff}}$$

All of the above results also hold in processes with LCFS and SIRO disciplines.

Example 7.2

> A bank has two tellers who serve customers waiting in a single line to talk to one of the tellers, whoever is available. Customers arrive according to a Poisson process with a rate of 38 customers per hour. The time it takes for a teller to serve a customer is an average of 3 minutes with standard deviation of 2 minutes. Due to the long waiting time of the customers in the line, bank managers are considering to hire another teller to reduce customer waiting time. However, they are not sure how much an additional teller would reduce waiting times. Find an estimate for customer average waiting time in the line if the third teller is hired.

Note that the queueing model of the two tellers is a $G/G/2$ system—an $M/G/2$ to be exact. In this system the interarrival times have an average of $T_{\text{arv}} = 60/38 = 1.579$ minutes. Since the arrivals are according to a Poisson process, the interarrival times follow exponential distribution, which has coefficient of variation $CV_{\text{arv}} = 1$. Serving a customer by a teller has effective mean process time $T_{\text{eff}} = 3$ minutes, standard deviation $\sigma_{\text{eff}} = 2$ minutes, and thus coefficient of variation $CV_{\text{eff}} = 2/3 = 0.667$. Considering the current number of servers $s = 2$, the utilization is

$$u = \frac{T_{\text{eff}}}{s\, T_{\text{arv}}} = \frac{3}{2(1.579)} = 0.95$$

Using Eq. (7.2), we can find an estimate for the average waiting time in line—the average in-buffer flow time—as follows:

$$T_b = \left( \frac{CV_{\text{arv}}^2 + CV_{\text{eff}}^2}{2} \right) \left( \frac{u^{\sqrt{2(s+1)}-1}}{s(1-u)} \right) T_{\text{eff}} = \left( \frac{1^2 + (0.667)^2}{2} \right) \left( \frac{0.95^{\sqrt{2(2+1)}-1}}{2(1-0.95)} \right) 3$$

$$= 20.1 \quad \text{minutes}$$

When an additional teller is hired, the system becomes $G/G/3$, with utilization

$$u = \frac{T_{\text{eff}}}{s\, T_{\text{arv}}} = \frac{3}{3(1.579)} = 0.633$$

Hence, with three tellers, the average customer waiting time in the line is

$$T_b = \left( \frac{CV_{\text{arv}}^2 + CV_{\text{eff}}^2}{2} \right) \left( \frac{u^{\sqrt{2(s+1)}-1}}{s(1-u)} \right) T_{\text{eff}} = \left( \frac{1^2 + (0.667)^2}{2} \right) \left( \frac{0.633^{\sqrt{2(3+1)}-1}}{3(1-0.633)} \right) 3$$

$$= 0.85 \quad \text{minutes} \times 60 = 51 \text{ seconds}$$

In conclusion, hiring an additional teller will reduce customer average waiting time from 20.1 minutes to 51 seconds. Considering the salary of the new teller and the benefits of the improvement in customer waiting time, bank managers can decide whether to hire a new teller or not.

While Eq. (7.2) looks a bit complex, it provides useful managerial insights into the relationship between in-buffer flow time, variability, utilization, and process capacity. Consider

Eq. (7.2) for the case of $s = 1$ server, for example. Since the effective capacity of the process is $C_{\text{eff}} = 1/T_{\text{eff}}$, we can rewrite Eq. (7.2) for $s = 1$ as follows:

$$T_b = \underbrace{\left( \frac{CV_{\text{arv}}^2 + CV_{\text{eff}}^2}{2} \right)}_{\text{VARIABILITY}} \underbrace{\left( \frac{u}{1-u} \right)}_{\text{UTILIZATION}} \underbrace{\left( \frac{1}{C_{\text{eff}}} \right)}_{\text{CAPACITY}} \tag{7.3}$$

Thus, as the equation shows, average in-buffer flow time is the product of three terms:

- The first term corresponds to the variability in the process, consisting of variability in the interarrival times and in the effective process times. The equation shows that the variability in arrivals or effective process times has similar impact on the average flow time. Any increase in either variabilities (keeping capacity and utilization unchanged) increases the average in-buffer flow times. We observed the same phenomenon in Sec. 7.1 using simulation models of Cases 3 and 4 of our packaging station example.

- The second term relates to process utilization ($u$) which implies that, as system utilization increases (keeping variability and capacity unchanged), the second term increases. This results in an increase in the average in-buffer flow time.

- The third term relates to the inverse of the effective capacity of the process. Hence, increasing the capacity of a process (keeping variability and utilization constant) reduces the average in-buffer flow time.

Since the impact of variability and utilization on flow times has significant implications on managing flow times of a process, we discuss them in more detail in the following sections.

### 7.5.1 Impact of Utilization on Flow Times

To analyze the effects of utilization on flow times of a process, we study the case of a furniture company in Example 7.3.

**Example 7.3**

Diorucci Furniture, a small manufacturer of luxury furniture, receives orders for a wide range of living room and bedroom furniture. Furniture is built in the company's small factory that operates in a produce-to-order mode and makes the orders on a first-come-first-served basis 7 days a week. Data shows that the time to complete a customer's order—which is the time that the factory starts working on an order until it finishes the entire order—is variable and has an average of 3 days and standard deviation of 6 days. One reason for the large standard deviation is the variety in the number of pieces of furniture in an order as well as the complexity of the design of those pieces. The company's furniture has become very popular and managers expect that their demand will increase by 12.5 percent from eight orders per month to nine orders per month. The demand, that is, the number of orders received per month, appears to be completely random, so a Poisson distribution is shown to well fit the demand distribution. The company's policy has been to keep the average time the customers wait until they get their order—the average order lead time—less than 50 days. Managers are worried that the expected increase in the demand can result in the average order lead time becoming significantly larger than 50 days, so they want to take necessary actions to prevent that. In a meeting, David, the head of marketing and operations, argues that there is no need to take any action. He mentions that he checked the sales data and has computed the *current* average order lead time to be around 31 days. He thinks that one unit increase in monthly demand will still keep the average order lead time below 50 days. He reminds everyone about 2 years ago when their average monthly demand also increased by one unit from 7 to 8. He mentions that, at that time, the order lead time increased by 13 days. So he concludes that they should expect about 13 days increase in average order lead time, if the monthly demand (as before) increases by one unit. This would increase the average order lead time from 31 days to $31 + 13 = 44$ days, which is still lower than the company's goal of 50 days. Thus, he concludes that no action is needed to handle the new demand. What do you think?

To analyze the head of marketing's argument, let's first compute the current average order lead time. Note that the average order lead time is, $T$, the average flow time of an order in the process, which is equal to the average in-buffer flow time plus the effective mean process time of an order, that is, $T = T_b + T_{\text{eff}}$. Considering the factory as a single resource that processes orders, we will have a $G/G/1$ queueing model, or $M/G/1$ to be exact, since arrival process is Poisson. Thus, having $s = 1$ in Eq. (7.2), we have

$$T = \left( \frac{CV_{\text{arv}}^2 + CV_{\text{eff}}^2}{2} \right) \left( \frac{u}{1 - u} \right) T_{\text{eff}} + T_{\text{eff}}$$

Assuming 30 days in a month, the effective capacity of the factory is $C_{\text{eff}}(\text{factory}) = 30/3 = 10$ orders per month. Considering the current demand, the order arrival rate is $R_{\text{arv}}(\text{factory}) = 8$ orders per month. Since all orders are accepted, then $R_{\text{in}}(\text{factory}) = R_{\text{arv}}(\text{factory}) = 8$ per month. Thus, current utilization of the factory is

$$u = \frac{R_{\text{in}}(\text{factory})}{C_{\text{eff}}(\text{factory})} = \frac{8}{10} = 0.8$$

Since order arrivals is Poisson, interarrival times are exponentially distributed with $CV_{\text{arv}} = 1$. The effective mean process time is 3 days or $3/30 = 0.1$ month, with variability of $CV_{\text{eff}} = 6/3 = 2$. Hence, the current average order lead time will be

$$T = \left( \frac{1^2 + 2^2}{2} \right) \left( \frac{0.8}{1 - 0.8} \right) 0.1 + 0.1 = 1.1 \text{ months} \times 30 = 33 \text{ days}$$

which is close to what the head of marketing estimated using sales data, that is, 31 days. This gives us the confidence that our queueing model fits reasonably well with what occurs in the factory's shop floor.

If the demand increases by one unit from 8 to 9 order per month, the factory utilization will increase from 0.8 to $u = 9/10 = 0.9$. In this case, assuming that the demand still follows Poisson distribution, the average order lead time will be

$$T = \left( \frac{1^2 + 2^2}{2} \right) \left( \frac{0.9}{1 - 0.9} \right) 0.1 + 0.1 = 2.35 \text{ months} \times 30 = 70.5 \text{ days}$$

which is 20.5 days above the goal of 50 days. So, the argument made by the head of marketing is not valid. Specifically, one unit increase in monthly demand (from 8 to 9) increases the average order lead time by 37.5 days from 33 to 70.5 days, which is much larger than 13 days—the increase in order lead time when the monthly demand increased by one unit (from 7 to 8) in previous years. Why?

Note that increase in demand results in increase in utilization. Thus, if we plot the average order lead time for different utilization levels, we find our answer. Figure 7.7 shows how the average flow time changes as the demand for furniture increases and thus factory utilization increases. To show the complete structure of the average flow time, the figure is for all the

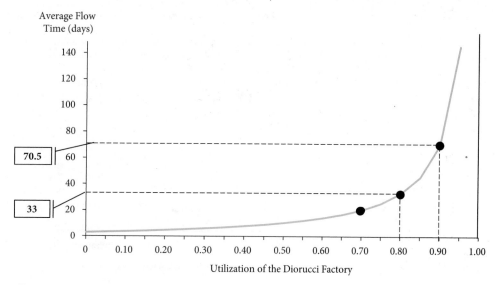

Figure 7.7    Average flow time at Diorucci Furniture in Example 7.3 for different factory utilizations.

utilization values between 0.1 and 1. The curve in Fig. 7.7 is also known as *Throughput-Delay* curve.

From the figure we observe that when process utilization increases, the average flow time increases nonlinearly with a slope that becomes larger and larger as the utilization gets closer to 1. This highly nonlinear behavior implies that, for example, one unit increase in monthly demand from 8 to 9 (i.e., increasing utilization from 0.8 to 0.9) will result in a much higher increase in flow time compared to that when the monthly demand is increased by one unit from 7 to 8 (utilization increases from 0.7 to 0.8).

Hence, the main issue with the argument made by the head of marketing is that we should not expect one unit increase in monthly demand to have the same impact on average flow time. The impact depends on the utilization level. The impact of changes in utilization on average flow time is more dramatic in processes with higher utilization.

Recall that utilization is the input rate (i.e., demand) divided by effective capacity of a process. Thus, an increase in utilization could be due to an increase in demand (as in the Diorucci Furniture example), or due to a decrease in effective capacity (e.g., more frequent machine failures or longer setup time). In either case, according to Little's law

$$I_b = \text{TH} \times T_b$$

any increase in average flow time results in an increase in average inventory. In other words, the average inventory also increases with utilization in a highly nonlinear fashion.

**PRINCIPLE**

**Impact of Utilization on Flow Time and Inventory**

The average flow time and the average inventory increase with utilization in a highly *nonlinear* fashion. Hence, if the demand for a process increases or the effective capacity of a process decreases, the average flow time and average inventory can increase significantly, especially in processes that are already highly utilized.

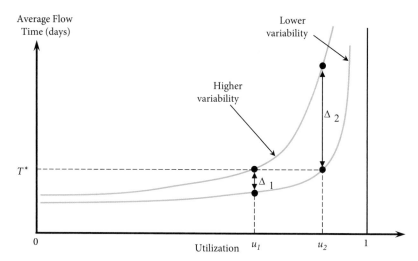

Figure 7.8   Impact of variability on average flow time.

### 7.5.2 Impact of Variability on Flow Times

Equation (7.3) shows that an increase in variability in the arrivals or in effective process times increases the average flow times. In other words, for any given effective capacity and arrival rate, increasing variability will push the average flow time curve up; see Fig. 7.8 as an example.

From Fig. 7.8, we can make the following observations:

- The average flow time curve with higher variability is always above that with lower variability. This, as expected, confirms that for any utilization level, processes with higher variability will have higher average flow times.

- For any given utilization level, the vertical distance between the two flow time curves is the reduction in average flow time (e.g., $\Delta_1$ and $\Delta_2$) if the variability in the process is reduced. As utilization increases (from $u_1$ to $u_2$), the benefit of variability reduction increases (i.e., $\Delta_2 > \Delta_1$). Hence, the impact of variability reduction in processes that work at higher utilization levels (e.g., bottleneck) is more significant than those that work at lower utilization levels.

- The average flow time curve with higher variability is always to the left of that with lower variability. This implies that, for example, to achieve a target average flow time (e.g., $T^*$), processes with higher variability must operate at a lower utilization level (e.g., $u_1$) than that in processes in which variability is lower (e.g., $u_2$). In other words, for any given demand, to achieve the same average flow times, processes with higher variability require a higher safety capacity. Larger safety capacity means larger capacity costs.

### Example 7.4

After realizing that an increase in monthly demand increases their customers' average order lead times up to 70.5 days, managers of Diorucci Furniture in Example 7.3 consider the following two actions to reduce their flow time below the target level of 50 days.

*Increasing Capacity:* Increasing capacity of the factory by 10 percent from 10 orders to 11 orders per month. This requires buying new equipment and hiring new workers. Increasing capacity is not expected to affect the variability of the effective process time; it will still be 2.

*Reducing Variability:* Hiring a consulting firm to reduce variability in their operations. A consulting team has already visited the factory and has given Diorucci a proposal and told them that by implementing lean manufacturing principles, the variability in the order processing time can be reduced by 40 percent from 2 to 1.2 or less.

Before making a final decision of which action to take, managers of Diorucci Furniture would like to know whether these actions can keep their target lead time below 50 days when the demand increases to 9 orders per month.

Under the first alternative, when the effective capacity is increased to $C_{eff}(\text{factory}) = 11$ orders per month, the factory utilization will be $u = 9/11 = 0.818$. On the other hand, the effective capacity of 11 orders per month implies effective mean process time $T_{eff}(\text{factory}) = 1/11 = 0.091$ months. Thus, the average order leadtime is

$$T = \left(\frac{1^2 + 2^2}{2}\right)\left(\frac{0.818}{1 - 0.818}\right)0.091 + 0.091 = 1.11 \text{ months} \times 30 = 33.3 \text{ days}$$

which is smaller than the acceptable average flow time of 50 days. This corresponds to Point A in Fig. 7.9 that shows how the average flow time curve changes for different values of utilization and variabilities.

If the second alternative is chosen, the variability of the effective process time is reduced from 2 to $CV_{eff} = 1.2$. However, factory utilization remains the same, that is, $u = 0.9$. Hence, the average order lead time will be

$$T = \left(\frac{1^2 + 1.2^2}{2}\right)\left(\frac{0.9}{1 - 0.9}\right)0.1 + 0.1 = 1.2 \text{ months} \times 30 = 36 \text{ days}$$

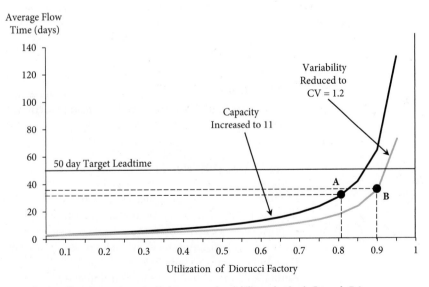

**Figure 7.9**  Average flow time under capacity increase and variability reduction in Example 7.4.

which also achieves management's goal of having an average order lead time below 50 days. This corresponds to Point B in Fig. 7.9.

In summary, with increased demand of nine orders per month, increasing capacity by 10 percent reduces the average order lead time by 37.2 days from 70.5 to 33.3 days. On the other hand, reducing variability has a similar effect. Specifically, reducing variability of effective process times from 2 to 1.2 reduces the average order lead time by 34.5 days from 70.5 to 36 days. This points out the important insight that variability reduction can act as a substitute for capacity expansion.

Which alternative should Diorucci Furniture choose? Well, it depends on the cost of each alternative. Expanding capacity (e.g., buying new equipment, hiring new workers, acquiring new space) is often more costly than efforts to reduce variability. Also, reducing variability has other advantages beyond just reducing flow time (e.g., higher throughput, better quality, and more efficient production and inventory control due to lower uncertainty). Thus, in general, variability reduction should always be a high priority goal of operations engineers and managers.

We summarize our observations about the interaction between variability and utilization and its impact on average flow times of a process in the following principles:

---

**PRINCIPLE**

**Impact of Variability on Flow Times and Inventory**

Processes with higher variability (in arrivals, in processing, or in both) will have a higher average flow time and inventory. Hence,

- To reduce average flow times and average inventory, variability reduction can be considered as a substitute for capacity expansion.

- The effects of variability reduction on reducing flow times and inventories are more significant in processes with higher utilization. Thus, variability reduction efforts should focus on highly utilized processes and resources (e.g., bottlenecks).

- To achieve a target average flow time, processes with higher variability must operate at a lower utilization than that in processes with lower variability. In other words, processes with higher variability require larger safety capacity to achieve a target flow time than what is required for processes with lower variability. Larger safety capacity results in larger capacity costs.

---

The above principles are for processes that have no limit on the number of flow units that can be held in the process. In the following section, we focus on processes with limited buffer size.

## 7.6 Single-Stage Processes with Limited Buffer

In processes with infinite buffer size, all arriving flow units enter the buffer and wait until they are processed. This is the reason that our analytics has $R_{in}(\text{process}) = R_{arv}(\text{process})$. In practice, as we mentioned, there are situations where the flow units in the buffer (i.e., the queue) cannot exceed above a certain level. This is often due to physical constraints that limit the number of units a buffer can hold. For example, a parking lot with a capacity of 100 cars cannot hold more than 100 cars and thus the arriving cars do not enter when the parking is full. The Michigan Yacht Club in Example 7.1 cannot hold more than four trailers in its buffer. The Easysurrance Call Center in the next example is another example of a process with limited buffer size. We use the next example to present the analytics required to estimate average flow times in such single-stage processes with limited buffer size.

### Example 7.5

Easysurrance is a large insurance company that offers a wide range of insurance products. All after-sales customer inquiries (e.g., questions about changing plans, submitting claims, etc.) are handled through the company's call center. The call center has two agents who answer the calls. The time that takes an agent to answer a caller's inquiry, called *Call Handling Time*, follows an exponential distribution with mean of 15 minutes. The call center has seven different phone lines, called *Trunk Lines*. If a customer calls and all seven lines are busy, the customer will hear a busy signal and will hang up. Calls that do not get busy signals (i.e., not blocked) and are connected to the call center, and wait on hold (if no agent is available). These calls are answered in the order they are received. Data shows that call arrivals are very random and a Poisson process with a mean of 11 calls per hour fits the call arrival pattern reasonably well. The call center performance is measured based on three metrics:

*Average Speed of Answer* (ASA), which is one of the metrics for evaluating the *speed* by which call centers offer service. Specifically, ASA is the average time a caller waits on hold before speaking to an agent.

*Fraction of Blocked Callers* (FBC), which is a metric for *accessibility* of the call centers. Specifically, FBC is the percent of callers who get busy signal and thus are blocked from getting connected to the call center.

*Occupancy* (OCC), which is a metric for *efficiency* of the workforce planning in call centers. Specifically, OCC is the percentage of time call center agents are busy answering calls. The smaller the OCC is, the more underutilized agents are.

Managers of call centers are told to improve the above service goals. The new service goals are to keep ASA below 4 minutes and FBC below 15 percent and occupancy above 75 percent. To achieve these goals, they are authorized to hire one additional agent if needed. Managers are wondering if hiring an additional agent can help them achieve the new service goals.

Easysurrance call center is an example of a process with limited buffer size. Customers (flow units) arrive at the process when they call. If they are not blocked, they enter the process and wait in the buffer (i.e., wait on hold) until they are connected to an agent (resource). They speak to the agent and leave the process when they hang up the phone. Because the call center has seven trunk lines, the maximum number of callers that are connected to the call center—either waiting on hold (in the buffer) or speaking to an agent (in process)—is 7. Thus, the maximum system size is 7. When the system reaches its maximum capacity (i.e., all seven lines are busy), the arriving calls are blocked and cannot enter the system (see Fig. 7.10). After hiring an agent, Easysurrance Call center will constitute an $M/M/3/7$ queueing system: interarrival and processing times are exponential, it will have three agents, the maximum number of customers (on hold or being served) is 7, and calls are answered in FCFS order. Customers arrive according to a Poisson process with a rate of 11 per hour and the effective mean process time is 15 minutes. Figure 7.10 shows the queueing model of the call center if Easysurrance hires one agent.

The bad news is that we cannot use Eq. (7.2) for $G/G/s$ queueing models to compute the average flow times, since the equation is for systems that have no limit on their buffer size. We need a new model. The good news is that since both interarrival times and processing times are exponentially distributed, we have a Markovian queue. Models for Markovian queues

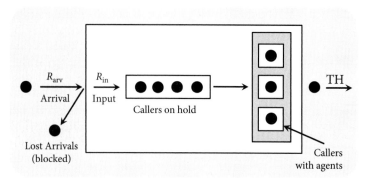

**Figure 7.10**   Queueing model of Easysurrance call center in Example 7.5: An $M/M/3/7$ queue.

provide a wide range of performance measures that can be used in managing and control of processes with limited buffer. In Markovian queueing models, both interarrival times and processing times are exponentially distributed.

**ANALYTICS**

**Estimating Performance Measures of a Single-Stage Process with Limited Buffer—The $M/M/s/K$ Model**

Consider a process with no rework and no yield loss and a pool of $s$ resources that process flow units independent of each other. The resources are identical, in that they all have the same effective mean process time $T_{\text{eff}}(\text{resource } i) = T_{\text{eff}}$, which is exponentially distributed. The process can hold a maximum number of $K$ flow units, that is, $s$ units with the resources, and $K - s$ units waiting in the buffer. If an arriving flow unit finds the process full, it is blocked and does not enter the process. Interarrival times of the flow units have a mean of $T_{\text{arv}}(\text{process})$ and are exponentially distributed. Hence, the process will have an *Arrival Rate* rate:

$$R_{\text{arv}}(\text{process}) = \frac{1}{T_{\text{arv}}(\text{process})}$$

To find the process-flow measures, we first need to compute the following:

$$r = R_{\text{arv}}(\text{process}) \times T_{\text{eff}}, \qquad \rho = r/s$$

where $r$ is called *Offered Load*, and

$$p_0 = \left( \sum_{j=0}^{s-1} \frac{r^j}{j!} + \sum_{j=s}^{K} \frac{r^j}{s! \, s^{j-s}} \right)^{-1} \tag{7.4}$$

where $p_0$ is the probability that the process is empty (i.e., all resources are idle). The probability that the process is full with $K$ flow units and thus an arriving flow unit is blocked is

$$P(\text{Blocking}) = \left( \frac{r^K}{s^{K-s}s!} \right) p_0 \tag{7.5}$$

- **Utilization:** Because some arrivals may be blocked, the input rate of the process will be less than or equal to the arrival rate of the process, that is, $R_{\text{in}}(\text{process}) \leq R_{\text{arv}}(\text{process})$. We can compute the input rate—the average number of arriving flow units that enter the process—using the probability of blocking as follows:

$$R_{\text{in}}(\text{process}) = \left[ 1 - P(\text{Blocking}) \right] R_{\text{arv}}(\text{process}) \tag{7.6}$$

consequently, the process utilization is

$$u = \frac{R_{\text{in}}(\text{process})}{C_{\text{eff}}(\text{process})} = \frac{\left[1 - P(\text{Blocking})\right] R_{\text{arv}}(\text{process})}{s/T_{\text{eff}}}$$
$$= \left[1 - P(\text{Blocking})\right] \rho \tag{7.7}$$

- **Throughput:** Since the process is always stable (i.e., its queue does not grow to infinity), the throughput of the process is equal to its input rate. Hence,

$$\text{TH} = R_{\text{in}}(\text{process}) = \left[1 - P(\text{Blocking})\right] R_{\text{arv}}(\text{process})$$

- **Average Flow Times:** The average in-buffer flow time is

$$T_b = p_0 \left(\frac{r^s \rho}{s!(1-\rho)^2}\right)\left[1 - \rho^{K-s+1} - (1-\rho)(K-s+1)\rho^{K-s}\right]\frac{1}{\text{TH}} \tag{7.8}$$

For processes with $\rho = 1$, the average in-buffer flow time can be obtained by applying L'Hopital's rule twice on the above equation. The average flow time is, therefore,

$$T = T_b + T_{\text{eff}}$$

- **Average Inventory:** The average in-buffer and in-process inventories, using Little's law, are as follows:

$$I_b = \text{TH} \times T_b \qquad I_p = \text{TH} \times T_p = \text{TH} \times T_{\text{eff}}$$

We can now use the above results to compute the performance measures of Easysurrance call center. After hiring an agent, Easysurrance call center would be an $M/M/3/7$ queueing system with

$$R_{\text{arv}}(\text{process}) = 11 \text{ calls per hour} \qquad T_{\text{eff}} = 15/60 = 0.25 \text{ hours}$$

To compute the average speed of answer (ASA)—which is $T_b$—and fraction of blocked callers—which is $P(\text{Blocking})$—we first need to compute $r$ and $\rho$ as follows:

$$r = R_{\text{arv}}(\text{process}) \times T_{\text{eff}} = 11 \times 0.25 = 2.75, \qquad \rho = r/s = 2.75/3 = 0.917$$

Hence,

$$p_0 = \left(\sum_{j=0}^{s-1}\frac{r^j}{j!} + \sum_{j=s}^{K}\frac{r^j}{s!\,s^{j-s}}\right)^{-1} = \left(\sum_{j=0}^{2}\frac{(2.75)^j}{j!} + \sum_{j=3}^{7}\frac{(2.75)^j}{3!\,3^{j-3}}\right)^{-1}$$
$$= \left(1 + \frac{2.75}{1} + \frac{2.75^2}{2} + \frac{(2.75)^3}{3!\,3^0} + \frac{(2.75)^4}{3!\,3^1} + \frac{(2.75)^5}{3!\,3^2} + \frac{(2.75)^6}{3!\,3^3} + \frac{(2.75)^7}{3!\,3^4}\right)^{-1}$$
$$= 0.045$$

Let's now check the fraction of blocked callers, which is the probability of blocking:

$$\text{FBC} = \text{P(Blocking)} = \left(\frac{r^7}{s^{7-s}s!}\right)p_0 = \left(\frac{(2.75)^7}{3^4\, 3!}\right)0.045 = 0.11 = 11\%$$

which satisfies the target level $\text{FBC} \leq 15\%$.

Having $p_0$, we can now compute the average in-buffer flow time as follows:

$$\begin{aligned}
\text{ASA} = T_b &= p_0\left(\frac{r^s\rho}{s!(1-\rho)^2}\right)\left[1 - \rho^{K-s+1} - (1-\rho)(K-s+1)\rho^{K-s}\right]\frac{1}{\text{TH}} \\
&= 0.045\left(\frac{2.75^3(0.917)}{3!(1-0.917)^2}\right)\left[1 - (0.917)^5 - (1-0.917)(5)(0.917)^4\right]\frac{1}{(1-0.11)17} \\
&= 0.123 \text{ hours} \times 60 = 7.4 \text{ minutes}
\end{aligned}$$

which is higher than target level of 4 minutes set by Easysurrance.

Finally, note that occupancy (OCC) measure of Easysurrance call center is in fact utilization $u$. Using, Eq. (7.7), we have

$$\text{OCC} = u = [1 - \text{P(Blocking)}]\rho = [1 - 0.11]0.917 = 0.816 = 81.6\%$$

Hence, after hiring an agent, Easysurrance can achieve two of its three goals, that is, the FBC = 11% which is below 15% and OCC = 81.6%, which is above 75%. However, it cannot reach the service goal of ASA $\leq$ 4 minutes, since the average speed of answer after hiring a new agent is 7.4 minutes.

But what if they hire two new agents? By hiring two agents, the underlying queueing system of Easysurrance call center would be an $M/M/4/7$. Repeating the above calculation for $s = 4$, we get ASA = 2.1 minutes, FBC = 4.6%, and OCC = 65.6%. While hiring two agents satisfies ASA and FBC metrics, it does not satisfy the occupancy goal of above 75%. After hiring two agents, agents will be idle 100% − 65.6% = 34.4% of time.

In conclusion, call center managers must inform the upper management that hiring one agent results in ASA of 7.4 minutes which is much larger than the target level of 4 minutes. On the other hand, hiring two agents achieves both the speed of answer and accessibility goals, but at the cost of lower agent occupancy. Thus, the upper management should revise their performance measures. Or, is there another option? We explore this in Example 7.6 in Sec. 7.6.1 where we analyze the impact of buffer size on flow time. But first, we return to Michigan Yacht Club in Example 7.1.

**Back to Example 7.1:** Our analytics about $M/M/s/K$ queues also helps us to estimate the increase in revenue of Michigan Yacht Club in Example 7.1, if the club adds one launching ramp. Note that the underlying queueing model for club's current operations is an $M/M/1/5$, and after adding a launching ramp it will be an $M/M/2/6$ queueing model. In both cases, the arrival process has a rate of $R_{\text{arv}}(\text{club}) = 12$ cars per hour, and effective mean process time $T_{\text{eff}} = 4$ minutes.

Under current operations, using our analytics for the $M/M/1/5$ queue we find $p_0 = 0.271$, and thus $\text{P(Blocking)} = 0.089$. Therefore, the number of boats launched in an hour—the throughput—is

$$\text{TH} = [1 - \text{P(Blocking)}]R_{\text{arv}}(\text{club}) = (1 - 0.089) \times 12 = 10.93 \text{ boats per hour}$$

Considering the revenue of $45 per boat, the current revenue of Michigan Yacht Club is $45 × 10.93 = $492.04 per hour.

If a new launching ramp is added and the process becomes an $M/M/2/6$ queueing model, then $p_0 = 0.43$, and $P(\text{Blocking}) = 0.004$, and throughput TH $= 11.96$ boats per hour. The revenue of the club after adding a new ramp will be $\$45 \times 11.96 = \$538.10$ per hour. Thus, the increase in revenue after adding a new ramp is $\$538.10 - \$492.04 = \$46.06$ per hour. Jack, the owner of the club, needs to do some calculation to see whether the hourly cost of renting the new space is lower or higher than $\$42.06$. If it is higher than $\$42.06$, then it is not beneficial for Jack to rent the property.

### 7.6.1 Impact of Limited Buffer

In Chap. 6 we discussed the impact of limited buffer size on the throughput of a process. We showed that, in processes with limited buffer, increasing the size of the buffer decreases the probability of blocking and thus increases the throughput of the process—reduces the throughput loss due to blocking. In this section, we examine the impact of limited buffer on another operational measure of a process—the average flow time. We return to the Easysurrance call center example to discuss this.

**Example 7.6**

> Managers of Easysurrance call center are not sure if they can get permission to hire more than one agent. Thus, before deciding to hire an additional agent, they are trying to find other ways to achieve their new service goals. One suggestion is to improve their per-formance by increasing the number of trunk lines. They would like to know how does increasing the number of trunk lines affect average speed of answer (ASA) and fraction of blocked callers (FBC) as well as occupancy (OCC) in their call center?

Before solving the above example, let's first study the impact of the limited buffer size on the utilization and the average flow time of the process.

**Impact on Utilization**
How does limiting the buffer size impact process utilization? Recall that utilization $u$ for processes with limited buffer size is computed as follows:

$$u = \left[1 - P(\text{Blocking})\right]\rho$$

where $\rho$

$$\rho = \frac{r}{s} = \frac{R_{\text{arv}}(\text{process}) \times T_{\text{eff}}}{s} = \frac{T_{\text{eff}}}{s\, T_{\text{arv}}(\text{process})}$$

and is independent of the buffer size.[7] However, in systems with larger buffer size, less arrivals are blocked, and thus $[1 - P(\text{Blocking})]$ increases as the buffer size increases. There-fore, all being the same, processes with larger buffers work at higher utilization levels. What is the implication for Easysurrance? This means that by increasing the number of trunk lines (i.e., maximum system size), Easysurrance should expect to see an increase in their agent OCC measure (i.e., utilization).

---

[7]In fact, $\rho$ is the utilization of the system with infinite buffer size, see Eq. (7.1).

**Impact on Average Flow Times**

For any given arrival rate, processes with smaller buffer will have less average flow time. The reason is that smaller buffer puts a tighter limit on the in-buffer inventory and prevents the in-buffer inventory from becoming very large. Smaller in-buffer inventory corresponds to smaller flow times. Thus, one way to reduce the average flow time of a process is to put limit on or reduce the size of its buffer.

These observations lead us to the following principle:

### PRINCIPLE

**Impact of Limited Buffer on Process Performance**

Processes with limited buffer size are always stable, since their in-buffer inventory cannot grow beyond the buffer size. In such processes, increasing the size of the buffer

- increases process utilization
- decreases blocking
- increases throughput
- increases average flow time

Hence, reducing buffer size can be considered as an option to reduce flow times.

What are the implications of the above principles for Easysurrance? Well, the most important implication is that, by increasing the number of trunk lines, managers will increase their ASA (i.e., the average in-buffer flow time). Increasing the number of trunk lines, however, will improve the other two performance measures: FBC and OCC. This is because, increasing the number of trunk lines will increase utilization (i.e., OCC) and decrease blocking (i.e., FBC). However, these two measures are not of concern, since after hiring one agent Easysurrance will achieve OCC $= 81.6\%$, and FBC $= 11\%$, which satisfies the goal of OCC $\geq 75\%$ and FBC $\leq 15\%$.

Hence, to decrease the ASA, Easysurrance can reduce its number of trunk lines. Table 7.4 shows the Easysurrance call center performance measures for different number of trunk lines.

As the table shows, if Easysurrance reduces the number of trunk lines from 7 to 5, its average in-buffer flow time (ASA) is reduced from 7.4 minutes to 3.5 minutes, which is below the goal of 4 minutes. As expected, decreasing the number of trunk lines decreases OCC (agent utilization) to 76%. The good news is that the new occupancy still satisfies the goal of OCC $\geq 75\%$. By decreasing the number of trunk lines, the FBC measure, however, is increased to 17.1%, which is 2.1% above the goal of FBC $\leq 15\%$.

Table 7.4   Performance Measures of Easysurrance in Example 7.6

| Number of trunk lines | Queueing model | ASA (*minutes*) | FBC | OCC |
|---|---|---|---|---|
| 5 | $M/M/3/5$ | 3.5 | 17.1% | 76.0% |
| 6 | $M/M/3/6$ | 5.4 | 13.5% | 79.3% |
| 7 | $M/M/3/7$ | 7.4 | 11.0% | 81.6% |
| 8 | $M/M/3/8$ | 9.3 | 9.2% | 82.3% |

In conclusion, there is a third option for Easysurrance: Hiring one agent and reducing the number of trunk lines from 7 to 5, resulting in an $M/M/3/5$ system:

- Compared to hiring one agent, this third option is less costly since the company does not need to pay the cost for maintaining the sixth and seventh lines. It also has much lower ASA—3.5 versus 7.4 minutes (with seven lines). The FBC, however, is 2.1 percent above the target level.

- Compared to hiring two agents, this third option has less labor cost—it has one less agent—and less costs with respect to trunk lines—it has two less lines. The only issue, as we mentioned, is that the FBC is 2.1 percent above target level 15 percent.

Which option should Easysurrance choose? Reducing the number of trunk lines—the third option—seems to be better from the cost perspective. The question is whether Easysurrance managers are willing to revise their FBC goal by 2% from FBC $\leq 15\%$ to about FBC $\leq 17.1\%$.

## 7.6.2 Impact of Variability

For processes with no limit on the buffer size, as we showed in Sec. 7.5.2, increasing variability results in an increase in average flow times. We also found that the resulting increase in average flow time is very significant in processes with high utilization. In this section, through Example 7.7, we examine whether these principles are also valid for systems with limited buffer.

 **Example 7.7**

> Managers of Easysurrance decided to hire one agent and reduce the number of trunk lines to 6. They have been happy with the performance of their call center after they made these changes. The arrival rate is still 11 calls per hour. Managers are now considering a training program for all their agents that can reduce the variability in (coefficient of variation of) call handling times from 1 to 0.5. They are wondering how this will impact the call center performance measures in the coming year. What do you think?

**Impact of Variability on Blocking**

Recall from Chap. 6 that for any given arrival rate, the throughput of a process with limited buffer decreases as variability in the process increases. On the other hand, for any given arrival rate, decreasing throughput implies that more flow units are blocked and cannot enter the process. This can also be concluded from Eq. (7.6):

$$\text{TH} = R_{\text{in}}(\text{process}) = \left[1 - \text{P(Blocking)}\right] R_{\text{arv}}(\text{process})$$

Hence, for any given call arrival rate $R_{\text{in}}(\text{process})$, if the variability in call handling time decreases, Easysurrance should expect a decrease in fraction of blocked callers (FBC).

Note that when Easysurrance reduces the variability of its call handling time from $\text{CV}_{\text{eff}} = 1$ to $\text{CV}_{\text{eff}} = 0.5$, the underlying queueing model of their call center changes from an $M/M/3/6$ to an $M/G/3/6$, for which the only analytical model to compute FBC is computer simulation. Using a computer simulation, we find that by reducing variability, the fraction of blocked callers would be reduced from FBC $= 13.5\%$ to FBC $= 7.8\%$. Hence, as we expected, variability reduction resulted in a decrease in blocking.

**Impact of Variability on Utilization**

Another performance measure of importance for Easysurrance is agents occupancy, which corresponds to agent (and process) utilization. The question is: How does decrease in variability affect process utilization?

In processes with infinite buffer, process utilization is independent of variability in the process. This is because the input rate of the process, $R_{in}(\text{process})$, does not change as the variability in the arrival or in effective process time changes. In processes with limited buffer, however, the input rate (and thus TH) of the process decreases as variability in the process increases. Since utilization of a process is defined as follows:

$$u = \frac{R_{in}(\text{process})}{C_{eff}(\text{process})}$$

for any given effective capacity, $C_{eff}(\text{process})$, a decrease in the input rate implies a decrease in process utilization. So, any time that the variability in call handling times increases, managers of Easysurrance call center should expect a decrease in their agent's occupancy. If the variability in call handling times decreases from $CV_{eff} = 1$ to $CV_{eff} = 0.5$, using computer simulation we find that the occupancy increases from $OCC = 79.3\%$ to $OCC = 84.1\%$. Thus, variability reduction will result in $4.8\% = (84.1\% - 79.3\%)$ increase in agent occupancy.

**Impact of Variability on Average Flow Time**
Equation (7.2) in Sec. 7.5 clearly shows that in processes with unlimited buffer size, variability reduction results in lower average flow time and thus lower average inventory. In processes with limited input buffer, however, variability reduction has two opposite effects. (i) Like in any process, for a given input (i.e., arrival) rate and a given system capacity, lower variability results in lower inventory and flow time—a positive effect. However, (ii) lower variability also decreases blocking, which in turn results in more arrivals entering the process. More inputs to the process result in higher inventory and thus higher flow times—a negative effect. Which effect—positive or negative—is stronger? The answer is not clear and it depends on the arrival rate, the effective capacity, and the variability in the process. In some processes the positive effect is stronger than the negative effect and thus variability reduction reduces flow times, and in some other processes the negative effect is stronger than the positive effect and thus the variability reduction results in an increase in flow times. Hence, to examine the impact of variability reduction on flow time of a process with limited input buffer, one should use some analytics to estimate the flow time after variability reduction and find whether the flow time is increased or decreased.

Using computer simulation we find that in Example 7.7, the variability reduction in call handling time from $CV_{eff} = 1$ to $CV_{eff} = 0.5$ results in reduction in in-buffer flow time from $ASA = 5.4$ minutes to $ASA = 4.9$ minutes—a 30-second improvement.

The results in this section lead us to the following principle:

**PRINCIPLE**

**Impact of Variability on Processes with Limited Input Buffer**

In processes with limited input buffer, variability reduction results in

- decrease in blocking
- increase in utilization
- increase in throughput
- increase or decrease in average flow time and average inventory, depending on the parameters of the process

## 7.7 Flow Time Service Levels

Average flow time of a process provides an overall process performance with respect to the time a flow unit spends in the process (waiting or being processed) before it leaves the process. But, does the average flow time give us a complete picture?

As an example, suppose that you need to take your car for oil change to one of the two locations that provide fast oil change to arriving customers. You go on the web and search for reviews of these two locations. You find reviews in which people mention how long they were waiting before their service started. Below is the waiting times of 20 customers (in minutes) in the queue that you get from the reviews of Location A:

8, 9.7, 8.9, 7.8, 7.8, 7.5, 9.8, 7.2, 7.4, 8.8, 8.9, 7.9, 9.5, 7.7, 7.4, 7.9, 9.9, 8, 7.2, 6.7

You also check the reviews for Location B, and are able to find the waiting time of 20 customers (in minutes) in the queue as follows:

14.9, 4.2, 6.2, 10.7, 3.2, 5.9, 5.1, 13.5, 10.5, 15, 8.5, 5.8, 4.9, 6.3, 13.6, 2.8, 9.8, 9.7, 11.1, 2.3

Both locations seem to have good reviews with respect to quality of service, and they are within the same distance from where you live. So, for you the decision boils down to which location has a smaller waiting time in queue (i.e., smaller in-buffer flow time).

To get a sense of the waiting time in the queue, you take the average of the 20 numbers for both locations and you find that both locations have an average in-buffer flow time of 8.2 minutes. Which location will you choose?

Well, while the *average* in-buffer flow times for both locations are the same, there is a big difference in the *actual* in-buffer flow times in these locations. According to the sample data, in Location A all 20 out of 20 (i.e., 100 percent of) customers waited less than 10 minutes to get their service. In Location B, however, only 13 out of 20 (i.e., 65 percent of) customers waited less than 10 minutes in the queue. So, it seems that if you go to Location A, there is a very high chance that your waiting time in the queue would be less than 10 minutes.

The probability that the flow time of a customer in a process is less than a predetermined target is called "flow time service level." Below we formalize this definition.

### CONCEPT

**Flow Time Service Level**

The *flow time service level* of a process, $SL^{flw}(t)$, is the probability that a flow unit that enters the process spends $t$ or less time in the process before it exits. In other words,

$$SL^{flw}(t) = P(\text{flow time of a unit} \leq t)$$

The *in-buffer flow time service level* of a process, $SL_b^{flw}(t)$, is the probability that a flow unit that enters the process waits for $t$ or less in the buffer of the process before its process starts. In other words,

$$SL_b^{flw}(t) = P(\text{in-buffer flow time of a unit} \leq t)$$

In call center industry, the in-buffer flow time service level is called *Telephone Service Fraction*.

Firms have realized the importance of service levels and are trying to improve their service levels and use it as a marketing tool. Mail carriers guarantee delivery within a day. Some restaurants guarantee to serve their food within 10 to 15 minutes of customers' arrivals, or the next meal is free. Suppliers of auto assembly plants guarantee delivery of parts to plants within couple of hours; otherwise, they will be fined for late delivery.

Computing service levels of a process is complex, even in single-stage processes, and thus simulation is often used to estimate it. However, in Markovian queues in which inter-arrivals and process times are exponentially distributed, there exists analytics that obtains flow time service levels. We use the case of Expresso in Example 7.8 to motivate and present the analytics.

### Example 7.8

Expresso is a manufacturer of espresso and cappuccino machines that produces a wide range of commercial and home coffee making machines. Due to the wide range of its features, the firm produces its gourmet-series in assemble-to-order mode, after it receives an order from a customer. Customers can choose from gourmet-series online catalogue. Orders from the customers are processed on first-come-first-served (FCFS) basis, and are produced in one of the three assembly cells allocated to gourmet-series products. When a cell finishes making an order, it then starts working on the next order. Products in gourmet-series are designed such that they all use a group of modular components, so that the cell's setup time to assemble the next order is negligible. A study has shown that the time a cell finishes an order follows an exponential distribution with a mean of 2 days. The study also shows that the firm receives orders for a product in gourmet-series according to a Poisson process with a rate of 1.4 orders per day. Factory and sales data revealed that it takes an average of around 10 days from the time a customer places an order until the order is shipped. It also shows that about 63 percent of the customers receive their orders within the 10 days; but most troubling is that about one of every five customers receives their orders in 17 days or more—about 1 week above the average of 10 days. While the average flow time seems low, the relatively large fraction of orders with longer than 17 days flow time is making managers at Expresso concerned. These are the customers who, most probably, will complain about their long order lead times and, in the long-run, can affect Expresso's reputation in the market. Thus, managers of Expresso want to take action to improve the situation. The goal is to increase their capacity such that (i) at least 90 percent of the customers receive their orders within a week of placing their orders, and (ii) the percent of customers who receive their orders later than 10 days would be less than 3 percent. Expresso is planning to invest on increasing the capacity of each cell by redesigning the cell and adding more workers. This will increase the capacity of each cell by 10 percent. Can this capacity investment help Expresso to achieve its new service goals?

The first goal of having at least 90 percent of the customers receiving their orders within $t = 7$ days (i.e., 1 week) of placing their orders translates into flow time service level $SL^{flw}(7) \geq 0.9$. The second goal of having at most 3% of the customers receiving their orders later than 10 days is equivalent to having at least 97 percent of the customers receive their orders in $t = 10$ days or less. This translates into flow time service level $SL^{flw}(10) \geq 0.97$.

The current effective capacity of each cell is one order every 2 days, or 0.5 orders a day. Increasing capacity by 10 percent will result in effective capacity of 0.55 orders per day. The question is whether this increase in the capacity can help Expresso achieve its goals. Considering one order as a flow unit, since order interarrival times and order assembly times in each cell follow exponential distribution, the underlying queueing model of Expresso production process after capacity expansion is an $M/M/3$ queueing model with arrival rate of 1.4 orders per day and service rate 0.55 orders per day for each server. To compute the flow time service level for such process, we need the following analytics.

**Flow Time Service Levels of a Single-Stage Process with Infinite Buffer—The $M/M/s$ Model**

Consider a process with no rework and no yield loss and a pool of $s$ resources that process flow units independent of each other. The resources are identical, in that they all have the same effective mean process time $T_{\text{eff}}(\text{resource } i) = T_{\text{eff}}$, and follow exponential distribution. Hence, each resource has effective capacity:

$$C_{\text{eff}}(\text{resource } i) = \frac{1}{T_{\text{eff}}}$$

Interarrival times of the flow units have mean $T_{\text{arv}}(\text{process})$ and are exponentially distributed. Hence, the process will have *Arrival Rate*:

$$R_{\text{arv}}(\text{process}) = \frac{1}{T_{\text{arv}}(\text{process})}$$

To simplify notation, we use $\lambda = R_{\text{arv}}(\text{process})$ and $\mu = C_{\text{eff}}(\text{resource } i)$. The underlying queueing system of the process is therefore an $M/M/s$ with arrival rate $\lambda$ and service rate $\mu$ for each server. As before, we define offered load as $r = \lambda/\mu$; hence, process utilization is $\rho = r/s$. The service-level measures are as follows:

- **Fraction of Delayed Customers (FDC):** Fraction of delayed customers (i.e., flow units) are the fraction of units that must wait in the buffer before being served by a server.

$$\text{FDC} = P(\text{customer waits in queue before being served})$$
$$= \left(\frac{r^s}{s!(1-\rho)}\right) p_0 \tag{7.9}$$

where $p_0$ is the probability of process being empty,

$$p_0 = \left(\frac{r^s}{s!(1-\rho)} + \sum_{j=0}^{s-1} \frac{r^j}{j!}\right)^{-1}$$

- **Flow Time Service Level:** Flow time service level $SL^{\text{flw}}(t)$ is

$$SL^{\text{flw}}(t) = \frac{s(1-\rho)-1+\text{FDC}}{s(1-\rho)-1}\left(1-e^{-\mu t}\right) - \frac{\text{FDC}}{s(1-\rho)-1}\left(1-e^{-(s\mu-\lambda)t}\right) \tag{7.10}$$

- **In-buffer Flow Time Service Level:** In-buffer flow time service level $SL_b^{\text{flw}}(t)$ is

$$SL_b^{\text{flw}}(t) = 1 - \text{FDC}\left[e^{-(s\mu-\lambda)t}\right] \tag{7.11}$$

Note that $\text{FDC} = 1 - SL_b^{\text{flw}}(t=0)$.

For the case of Expresso in Example 7.8, we have $r = \lambda/\mu = 1.4/0.55 = 2.55$, and $\rho = r/s = (1.4/0.55)/3 = 0.848$. To find the fraction of customers who receive their orders in 7 days or less, we need to compute $SL^{\text{flw}}(7)$, the flow time service level for $t = 7$ days. To do this, however, as Eq. (7.10) shows, we first need to compute the fraction of delayed customers as follows:

$$p_0 = \left( \frac{r^s}{s!(1-\rho)} + \sum_{j=0}^{s-1} \frac{r^j}{j!} \right)^{-1}$$

$$= \left( \frac{(2.55)^3}{3!(1-0.848)} + \frac{(2.55)^0}{0!} + \frac{(2.55)^1}{1!} + \frac{(2.55)^2}{2!} \right)^{-1} = 0.04$$

Therefore, the fraction of delayed customers is

$$\mathrm{FDC} = \left( \frac{r^s}{s!(1-\rho)} \right) p_0 = \left( \frac{(2.55)^3}{3!(1-0.848)} \right) 0.04 = 0.728$$

Having FDC, we can now compute the service level for $t = 7$ using Eq. (7.10) as follows:

$$SL^{\mathrm{flw}}(7) = \frac{3(1-\rho)-1+\mathrm{FDC}}{s(1-\rho)-1} \left( 1 - e^{-\mu t} \right) - \frac{\mathrm{FDC}}{s(1-\rho)-1} \left( 1 - e^{-(s\mu-\lambda)t} \right)$$

$$= \frac{3(1-0.848)-1+0.728}{3(1-0.848)-1} \left( 1 - e^{-0.55(7)} \right)$$

$$- \frac{0.728}{3(1-0.848)-1} \left( 1 - e^{-(3(0.55)-1.4)7} \right)$$

$$= 0.775$$

In other words, after investing in increasing the capacity by 10 percent, only 77.5 percent of customers will receive their orders within 1 week. To compute what fraction of customers receive their orders within 10 days, we repeat the above calculation for $t = 10$ days and we obtain $SL^{\mathrm{flw}}(10) = 0.892$. This implies that after capacity expansion, fraction $1 - 0.892 = 0.108 = 10.8\%$ of customers will receive their orders later than 10 days after they place their orders. In conclusion, a 10-percent increase in capacity does not achieve Expresso's service-level goals. How much capacity does Expresso need to reach its target service levels?

If we repeat the above computation for capacity $\mu = 0.55, 0.56, 0.57$, etc., we find that with capacity $\mu = 0.6$—a 20-percent increase in current capacity—Expresso achieves $SL^{\mathrm{flw}}(7) = 0.901$ and $SL^{\mathrm{flw}}(10) = 0.968$. This implies that 90 percent or more of Expresso's customers will receive their orders in a week or less, and only $1 - 0.968 = 0.032 = 3.2$ percent of the customers will receive their orders in more than 10 days.

## Want to Learn More?

### Flow Time Service Levels for Processes with Limited Buffer

In the online supplement of this chapter we extend the analytics for flow time service levels to processes with limited buffer. Specifically, we discuss a new case for Easysurrance call center in Example 7.7 with goal of reducing waiting times such that at least 80 percent of the callers wait less than 1 minute (on hold) to talk to an agent. The goal is to find the minimum number of new agents that must be hired to achieve the new service level goals.

### A Cost Model for Managing Flow Times

In addition to operational performance measures such as average flow times and flow time service levels, cost performance measures are also of interest to operations engineers and managers. The problem is that achieving better operational performance measures often results in higher costs. For example, one way to reduce the average in-buffer flow time is to invest in increasing the capacity by hiring more workers or buying more equipment to have higher safety capacity. Thus, one of the difficult tasks of operations engineers and managers is to achieve the target operational performance measures with minimum cost. In the online supplement of this chapter we use an example of an online retailer to illustrate a typical cost model that is useful in capturing the trade-offs between service levels and cost. Check it out.

### Flow Analysis in Multi-Stage Processes

The operations in manufacturing and service operations systems usually consists of several stages that perform different activities on flow units. For example, in auto assembly plants a car goes through several workstations (i.e., stages) and a different activity is performed on a car until the car assembly is finished. In Emergency Department (ED) of a hospital, a patient goes through several stages (e.g., registration, triage, lab, radiology, etc.). Depending on the level of details, one can model a multi-stage manufacturing or service system as a single-stage or a multi-stage process. For example, in Example 7.2 we modeled several stages of manufacturing of an order in Diorucci Furniture (e.g., cutting, assembling, painting, etc.) as a single-stage single-product process, since we were only interested in the *overall* average flow time of a *typical* order. However, one can perform a more detailed analysis of a manufacturing or service system by modeling every stage of its operation as a process. In this case, the entire operations constitutes a multi-stage process in which flow units of different types go through different single-stage processes before they leave the system. If you want to learn how to compute flow times in such processes see the online supplement of this chapter.

## 7.8 Summary

Large waiting time (flow times) of jobs waiting to be processed or large number of customers waiting in line to receive service have negative impact on firms' bottom line as well as on customer satisfaction. Unfortunately, these waiting times are inevitable in all operations due to the variability in demand and process times. To reduce these waiting times firms need to reduce the variability in their operations or increase their safety capacity (the capacity beyond demand). To determine how much safety capacity is needed to achieve an acceptable waiting time, one needs to understand how safety capacity and variability affect flow times. The main goal of this chapter is to present a series of analytical models—called queueing models—that can provide accurate estimates of flow times in different processes.

All analytical models illustrate that the flow time in a process increases with system utilization (in a nonlinear fashion). Therefore, if the demand for a process increases or if the effective capacity of a process decreases, flow times and average inventory in the process can increase significantly, especially in processes that are already highly utilized. Also, variability has negative impact on flow times. Hence, one approach to reduce flow time is to reduce variability in demand or in process times of a process. The effects of variability reduction on reducing flow times are more significant in processes with higher utilization. Thus, variability reduction efforts should focus on highly utilized processes and resources (e.g., bottlenecks). Processes with higher variability require larger safety capacity (and thus larger capacity cost) to achieve a target flow time than what is required for processes with lower variability.

Imposing limits on the size of buffers in a process has advantages and disadvantages. The advantage is that it reduces the average flow time (and inventory) in the process. The disadvantage is that it also reduces the throughput and process utilization. Strategies to reduce the flow times in processes with limited buffer include increasing safety capacity or reducing variability. Variability reduction has several advantages in such systems: It decreases blocking, it increases utilization, and it increases throughput.

## Discussion Questions

1. Explain why queues exist in practice and how does variability affect those queues.
2. What is safety capacity of a resource and how does it affect the flow time and inventory in the buffer of the resource?

3. What are the six main elements of queueing systems?

4. Explain how increasing the input buffer size of a process affects utilization, blocking, throughput, and average flow time of the process.

5. What is flow time service level? What are the two main flow time service levels?

## Problems

1. Tom and Joe work in two different workstations of a factory and both receive a job exactly every 5 minutes. Tom's utilization is about 90 percent, and Joe's utilization is about 85 percent. Jobs performed by Tom and Joe are routine jobs and thus have negligible (zero) variability. Which one of the following statements is true? Explain why.

   a. A 5-percent reduction in Tom's utilization results in a 5-percent reduction in in-buffer inventory in his station.

   b. A 5-percent reduction in both Tom's and Joe's utilizations results in a more significant reduction in in-buffer inventory in Tom's station than in Joe's station.

   c. Now suppose that there is some variability in tasks performed by Tom and Joe, while their utilization is still 90 percent and 85 percent, respectively. Which one of the following statements is true? Explain why.

   - A 5-percent reduction in both Tom's and Joe's utilizations results in a more significant reduction in the inventory in Tom's station than in Joe's station.

   - A 5-percent reduction in both Tom's and Joe's utilizations results in a more significant reduction in the inventory in Joe's station than in Tom's station.

   - More information is needed to compare the reduction in the inventory in Joe's and Tom's stations.

2. Jack and Jill work in the call center of an IT firm that produces commercial and technical software. Jack responds to calls regarding commercial software and Jill receives calls about technical software. Jack receives about 20 calls per hour and Jill receives about 30 calls per hour, both in a completely random fashion. Data shows that the call handling time for Jack to resolve an issue with a caller has mean of 170 seconds with coefficient of variation 1. Jill's average time to handle a call is 100 seconds with coefficient of variation 1. To reduce the waiting of their callers, Jack and Jill are considered for more training. The training is expected to reduce the variability in call handling time (i.e., coefficient of variation) from 1 to half. If one—Jack or Jill—should be chosen for training, which of the following statements are true? Explain your reason.

   a. Jill should be sent for training because she receives more calls to answer than Jack.

   b. Jack should be sent for training because, on an average, it takes him more time than Jill to handle a call.

   c. Since both Jack and Jill have the same variability in their call handling time, it does not matter which one is sent for training.

   c. None of the above statements are correct. If you chose this option, explain who should be sent for training and why?

3. Suppose in Problem 2, neither Jack nor Jill were sent for training due to major changes in the company. The company added many Help and Q&A pages on its website and customers are referred to the site before calling the company. This reduced the average number of calls, and the number of calls received per hour for commercial software is now the same as that for technical software. However, the data shows that the calls for commercial software has a larger variability in their arrivals than calls for technical software. Because of the Help and Q&A pages, both Jack and Jill are now responding to major issues which result in longer call handling time. Specifically, the call response time for both Jack and Jill now is an average of 5 minutes with coefficient of variation 1. A new study shows that both Jack and Jill have

10-percent idle time waiting for calls to arrive. Which of the following statements are true? Explain your reasons.

    a. The average waiting time for the two types of calls are different. To provide the same average waiting time for both call types, Jack's call handling time must be reduced.

    b. The average waiting time for the two types of calls are different. To provide the same service level for both call types, variability in Jack's call handling time must be reduced.

    c. Both calls for commercial and technical software experience the same average waiting time, because both Jack and Jill have the same utilization.

    d. More information is needed to make any statement.

4. Rossini's is a high-scale Italian restaurant that has 24 tables that can seat about 100 customers. On an average, the restaurant has about three tables available. When customers arrive and find no table available, they wait in the waiting area inside the restaurant. The waiting area can accommodate 20 customers. The restaurant has found that when their waiting area is full, arriving customers do not enter and go to other restaurants in the area. The restaurant is considering expanding its waiting area to accommodate 25 customers. Explain how expanding the waiting area affects the following. Specifically, whether the following measures increase, decrease, or stay the same.

    a. Restaurant's daily sales

    b. Customer's waiting time to get a table

    c. Average number of available tables in the restaurant

    d. Average number of customers in the restaurant

5. Consider Rossini's restaurant in Problem 4, and suppose that customers occupy tables for an average of 45 minutes with standard deviation 45 minutes. The manager of Rossini's has decided to eliminate one of the items in the menu that takes the kitchen a long time to prepare. It is believed that while this change does not affect the average time of 45 minutes that a table is occupied, it would reduce the standard deviation of this time. Explain how this change affects the following:

    a. Restaurant's daily sales

    b. Customer's waiting time to get a table

    c. Average number of available tables in the restaurant

    d. Average number of customers in the restaurant

6. You arrive at your favorite restaurant at 12:00 noon to have lunch and return to work. When you arrive you see a line with 10 people waiting to place order with the single cashier. After placing orders, customers wait until they are called to pick up their orders from the order pick-up counter. Based on your experience you know that the time that the cashier takes customers' orders is about 50 seconds, but the time is variable and has standard deviation of 50 seconds. While waiting in the line you feel that you need to go to the bathroom. If the arrivals to the restaurant are completely random with the rate of 60 customers an hour,

    a. If you decide to go to the bathroom, what is the chance that when you come back after 2 minutes, you find that the line has not moved and no one has taken your place in the line? (Assume arrival process is independent of cashier's queue length.)

Suppose that you decide to go to the bathroom after you place your order with the cashier.

    b. How long would you expect you need to wait before you can go to the bathroom?

    c. What is the chance that you will go to the bathroom in the next 3 minutes?

7. The ticketing counter of a major airline has four agents who serve its passengers. Passengers arrive at the rate of 110 per hour according to a Poisson process. They wait in the line to check in with the agents and get their boarding pass. The time it takes an agent to serve a passenger is about 2 minutes with standard deviation 3 minutes. The waiting line for ticketing has been long, resulting in some customers missing their flights. Thus the airline is considering

increasing capacity of the ticketing by adding one more agent. If the airline adds the fifth agent,

    a. How much, on an average, would this reduce the passengers' waiting times to get their boarding pass?

    b. How much, on an average, would this shorten the line for ticketing?.

    c. What fraction of time agents would be idle?

8. The airline in Problem 7, instead of adding an agent, is considering adding a self-service Kiosk that can check in passengers and issue boarding pass. The airline believes that most frequent flyers, which constitute about 30 percent of passengers, will use the kiosk. For these passengers, the time to get the boarding pass from the kiosk will have an average of 1.5 minutes with standard deviation 1.5 minutes. Adding a kiosk is expected to increase the average check-in times with the agents to 2.5 minutes, but reduces its standard deviation to 2 minutes. If the airline adds a kiosk,

    a. Which line—the line for Kiosk or the line for agents—would be shorter?

    b. What would be the average time a passenger spends to get a boarding pass?

    c. What fraction of time the kiosk is not used?

    d. What percent of passengers who use Kiosk must wait more than 2 minutes in line to use the kiosk?

9. A forklift truck is used to move engine blocks from the casting department to the machining department of a factory that produces engines. The throughput of the casting department is 12 engines per hour with inter-throughput standard deviation of 5 minutes. Engine blocks are stored in the output buffer of the casting department waiting for forklift truck to move them to the machining department. The forklift truck moves the engines one-by-one, and it takes an average of 4 minutes to move an engine from the casting to the machining department and come back. This transfer time is exponentially distributed. The cost of holding an engine block as work-in-process inventory (either waiting for the forklift truck or being transferred) is about $25 per hour.

    a. On an average, how many minutes of an hour the forklift truck is moving engines?

    b. How long it takes, on an average, for the throughput of the casting department to get to the machining department (including the waiting time for the forklift-truck)?

    c. What is the average hourly inventory holding cost of engine blocks?

10. Factory manager in Problem 9 is planning to purchase a new and larger forklift truck that is faster than the current truck. The new truck can move engines from the casting department to the machining department and come back in about 3 minutes, and this transfer time is exponentially distributed. While the operating cost (including capital cost) for the current forklift truck is about $30 per hour, the operating cost for the new forklift truck would be $45 per hour. Based on these operating costs and the inventory costs of engine blocks, do you recommend that the factory replace the current truck with the faster one?

11. You are hired by Big-Mart corporation to help its department stores improve their operations. Stacy, the manager of one of Big-Mart stores, is seeking your advice whether she should hire more checkout clerks than the three clerks she currently has. A clerk's salary is $12 per hour and customer satisfaction survey estimated that the store suffers approximately $2 in lost sales and goodwill for every hour a customer waits in checkout line. Data shows that about 120 customers per hour go to the checkout clerks in a completely random order (i.e., Poisson arrivals). It take a checkout clerk about 80 seconds with coefficient of variation 1 to serve a customer. How many more checkout clerks would you recommend Stacy to hire in order to minimize the total expected cost of her checkout operations?

12. Stacy in Problem 11 comes back to you in 1 month and mentions that she has been receiving many complaints about the long waiting time in checkout line. She is now thinking of setting the following two service goals for her checkout operations: (i) more than half of her

customers should not wait in checkout line at all, and (ii) more than 90 percent should wait less than 2 minutes in the line for checkout. How many clerks does Stacy need to achieve her service goals?

13. Quick Lube is a three-bay service facility next to a busy highway. It works 12 hours a day, 7 days a week (assume 30 days in a month). Besides the three bays (each serving one car), the facility has space for four vehicles lined up to wait for service in an alley that is the entrance of Quick Lube from the highway. There is no space for cars to line up on the busy highway, so if the waiting line is full, prospective customers go to another place for service. The owner of Quick Lube has noticed that he is losing customers due to lack of waiting space in the alley. A study of arrivals has shown that the arrivals of customers seeking lube is completely random with an average of 36 cars per hour. The study also shows that the mean time to perform the lube operation (in a bay) is 6 minutes per car. This time is found to be well-modeled by exponential distributions.
    a. How long, on an average, a customer waits in the line at Quick Lube before its service starts?
    b. What is the average number of idle bays at Quick Lube?
    c. The average profit of lube work is $8 per served car. What is the average potential profit that Quick Lube is losing each hour due to lack of waiting space in the alley?

14. Consider Quick Lube in Problem 13. To improve operations, the owner of Quick Lube is considering two alternatives for expansion.
    - Rent the adjacent land and expand the alley to create larger space for waiting cars to hold eight cars in the queue. The cost of rental is $1440 per month.
    - Invest in adding one bay to current facility. This results in cost of $2160 per month.
    A customer survey shows that the long-term customer satisfaction and loss of goodwill cost due to customer's waiting in queue is estimated to be $1 per 1 hour of waiting. Which alternative do you recommend to Quick Lube owner?

15. A fashion store in Downtown Chicago has a parking with a total capacity of 10 cars. Based on a time study, it was determined that car arrivals to the parking are completely random with an average of 10 cars an hour. Cars that find the parking full, go to another parking in the area. Cars are charged $1 for 10 minutes of parking (i.e., 10 cents a minute). Cars park in the parking for an average of 2 hours (assume exponential distribution). One suggestion to improve the revenue from parking fees is to charge a fix amount of $12 per car that enters the parking, regardless of how long the car is parked in the parking. Assuming that the demand (car arrivals) and the time a customer parks in the parking remain the same under the suggested pricing strategy, should the building adopt the fixed-charge strategy?

16. Rent-a-Van is a car rental company focusing only on renting vans. The company has a fleet of 20 vans. Rent-a-Van gets requests from customers randomly according to a Poisson process with rate of 12 per day. Customers who rent a van keep the van between 1 and 4 days, with an average of 2 days. Data shows that the duration of rental is well-approximated by an exponential distribution. The company makes a net profit of $30 for each day of rental. If a customer requests a rental and the company does not have any van available (all vans are already rented), the customer will rent a car from other rental companies—Rent-a-Van loses the customer.
    a. What fraction of customers who request a van from Rent-a-Van do not get it?
    b. What is the average number of vans available in Rent-a-Van each day for customers to rent?
    c. What is the fleet size that guarantees that least 90 percent of its customers who need a van get them.
    Rent-a-Van now has a contract with several local rental companies to share their fleet. Specifically, if a customer requests a van and Rent-a-Van does not have any van available

in its lot (all vans are already rented), it calls other rental companies and a van is delivered to Rent-a-Van. Rent-a-Van then gives the van to its customer and shares the profit of $30 with the other firm, that is, Rent-a-Van gets $15 for each day of rental.

   d. What is the average daily profit of Rent-a-Van?

17. Middle-Way is an airport that is the hub for a major airline. The airport has three runways for landing and three runways for takeoffs. Airplanes arrive randomly according to a Poisson process with a rate of 43 planes per hour to land at the airport. The time it takes an airplane to land and clear the runway is about 4 minutes. The landing time is well approximated by exponential distribution. If an airplane arrives and none of the runways are available for landing, the airplane must hover above the airport until a runway becomes available. Airplanes land according to first-come-first-served policy. There are also an average of 43 airplanes per hour that must take off from the airport. Time between two airplanes requesting a runway for takeoff is found to also follow exponential distribution. These airplanes use the three takeoff runways and the time it takes an airplane to take off is about 2.4 minutes (assume exponential distribution).

   To prevent accidents due to congestion, and to make sure that the planes do not run out of fuel while hovering above the airport, FAA (Federal Aviation Administration) has issued new safety rules. These rules state that (i) the average number of airplanes waiting for landing should not exceed two, and (ii) at least 98 percent of arriving planes must wait less than 25 minutes for landing?

   a. Does the airport satisfy the new FAA safety rules?
   b. The airport is planning to use one of the three takeoff runways for landing—the airport would have four runways for landing and two runways for takeoffs. Does this help airport satisfy the safety rules? How does this affect the waiting times of airplanes for takeoff?

18. Johnson Sealers (JS) is a large firm specialized in basement waterproofing. When a customer contacts JS, the firm sends one of its 12 advisors to customer location to evaluate the project and to provide a quotation about the cost of the project. It takes an advisor about 3 hours to go to the customer's location, evaluate the project and provide the initial quote, and write a contract for the customer to sign (assume exponentially distributed time). Data shows that only 20 percent of the customers sign the contract. After a contract is signed, the customer must wait until JS sends one of its 13 teams of technicians to customer's site to start the project. It takes an average of 20 hours for the team to finish the project (assume exponential distribution). Both advisors and technicians work from 8:00 am to 5:00 pm. JS receives about three calls per hour from its customers during 8:00 am to 5:00 pm according to a Poisson process requesting to make an appointment with JS's advisors.

   a. After requesting an appointment, how long, on an average, a customer must wait to meet with a JS advisor?
   b. What is the average time between when a customer signs a contract, until JS sends one of its 13 teams to work on customer's project?
   c. JS has noticed that if their advisors cannot make an appointment to meet with the customers within 5 days after a customer calls, most customers do not make an appointment and contact other firms. How many advisors does JS need, so it can make sure that at least 90 percent of its customers meet JS advisors in 5 days or less?
   d. If JS uses the number of advisors you obtained in Part (c), what fraction of customers who sign their contract must wait for more than 2 weeks for JS team to start their projects?

# CHAPTER 8

# Flow Time Improvement

## 8.0 Introduction

Flow time is the most critical operational performance measure in service systems, since it directly affects customers' experience with a firm. For example, a long waiting time to receive a computer that a customer orders, a long waiting time of a customer in a restaurant to get a table, or a long waiting time on hold to speak to an agent when a customer calls tech support will certainly affect customer's perception of a firm's overall quality of service. As we discussed in Chap. 1, speed and quick response time can be a competitive advantage.

In Chap. 7, we presented underlying reasons for long flow times and provided analytical tools to estimate flow time measures such as average flow times and flow time service levels in single-stage and multi-stage processes. In this chapter, we provide control and design strategies that can be used to improve flow time measures of a process. These strategies include

- Increasing capacity
- Reducing variability in demand and process time
- Synchronizing capacity with demand
- Using priority discipline, if appropriate
- Economies of scale and vertical pooling
- Horizontal pooling

- Reducing buffer size to block arrivals
- Changing batch size
- Redesigning flow and buffers

We also discuss customer psychology with respect to waiting times and suggest policies that a firm can implement to affect customers' perception of long waiting times. But, we first start with the strategy of increasing capacity.

## 8.1 Increasing Capacity

As shown in Eq. (7.2), increasing the effective capacity of a process reduces the average waiting time of a flow unit. In Sec. 6.5, we have presented several actions that one can take to increase the effective capacity of a process. These actions include adding more resources, increasing working hours, outsourcing activities, redesigning flow, eliminating or reducing supporting tasks such as setups and rework, reducing theoretical process time through work and time study, parallel working, better worker training, new technology, worker incentive, and making activities self-service. For more details see Chap. 6.

## 8.2 Reducing Variability in Demand and Process Time

We also showed in Chap. 7 that, for a given utilization, processes with higher variability will have longer flow times. Hence, another way to decrease flow times is to reduce variability in a process—variability in demand arrival and/or in process times.

To find ways to reduce variability in customer demand, we first need to recognize that the variability in demand is caused by: (i) variability due to randomness, and (ii) variability due to management decisions.

- *Variability Due to Randomness:* There is always randomness and uncertainty in customer demand. Customers make their purchasing decision independent of each other and at different times. Their decisions are also affected by external factors that are independent of management decisions. For example, changes in weather affect the daily demand in amusement parks, that is, there is lower demand on cold and rainy days and higher demand on nice and pleasant days. Other external factors that affect variability are state of economy, price of related goods, change in taste and preferences of customers, etc.

- *Variability Due to Management Decision:* Besides inherent randomness in demand, management decisions can also contribute to variability in demand. For example, a sales promotion—which is an internal decision made by a firm—can significantly increase demand during a short period, resulting in high variability in demand. Another example is an extensive advertising campaign that can increase demand and result in high variability. Frequent product launch and increase in product variety can also increase variability in demand.

In Sec. 5.6.1 of Chap. 5 we discussed several strategies for reducing variability in demand. In Chap. 11, we will discuss strategies to reduce the negative impact of variability in demand caused by management decisions. Thus, we refer readers to these chapters.

With respect to variability in process times, recall from Chap. 4 that processing time of a unit performed by a resource consists of a main task and some supporting tasks. Hence the variability in process times is due to: (i) variability in the main task, and (ii) variability caused by supporting tasks. A list of strategies to reduce variability in main and supporting tasks is presented in Sec. 5.7.2 of Chap. 5.

## 8.3 Synchronizing Capacity with Demand

Recall from Chap. 7 that the reason flow units wait in the buffer for their processing—the reason queues exist—is that the input rate to the process temporarily or permanently exceeds the effective capacity of the process. Thus, to decrease flow times of a process, we need to reduce the number of, or to shorten the length of, the periods in which input rate exceeds effective capacity. We can do this by synchronizing capacity with demand. Specifically, by *temporarily* increasing the capacity of the process in periods with high input rate, one can shorten the queue lengths in these periods and thus reduce the overall average in-buffer inventory and flow times. This practice has widely been used in many manufacturing and service operations systems. See Sec. 5.6.2 of Chap. 5 for strategies to synchronize capacity with demand.

## 8.4 Priority versus FCFS

First-come-first-served (FCFS) is the most common service discipline in business processes. The main reason is that it is considered fair, and it results in the same average flow times for every class of customers. However, we also observe many processes in which priority is given to a particular class of customers. For example, in Emergency Departments of hospitals priority is given to patients who are in more critical condition. In boarding an airplane, priority is given to business and first-class passengers. To study how priority discipline can be used to reduce flow times, we first start by introducing underlying queueing models of processes that use priority policies.

### 8.4.1  A Priority Queueing Model

We start this section with the case of Custom-Chrome Inc. that produces gear alloy wheels in produce-to-order fashion.

 **Example 8.1**

Custom-Chrome Inc. makes among other alloy products, alloy wheels (AW) for mid- and large-sized cars and trucks in produce-to-order mode. In its AW production facility, Custom-Chrome uses a base wheel that is machined and then it reshapes it based on customers' orders from the company's online catalog. Orders for custom-made AWs can be divided into two groups:

- *U-Chrome:* The firm receives an average of four orders every day for U-chrome wheels. The effective mean process time for this product is 3 hours with standard deviation of 4 hours.
- *Standard:* The firm receives an average of six orders for standard wheels every day. The effective mean process time of a standard wheel is 1.6 hours with standard deviation of 3 hours.

The orders for each product come from a large customer base and thus Poisson process is shown to be the best to model order arrivals for each type. Currently the firm processes the orders in a first-come-first-served basis. The processing of the next order starts only when the last order is completed. Due to large backlog of orders, Custom-Chrome has started to work three shifts, that is, 24 hours per day.

In the past 4 months, the firm has been receiving several complaints from its customers about the long waiting times for their orders. Most of the complaints are from customers who order U-chrome—the more profitable product. Based on a market study,

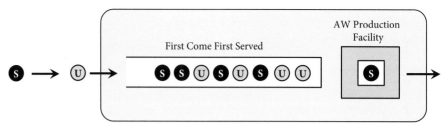

it is estimated that each day that a U-chrome customer is waiting for their order costs $10 in customer dissatisfaction and loss of future business. This number is $7 for standard orders.

The firm's managers are therefore thinking of giving priority to orders for U-chrome. Specifically, production of standard orders should start only when there is no outstanding U-chrome orders in the queue. The production process is such that when the process of making an order (standard or U-chrome) starts, it cannot be interrupted and must be completed before starting the next order. There are some concerns about implementing this new production policy. Custom-Chrome does not know whether or how much the new priority policy will improve customer satisfaction. Thus, before implementing the policy, the upper management would like to know how this new policy will affect customer waiting times and the overall cost associated with it.

To compute whether the priority policy improves performance by reducing customer waiting cost, we first need to estimate the average customer waiting cost under the current FCFS production policy. Let $C_U$ and $C_S$ be the waiting cost per unit time of a U-chrome and a standard customer, respectively. Hence, $C_U = \$10$ and $C_S = \$7$ per day.

Note that orders of U-chrome arrive at Custom-Chrome at the rate of $R_{in}^{(U)}$ per day. Each order will be in the system for $T^{(U)}$ days. Therefore, $R_{in}^{(U)} \times T^{(U)}$ is the total average number of days that $R_{in}^{(U)}$ orders wait in the system. Therefore, the total waiting cost of U-chrome is $c_U R_{in}^{(U)} T^{(U)}$. On the other hand, according to Little's law, $I^{(U)} = R_{in}^{(U)} T^{(U)}$, where $I^{(U)}$ is the average number (i.e., the average inventory) of U-chrome orders at Custom-Chrome. Therefore, the total average waiting cost of U-chrome orders in a day is $c_U I^{(U)}$. This is the same for standard wheels. Considering $TC(\text{FCFS})$ as the total expected waiting cost of U-chrome and standard orders per day under FCFS policy, we have

$$TC(\text{FCFS}) = \$10 \times I^{(U)} + \$7 \times I^{(S)}$$

Therefore, to compute the total expected cost $TC(\text{FCFS})$ we need to get an estimate for the average number of orders in Custom-Chrome shop, that is, $I^{(U)}$ and $I^{(S)}$.

To compute these inventories, we can either gather information about the number of outstanding orders in the Custom-Chrome at different times and take their average, or we can use an analytical model to estimate those inventories. Here we present the latter.[1] Note that the underlying queueing model of the current production process of Custom-Chrome is an $M/G/1$ queueing system. However, the queueing system has two different classes of customers: Class 1 (referring to U-chrome customers), and Class 2 (referring to standard customers). Figure 8.1 shows the queueing model for the current process.

---

[1] It is often recommended to do both, since the data gathered can help validate the analytical model.

As the figure shows, upon arrival, all customer classes join the end of the queue and thus are served on a FCFS basis. If all classes of customers had the same effective mean process times, we could have used Eq. (7.2) and Little's law to estimate the average flow times and inventory of the process. However, this is not the case here. Each class of customers has a different effective process time. The following analytics provides process-flow measures for such a system.

ANALYTICS

**Process-Flow Measures of a Single-Stage Process with Multiple Products—$M/G/1$ with FCFS**

Consider a process with one resource that processes $K$ different products in a make-to-order fashion. Demand for class-$k$ product ($k = 1, 2, \ldots, K$) arrives according to a Poisson process with arrival (input) rate $R_{\text{in}}^{(k)}$. The effective process time of class-$k$ product has mean $T_{\text{eff}}^{(k)}$ and standard deviation $\sigma_{\text{eff}}^{(k)}$. Products are processed according to first-come-first-served discipline. The buffer of the process has ample size and thus arriving demands are not blocked. The process-flow measures of the process can be obtained using the following steps:

**Step 1:**  For each class of product $k = 1, 2, \ldots, K$, compute $S_{\text{eff}}^{(k)}$ as follows:

$$S_{\text{eff}}^{(k)} = (\sigma_{\text{eff}}^{(k)})^2 + (T_{\text{eff}}^{(k)})^2 \tag{8.1}$$

**Step 2:**  Considering the total arrival rate $R_{\text{in}} = R_{\text{in}}^{(1)} + R_{\text{in}}^{(2)} + \cdots + R_{\text{in}}^{(K)}$, compute

$$p_k = \frac{R_{\text{in}}^{(k)}}{R_{\text{in}}} \qquad \text{for all } k = 1, 2, \ldots, K \tag{8.2}$$

Note that $p_k$ is the probability that an arriving demand is for class-$k$ products. Having $p_k$, compute $\mathbb{T}_{\text{eff}}$ and $\mathbb{S}_{\text{eff}}$ as follows:

$$\begin{aligned} \mathbb{T}_{\text{eff}} &= p_1 T_{\text{eff}}^{(1)} + p_2 T_{\text{eff}}^{(2)} + \cdots + p_K T_{\text{eff}}^{(K)} \\ \mathbb{S}_{\text{eff}} &= p_1 S_{\text{eff}}^{(1)} + p_2 S_{\text{eff}}^{(2)} + \cdots + p_K S_{\text{eff}}^{(K)} \end{aligned} \tag{8.3}$$

**Step 3:**  Process-flow measures can be obtained as follows:

- **Utilization:** Percent of time that the process is busy processing class-$k$ products, $k = 1, 2, \ldots, K$, is

$$u^{(k)} = R_{\text{in}}^{(k)} \times T_{\text{eff}}^{(k)}$$

Thus, the utilization of the process, that is, percent time that the process is busy processing all products, is

$$u = u^{(1)} + u^{(2)} + \cdots + u^{(K)} = R_{\text{in}} \times \mathbb{T}_{\text{eff}} \tag{8.4}$$

If utilization $u \geq 1$, then process is unstable.

- **Average Flow Times:** The average in-buffer flow time for all classes of products are the same and is

$$T_b^{(1)} = T_b^{(2)} = \cdots = T_b^{(K)} = \frac{R_{\text{in}} \times \mathbb{S}_{\text{eff}}}{2(1 - u)}$$

The average flow time for class-$k$ product, $k = 1, 2, \ldots, K$, is

$$T^{(k)} = T_b^{(k)} + T_{\text{eff}}^{(k)}$$

- **Average Inventory:** Using Little's law, the average in-buffer and in-process inventory for class-$k$ products, $k = 1, 2, \ldots, K$, are as follows:

$$I_b^{(k)} = R_{\text{in}}^{(k)} \times T_b^{(k)} \quad , \quad I_p^{(k)} = u^{(k)} \quad , \quad I^{(k)} = I_b^{(k)} + I_p^{(k)}$$

**Step 4:**    (**Overall Measures**) The overall process-flow measures of the process are as follows:

$$T_b = \frac{R_{\text{in}} \times \mathbb{S}_{\text{eff}}}{2(1 - u)} \quad , \quad I_b = \sum_{k=1}^{K} I_b^{(k)} \quad , \quad T = T_b + \mathbb{T}_{\text{eff}} \quad , \quad I = T \times R_{\text{in}}$$

### Custom-Chrome Operations Under FCFS Policy

Recall that Custom-Chrome has $K = 2$ classes of customers: (i) Class 1 are customers who order U-chrome, (ii) Class 2 are customers who order standard wheels. To obtain process-flow measures of the current operations at Custom-Chrome, as in Step 1 of the above analytics, we determine the arrival (input) rate of each class. Considering 1 hour as our time unit and the fact that the firm works 24 hours a day, we have: $R_{\text{in}}^{(1)} = 4/24 = 0.167$, and $R_{\text{in}}^{(2)} = 6/24 = 0.250$ customers per hour. Thus,

$$R_{\text{in}} = R_{\text{in}}^{(1)} + R_{\text{in}}^{(2)} = 0.167 + 0.250 = 0.417$$

and

$$p_1 = \frac{0.167}{0.417} = 0.4 \quad , \quad p_2 = \frac{0.250}{0.417} = 0.6$$

With respect to the processing time of customer orders, according to Step 1, we compute

$$S_{\text{eff}}^{(1)} = (\sigma_{\text{eff}}^{(1)})^2 + (T_{\text{eff}}^{(1)})^2 = 4^2 + 3^2 = 16 + 9 = 25$$

For Class 2 we get $S_{\text{eff}}^{(2)} = 3^2 + (1.6)^2 = 11.56$. Having these values, we move to Step 2 and compute

$$\mathbb{T}_{\text{eff}} = p_1 \, T_{\text{eff}}^{(1)} + p_2 \, T_{\text{eff}}^{(2)} = 0.4 \times 3 + 0.6 \times 1.6 = 2.16$$

and

$$\mathbb{S}_{\text{eff}} = p_1 \, S_{\text{eff}}^{(1)} + p_2 \, S_{\text{eff}}^{(2)} = 0.4 \times 25 + 0.6 \times 11.56 = 16.94$$

We can now compute process-flow measures of Custom-Chrome under the current FCFS production schedule. As shown in Step 3, the utilization of the process is

$$u = R_{\text{in}} \times \mathbb{T}_{\text{eff}} = 0.417 \times 2.16 = 0.90$$

Hence, we can find the average in-buffer flow time for all orders as follows:

$$T_b^{(1)} = T_b^{(2)} = \frac{R_{\text{in}} \times \mathbb{S}_{\text{eff}}}{2(1 - u)} = \frac{0.417 \times 16.94}{2(1 - 0.90)} = 35.32 \text{ hours} = 1.47 \text{ days}$$

Following Steps 3 and 4, we can compute other performance measures of the two classes of customers as follows:

Class 1:    $T^{(1)} = 1.59$ days,    $I_b^{(1)} = 5.88$,    $I^{(1)} = 6.36$

Class 2:    $T^{(2)} = 1.54$ days,    $I_b^{(2)} = 8.82$,    $I^{(2)} = 9.24$

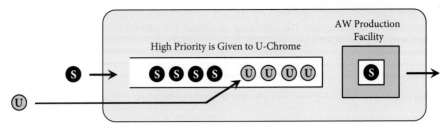

Figure 8.2 Queueing model of Custom-Chrome in Example 8.1 under priority production schedule.

The overall process-flow measures under FCFS production policy are $T_b = 1.47$ days, and

$$T = p_1 T^{(1)} + p_2 T^{(2)} = 0.4(1.59) + 0.6(1.54) = 1.56 \text{ days}$$
$$I = I^{(1)} + I^{(2)} = 6.36 + 9.24 = 15.6$$

Notice that while there are different average number of orders of each class in the queue (i.e., different in-buffer inventory), both classes of customers have the same average waiting time in queue (i.e., same average in-buffer flow time of 1.47 days). The reason is that, upon arrival, both classes join the end of the production queue, and thus experience the same average waiting time. The customers, however, have different average time to receive their orders (i.e., different average flow time), because each class has a different effective process time.

Having the average in-buffer inventory of each class, we can compute the average waiting cost per day

$$TC(\text{FCFS}) = \$10 \times I^{(1)} + \$7 \times I^{(2)} + = \$10 \times 6.36 + \$7 \times 9.24$$
$$= \$128.28 \text{ per day}$$

**Custom-Chrome Operations Under Priority Policy**
If Custom-Chrome decides to give priority to U-chrome (Class-1) customers, how much does this improve those customers' waiting times? Also, how does that affect the waiting times of standard orders? And finally, how do all these affect the overall customer waiting costs?

If Custom-Chrome gives higher priority to U-chrome wheels in their production schedule, then the queueing model of this new system is shown in Fig. 8.2. As the figure shows all arriving orders for U-chrome are put ahead of all standard orders in the production schedule, regardless of how long standard orders have been waiting for production.

To estimate the average waiting time for both standard and U-chrome customers, we need the following analytics for processes with priority policy.

ANALYTICS

**Process-Flow Measures of a Single-Stage Process with Multiple Products—$M/G/1$ with Priority**

Consider a process with one resource that processes $K$ different classes of products in a make-to-order fashion. Demand for class-$k$ product $(k = 1, 2, \ldots, K)$ arrives according to a Poisson process with arrival (input) rate $R_{\text{in}}^{(k)}$. The effective process time of class-$k$ product has mean $T_{\text{eff}}^{(k)}$ and standard deviation $\sigma_{\text{eff}}^{(k)}$. Products are processed according to a priority policy that gives priority to class $k - 1$ over class $k$. Specifically, at the end of any service completion, the server processes a product of Class 1. If there is no Class-1 product in the buffer, then the server processes a Class-2 product. If there is no Class-2 product, the server chooses a Class-3 product and so on. The buffer of the process has ample size and thus arriving demands for products are not blocked. The process-flow measures of the process can be obtained using the following steps:

**Step 1:**    Compute $u^{(k)} = R_{\text{in}}^{(k)} \times T_{\text{eff}}^{(k)}$ for all classes of products $k = 1, 2, \ldots, K$.

- **Utilization:.** The process utilization is

$$u = u^{(1)} + u^{(2)} + \cdots + u^{(K)}$$

  If utilization $u \geq 1$, then the process in unstable.

**Step 2:**    Set $a_0 = 0$ and compute $a_k$ for all $k = 1, 2, \ldots, K$ as follows:

$$a_k = u^{(1)} + u^{(2)} + \cdots + u^{(k)}$$

**Step 3:**    Process-flow measures of the process can be obtained as follows:

- **Average Flow Times:** As opposed to FCFS discipline, the average in-buffer flow time is different for each class of product. Considering the total arrival rate $R_{\text{in}} = R_{\text{in}}^{(1)} + R_{\text{in}}^{(2)} + \cdots + R_{\text{in}}^{(K)}$, the average in-buffer flow time for class-$k$ products is

$$T_b^{(k)} = \frac{R_{\text{in}} \; \mathbb{S}_{\text{eff}}}{2(1 - a_{k-1})(1 - a_k)} \quad : \quad \text{for all } k = 1, 2, \ldots, K$$

  where $\mathbb{S}_{\text{eff}}$ is the same as in Eq. (8.3). The average flow time for class-$k$ products is

$$T^{(k)} = T_b^{(k)} + T_{\text{eff}}^{(k)} \quad : \quad \text{for all } k = 1, 2, \ldots, K$$

- **Average Inventory:** Using Little's law, the average in-buffer and in-process inventory for class-$k$ products are as follows:

$$I_b^{(k)} = R_{\text{in}}^{(1)} \times T_b^{(k)} \quad , \quad I_p^{(k)} = u^{(k)} \quad , \quad I^{(k)} = I_b^{(k)} + I_p^{(k)} \quad \text{for all } k = 1, 2, \ldots, K$$

**Step 4:**    **(Overall Measures)** The overall process-flow measures for a randomly chosen product are as follows:

$$T_b = p_1 \, T_b^{(1)} + p_2 \, T_b^{(2)} + \cdots + p_K \, T_b^{(K)}$$

where $p_k$ are the same as in Eq. (8.2), and

$$I_b = I_b^{(1)} + I_b^{(2)} + \cdots + I_b^{(K)} \quad , \quad T = T_b + \mathbb{T}_{\text{eff}} \quad , \quad I = T \times R_{\text{in}}$$

Considering U-chrome (i.e., Class-1) orders having priority over standard (i.e., Class-2) orders, we can use the above analytics to compute process-flow measures of Custom-Chrome under the suggested priority. Following Step 1, we get

$$u^{(1)} = R_{\text{in}}^{(1)} \times T_{\text{eff}}^{(1)} = 0.167 \times 3 = 0.5$$

Similarly, $u^{(2)} = 0.25 \times 1.6 = 0.4$. Then, $u = 0.5 + 0.4 = 0.9 < 1$ and the process is stable. As in Step 2, we have $a_0 = 0$, and

$$a_1 = u^{(1)} = 0.5 \quad , \quad a_2 = u^{(1)} + u^{(2)} = 0.9$$

We can now compute the average in-buffer flow time (i.e., waiting time) for Class-1 customers using the equation in Step 3 as follows:

$$T_b^{(1)} = \frac{R_{in}\ \mathbb{S}_{eff}}{2(1-a_0)(1-a_1)} = \frac{0.417 \times 16.94}{2(1-0)(1-0.5)} = 7.06 \text{ hours}/24 = 0.294 \text{ days}$$

Similarly, for Class-2 customers

$$T_b^{(2)} = \frac{R_{in}\ \mathbb{S}_{eff}}{2(1-a_1)(1-a_2)} = \frac{0.417 \times 16.94}{2(1-0.5)(1-0.9)} = 70.64 \text{ hours}/24 = 2.943 \text{ days}$$

Following the remaining steps, we obtain process-flow measures of the two classes of customers as follows:

Class 1:    $T^{(1)} = 0.419$ days,    $I_b^{(1)} = 1.176$,   $I^{(1)} = 1.676$

Class 2:    $T^{(2)} = 3.010$ days,    $I_b^{(2)} = 17.658$,   $I^{(2)} = 18.060$

The overall operational performance measures under the priority policy are as follows:

$$T = p_1 T^{(1)} + p_2 T^{(2)} = 0.4(0.419) + 0.6(3.010) = 1.974 \text{ days}$$
$$T_b = p_1 T_b^{(1)} + p_2 T_b^{(2)} = 0.4(0.294) + 0.6(2.943) = 1.883 \text{ days}$$
$$I = I^{(1)} + I^{(2)} = 1.676 + 18.060 = 19.736$$

The above results lead us to the following points:

- After giving priority to Class-1 orders, that is, U-chrome, the average number of Class-1 orders in the queue is reduced from 5.88 to 1.176. However, the average number of Class-2 orders waiting for production is increased significantly from 8.82 under FCFS to 17.658 under the priority policy.

- Similarly, while the priority policy reduces the time a Class-1 customer waits to get the order, it results in a significantly larger waiting time for Class-2 customers. Specifically, the average flow time for Class-2 customers is doubled from 9.24 to 18.06 days.

- Under the priority policy, the average customer waiting cost per day would be

$$TC(\text{PRIOR}) = \$10 \times I^{(1)} + \$7 \times I^{(2)} = \$10 \times 1.676 + \$7 \times 18.06$$
$$= \$143.18 \text{ per day}$$

**Conclusion for Custom-Chrome**

As the analysis shows, implementing the priority policy does not decrease the overall waiting costs. It, in fact, increases the overall average waiting cost by $14.9 per day from $128.28 to $143.18. Hence, Custom-Chrome should not give priority to U-chrome, even though customers of U-chrome have a higher waiting cost than standard customers. But what about giving priority to standard? Does that reduce the overall average waiting costs? We discuss this in the following section.

## 8.4.2 Optimal Priority Policy

To provide a general principle of how to prioritize flow units to minimize waiting costs, we first need to introduce the "Conservation Identity Principle." Consider the finishing station of a factory that produces bowling balls covered with expensive metal. These balls are collective items that are in high demand. After a thin layer of gold (or titanium) is coated on

Table 8.1  Process-Flow Measures of the Bowling Ball Finishing Station

| | Class (1) | Class (2) | Class 1 $T^{(1)}_{\text{eff}}$ (minutes) | Class 1 $I^{(1)}$ (balls) | Class 2 $T^{(2)}_{\text{eff}}$ (minutes) | Class 2 $I^{(2)}$ (balls) | Overall Inventory (balls) | Overall Workload (minutes) |
|---|---|---|---|---|---|---|---|---|
| **FCFS** | | | | | | | | |
| | T-ball | G-ball | 4 | 2.275 | 2 | 5.188 | 7.463 | 19.475 |
| **PRIO** | | | | | | | | |
| | T-ball | G-ball | 4 | 0.713 | 2 | 8.313 | 9.025 | 19.475 |
| | G-ball | T-ball | 2 | 1.438 | 4 | 4.150 | 5.588 | 19.475 |

the balls using electroplating operations, the balls are sent to the finishing station in which a worker removes rough and bumpy material from the surface of the balls and sands and polishes them. The firm produces two types of balls: (i) titanium balls (T-balls), and (ii) golden balls (G-balls). It takes the worker 4 minutes with standard deviation of 3 minutes to finish a T-ball, and 2 minutes with standard deviation of 1 minute to finish a G-ball. T-balls and G-balls arrive at the finishing station according to the Poisson process with a rate of 6 and 15 balls per hour, respectively.

Considering balls as flow units, the sanding station would be a $M/G/1$ queue with two classes of products. Using our analytics we can compute the performance measures of this station under three different policies: (i) FCFS, (ii) priority to T-balls, and (iii) priority to G-balls. These measures are shown in Table 8.1.

As Table 8.1 shows, under FCFS policy, the average number of T-balls and G-balls in the finishing station are 2.275 and 5.188 balls, respectively, resulting in an overall inventory of $2.275 + 5.188 = 7.463$ balls. When priority is given to T-balls, the overall inventory is increased to 9.025 balls. However, when priority is given to G-balls, the overall average inventory is reduced to 5.588 balls. This points to the fact that different service policies result in different *overall* (total) average inventory in the process—nothing unexpected.

In the example above, we chose a ball as our flow unit and thus we measured the inventory of the station as the total number of balls in the station. Now let's take a different approach and measure the inventory of the process in terms of the average workload in the process. Specifically, let's define our flow unit as "a minute of work." Considering the effective mean process time of 4 minutes for T-balls, every unit of inventory of T-balls represents 4 minutes of work in the station. Similarly, every unit of inventory of G-balls represents 2 minutes of work in the station.

Measuring inventory using "minutes of work" instead of "number of balls," we can now compute the overall average inventory—overall minutes of work—in the finishing station under FCFS and priority policies. We call this overall workload in the process or simply *Overall Workload*. As Table 8.1 shows, under FCFS policy, we have $I^{(1)} = 2.275$ T-balls, and $I^{(2)} = 5.188$ G-balls. Considering the effective mean process times 4 minutes and 2 minutes for T-balls and G-balls, respectively, we have

$$\text{Overall (average) Workload} = T^{(1)}_{\text{eff}} I^{(1)} + T^{(2)}_{\text{eff}} I^{(2)} = 4 \times 2.275 + 2 \times 5.188 = 19.475 \text{ minutes}$$

When priority is given to T-balls, then $I^{(1)} = 0.713$ T-balls, and $I^{(2)} = 8.313$ G-balls, then the overall average workload would be

$$\text{Overall (average) Workload} = T^{(1)}_{\text{eff}} I^{(1)} + T^{(2)}_{\text{eff}} I^{(2)} = 4 \times 0.713 + 2 \times 8.313 = 19.475 \text{ minutes}$$

which is the same as that under FCFS. Finally, when G-balls are given higher priority—G-balls are Class-1 units—then $I^{(1)} = 1.438$ G-balls, and $I^{(2)} = 4.150$ T-balls. Thus,

$$\text{Overall (average) Workload} = T^{(1)}_{\text{eff}} I^{(1)} + T^{(2)}_{\text{eff}} I^{(2)} = 2 \times 1.438 + 4 \times 4.150 = 19.475 \text{ minutes}$$

Hence, when the inventory in the finishing station is measured by the minutes of work (instead of number of balls), the overall average workload is the same, regardless of the service policy being used (i.e., FCFS or priority). This is known as Conservation Identity Principle that we discuss below.

---

**PRINCIPLE**

### Work Conserving Policies and Conservation Identity Principle

Consider a single-server multi-class queueing system with a service discipline that is work conserving.

- *Work-Conserving* service disciplines are those under which the server never idles when there is work to do.
- *Conservation Identity Principle:* In processes with work-conserving disciplines, the overall (i.e., total) inventory measured in time units is independent of the service discipline being used. In other words, if the process has $K$ different classes of customers, then under any work-conserving service discipline

$$T_{\text{eff}}^{(1)} I^{(1)} + T_{\text{eff}}^{(2)} I^{(2)} + \cdots + T_{\text{eff}}^{(K)} I^{(K)} = \text{Constant} \qquad (8.5)$$

Note that FCFS, LCSF, PRIO, and SIRO are all work-conserving disciplines since they do not keep the server idle when there is a customer in the system. Hence, all these service disciplines result in the same overall workload (i.e., minutes of work in the process). But, why?

To understand this principle, let's return to our finishing station. Recall that we defined our flow unit as "one minute of work." Thus, a T-ball with effective mean process time of 4 minutes represented 4 minutes of work, or 4 flow units. To have a better visualization of the process, consider our new flow unit—1 minute of work—as a marble that requires an average 1 minute of sanding by the worker. We let a titanium marble represent 1 minute of sanding for a T-ball, and a golden marble to represent 1 minute of sanding for a G-ball. Thus, sanding a T-ball is analogous to sanding 4 T-marbles—which takes an average of 4 minutes—and sanding a G-ball is analogous to sanding 2 G-marbles—which takes an average of 2 minutes. On the other hand, arrival of a T-ball is analogous to simultaneous arrivals of 4 T-marbles and arrival of a G-ball is analogous to the simultaneous arrivals of 2 G-marbles. Hence, considering arrival rate of 6 and 15 for T-balls and G-balls, respectively, the process has arrival rate $4 \times 6 = 24$ T-marbles and $2 \times 15 = 30$ G-marbles, a total of 54 marbles per hour. Since the effective mean process time of a marble—titanium or golden—is 1 minute, the server has the capacity of serving 60 marbles an hour.

Note that in this system, the total average inventory of marbles is equal to the total average minutes of work—overall workload—in the finishing station. The question is why the total average number of marbles in the system is the same under different work-conserving service disciplines.

Notice that all work-conserving disciplines see the *same total arrival* of 54 marbles per hour. Furthermore, all arriving marbles—titanium or golden—have the *same effective mean process time* of 1 minute. Thus, regardless of what discipline (i.e., FCFS, LCSF, PRIO, and SIRO) is used to choose a marble (titanium or golden) from the buffer for sanding, the inventory of marbles is reduced by one every minute that a marble is finished. Thus, under all these work-conserving disciplines, the *total* average number of marbles in the process is the same, that is, 19.475 marbles. What is different under these policies, however, is the mix of titanium and golden marbles within 19.475 marbles. For example, under FCFS policy, among 19.475 marbles, the average number of T-marbles is $T_{\text{eff}}^{(1)}\, I^{(1)} = 4 \times 2.275 = 9.100$ and the average number of G-marbles is $T_{\text{eff}}^{(2)}\, I^{(2)} = 2 \times 5.188 = 10.375$, a total of $9.100 + 10.375 = 19.475$ marbles.

In conclusion, while each work-conserving service discipline results in different inventory of T-balls and G-balls as well as a different overall average inventory of balls, they all result in the same overall average workload (i.e., minutes of work) in the finishing station.

What are the implications of the Conservation Identity Principle for operations engineers and managers? Here we discuss two of them: (i) flow time trade-off among classes of customers, and (ii) optimal priority for minimizing overall waiting costs.

### Flow Time Trade-off among Classes

According to Little's law, for each class of customer we have $I^{(k)} = R_{\text{in}}^{(k)}\, T^{(k)}$. Thus,

$$T_{\text{eff}}^{(k)}\, I^{(k)} = T_{\text{eff}}^{(k)}\, R_{\text{in}}^{(k)}\, T^{(k)} = u^{(k)}\, T^{(k)}$$

Hence, Conservation Identity Principle in Eq. (8.5) can be rewritten as follows:

$$u^{(1)}\, T^{(1)} + u^{(2)}\, T^{(2)} + \cdots + u^{(K)}\, T^{(K)} = \text{Constant} \qquad (8.6)$$

The above equation leads to the following principle:

---

**PRINCIPLE**

**Flow Time Trade-Off Principle in Multi-Product Processes**

In multi-product processes, if the demand or the effective capacity of the process is not changed (i.e., utilization for classes of customers do not change), the improvement in flow time (or inventory) of one class of customers can be achieved only at the expense of increasing the flow time (or inventory) of some other class or classes of customers.

---

Consider the inventory of T-balls and G-balls in Table 8.1 under FCFS policy. Note that when priority is given to T-balls, the average inventory of T-balls is reduced by 1.562 from 2.275 to 0.713. This, according to Conservation Identity Principle, occurs at the expense of increasing the average inventory of G-balls by 3.125 from 5.188 to 8.313 balls. Giving priority to G-balls, on the other hand, decreases the average inventory of G-balls by 3.75 balls from 5.188 to 1.438 balls. This occurs at the expense of increasing the average inventory of T-balls by 0.875 balls from 2.275 to 4.150.

### Optimal Priority for Minimizing Waiting Cost—The c-mu Rule

Another important implication of Conservation Identity Principle is that it results in a simple rule to obtain the optimal queue discipline in order to minimize the overall waiting costs of flow units in the process. Consider a process with a single resource and two classes of flow units. Every unit of time that a Class 1 spends in the process (waiting or being processed) costs $c_1$. This waiting cost is $c_2$ for Class-2 flow units. Thus, the total average waiting cost per unit time is $TC = c_1\, I^{(1)} + c_2\, I^{(2)}$. If the arrival is Poisson, we can use our analytics for $M/G/1$ queues to find the optimal queue discipline among the following three: (i) FCFS, (ii) prioritizing Class-1, and (iii) prioritizing Class-2 flow units. The optimal queue discipline, however, can be easily determined without using the $M/G/1$ model and through a simple principle called "c-mu" rule, which we present below (see online Appendix B for details).

PRINCIPLE

**The c-mu Rule—Minimizing Total Average Waiting Cost**

Consider a multi-product process with $K$ classes of products (i.e., flow units). Class $k$ has an effective mean process time of $T_{\text{eff}}^{(k)}$. Holding one unit of class-$k$ product in the process (i.e., in the buffer or under process) for one unit of time costs $c_k$. Define $\mu_k = 1/T_{\text{eff}}^{(k)}$ for all $k = 1, 2, \ldots, K$. Consider the total waiting cost of all $K$ products per unit time as follows:

$$TC = c_1 I^{(1)} + c_2 I^{(2)} + \cdots + c_K I^{(K)}$$

Then, the c-mu rule that gives the first priority to the class of products with the largest $c_k \mu_k$, the second priority to the class with second largest $c_k \mu_k$, and so on will result in the lowest total average waiting cost (i.e., inventory cost) in the process.

Three important features of the c-mu rule are worth mentioning here:

- If $c_1 \mu_1 > c_2 \mu_2$, prioritizing Class-1 products not only is better than prioritizing Class-2 products, but also is better than FCFS, LCFS, and SIRO.

- The c-mu rule does not depend on the arrival process of each class of products. As long as $c_1 \mu_1 > c_2 \mu_2$, giving priority to Class-1 products is optimal, regardless of the distribution or the rate of the arrival of each class.

- The c-mu rule does not depend on the variability in arrivals or in process times. It only depends on the *average* processing times, that is, the effective mean process times of products.

We now use our finishing station to describe the c-mu rule. Suppose that the cost of holding one titanium ball, that is, a T-ball, in the finishing station is $c_T = \$3$ per minute, and the cost of holding a G-ball in the finishing station is $c_G = \$2$ per minute. On the other hand, $\mu_T$, the effective capacity of the station if it only produces T-balls, is $\mu_T = 1/T_{\text{eff}}^{(T)} = 1/4 = 0.25$ per minute. Similarly, $\mu_G = 1/2 = 0.5$ G-balls per minute. Since

$$\$2 \times 0.5 = \$1 = c_G \mu_G > c_T \mu_T = \$3 \times 0.25 = \$0.75$$

then giving priority to G-balls results in lower total average inventory holding cost than giving priority to T-balls.

The c-mu rule becomes clear if we use "unit of work"—or better yet a marble—as our flow unit instead of T-balls and G-balls and translate our holding cost from holding cost for a ball to holding cost for a marble. Note that the holding cost for a T-ball, that is analogous to 4 T-marbles, is $\$3$ per minute. Thus, the holding cost for a T-marble would be $\$3/4 = \$0.75$ per minute. On the other hand, the holding cost for a G-ball, which is analogous to 2 G-marbles, is $\$2$ per minute. Hence, the holding cost for a G-marble would be $\$2/2 = \$1$ per minute. Now assume that you are the worker in the finishing station with a total average of 19.475 marbles. Some of them are T-marbles and some are G-marbles. Both types of marbles have the same average process time of 1 minute. However, every minute you keep a T-marble in your station it costs you $\$0.75$, while every minute that you keep a G-marble in the station it costs you $\$1$. Which type of marble you process first in order to minimize the total average cost of holding marbles in your station?

Obviously, processing G-marbles that have a higher holding cost is the optimal strategy. Each time you process a G-marble you save $\$1$ in holding cost, while processing a T-marble saves you only $\$0.75$. This is exactly the rationale behind the c-mu rule. Note since $\mu_T = 1/T_{\text{eff}}^{(T)}$, then $c_T\mu_T = c_T/T_{\text{eff}}^{(T)} = \$3/4 = \$0.75$ is nothing but the average saving in holding cost for completing 1 minute of work on T-balls (i.e., completing one T-marble), and similarly, $c_G\mu_G = c_G/T_{\text{eff}}^{(G)} = \$2/2 = \$1$ is the average saving for completing 1 minute of work on G-balls (i.e., completing one G-marble).

Another application of the c-mu rule occurs when the goal is to minimize the total average inventory in a process (instead of minimizing the total average inventory cost). Recall that the c-mu rule minimizes $c_1 I^{(1)} + c_2 I^{(2)}$. If $c_1 = c_2 = 1$, then the c-mu rule minimizes $I^{(1)} + I^{(2)}$, which is the total average inventory. Thus, we can use the c-mu rule to obtain the optimal policy that minimizes the total average inventory. In this case $c_1\mu_1 = \mu_1$ and $c_2\mu_2 = \mu_2$. Hence, if $\mu_1 > \mu_2$, it is optimal to give priority to Class-1 products. Since $\mu_1 = 1/T_{\text{eff}}^{(1)}$, then $\mu_1 > \mu_2$ implies $T_{\text{eff}}^{(1)} < T_{\text{eff}}^{(2)}$. In other words, according to the c-mu rule, the optimal policy is to give priority to the class of products with shorter effective mean process time. This minimizes the total average inventory of products in the process. This special case of the c-mu rule is called *Shortest Expected Processing Time* or SPT Rule.

## PRINCIPLE

### The SPT Rule—Minimizing Total Average Inventory and Flow Time

Consider a multi-product process with $K$ classes of products (i.e., flow units). Class $k$ has an effective mean process time of $T_{\text{eff}}^{(k)}$. The *Shortest Expected Processing Time* or SPT rule that gives the first priority to the class of products with the smallest effective mean process time, the second priority to the class with the second smallest effective mean process time, and so on will result in the lowest overall average inventory in the process. According to Little's law, it will also result in the lowest overall average flow time in the process.

The rationale behind the SPT rule is also simple. Consider our finishing station and suppose that our goal is to minimize the total average number of balls in the station. Recall that the effective mean process times for T-balls and G-balls are 4 and 2 minutes, respectively. To reduce the overall number of balls in the station, what type of the balls would you process in the next 4 minutes? If you choose a T-ball, you would finish its processing in an average of 4 minutes and thus reduce the overall inventory of balls by one. However, if you decide to work on G-balls, you would be able to, on an average, process two G-balls and thus reduce the overall inventory in your station by two balls. Thus, obviously, working on units—balls—with shorter processing times first reduces your overall inventory at a faster rate, resulting in less overall average inventory in the process.

 Example 8.2

Recall Custom-Chrome Inc. in Example 8.1 that produces two types of wheels: (i) U-chrome with effective mean process time of 3 hours, and (ii) standard wheels with effective mean process time of 1.6 hours. Also, recall that it was estimated that each day that a U-chrome customer is waiting for their order it costs $10 in customer dissatisfaction and loss of future business. This number was $7 for standard orders. The company has decided to offer a third class of products called X-chrome wheels. X-chrome wheels is estimated to have an effective mean process time of 2 hours, and customer waiting cost of $9 per day.

- How should Custom-Chrome prioritize the production of these three products, if its goal is to minimize the total average customers' waiting cost?

> • If the goal is to minimize the overall average customer waiting time, how should the firm prioritize its production?

For U-chrome, we have $c_U = \$10$ per day, and $\mu_U = 8$ U-chrome per day; thus, $c_U \mu_U = \$10 \times 8 = \$80$. For X-chrome, we have $c_X = \$9$ per day, and $\mu_X = 12$ per day; thus, $c_X \mu_X = \$9 \times 12 = \$108$. Similarly, standard has $c_S = \$7$ per day, and $\mu_S = 15$ per day; thus, $c_S \mu_S = \$7 \times 15 = \$105$. Therefore, since X-chrome has the highest $c\mu$ of $\$108$, it should be given the first priority. Standard should be given the second priority since it has the second largest $c\mu$ of $\$105$. Finally, U-chrome should have the lowest (i.e., the third) priority. This will minimize the total average customer waiting cost, in other words, the lowest average inventory holding cost.

If the goal is to minimize the average customer flow time over all three classes of customers, considering the Little's law, this is equivalent to minimizing the overall average number of outstanding orders. Therefore, the SPT rule can be used to achieve this goal. Since

$$T_{\text{eff}}^{(S)} = 1.6 < T_{\text{eff}}^{(X)} = 2 < T_{\text{eff}}^{(U)} = 3$$

then the first, the second, and the third priorities should be given to standard, X-chrome, and U-chrome orders, respectively. This will minimize the total average number of customers waiting for their orders as well as minimize customer average waiting time over all three classes of orders.

In conclusion, one strategy that can be used to improve flow time is to prioritize flow units according to their effective mean process times. If the goal is to minimize the average flow time for a particular class of flow units, that class should be given the highest priority. If the goal, however, is to reduce the *overall* average flow time—the average over all classes of flow units—then flow units with smaller effective mean process times should be given priority over those with larger effective mean process times according to the SPT rule.

### 8.4.3 Implementing Priority Policy

One essential piece of information required for using SPT or c-mu rules in order to prioritize flow units is their effective mean process times. If the effective mean process times of flow units are not known, then SPT, or c-mu rule, or any policy that prioritizes units using effective mean process times cannot be implemented. Many business processes obtain this information upon arrival of a customer. For example, call centers of firms classify customers by using interactive voice response (IVR). The callers are asked to press numbers on their phones in response to a series of questions. These questions are designed to identify the class of customer and thus to give a better sense of the importance of the call and its effective mean process time.

Another important factor in choosing priority policy is its impact on customers' satisfaction. As we mentioned, FCFS policy is commonly accepted as a fair policy. Deviation from FCFS is shown to have a significant impact on customer dissatisfaction. There are several approaches that a firm can use to mitigate the negative impact of priority service.

**Make Priority Policy Seem Fair**

One strategy to implement priority policy and avoid customer dissatisfaction is to let the customers choose if they want to be of high or low priority. Why economy-class passengers never complain about business-class passengers given priority to board planes? Well, the economy-class passengers had the option of having higher priority—by buying a business-class ticket—but they chose not to do that. Then it seems fair that people who pay more for their tickets (i.e., business-class) get priority over those who pay less (i.e., economy-class). Financial institutions provide the option of private-client to their customers (if their account is above a certain level). This provides customers with discount and priority over regular customers (e.g., when they call the institution).

Finally, the Emergency Department of hospitals is another example of a process in which customers do not complain about the priority policy. In these processes, more critical patients are given higher priority over less critical patients. First, it seems fair to give priority to more critical patients, since that may save their lives. Second, while patients do not choose to be critical or non-critical, the fact that they will have higher priority when they are in a critical condition (if it happens) further implies the sense of fairness.

**Make the Queue Unobservable**

Another strategy to reduce or eliminate the negative impact of priority is to make the queue unobservable to customers. If the customers do not know their order in the queue, then they cannot determine whether customers are given priority, or they are served FCFS. How do firms make their queues unobservable? By using numbering systems and "Waiting Areas" instead of "Waiting Lines."

When customers are standing in a line, the order in the line is clear to everyone, and thus any deviation from FCFS is immediately noticed. If upon arrival, customers are given numbers or buzzers and are asked to wait in a "waiting area" and come back when their number is called or when their buzzers ring, then the order in the queue is less observable to customers. The reason is that customers spread out in the waiting area and sit in different seats. They do not worry about their orders in the queue, since they believe that the order in the queue is preserved by the numbering system. As long as numbers are called in order, for example, number 35 is called after number 34, then customers believe that the order is FCFS. This raises the following question: How can the firm implement priority policy, if every customer has a number corresponding to the order of the arrival?

The answer is to use letter/number systems. If 100 arriving customers are given numbers from 1 to 100, then if number 52 (i.e., a high-priority customer) is called before number 49 (a low-priority customer), it would not be acceptable to customers. Now, suppose that customer numbers start with a letter, for example, A21, B16, or C10 and high-priority customers get numbers starting with B and C, while low-priority customers' numbers start with A. While customers with A-numbers may complain if A21 is called before A20, they often do not complain if B21 is called before A20. Customers with numbers starting with letter A, for example, tend to compare their orders in the queue with all other numbers starting with letter A. The general impression is that numbers starting with letters other than A are probably requesting a different service or have different circumstances that have nothing to do with A-number customers. Hence, by using this numbering system the firm can implement its priority policy with less impact on customer satisfaction.

Some business processes already have unobservable queues. For example, suppose you have called an airline and they put you on hold. You and all customers who are put on hold (i.e., in the queue) do not know your positions and the order in the queue. Thus, it is easier for call centers to implement priority policies when serving customers.

Two important notes worth mentioning here:

- Under the letter numbering system, if the number of high-priority customers is large, then the firm should consider implementing partial priority scheme. Specifically, firm should occasionally call one or two low-priority customers even when there are high-priority customers waiting. This is easy using the numbering system mentioned above. For example, the firm calls two numbers starting with B and two numbers staring with C, and then two numbers starting with A. This results in serving two low-priority customer after serving four high-priority customers.

- If the waiting time is long, some curious customers may have enough time to learn about the service policy and observe that some customers who arrive after them (and are requesting the same service) are being called before them. This results in customer dissatisfaction, which can spread to other customers.[2] Thus, priority policy should be carefully implemented in highly utilized systems (i.e., systems with long waiting times), even when waiting areas and numbering system are used.

### Eliminate or Reduce Customer Physical Presence

Providing services online or through mail eliminates customers' physical presence in the queue and thus a firm can easily implement its priority policy which cannot be detected by customers. This is another example of having an unobservable queue. Other advantages of online services are saving customer travel time, reducing physical queues at the service center, saving labor time cost (e.g., eliminating customer interaction times with workers which often results in delay in processing), and improving worker scheduling (workers can work on the online applications when they are idle, or in the last 2 hours of their shifts, etc.).

## 8.5 Economies of Scale and Vertical Pooling

One interesting feature of queueing systems that can help managers reduce flow times is the phenomenon of economies of scale.

---

**CONCEPT**

**Economies of Scale**

Economies of scale refers to the cost advantages that a process obtains due to increase in size and scale of its operations. This translates into larger processes having smaller *cost per each unit* of their throughput.

---

Economies of scales exist in many different settings. For example, in production environments, producing in a larger batch size increases the effective capacity and reduces the setup cost associated with per unit produced. In logistics systems, ordering a larger quantity of items often results in lower shipment cost per ordered items. In marketing, cost of advertising per product becomes less for larger and wider range of outputs. In this section we first introduce this phenomenon in queueing environment and then describe how one can utilize this phenomenon in managing flow times and throughput of a process.

Consider the following two car wash facilities:

**SML Car Wash:**   SML car wash is a single-bay semiautomatic car wash that has a demand (arrival) rate of 16 cars per hour. The effective mean process time to wash a car is 3 minutes. The car interarrival times and car washing times follow exponential distributions.

**BG1 Car Wash:**   BG1 car wash, similar to SML car wash, has a single bay for washing cars. However, the bay is fully automated and can process a car in 1.5 minutes—half of that in SML car wash. On the other hand, the demand arrival rate is 32 cars per hour—twice of that in SML car wash.

Note that, while the effective capacity of BG1 (i.e., 40 cars per hour) is twice that of SML car wash (i.e., 20 cars per hour), it also serves a demand that is twice as large. Hence, both

---

[2]Unsatisfied customers often start talking to other customers and express their frustration.

SML and BG1 have same utilization of $u = R_{in}(process)/C_{eff}(process) = 16/20 = 32/40 = 0.80$—both facilities are idle 20 percent of the time. Which car wash is expected to have a lower average number of cars waiting in line (i.e., lower in-buffer inventory) and a lower average waiting time in line (i.e., lower in-buffer flow time)?

To answer this question, note that the underlying queueing model of SML car wash is an $M/M/1$ queue with $R_{in}(process) = 16$ and $C_{eff}(process) = 20$ units per hour. Thus, using Eq. (7.2) and Little's law we find the average queue length as follows:

$$I_b = T_b \times TH = \frac{u^2}{1-u} = \frac{(0.8)^2}{1-0.8} = 3.2 \text{ cars}$$

Since BG1 facility is also an $M/M/1$ queueing model with the same utilization $u = 0.80$, it will also have the same average queue length of 3.2 cars. Thus, the larger facility—BG1 facility—has the same average in-buffer inventory as that of the smaller process—SML facility. But, what about flow times?

In SML car wash with $TH = R_{in}(process) = 16$ units per hour, using Little's law, the average waiting time in queue is

$$T_b = I_b/TH = 3.2/16 = 0.2 \text{ hours} = 12 \text{ minutes}$$

For BG1 car wash, however, the average waiting time in queue is

$$T_b = I_b/TH = 3.2/32 = 0.1 \text{ hours} = 6 \text{ minutes}$$

Therefore, waiting times of cars in the queue in BG1—the larger facility—is about half of that in SML—the smaller facility. Why? The answer is clear. On an average, both facilities have the same queue length of 3.2 cars. However, since BG1 car wash has a higher effective capacity, it serves cars at a faster rate—twice faster to be exact. For example, if you are third in the queue of BG1 car wash, your average waiting time in queue is about half of that of someone who is third in the queue in SML car wash.

**BG2 Car Wash:**    BG2 car wash serves a demand that is twice as large as that of SML, that is, 32 cars per hour. The car wash uses a similar semiautomated bay as in SML, with effective mean process times of 3 minutes. However, to handle the demand, the car wash has two of these bays that can work independently on two different cars.

Thus, BG2 car wash has the same demand and the effective capacity as BG1 car wash. The only difference is that BG2 car wash has two (slower) bays, each can process 20 cars per hour, while BG1 has a single (twice faster) bay that can process 40 cars per hour. Does BG2 car wash also have smaller average flow time compared to SML car wash?

The underlying queueing model for BG2 is an $M/M/2$ with $R_{in}(process) = 32$ and $C_{in}(process) = 20$. Similar to BG1 and SML, the utilization of BG2 car wash is also $u = 80\%$. Using Eq. (7.2) we can compute the average waiting time in queue of BG2 to be $T_b = 5.43$ minutes, which is about half of that in SML car wash. The intuition is the same: Both larger systems (BG1 and BG2) have about the same average queue length as that of the smaller system (SML), but they can process units (cars) at a rate twice faster than that in the smaller system. This example shows that the economies of scale with respect to flow times holds in processes with larger capacity, regardless of whether the larger capacity is due to having faster servers or more servers.

Having economies of scale also improves flow time service levels. For example, if an acceptable waiting time in the queue is set to $t = 8$ minutes, then, using Eq. (7.11), we can compute the in-buffer flow time service levels at the three car washes as follows:

$$SML \text{ Car Wash} : SL_b^{\text{ftw}}(8) = 0.531$$
$$BG1 \text{ Car Wash} : SL_b^{\text{ftw}}(8) = 0.725$$
$$BG2 \text{ Car Wash} : SL_b^{\text{ftw}}(8) = 0.755$$

Hence, in SML car wash only 53.1 percent of customers wait in the queue for 8 minutes or less before their service starts. Due to economies of scale, however, BG1 and BG2 offer a much higher in-buffer flow time service level. Specifically, 72.5 and 75.5 percent of customers of BG1 and BG2, respectively, wait in the queue for 8 minutes or less.

Large processes benefit from their economies of scales in many ways, including the following two: (i) offering smaller flow times, and (ii) reducing capacity costs.

### Economies of Scale—Offering Smaller Flow Times

As shown in the above examples, for the same utilization level, larger processes have smaller flow times. This results in shorter queue (in-buffer inventory) and waiting times (in-buffer flow time). As discussed in Chaps. 2 and 7, lower inventory has many cost advantages, and lower flow times result in higher customer satisfaction, especially in service systems. Having more satisfied customers due to better service level leads to more demand for the firm, and thus more sales.

### Economies of Scale—Saving Capacity Costs

Large processes can provide the same service level as that of smaller processes, but with a lower capacity cost per throughput. For example, consider BG3 car wash as follows:

**BG3 Car Wash:**   BG3 car wash, similar to SML car wash, has a single bay for washing cars. However, the bay has an effective capacity of processing 36.4 cars per hour. Similar to BG1 and BG2, demand for BG3 is 32 cars per hour—twice that of SML car wash.

Using Eq. (7.2), with $u = 32/36.4 = 0.879$, we can compute the average waiting time in the queue at BG3 to be about 12 minutes—same as that in SML car wash. Considering the fact that BG3 is serving a demand that is twice as large as that of SML, then the question is what happened to the economies of scale advantage?

In the case of BG3, the economies of scale advantage manifests itself in BG3 providing the same average waiting time as that of SML but at a higher utilization ($u = 0.879$) than that of SML ($u = 0.8$). This higher utilized process results in less capacity costs associated with serving a customer. For example, let's assume that the costs associated with the capacity of serving 20 cars per hour (e.g., equipment depreciation cost, labor cost, utility cost, etc.) at SML translates to an hourly cost of $40 per hour. Since SML's throughput is 16 cars per hour, then the capacity cost per served car is $40/16 = $2.5. Now consider BG3 that has $36.4/20 = 1.82$ times larger capacity than SML. Let's assume that BG3's capacity cost is also 1.82 times larger than that of SML.[3] Thus, BG3's hourly capacity cost would be $40 \times 1.82 = $72.8 per hour. Considering the throughput of 40 customers per hour, the capacity cost associated with one served customer at BG3 is $72.8/32 = $2.27. Hence, while both SML and BG3 offer the same average waiting time in the queue, all other costs being the same, it costs BG3 about $2.5 - 2.27 = $0.22 less to serve each customer.

---

[3] This is an overestimation of the capacity cost. There is also economies of scale in capacity investment. For example, a machine that is twice faster usually costs less than twice of the cost of a machine that has half of the speed of the faster machine.

We summarize our findings in the following principle:

**Economies of Scale in Flow Time**

- For the same level of utilization, larger processes that serve larger demands have shorter average flow times and better service levels than those of smaller processes.

- Compared to smaller processes, larger processes serving larger demand can provide the same average flow times and service levels with higher utilization and thus less capacity costs per each unit of throughput.

### 8.5.1 Vertical Pooling

How can firms take advantage of economies of scale and reduce their flow time measures? Well, the answer is to make their processes larger. One way is to redesign the process by implementing *Vertical Pooling*. As we described in Sec. 6.5.3 of Chap. 6, vertical pooling pools different resources that process different demands into one pool to process the combined demand of the resources. This creates one larger process that has larger capacity and serves larger demand. Due to economies of scale, it is expected that the flow times in the pooled process are smaller than that in each individual process. We describe this in the following example.

 **Example 8.3**

AdaptOne is a small company that, among other electrical accessories, produces 110-volt and 220-volt adapters for personal computers for North America and Asian countries, respectively. Packaging station of the adapters has two workers. Adapters of type 1 and type 2 arrive independent of each other from two different assembly lines to their corresponding buffers at the packaging station, each with a rate of 15 adapters per hour. The coefficient of variation of interarrival times is 1. Adapters are kept in two different buffers in the packaging station, since each type requires a different package and accessories. Worker 1 is in charge of quality control check and packaging of 110-volt adapters and Worker 2 does the same for 220-volt adapters. The effective mean process time for quality control and packaging of a 110-volt adapter is about 3 minutes. This time is 3 minutes and 20 seconds for 220-volt adapters. The coefficient of variation of both process times is 1. Production managers have decided to redesign the process as follows: (i) train both workers to do packaging of both types of adapters, (ii) combine the two buffers into one so all arriving adapters will be kept in one buffer. Under this new design, upon completion of processing current adapter, the worker picks the next adapter (110- or 220-volt adapter) from the combined buffer and processes it. It is easy to distinguish 110-volt from 220-volt adapters, since they have different color cords and plugs. How much this new setting would reduce the average in-buffer inventory and flow time in the packaging station?

Figure 8.3 depicts the current underlying queueing models of the packaging station as well as that under the new design when workers and the buffer (i.e., queue) are pooled.

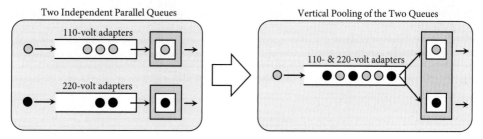

Figure 8.3    Vertical pooling of two workers in the packaging station of AdaptOne in Example 8.3.

**Current Operations:** As Fig. 8.3 shows, the current process consists of two $M/M/1$ queueing systems[4]: (i) the system for 110-volt adapters that has input rate of 15 adapters per hour and the effective capacity of 20 adapters per hour, and (ii) the system for 220-volt adapters that has an input rate of 15 adapters per hour, but the effective capacity of 18 adapters per hour. Using Eq. (7.2) and Little's law, we find the average in-buffer inventory and average in-buffer flow time for 110-volt adapters at the packaging station to be $I_b(110) = 2.25$ adapters and $T_b(110) = 9$ minutes. On the other hand, for 220-volt adapters we have $I_b(220) = 4.17$ adapters and $T_b(220) = 16.67$ minutes. Hence, under current operations, on an average, there is a total $I_b = 2.25 + 4.17 = 6.42$ adapters in the packaging station. And since the TH of the packaging station is 30 adapters per hour, using Little's law, the average in-buffer flow time of an adapter at the packaging station is $T_b = I_b/\text{TH} = 6.42/30 = 0.214$ hours $\times 60 = 12.84$ minutes.

**New Design (The Vertically Pooled System):** Under the new design, however, the underlying queueing model is a $G/G/2$ queueing system—an $M/G/2$ to be exact—with two different classes of customers (i.e., two types of adapters). Since our analytical model for $G/G/s$ queueing model is for a single class of customers all having the same effective process times, we aggregate the two products into one product that we simply call "adapter." The input rate for the adapter is therefore $R_{\text{in}}^{(agg)} = 15 + 15 = 30$ per hour according to a Poisson process.[5] Thus, the coefficient of variation of adapters interarrival times is $\text{CV}_{\text{arc}}^{(agg)} = 1$.

We now need to find the effective mean process time of an adapter—the aggregate product—and its coefficient of variation. Since half of arriving adapters are 110-volt adapters and the other half are 220-volt adapters, then half of the adapters that each worker processes would be 110-volt and the other half would be 220-volt adapters. Thus, the effective mean process time of an adapter would be

$$T_{\text{eff}}^{(agg)}(\text{adapter}) = 0.5 \times (3 \text{ minutes}) + 0.5 \times (3.33 \text{ minutes}) = 3.167 \text{ minutes}$$

To obtain the coefficient of variation of the effective process time of an adapter, we need to find the standard deviation of the effective process time, which is the square root of its variance. In probability and statistics, it is shown that if $X_1$ and $X_2$ are independent exponential random variables with average $1/\mu_1$ and $1/\mu_2$, respectively, and if $Y = p_1 X_1 + p_2 X_2$, where $p_1$ and $p_2$ are probabilities such that $p_1 + p_2 = 1$, then random variable $Y$ has variance:

$$\text{Var}(Y) = 2\left(\frac{p_1}{\mu_1^2} + \frac{p_2}{\mu_2^2}\right) - \left(\frac{p_1}{\mu_1} + \frac{p_2}{\mu_2}\right)^2$$

---

[4]Since interarrival and processing times have coefficient of variation 1, both interarrival and process times follow exponential distribution.

[5]Adding two independent Poisson arrival streams results in an arrival stream that is also Poisson.

Since effective processing time of 110- and 220-volt adapters are exponentially distributed with mean $1/\mu_1 = 3$ minutes and $1/\mu_2 = 3.33$ minutes, respectively, and since half of the adapters are 110-volt ($p_1 = 0.5$) and the other half are 220-volt adapters ($p_2 = 0.5$), we can find the variance of the effective process time of the aggregate product—the adapter—using the above equation as follows:

$$\text{Var}(Y) = 2\left(\frac{p_1}{\mu_1^2} + \frac{p_2}{\mu_2^2}\right) - \left(\frac{p_1}{\mu_1} + \frac{p_2}{\mu_2}\right)^2$$

$$= 2\left(0.5 \times (3)^2 + 0.5 \times (3.33)^2\right) - (0.5 \times 3 + 0.5 \times 3.33)^2 = 10.083$$

Thus, the coefficient of variation of effective process time of an adapter is $\text{CV}_{\text{eff}}^{(agg)} = \sqrt{10.083}/3.167 = 1.003$. Having the input $R_{\text{in}} = 30$ per hour and interarrival variability $\text{CV}_{\text{arv}}^{(agg)} = 1$ as well as effective mean process time $T_{\text{eff}}^{(agg)} = 3.167$ minutes and variability of $\text{CV}_{\text{eff}}^{(agg)} = 1.003$, we can use Eq. (7.2) for the $G/G/2$ queueing model of the new design. We find the average in-buffer flow time and the average in-buffer inventory for the new design as $T_b^{(agg)} = 5.43$ minutes and $I_b^{(agg)} = 2.72$ adapters, respectively.

**Current Operations versus New Design:** As these numbers show, by implementing the new design—implementing vertical pooling—the average in-buffer inventory will be reduced from a total of 6.42 adapters to 2.72 adapters. Also, the average in-buffer flow time for an adapter would be reduced from 12.84 minutes to 5.43 minutes. This is around a 60-percent improvement (i.e., decrease) in both in-buffer inventory and in-buffer flow times, which is very significant.

Note that, under the new design, the packaging station faces the same demand of packaging a total of 30 adapters per hour as in current operations. Also, the time to package one adapter is the same 3 minutes for 110-volt and 3.33 minutes for 220-volt adapters as in current operations. Then, where does the significant improvement of 60 percent come from? The improvement is the result of the following three factors:

- *Higher Effective Capacity:* As we described in Eq. (6.6) in Chap. 6, vertical pooling can increase the effective capacity. In this case, when we pool $n = 2$ workers with $p_o = 0.5$, $T_o = 3$, $p_b = 0.5$, and $T_b = 3.33$, the percent increase in capacity due to vertical pooling is as follows:

$$\text{Percent Increase in Capacity} = \frac{2}{1 + p_o T_o/p_b T_b} - 1$$

$$= \frac{2}{1 + [0.5(3)]/[0.5(3.33)]} - 1 = 0.052$$

In other words, implementing the new design will increase the effective capacity of the packaging station by about 5 percent. While not that significant, it will still have some impact in reducing inventory and flow times. The increase in the effective capacity would have been higher if the difference between the effective mean process times of 110-volt and 220-volt adapters was larger.

- *Economies of Scale:* The new design is a larger process that handles a demand (30 adapters per hour) twice as large as that in each individual queue in current operations. It also has a capacity that is about twice as large. Economies of Scale, as mentioned before, further decreases inventory and flow time.

- *More Efficient Use of Capacity:* Note that in the current operations, due to the variability in input and processing time, there are times that one of the workers, Worker 1 for example, is idle since there are no 110-volt adapters in the worker's station, while there are 220-volt adapters waiting in the buffer of Worker 2. In this case, current operations does not allow Worker 1 to package 220-volt adapters. Thus, in these situations, the capacity of Worker 1 will not be utilized to help reduce the overall inventory in the packaging station. This is also the case when Worker 2 becomes idle and there is an inventory of 110-volt adapters in the packaging station. This, however, does not occur in the vertically pooled process of the new design. Under the new design a worker is only idle if there is no adapter (100- or 220-volt) or there is only one adapter (being processed by the other worker). Thus, the capacity of the workers will not be wasted.

This example illustrates the significant impact of vertical pooling strategy in improving flow time measures such as average flow times and flow time service levels.[6] Firms take advantage of economies of scale by pooling their resources. For example, McDonald's has established call centers to receive orders from drive-thru customers in different states. It has a call center in California with about 35 workers who get orders remotely from 40 McDonald's restaurants in different states in the United States. After getting the order from a customer, the orders are sent to the restaurant via Internet.[7] Next time you order a sandwich from McDonald's drive-thru, you might be talking to a person hundreds of miles away.

### 8.5.2  To Pool or Not to Pool?

If vertical pooling of parallel queues has such a significant impact on improving flow time measures of a process, why do we still see separate parallel queues in many businesses, for example, separate lines for cashiers at retailers, or separate lines for security checks at the airports? There are several reasons for that, some of which are discussed below.

**1. Jockeying:** One fact is that some of the parallel unpooled queues we observe in practice are indeed pooled queues. Why? Because customers are allowed to move to another queue while waiting in one queue. Recall that one major inefficiency of having separate queues is that it prevents efficient use of server capacity, that is, there would be situations when one server is idle while there are customers waiting in the queues of other servers. Parallel queues in which customers in a queue are allowed to move to another queue—a phenomenon called *Jockeying*—does not result in a server being idle when there are still customers waiting in the system. When we are waiting in the queue of a cashier in a supermarket and we see the queue of another cashier is empty, we all move to that cashier. One main advantage of having a single pooled queue versus several parallel queues with jockeying is the sense of fairness. In a single pooled queue customers are served in FCFS order, while in separate parallel queues this is not guaranteed.[8]

**2. Physical Constraints:** Often times physical constraints do not allow vertical pooling of several resources. While it is easy to pool two call centers into one—as for McDonald's order-taking process for its drive-thru—it may not be economical for a car wash to pool two small car wash facilities located in the north and the south part of a town into one facility in the center of town. The firm may lose some of its customers who live in the north or the south due to long travel time to the center of town.

**3. Queue Looks Longer:** Suppose you would like to buy ice cream from two stores on the opposite side of a street. Both stores have three cashiers, but one has a single line with seven customers, and the other has three lines with three customers waiting in each line. Assuming that the quality and prices of both stores are comparable, which one do you go to? Most of us join the store with shorter individual line, because the line seems shorter there, implying that

---

[6]Note that smaller average flow times, in almost all cases, correspond to improvement in service levels.
[7]"The long-distance journey of a fast-food order," *The New York Times*, April 11, 2006.
[8]We further discuss fairness in later sections when we discuss psychology of waiting time.

the waiting time would be shorter. This is an issue in some vertically pooled processes, since their queues look longer. Even though the single queue in the vertically pooled processes moves faster (which customers realize *after* joining the queue), the first impression of the long queue may prevent customers from entering the process, resulting in lost sales.

**4. Cost of Pooling:** To implement vertical pooling of two (or more) resources (workers or machines), all resources must be made flexible to be able to do the two activities. This may not be feasible or economical. For example, while it might be easy to train the workers in the packaging station in Example 8.3 to perform packaging tasks of two types of adapters, it is not feasible or economical to train a worker doing assembly to also do welding; or it is costly to replace a simple drilling machine with a milling machine that can do drilling, cutting, and other machining tasks.

**5. Setups:** If a setup is required when a resource switches from performing one activity to another, then vertical pooling may not improve performance. The reason is that setup will reduce the effective capacity of the pooled resources. Setup was not an issue in the packaging station in Example 8.3, since no setup time was needed for the two workers to switch from packaging one type of adapter to the other.

**6. Benefit of Specialization:** Specialization and division of labor are known to have several advantages. First, having a worker to perform only one task improves the worker's learning, and the worker becomes more efficient in performing the task. This results in higher effective capacity and lower variability in effective process time. Both will result in lower average flow times. Second, workers who are specialized in one activity provide better quality outputs than those who perform multiple activities (e.g., in pooled processes). Depending on the complexity of the pooled activities, one disadvantage of pooling is that it does not utilize the above benefits of specialization.

**7. Priority versus FCFS:** As we mentioned earlier, allowing customers to jockey among parallel queues of cashiers in supermarket is equivalent to having a single pooled queue. In some supermarkets, however, some cashiers are assigned to serve only a particular class of customers—customers who have 10 items or less. This prevents other customers to go to that server and unpools the system into two queues. But, does this diminish the impact of pooling? Yes, but it provides other advantages. To illustrate this, note that customers with 10 items or less are those who have a shorter process time. Assigning a certain number of servers (cashiers) to the class of customers with shorter expected process times—while depools the process—acts like giving priority to customers with shortest processing time. Serving customers with shorter process times first (the SPT rule), as we mentioned in Sec. 8.4.2, helps reduce the overall in-buffer inventory and flow times. While this is not a complete priority discipline—customers with 10 items or less have priority over others in only one queue among several—this partial priority can improve overall inventory and flow time measures. The improvement becomes more significant as the class of customers with shorter process time becomes larger (assuming that the right number of servers are assigned to them). Other examples of assigning different servers to customers with shorter process time are assigning different call center agents to callers that require less service times, or assigning different lines for security check for families versus business travelers (who have lower service time).

**8. Variability Reduction:** Consider a similar packaging station as in Example 8.3 in which two workers work on packaging two different products independently. Type-1 products arrive according to a Poisson process with a rate of 18 per hour. It takes Worker 1 exactly 3 minutes to package type-1 products, that is, no variability in process times. Type-2 products, on the other hand, arrive at the rate of 2 per hour—also according to a Poisson process—and have effective process time of exactly 25 minutes, also with no variability.

Before these two workers are pooled, the underlying queueing model of this packaging station consists of two independent $M/D/1$ queues. Using Eq. (7.2) and Little's law, for the first worker, we find the average in-buffer flow time for type-1 product is 13.5 minutes. This number is 62.5 minutes for type-2 products. Since $90\%(= 18/(18 + 2))$ of products entering the packaging station are type 1 and the remaining are type 2, the overall average in-buffer flow time for a product at the packaging station would be $0.9(13.5) + 0.1(62.5) = 18.4$ minutes.

Now consider the pooled system, which would be an $M/G/2$ queueing system. Specifically, the arrival process to the pooled system would be a Poisson distribution with a rate of $18 + 2 = 20$ products per hour. On the other hand, the product mix is $(0.9, 0.1)$. Specifically, 90 percent of products processed by each worker are type 1 and 10 percent are type 2. Consequently, the effective mean process time of a product (the aggregate product) would be $0.9(3) + 0.1(25) = 5.2$ minutes. The variance of the effective process time of a product would be $0.9(3 - 5.2)^2 + 0.1(25 - 5.2)^2 = 43.56$. The variability of the effective process time is therefore $\mathrm{CV}_{\mathrm{eff}}^{(agg)} = \sqrt{43.56}/5.2 = 1.27$. Using Eq. (7.2) and Little's law we find the average in-buffer flow time of a job in the pooled system to be 20.69 minutes, which is 2.29 minutes more than that before pooling.

The reason for lower performance of the pooled system is that the pooled system has more variability. Before pooling, there was no variability in effective process times of each server; however, after pooling the effective process time has variability of 1.27. The negative impact of this increase in variability on flow times is larger than the positive impacts of pooling (e.g., increase in the effective capacity). Hence, when there is a significant difference between the effective process times of queues, one should caution about pooling the queues, since pooling queues may degrade the flow time performance measures of the process.

**9. Customer Preference:** In pooled systems, a flow unit may be processed by either one of the pooled resources. In service operations systems in which customers use the system periodically, customers often prefer to be served by one specific server. For example, in hair salons or tax services, customers often prefer to receive service from one particular server. Pooling the queues in these systems will not allow this customer preference.

## 8.6 Horizontal Pooling and Flow Times

As described in Chap. 6, horizontal pooling corresponds to the pooling of $n$ resources—each performing one of the $n$ consecutive activities on each flow unit—into a pool of $n$ resources, each performing all $n$ activities on a flow unit independent of others. We have also shown in that chapter that horizontal pooling can increase the capacity of a process. In the following example we show that horizontal pooling also improves flow times.

 **Example 8.4**

Easy Mortgage is a financial institution that offers loans and mortgages for home and car purchases. To evaluate an application for home mortgage, Easy Mortgage needs two documents, namely property survey and property appraisal. Property survey and property appraisal are prepared by Tom and Frank, respectively. Specifically, first the property survey for each application is prepared by Tom, which takes an average of 90 minutes with standard deviation of 90 minutes. Then Frank prepares property appraisal, which takes an average of 50 minutes with standard deviation of 50 minutes, independent of the time needed for property survey. One action to improve the throughput is to implement horizontal pooling and ask Tom to do both the survey and appraisal, and Frank to do the same independent of Tom (see Fig. 8.4). If Tom receives an application, on an average, every 100 minutes according to a Poisson process, how long an application waits in Easy Mortgage for survey and appraisal before and after horizontal pooling?

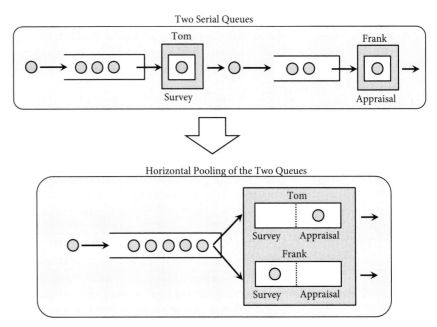

Figure 8.4    Horizontal pooling of Tom and Frank in Easy Mortgage in Example 8.4.

**Before Horizontal Pooling:** Before horizontal pooling, as shown in Fig. 8.4, the process of survey and appraisal is a two-stage process consisting of two queues in series. The underlying queueing model of survey is a $G/G/1$ model—an $M/M/1$ to be exact—with $R_{in} = 0.6$ applications per hour, $CV_{arv} = 1$, $T_{eff} = 90/60 = 1.5$ hours, and $CV_{eff} = 90/90 = 1$. Thus, survey process has utilization $u = R_{in} \times T_{eff} = 0.6 \times 1.5 = 0.9$. Using Eq. (7.2) and Little's law, the average in-buffer inventory and in-buffer flow times in the survey process are obtained as $I_b = 8.1$ and $T_b = 810$ minutes, respectively.

The underlying queueing model of appraisal, performed by Frank, is also a $G/G/1$ model—a $G/M/1$ to be exact—with effective mean process time of $T_{eff} = 50$ minutes with $CV_{eff} = 50/50 = 1$. The input to this queue is the output of the survey process. Using Eq. (5.11), we can find the variability of the output of the survey process as follows:

$$CV_{out}^2 = 1 + (1 - u^2)(CV_{arv}^2 - 1) + \frac{u^2}{\sqrt{s}}(CV_{eff}^2 - 1)$$

$$= 1 + (1 - (0.9)^2)((1)^2 - 1) + \frac{(0.9)^2}{\sqrt{1}}((1)^2 - 1) = 1$$

Thus, the input of the appraisal queue has a rate of $R_{in} = 0.6$ per hour with squared coefficient of variation $CV_{arv}^2 = 1$. Since utilization of the appraisal process is $u = 0.5$, using Eq. (7.2), we find the average in-buffer inventory and in-buffer flow times as $I_b = 0.5$ and $T_b = 50$ minutes, respectively.

Hence, overall, there are a total of $8.1 + 0.5 = 8.6$ applications in the queues of survey and appraisal. Furthermore, each application waits a total average of $810 + 50 = 860$ minutes in the queues of survey and appraisal.

**After Horizontal Pooling:** After horizontal pooling, the survey and appraisal process is a single-stage process with two resources. The underlying queueing model of the process is a $G/G/2$ queue—an $M/G/2$ to be exact. The queue has an input with rate $R_{in} = 0.6$ applications per hour with variability $CV_{arv} = 1$. The effective mean process time of a resource

(Tom or Frank) is $T_{\text{eff}} = 90 + 50 = 140$ minutes with standard deviation $\sqrt{(90)^2 + (50)^2} = 102.95$. Hence, in the horizontally pooled system, the effective process time has variability $\text{CV}_{\text{eff}} = 102.95/140 = 0.735$. Using Eq. (7.2), we can compute the average in-buffer inventory and the average in-buffer flow time for an application, respectively, as $I_b = 1.07$ applications and $T_b = 107.2$ minutes.

**Impact of Horizontal Pooling:** Thus, horizontal pooling has reduced the average number of applications waiting for survey or appraisal from about 8.6 to 1.07, which is about 87 percent improvement. Furthermore, the total time an application waits for survey and appraisal is reduced from 860 to 107.2 minutes—an 87-percent improvement. This example shows that, similar to vertical pooling, horizontal pooling can also significantly improve flow times. The reasons are as follows:

- *Higher Effective Capacity:* As we described in Eq. (6.4) of Chap. 6, horizontal pooling can increase the effective capacity of a process. In Easy Mortgage in Example 8.4, pooling $n = 2$ resources—Tom and Frank—with $T_o = 50$ and $T_b = 90$, the percent increase in capacity due to horizontal pooling is as follows:

$$\text{Percent Increase in Capacity} = \frac{2}{1 + T_o/T_b} - 1 = \frac{2}{1 + 50/90} - 1 = 0.29$$

  In other words, after horizontal pooling, the effective capacity increases by about 29 percent. This increase in capacity results in reduction in inventory and flow time.

- *More Efficient Use of Capacity:* Notice that, before horizontal pooling, the appraisal process—the second stage— has a utilization of 0.5, which implies that Frank is idle 50 percent of the time with no applications to process. On the other hand, the utilization of the survey process is $u = 0.9$, implying that Tom is working 90 percent of the time. Therefore, there will be many occasions that Frank is idle, but there is a queue of applications waiting for Tom to process them. In these situations, while Frank has an idle capacity, he cannot do survey reports. This, however, does not occur after horizontal pooling. Horizontal pooling better utilizes Frank's (and Tom's) capacity by allowing them to do both survey and appraisal. Thus, similar to vertical pooling, Tom or Frank will be idle only when there is no application in the queue.

- *Lower Variability in Pooled Activities:* Note that before the horizontal pooling, the variability of the tasks performed by Tom and Frank was $\text{CV}_{\text{eff}} = 1$. However, after horizontal pooling, the variability of the task performed by Tom and Frank was reduced to $\text{CV}_{\text{eff}} = 0.735$. Hence, another reason for horizontally pooled systems to have a lower flow time is that they have lower variability in the pooled tasks.[9]

Similar to vertical pooling, the advantage of horizontal pooling may not be as significant in some environments. This is one reason that, in practice, we see many queues in series. For example, consider a small gourmet sandwich restaurant with two workers that offers a wide range of sandwiches and gourmet dishes. Worker 1 works at cash register and gets customers' orders and payments. The worker then gives a buzzer to the customer; the buzzer will buzz when the customer's sandwich is ready. This ordering process takes an average of 1 minute. Customers then take the buzzers and sit on a table in the restaurant waiting for their orders. Orders appear on a monitor in the kitchen where Worker 2—a chef specialized in gourmet sandwiches—makes the sandwiches and activates customer's buzzer when the order is completed. The process of making a sandwich takes an average of 5 minutes.

---

[9]When tasks are horizontally pooled, the CV of the pooled task is usually lower than the CV of each individual task, unless there is a large difference between the CV of individual tasks.

The above restaurant is a process consisting of two consecutive stages—taking orders and making sandwiches. The bottleneck resource (i.e., the chef) has an effective mean process time that is five times as large as that of Worker 1 (i.e., the cashier). Thus, $T_o/T_b = 0.2$. If we implement horizontal pooling, the effective capacity of the process will be improved by $[2/(1 + 0.2)] - 1 = 0.667 = 66.7\%$, which can decrease the average flow time significantly. With such improvement, then why we see many restaurants and Cafes that operate similar to our gourmet sandwich restaurant not implement horizontal pooling. Some reasons are as follows:

**1. Cost of Pooling:** Similar to vertical pooling, horizontal pooling is not feasible or economical in some processes. For example, while it might be easy to train the chef how to process an order in the cash register, it is certainly not easy to train the cashier to become a chef, and the training cost is high. Also, a pooled process incurs other costs such as the cost of one more cash register, and additional space in the kitchen so that two workers can work there.

**2. Better Utilization of More Expensive Resource:** If it is not feasible to train the cashier to do the chef's jobs, it is certainly easy to train the chef to do the cashier's job. Should the restaurant do that? While this is not a complete horizontal pooling, it still pools the capacity of the chef (the second stage) to help the cashier (the first stage), but not vice versa. There are two issues with this approach. First, the chef is the bottleneck of the process, so it is better to reduce the chef's workload rather than adding to it. Second, it does not make economical sense to use a more expensive resource (i.e., the chef) to perform an activity that can be done by a much less expensive resource (i.e., the cashier). This is one of the reasons we do not see horizontal pooling in a doctor's office. The simple task of taking a patient's blood pressure and temperature and entering those in the computer is done by a nurse—a much cheaper resource—not by the doctor.

**3. Setups:** Similar to vertical pooling, setups play critical roles in horizontal pooling decisions. For example, let's assume that sandwiches and other meals that are offered by our restaurant are so simple that the cashier can be trained to also do the work at the second station. Hence, we can implement the horizontal pooling, where both the chef and the cashier take the order and then go to the kitchen and make the sandwich. This implies that when a worker (chef or cashier) finishes with taking the order of a customer, the worker needs to walk to the kitchen and wash hands before starting to make a sandwich. When the sandwich is ready, the worker needs to walk back to the cash register and take the order of the next customer. The walking times (back and forth) and hand washing times are setup times that increase the effective process time of the customers, and thus reduce the effective capacity in the pooled system. This, however, is not the case in pooling Tom and Frank in Easy Mortgage in Example 8.4. There is not a significant setup time that is required to work on property appraisal after finishing property survey. On the contrary, the time for appraisal report may take shorter if an agent (Tom or Frank) is doing the appraisal after finishing the survey report of an application. This is because, after doing the survey report, the application file is already uploaded in the computer and the agent has already studied the application (when doing survey).

**4. Benefit of Specialization:** Similar to the case of vertical pooling, separate processes in series may be preferred over a horizontally pooled process due to the benefit of specialization (e.g., better quality, lower process time, lower variability).

**5. Queues Look Shorter:** In multi-stage processes in series, when the first stage has shorter process time than the second stage, the queue in the process seems shorter to the customers. One reason is that the queue in the first stage is small (due to smaller process time), and the queue in the second stage—the longer queue—is often not visible. Consider our gourmet sandwich restaurant in which the effective mean process time of the cashier is five times smaller than the effective capacity of the chef. Thus, the cashier's utilization is much smaller than the chef's utilization—one-fifth to be exact. This in turn implies that the queue for the cashier is much shorter. Customers who enter the restaurant first see the cashier's line, which is very short. They do not see the second queue (which might be long), since customers in that queue are spread out—sitting on the tables—awaiting their food.

The second reason is that while waiting in the cashier's line is perceived by customers as waiting for service, waiting for their food (waiting in the second queue) is perceived as part of service, since it gives a feeling that their service has already started. Holding the buzzer in hand gives the customer the feeling that they are not forgotten and their service is in progress. These are all psychological factors that make customer's perception of their waiting times being shorter than what they actually are. We discuss these psychological factors in Sec. 8.11.

**7. Priority in the Second Stage:** In some processes, unpooled queues can implement priority discipline while horizontally pooled processes cannot. Recall from Sec. 8.4 that, by giving priority to customers with shorter effective mean process times, the overall average in-buffer inventory and flow times of the process decrease. One issue with priority discipline is that it is considered unfair by customers and can result in customer dissatisfaction. However, comparing to horizontally pooled queues, implementing priority policy is easier in the unpooled queues in which the queue in the second stage is not observable to the customers. For example, the chef in our restaurant example can prioritize orders which takes less time to prepare. This helps to decrease the number of customers waiting for their orders in the second queue faster.

## 8.7 Reducing Buffer Size to Block Arrivals

Reducing buffer size, as we discussed in Sec. 7.6.1 of Chap. 7, reduces the average flow time. Hence, one way to improve flow time measures is to reduce the buffer size. On the other hand, for a given demand, reducing the size of the buffer results in more blocking and thus decreases the throughput of the process, which in turn can result in lost sales in both short and long terms. Therefore, the decision to reduce the buffer size in order to reduce flow times must be made very carefully by considering its impact on the throughput of the process.

One factor that affects the decision of reducing the buffer size is the size of the system. To illustrate this, consider a small call center, that is, Call Center A.

**Call Center A:**    Call Center A has one agent. Calls arrive according to a Poisson process with a rate of eight calls per hour. The agent has the effective process time that is exponentially distributed with mean of 6 minutes to respond to a call. The call center has ample trunk lines, so no caller is blocked (i.e., no caller gets a busy signal).

Using the underlying $M/M/1$ queueing model for the call center, we find the average waiting time on hold (i.e., in-buffer flow time) to be $T_b = 24$ minutes. The number of calls served at the call center—the throughput—is equal to call arrival rate of eight per hour, since there is no throughput loss.

Now, suppose that the manager of the call center decides to reduce the average waiting time on hold to zero. Specifically, all customers who get connected should speak to an agent without waiting on hold. One way to achieve this is to reduce the buffer size to zero, that is, have only one trunk line. How does this affect call center's performance?

Well, reducing the buffer size to zero is a great idea if one ignores the loss of throughput. Reducing buffer size to zero leads to probability of blocking reaching to its maximum level. This in turn results in the process losing its throughput due to blocking. How much of the throughput of the call center is lost? With zero buffer, the queueing model of the call center would be $M/M/1/1$ model with arrival rate of eight calls per hour and effective mean process time of 6 minutes. Using the analytics for processes with limited buffer, we find the throughput of the call center to be 4.444 calls per hour. In other words, the center loses about $8 - 4.444 = 3.556$ calls per hour.

Now consider Call Center B, which is ten times larger than Call Center A.

**Call Center B:**    Call Center B receives 80 calls per hour—10 times larger than what Call Center A receives—and has 10 agents. Similar to Call Center A, the call arrivals are Poisson and the process times are exponentially distributed with an average of 6 minutes.

How much throughput of Call Center B is lost if its buffer size is reduced to zero? Using the underlying queueing model $M/M/10/10$, we find the throughput to be 70.27 calls per hour, and thus the throughput loss would be $80 - 70.27 = 9.73$ calls per hour, which is larger than throughput loss of Call Center A.

What happened to economies of scale? Shouldn't Call Center B, which is 10 times larger than Call Center A, have better performance measures? Why does the larger system lose more customers than the smaller system? Well, economies of scale is still there, but it manifests itself in different ways.

*Economies of Scale—Impact on Throughput Loss:* First, we should be careful how we compare the throughput of large and small systems. If we compute the probability of blocking a call, we find that in Call Center A $P(\text{Blocking}) = 0.4444$ and in Call Center B $P(\text{Blocking}) = 0.1217$. Therefore, Call Center B—the larger system—loses about 12.17 percent of its callers, while Call Center A—the smaller system—loses about 44.44 percent of its callers. In other words, Call Center A loses about 44 customers out of every 100 of its customers, while Call center B loses about 12 customers out of every 100 of its customers. This clearly shows the impact of economies of scale on throughput loss. As the system becomes larger, the throughput loss becomes smaller. Consider Call Center C, as another example.

**Call Center C:**    Call Center C receives 800 calls per hour—100 times larger than Call Center A—and has 100 agents. Similar to Call Center A, call arrivals are Poisson and the process times are exponentially distributed with an average of 6 minutes.

If we reduce the buffer size of Call Center C to zero, then we have an $M/M/100/100$ queueing model for which $P(\text{Blocking}) = 0.004$. Like Call Centers A and B, Call Center C also has a zero waiting time on hold for customers who do not get a busy signal. However, it only loses 0.4 percent of its throughput—losing about 4 customers out of every 1000 customers.

*Economies of Scale—Impact on Utilization:* Economies of scale also impacts resource utilization in systems with finite buffer. Using our analytics we find agent utilization (i.e., occupancy) to be 44.44 percent in Call center A and 70.27 percent in Call Center B. In other words, agents in Call Center B are answering calls 70.27 percent of the time, and are idle 29.73 percent of the time; while the agent in Call Center A is answering calls only 44.44 percent of the time and is idle about 55.56 percent of the time. As this simple example shows, economies of scale in systems with finite input buffer results in higher utilization of resources.

*Economies of Scale—Impact on Profit:* Suppose that both Call Centers A and B sell the same product to their customers at the same selling price of $15. In other words, each call that is answered by an agent generates a revenue of $15 for the call center. Also suppose that agents have the same hourly salary of $20 per hour. Which call center—A or B—is running a more profitable operations?

Obviously, Call Center B is 10 times larger than Call Center A, and thus it is expected to generate more profit. Hence, comparing the *total profits* of the two centers would not be a fair comparison. Note that in both call centers, agents are the main source of generating revenue. The more calls they answer in an hour, the more revenue is generated. Thus, one way to evaluate the profitability of the operations in a call center is to compute how much profit is generated by an agent in each call center.

The throughput of Call Center A is 4.444 calls per hour. Thus, considering the revenue of $15 per call, the total revenue of Call Center A in an hour is $4.444 \times \$15 = \$66.66$, all generated by its single agent. Now consider Call Center B that has throughput of 70.27 calls per hour. The total revenue generated in this call center is $72.27 \times \$15 = \$1054.05$ per hour. Since Call Center B has 10 agents, then the revenue generated by each agent is about $\$1054.05/10 = \$105.41$ per hour. Considering an agent's hourly salary of $20, the profit generated by an agent in Call Center B—the larger system—is $\$105.41 - \$20 = \$85.41$ per hour, while the profit generated by an agent in Call Center A—the smaller system—is $\$66.66 - \$20 = \$46.66$ per hour.[10] In summary, both Call Centers A and B are selling the same product at the same price and paying the same salary to their agents; both call centers face the same workload—they both have a demand that is 80 percent of their capacities; and agents in both Call Centers A and B have the same capacity of answering 10 calls an hour. Nevertheless, because of economies of scale, an agent in the larger system has higher utilization and thus answers more calls and generates more revenue.

We summarize our findings in the following principle:

**PRINCIPLE**

**Buffer Size Reduction and Economies of Scale**

Reducing the size of the buffer has a positive impact on flow time—it reduces flow times—but has negative impact on throughput—it increases throughput loss. However, the negative impact on throughput is lower in larger processes. Specifically, in systems with limited input buffer, for the same ratio of demand to system capacity,

- larger systems work in higher resource utilization and thus generate higher revenue per resource.
- larger systems have lower throughput loss than smaller systems.

The above principle is one of the reasons that some large call centers (e.g., call centers with more than 100 agents) reduce their buffer size to zero by setting the number of trunk lines equal to their number of agents. This results in zero waiting time for callers, and only a very small fraction of callers receive busy signals (or asked to call later). Even relatively smaller call centers that do not sell products (e.g., human resource department, tech support) reduce their buffer size to reduce customer waiting time on hold. In these call centers, (i) most customers call back when they get busy signal (i.e., the throughput loss is small), and (ii) they call back in non-peak hours. The latter reduces the variability in call arrivals, which in turn improves operational performance.

In conclusion, we must emphasize that the decision of limiting the buffer of a process to block arrivals is a critical decision that must be made with caution. It depends on the benefits gained by reducing customers' waiting time and the losses due to blocking and turning away customers. The trade-off between these gains and losses depends on many factors, including current buffer size, system utilization, number of servers, the cost of losing one unit of throughput, and the benefit of reducing 1 minute (or 1 hour) of customer waiting time.

---

[10]We do not include other operating costs such as overhead costs, since they do not affect our conclusion. In fact, it would further confirm our conclusion, since larger systems also benefit from economies of scale with respect to overhead costs.

Analytics provided in this chapter and the previous chapter can help make more informed decision of setting an appropriate buffer size.

## 8.8 Redesigning Process Flow

In Sec. 6.5.3 of Chap. 6 we showed that the capacity of a process can sometimes be increased by redesigning the process flow. Can redesigning process flow also help in reducing flow times? Let's return to Process A in Sec. 6.5.3 in which three workers perform three different activities of stamping, assembly, and inspection, with effective mean process times of 15, 30, and 5 minutes per product, respectively, with coefficient of variation 1. Process A is shown in Fig. 8.5. Worker who does the inspection detects the quality issues of the final product, which results in scraping of the product. Recall also that 10 percent of products are scraped due to quality problems resulting from the stamping activity, while the assembly activity results in almost no quality issues. Now, suppose that jobs are released to the first station—the stamping station—according to a Poisson process with a rate of 1.9 jobs per hour.

Note that Process A constitutes a multiple stage queueing process and, using computer simulation or the analytical model in the online supplement of Chap. 7 for multi-stage processes, we can obtain the average inventory in stamping, assembly, and inspection, respectively, as 0.90, 19, and 0.19—a total of 20.09 jobs. Using Little's law, the average flow time in Process A is $T = 20.09/1.9 = 10.57$ hours.

If the process flow is redesigned as shown in Process B in Fig. 8.5 such that the inspection activity is performed immediately after stamping and before assembly activities, then the average inventory in stamping, inspection, and assembly can be obtained as 0.905, 0.188, and 5.897 jobs, respectively—a total of 6.99 jobs. Using Little's law the average flow time for a job in Process B is $6.99/1.9 = 3.68$ hours. What is the underlying reason for this reduction in flow time?

The reduction was mainly due to decreasing the input rate at assembly. This is because the assembly station in Process B serves 10 percent less jobs than that in Process A. By pushing the scrap units out in the earlier stages of the process, we reduced the input rate and thus the utilization of later processes resulting in lower average flow time. So, as this example shows, one strategy to improve the flow time in a process is to (if possible) redesign flow such that flow units exit the process at the earlier stage and do not go through unnecessary processing in later stages. In manufacturing processes this points to the important practice

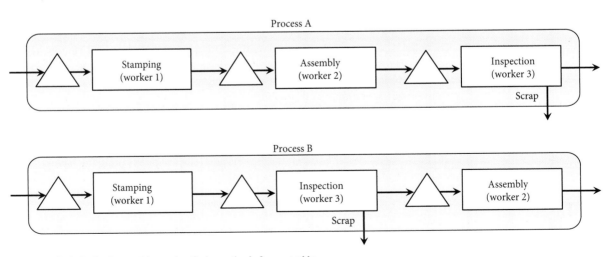

**Figure 8.5    Redesigning Process A by moving the inspection before assembly.**

of checking the quality of a job before sending the job to the later stages. This practice is known as "Quality at the Source," which we will discuss in detail in Chap. 15.

## 8.9 Changing Batch Size and Eliminating Setups

In Chap. 1 we described Batch Processes as production processes in which products are moved through work centers of the process in batches. A setup time or a changeover time is often required in each work center before the process of the next batch starts. We also mentioned that Batch Processes are suited for low-volume multi-product environments.

One main reason for moving or processing units in batches is the cost/time of moving units from one stage of process to another, and the other reason is that there is a need for setting up a resource after processing several units. The impact of batch size on flow times depends on the setup/transportation times. Below we consider two cases, one without setup and one with setup.

### Systems Without Setup Times/Costs

Suppose you work in a drilling station drilling a hole on jobs sent to your station. Drilling a job takes exactly 2 minutes, and thus you have an effective capacity of 30 jobs an hour with no variability. Also suppose that jobs are sent to your station in a pallet that contains three jobs. Specifically, you receive a batch (i.e., a pallet) of size 3 at exactly every 6 minutes—job arrival rate of 30 per hour with no variability. When you receive a batch, you start drilling the units one-by-one and put them back in the pallet. When all three jobs are completed the pallet leaves your station. What is the flow time of a job in your station?

The total flow time of a job in your station is the summation of the following three times:

- *Wait-In-Batch-Time* (WIBT): Consider time $t = 0$ when the first batch arrives. You immediately take the first job in the batch to drill. So, the first job waits in the batch for zero minutes before being drilled. The second job in the batch waits in the batch for 2 minutes for its drill to start. The third job waits in the batch for 4 minutes— waits for the completion of the first two jobs—before its drilling starts. So, the average wait-in-batch-time for the three jobs is WIBT $= (0 + 2 + 4)/3 = 2$ minutes.

- *Drilling (i.e., Processing) Time:* The drilling time for each of the three jobs is 2 minutes.

- *Wait-To-Batch-Time* (WTBT): In batch processes, jobs move in batches, so jobs do not leave your station until all three of them are complete. Hence, the first drilled job must wait for the completion of the other two jobs to leave your station. This takes 4 minutes. The second job must wait in the batch for 2 minutes, before the completion of the third job, and the third job waits for zero minutes, since the batch leaves your station immediately after the third job is completed. Therefore, for each job the average time to form a complete batch before leaving your station, known as wait-to-batch-time is WTBT $= (4 + 2 + 0)/3 = 2$ minutes.

Consequently, with a batch of size 3, the average flow time of a unit in your station is $2 + 2 + 2 = 6$ minutes. Each jobs waits an average of 2 minutes in the batch for its turn, then 2 minutes of drilling, and another 2 minutes of waiting to form a batch and leave the station. What would be the average flow time of a job, if you reduce your batch size to 2?

Assuming that a batch of size 2 is sent to your station at exactly every 4 minutes (same arrival rate of 30 per hour with no variability), you will have WIBT $= (0 + 2)/2 = 1$, and WTBT $= (2 + 0)/2 = 1$. Hence, the average flow time of a job in your station would be $1 + 2 + 1 = 4$ minutes. Therefore, in your case, reducing batch size from 3 to 2 reduces the flow time in your station from 6 minutes to 4 minutes. In fact, the optimal batch size is 1. If a batch of size 1 job is sent to your station every 2 minutes, then the flow time in your station is reduced to only 2 minutes—its minimum possible. The reason is that for batch of size 1 both wait-in-batch-time and wait-to-batch-time are zero (WIBT = 0 and WTBT = 0). This raises the following important question: If batch of size 1 has WIBT = 0 and WTBT = 0, then why do we see batches of larger size in most processes? Well, the reason is the setup and transportation times (and costs). We discuss this below.

### Systems with Setup Times/Costs

To illustrate the impact of setups on flow times, suppose that you must perform a setup, which takes about 3 minutes, each time you finish a batch. The setup does not depend on the batch size (e.g., after finishing a batch you must deliver it to another station and come back, which takes 3 minutes). What is the impact of this setup time on flow times of jobs in your station?

- *The positive impact of large batches on flow time:* For a batch of size $N$, and setup time $T_s = 3$ minutes, your effective process time is increased to $T_{eff} = T_0 + T_s/N = 2 + 3/N$. For a batch of size 1, for example, you have $T_{eff} = 2 + 3/1 = 5$ minutes, reducing your effective capacity to 12 jobs per hour (instead of 30 per hour for the case without setup). The effective capacity for batch of size $N = 3$, on the other hand, is 20 jobs per hour. This has two implications:
  - If, for example, you need to drill at least 18 jobs in an hour, the batch of size 1 and size 2 are not feasible. Formally stated, in processes with setup times, some smaller batch sizes may not be feasible to achieve a certain capacity.
  - With setup times, larger batch sizes result in higher effective capacity. For any given arrival rate, higher effective capacity results in lower flow times.
- *The negative impact of large batches on flow time:* The negative impact of a larger batch size is that it results in higher WIBT and WTBT, which in turn results in larger flow times.

So, while in systems with zero setup times reducing batch size to 1 is the optimal action, this may not be true when there is setup between processing two consecutive batches. In this case, two conflicting forces are in play. On one hand, reducing batch size decreases the effective capacity of the process and results in higher flow times. On the other hand, reducing batch size decreases wait-in-batch-time and wait-to-batch-time and results in lower flow time. Therefore, the batch size that results in the minimum flow time may not be 1. The optimal batch size would depend on the length of the setup time, the effective capacity of the process, and the variability in arrivals and process times. The following principle summarizes our findings.

**PRINCIPLE**

**Batching Principles**

When making decision about the batch size, one must consider the following:

- *Stability:* When setup times between batches are significant, the minimum batch size to make the process stable might be greater than 1.

- *Optimal Batch Size:* Depending on the current batch size and setup time, reducing the batch size may increase or decrease the average flow time. However, there exists an optimal batch size that minimizes the average flow time. For processes with zero setup time, the optimal batch size is 1.

In Chap. 15 we will show that small batch sizes have several advantages such as smoothing the flow of operations and improving quality. Thus, the goal of operations engineers and managers must be to reduce batch size. The main obstacle to batch size reduction is the setup time. Therefore, the first step for reducing the batch size is to reduce, or better yet, eliminate setups. With shorter setup times, smaller batch sizes become optimal. The goal must be to reduce the batch size to 1, if possible. Chapter 15 also illustrates how setup times can be reduced.

### Want to Learn More?

**Finding the Optimal Batch Size**

As illustrated above, when a setup is required between processing two consecutive batches, the optimal batch size that minimizes the average flow time may not be 1. What makes finding the optimal batch size difficult is the variability in arrival and process times. If you want to learn how to find the optimal batch size, check the online supplement of this chapter. The supplement develops a queueing model to obtain the optimal batch size.

## 8.10 Redesigning Buffers

You are going to City Hall to get a construction permit for the addition you are planning to build at your home. You arrive at the parking lot of City Hall and see that it is packed with cars. After 5 minutes of searching, you finally find a place to park your car. As you are getting out of your car, another driver who is also looking for a place to park, asks you if you are leaving. You tell him that you have just parked your car and are going to City Hall. The driver looks disappointed and continues looking for a parking spot. You feel sorry for him and think how long he would be searching for a place to park.

When you enter City Hall, you see people lining up in the different lines and you wonder which line you should join. After checking all the signs and directions posted on several places—which takes you a while—you find the line "Construction Permit for Residential." You are still not 100 percent confident that this is the right line, until after you ask the people in the line. They all, like you, are applying for a permit. When you join the line, to your surprise, you find the person you saw in the parking lot to be standing ahead of you in the line. That does not feel good, since you entered City Hall before him, but he is now ahead of you in the line. It seems that it did not take him much time to find which line to join. You start chatting with him about how difficult it was to find the line for construction permit. His name is John. John replies that he has been here before, so he knew which line to join. Frank, the person ahead of John, joins the conversation.

You wait in the line for about 20 minutes when Frank's turn comes up and he goes to one of the agents. To your surprise, Frank's application process takes less than a minute. When he is leaving, he tells you that he was missing a document and was asked to come back another day with complete documents. Frank was very unhappy that he waited for 30 minutes in the line just to find out that he is missing a document. You beginning to worry—what if you are missing a document?

When Frank leaves, John is called and he immediately goes to the available agent. You—while feeling tired standing in the line for 20 minutes—wonder what is going to happen to John. When John gets to the agent he starts taking documents out of his bag and searching for something in his bag while the agent is waiting for John to find what he wants. You are thinking: this is a waste of time. John should have had his documents ready to save his time, agent's time, your time, and everybody else's time. About to get frustrated, the person behind you taps your shoulder and tells you that it is your turn now.

You try to find which agent to go to when one of the agents waves at you and signals to go to that counter. You go to the agent and apologize for keeping the agent waiting. The agent asks for the documents. You give the documents to the agent who then gives you a form to fill out. When you ask what the form is for, the agent responds that this is a new form that is required for all construction permits. The agent waits for you to fill out the form, which takes you about 1 minute to do. The agent then checks your application and enters your information in to the computer and tells you that they will mail you the permit in about 10 days.

You are happy that you are not missing any documents and that you do not need to come back again. Before you leave they ask you to fill out a survey and tell them about your experience with customer service. The survey asks you to give them a number between 0 and 10—10 being excellent service—to evaluate the quality of service you received.

You think about your experience, which was not very pleasant. First, it took you some time to find which queue to join. Second, you were tired standing in the line for 20 minutes. Third, when you were called, you did not know which agent you need to go to. Fourth, they did not inform you about the new form, which took everybody about 1 minute to fill out while being with the agent. This made everyone's service time longer and thus increased waiting time in line. Fifth, you thought about Frank who waited in the line for more than 30 minutes just to find out that he is missing a document. On the positive side, you found the agent pleasant and respectful. Overall, you give them a five and leave City Hall, while thinking that Frank would have definitely given them a zero.

How can City Hall redesign its operations to improve customer waiting time and experience? Well, one way is to redesign customer flow and the buffers to eliminate some of the non-value-added activities. We discuss some of these strategies below.

**Directing Customers to the Right Queue**

Let's go back to the day you went to City Hall and consider a second scenario in which you have a completely different experience. When you arrive, you see a line close to the entrance with a big sign saying all inquiries should first join this line. The line is for a reception desk. You join the line, and when you get to the agent, the agent checks your documents for completeness and then gives you a number and directs you to the right waiting area. How do you feel about this? How your experience upon arrival is different from what you had in the first scenario?

Much better! First, you like the sense of fairness in the process. John, who arrives after you will be behind you in the line for reception desk. Second, having a number in hand, you also feel confident about the fairness throughout the rest of the process. Again, John would be behind you in all of the remaining steps of the process. Third, you find out about whether you are missing a document soon before waiting for a long time for the agent to tell you that. This certainly would have saved Frank a lot of time and reduced your worries while waiting to submit your application. Fourth, you do not need to spend time looking for the right line to join. The receptionist directs you to the right line. Fifth, this also eliminates your anxiety of whether you were in the right line.

As described above, reception desks improve customer experience significantly. They eliminate non-value-added activities such as customer searching for the right queue and improve customer experience upon arrival. As we will discuss in Sec. 8.11, customer's first impression of the service has a significant impact on the quality of service perceived by the customer. Automated kiosks can also be used to replace receptionists. After selecting the service type on the screen, the kiosk informs customers about the required documents. When the customer acknowledges that he has all the documents (by pressing a button or touching the screen), the machine prints a number with the instruction of which queue to join. Another advantage of a reception desk is that it provides managers with data about the arriving customer (e.g., the number of customers for each type of service and their arrival times). This provides the opportunity to track the customer throughout the process and evaluate the workload (demand) for each type of service at any time. Recording the time that the service of a customer ends, the system can obtain the flow time of the customers of each type, which can be used in performance evaluation and improvement.

### Making the Waiting in Line "Active Waiting"

Let's go back to our second scenario when you get a number from the reception desk and are going to the waiting area. You are also holding a form that the receptionist gave you and asked you to complete. Going to where the receptionists directed you, you see a waiting area—instead of a waiting line—with chairs and benches. You see people sitting with their numbers in hands and waiting. You also find desks and pens in the waiting areas, so you go there and fill out the form. You put the form in your document file, and then find a place to sit. When you sit down you notice several monitors in the waiting area that show the numbers of the customers currently being served by each agent. When someone's turn comes up, the monitor shows the number being called and an announcement is made: "Now serving Number … ." Noticing that they are serving Number B35 and your number is B52, you think it would take a while before they call you. Hence, you take out your laptop and start working on the report which is due tomorrow. While in the middle of your report, you hear "Now Serving Number B52." The monitor also shows that you need to go to Window 12. You put your laptop in your bag and go to Window 12.

Let's compare your experience in this scenario with your previous experience. First, obviously waiting in a waiting area with chairs is much more comfortable than standing in a line for 20 minutes. Second, you and other applicants get a chance to fill out the required form while waiting. This reduces the time with agents which in turn reduces waiting times (flow times) for you and for other customers. Third, sitting in the waiting area allowed you to get engaged in another activity—writing a report—which makes your waiting time a productive time, and thus does not annoy you. In fact, you did not even know how long you were waiting. Fourth, having audio announcement assured you that you would not miss your turn, so you could focus on your other activity. Fifth, showing which window number you should go to eliminated your confusion of which agent to go to. This also eliminates the agent's idle time waiting for the next customer.

**CONCEPT**

### Active Waiting

Utilizing customers' waiting times for service or products and engaging them in some activities is called *Active Waiting*.

In the second scenario, the waiting time is used to perform some tasks corresponding to customer service—filling a form—which improves flow time measures. In some commercial settings, firms use active waiting for advertising and marketing their products. They put printed catalogs of their goods and services in the waiting areas or use TV to promote those products.

In conclusion, we must mention that having a waiting area is not always preferred over having a waiting line. When service times are short and waiting line moves fast, for example, lines for ordering food in fast-food restaurants, it is more efficient to have a line instead of a waiting area. One reason is that the time from when the customer in head of the line is called until the service starts—non-value-added time—is shorter in lines than in waiting areas. This is because the customer ahead of the line is closer to the agent and is more alert since the line is constantly moving. This is the reason we see waiting lines (instead of waiting areas) in places such as banks and cash registers and post office. Waiting areas are more suited in processes in which the service times and waiting times in the queue are long. Examples are public service areas such as government offices and hospitals.

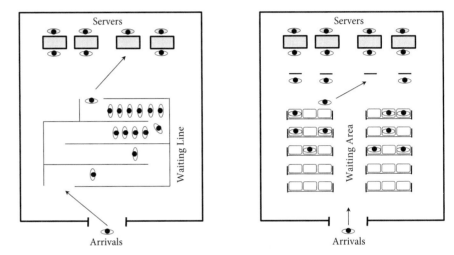

**Figure 8.6**    *Left:* A queueing system with waiting line. *Right:* A queueing system with waiting area and stand-by buffers.

### Establish Stand-by Buffers

Back to the second scenario. When your number is called, you rush to Window 12, but you see that the agent in Window 12 is not yet finished with its current customer. You also see a line on the floor and a sign which says: "Please stay behind the line, make your documents ready, and wait to be called." A few seconds after you have your documents ready, the agent calls you. You go to the window and give the agent your documents. The agent reviews your documents, including the new form, and processes your application. The agent then tells you that you will receive your permit by mail in about 10 days.

Let's compare the two scenarios after you were called. In the first scenario there were four different non-value-added activities: (i) looking for which agent to go to, (ii) walking to the agent, (iii) getting prepared to talking to the agent (i.e., taking out the documents from the bag), and (iv) filling out the new form while being with the agent. The numbering systems and active waiting eliminated the non-value-added activities in cases (i) and (iv) for both you and the agent. Waiting behind the line after you were called and preparing your documents eliminated the agents' waiting for the non-value-added activities in cases (ii) and (iii). While the agent was still with the customer, you walked to the window, and got ready to talk to the agent.

Note that in the second scenario, the buffer of the process for construction permit is divided into two: (i) waiting area before you were called, and (ii) waiting area (standing behind the line) after you were called. The latter is called "Stand-by Buffer."

**Stand-by Buffer**

Stand-by buffer is an area assigned to the person in the head of the queue. It is between the waiting area or line (i.e., the main buffer) and the server. The stand-by buffer is close to the server. Figure 8.6 depicts examples of a waiting line (the left figure) and a waiting area with stand-by buffers (the right figure).

As described above, the advantage of having a stand-by buffer is that it makes the customer more alert, closer to the server, and it reduces customer service time by eliminating agent's waiting time for the customer to come to the agent and to get ready for service.

The main question in processes with stand-by buffers is when to call the next customer to come to the stand-by buffer. One strategy is to call the customer 1 or 2 minutes before the service of the current customer ends. This can be done by the agent pushing the call button when wrapping up with the current customer service. The main concern is not to make the customer tired of standing for a long time in the stand-by buffer.

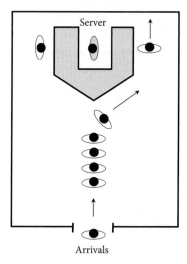

Figure 8.7    A single-server queue with two-sided service desk.

Stand-by buffers can also be created by having two-sided service desks, see Fig. 8.7. As the figure shows, a customer waits on one side of the desk while the server is processing the customer on the other side. When the server finishes processing a customer on one side, the server turns to the customer on the other side and starts the customer's service. Meanwhile a customer moves from the line to the empty side of the desk and waits there. In addition to the benefits of stand-by buffers, two-sided service desks also help customers to learn about service activities while watching the service being performed to the customer on the other side of the desk. This makes the customer to get better prepared for the service, which can save some time. Examples of two-sided service desks are cash registers in supermarkets and cafeterias. Note, however, that two-sided service desk cannot be used for services in which privacy of customers is important, for example, customer service requiring information such as social security and other personal information.

## 8.11 Customer Psychology and Flow Times

So far in this chapter our focus has been to find ways to *reduce the actual flow times* in a process. When flow units are humans who are physically or virtually (e.g., through Internet or phone calls) present in the process, human psychology can be used to make the customer *waiting time in the process feel shorter* than what it actually is. This is often referred to as Psychology of Queues or Psychology of Waiting Lines. In a paper with the same title David Maister[11] has introduced the first Law of Service as follows:

$$\text{Satisfaction} = \text{Perception} - \text{Expectation}$$

and he writes: "If you expect a certain level of service, and perceive the service received to be higher, you will be a satisfied customer." With respect to customer waiting time in a process, this law suggests two directions for improving customer satisfaction:

- Managing customers' *expectations* about their waiting times. According to Prospect theory,[12] people tend to experience losses more intensely than gains. Also, people determine their losses and gains by comparing it to a reference point. Thus, managing customers' expectation is mainly setting customer expectation about their waiting times such that the actual waiting time be shorter than expected (i.e., a gain for the customer).

- Affecting customers' *perceptions* of their waiting time. Researchers have found that people perceive their waiting time in a line to be much longer than what it actually is. Customer perception of waiting time is affected by many factors such as boredom, anxiety, confusion, stress, physical environment, uncertainty, and unknowns. The more anxious, stressed out, uncomfortable, bored, and uncertain the customer is, the longer the waiting time *feels* to the customer.

In this section we introduce several strategies that can be used to manage expectation and affect perception of customers about their waiting time.

### 8.11.1 Customer Arrival

It is known by researchers in human psychology that people remember the beginning and the end of an experience more than the middle. Thus, operations engineers and managers should pay careful attention to customers' experience upon arrival. Here are some tips that improve customer experience upon arrival.

**Avoid Customer Confusion Upon Arrival**

Recall the example in Sec. 8.10 about your experience in City Hall. One strategy introduced in that section that can improve flow time was "Directing Customers to the Right Queue." This, as we discussed, reduced customer's actual waiting time in the process by eliminating the additional time that the customer spends looking for which queue to join. This strategy also reduces customer's perception of waiting by reducing customer's confusion and anxiety while looking for the right queue and while waiting in the queue wondering whether she joined the right queue. After all, the only thing worse than waiting in line is waiting in the wrong line. Some retailers have their staff at their entrance that greet customers and direct them to the right departments. Hotels also use their staff to greet arrivals and help them with

---

[11] Maister, D.H., *The Psychology of Waiting in Lines.* Harvard Business School, Boston, MA, Note 9-684-064, Rev May 1984.

[12] Kahneman, D. and A. Tversky, "Prospect Theory: An Analysis of Decision under Risk." *Econometrica*, 47 (2) (1979), 263.

their inquiries. Some restaurants have their staff at the end of the line to welcome arriving customers. These all tend to give customers a good first impression, which as we said, has a major impact on customers' experience in the process, including the impact on perceived waiting time.

### Set Customer Expectation and Exceed It

Research has shown that people who have an estimate of their waiting time in line are less anxious than those who do not. As we mentioned, anxiety makes waiting times feel longer. Thus, it is important to tell customers what they should expect about their waiting time in the line. Pilots announce the arrival time of their flights, call centers provide an estimated waiting time on hold to their customers, restaurants give the estimated waiting time for customers in line, Disney provides the estimated waiting time for different rides in their parks. The new technology allows firms to inform customers about the expected waiting time even before they arrive. For example, Hartsfield-Jackson Atlanta International Airport has a service that sends emails to passengers and informs them about the waiting times in their security screening line.

While setting customer expectation reduces customer anxiety and uncertainty about their waiting times in the process, it can backfire if it is not set right. Suppose, for example, you are waiting in the line of a popular restaurant and the host tells you that the waiting time would be around 15 minutes. Now consider the following two cases: (i) your table is ready after 10 minutes, and (ii) your table is ready after 20 minutes. How do you compare your satisfaction about your waiting time in the two cases? In case (i) you got the table 5 minutes earlier than what the host said—a gain of 5 minutes—and in case (ii) you got the table 5 minutes later than what the host said—a loss of 5 minutes. Is the positive impact of gaining 5 minutes the same as the negative impact of losing 5 minutes. Behavioral economists have shown that for the same value (e.g., 5 minutes) the negative psychological value of loss is much more (i.e., an average of 2.5 times larger) than the positive psychological value of gains.[12] How does that affect setting customer expectations?

Operations engineers and managers should set customer expectations such that they can exceed it. That is the main reason hosts in the restaurants or pilots announce an overestimation of the waiting times or length of a flight. This achieves two goals. First, it sets the customer expectation and thus releases customer anxiety with respect to the waiting time. Second, it gives customers a positive psychological value when the firm exceeds expectation and the customers receive their services earlier than what was announced.

It is worth mentioning here that firms must pay careful attention to how much they overestimate the waiting time. If the overestimation is much larger than the actual expected time, the firm may lose customers. For example, if the restaurant host tells you that the waiting time for a table is about 45 minutes, then you may decide to go to another restaurant. Hence, the firm should set customer expectations realistically such that it is not too long and the firm can achieve it in most cases.

### 8.11.2 Customer Waiting for Service

By routing customers to the right queue and setting customer expectation, firms can reduce customers' confusion and anxiety. This creates a good first impression, which has a significant impact on customer experience. In this section, we focus on the time after customer arrival and before the service starts, that is, customer's waiting time in the buffer.

### Provide a Comfortable and Pleasant Environment for Customer Waiting

Obviously, a 10-minute standing in a line in a noisy, hot, and humid environment feels much longer than a 10-minute sitting on a comfortable chair in areas with air condition listening to a relaxing music. Several studies have shown the impact of environment on people's perception of service quality. For example, in a study, Lavender scent, which was known to have a calming effect, was released in the air in a DMV office in Australia. The effect was that it improved customer rating of service for those customers who waited for more than 10 minutes. A comfortable environment is not limited to the physical surroundings. It also includes a firm's staff. Waiting in an environment with cheerful and smiley staff also makes the waiting more comfortable.

**Make Customer Occupied During Waiting**

Several years ago passengers of Houston airport were complaining about the long waiting times to get their baggage after their flight landed. The airport managers increased the number of baggage handlers and the average wait was reduced to 8 minutes—the industry benchmark. To their surprise, passengers continued to complain. Doing a time study, the managers found that the 8 minutes consist of 1-minute walk from the gate to baggage claim and 7 minutes waiting for their baggage. They then decided to solve the problem in another way. Instead of further reducing the waiting time, they moved the arrival gates away from the main terminal and sent bags to the furthest carousel. Passengers now had to walk six times longer to get their bags. The passengers' complaints stopped. Why? During a large portion of waiting to get their baggage, passengers were occupied—walking to the carousel.

The psychological perception of time is influenced by mental activity. Thus, periods of time with events that occupy customers feel shorter than the same time with no events. When high rises started to grow in large metropolitan areas, people started complaining about elevator delays. After mirrors were installed outside the elevators, the complaints stopped. Why? Similar to the case of Houston airport, people were now occupied checking their looks in the mirror and were distracted from the passage of time.

The following are strategies that can make customers occupied and thus distract them from their waiting times to receive service.

*Active Waiting:* As we discussed in Sec. 8.10, Active Waiting can shorten the actual waiting time of customers in a process. Active waiting also makes the waiting time feel shorter, since it makes the customers occupied with activities.

*Group Waiting:* When you are waiting in line for a restaurant or to board a plane, the waiting time is perceived shorter when you are with a group of people than when you are by yourself. One reason is that, with solo waiting, you are more focused on the passage of time, while in group waiting, you are occupied interacting with a group of people. The more the customers are engaged with each other in the line, the less they notice their waiting time. In some situations, the group waiting is the part of the experience. Waiting in the line to buy a concert or a football ticket soon turns into a group of people talking about music and sports and having fun.

Another reason that group waiting feels shorter is that it creates a sense of comfort to be a part of a community. This reduces anxiety and stress of waiting. For example, if you are the only person whose flight is canceled and waiting for the next flight, your waiting time feels longer than if you are a part of a group in the same canceled flight.

Thus, to reduce the perceived waiting time, firms should promote the sense of community and group waiting in the line. This can be done by creating opportunities for customers to socialize and interact with each other. One example is to design the waiting area such that people sitting in groups face each other. This will encourage conversation among customers. Sometimes a firm's staff can initiate conversation among customers through different means. For example, in some concert clubs or amusement parks, the staff ask questions from the customers in line that requires them to talk to the person next to them. This initiates the conversation and creates a sense of community. In some amusement parks the customers are sent to different waiting stages in groups in order to maintain the sense of community among them.

*Advertising:* Firms can also use customers waiting area for advertisement. For example, some firms locate their products in places that waiting customers can see. This creates opportunities for the firm to sell more products. For example, the waiting area for cashiers in supermarkets have many items such as candies, chewing gums, chocolate, magazines, drinks, etc. Customers waiting in line often purchase one or two items while waiting. If products cannot be located around the waiting area, firms use media to advertise those products. For example, customers who are calling the insurance company to talk about their car insurance, while on hold, are exposed to other insurance products (e.g., home insurance) and are told how much they can save if they upgrade their car insurance to include home insurance.

*Entertainment:* Another, and perhaps more pleasant, way to occupy customers is to entertain them. Call centers play music for customers waiting on hold. Airports use TVs in their waiting areas that broadcast news and other programs. In Nickelodeon Studios customers can play video games while waiting in line. Some theme parks have staff dressed as cartoon characters that entertain and engage customers waiting in lines. At Disney, customers in line are asked to text message their jokes. They are told that their jokes may be used by actors during the show.

### Reduce Customer Uncertainty

Uncertainty magnifies the stress of waiting and makes waiting feel longer. Suppose you are a stand-by passenger for a flight. You are told to wait at the gate in the airport for your name to be called. After they call several passengers names you begin to worry: "Did they forget that I am also standing by? Why aren't they calling my name? Will I be able to get this flight?" These all add to your stress and make your waiting time feel long. How to prevent this? Here are some strategies aimed at reducing customer uncertainty.

*Explain the Process and Show the Line Is Moving:* It is known in queueing psychology that unexplained waits feel longer than explained waits. Thus, one way to reduce customers' uncertainty is to make sure customers understand how the process works and why they need to wait. For example, airlines now use TV monitors at the gate that show the list of all stand-by passengers for a flight. Airline staff explain to their stand-by passengers that their names will be in the list shown on the TV monitors and that they will be called according to the list. Thus, passengers understand how the process works and how passengers are chosen from the list. They also observe their progress in the list, as more names are called. This further reduces customer uncertainty that the process is moving forward, the passenger is not forgotten, and their turn will come soon.

*Explain Reasons for Unexpected Delay:* Unexplained waits feel longer than explained waits. For example, informing patients that, due to an emergency, the doctor will be 20 minutes behind the schedule makes the patients' waiting time more tolerable than leaving patients to wait for 20 minutes for some unknown reason. Airlines and subway systems announce the reason and the estimated time for delays to their passengers. One important point is that the reason for delay should be justifiable and unavoidable. How do you feel about the 20-minute delay if it is announced that the delay is because the doctor woke up 20 minutes late in the morning, or the pilot of your flight needs 20 minutes to finish his breakfast before he starts takeoff! Ok, these reasons for delay are not actually announced, but they make the point that unjustified and avoidable delays are not accepted by the customers.

### Use Stand-by Buffers

Another factor that makes waiting more frustrating is to see the time wasted by other customers resulting in longer waiting times for everyone, including you. Which one of the following times feel longer? A 1-minute wait behind a red traffic light, or a 20-second wait behind a green traffic light because the car in front of you did not notice the green light and its driver is texting on his phone? We all experience the frustration and irritation when the time is wasted by other cars ahead of us in traffic. The same is true in businesses. One example of time wasted by other customers is when the server is ready to serve a customer, but the customer in the head of the line is not aware of that. Often other customers in the line inform him to go to the server. Another commonly observed situation is the time the customer spends (wastes) preparing for his service when he is with the server (e.g., searching for a coupon in his bag in a supermarket cash register). Recall that stand-by buffers reduce or eliminate these wasted times. Therefore, in addition to reducing the actual waiting times, stand-by buffers also reduce the perceived waiting time, as we discussed above.

Another psychological advantage of stand-by buffers is that they create the feeling that your service has already been started. For example, waiting for 2 minutes on one side of the two-sided service counter feels shorter than waiting for 2 minutes in the head of the queue. This is because, when waiting on a two-sided service desk, you do not feel that you are waiting in a line anymore. In most doctor offices, the patients are taken from the waiting area (i.e., the waiting line) to a visit room (the stand-by buffer) and are told that the doctor will be with them shortly. The waiting in the visit room feels shorter than waiting in the waiting area, since patients have the feeling that their service has already started. Restaurants also use this approach and ask people in the line if they would like to wait at the bar (i.e, stand-by buffer). In addition to making the perceived waiting time shorter, restaurants also benefit from selling some items in the bar.

### Promote the Sense of Fairness

Perhaps the most critical factor that influences our experience of waiting in line is our perception of fairness. First-come-first-served (FCFS) is a universally accepted service discipline to ensure fairness; people who arrive after you should not be served before you. The desire for fairness is so strong that, in some cases, it can result in violence. In July 2012 a man was stabbed at a Maryland post office by a fellow customer who mistakenly thought he was violating FCFS by cutting in line. Thus, FCFS is considered as one of the key factors in managing queues in service operations.

While FCFS is an important practice from customers' perspective, it may not be the most profitable discipline in some businesses. As we showed in Sec. 8.4, in some cases, priority policy and the c-mu rule can result in higher profit. Thus, the question is how to implement priority policy while not resulting in customer dissatisfaction? Well, there are situations where priority policy is implemented and does not have much impact on customer's negative impression of the process. We have discussed these situations in Sec. 8.4.3 in detail and we refer the readers to that section.

Even when FCFS discipline is used, the issue of fairness arises in systems with multiple lines. How many times you enter a line for a cashier in a supermarket or in a fast-food restaurant and then realize the person who arrived after you and joined a different line was served before you. In other words, after joining a line we find that the other line moves faster, which affected our perception of fairness in the system.

But is it really true? Do other lines move faster? Well, no. The reason for us to feel that way is that we are affected more by the bad feeling of losing to the line on our right than by the good feeling of winning against the line on our left. Regardless of the underlying reason, systems with multiple queues will have this issue with fairness. How to remedy this? Pool the multiple queues into a single queue. A single pooled queue that implements FCFS is as fair as it can be. In Sec. 8.5 we presented a list of operational advantages of pooled queues over multiple queues. Here we add fairness to that list as a psychological advantage of pooled queues.

### Make Lines Appear Shorter Than They Are

Researchers have found that most people are concerned about how long a line is than how fast the line is moving. In other words, between a short slow-moving line and a long fast-moving line, most people join the short line. How can operations engineers and managers use this to affect customer's perception? One strategy is to make the line appear shorter that what it really is. How?

Amusement parks such as Disney hide their lines by turning them around and let them pass through tunnels so the length of the line is not clear to the customers. Customers feel that their line is as long as their distance to the next corner. While this practice may have a negative impact on the customers—since the customers may think that they are deceived—this negative impact seems to be less than the negative impact of facing a very long queue.

A similar approach to make waiting lines appear shorter is to make them zig-zag, as shown in Fig. 8.6-left. A zig-zag line with 40 customers looks much shorter than a straight line with 40 customers. Another advantage of zig-zag line is that it can be confined in a smaller space than a straight line.

Another strategy to make long lines appear shorter is to use waiting areas (instead of a single line) with appropriate numbering systems. In a system with waiting area (e.g., Fig. 8.6-right) customers do not see a long queue. The only indication of how long the queue is—how many customers are ahead of them in the line—is the number that customers have in hand. If your number is 76 and currently number 56 is being served, you infer that there are 20 customers ahead of you in the line. One way to hide this from customers is to use numbers starting with a letter. For example, consider a system in which numbers are given to the next 12 customers in the following sequence: A101, B301, C601, A102, B302, C602, A103, B303, C603, A104, B304, C604 and customers are served in the same order. If your number is B304 and number B301 is currently being served, you think that your queue is much shorter than the eight customers that are indeed between you and number B301.

### Pay Attention to the End of the Line

Customers who are waiting at the end positions of a line are more likely to leave if the line is not moving, or if it is moving very slowly. This, in the short term, results in loss of sales, and in the long term may result in losing customer's future business. Thus, firms should pay special attention to the tail of the line. How? Some fast-food or fast-casual restaurants start offering food samples to customers in the line when the line becomes large. Others, send an employee to take the orders from the customers in the line and send the order electronically to the people who prepare orders. When a customer at the end of a long line places an order with the employee, the possibility of leaving the line is reduced. Other commonly used approach is to put signs in different locations in the line that show the estimated waiting time from that point in line until getting to the head of the line, or until getting their food.

### Avoid Idling Servers When Customers Are Present

While the perception of waiting time is affected by observing time wasted by other customers, it is also affected by observing time being wasted by the servers. Suppose you are waiting in a long single line in a supermarket for cash registers. There are three cash registers open and working. You notice that one of the cashiers stopped serving customers and started talking to the other cashier about last night's football game. How do you feel about your waiting time? It has been found that observing idle servers or slow servers makes the waiting more irritating and results in longer perceived waiting time. People tolerate waiting in long lines where all servers are working (no idle server or idle service station) more than shorter lines where some of the servers are not working. What is the implication in managing waiting lines? If it is required that a server be idle for a period of time, the server's idleness should not be observable by the customer. That is one reason McDonald's asks its employees to take their breaks outside the store.

### 8.11.3 Customer Being Served

Flow time in a process consists of in-buffer flow time (time waiting in a line) and in-process flow time (time being served). While long waiting time in line degrades customer experience, in most cases, long service times also have negative impact on perceived waiting. Even if we wait for a short time in line to get to a server, we still get annoyed if our service takes longer than what we believe it should take. In this section we discuss strategies that can reduce or eliminate the negative impact of long service times on customers' perceived waiting time.

**Set Expectations and Exceed It**

As we mentioned, customers' perceived time at service is mainly affected by their expectation about how long the service would take. If service time takes longer than what the customer expected, then the service time feels longer, even if the actual service time is relatively short. Thus, service managers should set expectation of service time at the beginning or during each stage of service of a customer. Also, similar to waiting in line, the expectation should be met, or even better, should be exceeded. For example, when you take your car to a dealership for some repair, your waiting time feels shorter if the repair staff give you an estimated time for repairing your car (i.e., your service time). If they do not inform you about the time it takes to repair your car, uncertainty about your waiting time (as we mentioned in case of in-buffer flow time) makes you have the perception of a long waiting time.

**Show Progress in Service**

Recall that one way to reduce customer uncertainty about waiting time in line is to explain the process and show that the line is moving. This is also an effective strategy with respect to customer service time. Customers find waiting more tolerable when they can see their service being done. In an experiment, participants used two different websites to book a ticket for a flight. One website delivered a list of options almost immediately, but invisibly. The other website took around 30 to 60 seconds to list the options, but participants could see the search process for the ticket in different airlines. Which website was preferred by the participants? The slow one or the fast one? Study showed that the majority of participants preferred the slower website that showed the search progress (i.e., progress in service).

Most firms now show the progress in customer service times through different means. Travel site Kayak shows each airline it searches. Some restaurants use glass windows, so that customers can see the progress of their orders. Car washes show customers how their cars go through different stages of service. Amazon shows the progress of customers' orders and shipment. Domino's Pizza website shows several stages of making and delivering an order for pizza (e.g., prep, bake, quality check, out for delivery). Customers can trace the progress of their orders using the website (including the name of the person who delivers the pizza).

In some situations where it is not possible to show the service progress to customers, some firms use what is known as "Labor Illusion." Labor illusion is a fictitious demonstration of service progress to the customers. For example, Spanish Bank BBVA's ATM machines show an animation of bills being counted as customers wait for the machine to spit out their cash. Apple call centers added the pre-recorded sound of typing to their automated voice response (AVR) when customers push buttons to get to the right agent.

**Make the Waiting in the Last Stage of Service Short**

Which one feels a longer service time: a 2-minute wait to check in with a hotel staff, or a 2-minute wait to check out? A 10-minute wait to board a plane, or a-10 minute wait to get off the plane after the plane arrives at the gate? The more valuable a service is, the longer a customer is willing to wait. We board a plane or check in to a hotel to receive service. However, when the plane arrives at the gate, or when we are checking out of a hotel, we have already received our service and there is no value in the remaining service. Thus, every minute waiting in the system after our service is completed feels longer than every minute we are waiting for our service to start. What are the implications for service managers?

To make the service time perceived shorter, the last stages of service should be done faster. For example, hotels have shortened the last stage of their service—checking out. They put the receipt under your door the night before your departure. If you agree with the charges, you just need to drop your room key in a box when you leave the hotel. Some hotels let you check out using the TV in your room. Another example is the progress bars on our monitor when we download a file or install a software. In most cases, the bars are designed to move faster at the last stage of the process.

In conclusion, we summarize customer psychology principles that influence customers' perception of waiting times in processes as follows:

---

**PRINCIPLE**

**Queueing Psychology Principles**

The following strategies can be used to improve customer's perception of waiting times in the queue and in service:

- Avoid customer confusion upon arrival (lead customer to the right queue)
- Set customer expectation about the wait time in line (and service time) and exceed it
- Provide a comfortable and cheering environment for customer waiting
- Make customer occupied during waiting (e.g., active waiting, group waiting, advertising)
- Explain the process and show the line is moving
- Explain reasons for unexpected delays
- Use stand-by buffers
- Promote the sense of fairness
- Make lines appear shorter than they are
- Pay attention to the end of the line
- Avoid idling servers when customers are present
- Show customer progress in service
- Make the waiting in the last stage of service short

---

## 8.12 Summary

Flow time is the most critical operational measure in service systems in which customers directly experience waiting to get service. In Chap. 7, the underlying reasons for long flow times were discussed and analytical tools were presented to estimate flow time measures such as average flow times and flow time service levels. The main focus of this chapter was to present strategies that can improve flow time measures of a process.

As also discussed in the previous chapter, increasing capacity and reducing variability improve flow time measures of a process. Another strategy is to synchronize capacity with demand. This includes temporarily increasing capacity of the process (e.g., by using part-time workers) in periods with high demand. In systems with customers having different service times, managers can give priority to one group of customers (over others) to reduce their waiting times. The improvement in waiting time of one group of customers, however, can be achieved only at the expense of increasing the waiting time of some other class or classes of customers. One way to reduce the overall average waiting time in a process is to give priority according to shortest expected processing time (SPT) rule. SPT gives the first priority to the class of customers with the smallest effective mean process time, the second priority to the class with the second smallest effective mean process time, and so on.

Pooling the capacity of resources is another way to reduce customer waiting time. Pooling can be very effective in improving process performance, since it can increase the effective capacity of the process, provide the benefits of economies of scale, and allow more efficient use of resource capacities. Reducing the size of the buffer can also reduce flow time, but it increases throughput loss. Another approach that should be carefully considered for reducing flow times is changing the batch size in the process. Depending on the current batch size and setup time, reducing the batch size may increase or decrease the average flow time. However, there exists an optimal batch size that minimizes the average flow time.

In addition to above strategies that can reduce the actual flow time, there are human psychology principles that can be used to manage customers' waiting experience and improve customer satisfaction. Some of these principles include avoiding customer confusion upon arrival, setting customer expectation about the wait time and exceeding it, providing a comfortable and cheering environment for customer waiting, making customer occupied during waiting (e.g., active waiting, group waiting, advertising), explaining the process and showing that the line is moving, explaining reasons for unexpected delays, using stand-by buffers, promoting the sense of fairness, making lines appear shorter than they are, paying attention to the end of the line, avoiding idling servers when customers are present, showing customer progress in service, and making the waiting in the last stage of service short.

## Discussion Questions

1. What are c-mu and SPT rules, and how do they affect flow time and inventory in a process?

2. Describe three approaches that can mitigate the negative impact of implementing priority service on customer experience.

3. Explain how economies of scale of a process affects flow time and inventory of the process.

4. Describe three reasons why pooling resources of a process result in lower flow time and inventory in the process.

5. Present eight factors that must be considered when one decides whether to vertically pool resources of a process.

6. Present seven factors that must be considered when making decision whether to horizontally pool resources of a process.

7. Explain how the negative and positive impacts of reducing the input buffer of a process are affected by economies of scale.

8.  What is the impact of increasing batch size on flow time of a process?

9.  What is active waiting? Provide three examples of active waiting and their impact on process performance.

10. Describe two practices that can improve customer experience upon arrival to a service system. Provide examples for each practice.

11. Describe four strategies that firms can use to make customers waiting in line less annoying? Provide an example for each case.

12. What are the operational and psychological advantages of stand-by buffers?

13. What are the operational and psychological factors that one should consider when deciding whether to pool several lines into one line?

14. Describe three strategies that can be used to make the time a customer spends receiving service less annoying, resulting in higher customer satisfaction.

15. What is labor illusion and how does it help improve customer experience with waiting?

## Problems

1.  Sea Dock, a private firm, operates an unloading facility in the Gulf of Mexico for super-tankers delivering crude oil for refineries in the Port Arthur area of Texas. The arriving supertankers are either from the Middle East or from Africa. They arrive according to Poisson process with arrival rates of 4 and 16 tankers per hour, respectively, for those coming from the Middle East and Africa. Tankers form a single queue and are served (unloaded) in first-come-first-served basis. After unloading, tankers leave the port.

    Due to larger number of tankers from Africa, Sea Dock has decided to give these tankers higher priority over those coming from the Middle East. Specifically, the firm serves a tanker from the Middle East, only if there are no tankers from Africa in the queue. After implementing this new priority policy for 2 months, Sea Dock performs a time study and finds that the overall average time that a tanker waits in the port before its unloading starts is 4 days. On the other hand, this average time for a tanker from Africa is 2 days. What is the average number of tankers from the Middle East that are waiting in the port for their unloading to start?

2.  Precision Machine Tools (PMT) is a firm that makes a wide range of machine tools. The company has an after-sales department that provides services to its customers who purchase its products. These services include repair, maintenance, operator training, and so on. After-sales department in Chicago has only one team that provides these services to the customers in Illinois. When a customer calls and requests service, PMT sends its team to the customer's site. If there are other customers who called before this customer and are still waiting for PMT's team, then the team is sent to the customer who has been waiting the longest.

    PMT's products can be divided into two groups: (i) computer numerical control (CNC) machine tools and (ii) standard machine tools. After-sales department receives about one call every 10 days from its CNC customers and one call every 8 days from its standard customers. Number of calls are random and follow Poisson distribution. The time it takes for the PMT team to go to the customer's site and fix the issue and come back is random—following exponential distribution—and depends on the type of the machine. This time is, on an average, 1 day for standard machines and 2 days for CNC machines. Since PMT's machines are used in production processes of its customers, it is essential that PMT sends its team to the customer's site in a short time.

    a. What are the average waiting times for standard and CNC customers? What is the average number of customers waiting for PMT team?

    b. What fraction of time PMT team is idle?

    c. Due to an increase in CNC machine sales, PMT is expecting that it will receive more calls from its CNC customers. Specifically, they are expecting to get about one call from CNC customers, on an average, every 7 days. PMT knows that this will increase its customers' waiting times for after-sales services. Hence, it has set the following service goals: Keeping the average waiting time of CNC customers below 1 day, and the average waiting time of standard customers below 2 days. Can PMT achieve these service goals with its current team, or it needs to hire more teams?

3. The call center of a financial institution has 10 agents responding to customers' requests about its products. Customers' calls are routed to different agents who are trained to address customers' requests. One of the agents is in charge of addressing questions about mortgage and loans. Calls to this agent can be divided into three groups: (i) calls regarding home mortgage, (ii) calls regarding business loans, and (iii) calls regarding car loans. The agent receives about two calls per hour for home loans, one call per hour for business loans, and four calls per hour for car loans. Call arrivals are random and is well-approximated by Poisson distribution.

    It takes the agent about 4 minutes with standard deviation of 2 minutes to respond to calls about home mortgage. These numbers are 6 and 6 for business loans and 5 and 3 for car loans, respectively. Calls are answered in the orders they are received.

    a. What is the average time a customer of home mortgage must wait on hold before talking to the agent? What is this average waiting time for customers of business loans and car loans?

    b. To reduce the average waiting time for customers of business loans, call center managers are planning to implement the following policy: After finishing talking to a customer, the agent should only talk to a customer of home mortgage, if there are no customers of business loan waiting on hold. Also, the agent should only talk to customers of car loans, if there are no customers of home mortgage and business loans on hold. How would this policy change the waiting time of the three groups of customers on hold?

    c. If the goal is to minimize the total average number of customers waiting to speak with this agent, should the managers change their priority policy in Part (b)? How?

    d. Studies have shown that keeping a customer of home mortgage on hold for 1 minute costs about $10 in customer dissatisfaction and future loss of business. This number is $12 and $7 for business loan and car loan customers, respectively. How should the managers change their priority policy in Part (b) to minimize the total average customers' dissatisfaction cost?

4. In addition to the one in Chicago, Precision Machine Tools (PMT) in Problem 2 has two other after-sales departments: one in Los Angeles (LA), serving West Coast, and one in New York (NY), serving East Coast. Compared to the department in Chicago, departments in LA and NY serve a larger number of customers and thus operate differently. Specifically, they assign dedicated teams to each group of customers. There are teams in LA and NY that only serve standard customers and there are other teams that only serve CNC customers. Each team serves its customers first-come-first served. The number of calls for standard service in LA in a day is three times larger than that in NY. On the other hand, the number of calls for CNC service in LA in a day is half of that in NY. The service times for standard and CNC customers are the same as those in Chicago. Since the demand for standard service is three times larger in LA than in NY, to provide the same average waiting times for its standard

customers in LA and NY, management is planning to set the number of teams in LA that are assigned to standard calls to be three times larger than that in NY. This will also create a balanced workload between the teams in two locations. Similarly, the number of teams that are assigned to CNC calls in NY would be twice larger than that in LA. If this team allocation policy is implemented:

    a. Would the average waiting time of standard customers in LA be less than, more than, or equal to that of standard customers in NY?

    b. Would the standard teams in LA be idle more, less, or the same fraction of time as the standard teams in NY? What about CNC teams in LA and NY?

5. Davidson Realty is a firm that owns several parking structures in large metropolitan areas. The firm is planning to open two new structures in downtown Chicago. One in the center of downtown and the other in the north of downtown. It is estimated that the demand in the central location would be twice as large as that in the north. Therefore, the firm is planning to build a parking structure with 400 spaces in central location and a smaller structure with 200 spaces in the north. It is expected that the time a customer parks in either of these structures would be the same, ranging from 1 hour to 8 hours. If a customer arrives and the parking is full, customer goes to another parking structure in the same area. The firm is planning to charge the same rate, that is $1 for every 15 minutes in both locations. Which of the following statements are true and which ones are false? Explain your reasons:

    a. Although the demand in the central location is twice larger than that in the north location, because the central location also has twice number of spaces as that in the north location, it is expected that the fraction of time a parking space is empty would be the same in both locations.

    b. The revenue generated by each parking space in a day is the same in both parking structures.

    c. The average time in a day that the structure in central location is full is less than that for the structure in the north location.

6. A car rental company has two locations, one in airport and the other in downtown. Customers make reservations the day before and pick up their cars the next day. If the firm does not have a car available, customers call other car rental companies in the same area. Customers who rent a car from a location must return the car to the same location. The demand in the airport location is three times larger than the demand in the downtown location. The firm has been losing customers in its airport location, because they often do not have cars available for rental—cars are all rented. To keep the availability the same in both locations, the firm is planning to assign its fleet of 400 cars to the two locations in proportion to the demands in these locations. Specifically, it will assign 300 cars to the airport location and 100 cars to the downtown location. With this fleet allocation strategy, answer the following questions:

    a. Would the fraction of time that airport location has no cars (i.e., all of its cars are rented) be larger, smaller, or the same as that in downtown location?

    b. Cars are taken for overall repair and maintenance after every 2000 miles of usage. Assuming that, on an average, customers of airport and downtown drive a rental car for about the same mileage, which location would have a higher annual repair and maintenance cost per car?

To reduce the chance of losing customers in each location and to provide a better service, the firm is considering the following strategy: (i) allowing customers to return their cars to any of the two locations, regardless of where they picked up the car, and (ii) moving a car from one location to another, if the other location does not have a car that a customer wants.

    c. How does implementing this new strategy impact the current fraction of customers that the firm loses because all of its cars are rented out?

    d. How does implementing this new strategy impact the annual repair and maintenance cost of each car?

7. Explain why each of the following statements is true or false:

    a. Reducing setup times between batches always reduces the average inventory and flow time in a process.

    b. Increasing batch size always increases the average inventory and flow times in a process.

# CHAPTER 9

# Inventory Management for Deterministic Demand

## 9.0 Introduction

Inventory is a process-flow measure that affects several main functions of a firm. Operations wants to maintain enough inventory to create a smooth flow of production, allowing for more efficient production control. Finance, on the other hand, wants to minimize inventory due to capital needed to invest on inventory and projected cash flows. Marketing wants large inventory to prevent lost sales and provide high levels of service to customers.

All firms keep some kinds of inventory. In fact, the total inventory of the United States is estimated to be more than a trillion dollars, which is about $4000 for every man, woman, and child in the country.[1] Keeping inventory is costly. As we will discuss later in this chapter, buying, storing, insuring, managing, and controlling inventory require money. On the other hand, shortage of inventory is also costly. For example, when a customer demands a product and the product is not available in inventory, the firm may lose the customer. In some cases,

---

[1] Hillier, F.S., M.S. Hillier, and G.J. Lieberman, *Introduction to Management Science, A Modeling and Case Studies Approach with Spreadsheets.* Irwin/McGraw-Hill, Burr Ridge, IL, 2000.

in order to keep the customer, the firm offers to ship the item to the customer free of charge. In either case, the firm loses money due to shortage of inventory. Thus, keeping the right level of inventory—not too low and not too high—is one critical task of Inventory Management, and is the focus of this and the next two chapters.

In this chapter we discuss reasons why operations systems need to keep inventory and describe different types of inventories that firms hold. The main focus of this and the next chapter is on developing *inventory control* models to reduce the cost of inventory or to increase customer service level. While this chapter focuses on models of inventory management in which the variability in demand is negligible (i.e., deterministic demand), Chap. 10 focuses on inventory management models for systems in which the variability in demand is significant (i.e., stochastic demand).

## 9.1 Types of Inventory

Inventory management models are broad and have many applications in production and inventory systems. To be able to use these models, we first need to have a good understanding of the foundations of inventory management and control. In this section, we start by having a closer look at the types of inventory that production and service operations systems may carry.

Inventory of an operations can be classified in the following groups:

- *Raw Material Inventory:* Raw material inventory refers to the purchased (often unprocessed) items that are used in the early stages of manufacturing to be transformed into parts and components to make the final product. For example, wood and nail are raw material for furniture manufacturers. The wood is cut and shaped into different parts such as legs, back, and seat for chairs.

- *Finished Goods Inventory (FGI):* Finsihed goods inventory refers to the inventory of final products that are ready to sell. Finished chairs or tables are, for example, finished goods inventory of the furniture industry. Note that finished goods inventory of one manufacturer may be raw material for another. For example, a company that produces nails has finished goods inventory of nails. On the other hand, the company that produces furniture has nails as raw material inventory.

- *Work-in-Process (WIP) Inventory:* Work-in-process inventory refers to the inventory of items that are partially processed. This includes items between the beginning of the process (i.e., raw material inventory) and the end of the process (i.e., finished goods inventory). For example, in the furniture industry, all different parts that are made from wood (e.g., legs, back, seats) and are used in assembling the final product are WIP inventories.

- *Maintenance, Repair, and Operations Supply Inventory:* These refer to inventory of items that are not a part of the final product. They are used to facilitate the manufacturing of the product. Examples are tools, lubricants, spare parts for equipment, and cleaning material.

## 9.2 Why Do Firms Keep Inventory?

There are several reasons firms decide to keep inventory in their processes, some of which we explain in this section.

### 9.2.1 Achieving a Certain Throughput

The first reason for having inventory in a process is that, without enough inventory, firms cannot generate their desired throughput. Why? Because of Little's law that governs the relationship between inventory and throughput. As an example, consider a serial production line with five workstations. Jobs that are processed at Station $i$ are sent to Station $i + 1$ ($i = 1, 2, 3, 4$). Effective process times in all five stations is 1 minute—a balanced line. What is the minimum inventory required for the process to achieve throughput TH $= 50$ jobs per hour?

By Little's law, we have

$$\text{Inventory} = \text{Throughput} \times \text{Flow Times}$$

The minimum possible flow time for a job in line is when it does not wait in the buffers of the five stations (i.e., in-buffer flow time is zero in all five stations). In that case, the flow time in the serial line is equal to the sum of effective process times in all five stations, which is 5 minutes (i.e., 5/60 hours). Thus, the minimum inventory needed in this line to achieve throughput TH $= 50$ per hour is

$$\text{Minimum Inventory} = 50 \times (5/60) = 4.17 \text{ jobs}$$

In other words, the line cannot generate throughput of 50 jobs per hour if it does not have an inventory of *at least* 4.17 jobs (i.e., 5 jobs) in the line.

### 9.2.2 Taking Advantage of Economies of Scale

Another reason that processes hold inventory is to reduce the cost of their items through economies of scale. We illustrate this using the following case:

**Case I:**    Suppose that you like to drink your favorite imported iced coffee every day when you get home from work around 5:00 pm. The coffee is sold in 12-ounce bottles and the price is $3. You buy one bottle on your way home, put the bottle in the fridge, and drink it when you watch the 6:00 pm news on TV. In this case, you have the inventory of one bottle in your fridge for about 1 hour for your immediate use.

Now consider the following case:

**Case II:**    Suppose the store that sells the coffee does not carry the item anymore due to low demand. After searching the web, you find that you can buy the coffee for the same price of $3 from an online grocery firm. The firm, however, charges a fixed shipment cost of $5, regardless of how many bottles you order. Orders are delivered the next day. How many bottles of iced coffee will you order? Most probably you order more than one. If you order one each day (for use the next day), considering the shipment cost of $5, each bottle will cost you $8. If you order two every other day, each bottle will cost you ($5 + 2 \times$ $3)/2 = $5.5. Hence, the larger the order size, the cheaper the cost of each bottle, that is, Economies of Scale. Considering the limited space in your fridge, and the fact that you do all your shopping on the weekend, you decide to order seven bottles—1 week of supply. You place the order and you receive it the next day and put them all in your fridge.

As opposed to Case I where you had an inventory of one bottle in your fridge for *immediate use*, in Case II you are also carrying additional inventory of bottles for *future use*. This inventory is called cycle inventory.

---

**CONCEPT**

**Cycle Inventory**

*Cycle Inventory*, also known as *Cycle Stock*, is the excess inventory created when firms order (or produce) items beyond their immediate use and keep them for future use.

As mentioned above, the main reason for having cycle inventory is economies of scale. Specifically, a fixed cost (or time) related to the ordering process (e.g., shipment cost) or production process (e.g., setup costs) makes it more economical for firms to order or produce more than their immediate need.

### 9.2.3 Protecting Against Uncertainty in Demand and Supply

The third reason that firms keep inventory is to protect themselves against the possibility of stockout.

**Case III:**   It is Sunday night and you are about to place your weekly order for iced coffee, when you remember the football game the next Sunday afternoon. You always watch football games on TV together with five of your friends. You and your friends have not decided where you will watch the game, but there is a small possibility that your friends will come to your place to watch it. You know that your friends also like the imported iced coffee. Hence, you decide to order 12 bottles—5 more than your usual order of 7 bottles, in case everyone comes to your place to watch the game.

In Case III, when the order arrives, you will have an inventory of 12 bottles in your fridge. The additional five bottles are for the possibility that your friends might come to your place next Sunday to watch the game. This additional inventory is called Safety Inventory.

**CONCEPT**

**Safety Inventory**

*Safety Inventory* or *Safety Stock* is the excess inventory that is built up to protect against possible increase in demand or possible shortage in supply that may result in stockout.

To deal with uncertainty in supply, firms also build safety inventory to protect themselves against shortage in supply (e.g., delays in shipment from the supplier because of fire or earthquake or labor strike, etc.). In your iced coffee example, if you hear that the producer of your favorite iced coffee has some labor issues and may face labor strike in the coming month, you most probably order more than seven bottles to protect yourself against shortage of your favorite iced coffee in the coming weeks.

### 9.2.4 Protecting Against Uncertainty in Price

Firms also hold inventory to protect themselves against changes in the prices in the market. Let's return to our iced coffee example.

**Case IV:**   You are about to place an order for next week when you see that the company has its iced coffee for discounted sales price of $2 per bottle. Since you think that the sales discount will not continue until next week and the price will increase to its original $3, you decide to order 14 bottles (i.e., 2 weeks of supply) instead of 7 bottles.

The additional inventory of seven bottles beyond the weekly supply is the result of you speculating that the price will increase to its original price of $3. If you know for sure that the price in coming months will remain at $2, you would have not ordered more than 1 week of supply. Firms also buy more items and accumulate more inventory if they speculate that the price will increase in near future.

**Speculative Inventory**

*Speculative Inventory* or *Speculative Stock* is the excess inventory built in anticipation of changes in prices in the market.

Note that, while safety stock hedges against changes in the *amount* of inputs (i.e., supply) and outputs (i.e., demand) of a process, speculative inventory hedges against changes in the *price* of inputs and outputs of the process.

### 9.2.5 Maintaining Level Production Throughout the Year

Seasonal products are products that have periods of low demand followed by periods of high demand (or vice versa). For example, demand for items such as snowplowers or ski equipment is higher in fall and winter and lower in spring and summer, and demand for toys is higher in December. On the other hand, the capacities of manufacturers of these items are often fixed and are lower than the demand in high seasons and higher than the demand in low seasons. One strategy to deal with such seasonal demand pattern is to maintain a constant throughput—level production—throughout the year. This results in producing more products than what is needed in low seasons, carrying inventory, and using it to satisfy the demand later in high season. This inventory is called Seasonal Inventory.

**Seasonal Inventory**

*Seasonal Inventory* is the excess inventory that processes build up in low season to satisfy the demand in high season. Seasonal Inventory is the result of *Level-Production* strategy that maintains a constant throughput throughout the year.

## 9.3 A Basic Inventory Model

One main factor that affects the choice of modeling inventory problems is whether the product is make-to-order (MTO) or make-to-stock (MTS). Figure 9.1 shows the process flow in basic MTO and MTS systems.

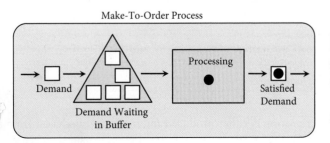
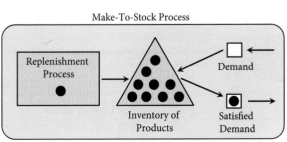

**Figure 9.1    Process-flow models of make-to-order and make-to-stock systems.**

Recall that in MTO systems, the processing of an item (e.g., an order, a job, a customer) does not start until the process receives a demand for the unit. Any new demand must join the queue of other demands before its process starts. For MTO systems the critical process-flow measure is flow time, that is, how long customers—demand—have to wait to receive their products after they place the order. For example, if you order a pizza, the key performance measure is when the pizza will be delivered to you. Chapters 7 and 8 provide models to analyze flow time as well as inventory (i.e., through Little's law) in MTO systems.

MTS processes, on the other hand, have a different dynamics, as shown in Fig. 9.1-right. Recall that in MTS systems, inventory of products is kept prior to receiving any demand. So, when a new demand arrives, it is immediately satisfied by the available inventory. For example, supermarkets hold the inventory of frozen pizzas in their freezers. If a customer wants to buy a frozen pizza, they would be able to take it from the freezer without any waiting or delay. In MTS systems, products are replenished through a replenishment process (e.g., by making it within the firm or by ordering it from a supplier). For example, when needed, supermarkets order frozen pizzas from the supplier of frozen pizza.

Because the flow time is zero for most customers in MTS systems, the key process-flow measure in MTS systems is inventory. Specifically, the main question is how much inventory should the process keep to make sure that the product is available when a demand arrives, so the demand (i.e., sales) is not lost. Larger inventory results in less number of lost sales, but larger inventory holding cost. Therefore, the main trade-off in MTS systems is between the cost of holding a large inventory and the cost of losing customers. We discuss this further in Sec. 9.7.

Many firms manage their inventory according to MTS process described above. All retailers (e.g., Walmart, Home Depot, Gap) acquire inventory and put them on their shelves prior to customer demand. Even firms that use MTO processes keep the inventory of some of their raw material according to an MTS process. For example, while a furniture manufacturer does not start to make a dining table until it receives an order—an MTO system—it still keeps inventory of several items such as standard screws, nails, and woods of different kinds as in MTS systems, so that they are immediately available when an order arrives. This is also true for service industry. For example, hotels have stock of soap and shampoo used in their rooms; and hospitals keep inventories of gauze and pads to be available when needed.

The goal of this chapter and Chaps. 10 and 11 is to provide tools and principles to manage inventory in such MTS systems. Specifically, we focus on systems in which an item is kept in inventory in anticipation of demand. Items in inventory are replenished—either by ordering them from outside of the process or by producing them within the process. The main question of how much inventory to keep and when to replenish inventory depends on several factors such as: (i) what the demand for the item is, (ii) how long it takes and how much it costs to replenish an item, and (iii) how much it costs to hold the item in inventory.

In the following section we start by identifying costs corresponding to holding inventory.

## 9.4 Costs of Inventory

The four main categories of costs associated with acquiring and holding inventory are: (i) unit cost of an item in inventory, (ii) inventory holding cost, (iii) inventory ordering or setup cost, and (iv) inventory shortage cost.

**Unit Cost of Items**
Each unit in the inventory of a process is either produced internally, or purchased from outside of the process. The *Unit Cost* of an item in inventory is the price paid to purchase the item or the internal cost of producing the item.

### Inventory Holding Cost

Inventory holding cost consists of several components as follows:

- *Opportunity Cost:* Inventory is money tied up in material and goods sitting in a process without generating any profit. Money invested in inventory could have been invested in projects with high returns. This is called the opportunity cost of capital (i.e., inventory).

- *Physical Holding Cost:* Keeping inventory requires physical space such as warehouses and distribution centers. The cost of having a warehouse or distribution center can be very significant. The cost includes depreciation cost of buildings, material handling equipment, and corresponding utility costs.

- *Manpower Cost:* Running a warehouse, for example, requires people who manage and control the inventory. Specifically, to carry inventory, we need people to keep record of the arriving and departing inventory, people who move and store inventory, security personnel to protect inventory from theft, etc.

- *Risk Cost:* Inventory may perish (e.g., foods and drugs passing their expiration date), may get stolen or become obsolete (e.g., VHS players became obsolete after DVD players entered the market). Thus, there is a risk of losing money spent in inventory.

- *Insurance Cost:* To safeguard their investment in inventory, firms may need to pay money to insure it. In fact, in some cases, it is mandatory to insure inventory (e.g., insurance for inventory of crude oil on tankers).

Considering the above costs associated with holding inventory, the question is how to compute the cost of holding, for example, 100 laptops in inventory for 1 month?

Although they believe that inventory holding cost is a real cost, the accounting department of firms do not measure it and do not have it in their reports. There are two ways one can compute the inventory holding cost. One way is to sum up all the above costs (and possibly other costs) associated with holding inventory and compute the cost per unit inventory for a specific period (e.g., $5 per item per year). Another way is to determine holding cost using Inventory Carrying Rate.

---

**CONCEPT**

### Inventory Carrying Rate

*Inventory Carrying Rate* is the cost of holding $1 of inventory for 1 year. Hence, if an item is purchased for $c$ dollars and the Inventory Carrying Rate is $i$, then the cost of holding one item in inventory for 1 year is $C_h$ and is computed as

$$C_h = i \times c$$

---

The typical inventory carrying rate is between 20 and 40 percent. The percentage depends on the type of item. For example, a 2-pound fresh lobster with the value of $20 requires refrigeration and is subject to rapid deterioration. Canned lobster with the same value of $20, however, is cheaper to hold, since it does not require refrigeration and can be kept for a long time. Thus, canned lobster will have a lower inventory carrying rate compared to that of fresh lobster.

### Inventory Order Setup Cost

There are also costs associated with ordering and purchasing items. These costs are often *fixed* and do not depend on the size of the order. They include cost of preparing the order, following the order, releasing the order, receiving the order, etc. Production systems also have fixed setup costs associated with configuring machines, tools, and equipment to produce items (to generate inventory).

### Inventory Shortage Cost

Inventory shortage cost corresponds to cost of running out of stock. There are two types of shortage costs: (i) backorder cost and (ii) lost sales cost.

- *Backorder cost* corresponds to cases where the firm is out of stock for a particular product; however, the arriving customer who wants that product is willing to wait until the product arrives, or the product is shipped to him (most probably for free). Hence, backorder costs include the cost of shipment and, in some cases, the loss of customer goodwill, which can result in losing the customer's future business.

- *Lost sales cost*, on the other hand, corresponds to cases where customers are not willing to wait for an out-of-stock item. Thus, lost sales costs include the immediate loss of profit from the missed sales as well as the possible future loss of business.

## 9.5 Demand

One of the key factors that affects inventory management decisions and determines which analytics can be used to make a good decision is the nature of demand. Two aspects of demand are crucial in inventory management: (i) whether demand is deterministic or stochastic, or (ii) whether demand is dependent or independent.

### Deterministic versus Stochastic Demand

Demand is *Deterministic* when it has no variability, that is, the future demand for an item is known with certainty. Consider a manufacturer of car audio system who has a contract to deliver 600 audio systems to a car assembly plant every day. Since the production rates of car assembly plants are very stable with almost no variability in their throughput, the demand that the audio system manufacturer faces is deterministic: exactly 600 systems per day.

*Stochastic* demand, on the other hand, is the demand that is uncertain and is not known for sure. Example of stochastic demand is the market demand for goods and services. For example, the daily demand for pasta sauce in a store or for tickets of a particular flight are stochastic (i.e., variable).

### Independent versus Dependent Demand

Demand for an item is called *independent demand* if it does not depend on the demand for other items, and only depends on the market demand. Examples are the demand for final products such as toys, appliances, light bulbs, etc. The demand for an item is called *dependent demand,* if it depends on the demand of other items. Consider a manufacturer of electric fans. While the demand for the electric fan (the final product) is independent demand—it only depends on the market—the demand for the fan blades is dependent demand since it depends on the demand for electric fans.

Since independent demand directly relates to consumer needs, it is mostly *stochastic* demand and needs to be forecasted. The dependent demand, however, is *deterministic*, since it is generated by a firm's production schedule. To illustrate this, suppose that the electric fan has four blades. Also suppose that the independent demand for electric fan is forecasted and the production schedule for the next 6 weeks is set to be: 400, 400, 350, 400, 450, and 350. Considering that each fan requires four blades, the demand for blades in the next 6 weeks is known with certainty (i.e., deterministic demand) to be: 1600, 1600, 1400, 1600, 1800, 1400 blades. We discuss the inventory management of items with dependent demand in Sec. 13.5 of Chap. 13. The focus of this chapter and Chaps. 10 and 11 is on inventory management of items with independent demand.

## 9.6 Replenishment Lead Time

There is a difference between how the inventory is replenished when the items are purchased from outside of a process versus when items are manufactured inside the process. In the former, the items are replenished simultaneously, where as in the latter, they are replenished constantly.

### Simultaneous Replenishment
As an example, consider the inventory of light bulbs (i.e., raw material inventory) of a manufacturer of desk lights. To purchase light bulbs, the manufacturer places an order for 5000 bulbs every week to the supplier of the bulbs. Two days after placing the order, the manufacturer receives the 5000 bulbs simultaneously in one shipment, and its raw material inventory of bulbs increases by 5000 instantly. This is called *Simultaneous Replenishment*.

### Continuous Replenishment
Now consider the inventory of desk lights at the end of desk light assembly line (i.e., finished goods inventory, FGI). Assuming that the line produces a desk light every 1 minute, the inventory of desk lights at FGI gradually increases at a constant rate of one per minute (or 60 per hour). This is called *Continuous Replenishment*.

### Replenishment Lead Time
In both continuous and simultaneous replenishment, there is a time lag between the time an order for purchase or an order for production is issued until the time the item is received. This time is called "Replenishment Lead Time," or simply "Lead Time."

---

**CONCEPT**

**Order and Production Lead Time**

*Order Lead Time* is the time between the placement of an order until the order is delivered. *Production lead time* is the time between releasing a job for production until the production is completed.

---

Replenishment lead time mainly consists of four time intervals: (i) The time that a firm decides to place an order until the order is actually placed. This includes going through the process of placing the order (preparing the documents, communication with the supplier, etc.). (ii) The time it takes for the supplier to have the order ready for shipment. If the supplier does not have the item in its inventory, this time includes production time of the order (if the supplier makes the item) or time to acquire the items (if the supplier orders the item from another supplier). If the supplier has the items in its inventory, this time includes picking items from inventory, packing them, and making them ready for shipment. (iii) Transshipment time, that is, the time it takes to ship the order from supplier to the firm. (iv) Receiving time, that is, the time from when the firm receives the order until the item is ready to be used by the firm. This time includes, for example, unloading the items, inspection, signing off the documents, and entering the order in the computer system.

Lead time plays an important role in inventory management decisions, especially in systems with stochastic demand. Consider a retailer who orders a particular energy drink from a supplier with 1-week lead time. During the lead time of 1 week when retailer's order has not yet arrived, the retailer faces a possibility of running out of stock if he does not have enough inventory to satisfy the demand during lead time. If the demand is highly variable, the possibility of stockout becomes larger. Thus, the decision of when to place an order for energy drinks depends on the demand pattern during the lead time. We will further discuss this in the next chapter.

## 9.7 Goals of Inventory Management

When making inventory decisions operations engineers and managers have two main goals in mind: (i) minimizing the cost of acquiring and holding inventory, and (ii) providing a high level of customer service.

### Minimizing Inventory Costs

As mentioned earlier, acquiring and holding inventory is costly. The costs include cost of purchasing items, administrative costs of ordering and shipment, cost of holding inventory, and costs associated with not satisfying a customer order (e.g., loss of profit, loss of goodwill, and costs of having backorder customers). Regardless of the type of inventory control system, one important goal of inventory management is to minimize the sum of all these costs.

### Providing High Level of Customer Service

The core idea behind customer service in an (MTS) inventory system is to satisfy customers' demand, that is, to have the product available for the customer when the customer demands it. Two commonly used metrics to measure service provided by an inventory system are: (i) Cycle-Service Levels and (ii) Fill Rate.

To understand the difference between the two service levels, suppose that you are in charge of managing inventory at a retail store. One of the items is a coffee-making machine that you order from a manufacturer in Italy. Your ordering policy is to order 50 machines when your inventory at store reaches 10 machines. When you place an order, it takes 2 weeks for the manufacturer to ship your order—lead time is 2 weeks. The demand for the machine is variable and thus you sometimes face situations when a customer asks for a coffee maker, but you are out of stock, waiting for the next shipment to arrive. Table 9.1 shows the demand for, and the shortage of, coffee maker in the last 10 times you placed an order (i.e., last 10 order cycles).

Table 9.1    Demand and Shortage in 10 Order Cycles

| Order cycle | Demand | Shortage |
|:---:|:---:|:---:|
| 1 | 44 | 0 |
| 2 | 35 | 0 |
| 3 | 68 | 8 |
| 4 | 46 | 0 |
| 5 | 58 | 2 |
| 6 | 33 | 0 |
| 7 | 54 | 1 |
| 8 | 22 | 0 |
| 9 | 60 | 5 |
| 10 | 41 | 0 |
| Total | 461 | 16 |

**Order Cycle**

*Order Cycle* is the time between receiving two consecutive orders.

For example, in the third order cycle, you had 68 customers who wanted the coffee maker, but 8 of those customers faced shortage and only $68 - 8 = 60$ customers got the machine. What services level are you offering to your customers? Well, there are two ways to look at it.

- What is the chance of not having a shortage (i.e., satisfying all demands) before the next order arrives? This is the probability of having no shortage in an order cycle.
- Overall, what fraction of customers do not face shortage?

The first question corresponds to type-1 service level also known as *Cycle-Service Level*. We can estimate cycle-service level using Table 9.1. As the table shows, in 6 out of 10 order cycles, you did not have shortage. Thus, the chance of not having a shortage in an order cycle—cycle-service level—can be estimated follows:

$$\text{Cycle-Service Level} = \frac{6}{10} = 0.6 = 60\%$$

The second question corresponds to type-2 service level, also known as *Fill Rate*. Using the data in Table 9.1, we can also find an estimate for the fill rate. Note that out of a total of 461 customers, 16 customers faced shortage. Hence, the fraction of customers who face shortage is $16/461 = 0.035$, and therefore the fraction of customers whose demand is served from the inventory (i.e., did not face shortage) is

$$\text{Fill Rate} = 1 - 0.035 = 0.965 = 96.5\%$$

Thus, under your current inventory policy, 96.5 percent of your customers of the coffee maker are served from the inventory at store; the remaining 3.5 percent face shortage. Also 60 percent of the time you would not have shortage before the next shipment arrives.

**Inventory Service Levels**

Inventory service levels correspond to the possibility of demand for an item not facing shortage and being satisfied by the inventory in the process. There are two types of inventory service levels:

- *Cycle-Service Level:* Cycle-service level, also known as "Type-1 Service Level" and denoted by $\alpha$, is the probability of not having shortage in an order cycle, that is,

$$\alpha = P(\text{having no shortage in an order cycle})$$

- *Fill Rate:* Fill rate, also known as "Type-2 Service Level" and denoted by $\beta$, is the fraction of overall demand that do not face shortage, that is,

$$\beta = 1 - \frac{\text{Demand facing shortage}}{\text{Total demand}}$$

Which service level to choose when making inventory management decision depends on the situations. For example, your inventory management (i.e., ordering) policy for the coffee maker results in low cycle-service level of 60 percent. Should you change your inventory policy to increase cycle-service level? Well, while your cycle-service level is low, 96.5 percent of your customers are served from your inventory and only 3.5 percent face shortage. This is because, although the number of cycles with shortage is high, the number of shortages in a cycle is small. So, your shortage cost is low and there is only a small fraction of customers who are unhappy. Thus, fill rate is a more appropriate service level (than cycle-service level) when the amount of shortage is more important than having or not having shortage.

On the other hand, cycle-service level is more appropriate than fill rate when having a shortage has the same consequence regardless of whether the amount of shortage is small or large. For example, suppose that you are a supplier of a component and your main customer is a large manufacturer that uses this component in its assembly line. In that case, a cycle-service level of $\alpha = 60\%$ means that 40 percent of times the large manufacturer does not receive *all* of the items in its order. This may result in occasional interruptions in the manufacturer's production, and hence the manufacturer faces costly adjustments in its production schedule and order shipments every time it faces shortage. Therefore, in this case, cycle-service level is of great importance.

**Trade-off between Costs and Customer Service**
After the appropriate type of service level (i.e., cycle-service or fill rate) is chosen, the question becomes what service level is appropriate for an item? Is service level of 95 percent good enough? Of course, higher (type-1 and type-2) service levels are better, implying lower possibility of shortage. The issue, however, is that providing higher service level is also more costly. For example, you can change your inventory management policy for coffee makers and order 50 machines when your inventory reaches 30 (instead of reaching 10). Thus, you will have $30 - 10 = 20$ more inventory during the 2 weeks that you are waiting for your next order to arrive. This reduces the chances of shortage and therefore increases both types of service levels. However, holding this additional inventory results in higher inventory cost.

Having the conflict between the goal of minimizing cost and the goal of maximizing service levels, the question is: which goal should managers choose to set their inventory policy? Cost or customer service? Well, the answer is both. When setting an inventory policy, operations engineers and managers should choose a policy that makes an acceptable trade-off between the two goals. In other words, they should decide how much increase in cost they are willing to accept to increase their service level. For example, are you willing to hold more inventory that costs you $20,000 more to increase your fill rate of the coffee makers from 96.5 to 98 percent?

In this chapter and in Chap. 10, we present analytics that obtain both cost and service levels of inventory policies, allowing managers to choose policies that result in their desired trade-off between the two.

## 9.8 Continuous Review Systems and (Q,R) Policies

Manufacturing and service firms carry a large number of different items in their inventories. A manufacturer of washing machines, for example, holds inventories of hundreds of items ranging from main components such as the electrical motors and switches to items such as tools, sand papers, pens, cleaning towels, and toilet papers. These items have different values and different levels of importance. Obviously, to plan production, it is very important to know exactly how many motors or switches the firm has at any time. But, it is not that critical for the manufacturer to know the exact number of pens and pencils in the plant at any time. More important items are monitored and controlled using "Continuous Review Systems," and less important items are often controlled using "Periodic Review Systems" which we will discuss in Sec. 10.7 of Chap. 10.

In a continuous review system, the inventory of items is constantly monitored, so at any point in time the inventory of each item is known. Computers and bar codes have made continuous review systems easy to implement. When you buy a TV set from a retailer, the cashier scans the bar code on the TV box and you pay for the item. At the same time, the computer system that keeps records of the items at the store reduces the inventory of that particular TV by one. As a result, the updated inventories of this TV (and all other items) in the store are constantly available to store managers at any time.

The main inventory control policy used in continuous review system is $(Q, R)$ policy. We use the following example to illustrate how $(Q, R)$ policy works.

### Ordering Bookshelves for Big-Mart

Mike works at the furniture department of Big-Mart, a large retail store that carries a wide range of products. One item in the furniture department is a particular bookshelf that is bought for $80 and sold for $120—a $40 profit margin. This is a high-profit item and thus its inventory is closely monitored using a continuous review system. There is variability in demand for bookshelves (i.e., stochastic demand), but the average demand seems to be five per week. A shipment of the bookshelves has just arrived and the computer shows that its inventory has increased to 25 bookshelves. To make sure that there is always enough inventory of bookshelves in the store, Mike needs to make two decisions:

- *When should the next order for the bookshelves be placed?* One month from now? Two months from now? What if the store sells all of its bookshelves in the first 2 weeks of the month? So, Mike soon realizes that ordering decision should not be made based on the time of the month. It should be based on the available inventory of bookshelves. If he has enough inventory, then he should not place an order, but if he does not have enough inventory (regardless of the time of the month), he needs to place an order. But, how much inventory is enough for not placing an order? Considering the demand of about 5 bookshelves a week, and the fact that the order lead time for bookshelves is 2 weeks, Mike expects to sell $2 \times 5 = 10$ bookshelves while he is waiting for the next shipment to arrive. So, Mike thinks 12 bookshelves is enough inventory to hold during the lead time. Mike considers the additional two bookshelves above the average demand of 10 to protect the store against possible stockout in case the demand is higher than its average of 10. This, as we mentioned, is Safety Stock. Thus, using their continuous review system, Mike can constantly monitor the inventory of bookshelves and when it reaches 12, he places the next order. But, Mike is wondering whether it is a good idea to always place an order in every order cycle when the inventory reaches 12. May be he should choose a different number (instead of 12) in each order cycle? His intuition tells him that, as long as nothing changes—no change in demand, lead time, supplier's price—there is no need to change the threshold of 12. So, Mike asks their IT staff to update their computer system such that any time the inventory reaches 12, the computer places an order with the supplier of the bookshelves.

- *How many bookshelves should be ordered?* Knowing when to place an order, Mike now needs to make a decision of how many bookshelves to order. Considering the supplier's price, the delivery cost of $400, the inventory space limitation in their warehouse, and the low demand for bookshelves, Mike decides to order 15 bookshelves each time.

So, Mike has determined their inventory policy for the bookshelves: Order 15 bookshelves any time its inventory reaches 12 or below. Following this policy for several months, Mike observes that the time between placing two orders varies. After placing an order, sometimes it took as short as 3 weeks to place the next order (store sold a lot during those weeks), and sometimes it took as long as 8 weeks to place the next order (the store did not sell many bookshelves).

The policy that always orders the same fixed quantity (e.g., 15 bookshelves) every time that the inventory reaches a certain level (e.g., 12 bookshelves) is called *Lot-Size Reorder-Point* or $(Q, R)$ policy.

---

**CONCEPT**

### (Q,R) Policy

The $(Q, R)$ policy, also known as *Lot-size Reorder-Point* policy, is used in continuous review systems in which the inventory of items is constantly monitored and an order can be placed at any time. The $(Q, R)$ policy works as follows:

- *When to place an order?* Place an order when the inventory reaches to or falls below $R$.

- *How much to order?* Always order fixed number of $Q$ units.

After understanding the $(Q, R)$ policy, in the rest of this chapter we provide a series of inventory management models for items with deterministic demand. We start with EOQ model.

## 9.9 Economic Order Quantity (EOQ)

Economic Order Quantity or EOQ model is perhaps the oldest and one of the most fundamental inventory management model that goes back to 1915. Even after 100 years, the model is still used in many inventory management situations to find the optimal order quantity $Q$ and reorder point $R$ for $(Q, R)$ policies in continuous review inventory systems. The model captures the best (the optimal) trade-off between the cost of holding inventory and the cost of placing an order. We use the case of Conphone, a producer of office phones, to illustrate the model.

**Managing Inventory of Batteries at Conphone**
Conphone is a producer of single-line and multi-line cordless phones with a high and stable demand. Its main manufacturing plant that produces 80 percent of its products is in Texas. The plant works 24 hours a day in three consecutive shifts, producing at the constant rate of 720 phones per day—its maximum production capacity. All models of single-line and multi-line phones use the same rechargeable battery packs that is purchased from a supplier in Illinois. The cost of placing an order to the supplier, including the transportation cost, is about \$200, regardless of the size of the order. The supplier requires a 1 week lead time for delivering an order. Conphone purchases each battery for \$6. The inventory carrying rate for batteries is estimated to be 25 percent. Thus, the cost of holding one battery in inventory for 1 year is $0.25 \times \$6 = \$1.5$. The supplier sells batteries in boxes that contain 600 batteries. Because the battery is one of the expensive components of the phone, Conphone uses a continuous review system to keep the records of the inventory of the batteries in its warehouse. Conphone uses a $(Q, R)$ policy with $Q = 21{,}600$ batteries (i.e., 36 boxes) and $R = 5040$ to manage its inventory. The order quantity is set based on Conphone's monthly requirement for batteries, that is, 720 per day $\times$ 30 days = 21,600 batteries (assuming 30 days in a month).

Managers of Conphone are not happy with the large money tied up in the inventory of batteries. Data shows that the plant carries an average inventory of around 10,500 batteries in its warehouse. Considering the price of $6 paid for each battery, the value of the inventory is therefore about $6 × 10,500 = $63,000. John, the plant manager, is asked to reduce the inventory of batteries while making sure that the production of the phones is not interrupted.

John realizes that to reduce inventories of batteries, he needs to order less than 21,600 batteries each time. But he knows that by ordering less, the cost per ordered batteries increases. Specifically, because of the fixed order setup cost of $200, the larger the order size, the less the total cost per battery would be. On the other hand, the larger the order size, the more the inventory of batteries they will hold. So, there is a trade-off between order setup cost and inventory holding cost. The question is then whether the current policy of ordering 21,600 phones captures the best trade between the two costs.

To compute the total cost, John uses the inventory data of batteries in the last year and draws the inventory curve as shown in Fig. 9.2.

As the figure shows, when the inventory of batteries reaches 5040, an order for $Q = 21,600$ batteries is placed. The order is received 1 week later when the inventory of batteries is about to become zero. When the order arrives, Conphone's inventory of batteries is raised to 21,600. This inventory is used at the constant rate of 720 batteries per day, which is the phone production rate at Conphone. In other words, the demand for batteries is deterministic (no variability) with constant rate of 720 per day.

Note that the inventory curve in Fig. 9.2, which resembles a saw tooth, exhibits the following structure:

- Following $(Q = 21,600, R = 5040)$ policy results in identical order cycles (i.e., triangles) that repeat every month. This is because there is *no variability* in the supplier's lead time and in Conphone's demand for batteries.

- Just when the inventory is about to reach zero, an order arrives and increases the inventory. The reason is that the orders always arrive exactly at the end of each month and the production rate (i.e., the demand for batteries) has a constant rate of 720 per day (i.e., $720 \times 30 = 21,600$ per month) with no variability. Hence, all 21,600 batteries will be gone exactly at the end of each month, when the next order arrives.

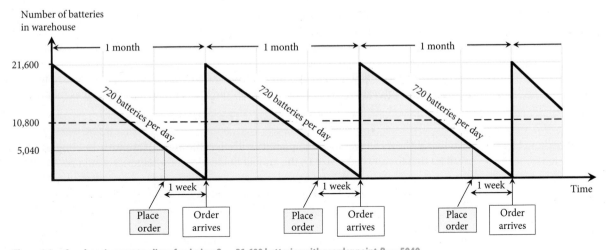

**Figure 9.2** Conphone's current policy of ordering $Q = 21,600$ batteries with reorder point $R = 5040$.

- When an order arrives, the inventory jumps to 21,600 in a straight line. The reason is that all 21,600 batteries are delivered in one shipment, that is, *simultaneous replenishment*.

Having the inventory curve, John decides to compute the total inventory, ordering and purchasing cost of batteries in a year.

**Annual Purchasing Cost**

Since Conphone uses 720 batteries per day, its annual demand for battery is 720 × 30 days × 12 months = 259,200 batteries. Considering the price of $6 that Conphone pays for each battery, the total annual purchasing cost is

$$\text{Annual Purchasing Cost} = \$6 \times \text{Annual Demand} = \$6 \times 259{,}200 = \$1{,}555{,}200$$

John realizes that, while this is a large amount, it is independent of the size of the order Conphone places with the supplier of the battery. Regardless of how often they order or how many batteries they order each time, they must purchase 259,200 batteries each year and pay $6 for each. The only way to reduce this cost is to negotiate with the supplier to get a price lower than $6.

**Annual Inventory Holding Cost**

The inventory of batteries changes throughout the year. It is at its maximum in the beginning of each order cycle (each month), and it is close to zero just about the end of the month. Thus, to compute the *annual* inventory holding cost, John needs to compute the *average inventory* in a year. To do that, John first computes the average inventory in a month, that is, in an order cycle. Recall from Chap. 2 that the average inventory in a period is equal to the area under inventory curve within that period divided by the length of the period. The area under inventory curve in each month is actually the area inside the triangle. Hence,

$$\text{Average inventory in a month} = \frac{(21{,}600 \times 1)/2}{1 \text{ month}} = \frac{21{,}600}{2} = 10{,}800 \text{ batteries}$$

John notices that the average inventory in a month is actually half of the order size 21,600. This makes sense to him since at the beginning of each month the inventory is at its maximum of 21,600 and it is zero at the end of the month. Since inventory is *linearly* decreasing at a constant rate, then the average inventory would be $(21{,}600 + 0)/2 = 10{,}800$.

Since all 12 months in a year have the same average inventory of 10,800 batteries, the average inventory in a year is also 10,800 batteries, shown by the dashed line in Fig. 9.2.[2] Also, John notices that the average inventory of 10,800 obtained in his calculation is very close to what inventory data shows, that is, 10,500. This makes him more confident about his calculations.

Having the annual average inventory of 10,800, and the holding cost of $1.5 to hold one battery in inventory for 1 year, John computes the total annual holding cost as follows:

$$\textit{Annual} \text{ holding cost} = \$1.5 \times 10{,}800 = \$16{,}200$$

Wondering how changes in order quantity will affect the annual inventory holding cost, John realizes that for any order size Q, the annual holding cost for batteries would be

$$\textit{Annual} \text{ holding cost} = \$1.5 \times \frac{Q}{2}$$

---

[2] If each month has a different average, then we should get the average of the 12 months by summing the 12 different averages and dividing by 12. But this does not happen in the EOQ model.

For example, if he changes the order quantity to $Q = 7200$ (i.e., covering 10 days of demand), then Conphone will have an average inventory of $7200/2 = 3600$ throughout the year resulting in the annual inventory holding cost of $\$1.5 \times (7200/2) = \$5400$, which is much less than the current cost of $\$16,200$. Note that $Q/2$ is the average *cycle inventory*, that is, inventory kept for future demand.

**Annual Order Setup Cost**

To compute the total annual order setup cost, John recalls that the fixed cost of placing an order (including shipment cost) is $\$200$. Thus, considering that they order every month, that is, 12 times a year, then

$$\textit{Annual} \text{ order setup cost} = \$200 \times \text{Number of orders in a year}$$
$$= \$200 \times 12 = \$2400$$

John is curious to see what happens to the annual order setup cost if he reduces the order size from 21,600 to $Q = 7200$. Since the annual demand for battery is 259,200, if John orders $Q = 7200$, then the number of orders he must place in a year is

$$\text{Number of orders per year} = \frac{\text{Demand in a year}}{Q} = \frac{259,200}{7200} = 36$$

Thus, if John orders $Q = 7200$, the annual order setup cost will be

$$\text{Annual order setup cost} = \$200 \times 36 = \$7200$$

which is much larger than the current total annual order setup cost of $\$2400$.

Hence, by switching order size from 21,600 to 7200, the total annual purchasing cost of $\$1,555,200$ will not change. However, in a year, Conphone will save $\$16,200 - \$5400 = \$10,800$ in inventory holding cost, while losing $\$7200 - \$2400 = \$4800$ in order setup cost. This results in a net annual saving of $\$10,800 - \$4800 = \$6000$. John realizes that there is an opportunity to reduce the inventory cost. The main question is: How many batteries should be ordered each time to reduce the cost to its minimum?

### 9.9.1 How Much to Order?

To find the order size that results in the lowest total annual cost, John summarizes his cost components for any given order size $Q$ as follows:

$$\text{Total annual cost of ordering } Q = \text{annual purchasing cost}$$
$$+ \text{ annual holding cost} + \text{annual order setup cost}$$

where annual purchasing cost is always $\$1,555,200$, regardless of the order size, but

$$\text{Annual Holding Cost} = \$1.5 \times \frac{Q}{2}$$

$$\text{Total Annual Order Setup Cost} = \$200 \times \frac{\text{Demand in a year}}{Q}$$

John develops an Excel spreadsheet model and computes the Annual Holding Cost and the Annual Order Setup Cost for different order sizes. Since the supplier sells batteries in boxes

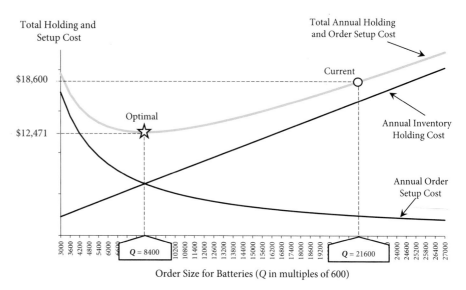

Figure 9.3    Finding the optimal order size for batteries at Conphone.

of 600 batteries, John considers $Q = 600, 1200, 1800, ...., 27,000$ in his spreadsheet—a total of 45 different order sizes. He limits his order size to 27,000, since the warehouse has a limited capacity for storing a maximum of 27,000 batteries. He then graphs the results as shown in Fig. 9.3.

As the figure shows, the order quantity that results in the lowest total cost—the optimal order quantity—is $Q = 8400$ (i.e., 14 boxes). It has a total annual holding and setup cost of $12,471, which is $6129 less than $18,600—the total annual holding and setup cost under the current policy $Q = 21,600$. Ordering less or more than 8400 will result in higher total cost.

While using spreadsheet model can help finding the optimal policy, it is a time-consuming task, especially when the number of orders to consider is large. Recall that, since the batteries are sold in boxes of 600 batteries, John's spreadsheet included the cost computation for 45 different order sizes. Now suppose that the supplier sells batteries in any quantity that Conphone wants (instead of only in multiples of 600). In this case, John's spreadsheet should compute the total holding and setup cost for 27,000 different order sizes of $Q = 1, 2, ..., 27,000$. Is there any easier way to do this? Yes, the following analytics provides a simple formula, known as EOQ formula, that yields the optimal order quantity in one simple calculation. Before we present the formula, we must first present the underlying assumptions of the EOQ model.

**Economic Order Quantity Assumptions**
Consider the buffer of an MTS process that is controlled with a $(Q, R)$ policy. Also, consider the following assumptions:

- *Constant and Deterministic Demand:* The demand occurs at a constant rate with no variability, and

$$D = \text{Total demand in a year}$$

- *Simultaneous Replenishment:* Units are ordered or produced in batches and thus all units of a batch are placed in inventory at once.

- *Constant and Deterministic Lead Time:* The production or order lead time is constant and known with no variability. Specifically, it takes $L$ units of time to produce a batch or to receive an order after the order is placed.

- *No Shortage:* Demand for units should be satisfied immediately with no shortage.

The costs associated with managing the buffer are as follows:

$c$ = Cost of producing or purchasing one unit

$K$ = Fixed production or order setup cost of a batch (independent of the batch size)

$i$ = Annual inventory carrying rate

Thus, the annual inventory holding cost is

$$C_h = i \times c$$

The following analytics presents the optimal order quantity in EOQ models.

---

**ANALYTICS**

### Finding the Optimal Order Quantity $Q^*$ in EOQ Models

**Total Annual Cost:** If the inventory is managed using a $(Q, R)$ policy, then the *total annual cost of purchasing, ordering, and holding inventory* for given order size $Q$ is ToT-Cost$(Q)$ is

$$\text{ToT-Cost}(Q) = cD + TC(Q) = cD + \left( C_h \frac{Q}{2} + K\frac{D}{Q} \right) \tag{9.1}$$

where $TC(Q)$ is the total annual order setup and inventory holding cost, which does not include purchasing cost.

**Optimal Order Quantity:** Note that Eq. (9.1) is independent of the reorder point $R$. To find the optimal $Q$ that minimizes the total annual cost, we take the derivative of Eq. (9.1), and set it equal to zero:

$$\frac{d}{dQ}\text{ToT-Cost}(Q) = \frac{d}{dQ}TC(Q) = \frac{C_h}{2} - \frac{KD}{Q^2} = 0$$

Solving for the optimal $Q^*$ that satisfies the above equation gives us[3]

$$\frac{C_h}{2} = \frac{KD}{(Q^*)^2}$$

Hence,

$$Q^* = \sqrt{\frac{2KD}{C_h}} \tag{9.2}$$

Note that,

$$\frac{C_h}{2} = \frac{KD}{(Q^*)^2} \quad \Longrightarrow \quad C_h\left(\frac{Q^*}{2}\right) = K\left(\frac{D}{Q^*}\right)$$

which implies that at the optimal order quantity $Q^*$, the annual inventory holding cost (the left-hand side) is equal to the annual order setup cost (the right-hand side). That is the reason why in Fig. 9.3, the optimal order quantity 8400 is the intersection between the annual order cost curve and the annual inventory cost curve.

**Optimal Number of Orders in a Year:** Considering the annual demand of $D$, by ordering $Q^*$ units each time, the total number of orders placed in a year is

$$\text{Number of orders in a year} = \frac{D}{Q^*}$$

**Length of an Order Cycle:** Define

$$T_{cy} = \text{Time between two consecutive order arrivals}$$

---

[3] Since second derivative of the total cost is always positive, then total cost is convex and $Q^*$ indeed minimizes (not maximizes) the cost.

then

$$\text{Optimal Length of an Order cycle} = T_{cy}^* = \frac{1}{\text{number of orders in a year}} = \frac{Q^*}{D} \text{ years}$$

**Flow time and Inventory Turn:** Since the average throughput of the inventory system is $D$ per year and the average inventory is $Q^*/2$, then using Little's law

$$\text{Average time a unit stays in inventory} = \frac{Q^*}{2D} \text{ years}$$

and the Inventory Turn is therefore

$$\text{Inventory Turn} = \frac{2D}{Q^*} = \sqrt{\frac{2C_h D}{K}} \tag{9.3}$$

**Total Optimal Annual Cost:** Substituting the optimal order quantity Eq. (9.2) in Eq. (9.1), we get

$$\text{ToT-Cost}(Q^*) = cD + C_h \frac{Q^*}{2} + K\frac{D}{Q^*} = cD + C_h \frac{\sqrt{2KD/C_h}}{2} + K\frac{D}{\sqrt{2KD/C_h}}$$

$$= cD + \sqrt{2KDC_h}$$

in which $\sqrt{2KDC_h}$ is the "total optimal annual order setup and holding cost" when ordering $Q^*$; in other words,

$$TC(Q^*) = \sqrt{2KDC_h} \tag{9.4}$$

**Inventory Service Levels:** Since all demands are satisfied from inventory and there is no shortage in any cycle, the EOQ model has cycle-service level of $\alpha = 100\%$ and fill rate $\beta = 100\%$.

Let's use Eq. (9.2) to compute Conphone's optimal order size for the batteries. Before we do that, we first need to make sure that inventory problem of batteries matches with the situation in the above analytics.

First, the demand for batteries indeed occurs at a constant and deterministic rate of 720 batteries per day. This is because the plant has a constant production rate of 720 phones per day and each phone requires one battery. Second, the replenishment of batteries is indeed simultaneous, because supplier ships all the batteries in one order. Third, the order lead time is constant with no variability. The supplier always delivers the batteries $L = 7$ days after Conphone places an order. Finally, shortage is not allowed, since managers do not want the production of phones to be interrupted because of shortage of batteries.

Since the situations in Conphone match with those required for our analytics, we can now use Eq. (9.2) to obtain the optimal order quantity. Recall that the *annual* demand for batteries is $D = 720 \times 30 \times 12 = 259{,}200$. Conphone pays $c = \$6$ for each battery in an order, with order setup cost of $K = \$200$. Also, it costs Conphone $C_h = \$1.5$ to keep one battery in inventory for 1 year. Having these numbers we can compute the optimal order quantity as follows:

$$Q^* = \sqrt{\frac{2KD}{C_h}} = \sqrt{\frac{2 \times 200 \times 259{,}200}{1.5}} = 8313.8 \text{ batteries}$$

If Conphone could order any amount (not just in boxes of 600 batteries), then the optimal order quantity would have been 8313 or 8314. However, since orders must be for boxes of 600, how should we go from the optimal order for batteries to the optimal order for boxes of batteries?

Actually, it is very simple. The optimal order quantity 8313.8 for batteries is equal to $8313.8/600 = 13.8$ boxes. Thus, the optimal order for boxes should be either 13 or 14 boxes.

All we need to do is to compare the cost of ordering 13 boxes with that of ordering 14 boxes. Note that, since the annual purchasing cost of $c \times D$ is independent of the order size, we only need to compare $TC(Q)$, the total annual inventory holding and order setup cost of both orders. For ordering 13 boxes, that is, ordering $Q = 13 \times 600 = 7800$ batteries, we have

$$TC(Q=7800) = C_h \frac{Q}{2} + K\frac{D}{Q} = \$1.5\frac{7800}{2} + \$200\frac{259,200}{7800} = \$12,496.15$$

For ordering 14 boxes, that is, ordering $Q = 14 \times 600 = 8400$ batteries, we have

$$TC(Q=8400) = C_h \frac{Q}{2} + K\frac{D}{Q} = \$1.5\frac{8400}{2} + \$200\frac{259,200}{8400} = \$12,471.43$$

which is lower than the cost of ordering 13 boxes. So, the optimal order quantity is $Q^* = 8400$ batteries. This is what John also found, except that we could get the answer by using EOQ formula in Eq. (9.2) and comparing only two order sizes of 7800 and 8400—John's spreadsheet compares 45 different order sizes to get to the same result.

The advantage of the EOQ formula is not just that it makes computing the optimal order quantity in $(R, Q)$ policy simple. The formula also reveals interesting insights and management principles that can be used to improve inventory costs. We discuss those in Chap. 11. We now present another example that shows the application of EOQ formula in a retail setting.

 **Example 9.1**

> H-Food is a supermarket that also sells a variety of dried fruits. One of its items that has a stable sales is its dried mango. The store orders the item from a supplier in West Coast with an order lead time of 5 days. H-Food pays $2 for each pound of dried mango. However, the supplier charges a shipment cost of $20 for each order, regardless of the size of the order. The sales data shows that the demand for dried mango has been steady with an average of 20 pounds per day. The store charges the inventory carrying rate of 40 percent when computing the annual inventory holding cost for 1 pound of dried mango. The store's inventory is constantly monitored—a continuous review inventory system.
>
> a. How much dried mango should H-food order each time?
> b. How many orders should H-Food place in a year?
> c. What is the time between two consecutive deliveries from the supplier in West Coast?
> d. How long, on an average, a pound of dried mango stays in H-Food before it is sold?
> e. What would be the inventory turn of dried mango?
> f. How much is the total annual inventory cost under the optimal order size?

*Part (a):* To find the optimal order quantity using the EOQ model, we need to first check whether the inventory dynamics for dried mango matches with that of the EOQ model. With a quick check we find that it does indeed satisfy the assumptions of EOQ model (i.e., deterministic and constant demand, simultaneous replenishment, etc.). Hence, we can use EOQ model to find the optimal order quantity.

Assuming 365 days in a year, the annual demand for dried mango is $D = 20 \times 365 = 7300$. The inventory cost for holding 1 pound of dried mango in the store is $C_h = i \times c = 40\% \times \$2 = \$0.8$. The order setup cost is $K = \$20$. Hence, the optimal order quantity will be

$$Q^* = \sqrt{\frac{2KD}{C_h}} = \sqrt{\frac{2 \times 20 \times 7300}{0.8}} = 604.15 \text{ pounds of dried mango}$$

*Part (b):* Since the annual demand is $D = 7300$, with order size of $Q^* = 604.2$, the total number of orders that H-Food places in a year is

$$\text{Number of orders placed in a year} = \frac{D}{Q^*} = \frac{7300}{604.15} = 12.1$$

*Part (c):* The time between two consecutive deliveries of the supplier is the length of an order cycle. Therefore,

$$\text{Length of a cycle} = \frac{1}{\text{number of orders in a year}} = \frac{1}{12.1} = 0.083 \text{ years} \times 365 \text{ days} = 30.2 \text{ days}$$

*Part (d):* The average flow time of dried mango at H-Food is

$$\text{Average time dried mango stays in inventory} = \frac{Q^*}{2D} = \frac{604.15}{2(7300)} = 0.041 \text{ years} \times 365 = 15 \text{ days}$$

*Part (e):* Inventory Turn for dried mango is therefore

$$\text{Inventory Turn} = \frac{2D}{Q^*} = \frac{2(7300)}{604.15} = 24.2 \text{ times in a year}$$

*Part (f):* The total annual cost when ordering $Q^*$ is

$$
\begin{aligned}
\text{ToT-Cost}(Q^*) &= cD + TC(Q^*) = cD + \sqrt{2KDC_h} \\
&= \$2 \times 7300 + \sqrt{2 \times \$20 \times 7300 \times \$0.8} \\
&= \$15{,}083.32
\end{aligned}
$$

**The Square Root Relationship**
Before we conclude this section we must refer to the square root relationship that holds between the optimal order quantity $Q^*$, inventory turn, cost, and demand, if orders are placed optimally according to the EOQ model. Rewriting the optimal order quantity in Eq. (9.2) we get

$$Q^* = \sqrt{\frac{2K}{C_h}} \times \sqrt{D}, \qquad \text{Inventory Turn} = \sqrt{\frac{2C_h}{K}} \times \sqrt{D}$$

Notice that both the optimal order quantity and Inventory Turn are multiple of two terms: the first term relates to the cost structure, and the second term relates to the demand. With this, we can clearly realize the following insights:

- *Order quantity and demand:* If ordered optimally, the order size increases proportionally with the *square root* of demand. For example, if demand is doubled (i.e., demand becomes $2D$—a 100-percent increase in demand), then the order quantity should not be doubled. It is enough to increase order quantity by factor $\sqrt{2}$ (i.e., by 41.4 percent) to keep the annual cost at its minimum.
- *Inventory and demand:* Recall that the average annual inventory is $Q^*/2$. Hence, if orders are placed optimally, then increase in average (cycle) inventory is also proportional to the square root of demand. Again, if the demand is doubled, the inventory should not be doubled. An increase in inventory by factor $\sqrt{2}$ (which is

about 41.4 percent) is enough to keep the optimal balance between inventory and order setup cost. Furthermore, if the demand is doubled, ordering optimally results in an increase in Inventory Turn by factor $\sqrt{2}$, that is, increasing by 41.4 percent.

- *Impact of holding and setup costs:* Order setup cost $K$ and inventory holding cost $C_h$ also have a square root relationship with order quantity and Inventory Turn. For example, if order setup cost is doubled, then optimal order quantity increases by factor $\sqrt{2}$ (i.e., by 41.4 percent) and Inventory Turn decreases by the same factor. In contrast, if inventory holding cost is doubled, then optimal order quantity decreases by factor $\sqrt{2}$ (i.e., by 41.4 percent) and Inventory Turn increases by the same factor.

- *Total cost and demand:* Since total optimal annual inventory holding cost is $C_h Q^*/2$, then it increases proportionally with $Q^*$, which in turn increases proportionally with the *square root* of demand. For example, if the demand is doubled, then the total optimal annual inventory holding cost would increase by a factor of $\sqrt{2}$ if orders are placed optimally. Also, since the optimal annual order setup cost (i.e., $KD/Q^*$) is equal to the optimal annual inventory holding cost, the optimal annual order setup cost also increases by factor $\sqrt{2}$, which is about 41.4-percent increase.

---

**PRINCIPLE**

**The Square Root Principle**

In inventory processes with order setup and inventory holding costs, if orders are placed optimally, then

- The total annual cycle inventory holding cost is the same as the total annual order setup cost.
- *Cycle inventory* and *Inventory Turn* increase with the square root of demand.
- Both total annual cycle inventory holding cost and the total annual order setup cost increase with the square root of demand.

---

### 9.9.2  When to Order?

Let's return to the case of batteries in Conphone. John decides to revise the current ($Q = 21{,}600, R = 5040$) policy to the optimal ($Q^*, R^*$) policy in which $Q^* = 8400$. But what is the optimal reorder point $R^*$? In other words, *when* should John place an order for 8400 batteries?

To find the answer, John draws the inventory curve for the new order quantity of $Q = 8400$ as in Fig. 9.4. Looking at the figure, he realizes that the answer is simple. As the figure shows, when an order of 8400 batteries arrives, considering the demand rate of 720 per day, it takes $8400/720 = 11.67$ days for the inventory to reach to zero when the next order arrives. Hence, time between arrivals of two consecutive orders is $T_{cy} = 11.67$ days.

John recalls the order lead time for batteries is 7 days. Hence, to receive an order of 8400 batteries at the end of every cycle, he needs to place an order 7 days in advance—7 days before its inventory reaches zero. Therefore, he needs to place the next order $11.67 - 7 = 4.67$ days after he receives the last order. While this is correct, John realizes that their computer system initiates an order based on inventory of batteries, not based on the time when the last order was placed. Specifically, the computer initiates an order when the inventory reaches the reorder point $R$. The question is, how much inventory does John have 4.67 days after he receives an order (which is 7 days before its inventory reaches zero, i.e., end of an order cycle)?

Number of batteries
in warehouse

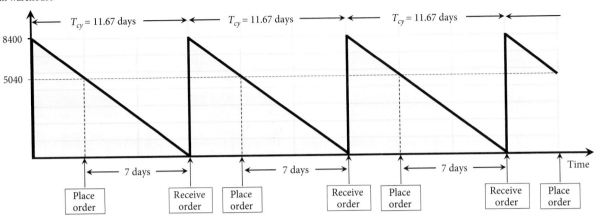

Figure 9.4   Conphone's inventory curve under the optimal policy of ordering $Q^* = 8400$.

The answer becomes obvious to John when he looks at the inventory curve in Fig. 9.4. Since they use batteries at a constant rate of 720 per day, this means that they use $7 \times 720 = 5040$ batteries in 7 days. Thus, if they place an order when the inventory of batteries reaches 5040, then 7 days later their inventory reaches zero, and that is exactly when the supplier delivers the next order. What puzzles John is that 5040 is exactly the same reorder point in the current $(Q = 21,600, R = 5040)$ policy. If he changes the order quantity to $Q^* = 8400$, shouldn't he also change the reorder point $R$? This makes him think for a while. However, after another careful look at the inventory curve in Fig. 9.4, John realizes that:

- The optimal reorder point $R^*$ is independent of the optimal order quantity $Q^*$. Its value depends only on the demand during lead time of 7 days.
- In fact, the optimal reorder point is indeed equal to the demand during lead time. Considering $D_L$ as demand during lead time, we have

$$
\begin{aligned}
R^* &= D_L \\
&= \text{Lead time in days} \times \text{demand in a day} = 7 \times 720 = 5040
\end{aligned}
$$

John concludes his analysis and makes his final decision to change the current $(Q, R)$ policy to the optimal one, that is, ordering $Q^* = 8400$ when the inventory of batteries reaches $R^* = 5040$.

### 9.9.3  Role of Inventory Position in Implementing $(Q, R)$ Policy

One advantage of the continuous review system is that the inventory is monitored constantly by computers. This makes it easy to implement the $(Q, R)$ policy. All is needed is to program the computer to initiate an order when the inventory of an item reaches or falls below its reorder point. Initiating an order might be just sending a message to the people in charge of placing an order, or directly sending an order to the supplier.

The concepts of "on-hand inventory," "pipeline inventory," "backorder inventory," and "inventory position" are essential in the implementation of $(Q, R)$ policy.

CONCEPT

**Inventory Position**

Before we introduce the concept of Inventory Position, we need to define the following:

- *On-Hand Inventory* of an item is the number of items that are physically available in inventory.
- *Pipeline Inventory* of an item is the number of items that are not physically available in inventory, but they are in transit (e.g., in trucks, in ships, in planes) and will arrive later.
- *Backorder (or Backlog) Inventory* of an item is the number of items that are not physically available in inventory, but have been promised to be shipped to the customers as soon as they arrive.

Thus, *Inventory Position* that presents the net inventory of a process is:

$$\text{Inventory Position} = \text{On-hand Inventory} + \text{Pipeline Inventory} - \text{Backorder Inventory}$$

To illustrate the concept of inventory position, let's return to the case of batteries in Conphone. Suppose that the new $(Q, R)$ policy has been implemented, that is, when "inventory" of batteries reaches or falls below reorder point $R = 5040$, Conphone places an order. The question is, which "inventory" of batteries: on-hand inventory, or inventory position?

What happens if we set Conphone's computer system to place an order when the number of batteries in its warehouse (i.e., on-hand inventory) reaches or falls below reorder point 5040? Consider the first time that this occurs. The computer will place an order for $Q = 8400$ batteries, which will arrive 7 days later. Some time later when the on-hand inventory reaches 5039, the computer will place another order for 8400, since the on-hand inventory is still below 5040. This goes on and on. Hence, obviously, the on-hand inventory should not be the determinant for when to place an order.

Now suppose that we set the computer system to place an order when the inventory position reaches or falls below reorder point $R = 5040$. This is shown in Fig. 9.5. Recall that backorders are not allowed in Conphone case, so the inventory position is the sum of

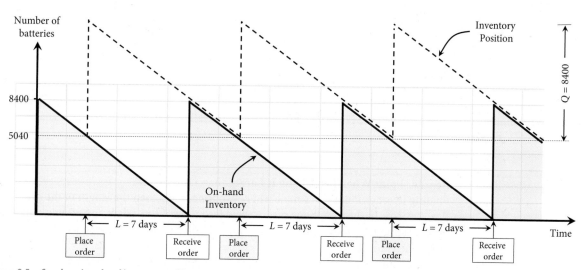

**Figure 9.5**   Conphone's on-hand inventory and inventory position under $(Q^*, R^*) = (8400, 5040)$.

on-hand inventory and pipeline inventory. Consider the first time in the figure that inventory position reaches $R = 5040$. The computer will place an order and increase the inventory position to $5040 + 8400 = 13{,}440$. The inventory position (and on-hand inventory) then decreases at the constant rate of 720 per day during the 7-day lead time. While on-hand inventory is below reorder point $R = 5040$ during the 7 days, inventory position is above the reorder point $R = 5040$, preventing system to place another order, since an order (i.e., pipeline inventory of 8400 batteries) is on the way. At the end of the 7-day lead time when the on-hand inventory reaches zero, the order arrives, and increases the on-hand inventory to 8400. This is when inventory position and on-hand inventory become equal, and this cycle repeats itself.

Before we summarize the analytics of finding the reorder point, two points worth mentioning are as follows:

- The inventory position and the on-hand inventory, as shown in Fig. 9.5, have an identical pattern. The lag between them is due to the lead time. If lead time is zero (i.e., pipeline inventory does not exist), then inventory position and on-hand inventory are the same in the EOQ model.

- Inventory position helps implementing the $(Q, R)$ policy, but, we use on-hand inventory to compute the optimal order quantity and the optimal reorder point. This is because, inventory position includes pipeline inventory. Firms usually do not incur inventory holding cost for pipeline inventory. Also, they often pay the supplier after they receive the pipeline inventory. Hence, in computing the annual holding cost under a $(Q, R)$ policy, only holding cost for on-hand inventory is considered.

---

**ANALYTICS**

**Finding the Optimal Reorder Point $R^*$ in $(Q, R)$ Policy for Systems with Deterministic and Constant Demand**

Consider the buffer of a make-to-stock (MTS) process that is managed with a continuous review system under the $(Q, R)$ policy. If the annual demand is $D$ and the lead time is $L$ (measured in years), and

$$D_L = \text{demand during lead time}$$

then the optimal Reorder Point $R^*$ is

$$R^* = D_L \tag{9.5}$$
$$= L \times D \tag{9.6}$$

The system should place an order when the *inventory position* reaches or falls below $R^*$.

**Reorder Point When Lead Time Is Larger than Order Cycle**

Finding Reorder Point $R$ is simple when lead time is less than the length of the order cycle (i.e., $L < T_{cy}$). In the Conphone case, we had lead time of $L = 7$ days and order quantity $Q = 8400$, resulting in order cycles of length $T_{cy} = 11.67$ days. Since demand per day is 720 batteries, then reorder point was simply demand during lead time, that is, $R = 7 \times 720 = 5040$.

Now suppose that the lead time for the supplier of batteries is 2 weeks, that is, $L = 14$ days. What would be the reorder point? The answer is in Fig. 9.6.

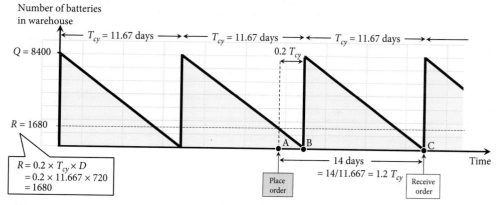

Figure 9.6    Finding reorder point when lead time is larger than order cycle.

To get the shipment of $Q = 8400$ batteries at the beginning of each other cycle of length $T_{cy} = 11.67$ days (e.g., at point C in Fig. 9.6), we must place an order $L = 14$ days in advance (i.e., at point A in the figure). Measuring lead time in term of the order cycle, this implies that we should place an order $L/T_{cy} = 14/11.67 = 1.2$ order cycles in advance. What would be the on-hand inventory at that time? As Fig. 9.6 shows, on-hand inventory at 1.2 order cycles before the arrival of the corresponding shipment at point C is exactly the same as on-hand inventory at 0.2 order cycles before the next arrival of shipment at point B. Thus, when we compute the ratio of lead time over length of the order cycle (i.e., ratio $L/T_{cy}$), what matters is the fraction value of the ratio (e.g., 0.2) not the ratio itself (i.e., 1.2).

What is on-hand inventory when fraction 0.2 of an order cycle is remaining? On-hand inventory would be equal to demand during a period of length $0.2T_{cy}$ days. Considering daily demand of 720, then demand during that period is $0.2T_{cy} \times D = (0.2)(11.67$ days$)(720$ per day$) = 1680$ batteries. Hence, each time on-hand inventory reaches or falls below 1680, Conphone should place an order for 8400 batteries.

But, this reorder point 1680 does not match with the optimal reorder point in Eq. (9.6) of our analytics for finding the reorder point. Specifically, if we use Eq. (9.6), we find the reorder point to be $R^* = L \times D = 14(720) = 10{,}080$. So, which one is the correct reorder point: 1680 or 10,080?

Well, both are correct. Note that the analytics states that we should place an order when the "Inventory Position" reaches reorder point $R = L \times D$. On the other hand, Fig. 9.6 shows that we need to place an order when "On-hand Inventory" reaches 1680. What is the inventory position when on-hand inventory is 1680? This is shown in Fig. 9.7.

Notice that, each time on-hand inventory reaches 1680, Conphone already has one order of size $Q = 8400$ that has not been delivered yet (i.e., pipeline inventory is 8400). Thus, when on-hand inventory is 1680 (just before placing an order), inventory position is $1680 + 8400 = 10{,}080$. Hence, placing an order when on-hand inventory reaches 1680 is equivalent to placing an order when inventory position reaches 10,080, which is demand during lead time of 14 days, that is, $R = L \times D = 14 \times 720 = 10{,}080$. Therefore, Eq. (9.6) also holds for finding the reorder point $R^*$ in systems in which lead time is longer than order cycles, as long as orders are placed based on inventory position (not based on on-hand inventory). If the orders are placed based on on-hand inventory, then the approach in Fig. 9.6 can be used to find the reorder point.

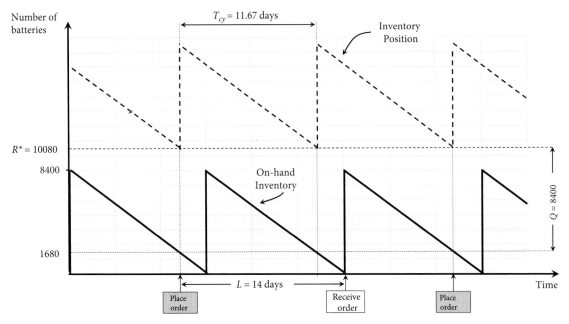

**Figure 9.7    On-hand inventory and inventory position when lead time is larger than order cycle.**

## 9.10  Robustness of the EOQ Model

The credibility of a solution obtained by modeling a management situation depends critically on how modeling assumptions represent the reality of the situations and how accurately the parameters of the model (e.g., costs, demand) are estimated. This is also true for our EOQ model. If parameters of the model such as setup costs $(K)$, holding cost $(C_h)$, or the annual demand $(D)$ are not accurate, or if the demand is not really constant and deterministic, the optimal order quantity obtained by EOQ formula in Eq. (9.2) is indeed not optimal. If that happens, the question then becomes how much using an order quantity that is not optimal increases the cost beyond the optimal cost. In other words, how sensitive is the total annual cost with respect to the deviation from the optimal order quantity?

To explore this, let's go back to the case of H-Food in Example 9.1. Recall that the demand for dried mango had a steady rate of 20 pounds per day, resulting in an optimal order quantity 604.15 pounds. Now suppose that while H-Food knows it faces a steady and constant demand, they make a 25-percent error in estimating their annual demand $D$. Specifically, to compute its order quantity, H-Food uses annual demand of $D = 9125$ pounds, which is 25-percent higher than the actual annual demand $D = 7300$ pounds. This leads to

$$Q^* = \sqrt{\frac{2KD}{C_h}} = \sqrt{\frac{2 \times 20 \times 9125}{0.8}} = 675.46 \text{ pounds of dried mango}$$

So, with 25-percent error in forecasting demand, the optimal order quantity would be 675.46 pounds. This is significantly (i.e., about 70 pounds) more than the true optimal quantity 604.15 for the actual annual demand of 7300. The question is, how much does H-Food's cost increase due to using the wrong order quantity?

To answer this, we first need to compare the annual cost for actual demand of $D = 7300$ when (i) H-Food orders optimal quantity $Q^* = 604.15$, and (ii) H-Food orders $Q = 675.46$.

Since, the annual purchasing cost is independent of the order size, we only compare the total annual setup and holding cost.

Under optimal order quantity $Q^* = 604.15$, the total annual setup and inventory holding cost is

$$TC(Q^* = 604.15) = C_h \frac{Q}{2} + K\frac{D}{Q} = \$0.8 \frac{604.15}{2} + \$20 \frac{7300}{604.15} = \$483.32$$

However, for order quantity $Q = 675.46$, the total annual setup and ordering cost under actual demand $D = 7300$ is

$$TC(Q = 675.46) = C_h \frac{Q}{2} + K\frac{D}{Q} = \$0.8 \frac{675.46}{2} + \$20 \frac{7300}{675.46} = \$486.33$$

which is only $3.01 (= \$486.33 - \$483.32)$ higher than that for the optimal order quantity. Interesting! Let's reiterate what happened. H-Food makes a 25-percent error in estimating its annual demand and places an order size which is significantly larger than the optimal; however, this results in only $\$3.01/\$483.32 = 0.6\%$ higher cost than the optimal cost. The $3.01 seems to be a very small penalty for making a 25-percent error in estimating annual demand. Is this specific to the case of dried mango in H-food, or is this true for all cases? Does this hold only with respect to the errors in estimating demand, or does it also hold for cases where other parameters (e.g., holding or setup cost) are not accurately estimated?

We answer these equations in the following analytics:

## ANALYTICS

### Robustness of the EOQ Model

Consider an Economic Order Quantity (EOQ) model with annual demand of $D$, inventory holding cost of $C_h$ and order setup cost of $K$. For this model, the optimal order quantity is $Q^*$, which is obtained using the EOQ formula in Eq. (9.2). Using Eq. (9.4), the optimal "Total Order Setup and Inventory Holding Cost" when ordering the optimal quantity $Q^*$ is $TC(Q^*) = \sqrt{2KDC_h}$.

Now suppose that for some reason (e.g., errors in estimating costs or demand) the order quantity is chosen to be $Q$, which is not optimal. Then,

$$\frac{TC(Q)}{TC(Q^*)} = \frac{C_h Q/2 + KD/Q}{\sqrt{2KDC_h}}$$

$$= \frac{C_h Q/2}{\sqrt{2KDC_h}} + \frac{KD/Q}{\sqrt{2KDC_h}}$$

$$= \frac{C_h Q/2}{C_h \sqrt{2KD/C_h}} + \frac{KD/Q}{2KD\sqrt{C_h/2KD}}$$

Considering that $\sqrt{2KD/C_h} = Q^*$ and $\sqrt{C_h/2KD} = 1/Q^*$, the above is simplified to

$$\frac{TC(Q)}{TC(Q^*)} = 0.5 \left\{ \frac{Q}{Q^*} + \frac{Q^*}{Q} \right\} \tag{9.7}$$

The above analytics leads to an important managerial insight about setting order quantity in EOQ situations. Equation (9.7) implies that, regardless of the values of holding or setup costs or demand, choosing an order quantity that is not optimal does not result in a significant increase in profit. For example, if one sets the order quantity $Q$ twice larger than the optimal order quantity $Q^*$ (i.e., $Q = 2Q^*$), then

$$\frac{TC(Q)}{TC(Q^*)} = 0.5 \left\{ \frac{Q}{Q^*} + \frac{Q^*}{Q} \right\} = 0.5 \left\{ 2 + \frac{1}{2} \right\} = 1.25$$

This implies that while the optimal annual purchasing cost remains the same (i.e., $pD$), ordering twice larger than the optimal quantity increases the annual setup and holding cost by only 25 percent.

It is important to note the underlying reason for this important feature of the EOQ formula, which is the fact that the total annual order setup and holding cost is very flat around its minimum. As an example, consider the total annual order setup and holding cost in Fig. 9.3 for batteries in Conphone. As the figure shows, while the curve's minimum occurs at $Q* = 8400$ batteries, the slope of the curve around the minimum is not very sharp. This implies that, for example, if one chooses an order quantity larger or smaller than the optimal, the impact on the optimal cost will not be that significant.

---

**PRINCIPLE**

### Robustness of EOQ Systems

In inventory processes in which items are ordered according to optimal EOQ policy, errors in estimating inventory holding cost or order setup cost, or errors in forecasting demand, do not have a significant impact on inventory management costs of the process.

Because of its robustness, the EOQ formula is used to approximate the optimal order quantity in the $(Q, R)$ policy in a variety of situations. As we will show in Chap. 10, even when the demand is stochastic, the EOQ formula provides a good estimate for the optimal order quantity for $(Q, R)$ policies.

---

**Want to Learn More?**

### Applications of EOQ Model Beyond Inventory Management

Have you taken a bus tour inside an amusement park? If you have, you have observed that buses leave according to a fixed schedule (e.g., every 30 minutes). On the other hand, visitors arrive at the bus station gradually and when the bus arrives, it takes all visitors (if it has enough space) for the next tour. There is a fixed cost of each bus trip and also there is a cost associated with keeping visitors too long waiting for the bus. Both of these costs are affected by the number of buses used and the length of the tour. How can one find the optimal number of buses and develop a bus schedule that results in minimum cost? Find the answer in the online supplement of this chapter that shows how the EOQ model can solve this problem as well as the problem of scheduling workforce training in a large engineering firm.

## 9.11 Economic Order Quantity with Quantity Discount

As we have shown in the previous sections, for a given price $c$, the total annual purchasing price $cD$ is independent of order quantity $Q$. However, there are situations where suppliers charge different prices for different order quantities or in different times of the year. In those cases, the total purchasing cost $cD$ depends on the order quantity. Thus, the optimal order quantity might be different from what Eq. (9.2) determines. In this section we consider the case called All-Unit Quantity Discount.

*Under All-Unit Quantity Discount*, the supplier charges different purchase prices for different order quantities. Specifically, the supplier is willing to charge lower price all the time for any order that is placed above a certain amount. We have discussed this in the following example.

### Example 9.2

Circuit Village is a retailer that sells electronic goods such as TVs, computers, cameras, phones, etc. One of its items is a smart wireless keyboard and mouse for computers, and its inventory is managed using a continuous review system. Its sales have been very steady at the rate of 6500 a year. The firm buys the item from a supplier who charges $20 for each set with a shipment cost of $100 for each order, regardless of the order size. Circuit Village charges an inventory carrying rate of 25 percent for computing its annual inventory holding cost. Currently Circuit Village changed its order size to 500 sets each time, which is believed to reduce the total inventory cost of keyboard to its minimum. However, they still would like to further reduce the inventory cost. They hope they would be able to negotiate for a lower price with the supplier of the keyboards. After a long phone conversation with the supplier, Jane—who is in charge of purchasing for Circuit Village—is able to get a lower price from the supplier, but only under one condition: Circuit Village must order more. Specifically, if Circuit Village orders 2000 or more each time, the supplier will charge a price of $19.2 per set. On the other hand, if Circuit Village orders more than 1000, but less than 1999, the price would be $19.5 per set. Otherwise, orders less than 1000 sets will have the same price of $20 per set. While happy that she was able to negotiate the price, Jane wonders whether this will help them reduce their costs. Obviously, lower prices such as $19.5 and $19.2 save them money, since it reduces cost of purchasing. However, increasing order size to 1000 or 2000 will result in higher inventory of keyboards and thus increases their inventory holding cost. Bottom line is that Jane does not know if the new quantity discount offered by the supplier will save them any money. What do you think? Should Circuit Village increase its order size to take advantage of the lower price, or continue with ordering 500 sets each time?

Before we find the answer, let's first formalize the supplier's quantity discount plan. The supplier is offering three different prices $c_1 = \$20$, $c_2 = \$19.5$, and $c_3 = \$19.2$, if Circuit Village's order quantity $Q$ is within Regions $\mathcal{R}_1$, $\mathcal{R}_2$, and $\mathcal{R}_3$, respectively. Then

$$c = \begin{cases} c_1 = \$20 & : \text{ if } Q \in \mathcal{R}_1 = [1, 999] \\ c_2 = \$19.5 & : \text{ if } Q \in \mathcal{R}_2 = [1000, 1999] \\ c_3 = \$19.2 & : \text{ if } Q \in \mathcal{R}_3 = [2000, \infty) \end{cases}$$

We can find Circuit Village's new optimal ordering policy following three simple steps as follows:

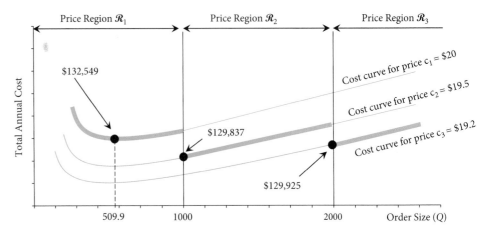

**Figure 9.8**   Total annual cost for each price $c_1 = \$20$, $c_2 = \$19.5$, and $c_3 = \$19.2$.

### Step 1—For each price, find the optimal order quantity and check if it is feasible

We find the optimal order quantity for each offered price, that is, $20, $19.5, and $19.2 using the EOQ formula as follows:

$$\text{For price } c_1 = \$20 \quad : \quad Q_1^* = \sqrt{\frac{2KD}{i \times c}} = \sqrt{\frac{2(100)(6500)}{0.25(\$20)}} = 509.9 \text{ sets}$$

$$\text{For price } c_2 = \$19.5 \quad : \quad Q_2^* = \sqrt{\frac{2KD}{i \times c}} = \sqrt{\frac{2(100)(6500)}{0.25(\$19.5)}} = 516.4 \text{ sets}$$

$$\text{For price } c_3 = \$19.2 \quad : \quad Q_3^* = \sqrt{\frac{2KD}{i \times c}} = \sqrt{\frac{2(100)(6500)}{0.25 \times \$19.2}} = 520.4 \text{ sets}$$

Figure 9.8 shows the Total Annual Cost curves for each price.

We define the optimal order quantity being *feasible* if it is within the interval corresponding to the purchase price we used to compute the optimal quantity. For example, the price $c_1 = \$20$ is valid for any order quantity in interval $\mathcal{R}_1 = [1, 999]$. For this price, the optimal quantity $Q_1^* = 509.9$ belongs to its corresponding region, that is, $Q_1^* \in \mathcal{R}_1$. Thus, order quantity $Q_1^* = 509.9$ is feasible. This is shown in Fig. 9.8. As the figure shows, the optimal quantity $Q_1^* = 509.9$ is within Region $\mathcal{R}_1$.

Now consider price $c_2 = \$19.5$. For this price, the optimal order quantity $Q_2^* = 516.4 \notin \mathcal{R}_2 = [1000, 1999]$. Hence, $Q_2^* = 516.4$ is not feasible. Similarly, for price $c_3 = \$19.2$, we have $Q_3^* = 520.4 \notin \mathcal{R}_3 = [2000, \infty)$, and thus $Q_3^* = 520.4$ is not feasible either.

### Step 2—Adjust the infeasible order quantities

In this step we only consider the orders that are not feasible and adjust them so that they become qualified for their corresponding prices. Let's start with optimal order quantity $Q_2^* = 516.4$ which is not feasible for price $c_2 = \$19.5$. Since Circuit Village cannot order $Q_2^* = 516.4$ at price $c_2 = \$19.5$, then the question is the following: At price $c_2$, what order size will have the lowest total annual cost? In other words, what order size within Region $\mathcal{R}_2$ will have the lowest total annual cost?

This can be easily answered using Fig. 9.8. As the figure shows, the total cost curve corresponding to price $c_2 = \$19.5$ is increasing within interval $\mathcal{R}_2$. This is because the minimum of the curve occurs on the left of Region $\mathcal{R}_2$—it occurs in Region $\mathcal{R}_1$. Thus, the order quantity within Region $\mathcal{R}_2$ that is qualified for price $c_2 = \$19.5$ and has the lowest total annual cost is 1000, that is, the smallest order quantity in Region $\mathcal{R}_2 = [1000, 1999]$. Hence, for price $c_2$ we adjust the optimal order quantity $Q_2^* = 516.4$ (which is not feasible) to $Q_2^{\text{adj}} = 1000$ which qualifies for price $c_2$ and has the lowest total annual cost among all orders qualified for price $c_2$. Therefore, besides $Q_1^* = 509.9$ which is feasible, we must consider $Q_2^{\text{adj}} = 1000$ as a candidate when we look for the best order quantity among candidates in Step 3.

Now consider price $c_3 = \$19.2$ for which the optimal order quantity $Q_3^* = 520.4$ is not feasible, that is, $Q_3^* = 520.4 \notin \mathcal{R}_3$. Similar to the previous case, we need to adjust this order to a quantity for price $c_3$ such that: (i) it is within Region $\mathcal{R}_3$, and (ii) it has the lowest total annual cost among all orders in that region. Again, Fig. 9.8 can do the job. As the figure shows, the total annual cost curve for price $c_3$ is increasing in Region $\mathcal{R}_3$. This, again, is because the minimum of this curve occurs in Region $\mathcal{R}_1$ (the region on the left of $\mathcal{R}_3$). Therefore, the best order quantity with minimum annual cost within region $\mathcal{R}_3 = [2000, \infty)$ is 2000. So, we adjust the optimal (infeasible) order quantity $Q_3^* = 520.4$ to the adjusted (feasible) quantity $Q_3^{\text{adj}} = 2000$.

Before we move to Step 3, we need to point out a general approach for changing the infeasible optimal order quantities to the adjusted ones. The approach is simple. For each price range, for example, $c_2$ which is a qualified price within Region $\mathcal{R}_2$, one of the following three cases occurs for the optimal order quantity $Q_2^*$ obtained in Step 1:

- **Case 1:** $Q_2^* \in \mathcal{R}_2$. In this case, $Q_2^*$ is feasible and there is no need for adjusting it. But that did not happen in the case of Circuit Village.

- **Case 2:** $Q_2^* \in \mathcal{R}_1$. In this case, $Q_2^*$ is *not* feasible, and thus we need to adjust the order size. Since $Q_2^*$ occurs in Region $\mathcal{R}_1$—on the *left* of Region $\mathcal{R}_2$—the total cost curve is *increasing* in region $\mathcal{R}_2$, and thus $Q_2^{\text{adj}}$ would be the *smallest* order size in Region $\mathcal{R}_2$. This is exactly what happened in the case of Circuit Village. We had $Q_2^* \in \mathcal{R}_1 = [1, 999]$ and thus we adjusted it to the smallest order size in $\mathcal{R}_2 = [1000, 1999]$, that is, we set $Q_2^{\text{adj}} = 1000$.

- **Case 3:** $Q_2^* \in \mathcal{R}_3$. Hence, $Q_2^*$ is *not* feasible, and thus we need to adjust its size. However, in this case, even if we adjust $Q_2^*$, the adjusted value will not be a good candidate for the final solution. For example, consider a hypothetical situation in Circuit Village where the optimal order quantity $Q_2^* = 2500$, which is infeasible and belongs to Region $\mathcal{R}_3$. To adjust it, note that $Q_2^*$ occurs in Region $\mathcal{R}_3$—on the *right* of Region $\mathcal{R}_2$. This implies that the total cost curve for price $c_2$ is *decreasing* in Region $\mathcal{R}_2$, and thus $Q_2^{\text{adj}}$ should be the *largest* order size in Region $\mathcal{R}_2$. In other words, $Q_2^{\text{adj}} = 1999$. However, the adjusted $Q_2^{\text{adj}} = 1999$ will never be a good solution because ordering one unit more, that is, $Q = 2000$ will result in a lower total annual cost. The reason is that for any order quantity $Q$, the *total annual cost* curve for price $c_3$ is always below that for price $c_2$. Thus, by ordering $Q = 2000$ (which is eligible for price $c_3 = \$19.2$), Circuit Village will have a lower total cost. Therefore, if Case 3 occurs for a region, we should not adjust the corresponding optimal order, and simply omit that region from our computation in Step 3.

### Step 3—Compare the total cost of all feasible and adjusted order quantities

Note that the feasible and the adjusted order quantities have the minimum total annual costs in their corresponding prices (i.e., regions). Hence, in this step, we compare the total cost of feasible and adjusted order quantities and choose the one with the lowest cost. We start with price $c_1 = \$20$, which has a feasible order quantity $Q_1^* = 509.9$. For this quantity, the total annual cost can be obtained as follows:

$$\text{ToT-Cost}(Q_1^* = 509.9) = c_1 D + \sqrt{2KDic_1}$$
$$= \$20(6500) + \sqrt{2(100)(6500)(25\%)(\$20)}$$
$$= \$132{,}549.5$$

For price $c_2 = \$19.5$, the optimal order quantity is not feasible and the adjusted order is $Q_2^{\text{adj}} = 1000$. Thus, the total annual cost is

$$\text{ToT-Cost}(Q_2^{\text{adj}} = 1000) = c_2 D + (i \times c_2)\frac{Q_2^{\text{adj}}}{2} + K\frac{D}{Q_2^{\text{adj}}}$$
$$= \$19.50(6500) + (25\% \times \$19.50)\frac{1000}{2} + \$100\frac{6500}{1000}$$
$$= \$129{,}837.50$$

Finally, for price $c_3 = \$19.2$, the adjusted order is $Q_3^{\text{adj}} = 2000$. Hence, we have

$$\text{ToT-Cost}(Q_3^{\text{adj}} = 2000) = c_3 D + (i \times c_3)\frac{Q_3^{\text{adj}}}{2} + K\frac{D}{Q_3^{\text{adj}}}$$
$$= \$19.20(6500) + (25\% \times \$19.20)\frac{2000}{2} + \$100\frac{6500}{2000}$$
$$= \$129{,}925$$

As the above numbers show, ordering 1000 units at price $c_2 = \$19.5$ has the lowest total annual cost. This is also clear in Fig. 9.8. The adjusted order quanity $Q_2^{\text{adj}} = 1000$, which is on the lower end of interval $1000 \leq Q \leq 1999$, has a lower total annual cost than the other two alternatives $Q_1^* = 509.9$ and $Q_3^{\text{adj}} = 2000$.

In conclusion, Circuit Village should change its order quantity from current order size of 500 to 1000 at the price of $19.5 per set. This will reduce its total annual cost from $132,550 to $129,837.5—a saving of $2712.50 a year.

Before we conclude this section, we formalize our algorithm for finding the optimal order quantity in EOQ models with All-Unit Quantity Discount.

---

**ANALYTICS**

### Finding the Optimal Order Size for EOQ Models with All-Unit Quantity Discount

Consider the buffer of a make-to-stock (MTS) process that is managed with a continuous review system under $(Q, R)$ policy. Items are ordered and are put in the buffer to satisfy the demand. The item has an annual demand $D$ and order setup cost $K$. The process charges inventory carrying rate $i$. The purchase price for the item depends on the quantity ordered, and follows the following quantity discount contract that offers $S$ different prices:

$$c = \begin{cases} c_1 & : \text{ if } Q \in \mathcal{R}_1 = (0, u_1] \\ c_2 & : \text{ if } Q \in \mathcal{R}_2 = [l_2, u_2] \\ \cdots & : \quad\quad\quad \cdots \\ c_S & : \text{ if } Q \in \mathcal{R}_s = [l_S, \infty) \end{cases}$$

where $l_j$ and $u_j$ are the lower and upper quantities in discount Region $\mathcal{R}_j$, such that $l_{j+1} > u_j$, and $\mathcal{R}_1 \cup \mathcal{R}_2 \cup \cdots \cup \mathcal{R}_S = \mathbb{R}$, and $\mathbb{R}$ is the set of real numbers.

The following steps yield the optimal order quantity that results in the minimum total annual (purchasing, setup, and holding) cost:

**Step 1:**   For each price $c_j$ compute the corresponding optimal order quantity $Q_j^*$ as follows:

$$Q_j^* = \sqrt{\frac{2KD}{i \times c_j}} \qquad \text{for } j = 1, 2, \ldots, S$$

If $Q_j^* \in \mathcal{R}_j$, then mark it as *Feasible* order quantity, and go to Step 2.

**Step 2:**   For each $Q_j^*$ that is not marked as feasible, that is, $Q_j^* \notin \mathcal{R}_j$, check the following:

- If $Q_j^*$ occurs on the left of Region $\mathcal{R}_j$ (i.e., if $Q_j^* \in \mathcal{R}_1 \cup \mathcal{R}_2 \cup \cdots \cup \mathcal{R}_{j-1}$), then set order quantity to $Q_i^{\text{adj}} = l_j$ and mark it as *Adjusted* order quantity.
- If $Q_j^*$ occurs on the right of Region $\mathcal{R}_j$ (i.e., if $Q_j^* \in \mathcal{R}_{j+1} \cup \mathcal{R}_{j+2} \cup \cdots \cup \mathcal{R}_S$), then no adjustment is needed. Omit $Q_j^*$ and discount Region $\mathcal{R}_j$ from further consideration.

**Step 3:**   For *feasible* order quantities $Q_j^*$ obtained in Step 1 compute the total annual cost as follows:

$$\text{ToT-Cost}(Q_j^*) = c_j D + \sqrt{2KD(i \times c_j)}$$

For all *adjusted* order quantities $Q_j^{\text{adj}}$ obtained in Step 2 compute the total annual cost as follows:

$$\text{ToT-Cost}(Q_j^{\text{adj}}) = c_j D + (i \times c_j)\frac{Q_j^{\text{adj}}}{2} + K\frac{D}{Q_j^{\text{adj}}}$$

The optimal order quantity is the one that has the lowest total annual cost among all feasible and adjusted quantities.

---

**Want to Learn More?**

**Trade Promotion**

*Fortune* magazine[4] reported that in the grocery industry, about 80 percent of the ordering made to manufacturers of grocery products is done through forward buying and trade promotion, corresponding to about $100 billion of inventory. As opposed to All-Unit Discount that is offered all the time, *Trade Promotion* is a price discount offered only for a short period of time (e.g., a week) during which a buyer (e.g., a grocery store) can place only one order at a discounted price. Should a firm change its order quantity when a trade promotion is offered? How does the order size depend on the discount price and the length of trade promotion? Well, you can find the answers in the online supplement of this chapter.

## 9.12 Economic Production Quantity

Recall that there are two types of inventory replenishments: (i) Simultaneous Replenishment, and (ii) Continuous Replenishment. Simultaneous replenishment often represents situations where items are purchased from outside of the process. Under simultaneous

---

[4] Sellers, P. "The Dumbest Marketing Ploy." *Fortune*, volume 126, 5, October 1992, 88-93.

replenishment after placing an order of size $Q$, all items of the order arrive simultaneously in one shipment. Hence, upon arrival of the order, the inventory instantly increases by the order size. The EOQ model discussed so far was developed to model such a replenishment.

Under continuous replenishment, however, all items do not arrive at the same time. When an order of size $Q$ is placed, items arrive at a constant rate (i.e., one-by-one) over a period of time (a day, a week, a month). Continuous replenishment often represents situations where items are produced within the process. Consider the finished goods inventory (FGI) of an auto assembly plant. The assembly line produces cars at a constant rate of about one per minute and cars are placed in the FGI at a constant rate (one per minute). We use the case of ABC Surround to illustrate this.

### Setting Production Schedule at ABC Surround

ABC Surround is a manufacturer of wireless surround sound (WSS) for home-theater systems. Its line of products includes Type-A, Type-B and Type-C series. The demand for Type-A series seems to be stable with low variability at a rate of 50 systems per day. ABC buys components of its systems from different suppliers and assembles them in its assembly plant. The most expensive and the most profitable system is Type-A series, and thus ABC schedules its production around this type. It costs ABC an average of $800 to produce a Type-A series system and it sells it for an average of $1200.

ABC uses a single assembly line to produce all different product series. A setup time of 1 day, however, is required to prepare the line to switch from producing one type of sound system to another. The setup operations cost ABC a total of $10,000. This includes the costs such as loss of production during setup and the cost of setup operations. After setting up the line to produce Type-A series, the line has the capacity of producing one Type-A series system every 6 minutes—150 per day, in a 15-hour work day. Assembled systems are moved (one-by-one) through a conveyor line to FGI. The inventory cost for Type-A series is charged at the 30 percent inventory carrying rate.

To meet the demand of 1500 in each month, ABC finishes setting up the line at the beginning of each month and produces a batch of 1500 Type-A series in the next 10 days. It then uses the line in the remaining 20 days of the month—assuming 30 days in a month—to produce other product series. ABC would like to decrease the production cost of Type-A series. One suggestion is to take advantage of economies of scale with respect to setup cost, and hence produce more than 1500 of Type-A series before switching to another type. While this reduces the setup cost per each assembled unit, it however increases the inventory cost. This is because producing a larger batch means having a larger finished goods inventory. Considering the high inventory cost of a Type-A system, i.e., $30\% \times 800 = \$240$ per year, ABC managers are not sure if producing a larger batch reduces the total inventory and setup costs. So, they are looking for the batch size that minimizes the total cost.

Before we show how to find the optimal batch size, let's first compute the total *annual* cost under the current production policy, which includes the production cost of $800 per Type-A series, the cost of $10,000 each time the line is set up for production of Type-A series, and the inventory cost of holding Type-A series in FGI. We compute these costs using the inventory curve of Type-A series in FGI as shown in Fig. 9.9.

As the figure shows, the inventory curve consists of production cycles that repeat themselves every month. Each cycle consists of two phases:

**Phase I:**  This phase corresponds to the first 10 days of each cycle. During this phase, ABC produces at the rate of 150 sets per day and puts them in FGI. This adds up to a total production of $150 \times 10 = 1500$ units. At the same time that units are produced and added to FGI, demand is depleting the FGI at the rate of 50 sets per day. That is the reason the inventory curve increases at the rate of $150 - 50 = 100$ sets per day. At

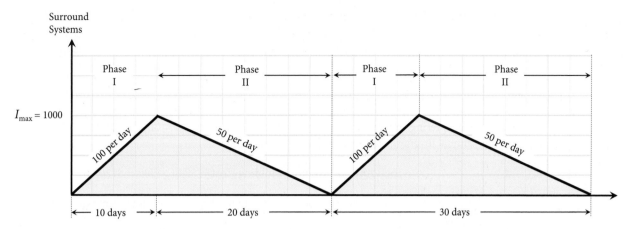

**Figure 9.9**   Inventory curve for Type-A Series in FGI under the current production schedule.

the end of this phase, production stops and the inventory reaches its maximum of $I_{max} = 10$ days $\times\ 100 = 1000$ sets.

**Phase II:**   This phase corresponds to the last 20 days of each production cycle. It starts with the maximum inventory of $I_{max} = 1000$ which is depleted at the rate of 50 per day—the demand—until the inventory reaches zero at the end of the 20 days.

The above inventory curve corresponds to a well-known production/inventory management model called *Economic Production Quantity* or EPQ model. The assumptions of the EPQ model are similar to those of the EOQ model, with one exception: In EOQ model all items in an order are added to inventory at the beginning of a cycle (i.e., simultaneous replenishment), while the production quantity in the EPQ model are added to inventory continuously during Phase I of the cycle (i.e., continuous replenishment). The production quantity in the EPQ model is called "*Production Lot Size*," or simply "*Lot Size*." We use $Q_\ell$ to denote the lot size in the EPQ model.

Using Fig. 9.9, we can find the total annual production, setup, and inventory cost under the current lot size of $Q_\ell = 1500$.

**Total Annual Production Cost**

Since the monthly demand is 1500, the annual demand would be $D = 1500 \times 12 = 18,000$. On the other hand, the production cost for one set is $c = \$800$. Since ABC must produce 18,000 sets to satisfy the annual demand, the total production cost in a year is

$$\text{Total } Annual \text{ Production Cost} = c \times D = \$800 \times 18,000 = \$14,400,000$$

**Total Annual Setup Cost**

The total annual setup cost depends on the number of setups performed in a year. Considering annual demand of $D = 18,000$, and the current production lot size $Q_\ell = 1500$, the number of setups in a year would be $D/Q_\ell = 18,000/1500 = 12$. Since the cost of a setup is $K = \$10,000$, then

$$\text{Total } Annual \text{ Setup Cost} = K\left(\frac{D}{Q_\ell}\right) = \$10,000\left(\frac{18,000}{1500}\right) = \$120,000$$

**Total Annual Inventory Cost**

Because all cycles in the inventory curve are identical, we can use the average inventory in a cycle to compute the average inventory during a year. Recall from Chap. 2 that the average inventory in a cycle is equal to the area under inventory curve during the cycle divided by the length of the cycle (i.e., $T_{cy}$). The area under inventory curve in a cycle is actually the area inside each triangle in Fig. 9.9, which is $(I_{max} \times T_{cy})/2$. Hence,

$$\text{The Average Inventory in a Cycle} = \frac{(I_{max} \times T_{cy})/2}{T_{cy}} = I_{max}/2 = 1000/2 = 500 \text{ sets}$$

Considering the annual inventory holding cost $C_h = i \times c = 30\% \times \$800 = \$240$, the total annual inventory holding cost is

$$\text{Total } \textit{Annual} \text{ Inventory Holding Cost} = \$240 \times 500 = \$120{,}000$$

Hence, the total annual production, setup, and holding cost under current policy of producing $Q_\ell = 1500$ units is

$$\text{ToT-Cost}(Q_\ell) = \$14{,}400{,}000 + \$120{,}000 + \$120{,}000 = \$14{,}640{,}000$$

The question that ABC is facing is whether increasing the lot size $Q_\ell$ decreases the total cost. More importantly, what value of $Q_\ell$ results in the lowest total annual cost? We can answer this question using the following analytics:

---

**ANALYTICS**

### Finding the Optimal Lot Size $Q_\ell$ in the EPQ Model

Consider a production process with annual effective capacity of $P$ of producing an item. Demand for the item is deterministic with the constant annual rate $D$, where $D < P$. After the process is set up for production, it starts producing the item for $T_1$ units of time and produces a lot of size $Q_\ell$. Produced items are put in the inventory from which the demand is satisfied. The process then stops production for $T_2$ time units (e.g., process is used to produce other products) until the inventory of the item reaches zero. The process then starts another cycle of producing lot size $Q_\ell$ and this schedule continues.

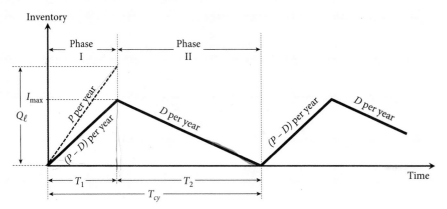

The cost of producing an item is $c$ and the setup cost is $K$. The process charges inventory carrying cost $i$ and thus the annual holding cost per item is $C_h = i \times c$. The goal is to obtain the optimal lot size $Q_\ell$ that minimizes the total annual production, setup, and holding cost.

Since the process must produce $D$ units to satisfy demand $D$ in a year, the total annual production cost is $c \times D$. Also, producing in lot size $Q_\ell$ in each cycle and the fact that each production run requires a setup means that the total number of setups in a year is $D/Q_\ell$. Considering setup cost $K$, the total annual setup cost is $K(D/Q_\ell)$.

As the above figure shows, under lot size $Q_\ell$ the maximum inventory in a cycle is $I_{max}$, which is accumulated during $T_1$ units of time. Since the inventory accumulation rate is $P - D$, then,

$$I_{max} = (P - D)T_1$$

On the other hand, all items in lot size $Q_\ell$ are produced during $T_1$ at the rate of $P$, which implies that

$$Q_\ell = PT_1 \quad \Rightarrow \quad T_1 = \frac{Q_\ell}{P}$$

Hence,

$$I_{max} = (P - D)T_1 = (P - D)\frac{Q_\ell}{P} = \left(1 - \frac{D}{P}\right)Q_\ell$$

The average inventory in a cycle is

$$\text{Average Inventory in a Cycle} = \frac{I_{max}}{2} = \left(1 - \frac{D}{P}\right)\frac{Q_\ell}{2}$$

Hence, the total annual inventory holding cost is

$$\text{Total Annual Inventory Holding Cost} = C_h\left(1 - \frac{D}{P}\right)\frac{Q_\ell}{2}$$

**Total Annual Cost:** Putting all these costs together, the total annual cost of producing in lot size $Q_\ell$ is

$$\text{ToT-Cost}(Q_\ell) = cD + C_h\left(1 - \frac{D}{P}\right)\frac{Q_\ell}{2} + K\frac{D}{Q_\ell} \tag{9.8}$$

The goal is to find the optimal value of $Q_\ell$ that minimizes the above total cost. With a closer look, we find that the above total cost for the EPQ model is exactly the same as that for the EOQ model (see Eq. (9.1)), with only one difference, that is, the inventory holding cost charged for the average inventory. Based on this similarity, we take a short cut and conclude that the optimal lot size in the EPQ model is

$$Q_\ell^* = \sqrt{\frac{2KD}{C_h(1 - D/P)}} = \sqrt{\frac{2KD}{C_h}}\sqrt{\frac{P}{P - D}} \tag{9.9}$$

**Service Levels:** Since all the demand for the item is satisfied in every production cycle, cycle-service level $\alpha = 100\%$ and fill rate $\beta = 100\%$.

Using Eq. (9.9), we can compute the optimal lot size for ABC Surround. Considering setup cost $K = \$10,000$, annual demand $D = 18,000$, annual holding cost $C_h = \$240$, and annual production capacity $P = 150 \times 30$ (days) $\times 12$ (months) $= 54,000$ units, we have

$$Q_\ell^* = \sqrt{\frac{2KD}{C_h}}\sqrt{\frac{P}{P - D}} = \sqrt{\frac{2(10,000)(18,000)}{240}}\sqrt{\frac{54,000}{54,000 - 18,000}} = 1500 \text{ units}$$

So, it seems that ABC Surround is already producing at optimal lot size. Changing the current lot size will increase their costs.

## 9.13 Summary

Operations systems cannot work without inventory such as raw material inventory, finished goods inventory (FGI), work-in-process (WIP) inventory, and maintenance, repair, and operations supply inventory. Inventory of these items is kept for several reasons. Firms keep cycle inventory, since it allows them to benefit from economies of scale in ordering items. Safety inventory is kept to safeguard against possible increase in demand or possible shortage in supply that may result in stockout. Speculative inventory is the excess inventory that firms build in anticipation of changes in prices in the market, and seasonal inventory is the inventory that firms build up in low season to satisfy the demand in high season.

Acquiring and holding inventory incur costs, including purchasing cost, inventory holding cost, inventory ordering or setup cost, and inventory shortage cost. The goal of inventory management is to develop inventory control and management policies that minimize these associated costs and provide a high level of customer service. When inventory levels are continuously monitored using firm's information technology, $(Q, R)$ policy, also known as Lot-size Reorder-Point policy, can be used to control and manage inventory effectively. Under this policy, a firm orders $Q$ units when its inventory position reaches $R$.

When variability in demand is not significant, Economic Order Quantity (EOQ) model yields the optimal order quantity $Q$ that achieves the best trade-off between order setup cost and inventory holding cost. Errors in estimating inventory holding cost or order setup cost, or errors in forecasting demand, do not have a significant impact on inventory management costs of the process in EOQ models.

Finding the optimal order quantity is also influenced by purchase price discounts offered by the supplier. In All-Unit Quantity discount, the supplier offers price discount every time an order is placed above a certain level.

While EOQ model is for situations where all items in an order are added to inventory at once (e.g., delivered by a supplier), Economic Production Quantity (EPQ) model is for cases where items are added to inventory one-by-one (e.g., produced by a recourse and added to inventory). The EPQ formula provides the production lot size that achieves the best trade-off between production setup and inventory holding costs. Both EOQ and EPQ models are suited for cases in which the variability in demand is insignificant, that is, deterministic demand. The next chapter illustrates how the EOQ model can be used in the analysis of inventory systems that face variable (i.e., stochastic) demand.

## Discussion Questions

1. Explain five reasons why firms hold inventory.
2. Describe the main reasons why holding inventory is costly.
3. Describe the differences between simultaneous replenishment and continuous replenishment and provide examples for each case.
4. What are order lead time and production lead time and what do they consist of?
5. Describe the main goals of inventory management and explain the trade-off between them.
6. Describe the characteristics of continuous review inventory systems and explain the class of policies that are appropriate for inventory control in these systems.
7. What is the difference between inventory position and on-hand inventory?
8. What are the differences between Economic Order Quantity (EOQ) and Economic Production Quantity (EPQ), and when should each one be used to manage and control inventory?

## Problems

1. Jack and Jill run two large warehouses of a retailer that sells electronics such as TV and computers. They purchase these items and keep them in their warehouses to satisfy the orders they receive from their retail stores. They purchase items from the same manufacturers and thus have the same order transaction (i.e., order setup) cost per purchase. They also have the same annual inventory holding cost per unit in their warehouses. The total number of units that Jack purchases from the manufacturers is about 180,000 a year, and his total annual inventory transaction and cycle inventory holding cost is about $450,000. Jill, on the other hand, purchases 45,000 units from the manufacturers in a year and her total annual transaction and cycle inventory cost is $150,000. Which one—Jack or Jill—is running a more efficient inventory system?

2. You work in the Distribution Center (DC) of a retailer in Chicago and have been in charge of ordering paper towels for 10 stores in south Chicago. You order towels from a producer in Ohio that delivers them to your DC. You then ship them to stores when stores place orders to your DC. The demand for the towel at your DC has been steady with no significant variability. Hence, you have been ordering optimally according to EOQ formula. The order setup cost for towels consists of only the shipment cost charged by the producer of towels, which is independent of your order size. How much additional cost above the optimal cost you would have, if
   a. The demand for towels doubles, but you do not change your order size; you instead order twice more in a year, cutting time between placing two orders to half?
   b. The shipment cost is cut to half, and therefore you cut your order size to half to save cycle inventory holding cost?
   c. You realize that you have overestimated the inventory holding cost $C_h$ by 30 percent, but you decide not to change your order size.

3. Consider Problem 2 and assume that your firm is consolidating its purchases, and they ask you to also order towels for the 40 stores in north Chicago. Now you have to place orders for 50 stores with the total demand that is three times larger than your current demand for 10 stores. The towels are all ordered from the same producer in Ohio with the same order setup cost. Explain which one of the following statements is true if you order optimally for the 50 stores:
   a. You should triple your order size.
   c. You should double your order size.
   b You should keep the order size the same, but place three times more orders in a year, cutting the time between orders to one-third.
   c. Information such as order setup cost and inventory holding cost is needed to make any conclusion.

4. H-Food, the supermarket in Example 1 in this chapter, also purchases dried peach and dried apple from two different suppliers in California. H-Food pays $4 per pound for dried peach and $2 per pound for dried apple to the suppliers. The order setup cost, which is only the shipment cost, is the same for both products, and these products are stored next to each other in the store. Sales have been steady for the two products and the total annual sales are almost the same for the two products. If H-Food orders these products optimally, then explain why each of the following statements is true or false:
   a. Dried peach has a higher inventory turn than dried apple.
   b. On an average, dried peach stays longer than dried apple at H-Food before it is purchased by the customers.
   c. Compared to dried peach, H-Food carries less inventory of dried apples at store.
   d. Dried apple has a larger total annual shipment cost than dried peach.

5. Consider Problem 4. How your answers to Parts (a) to (d) change, if the annual sales of dried peach is twice that of dried apple?

6. The optimal order quantity that H-Food in Problem 4 obtains for dried peach using economic order quantity implies that H-Food must place an order for 85 pounds each time. However, the supplier sells dried peach in packages that contain 10 pounds of the product. Thus, H-Food must decide whether to increase its order by 5 pounds to 90 pounds, or decrease it by 5 pounds to 80 pounds. Which one of the following statements is true? Explain why?
    a. Ordering 90 pounds results in lower total annual cost.
    b. Ordering 80 pounds results in lower total annual cost.
    c. Ordering 90 or 80 pounds have the same total annual cost.
    d. More information such as shipping and inventory holding cost is needed to make any conclusion.

7. The Distribution Center (DC) of Green Farms—a supermarket chain—buys organic milk in 1-gallon bottles from a supplier in Michigan at the price of $6.25 per bottle, and distributes it to its 50 stores in Michigan. The total demand of the 50 stores is about 1000 bottles in a week (assume 52 weeks in a year). The demand is steady with no significant variability. Order setup cost of the supplier in Michigan is $200, and is independent of the order size. The supplier has the order lead time of 2 days. The inventory carrying rate at DC is 36 percent.
    a. How many bottles of milk should DC order from its supplier in Michigan?
    b. The supply chain department has identified another supplier in Wisconsin that offers the same product but with the cheaper price of $6.00 per bottle. However, the supplier has order setup cost of $400 with order lead time of 4 days. Should Green Farms replace its current supplier with the one in Wisconsin?

8. Suppose shelf-life of a bottle of milk (i.e., the time from when the bottle leaves the supplier until it expires) in Problem 7 is about 5 weeks. Recently, Green Farms has been receiving complaints from some of its stores that the milk bottles that they receive from the DC are close to their expiration dates. This resulted in stores losing money because they had to discard the expired bottles or donate them 1 or 2 days before their expiration dates. The stores are demanding to get bottles that have at least 2 weeks before they expire. If it takes 1 day to ship milk from DC to its stores, what would be the answers to Parts (a) and (b) in Problem 7?

9. When a product is unique and popular, customers are willing to wait to receive the product, if the product is out of stock. Firms can take advantage of this to reduce their inventory cost and plan their inventory such that a certain fraction of customers is backordered. While having backordered customers incurs costs, in some cases, saving in inventory cost due to backorders is higher than the backorder cost. Consider the standard EOQ model with annual demand $D$, order setup cost $K$, and inventory holding cost $C_h$ per year. Suppose a firm decides to allow a maximum of $B$ backordered customers in an order cycle. The cost of each backorder customer for 1 year is $C_b$. Figure 9.10 shows the inventory curve for the EOQ model with planned backorders.

   As Fig. 9.10 shows, the order quantity in EOQ with planned backorders is $Q_b$ and the maximum inventory is $I_{max}$. Note that backorders are represented by negative inventory in the inventory curve in the figure.
    a. Show that the annual inventory holding cost is

$$C_h \frac{(Q_b - B)^2}{2Q_b}$$

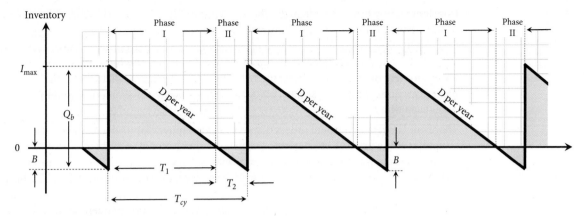

Figure 9.10    Standard EOQ models with planned backorders in Problem 9.

    b. Show that the annual order setup cost is

$$K\left(\frac{D}{Q_b}\right)$$

    c. Show that the annual backorder cost is

$$C_b\left(\frac{B^2}{2Q_b}\right)$$

    d. Show that the optimal order quantity that minimizes the total annual order setup, inventory holding, and planned backorder costs is

$$Q_b^* = \sqrt{\frac{2KD}{C_h}}\sqrt{\frac{C_h + C_b}{C_b}}$$

    and the optimal maximum number of backorders in each cycle is

$$B^* = Q_b^*\left(\frac{C_h}{C_h + C_b}\right)$$

    e. If the optimal order size in the above equation is used, what should be the optimal reorder point $R_b^*$?

10. Coby Inn is a hotel chain that offers low-price rooms with free breakfast. The hotel purchases small bottles of shampoos from a large distributor at the price of $4 per bottle. Shampoo bottles are put in rooms for daily use. Placing an order to the distributor costs $250 (which includes shipment cost), regardless of the size of the order. The distributor of shampoo delivers the order 10 days after the order is placed. Coby Inn uses 50 shampoos per day (assume 360 days in a year), and currently is ordering 4500 bottles every 3 months. The bottles are stored in the hotel's warehouse with inventory carrying rate of 20 percent.

    a. What is the hotel's current annual inventory holding and order setup cost for the shampoo bottles?

    b. What is the order size that minimizes the hotel total annual cost of ordering and holding shampoos?

    c. What is the optimal reorder point for the ordering policy in Part (b)?

    d. What is the time between two consecutive orders placed under the policy in Part (b)?

11. Consider Cobby Inn in Problem 10. The distributor of the shampoo has increased its price from \$4.00 to \$4.10. However, it informs Coby Inn about their new quantity discount plan. Specifically, the price will be reduced from \$4.10 to \$3.80 per bottle, if Coby Inn's order size is between 4000 and 5999. The price would be \$3.6, if Coby Inn orders 6000 or more. Orders for less than 4000 bottles would have the original price of \$4.10 per bottle. The shipment cost, however, would still be \$250. How many bottles of shampoo should Coby Inn order each time?

12. Thomson University hospital orders disposable exam gloves for its doctors and nurses from a supplier at the price of \$12 per box. Data shows that hospital uses about 20 boxes per day (assume 360 days in a year) at a steady rate with no significant variability. The hospital has a large supply room that keeps these gloves along with other medical supplies. The hospital has an inventory carrying rate of 25 percent. The only order setup cost associated with ordering gloves is its shipment cost which is about \$50 and is independent of the order size. Supplier's order lead time is 2 weeks.
    a. How many boxes should the hospital order each time in order to minimize its total cost of shipment and inventory holding cost? When should the hospital place its orders?
    b. What is the total annual cost, including the purchasing cost, of the gloves under the policy in Part (a)?
    c. What is the time between two consecutive orders placed by the hospital?
    d. What is the average inventory of gloves in the hospital's supply room? How long, on an average, a glove is kept in the supply room before it is used?
    e. What is inventory turn of the gloves if hospital implements the policy obtained in Part (a)?

13. Bright Desk (BD) is a manufacturer of a variety of desk lamps, all produced on two assembly lines in its Seattle plant. The plant uses 180 LED lamps in its assembly lines each day (assume 360 days in a year). BD orders the lamps from a supplier of electronics at the price of \$2.00 per lamp. The supplier has a lead time of 2 months and charges shipment cost of \$200 per order. BD charges inventory carrying rate of 30 percent for the LED lamps. The supplier has recently offered price discount depending on the size of the order. If BD orders 6000 lamps or less each time, the price is still \$2. However, for orders between 6001 and 8000 lamps, the price is reduced to \$1.75 per lamp. Finally, for orders larger than 8000 units, the price is \$1.50. How many lamps should BD order each time? What is the reorder point?

14. Consider the Bright Desk (BD) in Problem 13 and assume that the supplier does not offer any quantity discount and its price is \$2.00 per lamp for any order size. However, the supplier is increasing its shipment cost for large orders. Specifically, for orders of size 6000 or less, the order setup cost is the same as before, that is, \$200 per order. However, for orders with size larger than 6000 lamps, the order setup cost would be \$500. This is mostly due to the fact that the supplier needs to send two trucks for orders larger than 6000. How many lamps should BD order from the supplier?

15. CCAT is a manufacturer of construction vehicles such as loaders and bulldozers. Its transmission plant has the capacity of producing 10,000 transmissions each year. CCAT needs 4000 transmissions each year. Each transmission is valued about \$6000. The fixed cost of setting up to produce a batch is \$50,000. The annual inventory carrying rate for the transmission is 20 percent.
    a. How many transmissions should CCAT produce in each run in order to minimize its annual production setup and inventory holding costs of transmissions?
    b. What is the length of a production run?

      c. What is the maximum and average inventory of transmissions under the production run in Part (a)?

      d. What is the total annual productions setup and inventory holding cost of transmissions?

16. CCAT in Problem 15 uses a stamping press to make a metal fastener used to fasten hydraulic pipes in its vehicles. CCAT needs 200 fasteners each day and the press has the capacity of producing 500 fasteners a day (assume 360 days in a year). It costs CCAT $2 to make a fastener. This cost includes the overhead and variables costs such as material, labor, and press depreciation costs. It costs CCAT about $1200 to set up the process for producing fasteners. CCAT is now considering to outsource the fastener and purchase it from a supplier at the unit price of $2.20. The order setup and shipment cost of ordering from the supplier is $300, and is the same for any order size. CCAT charges inventory carrying rate of 20 percent for the fasteners in its inventory. Focusing only on inventory holding and production and order setup cost, should CCAT outsource the production of the fastener and buy it from the supplier?

# CHAPTER 10

# Inventory Management for Stochastic Demand

## 10.0 Introduction

Chapter 9 focused on inventory systems with deterministic demand—demand that has no or very small variability. In such inventory systems, the main trade-off is between two costs: order setup cost and inventory holding cost. Ordering in small quantities, but more frequently, results in lower inventory but in higher order setup cost (e.g., transactions costs, shipment costs, etc.). On the other hand, ordering in large quantities, but less frequently, results in lower order setup cost, but in higher inventory cost.

When demand is stochastic, finding the optimal inventory management policy is more difficult. The main reason is that, because the demand is uncertain, shortages may occur in each order cycle. Thus, the trade-off is among three costs: shortage cost, inventory holding cost, and order setup cost. However, the number of shortages is not predictable due to random nature of the demand. This adds another layer of complexity in finding the optimal inventory management policy.

Although finding the optimal inventory management policies in systems with stochastic demand is difficult, approximation models have been developed that result in simple policies with costs close to the optimal cost. This chapter focuses on these approximation models. We first start with the most basis inventory model that has a large range of applications—the single-period inventory system with stochastic demand.

## 10.1 Single-Period Inventory Problems with Stochastic Demand

The single-period inventory management involves a single decision of how much inventory should be purchased (or produced) to satisfy the demand during a single period. The model is also known as *Newsvendor* problem. Newsvendor problem deals with situations that involve only one order cycle. We explain the model in the following sections. We start with the cases with discrete demand distribution.

### 10.1.1 Newsvendor Problem with Discrete Demand Distribution

In this section we use a small-scale version of the newsvendor problem to explain the inventory management situation and the solution methodology of single-period inventory problems with stochastic demand. Consider a person, whom we call Newsvendor, running a newsstand in Chicago known for having newspapers and magazines from all different countries and cultures. One newspaper that has a very low demand is *WorldNews*. *WorldNews* is published daily by a publisher in Chicago. At the end of each day, Newsvendor places an order for *WorldNews* to the publisher for the next day. Newsvendor buys each paper for $1.10 and sells it for $1.50. When delivering the order each day, the publisher buys back previous day's newspapers from Newsvendor at the salvage price of $0.40 (as recycled paper).

It is the end of the day today and Newsvendor is deciding how many *WorldNews* to order for tomorrow. He is facing a trade-off between ordering too many or ordering too little.

- If he orders too many, then he might not be able to sell all and he will lose $0.70 ($= \$1.10 - \$0.40$) for every unsold *WorldNews*. This is called *Overstocking Cost*, and we show it by $C_o$.
- If he orders too little, then he may lose profit of $0.40 ($= \$1.50 - \$1.10$) for each customer that he would turn away tomorrow. This is called *Understocking Cost*, and we denote it by $C_u$.

Thus, Newsvendor must find the order quantity that strikes the optimum balance between the two costs.

Newsvendor's ordering decision is difficult because he does not know what the exact demand for *WorldNews* will be tomorrow. However, our Newsvendor knows the value of having data and thus he has been keeping records of the demand ($D$) for *WorldNews* in the last 100 days as shown in Table 10.1.

As the first column of the table shows, the daily demand for *WorldNews* is between 0 and 10. The second column shows the number of days (out of 100 days) that any particular demand was observed. For example, 12 days out of 100 days, the demand for *WorldNews* was 5. The last column of the table translates the frequency of the observed demand in the second column to the empirical probability distribution of the demand. For example, the probability of demand for *WorldNews* being 5 is $P(D = 5) = 12/100 = 0.12$. Hence, we can

Table 10.1    Demand for *WorldNews* in the Last 100 Days and its empirical demand distribution

| Demand (x) | Observed demand (days) | Probability $P(D = x)$ |
|---|---|---|
| 0 | 2 | 0.02 |
| 1 | 2 | 0.02 |
| 2 | 4 | 0.04 |
| 3 | 7 | 0.07 |
| 4 | 10 | 0.10 |
| 5 | 12 | 0.12 |
| 6 | 15 | 0.15 |
| 7 | 18 | 0.18 |
| 8 | 13 | 0.13 |
| 9 | 10 | 0.10 |
| 10 | 7 | 0.07 |

compute the average demand $\overline{D}$ as follows:

$$\overline{D} = \mathsf{E}(D) = 0(0.02) + 1(0.02) + 2(0.04) + \cdots + 9(0.10) + 10(0.07) = 6.11$$

Now, back to our Newsvendor's dilemma. Considering the demand information in Table 10.1 and purchase and selling prices of *WorldNews*, how many papers should Newsvendor order for tomorrow in order to maximize his profit? Considering that the average demand is 6.11 newspapers, should he order 6 newspapers, more than 6, or less than 6? How much more or how much less? Well, Newsvendor's options are to order $Q = 0, 1, 2, \ldots, 10$, since the demand is between 0 and 10. All he needs to do is to compute his profit for each of these 11 order quantities and choose the order quantity with the maximum profit. As an example, let's compute the profit, if Newsvendor orders $Q = 4$.

The important fact to point out is that, when ordering $Q = 4$ papers, Newsvendor's profit depends on how many customers will buy the paper tomorrow—it depends on the demand that can change between 0 and 10. If the demand, for example, turns out to be $D = 2$, then Newsvendor will sell only 2 of his papers at the price of $1.50 and will have 2 leftover papers that he will sell at salvage price of $0.40. In this case, there is no shortage of newspapers. Hence, the profit of Newsvendor if he orders $Q = 4$ *and* if the demand is $D = 2$ is

$$\begin{aligned} \mathsf{E}(\text{Profit}|\ Q = 4,\ D = 2) &= \$1.5 \times \text{sales} + \$0.4 \times \text{leftover} - \$1.1 \times \text{order quantity} \\ &= \$1.5 \times 2 + \$0.4 \times 2 - \$1.1 \times 4 \\ &= -\$0.6 \end{aligned}$$

Table 10.2 shows Newsvendor's profit for all other possible values of demand, if Newsvendor orders $Q = 4$.

However, according to Table 10.2, the demand of $D = 2$ only occurs with probability $P(D = 2) = 0.04$. If the demand turns out to be $D = 8$ with probability $P(D = 8) = 0.13$, the Newsvendor's profit will be $1.60. Considering all demand scenarios between 0 and 10 and their corresponding probabilities, we can compute Newsvendor's *Expected* (i.e., *Average*) profit if he orders $Q = 4$ as follows:

$$\begin{aligned} \mathsf{E}(\text{Profit}|\ Q = 4) &= (-2.8)\,P(D = 0) + (-1.7)\,P(D = 1) + (-0.6)\,P(D = 2) \\ &\quad + (0.5)\,P(D = 3) + \cdots + (1.6)\,P(D = 9) + (1.6)\,P(D = 10) \\ &= \$1.28 \end{aligned}$$

**Table 10.2    Computing Newsvendor's Profit for any Possible Demand when $Q = 4$**

| Demand $(x)$ | Probability $P(D=x)$ | Sales $\min(Q, x)$ | Shortage $(x - Q)^+$ | Leftover $(Q - x)^+$ | Profit |
|---|---|---|---|---|---|
| 0 | 0.02 | 0 | 0 | 4 | −2.80 |
| 1 | 0.02 | 1 | 0 | 3 | −1.70 |
| 2 | 0.04 | 2 | 0 | 2 | −0.60 |
| 3 | 0.07 | 3 | 0 | 1 | 0.50 |
| 4 | 0.10 | 4 | 0 | 0 | 1.60 |
| 5 | 0.12 | 4 | 1 | 0 | 1.60 |
| 6 | 0.15 | 4 | 2 | 0 | 1.60 |
| 7 | 0.18 | 4 | 3 | 0 | 1.60 |
| 8 | 0.13 | 4 | 4 | 0 | 1.60 |
| 9 | 0.10 | 4 | 5 | 0 | 1.60 |
| 10 | 0.07 | 4 | 6 | 0 | 1.60 |
| | Average | 3.71 | 2.40 | 0.29 | 1.28 |

Hence, ordering $Q = 4$ each day, Newsvendor should expect an average profit of $1.28 each day. To find the optimal order quantity that results in the maximum expected profit, we can repeat the above calculations and compute the expected profit for all other possible orders sizes $Q = 0, 1, 2, \ldots, 10$. The result is shown in Fig. 10.1.

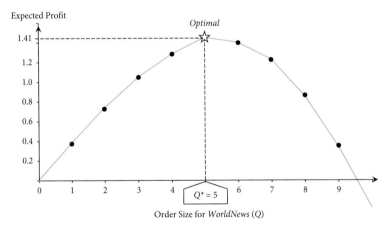

**Figure 10.1    Finding the optimal order size for *WorldNews*.**

As the figure shows, the expected profit curve is maximized at optimal order quantity $Q^* = 5$, with expected profit of $E(\text{Profit}|\ Q^* = 5) = \$1.41$.

From Table 10.2, which is for order quantity $Q = 4$, we can also observe the following:

- There is a relationship among the order quantity $(Q)$, average sales, and average leftover inventory as follows:

$$\text{Order Quantity } Q = E(\text{sales}|\ Q=4) + E(\text{leftover}|\ Q=4)$$
$$4 = 3.71 + 0.29$$

This is clear, since at the end of a day, any newspaper that was ordered is either sold (i.e., sales) or not (i.e., leftover). In fact, this relationship holds for all values of $Q$.

- There is also a relationship among average demand, average shortage, and average sales as:

$$\text{Average Demand} = \text{E(sales| } Q=4) + \text{E(shortage| } Q=4)$$
$$6.11 = 3.71 + 2.40$$

This is also clear, since at the end of a day, any demand for newspaper is either satisfied (i.e., sales) or not (i.e., shortage). This relationship also holds for all values of $Q$.

- Finally, for any value of $Q$, Newsvendor's expected profit can also be found based on average sales and average leftover inventory. For example, for $Q=4$ we have

$$\text{E(Profit| } Q=4) = \text{Average revenue} - \text{Average cost}$$
$$= \$1.50 \text{ E(sales| } Q=4) + \$0.40 \text{ E(leftover| } Q=4) - \$1.10 \times Q$$
$$\$1.28 = \$1.50 \times 3.71 + \$0.40 \times 0.29 - \$1.10 \times 4$$

Note that the revenue in the above comes from selling newspapers to customers (at $1.5) and selling the leftover newspapers to the publisher (at $0.4).

### Marginal Analysis

In our illustrative small-scale Newsvendor problem we had to compute the expected profit for 11 possible order quantities $Q = 0, 1, 2 \ldots, 10$. Computing the expected profit for each order quantity also requires some calculation as shown in Table 10.2. For realistic Newsvendor problems in which the demand for newspapers is in the order of hundreds, computing expected profit for each order size becomes tedious. Is there any easier way to find the optimal order quantity in Newsvendor problems? Yes. It is called Marginal Analysis.

The main challenge in Newsvendor problem is that both ordering too much and ordering too little are costly. Ordering too much (i.e., overstocking) results in leftovers that are sold with loss at their salvage value. Ordering too little (i.e., understocking) results in lost sales. Marginal Analysis approach is based on comparing the marginal gain of reducing the cost of understocking resulting from ordering one additional unit with the marginal loss of increasing overstocking cost resulting from ordering that additional unit. If the reduction in understocking cost is higher than the increase in overstocking cost, then the order quantity should be increased by one.

To describe the Marginal Analysis approach in detail, we first need to establish the following notation:

$$c = \text{Cost of buying a newspaper from the publisher}$$
$$p = \text{Selling price of a newspaper}$$
$$s = \text{Salvage value for an unsold newspaper}$$

Now suppose Newsvendor is thinking of ordering $Q=4$, and let's ask the question of whether he should increase his order by one unit to $Q+1=5$. If he does that, then two things can happen next day:

- $D > Q$, that is, *demand D is larger than order quantity Q:* In this case, the Newsvendor is understocked, and by increasing order quantity by one unit from $Q=4$ to $Q+1=5$, the Newsvendor will be understocked by one less unit. How much does this additional unit reduce the understocking cost? Easy! The Newsvendor can sell the additional newspaper and gain $p - c = \$1.50 - \$1.10 = \$0.40$. Thus, reducing understocking by one unit reduces the cost of understocking by $p - c$. This is called *understocking cost* of one unit and is shown by

$$C_u = p - c$$

- $D \leq Q$, that is, *demand D is smaller than or equal to order quantity Q*: In this case, the Newsvendor is overstocked, and by increasing order quantity by one unit from $Q = 4$ to $Q + 1 = 5$, the Newsvendor will be overstocked by one more unit. How much does this additional unit increase the overstocking cost? Since the additional newspaper will not be sold, the Newsvendor must sell it at salvage value $s$. This means that the additional unit that was bought for $c = \$1.10$ must be sold at salvage value $s = \$0.40$, resulting in $c - s = \$1.10 - \$0.40 = \$0.70$ less profit than if the Newsvendor would have not increased his order by one unit. In other words, one additional overstocked unit costs the Newsvendor $\$0.70$. This is called *overstocking cost* of one unit and is shown by

$$C_o = c - s$$

Hence, ordering one additional unit decreases the cost by $C_u = \$0.40$, if the Newsvendor is understocked. In contrast, ordering the additional unit increases the cost by $C_o = \$0.70$, if the Newsvendor is overstocked. So, should he or should he not increase his order quantity $Q = 4$ by one?

The answer depends on the probability that the Newsvendor is overstocked or understocked when he orders $Q = 4$. The probability of being understocked is the probability of demand being larger than $Q = 4$, that is, $P(D > 4)$, and the probability of being overstocked is the probability of demand being less than or equal to $Q = 4$, that is, $P(D \leq 4)$. Hence, the *marginal increase* in the expected profit (i.e., decrease in expected cost) from ordering one additional unit is

$$\Delta E[\text{Profit}(Q = 4 \rightarrow Q = 5)] = C_u P(D > 4) - C_o P(D \leq 4)$$

Thus, our Newsvendor should increase his order by one unit from $Q = 4$ to $Q = 5$, only if the marginal increase in his expected profit is positive, that is, only if

$$\Delta E[\text{Profit}(Q = 4 \rightarrow Q = 5)] \geq 0$$
$$C_u P(D > 4) - C_o P(D \leq 4) \geq 0$$
$$C_u [1 - P(D \leq 4)] - C_o P(D \leq 4) \geq 0$$

or if

$$P(D \leq 4) \leq \frac{C_u}{C_u + C_o} \tag{10.1}$$
$$\leq \frac{0.40}{0.40 + 0.70} = 0.364$$

In conclusion, Newsvendor should increase his order quantity from $Q = 4$ to $Q = 5$ only if $P(D \leq 4)$, that is, the cumulative distribution of the demand at $Q = 4$ is less than or equal to 0.364. Table 10.3 shows the cumulative distribution of demand for *WorldNews*.

As the table shows, $P(D \leq 4) = 0.25$ which is less than 0.364. Thus, if Newsvendor increases his order size by one unit from $Q = 4$ to $Q = 5$, there will be an increase in his expected profit.

Knowing that $Q = 5$ results in higher expected profit than $Q = 4$, the next question is whether Newsvendor should again increase his order size by one unit from $Q = 5$ to $Q = 6$? If

Table 10.3    Cumulative Distribution of Demand for *WorldNews*

| Demand (x) | Probability distribution $P(D=x)$ | Cumulative distribution $P(D \leq x)$ |
|---|---|---|
| 0 | 0.02 | 0.02 |
| 1 | 0.02 | 0.04 |
| 2 | 0.04 | 0.08 |
| 3 | 0.07 | 0.15 |
| 4 | 0.10 | 0.25 |
| 5 | 0.12 | 0.37 |
| 6 | 0.15 | 0.52 |
| 7 | 0.18 | 0.70 |
| 8 | 0.13 | 0.83 |
| 9 | 0.10 | 0.93 |
| 10 | 0.07 | 1 |

we repeat the above procedure we find that increasing order size $Q = 5$ by one unit increases the expected profit, only if

$$P(D \leq 5) \leq \frac{C_u}{C_u + C_o} = \frac{0.4}{0.4 + 0.7} = 0.364$$

However, from Table 10.3 we find that $P(D \leq 5) = 0.37$ which is *not* less than 0.364. Thus, if Newsvendor increases his order size by one unit from $Q = 5$ to $Q = 6$, there will be a *decrease* in his expected profit. Hence, Newsvendor should not increase his order from $Q = 5$ to $Q = 6$ (or any larger size). In other words, the optimal order quantity is indeed $Q^* = 5$. This is exactly what we found in Fig. 10.1.

Note that optimal order quantity $Q^* = 5$ is the smallest possible value of demand which has its cumulative distribution being larger than 0.364. This reveals the following simple property of the optimal order quantity in Newsvendor problems with discrete demand: The optimal order quantity $Q^*$ is the smallest value of demand, which we call $D_c$, for which we have

$$P(D \leq D_c) > \frac{C_u}{C_u + C_o}$$

Ratio $C_u/(C_u + C_o)$, which is called *Critical Ratio*, is the key to obtain the optimal order quantity in any Newsvendor problem.

---

### ANALYTICS

**Finding the Optimal Order Quantity Q\* in Newsvendor Problem with Empirical Demand Distribution**

Consider a Newsvendor who buys newspapers at price $c$ and sells the newspaper during the selling period (e.g., a day) at price $p$. Unsold newspapers at the end of the selling period are sold at the salvage value $s$ where $s < c$.[1] The demand $D$ for the newspapers is discrete and stochastic with probability distribution $P(D)$ and average $\overline{D}$.

**Optimal Order Quantity:** The optimal order quantity that maximizes Newsvendor's expected profit can be obtained as follows:

- **Step 1:** Compute understocking cost $C_u = p - c$ and overstocking cost $C_o = c - s$ and obtain critical ratio $\mathcal{R}_c$ as follows:

$$\mathcal{R}_c = \frac{C_u}{C_u + C_o}$$

---

[1] If $s \geq c$, then Newsvendor's problem is easy: order for the maximum possible demand, that is, order $Q = 10$ in our simple newsvendor example.

- **Step 2:** The optimal order quantity $Q^* = D_c$ is the smallest *possible* demand value that satisfies

$$P(D \leq D_c) > \mathcal{R}_c \tag{10.2}$$

**Average Shortage:** If an order of size $Q$ is placed, then the average shortage will be

$$E(\text{shortage}|\, Q) = \sum_{x=Q}^{\infty} (x - Q)P(D = x) \tag{10.3}$$

**Average Sales:** Using Eq. (10.3), the average sales during selling period can be computed as follows:

$$E(\text{sales}|\, Q) = \overline{D} - E(\text{shortage}|\, Q) \tag{10.4}$$

**Average Leftover Inventory:** Using Eq. (10.4), the average leftover inventory at the end of selling period will be

$$E(\text{leftover}|\, Q) = Q - E(\text{sales}|\, Q) \tag{10.5}$$

**Expected Profit:** Having average sales and average leftover inventory under order quantity $Q$, the expected profit can be computed as follows:

$$E(\text{Profit}|\, Q) = p\, E(\text{sales}|\, Q) + s\, E(\text{leftover}|\, Q) - c\, Q \tag{10.6}$$

### 10.1.2 Newsvendor Problems with Poisson Demand

As we mentioned in Chap. 3 Poisson distribution is commonly used to model demand for many products. Recall that Poisson probability distribution with an average demand $\overline{D}$ has the following form:

$$P(D = x) = \frac{e^{\overline{D}}\, (\overline{D})^x}{x!}: \qquad x = 0, 1, 2, \ldots$$

For Poisson distribution, the expected number of shortage in the Newsvendor problem for a given order $Q = x$, which is also called *Loss Function* $\mathcal{L}(x)$, is

$$\mathcal{L}(x) = E(\text{shortage}|\, Q = x) = \overline{D}\, P(D = x) + (\overline{D} - x) \left[1 - P(D \leq x)\right]$$

Excel spreadsheet provides simple commands to compute Poisson probability and its cumulative distributions as shown below:

$$P(D = x) = \text{POISSON}(x, \overline{D}, \text{False}) \tag{10.7}$$
$$P(D \leq x) = \text{POISSON}(x, \overline{D}, \text{True}) \tag{10.8}$$

Using the above Excel functions, the optimal order quantity for the Newsvendor problem is equal to the smallest possible demand $D_c$ that satisfies

$$\text{POISSON}(D_c, \overline{D}, \text{True}) > \mathcal{R}_c$$

The average shortage if an order of size $Q$ is placed when the demand follows Poisson distribution with mean $\overline{D}$, is, therefore

$$\begin{aligned}
\mathcal{L}(Q) &= E(\text{shortage}|\, Q) \\
&= \overline{D}\, \text{POISSON}(Q, \overline{D}, \text{False}) + (\overline{D} - Q)[1 - \text{POISSON}(Q, \overline{D}, \text{True})] \tag{10.9}
\end{aligned}$$

Having average shortage, we can compute average sales (using Eq. (10.4)), and average left-over inventory (using Eq. (10.5)). These quantities can be used in Eq. (10.6) to compute the expected profit.

 **Example 10.1**

> H-Food, a supermarket that sells organic food, buys challah bread made from organic flour from a bakery every other day at a price of $4.50 and sells it at $5.99. The store's policy is to sell fresh bread, so they do not keep the bread at the store for more than 2 days. The store orders 60 loaves of bread every 2 days. The breads that are not sold at the end of the 2 days are sold to nearby restaurants at the price of $2.00. The demand for challah during the 2 days is found to follow a Poisson distribution with an average of 50 loaves. The bakery has contacted the store manager and told him that it can offer a discount of $1.00 per loaf, if H-Food orders 75 or more loaves every other day. How many loaves of challah bread should H-Food order?

To answer this question, we first need to find the store's maximum expected profit under the current price $4.50. To do that, we first need to find the optimal order quantity under the current price.

**Expected Profit Under the Current Purchase Price:** The problem that H-Food managers are facing is a Newsvendor problem with selling period of 2 days with $c = \$4.50$, $p = \$5.99$, and $s = \$2.00$. The demand is Poisson with average of $\overline{D} = 50$ loaves. Thus,

$$C_u = p - c = \$5.99 - \$4.50 = \$1.49, \quad C_o = c - s = \$4.50 - \$2.00 = \$2.50$$

Hence, the critical ratio is computed as follows:

$$\mathcal{R}_c = \frac{C_u}{C_u + C_o} = \frac{1.49}{1.49 + 2.50} = 0.373$$

Considering average demand $\overline{D} = 50$ for the Poisson demand, the optimal order quantity is the smallest value of demand $D_c$ for which we have

$$\text{POISSON}(D_c, 50, \text{True}) > 0.373$$

Trying different values of $D_c$ in Excel function we find that

$$\text{POISSON}(47, 50, \text{True}) = 0.370, \quad \text{POISSON}(48, 50, \text{True}) = 0.425$$

Thus, the optimal order quantity is $Q^* = 48$ loaves.[2] So, the current order quantity of 60 does not maximize the store's expected profit.

To find the expected profit under the current price $4.50 with corresponding optimal order quantity $Q^* = 48$, we first need to compute the expected sales and leftover inventory under the order quantity $Q^* = 48$.

$$
\begin{aligned}
\text{E}(\text{shortage}| \, Q^* = 48) &= \mathcal{L}(Q^* = 48) \\
&= \overline{D}\,\text{POISSON}(48, \overline{D}, \text{False}) + (\overline{D} - 48)\left[1 - \text{POISSON}(48, \overline{D}, \text{True})\right] \\
&= 50 \times 0.055 + (50 - 48)[1 - 0.425)] \\
&= 3.9 \text{ loaves}
\end{aligned}
$$

---

[2] Cumulative Poisson distribution tables are also available that yield the cumulative probabilities for all values of $Q$. For different average demand, however, a different table is needed.

Having the average shortage, we can compute the average sales as follows:

$$E(\text{sales}|\, Q^* = 48) = \overline{D} - E(\text{shortage}|\, Q^* = 48) = 50 - 3.9 = 46.1$$

and having average sales, we can compute the average leftover inventory as follows:

$$E(\text{leftover}|\, Q^* = 48) = Q^* - E(\text{sales}|\, Q^* = 48) = 48 - 46.1 = 1.9$$

Hence, the expected profit under the current price $c = \$4.50$ is

$$
\begin{aligned}
E(\text{Profit}|\, Q^* = 48)) &= p\, E(\text{sales}|\, Q^* = 48) + s\, E(\text{leftover}|\, Q^* = 48) - c\, Q^* \\
&= \$5.99 \times 46.1 + \$2.00 \times 1.9 - \$4.50 \times 48 \\
&= \$63.94
\end{aligned}
$$

In other words, under the current price of $c = \$4.50$, H-Food should expect a profit of \$63.90 per every 2-day period if it orders the optimal values $Q^* = 48$ every 2 days.

**Expected Profit under Quantity Discount:** What would the maximum expected profit be if H-Food decides to take advantage of the discount offer? Under the discounted price of $c_1 = \$3.5$, we have

$$C_u = p - c_1 = \$5.99 - \$3.50 = \$2.49, \quad C_o = c_1 - s = \$3.50 - \$2.00 = \$1.50$$

Hence, the critical ratio is computed as follows:

$$\mathcal{R}_c = \frac{C_u}{C_u + C_o} = \frac{2.49}{2.49 + 1.50} = 0.624$$

Trying different values of $D_c$ in Excel function we find that

$$\text{POISSON}(51, 50, \text{True}) = 0.593, \quad \text{POISSON}(52, 50, \text{True}) = 0.646$$

Therefore, under discount offer, the optimal order quantity is $Q_1^* = 52$ loaves of bread. However, order size 52 does not qualify for the discounted price. To take advantage of the discount offer, H-Food should place an order of at least 75. The question is whether the expected profit under order quantity $Q = 75$ with purchase price $c_1 = \$3.50$ is higher than \$63.90—the expected profit of ordering optimally with $Q^* = 48$ at current price $c = \$4.50$.

Note that we do not need to compute the expected profit for orders larger than 75. This is because the optimal order quantity at the discounted price is $Q_1^* = 52$. In other words, the expected profit has its maximum at $Q^* = 52$. As the order size increases from 52, the expected profit decreases. Thus, any order size larger than 75 will have a lower expected profit than that for $Q = 75$.

Using Eqs. (10.3), (10.4), and (10.5), for $Q = 75$ and average demand $\overline{D} = 50$, we find

$$E(\text{shortage}|\, Q = 75) = 0.01, \quad E(\text{sales}|\, Q^* = 75) = 49.99, \quad E(\text{leftover}|\, Q^* = 75) = 25.01$$

Hence, the expected profit for the order size $Q = 75$ under the discount price $c_1 = \$3.50$ is

$$
\begin{aligned}
E(\text{Profit}|\, Q = 75) &= p\, E(\text{sales}|\, Q = 75) + s\, E(\text{leftover}|\, Q = 75) - c_1\, Q \\
&= \$5.99 \times 49.99 + \$2.00 \times 25.01 - \$3.50 \times 75 \\
&= \$86.96
\end{aligned}
$$

which is more than \$63.94. Thus, H-Food should increase its order from 60 to 75 to take advantage of the discount offered by the bakery.

### 10.1.3 Newsvendor Problem with Continuous Demand Distribution

As we also discussed in Chap. 3 sometimes the demand for products or services is modeled (or approximated) by continuous probability distribution (instead of Poisson distribution or an empirical distribution such as that in Table 10.1). There are several reasons for that. One reason is that some items are measured in continuous scale (e.g., pounds, tons, ounces, etc.). Another reason is that sometimes continuous distributions are better fit for modeling demand than discrete distributions. And, finally, working with discrete distribution becomes harder for cases with large demand. For example, there are no cumulative distribution tables for Poisson distribution with average demand of 5000 units. In fact, it is shown that Poisson distribution with large averages can be well-approximated by Normal distribution (which is a continuous distribution).

Thus, in this section we focus on Newsvendor problems in which the demand follows (or is well-approximated by) a continuous probability distribution with density function $f(x)$ and cumulative distribution function $F(x)$. We first present the following analytics. The details of the analytics can be found in online Appendix D.

---

**ANALYTICS**

**Finding the Optimal Order Quantity $Q^*$ in Newsvendor Problem with Continuous Demand**

Suppose demand for newspapers is approximated by a continuous probability distribution with density function $f(x)$ and cumulative distribution $F(x)$. Then, the optimal order quantity that maximizes Newsvendor's expected profit can be obtained as follows:

- **Step 1:** Compute understocking cost $C_u = p - c$ and overstocking cost $C_o = c - s$ and obtain critical ratio $\mathcal{R}_c$ as follows:

$$\mathcal{R}_c = \frac{C_u}{C_u + C_o}$$

- **Step 2:** The optimal order quantity is $Q^*$ that satisfies

$$F(Q^*) = P(D \le Q^*) = \mathcal{R}_c \tag{10.10}$$

If the item must be ordered in integer numbers, and $Q^*$ is not an integer, then the optimal order quantity is obtained by rounding $Q^*$ up to the next integer.

**Average Shortage:** If an order of size $Q$ is placed, then the average shortage will be

$$E(\text{shortage}|\, Q) = \int_Q^\infty (x - Q)f(x)dx \tag{10.11}$$

**Average Sales, Average Leftover Inventory, and Average Profit:** Average Sales, Average Inventory, and Average Profit can be found, respectively, using the same Eqs. (10.4), (10.5), and (10.6) that we had for the case of discrete demand distribution.

---

We use the above analytics in the following example.

 Example 10.2

> Consider the inventory management of challah breads at H-Food in Example 10.1. What is the optimal order quantity if the demand for challah bread is well-approximated by a Uniform Distribution between 40 and 100?

**Current Price:** Recall that the critical ratio for the current price was $\mathcal{R}_c = 0.373$. For this price the optimal order quantity would be $Q^*$ for which $F(Q^*) = 0.373$. Note that the cumulative probability distribution for uniform random variable with parameters $a$ and $b$ is

$$F(x) = \frac{x - a}{b - a}$$

Thus, the optimal order quantity would be

$$F(Q^*) = \frac{Q^* - a}{b - a} = \mathcal{R}_c \implies Q^* = (b - a)\mathcal{R}_c + a$$

Since for challah bread $a = 40$ and $b = 100$, we have

$$Q^* = (b - a)\mathcal{R}_c + a = (100 - 40)0.373 + 40 = 62.38 \text{ loaves}$$

Since challah breads are sold in loaves—an integer value—the optimal order quantity is obtained by rounding the value 62.38 up to the next integer $Q^* = 63$.

Why should we always round up and not round down? The answer comes from our analytics for discrete demand distributions. Note that since $F(Q^* = 62.38) = P(D \leq 62.38) = \mathcal{R}_c = 0.373$, then we have

$$P(D \leq 62) \leq \mathcal{R}_c, \qquad P(D \leq 63) > \mathcal{R}_c$$

Hence, according to marginal analysis, the optimal order quantity would be $Q^* = 63$.

**Discounted Price:** Recall that the critical ratio for the discounted price $c_1 = \$3.50$ is $\mathcal{R}_c = 0.624$. Thus, to find the optimal order quantity, we have

$$Q^* = (b - a)\mathcal{R}_c + a = (100 - 40)0.624 + 40 = 77.44 \text{ loaves}$$

Rounding up we get $Q^* = 78$. Since the discounted price is offered for any order quantity larger than 75, then H-Food should place an order for $Q^* = 78$ loaves of challah bread every 2 days.

### 10.1.4 Newsvendor Problem with Normal Demand Distribution

As mentioned in Chap. 3, one of the commonly used distributions to model demand is Normal distribution, which has the following density function:

$$f(x) = \frac{1}{\sigma_D \sqrt{2\pi}} \, e^{-\frac{1}{2}[(x - \overline{D})/\sigma_D]^2}$$

where $\overline{D}$ is the average demand and $\sigma_D$ is the standard deviation of the demand.

Normal distribution with mean 0 and standard deviation 1 is called *standard Normal distribution*, and random variable $Z$ is used to show it. Standard Normal distribution has the density and cumulative distribution functions as follows:

$$\phi(z) = \frac{1}{\sqrt{2\pi}} \, e^{-\frac{1}{2}(z)^2}$$

$$\Phi(z) = P(Z \leq z) = \int_{-\infty}^{z} \phi(u) du$$

The relationship between a Normal random variable $X$—with mean $\overline{D}$ and standard deviation $\sigma_D$—and standard normal random variable $Z$ is as follows:

$$Z = \frac{X - \overline{D}}{\sigma_D}$$

Thus, standard Normal random variable $Z$ will have mean of zero and standard deviation one. The above relationship between $Z$ and $X$ results in the following relationships between Normal and standard Normal distributions and their loss functions:

$$f(x) = \left(\frac{1}{\sigma_D}\right) \phi(z)$$

$$F(x) = P(D \le x) = \Phi(z) \quad \text{where} \quad z = \frac{x - \overline{D}}{\sigma_D}$$

$$\mathcal{L}(x) = \sigma_D \mathcal{L}(z)$$

where

$$\mathcal{L}(z) = \phi(z) - z[1 - \Phi(z)] \tag{10.12}$$

Standard Normal distribution plays an important role in computing cumulative distribution and loss function of any Normal distribution. Specifically, there exist tables such as the table in the Appendix that provides the values of probability density, cumulative distribution, and loss function of standard Normal distribution for any value of $z$. Also, Excel spreadsheet has the following functions that return the values of density and cumulative distributions for any value of $z$ for standard Normal distribution:

$$\phi(z) = \mathrm{NORMDIST}(z, 0, 1, \mathrm{False})$$
$$\Phi(z) = \mathrm{NORMDIST}(z, 0, 1, \mathrm{True})$$

Hence, considering Eq. (10.12), we have

$$E(\text{shortage}|\ Q) = \sigma_D \left(\mathrm{NORMDIST}(z, 0, 1, \mathrm{False}) - z[1 - \mathrm{NORMDIST}(z, 0, 1, \mathrm{True})]\right)$$

In the following example we show how the optimal order quantity and expected profit can be found if the Newsvendor orders products with the demand following a Normal distribution—or is well-approximated by a Normal distribution.

**Example 10.3**

Consider the inventory management of challah breads at H-Food in Example 10.1. What is the optimal order quantity under the current price if the demand for challah bread is well-approximated by a Normal distribution with mean 50 and standard deviation 15? What is H-Food's expected profit under the optimal policy?

Under the current price we have $R_c = 0.373$. Thus, the optimal order quantity $Q^*$ satisfies

$$F(Q^*) = P(D \le Q^*) = \mathcal{R}_c = 0.373$$

As Fig. 10.2 shows, the optimal order quantity $Q^*$ is the value of demand for which its cumulative probability—the gray area under the distribution—is equal to 0.373. We know that

$$F(Q^*) = \Phi(z), \quad \text{where} \quad z = \frac{Q^* - \overline{D}}{\sigma_D}$$

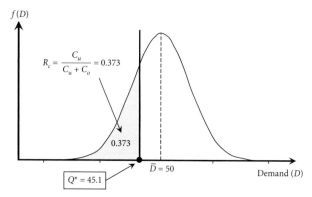

**Figure 10.2**    Finding the optimal order size for challah bread at H-Food in Example 10.3.

Hence, we use standard Normal table in the Appendix and we find that $\Phi(z = -0.325) = 0.373$. Since, $z = (Q^* - \overline{D})/\sigma_D$, then

$$Q^* = \overline{D} + z\,\sigma_D = 50 + (-0.325)15 = 45.1$$

which we round up to 46 loaves of bread.

Excel also provides an inverse of cumulative Normal distribution function that makes finding the optimal order quantity very easy. Specifically, for a Normal distribution with mean $\mu$ and standard deviation $\sigma$, $\mathsf{NORMINV}(\alpha, \mu, \sigma)$ yields the value of Normal random variable that has cumulative probability $\alpha$. Using this function, we can find the optimal order quantity for H-food as follows:

$$Q^* = \mathsf{NORMINV}(\mathcal{R}_c, \overline{D}, \sigma_D) = \mathsf{NORMINV}(0.373, 50, 15) = 45.1$$

**Expected Profit:** To compute the expected profit, we first need to compute the expected shortage. For Normal demand distribution the expected shortage if H-Foods orders $Q^* = 46$ units is

$$\mathsf{E}(\text{shortage}|\,Q^* = 46) = \mathcal{L}(Q^* = 46) = \sigma_D \mathcal{L}(z)$$

where

$$z = \frac{Q^* - \overline{D}}{\sigma_D} = \frac{46 - 50}{15} = -0.267$$

If we use the standard Normal table in the Appendix we find $\mathcal{L}(z = -0.267) = 0.546$. Thus,

$$\mathsf{E}(\text{shortage}|\,Q^* = 46) = \sigma_D \mathcal{L}(z) = 15 \times 0.546 = 8.20 \text{ loaves}$$

Instead of standard Normal table, we could have also used Excel commands to find the average shortage for an order quantity $Q$ as follows:

$$\mathsf{E}(\text{shortage}|\,Q) = \sigma_D\,(\mathsf{NORMDIST}(z, 0, 1, \text{False}) - z[1 - \mathsf{NORMDIST}(z, 0, 1, \text{True})])$$

For the optimal order Quantity $Q^* = 46$ corresponding to $z = -0.267$, we have

$$
\begin{aligned}
\text{E(shortage}|\, Q^* = 46) &= \mathcal{L}(Q^* = 46) \\
&= \sigma_D\,(\text{NORMDIST}(z, 0, 1, \text{False}) - z[1 - \text{NORMDIST}(z, 0, 1, \text{True})]) \\
&= 15\,(\text{NORMDIST}(-0.267, 0, 1, \text{False}) \\
&\quad + 0.267[1 - \text{NORMDIST}(-0.267, 0, 1, \text{True})]) \\
&= 15\,(0.385 + 0.267[1 - 0.395]) \\
&= 8.20 \text{ loaves}
\end{aligned}
$$

which is what we found using standard Normal table.

Using Eqs. (10.4) and (10.5), we can find the average sales and average leftover inventory as follows:

$$
\begin{aligned}
\text{E(sales}|\, Q^* = 46) &= \overline{D} - \text{E(shortage}|\, Q^* = 46) = 50 - 8.20 = 41.80 \\
\text{E(leftover}|\, Q^* = 46) &= Q^* - \text{E(sales}|\, Q^* = 46) = 46 - 41.80 = 4.20
\end{aligned}
$$

Hence, the expected profit under the optimal order quantity $Q^* = 46$ is

$$
\begin{aligned}
\text{E(Profit}|\, Q^* = 46) &= p\,\text{E(sales}|\, Q^* = 46) + s\,\text{E(leftover}|\, Q^* = 46) - c\,Q^* \\
&= \$5.99 \times 41.80 + \$2.00 \times 4.20 - \$4.5 \times 46 \\
&= \$51.78
\end{aligned}
$$

### 10.1.5  Service Levels in Newsvendor Problems

Recall from Chap. 9 that there are two commonly used service levels for inventory management problems: (i) cycle-service level $\alpha$, and (ii) fill rate $\beta$. Cycle-service level is the probability of not having shortage in an order cycle, and fill rate is the fraction of demand that is satisfied from inventory (without being backordered or lost). The following analytics shows how these service levels can be computed for Newsvendor problems.

---

**CONCEPT**

**Inventory Service Levels in Newsvendor Problem**

For any given order quantity $Q$, in Newsvendor problem with random demand $D$ following discrete or continuous distribution with mean $\overline{D}$, service levels are as follows:

- *Cycle-Service Level*: Cycle-service level $\alpha$ is the probability of not having shortage during the selling period (i.e., during the single period):

$$
\begin{aligned}
\alpha &= \text{P(having no shortage during selling period)} \\
&= P(D \leq Q) \tag{10.13}
\end{aligned}
$$

For the cases with continuous demand distribution in which an item can also be ordered in continuous units (e.g., in pound, ounces, etc.), the cycle-service level for the *optimal* order quantity $Q^*$ is equal to critical ratio, that is, $\alpha = \mathcal{R}_c$, see Eq. (10.10).

- *Fill Rate:* Fill rate $\beta$ is the fraction of overall demand that does not face shortage during the selling period:

$$\beta = 1 - \frac{\text{Average shortage during selling period}}{\text{Average demand}}$$

$$= 1 - \frac{E(\text{shortage}|\ Q)}{\overline{D}} \tag{10.14}$$

### Example 10.4

> Consider the inventory management of challah breads at H-Food in Example 10.3 where the demand was Normally distributed with mean 50 and standard deviation 15. What are the cycle-service level and fill rates under the optimal order quantity $Q^* = 46$?

For the optimal order quantity $Q^* = 46$, the cycle-service level is

$$\alpha = P(D \leq Q^*) = P(D \leq 46) = F(46)$$

Since demand is Normally distributed,

$$F(46) = \Phi\left(\frac{46 - \overline{D}}{\sigma_D}\right) = \Phi\left(\frac{46 - 50}{15}\right) = \Phi(-0.267)$$

using standard Normal table we find

$$\alpha = \Phi(-0.267) = 0.395 = 39.5\%$$

Therefore, $1 - 0.395 = 0.605 = 60.5\%$ of time (i.e., 60.5% of the 2-day periods) H-Food will face shortage of bread.

Recall from Example 10.2 that the expected shortage under the optimal order quantity $Q^* = 46$ is $E(\text{shortage}|\ Q^* = 46) = 8.20$. Thus, the fill rate for the challah bread if H-Food orders $Q^* = 46$ will be

$$\beta = 1 - \frac{E(\text{shortage}|\ Q = 46)}{\overline{D}} = 1 - \frac{8.20}{50} = 0.836 = 83.6\%$$

Hence, following the optimal policy during the 2-day period, about $1 - 83.6\% = 16.4\%$ of customers will find H-food's shelves empty of challah bread.

---

**Want to Learn More?**

**Applications of Newsvendor Model Beyond Inventory Management**

What should be the capacity of a new factory that a firm is planning to build? How many seats an airline should overbook in a flight from Chicago to Memphis, because some passengers may not show up? How many workers should an online retailer schedule in the 8-hour shift in the retailer's order processing center to process and ship orders to customers during Christmas peak days? These are critical decisions in capacity, revenue, and workforce management. Believe it or not, Newsvendor model can find the answers to these questions. How? See the online supplement of this chapter.

## 10.2 Multi-Period Inventory Problems

So far our focus has been on inventory systems with stochastic demand that have only a single order cycle—the Newsvendor problem. One important feature of Newsvendor problem is the following: although Newsvendor orders newspapers every day—a multi-period inventory problem—the decision in each day is independent of the decisions in the previous day. In other words, the multi-period inventory problem decouples into several independent single-period problems. The reason is that yesterday's newspapers cannot be used to fill the demand for today's newspaper. For instance, in Example 10.1 unsold challah bread in one (2-day) period was not sold the next period. However, there are many situations in which leftover inventory from last period—last order cycle—can be used in later periods. This includes standard products such as automobile tires, light bulbs, clothing such as socks and underwear, dried fruits, rice, among many other products.

In the following sections, we therefore focus on inventory management policies in Continuous and Periodic Review systems in multi-period problems. We first start with continuous review systems.

## 10.3 Continuous Review Multi-Period Inventory Systems and (Q, R) Policy

As we mentioned in Chap. 9 firms monitor their inventory either continuously or periodically. We showed that $(Q, R)$ policies are appropriate policies to manage inventory in continuous review systems. We developed models to compute the optimal order quantity $Q^*$ and optimal reorder point $R^*$ when demand was deterministic. In this section we show that these optimal values must be revised if demand is stochastic. We start by illustrating how variability in demand affects the performance of the models we derived for cases with deterministic demand.

### 10.3.1 Deterministic versus Scholastic Inventory Curves

We use the case of Auto Part Depot (APD) in the following example to show how the variability in demand affects the inventory curve and thus the optimal inventory policy.

 **Example 10.5**

Auto Part Depot (APD) sells a variety of automobile parts and accessories. One of their high selling item is a powerful polishing compound sold in 64-ounce cans. The compound is used for scratch removal on newly cured paint or older paints. APD orders this item from Germany and has the exclusive right to sell this item in the United States. It buys each can at the price of $80 and sells them at $87.99 per can. The administrative and shipment cost of ordering cans is about $187.5 per order, regardless of the size of the order. When an order is placed, it takes 10 days until APD receives the order. The store charges inventory costs at inventory carrying rate of 25%. The demand for the compound is highly variable. Some days it is as low as 2 or 3 cans and other days it is as high as 40 or 50 cans. A study of the sales shows that the daily demand for cans can be accurately estimated by a Normal distribution with mean 30 and standard deviation 10. If a customer demands the compound and the compound is out of stock, APD offers free shipment to deliver the item to the customer. All customers are willing to wait for the item since the compound cannot be bought from other stores. APD has a contract with a local delivery firm that does the same-day delivery of backorder items at a fixed cost of $10 per each order. The store ships the backordered items to the customers as soon as the next order

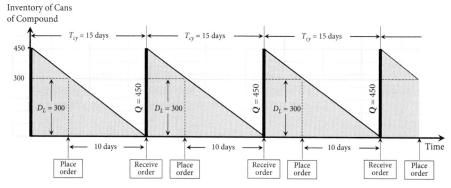

Figure 10.3    Polishing compound inventory curve under ($Q^* = 450, R^* = 300$) when demand is deterministic.

arrives. The inventory of cans is continuously monitored using the store's computerized inventory information system. How should the store manage its inventory of the polishing compound? Specifically, when should the store place an order and for how many cans of compound?

Obviously, this is a continuous review inventory system and the appropriate policy to manage and control inventory is $(Q, R)$ policy. However, as opposed to the $(Q, R)$ systems in Chap. 9 that face deterministic demand, the APD faces stochastic demand. To illustrate how the variability in demand impacts inventory management decisions, we compare the inventory curve in $(Q, R)$ policies with deterministic demand with that under stochastic demand.

**Deterministic Demand**
To understand the impact of variability on the $(Q, R)$ policy, let's assume that the demand for cans is *exactly* 30 per day—instead of being variable with an average of 30 per day. In this case, the problem becomes a simple EOQ model with annual demand $D = 30 \times 360 = 10,800$—assuming 360 days in a year—order setup cost $K = \$187.5$, and annual inventory holding cost of $C_h = \$80 \times 25\% = \$20$ per can. Thus, the optimal order quantity is

$$Q^* = \sqrt{\frac{2KD}{C_h}} = \sqrt{\frac{2(187.5)(10,800)}{20}} = 450 \text{ cans}$$

Reorder point $R^*$ is equal to the demand during lead time. Considering lead time of $L = 10$ days, we can compute the reorder point $R^*$ as follows:

$$R^* = D_L = D \times L = 30 \times 10 = 300 \text{ cans}$$

Figure 10.3 shows the inventory curve under the optimal policy ($Q^* = 450, R^* = 300$) if the daily demand for the compound was constant at the rate of 30 per day with no variability.

**Stochastic Demand**
Now what happens if APD uses the optimal ($Q^* = 450, R^* = 300$) when the daily demand is variable with mean 30 and standard deviation 10? Figure 10.4 illustrates an example of what can happen.

From Fig. 10.4 we observe the following changes in inventory curve compared with Fig. 10.3 in which the demand is deterministic. Note that both cases are using the same ($Q = 450, R = 300$) inventory management policy:

- *Variable demand results in order cycles with different lengths:* In the case of deterministic demand, all order cycles have the same length $T_{cy} = 15$ days. However, when

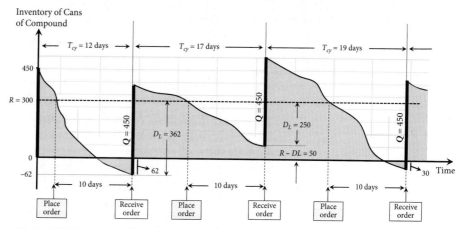

**Figure 10.4**    Polishing compound inventory curve under $(Q^* = 450, R^* = 300)$ when demand is stochastic.

demand is stochastic, each order cycle has a different length. For example, the first order cycle in Fig. 10.4 takes 12 days, while the third order cycle takes 19 days.

- **On-hand inventory** at the **end** of an order cycle is $R - D_L$: Since an order is placed when the inventory reaches $R = 300$, the system will use this inventory to satisfy the demand until the next order arrives, that is, satisfy demand during lead time $(D_L)$. Hence, the inventory at the end of an order cycle—just before an order arrives— is $R - D_L$. In the case with deterministic demand, the inventory at the end of all order cycles is zero, that is, $R - D_L = 0$. This is because the demand during lead time is exactly $D_L = 300$. However, in the case with variable demand, the inventory at the end of each order cycle can be zero, positive, or negative (i.e., shortage). This is because the demand during lead time in each order cycle may be different. For example, the demand during lead time in the first order cycle is $D_L = 362$. Hence, the on-hand inventory at the end of the first cycle—just before receiving the order— as shown in Fig. 10.4 is $R - D_L = 300 - 362 = -62$. However, the demand during the lead time in the second order cycle is $D_L = 250$, resulting in on-hand (positive) inventory of $R - D_L = 300 - 250 = 50$ at the end of the second order cycle. Note from the figure that when the order arrives at the end of the first cycle, APD ships the 62 backordered products to the customers and puts the remaining $450 - 62 = 388$ items in inventory.

- **Shortage** depends on reorder point $(R)$ and demand during lead time $(D_L)$: As we discussed above, negative on-hand inventory at the end of a period, that is, $R - D_L < 0$, corresponds to shortages. Thus, the number of shortages at the end of an order cycle is

$$\text{Shortage at the end of an order cycle} = (D_L - R)^+$$

where $(x)^+ = x$ if $x > 0$ and $(x)^+ = 0$, if $x \le 0$. Hence, as shown in the first order cycle, $D_L = 362$, then we have

$$\text{Shortage at the end of the first cycle} = (D_L - R)^+ = (362 - 300)^+ = 62$$

In contrast, in the second cycle, the demand during lead time is $D_L = 250$, and, therefore,

$$\text{Shortage at the end of the second cycle} = (D_L - R)^+ = (250 - 300)^+ = 0$$

Considering

$$\eta_L(R) = Average \text{ shortage in an order cycle with reorder point } R$$

then, we have

$$\eta_L(R) = \mathsf{E}(D_L - R)^+ = \int_{x=R}^{\infty} (x - R) f_L(x) dx \tag{10.15}$$

where $f_L(x)$ is the distribution of demand during lead time.

**Protection Interval**

Recall that safety stock is the inventory that is kept to protect the process against the variability in demand which can result in shortage. On the other hand, shortage $(D_L - R)^+$ can only occur during the interval when we placed an order and are waiting to receive it—the lead time. This leads us to the concept of protection interval.

---

**CONCEPT**

**Protection Interval**

*Protection Interval* is the time interval during which a firm must rely on its safety stock to "protect" against shortage.

---

Therefore, in $(Q, R)$ inventory control policy, the protection interval is the lead time.

### 10.3.2 Impact of *R* and *Q* on Costs

Having discussed how the variability in demand affects inventory of a process, the question now becomes: How should policy $(Q = 450, R = 300)$ be revised to deal with variability in demand? To answer this, we first need to understand how changing $Q$ and $R$ impacts inventory holding, shortage, and order setup cost.

**Impact of Reorder Point *R* on Costs**

First issue that must be resolved is how to deal with possible shortages in each order cycle.

- *Increasing reorder point R decreases the **possibility** of having shortage in an order cycle.* A shortage occurs when the demand during lead time becomes larger than reorder point $R = 300$. So, one way to reduce the possibility of shortage in an order cycle is to increase reorder point higher than 300. For example, if APD increases the reorder point by 50 and places an order when its inventory reaches $R = 350$, then its inventory curve will look something like the one in Fig. 10.5.

  Comparing the curve in Fig. 10.5 with that in Fig. 10.4, we see that the additional 50 units increase in reorder point eliminates the shortage in the third cycle. Thus, under a higher reorder point, there will be a less number of order cycles that face shortage, that is, lower probability of having shortage in an order cycle.

- *Increasing reorder point R decreases the **number** of shortages in an order cycle.* As Fig. 10.5 shows, increasing reorder point from $R = 300$ to $R = 350$ also reduces the number of shortages that occurs in an order cycle. For example, the shortage in the first cycle in the figure has been reduced from 62 to 12, and the number of shortages in the third order cycle is reduced from 30 to zero.

- *For any given order quantity Q, increasing reorder point R decreases the annual shortage cost, while increasing the annual inventory cost.* Note that by increasing reorder point from $R = 300$ to $R = 350$, APD will carry an additional 50 units of inventory throughout the year. Thus, it will have a higher annual inventory cost each year.

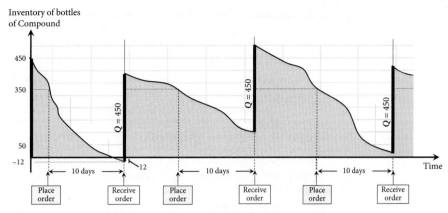

**Figure 10.5**    Polishing compound inventory curve under $(Q^*, R^*) = (450, 350)$ when demand is stochastic.

However, these additional 50 units will reduce the possibility of shortages in each order cycle as well as reduce the number of shortages in each order cycle, resulting in a lower annual shortage cost.

The additional 50 units of inventory—due to increasing reorder point—are to prevent or reduce the shortage caused by the variability in demand. Recall from Chap. 9 that the additional inventory that is built up to protect against stockout is called "Safety Stock" or "Safety Inventory." Obviously, holding more safety stock reduces the shortage costs, while it increases the inventory holding cost.

### Impact of Order Quantity Q on Costs
Let's keep the reorder point to its original value $R = 300$—no safety stock—and investigate what happens if we change order quantity $Q = 450$. Specifically, how does increasing $Q$ affect inventory curve in Fig. 10.4?

- *For any given reorder point R, increasing (or decreasing) order quantity Q does not affect the **possibility** of shortage in an order cycle.* Recall that shortage only occurs during lead time. The probability of having a shortage in an order cycle is the probability that the demand during lead time of 10 days exceeds reorder point $R = 300$. Thus, for the same reorder point and the same demand distribution during lead time, the possibility of shortage in an order cycle is independent of the order size $Q$.

- *For any given reorder point R, increasing (or decreasing) order quantity Q does not affect the **number** of shortages in an order cycle.* The number of shortages in an order cycle is the number of demand exceeding reorder point $R$. Thus, for the same reorder point $R = 300$ and the same demand distribution during lead time of 10 days, the number of shortages in an order cycle is independent of the order size $Q$.

- *For any given reorder point R, increasing order quantity Q decreases the **annual** shortage cost and the **annual** order setup cost, while increasing the **annual** inventory costs.* When $Q = 450$, considering the annual demand of $D = 10,800$, APD will place about $10,800/450 = 24$ orders in a year. Thus, on an average, there are about 24 order cycles in a year. Each order cycle corresponds to: (i) one order setup cost of $K = \$187.5$, and (ii) the possibility of having shortage that results in cost of $\$10$ per shortage. The total annual shortage and setup cost is equal to the sum of the shortages and setup costs in 24 cycles. Now, consider the extreme case of increasing the order size to $Q = 10,800$—equal to the 1-year demand. In this case, there is about one order cycle in each year, resulting in one order setup cost of $K = \$187.5$ per year. Also, the total shortage cost in a year is equal to the expected shortage cost in only one order cycle that has the same lead time of 10 days. Thus, for a given reorder point $R$, increasing order size $Q$ decreases the number of order cycles in a year and hence decreases the *annual* order setup and shortage cost. Obviously, order size $Q = 10,800$ results in a large cycle inventory throughout the year, which is much larger than cycle inventory when ordering $Q = 450$ units. Thus, increasing order size $Q$ increases the annual inventory holding cost.

All these observations indicate that we can control the total annual shortage and inventory costs by changing either $R$ or $Q$ or both. Changing $Q$, however, also affects the annual order setup cost. The question is, therefore, how to find the best values for order quantity $Q$ and reorder point $R$?

The answer depends on the management goals. Specifically, the appropriate values of $Q$ and $R$ depend on whether the management goal is to reduce costs, to provide a certain service level, or a balance of both. In the following section, we start with finding the optimal values for $Q$ and $R$, if the goal is to minimize the total average cost. Later in Sec. 10.6, we show how these values change if the goal is to provide a certain level of service.

Finding the optimal $(Q, R)$ policy that minimizes the total annual cost depends on whether the customers who face shortage (i) are backordered, or (ii) are lost. We start with the case of backorders.

## 10.4 Minimizing Cost in $(Q, R)$ Systems with Backorders

Recall that in processes with backorders, customers who face stockout are willing to wait until the product arrives. This is the case in APD store in Example 10.5. Customers who face stockout are willing to wait because the item will be delivered to them for free and they cannot buy the item from other stores. However, APD pays a cost of $10 to ship each backordered compound to its customers. What is the optimal value for order quantity $Q$ and reorder point $R$ that minimizes the total *average* annual backorder, holding, and order setup cost?

Note that we emphasize on "average" because, due to the variability in demand, the on-hand inventory, number of order cycles in a year, and number of shortages in a year are all random. Because of this randomness, finding the exact optimal values for $Q$ and $R$ is very complex. The good news, however, is that there are simple approximation methods that can result in $(Q, R)$ policies with close-to-minimum cost. One of these approaches is to break the total average annual cost into different components and try to estimate each component separately.

The approximation approach makes the following two assumptions to make the analysis simpler:

- *Assumption 1:* The average number of shortages is very small compared to the average on-hand inventory. This is not an unreasonable assumption. In practice, the short-term (e.g., loss of sales) and long-term (e.g., loss of future customer business) consequences of shortage are very high. Thus, inventory systems in real world have small shortages.

- *Assumption 2:* In every order cycle, at most one order is outstanding (i.e., is in the pipeline). This assumption is also not a restrictive assumption in practice. Having more than one outstanding order indicates that the inventory position reached below reorder point at least once during lead time (i.e., another order was placed while waiting for previous order to arrive). This implies that the system is facing an unusually large demand that can result in a large shortage. As we mentioned, in practice, systems are not run with large shortages.

While the approximation models are based on the above assumptions, it has been shown that these approximations give reasonable answers even in cases where the assumptions are violated.

We illustrate these approximation approaches using the case of APD polishing compound in Example 10.5. We assume that APD uses the $(Q, R)$ policy to manage the inventory of the polishing compound and we find the optimal values for both order quantity $Q$ and reorder point $R$. We also assume that the demand for the item is well-approximated using a *continuous probability distribution*.

Below, we first compute the total average annual cost under the $(Q, R)$ policy.

### 10.4.1 Computing the Total Average Annual Cost

Before we show how to compute the annual average cost, we first show that minimizing total cost is equivalent to maximizing profit. Consider a process that buys an item for $c$ and sells it for $p$. The annual demand has an *average* of $\overline{D}$ and the inventory of the item is managed using a $(Q, R)$ policy. Other costs include inventory holding cost, order setup cost, and shortage cost.

Since all backorders are eventually satisfied, the annual expected (i.e., average) profit is $(p - c)\overline{D}$. This, however, does not include the inventory holding and setup cost as well as the cost of shortage. Thus,

$$\begin{aligned} \text{Average annual profit} = (p - c)\overline{D} &- \text{Average annual shortage cost} \\ &- \text{Average annual holding cost} \\ &- \text{Average annual order setup cost} \end{aligned} \quad (10.16)$$

While $(p - c)\overline{D}$ does not depend on the $(Q, R)$ policy, the annual average inventory, order setup, and shortage cost do. Hence, considering $TC(Q, R)$ as the total average annual holding, setup, and shortage cost under $(Q, R)$ policy, we have

$$\text{Average annual profit under } (Q, R) \text{ policy} = (p - c)\overline{D} - TC(Q, R)$$

Since $(p - c)\overline{D}$ is constant, maximizing total average annual profit is equivalent to minimizing total average annual cost $TC(Q, R)$. Thus, in the following sections we focus on computing the elements of $TC(Q, R)$.

To compute the total average annual cost $TC(Q, R)$, we compute it in one order cycle and then multiply that by the average number of order cycles in a year. We start with average annual order setup cost.

**Average Annual Order Setup Cost**
Recall that in each order cycle one order is placed that incurs the order setup cost $K$. To find the average annual order setup cost we have

$$\textit{Average annual setup cost} = K \times \textit{Average number of order cycles in a year}$$

since exactly one order is placed in every order cycle (i.e., Assumption 2). Because there is variability in annual demand, the number of orders placed in a year—and thus the number of order cycles—varies from year to year. However, finding the *average* number of order cycles in a year is easy. If an order of size $Q$ is placed in each order cycle, to satisfy the average

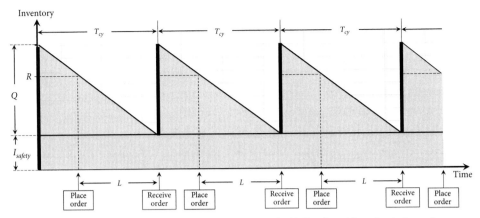

Figure 10.6    Continuous approximation of the inventory curve under $(Q, R)$ policy with stochastic demand.

annual demand $\overline{D}$, the average number of orders placed in a year must be

$$Average \text{ number of order cycles in a year} = \frac{\overline{D}}{Q}$$

Hence,

$$Average \text{ annual setup cost} = K \times \frac{\overline{D}}{Q}$$

### Average Annual Inventory Holding Cost

Recall that in EOQ model with deterministic demand, we computed the average inventory throughout the year by computing the average inventory in one order cycle. The reason was that all order cycles in the EOQ models were identical. For the case of stochastic demand, however, as shown in Fig. 10.4, order cycles have different lengths. Also, computing the area under the inventory curve in Fig. 10.4 to find the average inventory is difficult. But, here is where the approximation comes in.

One approach is to use the continuous approximation of the inventory curve—illustrated in Sec. 2.4 of Chap. 2—to compute the average inventory in our system with stochastic demand. Recall that the continuous approximation approach assumes that the demand is deterministic with a constant rate—in our case, the rate is the average demand of $\overline{D}$ per year. Thus, similar to the deterministic EOQ, the continuous approximation of the system will have *cycle inventory* through the year. On the other hand, because of variability in demand, we must also consider having *safety inventory* $I_{\text{safety}}$. Thus, the resulting inventory curve will look like the one in Fig. 10.6.

Considering annual inventory holding cost $C_h$ per item per unit time, then

$$Average \text{ annual holding cost} = C_h \times (\text{Cycle Inventory} + \text{Safety Inventory})$$

*Cycle Inventory:* Under the continuous approximation (see Fig. 10.6), finding the cycle inventory is easy. Note that in each order cycle, the inventory decreases linearly from the maximum of $Q + I_{\text{safety}}$ to the minimum of $I_{\text{safety}}$. Thus,

$$\text{Cycle Inventory} = \frac{Q + I_{\text{safety}} - I_{\text{safety}}}{2} = \frac{Q}{2}$$

*Safety Inventory:* Safety inventory depends on the reorder point $R$. Since we are looking for the optimal value of the reorder point $R$, we need to determine the exact relationship between reorder point $R$ and safety inventory $I_{\text{safety}}$. This requires a formal definition of safety inventory under $(Q, R)$ policy.

### Safety Inventory in $(Q, R)$ Models

Recall that *Safety Inventory* is the inventory kept to protect against variability in demand (or in supply). In the $(Q, R)$ model, safety inventory is defined as the *average* net inventory just before an order arrives, that is,

$$I_{\text{safety}} = \text{E}(\text{Net inventory just before an order arrives}) \qquad (10.17)$$

Three key features in the above definition are important to question:

- Why safety inventory is the inventory *just before* the arrival of the next order? The reason is that "just before" an order arrival is when the *total number* of shortages in a cycle is realized, and safety stock is designed to deal with that shortage.

- Why safety inventory is the *net* inventory? Net inventory refers to the fact that safety inventory can be positive or negative. For example, if shortage cost is much smaller relative to the inventory holding cost, then it may make economical sense to have some shortages (i.e., negative inventory). In other words, keeping some customers backordered is less costly than keeping items in inventory.

- Why safety inventory is the *average* net inventory? Because, due to the variability in demand, the number of shortages in different order cycles is different. Hence, the inventory just before the next order arrives (i.e., safety inventory) will also be different in different order cycles. Taking the average of those inventories is one way to quantify those inventories into one number, which is defined as "Safety Inventory."

Having defined safety inventory, we can now determine how reorder point $R$ and safety inventory $I_{\text{safety}}$ relate.

Note that for any given reorder point $R$, the net inventory just before the next order arrives is the difference between reorder point $R$ and demand during lead time, that is, $R - D_L$. Assuming $D_L$ follows probability density function $f_L(x)$, we can compute safety inventory as follows:

$$
\begin{aligned}
I_{\text{safety}} &= \text{E}(R - D_L) \\
&= \int_0^\infty (R - x) f_L(x) dx \\
&= \int_0^\infty R f_L(x) dx - \int_0^\infty x f_L(x) dx \\
&= R - \overline{D}_L
\end{aligned}
$$

where $\overline{D}_L$ is the average demand during lead time $L$. In other words, in the $(Q, R)$ policy with backorders, the safety inventory is the amount of inventory kept beyond the average demand during lead time. Thus,

$$
\begin{aligned}
\textit{Average} \text{ annual holding cost} &= C_h \times (\text{Cycle Inventory} + \text{Safety Inventory}) \\
&= C_h \times (Q/2 + R - \overline{D}_L)
\end{aligned}
$$

For cases that the demand during lead time is a discrete random variable, safety inventory is also found to be $R - \overline{D}_L$.

**Average Annual Backorder Cost**

Let's define average shortage cost as follows:

$$\Psi_s(R) = \text{Total average shortage cost in one } order\ cycle \text{ with reorder point } R$$

Since demand is random, each order cycle may have a different number of shortages. Thus, $\Psi_s(R)$ represents the total *average* shortage cost in one order cycle. Hence, for the total average annual shortage cost we have

$$Average \text{ annual shortage cost} = \Psi_s(R) \times Average \text{ number of order cycles in a year}$$
$$= \Psi_s(R) \times \frac{\overline{D}}{Q}$$

Putting all these cost components together, we get the following expression for the total average annual holding, setup, and shortage (i.e., backorder) cost under a $(Q, R)$ policy:

$$TC(Q, R) = K\left(\frac{\overline{D}}{Q}\right) + C_h\left(\frac{Q}{2} + R - \overline{D}_L\right) + \Psi_s(R)\left(\frac{\overline{D}}{Q}\right) \tag{10.18}$$

### 10.4.2 Finding the Optimal (*Q, R*) Policy in Systems with Backorders

Having the total average annual cost as in Eq. (10.18), we can now compute the optimal order quantity $Q^*$ and optimal reorder point $R^*$. The necessary conditions for values $Q$ and $R$ to minimize the above cost function are as follows:

$$\frac{\partial\ TC(Q, R)}{\partial Q} = 0 \quad \text{and} \quad \frac{\partial\ TC(Q, R)}{\partial R} = 0$$

In other words,

$$\frac{\partial\ TC(Q, R)}{\partial Q} = -K\left(\frac{\overline{D}}{Q^2}\right) + C_h\left(\frac{1}{2}\right) - \Psi_s(R)\left(\frac{\overline{D}}{Q^2}\right) = 0 \tag{10.19}$$

$$\frac{\partial\ TC(Q, R)}{\partial R} = C_h + \left(\frac{\overline{D}}{Q}\right)\frac{\partial\ \Psi_s(R)}{\partial R} = 0 \tag{10.20}$$

By solving Eq. (10.19) we get

$$Q = \sqrt{\frac{2\overline{D}[K + \Psi_s(R)]}{C_h}} \tag{10.21}$$

By solving Eq. (10.20), we get

$$\frac{\partial}{\partial R}\Psi_s(R) = -\frac{C_h Q}{\overline{D}} \tag{10.22}$$

Therefore, the optimal $Q^*$ and $R^*$ are the values of $Q$ and $R$ that satisfy both Eqs. (10.21) and (10.22). However, before we use these equations, we first need to find the value of $\Psi_s(R)$ which appears in both the equations.

The value of $\Psi_s(R)$, which is the average shortage cost in an order cycle with reorder point $R$ depends on how shortage cost—the backorder cost—incurs. Usually, a fixed cost incurs for each backorder demand. This is the case for APD in Example 10.5 that pays $10 for the shipment of each backordered demand. Another example of fixed cost per backorder demand is when firms offer discount to convince customers who face stockout to wait for their products.

### 10.4.3 Optimal $(Q^*, R^*)$ in Systems with Fixed Cost per Backorder Demand

In Example 10.5, the cost of having a backorder customer is \$10. This cost corresponds to the cost of free shipping of the product to the customer. In this case, the average shortage cost in an order cycle depends on the number of backordered demands in the order cycle. To find the optimal $(Q, R)$ policy for this case, we need the following analytics that modifies Eqs. (10.21) and (10.22) for the case of fixed cost per backordered demand.

---

**ANALYTICS**

**Finding the Optimal (Q, R) Policy for Minimizing Cost in Systems with Costs per Backorder Demands**

Consider a continuous review make-to-stock (MTS) inventory system managed by a $(Q, R)$ policy in which all demands that face shortage are backordered. Also, suppose that

$$C_b = \text{Backorder cost per backorder demand}$$
$$\eta_L(R) = \text{Average number of backorders in an order cycle with reorder point } R$$

Hence, for any given reorder point $R$, the average backorder cost in a cycle is

$$\Psi_s(R) = C_b\,\eta_L(R) \tag{10.23}$$

where

$$\eta_L(R) = \mathsf{E}(x - R)^+ = \int_R^\infty (D_L - R)f_L(x)dx$$

Finding $\eta_L(R)$, the average number of backorders in an order cycle with reorder point $R$ is indeed analogous to finding the average number of shortages in a Newsvendor problem with order quantity $Q$. The lead time is analogous to the selling period in the Newsvendor problem. In the Newsvendor problem the selling period starts with $Q$ units, while in an order cycle it starts with $R$ units.[3] Thus, the average number of shortages (i.e., backorders) at the end of an order cycle is equivalent to the average shortage at the end of the selling period in the Newsvendor problem.

When demand during lead time is Normally distributed with mean $\overline{D}_L$ and standard deviation $\sigma_L$, we have

$$\eta_L(R) = \sigma_L \mathcal{L}(z), \quad \text{where } z = \frac{R - \overline{D}_L}{\sigma_L}$$

and $\mathcal{L}(z)$ is the standard Normal loss function and is available in standard Normal table in the Appendix.

The optimal order quantity $Q^*$ and reorder point $R^*$ are those that *simultaneously* satisfy the following two equations (see online Appendix E for details):

$$Q^* = \sqrt{\frac{2\overline{D}[K + C_b\,\eta_L(R)]}{C_h}} \tag{10.24}$$

$$F_L(R^*) = 1 - \frac{C_h}{C_b\overline{D}}Q^* \tag{10.25}$$

where $F_L(R)$ is the value of cumulative probability distribution function of demand during lead time for reorder point $R$.

The total average annual holding, setup, and backorder cost under a $(Q, R)$ policy, including the optimal policy, is as follows:

$$TC(Q, R) = K\left(\frac{\overline{D}}{Q}\right) + C_h\left(\frac{Q}{2} + R - \overline{D}_L\right) + C_b\,\eta_L(R)\left(\frac{\overline{D}}{Q}\right) \tag{10.26}$$

---

[3] Recall that our approximation method assumes that there is at most one order outstanding in each order cycle. Thus, no orders arrive during the lead time and the system is using the $R$ units in inventory to satisfy demand during lead time.

While the set of Eqs. (10.24) and (10.25) is a set of two equations with two unknowns, solving it is not easy. This is because of the nonlinearity in both formulas and also because it includes cumulative probability distribution $(F_L(R))$ and loss function $(\eta_L(R))$ of demand during lead time. However, there exists a solution methodology that solves these equations iteratively, which we present below.

### ANALYTICS

**Iterative Approach for Finding the Optimal $(Q, R)$ Policy when Minimizing Cost**

**Step 0:** Set Iteration $i = 0$ and obtain the initial value of $Q_0$ using the optimal order quantity in EOQ model with deterministic demand as follows:

$$Q_0 = \sqrt{\frac{2\overline{D}K}{C_h}}$$

Then use $Q_0$ in the following equation and compute $R_0$,

$$F_L(R_0) = 1 - \frac{C_h}{C_b\overline{D}}Q_0$$

Having $R_0$, compute $\eta_L(R_0)$ and go to Step 1.

**Step 1:** Set $i = i + 1$ and compute $Q_i$ as follows:

$$Q_i = \sqrt{\frac{2\overline{D}[K + C_b\,\eta_L(R_{i-1})]}{C_h}}$$

Then using $Q_i$, compute $R_i$ that satisfies the following:

$$F_L(R_i) = 1 - \frac{C_h}{C_b\overline{D}}Q_i$$

Having $R_i$ compute $\eta_L(R_i)$ and go to Step 2.

**Step 2:** If $Q_i = Q_{i-1}$ and $R_i = R_{i-1}$, stop, the optimal policy is: $Q^* = Q_i$ and $R^* = R_i$. Otherwise, return to Step 1.

Using the above analytics, we can obtain the optimal $(Q, R)$ policy for the APD in Example 10.5. Recall that in Example 10.5 we have $\overline{D} = 10{,}800$, $K = \$187.5$, and $C_h = \$20$. Also, the delivery firm delivers each backorder demand at the cost of $C_b = \$10$. Thus, the optimal $Q^*$ and $R^*$ must satisfy Eqs. (10.24) and (10.25) as follows:

$$Q^* = \sqrt{\frac{2(10{,}800)[187.5 + 10\eta_L(R)]}{20}} \tag{10.27}$$

$$F_L(R^*) = 1 - \frac{20}{10(10{,}800)}Q^* \tag{10.28}$$

We can now use the iterative approach to obtain the optimal values of $Q^*$ and $R^*$.

**Step 0:** We set Iteration $i = 0$ and compute $Q_0$ using the standard EOQ formula for the case of deterministic demand as follows:

$$Q_0 = \sqrt{\frac{2(10{,}800)(187.5)}{20}} = 450$$

Hence,

$$F_L(R_0) = 1 - \frac{20}{10(10,800)} Q_0 = 1 - \frac{20}{10(10,800)}(450) = 0.9167$$

We now need to find $R_0$ such that $F_L(R_0) = 0.9167$. Recall that the daily demand follows a Normal distribution with mean 30 and standard deviation 10. Since lead time is 10 days, then the demand during lead time follows Normal distribution with mean $\overline{D}_L = 10 \times 30 = 300$ and standard deviation $\sigma_L = \sqrt{10} \times 10 = 31.62$. Recall also that $F_L(R_0) = \Phi(z)$, where $z = (R_0 - \overline{D}_L)/\sigma_L$. What value of $z$ gives $\Phi(z) = 0.9167$? We find from standard Normal table in the Appendix that $\Phi(z = 1.38) = 0.9167$. Thus,

$$R_0 = \overline{D}_L + z\sigma_L = 300 + 1.38(31.62) = 343.64$$

Having $R_0 = 343.64$, we can compute the average number of shortages in an order cycle with reorder point $R_0$ as follows:

$$\eta_L(R_0) = \sigma_L \mathcal{L}(z) = (31.62)\,\mathcal{L}(z = 1.38)$$

Using standard Normal table in the Appendix, we find $\mathcal{L}(1.38) = 0.0383$. So, at the end of initial iteration (i.e., Iteration 0), we computed

$$Q_0 = 450, \quad R_0 = 343.64, \quad \eta_L(R_0) = 31.62(0.0383) = 1.21$$

**Step 1:** We set Iteration $i = i + 1 = 0 + 1 = 1$. Having $\eta_L(R_0) = 1.21$, we can now compute $Q_1$ as follows:

$$Q_1 = \sqrt{\frac{2(10,800)[187.5 + 10\eta_L(R_0)]}{20}} = \sqrt{\frac{2(10,800)[187.5 + 10(1.21)]}{20}} = 464.30$$

We then compute $R_1$ as follows:

$$F_L(R_1) = 1 - \frac{20}{10(10,800)} Q_1 = 1 - \frac{20}{10(10,800)}(464.30) = 0.914$$

Using standard Normal table we find $\Phi(z = 1.37) = 0.914$. Thus, $R_1 = \overline{D}_L + z\sigma_L = 300 + 1.37(31.62) = 343.32$. Having $R_1 = 343.32$, we can compute the expected shortage cost in an order cycle with reorder point $R_1$ as follows:

$$\eta_L(R_1 = 343.32) = \sigma_L \mathcal{L}(z) = (31.62)\mathcal{L}(z = 1.37) = (31.62)(0.0392) = 1.24$$

So, at the end of Iteration 1 we have

$$Q_1 = 464.30, \quad R_1 = 343.32, \quad \eta_L(R_1) = 1.24$$

**Step 2:** Since $Q_0 \neq Q_1$ and $R_0 \neq R_1$, we return to Step 1. Results of the remaining iterations are shown in Table 10.4.

In general, the iterative procedure converges to the solution after three or four iterations—as also is the case in Table 10.4. According to the table, $Q^* = 464.64$ and $R^* = 343.32$. Since the volume of cans are integers, we can round the values to $Q^* = 465$ and $R^* = 343$. Hence, to minimize the total inventory and shortage cost of polishing compounds, APD must order 465 cans when its inventory position of cans reaches or falls below 343 cans. Under this policy, APD will carry a safety inventory of

$$I_{\text{safety}} = R - \overline{D}_L = 343 - 300 = 43 \text{ cans}$$

Table 10.4    Iterations to Find Optimal $Q^*$ and $R^*$ for Example 10.5

| Iteration $(i)$ | $Q_i$ | $F(R_i)$ | $z$ | $R_i$ | $\mathcal{L}(z)$ | $\eta_L(R_i)$ |
|---|---|---|---|---|---|---|
| 0 | 450 | 0.917 | 1.38 | 343.63 | 0.0383 | 1.21 |
| 1 | 464.30 | 0.914 | 1.37 | 343.32 | 0.0392 | 1.24 |
| 2 | 464.64 | 0.914 | 1.37 | 343.32 | 0.0392 | 1.24 |
| 3 | 464.64 | 0.914 | 1.37 | 343.32 | 0.0392 | 1.24 |

and should expect, on an average, to have $\eta_L(R = 343) = 1.24$ backordered customers in each order cycle.

---

**Want to Learn More?**

**Systems with Backorder Cost per Unit Time**

In the above case, the shortage cost is fixed and is independent of how long backorder customers must wait to receive their items. In general, the longer a customer waits, the less satisfied the customer is, and the more chance of losing customer's future business. If the waiting times of backordered customers are long and these long waiting times affect a firm's profit, a new model in which shortage cost is incurred per unit time (e.g., each day) is needed. The online supplement of this chapter presents analytics that yield optimal order quantity and optimal reorder point for such cases.

## 10.5 Minimizing Cost in $(Q, R)$ Systems with Lost Sales

In the previous sections we discussed how the optimal $(Q, R)$ policy can be determined if all demands facing stockout are *backordered* at a fixed cost of $C_b$ per backordered customers. This covers systems in which all customers who face stockout can be convinced to wait, for instance, by giving them discount on price and/or offering free shipping.

In this section, we focus on the case of *Lost Sales*, in which all customers facing stockout are not willing to wait and thus the firm loses their sales. This is very common in competitive business environments such as retail industry. For example, when we run out of milk and go to Supermarket A and find that the milk is sold out (i.e, we face stockout), we will not wait until the supermarket receives its next shipment of milk from its supplier and then ships it to us—we are not backordered. What we usually do is to go to Supermarket B and get the milk there—we are lost sales for Supermarket A.

Similar to the backorder case, losing sales is costly. The *Lost Sales* cost include immediate loss of profit from the missed sales as well as the possible future loss of business. We use $C_s$ to represent the lost sales cost. Specifically, we define lost sales cost as follows:

$$C_s = \text{Cost of losing one unit of demand when the item is out of stock}$$

### 10.5.1 Computing Total Average Annual Cost

Similar to the case with backorders, it can be shown that minimizing average annual cost is equivalent to maximizing average annual profit, as long as the lost sales cost includes the immediate loss of profit from missed sales, which is almost always the case. Thus, in this section we compute the total average annual cost in $(Q, R)$ models with lost sales, and in the following section we find the optimal $(Q^*, R^*)$ policy that minimizes this cost.

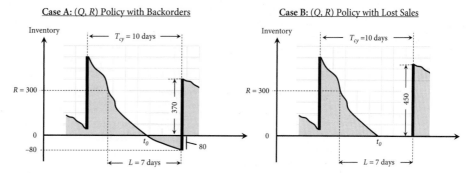

**Figure 10.7**    Inventory curves under $(Q, R) = (450,300)$ in a cycle with backorder (Case A) and lost sales (Case B).

To better understand the difference between backorder and lost sales systems, we discuss how—for the same $Q$ and $R$—the inventory curve when demand is lost is different from that when demand is backordered. Figure 10.7 compares a typical order cycle in systems with backorder (Case A) with a typical order cycle in systems with lost sales (Case B). Both systems use reorder point $R = 300$ and order quantity $Q = 450$ to manage inventory. The two graphs become different when on-hand inventory reaches zero (i.e., time $t_0$).

- *On-hand Inventory During Shortage:* From the time when on-hand inventory reaches zero at time $t_0$ until the next order arrives both systems face shortage. However, the on-hand inventory in the backorder case continues to decrease—it becomes negative—while the on-hand inventory in the lost sales case remains at zero until the next order arrives. As the figure shows, the on-hand inventory just before the next order arrives is $-80$ in the backorder cases and is zero in the lost sales case.

- *On-hand Inventory After Order Arrival:* Therefore, just after the next order arrives, the on-hand inventory in the lost sales case is higher than that in the backorder case. As the figure shows, after receiving an order of size 450, the on-hand inventory in the lost sales case increases to 450, while in the backorder case it increases to $(Q - \text{backorders}) = (450 - 80) = 370$. Thus, in the lost sales case, the next order cycle starts with higher inventory than that in the backorder case. The additional inventory in the lost sales case, that is, 80 units, is equal to the number of backorders in the backorder case, that is, $\eta_L(R)$.

From the above two observations we can conclude the following: For any given $(Q, R)$ policy we should expect that, on an average, order cycles in a system with lost sales carry more inventory than order cycles in an identical system with backorders. The additional inventory is equal to the average number of shortages in an order cycle, that is, $\eta_L(R)$. Thus, the components of the annual cost for lost sales case are the same as those for backorder case, except for the average annual inventory. Recall that

$$\text{Average Annual Inventory} = \text{Average Cycle Inventory} + \text{Average Safety Inventory}$$

Using continuous approximation of the inventory curve, similar to the backorder case, the average cycle inventory can be approximated by $Q/2$. For the average safety inventory, however, things would be different. Because in lost sales case the inventory never becomes negative, for the safety inventory we have

$$I_{\text{Safety}} = E(R - D_L)$$

$$= \int_0^R (R - x) f_L(x) dx + 0 \times \int_R^\infty (R - x) f_L(x) dx$$

$$= \int_0^\infty (R - x) f_L(x) dx + \int_R^\infty (x - R) f_L(x) dx$$

$$= R - \overline{D}_L + \eta_L(R)$$

Therefore, for the case with lost sales, the total annual average cost is

$$TC(Q, R) = K\left(\frac{\overline{D}}{Q}\right) + C_h\left(\frac{Q}{2} + R - \overline{D}_L + \eta_L(R)\right) + \Psi_s(R)\left(\frac{\overline{D}}{Q}\right) \qquad (10.29)$$

Considering the lost sales $C_s$, the average lost sales cost in an order cycle would be

$$\Psi_s(R) = C_s \, \eta_L(R)$$

where, as in the case of backorders, $\eta_L(R)$ is the average number of shortages in an order cycle—in this case, the average lost sales in an order cycle.

### 10.5.2 Finding the Optimal $(Q, R)$ Policy

When demands that face shortage are lost, we can use the following analytics to compute the optimal $(Q, R)$ policy.

---

**ANALYTICS**

**Finding the Optimal $(Q, R)$ Policy for Minimizing Cost in Systems with Lost Sales**

Consider a continuous review make-to-stock (MTS) inventory system managed by a $(Q, R)$ policy in which all customers who face shortage are lost. Suppose

$$C_s = \text{Cost per one unit lost demand}$$

which includes the immediate loss of profit as well as the potential future sales and loss of goodwill. For any given reorder point $R$, the average lost sales cost in an order cycle is

$$\Psi_s(R) = C_s \, \eta_L(R) \qquad (10.30)$$

When demand during lead time is Normally distributed with mean $\overline{D}_L$ and standard deviation $\sigma_L$, we have

$$\eta_L(R) = \sigma_L \mathcal{L}(z), \quad \text{where } z = \frac{R - \overline{D}_L}{\sigma_L}$$

and $\mathcal{L}(z)$ is found in standard Normal table in the Appendix.

The optimal order quantity $Q^*$ and reorder point $R^*$ that minimize the above cost are those that simultaneously satisfy the following two equations (see online Appendix G for details):

$$Q^* = \sqrt{\frac{2\overline{D}[K + C_s\eta_L(R)]}{C_h}} \qquad (10.31)$$

$$F_L(R^*) = 1 - \frac{C_h}{C_hQ^* + C_s\overline{D}}Q^* \qquad (10.32)$$

where $F_L(R)$ is the value of cumulative probability distribution function of *demand during lead time* for reorder point $R$. Both optimal $R^*$ and $Q^*$ can be found using a similar iterative approach as for the case of costs per backordered demand.

The total average annual holding, setup, and backorder cost under a $(Q, R)$ policy, including the optimal policy, is as follows:

$$TC(Q, R) = K\left(\frac{\overline{D}}{Q}\right) + C_h\left(\frac{Q}{2} + R - \overline{D}_L + \eta_L(R)\right) + C_s\, \eta_L(R)\left(\frac{\overline{D}}{Q}\right) \tag{10.33}$$

We use the following example to show how the above analytics is used to obtain the optimal $(Q, R)$ policy in systems with lost sales.

**Example 10.6**

Consider the Auto Part Depot (APD) in Example 10.5 which is now facing competition from an online firm that offers similar polishing compound product. A recent study has shown that almost all customers who come to APD to buy the compound and face stock-out do not wait and buy the similar product from the online firm. APD estimates that each lost customer costs them about $15, which includes the immediate loss of profit and loss of goodwill. How should APD revise its inventory policy for the polishing compound in order to minimize its total inventory costs?

We can obtain the optimal $(Q, R)$ policy for the APD with lost sales using the above analytics. Considering the lost sales cost of $C_s = \$15$, the optimal $Q^*$ and $R^*$ must satisfy Eqs. (10.31) and (10.32) as follows:

$$Q^* = \sqrt{\frac{2(10{,}800)[187.5 + 15\eta_L(R)]}{20}} \tag{10.34}$$

$$F_L(R^*) = 1 - \frac{20}{20Q^* + 15(10{,}800)}Q^* \tag{10.35}$$

We can now use the iterative approach to obtain the optimal values of $Q^*$ and $R^*$.

**Step 0:** We set Iteration $i = 0$ and compute $Q_0$ using the standard EOQ formula for the case of deterministic demand as follows:

$$Q_0 = \sqrt{\frac{2(10{,}800)(187.5)}{20}} = 450$$

Hence,

$$F_L(R_0) = 1 - \frac{20}{20Q_0 + 15(10{,}800)}Q_0 = 1 - \frac{20}{20(450) + 15(10{,}800)}450 = 0.947$$

Recall that demand during lead time follows Normal distribution with mean $\overline{D}_L = 300$ and standard deviation $\sigma_L = 31.62$. Recall also that $F_L(R_0) = \Phi(z)$, where $z = (R_0 - \overline{D}_L)/\sigma_L$. We find from standard Normal table that $\Phi(z = 1.62) = 0.947$. Thus,

$$R_0 = \overline{D}_L + z\sigma_L = 300 + 1.62(31.62) = 351.22$$

Table 10.5    Iterations to Find Optimal $(Q^*, R^*)$ Policy in Example 10.6

| Iteration $(i)$ | $Q_i$ | $F(R_i)$ | $z$ | $R_i$ | $\mathcal{L}(z)$ | $\eta_L(R_i)$ |
|---|---|---|---|---|---|---|
| 0 | 450 | 0.947 | 1.62 | 351.22 | 0.0222 | 0.7020 |
| 1 | 462.46 | 0.946 | 1.61 | 350.91 | 0.0227 | 0.7178 |
| 2 | 462.74 | 0.946 | 1.61 | 350.91 | 0.0227 | 0.7178 |
| 3 | 462.74 | 0.946 | 1.61 | 350.91 | 0.0227 | 0.7178 |

Having $R_0 = 351.22$, we can compute the average number of lost sales in an order cycle with reorder point $R_0$ as follows:

$$\eta_L(R_0) = \sigma_L \mathcal{L}(z) = (31.62) \, \mathcal{L}(z = 1.62)$$

Using standard Normal table in the Appendix, we find $\mathcal{L}(1.62) = 0.0222$. So, $\eta_L(R_0) = (31.62) \, 0.0222 = 0.7020$. Thus, at the end of initial iteration (i.e., Iteration 0), we computed

$$Q_0 = 450, \quad R_0 = 351.22, \quad \eta_L(R_0) = 0.7020$$

**Step 1:** We set Iteration $i = i + 1 = 0 + 1 = 1$. Having $\eta_L(R_0) = 0.7019$, we can now compute $Q_1$ as follows:

$$Q_1 = \sqrt{\frac{2(10{,}800)[187.5 + 15\eta_L(R_0)]}{20}} = \sqrt{\frac{2(10{,}800)[187.5 + 15(0.7020)]}{20}} = 462.46$$

We then compute $R_1$ as follows:

$$F_L(R_1) = 1 - \frac{20}{20Q_1 + 15(10{,}800)} Q_1 = 1 - \frac{20}{20(462.46) + 15(10{,}800)} 462.46 = 0.946$$

Using standard Normal table we find $\Phi(z = 1.61) = 0.946$. Thus, $R_1 = 300 + 1.61(31.62) = 350.91$. Having $R_1 = 350.91$, we can compute the expected lost sales in an order cycle with reorder point $R_1$ as follows:

$$\eta_L(R_1) = \sigma_L \mathcal{L}(z) = (31.62)\mathcal{L}(z = 1.61) = (31.62)(0.0227) = 0.7178$$

So, at the end of Iteration 1 we have

$$Q_1 = 462.46, \quad R_1 = 350.91, \quad \eta_L(R_1) = 0.7178$$

**Step 2:** Since $Q_0 \neq Q_1$ and $R_0 \neq R_1$, we return to Step 1. Results of the remaining iterations are shown in Table 10.5.

According to the table, we have $Q^* = 462.74$ and $R^* = 350.91$. Rounding these values we get $Q^* = 463$ and $R^* = 351$. Hence, to minimize the inventory management cost of polishing compounds, APD must order 465 cans when its inventory position reaches or falls below 351 cans. Under this policy, APD will carry a safety inventory of

$$I_{safety} = R - \overline{D}_L = 351 - 300 = 51 \text{ cans}$$

and should expect, on an average, to have $\eta_L(R = 351) = 0.7178$ lost sales in each order cycle.

## 10.6 Setting Service Levels

Recall from Sec. 9.7 of Chap. 9 that one of the goals of inventory management is to provide a certain level of service, that is, to make sure that the item is available to most customers who demand it. Two main reasons for setting a service level are as follows:

- *Competitive Market:* Even when total expected annual cost is minimized (using our analytics), due to the variability in demand, some customers will still face shortage. For example, if the cost of holding one item in inventory is very high and the cost of shortage is low, then the optimal policy that minimizes total cost may result in low inventory and large shortage. Hence, in competitive markets where customers can easily switch to another firm, in addition to minimizing cost, it is also important to set a limit on the number of shortages, that is, to set a certain service level.

- *Unknown Shortage Cost:* Recall that shortage cost includes two components: (i) short-term costs, and (ii) long-term costs. Short-term costs include the immediate costs associated with customer delay and loss of profit, while long-term costs are associated with the costs of possibly losing future business or loss of goodwill. While short-term costs are usually easy to estimate, in some cases, finding good estimates for long-term costs of shortage are not easy, if not impossible. For example, how much is the long-term cost of a customer who faces a shortage of a best-selling novel in a bookstore? When an accurate estimate of shortage cost is not available, then setting a service level provides an effective means to control the number of shortages.

As we discussed earlier, there are two main service levels: (i) Cycle-Service Level, and (ii) Fill Rate. In the following section, we first start by showing how one can set the cycle-service to a certain level, while still minimizing the total expected annual inventory costs.

### 10.6.1 Optimal (Q, R) Policy When Setting Cycle-Service Level

Cycle-service level $\alpha$, also known as type-1 service level, is the probability of not having shortage in an order cycle, that is,

$$\alpha = P(\text{having no shortage in a cycle})$$

Under a $(Q, R)$ policy, shortage occurs if the demand during lead lime becomes larger than reorder point $R$. Hence, the cycle-service level for any given $(Q, R)$ policy will be

$$\alpha = P(D_L \leq R) = F_L(R) \tag{10.36}$$

Two points worth mentioning here are as follows:

- As Eq. (10.36) shows, cycle-service level $\alpha$ does not depend on order quantity $Q$ or annual demand. It only depends on reorder point $R$ and demand during lead time $D_L$. This is obvious, since shortage only occurs during lead time when demand during lead time becomes larger than reorder point $R$.

- Cycle-service level is also independent of whether demands facing shortage are backordered or lost. This is because cycle-service level only captures the possibility of not having a shortage in an order cycle. For a given reorder point $R$ this possibility is the same in $(Q, R)$ policies with backorders and lost sales.

Cycle-service level $\alpha$ has different interpretations besides being the probability of not having shortage in an order cycle. It can also be interpreted as percentage of order cycles that does not have shortage, or the probability that the demand during lead time is immediately met from the available inventory.

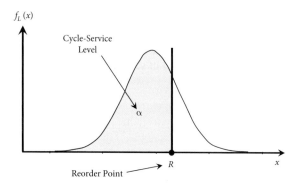

Figure 10.8    The relationship between cycle-service level and reorder point $R$ with demand distribution during lead time.

 **Example 10.7**

Consider the optimal inventory policy $(Q^*, R^*) = (463, 351)$ that we obtained for Auto Part Depot (APD) in Example 10.6 in order to minimize APD's total costs. Also, recall that customers who face shortage are lost. What would be the cycle-service level if APD implements this policy?

Under reorder point $R = 351$, the cycle-service level for APD is

$$\alpha = \mathsf{P}(D_L \leq 351) = F_L(351)$$

Since demand during lead time is Normally distributed with mean $\overline{D}_L = 300$ and standard deviation $\sigma_L = 31.62$, then

$$F_L(R = 351) = \Phi\left(z = \frac{351 - 300}{31.62}\right) = \Phi(z = 1.613) = 0.947$$

Hence, the optimal policy $(Q^*, R^*) = (463, 351)$ that minimizes the expected cost will result in cycle-service level $\alpha = 94.7\%$.

While Eq. (10.36) can be used to compute the resulting cycle-service level for a given reorder point, it can also be used to find the reorder point to achieve a certain cycle-service level. Specifically, if one desires to achieve a cycle-service level of $\alpha$ under a $(Q, R)$ policy, one should set the optimal reorder point $R$ such that the cumulative distribution of demand during lead time at point $R$ be $\alpha$ (see Fig. 10.8).

Hence, considering reorder point $R_\alpha^*$ as the optimal reorder point that results in cycle-service level of $\alpha$, then it must satisfy

$$\alpha = F_L(R_\alpha^*) \tag{10.37}$$

If the demand is Normally distributed, then the reorder point will be

$$R_\alpha^* = \overline{D}_L + z\sigma_L \quad \text{where} \quad \Phi(z) = \alpha$$

or using Excel,

$$R_\alpha^* = \mathsf{NORMINV}(\alpha, \overline{D}_L, \sigma_L)$$

But, what about order quantity $Q$? What is the optimal order quantity $Q_\alpha^*$ that results in cycle-service level $\alpha$, but with minimum total cost? To find the answer, first note that order quantity does not affect cycle-service level. Second, recall that Eq. (10.21) in Sec. 10.4.2 gives the optimal order quantity that minimizes the total average annual cost for a *given* reorder point $R$. Hence, if we set our reorder point to $R_\alpha^*$ that achieves cycle-service level $\alpha$, then the order quantity $Q_\alpha^*$ that minimizes the total average annual cost will be

$$Q_\alpha^* = \sqrt{\frac{2\overline{D}[K + \Psi_s(R_\alpha^*)]}{C_h}}$$

The average shortage cost $\Psi_s(R_\alpha^*)$ in an order cycle, as shown before, depends on whether shortages are lost or backordered, and also is a function of costs of backorders and lost sales. If, however, an accurate estimate of backorder cost $C_b$ or lost sales cost $C_s$ is not available, it has been shown that using the EOQ formula in Eq. (9.2) of Chap. 9 gives a good approximation for the optimal order quantity $Q_\alpha^*$. This leads us to the following analytics.

## ANALYTICS

### Finding the Optimal $(Q_\alpha^*, R_\alpha^*)$ Policy to Achieve a Cycle-Service Level of at Least $\alpha$

Consider a continuous review make-to-stock (MTS) inventory system managed by a $(Q, R)$ policy. The goal is to find the optimal reorder point $R_\alpha^*$ and order quantity $Q_\alpha^*$ that achieve cycle-service level $\alpha$ with minimum total expected annual holding, setup, and shortage cost.

- **Step 1:** (*Finding Reorder Point $R_\alpha^*$*) The optimal reorder point $R_\alpha^*$ that achieves cycle-service level $\alpha$ satisfies

$$\alpha = P(D_L \leq R_\alpha^*) = F_L(R_\alpha^*)$$

where $F_L(x)$ is the cumulative distribution of demand during lead time. If the demand during lead time is Normally distributed with mean $\overline{D}_L$ and standard deviation $\sigma_L$, then the reorder point will be

$$R_\alpha^* = \overline{D}_L + z\sigma_L \quad \text{where} \quad \Phi(z) = \alpha$$

and $\Phi(z)$ is the cumulative distribution of standard Normal distribution and is available in the Appendix.

- **Step 2:** (*Finding Order Quantity $Q_\alpha^*$*) The optimal order quantity for backorder case and lost sales case can be found as follows:
  - *Systems with Backorders:* If demands that face shortage are backordered with backorder cost $C_b$, then the optimal order quantity is

$$Q_\alpha^* = \sqrt{\frac{2\overline{D}[K + C_b\, \eta_L(R_\alpha^*)]}{C_h}}$$

  - *Systems with Lost Sales:* If demands that face shortage are lost with lost sales cost of $C_s$, then the optimal order quantity is

$$Q_\alpha^* = \sqrt{\frac{2\overline{D}[K + C_s\eta_L(R_\alpha^*)]}{C_h}}$$

If accurate estimates of backorder or lost sales costs are not available, then the optimal order quantity $Q_\alpha^*$ can be approximated by EOQ model as follows:

$$Q_\alpha^* = \sqrt{\frac{2K\overline{D}}{C_h}} \tag{10.38}$$

**Example 10.8**

Cool Fans is a firm that produces standard and customized ceiling fans, where customized fans are made-to-order after receiving orders from the customers through the firm's website. Customized fans require a particular three-speed electric motor called 3SPD motors that are purchased from a supplier at the price of $36. The supplier has a lead time of 2 weeks and the total order setup cost that also includes shipment cost is $120 per order. Cool Fans has an inventory carrying rate of 25 percent. The aggregated demand for 3SPD motors is about 10 per week. The demand during lead time is found to be Normally distributed with mean 20 and standard deviation 7. The inventory of the motors is monitored continuously and the firm orders $Q = 120$ motors when the inventory of motors reaches reorder point $R = 25$. John, the production manager, schedules a meeting with his shopfloor manager, Bill, to discuss recent delays in processing customer orders. In the meeting John says:

**John:** I have been looking at our production data and found that about 15 out of last 50 times—about 30 percent of times—that we placed an order for motors, we ran out of motors before the new shipment arrived. What do you do when we run out of motors?

**Bill:** As you know we produce both standard and customized fans on our two assembly lines. Motors are installed at the first station of our second assembly line in which we assemble *electrical* parts to fans coming from our first assembly line where *mechanical* parts are already assembled. When we run out of 3SPD motors, we stop the production of the customized orders in our second line and work on the standard fans that have lower priority. Meanwhile, we store the customized fans coming from the first line in a temporary storage until we receive the next shipment of 3SPD motors. When the new shipment arrives, we move the customized fans from the storage to the second line and start installing motors and complete the order.

**John:** How does this affect your production schedule?

**Bill:** It adds more inventory and more work to our process and delays customers' orders. We did a study and we estimated that the additional cost of labor and inventory due to shortage of motors add up to an average of $5.50 to our production cost per delayed order. There is also a cost associated with having an unsatisfied customer (with delayed order) which we estimated be about $2.00. So, each delayed order will cost us a total of $7.5.

**John:** Any suggestions what to do?

**Bill:** I think we need to either talk to the supplier of 3SPD and see if they can deliver our orders sooner than their current lead time of 2 weeks; or, ....

John interrupts Bill and says:

**John:** We have been trying to do that, but we found that due to their long production lead time, the supplier cannot deliver our orders in less than 2 weeks. I think we should change our inventory policy such that we reduce the chance of running out of stock when we are waiting for the next shipment from 30 percent to about 2 percent. Of course, we still want to minimize our inventory costs.

**Bill:** I need some time to look into this and will get back to you soon.

How should Bill revise the current ordering policy in order to achieve what John suggested?

Note that "the chance of running out of stock when waiting for the next shipment" refers to the probability of having a shortage in an order cycle. This, as we discussed earlier, corresponds to the cycle-service level. Specifically, setting cycle-service level $\alpha = 98\%$ guarantees that the chance of running out of stock while waiting for the next shipment of 3SPD motors is reduced to $100\% - 98\% = 2\%$.

Before finding the optimal $(Q_\alpha^*, R_\alpha^*)$ policy that achieves cycle-service level $\alpha = 98\%$, let's first compute the cycle-service level $\alpha$ under the current policy $(Q, R) = (120, 25)$. Recall that the demand during lead time $D_L$ is Normally distributed with mean $\overline{D}_L = 20$ and standard deviation $\sigma_L = 7$. Hence, since $R = 25$, using Eq. (10.36), we have

$$\alpha_{\text{current}} = P(D_L \leq 25) = \Phi\left(\frac{25 - 20}{7}\right) = \Phi(0.7143) = 0.762$$

Therefore, under the current ordering policy in $100\% - 76.2\% = 23.8\%$ of order cycles the firm faces shortage. Recall that John wants to reduce it to 2 percent by setting its cycle-service level to 98 percent. We use our analytics to find $(Q_\alpha^*, R_\alpha^*)$ that achieves this with minimum total average annual cost.

**Step 1:** The reorder point $R_\alpha^*$ that achieves cycle-service level $\alpha = 98\%$ must satisfy

$$\alpha = 98\% = P(D_L \leq R_\alpha^*)$$

Hence,

$$R_\alpha^* = \overline{D}_L + z\sigma_L, \quad \text{where} \quad \Phi(z) = 0.98$$

Using standard Normal table, we find $\Phi(z = 2.055) = 0.98$. Hence,

$$R_\alpha^* = \overline{D}_L + z\sigma_L = 20 + 2.055(7) = 34.4$$

which we round up to 35. Thus, to achieve cycle-service level of $\alpha = 98\%$, John must increase its current reorder point $R = 25$ to $R_\alpha^* = 35$ units.

**Step 2:** Recall that the average demand during lead time of 2 weeks is 20. Considering 52 weeks in a year, the average annual demand would be $\overline{D} = 520$. Following Step 2, since the shortages are backordered with the backorder cost \$7.5 per backorder demand, then the optimal order quantity would be

$$Q_\alpha^* = \sqrt{\frac{2\overline{D}[K + C_b' \, \eta_L(R_\beta^*)]}{C_h}} = \sqrt{\frac{2(520)[120 + 7.5\eta_L(35)]}{9}}$$

in which $C_h = 0.25 \times 36 = 9$. Computing $z = (35 - \overline{D}_L)/\sigma_L = (35 - 20)/7 = 2.14$, then

$$\eta_L(R_\alpha^*) = \sigma_L \mathcal{L}(z) \implies \eta_L(35) = 7\mathcal{L}(2.14) = 7(0.0058) = 0.0406$$

Hence,

$$Q_\alpha^* = \sqrt{\frac{2(520)[120 + 7.5\eta_L(35)]}{9}} = \sqrt{\frac{2(520)[120 + 7.5(0.0406)]}{9}} = 117.9 \simeq 118$$

Therefore, to achieve service level $\alpha = 98\%$, while minimizing the total annual expected costs, John must decrease its current order quantity from 120 to $Q_\alpha = 118$ and increase its current reorder point from $R = 25$ to $R_\alpha = 35$ units.

## 10.6.2 Optimal (Q, R) Policy When Setting Fill Rate

Recall from Sec. 9.7 of Chap. 9 that fill rate $\beta$, also known as type-2 service level, is the fraction of overall demand that does not face shortage, or

$$\beta = 1 - \frac{\text{Demand facing shortage}}{\text{Total demand}}$$

When demand is stochastic, under a $(Q, R)$ policy, each order cycle will have different demand and different number of shortages. Hence, we use the *average* demand in an order cycle as well as the *average* shortages in an order cycle to compute Fill Rate. Recall that for any given reorder point $R$, the average shortage (backordered or lost sales) is $\eta_L(R)$. Considering $\overline{D}_{cy}$ as the average demand during a cycle, we will have

$$\beta = 1 - \frac{\text{Average shortage in } \textit{an order cycle}}{\text{Average demand in } \textit{an order cycle}}$$

$$= 1 - \frac{\eta_L(R)}{\overline{D}_{cy}} \qquad (10.39)$$

But, what is the average demand during an order cycle? The answer depends on whether demands facing shortage are backordered or lost:

- *Systems with Backorders:* Since all demands (including backorders) are eventually satisfied by ordering $Q$ in each order cycle, to have a stable system under a $(Q, R)$ policy, the average demand during a cycle must be equal to $Q$, that is, $\overline{D}_{cy} = Q$. If for example, $\overline{D}_{cy} < Q$, this means that, in the long run, the inventory in the system will increase to infinity. Also, if $\overline{D}_{cy} > Q$, then in the long run, the number of backordered demands increases to infinity. Both of these cases cannot occur in a stable process. Thus,

$$\overline{D}_{cy} = Q$$

  Hence, under $(Q, R)$ policy, fill rate $\beta$ in systems with backorders can be computed as follows:

$$\beta = 1 - \frac{\eta_L(R)}{Q} \qquad (10.40)$$

  In other words, for any given order quantity $Q$, the reorder point $R$ that achieves fill rate $\beta$ must satisfy

$$\eta_L(R) = (1 - \beta)Q \qquad (10.41)$$

- *Systems with Lost Sales:* As opposed to systems with backorders, in systems with lost sales demands facing shortage are not satisfied. Thus, the demand during a cycle consists of satisfied demand and lost demand. Again, to have a stable system, the average demands that are satisfied must be the same as order quantity $Q$. The average lost demand, however, is $\eta_L(R)$. Thus,

$$\overline{D}_{cy} = Q + \eta_L(R)$$

  Hence, under $(Q, R)$ policy, fill rate $\beta$ in systems with lost sales can be found as follows:

$$\beta = 1 - \frac{\eta_L(R)}{Q + \eta_L(R)} \qquad (10.42)$$

  In other words, for any given order quantity $Q$, the reorder point $R$ that achieves fill rate $\beta$ must satisfy

$$\eta_L(R) = \frac{1 - \beta}{\beta} Q \qquad (10.43)$$

 **Example 10.9**

Consider the optimal inventory policy $(Q^*, R^*) = (463, 351)$ that we obtained for Auto Part Depot (APD) in Example 10.6 in order to minimize APD's costs with lost sales. We found, in Example 10.7, that this policy results in cycle-service level of $\alpha = 94.7\%$. But, if APD implements this policy, what fraction of its customers will be lost?

To compute fill rate $\beta$ using Eq. (10.39), we need to compute the average number of lost sales in an order cycle under policy $(Q^*, R^*) = (463, 351)$, that is, $\eta_L(351)$, and the average number of demand during a cycle, that is, $\overline{D}_{cy}$. Since the demand during lead time is Normal with mean $\overline{D}_L = 300$ and standard deviation $\sigma_L = 31.62$, then

$$\eta_L(351) = \sigma_L \mathcal{L}(\frac{351 - 300}{31.62}) = 31.62\mathcal{L}(1.613) = 31.62(0.0225) = 0.71$$

since using standard Normal tables we find $\mathcal{L}(1.613) = 0.0225$.

On the other hand, under policy $(Q^*, R^*) = (463, 351)$, the average demand during a cycle for systems with lost sales is

$$\overline{D}_{cy} = Q + \eta_L(R) = 463 + 0.71 = 463.71$$

Hence, the fill rate is

$$\beta = 1 - \frac{\eta_L(R)}{\overline{D}_{cy}} = 1 - \frac{0.71}{463.71} = 99.85\%$$

which implies that under $(Q^*, R^*) = (463, 351)$, APD should expect to lose $100\% - 99.85\% = 0.15\%$ of its customers.

Note that there may be several $(Q, R)$ policies that satisfy Eq. (10.40) and thus result in the same fill rate $\beta$. This is because, for different reorder point $R$, we can solve Eq. (10.40) for systems with backorders and obtain a different order quantity $Q$. This also is the case with Eq. (10.42) that corresponds to the case of lost sales. While all these policies achieve the same fill rate $\beta$, the question is which one of these policies has the minimum inventory costs? In other words, how can we find the optimal $(Q, R)$ policy that achieves a certain fill rate $\beta$ with minimum total expected cost?

In general, finding the answer to this question is very complex. However, it has been shown that the optimal order quantity $Q$ for the deterministic EOQ model provides good estimates for their stochastic counterparts. This leads us to the following analytics.

## ANALYTICS

### Finding the Optimal $(Q_\beta^*, R_\beta^*)$ Policy to Achieve Fill Rate $\beta$ with Minimum Cost

Consider a continuous review make-to-stock (MTS) inventory system managed by a $(Q, R)$ policy. The goal is to find the optimal reorder point $R_\beta^*$ and order quantity $Q_\beta^*$ that minimize the total expected annual cost, while achieving fill rate $\beta$.

- *Systems with Backorders:* If demands that face shortage are backordered, the optimal order quantity $Q_\beta^*$ can be estimated by

$$Q_\beta^* = \sqrt{\frac{2K\overline{D}}{C_h}} \qquad (10.44)$$

and the optimal reorder point $R_\beta^*$ satisfies

$$\eta_L(R_\beta^*) = (1 - \beta)Q_\beta^* \qquad (10.45)$$

- *Systems with Lost Sales:* If demands that face shortage are lost, the optimal order quantity can be estimated by

$$Q_\beta^* = \sqrt{\frac{2K\overline{D}}{C_h}} \qquad (10.46)$$

and the optimal reorder point $R_\beta^*$ satisfies

$$\eta_L(R_\beta^*) = \frac{1-\beta}{\beta}\, Q_\beta^* \tag{10.47}$$

 **Example 10.10**

Consider Cool Fans in Example 10.8 that produces standard and customized ceiling fans and orders 3SPD electric motors using $(Q, R) = (120, 25)$. After finding the new policy $(Q, R) = (118, 35)$ that achieves cycle-service level $\alpha = 98\%$, Bill calls John and says:

**Bill:** I think by increasing our reorder point from $R = 25$ to $R = 35$ and decreasing our order quantity from $Q = 120$ to $118$, we will be able to reduce the chance of facing a shortage of motors while waiting for the next shipment to about 2 percent. However, I am not sure if this is what you wanted.

**John:** What do you mean?

**Bill:** By implementing this new policy, we will indeed reduce the chance of facing a shortage—while waiting for our next shipment—from about 23.8 percent to 2 percent. However, when we face shortage in those 2 percent cases—causing delays in customer orders—the number of delayed customers might be very high. In fact, we will have no control over how many customer orders will be delayed. Considering the cost of shortage, I am not sure we are setting the right service level.

**John:** I am confused, could you elaborate?

**Bill:** Recall that, our second assembly line that installs 3SPD motors is very flexible and we can switch from assembling a customized order to a standard order with almost no cost. Thus, the number of times that we switch the assembly line to produce standard products—which is equal to the number of order cycles with shortage—does not cost us much. The main cost of the shortage, as you know, is for the number of delayed orders, which include the inventory and labor cost as well as customer dissatisfaction. Thus, I suggest that we revise our inventory policy in order to reduce the *fraction of delayed orders* to 2 percent instead of reducing the *fraction of order cycles with shortage* to 2 percent. Does that make sense?

**John:** Yes, I think you are right. But, I need to clarify one thing. After talking to our marketing people, I found that our estimate of $2.00 per dissatisfied customer is not that accurate. Hence, the overall cost of having a delayed order is higher than $7.5, but I do not know the exact amount. This, I believe, further emphasizes your point. Since we do not know what the exact cost of shortage is, it makes sense to make sure the fraction of delayed orders is small, that is, 2%. But, do you think by doing this we will need more inventory or less inventory?

**Bill:** Actually, our data shows that our old policy of $(Q, R) = (120, 25)$ was already achieving this, and we do not need to change it. But I am going to check that to make sure and will let you know.

How should Bill revise the policy to reduce the fraction of delayed orders to 2 percent while also minimizing cost?

The new 2-percent goal is, in fact, fill rate of $\beta = 98\%$. Specifically, the new goal is to make sure that $100\% - 2\% = 98\%$ of the orders will not be delayed due to shortage of 3SPD motors. Obviously, by keeping lots of inventory (high cycle inventory and safety inventory), one can reduce the fraction of delayed orders. However, what John also wants is to minimize the total inventory cost and order setup cost. We follow our analytics to find the $(Q_\beta^*, R_\beta^*)$ policy that achieves fill rate $\beta = 98\%$ at minimum total expected annual cost.

Since all shortages are backordered, according to Eq. (10.44), to achieve fill rate $\beta$, the order quantity is

$$Q_\beta^* = \sqrt{\frac{2K\overline{D}}{C_h}} = \sqrt{\frac{2(120)\,(520)}{9}} = 117.7 \simeq 118$$

And the optimal reorder point $R_\beta^*$ must satisfy

$$\eta_L(R_\beta^*) = (1 - \beta)Q_\beta^* = (1 - 0.98)(118) = 2.36$$

In other words, under the reorder point $R_\beta^*$ that achieves fill rate of 98%, we should expect to see an average shortage of $\eta_L(R_\beta^*) = 2.36$ units in each order cycle. But, what is the reorder point $R_\beta^*$? Since the demand during lead time is Normally distributed with mean $\overline{D}_L = 20$ and standard deviation $\sigma_L = 7$, we have

$$\eta_L(R_\beta^*) = \sigma_L \mathcal{L}(z)$$
$$2.37 = 7\mathcal{L}(z) \quad \Longrightarrow \quad \mathcal{L}(z) = 2.36/7 = 0.3373$$

Using standard Normal table we find $\mathcal{L}(z = 0.13) = 0.3373$. Thus,

$$R_\beta^* = \overline{D}_L + z\sigma_L = 20 + 0.13(7) = 20.91$$

which we round up to 21. Therefore, to achieve fill rate $\beta = 0.98$, John must order $Q_\beta^* = 118$ 3SPD motors each time its inventory position reaches $R_\beta^* = 21$.

## 10.7 Periodic Review of Multi-Period Inventory Systems

In continuous review systems, inventory of items is constantly monitored using computers. Hence, managers place an order of size $Q$ any time the inventory reaches (or falls below) reorder point $R$. In periodic review systems, however, an accurate review of inventory is obtained at the end of each period (e.g., end of each day, each week, or each month). The decision is then made about whether and how many items must be ordered for the next period. Thus, in periodic review systems, orders cannot be placed at any time; they can only be placed at the end of each period.

**Ordering Soy Sauce for Big-Mart**
Remember Mike in Chap. 9 who was working in Big-Mart. Due to Mike's success in developing a cost-effective inventory policy for bookshelves, he has been sent to other departments of Big-Mart to help them improve their inventory. At the grocery department, Mike is asked to help with setting an ordering policy for a particular Japanese soy sauce distributed by a large distributor of Asian food.

Checking the computer, Mike can see the inventory of the soy sauce. Happy that the inventory is continuously monitored (i.e., a continuous review system), he decides to manage the inventory of the sauce according to the $(Q, R)$ policy. All he needs to do is to find the value of $Q$ and $R$ for the soy sauce.

Talking to Lisa, the manager of the grocery department, however, he finds that the situation is completely different from what he had with the bookshelves. With the bookshelves, he could place an order any time he wanted (e.g., when its inventory reaches $R = 12$ or below).

With the sauce, however, as Lisa explains, Mike can only place an order on Friday afternoons. If he does not order any soy sauce on Friday, then he needs to wait until next Friday to do so. When he asks for the reason, Lisa explains:

> We order the sauce from a distributor that delivers Asian food products to all stores in different regions of the city. Their trucks deliver products to each region in a different day of a week. We are in the region that gets deliveries on Fridays. So every Friday afternoon their truck arrives at our store and asks if we need any products, including the soy sauce. Hence, each Friday morning, Eric counts the number of soy sauce bottles on the shelves and in the storage to get an accurate count. Confirming his counts with computer records, he then decides how many bottles we should order. We buy each bottle of sauce for $3 and the supplier does not charge us a delivery cost when it delivers them on Fridays. We also have the option of placing an order on other days of the week, but that will cost us $60 for rush delivery. The profit we make from the sauce is low and we do not have a large demand for it. So it is not worth to order on days other than Fridays.

Mike realizes that he is dealing with a periodic review system where he can place an order only at the end of each period (i.e., each week). He thinks about how to set an inventory management policy for the soy sauce. The big question is: *How many bottles should he buy (i.e., order) this Friday?* To get an idea of how many bottles to order for the next week, Mike checks the weekly sales in the last 20 weeks and he finds that sales range from 10 to 60 bottles per week, with an average of 40 bottles. Considering the average of 40, and to protect against possible shortage, Mike thinks that starting next week with 55 bottles of sauce is a good policy. This is because only in 2 weeks (of the past 20 weeks) the demand was above 55. So, he feels that starting a week with 55 bottles, the chance of having a shortage—demand being higher than 55—would be about $2/20 = 10\%$. Mike feels good about this number and sets his ordering policy as follows: Each Friday, place an order to raise the inventory of soy sauce up to 55. For example, if the count shows that the store has only 20 bottles of soy sauce on Friday noon, then order 35 bottles to raise the inventory to $20 + 35 = 55$ bottles for the next week's demand.

The policy that Mike came up with is a well-known inventory management policy for periodic review systems. It is called "order-up-to policy."

## CONCEPT

### Order-Up-To Policy and Order-Up-To Level

The *order-up-to policy* with *order-up-to level S* is used in periodic review systems in which the inventory of items is checked at the end of each period. The policy works as follows: If at the end of a period the inventory position is $I$:

- If $I < S$, then place an order for $S - I$ units to raise the inventory position to $S$.
- If $I = S$, then do not place an order.

Order-up-to policy is the optimal policy in periodic review systems when order setup cost $(K)$ is zero.

As mentioned above, the order-up-to policy is the optimal policy if the fixed cost of placing an order is zero. In the case of soy sauce, for example, supplier does not charge for shipping Big-Mart's order $(k = 0)$. If placing an order has a significant cost, however, order-up-to policy is not optimal; a more general version of the order-up-to policy, called $(s, S)$ policy, is optimal. We will discuss this later in Sec. 10.7.3.

Before we end this section, we must mention that counting the physical inventory to confirm the accuracy of inventory records before placing an order (as Eric does before ordering soy sauce) is not only for periodic review systems. Continuous review systems also have issues such as misplaced or stolen inventory, errors in entering data, and damaged inventory that is not known. Hence, all firms perform a physical counting of their inventories some time during their operations. The most common is the annual counting of inventory to satisfy auditors that inventory records are consistent with the actual value of inventory.

### 10.7.1 Order-Up-To Policy and Lead Times

To illustrate how order-up-to policy works, we use the inventory management of a product at Health-NS in the following example.

**Example 10.11**

> Health-NS is a retail store that sells health- and nutrition-related products, including vitamins, supplements, minerals, etc. One of its fast moving product is a chewable multivitamin tablet sold in boxes that contain 100 tablets. The demand for the product in a day has an average of 30 with standard deviation of 12 boxes, and is independent of the demand in other days. Health-NS buys each box for $12 and sells it for $16. The distributor of the vitamins delivers orders every 20 days, with order lead time of 10 days. Specifically, 10 days after receiving the last order, the store (if needed) must place an order. It then receives the order in 10 days. If the store does not place an order, then the next ordering opportunity is 20 days later. The order that is placed at that time will be received in 10 days (the lead time). The supplier does not charge any shipment cost for an order and the transaction cost of placing an order is insignificant. The store charges an inventory carrying rate of 20 percent. Mike, the store manager, has realized that the customers who face shortage are lost, which is estimated to cost the store about $8 for profit loss and potential loss of future business. Mike believes that they should always try to keep an inventory of 800 boxes at store. He reasons that, considering the average daily demand of 30 boxes, 600 of those units would take care of the demand during the 20-day period and the addition of 200 units is enough to protect against a possible spike in demand. Is this a good policy? If not, how should the store manage its inventory of the multivitamins if the store's objective is to minimize total cost; specifically, how many boxes should they order every 20 days?

The policy suggested by the store manager—trying to keep the inventory of 800 boxes—corresponds to the order-up-to policy with $S = 800$. While the store manager is correct that the order-up-to policy is the best policy (as we discussed in the previous sections), the value of the order-up-to level $S = 800$ may not be the optimal value. Before we find the optimal value of the order-up-to level, let's look at the inventory curve of Health-NS under store's order-up-to policy with $S = 800$. Figure 10.9 depicts an example.

We define a period (or an order cycle) as before, that is, the time between delivery of two consecutive orders.

Consider

$$D_P = \text{Demand during a period in periodic review systems}$$

which is a random variable. Also, recall that $D_L$ represents the demand during the lead time. From Fig. 10.9 we make the following observations:

- *All periods have the same length:* With stochastic demand, as opposed to $(Q, R)$ policy under which order cycles (i.e., periods) have different lengths, all order cycles

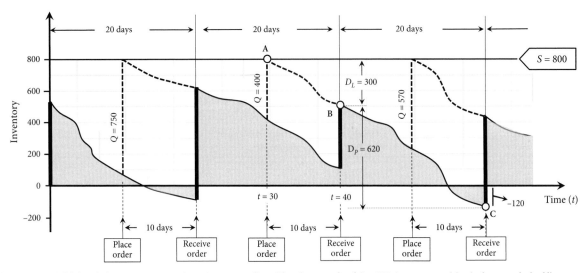

Figure 10.9    Multivitamin inventory curve under order-up-to policy with order-up-to level $S = 800$. Inventory position is shown as dashed line.

in periodic review systems have the same length.[4] For Health-NS, the length of all order cycles is 20 days.

- *Order quantities are different in different periods:* Recall that in continuous review systems under $(Q, R)$ policy, an order of the same size $Q$ is placed in all order cycles. This is because, in all order cycles, the order is placed when the inventory position reaches reorder point $R$. This, however, is not the case in periodic review systems. As shown in Fig. 10.9, in a periodic review system the order quantity in each period may be different. The reason is that, when an order is placed, its size depends on the inventory position at that time. In the first period in Fig. 10.9, for example, an order of size $Q = 750$ is placed to raise the inventory position to $S = 800$, since the inventory position was 50. On the other hand, the inventory position when the second order was placed was 400, and thus an order of size $Q = 400$ was placed to raise inventory position to order-up-to level $S = 800$.

- *On-hand inventory at the **beginning** of a period is $S - D_L$:* Note that, if lead time is zero, then on-hand inventory increases to 800 immediately after an order is placed. Thus, all periods start with on-hand inventory of $S = 800$. However, if lead time is not zero, after an order is placed, demand $D_L$ occurs before the order is received. Hence, immediately after an order is received, on-hand inventory increases to $S - D_L$. In other words, each period starts with on-hand inventory $S - D_L$. For example, the third period in Fig. 10.9 starts at time $t = 40$ with on-hand inventory of 500 items (i.e., Point B in the figure). This is because, an order of size $Q = 400$ was placed at time $t = 30$ when the inventory position was 400. This raised the inventory position to $S = 800$ at time $t = 30$ (i.e., Point A in the figure). The demand during lead time (the 10 days between $t = 30$ and $t = 40$) was $D_L = 300$. Consequently, inventory is reduced during lead time and the on-hand inventory at the beginning of the third period (i.e., Point B) was $S - D_L = 800 - 300 = 500$.

- *On-hand inventory at the **end** of a period is $S - D_L - D_P$:* If a period starts with on-hand inventory $S - D_L$ and if the demand during a period is $D_P$, then obviously the on-hand inventory at the end of a period (just before the next order arrives)

---

[4]When the demand is zero during a period, then there will be no delivery for the next period. We consider this event as the event of placing an order for zero units.

will be $S - D_L - D_P$. For example, consider the third period in Fig. 10.9 that starts at $t = 40$ with on-hand inventory of $S - D_L = 800 - 300 = 500$. Since the demand during the third period is $D_P = 620$, the on-hand inventory at the end of the period is $S - D_L - D_P = 800 - 300 - 620 = -120$.

- *Shortage depends on order-up-to level S, demand during lead time, and demand during the period:* A shortage occurs if the inventory at the end of a period is negative, that is, when $S - D_L - D_P < 0$. Hence,

$$\text{Shortage at the end of a period} = (D_L + D_P - S)^+$$

where $(x)^+ = x$ if $x > 0$ and $(x)^+ = 0$, if $x \leq 0$. Thus, as shown in the third order cycle of Fig. 10.9, we have $D_L = 300$ and $D_P = 620$, then we have

$$\begin{aligned}\text{Shortage at the end of second period} &= (D_L + D_P - S)^+ \\ &= (300 + 620 - 800)^+ \\ &= 120\end{aligned}$$

### Protection Interval and Average Shortage in a Period

Recall that the protection interval is the interval during which a firm must rely on its safety inventory to protect against a shortage. The expression for shortage, that is, $(D_L + D_P - S)^+$, clearly shows that the shortage depends on $D_L + D_P$, namely demand during lead time *and* during the period. Thus, the protection interval in order-up-to policies is the interval that consists of lead time followed by a period.

We define

$$\begin{aligned}D_{L+P} &= \text{Total demand during lead time } and \text{ period} \\ &= D_L + D_P\end{aligned}$$

and suppose $f_{L+P}(x)$ is the probability density function of $D_{L+P}$. Then, the average number of shortages in a period (just before an order arrives) will be

$$\eta_{L+P}(S) = E(D_{L+P} - S)^+ = \int_S^\infty (x - S)f_{L+P}(x)dx$$

If the demand during lead time and period is Normally distributed, then demand $f_{L+P}(x)$ follows a Normal distribution and we can use the standard Normal tables to compute $\eta_{L+P}(S)$.

## 10.7.2 Optimal Order-Up-To Policy—Minimizing Cost

Finding the optimal order-up-to level $S^*$ that minimizes total annual expected cost is similar to that of finding the optimal order quantity $Q^*$ and reorder point $R^*$ for $(Q, R)$ policy. Specifically, we break the components of the annual cost into inventory, order setup, and shortage costs.

### Computing Total Average Annual Cost in Systems with Backorders

We first focus on the case where shortages are *backordered*.

**Average Annual Order Setup Cost:** Recall that if placing an order in each period does not incur a fixed cost, then order-up-to policy is optimal. Hence in this case the average annual order setup cost is zero.

**Average Annual Inventory Review Cost:** In most cases, under periodic review system, the on-hand inventory is reviewed before an order is placed. Sometimes reviewing inventory results in additional cost. If

$$K_P = \text{Fixed cost of reviewing inventory in a period}$$
$$T_P = \text{Length of the period (}measured\ in\ years\text{)}$$

Then the number of periods in a year is $1/T_P$, and

$$\text{Total average annual inventory review cost} = K_P \left( \frac{1}{T_P} \right)$$

**Average Annual Inventory Cost:** Considering annual inventory holding cost $C_h$ per item, then

$$Average \text{ annual holding cost } = C_h \times (\text{Cycle Inventoy} + \text{Safety Inventory})$$

*Cycle Inventory:* Similar to the $(Q, R)$ policy, we use continuous approximation of the inventory curve under order-up-to policy and approximate the cycle inventory assuming that the demand during a period is deterministic at constant rate $\overline{D}$. Under this approximation, finding the cycle inventory is easy. Since in continuous approximation model the inventory in each period decreases linearly from $S - D_L$ at the beginning of the period to $S - D_L - D_P$ at the end of the period, then

$$\text{Cycle Inventory} = \frac{(S - D_L) - (S - D_L - D_P)}{2} = \frac{D_P}{2}$$

Considering
$$\overline{D}_P = \text{Average demand during a period}$$

then the *average cycle inventory* in a *period* is $\overline{D}_P/2$. Since under our approximation all periods are similar, the average cycle inventory during a *year* is also $\overline{D}_P/2$.

*Safety Inventory:* Safety inventory is the average net inventory just before an order arrives. Since we know that the inventory at the end of a period is $S - D_L - D_P$, then safety inventory is

$$I_{\text{safety}} = \mathrm{E}(S - D_L - D_P) = \mathrm{E}(S - D_{L+P})$$

Using $f_{L+P}(x)$, the probability density function of $D_{L+P}$, we can compute safety inventory as follows:

$$
\begin{aligned}
I_{\text{safety}} &= \int_0^\infty (S - x) f_{L+P}(x) dx \\
&= S - \int_0^\infty x f_{L+P}(x) dx \\
&= S - \overline{D}_{L+P}
\end{aligned}
$$

where $\overline{D}_{L+P}$ is the total average demand during lead time and period. Thus,

$$
\begin{aligned}
Average \text{ annual holding cost} &= C_h \times (\text{Cycle Inventory} + \text{Safety Inventory}) \\
&= C_h \times (\overline{D}_P/2 + S - \overline{D}_{L+P})
\end{aligned}
$$

**Average Annual Shortage Cost:** Similar to that for $(Q, R)$ policy, we define

$$\Psi_P(S) = Total \text{ average shortage cost in a period under order-up-to level } S$$

Hence, for the total average annual shortage cost we have

*Average* annual shortage cost $= \Psi_P(S) \times$ *Average* number of periods in a year

$$= \Psi_P(S) \times \frac{1}{T_P}$$

Putting all these cost components together, we get the following expression for the total average annual holding, inventory review, and backorder cost in systems with *backorders* under order-up-to policy with order-up-to level $S$:

$$TC(S) = K_P \left( \frac{1}{T_P} \right) + C_h \left( \frac{\overline{D}_P}{2} + S - \overline{D}_{L+P} \right) + \Psi_P(S) \left( \frac{1}{T_P} \right) \qquad (10.48)$$

### Computing Total Average Annual Cost in Systems with Lost Sales

Using a similar approach we can find the total annual average cost for systems with *lost sales* in periodic review systems with order-up-to level $S$ as follows:

$$TC(S) = K_P \left( \frac{1}{T_P} \right) + C_h \left( \frac{\overline{D}_P}{2} + S - \overline{D}_{L+P} + \eta_{L+P}(S) \right) + \Psi_P(S) \left( \frac{1}{T_P} \right) \qquad (10.49)$$

### Finding Optimal Order-Up-to Level $S$

When the review period $T_P$ is fixed (e.g., for Health-NS in Example 10.11, $T_P = 20$ days), finding the optimal order-up-to level is easy. The optimal order-up-to level $S$ must satisfy the following condition:

$$\frac{\partial \, TC(S)}{\partial S} = 0$$

$$C_h + \left( \frac{1}{T_P} \right) \frac{\partial \, \Psi_P(S)}{\partial S} = 0 \implies \frac{\partial \, \Psi_P(S)}{\partial S} = -C_h \, T_P$$

The above condition is similar to the optimality condition for reorder point $R$ in $(Q, R)$ policy, see Eq. (10.20). In Eq. (10.20), the average shortage cost $\Psi_P(R)$ is the average shortage during protection interval when the protection interval starts with $R$ units in inventory. While the protection interval is the lead time in $(Q, R)$ model, in order-up-to model, the protection interval is the lead time and period and the protection interval starts with $S$ units. So, to find the optimal order-up-level $S$ we can use the same analytics for reorder point $R$ by: (i) replacing $R$ with $S$, (ii) replacing $Q$ with $\overline{D}T_P$ (since $Q$ is equal to the average demand during order cycle), and (iii) replacing demand distribution $f_L(x)$ with $f_{L+P}(x)$. This leads us to the following analytics.

---

**ANALYTICS**

### Finding the Optimal Order-Up-To Level $S^*$ to Minimize Cost

Consider a period review make-to-stock (MTS) inventory system managed under an order-up-to policy with order-up-to level $S$. Inventory is reviewed at fixed intervals of length $T_P$ (measured in years) and an order is placed to raise the inventory to $S$. Costs include inventory holding cost $C_h$ per item per year, inventory review cost $K_P$ per period, and shortage cost. The optimal order-up-to level $S^*$ that minimizes the total expected annual cost depends on whether shortages are lost or backordered.

● **Systems with Backorders:** If demands that face shortage are backordered with backorder cost $C_b$ per backorder demand, then the optimal order-up-to level $S^*$ satisfies

$$F_{L+P}(S^*) = 1 - \frac{C_h}{C_b} T_P \qquad (10.50)$$

where $F_{L+P}(x)$ is the cumulative probability distribution of total demand during lead time and period. If the demand during protection interval (i.e., during lead time and period) is Normally distributed with mean $\overline{D}_{L+P}$ and standard deviation $\sigma_{L+P}$, then the optimal order-up-to level is

$$S^* = \overline{D}_{L+P} + z\sigma_{L+P}, \quad \text{where} \quad \Phi(z) = 1 - \frac{C_h}{C_b} T_P$$

and $\Phi(z)$ is the cumulative distribution of standard Normal distribution and is available in the Appendix.

● **Systems with Lost Sales:** If demands that face shortage are lost with lost sales cost $C_s$, then the optimal order-up-to level $S^*$ satisfies

$$F_{L+P}(S^*) = 1 - \frac{C_h}{C_h T_P + C_s} T_P \qquad (10.51)$$

Again, if the demand during protection interval (i.e., during lead time and period) is Normally distributed with mean $\overline{D}_{L+P}$ and standard deviation $\sigma_{L+P}$, then the optimal order-up-to level is

$$S^* = \overline{D}_{L+P} + z\sigma_{L+P}, \quad \text{where} \quad \Phi(z) = 1 - \frac{C_h}{C_h T_P + C_s} T_P$$

We now return to Health-NS in Example 10.11. Recall that the supplier delivers multivitamins every 20 days; thus, the length of the period is $T_P = 20$ days. Considering 30 days in a month, $T_P = 20/(30 \times 12) = 0.056$ years. The annual inventory holding cost is $C_h = 0.20(\$12) = \$2.40$, and lost sales cost is $C_s = \$8$.

To find the optimal order-up-to level, we need to find $f_{L+P}(x)$, the distribution of demand during the protection interval, that is, during lead time plus period. Since the lead time is 10 days and length of the period is 20 days, we need to find the distribution of the total demand during 30 days. Recall that the demand in a day has mean 30 and standard deviation 12, and is independent of the demand in other days. Using Central Limit Theorem, the demand during 30 days can be well-approximated by a Normal distribution with mean $30 \times 30 = 900$ and standard deviation $\sqrt{30} \times (12) = 65.7$. Hence, the optimal order-up-to level $S^*$ that minimizes the total expected annual cost satisfies the following:

$$F_{\ell+P}(S^*) = 1 - \frac{C_h}{C_h T_P + C_s} T_P = 1 - \frac{2.4}{2.4(0.056) + 8}(0.056) = 0.984$$

Using standard Normal table, we find that $\Phi(2.145) = 0.984$. Thus,

$$S^* = \overline{D}_{L+P} + z\sigma_{L+P} = 900 + 2.145(65.7) = 1040.92$$

which we round up to 1041. Hence, to minimize the annual expected inventory and lost sales cost, Health-NS should use the order-up-to policy with order-up-to level 1041.

### 10.7.3 Optimal (s, S) Policy—Minimizing Cost

Order-up-to policy is the optimal policy in periodic review systems in which placing an order does not incur a fixed cost, that is, the order setup cost $K$ is zero. For example, in the case of Health-NS in Example 10.11, the supplier of multivitamins delivers each order free of charge and the transaction cost for placing an order is also insignificant. Because of the zero setup cost, it is optimal to always place an order in each period and raise the inventory to order-up-to level $S$. But, when the order setup cost is not negligible, a different version of the order-up-to policy, called $(s, S)$ policy, becomes optimal.

 **Example 10.12**

Mike, the manager of Health-NS store in Example 10.11, receives a phone call from the supplier of the multivitamins who says:

> Hi Mike, just wanted to let you know about changes in our delivery policy. The trucking company that delivered our orders went out of business, and we hired a new delivery firm. The new firm charges higher fee for delivery. Hence, it is no longer economical for us to deliver our shipments for free. Our company's new policy is to charge about $80 to deliver an order, regardless of the order size. We still make deliveries every 20 days, and as before, we need lead time of 10 days to prepare and deliver your order. You also need to tell us if you decide not to order any items, so, we do not send the truck to your store and do not charge you the delivery cost.

Considering the new delivery cost of $80, how should Health-NS revise its ordering policy to minimize the total expected annual holding, delivery, and lost sales costs?

Because placing orders incurs costs, it might not be optimal for Health-NS to place an order each time. For example, consider a period in which the store has an on-hand inventory of 799 boxes of multivitamins. Should the store place an order for one box and raise the inventory to order-up-to level 800? Obviously, it does not make sense to pay the delivery cost of $80 to order only one box that—if not available—would cost the store about $8 of lost sales. So, with on-hand inventory of 799 boxes the store should not place an order for that period. But what if on-hand inventory is 798 or 797? If we continue with this line of reasoning, we realize that there would be a threshold (say $s$) on on-hand inventory above which the store should not place an order. However, if on-hand inventory is at or below that threshold, then the store should place an order and raise its inventory to order-up-to level $S = 800$.

This new version of the order-up-to policy is called $(s, S)$ policy and is shown to be the optimal policy in periodic review systems in which the order setup cost is not zero.

---

**CONCEPT**

**The (s, S) Policy**

The $(s, S)$ policy is used in periodic review systems in which the inventory of items is checked at the end of each period and there is a significant order setup cost if an order is placed. The policy includes *Reorder Point s* and *Order-Up-To Level S* and works as follows: If at the end of a period the inventory position is $I$:

- If $I \leq s$, then place an order for $S - I$ units to raise the inventory position to $S$.
- If $I > s$, then do not place an order.

Note that the order-up-to policy with order-up-to level $S$ is a special case of $(s, S)$ policy in which $s = S - 1$. When $s = S - 1$, we always place an order in each period, except when the demand in a period is zero (i.e., inventory position is $S$). As we mentioned earlier, this is optimal only if the order setup cost $K$ is zero or negligible. Also, note that order setup cost $K$ is different from Inventory Review Cost $K_P$. Inventory review cost $K_P$ occurs in every period to review inventory, regardless of whether an order is placed or not. The order setup cost $K$, however, occurs only when an order is placed.

How to find the optimal values of reorder point $s$ and order-up-to level $S$ to minimize total expected annual cost? The bad news is that the analytics to obtain these values is very complex. The good news, however, is that several approximations have been developed that yield reasonable results. One of these approximations uses the relationship between $(s, S)$ policy in periodic review systems and $(Q, R)$ policy in continuous review systems which we describe below.

## ANALYTICS

**Finding the Optimal $(s^* \; S^*)$ Policy to Minimize Cost**

Consider a make-to-stock (MTS) inventory system with $(s, S)$ policy in which the length of the period is $T_P$. Costs include inventory holding cost $C_h$ per item per year, order setup cost $K$, and shortage cost. The optimal values of $s$ and $S$ that minimize the total annual expected cost can be obtained as follows:

**Step 1:**   Assume that the problem is a continuous review system and find the optimal $(Q^*, R^*)$ policy that minimizes the total annual expected cost.

**Step 2:**   The optimal reorder point $s^*$ and order-up-to level $S^*$ that minimize the total annual expected cost in the periodic review system can then be approximated as follows:

$$s^* = R^*$$
$$S^* = R^* + Q^*$$

Similar to continuous review systems, customers who face shortages in periodic review systems may be backordered or lost. So, in Step 1 of the above analytics we must choose the corresponding $(Q, R)$ model. For example, if the customers who face shortage in the periodic review system are lost, then in Step 1 we should use the analytics for $(Q, R)$ policy with lost sales to obtain the optimal values of $Q^*$ and $R^*$.

Having been equipped with these analytics, we can now find Health-NS's new inventory control policy when the supplier of multivitamins starts charging delivery cost of $80. All we need to do is to find the optimal $(Q^*, R^*)$ policy for the multivitamins. Since the daily demand has a mean and standard deviation 30 and 12, respectively, demand during the lead time of 10 days will have a mean and standard deviation $30 \times 10 = 300$ and $\sqrt{10} \times 12 = 37.95$. Considering holding cost $C_h = \$2.4$, lost sales cost $C_s = \$8$, and order setup cost $K = \$80$, we can use the iterative approach to find the optimal $(Q^*, R^*)$ for the system with lost sales as shown in Table 10.6.

According to the table $Q^* = 862.42$ and $R^* = 375.52$. Hence, the optimal $(s^*, S^*)$ is as follows: $s^* = R^* = 375.52$ and $S^* = R^* + Q^* = 375.52 + 862.42 = 1237.94$. Therefore, after the supplier starts charging the delivery cost of $80, Health-NS should revise its current inventory policy to the following: In each period, if the inventory position is at $s^* = 375$ or lower, Health-NS should place an order to increase the inventory to order-up-to level

Table 10.6    Finding Optimal $Q^*$ and $R^*$ for Health-NS in Example 10.12

| Iteration $(i)$ | $Q_i$ | $F(R_i)$ | $z$ | $R_i$ | $\mathcal{L}(z)$ | $\eta_L(R_i)$ |
|---|---|---|---|---|---|---|
| 0 | 848.53 | 0.9770 | 2.00 | 357.90 | 0.0085 | 0.3226 |
| 1 | 862.11 | 0.9766 | 1.99 | 375.52 | 0.0087 | 0.3302 |
| 2 | 862.42 | 0.9766 | 1.99 | 375.52 | 0.0087 | 0.3302 |
| 3 | 862.42 | 0.9766 | 1.99 | 375.52 | 0.0087 | 0.3302 |

$S^* = 1238$. However, if the inventory position is above $s^* = 375$, Health-NS should not place an order for that period.

### 10.7.4 Setting Service Levels in Periodic Review Systems

Recall that, besides minimizing costs, providing a certain service level to the customers is another goal of inventory managers. In this section we show how the inventory should be managed in periodic review systems if the goal is to achieve a certain cycle-service level or a certain fill rate.

We described that when order setup costs are zero, then order-up-to policy minimizes the total expected annual costs. On the other hand, when order setup costs are not zero, then $(s, S)$ policy minimizes the total expected annual costs. Now, the question is what type of policy is optimal if the goal is to achieve a service level (instead of minimizing total expected cost)?

The answer turns out to be simple. It has been shown that if the goal is to set a certain service level (cycle-service level or fill rate), then order-up-to policy becomes optimal again. Of course, the optimal value of order-up-to level $S$ when setting a service level may be different from that when costs are minimized. Below we show how to obtain these values.

#### Setting Cycle-Service Level

Cycle-service level is the probability of not having shortage in an order cycle. In periodic review systems, every period is an order cycle. Now consider an order-up-to policy with order-up-to level $S$, and assume that we would like to find $S$ such that we set the cycle-service level to $\alpha$. Recall that the inventory at the end of each period (just before the next order arrives) under an order-up-to policy is $S - D_L - D_P$. Hence, if $S - D_L - D_P \geq 0$ or if $D_L + D_P \leq S$, then the period will not have any shortage. Considering $D_{L+P} = D_L + D_P$, to have the probability of no shortage being $\alpha$, we need to set order-up-to level $S$ such that

$$\alpha = P(D_{L+P} \leq S) = F_{L+P}(S)$$

As in the case of cost minimization, here also the optimal order-up-to level $S$ depends on the total demand during the protection interval of lead time and the period, that is, $D_{L+P}$.

#### Setting Fill Rate

Fill rate $\beta$ is the fraction of overall demand that does not face shortage, or

$$\beta = 1 - \frac{\text{Demand facing shortage}}{\text{Total demand}}$$

Now consider an order-up-to policy with order-up-to level $S$, and recall that the average shortage at the end of a period is $\eta_{L+P}(S)$. However, this shortage corresponds to the total demand $D_{L+P}$. In other words, out of a total average demand of $\overline{D}_{L+P}$, an average of $\eta_{L+P}(S)$

face shortage (backorder or lost sales), thus,

$$\beta = 1 - \frac{\eta_{L+P}(S)}{\overline{D}_{L+P}}$$

Hence, the order-up-to level $S$ that results in fill rate $\beta$ should satisfy

$$\eta_{L+P}(S) = (1 - \beta)\overline{D}_{L+P}$$

We summarize the above in the following analytics.

## ANALYTICS

### Setting Service Levels in Periodic Review Systems

Consider a make-to-stock (MTS) inventory system managed under a period review system in which the inventory is reviewed every $T_P$ units of time. The average demand during a period is $\overline{D}_P$ and the average demand during lead time is $\overline{D}_L$. The total demand during protection interval (i.e., lead time and period) follows probability density function $f_{L+P}(x)$. For such systems, order-up-to policy is optimal for setting cycle-service level or fill rate.

- **Cycle-Service Level:** The optimal order-up-to level $S_\alpha$ that achieves cycle-service level $\alpha$ satisfies

$$\alpha = F_{L+P}(S_\alpha)$$

where $F_{L+P}(x)$ is the cumulative probability distribution of total demand during protection interval. If the demand during protection interval (i.e., during lead time and period) is Normally distributed with mean $\overline{D}_{L+P}$ and standard deviation $\sigma_{L+P}$, then the optimal order-up-to level is

$$S_\alpha = \overline{D}_{L+P} + z\sigma_{L+P}, \quad \text{where} \quad \Phi(z) = \alpha$$

and $\Phi(z)$ is the cumulative distribution of standard Normal distribution and is available in the Appendix.

- **Fill Rate:** The optimal order-up-to level $S_\beta$ that achieves fill rate $\beta$ in systems with backorders or lost sales satisfies

$$\eta_{L+P}(S_\beta) = (1 - \beta)\overline{D}_{L+P}$$

where $\eta_{L+P}(x)$ is the average shortage during protection interval.

 **Example 10.13**

Consider the Health-NS in Examples 10.11 and 10.12. How should the store change its inventory management policy of multivitamins if the store decides to set a fill rate of 98 percent?

If the store changes its goal from minimizing costs to achieving a service level, then it should change its inventory management policy from a $(s^*, S^*) = (375, 1238)$ policy to order-up-to policy with order-up-to level $S_\beta$.

Recall that the demand during protection interval of 30 days has average $\overline{D}_{L+P} = 900$. Thus, according to our analytics, the optimal order-up-to level $S_\beta$ that achieves fill rate $\beta = 98\%$ satisfies the following:

$$\eta_{L+P}(S_\beta) = (1 - \beta)\overline{D}_{L+P} = (1 - 0.98)(900) = 18$$

Since demand during protection interval is Normally distributed with mean 900 and standard deviation 65.7, we have

$$\eta_{L+P}(S_\beta) = \sigma_{L+P}\mathcal{L}(z)$$
$$18 = 65.7\mathcal{L}(z) \implies \mathcal{L}(z) = 18/65.7 = 0.274$$

Using standard Normal table we find $\mathcal{L}(z = 0.28) = 0.274$. Thus, $z = 0.28$ and

$$S_\beta = \overline{D}_{L+P} + z\sigma_{D_{L+P}} = 900 + 0.28(65.7) = 918.40$$

which we round up to 919. Therefore, to achieve fill rate $\beta = 98\%$, Health-NS must order each time such that its inventory position is raised to $R_\beta = 919$.

## 10.8 Lead Time Variability

Replenishment lead time, as we described in Sec. 9.6 of Chap. 9, consists of four major times: (i) time to place an order to a supplier, (ii) time at the supplier, (iii) shipment time, and (iv) time needed to process the order after the order is received. While items (i) and (iv) are often short with small variability, items (ii) and (iii) are responsible for most of the variability in the lead time. The sources of variability in (ii) and (iii) are interruptions in the production process (e.g., machine failure, lack of raw material), or because of unexpected delays in transportation of the order (e.g., delay in customs when an order enters the United States).

The analytics presented in the previous sections are for cases where lead times have no variability. Can these analytics still be used to obtain the optimal $(Q, R)$ in continuous review systems or optimal order-up-to levels in periodic review systems when lead times are variable? The answer is yes, as long as our modeling assumptions still hold, that is, as long as the variability in demand does not result in having more than one outstanding order in each order cycle.

The good news is that, in practice, most systems do not usually face more than one outstanding order. The reason, as we discussed before, is that having more than one outstanding order is often an indication of low service levels, corresponding to a large shortage of items, which firms try to avoid. Thus, even in the case of variable lead times, the inventory policies of interest are those with reasonably high service levels that rarely result in having more than one outstanding order in a cycle.

How to revise our analytics to be able to incorporate the variability in lead time? All we need to do is to incorporate the variability in lead time into the demand distribution during the protection interval. There are two ways to do this:

- Collect the actual demand data during the protection interval. In a continuous review system, get a sample of the demand during the lead time for a large number of order cycles. In periodic review system, get a sample of the demand during the lead time and period. After the data is gathered, plot the histogram of the data and fit a distribution on them. Then use the fitted distribution as the demand distribution during protection interval.

- Often times the average and variance of daily demand and the average and variance of the lead time are known. One can combine this information and obtain the average and variance of demand during protection interval. The main assumption is that the demand in a day is independent of the length of the lead time. If these are indeed independent random variables, and if the successive lead times are independent of each other, then the following analytics can be used to compute the mean and variance of the demand during the protection interval in continuous and periodic review systems.

### Mean and Variance of Demand during Protection Interval

Suppose that the demand in one time unit (e.g., 1 day, 1 week, 1 month) has mean $\mu$ and variance of $\sigma^2$. Also, suppose the lead time is random variable $L$, which is independent of demand in one time unit, and has mean $E(L)$ and variance $Var(L)$.

- **Continuous Review:** In a continuous review system, the demand during protection interval (i.e., during lead time) has mean $\overline{D}_L$ and standard deviation $\sigma_L$ as follows:

$$\overline{D}_L = \mu \; E(L)$$
$$\sigma_L = \sqrt{\sigma^2 \, E(L) + \mu^2 \, Var(L)}$$

- **Periodic Review:** In a periodic review system with period of length $T_P$, the demand during protection interval (i.e., during lead time and period) has mean and standard deviation

$$\overline{D}_{L+P} = \mu \; (E(L) + T_P)$$
$$\sigma_{L+P} = \sqrt{\sigma^2 \, (E(L) + T_P) + \mu^2 \, Var(L)}$$

 **Example 10.14**

Consider Cool Fans in Example 10.8 that produces ceiling fans. Recall that the supplier of 3SPD motors had a lead time of 14 days (i.e., 2 weeks). However, the supplier has recently expanded its market to Europe. Due to the higher demand, the supplier sometimes falls behind its production schedule and delivers the motors to Cool Fans later than 14 days—the promised lead time. In a phone conversation with the Cool Fans manager, the supplier acknowledges the late deliveries, but promises that he will do his best to make his promised lead time of 14 days. Recently, Cool Fans purchasing team has reviewed the last deliveries of motors and concluded that, while the average lead time is indeed 2 weeks, there is a considerable variability in lead time. Specifically, lead time has standard deviation of 4 days. The demand during the variable lead time, however, still has a unimodal symmetric distribution which can be approximated by Normal distribution. Considering these new developments, how should Cool Fans revise its inventory control policy in order to maintain its goal of having less than 2 percent of its customers face late deliveries?

Recall that the demand during 2 weeks had an average of 20 and standard deviation of 7. Hence, the demand in a day has an average of $\mu = 20/14 = 1.43$ and standard deviation of $\sigma = 7/\sqrt{14} = 1.87$. Considering that lead time is variable with mean $E(L) = 14$ days and variance $Var(L) = (4)^2 = 16$, the demand during lead time will have mean and variance as follows:

$$\overline{D}_L = \mu \; E(L) = 1.43(14) = 20$$
$$\sigma_L = \sqrt{\sigma^2 \, E(L) + \mu^2 \, Var(L)} = \sqrt{(1.87)^2 \, (14) + (1.43)^2 \, 16} = 9.04$$

Since shortages are backordered, according to Eq. (10.44), to achieve fill rate $\beta$, the order quantity is

$$Q_\beta^* = \sqrt{\frac{2K\overline{D}}{C_h}} = \sqrt{\frac{2(120)\,(520)}{9}} = 117.7 \simeq 118$$

And the optimal reorder point $R_\beta^*$ must satisfy

$$\eta_L(R_\beta^*) = (1-\beta)Q_\beta^* = (1-0.98)(118) = 2.36$$

Since the demand during lead time can still be approximated by Normal distribution, we have

$$\eta_L(R_\beta^*) = \sigma_L \mathcal{L}(z)$$
$$2.36 = 9.04\mathcal{L}(z) \quad \Longrightarrow \quad \mathcal{L}(z) = 2.36/9.04 = 0.2611$$

Using standard Normal table we find $\mathcal{L}(z=0.315) = 0.2611$. Thus,

$$R_\beta^* = \overline{D}_L + z\sigma_{D_L} = 20 + 0.315(9.04) = 22.85$$

which we round up to 23. Therefore, to achieve fill rate $\beta = 98\%$, Cool Fans must order $Q_\beta^* = 118$ motors each time its inventory position reaches $R_\beta^* = 23$.

## 10.9 Trade-off between Costs and Service Levels

Finding an accurate estimate of backorder cost or lost sales cost is very difficult. Even when these values are accurately estimated, changes in market and customer behavior can affect these values and make them inaccurate. Setting service level is then a more reliable approach to control shortage. Obviously higher service levels result in higher inventory and therefore in higher holding costs. Thus, the decision of which service goals to choose depends on the costs of achieving those service goals. One useful tool that can help managers make more informed decision about setting service goals is the Cost-Service Trade-off Curve.

### CONCEPT

**Cost-Service Trade-off Curve**

The Cost-Service (or Profit-Service) Trade-off Curve shows the relationship between the cost (or profit) of a process and the service level offered by the process. The curve shows how much the total cost (or profit) changes if the service level is increased or decreased.

 **Example 10.15**

Recall Auto Part Depot (APD) in Example 10.6 that was facing competition from an online firm. APD used lost sales cost $C_s = \$15$ and computed its optimal inventory policy as $(Q^*, R^*) = (463, 351)$ that minimizes the total expected cost. APD managers, however, now believe that their estimate of the lost sales cost might not be accurate. Therefore, they have decided to change their inventory policy to achieve certain fill rate. While they would like to have a high fill rate, they know that a high fill rate requires holding a large inventory of the polishing compound. What makes the problem even more difficult is that new regulations are expected to be implemented by the United States Customs and Border Protection that would increase the variability of the lead time of the compound. Specifically, the lead time will have an average of 10 days with standard deviation of 3 days. Develop a Cost-Service Trade-off Curve to help managers in setting an appropriate fill rate.

Before we develop the Cost-Service Trade-off Curve, let's first compute the current variability in demand during lead time and that if the new regulations are adopted.

Recall that the daily demand has mean $\mu = 30$ and standard deviation $\sigma = 10$. Also, the current lead time has mean $E(L) = 10$ days and variance $Var(L) = 0$. Thus, the demand during lead time will have mean $\overline{D}_L = \mu \; E(L) = 30(10) = 300$ and standard deviation

$$\sigma_L = \sqrt{\sigma^2 \; E(L) + \mu^2 \; Var(L)} = \sqrt{(10)^2 \; (10) + (30)^2 \; (0)} = 31.62$$

Therefore, currently the demand during lead time has variability (i.e., coefficient of variation) $CV_L = 31.62/300 = 0.1$.

If the new regulation is adopted and implemented, we have $E(L) = 10$ days and variance $Var(L) = (3)^2 = 9$. Thus, demand during lead time will still have mean $\overline{D}_L = 300$, but with standard deviation

$$\sigma_L = \sqrt{\sigma^2 \; E(L) + \mu^2 \; Var(L)} = \sqrt{(10)^2 \; (10) + (30)^2 \; (9)} = 95.39$$

and variability $CV_L = 95.39/300 = 0.32$. Hence, new regulations will increase the variability in demand during lead time by a factor of 3.

To develop a Cost-Service Trade-off Curve, we first need to determine the cost and the service level for which we want to construct the curve. Obviously, APD is considering fill rate for its service goal and the inventory holding cost as its cost—they do not consider shortage cost, since they do not have an accurate estimate for it. Hence, we need to compute the expected inventory holding cost (the Y-axis) for each value of fill rate (the X-axis).

Based on our analytics, since the exact lost sales cost of the polishing compound is not known, for any fill rate $\beta$, the optimal order quantity is

$$Q_\beta^* = \sqrt{\frac{2K\overline{D}}{C_h}} = \sqrt{\frac{2(187.5)(10,800)}{20}} = 450$$

On the other hand, for any given fill rate $\beta$, the optimal reorder point $R_\beta^*$ must satisfy the following:

$$\eta_L(R_\beta^*) = \left(\frac{1-\beta}{\beta}\right) Q_\beta^* = \left(\frac{1-\beta}{\beta}\right) 450$$

We construct Cost-Service Trade-off Curve for values of fill rate $\beta = 0.9, 0.91, 0.92, \ldots,$ $0.98, 0.99, 0.999$. For each value we obtain its corresponding reorder point $R_\beta^*$ using the above formula. Finally, for the resulting policy $(Q_\beta^*, R_\beta^*) = (450, R_\beta^*)$, we compute the average annual inventory holding cost as follows (see Eq. (10.30)):

$$
\begin{aligned}
\text{Average Annual Inventory Holding Cost} &= C_h \left(\frac{Q_\beta^*}{2} + R_\beta^* - \overline{D}_L + \eta_L(R_\beta^*)\right) \\
&= 20 \left(\frac{450}{2} + R_\beta^* - 300 + \eta_L(R_\beta^*)\right) \\
&= 20[R_\beta^* + \eta_L(R_\beta^*)] - 1500
\end{aligned}
$$

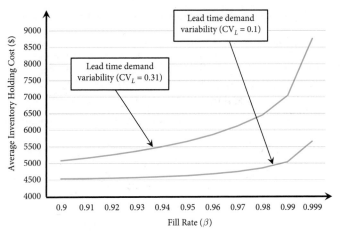

Figure 10.10 shows the values of the above cost for any given fill rate $\beta$ for two cases: (i) Regulations are not adopted: In this case, lead time $L$ of the compound is 10 days with no variability, and (ii) Regulations are adopted: In this case, lead time $L$ has an average of 10 days with standard deviation of 3 days.

Having the Cost-Service Trade-off Curve in Fig. 10.10, managers of APD can clearly see the impact of different fill rates on the inventory holding cost of the compound. They can also see how their inventory cost changes if the new regulations are adopted. Based on how much inventory cost they are willing to incur, they can choose their appropriate fill rate.

Figure 10.10 also provides the following general insights:

- The cost of achieving higher fill rates increases nonlinearly. The implication is that it is more costly to increase the fill when the fill rate is already high. For example, it is more costly to increase fill rate by 1 percent from 98 percent to 99 percent, than from 90 percent to 91 percent.

- Variability can have a significant impact on choosing a fill rate and its corresponding cost. Systems with lower variability can achieve higher service levels with lower inventory (and thus lower inventory holding cost). Consider fill rate $\beta = 0.98$ as an example. When there is no variability in lead time,[5] the average inventory cost in a year is about \$4840. Considering the holding cost of $C_h = \$20$ for the compound, this corresponds to an average inventory of $4840/20 = 242$ polishing compounds. However, when there is availability in lead time (resulting in higher variability in demand during lead time), the inventory cost is about \$6444, corresponding to an average inventory of $6444/20 = 322.2$ compounds. Hence, the system with lower variability will have the same fill rate $\beta = 0.98$, but it keeps $322.2 - 242 = 80.2$ less inventory of compounds throughout the year.

To conclude, we reiterate that one difficult decision in operations is to find the best balance between costs and service levels. The Cost-Service Trade-off Curve can help managers make such decisions by showing how much it costs to offer any particular service level.

---

[5]Note that when there is no variability in the lead time, there is still variability in demand during lead time due to variability in daily demand.

We conclude this section with the following principle:

---

**PRINCIPLE**

**Variability and Inventory Service Level**

Systems with lower variability can offer the same service level with lower inventory (and thus lower inventory costs) than systems with higher variability. In other words, for any given inventory control policy (e.g., any give $(Q, R)$ or $(s, S)$ policy), systems with lower variability have higher service level than systems with higher variability.

## 10.10 Summary

When demand is deterministic and known with high certainty, the challenge is to find the optimal trade-off between two costs: order setup cost and inventory holding cost. When demand is stochastic, however, shortages may occur in each order cycle. Thus, the optimal trade-off should be made among three costs: shortage cost, inventory holding cost, and order setup cost. The focus of this chapter was on finding the optimal inventory management and control policies that achieve the optimal trade-off.

The most fundamental inventory management problem with stochastic demand is Newsvendor problem. Newsvendor problem models situations in which the inventory manager has only one chance to place an order for a product. If the manager orders too much, there will be overstocking costs (e.g., product salvage cost); if the manager orders too little, there will be understocking cost (e.g., lost sales cost). This chapter provides a series of analytics that yields the optimal order quantity and its corresponding expected costs and service levels. Cycle-service level (type-1 service level) is the probability of not having shortage during the selling season and fill rate (type-2 service level) is the fraction of demand not facing shortage.

For inventory systems in which the inventory manager can place orders at any time (i.e., continuous review inventory systems), this chapter provides analytics to compute the optimal reorder point $R$ and optimal order quantity $Q$ in the $(Q, R)$ policy for two different cases: (i) backorder cases in which the customers who face shortage are willing to wait for the product to arrive, and (ii) lost sales cases in which the customers facing shortage do not wait and are lost (e.g., buying the product from the competitor). In all these models firms keep safety inventory to protect themselves against shortage costs and customer dissatisfaction.

Under $(Q, R)$ policy, increasing reorder point $R$ increases inventory holding cost, but it decreases shortage cost and results in a higher service level. For any given reorder point $R$, increasing (or decreasing) order quantity $Q$ does not affect the number of shortages in an order cycle. This is because only the demand during lead time and reorder point $R$ affect shortage in an order cycle. However, increasing order quantity $Q$ decreases the annual shortage cost and the annual order setup cost, while increasing the annual inventory cost.

Beside continuous review systems, another inventory control system that is commonly used is periodic review systems. As opposed to continuous review system in which an order can be placed at any time, in a periodic review system orders must be placed only at particular times, for example, beginning of each week or end of each month. In such systems, if the order setup cost is insignificant, then order-up-to policy is optimal. According to order-up-to policy, in each period firms order to increase their inventory to a certain amount, known as order-up-to level. This chapter provides analytics to find the optimal order-up-to levels in order to minimize total expected cost or achieve a certain service level. If there is indeed a significant cost associated with placing an order, then the optimal policy that minimizes the total expected cost is called $(s, S)$ policy. According to $(s, S)$ policy, an order is placed in each period to raise the inventory to order-up-to level $S$, only if the inventory position is less than or equal to reorder point $s$. The chapter shows how the optimal $(s, S)$ policy can be determined using the analytics for $(Q, R)$ policy.

## Discussion Questions

1. What are the main features of a Newsvendor decision setting? Provide three different examples of Newsvendor problem beyond inventory management.

2. What is protection interval in $(Q, R)$ inventory management policy, and why it is the key factor in managing and controlling inventory?

3. What is the impact of increasing reorder point $(R)$ on average shortage in an order cycle, average annual shortage cost, and average annual inventory holding cost?

4. What is the impact of increasing order quantity $(Q)$ on average shortage in an order cycle, average annual shortage cost, and average annual inventory holding cost?

5. What is periodic review inventory system, and what is protection interval in such systems?

6. What type of policy is appropriate for periodic review inventory systems when order setup cost is zero? How does the policy change if order setup cost is non-zero?

## Problems

1. Ravi, the manager of a retail store, has placed an order for a limited edition of a sports jacket that is only offered once by the manufacturer this summer. Ravi estimated that the demand for the jacket in summer will follow a Normal distribution with mean 2000. He uses the Newsvendor formula and orders optimally for 1600 jackets. Two weeks later he realizes that he made a mistake in estimating demand. Specifically, while the demand is still Normally distributed with mean 2000, the variability in demand is much less than what he estimated. Which one of the following statements is true? Explain why?
    a. Ravi should not change his order size.
    b. Ravi should decrease his order size.
    c. Ravi should increase his order size.
    d. More information such as overstocking and understocking costs are needed to make any conclusion.

2. Suppose Ravi in Problem 1 calls the manufacturer of the jacket to change his order size. The manufacturer informs Ravi that they have already started making his jackets and it is too late for Ravi to change his order size. Ravi knows that his order size 1600 is not optimal (because demand has lower variability), but decides not to change his order size. Explain why each of the following statements is true or false:
    a. Ravi will have a lower chance of stockout now that he did not change his order size to the optimal size.
    b. Ravi should expect to have less unsold jackets now that he did not change his order size to the optimal size.
    c. A larger fraction of his customers will face shortage now that he did not change his order size to the optimal size.
    d. His cycle-service level $\alpha$ does not change because the critical ratio $\mathcal{R}_c$ has not changed.

3. Tom and Stacy work together to order items for a large clothing retailer. They disagree about ordering a popular jeans brand from Italy. The demand for the jeans is variable and follows Normal distribution. They use $(Q, R)$ policy to manage their inventory. Stacy suggests that they should order 200 pairs when their inventory reaches or falls below 50. While Tom agrees that the order size should be 200, he thinks that they should place an order when their inventory reaches or falls below 80. Explain why each of the following statements is true or false.

   a. Stacy's policy results in higher service levels than those under Tom's policy.
   b. Stacy's policy is expected to result in more number of orders placed in a year than that under Tom's policy.
   c. Tom's policy results in lower average shortage cost in a year than that under Stacy's policy.
   d. Tom's policy results in lower total expected annual cost than that under Stacy's policy.

4. Tom in Problem 3 argues with Stacy that, if they place an order when the inventory reaches or falls below 50 (as Stacy suggests), then they need to increase their order size to 230. Explain why each of the following statements is true or false.

   a. Under Tom's new policy (i.e., ordering 230 when inventory reaches 50) the likelihood of shortage is less than that under Stacy's policy.
   b. The fraction of customers who face shortage under Tom's new policy is the same as that under Stacy's policy.
   c. Stacy's policy is expected to result in more number of orders placed in a year.
   d. Tom's new policy results in lower average shortage cost in a year.
   e. Tom's policy results in lower expected annual inventory holding cost in a year.

5. Finally, Tom and Stacy in Problem 3 come to an agreement and decide to place an order for 200 pairs of jeans when their inventory reaches or falls below 60. This results in cycle-service level of 90 percent. However, they get a call from the supplier of jeans in Italy, informing them that the demand for the jeans has increased throughout Europe. Because the supplier does not have enough capacity to satisfy the new demand, future orders will take twice the time to deliver. Specifically, the supplier's order lead time is doubled from 1 week to 2 weeks. Tom and Stacy decide not to change the new ordering policy they agreed upon, since *their* demand has not changed. If the new policy is implemented, explain why each of the following statements is true or false.

   a. The cycle-service level is reduced, but fill rate remains the same.
   b. Total average annual inventory holding cost is decreased.
   c. Total average annual order setup cost is decreased.
   d. Total average annual shortage cost is increased.

6. Tom and Stacy in Problem 5 come to the conclusion that since the supplier lead time is doubled, they should double their reorder point from 60 to 120 to maintain cycle-service level of 90 percent. The order size remains at 200 pairs. Explain why each of the following statements is true or false.

   a. Doubling reorder point results in higher cycle-service level than 90 percent.
   b. Doubling reorder point increases expected annual inventory holding cost compared to current annual holding cost.
   c. Doubling reorder point results in a safety inventory smaller than what is needed to achieve cycle-service level of 90 percent.
   d. Doubling reorder point makes the total average annual shortage cost remain the same.

7. Suppose a firm orders a product for which the demand is well-approximated by Normal distribution. To reduce the possibility of stockout to 5 percent or less, the firm orders 2500 units when its inventory of the product reaches or falls below reorder point 1200. If the

average demand is doubled but standard deviation remains the same, explain why each of the following statements is true or false.

    a. Doubling both order size and reorder point will keep the possibility of stockout at the 5-percent level.

    b. Doubling the reorder point is enough to keep the possibility of stockout at the 5-percent level.

    c. Only doubling the order size is enough to keep the possibility of stockout at the 5-percent level.

    d. Only doubling safety stock is enough to keep the possibility of stockout at the 5-percent level.

8.   A hardware store in a small town orders mold-resistant drywall panels from a supplier out of state. The supplier delivers drywall panels every 4 weeks and the store must place its orders 2 weeks before the delivery date (i.e., order lead time is 2 weeks). The supplier does not charge any shipment cost and the order setup cost is insignificant. Since the store is the only one in town that carries mold-resistant drywalls, customers who face shortage wait until the store receives its next delivery. The store manager orders drywalls according to order-up-to policy with order-up-to level of 500 panels, which minimizes the total average annual inventory and backorder cost. The store manager receives good news and bad news from the supplier. The bad news is that the supplier is changing its delivery schedule from every 4 weeks to every 5 weeks. The good news is that the supplier has reduced its order lead time from 2 weeks to 1 week. Which one of the following statement is true, explain your reasons. Assume weekly demand is independent and follows Normal distribution.

    a. The store manager should increase his order-up-to level to keep the total cost at its minimum.

    b. The store manager should decrease his order-up-to level to keep the total cost at its minimum.

    c. No need to change, current order-up-to level results in minimum costs.

    d. More information is needed to make any conclusion.

9.   Consider the hardware store in Problem 8 and assume that the order-up-to level of 500 was set to reduce the possibility of stockout to 10 percent and lower. Considering the new 5-week delivery schedule and lead time of 1 week, which one of the following statements is true. Explain your reasons.

    a. The store manager should increase his order-up-to level to maintain the same service level.

    b. The store manager should decrease his order-up-to level to maintain the same service level.

    c. No need to change, current order-up-to level yields the same service level.

    d. More information is needed to make any conclusion.

10.  Answer questions in Problem 9, if the order-up-to level of 500 was set to reduce the fraction of customers facing stockout to 10 percent or less.

11.  A bookstore is planning to order mugs that are designed for Valentine's day. The store buys a mug for $7.50 and sells it for $8.99. Because the store asked the manufacturer of the mug to print the bookstore's logo at the bottom of the mugs, the store needs to place its order 6 weeks in advance. The bookstore, therefore, has decided to place only one order 2 months before Valentine's day. All the mugs that are not sold by Valentine's day will be sold at the price of $3.99 to a discount store. The demand for the mugs is believed to follow a Poisson distribution with mean 160. The store manager has decided to order 180 mugs.

    a. How many mugs should the manager expect to sell to its customers and how many mugs should the manager expect to sell to the discount store?

    b. What is the bookstore's expected profit?

c. How many mugs should the manager order to maximize their expected profit?

d. What is the probability of not having shortage of mugs, if the manager orders optimally to maximize expected profit?

e. What fraction of customers will face shortage if the manager orders optimally?

12. Answer questions in Problem 11, if the demand for mugs follows a Normal distribution with mean 160 and standard deviation 50.

13. T.S. Fashion produces fashion clothing and sells it to high-scale fashion retailers. The firm must decide about the production schedule of a newly designed vest that is expected to be the fashion trend in the coming fall season. The total cost of producing a vest is $55 and the firm can sell it during the fall season for $95. However, there is not much demand for the vest after the fall season. The marketing department has found a discount retailer that is willing to buy the unsold vests at the end of the fall season for $25. The marketing department has estimated that the demand for the vest would be between 500 and 800 with the following probability distribution:

| Demand $(x)$ | $P(D=x)$ |
| --- | --- |
| 500 | 0.05 |
| 550 | 0.08 |
| 600 | 0.12 |
| 650 | 0.15 |
| 700 | 0.25 |
| 750 | 0.20 |
| 800 | 0.15 |

The firm sells the vests in batches of size 50 to department stores, which include five different sizes (small, medium, large, X-large, and XX-large).

T.S. Fashion has the following constraints with respect to the production schedule of the vests. Retailers start placing their orders for the vest when the fall season starts. Their orders are for boxes of the vest. However, the production of the vests should be completed before the fall season. This is because the entire production capacity of T.S. Fashion in the fall season is devoted to producing new fashion products for the winter season. Hence, T.S. Fashion must make a decision about how many vests to produce before receiving any orders from its retailers.

a. How many vests should the firm produce in order to maximize its expected profit?

b. How much would the expected profit be under the optimal production quantity?

c. How many leftover vests should T.S. Fashion expect to have at the end of the fall season?

14. K.G. Electronics is planning the production of its winter Olympic edition of one of its popular smart watches. The watch is expected to be a big hit during winter Olympics. K.G. is planning to produce 50,000 of these watches before winter Olympics. It costs K.G. $120 to make a watch and K.G. has two markets to sell them: (i) the U.S. market and (ii) overseas market. In the United States, the watch will be sold online through K.G.'s website at the price of $300. The sales in overseas will be through retailers in those countries, who will purchase the watch from K.G. at $220 and sell them in their stores. Watches must be shipped to overseas 1 month before winter Olympics start.

The overseas demand is so large that K.G. can sell all its 50,000 watches overseas, if it wants to. However, K.G. wants to sell as many watches as possible in the United States, since it can charge higher price for each watch.

The demand for the watch in the United States during Olympics is estimated to follow Normal distribution with mean 20,000 and standard deviation 5000. K.G. knows that watches that are not sold during the Olympics will not have demand overseas after Olympics. However K.G. is confident that it can sell its unsold watches in the United States at a discount price of $180. How many watches should K.G. sell in the United States and how many watches should it sell overseas in order to maximize its profit?

15. Your company has just bought a welding robot to install in its assembly line. One set of components of the robot that are subject to failure and would take a long time to order and install are pneumatics sensors. The manufacturer of the robot is offering a discount on these sensors (as spare parts). You are thinking of purchasing a number of sensors to replace the broken sensors during the lifetime of the machine. This is particularly important, since the manufacturer will stop the production of this robot in 1 year and produce the next generation of its welding robot that uses different sensors. You can purchase each sensor for $250. Holding a sensor during the lifetime of the robot costs about $150 and sensors have no salvage value if they are not used by the end of robot's lifetime. A shortage of a sensor, on the other hand, costs you about $45,000, which includes loss of production (due to robot breakdown), cost of sending the sensor for repair, and installing the repaired sensor. According to the machine maintenance catalog, and the working condition of the robot, it is expected that the number of sensors needed to replace the broken ones during the lifetime of the robot follow a Poisson distribution with mean 10. How many sensors would you purchase?

16. Sonbox Coffee orders paper cups from a supplier at the price of 20 cents per cup. It takes 3 weeks for the supplier to deliver cups after Sonbox places its order. Order setup cost is $80 which is mainly the shipment cost charged by the supplier and is independent of order size. Data shows that the weekly demand for to-go coffee that requires a paper cup follows a Normal distribution with mean 1400 and standard deviation 300. The inventory carrying rate for the cups at Sonbox is 25 percent. When Sonbox runs out of paper cups, customers who want the paper cup for to-go do not buy the coffee. This costs Sonbox about $1.5 in lost sales and loss of goodwill. Sonbox currently orders 5000 cups when its inventory of cups reaches 500.

   a. Under the current ordering policy, what is Sonbox's total average annual inventory holding, order setup, and lost sales cost?
   b. What percent of time does Sonbox face shortage under its current ordering policy?
   c. Under its current policy, what fraction of Sonbox customers are lost?
   d. If Sonbox keeps its order size of 5000 (to take advantage of full-truckload), but decides to change its reorder point, what reorder point should it choose to minimize its total average annual cost?
   e. If Sonbox decides to change both its order size and reorder point, how should they be changed in order to minimize the total average annual cost?
   f. If Sonbox wants to reduce the fraction of its lost customers below 5 percent, how should it change its ordering policy?

17. The supplier of paper cups in Problem 16 offers another ordering option to Sonbox. The supplier offers free shipment if Sonbox agrees to fixed delivery schedule. The supplier would like to deliver cups to Sonbox every 5 weeks and Sonbox must place its order 2 weeks before the delivery day (i.e., lead time is 2 weeks).

   a. If Sonbox agrees with the new delivery schedule, what would be Sonbox's optimal ordering policy that minimizes its total average cost?
   b. What fraction of customers are lost under your ordering policy in Part (a)?
   c. Should Sonbox accept the new delivery schedule?

18. DDA is a manufacturer of office furniture that sells its customized adjustable office desk after it receives an order from a customer. Because desks are made in an assemble-to-order process, DDA can finish orders and ship them on the same day, unless one of the components of the desk is not available. When a component is not available, the order is delayed until the

component is delivered from the supplier. All components of the desk are produced in DDA's factory, except for one component: an electric motor that is ordered from a supplier with order lead time of 2 weeks. DDA buys the motors at a unit price $110 and with order setup cost (including shipment cost) of $450, which is independent of the order size. Delivered motors are kept in the inventory with inventory carrying rate of 30 percent.

Sales data shows that weekly demand for adjustable desks follows a Normal distribution with mean 40 and standard deviation 12. To prevent dissatisfaction of delayed customers, DDA has decided to offer free delivery—which costs about $15 per order—to customers whose orders are delayed. Because all other components are produced in-house, the only component that often results in customer delay is the electric motor.

   a. What is the optimal ordering policy that minimizes the total average annual holding, setup, and customer delay cost of ordering the motor? What is the optimal average cost?
   b. What are cycle-stock service level and fill rate under the optimal policy obtained in Part (a).
   c. How should DDA change its inventory policy to guarantee that less than 2 percent of its orders are delayed?

19. Coby Inn is a hotel that offers low-price rooms with free breakfast. One of the popular items in its dinner menu is barbecue ribs. Coby Inn orders frozen ribs from a major distributor at the cost of $9 per pound. The distributor charges the hotel $60 for delivery and shipment cost. Coby Inn stores frozen ribs in its freezer with the inventory carrying rate of 20 percent. Coby Inn estimates that telling a guest that they are out of ribs and the guest needs to order another item from the menu results in $3 loss of profit. The data shows that the daily demand for ribs is Normally distributed with mean 50 pounds and standard deviation 15 pounds. The hotel manager realized that, while she has the flexibility of placing an order to distributor of rib at any time, the distributor often misses its delivery schedule. Looking at the past deliveries, the manager finds that the time duration from when the order was placed until the order was delivered has mean of 5 days with standard deviation of 2 days (assume 360 days in a year).

   a. How much ribs the hotel should order each time and when should it place its orders to minimize the total average annual holding, shipment, and shortage cost?
   b. What is the average time a pound of rib stays in the freezer before it is cooked and served to a guest?
   c. What is the optimal ordering policy if Coby Inn wants to reduce the fraction of its guests who must choose another dish (because rib is not available) to 5 percent? What is the safety inventory under this policy?
   d. Suppose Coby Inn has a limited space in its freezer and thus it cannot order more than 500 pounds each time. How does this change your answer to Part (a)?

20. The Coby Inn manager in Problem 19 calls the distributor and complains about their unreliable delivery schedule. The reason for delay, the distributor explains, is that Cobby Inn places its orders at different times each month. This interrupts the distributor's delivery schedule to other customers. The distributor suggests a fixed schedule of delivering ribs every 3 weeks on Mondays. Coby Inn needs to place its orders 5 days in advance and pay the shipment cost.

   a. If Cobby Inn accepts the distributor's delivery schedule, what would be its optimal ordering policy that minimizes the average annual cost?
To persuade Coby Inn, the distributor informs Coby Inn that if they accept the 3-week delivery schedule, the distributor will not charge shipment cost $60 per order.
   b. If Coby Inn accepts the free delivery and orders every 3 weeks, how much should it order each time?

    c. How long a pound of ribs stays in Coby Inn's freezer before it is cooked and served to a guest, if Coby Inn accepts the free delivery offer?

    d. Should Coby Inn accept the free delivery offer?

21. A pharmacy in Evanston orders a moisturizing cream from a pharmaceutical firm that has a distribution center in Michigan. The pharmacy purchases each cream for $12 and sells it at $20. The firm makes its deliveries to all pharmacies in Evanston every 10 days. Pharmacies must place their orders for the cream 2 days before the delivery day. The firm charges delivery cost of $80 if the firm places an order. To keep their customer base, the pharmacy gives 15 percent discount to those customers who wait for the cream if the pharmacy is out of stock. This has prevented the pharmacy from losing customers who left and bought the cream from other pharmacies. The pharmacy takes the customers' phone numbers and calls them when the cream arrives. The pharmacy has an inventory carrying rate of 35 percent. The demand for the cream in a day follows Normal distribution with mean 5 and standard deviation 2.

    a. How many creams should the pharmacy order each time in order to minimize its total average annual cost?

    b. How many creams should the pharmacy order each time if the pharmacy wants to reduce the possibility of stockout to less than 2 percent? What would be the total average annual cost under this ordering policy?

    c. How many creams should the pharmacy order each time if the pharmacy wants to reduce the percent of its customers getting discount (due to stockout) to less than 2 percent? On an average, how many customers receive discount from the pharmacy in a year?

# CHAPTER 11

# Inventory Improvement

## 11.0 Introduction

In Chap. 9 we discussed the basic concepts of inventory management problems and presented several analytics to develop effective inventory control policies for cases where the demand has no or very low variability—deterministic demand. In Chap. 10, on the other hand, we focused on inventory management in processes with significant variability in demand—stochastic demand. The goal of these two chapters was to introduce *inventory control* policies for continuous and periodic review systems to either minimize costs or achieve a certain service level.

Now suppose that you used the analytics in Chaps. 9 and 10 and found the optimal control policy that minimizes the total expected annual cost. This resulted in an improvement in your system (i.e., reduction in cost or increase in service level). However, the upper management is still not satisfied with the improvement and would like you to find ways to further improve the performance. Since you have already found the best—the optimal—control policy to minimize the cost, you are wondering if there are other ways to reduce costs even more. Fortunately, the answer is Yes, and this is what this chapter is all about.

This chapter provides a list of strategies and practices that can further improve the performance of inventory systems. These strategies range from tactical decisions such as reducing supplier lead time to strategic decisions such as redesigning the supply chain.

We start with simple, but important, task of identifying critical items that we should focus our improvement effort on.

## 11.1 Pareto Your Inventory—The ABC Analysis

Firms carry hundreds or thousands of different items, ranging from raw material, work-in-process, and finished goods inventory. The first step to make any improvement in inventory is to identify important and critical items that have significant impact on the overall inventory costs.

Many studies of inventory in manufacturing and retail industry have revealed an interesting and useful phenomenon regarding the usage rate of items: a few items (about 20 percent of the total items) account for most (about 80 percent) of the total annual dollar-demand or annual dollar-usage.[1] Therefore, by managing these few items more effectively, one can make a significant improvement on overall inventory costs.

The above phenomenon is known as *Pareto's Law*, named after the 19th century Italian scientist, Vilfredo Pareto, who noticed that 80 percent of Italy's land was owned by 20 percent of the population. Today, the Pareto's law has been observed in many different situations: 80 percent of your sales come from 20 percent of your clients, or 80 percent of purchased cost of items (e.g., material, parts) in manufacturing corresponds to about 20 percent of total number of purchased items.

How can one identify those few critical and important items that have significant impact on a firm's inventory costs? The answer is the *ABC Analysis*. ABC analysis uses Pareto's law and divides items into three classes of A, B, and C.

- *Class A:* Class A (most important) are about 10 to 20 percent of items that often account for 70 to 80 percent of the dollar-usage.
- *Class B:* Class B (intermediate in importance) are about 30 percent of the items that often account for about 15 to 25 percent of the dollar-usage.
- *Class C:* Class C (least importance) are about 50 percent of the items that often account for about 5 to 10 percent of the dollar-usage.

To show how ABC analysis works, consider a sample of 15 items used in a factory that produces a wide range of products. This is shown in Table 11.1. The first column of the table shows the items' names; the second and third columns show the annual usage and the dollar value of each item, respectively. As the table shows, for example, Item B11, which is a part bought from a supplier, costs $44 each, and the factory uses 12 of this item in a year. Hence, the total dollar-usage for this item is $12 \times \$44 = \$528$. This is shown in the fourth column of the table for this and all other items.

What percent of the total dollar-usage corresponds to Item B11? As Table 11.1 shows, the total annual dollar-usage for all 15 items is $65,894. Thus, Item B11 constitutes $(528/65894) = 0.008$ or 0.8% of the total annual dollar-usage. This is shown in the fifth column of the table.

---

[1] Annual dollar-usage is often used, since in addition to finished goods inventory, it also includes items such as raw material, work-in-process, etc., that are used in manufacturing systems to produce the final product.

Table 11.1    The ABC Analysis—Finding the Percent Usage

| Item | Annual demand (*units*) | Unit value ($) | Annual dollar-usage ($) | Percent of total dollar-usage |
|------|------|------|------|------|
| S201 | 85 | 9.00 | 765 | 1.16 |
| B11 | 12 | 44.00 | 528 | 0.80 |
| TS21 | 51 | 18.00 | 918 | 1.39 |
| BB13 | 43 | 60.00 | 2,580 | 3.92 |
| YT2 | 78 | 44.00 | 3,432 | 5.21 |
| BB23 | 300 | 80.00 | 24,000 | 36.42 |
| SC123 | 248 | 3.00 | 744 | 1.13 |
| A165 | 510 | 3.00 | 1,530 | 2.32 |
| DP222 | 188 | 1.50 | 282 | 0.43 |
| CV111 | 490 | 40.00 | 19,600 | 29.74 |
| B987 | 440 | 8.00 | 3,520 | 5.34 |
| N23 | 75 | 5.00 | 375 | 0.57 |
| Q234 | 280 | 4.00 | 1,120 | 1.70 |
| S12 | 1250 | 1.20 | 1,500 | 2.28 |
| S233 | 250 | 20.00 | 5, 000 | 7.59 |
| | | **Total:** | $65,894 | |

The percentages in the fifth column of Table 11.1 give useful information to classify the 15 items into A, B, and C classes. Note that Class-A items are the most important items that have the highest annual dollar-usage, and Class-C are the least important items with the lowest dollar-usage. Hence, to make the classification procedure easy, we can sort the 15 items in a descending order of their annual dollar-usage. This is shown in Table 11.2.

Table 11.2    The ABC Analysis—Ordering and Classification of Items

| Item | Percent dollar-usage | Item classification |
|------|------|------|
| BB23 | 36.42% | A |
| CV111 | 29.74% | A |
| S233 | 7.59% | A |
| | Total = 73.75% | |
| B987 | 5.34% | B |
| YT2 | 5.21% | B |
| BB13 | 3.92% | B |
| A165 | 2.32% | B |
| S12 | 2.28% | B |
| | Total = 19.07% | |
| Q234 | 1.70% | C |
| TS21 | 1.39% | C |
| S201 | 1.16% | C |
| SC123 | 1.13% | C |
| B11 | 0.80% | C |
| N23 | 0.57% | C |
| DP222 | 0.43% | C |
| | Total = 7.18% | |

The first column of the table is the list of items arranged from the one with the highest percent dollar-usage at the top to the item with the lowest at the bottom, with the corresponding percent dollar-usage in the second column.

The third column of Table 11.2 shows one classification of the 15 items according to the ABC analysis classification scheme. Note that the first three items, classified as Class-A items, are $3/15 = 20\%$ of the items and constitute 73.8 percent of the total dollar-usage. The next five items—Class-B items—are $5/15 = 33\%$ of the items and make up about 19.1 percent of the total dollar-usage. Finally, Class-C items are the last seven items that are $7/15 = 46.6\%$ of the items and make up only about 7.2 percent of the total dollar-usage.

While the total dollar-usage is a common criteria that corresponds to the importance of an item, ABC analysis can also be used to classify items according to other criterion such as shortage cost, storage requirement of an item, shelf life, etc.

How can ABC analysis help operations engineers and managers improve their inventory? By classifying items based on their importance, they can allocate their efforts more effectively to develop different inventory planning and control policies for each class:

- *Class-A Items:* Most specialized management attention must be paid to Class-A items. They should be watched closely using continuous review systems. The accuracy of the demand forecast for these items should be evaluated frequently, and sophisticated forecasting methods should be developed. Also, cost parameters such as inventory holding cost, shortage cost, order setup cost used in obtaining inventory control policies for Class-A items should be accurately estimated. The resulting policy (e.g., order quantity, reorder point, safety stock) should be evaluated frequently. The accuracy of the computer records should also be checked frequently. Significant effort must be made to reduce lead times for Class-A items.

- *Class-B Items:* Class-B items are of secondary importance compared to Class-A items. They require normal control effort. While they are not the most important items, they constitute a large portion of dollar-usage. Hence, they also need a significant attention, but not at the level of Class-A items. These items can be reviewed periodically using periodic review systems. Less sophisticated forecasting methods can be used for these items. They can be ordered in groups along with other Class-B items, instead of being ordered individually.

- *Class-C Items:* While they are of least importance, Class-C items may potentially consume a large amount of data and management time. One objective of the ABC analysis is to identify Class-C items in order to prevent this. Class-C items should receive minimum attention and minimum degree of control. The inventory of these items can be counted and reviewed physically. Inexpensive Class-C items should be ordered in large order quantities with large safety inventory. This reduces the order setup costs. For expensive Class-C items with very low demand, on the other hand, minimum inventory (or no inventory) should be kept.

## 11.2 Avoiding Ad Hoc Decisions and Using Analytics

As shown in all the chapters of this book, analytics are powerful decision tools that help operations engineers and managers make informed decisions, including inventory management decisions.

*Train Employees to Use Analytics:* Having employees with sufficient knowledge and training in inventory modeling and analytics is the first step in reducing inventory costs and providing better service.

Analytical models use data to generate cost-effective inventory management policies. Hence, one key factor for developing a good policy is to get accurate data. Some of the data used in analytics can be estimated with a reasonable degree of accuracy (e.g., order setup cost, inventory holding cost, purchase cost, etc.). The two types of data that are often not easy to accurately estimate are: (i) demand, and (ii) shortage cost.

*Improve Demand Forecast:* Information about future demand (e.g., average, standard deviation, probability distribution) is one of the critical inputs to all analytical and even ad-hoc inventory management models. Therefore, a good demand forecasting model is essential for developing effective inventory control and improvement policy. Chapter 3 presents several techniques for forecasting demand.

*Use Imputed Shortage Cost when Setting Service Levels:* As we mentioned in Chap. 10, in some cases having an accurate estimate of shortage cost is difficult, if not impossible. Hence, to have a control on the number of shortages, managers set their inventory policy to achieve a certain service level. The question in such cases is how high or how low the service level should be set. In other words, what is the appropriate service level for a process?

The answer depends on how severe the consequences of shortages are. Several factors such as product type, product's profit margin, and market condition should be considered when setting a service level for a product. Consider the product type as an example. Shortage of blood in blood banks and shortage of a particular part used in maintaining commercial aircraft can have devastating and costly consequences. In the former, the shortage can cause serious consequences for a patient who is in need for blood transfusion, and in the latter, shortage can result in delaying a flight, which often costs airlines a lot of money. For these two products, the process needs to maintain a very high service level. Now consider a different product: a pencil in a store that sells office products. Shortage of pencils does not have a severe consequence and thus stores do not need to have a very high service level for pencils.

Market condition also affects the service level of a product. In a competitive market, shortage of an item often results in immediate lost sales as well as a very high chance of losing customers to the competitor. If the number and frequency of shortage are high, firms may lose market share resulting in significant reduction in sales. In this case, setting the appropriate service level is of great importance.

In conclusion, one must be very careful in setting the right service level and should avoid ad hoc decision without any analytical or experimental support. The key point is that not having an accurate estimate for the shortage cost does not mean that it does not exist. Thus, to set a target service level, one should also consider its impact on the shortage cost. One concept that can help managers to achieve this and make a more informed decision about the service level is the "Imputed Shortage Cost." We use the following example to illustrate it.

 **Example 11.1**

Consider the Auto Part Depot (APD) in Example 10.6 of Chap. 10 in which the lost sales cost is estimated to be $C_s = \$15$. Using our analytics, we found that for the polishing compound, the optimal $(Q^*, R^*) = (463, 351)$ minimizes the total average annual shortage, inventory, and order setup cost. However, APD managers realized that the estimate of their lost sales cost was not accurate. While they have not been able to find an exact estimate, new data shows that the lost sales cost is higher than $20 per lost customer. Since they do not know the exact lost sales cost, managers of APD have decided to change their inventory policy and reduce the possibility of having a shortage to a certain level, that is, set a cycle-service level. The question they now face is what cycle-service level to choose? After some discussion they decide to set the cycle-service level for the polishing compound at $\alpha = 95\%$, reducing the chance of a shortage to only 5 percent. What do you think? Do you agree with the cycle-service level of 95 percent? Should they choose a higher or a lower cycle-service level?

To answer this question, let's first find the policy that achieves cycle-service level $\alpha = 95\%$. As our analytics states, since we do not have an estimate for the shortage cost, we use the EOQ formula to find the optimal order quantity:

$$Q_\alpha^* = \sqrt{\frac{2\overline{D}K}{C_h}}\sqrt{\frac{2(10800)(187.5)}{20}} = 450$$

Considering $\Phi(z = 1.65) = 0.95$, we have

$$R_\alpha^* = \overline{D}_L + z\sigma_L = 300 + (1.65)31.62 = 352.2$$

Thus, the new policy $(Q_\alpha^*, R_\alpha^*) = (450, 353)$ achieves a cycle-service level $\alpha = 95\%$. This policy does not minimize the total lost sales, inventory, and order setup cost, because APD does not know what the actual lost sales cost is. While the exact value of the lost sales cost is not known, it is interesting and useful to know for what value of lost sales cost policy $(Q_\alpha^*, R_\alpha^*) = (450, 353)$ minimizes the total average annual cost (including lost sales cost)?

Recall that, for the case of lost sales, the optimal $(Q, R)$ policy that minimizes the total average annual cost satisfies (see Eq. (10.33)):

$$F_L(R) = 1 - \frac{C_h}{C_h + C_s\overline{D}}Q \quad \text{or} \quad C_s = \frac{C_h}{\overline{D}}\left(\frac{Q}{1 - F_L(R)} - 1\right)$$

Hence, policy $(Q_\alpha^*, R_\alpha^*) = (450, 353)$ minimizes the total average cost, if the value of lost sales cost $C_s$ is

$$C_s = \frac{C_h}{\overline{D}}\left(\frac{450}{1 - F_L(353)} - 1\right) = \frac{20}{10800}\left(\frac{450}{1 - 0.95} - 1\right) = 16.66$$

Therefore, the policy that achieves cycle-service level 95 percent also minimizes the total average annual cost if the lost sales cost was $C_s = 16.66$. In other words, if lost sales cost was $C_s = 16.66$, then under policy $(Q_\alpha^*, R_\alpha^*) = (450, 353)$ that yields $\alpha = 95\%$, the optimal balance between the shortage cost and inventory cost is achieved. The lost sales cost $C_s = 16.66$ is called *imputed shortage cost* (or imputed lost sales cost).

---

**CONCEPT**

**Imputed Shortage Cost**

For any given policy that is not cost-optimal, that is, does not minimize the total average annual holding, shortage, and setup cost, the *imputed shortage cost* is the value of the shortage cost for which the given policy becomes the cost-optimal policy.

---

The imputed shortage cost $C_s = 16.66$ can help us decide whether the APD's decision of setting $\alpha = 95\%$ was an appropriate decision. Note that only when $C_s = 16.66$ then APD achieves the optimal balance between the shortage and inventory costs. Considering that the actual shortage (i.e., lost sales) cost is known to be more than \$20, should APD revise its cycle-service level? Well, since for cycle-service level $\alpha = 0.95$, imputed shortage cost is lower than the actual shortage cost, then implementing the policy $(Q_\alpha^*, R_\alpha^*) = (450, 353)$ will result in higher expected annual cost than that in a cost-optimal policy with shortage cost above \$20. Again, not knowing the exact value of shortage cost does not mean that shortage does not exist. There is a shortage cost and it is above \$20. To reduce this shortage cost, APD should increase its cycle-service level above $\alpha = 95\%$.

Before we conclude this section, it is important to note that the concept of imputed shortage cost does not determine the best service level for a system. It, however, can be used to rule out service levels that are not appropriate. In case of the APD, for example, while we could not suggest a cycle-service level, using the imputed shortage cost, we were able to determine that the cycle-service level 95 percent was low.

## 11.3 Reducing Order Setup Costs

Order setup cost corresponds to the cost associated with placing an order. Consider the optimal order quantity in continuous review system

$$Q^* = \sqrt{\frac{2KD}{C_h}}$$

It is clear that the higher the order setup cost $K$, the larger the optimal order quantity $Q^*$. On the other hand, the average cycle inventory in continuous review systems is about half of the order quantity, that is, $I_{cycle} = Q^*/2$. Hence, any reduction in order setup cost results in reduction in cycle inventory.

This is also true in periodic review systems. Recall that the optimal policy in periodic review system with non-zero order setup cost is $(s, S)$ policy. Also note that the optimal values for this policy are $s^* = R^*$ and $S^* = R^* + Q^*$, where $R^*$ and $Q^*$ are the optimal reorder point and order quantity, respectively, in corresponding continuous review system (see Sec. 10.7.3 of Chap. 10). Therefore, reduction in order setup cost will result in reduction in order quantity $Q^*$ and thus results in reduction in order-up-to level $S^*$, which in turn results in lower cycle inventory.

As shown above, both continuous review and periodic review systems benefit from reduction in order setup cost. But, how can one reduce the order setup cost? To answer this question, note that the main two components of order setup cost are: (i) the transactions costs of placing an order, and (ii) transportation cost of the order.

**Reducing Order Transaction Costs**

Transaction costs of placing an order include costs such as cost of preparing the order, following the order, invoice and payment processing, etc. One way to reduce these costs is to simplify the process of placing an order. This can be done using information technologies such as computer-assisted ordering (CAO), electronic data interchange (EDI), and enterprise resource planning (ERP) systems. These systems reduce the cost of paperwork and automate the ordering process (thus reducing the labor cost). For example, by using information technology and matching their suppliers and buyers throughout the company, General Electric (GE) was able to reduce its paper purchase order that cost \$50 to as low as \$5.[2]

**Reducing Transportation Costs**

Consider a supermarket that orders milk from a producer of dairy products. The supplier charges about \$100 for delivery of milk. How can the supermarket reduce this transportation cost?

---

[2]Smart, T., "Jack Welsh's Cyber-Czar." August 5, 1996, 82–83.

- *Negotiation:* By negotiating with the supplier, the supermarket might be able to reduce the delivery cost to less than $100. For example, the supplier might be willing to reduce the delivery cost if the supermarket places an order in certain periods that make production planning and delivery easier for the supplier.

- *Order Assortment:* Assume that the supermarket also orders other dairy products such as cheese and yogurt from the same supplier, but on different days. This results in order setup cost of $100 for each order. However, if the supermarket places one order for an assortment of different products, the supplier may be able to deliver all products in one truck, resulting in lower delivery cost. For example, if the supermarket orders milk, cheese, and yogurt, which are delivered in one truck, then the order setup cost (i.e., the transportation cost) for each product is cut to third.

- *Third-Party Logistics:* Now suppose that our supermarket orders only milk from the supplier, so it cannot use order assortment to reduce its order setup cost. However, there are other supermarkets (of the same firm or of different firms) in a relatively short distance to our supermarket. How can our supermarket reduce its order transportation cost of $100? Through Third-Party Logistics firms. *Third-party logistics (3PL) firms* are firms that provide outsourced logistics services to other firms. For example, a 3PL firm may accept to deliver all the orders for milk from the supplier of dairy product to the supermarkets (including our supermarket). Again, the 3PL might be able to put all the orders in one truck and deliver that to the supermarkets with lower transportation cost.

In production systems that produce items in batches, the order setup cost corresponds to the fixed setup cost associated with configuring machines, tools, and equipment to produce the next batch. Chapter 15 provides a list of strategies that can reduce the time and cost of implementing setup operations.

## 11.4 Reducing Length of the Period in Periodic Review Systems

Reducing length of the period in periodic review systems reduces the safety stock and thus reduces the average annual inventory cost. To show this, we use a simple example of a period review system in which the demand during protection interval is Normally distributed. If the goal is to achieve cycle-service level $\alpha$, for example, then the safety inventory under the order-up-to policy with order-up-to level $S_\alpha^*$ is

$$I_{\text{safety}} = S_\alpha^* - \overline{D}_{L+P}$$

Since demand during protection interval is Normally distributed, then to achieve cycle-service level $\alpha$, the optimal order-up-to level would be

$$S_\alpha^* = \overline{D}_{L+P} + z\sigma_{L+P}, \quad \text{where} \quad \Phi(z) = \alpha$$

Hence,

$$
\begin{aligned}
I_{\text{safety}} &= S_\alpha^* - \overline{D}_{L+P} = \overline{D}_{L+P} + z\sigma_{L+P} - \overline{D}_{L+P} \\
&= z\,\sigma_{L+P}
\end{aligned}
\tag{11.1}
$$

Therefore, as the above equation shows, lower standard deviation of demand during protection interval—lower $\sigma_{L+P}$—results in lower safety inventory. But how does that relate to length of a period? The answer is in the following equation that was presented in Sec. 10.7.4 of Chap. 10:

$$\sigma_{L+P} = \sqrt{\sigma^2 \left[ \mathsf{E}(T_L) + T_P \right] + \mu^2 \, \mathsf{Var}(T_L)} \tag{11.2}$$

As the equation shows, the standard deviation of demand during protection interval decreases as the length of the period ($T_P$) decreases. Hence, by decreasing the length of the period, periodic review systems can maintain the same cycle-service level with lower safety inventory. This is also true for cases where the goal is to maintain a fill rate or to minimize cost. This leads us to the following principle:

**PRINCIPLE**

**Impact of the Review Period on Inventory**

Reducing the length of the review period in periodic review systems results in reduction in inventory. Hence, systems with shorter review periods can maintain the same service level with less inventory.

Therefore, firms that use periodic review systems to manage their inventories should try to shorten the length of their review period. But, how much the length of the period can be reduced depends on physical constraints. For example, in some cases the length of the period is decided based on the supplier's delivery schedule (e.g., supplier delivers every week or every month). Decision on how much to shorten the review period may also depend on the cost structure. Recall that some systems incur a cost $K_P$ each time the inventory is reviewed (see Sec. 10.7.2 of Chap. 10). In such systems, while reducing the length of the period reduces the annual inventory cost, it will increase the annual inventory review cost. The analytics in Chap. 10 can be used to find the optimal review period that minimizes the total annual cost.

## 11.5 Reducing Mean and Variability of Lead Time

Reducing mean and variance of lead time results in reducing safety inventory in both continuous and periodic review systems. Consider Eqs. (11.1) and (11.2) corresponding to the safety inventory in a periodic review system to achieve cycle-service level $\alpha$. As Eq. (11.2) shows, the standard deviation of demand during protection interval $\sigma_{L+P}$ depends on the lead time. Specifically, any reduction in average lead time $E(T_L)$ or variance of lead time $\text{Var}(T_L)$ results in reduction in $\sigma_{L+P}$, leading to reduction in safety inventory.

This is also true for continuous review systems. Recall that under $(Q, R)$ policy the safety inventory is

$$I_{\text{safety}} = R - \overline{D}_L$$

Now suppose the demand is Normally distributed. Thus, if, for example, one would like to set a cycle-service level $\alpha$, then the optimal reorder point would be

$$R_\alpha^* = \overline{D}_L + z\sigma_L, \quad \text{where} \quad \Phi(z) = \alpha$$

Hence, for the policy that achieves cycle-service level $\alpha$, we have

$$\begin{aligned} I_{\text{safety}} &= R_\alpha^* - \overline{D}_L = \overline{D}_L + z\sigma_L - \overline{D}_L \\ &= z\,\sigma_L \end{aligned}$$

showing that the safety inventory depends on the standard deviation of demand during lead time. On the other hand, for $(Q, R)$ policy we have

$$\sigma_L = \sqrt{\sigma^2\, E(T_L) + \mu^2\, \text{Var}(T_L)} \tag{11.3}$$

As the equation shows, $\sigma_L$ depends on both the average length of lead time, that is, $\mathsf{E}(T_L)$ and the variance of lead time $\mathsf{Var}(T_L)$. Any reduction in mean or variance of lead time results in reduction in standard deviation of demand during lead time, which in turn leads to a smaller safety inventory.

Using similar examples one can show that lead time reduction can also reduce safety inventory when the goal is to achieve fill rate or to minimize cost. We summarize this in the following principle:

**PRINCIPLE**

**Impact of Lead Time on Inventory**

Reducing average lead time or variance of lead time results in reduction in inventory in both continuous and periodic review systems. Hence, systems with smaller average lead times and/or smaller variance in their lead times can maintain the same service level with less inventory, and hence with less costs.

How the length of lead time or its variance can be reduced? To answer this, recall that replenishment lead time consists of four major times: (i) time to place an order to a supplier, (ii) time at the supplier, (iii) shipment time, and (iv) time needed to process the order after the order is received.

### Use Information Technologies

The time to place an order (i.e., item (i)) and the time needed to process the order after it arrives (i.e., item (iv)) are often not that long. Nevertheless, these times and their variance can be reduced by placing orders and processing their receipts electronically (e.g., using EDI or ERP systems).

### Implement Lean Operations Practices

Strategies to reduce item (ii)—time at the supplier—depends on whether the supplier produces the item in-house or orders it from another supplier. If the supplier is a manufacturer, the time at the supplier corresponds to the flow time of the item in the manufacturer's process. In Chap. 8 we presented strategies that can be used to reduce flow times. Lean manufacturing principles can also help manufacturers reduce the mean and the variability in production lead times (see Chap. 15).

### Reduce Transportation Time

The shipment time and its variance—item (iii) above—can be reduced in a variety of ways.

- *Use Faster Transportation:* Obviously using faster modes of transportation can reduce the shipment time. One issue, however, is that it is also more costly. So, the mode of transportation should be chosen carefully, taking into account its cost, its speed, and its variability.
- *Cross-Ducking:* Under cross-docking, items are moved directly from inbound unloading docks in a warehouse to the outbound loading docks without being placed in the storage. Walmart has used cross-ducking practices and has significantly reduced its lead times to deliver orders from manufacturers to its stores.
- *Supplier Proximity:* Another way to reduce shipment time and its variability is to choose suppliers that are in close proximity. For example, Toyota has 12 plants in Toyota City. Suppliers of these plants are located within 50-mile radius of these plants.

**Redesign Product or Supply Chain**

If the long lead times and their variability are serious concern, especially for some critical items, one last option is to make those items in-house instead of ordering them from another supplier. This gives firms more control over the length and variability of lead times. Of course, this is a strategic decision with long-term consequences and therefore it should be made carefully. Another strategic decision that can reduce lead time and its variability is to redesign the products such that it requires less parts or requires parts that can be acquired from reliable suppliers with shorter lead times.

## 11.6 Aggregate Demand—Inventory Pooling

There is always variability in the number of customers who demand a product and in the time until the next order arrives (i.e., demand interarrival times). In Sec. 5.6 of Chap. 5 we discussed the overall strategies to reduce variability in demand. In this section, we discuss in more detail how those strategies can be used to reduce inventory. We start with Inventory Pooling.

Firms save significant money by pooling their inventory in several locations into one location. Auto manufacturers such as Ford often close several of their dealerships and pool their inventories into a larger dealership. Retailers such as Walmart pool their inventory in several small stores into one superstore. Besides saving in fixed costs such as staffing and equipment, physical pooling of inventory into one location also results in saving in inventory costs. To illustrate this, we use the Case of W&D below.

**Pooling Inventory at W&D**

Windows & Doors (W&D) is a large retailer in New York that specializes in interior and exterior home improvement products. To save overhead costs such as space rent and employee salary, W&D is planning to close five of its small stores in New York and instead open a large superstore there. To get an idea of how much the firm can save in its inventory costs, managers of W&D are analyzing the savings in one of their products: a standard 34 inch-by-41 inch double hung window with grilles, called S-200. The firm buys the window from a manufacturer at $160, and sells it for $220. Considering the firm's overhead costs of $10 per window, W&D is making a net profit of $220 − $160 − $10 = $50 per window sold. This does not include the inventory-related costs (i.e., inventory holding cost and order setup cost) at each of its retail stores.

*Current System with Five Stores:* The daily demand for S-200 at each store $i$ is Normally distributed with mean $\mu_i = \mu = 10$ and standard deviation $\sigma_i = \sigma = 5$, for all $i = 1, 2, \ldots, 5$ stores (see Fig. 11.1–left). While demands at stores have the same mean and standard deviation, there is a correlation among the daily demands at the stores. Formally stated, $\rho_{ij}$ is the correlation between the daily demands at stores $i$ and $j$ (where $i \neq j$). Data shows that the correlation coefficient between the daily demand of each pair of store is 0.3, that is, $\rho_{ij} = 0.3$ for all $i, j$, and $i \neq j$.

All five stores order S-200 windows from the same manufacturer in California at price $c = \$160$ with order lead time of $L = 20$ days (with no variability) and order setup cost $K = \$5000$. They all have the same annual inventory holding cost $C_h = \$40$. Shortages are backordered, but an accurate estimate of backorder cost is not available. Hence, to keep backorders under control, W&D has limited the probability of having a backorder customer to less than 5 percent. In other words, W&D has been using a $(Q, R)$ policy that achieves the cycle-service level $\alpha = 95\%$.

*Proposed System with One Superstore:* The superstore will buy S-200 from the same manufacturer with the same lead time $L = 20$ days at the same price $c = \$160$ and order setup

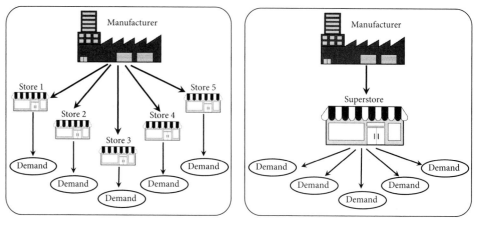

Figure 11.1    *Left:* Current system with five stores. *Right:* Proposed system with one superstore.

cost $K = \$5000$ (see Fig. 11.1–right). The superstore will also have the same annual inventory holding cost $C_h = \$40$. Since there is no major competition, W&D expects that the demand of all five stores would go to the superstore. The managers would like to maintain the same cycle-service level $\alpha = 95\%$ at the superstore.

Note that in the proposed system, all of W&D's inventory is physically pooled at the superstore. To study the impact of inventory pooling of S-200 window on W&D's costs, we focus on answering the following questions: (i) whether or how much does pooling inventory at the superstore save inventory cost? and (ii) What factors impact this saving?

Does pooling inventory of several locations into one location save inventory costs? The answer is yes. The underlying reason for the saving comes from "Demand Aggregation" for S-200 windows—*the demand at the superstore is the aggregated demand of all five stores.* Figure 11.2 shows the demand distribution in each of the five stores and the aggregated demand distribution at the superstore.

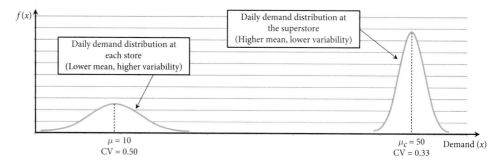

Figure 11.2    Daily demand distribution for S-200 windows at each store and at the superstore.

As the figure shows, demand aggregation results in the following two advantages for the superstore:

- *Economies of Scale:* The mean of the aggregated daily demand at the superstore (i.e., $\mu_c = 50 = 5 \times 10$) much larger than the mean daily demand at each one of the five stores (i.e., $\mu = 10$).

- *Variability Reduction:* The daily demand at the superstore has a lower variability than that of the five stores. As discussed in Chap. 5, the variability of the aggregated demand at the superstore (i.e., CV = 0.33) would be smaller than the variability of the demand at each of the five stores (i.e., CV = 0.5).

How do economies of scale and variability reduction help W&D improve its inventory (e.g., reduce inventory cost or increase service level)? That is what we discuss in the following section. We first start with Economies of Scale.

### 11.6.1 Pooling Inventory—Economies of Scale

We first compare W&D's cycle inventory in both the current and the proposed systems.

**Saving in Cycle Inventory**
According to analytics in Sec. 10.6.1 of Chap. 10, the optimal order quantity at each store can be approximated using the EOQ formula. Considering 30 days in a month, the average annual demand at each store is $\overline{D} = 10 \times 30 \times 12 = 3600$. Hence,

$$Q^* = \sqrt{\frac{2K\overline{D}}{C_h}} = \sqrt{\frac{2(5000)(3600)}{40}} = 948.7 \simeq 949 \tag{11.4}$$

Thus, each store carries an average cycle inventory of $I_{\text{cycle}} = Q^*/2 = 949/2 = 474.5$ and the total average cycle inventory of the firm in its $n = 5$ stores is

$$\underline{\text{Current System}} \implies \text{Total Cycle Inventory} = n \times I_{\text{cycle}} = 5 \times 474.5 = 2372.5$$

Considering the annual inventory holding cost $C_h = \$40$, the total annual holding cost of the cycle inventory is

$$\underline{\text{Current System}} \implies \text{Average Annual Cycle Inventory Holding Cost}$$
$$= C_h\, I_{\text{cycle}} = (\$40)\, 2372.5 = \$94,900$$

What is the annual holding cost of the cycle inventory at the superstore? Note that the superstore faces an average annual demand $\overline{D}_c = n \times \overline{D}$, where $\overline{D}$ is the average annual demand in each small store. Thus, its optimal order quantity would be

$$Q_c^* = \sqrt{\frac{2K(n\overline{D})}{C_h}} = \sqrt{n}\left(\sqrt{\frac{2K\overline{D}}{C_h}}\right) = \sqrt{n}\, Q^* = \sqrt{5}\,(948.7) = 2121.3 \simeq 2121$$

resulting in an average cycle inventory

$$\underline{\text{Superstore}} \implies I_{\text{cycle}} = \frac{Q_c^*}{2} = \frac{2121}{2} = 1060.5$$

In other words, while the average annual demand at the superstore is $n = 5$ times larger than that at each store, the optimal order quantity and the average cycle inventory at the superstore are only $\sqrt{n} = \sqrt{5} = 2.24$ times larger than that in each store. This is the underlying reason for economies of scale advantage of the superstore. Specifically, considering that the superstore sells an average of $\overline{D}_c = 5 \times 3600 = 18,000$ S-200 windows, and considering the annual holding cost $C_h = \$40$, W&D incurs a cycle inventory holding cost per sold item at the superstore as follows:

$$\text{Average cycle inventory holding cost } \textit{per sold unit} = \frac{C_h I_{\text{cycle}}}{\overline{D}_c} = \frac{40(1060.5)}{18,000} = \$2.36$$

However, in the current system, the firm incurs an average cycle inventory holding cost of $5.27 for each item sold in each of its five stores, that is,

$$\text{Average cycle inventory holding cost } per \text{ sold unit} = \frac{C_h I_{\text{cycle}}}{\overline{D}_c} = \frac{40(2372.5)}{18{,}000} = \$5.27$$

In other words, by pooling inventory of its five stores into one location—the superstore— W&D would have a smaller cycle inventory holding cost per sold unit, that is, Economies of Scale.

Note that while the saving in cycle inventory holding cost is small per sold item, the reduction in average annual holding cost is large. Specifically, considering the average annual demand of $5 \times 3600 = 18{,}000$ for S-200 windows, the total average annual holding cost of the cycle inventory in current system is $\$5.27(18{,}000) = \$94{,}860$, while at the superstore it will be $\$2.36(18{,}000) = \$42{,}480$, which is $\$94{,}860 - \$42{,}480 = \$52{,}380$ lower than that in the current system.

### Saving in Ordering Cost

Every time each of the five stores or the superstore places an order, W&D incurs a fixed order setup cost $K = \$5000$, regardless of the order size. Does pooling inventory into one location also save the ordering cost?

Before pooling inventory in one location, each of $n = 5$ stores places an order of size $Q^* = 949$, for which it pays the order setup cost $K = \$5000$. Thus,

$$\text{Average order setup cost } per \text{ sold unit} = \frac{K}{Q^*} = \frac{5000}{949} = \$5.27$$

In the proposed system, the superstore orders $Q_c^* = \sqrt{n}\, Q^* = 2121$, for which it pays the order setup cost $K = \$5000$. Thus, for the proposed system

$$\text{Average order setup cost } per \text{ sold unit} = \frac{K}{\sqrt{n}Q^*} = \frac{1}{\sqrt{n}}\left(\frac{K}{Q^*}\right) = \frac{1}{\sqrt{5}}(5.27) = 2.36$$

So, while the average demand in the superstore is larger than that in each store by a factor of 5, the order setup cost per sold unit at the superstore is smaller by a factor of $\sqrt{5}$. This again points to the economies of scale. Specifically, the average order setup cost per sold unit is smaller in the larger system facing larger demand (i.e., the superstore) than in smaller systems (i.e., each of the five stores).

An interesting observation is that in the current system the average cycle inventory holding cost per sold unit (i.e., $5.27) is the same as the average order setup cost per sold unit (i.e., $5.27). This is also true for the proposed system where the average cycle inventory holding cost per sold unit (i.e., $2.36) at the superstore is the same as the average order setup cost per sold unit (i.e., $2.36). The reason is that both systems are placing their orders optimally according to the EOQ formula. As we mentioned in Sec. 9.9.1 of Chap. 9, if ordered optimally, the annual order setup cost and the annual (cycle) inventory holding cost are equal. This implies that the total average annual order setup cost in the current system is $94,860, while in the proposed system it would be $42,380, resulting in an additional saving of $94,860 − $42,480 = $52,380 due to inventory pooling.

### Possible Saving in Purchasing Costs

Each of the five stores in the current system orders $Q^* = 949$ units from the same manufacturer at the unit price $c = \$160$. The superstore, on the other hand, would place a larger order of size $Q_c^* = 2121$. In some cases where the manufacturer offers quantity discount (see Sec. 9.11 of Chap. 9), the superstore might be able to get a discount on purchase price $c = \$160$, because its order size is more than twice larger than that of each of the five stores. Hence, larger order size may result in smaller purchase cost per unit sold—Economies of Scale.

As the above cases show, by pooling inventory of several locations into one location, firms can reduce their inventory costs.

### Pooling Inventory—Economies of Scale

When the inventory at several locations is pooled into one location, then the pooled location benefits from economies of scale. Specifically, under the same holding and order setup cost structure, the pooled system will incur lower cycle inventory holding cost and lower order setup cost per unit demand than those in unpooled system.

The economies of scale advantage of the pooled location comes from the relationship between inventory holding cost and order setup cost and the annual demand in the EOQ formula in Eq. (9.2). If ordered optimally according to the EOQ formula, then while the larger (pooled) system has $n$ times larger demand, its average cycle inventory and the order setup cost per sold unit is smaller than those in the unpooled system by a factor of $\sqrt{n}$.

### 11.6.2  Pooling Inventory—Variability Reduction

While the saving in cycle inventory and order setup cost originated from economies of scale, saving in safety inventory comes from lower variability due to demand aggregation. Specifically, as we discussed in Sec. 5.6.1 of Chap. 5, demand variability measured by the coefficient of variation is lower when the demands of several items are aggregated. This has a direct impact on safety inventory, as we show below.

### Saving in Safety Inventory

Besides savings in cycle inventory and order setup costs, pooling inventory also results in saving in safety inventory cost. The smaller safety inventory in the pooled system is due to lower variability in the aggregated demand distribution (as shown in Fig. 11.2).

To show this, we first compute the safety inventory needed in each of the five stores to maintain cycle-service level $\alpha = 0.95$. Note that demand during lead time at each of the five stores is Normally distributed with mean $\overline{D}_L = L\,\mu = 20 \times 10 = 200$ and standard deviation $\sigma_L = \sqrt{L}\,\sigma = \sqrt{20} \times 5 = 22.36$. Since for $\Phi(z) = 0.95$, we have $z = 1.645$, then the safety inventory at each of the five stores would be $I_{\text{safety}} = z\,\sigma_L = 1.645(22.36) = 36.78$, which we round up to 37. Then the total safety inventory at all the five stores to maintain cycle-service level $\alpha = 0.95$ is

$$\underline{\text{Current System}} \quad \Longrightarrow \quad \text{Total safety inventory} = n \times I_{\text{safety}} = 5(37) = 185$$

But, what about the safety inventory if W&D pools the inventory of its five stores into one superstore? Using Eq. (5.3), and considering $\rho_{ij} = 0.3$, we can compute the standard deviation of the aggregated daily demand at the superstore as follows:

$$\sigma_c = \sigma \times \sqrt{n + 2\sum_{j>i} \rho_{ij}} = 5 \times \sqrt{5 + 2\sum_{j>i} 0.3} = 16.58$$

Recall that the order lead time is $L = 20$ days. Hence, the standard deviation of demand during lead time at the superstore is

$$\sigma_L = \sqrt{L} \times \sigma_c = \sqrt{20}\,(16.58) = 74.15$$

Hence, the safety inventory at the superstore to maintain cycle-service level $\alpha = 0.95$ can be computed as follows:

$$\text{Superstore} \quad \Longrightarrow \quad I_{\text{safety}} = z\,\sigma_L = (1.645)(74.15) = 122$$

which is lower than the safety inventory in the current system. Why? Because of variability reduction due to demand aggregation. Note that the daily demand at each store in the current system has a coefficient of variation $CV = 5/10 = 0.5$. However, aggregated daily demand in the pooled system—the superstore—has coefficient of variation $CV_c = 16.58/50 = 0.33$. Thus, through demand aggregation, the superstore faces a lower variability in its daily demand, resulting in a lower safety inventory to maintain the service level of $\alpha = 0.95$.

This leads us to the following principle:

---

**PRINCIPLE**

**Pooling Inventory—Safety Inventory and Service Level**

Systems with pooled inventory can offer the same service level as that of an unpooled system, but with less safety inventory; or said another way, for any given safety inventory, systems with pooled inventory offer larger inventory service levels (e.g., larger cycle-service level or fill rate) than those of the systems in which the inventory is not pooled.

---

Considering the annual inventory holding cost $C_h = \$40$, the annual saving in safety inventory holding cost would be $\$40(185 - 122) = \$2520$.

### Impact of Pooling on Inventory Turn

One concluding remark is that, since the superstore serves the same demand (i.e., has the same throughput) as that of the five stores, but with lower average (cycle and safety) inventory, then the inventory turn for the S-200 windows is higher in the superstore than that in each of the five stores. This is another potential benefit of inventory pooling.

### 11.6.3  Which Systems Benefit More from Inventory Pooling?

By pooling the inventory of five stores into one superstore, W&D can reduce its total average annual inventory holding and order setup cost by

$$\text{Total average annual saving} = \$52{,}380 + \$52{,}380 + \$2520 = \$107{,}280$$

One interesting question is, how this saving is affected by W&D's cost structure, its daily demand, and its supplier's lead time?

To gain insights into this, we derive expressions for cost savings per demand (i.e., per unit sold) when the inventory of the $n$ stores is pooled into one location. This is shown in the following analytics.

---

## ANALYTICS

### Pooling Inventory—Cost Saving per Unit Demand

Consider $n$ locations, each having an average annual demand of $\overline{D}$. The demand in each day at each location for a single item is independent of the demand at the other days and follows a Normal distribution with mean $\mu$ and standard deviation $\sigma$. Also, there is a correlation $\rho_{ij}$ between the daily demand at locations $i$ and $j$, where $i, j = 1, 2, \ldots, n$. Each location has an annual inventory holding cost $C_h$ and orders from the same manufacturer with order setup cost $K$ and order lead time $L$ days. Each location maintains a cycle-service level $\alpha$ with minimum cost.

Now suppose that the inventory of all $n$ locations are pooled into one central location, serving the aggregated demand of all $n$ locations. The central location orders from the same manufacturer with the same order setup cost $K$ and order lead time $L$.

*Saving in Cycle Inventory Holding and Order Setup Costs:* Pooling the inventory of the $n$ locations into one central location results in savings in order setup cost as well as in cycle inventory holding cost. The total cost saving per unit demand (see Appendix H for details) is

$$\text{Total average saving in } \textit{order setup} \text{ and } \textit{cycle inventory} \text{ holding costs per unit demand} = \left( \sqrt{\frac{2C_h \, K}{\overline{D}}} \right) \left( 1 - \frac{1}{\sqrt{n}} \right) \tag{11.5}$$

*Saving in Safety Inventory Holding Cost:* Pooling inventory also results in saving in safety inventory holding cost as follows:

$$\text{Average saving in } \textit{safety inventory} \text{ holding cost per unit demand} = \left( \frac{C_h}{d_o} \right) z_\alpha \sqrt{L} \, \text{CV} \left( 1 - \frac{\text{CV}_c}{\text{CV}} \right) \tag{11.6}$$

where $\text{CV} = \sigma / \mu$ is the coefficient of variation of daily demand at each location and $\text{CV}_c$ is the coefficient of variation of daily demand in the central location, where

$$\text{CV}_c = \frac{\sigma \sqrt{n + 2 \sum_{j>i} \rho_{ij}}}{n\mu}$$

and $d_o$ is the number of days in a year.

The above analytics provides several useful insights about how the benefit (i.e., cost saving) of pooling is affected by demand and cost structure of a system.

- *Impact of Number of Locations:* As Eqs. (11.5) and (11.6) show the cost reduction in cycle inventory, order setup cost, and safety inventory increases with the number of pooled location $n$. In other words, the more the inventory locations are pooled, the more saving a firm can get by pooling inventory.

- *Impact of Holding Cost:* Both Eqs. (11.5) and (11.6) show that the saving in cycle and safety inventory holding cost increases with annual inventory holding cost $C_h$.

Hence, pooling inventory of more expensive items (i.e., having higher inventory holding cost) results in higher cost saving per unit demand.

- *Impact of Order Setup Cost:* Equation (11.5) shows that the saving in both cycle inventory holding cost and order setup cost increases with order setup cost $K$. Therefore, pooling inventory of items with larger order setup costs (e.g., larger transportation cost) results in higher cost saving per unit demand.

- *Impact of Service Level:* The cost saving in safety inventory, as shown by Eq. (11.6), increases with $z_\alpha$. Note that $z_\alpha$ increases with cycle-service level $\alpha$. Hence, pooling inventory of items with higher cycle-service levels results in more cost saving per unit demand.

- *Impact of Lead Time:* The cost saving in safety inventory also increases with the supplier's lead time $(L)$. Thus, pooling items with higher lead times results in more cost saving in safety inventory.

- *Impact of Variability:* The variability reduction due to demand aggregation (i.e., $CV_c/CV$) also impacts the cost saving, as shown in Eq. (11.6). Specifically, systems benefit more from pooling inventories of items for which demand aggregation results in a higher variability reduction.

- *Impact of Demand Correlation:* Increase in $\sum_{j>i} \rho_{ij}$ results in increase in $CV_c$, that is, the variability of the aggregated demand. This results in decreasing the cost saving in safety inventory as shown in Eq. (11.6). Hence, as the daily demand in $n$ locations becomes more negatively correlated (i.e., $\sum_{j>i} \rho_{ij}$ becomes smaller), the variability of the aggregated demand becomes smaller and thus the cost saving increases. Negatively correlated demand implies that increase in demand in one location is associated with decrease in demand in another location (e.g., customers of one location may buy from another location).

We summarize the above observations into the following principle:

---

**PRINCIPLE**

**Pooling Inventory—When Is the Benefit of Pooling High?**

Benefit of inventory pooling is higher in systems with more expensive products (i.e., higher inventory holding cost), higher order setup costs, higher lead times, higher service levels, higher variability (coefficient of variation) in demand and with more negative correlation among demands. The benefit also increases as more number of inventory locations are pooled.

### 11.6.4  Pooling Inventory—Diminishing Return

While pooling inventory of more locations results in lower costs per unit sold, there is a limit on the cost saving achieved by pooling. To illustrate this, let's return to our example of W&D. We found that by pooling inventory of its five locations, W&D saves a total of $107,280. Considering 30 days in a month, the total average demand (unit sold) of the five stores in a year would be $5 \times (10 \times 30 \times 12) = 18,000$. Hence, the average cost saving per unit sold is $107,280/18,000 = \$5.96$. We know that the more the number of stores pooled, the higher this cost saving would be. How many stores should W&D pool to double this cost saving from $5.96 to about $12.00?

To answer this question, we need to repeat our calculation and compute the cost saving per unit sold for all different number of pooled locations, that is, for different values of $n$.

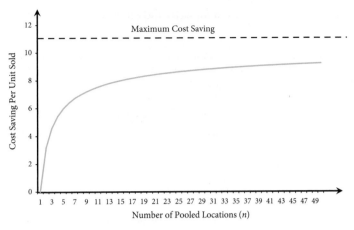

Figure 11.3    Cost saving per unit sold for S-200 windows by pooling *n* locations.

This is shown in Fig. 11.3 for values of $n = 1, 2, \ldots, 50$ locations. All locations are assumed to have the same demand distributions and correlation coefficients as those of the five locations.

Figure 11.3 provides some interesting insights about the benefits of pooling in reducing inventory costs:

- *Maximum Cost Saving:* There is a maximum limit in the cost saving achieved by pooling locations. As the figure shows, as we increase the number of locations, the cost saving per unit sold converges to a maximum limit. This maximum limit is different in different cases. For our W&D example with the goal to maintain cycle-service level $\alpha$, we can obtain the maximum cost saving by letting $n$ go to infinity in Eqs. (11.5) and (11.6) and summing them up. Assuming 30 days in a month, that is, $d_o = 360$, this results in[3]:

  Maximum average cost saving per unit sold

  $$= \sqrt{\frac{2C_h\,K}{\overline{D}} + \left(\frac{C_h}{d_o}\right)}\,z_\alpha\sqrt{L}\;(\mathrm{CV})$$
  $$= \sqrt{\frac{2(40)(5000)}{3600} + \left(\frac{40}{360}\right)}(1.645)\sqrt{20}\left(\frac{5}{10}\right)$$
  $$= \$10.95$$

  So, W&D cannot get the cost saving of $12.00 by pooling the inventory of any number of its stores, since the maximum cost saving per unit sold is $10.95, as shown above.

- *Diminishing Return:* Figure 11.3 also shows that the increase in cost saving as the number of locations increases has a diminishing return. In other words, adding a new location to the current pool results in a smaller cost saving than that when the last location was added to the pool. Diminishing return has an important implication: By pooling a small number of locations, one could get a large portion of the maximum cost saving.

---

[3]Note that as $n$ gets larger and goes to infinity, the variability of the aggregated demand (i.e., $CV_c$) becomes smaller and approaches zero.

We summarize the above observations into the following principle:

**Pooling Inventory—Diminishing Return**

There is a limit on how much inventory pooling can improve costs. The benefit of pooling more locations has diminishing return. Hence, a large portion of the maximum benefit may be achieved by pooling a small number of locations.

### 11.6.5 Pooling Inventory—Offering More Variety

Another advantage of pooling inventory is that it can make carrying the items with low demand profitable. This allows firms to increase the number of items that they carry and thus provide a wide variety of products. Consider, for example, another product, that is, S-500 window that W&D sells in its stores. For sake of presentation, suppose that S-500 has the same cost and lead time as those for S-200 window, except for one: Daily demand for S-500 windows is well-approximated by a Normal distribution with mean of 7 per month and standard deviation 3.5.

S-200 window that has a large demand is called a *fast-moving* product and S-500 window that has a small demand is called *slow-moving* product. To study the impact of inventory pooling on slow-moving products, we would now compare the net profit of the product.

*S-200 Windows—The Fast-Moving Product:* Recall that each S-200 window was bought for $160 and sold for $220. Also, before pooling inventory, the total average cycle inventory and order setup cost per sold unit at each store was $5.27 + $5.27 = $10.54. Carrying a safety inventory of 37, the safety inventory holding cost per unit sold at each store is $40(37)/3600 = $0.41. Hence, the net profit of selling a S-200 window at each store is $220 − $160 − $10 − ($10.54 + $0.41) = $39.05 (recall that W&D had an overhead cost of $10 per sold window). Thus, without pooling inventory the firm can still make a profit of about $39 per sold S-200 window.

*S-500 Windows—The Slow-Moving Product:* For S-500 windows with average annual demand of $\overline{D} = 7 \times 12 = 84$, the optimal order quantity at each store is $Q^* = \sqrt{2K\overline{D}/C_h} = \sqrt{2(5000)84/40} \simeq 145$. Hence, the average cycle inventory per unit sold is $(\$40(145/2))/84 = \$34.52$. The order setup cost per sold unit is also $34.52. Thus, the total cycle inventory and order setup cost per unit sold is $34.52 + $34.52 = 69.04. The safety inventory can be found to be one unit that results in the safety inventory holding cost per unit sold of $40(1)/84 = $0.48. Consequently, with the selling price of $220, the net profit of selling an S-500 window at each store is $220 − $160 − $10 − $69.04 − $0.48 = −$19.52. In other words, with an average demand of seven windows in a month, each store loses about $19.52 for each S-500 window that it sells. So, it might be in W&D's best interest to stop selling S-500 windows—its slow-moving product.

Now consider pooling inventory of S-500. Figure 11.4 represents the cost saving from pooling the inventory of S-500 windows of $n$ locations. As the figure shows, pooling five locations results in cost saving of about $38 per unit sold. This will turn the loss of −$19.52 to profit of $38 − $19.52 = $18.48 and make S-500 windows profitable for W&D. Hence, while before pooling inventory it is not profitable to offer S-500 windows, after pooling inventory at the superstore, this slow-moving product becomes profitable for W&D.

The above example explains why online retailers—also called e-tailers—such as Amazon that pools all its inventory into a few locations can offer a wider range of products (including slow-moving products) than traditional retailers such as Target, Home Depot,

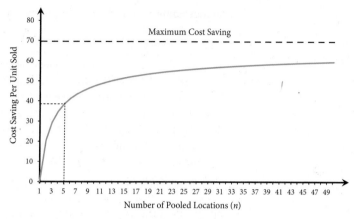

**Figure 11.4    Cost saving per unit sold for S-500 windows by pooling *n* locations.**

Barns & Noble, and Gap. This is also the reason why these retailers started offering some of their products (i.e., products with low demand) only through their websites.

We summarize the above observations into the following principle:

**Pooling Inventory—Slow Moving Products and Large Variety**

Pooling inventory reduces the inventory costs per unit sold. Hence, pooling inventory can make slow-moving products (i.e., products with low demand) profitable, which allows firms to offer more variety of products.

### 11.6.6  To Pool or Not to Pool?

Pooling inventory into a central location provides several benefits. First, centralized systems have lower overhead costs such as costs of facilities, building, and employees per unit sold. Second, with respect to inventory management costs, physical pooling of inventory into one location results in lower cycle inventory, lower order setup, and lower safety inventory costs per unit sold. Third, it allows firms to offer a wide range of products, including slow-moving products. One question that comes up, however, is the following: If inventory pooling has all these advantages, why firms are not pooling all their inventories into one location?

Well, the answer is that there are also some disadvantages of pooling inventories of several locations into one, some of which are as follows:

- *Losing Market Share:* In our example of W&D, we assumed that all customers of the five stores will go to superstore after those stores are closed. However, this may not always be the case. After closing the five stores, some customers who were close to those stores may decide not to go to the superstore, because it is far from them. This may result in losing customers to the competitor who has stores closer to those customers.

- *Response Time:* Suppose that the five stores in our W&D example are Distribution Centers (DCs) of an online retailer that buys S-200 windows from the manufacturer, keeps them in the five DCs, and ships them to the customers. While in this case the firm may not lose market share, it may degrade its response time to customer orders. Specifically, since DCs in the decentralized system are much closer to most customers than the large DC in centralized system, it would take the firm a longer time to ship the product to the customers from centralized DC than from five decentralized DCs. For example, the firm may not be able to offer next-day or 2-day delivery guarantee any more.

- *Transportation Cost:* Having a longer distance between the pooled inventory (e.g., centralized DC) and customers results in higher cost of shipping customer orders. It is important to mention, however, that the per-unit transportation cost of items from the supplier/manufacturer to the centralized location is lower than that in the decentralized system. Hence, the impact of transportation cost in pooling location depends on the location of the supplier as well as inventory locations in decentralized and centralized systems.

- *Risk of Disruption:* Natural disasters (e.g., earthquakes, tsunami), accidents (e.g., fire), and political incidents can cause serious disruptions in supply chains. Pooling all inventory into one large warehouse increases the risk of disruption. If the operations in the warehouse is disrupted (e.g., due to flood), then the entire supply chain will be disrupted. To protect a supply chain from serious and costly disruptions, firms often hold inventory in different locations and use multiple suppliers.

The diminishing return principle of pooling as well as the above factors prevent firms from pooling all of their inventories into one location. The optimal strategy is to pool small number of locations that are carefully chosen such that a firm does not lose market share and is able to provide a reasonable response time. For example, while Amazon started with one large warehouse (distribution center) to serve all its customers, it now has several warehouses that are strategically located throughout the United States to better serve its customers.

## 11.7 Aggregate Demand—Centralized Distribution System

While not benefiting from inventory pooling, decentralized inventory systems have their own advantages, for example, being closer to the customers, maintaining market share, having faster response time, and smaller outbound transportation cost. But, is it possible to have the benefits of inventory pooling without pooling inventory to one location? In the case of W&D, is it possible that the firm keeps its stores open but still benefits from economies of scale and variability reduction?

Yes. There are several strategies that W&D can take to achieve that. We discuss some of these strategies in this and in the following sections. One strategy is to open a Distribution Center (DC) that orders S-200 windows for all the five stores from the manufacturer. Each store would then order from the DC (instead of ordering from the manufacturer) and sell them at the store. To guarantee that DC always has items to satisfy stores' orders, DC will work in high cycle-service level $\alpha = 0.99$. Currently there are three locations that the firm is considering to open its DC: Arizona, Kansas, and Virginia. The DC will be used to order several items (not just S-200 windows) from the manufacturer in California and distribute them to its stores in New York. Figure 11.5 shows these options.

In the figure

$$K_c = \text{Order setup cost of DC to deliver to a store}$$
$$L_c = \text{Order lead time of DC to deliver to a store}$$
$$K_m = \text{Order setup cost of the manufacturer to deliver to DC}$$
$$L_m = \text{Order lead time of the manufacturer to deliver to DC}$$

Table 11.3 shows the order setup cost and order lead time for the three locations.

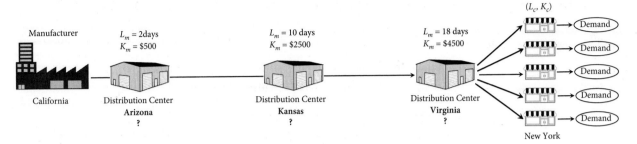

Figure 11.5   Three alternative locations to establish a distribution center.

Notice that locations that are closer to the stores in New York (e.g., Virginia) have smaller $K_c$ and $L_c$, because it takes less time with less cost for DC to deliver orders to stores. In contrast, locations that are closer to the stores in New York have larger $K_m$ and $L_m$, because they are farther away from the manufacturer in California.

Table 11.3   Order Setup Costs and Order Lead Times at Three Alternative Locations

| Location | $L_m$ | $K_m$ | $L_c$ | $K_c$ |
|---|---|---|---|---|
| Virginia | 18 | 4500 | 2 | 500 |
| Kansas | 10 | 2500 | 10 | 2500 |
| Arizona | 2 | 500 | 18 | 4500 |

Of course, opening a DC is a strategic decision that depends on several factors such as cost of establishing the DC and its overall impact on the supply chain. Here we only focus on one factor, that is, saving in inventory costs, and we use only one item to illustrate this approach, that is, S-200 window.

**Saving in Cycle Inventory Holding Cost**
Recall from Sec. 11.6.2 that the decentralized system without DC has total cycle inventory cost per sold unit of $\$94,900/(5 \times 3600) = \$5.27$. Here we compute this number if the firm opens a DC in one of the locations. In the system with DC, the cycle inventory is kept in all $n$ locations and in DC. Each location orders $Q_c^*$ from DC with order setup cost $K_c$ and DC orders $Q_m^*$ from the manufacturer with order setup cost $K_m$. Hence, the total average cycle inventory at the system with DC is

$$I_{cycle} = n\left(\frac{1}{2}\right)Q_c^* + \left(\frac{1}{2}\right)Q_m^* = n\left(\frac{1}{2}\right)\sqrt{\frac{2K_c\overline{D}}{C_h}} + \left(\frac{1}{2}\right)\sqrt{\frac{2K_m(n\overline{D})}{C_h}} \qquad (11.7)$$

Assuming that DC has the same inventory holding cost as stores, the total cycle inventory cost if the DC is opened in Arizona, for example, will be

$$\text{Total Cycle Inventory } = 5\left(\frac{1}{2}\right)\sqrt{\frac{2(4500)(3600)}{40}} + \left(\frac{1}{2}\right)\sqrt{\frac{2(500)(5)(3600)}{40}} = 2585.4$$

Considering the total annual demand of $5 \times 600 = 18,000$ and annual inventory holding cost $C_h = \$40$, the total cycle inventory holding cost per sold unit is

$$\text{Total cycle inventory holding cost } per\ sold\ unit = \frac{\$40(2585.4)}{18,000} = \$5.74$$

Thus, if DC is opened in Arizona, then the saving in cycle inventory holding cost would be

$$\text{Saving in cycle inventory holding cost } \textit{per sold unit} = \$5.27 - \$5.74 = -\$0.47$$

### Saving in Order Setup Cost

Recall from Section 11.6.2 that the decentralized system without DC has total order setup cost per sold unit of $5.27, which is equal to the total cycle inventory holding cost per sold unit (since stores are ordering optimally according to the EOQ model).

Also, in the system with DC, since both stores and DC order optimally according to the EOQ formula, then their annual cycle inventory holding cost would be the same as their annual order setup cost. Therefore, the saving in order setup cost per sold unit would be the same as the saving in cycle inventory holding cost per sold unit. Hence, if a DC is opened in Arizona, we have

$$\text{Saving in order setup cost } \textit{per sold unit} = \$5.27 - \$5.74 = -\$0.47$$

### Saving in Safety Inventory Holding Cost

We have already computed in Sec. 11.6.3 that the total safety inventory in the decentralized system without DC is 185 units. Considering holding cost $C_h = \$40$ and total demand of 18,000, the total safety inventory holding cost per sold unit is

$$\text{Total safety inventory holding cost } \textit{per sold unit} = \frac{\$40(185)}{18,000} = \$0.41$$

For the system with DC, the safety inventory is kept in all $n$ locations and in DC. Considering order lead time $L_c$, standard deviation of daily demand at each store $\sigma = 5$, and cycle-service level $\alpha = 0.95$, the safety inventory in each location that orders from DC is $z_\alpha \sqrt{L_c}\sigma$: If a DC is opened in Arizona that has $L_c = 18$ days, the safety inventory in each store will be[4]

$$z_\alpha \sqrt{L_c}\sigma = (1.645)\sqrt{18}\,(5) = 34.89$$

which we round up to 35. Recall from Sec. 11.6.3 that the aggregated daily demand of the five stores—which is now the daily demand at DC—has mean 50 and standard deviation $\sigma_c = 16.58$, and thus coefficient of variation $CV_c = 16.58/50 = 0.33$. If the DC is opened in Arizona and operates with cycle-service level $\alpha = 0.99$, the safety inventory at DC, which has order lead time $L_m = 2$ days is, therefore,

$$z_\alpha \sqrt{L_m}\sigma_c = (2.326)\sqrt{2}(16.58) = 54.54$$

which we round up to 55. Thus, considering the inventory holding cost $C_h = \$40$ and total demand of 18,000, the total safety inventory holding cost per sold unit in the system with DC is

$$\text{Total safety inventory holding cost } \textit{per sold unit} = \frac{40(5 \times 35 + 55)}{18,000} = \$0.51 \qquad (11.8)$$

Therefore, the saving in holding cost of safety inventory per sold unit if DC is opened in Arizona is

$$\text{Saving in safety inventory holding cost } \textit{per sold unit} = \$0.41 - \$0.51 = -\$0.10$$

---

[4]Because of the high cycle-service level $\alpha = 0.99$ at DC, we are making the assumption that items are always available at DC to be shipped to stores. This simplifies the computation of safety inventory at each store.

**Total Saving**

In summary, if W&D opens a DC in Arizona, its cycle inventory holding cost per sold unit increases by $0.47, its order setup cost per sold unit increases by $0.47, and its safety inventory holding cost increases by $0.10. Thus, overall, W&D will lose a total of $0.47 + $0.47 + $0.10 = $1.04 per sold unit in inventory-related costs if it opens a DC in Arizona. Hence, opening a DC in Arizona to consolidate orders from the five stores in New York is not beneficial.

This points to the following important observation. As opposed to aggregating demand through inventory pooling that always results in a decrease in inventory costs, aggregating demand through a DC does not always reduce costs. To further illustrate this, let's check whether opening a DC in Kansas or Virginia can reduce costs. Repeating our analysis for Kansas we find that opening a DC in Kansas results in an increase of $0.42 in inventory-related costs per sold unit. Opening a DC in Virginia, in contrast, saves W&D about $2.63 in inventory-related costs per sold unit. Hence, from inventory holding and order setup cost's perspective, opening a DC in Virginia is a profitable option.

But what is it that makes Virginia a profitable option? The answer is that Virginia is closer to the stores in New York than the other two options, and thus stores have shorter lead time and smaller order setup costs when ordering from Virginia.

To understand this, consider the extreme case in which DC is so close to stores that the order setup cost $(K_c)$ and order lead times $(L_c)$ are zero, that is, stores can get an item from the DC in no time and with zero setup cost. In this case, stores do not keep any inventory and directly use items at the DC. This system with DC is then equivalent to pooling all inventory in one location. We have already shown that pooling inventory in one location always saves cycle and safety inventory cost as well as order setup cost. As we move the DC away from the stores, the savings in inventory costs decrease because we now are keeping inventory in both stores and DC. If the DC is moved further away from the stores, there will be a point where the savings become negative and opening a DC results in increase in inventory-related costs. That is why opening a DC in Virginia (which is close to New York) saves $2.63 in inventory costs per sold unit, while opening a DC in Kansas (which is farther away from New York) results in loss of $0.42 per unit sold. The DC in Arizona has the most loss, that is, $1.04, because it is the farthest location from the stores.

One important point to emphasize here is that aggregating demand through a DC does not always result in variability reduction. In other words, the demand at the DC, which is the total demand of all the five stores, may have a variability (coefficient of variation) larger than the variability of demand at each store. This is called Bullwhip Effect. There are several reasons for bullwhip effect which we discuss in detail in Sec. 11.9. In our analysis of the W&D case, we assumed that the bullwhip effect did not exist.

Finally, while aggregating demand through a DC may not always result in saving in inventory costs, when it does save inventory cost, the saving is larger in systems with larger number of stores $(n)$, higher inventory holding cost $(C_h)$, and lower demand variability at DC (i.e., lower $CV_c$).

We summarize the above observations in the following principles:

## PRINCIPLE

### When Is the Benefit of a Centralized Distribution High?

- Consolidating orders of $n$ inventory locations through a centralized Distribution Center (DC) does not always result in inventory holding and order setup cost reduction. The closer the DC to the inventory locations, the higher the benefit of having the DC.

- If having a centralized distribution results in cost saving, then, in general, the cost saving is higher in systems with more expensive products (i.e., higher holding cost), higher order setup costs ($K$), higher lead time ($L$), higher service levels, higher variability (coefficient of variation) in demand at each location and with more negative correlation among demands. The benefit also increases as more locations receive their items from the centralized DC.

In summary, one of the disadvantages of closing $n$ locations (e.g., several stores) and pooling their inventories into a central location (e.g., a superstore) is that it moves product away from the customers and may result in losing customers. It also increases the response time to customer orders and the transportation cost of shipping items to the customers. By keeping the inventory locations and opening a DC, a firm can avoid these disadvantages and still capture some benefits of inventory pooling. The closer the DC to the locations, the higher the benefit of inventory pooling captured by the DC. Of course, additional costs of establishing and operating the DC should be considered along with its reduction in inventory costs.

Before we discuss other ways of aggregating demand, the following points are worth mentioning here:

- While we used cycle-service level as the inventory management goal in the case of W&D, the fundamental insights obtained also hold when the goal is to set a fill rate or to minimize costs.

- In our example, the centralized distribution center (DC) was only used to consolidate retailers' orders. The DC, however, can also be used for pooling inventory of some items. Specifically, the firm can pool the inventory of its slow-moving products at the DC, stop offering those items at retail stores, and only offer them online.

## 11.8 Other Strategies for Aggregating Demand

Physically moving items from several locations to one centralized store or consolidating the orders of all locations through a central Distribution Center (DC) are two different ways of aggregating demand. As we mentioned, demand aggregation can utilize economies of scale as well as variability reduction, both resulting in saving in inventory costs.

But, there are several other ways that firms can utilize the benefits of demand aggregation: (i) Virtual Pooling, (iii) Specialization, (iii) Substitution, and (iv) Postponement. We discuss these in detail in this section.

### 11.8.1 Virtual Inventory Pooling

One disadvantage of decentralized inventory is that a customer in one location—Location A—can face shortage while the product is available in another location—Location B. Inventory Pooling prevents this by pooling the inventories of all locations into one. But, what if: (i) the customer who faces shortage at Location A is willing to go to Location B to get the product, or (ii) the customer is willing to wait for Location B to ship the product.

If one or both of the above conditions hold, the inventory at each of the locations can be used to satisfy the demand of all other locations. Hence, while the inventory is not physically centralized, the system still benefits from inventory pooling implications such as reduction in inventory. This is called *Virtual Centralization of Inventory* or *Virtual Pooling*, see Fig. 11.6 for virtual pooling of W&D example. Of course, virtual pooling would not

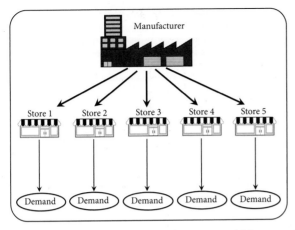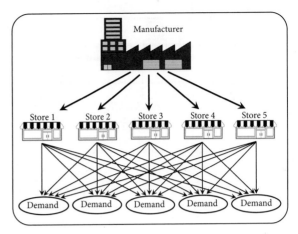

Figure 11.6    *Left:* W&D's current decentralized inventory with five stores. *Right:* Virtual pooling of the five stores.

be possible without a centralized information system that provides products' demand and inventory information to all inventory locations of the firm and to the customers.

Many retail stores such as Walmart, Macys, Gap, Target, and Home Depot use virtual pooling in managing their inventories. For example, customers of Home Depot can search Home Depot's website and find the products they need. The website finds the products in several of its stores and gives the customers the following two options: (i) picking up the product at their closest store, or (ii) shipping the product to them. Home Depot offers free shipping on orders above a certain amount. Even if a customer does not check the website and goes to a Home Depot store and finds that the item is out of stock, Home Depot staff can locate the item in the closest store and offer the customer the same two options.

Another type of virtual inventory pooling that has been used by firms is to avoid stockouts by moving items from stores to stores. For example, through Walmart's efficient transportation system, managers of each Walmart store can exchange products from stores with excess inventory with those in stores with low inventory. This reduces the chance of stockout in all stores and allows Walmart to reduce its safety inventory while maintaining a desired service level.

### 11.8.2 Specialization

Specialization is another way to benefit from the advantages of physical centralization of inventory in systems with multiple inventory locations. Under specialization, the safety inventory of each product is pooled into only a few stores that have large demand for that product—the location specializes in handling that product. To illustrate the concept of Specialization, consider the example of an inventory system with two stores and a DC as shown in Fig. 11.7-left. Both stores have demand for three products A, B, and C. At Store 1, the demand for Product A is high, the demand for Product B is low, and the demand for Product C is very low. At Store 2, however, the demand for B is high, demand for Product A is low, and the demand for Product C is very low. Both stores order items from the DC and keep safety inventory of all three items.

Figure 11.7-right shows one example of specialization. In that figure, since Store 1 has a large demand for Product A, it is specialized in that product and thus holds all the safety inventory of Product A (for Stores 1 and 2). Product A is shipped to Store 2 (if needed) to prevent stockouts. Store 2, on the other hand, has a large demand for Product B and thus holds all the safety inventory of Product B (for Stores 1 and 2)—Store 2 is specialized in

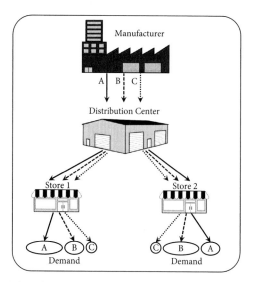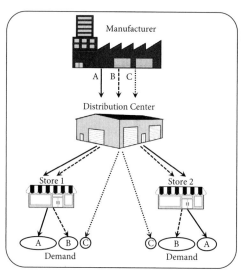

Figure 11.7    *Left:* **Both stores hold safety inventory for all three products.** *Right:* **Specialization: Store 1 specializes in Product A, Store 2 specializes in Product B, and distribution center specializes in Product C.**

Product B. Both stores have a very low demand for Product C; thus, they do not offer Product C. Product C is only offered online and is shipped to customers from the DC, which is now specialized in Product C.

How does specialization improve performance?

- *Product A:* By keeping the decentralized inventory locations (Stores 1 and 2), the system can maintain the advantages of having multiple inventory locations (e.g., having products closer to the customers, lower transportation costs, etc.). By pooling the safety inventory of Product A in Store 1, the firm can provide the same service level with lower safety inventory.

- *Product B:* The improvement in safety inventory and service level for Product B is similar to those for Product A. Since Store 2 pools the safety inventory of both stores, the firm can provide the same service level for Product B with lower inventory.

- *Product C:* By specializing the DC on Product C, the inventory of Product C at both stores is physically centralized in one location—the DC. This allows the firm to benefit from both economies of scale and variability reduction, including reduction in safety inventory of Product C.

The decision of which inventory location should be specialized in which product, or which product should or should not be offered in a location, is an important decision that requires a careful study of market as well as inventory costs. In our example in Fig. 11.7, the decision of offering Product C only through DC (e.g., selling online) is a reasonable decision, since Product C has a very low demand. As we discussed in Sec. 11.6.6, pooling inventory of slow-moving products can significantly reduce its inventory costs and make them more profitable. This is one of the main reasons that most retailers such as Home Depot or Target offer some of their products exclusively online.

### 11.8.3 Substitution

Why at the supermarkets same products of different brands are located in shelves next to each other? Well, one obvious reason is that supermarkets would like to make it easier for customers to find what they need. But another important fact is that supermarkets also want to make it easier for customers to find substitute for their products, if their products are out of stock. For example, what would you do if you go to a supermarket to buy a can of chicken noodle soup of your favorite brand X, but the store has run out of it? Well, some of us may not buy the soup but others buy the soup of a different brand—brand Y. This is an example of product substitution driven by a customer.

Product substitution corresponds to satisfying demand of one product, using inventory of another product. If customers of brand X products are willing to buy brand Y products and vice versa—a two-way substitution—then the safety inventory of these two products can be pooled, resulting in lower inventory. Even one-way substitution where customers of brand X are willing to buy brand Y, but not the other way around, the safety inventory of both products can be pooled where the safety inventory of brand Y serves customers of both demands.

In contrast to customer-driven substitution in which customers make the decision to substitute one product for another, manufacturer-driven substitution refers to the case where a manufacturer makes the decision to substitute one product for another. In this situation, the substitute product should be the one that the customer would certainly accept. This is often accomplished by offering a higher value product to substitute a lower value product that is out of stock. Computer manufacturers, for example, install higher-performance components (e.g., large RAM or hard drive) if lower-performance components are out of stock (without charging the customer).

The decision of pooling safety inventory of substitutable products requires a good understanding of the customers' substitution patterns, variability of product demands, and correlation among them. As discussed earlier, larger demand variability and more negatively correlated demand result in higher benefit of pooling inventory—in this case, pooling safety inventory through substitution. Also, there should be mechanisms in place that can persuade customers to substitute their products. For example, a retailer such as Best Buy or car dealers rely on the sales force at their stores to convince customers to buy another similar product if what customer wants is not available. Online retailers such as Amazon also provide the information about comparable (i.e., substitutable) products when a customer searches for a particular product.

### 11.8.4 Component Commonality

Another way that demand aggregation can reduce inventory is "Component Commonality." The idea is to design products such that several versions of a product use the same common component. For example, suppose that a firm produces Product A by assembling two components $C_1$ and $C_2$. Product B is produced by assembling two components $C_3$ and $C_4$, see Fig. 11.8-left. Both products are made to order. Components $C_1$, $C_2$, $C_3$, and $C_4$ are ordered

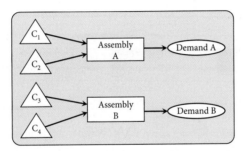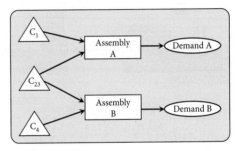

**Figure 11.8**  *Left:* Products A and B made from different components. *Right:* Products A and B use a common component.

from different suppliers. To protect against stockouts of the components, the firm keeps a high cycle-service level for all four components. Since each Product A uses one Component $C_1$ and one Component $C_2$, the demand for Components $C_1$ and $C_2$ is the same as that for Product A. Similarly, the demand for Components $C_3$ and $C_4$ is the same as that for Product B.

Now suppose the firm designs a new component—Component $C_{23}$—that can be used in both products A and B, see Fig. 11.8-right. Specifically, Component $C_{23}$ is a common component that is used instead of Components $C_2$ and $C_3$ in Products A and B, respectively. Component $C_{23}$ is supplied by a new supplier.

How does component commonality—Component $C_{23}$—result in saving inventory costs? The answer should be clear by comparing the two systems in Fig. 11.8. The inventory of Components $C_1$ and $C_4$ do not change in the new system, since they still face about the same demand in both the systems. However, Component $C_{23}$ faces the aggregated demand of Products A and B. In fact it can be considered as the physical centralization of inventories of Components $C_2$ and $C_3$. Therefore, with respect to the new Component $C_{23}$, the new system benefits from both reduction in cycle inventory and order setup cost (economies of scale) as well as reduction in safety inventory (variability reduction).

Dell Computers was pioneer in using component commonality in its operations. They use many common modular components (hard drives, random access memory) in making a wide variety of standard and customized products. We will discuss component commonality and modular product design in detail in Chap. 14.

## 11.9 Reducing Bullwhip Effect

*Supply chain* of a product is a group of two or more entities that are linked through the production and delivery of the product to the customers. For example, Procter & Gamble's (P&G) supply chain consists of suppliers that provide raw material and parts to P&G's manufacturers. Manufacturers then produce the product and ship them to Distribution Centers (DCs). DCs send the products to retailers that order those products. And finally, retailers sell the products to the customers. See Fig. 11.9 as an example of a supply chain.

As the figure shows, while each member of supply chain produces or delivers the same product, each faces a different demand. For example, while retailers face direct customer demand (e.g., customers buying one or more items), the DC faces the demand in form of orders received from one or more retailers. The orders are often for a large number of items. Similarly, the demand that the manufacturer faces are the orders it receives from one or more DCs.

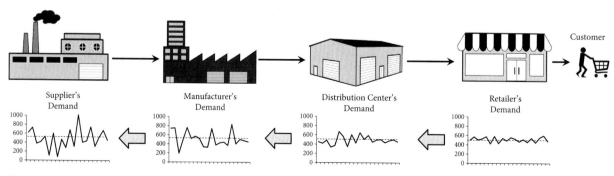

Figure 11.9    An example of a supply chain consisting of a supplier, a manufacturer, a distribution center, and a retailer.

Strategies in reducing variability in demand of a firm depend on where in the supply chain the firm operates. If the firm is a retailer—the last stage of the supply chain—the variability in the firm's demand is often affected by uncontrollable factors such as state of economy, competitor's price, and customers' preference. However, if the firm operates in upstream of the supply chain (e.g., a manufacturer), in addition to those uncontrollable factors, the firm's demand is also affected by the decision (e.g., order size) made by the downstream members of the supply chain. But, how do downstream members' decisions affect upstream members' demand?

In the 1990s, P&G observed this firsthand. The company found that while the variability in demand for its best-selling product—Pampers—was low, the variability in demand at P&G distribution center was significantly higher. To their surprise, when they checked the variability in demand that their suppliers face, they found that it was even higher than the variability in demand at the DC. In summary, they found that the variability in demand was amplified as they moved up the supply chain (see Fig. 11.9). P&G called this phenomenon the "Bullwhip Effect."

### CONCEPT

**Bullwhip Effect**

When the bullwhip effect is present in a supply chain, the demand variability is higher at upstream stages of the supply chain (e.g., manufacturers or suppliers) than that at downstream stages (e.g., retailers).

The bullwhip effect has been observed by many industries such as auto industry, computer and electronics, apparel, food, and grocery industries. The presence of the bullwhip effect—higher variability in upstream stages—results in high inventory-related costs. To reduce or to remove the bullwhip effect, one should first identify the causes of this unwanted phenomenon. Lee et al (1997)[5] identified four main causes of the bullwhip effect as follows:

- Demand Forecast Updating
- Order Batching
- Price Fluctuations
- Rationing and Shortage Gaming

In the following sections, we discuss these causes in more detail and provide strategies to mitigate the bullwhip effect.

### 11.9.1 Demand Forecast Updating

Forecasting demand is essential for any production and inventory planning and control. Forecasting methods such as time series forecasting use the demand in previous periods to generate a forecast for demand in the next period. Each time a new demand is observed, it is used to update the demand forecast (i.e., to generate a new forecast) for the next period. The *demand forecast updating* refers to the circumstances where a small increase (or decrease) in forecasted demand can result in a large increase (or decrease) in orders placed by the process.

---

[5]Lee H.L., V. Padmanabhan, and S. Whang, "The Bullwhip Effect in Supply Chains." *Sloan Management Review,* Spring 1997.

To show how demand forecasting can cause bullwhip effect in a supply chain, we use the following simple example. Consider the retailer in Fig. 11.9 that orders a product from the Distribution Center (DC) of a manufacturer. The retailer places an order every week on Monday according to a periodic review system. The DC has a lead time of $L = 1$ week. The retailer's forecast indicates that the demand in each week follows a Normal distribution with mean 100 and standard deviation 30. The retailer's goal is to reduce its chance of a shortage to less than 1 percent, achieving a cycle-service level of $\alpha = 0.99$. Hence, the retailer uses an order-up-to policy with order-up-to level $S_\alpha^*$ that satisfies

$$\alpha = F_{L+P}(S_\alpha^*)$$

where $F_{L+P}(x)$ is the cumulative demand distribution during protection interval (period plus lead time) of $1 + 1 = 2$ weeks. Since the weekly demand has a mean of 100 and standard deviation of 30, the demand during the protection interval is Normally distributed with mean $\overline{D}_{L+P} = 2(100) = 200$ and standard deviation $\sigma_{L+P} = \sqrt{2}(30) = 42.43$. Hence, considering $\Phi(2.33) = 0.99$, the optimal order-up-to level $S_\alpha^*$ is

$$S_\alpha^* = \overline{D}_{L+P} + z\sigma_{L+P} = 200 + 2.33(42.43) = 299$$

Thus, the retailer places an order every week to raise its inventory position to 299.

Consider several weeks in which the demand during the week was about 100 units. Hence, at the end of the week the retailer's inventory position is about $299 - 100 = 199$, and the retailer orders about 100 units each week to raise its inventory position to order-up-to level 299. Consequently, DC observes the average demand of 100 units per week. Now suppose that the retailer observes a high demand of 150 during a week. Incorporating this new demand into its forecasting method, and after several studies, the retailer concludes that the mean weekly demand has increased from 100 to 110—a 10-percent increase.

Using this updated forecast, the retailer then revises its order-up-to level considering the new demand distribution during protection interval, which now has a mean of $\overline{D}_{L+P} = 2(110) = 220$ and standard deviation $\sigma_{L+P} = \sqrt{2}(30) = 42.43$. This results in new order-up to level:

$$S_\alpha^* = \overline{D}_{L+P} + z\sigma_{L+P} = 220 + 2.33(42.43) = 319$$

To raise the inventory position to the new order-up-to level 319, and considering the current inventory position of $299 - 150 = 149$, the retailer then places an order of size $319 - 149 = 170$. Since the retailer has been ordering about 100 units each week—about the average weekly demand—this correspond to a 70-percent increase in order size. In other words, a 10-percent increase in the mean demand resulted in a 70-percent increase in order size. Considering that retailer's orders are the DC's demand, this implies that a 10-percent increase in retailer's demand translated into a 70-percent increase in the DC's demand.

If the mean demand was not actually increased by 10 percent, then the retailer will eventually realize it after some period and then decrease its mean demand by 10 percent. Similarly, this decrease will result in a larger decrease in retailer's order size and thus in DC's demand.

This simple example shows how small changes (small variations) in demand of a downstream member of a supply chain—the retailer—can result in large changes (large variations) in the demand of an upstream member—the DC. This can cascade in a similar way from the DC to the manufacturer in Fig. 11.9. Specifically, the DC can also update its forecast of the retailer's order by a 70-percent increase. This can result in an even larger increase in DC's orders placed to the manufacturer, which in turn leads the manufacturer to update its forecast. So, each time a downstream member updates its forecast that affects its ordering policy, it can lead to an increase in variability of the demand of the upstream members—the bullwhip effect.

One important factor is that the impact of demand forecast updating on amplifying variability is larger for larger lead times. For example, if the DC has a lead time of $L = 2$ weeks, then the protection interval will have an average demand of $\overline{D}_{L+P} = (2 + 1)100 = 300$. Consequently, the retailer will increase its order-up-to level to $300 + 2.33(42.43) = 399$. After updating the forecast of the mean demand to 110, the new order-up-to level would be $330 + 2.33(42.43) = 429$. Hence, considering the current inventory position of $399 - 150 = 249$, the retailer would place an order for $429 - 249 = 180$. This is about an 80-percent increase in previous order size of 100 and is 10 percent higher than when the lead time was 1 week. Therefore, as the lead time increases, the impact of demand forecast updating is intensified. This emphasizes the importance of lead time reduction: lead time reduction not only decreases the safety inventory (as we discussed in Sec. 11.5), but also mitigates the bullwhip effect.

### Demand Forecast Updating—Remedy

How to prevent demand forecast updating in order to eliminate the bullwhip effect? Note that except for the retailer that directly observes the customer demand for the product, other supply chain members such as the DC or the manufacturer observe the demand in form of orders placed by downstream supply chain members—a distorted information about customer demand. Hence, one way to prevent the bullwhip effect is to share the actual customer demand (e.g., the point-of-sale (POS)) information with all supply chain members. This allows all supply chain members to forecast their future demand based on the actual customer demand and plan their operations accordingly. For example, in our above example, if the DC had access to the POS data, it would have realized that there is only a 10-percent increase in customer demand not 70 percent. This would allow the DC to better manage its inventory and ordering process. Many firms such as Walmart, P&G, and Dell share their POS data as well as their inventory position data with the suppliers and manufacturers in their supply chains.

Even when the customer demand information is shared, the human tendency to over-react to changes in demand can still impact the variability in demand throughout the supply chain. The underlying reason for overreacting is rooted in human psychology. Specifically, humans have the tendency to be most affected by what they have last seen or heard, because humans tend to retain most information about the most recent events. Thus, in the context of our example, even though the retailer has observed several weeks with a demand of about 100 units, the most recent demand of 150—a large increase—has a significant impact on retailer's decision to conclude that the mean demand has increased. How can this be prevented? Statistical methods can be used to identity whether a large increase in most recent demand is an outlier, or it is due to a real increase in demand.

### 11.9.2 Order Batching

Another reason for supply chain members to inflate or deflate their orders—and thus distort demand information transferred upstream—is ordering in batches. The main reason for ordering in batches is economies of scale with respect to order setup cost (i.e., order transaction cost and order shipment cost), which is usually a fixed cost. But how does ordering in batches result in the bullwhip effect?

The main reason is that ordering in batches results in larger and more infrequent orders. To show this, we utilize the simple example of a supply chain of a product in Fig. 11.10 that consists of one retailer and one manufacturer. The daily demand for the product at the retailer is equally likely between 60 and 100 units. The retailer uses a periodic review system with a period of length 3 days and an order-up-to policy with order-up-to level $S_\alpha = 300$

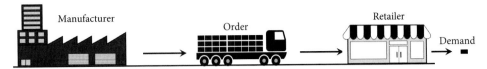

**Figure 11.10**   Retailer demand is in single units, but manufacturer's demand is in batches.

to maintain a certain cycle-service level. Hence, the retailer places an order at the end of every third day to raise its inventory position to $S_\alpha = 300$. The manufacturer charges a fixed shipment cost of $5000 for every truck that delivers its orders, regardless of the order size. Each truck has a capacity of carrying 100 units. It takes the manufacturer $L = 1$ day to deliver retailer's order.

**Table 11.4**   Retailer's Daily Demand and Submitted Orders

|  | Retailer's demand | Retailer's inventory position | Required order size | Submitted order size |
|---|---|---|---|---|
| 1 | 79 | 221 | 0 | 0 |
| 2 | 99 | 122 | 0 | 0 |
| 3 | 99 | 23 | 277 | 300 |
| 4 | 78 | 245 | 0 | 0 |
| 5 | 67 | 178 | 0 | 0 |
| 6 | 86 | 92 | 208 | 200 |
| 7 | 68 | 224 | 0 | 0 |
| 8 | 99 | 125 | 0 | 0 |
| 9 | 81 | 44 | 256 | 300 |
| 10 | 74 | 270 | 0 | 0 |
| 11 | 95 | 175 | 0 | 0 |
| 12 | 98 | 77 | 223 | 200 |
| 13 | 76 | 201 | 0 | 0 |
| 14 | 63 | 138 | 0 | 0 |
| 15 | 82 | 56 | 244 | 200 |
| 16 | 83 | 173 | 0 | 0 |
| 17 | 73 | 100 | 0 | 0 |
| 18 | 89 | 11 | 289 | 300 |
| 19 | 95 | 216 | 0 | 0 |
| 20 | 65 | 151 | 0 | 0 |
| 21 | 78 | 73 | 227 | 200 |
| 22 | 85 | 188 | 0 | 0 |
| 23 | 91 | 97 | 0 | 0 |
| 24 | 76 | 21 | 279 | 300 |
| 25 | 94 | 227 | 0 | 0 |
| 26 | 67 | 160 | 0 | 0 |
| 27 | 96 | 64 | 236 | 200 |
| 28 | 76 | 188 | 0 | 0 |
| 29 | 78 | 110 | 0 | 0 |
| 30 | 70 | 40 | 260 | 300 |

Table 11.4 shows the daily demand during the period of last 30 days. The period starts with inventory position of 300. The table also includes the retailer's inventory position at the end of each day, the order size *required* to raise the inventory to order-up-to level, and the actual order *submitted* to the manufacturer.

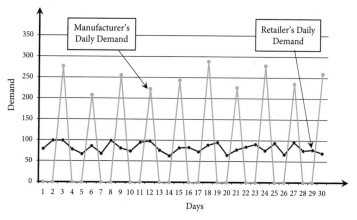

Figure 11.11    The bullwhip effect due to order batching.

To illustrate how daily demand resulted in orders submitted to the manufacturer, consider day 1 in the table that starts with inventory position of 300. Since the demand during that day is 79, the inventory position at the end of the day is $300 - 79 = 221$. The inventory at the end of the second and the third day is $221 - 99 = 122$ and $122 - 99 = 23$, respectively, because the daily demand at those days were 99 and 99. Since the retailer places an order every 3 days, the orders placed to the manufacturer at days 1 and 2 are zero. With inventory position of 23 at the end of the third day, the retailer needs to place an order of size $300 - 23 = 277$ units, that is, the required order size is 277.

Now consider the required order of 277 units. Delivering this order requires three trucks—two trucks delivering their full load of 100 units and one truck delivering only 77 units. Since the third truck costs the retailer $5000 and is only carrying 77 units, to take advantage of the full truckload, the retailer can revise its order size in two different ways: (i) reduce its order size to 200 and pay $10,000 for two full truckloads, or (ii) increase its order size to 300 and pay $15,000 for three full truckloads. In both cases, the shipment cost per unit delivered is $5000/100 = $50$. Suppose the retailer would like to stick as close as possible[6] to the required order size 277. Thus, the retailer revises its required order of size 277 to 300, which is 23 units larger than what it needed (as opposed to ordering 200, which is 77 units smaller than what it needed). The retailer therefore submits an order of size 300 and raises its inventory position to $23 + 300 = 323$ at the end of day 3. Hence, considering the demand of 78 during day 4, the inventory position at the end of day 4 will be $323 - 78 = 245$. Continuing this way we can see how the inventory positions and order sizes in Table 11.4 are computed.

Why does order batching result in the bullwhip effect in our example? Two reasons: (i) the retailer orders in batches, and (ii) the retailer utilizes the full truckload. We explain each in detail below.

### Impact of Batching

To see how the order batching affects the variability of manufacturer demands, let's compare the daily demand of the retailer with that of the manufacturer in Table 11.4, which is shown in Fig. 11.11. The figure clearly shows that the variability in orders submitted to the manufacturer (i.e., manufacturer's demand) is significantly higher than the variability in retailer's daily demand. In fact, using the data in Table 11.4, we find that the retailer's daily demand has

---

[6]The retailer revises the required orders of size 249 or less to 200 and the orders of size 250 or higher to 300.

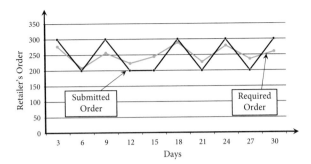

**Figure 11.12    The impact of full truckload on retailer's order quantity (i.e., manufacturer's demand).**

CV = 0.14, while the manufacturer's demand has a CV = 1.46. This is because, due to ordering a large batch every 3 days, the manufacturer's daily demand varies between zero and a large batch size which is about three times larger than the retailer's average daily demand. For example, while the retailer's demand on the first 3 days varies between 79 and 99—a small variation—the daily demand that the manufacturer faces (i.e., the orders received from the retailer) are 0, 0, and 300—a large variation.

### Impact of Full Truckload

To see how using full truckload increases the variability of the manufacturer's demand, let's compare the retailer's *required* order size (the fourth column in Table 11.4) with the *submitted* order size (i.e., the fifth column in the table), see Fig. 11.12. As the figure shows, increasing and decreasing the required order size in order to minimize shipment cost per units shipped (by ordering in full truckloads) results in an increase in variability in the orders submitted to manufacturer—increase in the variability in manufacturer's demand.

### Impact of Multiple Retailers

Besides utilizing full truckload, having multiple retailers also intensifies the bullwhip effect. Suppose that the manufacturer in our example serves two retailers. The second retailer faces the same demand as the first retailer and places an order every 3 days like the first retailer, that is, placing orders at the end of days 3, 6, 9, etc. In this case, the manufacturer's demand in each period of length 3 days consists of 2 days with zero demand and the third day with a demand which is about twice as large as that when serving only one retailer. This results in a much higher variability in the manufacturer's demand—a more intensified bullwhip effect.

### Order Batching—Remedy

How to reduce the impact of order batching in creating the bullwhip effect? The obvious answer is to reduce the order size placed at upstream stages of the supply chain. Several strategies can be taken to reduce the order size:

- *Reduce the Order Setup Cost:* As we discussed, the order size can be reduced by reducing the order setup cost (i.e., order transaction cost, order shipment cost, etc.). How to reduce order setup cost? You can find the answers in Sec. 11.3.

- *Reducing the Impact of Full Truckload:* As was illustrated in Fig. 11.12, tendency to take advantage of economies of scale by ordering in full truckload intensifies the bullwhip effect. Hence, reducing this tendency will diminish the bullwhip effect. One way to order in smaller batches, but still take advantage of the full truckload, is to fill the truck by ordering small batches of different products. For example, P&G encourages its retailer to order a variety of products—in small batches—to fill a truck. But what if a retailer only orders one product from a certain supplier? In those cases, the supplier can combine the orders of several retailers in one full truck. Thus, the shipment cost for each retailer is reduced, allowing them to order in smaller batches.

- *Coordinating Orders with Several Retailers:* Recall that receiving orders from multiple retailers can amplify the bullwhip effect. Consider an example of seven retailers, each ordering an average of 100 units from the same manufacturer once a week. If all retailers place their orders on Monday, then the manufacturer will see an average demand of 700 on Monday, with the demand of zero in the rest of the week—a very large demand variability. However, if the seven retailers place their orders in different days of the week, the manufacturer will have an average daily demand of 100 units throughout the week—a very low demand variability. The good news is that while it matters a lot to the manufacturer, it does not matter much to the retailers which day of the week they place an order, as long as they are placing an order once a week. Therefore, it is not a difficult task to coordinate the timing of the orders placed by different retailers.

### 11.9.3 Price Fluctuations and Speculation

Prices change at every stage of a supply chain, because of the inherent randomness in price or because of the management decisions. For example, prices of commodities such as farm products, oil, and steel are very random and change constantly due to market and economic situations. This results in price fluctuations at upstream stages of the supply chain. At downstream or middle stages of a supply chain, on the other hand, the change in price is due to decisions made by the management. For example, retailers that sell products to customers frequently offer lower prices (e.g., coupons or sales) to boost their sales. Manufacturers who sell their products to retailers also offer lower price through trade promotion (see the online supplement of Chap. 9). But, how do such price fluctuations cause the bullwhip effect?

To answer this question, consider a retailer who buys a product from a manufacturer and sells it to the customers. If the retailer offers a discount on a particular product, the demand for that product will most probably increase. Because of the lower price, customers may buy more than they actually need and keep the product for future use. On the other hand, when the price returns to its original price, customers stop buying until they finish what they bought before. Hence, the variation in customer buying pattern—the demand observed by the retailer—is higher than the variation of actual customer needs. This is the bullwhip effect.

But that is not the end of the story. Increase in retailer's demand may result in placing a larger order size to the manufacturer, which in turn results in an increase in the variability in manufacturer's demand—again, the bullwhip effect.

The price fluctuations at the manufacturer can also cause the bullwhip effect. Consider the manufacturer in our example and suppose that the manufacturer offers a trade promotion for a period of 2 weeks with 10-percent discount in price. As we have shown in the online supplement of Chap. 9, it is optimal for the retailer to increase the order size to take advantage of this price discount. Consequently, the manufacturer will observe an unusually large order size, contributing to an increase in the demand variability—the bullwhip effect.

It is worth mentioning that sometimes placing a larger order is due to rumor and speculation of a price increase that makes a member of the supply chain to increase its order size beyond its actual needs. Regardless of whether the fluctuations in price are real or are speculated, they will cause bullwhip effect.

### Price Fluctuations and Speculations—Remedy

As the above examples show, price fluctuations at any stage of the supply chain can contribute to the bullwhip effect. Hence, one way to diminish or eliminate the bullwhip effect is to stabilize the price.

- *Limit the Maximum Purchase under Discounted Price:* Limiting the number of units a retailer can buy from a manufacturer under a trade promotion restricts the size of the order placed by the retailer, and thus reduces the variability in manufacturer's demand.

- *Offer Volume-Based Not Order-Based Quantity Discount:* Trade promotions that offer lower price for only a single order (i.e., an order-based discount) result in retailers placing very large orders. The volume-based discount, however, offers a lower price for the total number of units purchased in a certain period (e.g., in 3 months). Thus, retailers can place smaller order sizes during the discount period and still benefit from lower prices.

- *Everyday Low Price (EDLP):* Perhaps the best way to stabilize price is to eliminate trade promotion and discounted prices. Keeping prices at a low price and not offering promotions prevent large variations in order sizes at different stages of a supply chain. It also reduces the distortion in demand information, allowing stages of supply chain to have a better forecast for their demand. P&G, which was among the first to discover the bullwhip effect, has moved away from practice of trade promotion and follows an everyday low pricing strategy. After starting to offer lower prices every day, P&G reported an increase in its profits as well as in its market share. Other firms such as Kraft, Pillsbury, and Walmart adopted the EDLP strategy. In fact, Walmart made the EDLP into its slogan. By offering the same low price every day, the firm also saves on promotion-related costs such as advertising and transaction costs.

### 11.9.4 Rationing and Shortage Gaming

When demand for a product surpasses its supply, the manufacturer of the product often rations the product among its customers according to their order sizes. For example, consider a manufacturer that has the production capacity of 125 units per month. It sells its product to two retailers who order every month. Now suppose that due to some equipment failure, the manufacturer's production capacity reduces to 100 units per month. Consider this month when the manufacturer receives orders of size 50 and 75, respectively, from Retailers 1 and 2. Note that the total demand is $50 + 75 = 125$; hence, the capacity is only $100/125 = 80\%$ of demand. To be fair, the manufacturer then rations its product and each retailer receives 80 percent of what they ordered, that is, Retailer 1 receives $80\% \times 50 = 40$ and Retailer 2 receives $80\% \times 75 = 60$ units.

Now suppose that retailers are aware of the manufacturer's capacity shortage and know that the manufacturer will ration the product. In this case, to get their actual need, retailers game the system and inflate their order size so they could get what they actually need. When the period of product shortage ends, retailers' orders reduce to the original amount and some times even lower. These increase and decrease in order sizes that do not reflect the actual retailers' demands increase the variability in manufacturer's demand—the bullwhip effect.

The rationing and shortage gaming in our simple example is a real phenomenon that many firms face during periods of product shortage. For example, there was a period in which Hewlett-Packard (HP) was not able to meet the demand for its laserJet III printer and started rationing the product. Soon after, HP realized a significant increase in orders that it received. After HP increased its production to overcome the shortage, many of HP customers canceled their orders. This left HP with a significant cost related to increase in production capacity and excess inventory.

Sometimes not the actual shortage, but the speculation of a shortage can cause the bullwhip effect. In 1994, IBM was very successful in selling its apriva personal computer. The sales were going extremely well, to the point that IBM's customers (i.e., retailers and resellers) speculated that there would be a shortage of apriva before the Christmas season. Retailers then increased their orders with IBM in the fear of rationing by IBM. This resulted in a spike in IBM's demand and increased variability in demand for apriva.

### Rationing and Shortage Gaming—Remedy

There are several strategies that firms can adopt to dampen the impact of rationing and shortage gaming that cause the bullwhip effect

- *Choosing an Appropriate Rationing Policy:* There are several rationing policies that prevent retailers to place large orders beyond what they actually need. One policy called *turn-and-earn* allocates the available units based on past retailers' sales—not the current order sizes. This policy has two advantages. First, it eliminates retailers' incentives to inflate their orders. Second, it encourages retailers to sell more during low-demand periods, so they can place larger orders during high-demand periods. Firms such as GM and HP have been using this strategy for allocating their products to their retailers. Other rationing policies that have been used are policies that ration based on previous order sizes. For example, one can allocate the available units proportionally to the last period's order size, or limit the retailers' orders not to increase beyond a certain percent of their last period's orders.

- *Restrict Order Cancellations:* Restricting order cancellation removes retailers' incentive to order beyond what they actually need. Manufacturers use several strategies to do that. One strategy is to use time fences that restrict retailers' ability to cancel their orders after a particular date. The goal is to make it more difficult for retailers to cancel their orders as they get closer to the delivery date. Another strategy is to set a penalty and charge retailers if they cancel their orders. One practice is to charge retailers a reservation price (also called reservation payment) when retailers place an order. The reservation price is a prepayment of a fraction of the actual price of the product. When retailers receive the product, they will then be charged for the remaining fraction of the price. However, if a retailer cancels the order or decreases its order size, the reservation price will not be returned.

- *Sharing Capacity/Availability Information:* The shortage gaming intensifies when retailers have little information about the availability of the product. Therefore, one way to diminish the impact of shortage gaming is to share the information about the capacity or available inventory of the product. This is a good strategy when there is an untrue perception of the shortage or when the shortage is not significant. However, when there is a significant shortage, sharing the capacity/availability information may intensify the shortage gaming.

- *Preventing Shortage:* Perhaps the best remedy for shortage gaming and rationing is to prevent shortage in the first place. One approach used by many firms is to give retailers incentive to place their orders, or at least part of their orders, way in advance. This allows manufacturer to plan its resources (labor, material) across products accordingly and schedule its production to prevent shortage. Having a flexible production capacity that can shift capacity from products with low demand to products with high demand also helps manufacturer to prevent shortage.

## 11.10 Improving Coordination in Supply Chains

Lack of coordination among supply chain stages plays an important role in increasing costs and therefore decreasing profit of the supply chain. In case of bullwhip effect, for example, we saw how pricing or batching decisions at one stage of the supply chain can increase operating costs in upstream stages. Finding mutually acceptable ways to coordinate decisions at different stages of a supply chain with the common goal of improving total supply chain's profit has been of great interest to many firms in the past decades. This is referred to as Supply Chain Coordination.

---

**CONCEPT**

**Supply Chain Coordination**

Supply chain coordination refers to coordinating activities at different stages of a supply chain—directly or indirectly—to increase the *total* supply chain profit.

---

Why a firm (e.g., a manufacturer) which is only one part of a supply chain should care about increasing the *total profit*—the sum of all supply chain members'—profits? Well, if the total supply chain profit increases, the firm may get some portion of the surplus.

One reason behind the lack of coordination in supply chains is the conflicts among incentives of supply chain members, also called *Misalignment of Objectives*. Specifically, each stage of the supply chain optimizes its own profit, regardless of how its decision affects the profits of other stages of the supply chain. These locally optimized decisions dictate the dynamics of the relationship between stages of a supply chain that might not result in the maximum profit for all supply chain members. Coordinating a supply chain refers to designing mechanisms that govern the relationship among stages of supply chains such that all stages see an increase in their profits—a win-win situation. We illustrate this using the following simple example of a supply chain consisting of a manufacturer and a retailer.

### You and the Laser Projectors

Suppose that you work as the sales and inventory manager of a firm that produces a variety of light fixtures, including a laser projector with motion capability that is used in Christmas season for decoration. The projector creates an illusion of dropping snowflakes on the exterior of a house. It costs you $60 to produce each laser projector and your main customer for the projector is a retailer of home appliances called Dave Homes. Due to high demand for your other products, and to set your production schedule, you have asked Dave Homes to place its order 3 months before Christmas. And you promised to deliver their order 1 month before Christmas when the holiday shopping season starts. Your firm's policy is to set a large penalty for canceling or changing order sizes, and your retailer is aware of that.

As the sales manager, you need to make a decision about how much to price the projector in order to maximize your profit. To get more information, you call Dave Homes' supply chain manager, Jane, whom you have been working with for the last 8 years. Jane tells you that, due to low demand, they decided not to offer the projector in their retail stores, and only offer them online. She also tells you that they made a decision to price it at $110, which is about $5 less than the competitor's price of $115. With this price, Dave Homes' marketing manager believes that there will be an average demand of about 450 for the projector during the holiday season. Jane also mentions that they expect a moderate variability in demand with a standard deviation of about 150. She tells you that they believe the demand for the projector would well-fit a Normal distribution, as they have seen it for their other similar products. When you ask about their pricing strategy after the holiday season when there is almost no demand for the projector, Jane informs you that they have already signed a contract with a discount store that buys unsold projectors after the holiday season at a discount price of $45. Jane asks you how much you are planning to price your projectors. You tell Jane that your firm has not yet decided how to price the projector, and you will contact her as soon as the decision is made. Jane then tells you that she will place her order after she knows your price.

Well, with this information, you must now identify a price that maximizes your expected profit.

### 11.10.1 Wholesale Price Contract

Let's focus on the supply chain of this projector. It consists of your firm (i.e., the manufacturer of the projector) and Dave Homes (i.e., the retailer). You must make a decision of what wholesale price you should announce to Jane, and Jane must decide how many projectors to order from you. Your goal is to maximize your firm's profit, and Jane's goal is to maximize Dave Homes' profit. This is called Wholesale Price Contract.

---

**CONCEPT**

**Wholesale Price Contract**

Under a wholesale price contract, the seller (e.g., a supplier or a manufacturer) sets a wholesale price $w$ for its product, and the buyer (e.g., a retailer) then decides how many of the products to order based on the price set by the seller.

---

**Your Optimal Wholesale Price Contract**

What wholesale price should you set for the projector in order to maximize your profit? To answer this question, you realize that you need to do some math. If you set your wholesale price, say at $w$, and Jane places an order of size $Q$, then considering that Jane places only a single order and cannot change her order, your expected profit would be $(w - 60)Q$, where $60 is your cost of producing one projector.

Obviously, your wholesale price $w$ should be larger than the production cost $60; otherwise, you do not make any profit selling the projector. Also, your wholesale price $w$ should not be larger than Jane's selling price $110; otherwise, she would not order any. Therefore, your wholesale pricing decision is in fact the following optimization problem with the goal of finding the optimal value of $w$ that maximizes your expected profit:

$$Max \ \mathsf{E}(\text{Your Profit}) = (w - 60)Q \qquad (11.9)$$
$$\text{subject to:}$$
$$60 \leq w \leq 110$$

The optimization problem seems simple, except that you do not know the order size $Q$ in the objective function. In fact, the order size $Q$—Jane's decision—depends on your wholesale price $w$. If you set $w$ very high, Jane would not order much, and your profit goes down because you are not selling much. On the other hand, if $w$ is too small, Jane would order a lot, but your profit would still be low, because you are not making much money on each projector. You soon realize that you need to find the relationship between Jane's order quantity $Q$ and your wholesale price $w$.

What would you do if you were in Jane's situation? After some thinking you realize that Jane is facing a Newsvendor problem, since she can only place one order before the selling season and demand during selling season is random. Specifically, Jane buys a projector at price $w$ and sells it during the selling season at $p = \$110$. Unsold projectors would be sold to the discount store at salvage price $s = \$45$. Hence, the understocking and overstocking costs for the projector are $C_u = p - w = 110 - w$ and $C_o = w - s = w - 45$, respectively. Therefore,

$$\mathcal{R}_c = \frac{C_u}{C_u + C_o} = \frac{110 - w}{(110 - w) + (w - 45)} = \frac{110 - w}{65}$$

Recall that demand for projectors is Normally distributed with mean 450 and standard deviation 150. Thus, if you set your wholesale price at $w$, Jane would place her optimal order of size $Q^*$ that maximizes her expected profit as follows:

$$Q^* = \text{ROUNDUP( NORMINV}((110 - w)/65, 450, 150)\,,\,0\,) \tag{11.10}$$

where Excel function $\text{ROUNDUP}(x, 0)$ rounds up the value of $x$ to the next integer number. Therefore, for any given value of $w$ you can easily compute your expected profit using Excel function as follows:

$$Max\ \text{E(Your Profit)} = (w - 60) \times \overbrace{\text{ROUNDUP( NORMINV}((110 - w)/65, 450, 150)\,,\,0\,)}^{Q^*}$$

So, all you need to do is to evaluate the above function for different values of wholesale price $w$ between 60 and 110. This is shown in Fig. 11.13.

As the figure shows, your expected profit is maximized at wholesale price $w = \$97.80$. Under this wholesale price you should expect a profit of about \$12,020.40.

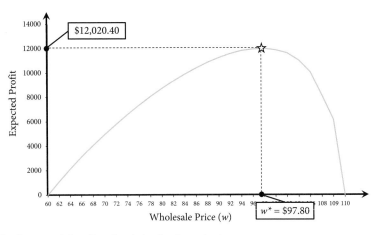

**Figure 11.13    Your expected profit under wholesale price contract.**

**Jane's Optimal Ordering Decision**

If you set your wholesale price at $w = \$97.80$, then using Eq. (11.10), Jane's optimal order quantity would be

$$Q^* = \text{ROUNDUP}(\ \text{NORMINV}((110 - 97.80)/65, 450, 150)\ , 0\ )$$
$$= 318$$

You are curious to know how much Jane's expected profit would be if she orders her optimal order size $Q^* = 318$. To compute her expected profit, you first compute the expected shortage if Jane orders 318 units. For Normal demand distribution with mean 450 and standard deviation 150, order $Q^* = 318$ corresponds to $z = (318 - 450)/150 = -0.88$, and thus Jane's expected shortage is

$$E(\text{shortage}|\ Q^* = 318) = \mathcal{L}(Q^* = 318) = \sigma_D \mathcal{L}(z = -0.88) = 150(0.9842) = 147.63$$

Using Eqs. (10.4) and (10.5), you then find the average sales and average leftover inventory as follows:

$$E(\text{sales}|\ Q^* = 318) = \overline{D} - E(\text{shortage}|\ Q^* = 318) = 450 - 147.63 = 302.37$$
$$E(\text{leftover}|\ Q^* = 318) = Q^* - E(\text{sales}|\ Q^* = 318) = 318 - 302.37 = 15.63$$

Hence, Jane's expected profit under the optimal order quantity $Q^* = 318$ is

$$E(\text{Jane's profit}) = p\,E(\text{sales}|\ Q^* = 318) + s\,E(\text{leftover}|\ Q^* = 318) - c\,Q^*$$
$$= \$110 \times 302.37 + \$45 \times 15.63 - \$97.80 \times 318$$
$$= \$2864.75$$

While you feel good that your profit of $12,021.40 is larger than Jane's expected profit, you still are not satisfied and would like to find a way to further increase your profit. Of course, if Jane orders more than 318, your profit goes up. The problem, however, is that Jane has no incentive to deviate from her optimal order quantity of $Q^* = 318$, unless you reduce your wholesale price. But, you do not want to deviate from your optimal wholesale price $w = \$97.80$, since it maximizes your expected profit. So, it seems that you and Jane are making the best decisions that maximize your expected profits and thus maximizes the total supply chain (sum of yours and Jane's) profit, right?

Wrong! There are still ways to further increase your profit as well as Jane's profit. Before we discuss those ways, we first need to understand the concept of a centralized supply chain.

## 11.10.2 Centralized Supply Chain

When both you (the manufacturer) and Jane (the retailer) maximize your own expected profits, you and Jane earn $12,020.40 and $2,864.75, respectively. Therefore, your supply chain—which we use notation SC to refer to—makes the total expected profit of

$$E[\text{SC profit}] = \$12,020.40 + \$2,864.75 = \$14,885.15$$

To find ways to further increase the supply chain members' profits, we first need to identify factors that impact profit. There are obvious factors such as demand, production cost $c = \$60$, and selling price $p = \$110$ that affect profits. Lower production cost, higher selling price, higher demand, and lower demand variability, obviously, increase the expected profit. But there is one—not so obvious—factor that also impacts profit, and that is Misalignment of Objectives of supply chain members. Specifically, you and Jane have two different objectives that are in conflict. To increase her profit, she would like you to reduce your wholesale price from its optimal value of $w = \$97.80$. But that conflicts with your goal of maximizing your profit. To maximize your profit, on the other hand, you would like Jane to increase her order quantity above her optimal $Q^* = 318$, which conflicts with her objective of maximizing her profit. These conflicting objectives reduce the *total* supply chain profit—which includes yours and Jane's expected profit.

But what would be the expected profit if one can eliminate the conflict between the objectives of supply chain members? To answer this question, suppose that your company and Jane's company are owned by the same firm, called Central Light. You and Jane now work under Andrew—the vice president of the supply chain at Central Light. Andrew is now in charge of making the decision of how many projectors must be produced in your factory and shipped to Jane's DC. Note that there is no misalignment of objectives in this new supply chain, since only one person—Andrew—makes the decision for you and Jane with the single goal of maximizing the total supply chain's profit.

This new supply chain in which one decision maker (i.e., Central Light) makes the decision for all supply chain members with the goal of maximizing the total supply chain profit is called *Centralized Supply Chain*. In contrast, your supply chain in which two decision makers (i.e., you and Jane) make decisions to maximize your own expected profits is called *Decentralized Supply Chain*.

To find the optimal production quantity (i.e., the order quantity shipped to Dave Homes), Andrew must solve a new Newsvendor problem with production cost $c = \$60$, selling price $p = \$110$, and salvage value of $s = \$45$. Note that there is no wholesale price decision, since Central Light owns both yours and Jane's companies. Hence, $C_u = p - c = 110 - 60 = 50$ and $C_o = c - s = 60 - 45 = 15$.

$$ \mathcal{R}_c = \frac{C_u}{C_u + C_o} = \frac{50}{50 + 15} = 0.769 $$

Thus, Central Light's optimal production quantity in this centralized supply chain is

$$ Q_c^* = \text{ROUNDUP}( \text{NORMINV}(0.769, 450, 150) , 0 ) $$
$$ = 561 $$

For order quantity $Q_c^* = 561$, we can compute the average shortage, average sales, and average leftover inventory to be 20.02, 429.98, and 131.02, respectively. So the total supply chain profit—now owned by Central Light—is

$$ E(\text{Central Light profit}) = p\, E(\text{sales}|\, Q_c^* = 561) + s\, E(\text{leftover}|\, Q_c^* = 561) - c\, Q_c^* \quad (11.11) $$
$$ = \$110 \times 429.983 + \$45 \times 131.017 - \$60 \times 561 $$
$$ = \$19{,}533.89 $$

### 11.10.3 Decentralized versus Centralized Supply Chain

By comparing the order quantities and the expected profits of centralized and decentralized supply chains of the projector, we observe the following:

*Order quantity is higher in the centralized supply chain (i.e., $Q_c^* > Q^*$).* Note that the optimal order quantity in the centralized supply chain $Q_c = 561$ is higher than in decentralized supply chain, $Q = 318$. But why? The reason is that the understocking cost $C_u = p - c = 110 - 60 = 50$ is higher in the centralized supply chain than understocking cost $C_u = p - w = 110 - 97.80 = 12.20$ in the decentralized supply chain. In other words, profit margin of selling one more unit in the centralized supply chain is about $50, while it is only about $12.20 in the decentralized system. This higher profit margin results in the centralized system having a larger order size.

The higher profit margin in the centralized system refers to the concept of *Double Marginalization*. Note that in the centralized system, there is only one decision maker—Central Light. Central Light produces the product at the cost of $60 and sells it at the price of $110, a single margin of $p - c = \$110 - \$60 = \$50$. In the decentralized system, this single margin is divided between two decision makers with conflicting interests: Your margin is $w - c = \$97.80 - \$60 = \$37.80$, and Jane's margin is $p - w = \$110 - \$97.80 = \$12.20$. So, the profit margin of $12.20, which is the basis of making ordering decision in the decentralized system, is one of the two margins in the decentralized supply chain.

*Centralized supply chain has a higher total supply chain profit.* Recall that the total expected profit in the centralized supply chain is $19,533.89, which is higher than that in the decentralized supply chain, that is, $14,885.15. In fact, this is true for all supply chains. To show this, note that, for any order of size $Q$, your expected profit is $(w - c)Q$. Jane's expected profit is

$$E(\text{Jane's profit}) = p \, E(\text{sales} | \, Q) + s \, E(\text{leftover} | \, Q) - w \, Q$$

Hence, for any order quantity $Q$, the expected *total* supply chain's profit in the decentralized system is

$$
\begin{aligned}
E(\text{SC profit}) &= (w - c)Q + p \, E(\text{sales} | \, Q) + s \, E(\text{leftover} | \, Q) - w \, Q \\
&= p \, E(\text{sales} | \, Q) + s \, E(\text{leftover} | \, Q) - c \, Q
\end{aligned}
$$

which is also the total expected profit[7] in the centralized supply chain for the order size $Q$, see Eq. (11.11). Hence, order quantity $Q_c^* = 561$ that maximizes Central Light's expected profit in the centralized system also maximizes the total supply chain (sum of yours and Jane's) profit in the decentralized system. Since under wholesale price contract the order quantity in the decentralized system $Q^* = 318$ is smaller than $Q_c^* = 561$, the total supply chain profit in the decentralized system is lower than that in the centralized system. This leads us to the concept of efficiency of contracts in coordinating supply chains.

**CONCEPT**

**Efficiency of Contracts**

The efficiency of a contract in coordinating a decentralized supply chain can be measured by how much the contract can achieve the maximum possible total supply chain profit. Specifically, considering $E(\text{SC profit}(j))$ as the total expected profit of all supply chain members under Contract $j$ and $E(\text{Cent-SC profit})$ as the total expected profit in the corresponding centralized supply chain, then

$$\text{Efficiency of Contract } j = \frac{E(\text{SC profit}(j))}{E(\text{Cent-SC profit})}$$

So, your wholesale price contract has an efficiency of

$$\text{Efficiency of wholesale price contract} = \frac{E(\text{SC profit}(wholesale))}{E(\text{Cent-SC profit})} = \frac{\$14,885.15}{\$19,533.88} = 76.2\%$$

---

[7]Note that since both centralized and decentralized systems face the same demand, for any given order quantity $Q$, the average sales and the average leftover inventory are the same in both systems.

Hence, although you use your optimal wholesale price, your wholesale price contract is only capturing 76.2 percent of the maximum possible supply chain's expected profit. In other words, your wholesale price contract, even when optimized, cannot coordinate your supply chain.

### 11.10.4 Supply Chain Coordinating Contracts

You now know the following fact: There is opportunity to increase the total supply chain by $4,649.50 from $14,885.32 to $19,528.50. However, this can only occur if Jane increases her order from 318 to 561. You become curious to know what happens to your profit and Jane's profit if Jane orders $Q = 561$, instead of her optimal order $Q^* = 318$?

Well, an order of size $Q = 561$ results in an expected profit of $(w - c)Q = (97.80 - 60)561 = 21,205.80$ for you, which is about $21,205.80 - 12,020.40 = \$9185.40$ higher than when Jane orders the optimal quantity $Q^* = 318$. So in this case, placing an order of size 561 not only increases the total profit of the supply chain, but also increases your expected profit. But what about Jane's profit?

As we computed in Sec. 11.10.2, for an order of size $Q = 561$, the average sales and average leftover inventory are 429.98, and 131.02, respectively. So Jane's expected profit will be

$$
\begin{aligned}
\text{E(Jane's profit)} &= \$110 \times 429.98 + \$45 \times 131.02 - \$97.80 \times 561 \\
&= -\$1672.10
\end{aligned}
$$

In other words, Jane loses about $1672.10 if she orders $Q = 561$ that maximizes the total supply chain profit. So, there is no way you could convince Jane to change her order size. Then it hits you. May be you can use some portion of the increase in supply chain profit (i.e., some portion of $4649.50) to cover Jane's loss. But how?

Well you realize that you need a mechanism that achieves the following three goals:

(i)   It should result in Jane placing an order of size $Q_c^* = 561$ that maximizes the total supply chain profit,

(ii)   The order of size $Q_c^* = 561$ should maximize[8] Jane's expected profit; otherwise, she will not order $Q_c^* = 561$, and

(iii)   The order of size $Q_c^* = 561$ should also increase your expected profit; otherwise, you will stick with your wholesale price contract with the optimal price $w^* = \$97.80$.

A contract that achieves the above three objectives is called a supply chain coordinating contract.

---

**CONCEPT**

**Supply Chain Coordinating Contract**

A contract is called a supply chain coordinating contract if it maximizes the total supply chain's expected profit to that of the centralized system. In other words, it has an efficiency of 100 percent. The expected profits of all supply chain members under a coordinating contract are higher (or the same) as those in the decentralized system.

---

[8]Jane's maximum expected profit should not be negative; otherwise, she will not buy the projectors at all.

In the following sections we introduce several supply chain coordinating contracts, starting with the buyback contracts.

### 11.10.5 Buyback Contracts

To illustrate how buyback contracts coordinate a supply chain, we first need to understand what is preventing Jane from ordering higher than 318 projectors. Note that, because of randomness in demand, Jane faces two different risks when she places an order. Ordering too little may result in losing profit of $p - w = \$110 - \$97.80 = \$12.20$ per lost demand. On the other hand, ordering too many results in leftover inventory at the end of the holiday season that will cost Jane $w - s = 97.80 - 45 = 52.80$ per unsold projector. The optimal order quantity $Q^* = 318$ strikes the optimal balance between the costs corresponding to these two risks.

So, for Jane to increase her order size, the cost of unsold projectors (i.e., $w - s = \$52.80$) must be reduced. So, if you reduce your wholesale price $w = 97.80$, or if Jane could salvage her unsold projector higher than the salvage price $s = 45$, or both, then Jane will increase her order size. This is exactly what buyback contracts do.

---

**CONCEPT**

**Buyback Contract**

Under a buyback contract, the seller (e.g., a supplier or a manufacturer) sets a wholesale price $w_b$ for its product, and agrees to buy *all* of the buyer's (e.g., retailer's) unsold products at a buyback price $s_b$. Based on these two prices, the buyer then decides how many products to order from the seller.

---

If you want to offer Jane a buyback contract, you should sell the projector to Jane at wholesale price $w_b$, and promise her that you will buyback any unsold projector from her at price $s_b$, which is higher than the salvage value of $s = \$45$ (i.e., $s_b > s$). By setting the right wholesale price $w_b$ and right buyback price $s_b$, you can maximize the total supply chain's profit, that is, you can coordinate your supply chain. How? Very easy!

Buyback contracts eliminate the misalignment of objectives in your supply chain. Recall that your objective of maximizing your profit is in conflict with Jane's objective of maximizing her own profit. Buyback contract provides a *mechanism under which maximization of yours and Jane's profits becomes aligned with the maximization of the total supply chain profit.* How? It finds the right pair $(w_b, s_b)$ that makes Jane's expected profit to be a fraction $\pi$ (e.g., 30 percent) of the total supply chain profit and makes your expected profit to be the remaining fraction $(1 - \pi)$ (e.g., 70 percent) of the total supply chain profit. Now when you and Jane are maximizing your own profits, you are indeed maximizing the total supply chain profit. Think of the total supply chain profit as an apple pie. Jane is promised to get $\pi = 30\%$ of the pie and you are promised to get $(1 - \pi) = 70\%$ of the pie. Of course, as the size of the pie is increased, your 70 percent share and Jane's 30 percent share also increase. Hence, it is in yours and Jane's benefit to increase the size of the apple pie. The size of the apple pie would be its maximum if Jane orders 561—and under a buyback contract, she will.

But what are the values of $(w_b, s_b)$ that achieve all of these? To get those values, all we need to do is to set Jane's expected profit to be fraction $\pi$ of the total supply chain profit. Under buyback contract, if Jane orders $Q$, her expected profit would be

$$E(\text{Jane's profit}) = p\, E(\text{sales}|\, Q) + s_b\, E(\text{leftover}|\, Q) - w_b\, Q$$

since $E(\text{leftover}|\ Q) = Q - E(\text{sales}|\ Q)$, the above can be rewritten as follows:

$$E(\text{Jane's profit}) = (p - s_b)\ E(\text{sales}|\ Q) + (s_b - w_b)\ Q$$

Your expected profit under buyback contract, if Jane orders $Q$, would be

$$E(\text{Your profit}) = w_b\ Q - c\ Q - (s_b - s)E(\text{leftover}|\ Q)$$

Again, since $E(\text{leftover}|\ Q) = Q - E(\text{sales}|\ Q)$, the above can be rewritten as follows:

$$E(\text{Your profit}) = (s_b - s)\ E(\text{sales}|\ Q) + [(w_b - c) - (s_b - s)]\ Q$$

Adding your expected profit to Jane's expected profit, after some simple algebra, we get the total expected profit in your supply chain as follows:

$$E(\text{SC profit}) = (p - s)\ E(\text{sales}|\ Q) + (s - c)\ Q$$

Now, we set Jane's expected profit to be a fraction of $\pi$ of the total supply chain as follows:

$$E(\text{Jane's profit}) = \pi \times E[\text{SC profit}] \tag{11.12}$$

$$(p - s_b)\ E(\text{sales}|\ Q) + (s_b - w_b)\ Q = \pi(p - s)\ E(\text{sales}|\ Q) + \pi(s - c)\ Q \tag{11.13}$$

From the above observations we get the following insights:

- First, from Eq. (11.12), we find that *Max* $E(\text{Jane's profit}) = \pi \times Max\ E(\text{SC profit})$. Therefore, Jane's maximum expected profit is aligned with the supply chain's maximum expected profit. Hence, to maximize her profit, Jane must choose the order quantity that maximizes the total supply chain profit, that is, $Q_c^* = 561$.

- Second, from Eq. (11.13) above we can see that, to have Jane's expected profit being fraction $\pi$ of the total supply chain profit, we must have

$$\begin{cases} (p - s_b) = \pi(p - s) \\ (s_b - w_b) = \pi(s - c) \end{cases}$$

By solving the above two equations with two unknowns (i.e., $w_b$ and $s_b$), we find

$$w_b = p - \pi(p - c) \quad \text{and} \quad s_b = p - \pi(p - s) \tag{11.14}$$

So, suppose you want to have 80 percent of the total supply chain's profit and give Jane the remaining 20 percent (i.e., $\pi = 20\%$). Then all you need to do is to offer Jane a buyback contract with wholesale price $w_b = p - \pi(p - c) = 110 - 0.2(110 - 60) = \$100$ and buyback price $s_b = p - \pi(p - s) = 110 - 0.2(110 - 45) = \$97$. This will guarantee that Jane will order 561 projectors and the total supply chain's profit reaches its maximum of \$19,533.89. Also, you will get $80\% \times \$19,533.89 = \$15,627.11$ of the profit—which is \$3606.71 higher than what you get under your wholesale price contract. Jane, on the other hand, gets $20\% \times \$19,533.89 = \$3906.78$ of the profit—which is \$1042.80 higher than what she gets under your wholesale price contract. A win-win situation!

Buyback contracts are commonly used in different supply chains, especially for products that have low variable costs, for example, newspapers, books, software, music, and fashion apparel. Another contract that can coordinate supply chains is Revenue Sharing Contract that we discuss next.

### 11.10.6 Revenue Sharing Contract

You decide to give Jane a call and tell her about your buyback contract. You are certain that she will accept it. Your wholesale price in the buyback contract, i.e., $w_b = \$100$ is only $2.2 higher than that in your wholesale price contract ($w^* = \$97.80$). However, Jane can get $s_b = \$97$ for each unsold projector from you, rather than getting $s = \$45$ from the discount store.

You call Jane to propose your buyback contract, but Jane interrupts you and says that she has a proposal for you. She asks you if you are willing to sell your projectors to her for $30 each; in return, Jane gives you 50 percent of the selling price of every projector that she sells. That means $50\% \times \$110 = \$55$ of every sales during the holiday season, and $50\% \times \$45 = \$22.5$ of every sales after the holiday season. You pause for a second to comprehend. Her suggested wholesale price $30 is half of what it costs you to produce a projector, that is, half of $c = \$60$. However, getting half of Jane's total sales seems an attractive proposal. Of course, it all depends on how many projectors Jane will order and sell, if you accept her proposal. For you, the main question is whether you make more profit under Jane's new proposal, or under your buyback contract. You tell Jane that you need to think about this and you will get back to her tomorrow with an answer.

What Jane is suggesting is called Revenue Sharing Contract.

---

**CONCEPT**

**Revenue Sharing Contract**

Under a revenue sharing contract, the seller (e.g., a supplier or a manufacturer) sets a wholesale price $w_r$, and the buyer (e.g., a retailer) agrees to keep only fraction $\delta$ of the *sales* of each product and gives the seller the remaining fraction $(1 - \delta)$. Based on the wholesale price $w_r$ and fraction $\delta$, the buyer then decides how many products to order from the seller.

---

Now that you know what revenue sharing contract is, the questions that come to your mind are: (i) Can revenue sharing contract coordinate your supply chain? (ii) If it can, does it have a mechanism to allocate the total supply chain to you and Jane according to a certain percentage? And most important of all, (iii) Does Jane's proposed revenue sharing contract result in a higher expected profit for you than your buyback contract?

Can revenue sharing contract coordinate your supply chain (of projectors)? To answer this, you need to find if there exists a pair of $(w_r, \delta)$ that can align Jane's and your objectives with the objective of maximizing the total supply chain profit. You now know that one way to do this is to set Jane's expected profit to be a fraction, say $\pi$, of the total supply chain profit. But what is Jane's expected profit under a revenue sharing contract? Not that complicated; if Jane orders $Q$, then her expected profit under revenue sharing with parameters $(w_r, \delta)$ is

$$E(\text{Jane's profit}) = \delta \left[ p\, E(\text{sales}|\, Q) + s\, E(\text{leftover}|\, Q) \right] - w_r\, Q$$

Considering $E(\text{leftover}|\, Q) = Q - E(\text{sales}|\, Q)$, and using it in the above equation, after some algebra we get

$$E(\text{Jane's profit}) = \delta(p - s)\, E(\text{sales}|\, Q) + (\delta s - w_r)\, Q$$

The total supply chain's expected profit under an order of size $Q$ under revenue sharing contract is $E(\text{SC profit}) = (p - s)\, E(\text{sales}|\, Q) + (s - c)\, Q$ (you can check that). Thus, to set Jane's profit to be fraction $\pi$ of the supply chain profit, we have

$$E(\text{Jane's profit}) \;=\; \pi \times E(\text{SC profit})$$
$$\delta(p - s)\, E(\text{sales}|\, Q) + (\delta s - w_r)\, Q \;=\; \pi(p - s)\, E(\text{sales}|\, Q) + \pi(s - c)\, Q \quad (11.15)$$

From the above observations we get the following insights:

- First, from Eq. (11.15) we can see that, to have Jane's expected profit being fraction $\pi$ of the total supply chain profit, we must have

$$\begin{cases} \delta(p - s) = \pi(p - s) \\ \delta s - w_r = \pi(s - c) \end{cases}$$

  By solving the above two equations with two unknowns $w_r$ and $\delta$ we find

$$w_r = \pi c \quad \text{and} \quad \delta = \pi \qquad (11.16)$$

  In other words, there exists a revenue sharing contract with the above parameters that can coordinate the supply chain resulting in Jane ordering $Q_c^* = 561$ and maximizing the total supply chain profit to \$19,533.89.

- Second, the revenue sharing contract can allocate fraction $\pi$ of the total profit to Jane (the retailer) and the remaining fraction $(1 - \pi)$ to you (the manufacturer). This is simply done by setting the parameter $\delta$ of the contract to $\pi$ as shown in Eq. (11.16).

Knowing the above calculations, the first question comes to your mind is whether the revenue sharing contract that Jane suggests with wholesale price $w_r = \$30$ and Jane's revenue sharing factor $\delta = 50\%$ is the coordinating contract. Considering Eq. (11.16), you find that her proposed contract is indeed the coordinating contract. Specifically, under her contract $\pi = \delta = 50\%$ and $w_r = \pi c = 50\% \times \$60 = \$30$. This implies that, if you accept Jane's proposal, your expected profit would be $(1 - \pi) = (1 - 0.5) = 50\%$ of the total supply chain (i.e., $50\% \times \$19,533.88 = \$9766.94$) and Jane will get the remaining 50% (i.e., \$9,766.94). This would be less than what you had in mind when you designed your buyback contract. You wanted to get 80 percent of the total profit.

To get 80 percent of the total profit under a revenue sharing contract, you must set $\pi = 0.2$. The wholesale price and revenue sharing factor would then be $w_r = \pi c = 0.2(\$60) = \$12$, and $\delta = \pi = 20\%$. So, you decide to call Jane and offer the following two options as staring point of negotiation: (i) a buyback contract with $w_b = 100$ and $s_b = 97$, or a revenue sharing contract with $\delta = 20\%$ and $w_r = \$12$. You expect that you would have a long negotiation with Jane and finally come up with a mutually acceptable contract. You are confident that the final contract will give you more than 50 percent of the total profit, since you know that your brand is well known and Jane would do everything to keep its relationship with your company. In other words, your firm has more power in the supply chain than Jane's firm.

### 11.10.7 Other Coordinating Contracts

All coordinating contracts have one thing in common: they have a mechanism that allows members of the supply chain to share the risk. To be specific, let's consider the buyback contract as an example.

Under the wholesale price contract, Jane—the retailer—faces an uncertain demand. Due to the variability in the demand, Jane faces the risk of having unsold projectors at the end of the holiday season that costs her about $w - s = \$97.80 - \$45 = \$52.80$ per unsold projector. On other hand, you—the manufacturer—do not face any risk of unsold projector (i.e., leftover inventory), since Jane places an order of $Q^* = 318$ three months in advance and you produce exactly the same amount and sell it to her. The high cost of leftover inventory (i.e., \$52.80) prevents Jane to order $Q_c^* = 561$ that coordinates the supply chain.

With buyback contract $(w_b = \$100, s_b = \$97)$, you share some of Jane's risk of unsold projectors as follows. Each unsold projector will cost Jane about $w_b - s_b = \$100 - \$97 = \$3$ which makes it optimal for Jane to increase her order size to $Q_c^* = 561$. You now also face the risk of unsold projectors that cost you about $s_b - s = 97 - 45 = 52$, because you buy them from Jane at \$97 and sell them to the discount store for \$45. In fact, an unsold projector costs you much more than it costs Jane. However, you cover this cost by charging a higher wholesale price $w_b = \$100$ (instead of \$97.80) and receiving a higher order of size 561 instead of 318.

Revenue Sharing contract also increases the supply chain profit by letting you share some of Jane's risk of unsold projectors. This is done by tying your profit to Jane's sales. Specifically, by sharing fraction $(1 - \delta)$ of Jane's expected sales, you also accept fraction $(1 - \delta)$ of the cost associated with the risk of unsold projectors. This results in Jane increasing her order size to supply chain coordinating order $Q_c^* = 561$.

There are several other contracts that facilitate risk sharing among supply chain members and can coordinate supply chain, some of which we introduce below:

- *Quantity Flexibility Contract:* Quantity flexibility contract is similar to the buyback contract with a small twist. Under quantity flexibility contract the seller (e.g., a supplier or a manufacturer) allows the buyer (e.g., a retailer) to return the unsold item at full wholesale price up to some limit. While, under the buyback contract the seller buys back *all* unsold units at lower-than-full wholesale price, under quantity flexibility contract the seller accepts only a *certain number* of unsold units, but pays the buyer the full wholesale price. With the quantity flexibility contract, the seller shares some risk of having unsold units by fully refunding the buyer for some of its purchased order.

- *Sales Rebate Contract:* With a sales rebate contract, the seller offers a rebate to the buyer for a certain amount for each unit that buyer sells beyond a certain threshold. This motivates the buyer to order more, and if the contract is properly designed, the contract can coordinate the supply chain.

- *Option Contracts:* Under an option contract, the seller (e.g., a manufacturer) divides its wholesale price $w$ into two parts: (i) the option price $w_o$, also called the reservation price, and (ii) an exercise price $w_e$, which is paid to exercise the option, where $w_o + w_e = w$. Before the selling season starts, the retailer purchases options for $Q$ units from the manufacturer at per unit option price of $w_o$. Each option gives the retailer the right (not the obligation) to buy each unit of $Q$ at the exercise price of $w_e$, after the retailer realizes its demand. When retailer realizes its demand, its order size is equal to the observed demand. Therefore, the retailer pays the exercise price only for those products that it actually sells. Note that since the retailer does not pay the exercise price for the options it does not use, it will order higher than the order size in a wholesale price contract. By finding the right option and exercise prices, one can coordinate the supply chain in many cases. Option contracts are common practice in *Contract Manufacturing* in which retailers pay reservation price to reserve $Q$ units of the manufacturer's capacity for production of their items.

## 11.11 Centralized Control and Vendor-Managed Inventory

As mentioned in Sec. 11.10.3, one source of inefficiency in the decentralized supply chains is due to the fact that several decision makers with conflicting incentives are controlling the activities in different stages of the supply chain. Coordinating contracts discussed in the previous section was one approach to align decision makers' incentives to increase each member's and the supply chain's profits. Another approach is to give authority to one decision maker to make decision for all stages of the supply chain—creating a centralized control system.

Walmart and P&G were among the first that developed several new approaches to coordinate their supply chains by centralizing decision-making process and sharing information. These new innovations not only align supply chain members' incentives, but also mitigate the bullwhip effect. We discuss some of these innovations here.

### Vendor Managed Inventory (VMI)

Under Vendor Managed Inventory, an upstream supply chain member—a manufacturer or a supplier—makes decisions regarding the inventory levels at a downstream member—a retailer. Specifically, the supplier decides when and how much of an item should be shipped to the retailer. Supplier's decision is made based on an objective (e.g., fill rate) that both parties agreed upon. In most cases, the supplier actually owns the inventory until it is sold at the retailer. Implementing VMI requires that the retailer shares its demand and inventory information with the supplier. To eliminate the bullwhip effect both parties eliminate price fluctuations such as sales and promotions.

### Collaborative Planning Forecasting and Replenishment (CPFR)

While in VMI the inventory decisions are often made by a single member of the supply chain, in CPFR supply chain members collaborate in making such decisions. Specifically, under CPFR members of a supply chain collaborate in some or all of different activities including production and inventory planning, sales and promotions, sales forecast, order plans, shipping, and stocking products. This close collaboration allows members of the supply chain to better align their objectives and prevent or mitigate the bullwhip effect.

## 11.12 Summary

This chapter provides a list of strategies and practices that can improve the performance of inventory systems, ranging from tactical decisions such as reducing supplier lead time to strategic decisions such as redesigning the supply chain.

Classifying inventory to three classes according to ABC analysis allows firms to reduce their inventory costs by allocating more effort to more critical class of items. Reducing order setup cost reduces cycle inventory in both continuous and periodic review systems. Order setup cost can be reduced by reducing the transaction costs (e.g., using information technology) and reducing transportation costs (via negotiation, order assortment, or third-party logistics). Reducing both lead time and its variability also results in a decrease in safety inventory. This can be done using lean operations tools, cross-ducking, and choosing suppliers that are in close proximity.

Aggregating customer demand is a powerful strategy to improve supply chain inventory costs. Demand aggregation provides benefits of both economies of scale and variability reduction. One way to aggregate customer demand is to pool inventory of several locations into one location. This reduces inventory costs per unit sold and allows firms to offer more variety of products at lower costs. Another approach to aggregate demand is to open a centralized distribution center. Centralized distribution reduces inventory costs more if it is located closer to the demand (e.g., closer to stores). Other strategies for demand aggregation are (i) virtual inventory pooling where available inventory in one store can be used to satisfy the demand of a customer who faces stockout in another store, for example, by shipping the product to the customer; (ii) specialization in which the safety inventory of each product is pooled into only a few stores that have large demand for that product—the location specializes in handling that product; (iii) product substitution where the demand for one product (which is out of stock) is satisfied using inventory of another product; and (iv) component commonality in which several versions of a product use the same common component.

One source of inefficiency in supply chain is the bullwhip effect. Bullwhip effect is present in a supply chain, if demand variability is higher at upstream stages of the supply chain (e.g., manufacturers or suppliers) than that in downstream stages (e.g., retailers). The increase in demand variability results in increase in production and inventory costs. The main causes of bullwhip effect are (i) demand forecast updating, (ii) order batching, (iii) price fluctuations, and (iv) rationing and shortage gaming. Sharing demand information with supply chain members eliminates or reduces the negative impact of demand forecast updating. Strategies to reduce order batching is to reduce order setup cost and to coordinate orders with several retailers. To reduce variability in demand caused by price fluctuations, firms can limit the maximum purchase under discounted price, offer volume-based (not order-based) discount, or implement the everyday low price strategy. Strategies to prevent rationing and shortage gaming include choosing an appropriate rationing policy, restricting order cancellations, sharing capacity/availability information, and preventing shortage. Vendor managed inventory (VMI) and collaborative planning forecasting and replenishment (CPFR) systems improve supply chain performance by improving information sharing among supply chain members as well as facilitating supply chain members to collaborate in activities such as production, inventory, sales forecast and planning, shipping, and stocking products.

Another cause of inefficiency in the supply chains that results in loss of profit is the conflicting incentives of supply chain members. A supplier, for example, makes more profit by increasing the price to a manufacturer, while this reduces manufacturer's profit. Supply chain coordination refers to strategies to overcome this source of inefficiency by coordinating activities at different stages of a supply chain—directly or indirectly—to increase the total supply chain profit. One mechanism for supply chain coordination is to design contracts that can align incentives of supply chain members and increase supply chain efficiency. Contracts such as buyback contract, revenue sharing contracts, quantity flexibility contracts, sales rebate contracts, and options contracts—if properly designed—can increase the total profit of the supply chain as well as the profit of each member of supply chain.

## Discussion Questions

1. Describe three strategies that firms can use to reduce their transportation costs of receiving orders from the suppliers.

2. What is "cross ducking" and how can it impact inventory costs in supply chains?

3. Explain why pooling inventory allows firms to offer a wide range of products.

4. Besides saving in inventory, describe four other factors that should be considered when making decision to pool inventory.

5.  What is "virtual inventory pooling"? Provide examples of virtual pooling and explain how it helps reducing inventory costs in supply chains?

6.  What is "specialization" in the context of demand aggregation? How does it help improving inventory costs?

7.  What is "customer-driven substitution" and how does it differ from "manufacturer-driven substitution"? Provide an example for each one.

8.  What is "component commonality" and how can it decrease production and inventory costs?

9.  What is "bullwhip effect" and what are the four main causes of bullwhip effect?

10. What is "demand forecast updating" and what are the actions that can reduce its negative impact on supply chain inventory costs?

11. Describe three actions that a firm can take to mitigate the impact of order batching that results in bullwhip effect.

12. Describe three strategies that can diminish or eliminate the negative impact of price fluctuations and speculation on supply chain performance?

13. Explain how shortage gaming and rationing intensifies bullwhip effect and describe four strategies that can be implemented to diminish their negative impact.

14. What is supply chain coordination and what are the two major causes of lack of coordination in supply chains?

15. What is VMI and how can it improve supply chain performance?

## Problems

1.  Suppose a supplier delivers a product to your facility every 4 weeks and you need to place your orders 2 weeks before the delivery date. You place orders according to order-up-to policy with order-up-to level 350 to achieve cycle-service level of 95 percent. You would like to reduce your inventory costs while still keeping your cycle-service level at 95 percent. Explain why each one of the following statements is true or false.
    a.  If you could place your order 1 week before the delivery date, you could decrease your inventory costs.
    b.  If the supplier delivers every 3 weeks (instead of every 4 weeks), you could decrease your inventory costs.
    c.  Parts (a) and (b) do not reduce inventory unless the variability in demand is also decreased.

2.  Mabt, a large home appliance retailer, has a store in which the daily demand for a refrigerator is Normally distributed with mean 5 and standard deviation 2. The store orders refrigerator from a manufacturer at the price of $1200 and sells it at $1300. The manufacturer has lead time which is variable with mean of 5 days and standard deviation of 4 days. The order setup cost is $80 and is independent of the order size. The store has an inventory carrying rate of 35 percent for refrigerators. The store manager knows that a shortage of a refrigerator results in an immediate loss of profit of $100, because customers who face shortage are lost. The store manager also knows that the frequent shortage impacts the profit in the long term, but is not able to estimate the long-term cost. Therefore, the manager decides to set its ordering policy to reduce the percent of its customers who face shortage to 5 percent or less. What do you think about the manager's choice of 5 percent? Support your answer with numbers.

3.  Precision Golf (PG) is a manufacturer of golf clubs. The production manager of PG is struggling with its inventory of parts and material used to make clubs. He often faces the shortage of some parts that interrupts production, while he has too many of noncritical parts that is

increasing his inventory cost. The following is the list of the main parts used in production of golf clubs, the numbers needed for production in a month, and their costs.

| Part no. | Monthly demand | Unit cost ($) |
|---|---|---|
| GC-01 | 107 | 22.75 |
| GC-02 | 15 | 112 |
| GC-03 | 64 | 45.25 |
| GC-04 | 54 | 145.25 |
| GC-05 | 98 | 108.25 |
| GC-06 | 378 | 195 |
| GC-07 | 312 | 8.5 |
| GC-08 | 643 | 6.75 |
| GC-09 | 237 | 3.75 |
| GC-10 | 617 | 98.2 |
| GC-11 | 554 | 19.75 |
| GC-12 | 95 | 11.59 |
| GC-13 | 353 | 10.75 |
| GC-14 | 1575 | 4.25 |

Use ABC analysis and classify parts to help PG better allocate effort in managing their inventory costs. What are your suggestions for each class of items?

4. Mabt, the retailer in Problem 2, has 10 stores in Chicago area. One of their product is Oven-SX, which is a gas oven that is ordered from a manufacturer in Los Angeles. Each store faces similar demand and has been ordering the oven according to a $(Q, R)$ policy to achieve cycle-service level of 95 percent. Due to low demand in each store, Mabt is planning to hold inventory of the oven only in 1 of the 10 stores—the store in downtown. The store in downtown is larger than other stores and can hold more inventory. It is also close to other nine stores. The idea is to keep only one oven in the other nine stores for display. When a customer orders the oven, the oven will be shipped from the store in downtown. To keep the availability the same, the plan is to keep the cycle-service level for the oven in downtown store at 95 percent. If this plan is implemented, explain why each of the following statements is true or false.
    a. The safety stock of the oven in the downtown store would be smaller than the safety stock in each stores before the plan is implemented.
    b. The safety stock of the oven in the downtown store would be smaller than the total safety stock of the 10 stores before the plan is implemented, but the chance of stock-out would be larger in the store in downtown than that of the 10 stores before the plant is implemented.
    c. The inventory turn of the oven in the downtown store would be larger than that before the plan is implemented.
    d. More information is needed to make any conclusion.

5. Before implementing their plan for Oven-SX, one of the Mabt managers in Problem 4 recommends implementing the plan for another oven, that is, Oven-BT, which has lower sales in the 10 stores and is ordered from a manufacturer in Atlanta. All stores purchase Oven-SX at price $800 with order lead time of 3 weeks. Their order setup cost is $1200. The weekly

demand for the oven at each of the 10 stores is Normally distributed with mean 8 and standard deviation 2. On the other hand, all stores purchase Oven-BT at price $1400 with order lead time of 1 week. Their order setup cost is $850. The weekly demand for the oven at each of the 10 stores is Normally distributed with mean 3 and standard deviation 1. Stores have inventory carrying rate of 25 percent and their demands are independent of each other. Which one of the two ovens would you recommend to be kept in the store in downtown? Assume 52 weeks in a year.

6. Billy Bookstore would like to centralize the ordering process of its 10 stores in Seattle for ordering marble bookends through its Distribution Center (DC) in Columbus Ohio. Currently, each store orders bookends separately from a supplier in Alabama at the price of $28 with order lead time of 4 weeks. The order setup cost (including shipment cost) is $320, which is independent of the order size. Stores have an inventory carrying rate of 25 percent. The weekly demand for the bookends at each store is almost the same and is independent of each other and is well-approximated by a Normal distribution with mean 12 and standard deviation 4 (assume 52 weeks in a year). Each bookstore orders according to $(Q, R)$ policy such that the chance of stockout is reduced to 3 percent. The plan is that DC would order the bookends from the supplier according to a $(Q, R)$ policy and then stores place their orders to the DC. The supplier's lead time and order setup cost to deliver bookends to the DC are 3 weeks and $250, respectively. The DC's lead time and order setup cost to deliver bookends to each store are 1 week and $150, respectively. To have enough bookends at the DC to make sure it can always satisfy stores' orders without delay, the DC is run with cycle-service level of 99 percent. Based on past data of orders sent from all stores to the supplier, it is expected that the DC will face the weekly demand from stores that is Normally distributed with mean of 120 and standard deviation of 30. Should Billy Bookstore use its DC to order bookends, or allow bookstores to order directly from the supplier in Columbus?

7. You work in the marketing department of a clothing company. A large movie studio is planning to release its highly anticipated movie in the beginning of summer—3 weeks before the release of the new Star Wars movie. One of the movie's promotional product is a T-shirt with the picture of the main character. The movie studio has contacted you and asked you how much you charge for a T-shirt, if they sign a contract with you to make the T-shirts for them. You ask for more information, and the movie studio provides the following: (i) they are planning to offer the T-shirts online for $24.99, and if the T-shirts are not sold before the release of the Star Wars movie, their market value is reduced. However, all unsold T-shirts can be sold at discount price of $10. (ii) It is estimated that the demand for the T-shirts be about 20,000 with standard deviation 6000 (assume Normal distribution). You inform the movie studio that because you already made your production plan, you can only set up your operations once to produce T-shirts for the movie studio. That means that the studio can place only one order 2 months before summer. The studio accepts your condition and awaits you to tell them what price you charge for a T-shirt. If it costs you $12 to make a T-shirt,
   a. What wholesale price would you sell the T-shirt to the studio in order to maximize your expected profit?
   b. What is the efficiency of the wholesale price contract in Part (a)?

8. The movie studio in Problem 7 informs you that another firm has offered to make the T-shirt for $20. However, they are willing to work with you, if you sell the T-shirt for $20.
   a. What would be your expected profit, if you sell your T-shirts for $20 to get the movie studio's contract?
   b. Design a buyback contract that you can offer to the movie studio to maximize your profit.
   c. Design a revenue sharing contract that you can offer to the movie studio to maximize your profit.

# Aggregate Planning

## 12.0 Introduction

One important source of poor performance of an operations is that supply and demand of the operations do not match. When a firm produces a product more than the market demand, the unsold products often become obsolete or have to be sold with deep discount. In contrast, when a firm produces less than the market demand, it faces costs related to unsatisfied demand. This includes the direct loss of profit due to stockout and the lost sales cost of unhappy customers who may switch to a different brand in their future purchases. Hence, one important and difficult task of operations engineers or managers to improve the process is to find ways to match supply with demand. What are the levers that one can use to match supply with demand? Well, there are mainly two options: (i) either adjust demand to match with supply, or (ii) adjust supply to match with demand, or both.

The focus of this chapter is to provide cost-effective strategies and analytics that can be used to either adjust the demand or the supply. After introducing ways to match demand with supply, this chapter focuses on "Aggregate Planning," which is a very effective method to match supply with demand.

## 12.1 Adjusting Demand

How can a firm affect demand so that it matches with its capacity (i.e., supply)? In general, this is a very difficult task and—if it is not done carefully—can have a significant impact on a firm's profits. There are, however, several strategies that have shown to be successful. A firm can influence customer demand by providing *incentive* such as price incentive. For example, airlines and hotels offer cheaper price for periods with low demand for travel (i.e., low season) in order to encourage customers to purchase their products (i.e., a seat in a flight or a room in a hotel). Another option is *backorder,* that is, postponing satisfying demand in the current period to future periods. The key factors in this approach are the customers' willingness to wait and the cost associated with late deliveries. If there are comparable products in the market, customers often do not wait and purchase the products from competitors. Also, it is difficult to accurately estimate the cost of backorders, since it includes the direct costs (e.g., additional transactions and shipment costs) and indirect costs (e.g., cost of unhappy customers and its possible impact on the customers' future purchases).

## 12.2 Adjusting Supply

The other option of matching supply with demand is to increase or decrease supply of the product to match with the demand in the market. There are several ways that a firm can change its supply of products.

### Increasing Capacity by Working Overtime

Not many companies work three shifts (i.e., 24 hours a day); therefore, it is possible to add additional overtime hours to a working day to increase the capacity of the process. Overtime is the most commonly used way of increasing capacity, but it has some limitations. First, the worker's wage for overtime hours is about 50 percent higher than that for working regular hours. Second, worker's fatigue due to working additional hours has a negative impact on worker's productivity, on product quality, and on worker's safety (e.g., increases the possibility of accidents). This will increase the labor and quality costs per unit of production. Finally, in some industries, firms have contracts with their workers' union allowing workers to refuse to work overtime, or limiting the number of hours of overtime in a day or in a week. Hence, while overtime is quick and easy to implement, it should only be considered as a short-term option for increasing the capacity.

### Increasing Capacity by Hiring More Workers

A long-term option for increasing capacity in labor-intensive operations is to hire more workers. When making decision on hiring new workers, several factors must be considered. Hiring incurs recruitment costs such as evaluating applicants, verifying employment, etc. After hiring, workers go through training, during which their productivity and product quality is low. After training, the main cost is their salary, including fringe benefits. Similar to overtime, union contracts may limit the number of hiring (and layoffs) that a firm can do. In industries with seasonal demand in which work is highly labor-intensive with low to medium skill required, hiring part-time workers is a good option. Examples of such cases are retail stores and restaurants. The salary and fringe benefits of part-time workers are often less than those for regular workers. The problem, however, is that the unions often consider part-time workers as threat (e.g., threat of replacing the full-time workers) and that part-time workers do not pay union dues.

### Increasing Supply Through Subcontracting

Subtracting is when a firm hires another firm to produce a part, a subassembly or even a product to perform a service. Subcontracting does not increase the capacity of a firm to produce a unit; it only increases the supply of the unit. The advantage of subcontracting is that the firm can increase the supply without investing in labor, equipment, or material. The disadvantage of subcontracting is that the firm has less control over the quality of the sub-contracted units. Also, subcontracting incurs additional cost of shipping parts or material and cost of receiving the finished units from the subcontractor. Finally, there is often a limit on the minimum or the maximum number of units that can be acquired through subcontracting. The minimum limit corresponds to the fact that it may not benefit the subcontractor to produce less than a certain amount. This is due to the setup cost that the subcontractor may incur for setting its operations to produce the required parts. The maximum limit corresponds to the maximum capacity of the subcontractor, or the maximum supply available in the market.

### Increasing Supply Using Inventories

Another way to increase the supply of a part (or a product) is to produce the part in a period and keep it in inventory to be used in future periods. This is a common strategy in firms with large fluctuations in their demands. This strategy allows these firms to keep a stable throughput during the year equal to the average demand. In periods that the demand is low (i.e., lower than the average), the unsold units are kept in inventory. These units will later be used in periods in which the demand is high (i.e., higher than the average). Holding inventory, of course, incurs costs including opportunity cost of investing in inventory, cost of holding inventory in warehouses, labor cost of handling and moving inventory, and insurance cost. See Sec. 9.4 of Chap. 9 for details.

There are also options to reduce supply to match it with demand, some that are discussed next.

### Decreasing Capacity Through Undertime

Undertime or slack time is time of a process during which the process remains idle and does not produce any item. This obviously decreases the throughput of the process and prevents the process from building inventory. While undertime does not increase the labor cost per unit time, it does increase the labor cost per unit produced, because less units will be produced during the regular working hours. Other issue with undertime is its impact on workforce morale; they may think that the firm is losing market and soon they may be laid off. Hence, undertime should be considered as a short-term solution to reduce throughput. If the firm decides to have undertime, then it can use those times for activities such as training, problem solving, and process improvement. These activities improve workforce morale and knowledge as well as improve process performance.

One principle of capacity management is that, because of the variability in production environments, processes should always maintain some level of undertime to react to unpredictable increase in demand. In other words, processes should always maintain an effective capacity larger than their average demand. The additional capacity beyond average demand, as introduced in Chap. 7, is called *Safety Capacity* or *Capacity Cushion*. With safety capacity, the utilization of the process will be less than 100 percent. How much safety capacity should a firm have? The answer depends on many factors including the variability in demand and processes, as well as the firm's competitive advantage. For example, firms that compete on cost will generally have a small safety capacity (i.e., higher utilization), while firms that compete on quality, speed, or flexibility employ larger safety capacity. See Chap. 7 for a more detailed discussion on Safety Capacity.

### Decreasing Capacity by Reducing Workforce

If the company is facing frequent idle times and the demand forecast indicates a decrease in demand, a firm may consider reducing its workforce by firing some of its workers. Similar to hiring, firing workers also costs firms. Costs include the severance pay that the firm must pay full-time workers to terminate their employment. Besides this tangible cost, the firm also incurs intangible costs such as the cost of losing knowledge and know-how of the fired workers. The worker who is assigned to do the task of the fired worker must go through training and it will take some time to acquire the know-how. The other intangible cost is the cost of morale: the remaining workforce may feel that the firm is not doing well and they may also be fired. This affects worker's productivity and product quality.

## 12.3 Goals of Aggregate Planning

As discussed above, an operations engineer or manager has several options to balance the supply and demand. However, the challenge is how to effectively use these options to develop production plans that match supply with demand with minimum cost. This is the goal of aggregate planning.

*Aggregate Planning* aims to develop a plan that uses firm's resources effectively to satisfy expected demand. This includes finding production mix and volumes, changes in workforce level, required overtime, backorders, and subcontracting. For example, if the demand in the next 2 months will be higher than the demand in this month, the planner faces questions of whether: (i) to produce more products this month to be used in the next 2 months, or (ii) to use overtime in the next 2 months to satisfy demand, or (iii) to hire some part-time workers next month to increase production, or (iv) to satisfy the demand by subcontracting the product to another manufacturer, or (v) a combination of these policies. The answer is not easy and depends on the cost and limitations of each of these options. The goal of aggregate planning is to find the most cost-effective solution to above questions.

Aggregate planning is also called macro planning. It is an intermediate capacity/supply planning for a time horizon of 2 to 12 months. This intermediate time horizon is not long enough to build a new production facility, but it is feasible to build inventory or increase (or decrease) capacity by overtime, subcontracting, hiring, and firing.

## 12.4 Aggregate Unit of Production

Aggregate planning is established based on the concept of aggregate unit of production. Specifically, planners group the large number of products that they offer into a few family of products. Consider a producer of dairy products that produces many different types of milk (e.g., whole, 2 percent, 1 percent, skimmed, lactose-free milk) in different sizes (e.g., 8 ounce, 1 gallon, 2 gallons) and different packaging (e.g., bottles or paper boxes), as well as many different types of yogurt (e.g., plain, 2 percent, whole milk, with fruit yogurt) with different sizes (e.g., small, medium, and large), and also produces different types of cheese (e.g., cream cheese, Swiss, American, provolone) in different packaging (e.g., small, large, and sliced).

The aggregate planning approach groups all these (tens or hundreds of) products, for instance, into three groups of milk, yogurt, and cheese. Each of these products is called an aggregate product. The production unit for the aggregate products milk and yogurt can be a gallon, and the unit for cheese can be a pound. Hence, instead of finding how many bottles of skimmed, whole, 1 percent, 2 percent, and lactose-free milk the firm should produce in the next 6 months, aggregate planning focuses on finding how many gallons of milk the firm should produce in the next 6 months.

Sometimes all products that a firm offers can be aggregated into one. For example, a bicycle manufacturer that produces many different types of bicycles (e.g., speed bikes, mountain bikes, hybrid bikes) and in different sizes may aggregate all these products into one aggregate product, that is, "a bike." Other examples are steel producers that can define their aggregate product to be "one ton of steel," or TV producers that can aggregate all models and sizes of their TVs into one aggregate product: "a TV."

Aggregate planning is a useful tool that helps managers make more informed production, inventory, and workforce planning decisions for intermediate and long-term future. First, developing such plans for tens or hundreds of products is very complex. Aggregating these products into a few aggregate products allows planners to develop analytics to find cost-effective plans. Second, as mentioned in Chap. 3, forecasting demand for a group of products is more accurate than that for each individual product. Hence, aggregate planning has the advantage of developing overall production and workforce plan based on a more accurate demand forecast.

The concept of product aggregation was discussed in Sec. 4.10 of Chap. 4. In that chapter it is shown how the effective process time of individual products can be used to compute the effective process time of the aggregate unit of production. After deciding how products should be aggregated into a few groups, the planner's task is to find the demand forecast for each aggregate product in each planning period—see Chap. 3 for demand forecasting methods. In most cases, the *planning period* is a month and the *planning horizon* is between 2 and 12 months. However, depending on the application, one may decide to choose a week or a quarter (i.e., 3 months) as the planning period and 2 or more years as the planning horizon.

## 12.5 Aggregate Resources

In a production system products are processed by several different resources. These resources include labor, tools, and machines, among others. For example, in an auto assembly plant, a car goes through tens and hundreds of different stations before it is completed. Incorporating these tens and hundreds of different resources into an aggregate planning model is very complex. Hence, similar to the aggregate product that represents a group of products, we can define an *Aggregate Resource* to represent a pool of several resources.

In processes that are very labor-intensive (such as call centers), the main aggregate resource is labor. Other resources such as phones and computers do not play a significant role in planning services and workforce. In processes that rely on machines and operators, where both can become a bottleneck, resources can be aggregated into two aggregate resources: Equipment and Labor. Another option, which is often used, is to aggregate based on their functions. For example, one can aggregate all the workers and equipment in a drilling department into one aggregate resource: Drilling Department.

The decision of how to aggregate resources depends on many factors such as the resource's contribution to the final product as well as the variable and fixed costs of resource, among others. When resources are aggregated into one or a few resources, for each resource the following must be determined. First, if the aggregate resource is a pool of resources (e.g., labor in the call center example), we must determine how many of the resources are in the pool (e.g., how many agents are in the call center). Second, it should also be determined how long it takes for the aggregate resource to process each unit of aggregate product (e.g., how long it takes on an average for an agent to respond to a phone call). Third, how many hours the aggregate resource is available during each period (e.g., what are the working hours of the agents in a week or in a month?).

Establishing the concepts of aggregate units of production and aggregate resources, the remaining of this chapter is devoted to presenting some basic and fundamental aggregate planning analytics that are very effective in finding cost-effective production, inventory, and workforce plans. However, while the general aggregate planning is concerned with developing those plans for multiple future periods, it is useful to first show how aggregate planning helps to develop such plans in single-period problems.

## 12.6 Aggregate Planning—Single-Period with Multiple Resources

There are several approaches to solve aggregate planning problems, most of which are ad hoc and are based on some simple rules of thumb. While these approaches lead to simple solutions with reasonably low cost, they do not result in the minimum costs. Mathematical Programming approach that is discussed in this chapter, however, guarantees to result in plans with minimum cost (or maximum profit). We illustrate the mathematical programming approach using the most basic aggregate planning problem called the "Product-Mix" problem.

### 12.6.1 The Product-Mix Problem

The product-mix problem arises when firms produce more than one product. Since each product takes a different time to make and has a different production cost and selling price (and thus different profit), the question is how many of each product should be produced in a period. We use the following example to illustrate this problem.

**Example 12.1**

Fancy Windows is a producer of 82 different types of high-end, expensive, and hand-crafted doors, windows, and sunroofs. Specifically, the firm produces 27 different types of windows, which include awning windows, casement windows, double-hung windows, single-hung windows, bay or bow windows, sliding windows, and windows with special shapes. The firm also produces 43 different types of doors and 12 different types of sunroofs. To develop an aggregate plan for the next month's production, the production manager has grouped the 82 products into three aggregate products "windows," "doors," and "sunroofs." These products are produced in three different plants, namely, fabrication plant, metalshop plant, and assembly plant. Figure 12.1 shows the process-flow chart of the three aggregate products.

As the figure shows, producing a sunroof requires processing in the metalshop plant and then the assembly plant. Producing a window, on the other hand, requires processing in the fabrication plant and then in the assembly plant. Finally, producing a door requires processing in all three plants. Table 12.1 shows the effective mean process time of a worker to process each aggregate product in each plant.

The production process in the three workshops is labor-intensive and thus workers are the main production resources. The metalshop plant has 16 workers, while the fabrication plant and the assembly plant have 25 and 22 workers, respectively. All plants work 8 hours a day, and 25 days in a month. The average production cost of a sunroof is estimated to be $100, which includes cost of material, labor cost, overhead costs, etc. This cost is $75 and $95 for windows and doors, respectively. The average selling prices for a sunroof, a window, and a door are $120, $100, and $135, respectively.

The marketing department has forecasted that the average demand for sunroofs, windows, and doors in the next month will be 5000, 4500, and 4000, respectively. How many doors, windows, and sunroofs should the firm produce next month in order to maximize its profit?

The problem that Fancy Windows faces can be modeled as the standard single-period aggregate planning model with three aggregate products (i.e., sunroofs, windows, and doors) and three aggregate resources (metalshop, fabrication, and assembly plants). This problem is also known as the *Product-Mix Problem*, and can be solved using mathematical programming model presented in the following analytics.

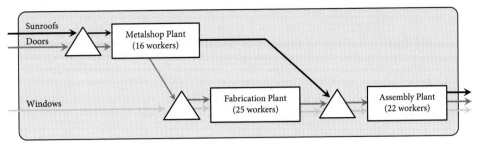

Figure 12.1    Process-flow chart for three aggregate products in Fancy Windows in Example 12.1.

Table 12.1    Effective Mean Process Times (in Hours) of a Worker in Example 12.1

| Resources (workers in) | Sunroof | Window | Door |
|---|---|---|---|
| Metalshop plant | 1 | 0 | 0.6 |
| Fabrication plant | 0 | 1 | 0.5 |
| Assembly plant | 0.5 | 0.8 | 0.7 |

---

## ANALYTICS

### Single-Period Aggregate Planning—The Product-Mix Problem

Consider a process in which resources are aggregated into $\mathcal{I}$ aggregate resources that process $\mathcal{K}$ different aggregate products. Each aggregate resource consists of a pool of workers. Also, suppose

*Period:*

$n_o$ = Number of days in the period

$\ell$ = Number of regular working hours in a day

*Resource Pool i:*

$w_i$ = Number of workers in Resource Pool $i$

$T_{\text{eff}}^{(k)}(i)$ = Effective mean process time (in hours) of one worker in Resource $i$ to process one product of type $k$

*Costs and Profit:*

$r^{(k)}$ = Selling price of Product $k$

$c^{(k)}$ = Cost of producing one unit of Product $k$

*Demand:*

$d^{(k)}$ = Demand for Product $k$ in the period

*Decision Variables:* The decision variables are the production plans for each product in the single period; specifically,

$$X_k = \text{Number of products of type } k \text{ produced in the period}$$

where $X_k \geq 0$, for all $k = 1, 2, \ldots, \mathcal{K}$.

*Mathematical Programming Model:* The mathematical model that finds the optimal production plan $X_k$ in order to maximize the total profit in the single period is the following linear programming:

$$Maximize \quad Z = \sum_{k=1}^{\mathcal{K}} (r^{(k)} - c^{(k)}) X_k$$

subject to:

$$\sum_{k=1}^{\mathcal{K}} T_{\text{eff}}^{(k)}(i) X_k \leq \ell n_o w_i \quad \text{for all} \quad i = 1, 2, \ldots, \mathcal{I} \quad \text{(Resource time constraints)}$$

$$X_k \leq d^{(k)} \quad \quad \text{for all} \quad k = 1, 2, \ldots, \mathcal{K} \quad \text{(Demand constraints)}$$

$$X_k \geq 0 \quad \text{for all} \quad k = 1, 2, \ldots, \mathcal{K}$$

We follow the above analytics to develop and solve Fancy Windows' single-period aggregate planning problem. The single period refers to 1 month.

**Aggregate Products and Resources**

Fancy Windows produces $\mathcal{K} = 3$ types of products, namely sunroofs, windows, and doors. We use $k = S$, $k = W$, and $k = D$ to refer, respectively, to sunroofs, windows, and doors. Products are processed through $\mathcal{I} = 3$ different aggregate resources, namely, metalshop, fabrication, and assembly plants. We use $i = M$, $i = F$, and $i = A$ to refer, respectively, to metalshop, fabrication, and assembly plants.

All three plants (i.e., resources) work $\ell = 8$ hours a day and $n_o = 25$ days a month. Hence, the total number of hours that each plant works in a month is $\ell n_o = 8 \times 25 = 200$ hours, for $i = M, F, A$. Metalshop plant has $w_M = 16$ workers, while fabrication and assembly plants have $w_F = 25$ and $w_A = 22$ workers, respectively. So, the total available worker-hours in a month in the metalshop plant is $\ell n_o w_M = 200 \times 16 = 3200$. For fabrication and assembly plants we have $\ell n_o w_F = 200 \times 25 = 5000$ and $\ell n_o w_A = 200 \times 22 = 4400$, respectively.

Also, $T_{\text{eff}}^{(k)}(i)$, the effective mean process time of a worker in Resource Pool (i.e., plant) $i$ to process a product of type $k$ is presented in Table 12.1. For example, it takes a worker in the metalshop an average of $T_{\text{eff}}^{(S)}(M) = 1$ hour to process a sunroof.

**Demand and Costs**

As the example states, the demands for sunroofs, windows, and doors are forecasted to be $d_s = 5000$, $d_w = 4500$, and $d_D = 4000$, respectively. Sunroofs, windows, and doors are sold, respectively, for $r^{(S)} = \$120$, $r^{(W)} = \$100$, and $r^{(D)} = \$135$, and their production costs are $c^{(S)} = \$100$, $c^{(W)} = \$75$, and $c^{(D)} = \$95$. Hence, the net profits for sunroofs, windows, and doors are $r^{(S)} - c^{(S)} = \$120 - \$100 = \$20$, $r^{(W)} - c^{(W)} = \$100 - \$75 = \$25$, and $r^{(D)} - c^{(D)} = \$135 - \$95 = \$40$, respectively.

**Decision Variables**

The decision variables are the number of products that must be produced in the single-period, that is, next month. Specifically,

$$X_S = \text{Number of } \textit{Sunroofs} \text{ to produce next month}$$
$$X_W = \text{Number of } \textit{Windows} \text{ to produce next month}$$
$$X_D = \text{Number of } \textit{Doors} \text{ to produce next month}$$

where $X_S \geq 0$, $X_W \geq 0$, and $X_D \geq 0$, because the firm cannot produce negative number of products.

Before, we present the mathematical programming model for Fancy Windows, let's have a brief analysis of the problem. Since doors are the most profitable product, it seems that the firm should produce and sell as many doors as it has demand for. However, there are several issues with this strategy. (i) Producing a door requires processing in all three plants and takes up more processing time than other products (i.e., it takes $0.6 + 0.5 + 0.7 = 1.8$ hours), while producing a sunroof, for example, requires processing in only two plants with a total processing time of $1 + 0.5 = 1.5$ hours. (ii) The demand for sunroofs (i.e., 5000) is higher than the demand for doors (i.e., 4000). In summary, while doors are the most profitable product, sunroofs require processing in less number of resources and can have higher sales (due to its higher demand). As this simple analysis shows, finding the best aggregate production plan is not simple, and hence we need the mathematical programming model, as shown below, to develop a plan:

$$\text{Maximize} \quad Z = 20X_S + 25X_W + 40X_D$$

$$\text{subject to :}$$

| | | |
|---|---|---|
| $X_S$ | $+ 0.6X_D \leq 3200$ | (Metalshop time constraint) |
| | $X_W + 0.5X_D \leq 5000$ | (Fabrication time constraint) |
| $0.5X_S + 0.8X_W + 0.7X_D \leq 4400$ | | (Assembly time constraint) |
| $X_S$ | $\leq 5000$ | (Sunroofs demand constraint) |
| | $X_W \leq 4500$ | (Windows demand constraint) |
| | $X_D \leq 4000$ | (Doors femand constraint) |

$$X_S \geq 0, \quad X_W \geq 0, \quad X_W \geq 0$$

## Objective Function

Objective function is the total profit of producing $X_S$, $X_W$, and $X_D$ sunroofs, windows, and doors, respectively, in the next month, which is $Z = 20X_S + 25X_W + 40X_D$.

## Resource Time Constraints

Time constraints for different aggregate resources of the process (i.e., workers in each plant) in the mathematical programming model were obtained as follows. As the process-flow chart in Fig. 12.1 shows, the metalshop processes sunroofs and doors. The effective mean process times of one worker in the metalshop is $T_{\text{eff}}^{(S)}(M) = 1$ hour to process a sunroof and $T_{\text{eff}}^{(D)}(M) = 0.6$ hours to process a door. Hence, the total time the workers in the metalshop will spend next month producing $X_S$ sunroofs and $X_D$ doors would be $(1)X_S + (0.6)X_D$. But, this total time cannot be more than the total available worker-hours in the metalshop, that is, 3200 worker-hours. Mathematically speaking,

$$X_S + 0.6X_D \leq 3200 \qquad \text{(Metalshop time constraint)}$$

Time constraints for fabrication and assembly plants in the mathematically programming model are established similarly.

## Demand Constraints

The demand constraints are simple. Since forecast for the next month's demand for sunroofs is 5000, the number of sunroofs to produce next month ($X_S$) should not be more than 5000. This can be stated mathematically as $X_S \leq 5000$. Considering the demand forecasts 4500

| | A | B | C | D | E | F | G |
|---|---|---|---|---|---|---|---|
| 1 | | | | | | | |
| 2 | Decision Variable Names: | $X_S$ | $X_W$ | $X_D$ | | | Objective |
| 3 | Decision Variable Values: | 800 | 1500 | 4000 | | | Function Value |
| 4 | | | | | | | (Z) |
| 5 | Objective Function Coefficients: | 20 | 25 | 40 | | | 213500 |
| 6 | | | | | | | |
| 7 | CONSTRAINTS: | | | | | LHS | | RHS |
| 8 | Metalshop Hours : | 1 | 0 | 0.6 | 3200 | < or = | 3200 |
| 9 | Fabrication Hours : | 0 | 1 | 0.5 | 3500 | < or = | 5000 |
| 10 | Assembly Hours : | 0.5 | 0.8 | 0.7 | 4400 | < or = | 4400 |
| 11 | Sunroof Demand : | 1 | | | 800 | < or = | 5000 |
| 12 | Window Demand : | | 1 | | 1500 | < or = | 4500 |
| 13 | Door Demand : | | | 1 | 4000 | < or = | 4000 |

**Figure 12.2**     Optimal production plan for Fancy Windows in Example 12.1.

and 4000 for windows and doors, the corresponding demand constraints, as shown in the mathematical programming model, are $X_W \leq 4500$ and $X_D \leq 4000$, respectively.

**Optimal Solution**

The solution of the above problem using Solver is shown in Figure 12.2 (see Appendix K that illustrates how Solver can be used to solve mathematical programming problems). As the solution shows, the optimal production schedule is to produce $X_S^* = 800$ sunroofs, $X_W^* = 1500$ windows, and $X_D^* = 4000$ doors. This results in the maximum profit of $Z^* = \$213,500$.

One important feature of aggregate planning is that it assumes there is no variability in demand. As in the example of Fancy Windows, aggregate planning uses the point forecast—the average demand—of the next month to develop production and workforce plans. Aggregate planning, therefore, does not provide any buffer against errors in demand forecasts. While this might be considered as a weakness, it is also a strength. First, the demand variability for aggregate products—which is the sum of the demand for a group of products—is lower than the demand variability of each product. Hence, in most cases, the variability in demand of an aggregate product is not very large and assuming no demand variability is not unreasonable. Second, planning for average demand allows the planner to use deterministic mathematical programming models (as in the above) that are less complex and are easier to solve. Third, aggregate planning is a medium- to long-term planning. Therefore, by using the average demand, aggregate planning takes into account the changes in base demand (not the noise), which is a better representation of the demand pattern for developing future production and workforce plans. Finally, while the solution of the aggregate planning is the basis for future decisions, these decisions can be adjusted later to include variability in demand when more information about demand becomes available.

### 12.6.2  The Product-Mix Problem with Overtime

One option to increase the capacity of a production process is to use overtime.

 **Example 12.2**

> Managers of Fancy Windows are planning to use overtime to produce more products. However, the problem is that the worker union does not allow overtime, since the plants are in a remote location and it takes a long time for workers to go home when the plants close. After a long negotiation, the union agreed with overtime under two conditions. First, the maximum overtime hours in a day should not exceed 3 hours. Second, Fancy

Windows should pay an additional $12 above the regular hourly wage for each hour of overtime. Should Fancy Windows accept the union's conditions? If it does, what would be the optimal number of overtime hours in each of its plants in order to maximize its total profit?

The product-mix problem discussed in the previous section can be easily extended to the following analytics that incorporates overtime.

---

**ANALYTICS**

### Single-Period Product-Mix Problem with Overtime

Consider the single-period product-mix problem and assume that the process can increase its available hours in the period by working overtime hours every day. Also suppose

$$\mathcal{O}^{\text{max}} = \text{Maximum allowed overtime hours for one worker in a day}$$
$$c_O = \text{Additional cost of 1 hour of overtime for one worker (beyond the regular hourly wage)}$$

Define new decision variables as follows:

$$O_i = \text{Total } \textit{worker-hours} \text{ of overtime for } \textit{all} \text{ workers in Resource Pool } i \text{ in the period}$$

The following mathematical programming model can be used to find the optimal production plan as well as the optimal overtime hours that result in the maximum profit in the period:

$$\text{Maximize} \quad Z = \sum_{k=1}^{\mathcal{K}} (r^{(k)} - c^{(k)}) X_k - c_O \sum_{i=1}^{\mathcal{I}} O_i$$

subject to :

$$\sum_{k=1}^{\mathcal{K}} T_{\text{eff}}^{(k)}(i) X_k - O_i \leq \ell \, n_o \, w_i \qquad \text{for all } i = 1, 2, \ldots, \mathcal{I} \quad \text{(Resource time constraints)}$$

$$X_k \leq d^{(k)} \qquad \text{for all } k = 1, 2, \ldots, \mathcal{K} \quad \text{(Demand constraints)}$$

$$O_i \leq w_i n_o \mathcal{O}^{\text{max}} \qquad \text{for all } i = 1, 2, \ldots, \mathcal{I} \quad \text{(Overtime constraints)}$$

$$X_k \geq 0, \quad O_i \geq 0 \quad \text{for all } i = 1, 2, \ldots, \mathcal{I}; \quad k = 1, 2, \ldots, \mathcal{K}$$

where $w_i n_o \mathcal{O}_i^{\text{max}}$ is the maximum total overtime worker-hours when $w_i$ workers in Resource Pool $i$ work for maximum of $\mathcal{O}_i^{\text{max}}$ hours of overtime in each of the $n_o$ days in the period.

For the three plants of Fancy Windows, we define the following decision variables to model the total overtime in each plant:

$$O_M = \text{Total worker-hours of overtime in the } \textit{Metalshop} \text{ plant next month}$$
$$O_F = \text{Total worker-hours of overtime in the } \textit{Fabrication} \text{ plant next month}$$
$$O_A = \text{Total worker-hours of overtime in the } \textit{Assembly} \text{ plant next month}$$

where $O_M \geq 0$, $O_F \geq 0$, and $O_A \geq 0$. Using the analytics, we incorporate these variables into our mathematical programming model, and we have

$$Maximize \quad Z = 20X_S + 25X_W + 40X_D - 12O_M - 12O_F - 12O_A$$

subject to:

$$
\begin{aligned}
X_S \qquad\qquad + 0.6X_D - O_M &\leq 3200 \\
X_W + 0.5X_D - O_F &\leq 5000 \\
0.5X_S + 0.8X_W + 0.7X_D - O_A &\leq 4400 \\
X_S \qquad\qquad\qquad &\leq 5000 \\
X_W \qquad\qquad &\leq 4500 \\
X_D \qquad &\leq 4000 \\
O_M &\leq 1200 \\
O_F &\leq 1875 \\
O_A &\leq 1650
\end{aligned}
$$

$$X_k \geq 0 , \; O_i \geq 0 \quad : \quad k \in \{S, W, D\} \quad i \in \{M, F, A\}$$

### Objective Function

Note that the overtime cost is an additional cost on top of production cost $c^{(k)}$. Therefore, the objective function now also includes the total cost of overtime hours in the three plants.

### Resource Time Constraints

When a plant works overtime, the overtime hours are added to the time the plant is available for production. For the metalshop, for example, if the plant works for a total of $O_M$ worker-hours of overtime next month, then its total available worker-hours next month would be $3200 + O_M$, and thus we have

$$X_S + 0.6X_D \leq 3200 + O_M \quad \Longrightarrow \quad X_S + 0.6X_D - O_M \leq 3200$$

The constraints for fabrication and assembly plants are obtained in the same manner. Also, demand constraints are the same as in the previous model.

### Overtime Constraints

Recall that the maximum overtime allowed for a worker in each day is $\mathcal{O}^{\max} = 3$ hours. Consider the metalshop plant that has $w_M = 16$ workers and works $n_o = 25$ days in a month. Thus, the maximum possible worker-hours of overtime in that plant would be $w_M n_o \mathcal{O}^{\max} = 16 \times 25 \times 3 = 1200$ hours. The maximum worker-hours of overtime in fabrication and assembly plants are $w_F n_o \mathcal{O}^{\max} = 25 \times 25 \times 3 = 1875$ hours and $w_A n_o \mathcal{O}^{\max} = 22 \times 25 \times 3 = 1650$, respectively. The last three constraints reflect these maximum limits.

### Optimal Solution

Figure 12.3 shows the spreadsheet model for our new Linear Programming model with overtime that contains the optimal solution found by Excel Solver. As the figure shows:

$$X_S^* = 800, \quad X_W^* = 3562.5, \quad X_D^* = 4000, \quad O_M^* = 0, \quad O_F^* = 562.5, \quad O_A^* = 1650$$

with the total profit of $Z^* = \$238{,}512.5$, which is the maximum possible profit.

The optimal solution shows that by having a total of 562.5 hours of overtime in the fabrication plant and 1650 hours in the assembly plant, the firm will be able to produce 3562.5 windows. In the previous plan with no overtime, the firm could produce only 1500

| | A | B | C | D | E | F | G | H | I | J |
|---|---|---|---|---|---|---|---|---|---|---|
| 1 | | | | | | | | | | |
| 2 | Decision Variable Names: | $X_S$ | $X_W$ | $X_D$ | $O_M$ | $O_F$ | $O_A$ | | | Objective |
| 3 | Decision Variable Values: | 800 | 3562.5 | 4000 | 0 | 562.5 | 1650 | | | Function Value (Z) |
| 4 | | | | | | | | | | |
| 5 | Objective Function Coefficients: | 20 | 25 | 40 | -12 | -12 | -12 | | | 238512.50 |
| 6 | | | | | | | | | | |
| 7 | CONSTRAINTS: | | | | | | | LHS | | RHS |
| 8 | Metalshop Hours : | 1 | 0 | 0.6 | -1 | | | 3200 | < or = | 3200 |
| 9 | Fabrication Hours : | 0 | 1 | 0.5 | | -1 | | 5000 | < or = | 5000 |
| 10 | Assembly Hours : | 0.5 | 0.8 | 0.7 | | | -1 | 4400 | < or = | 4400 |
| 11 | Sunroof Demand : | 1 | | | | | | 800 | < or = | 5000 |
| 12 | Window Demand : | | 1 | | | | | 3562.5 | < or = | 4500 |
| 13 | Door Demand : | | | 1 | | | | 4000 | < or = | 4000 |
| 14 | Metalshop Overtime Hours : | | | | 1 | | | 0 | < or = | 1200 |
| 15 | Fabrication Overtime Hours : | | | | | 1 | | 562.5 | < or = | 1875 |
| 16 | Assembly Overtime Hours : | | | | | | 1 | 1650 | < or = | 1650 |

**Figure 12.3**    Optimal production and overtime for Fancy Windows in Example 12.2.

windows. With the overtime, the firm's profit increases from $213,500 (with no overtime) to $238,512.5 (with overtime). Note that since the fabrication plant has 25 workers and works 25 days a month, the total 562.5 worker-hours of overtime translates into $562.5/(25 \times 25) = 0.9$ hours a day for each worker. In the assembly plant, however, the overtime hours in a day would be $1650/(25 \times 22) = 3$ hours for each worker. No overtime is needed in the metalshop plant.

### 12.6.3 The Product-Mix Problem with Setup Cost

Although Fancy Windows' new production plan with overtime increases production capacity, there are still 4200 unsatisfied demands for sunroofs and 937.5 unsatisfied demands for windows.[1] Therefore, the firm may want to find other ways to satisfy these demands. One strategy that firms use to increase the production capacity in the short term is to set up a temporary resource to produce more products, as in the following example.

**Example 12.3**

To maintain their current market share, managers of Fancy Windows are determined to find other ways to satisfy all of their demand. The production manager comes up with an idea. He says that he might be able to setup a temporary mixed production line capable of producing sunroofs, windows, and doors. This, however, will take 1 week and will cost about $10,000 to set up. Also, the cost of producing a sunroof, a window, and a door on the new line would be about $115, $80, and $120, respectively. Furthermore, the line will have the capacity of producing a total of 1000 products next month. While this does give Fancy Windows a new resource for production, it is not clear whether this would increase profit. First, it requires a setup cost of $10,000. Second, the production cost of the products on the new line is higher than what costs Fancy Windows to produce those products using existing resources (i.e., plants). So, the question is whether the new line should be set up, and if it should, how many of each product should be made next month?

So, in addition to the decision of production in the existing plants and overtime, the firm also faces two new decisions: (i) should the new line be set up or not? and (ii) if the line is set up, how many of each product should the line produce next month?

---

[1] You can find unsatisfied demand by subtracting left-hand side (LHS) of demand constraints from their right-hand side (RHS).

In aggregate planning, binary variables are used to model the decision of whether an event/decision occurs or not. Binary variables take only two values of zero or one. For the decision of whether the firm should set up the new line or not, we define binary variable $Y$ as follows:

$$Y = \begin{cases} 1 \ : \ \text{if the new line is set up} \\ 0 \ : \ \text{if the new line is not set up} \end{cases}$$

To incorporate the decision of how many products the line should produce next month, we need to define three new decision variables:

$$L_S = \text{Number of } \textit{Sunroofs} \text{ to produce on the new line next month}$$
$$L_W = \text{Number of } \textit{Windows} \text{ to produce on the new line next month}$$
$$L_D = \text{Number of } \textit{Doors} \text{ to produce on the new line next month}$$

where $L_S \geq 0$, $L_W \geq 0$, and $L_D \geq 0$.

There is an important relationship between the binary variable $Y$, production quantities $L_S$, $L_W$, and $L_D$, and the capacity of the new production line that is described in the following analytics.

---

**ANALYTICS**

**Incorporating Setup Costs in Aggregate Planning**

Consider a production resource that can be set up with setup cost $c_S$ to produce products $k = 1, 2, \ldots, \mathcal{K}$. The production cost of Product $k$ made by the resource is $c_L^{(k)}$. The resource has a total capacity of producing $\mathcal{L}$ products, regardless of the product mix. To incorporate this resource into the aggregate planning problem, we must define binary variable $Y$ as follows:

$$Y = \begin{cases} 1 \ : \ \text{if the resource is set up} \\ 0 \ : \ \text{if the resource is not set up} \end{cases}$$

We also must define decision variables regarding the production quantity of the resource, that is,

$$L_k = \text{Number of type-}k \text{ products produced by the resource}$$

where $L_k \geq 0$. Then the following must be added to the aggregate planning mathematical programming model:

**Objective Function:** Terms $-c_S Y$ and $-(c_L^{(1)} L_1 + c_L^{(2)} L_2 + \cdots + c_L^{(\mathcal{K})} L_\mathcal{K})$ must be added to the objective function representing the (fixed) cost of setting up the resource and the (variable) production costs of units produced by the resources, respectively.

**Capacity Constraint:** The following constraint should be added to the set of constraints to reflect the capacity of the resource:

$$L_1 + L_2 + \cdots + L_\mathcal{K} \leq \mathcal{L} Y$$

Note that,

- If the resource is not set up, that is, if $Y = 0$, then (i) the above constraints becomes $L_1 + L_2 + \cdots + L_\mathcal{K} \leq 0$, implying $L_1 = L_2 = \cdots = L_\mathcal{K} = 0$, that is, no products are produced by the resource, and (ii) $-c_S Y = 0$ and $-(c_L^{(1)} L_1 + c_L^{(2)} L_2 + \cdots + c_L^{(\mathcal{K})} L_\mathcal{K}) = 0$; thus, no costs are added to the objective function.
- If the resource is set up, that is, if $Y = 1$, then (i) the above constraint becomes $L_1 + L_2 + \cdots + L_\mathcal{K} \leq \mathcal{L}$, reflecting the capacity of the resource, and (ii) terms $-c_S$ and $-(c_L^{(1)} L_1 + c_L^{(2)} L_2 + \cdots + c_L^{(\mathcal{K})} L_\mathcal{K})$ will be added to the objective function.

Using the above analytics, the mathematical programming model for the single-period aggregate programming for Fancy Windows in Example 12.3 would be

$$Maximize \quad Z = 20X_S + 25X_W + 40X_D - 12O_M - 12O_F - 12O_A$$
$$- 10{,}000Y + 5L_S + 20L_W + 15L_D$$

*subject to:*

$$X_S \quad\quad + 0.6X_D - O_M \leq 3200$$
$$X_W + 0.5X_D - O_F \leq 5000$$
$$0.5X_S + 0.8X_W + 0.7X_D - O_A \leq 4400$$
$$X_S \quad\quad\quad + L_S \leq 5000$$
$$X_W \quad\quad + L_W \leq 4500$$
$$X_D + L_D \leq 4000$$
$$O_M \leq 122$$
$$O_F \leq 1875$$
$$O_A \leq 1650$$
$$- 1000Y + L_S + L_W + L_D \leq 0$$

$$Y = \{0, 1\}, \ X_k \geq 0, \ L_k \geq 0, \ O_i \geq 0 \ : \ k \in \{S, W, D\} \quad i \in \{M, F, A\}$$

## Objective Function

As the model shows, the objective function has two new components. One incorporates the cost of setting up the line (i.e., $-c_S Y = -10{,}000Y$) if the model decides to do so. The other new component is the net profit of the products made using the new line. Considering new line's production costs $c_L^{(S)} = \$115$, $c_L^{(W)} = \$80$, and $c_L^{(D)} = \$100$, for sunroofs, windows, and doors, respectively, the net profit for each product will be $r^{(S)} - c_L^{(S)} = \$120 - \$115 = \$5$, $r^{(W)} - c_L^{(W)} = \$100 - \$80 = \$20$, and $r^{(D)} - c_L^{(D)} = \$135 - \$120 = \$15$. This is shown by terms $5L_S + 20L_W + 15L_D$ in the objective function.

## Constraints

The constraints on the available regular hours and available overtime hours are the same as those in the product-mix model. The demand constraints, however, are revised to incorporate the products made on the new line. The last constraint is the capacity constraint for the new line, as described by the above analytics.

## Optimal Solution

Solving the mathematical programming model using Excel Solver, the optimal production, overtime, and the new line setup are obtained as in Fig. 12.4

As Fig. 12.4 shows, the optimal solution is

$$X_S^* = 900, \ X_W^* = 3500, \ X_D^* = 4000, \ O_M^* = 100, \ O_F^* = 500, \ O_A^* = 1650$$

$$Y^* = 1, \ L_S^* = 0, \ L_W^* = 1000, \ L_D^* = 0$$

which indicates that Fancy Windows should produce 900, 3500, and 4000 sunroofs, windows, and doors, respectively, in the existing plants. This requires a total overtime worker-hours of 100, 500, and 1650 in the metalshop, fabrication, and assembly plants, respectively. Considering the number of workers in each plant and 25 working days in the next month,

| | A | B | C | D | E | F | G | H | I | J | K | L | M | N |
|---|---|---|---|---|---|---|---|---|---|---|---|---|---|---|
| 1 | | | | | | | | | | | | | | |
| 2 | Decision Variable Names | $X_S$ | $X_W$ | $X_D$ | $O_M$ | $O_F$ | $O_A$ | $L_S$ | $L_W$ | $L_D$ | Y | | | Objective |
| 3 | Decision Variable Values: | 900 | 3500 | 4000 | 100 | 500 | 1650 | 0 | 1000 | 0 | 1 | | | Function Value (Z) |
| 4 | | | | | | | | | | | | | | |
| 5 | Objective Coefficients: | 20 | 25 | 40 | -12 | -12 | -12 | 5 | 20 | 15 | -10000 | | | 248500 |
| 6 | | | | | | | | | | | | | | |
| 7 | CONSTRAINTS: | | | | | | | | | | | LHS | | RHS |
| 8 | Metalshop Hours : | 1 | 0 | 0.6 | -1 | | | | | | | 3200 | < or = | 3200 |
| 9 | Fabrication Hours : | 0 | 1 | 0.5 | | -1 | | | | | | 5000 | < or = | 5000 |
| 10 | Assembly Hours : | 0.5 | 0.8 | 0.7 | | | -1 | | | | | 4400 | < or = | 4400 |
| 11 | Sunroof Demand : | 1 | | | | | | 1 | | | | 900 | < or = | 5000 |
| 12 | Window Demand : | | 1 | | | | | | 1 | | | 4500 | < or = | 4500 |
| 13 | Door Demand : | | | 1 | | | | | | 1 | | 4000 | < or = | 4000 |
| 14 | Metalshop Overtime Hours : | | | | 1 | | | | | | | 100 | < or = | 1200 |
| 15 | Fabrication Overtime Hours : | | | | | 1 | | | | | | 500 | < or = | 1875 |
| 16 | Assembly Overtime Hours : | | | | | | 1 | | | | | 1650 | < or = | 1650 |
| 17 | New Line Constraint : | | | | | | | | 1 | 1 | 1 | -1000 | 0 | < or = | 0 |

Figure 12.4   Optimal production, overtime, and new line setup for Example 12.3.

this corresponds to $100/(25 \times 16) = 0.25$, $500/(25 \times 25) = 0.8$, and $1650/(25 \times 22) = 3$ hours of overtime for each worker in the metalshop, fabrication, and assembly plants, respectively. Also, the firm should set up the line (i.e., $Y^* = 1$), but the line should only be used to produce $L_W^* = 1000$ windows. This plan results in a total profit of $Z^* = \$248,500$.

### 12.6.4 The Single-Period Product-Mix Problem with Subcontracting

Subcontracting is another option to increase supply of a product. We use Example 12.4 to show how subcontracting can be incorporated into the aggregate planning model of Fancy Windows in Example 12.1.

### Example 12.4

Fancy Windows contacted the supplier of material for its new line in Example 12.3, but got an unexpected answer. The supplier informed Fancy Windows that it cannot provide the additional material required for production on the new line until the end of next month. This means that Fancy Windows cannot use the new line to satisfy next month's demand. To maintain their current market share, managers of Fancy Windows are determined to find other ways to satisfy all of their demands. In the previous years, the firm hired another high-quality manufacturer, Quality Hardware (QH), to make some of their products. After contacting QH, Fancy Windows receives the following subcontracting offer from them. QH is willing to produce sunroofs and deliver them to Fancy Windows at a unit cost of $110. However, the contract should be for at least 1000 sunroofs. Producing less than 1000 sunroofs does not benefit QH, since its revenue does not cover QH's production, material, setup, and delivery costs. In the offer the unit cost and minimum order size for windows are $85 and 1000, and for doors are $105 and 1000, respectively. The question is: should Fancy Windows subcontract some of its production to QH? And for how many units? Some managers of Fancy Windows believe that, although the subcontracting costs $110, $85, and $105, for sunroofs, windows, and doors, respectively, are higher than the production costs, it still costs less than producing during overtime. Hence, they insist on signing the contract. What is your recommendation? Should the firm sign the contract? If they should, how many of each product should they order?

To see whether Fancy Windows should sign the subcontracting contract, we add the subcontracting decision variables and subcontracting costs to our aggregate planning model we developed in the previous sections. First, we need to define the following decision variables:

$$Q_S = \text{Number of } \textit{Sunroofs} \text{ to be subcontracted next month}$$
$$Q_W = \text{Number of } \textit{Windows} \text{ to be subcontracted next month}$$
$$Q_D = \text{Number of } \textit{Doors} \text{ to be subcontracted next month}$$

where $Q_S \geq 0$, $Q_W \geq 0$, and $Q_D \geq 0$.

There is a fundamental difference between decision variables $Q_k$ and $X_k$. Production value $X_k$ can be any positive value, and hence we used constraint $X_k \geq 0$ to model it. However, subcontracting variables $Q_k$ can either be zero or a positive value greater than 1000. Note that constraint $Q_k \geq 1000$ alone does not capture this, since it does not allow $Q_k$ to be zero. "Either-Or" constraints such as "either $Q_k = 0$ or $Q_k \geq 1000$" is not recognized by software that solves mathematical programming problems. However, the Either-Or constraints can be modeled as standard constraints using binary variables as shown in the following analytics.

**ANALYTICS**

### Incorporating Subcontracting with Minimum Order Size in Aggregate Planning

Consider a subcontractor that can supply the aggregate Product $k$. The cost of acquiring a product of type $k$ from the subcontractor is $c_Q^{(k)}$. For type-$k$ product, however, the subcontractor only accepts orders of size $Q_{\min}^{(k)}$ or higher (where $Q_{\min}^{(k)} > 0$). To incorporate this option into the aggregate planning problem, we must define binary variable $Y_k$ as follows:

$$Y_k = \begin{cases} 1 : \text{if Product } k \text{ is ordered from the subcontractor} \\ 0 : \text{if Product } k \text{ is not ordered from the subcontractor} \end{cases}$$

We must also define decision variable $Q_k$ as follows:

$$Q_k = \text{Number of type-}k \text{ products ordered from the subcontractor}$$

Then the following must be added to the aggregate planning mathematical programming model:

**Objective Function:** Term $-c_Q^{(k)} Q_k$ must be added to the objective function to incorporate the cost for each subcontracted unit of type $k$.

**Minimum Order Constraints:** The following constraints should be added to the set of constraints to reflect the minimum order size:

$$Q_k \leq MY_k$$
$$Q_k \geq Q_{\min}^{(k)} Y_k$$

where $M$ is a very large number, for example, $M = 1,000,000$. Note that,

- If the subcontractor is not used, that is, if $Y_k = 0$, then (i) the above constraints become $Q_k \leq 0$ and $Q_k \geq 0$, which imply $Q_k = 0$, which implies, sending no order to the subcontractor. (ii) Having $Q_k = 0$, we have $-c_Q^{(k)} Q_k = 0$ and no subcontracting cost is added to the objective function.

- If the subcontractor is used, that is, if $Y = 1$, and order of size $Q_k$ is placed, then (i) term $-c_Q^{(k)} Q_k$ will be added to the objective function, and (ii) the above two constraints become $Q_k \leq M$ and $Q_k \geq Q_{\min}^{(k)}$. Since

$M$ is a very large number, the former constraint becomes redundant (it will always hold, since $M$ is chosen such that it is always greater than $Q_k$); and the latter constraint guarantees the minimum order size on the subcontracted units.

Following the above analytics, we define binary variable as follows:

$$Y_k = \begin{cases} 1 : \text{if Product } k \text{ is subcontracted} \\ 0 : \text{if Product } k \text{ is not subcontracted} \end{cases}$$

for $k = S, W, D$, corresponding, respectively, to sunroofs, windows, and doors. Since the minimum order size for subcontracting is $Q_{\min}^{(k)} = 1000$, and considering $M = 1,000,000$, following the above analytics, the minimum subcontracting order constraints for type-$k$ product are

$$Q_k \leq 1,000,000 Y_k$$
$$Q_k \geq 1000 Y_k$$

Note that number 1,000,000 in the first constraint is an arbitrary, very large number. The key is that this number should be large enough such that it does not limit $Q_k$. For example, considering that the total demand for sunroofs is 5000 units, the value of $Q_S$ would not be above 5000 (the firm does not subcontract above what it can sell). Hence, for sunroofs, the first contract can be written as $Q_S \leq 5000 Y_S$ instead of $Q_S \leq 1,000,000 Y_S$.

For each of the three products, we need a pair of constraints to model the lower limit on the subcontracting products. The aggregate planning model will, therefore, be

$$\textit{Maximize} \quad Z = 20X_S + 25X_W + 40X_D - 12O_M - 12O_F - 12O_A$$
$$+ 10Q_S + 15Q_W + 30Q_D$$

$$\textit{subject to:}$$

$$
\begin{aligned}
X_S \qquad\quad + 0.6X_D - O_M &\leq 3200 \\
X_W + 0.5X_D - O_F &\leq 5000 \\
0.5X_S + 0.8X_W + 0.7X_D - O_A &\leq 4400 \\
X_S \qquad\qquad\quad + Q_S &\leq 5000 \\
X_W \qquad\quad + Q_W &\leq 4500 \\
X_D + Q_D &\leq 4000 \\
O_M &\leq 122 \\
O_F &\leq 1875 \\
O_A &\leq 1650 \\
Q_S - 5000Y_S &\leq 0 \\
Q_S - 1000Y_S &\geq 0 \\
Q_W - 4500Y_W &\leq 0 \\
Q_W - 1000Y_W &\geq 0 \\
Q_D - 4000Y_D &\leq 0 \\
Q_D - 1000Y_D &\geq 0
\end{aligned}
$$

$$X_k \geq 0, \ Q_k \geq 0, \ O_i \geq 0, \ Y_k = \{0, 1\} \ : \ k \in \{S, W, D\} \quad i \in \{M, F, A\}$$

| | A | B | C | D | E | F | G | H | I | J | K | L | M | N | O | P |
|---|---|---|---|---|---|---|---|---|---|---|---|---|---|---|---|---|
| 1 | | | | | | | | | | | | | | | | |
| 2 | Decision Variable Names: | $X_S$ | $X_W$ | $X_D$ | $O_M$ | $O_F$ | $O_A$ | $Y_S$ | $Y_W$ | $Y_D$ | $Q_S$ | $Q_W$ | $Q_D$ | | | Objective |
| 3 | Decision Variable Values: | 2600 | 4500 | 1000 | 0 | 0 | 1200 | 1 | 0 | 1 | 2400 | 0 | 3000 | | | Function Value (Z) |
| 4 | | | | | | | | | | | | | | | | |
| 5 | Objective Function Coefficients: | 20 | 25 | 40 | -12 | -12 | -12 | 0 | 0 | 0 | 10 | 15 | 30 | | | 304100 |
| 6 | | | | | | | | | | | | | | | | |
| 7 | CONSTRAINTS: | | | | | | | | | | | | | LHS | | RHS |
| 8 | Metalshop Hours : | 1 | 0 | 0.6 | -1 | | | | | | | | | 3200 | < or = | 3200 |
| 9 | Fabrication Hours : | 0 | 1 | 0.5 | | -1 | | | | | | | | 5000 | < or = | 5000 |
| 10 | Assembly Hours : | 0.5 | 0.8 | 0.7 | | | -1 | | | | | | | 4400 | < or = | 4400 |
| 11 | Sunroof Demand : | 1 | | | | | | | | | 1 | | | 5000 | < or = | 5000 |
| 12 | Window Demand : | | 1 | | | | | | | | | 1 | | 4500 | < or = | 4500 |
| 13 | Door Demand : | | | 1 | | | | | | | | | 1 | 4000 | < or = | 4000 |
| 14 | Metalshop Overtime Hours : | | | | 1 | | | | | | | | | 0 | < or = | 1200 |
| 15 | Fabrication Overtime Hours : | | | | | 1 | | | | | | | | 0 | < or = | 1875 |
| 16 | Assembly Overtime Hours : | | | | | | 1 | | | | | | | 1200 | < or = | 1650 |
| 17 | Setup Subcontract Sunroof : | | | | | | | -5000 | | | 1 | | | -2600 | < or = | 0 |
| 18 | Setup Subcontract Window : | | | | | | | | -4500 | | | 1 | | 0 | < or = | 0 |
| 19 | Setup Subcontract Door : | | | | | | | | | -4000 | | | 1 | -1000 | < or = | 0 |
| 20 | Minimum Subcontract Sunroof : | | | | | | | -1000 | | | 1 | | | 1400 | > or = | 0 |
| 21 | Minimum Subcontract Window : | | | | | | | | -1000 | | | 1 | | 0 | > or = | 0 |
| 22 | Minimum Subcontract Door : | | | | | | | | | -1000 | | | 1 | 2000 | > or = | 0 |

Figure 12.5   Optimal production, overtime, and subcontracting plans for Example 12.4.

### Objective Function

Considering the variable subcontracting costs $c_Q^{(S)} = \$110$, $c_Q^{(W)} = \$85$, and $c_Q^{(D)} = \$105$, for sunroofs, windows, and doors, respectively, the net profit for each subcontracted product will be $r^{(S)} - c_Q^{(S)} = \$120 - \$100 = \$10$, $r^{(W)} - c_Q^{(W)} = \$100 - \$85 = \$15$, and $r^{(D)} - c_Q^{(D)} = \$135 - \$105 = \$30$. This is shown by $10Q_S + 15Q_W + 30Q_D$ in the objective function.

### Constraints

The constraints on available regular and overtime hours are similar to the previous model. The only change is in the demand constraints, which now include both in-house production and subcontracted units.

### Optimal Solution

The optimal production, overtime, and subcontracting plan are shown in Fig. 12.5.

As the figure shows

$$X_S^* = 2600, \quad X_W^* = 4500, \quad X_D^* = 1000, \quad O_M^* = 0, \quad O_F^* = 0, \quad O_A^* = 1200$$

$$Y_S^* = 1, \quad Y_W^* = 0, \quad Y_D^* = 1, \quad Q_S^* = 2400, \quad Q_W^* = 0, \quad Q_D^* = 3000$$

which indicates that Fancy Windows should produce 2600, 4500, and 1000 sunroofs, windows, and doors, respectively, in the existing plants. This requires a total of 1200 worker-hours of overtime only in the assembly plant, that is, $1200/(25 \times 22) = 2.18$ hours a day for each worker. The firm should sign the subcontracting contracts only for sunroofs and doors, and should order 2400 sunroofs and 3000 doors. The plan results in the total profit of $Z^* = \$304,100$.

## 12.7 Aggregate Planning for Multi-Period Problems

As discussed in Sec. 12.2, inventory is another means to increase supply. Specifically, a firm can produce more than demand in one period, keep the items in inventory, and use them as supply in future periods. This option can be utilized when aggregate planning has planning horizon of more than a single period, that is, multi-period problems. Hence, in addition to overtime, subcontracting, and setting up new resources, aggregate planning in multi-period problems can also use inventory to match its supply with demand.

### 12.7.1 Single Aggregate Resource and Single Aggregate Product

The most basic multi-period aggregate planning is when all products are aggregated into one aggregate product and all resources are also aggregated into one aggregate resource. The following example presents such a case.

**Example 12.5**

Delish Soup is a firm that produces about 27 different types of chicken, beef, seafood, and vegetable soups. Their production system consists of two departments: kitchen and packaging. In the kitchen department, ingredients are washed and prepared for different recipes and are cooked in large pots. The soups are then sent to the packaging department where they are filled into cans, sealed, and cooked for another half an hour. The cans are then cooled down for 10 minutes and labeled. While almost all of the operations in the packaging department are automated, the main tasks in the kitchen department are done manually by workers.

*Demand for Aggregate Product:* The firm is in the process of planning its production and workforce for the next 6 months. Since all soups are sold in cans of the same size (i.e., cans that contain 14.75 ounces of soup), managers have aggregated all of their 27 different products into one, which we call "Soup." They also decided to use 100 cans as the unit of the aggregate product, that is, one unit of the aggregate product is 100 cans of soups. The forecasted demand for the aggregate product in the next 6 months is presented in Table 12.2.

*Profit and Costs:* The net profit of each aggregate unit, not considering worker wages and inventory holding costs, is about $50. The firm pays workers an hourly wage of $8 during regular working hours and $12 during overtime. Because of the change in demand from month to month, the firm often hires temporary workers to use for 1 or more months. Hired workers start their jobs in the beginning of a month. A hired worker is fired at the end of a month, if the worker's service is not needed in the next month. A hired worker will work for at least 1 month, and if needed can continue working for another month and so on. The costs of hiring and firing a worker are $3000 and $4000, respectively. The cost of holding one unit of the aggregate product (i.e., 100 cans) in inventory for 1 month is about $5.

*Resources:* The main resource used in the production is labor, which is the bottleneck. Equipment (e.g., can fillers and sealers) used in the process are semiautomated and have ample capacity to satisfy demand. A time study has shown that 6 minutes of a worker's time (mainly in the kitchen department) are required to process one aggregate product. The firm has 50 permanent workers who work 8 hours a day and 25 days a month. The firm does not want to fire its 50 permanent workers and wants to return to the workforce

**Table 12.2    Forecasted Demand for the Aggregate Product (Soup) in Example 12.5**

| Month  | 1      | 2       | 3       | 4      | 5       | 6       |
|--------|--------|---------|---------|--------|---------|---------|
| Demand | 60,000 | 240,000 | 192,000 | 90,000 | 210,000 | 120,000 |

level of 50 at the end of the 6 months. Also due to space and equipment limitations, the number of workers that can work in a month cannot be more than 70 workers.

To plan its workforce in the next 6 months, the firm needs to know if it needs to hire temporary workers, or use overtime instead (or do both). Furthermore, it would like to know how many cans of soup it must produce each month and how much inventory it should carry from month to month. The goal is to maximize the total profit in the next 6 months.

The problem that Delish Soup is facing is a standard multi-period aggregate planning problem that involves only one aggregate product (i.e., batch of 100 canned soups) and one aggregate resource (i.e., labor who mainly work in the kitchen department). The following analytics illustrates how this problem can be solved using mathematical programming models.

**Multi-Period Aggregate Planning—Single Resource and Single Product**

*Periods and Products:* Consider a process that uses one pool of "aggregate resource" (i.e., workers) to produce several products, which are all aggregated into one "aggregate product." The aggregate planning is for $T$ periods. Also, suppose

$$\ell = \text{Number of regular working hours in a day}$$
$$n_t = \text{Number of working days in Period } t$$

*Resources:* The process has one main aggregate resource—pool of $w$ workers—that processes the product, where

$$T_{\text{eff}} = \text{Effective mean process time (measured in hours) of a worker to process a product}$$

Temporary workers can be hired in the beginning of a period to work for at least one period. A temporary worker, if not needed, is fired at the end of a period.

*Demand and Costs:*

$$d_t = \text{Forecasted demand for the aggregate product in Period } t$$
$$r = \text{Net profit of one product (not including resource hourly cost and inventory costs)}$$
$$c_I = \text{Inventory holding cost of one unit of aggregate product in a period}$$
$$c_R = \text{Hourly cost of one resource used in regular working hours}$$
$$c_O = \text{Hourly cost of one resource used in overtime hours}$$
$$c_H = \text{Cost of hiring one worker}$$
$$c_F = \text{Cost of firing one worker}$$

Note that in this model (as opposed to the single-period model), net profit $r$ does not include the resources' (regular) hourly cost. Also, $c_O$ is the total cost of one overtime hour of a resource (not just the additional cost beyond regular hourly cost).

*Decision Variables:* Decisions are the production, inventory, and workforce levels in each Period $t = 1, 2, \ldots, T$, which are represented by the following decision variables:

<u>Production and Sales:</u>
$$X_t = \text{Number of products produced in Period } t$$
$$S_t = \text{Number of products sold in Period } t$$

*Inventory:*

$$I_t = \text{Number of products in inventory at the } \textit{beginning} \text{ of Period } t$$

*Workforce Level:*

$$H_t = \text{Number of workers hired in the } \textit{beginning} \text{ of Period } t$$
$$F_t = \text{Number of workers fired at the } \textit{end} \text{ of Period } t$$

*Resource Work Schedule:*

$$W_t = \text{Number of workers in Period } t$$
$$O_t = \text{Total number of overtime worker-hours in Period } t$$
$$\mathcal{O}^{\max} = \text{Maximum allowed overtime hours for a worker in a day}$$

**Mathematical Programming Model:** The mathematical programming model that finds the optimal production, overtime, inventory, and workforce levels in order to maximize the total profit in the next $\mathcal{T}$ periods is the following linear programming:

$$\textit{Maximize} \quad Z = \sum_{t=1}^{\mathcal{T}} rS_t - c_I I_t - c_R \ell\, n_t\, W_t - c_O O_t - c_H H_t - c_F F_t$$

$$\textit{subject to:}$$

$$
\begin{aligned}
T_{\text{eff}}\, X_t - \ell\, n_t\, W_t - O_t &\le 0 && \text{for all } t = 1, 2, \ldots, \mathcal{T} \quad \text{(Resource time constraints)}\\
S_t &\le d_t && \text{for all } t = 1, 2, \ldots, \mathcal{T} \quad \text{(Sales constraints)}\\
O_t - n_t\, \mathcal{O}^{\max}\, W_t &\le 0 && \text{for all } t = 1, 2, \ldots, \mathcal{T} \quad \text{(Maximum overtime constraints)}\\
W_t - W_{t-1} + H_t - F_{t-1} &= 0 && \text{for all } t = 1, 2, \ldots, \mathcal{T} \quad \text{(Workforce conservation constraints)}\\
I_t - I_{t+1} + X_t - S_t &= 0 && \text{for all } t = 1, 2, \ldots, \mathcal{T} \quad \text{(Production conservation constraints)}
\end{aligned}
$$

$$X_t, S_t, I_t, H_t, F_t, W_t, O_t \ge 0 \quad \text{for all } t = 1, 2, \ldots, \mathcal{T}$$

where $W_0 = w$ is the current number of workers at the beginning of the planning horizon and $F_0 = 0$. Also, $I_1$ is the available inventory of the aggregate product in the beginning of the planning horizon—beginning of Period 1, and the inventory at the end of planning horizon is zero, that is, $I_{\mathcal{T}+1} = 0$.

The period in the Delish Soup multi-period aggregate planning in Example 12.5 is a month and the firm is facing an aggregate planning for a 6-month planning horizon (i.e., $\mathcal{T} = 6$). Using the decision variables defined in the above analytics, we develop the objective function and constraints as follows.

**Objective Function**

In the Delish Soup example, we have net profit of selling a unit of the aggregate product (i.e., 100 cans of soup) to be $r = \$50$. The costs include inventory holding cost of $c_I = \$5$ per month, cost of regular hour $c_R = \$8$ per worker per hour, cost of overtime $c_O = \$12$ per worker per hour, cost of hiring $c_H = 3000$ per worker, and cost of firing $c_F = \$4000$ per worker. Thus, the objective function of the aggregate planning that maximizes the total profit in the 6 months would be

$$\textit{Maximize } Z = 50(S_1 + \cdots + S_6) - 5(I_1 + \cdots + I_6) - 8(25 \times 8)(W_1 + \cdots + W_6)$$
$$-12(O_1 + \cdots + O_6) - 3000(H_1 + \cdots + H_6) - 4000(F_1 + \cdots + F_6)$$

### Resource Time Constraint

Recall from the example that it takes a worker to process a unit of aggregate product about $T_{\text{eff}} = 6$ minutes $= 0.1$ hours. Also, the firm works $\ell = 8$ regular hours a day and $n_t = 25$ days a month. Thus, the total number of regular working hours in a month is $\ell n_t = 8(25) = 200$ hours. Using our analytics, the resource (i.e., labor) time constraints would be

$$0.1\, X_1 - 200\, W_1 - O_1 \leq 0$$
$$0.1\, X_2 - 200\, W_2 - O_2 \leq 0$$
$$0.1\, X_3 - 200\, W_3 - O_3 \leq 0$$
$$0.1\, X_4 - 200\, W_4 - O_4 \leq 0$$
$$0.1\, X_5 - 200\, W_5 - O_5 \leq 0$$
$$0.1\, X_6 - 200\, W_6 - O_6 \leq 0$$

Each constraint is similar to what we had in the single-period aggregate planning problem with overtime. However, we have six of them, one for each of the 6 months.

### Sales Constraints

The forecasts for demand in each month are shown in Table 12.2. The sales in each month $t$ (i.e., $S_t$) cannot be larger than the demand in that month (i.e., $d_t$); in other words, $S_t \leq d_t$. Therefore, following the analytics, the sales constraints corresponding to the aggregate planning model of Delish Soup for the 6 months would be:

$$S_1 \leq 60{,}000 \qquad S_2 \leq 240{,}000 \qquad S_3 \leq 192{,}000$$
$$S_4 \leq 90{,}000 \qquad S_5 \leq 210{,}000 \qquad S_6 \leq 120{,}000$$

### Overtime Constraints

The maximum allowed overtime for a worker in a day is $\mathcal{O}^{\max}$ hours. Hence, the maximum allowed overtime for a worker in a month is $n_t\, \mathcal{O}^{\max}$. If there are $W_t$ workers in month $t$, then the maximum total worker-hours of overtime for all workers will be $n_t\, \mathcal{O}^{\max}\, W_t$. Hence, when deciding for the total overtime worker-hours $O_t$ for month $t$, we must have it less than the maximum allowed; specifically, $O_t \leq n_t\, \mathcal{O}^{\max}\, W_t$ or $O_t - n_t\, \mathcal{O}^{\max}\, W_t \leq 0$, as shown in the analytics. For Delish Soup we have $n_t = 25$ days and $\mathcal{O}^{\max} = 3$ hours, then $n_t\, \mathcal{O}^{\max} = 25(3) = 75$. The corresponding maximum overtime constraints are therefore,

$$O_1 - 75\, W_1 \leq 0$$
$$O_2 - 75\, W_2 \leq 0$$
$$O_3 - 75\, W_3 \leq 0$$
$$O_4 - 75\, W_4 \leq 0$$
$$O_5 - 75\, W_5 \leq 0$$
$$O_6 - 75\, W_6 \leq 0$$

### Workforce Conservation Constraints

The main idea behind workforce conservation constraints is to model how hiring and firing of workers in a period impact the available workforce in the next period. Recall that Delish Soup hires workers to start in the *beginning* of a month and fires them, if needed, at the *end* of a month. Suppose that the current month is month $t$. The workforce conservation constraint implies that the number of workers that will be working this month (i.e., $W_t$) is equal to the number of workers that were working last month (i.e., $W_{t-1}$) minus the number of workers who were fired at the end of last month (i.e., $F_{t-1}$) plus the number of workers that are hired in the beginning of this month (i.e., $H_t$). In other words,

$$W_t = W_{t-1} + H_t - F_{t-1} \quad \Longrightarrow \quad W_t - W_{t-1} - H_t + F_{t-1} = 0$$

Considering that Delish Soup currently has $W_0 = 50$ permanent workers and since $F_0 = 0$, the above constraint for the 6 months are as follows:

$$
\begin{aligned}
W_1 \quad\quad - H_1 \quad\quad &= 50 \\
W_2 - W_1 - H_2 + F_1 &= 0 \\
W_3 - W_2 - H_3 + F_2 &= 0 \\
W_4 - W_3 - H_4 + F_3 &= 0 \\
W_5 - W_4 - H_5 + F_4 &= 0 \\
W_6 - W_5 - H_6 + F_5 &= 0
\end{aligned}
$$

The firm wants to keep its current 50 permanent workers throughout the planning horizon. Therefore, we must add the following constraints to make sure that the workforce level in each period does not go below 50:

$$
\begin{aligned}
W_1 \geq 50 \quad\quad W_2 \geq 50 \quad\quad W_3 \geq 50 \\
W_4 \geq 50 \quad\quad W_5 \geq 50 \quad\quad W_6 \geq 50
\end{aligned}
$$

On the other hand, as stated in the example, because of the limitation in space and equipment, the number of workers working in a month cannot exceed 70. This is modeled with the following constraints:

$$
\begin{aligned}
W_1 \leq 70 \quad\quad W_2 \leq 70 \quad\quad W_3 \leq 70 \\
W_4 \leq 70 \quad\quad W_5 \leq 70 \quad\quad W_6 \leq 70
\end{aligned}
$$

Finally, since the firm would like to end the planning horizon with the same 50 workers, we must have the following constraint:

$$W_6 - F_6 = 50$$

The above constraint is obtained by writing the workforce conservation constraint for period (i.e., month) 7. Note that the number of workers at the end of period 6 is equal to the number of workers at the beginning of period 7 (since no one is hired in period 7). Hence, the firm wants to have $W_7 = 50$. For period 7, the workforce conservation constraint is $W_7 = W_6 + H_7 - F_6 = 50$, which also guarantees that the number of workers at the beginning of period 7 is 50. Considering that the firm does not hire any workers in period 7, that is, $H_7 = 0$, then we get $W_6 - F_6 = 50$.

### Production Conservation Constraints

The production conservation constraints imply that a unit produced in a period will either be sold in that period, or will be carried as inventory to be sold in future periods. Consequently, there is a relationship among the inventory at the beginning of a period and production quantity and sales in the previous period. Suppose that the current month is month $t + 1$. The inventory at the beginning of this month (i.e., $I_{t+1}$) is equal to the inventory at the beginning of the last month (i.e., $I_t$) plus what was produced last month (i.e., $X_t$) minus what was sold last month (i.e., $S_t$); therefore, the production conservation constraint for month $t + 1$ will be

$$I_{t+1} = I_t + X_t - S_t \quad \Longrightarrow \quad I_t - I_{t+1} + X_t - S_t = 0$$

**Table 12.3    Optimal Production, Inventory, and Workforce Plans for Delish Soup in Example 12.5**

| Optimal solution | Month 1 | Month 2 | Month 3 | Month 4 | Month 5 | Month 6 |
|---|---|---|---|---|---|---|
| Sales ($S_t^*$) | 60,000 | 240,000 | 192,000 | 90,000 | 210,000 | 120,000 |
| Production ($X_t^*$) | 107,500 | 192,500 | 192,000 | 107,500 | 192,500 | 120,000 |
| Inventory ($I_t^*$) | 0 | 47,500 | 0 | 0 | 17,500 | 0 |
| Workforce level ($W_t^*$) | 54 | 70 | 70 | 70 | 70 | 60 |
| Overtime hours ($O_t^*$) | 0 | 5250 | 5200 | 0 | 5250 | 0 |
| Hiring ($H_t^*$) | 4 | 16 | 0 | 0 | 0 | 0 |
| Firing ($F_t^*$) | 0 | 0 | 0 | 0 | 10 | 10 |

Considering that the firm does not currently have any inventory (i.e., $I_1 = 0$) and that the firm does not want to have inventory at the end of the planning horizon (i.e., $I_7 = 0$), we have the following production conservation constraints for Delish Soup:

$$-I_2 + X_1 - S_1 = 0$$
$$I_2 - I_3 + X_2 - S_2 = 0$$
$$I_3 - I_4 + X_3 - S_3 = 0$$
$$I_4 - I_5 + X_4 - S_4 = 0$$
$$I_5 - I_6 + X_5 - S_5 = 0$$
$$I_6 \quad + X_6 - S_6 = 0$$

Finally, we should not forget important constraints $X_t \geq 0, S_t \geq 0, I_t \geq 0, H_t \geq 0, F_t \geq 0, W_t \geq 0, O_t \geq 0$ for all $t = 1, 2, \ldots, 6$.

**Optimal Solution**

Using Excel Solver we find the optimal solution for Delish Soup's multi-period aggregate planning problem as shown in Fig. 12.3.

From the optimal solution in Table 12.3 we conclude the following:

- *Production and inventory:* In each month, the firm should produce equal to the demand in that month, except for the first and fourth months, in which the firm should produce more than the demand. Specifically, in the first month the firm should produce 47,500 aggregate products more than the demand in that month and should carry them as inventory to satisfy demand in the second month. Also, in the fourth month, the firm should produce an additional 17,500 units more than demand in that month and use them to satisfy the demand in the fifth month.

- *Workforce:* To achieve the above production plan, the firm should hire 4 and 16 temporary workers in the first and the second month, respectively—a total of 20 workers. It should then fire 10 of those workers at the end of the fifth month and fire the rest at the end of the sixth month.

- *Overtime:* The aggregate plan calls for overtime in the second, third, and the fifth month. Specifically, in the second month the firm must have 5250 worker-hours of overtime. Having $W_2^* = 70$ workers working in the second month, and $n_2 = 25$ working days in a month, this translates into $O_2^*/(n_2 W_2^*) = 5250/(25 \times 70) = 3$ hours of overtime for each worker in that month. This number is 2.97 hours and 3 hours for the third and the fifth months.

The above plan results in a total profit of $44,316,200 in the 6 months.

One point worth mentioning here is the relationship between $H_t$ and $F_{t-1}$ in the optimal solution. As Table 12.3 shows, in all months we either have ($H_t^* \neq 0$ and $F_{t-1}^* = 0$), or ($H_t^* = 0$ and $F_{t-1}^* \neq 0$), or we have ($H_t^* = 0$ and $F_{t-1}^* = 0$). In the optimal solution we will never have ($H_t^* \neq 0$ and $F_{t-1}^* \neq 0$). The reason becomes obvious when we consider an example of ($H_3 = 5 \neq 0$ and $F_2 = 3 \neq 0$). This implies that the firm should fire three workers at the end of month 2, and then immediately hire five workers in the beginning of month 3. This will increase the workforce level by two workers from month 2 to month 3. Under this hiring/firing plan, the firm will pay firing cost for three workers and hiring cost for five workers. However, a cheaper way to increase workforce level by two workers is not to fire anyone at the end of month 2 (i.e., $F_2 = 0$) and just hire two workers in the beginning of month 3 (i.e., $H_3 = 2$). Under this plan, the firm only pays the hiring cost for two workers. Hence, the optimal solution of an aggregate planning always results in $H_t^* \times F_{t-1}^* = 0$.

### 12.7.2 Multiple Aggregate Resources and Multiple Aggregate Products

The Delish Soup case is an example of a basic multi-period aggregate planning problem with a single aggregate product and single aggregate resource. In this section, we extend our analytics to cases in which products are aggregated to more than one aggregate product and there is more than one aggregate resource that processes these products.

---

**ANALYTICS**

**Multi-Period Aggregate Planning with Multiple Resources and Multiple Products**

Consider the basic multi-period model and suppose that all products are aggregated into $\mathcal{K}$ different aggregate products processed by $\mathcal{I}$ different pools of resources. Aggregate Resource Pool $i$ is a pool of $w_i$ resources (i.e., workers) who work $n_{it}$ days in Period $t$. Also, let

$$T_{\text{eff}}^{(k)}(i) = \text{Effective mean process time of a worker in Resource } i \text{ to process a product of type } k$$

where $T_{\text{eff}}^{(k)}(i)$ is measured in hours.

*Demand and Costs:*

$$
\begin{aligned}
d_t^{(k)} &= \text{Forecasted demand for aggregate Product } k \text{ in Period } t \\
r^{(k)} &= \text{Net profit of one Product } k \text{ (not including resource hourly cost and inventory costs)} \\
c_I &= \text{Inventory holding cost of one unit of aggregate product in a period} \\
c_{iR} &= \text{Hourly cost of one resource in aggregate Resource } i \text{ used in regular working hours} \\
c_{iO} &= \text{Hourly cost of one resource in aggregate Resource } i \text{ used in overtime hours} \\
c_{iH} &= \text{Cost of hiring one worker for Resource } i \\
c_{iF} &= \text{Cost of firing one worker of Resource } i
\end{aligned}
$$

*Decision Variables:* Decisions are production, overtime, inventory, and workforce levels in each Period $t = 1, 2, \ldots, \mathcal{T}$, for aggregate Resources $i = 1, 2, \ldots, \mathcal{I}$ and aggregate Products $k = 1, 2, \ldots, \mathcal{K}$, which are represented by the following decision variables:

*Production and Sales:*

$$
\begin{aligned}
X_{kt} &= \text{Number of Products } k \text{ produced in Period } t \\
S_{kt} &= \text{Number of Products } k \text{ sold in Period } t
\end{aligned}
$$

*Inventory:*

$$I_{kt} = \text{Number of Products } k \text{ in inventory at the } beginning \text{ of Period } t$$

*Workforce Level:*

$$H_{it} = \text{Number of workers hired for Resource } i \text{ in the } beginning \text{ of Period } t$$

$$F_{it} = \text{Number of workers fired from Resource } i \text{ at the } end \text{ of Period } t$$

*Resource Work Schedule:*

$$W_{it} = \text{Number of workers in Resource } i \text{ in Period } t$$

$$O_{it} = \text{Total number of overtime worker-hours of Resource } i \text{ in Period } t$$

$$\mathcal{O}^{\max} = \text{Maximum allowed overtime hours for a worker in a day}$$

***Mathematical Programming Model:*** The mathematical programming model that finds the optimal production, overtime, inventory, and workforce levels in order to maximize the total profit in $\mathcal{T}$ periods is

$$\text{Maximize} \quad Z = \sum_{t=1}^{\mathcal{T}} \left( \sum_{k=1}^{\mathcal{K}} r^{(k)} S_{kt} - \sum_{k=1}^{\mathcal{K}} c_I I_{kt} - \sum_{i=1}^{\mathcal{I}} [c_{iR}\, \ell\, n_{it}\, W_{it} + c_{iO} O_{it} + c_{iH} H_{it} + c_{iF} F_{it}] \right)$$

*subject to :*

$$\sum_{k=1}^{\mathcal{K}} T_{\text{eff}}^{(k)}(i)\, X_{kt} - \ell\, n_{it}\, W_{it} - O_{it} \leq 0 \qquad \text{for all } t = 1, 2, \ldots, \mathcal{T} \ , \ i = 1, 2, \ldots, \mathcal{I}$$

$$S_{kt} \leq d_{kt} \qquad \text{for all } t = 1, 2, \ldots, \mathcal{T} \ , \ k = 1, 2, \ldots, \mathcal{K}$$

$$O_{it} - n_{it}\, \mathcal{O}^{\max}\, W_{it} \leq 0 \qquad \text{for all } t = 1, 2, \ldots, \mathcal{T} \ , \ i = 1, 2, \ldots, \mathcal{I}$$

$$W_{it} - W_{i,t-1} + H_{it} - F_{i,t-1} = 0 \qquad \text{for all } t = 1, 2, \ldots, \mathcal{T} \ , \ i = 1, 2, \ldots, \mathcal{I}$$

$$I_{kt} - I_{k,t+1} + X_{kt} - S_{kt} = 0 \qquad \text{for all } t = 1, 2, \ldots, \mathcal{T} \ , \ k = 1, 2, \ldots, \mathcal{K}$$

$$X_{kt}, S_{kt}, I_{kt}, H_{it}, F_{it}, W_{it}, O_{it} \geq 0 \ \text{ for all } \ k = 1, 2, \ldots, \mathcal{K}, \ i = 1, 2, \ldots, \mathcal{I}, \ t = 1, 2, \ldots, \mathcal{T}$$

Note that $W_{i0}$ is the number of workers in aggregate Resource Pool $i$ in the beginning of the planning horizon (time $t = 0$) and $F_{i0} = 0$. Also $I_{k1}$ and $I_{k,\mathcal{T}+1}$ are inventory of aggregate Product $k$ in the beginning and at the end of planning horizon, respectively.

The above aggregate planning model is essentially the same as the previous model with one aggregate product and one aggregate resource, except that: (i) each aggregate resource has its own set of available regular time, overtime, and workforce conservation constraints, and (ii) each aggregate product has its own set of sales and production conservation constraints.

### 12.7.3 Multi-Period Aggregate Planning with Lost Sales and Backorders

In some cases when the total demand is higher than supply, then different aggregate products compete for resources, as in product-mix problem. Hence, it is possible that in some periods, the process is not capable of satisfying the demand for one or some of the aggregate products. In these situations, the main question is: Which demand should the firm satisfy fully, and which demand should be partially satisfied? The answer is not easy to find, since it depends on several factors such as net profit of the products, the costs of unsatisfied demand of each type of product, and the amount of resources needed to produce each type. However, in this section we show how the analytics for aggregate planning can be extended to incorporate shortage of products and their corresponding costs. When there is a shortage of a product in a period, the unsatisfied demand in the period is either *lost* or *backordered.*

### Incorporating Lost Sales

Lost sales correspond to situations where the customers whose demands are not satisfied in a period are not willing to wait to receive their products in future periods. They buy the product from competitors. This, in addition to the immediate loss of profit, has long-term cost of losing customers' future business. The following analytics shows how lost sales can be incorporated to the aggregate planning models.

---

**ANALYTICS**

### Multi-Period Aggregate Planning—Incorporating Lost Sales

Consider the multi-period aggregate planning model with $\mathcal{I}$ aggregate resources and $\mathcal{K}$ aggregate products. Suppose that unsatisfied demands in a period are lost, and let

$$c_D^{(k)} = \text{Cost of losing a unit of demand for Product } k$$

and define decision variable

$$U_{kt} = \text{Total lost sales for Product } k \text{ in Period } t$$

where $U_{kt} \geq 0$. The multi-period aggregate planning model can be revised as follows to incorporate lost sales:

- *Objective Function:* The following term should be added to the objective function to include lost sales costs:

$$\sum_{t=1}^{\mathcal{T}} \sum_{k=1}^{\mathcal{K}} c_D^{(k)} U_{kt}$$

- *Constraints:* The sales constraints $S_{kt} \leq d_{kt}$ must be replaced with the following:

$$S_{kt} + U_{kt} = d_{kt} \quad \text{for all} \quad t = 1, 2, \ldots, \mathcal{T}, \quad k = 1, 2, \ldots, \mathcal{K}$$

---

Note that the new sales constraints in the analytics are straightforward, since they state that the demand for Product $k$ in Period $t$ (i.e., $d_{kt}$) is either satisfied by sales in that period (i.e., $S_{kt}$) or is lost (i.e., $U_{kt}$).

### Incorporating Backorders

Backorders correspond to the cases where customers whose demands are not satisfied in a period are willing to wait to receive their products in later periods. This often occurs when the product is popular and the firm has no competition. Hence, customers do not have a comparable option to the products offered by the firm. Nevertheless, there is a cost associated with postponing a demand from one period to another. This include costs such as transaction costs, price discount, and cost of free shipping to convince customers to wait for their orders. Other more long-term cost is the cost of unhappy customers corresponding to the potential loss of customers' future business.

Similar to the case of lost sales, a simple change in the aggregate planning model can incorporate the case of backorders. The key is to realize that backordered products are nothing but negative inventory. For example, if the demand in the first period is for 100 units and only 80 products are produced and sold in that period, then the inventory of products at the end of the first period is $-20$, representing a backorder (i.e., shortage) of 20 units. In other words, the total of 20 demands are postponed to the second period.

Hence, at first glance, it seems that all we need to incorporate backorders in our model is to allow the decision variable $I_{kt}$ (i.e., the inventory of Product $k$ in the beginning of Period $t$) to also take negative values. However, there are two issues with this. First, the corresponding term in the objective function is $-c_I I_{kt}$ representing the holding cost of having (positive) inventory of Product $k$. If $I_{kt}$ becomes negative, then the term $-c_I I_{kt}$ becomes positive and more shortages will increase the value of objective function (i.e., increase profit), which does not make sense. Second, the cost of positive inventory (i.e., holding cost) is different from the cost of negative inventory (i.e., backorder cost). Therefore, $c_I$ in the current aggregate planning model cannot present both.

These imply that we need to define another decision variable to represent backorders, as illustrated in the following analytics.

---

**ANALYTICS**

## Multi-Period Aggregate Planning—Incorporating Backorders

Consider the multi-period aggregate planning model with $\mathcal{I}$ aggregate resources and $\mathcal{K}$ aggregate products. Suppose that unsatisfied demands in a period are backordered, and let

$c_I^+ =$ Cost of holding one unit of (positive) inventory of one product for *one period*

$c_I^- =$ Cost of carrying one unit of backorder demand (i.e., negative inventory) for *one period*

with their corresponding decision variables

$I_{kt}^+ =$ Inventory of Product $k$ in the beginning of Period $t$

$I_{kt}^- =$ Backorders of Product $k$ in the beginning of Period $t$

where $I_{kt}^+ \geq 0$ and $I_{kt}^- \geq 0$.

The multi-period aggregate planning model can be revised as follows to incorporate lost sales:

- *Objective Function:* The term $-(\sum_{k=1}^{\mathcal{K}} c_I I_{kt})$ should be replaced by

$$-\left( \sum_{k=1}^{\mathcal{K}} c_I^+ I_{kt}^+ + c_I^- I_{kt}^- \right)$$

- *Constraints:* The following constraints should be added to the set of constraints:

$$I_{kt} = I_{kt}^+ - I_{kt}^- \quad \text{for all } k = 1, 2, \ldots, \mathcal{K}, \ t = 1, 2, \ldots, \mathcal{T}$$

Also, we should omit the non-negativity constraint $I_{kt} \geq 0$ and instead add $I_{kt}^+ \geq 0$ and $I_{kt}^- \geq 0$.

Note that in an aggregate planning model with backorders, while inventory $I_{kt}^+$ and backorder $I_{kt}^-$ are non-negative variables, variable $I_{kt} = I_{kt}^+ - I_{kt}^-$ can either be positive or negative. Also, it is clear that in the optimal solution, the inventory in the beginning of each period is either zero ($I_{kt}^+ = 0$ and $I_{kt}^- = 0$), positive ($I_{kt}^+ > 0$ and $I_{kt}^- = 0$), or negative ($I_{kt}^+ = 0$ and $I_{kt}^- > 0$). In other words, under optimal policy, we have $I_{kt}^+ \times I_{kt}^- = 0$.

Finally, notice that $I_{kt}$ is actually *Inventory Position*, as discussed in Chap. 9. Inventory position is used to make a decision whether to place an order (or start production) to increase inventory.

## 12.8 Summary

Matching supply (e.g., production) with demand is the main challenge in operations engineering and management. There are mainly two approaches to match supply with demand: (i) adjusting demand to match with supply, or (ii) adjusting supply to match with demand. A firm can influence customer demand by providing incentives such as price incentive to encourage customers to purchase the product at a different time. The supply, on the other hand, can be adjusted through working overtime, hiring or firing workers or buying new equipment, subcontracting, and seasonal and cycle inventory. While all these production resources are available to adjust supply, they are all costly. The goal of aggregate planning is to develop a plan that uses these resources effectively to satisfy expected demand. Aggregate planning is an intermediate capacity/supply planning for a time horizon of 2 to 12 months. This intermediate time horizon is not long enough to build a new production facility, but it is feasible to build inventory or increase (or decrease) capacity by overtime, subcontracting, hiring, and firing.

Aggregate planning is established based on the concept of aggregate unit of production. Each aggregate unit of production represents a family of a large number of products that have similar production requirements. Similarly, an aggregate resource represents a pool of several resources that are identical or performing activities of a certain production process (e.g., workers in assembly department). One of the effective analytical tools that yields the optimal aggregate plan is mathematical programming model. This chapter provides a series of mathematical programming models that can be used to develop aggregate production plans for single- and multiple-period planning problems. These models provide wide range of optimal decisions such as production level in each period, inventory to carry to the next period, number of required overtimes in each period, number of outsources and subcontracted units in each period, and optimal planned backorders in each period.

## Discussion Questions

1. Provide two strategies that are commonly used to adjust the demand of a product (goods or service) to match with the supply of the product.

2. Provide five strategies that are commonly used to adjust the supply of a product to match with its demand.

3. What are the goals of aggregate planning?

4. What are aggregate unit of production and aggregate resources? Provide examples for each one and explain why aggregate planning models are established based on these two concepts.

## Problems

1. Office Tools (OT) produces adjustable height desks in three different sizes of small, medium, and large. Its production process consists of three departments: welding, painting, and finishing. The main production resource in these three departments is labor. The welding department has 12 workers, the painting department has 6 workers, and the finishing department has 35 workers. All departments work 8 hours a day and 25 days in a month. Producing one small desk requires 1 worker-hour in the welding department, 0.5 worker-hours in the

painting department, and 2 worker-hours in the finishing department. These numbers are 1.5, 0.6, and 2 worker-hours for the medium desk, and 2, 0.9, and 3 worker-hours for large desks, respectively.

OT can sell a maximum of 2000 desks of each type next month and it has a contract with a large retailer to deliver 300 small, 300 medium, and 450 large desks next month. OT sells small, medium, and large desks for $250, $340, and $430, respectively. The production costs are $132, $187.20, and $252.80, respectively, for small, medium, and large desks. Production cost includes all costs, expect for the labor costs. OT pays $8 an hour to workers in its three departments. How many small, medium, and large desks should OT make next month in order to maximize its profit?

2. Consider Office Tool's (OT) production planning in Problem 1. Managers of OT would like to increase their production through overtime. The maximum overtime hours allowed per day is 3 hours. Worker's wage for overtime hour is $14. In order to maximize its profit,
   a. How many hours of overtime should OT ask each worker in the welding, painting, and finishing departments to do next month?
   b. How many desks of each type should OT produce next month? What would be the profit?

3. Office Tools (OT) in Problem 2 has the option of subcontracting the production (i.e., welding and painting) of the parts used in making each desk to another manufacturer. The manufacturer will deliver the parts to the finishing department and OT then makes the final product. It takes the finishing department 4 worker-hours to assemble the parts delivered by the manufacturer and make a small desk. This number is 4 worker-hours for the medium desk and 6 worker-hours for the large desk. The manufacturer charges $200, $270, and $350 for making the parts for one unit of small, medium, and large desks, respectively. The maximum number that the manufacturer can deliver next month is 800 desks of each type. Should OT use the subcontractor? What is the optimal production and subcontracting plan that maximizes OT's profit?

4. Creative Toys (CT) is a manufacturer of toys with an innovative product development team that designs new toys for children of 10 years old or younger. The team has just finished the design of three different remote control toy cars, that is, Car-TA, Car-TB, and Car-TC. There is an electronic component, called EX-30, that is used in all three cars. Making one Car-TA requires three EX-30 components, while making one Car-TB and Car-TC require four and five EX-30 components, respectively. CT has an inventory of 180,000 EX-30 components for production of these cars.

Cars' components are made in the fabrication department. The fabrication department then sends the finished parts to the assembly department, where production of cars is completed. To start producing each car model, CT needs to set up its fabrication and assembly departments for these new products. The setup operations costs are $10,000, $12,000, and $8,000 for Car-TA, Car-TB, and Car-TC models, respectively. If the operations is set up to produce a car, it should produce at least 8000 cars to reach its break-even point and become profitable.

The bottleneck resource in each department is workers. Car-TA requires 5 minutes of a worker time in the fabrication department, while Car-TB requires 6 minutes in the fabrication department. This number is 5 minutes for Car-TC. All three types of cars have the same assembly time of 3 minutes in the assembly department. The fabrication department has 16 workers and the assembly department has 6 workers. Both departments work 8 hours a day and 25 days a month.

CT's plan is to sell Car-TA, Car-TB, and Car-TC for $138, $169, and $146, respectively. The production cost of one Car-TA is $54, which includes all costs such as overhead cost, cost of material, and labor cost (working in regular hours). Production costs for Car-TB and

Car-TC are $73 and $62, respectively. It is estimated that the demands for Car-TA, Car-TB, and Car-TC are 10,000, 13,000, and 25,000 units, respectively. CT is planning to set up its operations to produce these cars next month. Answer the following questions if the goal is to maximize the profit CT gets by producing and selling these cars next month:

    a. Should CT make all three types of cars, or should it only produce one or two types?

    b. How many of each type should be produced under the optimal production plan next month? What is the total profit under this plan?

5. Consider the Creative Toys (CT) case in Problem 4. CT decides to use overtime to increase its production capacity next month. While workers are paid $12 per hour for working in regular hours, the hourly pay for overtime is $18. Also, CT does not want to ask workers to work more than 3 hours of overtime each day, since it may affect the quality of the products made at the end of the day.

    a. What is the optimal production plan if CT uses overtime to increase its production capacity? What is the profit under the optimal plan?

    b. How many hours of overtime each worker in the fabrication and assembly departments must do in each day under the optimal plan in Part (a)?

6. Ceramico is a firm that produces a variety of ceramic mugs, bowls, plates, and so on. Its customers are major retail stores that place orders for Ceramico's products months in advance. One of its factories that is devoted to producing mugs has 32 workers and works 8 hours a day and 25 days a month. Production of a mug requires 9 minutes of a worker time and a worker is paid $8 an hour at regular working hours and $12 an hour at overtime hours. The demand forecast for mugs in the next 4 months is determined to be 40,000, 50,000, 45,000, and 20,000. The net profit of selling a mug, not including workers' wage and inventory holding cost of finished goods, is $10. The cost of holding one mug for 1 month in inventory is about $2 and the maximum overtime allowed per day is 1 hour. Find the optimal production plan for the next 4 months that maximizes Ceramico's profit. How many hours of overtime does Ceramico need under the optimal plan?

7. Ceramico in Problem 6—in addition to overtime—can increase its production capacity by hiring new temporary workers to use for one or more months. Hired workers start their jobs at the beginning of a month. A hired worker is fired at the end of a month, if the worker's service is not needed in the next month. A hired worker will work for at least 1 month, and if needed can continue working for another month and so on. It costs Ceramico $500 to hire a new worker and the worker will be paid the same hourly regular and overtime wages as other workers. The cost of firing a worker is $200. Ceramico's plan is to use temporary workers, if needed, during the next 4 months; however, it wants to keep its workforce at the current level of 32 workers at the end of the fourth month. What is the optimal production plan for the next 4 months, if Ceramico goes with its plan of hiring temporary workers?

8. Because of the high quality of Ceramico's products, retail stores who ordered mugs in Problem 6 are willing to wait for their orders to be delivered later than delivery dates. In fact, Ceramico has told them that it will give them $1 discount for each month that a mug is delivered late. If Ceramico does not hire temporary workers, how should it change its optimal production in Problem 6?

# Operations Scheduling

## 13.0 Introduction

Chapter 12—Aggregate Planning—focused on finding a cost-effective production plan that determines how many products must be produced in each period (e.g., a day, a week, a month), and how many resources (e.g., workers, machines) are needed in each period to achieve the production plan. While Aggregate Planning is an effective tool to improve production planning, it is only one side of the story. The other side is how production plans must be executed in the process in an effective manner to achieve process improvement. This is called production "Scheduling."

In general, *Scheduling* refers to the process of *assigning* activities needed to produce a product (or to provide a service) to resources and specifying *when* each resource should start performing each activity in order to achieve a certain goal. All manufacturing and service operations involve scheduling. In manufacturing systems, operations engineers and managers face daily or weekly production scheduling problems such as deciding which job should be processed on which machine at what time as well as maintenance scheduling, delivery scheduling, to mention a few. Service systems such as hospitals, for example, face scheduling challenges such as scheduling patients admission, scheduling doctors and nurses working shifts, and scheduling surgery in operations rooms, among others. A bad operations scheduling can have a significant negative impact on any process performance. The main goal of this chapter, therefore, is to present a series of effective scheduling policies that can improve a wide variety of process-flow measures such as throughput, flow time, and inventory.

## 13.1 What Is Operations Scheduling?

Operations Scheduling is the final decision stage of producing goods or services, as shown in Fig. 13.1. As the figure shows, the first stage is planning, which requires a demand forecast. Having estimates for future demand, one can develop production and workforce plans. As discussed in Chap. 12, this is done through aggregate planning that aggregates all products into a few aggregate products and uses analytics to find a cost-effective production plan. For instance, a firm that produces five different types of office chairs can aggregate all five types into one aggregate product, "Chair." Using aggregate planning, the firm can determine the production plan of chairs for the next 12 months.

The next step, as shown in Fig. 13.1, is to disaggregate production plan of the aggregate products and develop Master Production Schedule for each individual product.

CONCEPT

**Master Production Schedule (MPS)**

Master production schedule (MPS) is a production schedule that determines the exact number of each product that must be produced in each of the future time periods (e.g., each day, each week, each month). MPS is the basis for scheduling the production of each individual product.

Master production schedule (MPS) translates the aggregate production plan for the aggregate product into the production schedule for each individual product. Recall that aggregate planning is long- or medium-range planning, for example, planning production and workforce in the next 2 to 18 months. However, MPS is developed when the firm gets close to actual production time, for example, MPS is developed for the next 10 days or the next 10 weeks. But as the firm gets closer to the actual production, new information

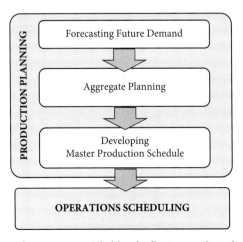

Figure 13.1   **Hierarchy of operations management decisions leading to operations scheduling.**

becomes available. This includes a better forecast (which now also includes the actual orders made by the customers), inventory consideration such as the level of safety stock needed for the product, labor availabilities, supplier's considerations, financial situations, and market prices. Operations engineers and managers consider these factors, along with the aggregate production plan, and develop MPS for each individual product.

Having MPS that determines how many of each type of products must be produced in each hour, each day, or each week, the next step is to allocate the activities required to produce each product to resources, and to determine when each resource should start performing each activity. This is called "Operations Scheduling." For example, if according to MPS a worker is supposed to process 200 item A, 400 item B, and 550 item C in this week, an operations schedule determines which item the worker should process first, second, or third, and how many of each item the worker should process each day of the week.

While operations scheduling is the last step of actually making the product or delivering a service, it is critical to a firm's success. From the production (supply) side, a good production or delivery schedule utilizes firm's resources in the most effective way that can result in higher throughput, and thus less cost per unit produced. From the customer (demand) side, a good production and delivery schedule can result in on-time delivery that improves customer satisfaction.

## 13.2 Goals of Scheduling

The main goal of operations scheduling is to improve process-flow measures as well as customer service. Recall from Chap. 2 that main process-flow measures that affect a firm's profit are process utilization, inventory, flow time, and throughput. A profitable process is the one that is highly utilized and achieves the maximum throughput with minimum inventory and minimum flow time. The goal of operations scheduling is to achieve that.

### Maximizing Utilization and Throughput

Higher utilization of a resource (e.g., a worker or a machine) means higher return on investment. Consider a work center that is supposed to finish processing 20 different jobs. Each job requires processing on Machine 1 and then on Machine 2. Different jobs have different process times on different machines. A production schedule determines when each job must be processed on each machine.

Consider Production Schedule I that schedules jobs on machines such that all 20 jobs are completed in 40 hours. Now consider Production Schedule II that completes all 20 jobs in 50 hours.

---

**CONCEPT**

**Makespan**

Makespan is the time that a process takes to finish a fixed number of jobs. It is the time from when the processing of the first job starts until when the processing of the last job ends. Define,

$$C_i = \text{Time that Job } i \text{ is completed}$$

and suppose all $N$ jobs are available now (at time zero), then Makespan $C_{max}$ is

$$C_{max} = \max\{C_1, C_2, \ldots, C_N\}$$

Hence, Production Schedule I has a makespan of 40 hours and Production Schedule II has a makespan of 50 hours. Which schedule is better? Obviously, the schedule with the smaller makespan. A smaller makespan results in:

(i) *Higher Utilization:* Note that processing times of a job on a machine is the same under both schedules. If all jobs can be processed in 40 hours (as in Schedule I), then Schedule II that finishes jobs in 50 hours must have some idle times (e.g., Machine 2 is idle while waiting for Machine 1 to finish processing of a job). More idle time implies less utilization.

(ii) *Higher Throughput:* Schedule I finished 20 jobs in 40 hours, implying a throughput of $20/40 = 0.5$ jobs an hour. Schedule II, however, finished 20 jobs in 50 hours, a throughput of $20/50 = 0.4$ jobs an hour.

Hence, maximizing utilization or maximizing throughput is equivalent to minimizing makespan. That is the reason why makespan is one of the main goals in operations scheduling.

### Minimizing Inventory and Flow Time

We have illustrated in Chap. 2 and will discuss in Chap. 15 that minimizing inventory results in lower costs as well as higher quality products. Therefore, another major goal of operations scheduling is to minimize inventory in the system. Minimizing flow time—the time a job spends in a process before it is completed—is also an important goal of operations scheduling, especially in make-to-order systems where customers are waiting for their orders.

The two main goals corresponding to inventory and flow time in operations scheduling are as follows:

- *Minimizing Total Flow Time:* Considering

$$F_j = \text{Flow time of Job } j$$

the total flow time of $N$ jobs is

$$\text{Total Flow Time} = F_1 + F_2 + \cdots + F_N$$

Hence, the average flow time for the $N$ jobs is

$$\text{Average Flow Time} = \frac{F_1 + F_2 + \cdots + F_N}{N}$$

Therefore, for a given number of jobs, minimizing the total flow time is equivalent to minimizing the average flow time. On the other hand, average inventory and average flow time are closely related through Little's law. Hence, for a given number of jobs, minimizing average flow time results in minimizing average inventory.

- *Minimizing Weighted Flow Time:* Suppose $w_j$ is the weight assigned to Job $j$. Then, the weighted flow time is

$$\text{Weighted Flow Time} = w_1 F_1 + w_2 F_2 + \cdots + w_N F_N$$

When $w_j$ is the inventory holding cost of Job $j$ per unit time, minimizing the weighted flow time is, in fact, minimizing the total inventory costs of all jobs.

### Meeting Due Dates

Another major goal of operations scheduling is to meet the due dates for each job or a group of jobs. Due dates are often originated from two sources: (i) customers, or (ii) other production processes in the firm. Example of the former is the due date agreed upon with a customer to deliver its product (e.g., delivering pizza). This mainly occurs in make-to-order systems. Example of the latter is the due date to send the door panels from the stamping plant of an auto manufacturer to its assembly plant. This usually occurs in make-to-stock systems.

When compared with its due date, a job is finished either before its due date (i.e., the job is early) or after its due date (i.e., the job is late). To formally define earliness and lateness, let

$$d_j = \text{Due date of Job } j$$

---

**CONCEPT**

---

### Lateness, Tardiness, and Earliness

*Lateness* of Job $j$, denoted by $L_j$, is defined as

$$L_j = C_j - d_j$$

Lateness can be positive or negative:

- *Tardiness:* When lateness is positive, that is, when $C_j > d_j$, then the job is finished after its due date. In this case, the job is called a *Tardy* job and $C_j - d_j$ is called *Tardiness*. Defining $T_j$ as the tardiness of Job $j$, we have

$$T_j = \max\{L_j,\ 0\}$$

- *Earliness:* When lateness is negative, that is, when $C_j < d_j$, then the job is finished before its due date. In this case, the job is called an *Early* job and $C_j - d_j$ is called *Earliness*.

---

There are usually more costs associated with tardiness than with earliness. Most contracts have severe penalty costs (in order of thousands or ten thousands of dollars) for a job being tardy or for each day a job is tardy. The cost of earliness, however, is not as significant. It is mainly associated with the cost of holding the finished job in the finished goods inventory until its delivery due date. In fact, cost of earliness is nothing but the holding cost of the finished goods inventory.[1] Hence, the focus of this chapter will be on improving performance measures corresponding to tardiness (not earliness).

Because lateness can be positive or negative, "Total Lateness" or "Average Lateness" is not a good indication of due date performance. For example, a total or average lateness of zero can occur when (i) all jobs are finished on time (no lateness or earliness), or (ii) some jobs are very tardy and some jobs are very early; but the sum of all tardiness and earliness is zero. Hence, the average lateness without its variance does not mean much. A schedule with small average lateness and small variance of lateness implies that jobs are finished around their due dates.

---

[1] In many cases, customers do not want their jobs to be delivered earlier than their due dates. For example, lean manufacturers with just-in-time practices do not want their suppliers to bring their orders earlier than what was planned.

Most due date performance measures in operations scheduling focus on tardiness measures that have clear meanings. Some of the commonly used operations scheduling goals associated with the due date are the following:

- *Minimizing Maximum Tardiness:* Considering $T_j$ to be the tardiness of Job $j$, then maximum tardiness (i.e., $T_{max}$) of $N$ jobs is

$$T_{max} = \max\{T_1, T_2, \ldots, T_N\}$$

  Minimizing the maximum tardiness tends to improve the worst-case scenario by making the largest tardiness among $N$ jobs as small as possible. This is an important goal when even one very late job can have a significant impact on a firm's profit. For example, a customer who receives the order much later than the quoted due date can damage a firm's reputation through social media. Hence, sometimes it is better to have some jobs being late for a maximum of 1 or 2 days than having only one job late for 3 weeks. Minimizing $T_{max}$ tends to prevent the latter.

- *Minimizing Weighted Number of Tardy Jobs:* Often times, there is penalty associated with missing due dates. Let $w_j$ be the penalty cost if Job $j$ misses its due date, that is, if Job $j$ is tardy. Let's define binary variable

$$U_j = \begin{cases} 1 : & \text{if Job } j \text{ is tardy} \\ 0 : & \text{otherwise} \end{cases}$$

then we have

$$\text{Weighted Number of Tardy Jobs} = w_1 U_1 + w_2 U_2 + \cdots + w_N U_N$$

  When minimizing the weighted number of tardy jobs, the goal is to find a schedule that minimizes the above function. When $w_j = 1$ for all jobs, then minimizing the weighted number (i.e., total cost) of tardy jobs reduces to simply minimizing the total tardiness of all jobs.

- *Minimizing Total Weighted Tardiness:* In some cases, the penalty associated with a tardy job is not about whether a job is tardy or not; it is about how long the job is tardy. Now suppose that $w_j$ presents the cost associated with each day (or each hour) Job $j$ is tardy. Then

$$\text{Total Weighted Tardiness} = w_1 T_1 + w_2 T_2 + \cdots + w_N T_N$$

  Therefore, minimizing total weighted tardiness minimizes the total cost for all the days of tardy jobs. When $w_j = 1$ for all jobs, then total weighted tardiness becomes total tardiness of all jobs.

Having discussed the main goals of operations scheduling, the rest of this chapter is devoted to introducing some common scheduling problems that occur in manufacturing and service operations. These scheduling problems are often different depending on how jobs flow within the system and the type of products or services. Recall from Chap. 1 that based on the structure of a process and the way units flow within the process, the operations processes can be divided into five main processes: (i) continuous-flow process, (ii) flow-line process, (iii) batch process, (iv) job shop process, and (v) project process. The scheduling in continuous-flow process is presented in the online supplement of this chapter. The remaining part of this chapter focuses on operations scheduling decisions that are prevalent in the remaining four processes.

## 13.3 Operations Scheduling in Flow-Line Processes

Flow-line processes are used to produce standard products in high volumes. Flow lines can be divided into two groups: (i) fabrication lines, and (ii) assembly lines. Fabrication lines produce a part or subassembly needed to make the final product, and assembly lines use those parts and subassemblies to make the final product.

Because of the main goal of achieving a high production rate, in flow lines (fabrication or assembly lines) the process of making the product is divided into its smallest indivisible activities. These activities are so small that they cannot be performed by more than one worker. However, in flow lines, it is common that a worker performs more than one activity in the line. Each activity (or a subset of activities) is performed in a workstation and the product moves from one workstation to the next on a conveyor or chain (if the product is heavy). The last workstation in the line performs the last activity and the completed product exits the line.

Since the analytics and management principles are the same for the fabrication and assembly lines, in the rest of this section we use "Assembly Line" to represent both lines. As a simple example, suppose that making a product requires performing three activities A, B, and C. The effective mean process time of activities A, B, and C is 2 minutes. Thus, producing a product requires a total of 6 minutes of work. Figure 13.2 shows two different flow lines that can be set up to produce the product.

In both lines one workstation with one worker is assigned to perform each activity. But there is a fundamental difference between the two lines, which relates to the way jobs are moved from one station to another.

- *Paced Flow Line:* In a paced flow line, as in Fig. 13.2-top, there is only one job in each workstation with no buffer between workstations. Jobs are sitting on a conveyor, and each worker has exactly 2 minutes to finish its tasks. At the end of every 2 minutes all three workers must be done with their tasks and the jobs in all three workstations are moved simultaneously by the conveyor to the next station. Hence, the line is capable of producing one product every 2 minutes, that is, an effective capacity of 30 jobs an hour.

- *Unpaced Flow Line:* In an unpaced line, as in Fig. 13.2-bottom, there are buffers between workstations. Also, jobs are not moved simultaneously between stations by a conveyor. Hence, workers do not have to finish their tasks in exactly 2 minutes. Workers who finish their tasks in less than 2 minutes, put their jobs in the buffer of

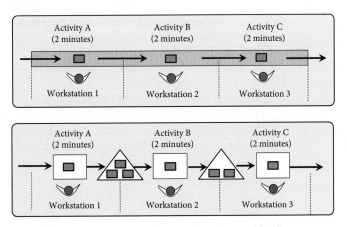

**Figure 13.2**   *Top:* A paced flow line with conveyor system. *Bottom:* An unpaced flow line.

the next workstation, regardless of whether other workers have finished their jobs or not. Similar to the paced line, this line also has an effective capacity of 30 jobs an hour.

Which flow line—paced or unpaced—is more efficient? The answer depends on the variability in activities, A, B, and C. Note that both lines have the same effective capacity of 30 jobs an hour. If there is no variability in effective process time of activities A, B, and C, then all three workers in the unpaced line will finish their jobs in exactly 2 minutes, and all will put their completed jobs in the buffer of the next workstation at the same time—every 2 minutes. In other words, the unpaced flow line would behave exactly like a paced flow line, and therefore there is no need for buffers between workstations. On the other hand, if there is a high variability in the process times, then paced lines would become very inefficient. For example, consider a case where—because of variability—a worker cannot finish his task in 2 minutes. Then the incomplete job will be moved to the next station by the conveyor. Therefore, the final product exiting the paced line would be defective.

Hence, paced lines are suited for systems in which activities have no variability; or if they do have variability, the speed of the conveyor is set such that the probability of one workstation not finishing its job before the conveyor moves the job is very small.

One advantage of paced assembly lines is the issue of fairness from workers' perspective who work on the line. Workers do not like to work more than others on the line for the same pay. In paced assembly lines (if they are not fully staffed by robots) each worker in each station will have the same workload, and the issue of fairness will not arise.

The main operations scheduling questions in flow lines are as follows:

- *Line Balancing:* Suppose that producing a product requires five activities A, B, C, D, and E, each having a different effective process time. Line balancing is concerned with answering the following questions: How many workstations are needed to achieve a certain throughput? Which activities should be performed in which workstation? Note that when process times are different, two tasks with small process times can be assigned to one workstation.

- *Final Assembly Schedule (FAS):* Workers on a flow line do not have the scheduling problem of which job they should do next, or when to start working on a job. Jobs enter each workstation one-by-one and workers in the workstation process jobs as soon as they enter the station. Therefore, the main scheduling question in flow lines is to match production with demand. Specifically, the production scheduling problem is to identify how many hours in a day the flow line should work and how many jobs should be released into the line in each hour to satisfy demand. If a line produces more than one type of products, for example, a mixed-model assembly line, then another scheduling question is the order by which different products should be fed to the line. This is called Final Assembly Schedule (FAS), and is discussed in Chap. 15 when we discuss Toyota's production scheduling method of Heijunka.

In this chapter, we focus on line balancing in paced assembly lines. The analytics described in Chaps. 7 and 8 can be used to solve the line balancing problem of unpaced lines. This is because each workstation in an unpaced line is a $G/G/m$ queueing system for which we can compute process-flow measures such as throughput, average inventory, and flow times.

### Assembly Line Balancing

Assembly line balancing problem is a classical and important industrial engineering problem since 1950s when firms started using flow lines to increase their throughput. The early application of the problem was in the design of assembly lines and their layout in the shopfloor.

Traditionally, once balanced line was designed (i.e., the number of workstations and assignment of activities to each workstation was determined), the layout of the facilities required was determined and the line was established. The line was then used for several years producing the same product (with some minor changes). The issue with traditional assembly lines was that it was very difficult to change their configurations. Therefore, the assembly line balancing was a one-time decision that was made in the beginning to design and set up an assembly line in a plant.

In today's plants, however, the assembly line balancing problem arises more often. This is because today's assembly lines are set up to be flexible enough to produce a wide range of products. With a short setup time, these flexible lines can be reconfigured (i.e., the sequence of activities on the line can often be changed) to produce another product. Therefore, the question of how to reconfigure the line—the line balancing problem—may come up each time the firm switches from one type of product to another.

The assembly line balancing is a very complex and difficult problem to solve. There are several reasons for that. First, as opposed to our simple example in Fig. 13.2 in which there are only three activities and all activities have the same process time of 2 minutes, real assembly line balancing problems involve more than three activities, and activities have different process times. Second, there are precedence relationships among activities—some activities cannot start until some other activities are completed. For example, you cannot seal a potato chips bag before the bag is filled with potato chips. There are also other constraints such as two activities must be performed in one workstation (because they require a common tool), or some activities must be done by two workers (because the item is heavy to move or difficult to assemble).

To illustrate the concepts and analytics in line balancing problems, we use the following example.

**Example 13.1**

FreshAir is a producer of a wide range of commercial and industrial humidifiers. One of their products that is designed for small rooms is called Humid-X10. The demand for the product has increased and is estimated to be 420 units per week. Since the assembly operations are mainly manual operations performed by workers using simple tools, the firm can easily reconfigure its assembly lines, if needed. Because of the increase in demand, FreshAir is planning to revise its current assembly line of Humid-X10 to be able to satisfy the demand. The main activities required for assembling Humid-X10 are presented in Table 13.1. The processing time of the activities has very low variability and the numbers in the table were inflated to ensure that activities are completed within those times.

Table 13.1    Activities for Assembling Humid-X10 in Example 13.1

| Activity $(i)$ | Description | Process time $(t_i)$ (*minutes*) | Predecessor activities |
|---|---|---|---|
| A | Set up the base and power knob | 5 | — |
| B | Install the heating device | 3 | A |
| C | Connect and test wirings | 5 | B |
| D | Assemble thermostat | 6 | B |
| E | Install demineralization cartridge | 6 | A |
| F | Test thermostat and power knob | 3 | C,D |
| G | Assemble the water tank | 3 | E |
| H | Final assembly and inspection | 4 | F,G |

The firm is considering to set up a paced assembly line that will work two shifts (i.e., 12 hours a day) and 7 days a week. Design an assembly line with the minimum number of stations that can satisfy the demand for Humid-X10.

Assembly line problems are very difficult problems, especially finding the optimal design. However, several heuristic algorithms have been developed that yield designs that may not be optimal, but are close to the optimal design. Below we illustrate one of these heuristics.

**Step 1: Finding Cycle Time**
The first step to design a balanced line is to find the capacity of the line. The capacity of the line is the maximum number of products that the line can produce per unit time. Obviously, to satisfy demand, the capacity of the line must be larger than or equal to demand. This implies that the inter-throughput of the line should be equal to or smaller than the demand interarrival times. Takt time is a concept that captures this.

---

**CONCEPT**

**Takt Time**

The inter-throughput of a process that enables the process to satisfy its demand is called *Takt Time*. "Takt" is the German word for the baton that an orchestra conductor uses to regulate the speed, beat, or timing at which musicians play. So takt time can be considered as time that tends to regulate the inter-throughput times, so the process can satisfy its demand. Takt time for a process can be computed as follows:

$$\text{Takt Time} = \frac{\text{Working time during a period}}{\text{Demand to be met during the period}}$$

While, inter-throughput time corresponds to the process—the supply side of operations—takt time corresponds to the demand side of operations. If the average inter-throughput time of a process is larger than its takt time, the process will not be able to satisfy its demand.

For Humid-X10 takt time would be:

$$\begin{aligned}
\text{Takt Time} &= \frac{\text{Working time during a week}}{\text{Demand to be met during a week}} \\
&= \frac{60 \text{ minutes} \times 12 \text{ hours} \times 7 \text{ days}}{420} \\
&= 12 \text{ minutes}
\end{aligned}$$

So, to satisfy the demand, the line must be able to produce one product every 12 minutes. Considering the way paced flow lines work, this means that the total time of all activities performed in each station of the line should be equal to or less than 12 minutes. This is known as "Cycle Time" of the line in line balancing literature, and we use $C_{\text{line}}$ to denote it, that is, $C_{\text{line}} = 12$ minutes.

**Step 2: Constructing Precedence Diagram**
One useful tool in line balancing is *Precedence Diagram* as shown in Fig. 13.3. The precedence diagram is constructed based on the precedence relationship in Table 13.1.

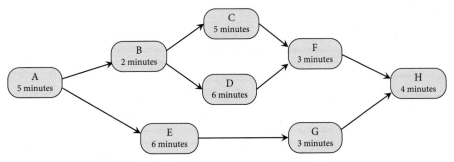

Figure 13.3    Precedence diagram for activities required for assembling Humid-X10 in Example 13.1.

As the figure shows, precedence diagram is a network in which activities are represented by nodes and precedence relationships are represented by arrows connecting the nodes. This is why the diagram is also called *Activity-on-Node Network* or AON Network.

### Step 3: Ranking Activities Based on their Total Number of Successors

Consider Activity A in the precedence diagram in Fig. 13.3. All other seven activities B, C, ..., H cannot be performed without completion of Activity A. Thus, Activity A has a total of seven successors. Activity B, on the other hand, has a total of four successors, namely, C, D, F, and H. Table 13.2 shows the ranked activities based on the decreasing order of the total number of successors. In case of a tie (e.g., activities C and D), the activity with larger processing time comes first—if there is a tie in processing times, then one is randomly ranked higher (e.g., activities D and E).

Table 13.2    Ranked List of Activities for Assembling Humid-X10 in Example 13.1

| Activity ($i$) (ranked)      | A | B | D | E | C | F | G | H |
|------------------------------|---|---|---|---|---|---|---|---|
| Total number of successors   | 7 | 4 | 2 | 2 | 2 | 1 | 1 | 0 |
| Processing time ($t_i$)      | 5 | 3 | 6 | 6 | 5 | 3 | 3 | 4 |

### Step 3: Assigning Activities to Stations, Starting with the First Station

In this step, activities are assigned sequentially to stations, starting from Station 1. To do that, we need to determine the set of eligible activities that can be assigned to Station 1. The set of *eligible activities* $\mathcal{E}$ includes activities that: (i) have not yet been assigned to any station, and (ii) all of their predecessors have already been assigned (or they do not have any predecessors). Since no activities are assigned to any stations yet, and based on the precedence diagram, our set of eligible activities includes only Activity A, that is, $\mathcal{E} = \{A\}$. We can then assign Activity A to Station 1. If set of eligible activities includes more than one activity, then the activity with the largest total number of successors (i.e., the activity with a higher rank in Table 13.2) must be assigned first.

Each time an activity is assigned, we must: (i) update the set of eligible activities, and (ii) compute the remaining time at the current workstation in which the last activity was assigned. After assigning Activity A, the set of Eligible Activities would be $\mathcal{E} = \{B, E\}$. Also, considering line cycle time $C_{\text{line}} = 12$ and $t_A = 5$, the remaining available time at Station 1 would be $C_{\text{line}} - t_A = 12 - 5 = 7$. Having updated these values, we choose the next activity from set $\mathcal{E} = \{B, E\}$ to assign to Station 1. Between two activities B and E, we choose Activity B that has a larger total number of successors (i.e., 4). Before we assign Activity B to Station 1, we must first make sure that its process time is not larger than the remaining available time in Station 1. Since $t_B = 3 \leq 7$, we can assign Activity B to Station 1. Thus, the new set of

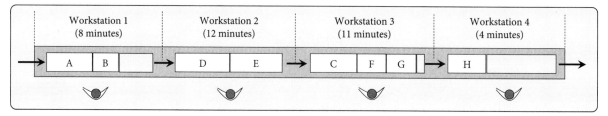

eligible activities would be $\mathcal{E} = \{D, E, C\}$ and Station 1 has the remaining available time of $C_{\text{line}} - t_A - t_B = 12 - 5 - 3 = 4$.

We can now assign the next activity from Set $\mathcal{E} = \{D, E, C\}$ to Station 1. Since all three activities C, D, and E have the same total number of successors (i.e., 2), we use the tie-breaking rule—choosing the activity with the largest process time. Both activities D and E have the same processing time 6 (i.e., another tie); hence, we use a third tie-breaking rule and arbitrarily choose one, say activity D. However, since processing time of Activity D, that is, $t_D = 6$, is larger than the remaining available time at Station 1 (i.e., 4), we cannot assign Activity D to Station 1. Thus, we try the next Activity in Set $\mathcal{E}$, that is, Activity E, with processing time $T_E = 6$, which is also larger than the remaining available time at Station 1. Thus, we cannot assign Activity E to Station 1. The same occurs with Activity C. Since we cannot assign any of the eligible activities to Station 1, we close Station 1 and open the next Station—Station 2.

With 12 minutes of available time at Station 2, and with set of Eligible Activities $\mathcal{E} = \{D, E, C\}$, we assign Activity D to Station 2 and update the set of Eligible Activities as $\mathcal{E} = \{E, C\}$ and remaining available time as $C_{\text{line}} - t_D = 12 - 6 = 6$. The next activity to assign to Station 2 would be Activity E. Assigning this activity, we update $\mathcal{E} = \{C, F, G\}$ with remaining time of $C_{\text{line}} - t_D - t_E = 12 - 6 - 6 = 0$. Since there is no remaining time, we close Station 2 and we open a new station—Station 3.

Following the same approach we assign activities C, F, and G to Station 3, resulting in remaining time $C_{\text{line}} - t_C - t_F - t_G = 12 - 5 - 3 - 3 = 1$. Since the last remaining activity, Activity H, has processing time $t_H = 4$, we close Station 3 and open Station 4 which will have only Activity H. The resulting assembly line is a four-station line with cycle time of 12 minutes as shown in Fig. 13.4.

With cycle time of 12 minutes, the line will be able to satisfy demand of 420 per week. The worker in Station 1 has $12 - 8 = 4$ minutes idle time, the worker in Station 3 has $12 - 11 = 1$ minute idle time, and the worker in Station 4 has $12 - 4 = 8$ minutes idle time—a total of $4 + 1 + 8 = 13$ minutes of idle time.

## Want to Learn More?

### Designing the Most Efficient Assembly Line

The line in Fig. 13.4 with four workstations has a total of 13 minutes of idle time. Is this the most efficient line? Is there another line with the same capacity (i.e., cycle time of 12 minutes) that has less than four workstations and less idle time? Indeed, there is one. The reason that our method was not able to find that line is that it is a heuristic method with no guarantee of finding the optimal line. The reason the method is used in practice is its simplicity of use, especially in designing lines with large number of workstations. If you would like to know how to find the most efficient line using mathematical programming models, you should check out the online supplement of this chapter. The online supplement of this chapter presents two main types of assembly line balancing problems, both of which find the optimal line design.

*Type-1 assembly line balancing* finds the line with the minimum number of stations that can achieve a certain throughput (i.e., a certain cycle time). Minimum number of stations means minimum investment cost in setting up the line. On the other hand, there are situations where a firm has already made investment and set up a paced assembly line with a given number of stations. In this case, the firm's goal is to assign tasks to the stations to maximize line's throughput (i.e., minimize line cycle time). This is known as *type-2 assembly line balancing* problem. Both Type-1 and Type-2 assembly line balancing problems are presented in detail in the online supplement of this chapter.

## 13.4 Operations Scheduling in Job Shop and Batch Processes

In both job shop and batch processes, different products move between several workstations in batches. Each family of products requires processing in different workstations. Because operations scheduling problems in these two processes are almost the same, we use job shop to refer to both systems in the remaining of this chapter.

Operations scheduling in job shops is more difficult than that in flow-line processes. It involves more layers of decisions as shown in Fig. 13.5. We use the case of Solid Furniture to explain each scheduling decision in the figure.

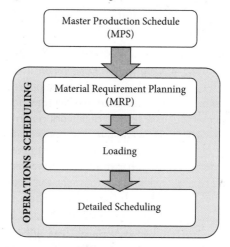

**Figure 13.5    Operations scheduling decisions in job shops and batch processes.**

### Operations Scheduling at Solid Furniture

Solid Furniture is a producer of home and office furniture. The firm produces four types of dining chairs called CC-12, CC-13, CC-14, and CC-15 chairs. After finding the aggregate production plan and new information about orders received from several retailers, the firm establishes master production schedule for one of the chairs—CC-12 chair—also called classic cafe chair. MPS for CC-12 chair requires the firm to produce these chairs in the next 10 weeks according to Table 13.3.

Table 13.3    Master Production Schedule (MPS) for CC-12 Chairs

| Week | 1 | 2 | 3 | 4 | 5 | 6 | 7 | 8 | 9 | 10 |
|---|---|---|---|---|---|---|---|---|---|---|
| CC-12 chairs | 0 | 0 | 0 | 0 | 0 | 200 | 600 | 900 | 800 | 700 |

### X-12 Chair and Its Parts

As Fig. 13.6 shows, CC-12 chair consists of three main assemblies and parts: (i) front assembly, (ii) back assembly, and (iii) seat. Front assembly is composed of two legs and

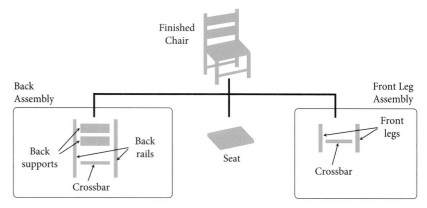

**Figure 13.6**    **Assembly diagram for a CC-12 model chair.**

one crossbar, while back assembly consists of two back rails, two back supports, and one crossbar. Seat is a cushioned seat that is ordered from a supplier.

**The Production Process**

The production process of Solid Furniture consists of several departments, two of which are used to make CC-12 chairs, namely the fabrication and assembly departments.

- *Fabrication Department:* The fabrication department produces front legs, crossbars, back rails, and back supports. To reduce setup times and costs, each of these parts are assigned to different work centers in the fabrication department. For simplicity, we refer to each work center by the name of the part it produces, for example, front leg center, back support center.

  Because each center in the fabrication department is also used in the production of other products, the center cannot immediately start producing parts for CC-12 chairs. For example, the front leg center requires a lead time of $L = 1$ week to produce the required number of front legs. Hence, if an order for producing 4000 front legs is released to the front leg center, the production of 4000 legs will be finished 1 week later. The lead times for back support, crossbar, and back rails are 2, 1, and 2 weeks, respectively.

- *Assembly Department:* There are three work centers in the assembly department that work on CC-12 chairs. Front leg assembly center makes the front leg assembly by assembling two front legs and one crossbar that it receives from the fabrication department. This takes a lead time of 2 weeks to finish a batch. The back assembly center makes the back assembly by assembling two back rails, two back supports, and one crossbar that it receives from the fabrication department. This takes a lead time of 2 weeks for each batch. The chair assembly center makes the finished product—CC-12 chair—by assembling front assembly and back assembly to the seat it receives from a supplier. This final assembly takes 1 week for each batch.

- *Seat Supplier:* Seats with soft cushions are not produced in Solid Furniture's manufacturing facility. They are ordered from an outside supplier with order lead time of 3 weeks. Specifically, when the firm places an order for seats, it receives its order in 3 weeks.

**Material Requirement Planning (MRP)**

Having MPS for CC-12 chair—the finished product—the first operations scheduling decision is how to translate the MPS for the product to production schedule of each of its parts in the fabrication and assembly departments. Specifically, considering production lead times in different centers of the fabrication department, when should each center start making their parts and how many of each part should they make each week? Also, when should Solid Furniture place orders for seats to the supplier and how many seats must be ordered each time? Finally, when should the assembly department start assembling the chair? All these must be scheduled such that CC-12 chairs are completed according to MPS in Table 13.3.

Material requirement planning provides answers to all the above questions. As shown in Fig. 13.5, material requirement planning uses MPS as an input and generates a production schedule for all parts and subassemblies in the fabrication department and develops schedules for when the assembly department needs to start assembling back, front, and the finished product.

### Loading

Production activities in a work center can often be performed by various resources. For example, assembling crossbars to front legs can be performed by different workers, some may be more experienced and can do the activities more efficiently. When there is enough time and enough capacity, then activities should be assigned to resources that have higher efficiency to perform those activities. However, when there is not enough capacity or when some resources are overloaded, then decisions must be made how to assign (or reassign) activities to resources to achieve a certain operational performance. The process of assigning activities to resources to achieve a certain goal is called *Loading*.

Suppose that MRP dictates that the fabrication department in our example must produce 4500 back supports, 2000 crossbars, 3400 front legs, and 5000 back rails in Week 4. There are several workers and machines in the fabrication department that can make these parts. The loading problem that the managers of fabrication department is facing is *which* of these products should be assigned to *which* worker in order to make sure that these products are finished before the end of Week 4.

### Detailed Scheduling

Detailed Scheduling, as shown in Fig. 13.5, is the last stage of operations scheduling that corresponds to detailed schedule of *when* each resource should start working on each job. Consider a work center in the fabrication department that is supposed to produce 3400 front legs for CC-12 chairs in Week 4. The center also produces front legs for other types of chairs that the manufacturer produces. Specifically, suppose that the center needs to make a batch of 1200 front legs of CC-13 chairs, a batch of 400 front legs of CC-14 chairs, and a batch of 3300 front legs of CC-15 chairs. The detailed scheduling question that the center faces is: when the center should start working on each batch. In other words, which one of these batches should the center finish first, second, and third?

After having discussed an overview of operations scheduling decision in Fig. 13.5, in the rest of this chapter we will present concepts and analytics corresponding to each individual decision shown in the figure. We start with Material Requirement Planning.

## 13.5 Material Requirement Planning (MRP)

We use our example of CC-12 chair to describe how MRP works. Imagine you are the production manager in charge of CC-12 chairs, and your job is to make sure that the number of chairs made meet the required numbers in the MPS in Table 13.3. The head of the fabrication department is asking you the following question: "*When* exactly should I start making legs, back rails, crossbars, and back supports for the CC-12 chairs, and *how many* of each should I make each week?" The head of the assembly department is also asking a similar

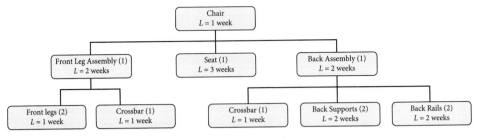

**Figure 13.7** Product structure diagram for CC-12 chairs.

question: "*When* should my work centers start assembling back and front subassemblies and the CC-12 chairs, and *how many* chairs should I expect to assemble each week?"

All the above questions correspond to the production schedule for each part of CC-12 chairs made, ordered, or assembled in different work centers in your production process. How do you schedule production in those centers to meet the MPS for CC-12 chairs?

One tool that helps you better understand the relationship between parts and subassemblies of a product and the lead time needed to make each part is product structure diagram, as shown in Fig. 13.7. *Product Structure Diagram* is the simplest form of what is known as Bill of Material. *Bill of Material* is a list of all assemblies, subassemblies, components, and raw material and their quantities needed to make *one* final product. In Fig. 13.7, *L* in each block corresponds to production or ordering lead time for a component or assembly. Another term that is often used to describe the levels in the product structure diagram is parent-child relationship. Each immediate higher level is called *Parent Level* and each immediate lower level is called *Child level*. Finally, *Usage* of a part or subassembly corresponds to the number of those parts or subassemblies needed to produce *one* unit of the item in its parent level. For each part, usage is shown in the parenthesis in front of the part's name.

### 13.5.1 Forward and Backward Scheduling

Consider the 900 chairs required in Week 8 in MPS in Table 13.3. To make sure that you have those 900 chairs ready when you need them, one approach is to immediately start making the required parts in all work centers in the fabrication department now, that is, in Week 1. This is called "Forward Scheduling."

**Forward Scheduling**
Considering the usage of each part in the final product (see Fig. 13.7), this means releasing a production order for $900 \times 2 = 1800$ front legs, $900 \times 2 = 1800$ crossbars, $900 \times 2 = 1800$ back rails, $900 \times 2 = 1800$ back supports, and also place an order for $900 \times 1 = 900$ seats to the supplier now. Considering the corresponding lead times, also shown in Fig. 13.7, the production schedule in each center of the fabrication and assembly departments can be presented as in Fig. 13.8.

In *Forward Scheduling* the production starts immediately when a job becomes available, regardless of the due date. With forward scheduling the product is often finished before its due date. Is that good? Yes, if there is a benefit for finishing the job before its due date. However, if the job is not leaving the process until its due date, forward scheduling results in work-in-process (WIP) and finished goods inventory (FGI).

Consider the schedule in Fig. 13.8. The firm finishes producing 900 chairs 2 weeks before they are needed in Week 8, and hence must carry 900 CC-12 chairs in its FGI for 2 weeks. Also, the firm finishes 1800 crossbars and receives 900 seats (from the supplier) 1 week before they are needed for assembly. This results in a total of $1800 + 900 = 2700$ WIP

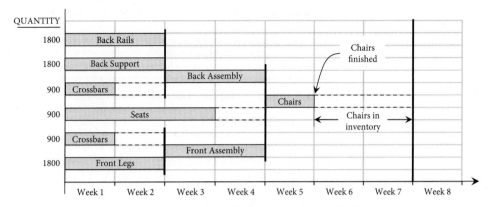

Figure 13.8    Forward scheduling of production for CC-12 chairs to satisfy demand in Week 8.

inventory that must be kept for 1 week. Keeping this inventory for 1 week results in waste in money and space, and can lead to quality issues, as we will discuss in Lean Operations in Chap. 15. How can we schedule production such that 900 CC-12 chairs are completed on their due date with the minimum inventory? Use Backward Scheduling.

### Backward Scheduling

*Backward Scheduling* schedules the last activity first, so the product is finished right at its due date. Backward scheduling then works backward and schedules the next-to-last activity and so on. Knowing each activity must be completed at its due date, the start of each activity is computed by subtracting lead time from the completion time of the activity (i.e., due date). Figure 13.9 shows a production schedule that has resulted from backward scheduling.

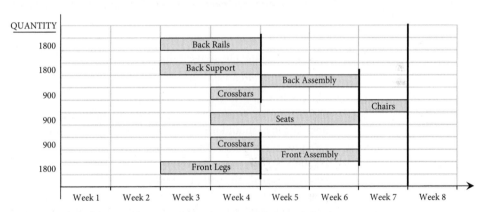

Figure 13.9    Backward scheduling of production for CC-12 chairs to satisfy demand in Week 8.

Comparing production schedule in Fig. 13.9 (backward scheduling) with that in Fig. 13.8 (forward scheduling), we observe the following: while both schedules satisfy demand in Week 8, the backward scheduling results in no WIP or FGI. Since there is no benefit for the firm to produce 900 CC-12 chairs earlier than Week 8, the production schedule developed by backward scheduling is preferred.

### Material Requirement Planning

The production schedule in Fig. 13.9 determines exactly how many of each part should be produced or ordered in each week to satisfy demand for 900 chairs in Week 8 with minimum inventory. This is what material requirement planning does.

*Material Requirement Planning* (MRP) is a computer-based information and decision support system that is designed to determine the production schedule (in-house) and ordering schedule (from suppliers) of all raw material, components, and subassembly used in making the finished product. This is done through backward scheduling. The main goal is the timely completion of the finished product to meet master production schedule (MPS), while keeping the WIP and FGI as low as possible.

### 13.5.2 Explosion Calculus

How does MRP translate MPS for the final product into production and ordering schedules for each individual part, components, and subassemblies of the product? MRP does it through what is known as "Explosion Calculus." *Explosion Calculus* is a set of rules that use information in product structure diagram, inventory records, and MPS and create the final production schedule in three steps: (i) Computing Net Requirement, (ii) Time Phasing, and (iii) Lots Sizing.

MRP first performs the above three steps for the final product, that is the highest level in the product structure diagram, see Fig. 13.7. It then repeats the three steps for each individual component, part, or subassembly in the next immediate lower level in the product structure diagram. MRP continues this backward scheduling process until it reaches the last level (i.e., the lowest level).

**Step 1—Computing Net Requirements**
The first step of MRP requires the following information:

- *Gross Requirements* (GR) for an item is the total quantity of the item needed to make the item at the next-higher level in the product structure diagram. For the end item (i.e., the finished product) gross requirements are those in the master production schedule. For other items (i.e., parts, components, subassemblies) gross requirements are calculated from planned-order releases of their immediate parents in the product structure diagram (we will show this in the later sections).
- *Scheduled Receipts* (SR) are the items that have already been ordered and are scheduled to arrive in the coming periods.
- *On-hand Inventory* (OI) of an item is the number of items available in the inventory. On-hand inventory, if not used, will be carried to the next period.

Calculating net requirements is based on a simple idea. To satisfy gross requirement for an item, the firm should first use its on-hand inventory and the arriving orders (i.e., scheduled receipts) of the item. If these do not satisfy gross requirements, the remaining needed item is called Net Requirement. Hence, *Net Requirements* (NR) of an item is the number of items actually needed. The firm must then produce the required items in its production facility, or order them from outside supplier to satisfy net requirements.

Now let's put the above idea in a formula. For each Period $t$,

$$\text{GR}_t = \text{Gross requirements in Period } t$$

$$\text{NR}_t = \text{Net requirements in Period } t$$

$$\text{OI}_t = \text{On-hand inventory at the } \textit{beginning} \text{ of Period } t$$

$$\text{SR}_t = \text{Scheduled receipts at the } \textit{beginning} \text{ of Period } t$$

To compute net requirements in Period $t$, we must first compute

$$x = \text{GR}_t - \text{SR}_t - \text{OI}_t \tag{13.1}$$

then

$$\begin{cases} \text{If } x \geq 0 & \text{then } \text{NR}_t = x \text{ and } \text{OI}_{t+1} = 0 \\ \text{If } x < 0 & \text{then } \text{NR}_t = 0 \text{ and } \text{OI}_{t+1} = |x| \end{cases} \tag{13.2}$$

To see how the above calculus works, consider a simple example. Suppose in Period $t = 8$, we have $GR_8 = 100$, $SR_8 = 40$, and $OI_8 = 50$. To compute net requirements in Period 8, we use the above formula:

$$x = GR_8 - SR_8 - OI_8 = 100 - 40 - 50 = 10$$

Hence, after using 40 units from scheduled receipts and 50 units in inventory to satisfy gross requirement of 100 units, there is still $x = 100 - 40 - 50 = 10 \geq 0$ units of gross requirement that are not satisfied. Thus, net requirement in Period 8 is $NR_8 = 10$. Since all scheduled receipts and on-hand inventory are used to satisfy gross requirement in Period $t = 8$, then no inventory is carried to the next period, that is, Period $t + 1 = 9$. In other words, $OI_9 = 0$.

Now consider the above example, and assume that gross requirement in Period 8 is $GR_8 = 70$. Then

$$x = GR_8 - SR_8 - OI_8 = 70 - 40 - 50 = -20$$

The negative $x = -20$ implies that the sum of schedule receipts and on-hand inventory is 20 units higher than what is needed in Period 8, that is, higher than gross requirement. Thus, gross requirement in Period 8 is satisfied and the net requirement is $NR_8 = 0$. The additional 20 units will be carried to Period 9 as on-hand inventory, that is, $OI_9 = |-20| = 20$.

When the number of periods is large, it is easier to compute the net requirements in a tabular form. To show this, let's return to our CC-12 chair example and suppose that the inventory and sales records indicate the following:

- *On-hand Inventory:* Currently, there are 250 finished CC-12 chairs in FGI that can be used to satisfy demand in Weeks 6 to 10. Also, there are 300 back assembly and 200 crossbars in WIP inventory that can be used to make CC-12 chairs.

- *Scheduled Receipts:* The firm will receive 150 and 80 seats in the beginning of Weeks 2 and 3, respectively. These are orders that the firm has already placed when the firm was consolidating its orders for seats for different types of chairs.

These information imply

$$
\begin{array}{lll}
\text{For chair} & : & OI_1 = 250 \\
\text{For back assembly} & : & OI_1 = 300 \\
\text{For crossbars} & : & OI_1 = 200 \\
\text{For seats} & : & SR_2 = 150, \quad SR_3 = 80
\end{array}
$$

With this information, we can establish the first table for calculating net requirements as shown in Fig. 13.10.

As we mentioned, MRP is a backward scheduling approach, so it starts from the top of the product structure diagram. Hence, Fig. 13.10 is for computing net requirements for

| Weeks | | 1 | 2 | 3 | 4 | 5 | 6 | 7 | 8 | 9 | 10 |
|---|---|---|---|---|---|---|---|---|---|---|---|
| **CHAIR** | Gross Requirement (GR) | 0 | 0 | 0 | 0 | 0 | 200 | 600 | 900 | 800 | 700 |
| Lead Time = 1 Week | Scheduled Receipts (SR) | | | | | | | | | | |
| Usage = 1 | On-Hand Inventory in the Beginning of the Week (OI) | 250 | 250 | 250 | 250 | 250 | 250 | 50 | 0 | 0 | 0 |
| | Net Requirements (NR) | 0 | 0 | 0 | 0 | 0 | 0 | 550 | 900 | 800 | 700 |
| | Time-Phased Net Requirements (T-NR) | 0 | 0 | 0 | 0 | 0 | 550 | 900 | 800 | 700 | 0 |
| | Planned-Order Release (POR) using Lot-for-Lot | 0 | 0 | 0 | 0 | 0 | 550 | 900 | 800 | 700 | 0 |

Figure 13.10 Tabular form of explosion calculus for final product, that is, CC-12 chair.

CC-12 chairs. The first row of the table shows the time periods—weeks. The second row is for gross requirement of chairs in each of the 10 periods. For the finished product, the gross requirement is the MPS, which is shown in Table 13.3.

The third row is scheduled receipts, which are all zeros, since the firm will not receive any chairs (e.g., returns from retailers) in the next 10 weeks. The fourth row is on-hand inventory. Recall that the firm already has on-hand inventory $OI_1 = 250$ chairs, which is carried over from Week 1 to Week 6. The fifth row is net requirements that are obtained using Eqs. (13.1) and (13.2). Consider the net requirement in Week 6, for which we have

$$x = GR_6 - SR_6 - OI_6 = 200 - 0 - 250 = -50 \implies NR_6 = 0, \quad OI_7 = 50$$

For Week 7, the net requirement is

$$x = GR_7 - SR_7 - OI_7 = 600 - 0 - 50 = 550 \implies NR_7 = 550, \quad OI_8 = 0$$

The net requirements for Weeks 8 to 10 are calculated in the same manner. Having net requirements for CC-12 chairs, the next step of the MRP is Time Phasing.

**Time Phasing**

Net requirements of CC-12 chairs in Week 7, that is, $NR_7 = 550$ indicate that the firm must have 550 finished chairs ready in the beginning of Week 7. Since the assembly department has a lead time of 1 week to assemble chairs, this implies that the assembly department should start assembling those 550 chairs 1 week before they are needed—start assembling in the beginning of Week 6. To incorporate the lead time into the production schedule, MRP transfers the net requirements back in time for $\ell$ periods, where $\ell$ is the lead time, that is, $L = \ell$. This is called Time Phasing. Therefore, considering

$$\text{T-NR}_t = \text{Time-phased net requirements in Period } t$$

then we have

$$\text{T-NR}_t = NR_{t+\ell}$$

Assembling chairs has a lead time of $L = 1$ week; thus, the time-phased net requirement in Week 6, for example, is $\text{T-NR}_6 = NR_{6+1} = NR_7 = 550$. This means that, to make sure that MPS for month 7 is satisfied, the assembly department must start assembling 550 CC-12 chairs in Week 6.

The time-phased net requirements for other weeks are shown in the sixth row in the table in Fig. 13.10. As the figure shows, all net requirements are phased-back in time for 1 week to incorporate the assembly department's lead time of $L = 1$.

**Lot Sizing**

Lot sizing is the last stage of MRP.

---

**CONCEPT**

### Lot Sizing

A *lot size* is the number of an item in an order that is either produced in an in-house production facility or purchased from an outside supplier. *Lot sizing* is the process of determining *when* to place an (production or purchasing) order and *how many* items to order to achieve a certain goal.

Setting up a production facility to produce a lot (i.e., a batch) or placing an order to an outside supplier often involves a setup cost. Thus, producing in small lots results in frequent setup cost. On the other hand, producing a large lot to satisfy demands (i.e., net requirements) of future periods results in inventory holding cost. What makes lot sizing problem difficult is finding the lot size that results in the optimal balance between order setup and inventory holding costs. Economic Order Quantity (EOQ) problem discussed in Chap. 9, for example, is a lot sizing problem that determines the order quantity (i.e., lot size) and the time to place an order to achieve the minimum total order setup and inventory holding cost.

When order setup costs are small, a simple lot sizing policy called Lot-for-Lot results in minimum total cost. Under a *Lot-for-Lot* lot sizing, in each period the order size (production or purchasing order) is set to be equal to the demand in that period. Because the demand in each period is satisfied by the production in that period, no inventory is carried to the next period, and hence no inventory cost is incurred. We use lot-for-lot policy in this step of MRP for our CC-12 chair problem.

The last row in the MRP tables, that is, planned-order release, corresponds to the lot sizing step of the MRP. *Planned-Order Release* (POR) specifies the date at which an order for the item should be released to the firm's production facility or to an outside supplier. Recall that time-phased net requirement $T\text{-}NR_6 = 550$ implies that, to make sure the MPS is satisfied, the assembly department must start assembling a lot of size 550 chairs in Week 6. The firm has the option of producing 550 chairs or more in Week 6 (to use in future periods). Under lot-for-lot policy, however, the firm releases a production order for producing exactly 550 chairs and avoids accumulating inventory for future periods. Hence, the Planned-Order Release for CC-12 chair in Week 6 is 550. Overall, under lot-for-lot policy

$$\text{Under lot-for-lot policy} \quad \Longrightarrow \quad POR_t = T\text{-}NR_t$$

Planned-Order Release for other weeks is obtained in a similar way using the above formula. As Fig. 13.10 shows, under lot-for-lot lot sizing, Planned-Order Release row in MRP table is the exact copy of time-phased net requirements.

After finding Planned-Order Release for CC-12 chair, MRP moves down one level in the product structure diagram and performs explosion calculus for items in that level. Consider seats that have a lead time of $L = 3$ weeks and usage = 1. What is gross requirement for seats?

To find gross requirement for seats, we need to find: (i) *How many* seats are required? and (ii) *When* are they required? The number of seats required depends on how many seats are used to make a CC-12 chair (i.e., usage) and the number of CC-12 chairs that are planned for production. The date that the seats are required is when they are needed for assembling CC-12 chairs (i.e., Planned-Order Release for CC-12 chairs). Consider the Planned-Order Release of 550 for CC-12 chairs in Week 6 in Fig. 13.10. For the assembly department to start assembling these 550 chairs, and since each chair requires only one seat, the assembly department must have 550 seats already available in that week. In other words, the gross requirement for seats in Week 6 is 550. Hence, gross requirement for seats is equal to the Planned-Order Release for chairs (i.e., the product) times number of seats used to make one chair (i.e., usage). Overall, in each Period $t$,

$$GR_t(\text{Item } i) = \text{Usage of Item } i \times POR_t(\text{Parent of Item } i)$$

Figure 13.11 shows the complete MRP calculations for CC-12 chair and all its parts and components. Consider gross requirements for "front leg assembly." It has usage = 1, implying that one unit of its parent item (i.e., finished CC-12 chair) requires one unit of "front leg assembly." Thus, gross requirement for "Front-Leg Assembly" in each week is one times Planned-Order Release for CC-12 chair in that week. As shown in the figure, Week 6 gross requirement for front leg assembly is $550 \times 1 = 550$. Now consider gross requirements for "front legs" that have usage = 2. Hence, two front legs are required to make one "front leg

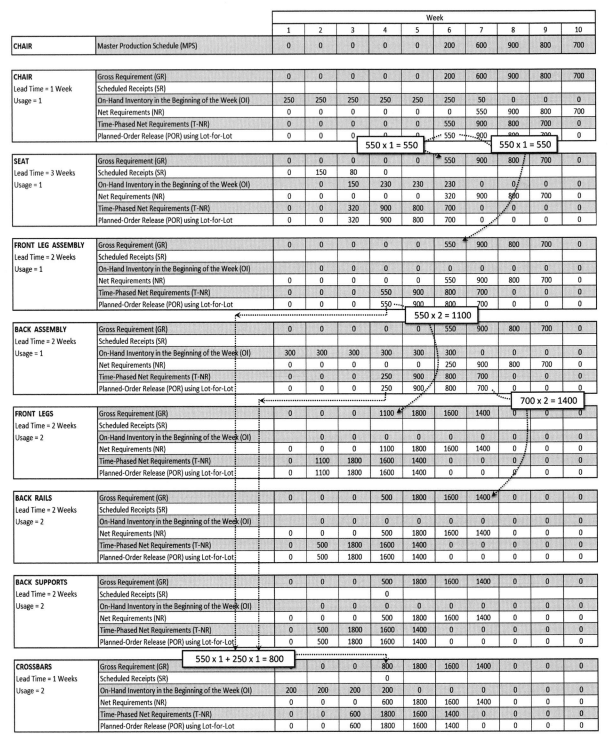

| | | Week | | | | | | | | | |
|---|---|---|---|---|---|---|---|---|---|---|---|
| | | 1 | 2 | 3 | 4 | 5 | 6 | 7 | 8 | 9 | 10 |
| **CHAIR** | Master Production Schedule (MPS) | 0 | 0 | 0 | 0 | 0 | 200 | 600 | 900 | 800 | 700 |

| **CHAIR** | Gross Requirement (GR) | 0 | 0 | 0 | 0 | 0 | 200 | 600 | 900 | 800 | 700 |
|---|---|---|---|---|---|---|---|---|---|---|---|
| Lead Time = 1 Week | Scheduled Receipts (SR) | | | | | | | | | | |
| Usage = 1 | On-Hand Inventory in the Beginning of the Week (OI) | 250 | 250 | 250 | 250 | 250 | 250 | 50 | 0 | 0 | 0 |
| | Net Requirements (NR) | 0 | 0 | 0 | 0 | 0 | 0 | 550 | 900 | 800 | 700 |
| | Time-Phased Net Requirements (T-NR) | 0 | 0 | 0 | 0 | 0 | 550 | 900 | 800 | 700 | 0 |
| | Planned-Order Release (POR) using Lot-for-Lot | 0 | 0 | 0 | 0 | 0 | 550 | 900 | 800 | 700 | 0 |

550 x 1 = 550    550 x 1 = 550

| **SEAT** | Gross Requirement (GR) | 0 | 0 | 0 | 0 | 0 | 550 | 900 | 800 | 700 | 0 |
|---|---|---|---|---|---|---|---|---|---|---|---|
| Lead Time = 3 Weeks | Scheduled Receipts (SR) | 0 | 150 | 80 | 0 | | | | | | |
| Usage = 1 | On-Hand Inventory in the Beginning of the Week (OI) | | 0 | 150 | 230 | 230 | 230 | 0 | 0 | 0 | 0 |
| | Net Requirements (NR) | 0 | 0 | 0 | 0 | 0 | 320 | 900 | 800 | 700 | 0 |
| | Time-Phased Net Requirements (T-NR) | 0 | 0 | 320 | 900 | 800 | 700 | 0 | 0 | 0 | 0 |
| | Planned-Order Release (POR) using Lot-for-Lot | 0 | 0 | 320 | 900 | 800 | 700 | 0 | 0 | 0 | 0 |

| **FRONT LEG ASSEMBLY** | Gross Requirement (GR) | 0 | 0 | 0 | 0 | 0 | 550 | 900 | 800 | 700 | 0 |
|---|---|---|---|---|---|---|---|---|---|---|---|
| Lead Time = 2 Weeks | Scheduled Receipts (SR) | | | | | | | | | | |
| Usage = 1 | On-Hand Inventory in the Beginning of the Week (OI) | | 0 | 0 | 0 | 0 | 0 | 0 | 0 | 0 | 0 |
| | Net Requirements (NR) | 0 | 0 | 0 | 0 | 0 | 550 | 900 | 800 | 700 | 0 |
| | Time-Phased Net Requirements (T-NR) | 0 | 0 | 0 | 550 | 900 | 800 | 700 | 0 | 0 | 0 |
| | Planned-Order Release (POR) using Lot-for-Lot | 0 | 0 | 0 | 550 | 900 | 800 | 700 | 0 | 0 | 0 |

550 x 2 = 1100

| **BACK ASSEMBLY** | Gross Requirement (GR) | 0 | 0 | 0 | 0 | 0 | 550 | 900 | 800 | 700 | 0 |
|---|---|---|---|---|---|---|---|---|---|---|---|
| Lead Time = 2 Weeks | Scheduled Receipts (SR) | | | | | | | | | | |
| Usage = 1 | On-Hand Inventory in the Beginning of the Week (OI) | 300 | 300 | 300 | 300 | 300 | 300 | 0 | 0 | 0 | 0 |
| | Net Requirements (NR) | 0 | 0 | 0 | 0 | 0 | 250 | 900 | 800 | 700 | 0 |
| | Time-Phased Net Requirements (T-NR) | 0 | 0 | 0 | 250 | 900 | 800 | 700 | 0 | 0 | 0 |
| | Planned-Order Release (POR) using Lot-for-Lot | 0 | 0 | 0 | 250 | 900 | 800 | 700 | 0 | 0 | 0 |

700 x 2 = 1400

| **FRONT LEGS** | Gross Requirement (GR) | 0 | 0 | 0 | 1100 | 1800 | 1600 | 1400 | 0 | 0 | 0 |
|---|---|---|---|---|---|---|---|---|---|---|---|
| Lead Time = 2 Weeks | Scheduled Receipts (SR) | | | | | | | | | | |
| Usage = 2 | On-Hand Inventory in the Beginning of the Week (OI) | | 0 | 0 | 0 | 0 | 0 | 0 | 0 | 0 | 0 |
| | Net Requirements (NR) | 0 | 0 | 0 | 1100 | 1800 | 1600 | 1400 | 0 | 0 | 0 |
| | Time-Phased Net Requirements (T-NR) | 0 | 1100 | 1800 | 1600 | 1400 | 0 | 0 | 0 | 0 | 0 |
| | Planned-Order Release (POR) using Lot-for-Lot | 0 | 1100 | 1800 | 1600 | 1400 | 0 | 0 | 0 | 0 | 0 |

| **BACK RAILS** | Gross Requirement (GR) | 0 | 0 | 0 | 500 | 1800 | 1600 | 1400 | 0 | 0 | 0 |
|---|---|---|---|---|---|---|---|---|---|---|---|
| Lead Time = 2 Weeks | Scheduled Receipts (SR) | | | | | | | | | | |
| Usage = 2 | On-Hand Inventory in the Beginning of the Week (OI) | | 0 | 0 | 0 | 0 | 0 | 0 | 0 | 0 | 0 |
| | Net Requirements (NR) | 0 | 0 | 0 | 500 | 1800 | 1600 | 1400 | 0 | 0 | 0 |
| | Time-Phased Net Requirements (T-NR) | 0 | 500 | 1800 | 1600 | 1400 | 0 | 0 | 0 | 0 | 0 |
| | Planned-Order Release (POR) using Lot-for-Lot | 0 | 500 | 1800 | 1600 | 1400 | 0 | 0 | 0 | 0 | 0 |

| **BACK SUPPORTS** | Gross Requirement (GR) | 0 | 0 | 0 | 500 | 1800 | 1600 | 1400 | 0 | 0 | 0 |
|---|---|---|---|---|---|---|---|---|---|---|---|
| Lead Time = 2 Weeks | Scheduled Receipts (SR) | | | | 0 | | | | | | |
| Usage = 2 | On-Hand Inventory in the Beginning of the Week (OI) | | 0 | 0 | 0 | 0 | 0 | 0 | 0 | 0 | 0 |
| | Net Requirements (NR) | 0 | 0 | 0 | 500 | 1800 | 1600 | 1400 | 0 | 0 | 0 |
| | Time-Phased Net Requirements (T-NR) | 0 | 500 | 1800 | 1600 | 1400 | 0 | 0 | 0 | 0 | 0 |
| | Planned-Order Release (POR) using Lot-for-Lot | 0 | 500 | 1800 | 1600 | 1400 | 0 | 0 | 0 | 0 | 0 |

550 x 1 + 250 x 1 = 800

| **CROSSBARS** | Gross Requirement (GR) | 0 | 0 | 0 | 800 | 1800 | 1600 | 1400 | 0 | 0 | 0 |
|---|---|---|---|---|---|---|---|---|---|---|---|
| Lead Time = 1 Weeks | Scheduled Receipts (SR) | | | | 0 | | | | | | |
| Usage = 2 | On-Hand Inventory in the Beginning of the Week (OI) | 200 | 200 | 200 | 200 | 0 | 0 | 0 | 0 | 0 | 0 |
| | Net Requirements (NR) | 0 | 0 | 0 | 600 | 1800 | 1600 | 1400 | 0 | 0 | 0 |
| | Time-Phased Net Requirements (T-NR) | 0 | 0 | 600 | 1800 | 1600 | 1400 | 0 | 0 | 0 | 0 |
| | Planned-Order Release (POR) using Lot-for-Lot | 0 | 0 | 600 | 1800 | 1600 | 1400 | 0 | 0 | 0 | 0 |

Figure 13.11    MRP calculations for CC-12 chair and its components.

assembly." Therefore, as shown in the figure, the Week 4 gross requirement for "front legs" is two times Planned-Order Release for "front leg assembly," that is, $2 \times 550 = 1100$.

Often times a part or a component is used in two different subassemblies, that is, it has two parents. For such components gross requirement is the sum of planned-order releases of

|  | Week1 | Week 2 | Week 3 | Week 4 | Week 5 | Week 6 | Week 7 | Week 8 | Week 9 | Week 10 |
|---|---|---|---|---|---|---|---|---|---|---|
| Assembling X-12 Chair | 0 | 0 | 0 | 0 | 0 | 550 | 900 | 800 | 700 | 0 |
| Placing Orders for Seats | 0 | 0 | 320 | 900 | 800 | 700 | 0 | 0 | 0 | 0 |
| Making Front Leg Assembly | 0 | 0 | 0 | 550 | 900 | 800 | 700 | 0 | 0 | 0 |
| Making Back Assembly | 0 | 0 | 0 | 250 | 900 | 800 | 700 | 0 | 0 | 0 |
| Making Front Legs | 0 | 1100 | 1800 | 1600 | 1400 | 0 | 0 | 0 | 0 | 0 |
| Making Back Rails | 0 | 500 | 1800 | 1600 | 1400 | 0 | 0 | 0 | 0 | 0 |
| Making Back Supports | 0 | 500 | 1800 | 1600 | 1400 | 0 | 0 | 0 | 0 | 0 |
| Making Crossbars | 0 | 0 | 600 | 1800 | 1600 | 1400 | 0 | 0 | 0 | 0 |

Figure 13.12    Output of MRP: production schedule for CC-12 chair and its components.

each parent times its corresponding usage. Consider crossbars in our CC-12 chair example that are used in both "front leg assembly" and "back assembly." One crossbar is used in Front Leg Assembly and one crossbar is used in Back Assembly. On the other hand, the planned-order-releases for front leg assembly and back assembly in Week 4, for example, are 550 and 250, respectively. Thus, gross requirement for crossbars in Week 4 is $550 \times 1 + 250 \times 1 = 800$. Gross requirements for crossbars in other weeks are shown in Fig. 13.11.

In summary, material requirement planning (MRP) translates master production schedule (MPS) for CC-12 chairs into the production schedule (i.e., Planned-Order Release) for CC-12 chairs and all of its components as summarized in Fig. 13.12. This table clearly shows the weekly production schedule in all work centers in the fabrication and assembly departments. The table also shows the timing and the sizes of the orders that must be placed to the suppliers of seats.

### 13.5.3 Lot Sizing with Setup Costs and Capacity Constraints

When setup cost of producing a lot is not significant and when the process has enough capacity, lot-for-lot lot sizing policy is an effective policy, since it minimizes inventory carried from one period to another. However, if setup cost for making a lot is large, then it might not be economical to incur setup costs and make a lot every period. Lot sizing problems in operations scheduling is concerned with finding the production schedule (i.e., planned-order release) that minimizes the total setup and inventory holding cost.

Example 13.2

(*Lot Sizing Crossbars at Solid Furniture*) Because of the long and costly setups, crossbars for CC-12 chairs are made in batches each week. Specifically, each week the crossbar work center is set up to produce a batch with setup cost of $4000. If a crossbar is not used in that week, it is kept in inventory to be used in future weeks. The cost of holding one crossbar in inventory for 1 week is $0.5. The crossbar work center has a capacity of producing a maximum of 3000 crossbars in a week.

The managers are concerned about the high setup cost of lot-for-lot policy. Considering the planned-order release for crossbars that requires making 600, 1800, 1600, and 1400 crossbars in weeks 3, 4, 5, and 6, respectively, the total setup cost in 4 weeks is $4 \times \$4000 = \$16,000$. On the other hand, holding a crossbar in inventory for 1 week is not that costly. Therefore, considering the large setup cost and the small inventory holding cost, using lot-for-lot policy does not make sense. Hence, the managers would like to use another lot sizing policy that minimizes the total setup and inventory holding costs of crossbars in the 6 weeks. What lot sizing policy has the minimum total cost?

Lot sizing problems with setup and inventory holding costs can be solved using the mathematical programming model presented in the following analytics.

**Lot Sizing—Minimizing Total Setup and Inventory Holding Cost**

Consider a process that produces one item to satisfy demand in periods $1, 2, \ldots, T$, where

$$d_t = \text{Demand in Period } t$$
$$b_t = \text{Process capacity in Period } t$$

Items are produced in lots (i.e., batches). Producing a lot in a period requires the process to be set up for production in that period. When the process is set up in Period $t$, it can produce a lot of any size with the maximum size of $b_t$—the capacity of the process. Produced items that are not used in one period are kept in inventory to be used in future periods. Costs include

$$c_I = \text{Inventory holding cost of one item per period}$$
$$c_P = \text{Setup cost of the process}$$

*Decision Variables:* Decisions variables in Period $t = 1, 2, \ldots, T$ are as follows:

$$X_t = \text{Size of the lot produced in Period } t$$
$$I_t = \text{Number of items in inventory at the } \textit{beginning} \text{ of Period } t$$

and binary variable for setup decision is

$$Y_t = \begin{cases} 1 : & \text{if process is set up for production in Period } t \\ 0 : & \text{otherwise} \end{cases}$$

Note that it might be optimal for the process not to produce any lots in a period and use inventory to satisfy demand in that period.

*Mathematical Programming Model:* The mathematical programming model that finds the optimal solution to lot sizing problem (i.e., optimal production, inventory, and setup decisions) in order to minimize the total setup and inventory holding cost in the next $T$ periods is

$$\text{Minimize} \quad Z = \sum_{t=1}^{T} c_I I_t + c_P Y_t$$

$$\text{subject to:}$$
$$X_t - b_t Y_t \leq 0 \qquad \text{for all} \quad t = 1, 2, \ldots, T$$
$$I_t - I_{t+1} + X_t = d_t \qquad \text{for all} \quad t = 1, 2, \ldots, T$$

$$X_t, I_t \geq 0 \quad \text{and} \quad Y_t = \{0 \text{ or } 1\} \quad \text{for all} \quad t = 1, 2, \ldots, T$$

where $I_1 = 0$ and $I_{T+1} = 0$ are the available inventory at the beginning of Period 1 and at the end of Period $T$ (i.e., beginning of Period $T + 1$), respectively.

The first set of constraints corresponds to the capacity of the process in Period $t$. To explain the constraint, we can rewrite it as follows:

$$X_t \leq b_t Y_t$$

If the process is set up in Period $t$ to produce, then $Y_t = 1$, and the above constraint becomes $X_t \leq b_t$, which implies that the size of the lot produced in Period $t$ should not be more than the capacity of the process in that period. If the process is not set up to produce in Period $t$, then $Y_t = 0$ and the above constraint becomes $X_t \leq 0$. Since $X_t$ cannot be negative (i.e., we set $X_t \geq 0$ in our model), then $X_t = 0$, and there will be no production in Period $t$.

The second set of constraints is the "Production Conservation Constraints" that ensures that an item produced in each period is either used to satisfy demand in that period, or carried as inventory to satisfy demand in future periods (see Sec. 12.7.1 of Chap. 12).

To find the optimal lot sizing policy for crossbars in Example 13.2, we have the demands (i.e., time-phased net requirements) as $d_1 = 0$, $d_2 = 0$, $d_3 = 600$, $d_4 = 1800$, $d_5 = 1600$, and $d_6 = 1400$. Using the above analytics and considering $I_0 = 0$ and $I_7 = 0$, the mathematical programming model for the lot sizing problem of crossbars would be as follows:

$$Minimize \quad Z = 0.5(I_1 + I_2 + \cdots + I_6) + 3000(Y_1 + Y_2 + \cdots + Y_6)$$

*subject to:*

$$X_1 - 3000Y_1 \leq 0$$
$$X_2 - 3000Y_2 \leq 0$$
$$X_3 - 3000Y_3 \leq 0$$
$$X_4 - 3000Y_4 \leq 0$$
$$X_5 - 3000Y_5 \leq 0$$
$$X_6 - 3000Y_6 \leq 0$$
$$-I_2 + X_1 = 0$$
$$I_2 - I_3 + X_2 = 0$$
$$I_3 - I_4 + X_3 = 600$$
$$I_4 - I_5 + X_4 = 1800$$
$$I_5 - I_6 + X_5 = 1600$$
$$I_6 \quad + X_6 = 1400$$

$$X_t, I_t \geq 0 \quad \text{and} \quad Y_t = \{0 \text{ or } 1\} \quad \text{for all} \quad t = 1, 2, \ldots, 6$$

Solving the above linear programming model with binary variables using Excel Solver, we find the optimal lot sizing solution as follows:

$$Y_1^* = 0, \quad Y_2^* = 0, \quad Y_3^* = 1, \quad Y_4^* = 0, \quad Y_5^* = 1, \quad Y_6^* = 0$$

which implies that only in Weeks 3 and 5 the work center must be set up to produce crossbars; in other weeks the work center should not produce any crossbars. The optimal lot sizes in Weeks 1 to 6 and the resulting inventory are given by

$$X_1^* = 0, \quad X_2^* = 0, \quad X_3^* = 2400, \quad X_4^* = 0, \quad X_5^* = 3000, \quad X_6^* = 0$$

$$I_1^* = 0, \quad I_2^* = 0, \quad I_3^* = 0, \quad I_4^* = 1800, \quad I_5^* = 0, \quad I_6^* = 1400$$

The above solution shows that the work center should not produce any lots in Weeks 1 and 2. In Week 3, the center must produce a lot of size $X_3^* = 2400$ crossbars and use 600 of them to satisfy demand in Week 3. The remaining $2400 - 600 = 1800$ units are carried as inventory to Week 4 (i.e., $I_4^* = 1800$). The demand in Week 4 is $d_4 = 1800$, so the inventory of 1800 is used to satisfy demand in Week 4 and thus there is no production in Week 4 (i.e., $X_4^* = 0$). However, the center is set up in Week 5 and produces a lot of size $X_5^* = 3000$ crossbars. This lot is used to satisfy demand $d_5 = 1600$ in Week 5 and the rest is carried as inventory to Week 6 (i.e., $I_6^* = 1400$) to satisfy demand $d_6 = 1400$ in that week.

One interesting observation that holds for all lot-sizing problems is that under the optimal policy we always have $X_t^* \times I_t^* = 0$. This implies that the entire demand in each Period $t$ is satisfied by either production in that period (i.e., $X_t^* \neq 0, I_t^* = 0$), or by inventory in that period (i.e., $X_t^* = 0, \ I_t^* \neq 0$), but not both; in other words, the optimal solution does not result in $X_t^* \neq 0, I_t^* \neq 0$.

### 13.5.4 Incorporating Variability in MRP Process

There are three underlying assumptions in MRP explosion calculus: (i) there is no variability in demand, (ii) there is no variability in production time, and (iii) lead time is independent of the size of the lot. How can one incorporate variability into MRP procedure?

**Variability in Demand**

Recall that MPS was established based on the results of aggregate planning, which in turn used demand forecasts to develop production plan for products. Demand forecasts, as mentioned in Chap. 3, have two sources of variability: one for demand itself and the other for errors in the forecast. One approach to protect against variability in demand is to use safety inventory. How much safety inventory? We can use the cost models or service level models in Chap. 10 to find the proper safety inventory for the finished product. A similar approach is to, instead of using the point forecast for demand (i.e., mean demand), use the upper limit of a two-sided prediction interval. Using the upper limit of a two-sided 90-percent prediction interval forecast results in demand being less than the upper limit with probability 0.95. Planning for that demand creates a safety inventory, which is inventory beyond the average demand.

The good news is that we only need to incorporate safety inventory for the finished product. This is because, through explosion calculus, the safety inventory of the finished product will be automatically transmitted down to all items in the lower levels.

**Variability in Lead Times**

There is also variability in production or purchasing lead times due to interruptions such as machine failures or interruptions in supplier's delivery. Similar to safety inventory, one can use *safety lead time* to protect against variability in lead time. Specifically, one can inflate lead time at each level by a safety factor to account for unexpected interruptions in production or ordering process. For example, a lead time of 2 weeks for back supports in our CC-12 chair example can be multiplied by a safety factor 1.5, resulting in lead time of $2 \times 1.5 = 3$ weeks. Then lead time of 3 weeks (instead of 2 weeks) is used to obtain time-phased net requirements for back supports. As opposed to safety inventory that is only needed for the finished product, safety lead time must be used for items at all levels. How large the safety factor should be depends on the variability in the production or ordering process. Lead times with larger coefficient of variation require larger safety factors.

Setting up safety inventory results in larger on-hand inventory in all levels and setting safety lead time results in having those inventories earlier than when they are actually needed—both result in increasing inventory costs. However, this is the cost of having variability in the process, which again emphasizes the importance of reducing variability in operations.

**Lot-Dependent Lead Times**

The explosion calculus in MRP assumes that the lead time required to produce or order a lot is independent of the size of the lot. In other words, it takes the same lead time of $L = 3$

weeks to make 800 front legs or to make 1800 front legs. While this might be true for when items are ordered from outside suppliers (e.g., seats in the CC-12 chair example), it is usually not the case for items produced in the firm's facility. It is often the case that larger lot sizes take more time to produce. Incorporating lot-dependent lead times in MRP calculation is a very complex process that is out of scope of our discussion here. To deal with this complexity and other similar challenges, different approaches have been developed and used in higher generations of MRP software, as discussed in the following section.

### 13.5.5 From MRP to Enterprise Resource Planning (ERP)

After successful implementation of MRP in planning and scheduling manufacturing activities, MRP became popular with manufacturers, and its features were expanded beyond finding the planned-order releases of parts, components, and subassemblies. The next generation of MRP developed in the early 1980s was called MRP II or *Manufacturing Resource Planning*. MRP II linked the manufacturing activities under MRP to firm's other functions, including demand management, capacity planning, master production scheduling, and capacity requirement planning.

With the advances in computer technology and the increasing need of information sharing and coordination among all functions of a firm, MRP II evolved into a much larger system called ERP, which stands for *Enterprise Resource Planning*. While MRP II linked some functional areas of firms, ERP links all functional areas. ERP includes several modules that are linked together using a central database. The main modules are as follows:

- *Manufacturing Module:* This module of ERP provides the same functions as MRP and MRP II, including forecasting, master production scheduling, product data management, capacity requirement planning, and shop floor control. The goal of this module is to improve efficiency of these manufacturing-related activities.
- *Supply Chain Management (SCM):* This module includes planning, coordination, and control of activities such as demand management, inventory and ordering management, procurement, distribution management, and sourcing, among others.
- *Customer Relationship Management (CRM):* This module is designed to improve customer service. It covers activities such as marketing campaigns, sales quotes, customers' orders and invoices, customer service needs, gathering and analyzing customers' information, substituting products for an order, and identifying profitable customer segments, among others.
- *Finance and Accounting:* This module includes all accounting and finance activities of a firm, including updating and compiling billing, account payable and account receivable, cash flow management, budgeting, risk management, tax management, etc.
- *Human Resource Management (HRM):* This module is about managing human resources, including activities such as tracking working times of workforce, benefits and compensation, payroll, workforce planning, and workforce performance evaluation.

With globalization of business activities, ERP provides decision support systems and improves efficiency of operation. A manufacturing manager sitting in the office in New York can see (in the laptop) a new order that has just been placed by supply chain group to a supplier in China. The firm's accountant can then immediately see any transactions related to this order. The supply chain manager can see when this order is supposed to be delivered to the warehouse in Los Angeles. There are more than 150 firms now providing ERP systems to businesses. The big players in the market are SAP SE (Germany), Oracle Corporation and Microsoft Corporation (U.S.), and The Sage Group plc. (U.K.).

ERP systems are not only for manufacturing firms. Service firms also use ERP systems. Because operations in service systems is different from those in manufacturing systems, the manufacturing module of ERP must be changed and customized to the firm's needs. Overall, ERP systems designed for services improve their efficiency by better matching supply with demand, providing backoffice support, time management, resource management, and project management. The Human Resource Management, Accounting and Finance, and Customer Relationship Management modules of ERP for services are similar to those for manufacturing systems.

## 13.6 Loading

After MRP, Loading is the next step in the hierarchy of decisions for operations scheduling in job shops and batch processes, see Fig. 13.5. *Loading* refers to assignment of activities or jobs to resources of a process in order to achieve a certain goal (e.g., minimizing costs, minimizing setups, completing jobs in minimum time). Below is a typical example of a loading problem in job shop scheduling.

**Example 13.3**

*(Loading in the Fabrication Department of Solid Furniture)* Consider the production scheduling of the CC-12 chair in Solid Furniture. Recall that the fabrication department makes front legs for CC-12 chairs in its front leg work center. According to the production schedule in Fig. 13.12, in Week 2, the fabrication department must produce 1100 front legs for CC-12 chairs. On the other hand, this work center also makes front legs for three other types of chairs, namely, CC-13, CC-14, and CC-15 chairs. The MRP procedure for these chairs resulted in planned-order releases of 1200, 2300, and 2800 front legs for CC-13, CC-14, and CC-15 chairs, respectively. Front leg work center has four specialized machines making front legs. Front legs for each type of chair have a different design and hence have different effective process times on these four machines. Also, some of the four machines are old and slow with longer setup times, while others are new, fast with shorter setup times. These machines are also used for other purposes, so their available times are different, as shown in Table 13.4.

Because of the setup and operator costs, producing each lot on each machine has different costs, as shown in Table 13.4. The time it takes (in days) for each machine to produce each required lot is different and is also shown in the table. The question for the manager of the fabrication department is the following: Considering the availabilities of machines in Week 2, how should each lot of four types of legs be assigned to the four machines in order to minimize the total cost? A lot cannot be assigned to different machines.

**Table 13.4    Processing Times and Costs of Making Batches of Front Legs on Four Machines**

| Machines | Costs of processing (*dollars*) | | | | Processing times (*days*) | | | | Availability (days) |
|---|---|---|---|---|---|---|---|---|---|
| | CC-12 | CC-13 | CC-14 | CC-15 | CC-12 | CC-13 | CC-14 | CC-15 | |
| Machine 1 | 2400 | 1800 | 2500 | 3200 | 4 | 3 | 1 | 2 | 5 |
| Machine 2 | 1920 | 2240 | 3250 | 4160 | 3 | 3.5 | 5.5 | 6.5 | 6 |
| Machine 3 | 1200 | 1300 | 2080 | 1560 | 1.5 | 2.5 | 4 | 3 | 4 |
| Machine 4 | 2700 | 3000 | 3600 | 2700 | 4.5 | 5 | 6 | 4.5 | 7 |

The above Loading problem is a variant of a problem called "Assignment Problem." In its general form, *Assignment* problem is about assigning members of one set to members of another set to minimize or maximize a certain performance measure (e.g., minimize total cost or total time, or maximize total profit). Examples of assignment problem include assigning jobs to machines, contracts to bidders, salespeople to sales territories, trucks to routes, repair jobs to repair crews, warehouse capacity to demand, etc.

The following analytics shows how a mathematical programming model can be used to solve the above loading problem.

---

### ANALYTICS

#### Loading—Minimizing Total Cost

There are $N$ jobs available in a process that must be assigned to $M$ resources. If Job $j$ is assigned to Resource $i$, then the time it takes Resource $i$ to complete Job $j$ is $t_{ij}$, resulting in cost of $a_{ij}$. Resource $i$ is available for $T_i$ units of time to process jobs. The goal is to assign jobs to resources in order to minimize the total cost of processing $N$ jobs during the available time of resources.

The following mathematical programming model with binary decision variables can be used to find the optimal assignment.

*Decision Variables:* The binary decision variables for all $i = 1, 2, \ldots, M$ and $j = 1, 2, \ldots, N$ are as follows:

$$Y_{ij} = \begin{cases} 1 & : \text{ if Job } j \text{ is assigned to Resource } i \\ 0 & : \text{ otherwise} \end{cases}$$

*Objective Function:* The goal is to minimize the total cost of processing $N$ jobs by $M$ resources

$$\text{Minimize} \quad Z = \sum_{i=1}^{M} \sum_{j=1}^{N} a_{ij} Y_{ij}$$

*Constraints:* There are two groups of constraints:

- *Resource Time Constraints:* The first group of constraints ensures that resources do not work beyond their available times. For Resource $i$ the constraint is

$$t_{i1} Y_{i1} + t_{i2} Y_{i2} + \cdots + t_{iN} Y_{iN} \leq T_i \quad \text{(Resource } i \text{ availability)}$$

There are $M$ such constraints, one for each resource.

- *Job Assignment Constraints:* The second group of constraints ensures that each job is completed by one of the resources. For Job $j$, the constraint is

$$Y_{1j} + Y_{2j} + \cdots + Y_{Mj} = 1 \quad \text{(Job } j \text{ assignment)}$$

There are $N$ such constraints, one for each job.

Hence, the above mathematical programming problem has $M \times N$ binary decision variables and $M + N$ constraints.

---

Let's use the above analytics to find the optimal solution. Consider producing the batch of 1100 front legs for CC-12 as Job 1; producing the batch of 1200 front legs for CC-13 as Job 2; producing the batch of 2300 front legs for CC-14 as Job 3; and producing 2800 front legs for CC-15 as Job 4. Therefore, the decision variables are binary variables $Y_{ij} = (0 \text{ or } 1)$, for $i = 1, 2, 3, 4$ machines and $j = 1, 2, 3, 4$ jobs. Hence, the optimization problem with 16 variables and 8 constraints is

$$Minimize \ Z = 2400Y_{11} + 1800Y_{12} + 2500Y_{13} + 3200Y_{14} + 1920Y_{21} + 2240Y_{22}$$
$$+ 3520Y_{23} + 4160Y_{24} + 1200Y_{31} + 1300Y_{32} + 2080Y_{33} + 1560Y_{34}$$
$$+ 2700Y_{41} + 3000Y_{42} + 3600Y_{43} + 2700Y_{44}$$

subject to :

$$4Y_{11} + 3Y_{12} + Y_{13} + 2Y_{14} \leq 5 \quad \text{(Machine 1 availability)}$$
$$3Y_{21} + 3.5Y_{22} + 5.5Y_{23} + 6.5Y_{24} \leq 6 \quad \text{(Machine 2 availability)}$$
$$1.5Y_{31} + 2.5Y_{32} + 4Y_{33} + 3Y_{34} \leq 4 \quad \text{(Machine 3 availability)}$$
$$4.5Y_{41} + 5Y_{32} + 6Y_{43} + 4.5Y_{44} \leq 7 \quad \text{(Machine 4 availability)}$$
$$Y_{11} + Y_{21} + Y_{31} + Y_{41} = 1 \quad \text{(Job 1 assignment )}$$
$$Y_{12} + Y_{22} + Y_{32} + Y_{42} = 1 \quad \text{(Job 2 assignment )}$$
$$Y_{13} + Y_{23} + Y_{33} + Y_{43} = 1 \quad \text{(Job 3 assignment )}$$
$$Y_{14} + Y_{24} + Y_{34} + Y_{44} = 1 \quad \text{(Job 4 assignment )}$$

$$Y_{ij} = (0 \text{ or } 1) \quad \text{for} \quad i = 1, 2, 3, 4, \ j = 1, 2, 3, 4$$

If Excel Solver is used to solve the above assignment problem, we find $Z^* = \$7700$. The optimal values for decision variables are $Y_{ij}^* = 0$ for all decision variables, except for the following:

$$Y_{31}^* = 1, \ Y_{32}^* = 1, \ Y_{13}^* = 1, \ Y_{44}^* = 1$$

The solution implies that producing the batch of 1100 legs for CC-12 chair (i.e., Job 1) must be assigned to Machine 3; producing the batch of 1200 legs for CC-13 chair (i.e., Job 2) must also be assigned to Machine 3; producing the batch of 2300 legs for CC-14 chair (i.e., Job 3) must be assigned to Machine 1; and producing the batch of 2800 legs for CC-15 chair (i.e., Job 4) must be assigned to Machine 4. Under this assignment, no jobs are assigned to Machine 2.

Minimizing the total cost is one performance measure for loading jobs to machines. The cost $a_{ij}$ can also represent the setup time required if Machine $i$ is set up for Job $j$, for which the goal would be to minimize the total time required to set machines up for jobs.

## 13.7 Detailed Scheduling

Loading decisions determine which job or group of jobs must be assigned to which work center. The loading decision, however, does not specify in what order the jobs waiting in a work center must be processed. This is determined by Detailed Scheduling. *Detailed Scheduling* is the final stage of operations scheduling before the product is completed and shipped to the customer (see Fig. 13.5).

Consider the loading of crossbars for CC-12, CC-13, CC-14, and CC-15 chairs in Example 13.3 of Solid Furniture. Recall that the optimal loading assigned two jobs (making crossbars for CC-12 and CC-13 chairs) to Machine 3. Suppose you are the manager of the fabrication department and the worker operating Machine 3 is asking which batch should be made first: the batch of 1100 crossbars for CC-12 chairs, or the batch of 1200 crossbars for CC-13 chairs? Which one would you recommend?

Your answer, of course, depends on several factors including when each batch is needed (i.e., due dates), how long it takes to finish each batch (i.e., process times), how long it takes to set up the machine to produce each batch, etc. Detailed scheduling analytics consider these factors and determine the best way to schedule jobs on machines in order to achieve a certain performance (e.g., meeting due dates with least costs).

The most commonly used detailed scheduling problem is for processing a set of jobs on a single resource. Although simple, this problem covers a wide range of applications in both manufacturing and service systems. In manufacturing, for example, bottleneck resource—a single worker, a single machine, or a single workstation—determines the effective capacity and hence the throughput of the entire system. Thus, scheduling this single resource efficiently is critical. Also, the single resource can be the entire plant that is supposed to schedule the completion of several orders it received from its customers. In service systems, the single resource can be the operations room of a hospital to be scheduled for surgery of patients, or an accountant who is scheduling the work of processing tax documents of several clients.

In operations scheduling terminology, there is a difference between "Scheduling" and "Sequencing."

**Sequencing and Scheduling**

- *Sequencing* is often used for the case of a single resource. A sequence determines the *order* by which the jobs are to be processed by a single resource.

- *Scheduling* is used for more complex systems with multiple resources. A schedule determines *when* each job must be processed by *which* resource.

Note that in a single resource case, a sequence also determines *when* a job must be processed, because the starting time of a job is the finishing time of the previous job in the sequence. Hence, Sequencing is a simpler version of Scheduling, mainly used in cases with a single resource processing a set of jobs.

### 13.7.1 Sequencing Jobs in a Process with a Single Resource

In this section we focus on the most common sequencing problems in processes in which a single resource (a worker, a machine, or entire factory) must process several jobs. A schematic of such a system is shown in Fig. 13.13.

A wide range of analytics have been developed for processing jobs in processes with single resource. The common assumptions for these analytics are the following:

- There are $N$ jobs available at time zero that must be processed one-at-a-time by a single resource. No jobs arrive after processing starts.

- The processing times of jobs are deterministic (i.e., have no variability). They are also independent of each other and independent of the sequence by which they will be processed.

Figure 13.13    Processing *N* jobs in a process with a single resource.

- The resource is continuously available, it does not breakdown, and it never idles when jobs are waiting.
- When processing of a job starts, the job is processed to completion without any interruptions.

While some sequencing problems may violate one or more of the above assumptions, some of the analytics developed for systems with the above assumptions are shown to result in optimal or close-to-optimal sequence, even when one or more assumptions are violated.

### Minimizing Flow Times

Minimizing flow times is an important goal of the operations engineers and managers. In a manufacturing setting, lower flow times result in lower response time and lower inventory, and in service setting, lower flow times result in lower customer waiting time and hence higher customer satisfaction.

**Example 13.4**

Safe-Belt (SB) is a small producer of safety belts that are used in children's car seats and strollers. The firm makes these belts in its plant in Iowa and supplies them to several manufacturers of those products inside and outside the United States. Currently, SB has eight outstanding orders from eight different manufacturers as shown in Table 13.5.

Because each order is for a different type of belt with different length and width, different color, and different buckles, SB must complete one order before starting work on another. SB has material and labor to start working on any of the orders. The question is how to sequence processing these orders. SB knows that under any sequence, some manufacturers whose orders are processed first will wait less and those whose orders are processed last will wait more. Is there a sequence that results in the minimum total waiting time for all manufacturers?

**Table 13.5    Process Times of Orders in Safe-Belt Plant in Example 13.4**

| Orders ($j$) | 1 | 2 | 3 | 4 | 5 | 6 | 7 | 8 |
|---|---|---|---|---|---|---|---|---|
| Process time ($t_j$) | 10 | 6 | 3 | 5 | 2 | 4 | 8 | 11 |

SB's sequencing problem has all the features of our basic single-resource sequencing problem. There is a single resource (i.e., SB plant) processing several jobs (i.e., eight orders) one-at-a-time. All orders are available at time zero (i.e., now). The processing times of the orders are known and are independent of each other. SB's plant is continuously available for processing orders and can process without interruptions.

Because all orders are available now, the waiting time of Manufacturer $i$ would be $F_i$—the flow time of its order in SB's plant. The SB's goal is to find a sequence that minimizes the total waiting time of all eight manufacturers. Formally stated, the goal is to minimize $F_1 + F_2 + \cdots + F_8$. On the other hand, average flow time of the eight orders is $(F_1 + F_2 + \cdots + F_8)/8$. Thus, minimizing total flow time also minimizes the average flow time.

How many options does DB have to sequence the eight orders? Well, a lot. There are $8! = 40{,}320$ different sequences that can be used to finish all eight orders. The good news, however, is that we do not need to check all these sequences. There is a simple rule, called "SPT Rule," that easily identifies the optimal sequence (see Appendix I that shows why SPT Rule minimizes the average flow time).

---

**PRINCIPLE**

**Minimizing Flow Time and Inventory—The SPT Rule**

Consider a resource that must process $N$ jobs that are all available at time zero. The process time of Job $j$ is $t_j$. Shortest Processing Time *or* SPT rule minimizes (i) total flow time of $N$ jobs, (ii) average flow time of $N$ jobs, and (iii) average inventory. Under SPT rule, jobs are sequenced from the smallest process times to the largest process times.

---

The rationale behind optimal performance of SPT rule is clear. By working on jobs with smaller processing times, these jobs leave the process earlier and thus the number of jobs in the process (i.e., inventory) reduces faster. Hence, processes that use SPT have lower inventory at any time compared to processes that do not follow SPT. Lower inventory, according to Little's law, results in lower flow time.

Let's now find the optimal sequence for processing eight orders at Safe-Belt (SB) in Example 13.4. Considering the process times of orders in Table 13.5, the SPT rule leads to sequence

$$\text{SPT Sequence} \quad \Longrightarrow \quad 5,\ 3,\ 6,\ 4,\ 2,\ 7,\ 1,\ 8$$

Under this sequence, the flow time of the first job, Job 5, is $F_5 = t_5$. The flow time of the second job, Job 3, is $F_3 = t_5 + t_3$. The flow time of the third job, Job 6, is $F_6 = t_5 + t_3 + t_6$. Continuing in this manner and adding the flow times of all eight jobs, the total flow time of the sequence would be

$$
\begin{aligned}
\text{Total flow time} &= 8t_5 + 7t_3 + 6t_6 + \cdots + 2t_1 + t_8 \\
&= 8(2) + 7(3) + 6(4) + \cdots + 2(10) + 11 \\
&= 165 \text{ days}
\end{aligned}
$$

The average flow time is therefore $(165/8) = 20.6$ days, which is the minimum possible average flow time.

**Minimizing Total Weighted Flow Time**

Firms often consider some customers more important than others, and hence they try to finish those customers' orders sooner than other orders. Therefore, for example, firms prefer a sequence with the total flow time of 100 days for two customers in which the important customer waits for 20 days and the unimportant customer waits for 80 days over another sequence that also has a total flow time of 100 days, but with important customers waiting for 80 days and unimportant customers waiting for 20 days. Total flow time cannot capture this, since it assumes customers are equally important.

Also flow time represents the time a job spends in the process as inventory (in-buffer or in-process inventory). Different jobs have different values and thus different holding costs. In these cases, minimizing total flow time is not an appropriate goal either, since it assumes that all jobs have the same holding cost. Then, what is an appropriate goal?

In the above situations one can use the total weighted flow time that considers the importance and/or the holding costs of jobs. As mentioned in Sec. 13.2, in "total weighted flow time" each Job $j$ has weight $w_j$ and the goal is to minimize

$$\text{Total Weighted Flow Time} = w_1 F_1 + w_2 F_2 + \cdots + w_N F_N$$

We use the following example to introduce the sequencing rule that minimizes the above measure.

Table 13.6   Process Times and Holding Costs of Jobs in Example 13.5

| Job | 1 | 2 | 3 | 4 | 5 | 6 | 7 | 8 |
|---|---|---|---|---|---|---|---|---|
| Process time | 10 | 6 | 3 | 5 | 2 | 4 | 8 | 11 |
| Holding cost | $240 | $480 | $320 | $500 | $175 | $650 | $600 | $750 |

### Example 13.5

Managers of Safe-Belt (SB) in Example 13.4 realize that under the SPT sequencing rule, the flow time of Job 8 in their plant would be 49 days. However, Job 8 consists of a batch of expensive high-end belts and is the most profitable product that is sold to a manufacturer in Texas. Hence, SB does not want to keep the manufacturer waiting a long time for its order. In fact, some of the other orders also belong to SB's major customers and SB would like to also shorten the waiting times of those customers as much as possible. A study of the orders estimates the cost of keeping each order in the plant for 1 day as shown in Table 13.6. This cost includes the work-in-process inventory holding cost of each order as well as long-term loss of goodwill cost of each manufacturer. Considering this new information, how should SB process these eight jobs in order to minimize its total cost?

If we denote $w_j$ as the daily cost of keeping Job $j$ in SB's plant (waiting or being processed), then $w_1 F_1 + w_2 F_2 + \cdots + w_8 F_8$ would be the total cost of eight jobs in SB's plant. Therefore, SB's goal now becomes minimizing the total weighted flow time. But, what sequence of jobs minimizes this new goal?

Obviously, SPT rule cannot be used since it ignores the costs associated with flow times. However, a modified version of SPT rule called WSPT results in the optimal sequence that minimizes the total weighted flow time.

---

**PRINCIPLE**

**Minimizing Total Weighted Flow Time—The WSPT Rule**

Consider a process with a single resource that must process $N$ jobs, all available at time zero. The process time of Job $j$ is $t_j$. Also, holding Job $j$ in the process (waiting or being processed) costs $w_j$ per unit time. To minimize the total weighted flow time, jobs must be processed according to WSPT rule. Under *Weighted Shortest Processing Time* (WSPT) rule, jobs are sequenced from the largest ratio of $w_j/t_j$ to the smallest. If $w_j$ is the inventory holding cost of Job $j$ per unit time, then WSPT minimizes the total inventory holding cost of all jobs.

The rationale behind WSPT rule is as follows. There are two ways to reduce costs: (i) processing jobs with *larger holding costs* ($w_j$) first. This reduces costs by completing those more expensive jobs earlier. (ii) Processing jobs with *shorter processing times* ($t_j$) first. This reduces costs by reducing the total waiting time (i.e., total flow time) of all jobs in the process. WSPT rule tends to capture these two factors through ratio $w_j/t_j$. Larger value of $w_j/t_j$ corresponds to: (i) larger holding costs $w_j$, or (ii) smaller processing time $t_j$. Hence, jobs with larger ratio $w_j/t_j$ should get priority over jobs with smaller ratio. This is proven to minimize the total weighted flow times.

We now use WSPT rule to find the optimal sequence for SB in Example 13.5. Using process times $t_j$ and costs $w_j$, we can compute ratio $w_j/t_j$ as in Table 13.7.

Hence, the optimal sequence for processing eight orders at Safe-Belt (SB) in order to minimize the total holding cost of all orders is

$$\text{WSPT Sequence} \implies 6, 3, 4, 5, 2, 7, 8, 1$$

Table 13.7   Computing Ratio $w_j/t_j$ for Jobs in Example 13.5

| Job | 1 | 2 | 3 | 4 | 5 | 6 | 7 | 8 |
|---|---|---|---|---|---|---|---|---|
| $t_j$ | 10 | 6 | 3 | 5 | 2 | 4 | 8 | 11 |
| $w_j$ | $240 | $480 | $320 | $500 | $175 | $650 | $600 | $750 |
| $w_j/t_j$ | 24 | 80 | 106.67 | 100 | 87.5 | 162.5 | 75 | 68.18 |

One interesting feature of WSPT rule is that it also works in systems in which job processing times are stochastic (i.e., have variability), and new jobs arrive randomly to the process. Such systems are $G/G/1$ queuing systems (see Chap. 7). Considering $E(t_j)$ as the mean effective process time of Job $j$, and defining $\mu_i = 1/E(t_j)$ as the processing rate for type-$j$ jobs, the WSPT ratio becomes

$$\frac{w_j}{E(t_j)} = w_j\,\mu_j$$

which is the c-mu rule presented in Sec. 8.4.2 of Chap. 8 (with holding cost $c_j$ replaced by $w_j$). Recall that according to the c-mu rule, jobs with higher values of $c_j\mu_j$ are processed before jobs with lower values of $c_j\mu_j$. Hence, WSPT rule is a special case of c-mu rule for systems with deterministic process times and no job arrivals.

**Minimizing Maximum Tardiness**

Jobs often have due dates that either originate from customers (e.g., delivery date for a sofa) or from other parts of the production process (e.g., due date for finishing the legs of the sofa needed in the assembly department). When there are firm due dates, the major goal of operations scheduling is to meet those due dates.

**Example 13.6**

Consider Safe-Belt's sequencing problem in Example 13.5, and now assume that the eight orders from the manufacturers have due dates that were negotiated by the sales department. The due dates are shown in Table 13.8.

SB realizes that, because of lack of coordination with the sales department when they negotiated the due dates, SB cannot meet the due dates of all jobs and some of the jobs may have delays. According to the sales department, manufacturers are not too sensitive about their delays as long as the delays are not very long. Therefore, the sales department has requested SB's plant manager to complete the orders such that, if a job is delayed, the delay is not too long. How should SB process those orders?

Table 13.8   Due Dates for Jobs in Example 13.6

| Job | 1 | 2 | 3 | 4 | 5 | 6 | 7 | 8 |
|---|---|---|---|---|---|---|---|---|
| Process time | 10 | 6 | 3 | 5 | 2 | 4 | 8 | 11 |
| Due date | 15 | 30 | 6 | 35 | 9 | 39 | 23 | 20 |

Recall that, $T_j$, tardiness of a Job $j$ is the lateness of the job when the job is completed after its due date $d_j$. Hence, the manager of SB in Example 13.6 is facing a different goal. The first goal is to complete jobs in an order that no jobs are delayed. If that is not possible, then the goal is to make sure those jobs that are delayed are not delayed for too long. One way to achieve this is to make the largest delay (i.e., largest tardiness) among eight jobs as small as possible. This, as we mentioned in Sec. 13.2, is to minimize the maximum tardiness $T_{\max}$. But, which sequence minimizes the maximum tardiness?

**Minimizing Maximum Tardiness—The EDD Rule**

Consider a process with a single resource that must process $N$ jobs, all available at time zero. Job $j$ has process time $t_j$ and due date $d_j$. Under *Earliest Due Date* (EDD) rule, jobs are sequenced according to their due dates from the earliest due date to the latest due data. EDD rule has the following properties:

- If there exists a sequence that results in all jobs being completed before their due dates (i.e., no tardy jobs), then EDD sequence also results in no tardy jobs. In other words, if a job is tardy in an EDD sequence, then there is no sequence under which all jobs are completed before their due dates.
- EDD rule minimizes the maximum tardiness $T_{\max}$ and maximum lateness $L_{\max}$.

Therefore, EDD rule can be used to check if it is possible to have all jobs completed before their due dates. This becomes useful when we minimize the number of tardy jobs.

We can now use the EDD rule in Safe-Belt sequencing problem to find the sequence with minimum $T_{\max}$. Using due dates in Table 13.8, we construct the EDD sequence as follows:

$$\text{EDD Sequence} \implies 3, 5, 1, 8, 7, 2, 4, 6$$

Under the above EDD sequence, the tardiness of eight jobs can be computed as in Table 13.9.

Table 13.9    Computing Tardiness Under EDD Sequence for Example 13.6

| EDD sequence | 3 | 5 | 1 | 8 | 7 | 2 | 4 | 6 |
|---|---|---|---|---|---|---|---|---|
| Process time ($t_j$) | 3 | 2 | 10 | 11 | 8 | 6 | 5 | 4 |
| Completion time ($C_j$) | 3 | 5 | 15 | 26 | 34 | 40 | 45 | 49 |
| Due date ($d_j$) | 6 | 9 | 15 | 20 | 23 | 30 | 35 | 39 |
| Lateness ($L_j$) | −3 | −4 | 0 | 6 | 11 | 10 | 10 | 10 |
| Tardiness ($T_j$) | 0 | 0 | 0 | 6 | 11 | 10 | 10 | 10 |

As the table shows, five out of eights jobs would be tardy with a maximum tardiness among five tardy jobs being 11 days.

**Minimizing other Tardiness Measures**

While minimizing $T_{\max}$ reduces the tardiness of the job with the most tardiness, it does not reduce the number of days that other jobs are tardy. However, in practice there are situations that a company must pay a penalty if a job is late, or pay a penalty for each day a job is late. Under the EDD sequence in Example 13.6, four jobs (i.e., jobs 2, 4, 6, and 8) are tardy in which jobs 2, 4, and 6 have a tardiness of 10 days and Job 8 has a tardiness of 6 days—a total tardiness of 47 days (including Job 7). What is the schedule that minimizes the total number of tardy jobs or the total number of tardy days? If you would like to know the answer, check out the online supplement of this chapter. The supplement shows how mathematical programming models and Excel spreadsheet can be used to find the sequence that minimizes the total weighted tardiness or the total weighted number of tardy jobs.

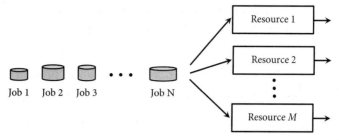

**Figure 13.14    Processing *N* jobs in a process with a pool of *M* resources.**

### 13.7.2  Scheduling Jobs in a Process with a Single Pool of Resources

Often times processes have a pool of multiple resources that are capable of performing same activities to make a product or to provide a service. Recall that in job shops, resources are physically grouped by their functions. Drilling department, for example, consists of several drilling machines, all capable of performing drilling operation. These machines can work independent of each other in parallel and process different jobs. Example of pool of resources working in parallel in service systems is accountants working on different projects in an accounting firm, or cashiers in fast-food restaurants taking orders from different customers.

A schematic of a process with a pool of $M$ resources working in parallel is shown in Fig. 13.14.

The basic problem of scheduling jobs for a pool of servers has the following features:

- There are $N$ jobs available at time zero that must be processed by one of $M$ resources. No jobs arrive after processing starts.

- The processing times of jobs are deterministic (i.e., no variability). They are also independent of each other, independent of the resource that processes them, and independent of the sequence by which they will be processed by resources.

- Each resource can work on only one job at a time.

- Resources are continuously available, they do not breakdown, and never idle when jobs are waiting.

- When processing of a job starts, the job is processed to completion without any interruptions.

While the analytics and principles of scheduling jobs in systems with one resource do not exactly apply to systems with a pool of resources, they provide insights into developing optimal schedules for such systems. In the following sections we present optimal schedules for minimizing total flow time and makespan in processes with a pool of resources.

**Minimizing Total Flow Time**

To minimize total flow time in processes with a pool of resources, we return to Safe-Belt manufacturer in Example 13.4.

 **Example 13.7**

Recall that Safe-Belt (SB) in Example 13.4 has eight orders from eight different manufacturers with the process times shown in Table 13.10. While the original decision was to process all these eight jobs in SB's factory in Iowa, some concerns were raised that the average waiting time of 20.6 days obtained under the optimal SPT rule is still large. The managers of SB came to the conclusion that they should make some of these orders in SB's other plant in Nebraska. This way, both plants can work simultaneously on different orders and finish orders earlier. The question is now which orders must be processed in which plant, and how orders in each plant must be sequenced to minimize the average waiting time of all eight manufacturers? The process times of orders in Nebraska plant are the same as those in Table 13.10.

Table 13.10    Process Times of Jobs in Safe-Belt Plant in Example 13.7

| Job ($j$) | 1 | 2 | 3 | 4 | 5 | 6 | 7 | 8 |
|---|---|---|---|---|---|---|---|---|
| Process time ($t_j$) | 10 | 6 | 3 | 5 | 2 | 4 | 8 | 11 |

Finding the optimal schedule that minimizes the total (and hence the average) flow times in the case with two plants is more difficult than that with one plant. In addition to finding the sequence of orders in each plant, we also need to find which orders must be processed in which plant. The good news, however, is that the SPT rule that minimizes the total flow time in processes with a single resource can be extended to processes with multiple resources.

ANALYTICS

**Minimizing Total Flow Time in a Process with a Pool of Resources—The SPT Rule**

Consider $N$ jobs that must be processed by a pool of $M$ resources. Each job requires processing by only one of the resources—whichever is available. The process time of Job $j$ is $t_j$. The following steps result in the optimal schedule of $N$ Jobs that minimizes total flow time of all jobs (and thus the average flow time):

- **Step 1:** Sequence all jobs in a list according to the SPT rule. Denote the list by $\mathcal{L}$.
- **Step 2:** Assign the first job (i.e., the job with the least process time) in the list $\mathcal{L}$ to the resource with the smallest total process time already assigned. If two or more resources have the smallest total process times already assigned, choose one arbitrary.
- **Step 3:** Omit the job from list $\mathcal{L}$ and return to Step 2.

Following Step 1, and arranging the jobs in list $\mathcal{L}$ according to SPT rule we have

$$\mathcal{L} = \{5, \ 3, \ 6, \ 4, \ 2, \ 7, \ 1, \ 8\}$$

Following Step 2, we must assign Job 5 with process time $t_5 = 2$ days to Iowa plant or Nebraska plant. Because both plants have zero total process time already assigned, we can assign Job 5 to any of them. Hence, we choose Iowa. Delete Job 5 from the list and return to Step 2. The first job in the new list is Job 3 with process time $t_3 = 3$ days. Since Nebraska has the least total process time already assigned (i.e., zero), we assign Job 3 to Nebraska plant. We omit Job 3 from list $\mathcal{L}$ and return to Step 2. The next job in the new list is Job 6 with process time $t_6 = 3$ days. Since Iowa has the least total process times already assigned (i.e., 2 days compared to 3 days for Nebraska), we assign Job 6 to Iowa plant. Continuing in the same manner, all eight jobs will be assigned to plants as shown in Fig. 13.15.

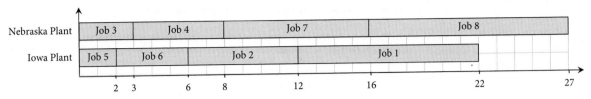

Figure 13.15    Optimal production schedule in Iowa and Nebraska plants in Example 13.7.

As the figure shows, jobs 1, 2, 5, and 6 must be assigned to Iowa plant and jobs 3, 4, 7, and 8 must be assigned to Nebraska plant. The optimal sequence in each plan is

Iowa plant   $\Longrightarrow$   3 , 4 , 7 , 8        Nebraska plant   $\Longrightarrow$   5 , 6 , 2 , 1

Figure 13.15 can also be used to compute the total flow times as follows:

$$\begin{aligned}\text{Total flow time} &= F_1 + F_2 + F_3 + F_4 + F_5 + F_6 + F_7 + F_8 \\ &= 22 + 12 + 3 + 8 + 2 + 6 + 16 + 27 \\ &= 96 \text{ days}\end{aligned}$$

Therefore, the average waiting time of the eight manufacturers is $96/8 = 12$ days, which is significantly lower than that when all orders were planned to be processed in Iowa plant (i.e., 20.6 days).

One interesting question that comes to mind is whether we can use the above analytics and establish list $\mathcal{L}$ according to WSPT rule (instead of SPT rule) to find the optimal schedule that minimizes the total weighted flow time. The answer is no. However, while this revised analytics is not guaranteed to result in the optimal schedule, it is shown to give good schedules with total weighted flow time close to the optimal.

**Minimizing Makespan**
Besides flow time, another goal in scheduling $N$ jobs to multiple resources is to finish all $N$ jobs as soon as possible, that is, minimize the makespan of all jobs. This is a common goal in job shops and batch process systems in which the entire batch of $N$ jobs cannot be moved to the next stage until all jobs in the batch are completed. Minimizing makespan is also an important goal when multiple orders of a customer cannot be shipped unless all orders are completed. Example 13.8 presents such a case.

 Example 13.8

> Fisher Dies has received an order for six stamping dies from an auto manufacturer. Fisher Dies has three numerically controlled (NC) work centers that can be used to make these dies. Each work center has different capabilities for performing a wide range of precision drilling, milling, and surface operations. Therefore, making the same die may require a shorter time in one center than the other. The time that it takes for each work center to complete each die is shown in Table 13.11.
>
> To minimize the shipment cost, all six stamping dies must be shipped to the manufacturer together in one shipment. How should Fisher Dies assign six dies to work centers in order to ship the manufacturer's order as soon as possible?

The problem that Fisher Dies is facing is to assign six jobs (dies) to a pool of three resources (work centers), each capable of making the dies independent of other centers. Since the dies cannot be shipped until all six of them are completed, shipping orders as soon as possible

Table 13.11    Process Times (in Days) of Dies in Work Centers of Example 13.8

|                | Die 1 | Die 2 | Die 3 | Die 4 | Die 5 | Die 6 |
|----------------|-------|-------|-------|-------|-------|-------|
| Work Center 1  | 1     | 2     | 7     | 5     | 8     | 2     |
| Work Center 2  | 2     | 4     | 7     | 8     | 9     | 4     |
| Work Center 3  | 6     | 5     | 3     | 5     | 7     | 9     |

corresponds to minimizing the time that the last die is completed—minimizing makespan of the six dies.

The following analytics describes how a mathematical programming model can be used to find the optimal schedule of dies in the three work centers of Fisher Dies.

## ANALYTICS

**Minimizing Makespan in a Process with a Pool of Resources—Mathematical Programming Model**

Consider $N$ jobs that must be processed by a pool of $M$ resources. Each job requires processing by only one of the resources—whichever is available. The time it takes for Resource $i$ to process Job $j$ is $t_{ij}$ for $i = 1, 2, \ldots, M$ and $j = 1, 2, \ldots N$. The goal is to minimize the time that all jobs are completed, that is, minimizing makespan.

*Decision Variables:* Decision variables include binary variables for all $i = 1, 2, \ldots, M$ and $j = 1, 2, \ldots, N$ as follows:

$$Y_{ij} = \begin{cases} 1 : & \text{if Job } j \text{ is processed by Resource } i \\ 0 : & \text{otherwise} \end{cases}$$

as well as a continuous variable

$$C_{\max} = \text{Time that all jobs are completed}$$

where $C_{\max} \geq 0$.

*Objective Function:* The goal is to minimize the time that all jobs are completed,

$$Minimize \quad Z = C_{\max}$$

*Constraints:* There are two groups of constraints:

- *Job Assignment Constraints:* This group of constraints ensures that each job is completed by one of the resources. For Job $j$, the constraint is

$$Y_{1j} + Y_{2j} + \cdots + Y_{Mj} = 1 \qquad \text{(Job } j \text{ assignment)}$$

There are $N$ such constraints, one for each job.

- *Makespan Constraints:* This group of constraints captures the relationship between makespan and the completion times of jobs by resources. Suppose jobs are assigned to resources. Some resources finish their jobs earlier than others, and one resource will finish its job last. Hence, the makespan of jobs would be the time that the last resource finishes its job. Consequently, the time that each resource finishes its jobs would be less than—and for the last resource would be equal to—the makespan. For Resource $i$ this relationship can be captured by the following constraint:

$$t_{i1} Y_{i1} + t_{i2} Y_{i2} + \cdots + t_{iN} Y_{iN} \leq C_{\max} \qquad \text{(Resource } i)$$

There are $M$ such constraints, one for each resource.

The above mathematical programming problem has $M \times N + 1$ decision variables and $M + N$ constraints.

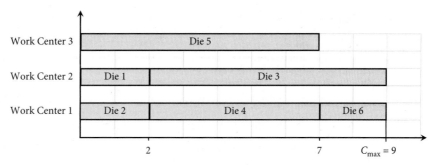

**Figure 13.16    Optimal schedule of making six dies in three work centers of Example 13.8.**

For Fisher Dies, the decision variables are binary variable $Y_{ij} = (0 \text{ or } 1)$, for $i = 1, 2, 3$ resources (i.e., work centers) and $j = 1, 2, 3, 4, 5, 6$ jobs (i.e., stamping dies). The other decision variable is $C_{max} \geq 0$, where $C_{max}$ is the time all six stamping dies are finished by Fisher Dies. Hence, the optimization problem is

$$Minimize \ Z = C_{max}$$

$$subject \ to:$$

$$Y_{11} + Y_{21} + Y_{31} = 1 \qquad (\text{Job 1 assignment})$$
$$Y_{12} + Y_{22} + Y_{32} = 1 \qquad (\text{Job 2 assignment})$$
$$Y_{13} + Y_{23} + Y_{33} = 1 \qquad (\text{Job 3 assignment})$$
$$Y_{14} + Y_{24} + Y_{34} = 1 \qquad (\text{Job 4 assignment})$$
$$Y_{15} + Y_{25} + Y_{35} = 1 \qquad (\text{Job 5 assignment})$$
$$Y_{16} + Y_{26} + Y_{36} = 1 \qquad (\text{Job 6 assignment})$$
$$1Y_{11} + 2Y_{12} + 7Y_{13} + 5Y_{14} + 8Y_{15} + 2Y_{16} - C_{max} \leq 0 \qquad (\text{Work Center 1})$$
$$2Y_{11} + 4Y_{12} + 7Y_{13} + 8Y_{14} + 9Y_{15} + 4Y_{16} - C_{max} \leq 0 \qquad (\text{Work Center 2})$$
$$6Y_{11} + 5Y_{12} + 3Y_{13} + 5Y_{14} + 7Y_{15} + 9Y_{16} - C_{max} \leq 0 \qquad (\text{Work Center 3})$$

$$Y_{ij} = (0 \text{ or } 1) \quad for \quad i = 1, 2, 3 \ , \ j = 1, 2, 3, 4, 5, 6 \ \text{ and } \ C_{max} \geq 0$$

If Excel Solver is used to solve the above assignment problem, we find $Z^* = 9$. The optimal values for decision variables are $C_{max}^* = 9$ and $Y_{ij}^* = 0$ for all decision variables except for the following:

$$Y_{12}^* = 1, \ \ Y_{14}^* = 1, \ \ Y_{16}^* = 1, \ \ Y_{21}^* = 1, \ \ Y_{23}^* = 1, \ \ Y_{35}^* = 1$$

The solution implies that the earliest time Fisher Dies can finish making all six stamping dies and ship them to the manufacturer is $C_{max} = 9$ days. This can be accomplished if stamping dies 2, 4, and 6 are made in Work Center 1, stamping dies 1 and 3 are made in Work Center 2, and stamping die 5 is made in Work Center 3. This schedule is shown in Fig. 13.16.

### 13.7.3 Scheduling Jobs in a Two-Stage Process

The detailed scheduling problem discussed so far assumes a single-stage process in which jobs are completed after being processed by only one resource. There are also cases in practice where jobs require processing by more than one resource of the process. For example, in manufacturing systems (e.g., assembly lines) jobs often go through a number of stages, one after another. The resources are set up in series and there is a single routing for all jobs. These

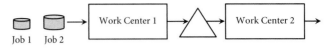

Figure 13.17    *Two work centers in series processing two jobs.*

systems are called *Flow Shops* or flow-line processes. Flow shops also exist in service system. For example, in Emergency Department of hospitals patients go through the same stages of registration, triage, and being seen by a doctor. The scheduling question in flow shops is the following: In what sequence jobs must be processed in the first stage, in the second stage, . . ., in the last stage of a flow shop?

Finding the optimal job schedule in flow shops is very complex and requires sophisticated mathematical programming models. However, when the flow shop has only two stages, finding the optimal schedule becomes easy. The main assumptions of a two-stage flow shop are as follows:

- There are *N* jobs available at time zero that must be processed in two stages. No jobs arrive after processing starts.
- The processing times of jobs in each stage are deterministic (i.e., no variability). They are also independent of each other, and independent of the sequence by which they will be processed at each stage.
- Each stage has a single resource that can work on only one job at a time.
- Resources are continuously available, they do not breakdown, and never idle when jobs are waiting.
- When processing of a job starts, the job is processed to completion without any interruptions.
- When the processing of a job at stage 1 is completed, the job waits in the buffer of stage 2 to be processed by the second resource.

To better understand the dynamics of flow shops, consider a simple example of two jobs that must be processed in two work centers as shown in Fig. 13.17.

The process times of Job 1 in Work Centers 1 and 2 are 2 days and 6 days, respectively, and the process times of Job 2 in Work Center 1 and 2 are 6 days and 2 day, respectively. The question is: how should these jobs be scheduled for processing in these two work centers in order to finish them as soon as possible (i.e., to minimize makespan)?

As shown in Fig. 13.18, there are a total of four different ways to process these jobs in the two centers.

From Fig. 13.18 we make the following observations:

- *Idle Time in Work Center 2:* Both Schedules III and IV result in less idle times in Work Center 2 than those in Schedules I and II. For example, under Schedule III, Work center 2 is idle for 6 days to start working on the jobs, while in Schedule I Work Center 2 is idle for 8 days to start working on the jobs.
- *Idle Time in Work Center 1:* Both Schedules III and IV also result in less idle times in Work Center 1. For example, under Schedule III, Work Center 1 is idle for 6 days during the time Work Center 2 is processing jobs. However, this number is 8 days under Schedule I.

Idling in both Work Centers 1 and 2 is important, because schedules in which the total idling in Work Centers 1 and 2 are lower result in completing both jobs earlier, leading to a smaller makespan. That is the reason Schedules III and IV have smaller makespan. Under

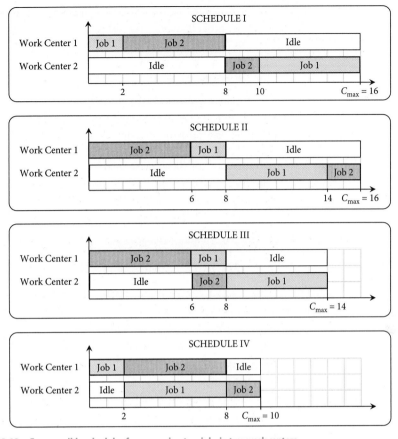

**Figure 13.18    Four possible schedules for processing two jobs in two work centers.**

Schedules III and IV both jobs are completed on day 14 and 10, respectively, but under Schedules I and II those jobs are completed on day 16.

But, is there anything special about Schedules III and IV that makes them perform better than the other schedules? The answer is Yes; they are Permutation Schedules. *Permutation Schedules* are schedules in which sequence of jobs is the same in all stages of the process. In Schedule IV, for example, both work centers process Job 1 first and then Job 2. In Schedule III, both work centers process Job 2 first and then Job 1. In Schedule I, however, Work Center 1 processes Job 1 and then Job 2, while Work Center 2 processes Job 2 and then Job 1. Hence, Schedule I is not a permutation schedule and so is not Schedule II.

Now let's compare two permutation Schedules III and IV. As shown in Fig. 13.18, Schedule IV has less idle times and smaller makespan compared to Schedule III. Why? More specifically, what makes a permutation schedule to have a lower makespan than other permutation schedules?

Taking a closer look at Schedules III and IV we find that Schedule IV has the following features. Recall that the sequence of jobs in Schedule IV is (1,2), but in Schedule III it is (2,1).

- The idling of Work Center 2 in both Schedules III and IV is equal to the process time of the first job in Work Center 1. Since in Schedule IV the first job in the sequence (i.e., Job 1) has a smaller processing time in Work Center 1 than that of the first job in Schedule III (i.e., Job 2), Schedule IV has a smaller idling time in Work Center 2.

- Also, note that the idling of Work Center 1 in both Schedules III and IV is equal to the process time of the last job in Work Center 2. Since in Schedule IV the last job in the sequence (i.e., Job 2) has a smaller processing time in Work Center 2 than that of the last job in Schedule III (i.e., Job 1), Schedule IV has a smaller idle time in Work Center 1.

These two observations lead us to the following general insights: Permutation Schedules in which the jobs in the beginning of the sequence have smaller process times in Work Center 1, and jobs at the end of the sequence have smaller process times in Work Center 2 result in smaller makespan. While we just concluded this using our small example, in 1954 Johnson proved that our conclusion is indeed true when jobs are processed through two stages. Based on his proofs, Johnson developed a simple algorithm that can find the optimal sequence of the permutation schedule as illustrated in the following analytics.

---

**ANALYTICS**

**Minimizing Makespan of $N$ Jobs in a Two-Stage Flow Shop—Johnson's Rule**

Consider $N$ jobs that must be processed in two work centers. All jobs must be processed in Work Center A before they are processed in Work Center B. The processing time of Job $j$ in Work Center $i$ is $t_{ij}$. The goal is to find the optimal schedule of jobs in work centers to minimize the makespan of the $N$ jobs.

The optimal schedule is a permutation schedule that can be obtained as follows:

- *Step 1:* List the jobs and their processing times in Work Centers A and B as shown in the following table:

| Job | Work Center A | Work Center B |
|-----|---------------|---------------|
| 1 | $t_{A1}$ | $t_{B1}$ |
| 2 | $t_{A2}$ | $t_{B2}$ |
| 3 | $t_{A3}$ | $t_{B3}$ |
| ... | ... | ... |
| $N$ | $t_{AN}$ | $t_{BN}$ |

- *Step 2:* Select the job with the shortest time in the above table. If the shortest time is at the first work center (i.e., it appears under Column A), then schedule that job first. If the shortest time is at the second work center (i.e., it appears under Column B), then schedule that job last.

- *Step 3:* Eliminate the scheduled job and its process times from the table and return to Step 2 until all jobs have been scheduled.

We use the following example to illustrate Johnson's rule.

  **Example 13.9**

Fisher Dies in Example 13.8 receives an order from a major auto manufacturer to make five different stamping dies used to produce door panels for different car models. The manager of Fisher Dies believes that this is a great opportunity to establish a good relationship with the manufacturer, and hence the manager would like to finish the five jobs and ship them to the manufacturer as soon as possible. To minimize the shipment cost, manufacturer has requested that all five dies are shipped together in one shipment.

Making a stamping die first requires machining operations and then sanding. Machining operations are performed using the shop's only CNC machine, and sanding is done

Table 13.12    Process Times of Dies on Two Machines

| Stamping die | CNC machine | Sanding machine |
|:---:|:---:|:---:|
| 1 | 8 | 3 |
| 2 | 10 | 8 |
| 3 | 6 | 4 |
| 4 | 10 | 7 |
| 5 | 5 | 7 |

using precision sanding machine. The process times (in days) of each die on CNC and sanding machines are shown in Table 13.12.

How should these jobs be scheduled on CNC and on sanding machine in order to ship these jobs in the least time?

Finishing all five stamps in least time refers to minimizing the makespan of the five jobs. For that, we use Johnson's rules. We do not need to do Step 1, since the process times of jobs are already arranged in two columns in Table 13.12. Considering the CNC machine as Work Center A and sanding machine as Work Center B, we move to Step 2.

Following Step 2, the shortest process time in the table is 3, which is the processing time of Die 1 on the second machine (i.e., sanding machine). Since this number appears under the second machine (i.e., under Column B), we assign it as the last job in the sequence and obtain

$$\text{Optimal Sequence} \implies ?, ?, ?, ?, 1$$

Following Step 3, we eliminate Die 1 from the table and return to Step 2.

After removing Die 1, the shortest process time in the remaining of the table is 4, which is the process time of Die 3 on the sanding machine—the second column. Thus, we assign Die 3 last, which leads us to

$$\text{Optimal Sequence} \implies ?, ?, ?, 3, 1$$

We eliminate Die 3 from the table and return to Step 2.

The shortest process time in the table after removing Dies 1 and 3 would be 5, which is the process time of Die 5 on the CNC machine—the first column. We, therefore, assign Die 5 first, which leads us to

$$\text{Optimal Sequence} \implies 5, ?, ?, 3, 1$$

We eliminate Die 5 from the table and return to Step 2.

The only remaining dies in the table would be Die 2 and Die 4. The shortest processing time in the table would be 7, which is the process time of Job 4 on the second machine, thus

$$\text{Optimal Sequence} \implies 5, ?, 4, 3, 1$$

Finally, the only remaining die is Die 2, which will be put in the only remaining position in the sequence. Hence, the optimal permutation schedule processes the five jobs on both machines according to the following sequence:

$$\text{Optimal Sequence} \implies 5, 2, 4, 3, 1$$

This sequence finishes all five dies in the least time—it minimizes makespan of five dies.

## 13.8 Operations Scheduling in Project Processes

As discussed in Sec. 1.1.6 of Chap. 1, project processes are used to produce highly customized and unique products (or services) in a very small quantity, often just one unit. Examples of project processes are construction projects, software development, medical procedures, and production process of large ships and aircrafts. As it was also mentioned in that section, the main management problem in project processes is the scheduling of activities and coordination among resources. The two major tools used in management and control of activities in project processes are *Critical Path Method (CPM)* and *Project Evaluation and Review Technique (PERT)*.

### 13.8.1 Critical Path Method (CPM)

CPM is the most widely used technique in project management. It coordinates resources used in projects and provides information that can help manage and control activities such that the project is completed on time. We describe different steps of CPM using the case of ComSoft, a software developing firm.

**Operations Scheduling at ComSoft**

ComSoft is a firm that develops software for business operations, focusing on sales through online channels. The firm has currently started negotiations with Flowers-For-You (FFY), a firm that sells flowers online. FFY would like to upgrade its current system to integrate its ordering, sales, and accounting processes into one system. FFY wants to have the new system up and running in no later than 60 days. ComSoft is currently working on a different project that is expected to finish in 10 days. Bill, ComSoft's operations manager, believes that after finishing their current project in 10 days, they can meet the deadline of 60 days to finish FFY's project. He, however, would like to do a more detailed analysis of the project before he signs a contract with FYY with the due date of 60 days.

Bill asks his assistant, Julie, to send him a report that includes all activities needed to finish FFY's system and also give him an estimate of how long it takes to do each activity. Julie summarizes her analysis as shown in Table 13.13 and sends it to Bill.

As Table 13.13 shows, completing the new software requires nine major activities. The table also shows the time required to finish each activity, each activity's predecessor, and the resources needed to perform each activity. Activities cannot start until all their predecessors are already finished. For example, "Design customer interface for ordering module" (Activity G) cannot start unless "Develop codes for ordering module" (Activity E) is completed. Activity G is done by Team D, while Activity E is done by Team O.

**Table 13.13    Activities Required for Completing FFY's Software Project**

| Activity | Activity description | Activity duration (*days*) | Immediate predecessor | Required resource |
|---|---|---|---|---|
| A | Identify teams to work on each module | 10 | — | Project manager |
| B | Determine the required features of ordering module | 7 | — | Team O |
| C | Develop codes for accounting module | 18 | A | Team A |
| D | Develop codes for sales module | 15 | A | Team S |
| E | Develop codes for ordering module | 20 | A,B | Team O |
| F | Integrate accounting and sales modules | 5 | C,D | Teams A and S |
| G | Design customer interface for ordering module | 8 | E | Team D |
| H | Develop training module for accounting and sales | 7 | F | Team A |
| I | Integrate sales, accounting, and ordering modules | 10 | F,G | Teams S, A, and O |

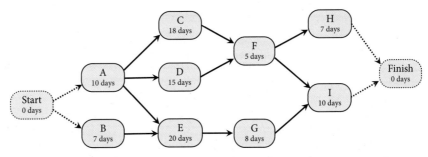

**Figure 13.19    Precedence diagram of the activities required for FFY software project.**

Figure 13.19 shows the precedence diagram, also known as *activity-on-node* network, of all activities of the project. Note that the figure includes two (dummy) activities "Start" and "Finish" corresponding to start and finish of the project. The time to perform these activities is zero, since they are not actually an activity. These dummy nodes make the analytical process of planning and scheduling the project easier, as we will show later.

To finish the project, ComSoft must finish all nine activities in Table 13.13. Thus, the project completion time is equal to makespan (i.e., $C_{max}$) of all activities (see Sec. 13.2). Consequently, finding the earliest time to finish FFY's project is equivalent to finding the minimum makespan of all nine activities. But what is the minimum makespan?

If ComSoft performs the nine activities in sequence A, B, C, D, E, F, G, H, I, then the total time to finish the project would be $C_{max} = 0 + 10 + 7 + \cdots + 7 + 10 + 0 = 100$ days. But is this the earliest time that ComSoft can finish the project?

Well, if there is only one resource (e.g., one person or one team) that performs all nine activities, then yes, it will take the resource a total of 100 days to finish all activities. In this case, the scheduling problem reduces to the scheduling problem of a single resource performing nine activities. Since all sequences will finish all nine activities in 100 days (i.e., $C_{max} = 100$ days), the goal is to find a sequence that does not violate the precedence constraints among activities.

ComSoft project management problem, however, is not a single-resource scheduling problem. As Table 13.13 shows, activities are performed by different teams that can work on the activities independent of each other and in parallel. For example, while Team O is developing a code for ordering module (Activity E), Team A can simultaneously work on developing the code for accounting module (Activity F). Therefore, having several teams working in parallel on different activities will able ComSoft to finish the project in less than 100 days. But, again, what is the earliest time that ComSoft can finish the project?

One way to answer this question is to compute the length of all paths in the network in Fig. 13.19 from start to finish. Note that each path represents the sequence of activities that follows the precedence relationship in precedence diagram. Also, each path can be done in parallel with other paths. Therefore, the time the project can be completed is equal to the time when all paths are completed. Since paths are performed in parallel, this is equal to the time when the path with the longest length is completed.

Let's use this approach to find the earliest time that ComSoft can finish the project. Note that there are six paths in Fig. 13.19 from the "Start" node to "Finish" node:

Path P1 : start $\rightarrow A \rightarrow C \rightarrow F \rightarrow H \rightarrow$ finish $\Longrightarrow$ Path Length $= C_{P1} = 0 + 10 + 18 + 5 + 7 + 0 = 40$

Path P2 : start $\rightarrow A \rightarrow D \rightarrow F \rightarrow H \rightarrow$ finish $\Longrightarrow$ Path Length $= C_{P2} = 0 + 10 + 15 + 5 + 7 + 0 = 37$

Path P3 : start $\rightarrow A \rightarrow C \rightarrow F \rightarrow I \rightarrow$ finish $\Longrightarrow$ Path Length $= C_{P3} = 0 + 10 + 18 + 5 + 10 + 0 = 43$

Path P4 : start $\rightarrow A \rightarrow D \rightarrow F \rightarrow I \rightarrow$ finish $\Longrightarrow$ Path Length $= C_{P4} = 0 + 10 + 15 + 5 + 10 + 0 = 40$

Path P5 : start $\rightarrow A \rightarrow E \rightarrow G \rightarrow I \rightarrow$ finish $\Longrightarrow$ Path Length $= C_{P5} = 0 + 10 + 20 + 8 + 10 + 0 = 48$

Path P6 : start $\rightarrow B \rightarrow E \rightarrow G \rightarrow I \rightarrow$ finish $\Longrightarrow$ Path Length $= C_{P6} = 0 + 7 + 20 + 8 + 10 + 0 = 45$

As mentioned earlier, the project completion time (i.e., $C_{max}$) is equal to the length of the longest path that is,

$$C_{max} = \max\{C_{P1}, C_{P2}, C_{P3}, C_{P4}, C_{P5}, C_{P6}\} = \max\{40, 37, 43, 40, 48, 45\} = 48$$

which corresponds to Path P5.

---

**CONCEPT**

**Critical Path and Critical Activities in a Project Process**

- *Critical Path* (or paths) of a project is the path (or paths) with the longest length from Start to Finish of the project. The minimum time to complete a project is equal to the length of its critical path.
- *Critical Activities* of a project are all activities on the critical path of the project.

Path P5 has the longest path length (i.e., 48 days) in FFY's project and thus is the Critical Path. The path contains activities A, E, G, I, which are critical activities of the project. Since ComSoft starts FFY's project on day 10, the earliest time it can finish the project is on day $10 + 48 = 58$.

Finding critical path of a project is often a tedious task, especially in projects that have a large number of activities. The good news, however, is that there is a method called *Critical Path Method* (or CPM) that provides a simple approach to identify the critical path and thus the time required to finish the project. CPM consists of the following steps: (i) computing earliest times that each activity can start and finish, (ii) computing latest times that each activity can start and finish, (iii) computing slack time of each activity, and (iv) identifying the critical path (or paths) and critical activities. We describe each of these steps, starting with computing the earliest times.

**Computing Earliest Times—Forward Scheduling**
CPM first computes the earliest time that each activity can start and the earliest time that each activity can be completed. This allows CPM to compute the earliest time that the project can be completed.

---

**CONCEPT**

**Earliest Start and Earliest Finish Times**

- *Earliest Start* time of Activity $i$, denoted by $ES_i$, is the earliest time that Activity $i$ can start.
- *Earliest Finish* time of Activity $i$, denoted by $EF_i$, is the earliest time that Activity $i$ can be finished.

Considering $T_i$ as the duration of Activity $i$, we have

$$EF_i = ES_i + T_i \tag{13.3}$$

Hence, to compute the earliest finish time of each Activity $i$ (i.e., $EF_i$) using Eq. (13.3), CPM needs the duration of the activity (i.e., $T_i$, available in Table 13.13), and the earliest start of the activity (i.e., $ES_i$), which is not available and must be determined.

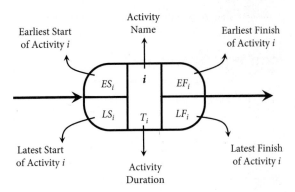

**Figure 13.20    Activity information in CPM.**

CPM uses a *Forward Scheduling* approach and computes the earliest start and finish times of all activities from the "Start" node to "Finish" node in the precedence diagram. Here is how it works. CPM starts with finding the earliest time ComSoft can start the project. Recall that the earliest time ComSoft can start FFY's project is day 10; hence, $ES_{start} = 10$. Using Eq. (13.3) and considering the duration of Activity "Start" being $T_{start} = 0$, the earliest time that Activity "Start" can be finished is $EF_{start} = ES_{start} + T_{start} = 10 + 0 = 10$.

Now consider the next activity—Activity A. What is the earliest time that Activity A can start? Because Activity "start" is the predecessor of Activity A (see Fig. 13.19), then Activity A cannot start until Activity "start" is completed. Hence, the earliest time that Activity A can start is the earliest time that Activity "start" is finished. In other words, $ES_A = EF_{start} = 10$. Having the earliest start of Activity A, we can use Eq. (13.3) to compute the earliest finish of Activity A as $EF_A = ES_A + T_A = 10 + 10 = 20$.

Now consider Activity C. Because Activity A is the predecessor of Activity C, then the earliest time that Activity C can start is the earliest time that Activity A is finished, that is, $ES_C = EF_A = 20$. Having the earliest start time of Activity C, we compute the earliest finish time of Activity C as $EF_C = ES_C + T_C = 20 + 18 = 38$. Using the same approach we can compute earliest start and earliest finish times for the remaining activities as shown in Fig. 13.21. Each node in Fig. 13.21 includes information that is described in Fig. 13.20.

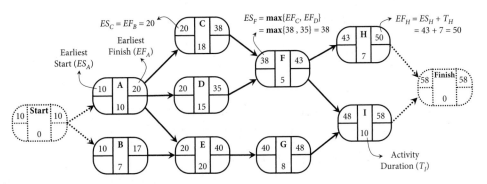

**Figure 13.21    Computing earliest start and earliest finish times for all activities of FFY project.**

When an activity has only one immediate predecessor (e.g., activities C, D, G), its earliest start time is equal to the earliest finish time of its predecessor. However, when an activity has more than one immediate predecessor, things are different. Consider Activity F that has two immediate predecessors—activities C and D. The earliest finish times for activities C and D are 38 and 35, respectively (see Fig. 13.21). When is the earliest time that Activity F can start? The answer is simple. Since Activity F can start only when both activities C and D are finished, then its earliest start time would be equal to the earliest finish time of the activity among C and D that finishes last. In other words, the earliest start of Activity F is equal to the maximum of the earliest finish times of activities C and D, that is, $ES_F = \max\{EF_C, EF_D\} = \max\{38, 35\} = 38$. This leads us to the following general rule:

$$ES_i = \max\{\text{Earliest finish times of all immediate predecessors of Activity } i\}$$

Using the above rule, as shown in Fig. 13.21, the earliest time that Activity "Finish" can start is $EF_{\text{finish}} = \max\{EF_H, EF_I\} = \max\{50, 58\} = 58$. Considering $T_{\text{finish}} = 0$, we have $EF_{\text{finish}} = ES_{\text{finish}} + T_{\text{finish}} = 58 + 0 = 58$. Thus, the earliest time ComSoft can complete FFY's project would be in 58 days.

### Computing Latest Times—Backward Scheduling

Finding the earliest start and finish times, Bill remembers that he may need to use one or more of his teams to work on other projects during the next 60 days. This may delay the start time or delay the finish time of some of the activities, and thus delay the completion of FFY's project. Hence, Bill would like to know when are the latest times that ComSoft can start or finish each activity such that the project is still completed on time in 58 days.

The question that Bill is facing is critical in management and control of project processes. CPM provides answer to this question since, in addition to earliest start and earliest finish times, CPM also computes the latest start and latest finish times for each activity of the project.

---

**CONCEPT**

### Latest Start and Latest Finish Times

- *Latest Start* time of Activity $i$, denoted by $LS_i$, is the latest time that Activity $i$ can be started without delaying the completion of the project.

- *Latest Finish* time of Activity $i$, denoted by $LF_i$, is the latest time Activity $i$ can be finished without delaying the completion of the project.

For any Activity $i$,

$$LS_i = LF_i - T_i \tag{13.4}$$

While earliest start and finish times of activities correspond to earliest time that a project can be completed, latest start and finish times correspond, respectively, to the latest time to start or finish each activity such that the project is still completed in its earliest possible time (i.e., the time obtained in Forward Scheduling).

Having latest finish time $LF_i$ of Activity $i$ and its duration $T_i$, CPM uses Eq. (13.4) to compute the latest start time $LS_i$ for the activity. This is done though a *Backward Scheduling* process. Backward scheduling process starts from the last activity (i.e., Activity "Finish") in the project and computes the latest (start and finish) times and moves backward until it finds the latest times for the first activity (i.e., Activity "Start") of the project. Here is how it works. To have the project still completed on day 58, the latest finish time for the last activity of the project—Activity "Finish"—should be $LF_{\text{finish}} = 58$. Since Activity "Finish" takes $T_{\text{finish}} = 0$ days, using Eq. (13.4), the latest start time of Activity "Finish" would be $LS_{\text{finish}} = LF_{\text{finish}} - T_{\text{finish}} = 58 - 0 = 58$.

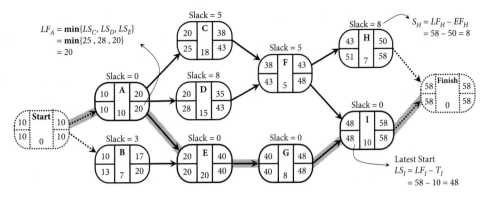

Figure 13.22   Earliest and latest times and slack times of all activities in FFY's project.

Moving backward, what would be the latest start and finish times of Activity H? If the latest time that we can start Activity "Finish" is 58, then the latest time its immediate predecessor activity—Activity H—can be finished should be 58, that is, $LF_H = LS_{finish} = 58$. If Activity H finishes later than 58, then the completion of the project is delayed beyond day 58. Having $LF_H = 58$ and the duration of Activity H to be $T_H = 7$ days, using Eq. (13.4) we find the latest start of Activity H as $LS_H = LF_H - T_F = 58 - 7 = 51$. Latest start and latest finish times of all other activities can be obtained in the same manner, as shown in Fig. 13.22.

When an activity has only one immediate successor (e.g., Activity D has Activity F as its only immediate successor), then its latest finish time is equal to the latest start of its immediate successor (e.g., $LF_D = LS_F = 43$). However, when an activity has more than one immediate successor, then things are different. Consider Activity A that has three immediate successors, that is, activities C, D, and E. Also, note that following the backward scheduling process we find $LS_C = 25$, $LS_D = 28$, and $LS_E = 20$ (shown in Fig. 13.22). This implies that, if activities C, D, and E start later than days 25, 28, and 20, respectively, then the project will take more than 58 days to complete. What is the latest time that Activity A should be completed in order to still finish the project by day 58?

Note that if Activity A finishes on or before day 20 (i.e., the minimum of $LS_C = 25$, $LS_D = 28$, and $LS_E = 20$), then all three activities C, D, and E can start no later than their latest start times, and thus the project will not be delayed. Therefore, $LF_A = \min\{LS_C, LS_D, LS_E\} = \{25, 28, 20\} = 20$. This is also shown in Fig. 13.22 and points to the following general rule:

$$LF_i = \min\{\text{Latest start times of all immediate successors of Activity } i\}$$

**Computing Slack Times of Activities**
Comparing the latest start and finish times of activities with their earliest start and finish times provides useful information about the flexibility that ComSoft has in managing and controlling each activity. Consider Activity D as an example. It has earliest start time $ES_D = 20$ and latest start time $LS_D = 28$. This implies that ComSoft can start Activity D as early as day 20 and as late as day 28, and the project will still be completed on day 58. In other words, ComSoft has the flexibility to delay the start of activity D by $LS_D - ES_D = 28 - 20 = 8$ days without affecting the project completion by day 58. The same conclusion can be made if we compare the earliest finish and the latest finish times of Activity D. Specifically, ComSoft has the flexibility to delay the completion of Activity D for $LF_D - EF_D = 43 - 35 = 8$ days without delaying the completion of the project. For example, if Activity D starts at its earliest time $ES_D = 20$, but it takes 8 days more than its original estimate $T_D = 15$ days to finish D, the project will still be completed by day 58. This time flexibility in start or finish times of an activity is called Slack Time of the activity.

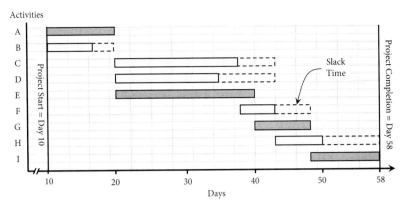

Figure 13.23   Start, finish, and slack times of activities to complete the project on day 58. Critical activities are shown in darker shades.

---

CONCEPT

### Slack Times of Activities in Project Processes

*Slack Time* of Activity $i$, denoted by $S_i$, is the number of time periods (e.g., days, weeks, months) that the start time or the finish time of Activity $i$ can be delayed without affecting the project completion time.

$$S_i = LS_i - ES_i = LF_i - LS_i$$

Slack times for all *critical activities* are zero.

Figure 13.22 also shows the slack times of all activities. As the figure shows, activities A, E G, and I have slack times of zero. For these activities the latest time they can be started (and still get project done on time) is also the earliest time they can be started. Consequently, any delay in start (or in finish) times of these activities will delay the completion time of the project beyond day 58. For example, if Activity G takes 1 day longer than its original estimate of $T_G = 8$ days, then the project will be delayed for 1 day and will be completed on day 59.

### Identifying Critical Path

As Fig. 13.22 shows, activities A, E, G, and I with zero slack times are the critical activities of the project. The critical path passing through these critical activities is also shown in Fig. 13.22 with shaded arrows.

Critical activities play an important role in managing and controlling project processes. Operations engineers and managers must pay close attention to critical activities and make sure that they are not delayed. In some cases, the resources allocated to non-critical activities can be allocated to critical activities. This can reduce the length of critical path and thus reduce the project completion time.

After earliest and latest times of all activities are found, start and finish times of activities can be scheduled. This also allows to schedule the resources that perform those activities (e.g., work schedules for teams in FFY project). Figure 13.23 depicts the Gantt chart of the activities showing when each activity can start and finish in order to meet the deadline of 58. The chart also shows the slack times for the activities.

We conclude this section by summarizing the critical path method in the following analytics:

**Critical Path Method for Scheduling Project Processes**

- ***Step 1:*** *(Activity Durations and Precedence Diagram)* Identify the activities required to finish the project and estimate the duration of each activity. Also, draw the precedence diagram of the activities, adding dummy activities "Start" and "Finish" with durations of zero.

- ***Step 2:*** *(Earliest Times through Forward Scheduling)* Set the earliest start time of Activity "Start" to the time the project can start. Then compute the earliest start and finish times of Activity $i$ by working forward through project precedence diagram using the following formulas:

$$ES_i = \max\{\text{Earliest finish times of all immediate predecessors of Activity } i\}$$
$$EF_i = ES_i + T_i$$

The earliest completion time of the project is the earliest finish time of Activity "Finish."

- ***Step 3:*** *(Latest Times through Backward Scheduling)* Set the latest finish time of Activity "Finish" equal to its earliest finish time. Then compute the latest start and finish times of Activity $i$ by working backward through project precedence diagram using the following formulas:

$$LF_i = \min\{\text{Latest start times of all immediate successors of Activity } i\}$$
$$LS_i = LF_i - T_i$$

- ***Step 4:*** *(Critical Path and Slack Times)* For each Activity $i$, compute its slack time as follows:

$$S_i = LS_i - ES_i \quad (\text{or } S_i = LF_i - EF_i)$$

Critical activities are those with slack time of zero. The path (or paths) passing through critical activities is the Critical Path.

### 13.8.2 Shortening the Project Completion Time

Often times firms need to shorten the project completion time due to contractual agreements, financial incentives, or because they need to start another (higher priority) project. Let's return to our ComSoft example. While confident that he can finish the project by day 60, as was discussed with FFY, Bill gets a phone call from FFY manager who informs him that they are willing to pay a bonus of $25,000, if ComSoft can install their system in 46 days (instead of in 60 days). This means that—to get the bonus—ComSoft must complete the project 12 days earlier than day 58. How can ComSoft accelerate the completion of the project?

Well, one way to finish FFY's project is to start it before day 10. But this is not possible, Bill realizes, since ComSoft has a contract with rigid penalty to finish its current project by day 10. Besides starting the project earlier (which is not possible for ComSoft), the other approach is to shorten the length of the critical path. This can be done in two ways: (i) by overlapping some of the critical activities, and (ii) by shortening the duration of the critical activities.

**Overlapping Critical Activities**

The precedence diagram shows the dependence among activities of a project. When Activity A is predecessor of Activity C, it implies that Activity C cannot start until Activity A is

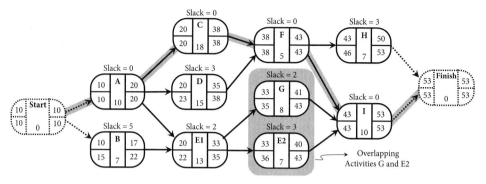

**Figure 13.24** Precedence diagram of FFY project after overlapping activities G and E2.

*completely* finished. However, in some cases it is possible to start an Activity before its predecessor activity is "completely" finished. Consider Activity E, "Develop codes for ordering module," which is the predecessor of Activity G, "Designing customer interface for ordering module." Activity E takes 20 days and is performed by Team O, while Activity G takes 8 days and is performed by Team D. These two activities are critical activities and take a total of 28 days. Now suppose that ordering module consists of two subactivities: (i) order processing, that is, Activity E1, and (ii) order tracing, that is, Activity E2. Activity E1 takes 13 days and Activity E2 takes 7 days. To start designing customer interface (i.e., Activity G) Team D only requires Activity E1 to be completed. Activity E2 (order tracing) is an internal process and does not relate to customer interface. Hence, Activity G can start after Activity E1 is completed—it does not require the completion of Activity E2. How does this affect project completion time?

Figure 13.24 shows the new precedence diagram. Note that activities G and E2 now overlap, that is, they can be performed simultaneously. Implementing CPM we find that the critical path is changed to activities $A \rightarrow C \rightarrow F \rightarrow I$. This reduces the completion time of the project by 5 days from day 58 to day 53. However, the reduction is still not enough to get the bounce of $25,000. ComSoft still needs to reduce the length of the project by 7 more days to complete the project on day 46. But how can ComSoft do that?

**Shortening the Duration of Critical Activities**

The duration of an activity can often be reduced by increasing the number of resource-hours used to perform the activity. In ComSoft project, for example, teams can work overtime and on weekends to shorten the duration of the activity. Also, ComSoft can allocate more staff to each team to speed up the activity. These options, however, increase costs. By performing a time and cost study of the activities, ComSoft estimates the cost of reducing duration of each activity by 1 day and the maximum possible reduction as shown in Table 13.14.

As the table shows, for example, Activity C has a duration of 18 days. This activity can be reduced by 1, 2, 3, 4 or a maximum of 5 days. The cost of reducing the duration of this activity for each day is $6000. The first question that ComSoft is facing is to decide which activities should be reduced and for how many days in order to finish the project in 46 days with minimum cost. The second question is whether the additional cost of activity time reduction is less than the bonus of $25,000, which makes time reduction a profitable decision.

The process of shortening project completion time by reducing the duration of project activities is called *Crashing the Project*. One method that is often used for crashing projects is Linear Programming (LP). We start by introducing the decision variables of the LP model.

**Table 13.14    Maximum Time Reduction of Each Activity and Its Corresponding Costs in FFY Project**

| Activity | Activity description | Duration (days) | Reduction cost (per day) | Maximum reduction (days) |
|---|---|---|---|---|
| A | Identify teams to work on each module | 10 | $5000 | 3 |
| B | Determine the required features of ordering module | 7 | $7000 | 2 |
| C | Develop codes for accounting module | 18 | $6000 | 5 |
| D | Develop codes for sales module | 15 | $2500 | 4 |
| E1 | Develop codes for ordering processing | 13 | $3000 | 2 |
| E2 | Develop codes for order tracing | 7 | $1500 | 1 |
| F | Integrate accounting and sales modules | 5 | $4500 | 1 |
| G | Design customer interface for ordering module | 8 | $3400 | 3 |
| H | Develop training module for accounting and sales | 7 | $5000 | 2 |
| I | Integrate sales, accounting, and ordering module | 10 | $4500 | 3 |

*Decision Variables:* Decision variables of the Linear Programming model are as follows:

$$X_i = \text{The earliest finish time for Activity } i$$
$$R_i = \text{Number of days that Activity } i\text{'s duration is reduced}$$

*Objective Function:* The objective is to minimize the total cost of crashing the project. Considering the costs in Table 13.14, and defining $Z$ as the total cost of crashing the activities of the project, the objective function can be written as follows:

$$\text{Minimize } Z = 5000R_A + 7000R_B + 6000R_C + 2500R_D + 3000R_{E1} + 1500R_{E2}$$
$$+ 4500R_F + 3400R_G + 5000R_H + 4500R_I$$

*Project Start Time Constraint:* Because the project starts on day 10, then the earliest time Activity "Start" will be finished is 10, or

$$X_{\text{start}} = 10$$

*Project Due Date Constraint:* The due date to complete the project (in order to get the bonus) is 46 days. Hence, the latest time that Activity "Finish" should be completed is day 46. Consequently, the earliest finish time for Activity "Finish" (i.e., $X_{\text{finish}}$) should be before or on day 46. This can be written as the following constraint for the LP model:

$$X_{\text{finish}} \leq 46$$

*Crash Time Constraints:* Recall that there is a limit on the maximum number of days each activity can be reduced. These maximum limits are given in Table 13.14. The following constraints guarantee that the reduction in activities do not exceed their maximum limits:

$$R_A \leq 3, \quad R_B \leq 2, \quad R_C \leq 5, \quad R_D \leq 4, \quad R_{E1} \leq 2$$

$$R_{E2} \leq 1, \quad R_F \leq 1, \quad R_G \leq 3, \quad R_H \leq 2, \quad R_I \leq 3$$

*Activity Precedence Constraints:* To make sure that an activity cannot start until its predecessor activities have already been finished, we need to add some constraints to our LP model to capture this precedence relationship among activities. To illustrate this, consider Activity F as an example. Activity F cannot start until its immediate predecessor activities C and D are completed. Let's focus on precedence relationship between activities F and C first, which corresponds to link $C \to F$ in the precedence diagram in Fig. 13.24. Recall that the duration of Activity F is $T_F = 5$ days. Therefore, the earliest time Activity F can finish is 5 days after Activity C is finished. In other words, $X_F \geq X_C + 5$. However, since Activity F is reduced for $R_F$ days (if needed), then the duration of Activity F is $(5 - R_F)$. Putting these together, we will have

$$X_F \geq X_C + (5 - R_F) \quad \implies \quad X_F - X_C + R_F \geq 5$$

Now consider the precedence relationship between activities D and F, corresponding to link $D \to F$ in the precedence diagram. Using a similar argument, we will have the following constraint:

$$X_F \geq X_D + (5 - R_F) \quad \implies \quad X_F - X_D + R_F \geq 5$$

Thus, for every link $(i \to j)$ in the precedence diagram we will have one constraint. Since the diagram has 16 links, we will have the following 16 activity precedence constraints (which includes the above two):

$$
\begin{aligned}
X_A - X_{\text{start}} + R_A &\geq 10 & &: \text{Precedence (start} \to A) \\
X_B - X_{\text{start}} + R_B &\geq 7 & &: \text{Precedence (start} \to B) \\
X_C - X_A + R_C &\geq 18 & &: \text{Precedence } (A \to C) \\
X_D - X_A + R_D &\geq 15 & &: \text{Precedence } (A \to D) \\
X_{E1} - X_A + R_{E1} &\geq 13 & &: \text{Precedence } (A \to E1) \\
X_{E1} - X_B + R_{E1} &\geq 13 & &: \text{Precedence } (B \to E1) \\
X_F - X_C + R_F &\geq 5 & &: \text{Precedence } (C \to F) \\
X_F - X_D + R_F &\geq 5 & &: \text{Precedence } (D \to F) \\
X_G - X_{E1} + R_G &\geq 8 & &: \text{Precedence } (E1 \to G) \\
X_{E2} - X_{E1} + R_{E2} &\geq 7 & &: \text{Precedence } (E1 \to E2) \\
X_H - X_F + R_H &\geq 7 & &: \text{Precedence } (F \to H) \\
X_I - X_F + R_I &\geq 10 & &: \text{Precedence } (F \to I) \\
X_I - X_G + R_I &\geq 10 & &: \text{Precedence } (G \to I) \\
X_I - X_{E2} + R_I &\geq 10 & &: \text{Precedence } (E2 \to I) \\
X_{\text{finish}} - X_H &\geq 0 & &: \text{Precedence } (H \to \text{finish}) \\
X_{\text{finish}} - X_I &\geq 0 & &: \text{Precedence } (I \to \text{finish})
\end{aligned}
$$

*Non-negativity Constraints:* Of course, all decision variables (i.e., the earliest finish times and activity crashing times) cannot be negative. Hence,

$$X_i \geq 0, \quad R_i \geq 0, \quad i = A, B, \ldots, H, I$$

*Solving the Linear Programming:* We can use Excel Solver to solve the LP. The total optimal cost of crashing the project to finish by day 46 is obtained to be $Z^* = \$33,000$. This is achieved if we reduce the duration of each activity by the following number of days:

$$R_A^* = 3, \quad R_B^* = 0, \quad R_C^* = 0, \quad R_D^* = 0, \quad R_{E1}^* = 0$$

$$R_{E2}^* = 0, \quad R_F^* = 1, \quad R_G^* = 0, \quad R_H^* = 0, \quad R_I^* = 3$$

In other words, to finish the project on day 46, ComSoft needs to only reduce the durations of activities A, F, and I by 3, 1, and 3 days, respectively. The optimal values for the earliest finish times for all activities are as follows[2]:

$$X^*_{\text{start}} = 10 \ , \ X^*_A = 17, \ X^*_B = 17, \ X^*_C = 35, \ X^*_D = 32, \ X^*_{E1} = 30,$$

$$X^*_{E2} = 39, \ X^*_F = 39, \ X^*_G = 38, \ X^*_H = 46, \ X^*_I = 46, \ X^*_{\text{finish}} = 46$$

In summary, as the solution shows, it costs ComSoft an additional \$33,000 to finish FFY's project on day 46 (instead of day 60). This is higher than the bonus of \$25,000 offered by FFY. Therefore, it does not make economical sense for ComSoft to crash the project.

Before we conclude this section, we summarize the LP model for crashing a project in the following analytics.

## ANALYTICS

### Crashing a Project Using Linear Programming

Consider a project that consists of N activities with a given precedence diagram that has dummy activities "Start" and "Finish." Activity $i$ has normal duration of $T_i$ periods (e.g., days, weeks, months). The project starts in time period $t_0 \geq 0$. The duration of Activity $i$ can be reduced at the cost of $c_i$ per period. However, the duration of Activity $i$ can be reduced for a maximum of $R^{\max}_i$ periods. The goal is to reduce the durations of the activities such that the project is completed by due date $T_{due}$ with the minimum cost of crashing the activities.

To find the optimal crashing times, define the following decision variables:

$$X_i = \text{The earliest finish time for Activity } i$$

$$R_i = \text{Number of periods that Activity } i\text{'s duration is reduced}$$

Considering the two dummy activities, the problem has $2N + 2$ decision variables.

The following linear programming model yields the optimal reduction in each activity:

$$Minimize \ Z = \sum_{i=1}^{N} c_i R_i$$

$$subject \ to:$$

$$\begin{aligned}
X_{\text{start}} &= t_0 \\
X_{\text{finish}} &\leq T_{\text{due}} \\
R_i &\leq R^{\max}_i &&: \quad \text{for all activities } i = 1, 2, \ldots, N \\
X_j - X_i + R_j &\geq T_j &&: \quad \text{for all precedence } (i \rightarrow j) \text{ in precedence diagram} \\
X_i \geq 0, \ R_i &\geq 0 &&: \quad \text{for all activities } i = 1, 2, \ldots, N
\end{aligned}$$

Considering $K$ as the number of links (i.e., arrows) in the precedence diagram, the above linear programming has $2 + N + K$ constraints.

### 13.8.3 Variability in Project Scheduling—PERT

Having realized that completing the project in 46 days is costly, Bill calls FFY and informs them that ComSoft cannot finish the project in 46 days. FFY manager then asks Bill if they can complete the project in 60 days, as they originally agreed. Bill responds: "Absolutely, we

---

[2] This LP problem has multiple optimal solutions, all resulting in the minimum total cost of \$33,000, but the earliest finish times of some activities might be different in alternative solutions.

can have your system up and running in 60 days." FFY manager says: "Great! Then I will send you the contract to sign tomorrow. The due date in the contract would be 2 months from now."

The next day, Bill receives the contract with the due date of 60 days, but the contract also has a penalty of $10,000 for each day the project is finished beyond the due date of 60. This makes Bill nervous. While he believes that they can finish the project in 60 days—especially after overlapping activities E2 and G—he would like to know the likelihood that the project is completed after the due date of 60. After all, many things can go wrong and activities can take longer than they expected. This makes Bill thinking about how Julie estimated the duration of each task and how reliable those estimates are? Bill calls Julie and asks her about her estimates.

Julie explains that she estimated the duration of each activity after she met with the team that is assigned to do the activity. In the meetings, she asked them to give her the following three estimates for activity durations:

- *Optimistic Estimate* ($a$): What would be the shortest possible time to finish the activity if everything progresses in an ideal manner?
- *Pessimistic Estimate* ($b$): What would be the longest possible time to finish the activity if everything progresses at the slowest possible pace?
- *Most Likely Estimate* ($m$): What would be the best estimate of the time to finish the activity, if everything progresses normally?

She then explains that she developed her own estimate for each activity by taking the weighted average of the three estimates—putting 4 times more weight on the most likely estimate—as follows:

$$\text{Estimate for Activity Duration} = \frac{a + 4m + b}{6}$$

Thus, all estimates shown in Table 13.13 are obtained from the above formula. She also mentions that, because of the variability in activity times, her estimates are the expected (i.e., the average) durations of the activities. To capture the variability in activity times, Julie explains that they can compute the variance of activity times using the following formula:

$$\text{Variance of Activity Duration} = \left( \frac{b - a}{6} \right)^2$$

Bill asks for Julie's detailed calculation of the average and variance of all activities (including new activities E1 and E2), and Julie sends Bill all the calculations as shown in Table 13.15.

The technique that Julie used to estimate the average duration of each activity as well as its variance is a part of what is known as *Project Evaluation and Review Technique* or PERT. PERT was developed in the 1950s by U.S. Department of Defense and Booz Allen Hamilton for planning, scheduling, and controlling the Polaris ballistic missile project. PERT assumes that the time to finish an activity is a random variable following a particular probability distribution. Probability distributions of activities are often not known, and thus it is not easy to compute their average and variance. One advantage of using the three estimates is its simplicity and intuitive way of computing mean and variance of the durations of activities. This is because, managers can usually determine the worst case, the best case, and the most likely case for duration of an activity.

The most common probability distribution associated with activity time is beta distribution. The reasons is that, similar to the three estimates, beta distribution has a start point (corresponding to the optimistic time), an end point (corresponding to the pessimistic

Table 13.15    Computing Mean and Variance of Activity Durations Using Three Time Estimates

| Activity (i) | Optimistic time $a_i$ | Most likely time $m_i$ | Pessimistic time $b_i$ | Mean $E(T_i)$ | Variance $VAR(T_i)$ |
|---|---|---|---|---|---|
| A | 8 | 9 | 16 | 10 | 1.78 |
| B | 3 | 6 | 15 | 7 | 4 |
| C | 11 | 19 | 21 | 18 | 2.78 |
| D | 12 | 15 | 18 | 15 | 1 |
| E1 | 12 | 13 | 14 | 13 | 0.11 |
| E2 | 7 | 7 | 7 | 7 | 0 |
| F | 1 | 5 | 9 | 5 | 1.78 |
| G | 6 | 7 | 14 | 8 | 1.78 |
| H | 4 | 6 | 14 | 7 | 2.78 |
| I | 5 | 10 | 15 | 10 | 2.78 |

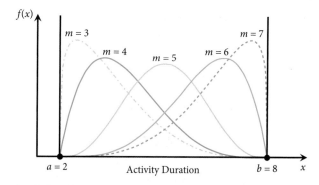

Figure 13.25    Examples of a beta distribution with $a = 2, b = 8$ with different values of $m$.

time), and a mode (corresponding to the most likely time). Furthermore, beta distribution is a very flexible distribution that can capture a wide range of variability in activity times (see Fig. 13.25).

Bill now understands that the time estimates for activities of FFY project are the *average* duration of each activity. Thus, the project completion date of 53 he obtained after overlapping activities E2 and G (shown in Fig. 13.24) is an average estimate. This means that the project can take longer and may be completed after day 53 or even after day 60—the due date of the project. Bill is now worried and wondering what is the probability that the project will be completed after day 60? Starting on day 10, this corresponds to the probability that the project takes more than 50 days?

PERT provides the answer to this question in a simple way, using central limit theorem. Note that the critical path in Fig. 13.24 has the length of 43 days. The probability that the project is completed after day 60 is equal to the probability that the length of the critical path exceeds 50. PERT assumes that the activities on the critical path are independent (i.e., delay in duration of one activity does not affect the duration of other activities) and their durations are identically distributed (i.e., follow beta distribution).

By the central limit theorem, the distribution of the summation of all activities on the critical path—the length of the critical path—can be approximated by a Normal distribution with a mean and variance, respectively, equal to the sum of the means and variances of all activities on the critical path.[3] The critical path in Fig. 13.24 includes activities A, C, F, and I.

---

[3]The approximation becomes more accurate as the number of activities on the critical path increases.

Considering $C_{\max}$ as the length of the project, then according to the central limit theorem:

$$E(C_{\max}) = E(T_A) + E(T_C) + E(T_F) + E(T_I)$$
$$= 10 + 18 + 5 + 10 = 43$$
$$VAR(C_{\max}) = VAR(T_A) + VAR(T_C) + VAR(T_F) + VAR(T_I)$$
$$= 1.78 + 2.78 + 1.78 + 2.78 = 9.11$$

and therefore a Normal distribution with mean 43 and variance 9.11 can be used to approximate the probability that the length of the project (i.e., $C_{\max}$) be smaller or larger than a certain time. Thus, we can compute the probability that FFY project takes longer than 50 days as follows:

$$P(C_{\max} > 50) = 1 - P(C_{\max} \le 50)$$
$$= 1 - P\left(Z \le \frac{50 - 43}{\sqrt{9.11}}\right)$$
$$= 1 - \Phi(2.32) = 1 - 0.99$$
$$= 0.01$$

which implies that the probability that FFY project takes more than 50 days, and thus the project is completed after the due date of 60, is only 1%. Hence, Bill can sign the contract with a high level of confidence that the project will be completed before its due date.

We summarize PERT's approach in estimating the probability of project completion time in the following analytics.

## ANALYTICS

**Computing Probability of Completing a Project by a Certain Time**

Consider a project that consists of $N$ activities with a given precedence diagram. Three estimates for duration of Activity $i$ are made, namely optimistic estimate $a_i$, pessimistic estimate $b_i$, and most likely estimate $m_i$. The project is supposed to start at time $t_0 \ge 0$. The goal is to estimate the probability that the project is completed by due date $T_{\text{due}}$.

- **Step 1:** Compute the average and variance of the duration of all activities as follows:

$$E(T_i) = \frac{a_i + 4m_i + b_i}{6}$$

$$VAR(T_i) = \left(\frac{b_i - a_i}{6}\right)^2$$

- **Step 2:** Set the earliest start time of dummy Activity "Start" to $t_0$ and consider $E(T_i)$ as the duration of Activity $i$. Then use CPM approach and identify the critical path of the project.
- **Step 3:** Compute the average and variance of the length of the project (i.e., length of critical path) as follows:

$$E(C_{\max}) = \sum_{j \in CR} E(T_j)$$

$$VAR(C_{\max}) = \sum_{j \in CR} VAR(T_j)$$

where set $CR$ is the set of all activities on the critical path.

- *Step 4:* Compute $z_o$ as

$$z_o = \frac{(T_{\text{due}} - t_0) - E(C_{\text{max}})}{\sqrt{\text{VAR}(C_{\text{max}})}}$$

If the project starts at $t_0$, then the probability that the project is completed by due date $T_{\text{due}}$ can be computed as follows:

$$P(C_{\text{max}} \leq T_{\text{due}} - t_0) = P(Z \leq z_o)$$
$$= \Phi(z_o)$$

where $\Phi(z)$ is the cumulative probability distribution of standard Normal distribution for value of $z$.

### 13.8.4 Difficulties with PERT

While PERT is a simple and useful method to evaluate the project completion time and its probabilities, it should be used with caution. The reason is that some of its assumptions may not hold in some projects.

**Independence Among Activities**

For the central limit theorem to be accurate, the duration of activities on the critical path must be independent of each other. In other words, the change in duration of one activity should not impact the duration of other activities. This, however, might not be the case in some projects. For example, in construction projects if an activity takes more time than its normal time due to bad weather, its subsequent activities are highly likely to also take longer times due to bad weather. Despite these issues, central limit theorem is shown to be robust and therefore its results should be reasonably accurate.

**Varying Critical Path**

Recall that the project completion time (i.e., $C_{\text{max}}$) is equal to the length of the path (from start to finish) with the longest path length among all paths. For example, if a project has $M$ paths from start to finish, the project completion time would be

$$C_{\text{max}} = \max\{C_{P1}, C_{P2}, C_{P3}, \ldots, C_{PM}\}$$

where $C_{Pj}$ is the length of Path $Pj$. According to the central limit theorem, length of each path (i.e., $C_{Pj}$) is approximately a Normal random variable. Therefore, project completion time $C_{\text{max}}$ would be the maximum of $M$ Normal random variables, which is not a Normal random variable as PERT assumes. This, however, does not create a big issue if there is only a single critical path, the variance of the critical path is small, and other non-critical paths with large variances have much smaller length. In this case, there is a low chance that any non-critical path becomes critical, and hence, project completion time is almost always equal to the length of the critical path. In other words, the maximum of all path lengths is almost always equal to the length of the critical path, which is what PERT assumes.

However, when there are non-critical paths that have lengths close to the (average) length of the critical path, then PERT may not be accurate, especially when non-critical paths have large variance. In such cases, there is a high chance that one of the non-critical paths becomes critical, and PERT results in underestimation of the project length and overestimation of the probability of completing the project by a certain time.

Despite the above concerns, PERT has been shown to be a useful method providing accurate estimates, and thus is broadly used in practice. For projects with severe consequences in missing due dates, an alternative methodology to PERT—the Monte Carlo Simulation—should be used to estimate the probability of project completion time. In Monte Carlo simulation, activity durations are generated for each activity according to its probability distribution and then lengths of all paths from start to finish are computed. The maximum of all these paths is then selected as the project completion time. This is repeated several (thousand) times resulting in several (thousand) project completion times. The empirical distribution of those project completion times can then be used to estimate the probability of project completion for any given time. Monte Carlo simulation procedure is out of scope of this text, and hence we do not discuss it here.

## 13.9 Workforce Scheduling

All operations scheduling models developed to find the production schedule of goods and services assume that the resources performing those tasks are always available. While this is true for resources such as equipment and tools, it is not true for workforce. Workers, as opposed to equipment, work in shifts, have lunch breaks, and have days off. Therefore, besides finding the production schedule, other main task of operations scheduling is to develop working schedules for the workforce in order to achieve the production goals.

Developing work schedule for workforce is time-consuming and is often a difficult problem, especially when the process works in multiple shifts and uses full-time and part-time workers, as is in many manufacturing and service systems. The work schedule must achieve multiple goals while satisfying demand constraints. To develop a work schedule for the next month, for example, the scheduler must consider number of working days in the month, number of shifts in each working day, number of working hours in a shift, and the number of workers required to satisfy demand in each hour. Having this information, the scheduler must assign full-time and part-time workers to each shift in order to satisfy demand in each hour of the shift.

### 13.9.1 *T*-Period Cycle Work Schedule

The most common workforce scheduling problem is assigning workers to a *T*-period cyclic schedule. Examples include scheduling workers in 7-day weekly schedule, or assigning workers to 3-shift (3-period) daily schedule. In each case, the schedule repeats itself every cycle. Mathematical programming models have been used by many firms (e.g., McDonald's, American Airlines) to develop cyclic schedules. We present two examples to illustrate how cyclic schedules can be designed using such models.

 **Example 13.10**

Neighborhood Bakery (NB) is a restaurant that is open for breakfast and lunch from 7:00 am to 3:00 pm 7 days a week. The restaurant has nine full-time workers who are trained to do all tasks including cash register and serving food. The demand in each

**Table 13.16**    Number of Workers Required in Each Day of the Week in Example 13.10

|  | Monday | Tuesday | Wednesday | Thursday | Friday | Saturday | Sunday |
|---|---|---|---|---|---|---|---|
| Workers needed | 5 | 6 | 6 | 5 | 6 | 7 | 6 |

7 days of the week (in terms of the minimum number of workers needed) is shown in Table 13.16.

Workers are working 40 hours a week. They work 5 days a week and have 2 days off. Finding which worker should have which day off has been very challenging. Currently some workers have complained that their off-days are not 2 consecutive days. For example, some workers get Mondays and Thursdays off, while some other workers have Saturdays and Sundays off. This has created a working environment in which some workers believe they are not treated fairly. Therefore, the manager would like to create a work schedule that is fair and gives every employee 2 consecutive days off each week. The manager however, is beginning to think that with the current number of nine workers, it might not be possible to create a schedule that everyone gets 2 consecutive days off and have the minimum required workers every day of the week. What do you think? Would it be possible to create such a work schedule?

NB manager is looking to find a 7-day cyclic schedule to satisfy its labor requirement in each day of the week and have 2 consecutive off-days (not necessarily the same 2 days) for all its employees.

The first step is to determine the cycle and the number of periods in the cycle. In the case of NB, the cycle is obviously 1 week and the number of periods in the cycle is $T = 7$ days.

The second step is to list all possible work schedules in a week that have 2 consecutive days off. We call each possible schedule a shift. Figure 13.26 shows that there are only seven shifts in a week (i.e., in a cycle) that allows for 2 consecutive off-days.

|          | Monday  | Tuesday | Wednesday | Thursday | Friday  | Saturday | Sunday  |
|----------|---------|---------|-----------|----------|---------|----------|---------|
| Shift 1: |         |         |           |          |         | Off Day  | Off Day |
| Shift 2: | Off Day |         |           |          |         |          | Off Day |
| Shift 3: | Off Day | Off Day |           |          |         |          |         |
| Shift 4: |         | Off Day | Off Day   |          |         |          |         |
| Shift 5: |         |         | Off Day   | Off Day  |         |          |         |
| Shift 6: |         |         |           | Off Day  | Off Day |          |         |
| Shift 7: |         |         |           |          | Off Day | Off Day  |         |

Figure 13.26    Shift patterns with 2 consecutive off-days for fast-food restaurant in Example 13.10.

As the figure shows, Shift 1, for example, has Monday to Friday as working days and Saturday and Sunday as the 2 consecutive off-days. Shift 4, on the other hand, has Thursday to Monday as working days, and Tuesday and Wednesday as 2 consecutive off-days.

Having identified the shift patterns, the question is whether it is possible to assign current nine workers to each shift such that the demand in each day is satisfied? If not, what is the minimum number of workers required for such a shift pattern to work? If we find the answer to the second question, then the first question is also answered.

The mathematical programming model that is used to solve cyclic workforce scheduling problem is a simple linear programming model with the following decision variable:

$$X_i = \text{Number of workers assigned to Shift } i$$

The number of decision variables is equal to the number of shifts; hence, the problem has seven decision variables.

The linear programming model is as follows:

$$Minimize\ Z = X_1 + X_2 + X_3 + X_4 + X_5 + X_6 + X_7$$

*subject to :*

$$X_1 \qquad\qquad + X_4 + X_5 + X_6 + X_7 \geq 6 \qquad \text{(Monday demand)}$$
$$X_1 + X_2 \qquad\qquad + X_5 + X_6 + X_7 \geq 7 \qquad \text{(Tuesday demand)}$$
$$X_1 + X_2 + X_3 \qquad\qquad + X_6 + X_7 \geq 7 \qquad \text{(Wednesday demand)}$$
$$X_1 + X_2 + X_3 + X_4 \qquad\qquad + X_7 \geq 6 \qquad \text{(Thursday demand)}$$
$$X_1 + X_2 + X_3 + X_4 + X_5 \qquad\qquad \geq 6 \qquad \text{(Friday demand)}$$
$$X_2 + X_3 + X_4 + X_5 + X_6 \qquad \geq 7 \qquad \text{(Saturday demand)}$$
$$X_3 + X_4 + X_5 + X_6 + X_7 \geq 6 \qquad \text{(Sunday demand)}$$

$$X_i \geq 0 \quad \text{and integer for} \quad i = 1, 2, \ldots, 7$$

**Objective Function**

The goal is to find the minimum number of workforce needed to satisfy daily demand with work schedule shown in Fig. 13.26. Therefore, the objective function is set to minimize the total number of workers assigned to all shifts. This would be the summation of all seven decision variables, as shown in the model.

**Constraints**

The problem has seven constraints corresponding to seven periods in the cycle (i.e., 7 days in a week). Consider the first constraint as an example:

$$X_1 + X_4 + X_5 + X_6 + X_7 \geq 6 \qquad \text{(Monday demand)}$$

The left-hand side of the constraint is the total number of workers who will be working on Monday, which must be at least six workers (i.e., the right-hand side). The left-hand side becomes clear when we see the shift patterns in Fig. 13.26. As the figure shows, only workers who are assigned to Shifts 1, 4, 5, 6, and 7 will be working on Monday (see the column under Monday). Workers assigned to Shifts 2 and 3 are in their off-days on Monday. The constraints for Tuesdays to Sundays are established in a similar manner.

**Optimal Cyclic Schedule**

Using Excel Solver, the optimal solution is found to be $Z^* = 9$, indicating that if NB wants to give its workers 2 consecutive days off (i.e., following the shift patterns in Fig. 13.26) and be able to satisfy demand in each day, it must have at least $Z^* = 9$ workers. With those nine workers, NB must assign them to each shift according to the following optimal solution (found by Solver):

$$X_1^* = 1, \quad X_2^* = 2, \quad X_3^* = 1, \quad X_4^* = 1, \quad X_5^* = 1, \quad X_6^* = 2, \quad X_7^* = 1$$

The solution implies that $X_1^* = 1$ worker must work in Shift 1 that gives the worker Saturday and Sunday off. The number of workers who would have Sunday and Monday off is $X_2^* = 2$ workers, and so on.

The good news is that NB already has nine workers and hence the current work schedule can be changed to the above in order to give everyone 2 consecutive days off. One issue that may arise with this schedule is who will be assigned to which shift. More specifically, Shift 1 with Saturday and Sunday off is probably the shift everyone would like to be scheduled for. Therefore, people who are not assigned to that shift may feel that they are treated unfairly.

A simple way to address the issue of fairness is to follow a rotation policy. One example of a rotation policy is to have some of the workers who work in Shift 1 this week, for example,

**Table 13.17   Minimum Number of Tellers Required in Each Hour of the Day in Example 13.11**

| Time periods | Tellers needed |
|:---:|:---:|
| 9:00–10:00 | 8 |
| 10:00–11:00 | 5 |
| 11:00–Noon | 7 |
| Noon–1:00 | 8 |
| 1:00–2:00 | 7 |
| 2:00–3:00 | 7 |
| 3:00–4:00 | 10 |
| 4:00–5:00 | 11 |

work in Shift 2 next week (or next month). This can ensure that everyone will have the same number of Saturdays and Sundays off each year.

### 13.9.2  Scheduling Part-Time Workforce

Another common problem of workforce scheduling arises when firms use part-time workers. Example 13.11 presents such a situation that also includes the problem of scheduling lunch breaks, which is common in most operations systems.

**Example 13.11**

> Second-Fifth (SF) National Bank is open Monday to Friday from 9:00 am to 5:00 pm. The bank currently has six full-time tellers, who come to work every day. Full-time tellers work from 9:00 am to 5:00 pm and have 1 hour lunch break, which is either from noon to 1:00 pm or from 1:00 pm to 2:00 pm.
>
> The bank manager was informed that another branch of SF National is closing next month, and the manager should expect to have more customers coming to the branch. The manager was also given an estimate of the minimum number of tellers the branch will need to be able to handle the current demand and the new demand starting next month. This is shown in Table 13.17.
>
> The good news is that the bank manager is authorized to hire both full-time and part-time tellers. Including fringe benefits, full-time employees are paid $18 per hour, including the lunch hour, and part-time tellers are paid $14 per hour and receive no fringe benefit. Part-time tellers, if hired, must work Monday to Friday. However, they work only for 3 consecutive hours each day. The bank is only authorized to hire a maximum of five part-time tellers. The bank manager now needs to determine how many part-time and full-time tellers should be hired to minimize the total personnel cost. The manager also needs to find a work schedule for part-time hires as well as a schedule for lunch break of full-time employees. Do you have any suggestion?

SF National Bank is facing a cyclic worker scheduling in which the cycle is 1 day with $T = 8$ one-hour periods starting from 9:00 am and ending in 5:00 pm. The possible shift patterns for full-time and part-time employees are shown in Fig. 13.27.

Note that there are no shifts for part-time tellers who start at 3:00 pm or 4:00 pm. This is because bank closes at 5:00 pm and part-time tellers must work for 3 hours.

SF National Bank would like to find how many tellers must be assigned to each of the shifts in order to satisfy hourly demand with minimum personnel cost. For each shift, we

| | | 9:00 --10:00 | 10:00 -- 11:00 | 11:00 -- Noon | Noon -- 1:00 | 1:00 -- 2:00 | 2:00 -- 3:00 | 3:00 -- 4:00 | 4:00 -- 5:00 |
|---|---|---|---|---|---|---|---|---|---|
| Full-Time Tellers | Shift 1 : | | | | Lunch Break | | | | |
| | Shift 2 : | | | | | Lunch Break | | | |
| Part Time Tellers | Shift 1 : | | | | Off | Off | Off | Off | Off |
| | Shift 2 : | Off | | | | Off | Off | Off | Off |
| | Shift 3 : | Off | Off | | | | Off | Off | Off |
| | Shift 4 : | Off | Off | Off | | | | Off | Off |
| | Shift 5 : | Off | Off | Off | Off | | | | Off |
| | Shift 6 : | Off | Off | Off | Off | Off | | | |

Figure 13.27   Shift patterns for full-time and part-time tellers in SF National Bank in Example 13.11.

define a decision variable as follows:

$$X_1 = \text{Number of full-time tellers with lunch break from noon to 1:00}$$
$$X_2 = \text{Number of full-time tellers with lunch break from 1:00 to 2:00}$$
$$P_i = \text{Number of part-time tellers who work in Shift } i$$

Using the above decision variables, the mathematical programming model that minimizes the daily personnel costs and satisfies demand is

$$\textit{Minimize } Z = 144(X_1 + X_2) + 42(P_1 + P_2 + P_3 + P_4 + P_5 + P_6)$$

$$\textit{subject to :}$$

$$
\begin{aligned}
X_1 + X_2 + P_1 &\geq 8 & \text{(9:00--10:00 demand)} \\
X_1 + X_2 + P_1 + P_2 &\geq 5 & \text{(10:00--11:00 demand)} \\
X_1 + X_2 + P_1 + P_2 + P_3 &\geq 7 & \text{(11:00--noon demand)} \\
X_2 + P_2 + P_3 + P_4 &\geq 8 & \text{(Noon--1:00 demand)} \\
X_1 + P_3 + P_4 + P_5 &\geq 7 & \text{(1:00--2:00 demand)} \\
X_1 + X_2 + P_4 + P_5 + P_6 &\geq 7 & \text{(2:00--3:00 demand)} \\
X_1 + X_2 + P_5 + P_6 &\geq 10 & \text{(3:00--4:00 demand)} \\
X_1 + X_2 + P_6 &\geq 11 & \text{(4:00--5:00 demand)} \\
P_1 + P_2 + P_3 + P_4 + P_5 + P_6 &\leq 5 & \text{(Part-time tellers hired)}
\end{aligned}
$$

$$X_i, P_j \geq 0 \quad \text{and integer for} \quad i = 1, 2 \quad \text{and} \quad j = 1, 2, \ldots, 6$$

**Objective Function**

Because a full-time teller is paid $18 an hour and works 8 hours a day, he or she costs the bank about $18 \times 8 = \$144$. For $X_1 + X_2$ number of full-time tellers, the daily personnel cost is therefore $144(X_1 + X_2)$. On the other hand, part-time tellers are paid $14 an hour and work for only 3 hours a day; hence, each part-time teller costs $14 \times 3 = \$42$. The total daily cost for part-time tellers is therefore $42(P_1 + P_2 + P_3 + P_4 + P_5 + P_6)$. The objective function is set to minimize the sum of part-time and full-time daily personnel cost.

### Constraints

For each time period, the left-hand side of each constraint is the total number of tellers (part-time and full-time) that are working in the time period, and the right-hand side is the minimum number of tellers required in that time period. Consider the constraint corresponding to time period 1:00 pm to 2:00 pm:

$$X_1 + P_3 + P_4 + P_5 \geq 7 \qquad (1\text{:}00\text{--}2\text{:}00 \text{ demand})$$

Note from Fig. 13.27 that the tellers who work from 1:00 pm to 2:00 pm are full-time tellers who have their lunch break from noon to 1:00 (i.e., $X_1$), part-time tellers who start their shift at 11:00 am (i.e., $P_3$), part-time tellers who start at noon (i.e., $P_4$), and part-time tellers who start at 1:00 pm (i.e., $P_5$). Therefore, the left-hand side of the above constraint is the total number of full-time and part-time tellers who are available in the bank from 1:00 pm to 2:00 pm. The right-hand side of the constraint is the minimum number of tellers required from 1:00 pm to 2:00 pm.

The last constraint ensures that the total number of part-time tellers hired does not exceed 5.

### Optimal Cyclic Schedule

After solving the mathematical programming using Solver, the optimal values for decision variables and objective function are found as follows:

$$Z^* = 1506, \quad X_1^* = 4, \quad X_2^* = 5$$

$$P_1^* = 0, \quad P_1^* = 0, \quad P_1^* = 3, \quad P_1^* = 0, \quad P_1^* = 0, \quad P_6^* = 2$$

The solution implies the following workforce schedule:

- *Full-time tellers:* SF National Bank needs a total of $X_1^* + X_1^* = 4 + 5 = 9$ full-time tellers. Since the bank already has six tellers, it needs to hire three more full-time tellers. Among those nine tellers, $X_1^* = 4$ should take their lunch break from noon to 1:00 pm, and $X_2^* = 5$ of them should take their lunch break from 1:00 pm to 2:00 pm.
- *Part-time tellers:* SF National Bank needs to hire part-time tellers for only two 3-hour shifts: specifically, hiring $P_3^* = 3$ tellers to work from 11:00 am to 2:00 pm, and hiring $P_6^* = 2$ tellers to work from 2:00 pm to 5:00 pm. There is no need to hire part-time tellers for other shifts.

This results in the total daily personnel cost of $Z^* = \$1506$, which is the minimum possible for satisfying demand.

## 13.10 Summary

Operation scheduling refers to the process of assigning activities needed to produce a product (or to provide a service) to resources and specifying when each resource should start performing each activity in order to achieve a certain goal. Main goals of operations scheduling include maximizing throughput (e.g., minimizing makespan), minimizing flow time and inventory (e.g., minimizing average flow time or weighted flow times), and meeting promised due dates (e.g., minimizing tardiness or number of tardy jobs). There are many analytics designed to find schedules that achieve these goals in different production processes. This chapter presents a summary of these analytics for the following processes: (i) continuous-flow process, (ii) flow-line process, (iii) batch process, (iv) job shop process, and (v) project process.

The main challenge in the scheduling of continuous processes is to determine the production cycle; specifically, to find: (i) how long should the process produce each product, and (ii) how should products be sequenced for production? The online supplement of this chapter illustrates how this scheduling problem can be modeled in a spreadsheet and how evolutionary algorithm of Excel Solver can be set up to find the optimal production cycle.

Flow-line processes achieve high production rate by breaking production activities into smaller tasks and performing them in several workstations arranged sequentially, for example, assembly lines. The main goal of operations scheduling in these processes is to achieve high throughput with minimum cost. Unfortunately, these two goals often contradict with each other, that is, higher throughput is usually achieved with more stations and more workers and thus leads to higher costs. One way to achieve a good trade-off between higher throughput and lower cost is to balance the flow line. This chapter presents models that can balance paced flow-line processes to achieve a certain throughput with minimum number of stations.

Operations scheduling in job shop and batch processes is more complex, because it involves several stages of planning. The scheduling starts by finding the master production schedule (MPS) that determines how many of each product must be produced in each of the future weeks or months. Material requirement planning (MRP) is then used to translate MPS into the production and ordering schedules for each part and subassembly used to make the final product. The next step, which is called loading, is to find how parts in each production batch must be assigned to resources (e.g., machines) in order to finish production as soon as possible. The final scheduling problem, called detailed scheduling, corresponds to finding the sequence by which each resource must process these parts.

This chapter presents several detailed scheduling policies designed to achieve different goals. It describes how shortest processing time (SPT) and weighted shortest processing time (WSPT) rules minimize total inventory and total inventory cost in systems with a single resource. Earliest due date (EDD) rule is shown to minimize the maximum tardiness and maximum lateness. The chapter also presents analytics to determine how jobs should be processed by a pool of resources or by a two-stage process in order to finish all jobs as soon as possible, that is, to minimize makespan.

Operations scheduling in project processes has its own techniques called critical path method (CPM) and project evaluation and review technique (PERT). CPM's goal is to determine when each activity of the project must be started or finished to make sure that the project is completed before its due date. Linear programming models are discussed and used to crash a project such that the length of the project is shortened with minimum crashing cost. Finally, the chapter illustrates how variability in activity times can be incorporated using PERT to obtain an estimate for the likelihood of finishing the project before a certain due date.

The last scheduling problem discussed in the chapter is workforce scheduling. Developing work schedule for workforce in manufacturing and service systems is a time-consuming and difficult problem, especially in processes with multiple shifts and full-time and part-time workers. This chapter illustrates how linear programming models can be used to find the best way to assign part-time and full-time workers to different working shifts with minimum labor costs.

## Discussion Questions

1. What is master production schedule (MPS) and how is it developed?
2. Describe three main goals of operations scheduling?
3. What is makespan and how does it relate to utilization and throughput of a resource?

4. When should a manager use the following criteria to evaluate a scheduling policy: (i) minimizing the maximum tardiness, (ii) minimizing weighted tardiness, and (iii) minimizing weighted number of tardy jobs?

5. What is the difference between paced flow lines and unpaced flow lines? What is the main scheduling problem in paced flow lines?

6. Describe the advantage and disadvantage of forward and backward scheduling?

7. What is material requirement planning (MRP)? How can one incorporate variability into MRP process? How does MRP differ from ERP?

8. What is the difference between sequencing and scheduling? What are the common assumptions in sequencing analytics for a single resource?

9. Explain when each of the following rules must be used to sequence jobs that require processing by a single resource?
    a. SPT
    b. WSPT
    c. EDD

10. Explain when Johnson's rule can be used to find optimal schedule in an operations?

11. What are the two main techniques used in scheduling activities in a project process?

12. What is a critical path in a project and why is it important in managing projects?

13. What are non-critical activities? Is it possible that they become critical activity as the project progresses?

14. What are the outputs of CPM analysis and why are they important in managing projects?

15. How can one estimate the probability that a project is completed by a certain due date? Describe the assumptions that are made to compute this probability.

16. What does it mean to crash a project? Explain three approaches to crash a project.

## Problems

1. Ocean Briz is a firm that produces a variety of ceiling and standing fans. Assembling their SF-32 fans requires the following activities:

| Activity | Process time (*minutes*) | Predecessor activities |
|---|---|---|
| A | 5 | — |
| B | 4 | A |
| C | 8 | A |
| D | 2 | C |
| E | 7 | B,D |
| F | 4 | E |
| G | 3 | F |
| H | 4 | F |

If the demand for SF-32 is about five per hour, design an assembly line with the minimum number of workstations that can achieve the demand of five per hour. How many workstations are needed and which activities should be assigned to which station?

2. A firm that makes accessories for electronics wants to redesign one of its assembly lines to produce a new TV wall mount brackets. The demand for the bracket is estimated to

be about 32 per day. Parts and components of the bracket will be made in the fabrication department and the bracket will be assembled in the assembly department. There are nine activities that are required to assemble the bracket. These activities, their process time, and their predecessor activities are shown in the following table:

| Activity | Process time (minutes) | Predecessor activities |
|---|---|---|
| A | 4 | — |
| B | 2 | — |
| C | 4 | A |
| D | 6 | A,B |
| E | 7 | C,D |
| F | 5 | D |
| G | 1 | E |
| H | 4 | F |
| I | 3 | H,G |

The assembly department works 8 hours a day. Design an assembly line with minimum number of workstations that can satisfy the demand for the brackets. How many workstations are needed? Which activities should be assigned to each workstation in the line?

3. A producer of home appliances is planning the production of its only electric tea kettle, known as Product Z. One unit of Product Z is made of 2 units of subassembly Y and 4 units of subassembly X. One unit of Y, on the other hand, requires 3 units of A and 3 units of B. One unit of A is made of 1 unit of B and 2 units of C. One unit of subassembly X is made of 5 units of A and 4 units of C. Production lead time to assemble subassemblies X and Y to make Product Z is 2 weeks. Production lead time to make subassemblies X and Y are 2 weeks and 1 week, respectively. Production lead time for Part A is 1 week. Parts B and C are ordered from outside suppliers with order lead times 2 and 3 weeks, respectively. The firm has pipeline inventory for these parts. Specifically, the firm will receive 200 Part A in week 2 and 400 Part A in week 5. The supplier is also scheduled to deliver 350 Part C in week 4. Master production schedule (MPS) shows a demand of 800, 1200, 1000, and 1500 units for Product Z in weeks 12, 13, 14, and 15, respectively.
    a. Draw product structure diagram for the electric kettle.
    b. Develop a weekly production and ordering schedule to satisfy the demand for the kettle. Specifically, determine how many of each part or subassembly must be produced or ordered each week to satisfy MPS. Use lot-for-lot as your lot sizing policy.
    c. Part C is an expensive component that is ordered from a supplier in Asia. There is a large order setup and shipment cost of $1500 associated with placing and receiving an order of Part C. This cost is independent of the order size. Therefore, the firm wants to use another lot sizing policy (instead of lot-for-lot) to minimize the cost of ordering and holding inventory of Part C. If holding one unit of Part C in inventory for 1 week costs $1, when should the firm place order for Part C to satisfy MPS? How many Part C should be ordered each time?

4. Times R Us is a producer of digital and analog clocks. To reduce inventory costs, Times R Us designed its clock such that they use many common components. The firm is planning the production of two of its popular digital clocks—clock C-33 and clock D-33—ahead of Christmas shopping season, which is 14 weeks from now. The product structure diagram for these products is shown in Fig. 13.28.

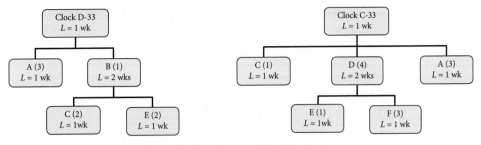

Figure 13.28    Product structure diagram for clocks C-33 and D-33 in Problem 4.

Parts A, C, E, and F are purchased from suppliers and are used to make clocks. Master production schedule (MPS) for these two clocks are set, indicating the demand of 300, 500, 900, and 200 for clock D-33 in weeks 16, 17, 18, and 19, respectively; and demand of 500, 800, 1200, and 400 for clock C-33 in weeks 16, 17, 18, and 19, respectively. Based on the previous orders placed, Times R Us is scheduled to receive 120 and 80 Part C, respectively, in weeks 5 and 6, and 200 Part E in week 3, and 180 Part F in week 4. Develop a weekly production and ordering schedule for all parts to satisfy MPS. Use lot-for-lot as your lot sizing policy.

5.  L&C, a manufacturer of automotive seats, has just received orders for leather seats from five different auto manufacturers. Each order consists of 7500 sets of seats used in different car models. L&C has five plants P1, P2, P3, P4, and P5 with unused capacities that can be used to make the seats. Since each plant is designed to make different types of seats, it takes each plant a different time to finish each order. The following table shows the time (in days) that takes each plant to finish each order. The table also shows the number of available days in each plant that can be used to make the orders.

| Order no. | Processing times (*days*) in plant | | | | |
|---|---|---|---|---|---|
| | P1 | P2 | P3 | P4 | P5 |
| 1 | 3 | 4 | 4 | 6 | 4 |
| 2 | 4 | 8 | 2 | 6 | 2 |
| 3 | 5 | 3 | 8 | 4 | 4 |
| 4 | 6 | 6 | 4 | 8 | 5 |
| 5 | 4 | 3 | 4 | 4 | 6 |
| Available days: | 7 | 8 | 5 | 5 | 4 |

Also, due to different degrees of automation and overhead costs, the cost (in $10,000) of making each order in each plant is different, as shown in the following table:

| Order no. | Production cost (*in $10,000*) in plant | | | | |
|---|---|---|---|---|---|
| | P1 | P2 | P3 | P4 | P5 |
| 1 | 27 | 10 | 15 | 19 | 14 |
| 2 | 10 | 24 | 7 | 15 | 8 |
| 3 | 13 | 7 | 26 | 13 | 10 |
| 4 | 19 | 15 | 13 | 21 | 9 |
| 5 | 14 | 8 | 11 | 7 | 20 |

How should L&C allocate these orders to its plants in order to minimize the total cost of making all the seats in the orders?

6. Collier Financial (CF) has a tax processing department that does tax document preparation for large corporations. CF needs to assign five of its tax teams to five projects that it got from its five major clients. Because of the size of each team and the type of the project, the time it takes each team to finish each project is different. These times (in days) are shown in the following table:

| Team | Processing times (*in days*) of project | | | | |
|------|---|---|---|---|---|
|      | 1 | 2 | 3 | 4 | 5 |
| 1 | 7 | 3 | 2 | 3 | 12 |
| 2 | 9 | 3 | 6 | 4 | 14 |
| 3 | 5 | 2 | 4 | 3 | 9 |
| 4 | 4 | 5 | 3 | 6 | 11 |
| 5 | 6 | 10 | 11 | 12 | 5 |

CF does not want to make its clients wait a long time for the completion of their projects.
   a. How should CF assign its teams to the five projects in order to minimize the average waiting time of its clients for the completion of their projects?
   b. The client of project 5 has requested that CF puts team 3 on its project. If CF warrants this request, how does that affect the assignment in Part (a)?

7. A firm that produces soft drinks has just finished making a 30-second TV commercial to promote its new flavored carbonated water. The plan is to show the commercial on Saturday prime time TV between 6:00 pm and 11:00 pm. The marketing department has contacted five major TV stations and got the following costs of playing the commercial during different time slots in prime time (numbers are in $10,000):

| Prime time | Cost (*in $10,000*) of playing on channel | | | | |
|------------|-----|-----|-----|-----|-----|
|            | CH1 | CH2 | CH3 | CH4 | CH5 |
| 6:00–7:00 | 21 | 13 | 19 | 24 | 18 |
| 7:00–8:00 | 28 | 30 | 9 | 19 | 10 |
| 8:00–9:00 | 16 | 9 | 33 | 16 | 13 |
| 9:00–10:00 | 24 | 19 | 16 | 26 | 11 |
| 10:00–11:00 | 18 | 10 | 14 | 9 | 18 |

The marketing department also gathered information about the expected number of viewers watching each channel during the above time slots as follows (numbers are in 1000 viewers):

| Prime time | Number of viewers (*in 1000*) | | | | |
|------------|-----|-------|-------|------|------|
|            | CH1 | CH2 | CH3 | CH4 | CH5 |
| 6:00–7:00 | 63 | 75.6 | 91.4 | 39.7 | 28.4 |
| 7:00–8:00 | 84 | 100.8 | 121.8 | 22.1 | 12.3 |
| 8:00–9:00 | 48 | 57.6 | 69.6 | 30.2 | 21.6 |
| 9:00–10:00 | 72 | 86.4 | 104.4 | 45.4 | 32.4 |
| 10:00–11:00 | 54 | 64.8 | 78.3 | 34 | 24.3 |

TV stations have informed the firm that they can play the commercial in only one of their prime time slots. The firm has a budget of $90,000 assigned to playing the commercial on Saturday night. If the firm's goal is to maximize the number of viewers of its commercials, which time slots of which TV stations should it choose for its commercial? How many viewers will the firm have under the optimal plan?

8. Bizzi is a small airport that has only one runway for landing and one runway for takeoff. Ten airplanes have finished boarding in six different terminals and are all ready for takeoff. Because of the location of the terminals and the type of the airplanes, the time it takes each airplane to takeoff (including taxiing to runway) is different. The following table shows the time needed for takeoff and the number of passengers in each plane:

| Flight | Takeoff time (*minutes*) | Number of passengers |
|---|---|---|
| U212 | 11 | 210 |
| SS23 | 8 | 450 |
| A43 | 4 | 340 |
| A67 | 5 | 480 |
| U10 | 3 | 210 |
| SS44 | 6 | 430 |
| U32 | 8 | 355 |
| U19 | 9 | 324 |
| A11 | 8 | 155 |
| SS22 | 15 | 480 |

   a. In what sequence should the air traffic control schedule the takeoff of these 10 airplanes in order to minimize the average waiting time of each plane? What is the minimum average waiting time?

   b. Answer Part (a) if the goal is to minimize the average waiting time of passengers of all 10 airplanes.

   c. Answer Part (a) if the air traffic control decides to temporarily use both runways for takeoffs of these 10 airplanes.

9. The air traffic control of Bizzi airport in Problem 8 needs to schedule landing of eight airplanes that have arrived earlier than scheduled and are hovering above the airport for their turn to land. The following table shows the scheduled arrivals of each airplane and the time each airplane uses a runway for landing. The scheduled arrival is the time that an airplane is supposed to finish landing.

| Flight | Scheduled arrival | Landing time (*minutes*) |
|---|---|---|
| U102 | 10:07 | 4 |
| A23 | 10:10 | 3 |
| U105 | 10:14 | 9 |
| SS02 | 10:20 | 10 |
| A12 | 10:25 | 9 |
| A23 | 10:31 | 7 |
| U342 | 10:37 | 6 |
| SS11 | 10:40 | 5 |

   Because of the large number of airplane arrivals in each hour, it is critical that airplanes stick to their scheduled arrivals. Delay in arrivals exacerbates the delays in landing of future arrivals.

   a. In what sequence should the air traffic control schedule the landing of these airplanes in order to minimize the maximum delay in landing among all airplanes? What would be the maximum delay?

   b. Answer Part (a) if the goal is to minimize the air pollution caused by these airplanes hovering above the airport. Assume all airplanes burn the same amount of fuel for every minute of flying.

10. Air traffic control of Bizzi airport in Problem 9 is informed by the pilots of the eight flights that they may not have enough fuel to hover above the airport for long time. Thus, air traffic control decides

to use two runways for landing of the eight airplanes. How should the airplanes be scheduled for landing in these runways such that the time the last airplane hovers above the airport before landing is minimized? What would be the hovering time of the last airplane in your schedule?

11. JD Precision Parts is a firm that supplies parts and components to auto manufacturers. Because orders received from manufacturers are for different parts, JD must set up its production process to make each order. The following table shows the list of orders that JD has received from eight different manufacturers. The table also depicts the number of parts in each order and the time to make each part.

| Order | Number of parts | Setup time (*days*) | Part process time (*minutes*) |
|-------|-----------------|---------------------|-------------------------------|
| 1     | 9883            | 1                   | 0.34                          |
| 2     | 5452            | 0.8                 | 0.81                          |
| 3     | 2000            | 0.5                 | 0.6                           |
| 4     | 6272            | 1.2                 | 0.75                          |
| 5     | 2400            | 0.4                 | 1.12                          |
| 6     | 12000           | 1.5                 | 0.14                          |
| 7     | 5648            | 1                   | 0.51                          |
| 8     | 3508            | 0.2                 | 0.52                          |

JD cannot set up its operations for next order until the production of the last order is completed. Also, the production system cannot be used during the set up operations. The plan is to process all these orders in JD's plant in Detroit.

a. In what order should JD make and deliver these orders to minimize the average waiting time of the manufacturers? What is the minimum average waiting time?

JD has already purchased the material used to make all eight orders. Thus, it will be facing a large inventory holding cost if it does not finish these orders fast. Holding each part in order 1 as inventory (in form of raw material, WIP, or finished goods) in each day costs JD about $0.01. This number is $0.09 for each part in order 2; $0.05 for order 3; $0.04 for order 4; $0.02 for order 5; $0.06 for order 6; $0.07 for order 7; and $0.03 for each part in order 8.

b. An order cannot be shipped to a manufacturer until all parts in the order are completed. In what sequence should JD process the orders to minimize its total inventory holding cost of these eight orders?

12. Besides its plant in Detroit, JD Precision Part in Problem 11 decides to also use its plant in Indiana to process the orders. Each order can only be allocated to one of the two plants. Which order should be allocated to each plant in order to minimize the average waiting time of the manufacturers?

13. You come back to school from Christmas and new year's break on January 3rd and you realize that you have eight homework assignments that you need to submit this month. The list of the homework, the time it takes (in days) to finish each one, and their due dates are in the following table:

| Homework | Required time to finish (*days*) | Due date |
|----------|----------------------------------|----------|
| H-1      | 4                                | Jan-8    |
| H-2      | 3                                | Jan-11   |
| H-3      | 2                                | Jan-6    |
| H-4      | 1                                | Jan-4    |
| H-5      | 5                                | Jan-16   |
| H-6      | 2                                | Jan-18   |
| H-7      | 2                                | Jan-14   |
| H-8      | 4                                | Jan-10   |

You realize that you will be late in submitting some of these assignments and you regret that you did not finish them in the last 2 weeks. The due date of a homework refers to the end of the day. For example, the due date of homework H-1 is the end of the day of January 8th—six days from now, including today, January 3rd. You realize that each homework has a penalty for late submission. Specifically, the instructor of each course deducts 10 percent of the homework grade for each day the submission is delayed. In what order should you finish and submit your homework in order to minimize the maximum penalty you get among the delayed homework?

14. Superior Games (SG), a video game design company, is in the process of designing a new video game that is expected to launch during the next 3 months. The game consists of seven software modules that can be developed independent of each other. After these modules are completed, the final step is to create the interface between these modules and complete the game. This final task takes about 1 month. SG has three teams of engineers who can work in any of the seven modules. Because of the expertise and experience of the teams, the time it takes a team to develop a module is different for different teams as shown in the following table:

| Team | Process time of modules (*days*) | | | | | | |
|------|----|----|----|----|----|----|----|
|      | 1  | 2  | 3  | 4  | 5  | 6  | 7  |
| 1    | 7  | 13 | 46 | 33 | 52 | 13 | 21 |
| 2    | 13 | 26 | 46 | 52 | 59 | 26 | 15 |
| 3    | 39 | 33 | 20 | 33 | 46 | 59 | 32 |

What is the earliest time that SG can finish the new game and introduce it to the market?

15. Solid Foundation and Concrete (SFC) is specialized in excavation and concrete services for commercial and residential construction. SFC has two teams: (i) excavation team and (ii) concrete team. For each project, first excavation team excavates the site, making it ready for the concrete team to pour the concrete foundation. SFC is about to sign a contract with a construction company that is making a housing complex consisting of 10 houses. The following table shows the time for excavation and concrete activities of each house.

| House | Excavation time (*days*) | Concrete time (*days*) |
|-------|--------------------------|------------------------|
| A     | 4                        | 8                      |
| B     | 3                        | 5                      |
| C     | 1                        | 7                      |
| D     | 5                        | 4                      |
| E     | 12                       | 5                      |
| F     | 10                       | 9                      |
| G     | 8                        | 3                      |
| H     | 7                        | 9                      |
| I     | 4                        | 2                      |
| J     | 8                        | 6                      |

Before signing the contract and committing to a due date, SFC wants to know the earliest time it can complete its work on all 10 houses. The head of the excavation team suggests that they should sequence the houses such that they work on the house with the smallest excavation time next (i.e., SPT sequence). This finishes excavating the houses at a faster rate and the concrete team can start their work earlier. The concrete team then follows the same sequence of the excavation team.

a. For the suggestion made by the head of the excavation team, develop a Gantt chart to show the schedule of the two teams working on the 10 houses. What is the time that all 10 houses are completed?

b. What is the schedule that results in the earliest time that SFC can finish all 10 houses?

16. A firm provides cleaning services to houses in a metropolitan area. The cleaning of a house consists of two main activities: (i) cleaning bathrooms and kitchen, (ii) vacuuming the entire house, which has to be done after cleaning bathrooms and kitchen. The firm has seven houses to clean today and two workers are available to clean these houses. There are two options to clean houses.

Option 1 is to assign worker 1 to do the cleaning and worker 2 to do vacuuming. Worker 1 cleans a house and then moves to the next house. Worker 2 cannot start vacuuming until after worker 1 is finished cleaning. The following table shows the time of cleaning and vacuuming of each house if the tasks are done by workers 1 and 2 independently:

| House | Cleaning time (*minutes*) | Vacuum time (*minutes*) |
|---|---|---|
| 1 | 25 | 72 |
| 2 | 24 | 48 |
| 3 | 12 | 60 |
| 4 | 48 | 36 |
| 5 | 108 | 48 |
| 6 | 96 | 84 |
| 7 | 72 | 24 |

Option 2 is that the two workers work as a team and first clean bathrooms and kitchen and then vacuum the house together. Under this option, cleaning and vacuuming of each house would take much shorter than those in Option 1. The following table shows the expected time for cleaning and vacuuming each house in Option 2:

| House | Cleaning time (*minutes*) | Vacuum time (*minutes*) |
|---|---|---|
| 1 | 13 | 42 |
| 2 | 13 | 21 |
| 3 | 6 | 30 |
| 4 | 27 | 18 |
| 5 | 68 | 24 |
| 6 | 44 | 49 |
| 7 | 38 | 11 |

Which option results in a shorter time to finish cleaning all seven houses?

17. Smaag Homes is a construction firm that builds houses for its clients. The following table presents the list of activities that Smaag Homes must complete for its current client's project:

| Activity | Duration (*days*) | Immediate predecessor |
|---|---|---|
| A : Foundation | 8 | — |
| B : Walls and ceilings | 10 | A |
| C : Roof | 9 | B |
| D : Electrical wiring | 8 | B,C |
| E : Windows and doors | 5 | B |
| F : Exterior sidings | 10 | E |
| G : Utilities | 2 | C,D,F |
| H : Landscaping | 3 | F |
| I : Painting | 10 | G,H |

a. What is the earliest time that Smaag Homes can finish the project?

b. What is the earliest time that each activity can start?

c. If the deadline to finish the project is day 50, what is the latest time that each activity can start and Smaag Homes can still finish the project before the deadline?

d. The company who does electrical wiring (Activity D) has just informed Smaag Homes that they cannot start their project until day 30. Would that cause Smaag Homes to miss its deadline?

e. What is the maximum time each activity can be delayed and Smaag Homes can still meet the deadline?

18. Smaag Homes in Problem 17 is informed that its client is willing to pay a bonus if Smaag Homes can finish the project in 40 days. The following table shows the cost of reducing the duration of each activity by a day and the maximum possible reduction:

| Activity | Duration (*days*) | Reduction cost (*per day*) | Maximum reduction (*days*) |
|----------|-------------------|----------------------------|----------------------------|
| A | 8 | $5000 | 2 |
| B | 10 | $9000 | 2 |
| C | 9 | $7000 | 4 |
| D | 8 | $4500 | 3 |
| E | 5 | $1500 | 1 |
| F | 10 | $3500 | 3 |
| G | 2 | — | 0 |
| H | 3 | $4500 | 1 |
| I | 10 | $9500 | 3 |

a. Develop a linear programming model to find the minimum bonus that Smaag Homes should accept to finish the project in 40 days.

b. How many days the duration of each activity must be reduced to finish the project in 40 days with minimum crashing cost?

19. Consider Smaag Homes in Problem 18 and assume that because of budget considerations, Smaag Homes decided not to spend money to reduce the durations of the tasks in its current project in Problem 17. However, Smaag Homes has just got a new project from another client who requests Smaag Homes to start its project on day 43. To compute the probability of finishing its current project before day 43, Smaag Homes asks its teams to provide the optimistic, pessimistic, and most likely estimates for finishing each task of its current project. The following table presents these estimates:

| Activity | Estimates (*days*) | | |
|----------|-----------|-------------|-------------|
| | Optimistic | Most likely | Pessimistic |
| A | 4 | 7 | 16 |
| B | 6 | 9 | 18 |
| C | 6 | 8 | 16 |
| D | 4 | 8 | 12 |
| E | 1 | 5 | 9 |
| F | 6 | 10 | 14 |
| G | 1 | 2 | 3 |
| H | 1 | 2 | 9 |
| I | 5 | 10 | 15 |

What is the probability that Smaag Homes can finish its current project by day 43?

20. A large online retailer is working with an information technology consulting firm to update its order processing and sales equipment and software. The transition from the old system to the new system consists of the following activities:

| Activity | Duration (days) | Immediate predecessor |
|---|---|---|
| A : System design | 7 | — |
| B : Hiring technicians | 9 | — |
| C : Updating equipment | 5 | — |
| D : Initial tests | 3 | A |
| E : Job training | 7 | B |
| F : Final test | 7 | C,E |
| G : Experimental run | 8 | D,E |
| H : System transition | 1 | G,F |

a. What is the earliest time that the retailer can use the new IT system?

b. What is the earliest time that each activity can start?

c. If the deadline for complete transition is in 1 month (i.e., in 30 days), what is the latest time that each activity can start and finish, but the firm can still make its transition to the new system within a month?

d. What is the maximum time each activity can be delayed and the retailer would still be able to make the transition within a month?

21. Consider the online retailer in Problem 20. Because the sales promotion campaign is scheduled to start on day 20, the retailer is willing to pay a bonus of $30,000 if the consulting firm can complete the transition by day 20. Suppose you work for the consulting firm and you have gathered the following information about the cost of reducing the duration of each activity by 1 day and the maximum number of days that each activity can be reduced:

| Activity | Duration (days) | Reduction cost (per day) | Maximum reduction (days) |
|---|---|---|---|
| A | 7 | $2000 | 3 |
| B | 9 | $5000 | 3 |
| C | 5 | $3000 | 1 |
| D | 3 | $4500 | 1 |
| E | 7 | $4000 | 3 |
| F | 7 | $2500 | 2 |
| G | 8 | $2000 | 4 |
| H | 1 | — | 0 |

a. Would you recommend that the consulting company accept the bonus and finish the project within 20 days?

b. How many days the duration of each activity must be reduced to complete the project in 20 days with minimum crashing costs?

22. Consider the online retailer in Problem 20 that is making a transition to a new system for order processing and sales. The consulting firm had a long negotiation and is planning to sign a contract to finish the project in 23 days. But, before doing that, they are asking your opinion about the possibility that the project takes more than 23 days. To estimate this possibility, you contacted the experts in your teams and asked them to give you their pessimistic, optimistic, and most likely estimates for the time it takes to finish each activity of the project. You get the following information:

| Activity | Estimates (*days*) | | |
|---|---|---|---|
| | Optimistic | Most likely | Pessimistic |
| A | 4 | 7 | 10 |
| B | 6 | 8 | 16 |
| C | 1 | 5 | 9 |
| D | 2 | 3 | 4 |
| E | 2 | 5 | 20 |
| F | 3 | 6 | 15 |
| G | 4 | 8 | 12 |
| H | 1 | 1 | 1 |

Using the above data, what would be your estimate for the probability that completing the project takes more than 23 days?

23. Call center of a financial institution works from 7:00 am to 5:00 pm. The firm has a plan to offer more of its services online and through its call center, and thus expecting to have significant increase in calls to its call center. When the plan is implemented, the number of agents required to provide low waiting time for callers during each hour of operations is expected to be as follows:

| Time periods | Number of agents needed |
|---|---|
| 7:00–8:00 | 5 |
| 8:00–9:00 | 7 |
| 9:00–10:00 | 12 |
| 10:00–11:00 | 18 |
| 11:00–Noon | 15 |
| Noon–1:00 | 12 |
| 1:00–2:00 | 20 |
| 2:00–3:00 | 15 |
| 3:00–4:00 | 12 |
| 4:00–5:00 | 7 |

Call center has three 8-hour working shifts with 1-hour lunch break: (i) first shift starts at 7:00 am with lunch break from 11:00 am to 12:00 pm, (ii) second shift starts at 8:00 am with lunch break from 12:00 pm to 1:00 pm, and (iii) third shift starts at 9:00 am with lunch break from 1:00 pm to 2:00 pm.

Currently, the call center has 15 agents working in different shifts. However, call center manager knows that more agents may need to be hired to maintain the required service level.

a. What is the minimum number of agents needed to satisfy the workforce requirement in each hour?

b. How many agents must be assigned to each working shift to maintain current service level?

24. Call center in Problem 23 pays its full-time agents who work in its three 8-hour shifts $20 per hour. In addition to hiring new full-time agents, the manager is also considering hiring part-time agents who work for 4 consecutive hours in a day. The part-time shifts include seven 4-hour shifts with no breaks. The first part-time shift starts at 7:00 am, the second shift starts at 8:00 am, and so on, and the last part-time shift starts at 1:00 pm. Part-time agents are paid $18 per hour. How many full-time and how many part-time agents should the call center hire in order to minimize its labor cost? How these agents should be assigned to the 8-hour and 4-hour (part-time) shifts? The manager would like to keep its current 15 full-time workers.

# CHAPTER 14

# Flexible Operations

## 14.0 Introduction

In Chap. 4 of his autobiography—*The Secret of Manufacturing and Serving*—Henry Ford wrote:

> *Therefore in 1909 I announced one morning, without any previous warning, that in the future we were going to build only one model, that the model was going to be "Model T," and that the chassis would be exactly the same for all cars, and I remarked:*
>
>> *"Any customer can have a car painted any colour that he wants so long as it is black."*
>
> *I cannot say that any one agreed with me. . . . The sales people had ground for their objections and particularly when I made the following announcement:*
>
>> *"I will build a motor car for the great multitude. It will be large enough for the family but small enough for the individual to run and care for. It will be constructed of the best materials, by the best men to be hired, after the simplest designs that modern engineering can devise. But it will be so low in price that no man making a good salary will be unable to own one and enjoy with his family the blessing of hours of pleasure in God's great open spaces."*
>
> *This announcement was received not without pleasure. The general comment was:*
>
>> *"If Ford does that he will be out of business in six months."*

But, Henry Ford did not go out of business in 6 months. Through his invention—the flow line—he was able to make Model T cars in one color and sell them at a very low price. Ford became the largest manufacturer of cars in the world. Henry Ford's idea of reducing the variety in the production into one and producing them in large quantities using flow lines revolutionized manufacturing and created what we know as *Mass Production.*

## Mass Production

Mass Production is the production of large quantities of standard products at a low cost (per unit). Mass production is based on several principles including division and specialization of labor, automated machines, and production using flow-line processes (e.g., assembly lines).

While mass production is a cost-efficient way to produce products in large quantities, it is mainly suited for producing standardized products and does not allow for product variety. But this is not what today's market wants. Today, customers can indeed have any color car they want. In fact, through websites of auto manufacturers, customers can build their own car, by selecting the exterior color, type of transmission (manual or automatic), engine type, interior color and type, audio systems, etc. This product customization is not just for car industry. Nowadays customers can custom design their goldfish crackers, their bikes, their audio speakers, their mugs, you name it.

Businesses are realizing the value of offering customized products. Offering a wide range of customized products and/or allowing customers to customize their products, not only increases a firm's market share but also strengthens its brand loyalty. Studies have shown that customers who had customized a product online were more loyal to the brand than those who bought the regular products from the same manufacturer. Higher loyalty translates to higher sales, more referrals, and lifetime customer value.

Considering market trend toward customized products, and firms' desire to utilize the benefit of offering customized products, the question is how firms can produce a large variety of customized product with low cost? The answer is *Mass Customization.*

## Mass Customization

*Customization* is the ability to make a product to match exactly to what customer wants. *Mass Customization* is the high-volume production of customized products—each made differently based on customer order. The key feature of mass customization is to produce products at almost the same cost of mass production that produces a large volume of products at a faster speed, but with much less variety.

In addition to advanced information technology (connecting customers with organizations), lean operations (reducing costs) and efficient supply chain management, mass customization requires operational flexibility. In fact, in a survey of executives of manufacturing companies, 9 out of 10 executives stated that operational flexibility is critical in their companies' success in achieving their revenue growth.[1]

[1]How US Manufacturers are Thriving (or Not) in a World of Ongoing Volatility and Uncertainty. The US Manufacturing Flexibility Index. *Accenture Strategy*, 2014.

But what is "operational flexibility"? Dictionaries define flexibility as the ability to bend or to change or to do different things. From operations perspective, the operational flexibility implies the ability of an operations to adapt quickly and rapidly to changes in its environment.

Operational flexibility relates to product mix or product volume flexibility. Product mix flexibility, as discussed in Chap. 1, is the operations' ability to provide a *wide range* of customized products to meet changing customers' demand in a fast and cost-effective manner. Product volume flexibility, on the other hand, refers to the operations' ability to—quickly and with minimum cost—respond to the changes in the *demand volume* for each product by changing its throughput. Both product mix flexibility and volume flexibility are critical for Mass Customization.

There are several factors that help achieve operational flexibility. They include efficient use of information technology to rapidly recognize changes in the market and operations and react accordingly, utilizing digital manufacturing technology such as robotics, computer-integrated manufacturing systems (CIMS), 3-D printing, and using flexible resources.

This chapter focuses on improving operations by achieving operational flexibility through two different strategies:

(i) *Using Flexible Resources.* Flexible resources are resources that can perform more than one activity on flow units. Examples are cross-trained workers who can work in more than one station in a line, or multi-functional machines (e.g., CNC machine tools) that can do more than one machining operations on a job, or plants that can produce more than one type of product.

(ii) *Postponement.* Postponement refers to making products in standard form in the earlier stages of a process and creating variety in later stages by adding unique components. For example, making standard computer memory, processors, and wireless boards and then assembling them later according to customer's order after receiving the order.

We start with flexible resources.

## 14.1  Flexible Resources

As mentioned above, flexible resources are those that can perform more than one activity. For example, agents in a call center are often cross-trained to respond to more than one type of calls. Advanced machine tools are capable of performing a wide range of machining operations, and most car assembly plants are designed to assemble more than one car model.

In this section, we present several simple examples to show how flexible resources can improve performance by improving effective capacity of a process as well as by mitigating the impact of variability, both of which are important in mass customization that deals with high demand of customized products. We first start by showing how flexible resources can improve process capacity.

## 14.2  Flexible Resources and Process Capacity

One reason that flexible resources can improve process performance is that they result in capacity pooling. We use the following case to illustrate this.

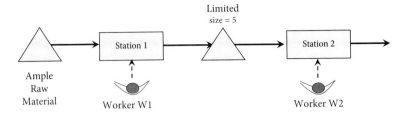

Figure 14.1　The two-station line in Case A.

**Case A:** Consider the two-station line in Figure 14.1 with effective process times of Worker 1 and Worker 2 to be exactly 1 and 0.6 minutes, respectively, with no variability. The buffer at Station 2 has a limited size of five jobs. There is no variability in the process times and there are ample jobs at the buffer of Station 1.

Obviously, Worker 1 is the bottleneck of the line and thus, the line has effective capacity of one job per minute, or 60 jobs per hour. Therefore, the throughput of the line will be 60 jobs per hour—equal to its effective capacity. The utilization of Worker 1 is 100 percent, while Worker 2's utilization is 60 percent, implying that the worker is idle 40 percent of the time.

### Improving Throughput by Balancing the Line
How can we improve the throughput of the line? One way is to redesign the tasks performed by the two workers in order to balance the line. *Line Balancing*, in its traditional sense, means to design (or allocate) tasks to stations such that all workers in all stations have the same effective mean process times (see Chap. 13 for details). Note that the total processing time of a job in the line is $1 + 0.6 = 1.6$ minutes. One way to balance the line is to redesign tasks (or product) such that the effective mean process times at both Stations 1 and 2 be 0.8 minutes— still a total of $0.8 + 0.8 = 1.6$ minutes. This results in effective capacity of both workers to be $60/0.8 = 75$ jobs per hour. Hence, the line will have a throughput of 75 jobs an hour, that is, a 25-percent improvement over the throughput of 60 jobs an hour. In this line both workers will be 100 percent utilized.

The problem, however, is that in most cases, dividing activities into smaller tasks in order to have a line with the same process times in all stations—a balanced line—is not always feasible or economical. For example, it is not often possible to divide a machining operation such as drilling or milling into smaller parts and perform it on different machines. One reason is that this requires additional setups (e.g., loading and unloading) and may result in quality issues.

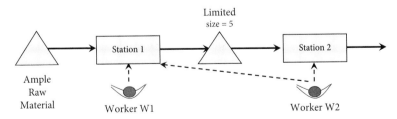

Figure 14.2　Using flexible worker in the two-station line of Case A.

### Improving Throughput Using Flexible Workforce
How can a flexible resource improve the throughput of the line in Fig. 14.1? Easy! Suppose we cross-train Worker 2, as shown in Fig. 14.2, to also work at Station 1 when the worker becomes idle.

Specifically, suppose we start the line at time $t = 0$ with five jobs in the buffer of Station 2. Three minutes later at time $t = 3$, Worker 2 will finish processing five jobs (0.6 minutes each). During the 3 minutes, Worker 1 processes three jobs (1-minute each) and thus there will be three jobs at the buffer of Station 2 at time $t = 3$. Now, suppose at this time Worker 2 switches to Station 1 and processes one job there (while Worker 1 is also processing a job there). One minute later at time $t = 4$ they both finish processing their jobs and put them in the buffer of Station 2. Worker 2 then switches back to Station 2. Hence, at time $t = 4$ the buffer at Station 2 will have five jobs—the same number of jobs it had at time $t = 0$. The workers can repeat this cycle and produce five jobs in every 4-minute cycle. This results in the line throughput of 75 jobs an hour, and both workers will have the utilization of 100 percent.

As this example shows, by cross-training Worker 2, the throughput of the line is increased by 25 percent from 60 to 75 jobs per hour without balancing the line, that is, without changing the tasks in Stations 1 and 2. All that needed is enough space at Station 1, so both workers can work on two different jobs there.

What is the reason for this significant improvement in throughput? The reason is the increase in the effective capacity of the line achieved by utilizing the idle time of Worker 2 to help the bottleneck worker at Station 1. This, as discussed in Chap. 6 is "horizontal pooling" of the capacities of the workers and is achieved by using cross-trained (i.e., flexible) Worker 2. In that chapter we showed that horizontal pooling can improve the effective capacity of a process, resulting in increase in throughput, as is in our example here.

One interesting observation is that while cross-training Worker 2 does not balance the line, it indeed balances the workers' utilizations. In the original line in Fig. 14.1 Worker 1's and Worker 2's utilizations were 100 percent and 60 percent, respectively. By cross-training Worker 2, the utilization of both workers became 100 percent. This points to the following flexibility principle:

**Flexible Resources and Balancing Utilization**

To achieve improvement in process-flow measures (through improving process capacity), use flexible resources and pool their capacities such that it results in the same utilization for all resources.

Vertical Pooling of resources discussed in Chap. 6 is also based on using flexible resources to work in more than one station. Compared to horizontal pooling, the difference is only in pooling the capacities of workers who work in two parallel stations.

## 14.3  Flexible Resources and Variability

One source of variability in processes is the variability in the effective process times of resources. One factor that causes this variability is when resources are processing different products, each requiring a different process time. The higher the variations in products (as it is in mass customization), the higher the variability in the effective process times of resources. In this section, we show that improvement in performance obtained by using flexible resources is not limited to the increase in the effective capacity of a process. Flexible resources, if properly used, can eliminate or reduce the negative impact of variability in the process.

As an example, consider the two-station line in Fig. 14.1 with no variability and suppose that we were able to balance the line by redesigning the tasks such that both stations in the line now have an effective mean process time of 0.8 minutes. In this situation, the line has its

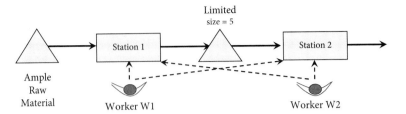

Figure 14.3    Using flexible workers in the balanced line in Case B.

maximum effective capacity of 75 jobs an hour and therefore cross-training workers cannot increase the capacity. In other words, resource flexibility does not provide any benefits.

Now suppose that there is variability in the process times of each station, as explained in Case B:

**Case B:**    Consider a two-station balanced line with effective mean process time of 0.8 minutes for both workers. Effective process times of both workers have coefficient of variation 1. Similar to Case A, the buffer at Station 2 has a limited size of five jobs and there are ample jobs at the buffer of Station 1.

Developing a computer simulation of this line we obtain the throughput of the line to be 62.77 jobs per hour and utilizations of Workers 1 and 2 be around 84 percent. Note that, although Station 1 has ample jobs in its buffer, the throughput of the line cannot reach its effective capacity of 75 jobs per hour. This, as we discussed in Chap. 6, is because of throughput loss due to limited buffer size at Station 2. Can cross-training workers improve throughput? Let's try it.

Suppose we cross-train Worker 2 to also work at Station 1 and Worker 1 to also work at Station 2 (see Fig. 14.3). Specifically, workers are asked to give priority to working in their own station and work in the other station if they become idle (due to starvation or blocking). Using a simulation model we find the throughput of the line to be 74.17 jobs per hour, which is an 18.2-percent increase in throughput and is close to the maximum possible throughput of 75 jobs per hour. Also both workers have utilization of about 99 percent. Thus, even in this case that resource flexibility does not increase process capacity, it still improves throughput by increasing resource utilization from 84 percent to 99 percent. The reason is as follows. Because of the *variability* in process times, the line with no flexibility will experience situations in which Worker 2 is idle (i.e., starved) because the buffer is empty, or Worker 1 is idle (i.e., blocked) because the buffer of Station 2 is full. This reduces worker utilization. Resource flexibility diminishes this negative impact of variability on worker utilization. Specifically, when cross-trained workers become idle, they switch to the other station and process a job. This increases the utilization of both workers (compared to those in the line with no flexibility). Higher utilization means workers are working more in an hour and thus produce more in an hour, resulting in higher throughput for the line.

**Improving Inventory and Flow Time Using Flexible Workforce**

While resource flexibility can increase resource utilization in some processes, there are other systems in which flexibility does not increase resource utilization. As an example, consider the two-station line in Case C.

**Case C:**    Consider Case B and assume that the buffer at Station 2 has unlimited space. Also assume that, instead of keeping ample jobs at the buffer of Station 1, jobs are released at that buffer at a rate of 60 jobs per hour.[2] The job interarrival times at Station 1 have coefficient of variation 1.

---

[2] If we provide ample jobs at Station 1, Worker 1 becomes 100 percent utilized and thus processes jobs at the rate of 75 jobs per hour (i.e., the worker's capacity). This results in job arrival rate of 75 per hour at Station 2, equal to Station 2's capacity, resulting in an unstable Station 2.

Can we increase the line's throughput using flexible workers? The answer is No. Note that this line is a demand-constrained process with no throughput loss. Thus, the only way to increase the throughput is to release more than 60 jobs per hour to the line.

But, what about other performance measures? Specifically, can we reduce the line's flow time or inventory using flexible workers? To check this, notice that with no flexible workers, the first station in Case C constitutes an $M/M/1$ queueing system with arrival rate of 60 jobs per hour and average process time of 0.8 minutes. Using Eq. (7.2), we can compute the average flow time and average inventory at Station 1 to be 4 minutes and four jobs, respectively. The second station is also an $M/M/1$ queueing system with arrival rate 60 jobs per hour and effective mean process time of 0.8 minutes. Thus, there are also an average of four jobs in Station 2 and each job spends an average of 4 minutes in that station. Overall, the line has an average inventory of eight jobs with an average flow time of 8 minutes. Can we improve the line's flow time and inventory by using flexible workers?

The answer may not be so clear, because in this new line: (i) the line is balanced, that is, both workers have effective capacity of 75 jobs per hour; hence, flexibility does not increase the effective capacity of the line. (ii) Both workers receive jobs at the rate of 60 per hour and thus have the same utilization of $u = 60/75 = 0.8$, which is the maximum utilization under job release rate of 60 per hour. Thus, resource flexibility will neither increase the capacity nor increase the utilization of the workers. So, the question remains: Can line performance be improved by making workers flexible?

While in Case C flexibility does not improve the line's capacity or worker utilization, it still improves performance by mitigating the impact of variability. Suppose we cross-train workers to work in both stations and ask workers to follow the following control policy: Work at your own station until you become idle with no jobs. At that point switch to the other station and process a job. Upon completion of the job, return to your own station only if there is a job there; otherwise, process another job and follow the control policy. Using a computer simulation we find the average inventory and average flow time in the line to be 4.44 jobs and 4.44 minutes, respectively. Compared to the line with no flexibility, this is about $(8 - 4.44)/8 = 44.5\%$ reduction in inventory and flow times of the line.

What is the reason for such significant improvement in flow time and inventory? The reason is that resource flexibility diminishes the negative impact of variability on flow time and inventory by using the capacities of resources more efficiently. In the line with no flexibility, due to *variability* in process times, there will be times that Worker 1 is idle when there are jobs at Station 2 and other times that Worker 2 is idle when there are jobs waiting at Station 1. However, when workers are cross-trained, variability cannot result in such situations. Thus, in addition to increasing capacity and worker utilization, resource flexibility diminishes the negative impact of variability by using the capacity of the workers more efficiently.

---

**PRINCIPLE**

**Flexible Resource and Variability**

Flexible Resources also improve process performance by reducing or eliminating the negative impact of variability.

---

## 14.4 Flexible Resources and Control Policy

As illustrated earlier, using flexible resources can improve process-flow measures such as throughput, inventory, or flow times, since it can increase the capacity of a process and diminish the negative impact of variability. One question that arises when using flexible resources is how those resources' times should be allocated to different tasks. A *worker control policy* determines how each worker's time should be allocated to the different tasks for which the worker is cross-trained. For example, consider the line in Case C with no flexible workers. In this case, all of Worker 1's time will be spent on processing jobs at Station 1 and all of Worker 2's time will be spent on processing jobs at Station 2. When both workers are flexible and thus can work in both stations, the question becomes: When should Worker 1 switch to Station 2 and process jobs there, and when should Worker 1 switch back to Station 1? Same questions arise for Worker 2.

Recall that when we used flexible workforce in Case C, we used a worker control policy under which workers give priority to their own stations and switch to the other stations only when their own station is empty. This control policy made an efficient use of worker capacities by preventing idleness of each worker as long as there is at least one job waiting in the buffers of the line. This, as the simulation showed, resulted in 44.5-percent improvement in average inventory and flow time of the line. Can we further improve line performance by using a different control policy?

**Improving Performance Using a Better Control Policy**
Note that the priority policy implemented by Worker 1 gives priority to jobs at Station 1. Is that a good policy that helps reduce inventory in the line? The answer is No. To understand the reason, consider the two-station line as one process with two different types of jobs: *Type-1 jobs* are jobs waiting in the buffer of Station 1, and *Type-2 jobs* are jobs waiting in the buffer of Station 2. The total effective mean process times of type-1 jobs in the line (before they exit the line) is $0.8 + 0.8 = 1.6$ minutes, while this is only 0.8 minutes for type-2 jobs. Which type of job should Worker 1 give priority to?

Recall from Chap. 8 that in processes with multiple jobs, giving priority to jobs with the shortest expected process time—the SPT rule—minimizes the inventory and flow times of the process. Thus, the inventory and flow times of the line in Case C can be further improved if—similar to Worker 2—Worker 1 also gives priority to Station 2 (i.e., type-2 jobs), since they have a shorter effective mean process times (to leave the line) than those jobs in Station 1. This new priority policy translates into the following worker control policy that we call *Sequential* policy: Both workers should process a job at Station 2 immediately after they finish processing a job at Station 1. They then return to Station 1 and repeat the cycle. Under this policy, the buffer of Station 2 is always empty since jobs do not wait in that station.

How much implementing this new control policy improves line performance? Using a simulation model, we find the average inventory in the line to be 3.74 jobs with an average flow time of 3.74 minutes. This is 53.3-percent improvement in inventory over the case with no flexible workforce and it is $53.3\% - 44.5\% = 8.8\%$ improvement over the line with flexible workforce in which both workers give priority to their own stations.

Now let's have a look at the line in Case B in which the buffer between the two stations has a limited size of five jobs and there are ample jobs at the buffer of Station 1. Note that, by using flexible workforce (as in Fig. 14.3) we showed that the throughput of the line can be increased from 62.77 to 74.17 jobs per hour. Recall that in that case the worker control policy was also the same: workers giving priority to their own stations and only switching to the other station when they become idle. What happens if we change the control policy to the sequential policy introduced above?

Under sequential policy, the buffer of Station 2 is always empty; thus, its limited size does not result in blocking of workers in Station 1. Consequently, under the sequential policy it takes each worker an average of 1.6 minutes to finish a job independent of the other worker and results in a throughput of $60/1.6 = 37.5$ jobs an hour. Having two workers, the throughput of the line becomes $2 \times 37.5 = 75$ jobs an hour—the maximum possible throughput. The utilization of both workers will be 100 percent. While this is only slightly better than that when workers do not use sequential policy and give priority to their own stations, it, however, still makes the following important point:

---

**PRINCIPLE**

**Flexible Resources and Control Policy**

The improvement in process-flow measures using flexible resources also depends on the control policy by which each resource time is allocated to different tasks. A good control policy can further improve the process performance achieved using flexible resource.

---

There are several commonly used control policies in practice. To describe some of these policies, consider a worker who is flexible to work in two workstations A and B.

- *Fixed Priority:* Under this control policy, the worker prioritizes the two stations as Priority 1 and 2. Upon completion of processing of a job at a station, the worker processes a job at the station with Priority 1. If there is no job there, the worker processes a job at the station with Priority 2. If there is no job there, the worker remains idle until a job arrives at one of the two stations. This is the policy we discussed in the above examples.

- *Largest In-buffer Inventory:* Under this control policy, upon completion of processing a job, the worker works in the station with the largest in-buffer inventory. The policy is also called "longest-queue" control policy. This policy is often used in manufacturing processes in order to balance inventories of stations in a line to prevent large inventory buildup in any station.

- *Largest In-buffer Flow Time:* Under this control policy, upon completion of processing a job, among all jobs at stations A and B, the worker processes the job with the largest in-buffer flow time. This policy is also called "longest-waiting time" control policy. A common application of this policy is in call centers, where a call center agent is trained to answer multiple types of calls. The agent chooses a customer among those calls that have been kept on hold for the longest time.

In summary, flexible resources can improve process-flow measures because: (i) they can balance and increase utilizations of the resources and thus increase the effective capacity of the process, and (ii) they can diminish the negative impact of variability and hence result in more efficient use of resource capacity. Using a good control policy to allocate resource times to tasks can further improve performance of a process.

## 14.5 Computing Capacity for Processes with Flexible Resources

Recall from Chap. 4 that the effective capacity of a process is the maximum long-run average number of outputs that the process can produce per unit time, if the resources of the process are not idle due to unavailability of input units or unavailability of other resources in the process. In that chapter we showed how the effective capacity of a process can be computed. The analytics developed in that chapter did not cover processes with flexible resources. Computing the effective capacity of processes with flexible resources is more complex and is the focus of this section. We use the case of Leather Office Furniture to illustrate this.

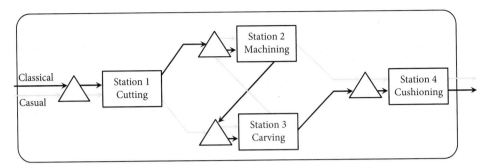

**Figure 14.4**    Process-flow chart of Leather Office Furniture for casual and classical office chairs.

**Leather Office Furniture (LOF)**

Leather Office Furniture (LOF) is a small company that produces a variety of office furniture, focusing on classical designs made of wood and leather. One of LOF's popular product—office chair—consists of several models that are grouped into two family of products: (i) casual and (ii) classical, which we call type-1 and type-2 products, respectively. Type-1 and type-2 products constitute 40 percent and 60 percent of the LOF's demand for chairs, respectively. Figure 14.4 depicts the process-flow chart of LOF's chair manufacturing shop.

Since chairs are made-to-order and are handmade products—the market that LOF is focusing on—the main resources of chair manufacturing process are workers using simple tools. Table 14.1 shows the effective mean process times of each activity for both types of products at LOF's shop. The shop works 8 hours per day.

**Table 14.1**    Effective Mean Process Times at LOF Chair Manufacturing Shop

| Workstation | Effective mean process time (*minutes*) | | Resource |
|---|---|---|---|
| | Casual | Classical | |
| 1. Cutting | 20 | 12 | Worker 1 |
| 2. Machining | 25 | 10 | Worker 2 |
| 3. Carving | 30 | 20 | Worker 3 |
| 4. Cushioning | 40 | 30 | Worker 4 |

The demand for chairs has recently increased, resulting in long waiting times for orders and a large inventory in LOF shop, especially in the bottleneck of the shop—the cushioning station. The demand is projected to increase by about 20 percent to around 17 chairs per day (with the same product mix). To deal with the new demand, managers of LOF are thinking of cross-training the workers in order to increase the effective capacity of the shop to at least 10 percent above the projected demand—to 18.7 per day—to safeguard against variability. One idea is to cross-train workers at cutting, machining, and carving to also work at the cushioning station. Before implementing this idea, they are wondering whether this will indeed increase the capacity of their chair manufacturing shop to 18.7 per day?

LOF operations is a multi-stage process producing multiple products. The product mix is $(p_1, p_2) = (0.4, 0.6)$. We have shown in Chap. 4 how product aggregation method can be used to find the effective capacity of such processes. When some of the resources of a process are flexible and perform different activities, product aggregation method cannot be used to find the effective capacity of the process; instead, linear programming (LP) models are used. Before we present the LP model for LOF with flexible workers, let's first develop an LP model to compute the capacity of LOF in its current process with no flexible workers.

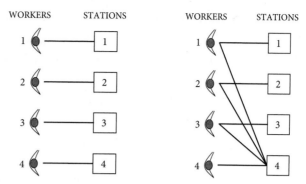

Figure 14.5    *Left:* LOF current worker assignment to stations (no worker flexibility). *Right:* Proposed worker assignment to stations (cross-training three workers).

### Linear Programming Model to Compute Effective Capacity of Processes with No Resource Flexibility

Figure 14.5-left shows the current resource flexibility structure. A link that connects a worker to a station implies that the worker is trained to perform the activity in that workstation. As the figure shows, currently workers are not flexible and each is only performing one task.

### Step 1—LP Input

The inputs of the LP model are: (i) the effective mean process time of each resource to process each product type, (ii) product mix, and (iii) working times of resources in a day. The resources and their effective mean process times to process type-1 (casual) and type-2 (classical) products are shown in Table 14.1. Product mix is $(p_1, p_2) = (0.4, 0.6)$, and resources (i.e., workers) work 8 hours, that is, 480 minutes, in a day.

### Step 2—Defining Decision Variables

Since we are interested in the effective capacity of the shop for producing each type of product (chairs), the decision variables would be

$$X_1 = \text{Number of products of } type \text{ 1 produced in a } day$$
$$X_2 = \text{Number of products of } type \text{ 2 produced in a } day$$

### Step 3—Objective Function

Considering $Z$ as the total number of chairs produced by LOF in a day, we will have $Z = X_1 + X_2$. Since the effective capacity of a process is the maximum number of products that the process can produce, our objective function becomes

$$C_{\text{eff}}(\text{LOF}) = Maximize \ Z = X_1 + X_2$$

### Step 4—Resource Time Constraints

Consider Worker 1 who works in Station 1 (cutting). The worker processes both type-1 and type-2 chairs. Since the worker's effective mean process times for type-1 and type-2 chairs are 20 and 12 minutes, respectively (see Table 14.1), the total time that the worker spends in a day processing $X_1$ type-1 chairs and $X_2$ type-2 chairs is $(20)X_1 + (12)X_2$ minutes. However, this time cannot exceed 480 minutes in a day; hence,

$$20X_1 + 12X_2 \leq 480 \qquad \text{(Worker 1 time constraint)}$$

Similar time constraints can be developed for Workers 2, 3, and 4 as follows:

$$25X_1 + 10X_2 \leq 480 \qquad \text{(Worker 2 time constraint)}$$
$$30X_1 + 20X_2 \leq 480 \qquad \text{(Worker 3 time constraint)}$$
$$40X_1 + 30X_2 \leq 480 \qquad \text{(Worker 4 time constraint)}$$

### Step 5—Product Mix Constraints

Product mix $(p_1, p_2) = (0.4, 0.6)$ implies that fraction 0.4 of the total number of products produced in a day (i.e., $X_1 + X_2$) must be of type 1. Thus,

$$X_1 = 0.4(X_1 + X_2) \qquad \Longrightarrow \qquad 0.6X_1 - 0.4X_2 = 0$$

### Step 6—Computing Effective Capacity

Putting the objective function and the above constraints together, we have the following LP model:

$$\textit{Maximize} \quad Z = X_1 + X_2$$
$$\textit{subject to}:$$
$$20X_1 + 12X_2 \leq 480 \qquad \text{(Worker 1 time constraint)}$$
$$25X_1 + 10X_2 \leq 480 \qquad \text{(Worker 2 time constraint)}$$
$$30X_1 + 20X_2 \leq 480 \qquad \text{(Worker 3 time constraint)}$$
$$40X_1 + 30X_2 \leq 480 \qquad \text{(Worker 4 time constraint)}$$
$$0.6X_1 - 0.4X_2 = 0 \qquad \text{(Product mix constraint)}$$
$$X_1, X_2 \geq 0$$

Solving this problem using Excel Solver, we find

$$X_1^* = 5.65, \quad X_2^* = 8.46, \quad Z^* = 14.12$$

Thus, under product mix (0.4,0.6), LOF has an effective capacity of producing 5.65 casual (type-1) chairs per day (i.e., 565 in 100 days), and 8.46 classical (type-2) chairs per day (i.e., 846 in 100 days)—a total capacity of 14.12 chairs a day (or 1412 chairs in 100 days).

### Linear Programming Model to Compute Effective Capacity of Processes with Flexible Resources

We now extend the above linear programming model to incorporate resource flexibility in computing the effective capacity of the chair manufacturing shop at LOF. As before, we call workers at cutting, machining, carving, and cushioning, workers 1, 2, 3, and 4, respectively, and we call their corresponding activities, activities 1, 2, 3, and 4. Hence, LOF's idea is to cross-train workers 1, 2, and 3 to also perform Activity 4. This is shown in Fig. 14.5-right.

#### Worker Time Allocation Decision Variables

In LOF's current situations where workers are not flexible and only perform one activity, each worker spends the entire time—480 minutes of a working day—performing the activity. However, when workers are flexible and perform several different activities, the key question becomes: what fraction of a working day the worker should spend performing each of those activities? For example, suppose Worker 1 who is currently doing the cutting activity is also trained to do cushioning. If this worker spends most of the time doing cutting and does not spend enough time doing cushioning (the bottleneck activity), then the capacity of the bottleneck may not be increased enough to satisfy the new demand. On the other hand, if the worker spends most of the time working in the bottleneck station, and not much time doing the task of cutting, then the cutting station may become the bottleneck and reduce the capacity of the process. Therefore, to find the capacity of the shop, we need to find the fraction of time each flexible resource should spend on each activity in order to process as many units as possible, that is, achieving maximum effective capacity.

Hence, we need to add the following new decision variable to our linear programming model:

$$Y_{ij} = \text{Fraction of time Resource } i \text{ performs Activity } j \text{ in a day}$$

where $0 \leq Y_{ij} \leq 1$. Thus, $Y_{14} = 0.2$, for example, implies that Worker 1 should spend 20 percent of the time in a day performing Activity 4—cushioning. For the LOF's plan of cross-training workers 1, 2, and 3 to also do Activity 4, we need to add the following variables to our LP model:

$$
\overbrace{Y_{11}\ ,\ Y_{14}}^{\text{Worker 1}}\ ,\ \overbrace{Y_{22}\ ,\ Y_{24}}^{\text{Worker 2}}\ ,\ \overbrace{Y_{33}\ ,\ Y_{34}}^{\text{Worker 3}}\ ,\ \overbrace{Y_{44}}^{\text{Worker 4}}
$$

### Worker Time Allocation Constraint

Since the fraction of time each worker allocates to performing different activities cannot exceed 100 percent, under LOF's proposed plan we must add the following constraints to our LP:

$$
\begin{aligned}
Y_{11} + Y_{14} &\leq 1 &&\text{(Worker 1 time allocation)}\\
Y_{22} + Y_{24} &\leq 1 &&\text{(Worker 2 time allocation)}\\
Y_{33} + Y_{34} &\leq 1 &&\text{(Worker 3 time allocation)}\\
Y_{44} &\leq 1 &&\text{(Worker 4 time allocation)}
\end{aligned}
$$

Consider Worker 1 as an example. Constraint $Y_{11} + Y_{14} \leq 1$ implies that the fraction of time during a day that the Worker 1 spends working on activities 1 and 4 should not exceed 100 percent.

### Activity Time Constraint

Finally, recall that the shop works for 480 minutes a day and consider Activity 4—cushioning. Also, recall that effective mean process time of activity cushioning for chairs of types 1 and 2 are 40 and 30 minutes, respectively. In the current situation, cushioning is performed by only one worker (Worker 4), allocating 100 percent of the worker's time (i.e., 480 minutes a day) performing that activity. This, as shown before, translates into the following constraint: $40X_1 + 30X_2 \leq 100\% \times 480$, which we called "Worker 4 Time Constraint."

Now, under LOF's process with flexible workers, in addition to Worker 4 that spends $Y_{44}$ fraction of its time at Work Station 4, Workers 1, 2, and 3 spend $Y_{14}$, $Y_{24}$, and $Y_{34}$ fractions of their times in a day, respectively, working at Station 4. Thus, the time constraint for Activity 4—cushioning—would be

$$40X_1 + 30X_2 \leq (Y_{14} + Y_{24} + Y_{34} + Y_{44}) \times 480$$

or

$$40X_1 + 30X_2 - 480(Y_{14} + Y_{24} + Y_{34} + Y_{44}) \leq 0 \quad \text{(Activity 4 time constraint)}$$

Note that the above constraint corresponds to the maximum time spent on Activity 4 by all workers. Hence, we call it "Activity 4 Time Constraint." Similar constraints for activities 1,

2, and 3 are as follows:

$$20X_1 + 12X_2 - 480Y_{11} \leq 0 \quad \text{(Activity 1 time constraint)}$$
$$25X_1 + 10X_2 - 480Y_{22} \leq 0 \quad \text{(Activity 2 time constraint)}$$
$$30X_1 + 20X_2 - 480Y_{33} \leq 0 \quad \text{(Activity 3 time constraint)}$$

Putting them all together, we have the following linear programming model that can determine the maximum capacity of the shop under LOF's proposed process with flexible workers:

$$\text{Maximize} \quad Z = X_1 + X_2$$

$$\begin{aligned}
\text{subject to:} \\
20X_1 + 12X_2 - 480Y_{11} & \leq 0 \quad \text{(Activity 1)} \\
25X_1 + 10X_2 - 480Y_{22} & \leq 0 \quad \text{(Activity 2)} \\
30X_1 + 20X_2 - 480Y_{33} & \leq 0 \quad \text{(Activity 3)} \\
40X_1 + 30X_2 - 480(Y_{14} + Y_{24} + Y_{34} + Y_{44}) & \leq 0 \quad \text{(Activity 4)} \\
Y_{11} + Y_{14} & \leq 1 \quad \text{(Worker 1 time allocation)} \\
Y_{22} + Y_{24} & \leq 1 \quad \text{(Worker 2 time allocation)} \\
Y_{33} + Y_{34} & \leq 1 \quad \text{(Worker 3 time allocation)} \\
Y_{44} & \leq 1 \quad \text{(Worker 4 time allocation)} \\
0.6X_1 - 0.4X_2 & = 0 \quad \text{(Product mix constraint)} \\
X_1, X_2, Y_{11}, Y_{14}, Y_{22}, Y_{24}, Y_{33}, Y_{34}, Y_{44} & \geq 0
\end{aligned}$$

Solving the above LP, we find the following optimal solution:

$$X_1^* = 8, \quad X_2^* = 12, \quad Z^* = 20$$

$$Y_{11}^* = 0.63, \quad Y_{14}^* = 0.37, \quad Y_{22}^* = 0.67, \quad Y_{24}^* = 0.05, \quad Y_{33}^* = 1, \quad Y_{34}^* = 0, \quad Y_{44}^* = 1$$

which indicates that by implementing the LOF's proposed worker flexibility plan, the effective capacity of the shop can be increased from 14.12 to 20 chairs per day. This capacity can be obtained when 63% and 37% of Worker 1's time during a day is spent, respectively, working at Station 2 and Station 4; 67% and 5% of Worker 2's time is spent, respectively, working at Station 2 and Station 4; 100% of Worker 3's time is spent working at Station 3; and 100% of Worker 4's time is spent working at Station 4.

Notice that, while LOF's plan is to train Worker 3 to also work at Station 4, the solution of the LP shows that there is no need to do that. In other words, the maximum capacity of 20 chairs per day is achieved when 100% of worker 3's time is spent working at station 3. The main reason is that, by allocating 37 percent of Worker 1's time to Station 4 (i.e., $Y_{14}^* = 0.37$), the capacity of Station 4—the bottleneck—increases and then Station 3 becomes the bottleneck.

The above example shows how a linear programming model can be developed to compute the effective capacity of complex processes with flexible resource. The following analytics formalizes the LP model.

---

**ANALYTICS**

### Finding Effective Capacity of a Process with Flexible Resources

Consider a process that produces $K$ different types of products by performing $J$ different activities. The process has $I$ resources and product mix is $(p_1, p_2, \ldots, p_K)$. The effective mean process time of performing Activity $j$ on Product $k$ is $T_{\text{eff}}^{(k)}(j)$. If Product $k$ does not require Activity $j$, then $T_{\text{eff}}^{(k)}(j) = 0$.

The process has flexible resources—some of the resources are trained to perform more than one activity. Specifically, suppose set $RES_j$ is the set of all resources that are trained to perform Activity $j$ and set $ACT_i$ is the set of all activities that Resource $i$ is trained for. Hence, for example, $RES_2 = \{1, 3\}$ indicates that Resources 1 and 3 are trained to do Activity 2, and $ACT_3 = \{3, 2, 4\}$ indicates that Resource 3 is trained to do Activities 3, 2, and 4.

To compute the effective capacity of the process during a period of length $T$ (e.g., an hour, a day, a week), we first need to define decision variables as follows: for all $i = 1, 2, \ldots, I$ and for all $j \in ACT_i$:

$$Y_{ij} = \text{Fraction of time Resource } i \text{ performs Activity } j \text{ during } T$$

If Resource $i$ is trained to perform Activity $j$ (i.e., if $j \in ACT_i$), then we need to define variable $Y_{ij}$. However, if Resource $i$ is not trained to perform Activity $j$ (i.e., if $j \notin ACT_i$), then we do not need to define the corresponding variable $Y_{ij}$.

As before, considering

$$X_k = \text{Number of products of type } k \text{ produced during } T$$

the following linear programming model finds the effective capacity of the process during a period of length $T$:

$$\text{Maximize } Z = \sum_{k=1}^{K} X_k$$

$$\text{subject to:}$$

$$\sum_{k=1}^{K} T_{\text{eff}}^{(k)}(j) X_k - T \sum_{i \in RES_j} Y_{ij} \leq 0 \qquad \forall j = 1, 2, \ldots, J \qquad \text{(Activity } j \text{ time constraint)}$$

$$(1 - p_k) X_k - p_k \sum_{j \neq k} X_j = 0 \qquad \forall k = 1, 2, \ldots K - 1 \qquad \text{(Product mix constraint)}$$

$$\sum_{j \in ACT_i} Y_{ij} \leq 1 \qquad \forall i = 1, 2, \ldots, I \qquad \text{(Resource } i \text{ time allocation constraint)}$$

$$X_k \geq 0, \ Y_{ij} \geq 0 \qquad \forall i, j, k$$

The optimal value of $Z^*$ is the effective capacity of the process during a time period of length $T$, and the optimal value of $Y_{ij}^*$ is the fraction of time Resource $i$ should spend in performing Activity $j$ during time Period $T$.

In the LOF example by making workers flexible the firm can increase its capacity from 14.12 to 20 chairs per day—a 42-percent increase. If resource flexibility can provide such an improvement, then why not make all resources fully flexible, so that all resources can perform all activities?

There are several reasons for that, some of which are as follows:

- *Costs:* Some activities are very specialized and thus it is very expensive (if not impossible) to make all resources capable of performing those activities. For example, it is not economical to send a nurse to medical school so she will be able to also perform doctors' activities; or it may be too costly to replace a simple lathe with a milling machine that can do drilling, cutting, and other machining operations. Another example is when performing different activities require different equipment. In those cases, making workers flexible would require buying additional equipment for flexible workers. This, again, increases the cost of resource flexibility.

- *Efficiency:* Specialization and division of labor is known to have several advantages including higher efficiency. For example, workers who are performing multiple tasks are not often as efficient in each task as those who perform only a single task.
- *Quality:* Another benefit of specialization is having higher quality. Thus, depending on the complexity of the activities, flexible workers may produce lower quality products than specialized (non-flexible) workers.
- *Setups:* Often times a setup is required when a flexible resource switches from performing one activity to another. Setups are non-value-added activities that reduce the effective capacity of a process. Hence, depending on the cost and the length of the setup operations, it might not be beneficial to make resources more flexible.

In summary, we might not be able to make all resources fully flexible because it may not be possible or it may be too expensive. The good news, however, is that there are some particular flexibility structures that result in low flexibility cost, but work almost as good as fully flexible structures. One of them is Chain Flexibility structure that we discuss next.

## 14.6 Chain Resource Flexibility Design

Let's assume for a moment that LOF does not have any constraints on worker flexibility, and hence, all four workers can be cross-trained to work in all four stations. What would be the capacity of the shop? For this case of full flexibility, the capacity of LOF's shop can be computed as 21.52 (see Problem 2 at the end of the chapter).

It is interesting that by training only two workers, each performing only one additional activity, LOF can increase its capacity significantly from 14.12 to 20 chairs per day, which is close to that under full flexibility design (i.e., 21.52). This points to the fact that a small amount of flexibility can go a long way. In fact, operations engineers and managers have identified resource flexibility designs with limited flexibility that can significantly improve effective capacity of a process and diminish the impact of variability. One of these designs is called "chain" resource flexibility design which we discuss in this section.

The chain resource flexibility design was first introduced by Jordan and Graves in 1995 through a simple simulation experiment as follows. Consider an auto manufacturer that has 10 assembly plants to satisfy demand for 10 different car models. Plants are not flexible and hence each plant produces only one car model. Figure 14.6 shows the resource flexibility structure of the firm with no flexibility. Each plant has a fixed capacity of producing 100 cars in each period with no variability and demand for each car model in each period is also 100 units with no variability. In this case, each plant produces 100 cars in each period and satisfies the demand of 100 in that period. Hence, the utilization of each plant is 100 percent and all demands are satisfied with no shortage. In other words, the total sales of the auto manufacturer in each period is $10 \times 100 = 1000$ cars. Now what happens if there is variability in demand, for example, if in each period the demand for each model is Normally distributed with mean 100 and standard deviation 40 independent of the demand for other models and in other periods?

Depending on the realized demand in each period, the plant capacity may not be fully utilized or the demand may not be fully satisfied. For example, if in one period the demand for Car Model 2 turns out to be 80, then Plant 2 will make 80 cars and satisfy all the demand in that period. Plant 2 is therefore 80% ($= 80/100$) utilized and has a sales of 80 cars in that period. Now consider a period in which the demand for Car Model 2 is 120 units. Due to its limited capacity of 100 units, Plant 2 produces 100 cars and the firm loses the remaining demand of 20 Car Model 2. In this case, the utilization of Plant 2 is 100 percent, while its sales is 100 with shortage of 20.

Plants                           Car Models

Figure 14.6    Ten plants producing 10 car models with no flexibility.

Hence, depending on the demand for each car model in each period, some plants will have different utilizations with different sales. But what are the average utilization of the 10 plants and the total average sales if plants work for several periods? To explore this, Jordan and Graves developed a computer simulation to compute the average utilization and total average sales in a period. They found the average utilization of 10 plants to be 86 percent and total average sales in a period to be 858 cars. Recall that with no variability, all plants were 100 percent utilized with total sales of 1000 cars per period. Thus, variability in demand reduces the utilization of plants from 100 percent to 86 percent and decreases the firm's sales from 1000 cars per period to 858 cars per period. Now the question is whether or how much making plants flexible improves utilization and sales?

Jordan and Graves did an experiment and made each plant flexible through a systematic approach as shown in Fig. 14.7. First, they made Plant 1 flexible to also produce Car Model 2. This corresponds to the plant flexibility structure with 11 links in Fig. 14.7. For this plant flexibility design, they computed the average utilization and average sales of plants in a period and plotted them as shown in Fig. 14.8. Then they made Plant 2 flexible to also make Car Model 3 (in addition to Car Model 2) and computed the average utilization and sales for this new flexibility design with 12 links. They continued in this manner and made Plant 9 flexible to make Car Model 10 resulting in a process flexibility with 19 links. At this step plants 1 to 9 produce two different car models, and each car models 2 to 10 are assembled in two different plants. The next natural step to add flexibility is to make Plant 10 flexible to make Car Model 1, resulting in a flexibility design with 20 links (Fig. 14.7) in which all plants make two cars and all cars are made in two different plants, in a particular way. They called this particular resource flexibility design a *Chain* flexibility design.

Fig. 14.8 shows how average utilization and average sales in a period improve as more flexibility is added to the process—process flexibility increased from 10 to 20. It also shows those values for the case of full plant flexibility design with 100 links in which all plants can produce all car models. Several interesting observations can be made from the figure:

- The chain structure with 20 links (i.e., 10 additional flexible links) performs close to the full-flexibility design with 100 links. This, again, points to the fact that a

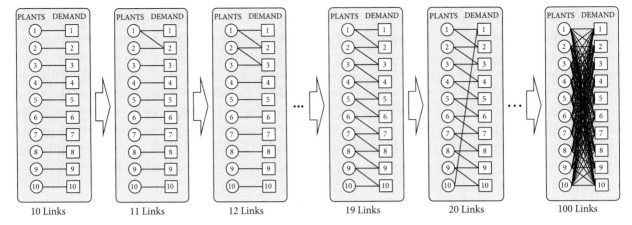

10 Links      11 Links      12 Links      19 Links      20 Links      100 Links

Figure 14.7   Adding flexibility to 10 plants.

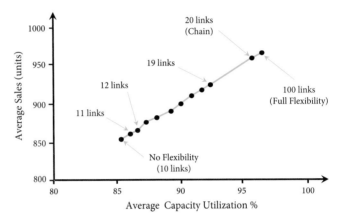

Figure 14.8   Average capacity utilization and average sales for structures in Fig. 14.7.

limited amount of flexibility (if properly designed) can capture most benefits of full-flexibility design.

- The last link that increases process flexibility level from 19 to 20 and creates a chain structure results in a significantly larger improvement in performance compared to those resulted from adding previous links. This points to the importance of how flexibility should be added to a process. Any link that increases flexibility level from 19 to 20 does not have as significant impact as the one that results in a chain structure.

**Why Chain Performs So Well?**
It is surprising how close the performance of chain flexibility design is to that of full flexibility. But, what is the reason? One reason is that chain structure allows the capacity of each of the 10 plants be used (directly or indirectly) to satisfy the demand for any of the 10 car models. To understand this, note that the main advantage of full-flexibility design is that all plants can produce all 10 car models. Chain flexibility design also has the same feature but in an indirect way. For example, consider a period in which the demand for all car models is 100, except for Car Model 1, which is 50, and Car Model 4, which is 150. So, Plant 1 will have 50 units of unused capacity, while Plant 4 has a shortage of 50 units. Under full flexibility, Plant 1 is flexible to produce both Car Models 1 and 4. Thus, it can use its remaining unused 50 units of capacity to produce 50 Car Model 4. But what about chain flexibility structure? In chain structure, Plant 1 is not flexible to produce Car Model 4.

While in chain structure the unused capacity of Plant 1 cannot be used *directly* to satisfy the 50 units shortage of Car Model 4, the particular structure of the chain allows the capacity of Plant 1 to be used *indirectly* to satisfy the 50 units shortage of Car Model 4. Under chain flexibility design, Plant 1 can use its 50 units unused capacity to produce Car Model 2. This releases 50 units of capacity of Plant 2 to be used to produce 50 units of Car Model 3. This in turn releases 50 units of capacity of Plant 3 that can then be used to make the 50-unit shortage of Car Model 4. Hence, similar to the full-flexibility design, chain flexibility design also allows the capacity of all resources (i.e., plants) to be used to perform all activities (i.e., making cars).

This raises the following question: If under both full flexibility and chain structure the capacities of all 10 plants can be utilized (directly or indirectly) to make all 10 car models, then why does full flexibility perform better than chain? The answer relates to the *amount* of capacity that each design can provide for each car model. Specifically, under full-flexibility design the total capacity of all 10 plants (1000 units per period) can satisfy all different demand scenarios as long as the total demand for all 10 cars in a period is 1000 or less. However, this is not the case for the chain flexibility structure. As an example, consider a period in which demand for all cars is 50, except the demand for Car Model 2, which is 200—a total demand of 650. In this case, all other plants can satisfy their own demand of 50 and have unused capacity of 50, while Plant 2 will make 100 cars and will have a shortage of 100 cars. Under full-flexibility design, the unused capacity of any two plants can be used to satisfy the shortage, since all plants are flexible to produce all car models. Under chain flexibility design, however, Car Model 2 can only be made by plants 1 and 2. Plant 2 has no remaining capacity to satisfy the shortage of Car Model 2, and Plant 1 has only 50 units of capacity to make Car Model 2. Thus, while the total demand of 650 units is less than the total capacity of 1000 per period, chain flexibility design cannot utilize the unused capacity of $1000 - 650 = 350$ units to satisfy the entire shortage of 100 units.

Similar demand scenarios can occur that make full-flexibility design work better than chain flexibility design. The main reason, as we discussed, is due to existences of situations when the demand for one car model is significantly higher than the total capacity of the two plants capable of producing that model. The large demand can occur because of high variability in demand, or because the (average) demand for one car model is significantly higher than the demand for other car models—asymmetric[3] demand. It is, however, worth mentioning that many studies have been performed to evaluate the performance of the chain structure in high variability and asymmetric demand cases in different settings (e.g., chaining workers in production lines, chaining repair crew in maintenance departments, chaining agents in call centers). The general finding is that chain flexibility design captures the most benefit offered by full resource flexibility, even in asymmetric systems. Also, chain flexibility design performs the best among flexibility designs with the same number of links; if it doesn't, its performance is very close to the best among those designs.

In summary, recall that using our linear programming models for LOF we found that by training only one or two workers, LOF can achieve a capacity close to full resource flexibility. In this section we also observed that the chain flexibility design in which each plant produces only two types of cars had a performance (utilization and sales) close to that of full-flexibility design. Both cases of LOF and the auto manufacturer example point to the following important flexibility design principle:

---

[3]In Jordan and Graves' experiment the demand for all car models has the same average of 100—a symmetric demand. It can be that the average demand for a car model is two or three times larger than the average demand of other models—asymmetric demand.

---

**Resource Flexibility Design**

- A limited amount of flexibility—if properly designed—can achieve the process performance close to that under a full-flexibility design.
- Chain resource flexibility design is a robust design that performs well in a wide range of environments.

---

In conclusion, it is worth mentioning that in our auto manufacturer example flexibility does not improve the effective capacity, since the firm has a total capacity of 1000 cars in systems with and without flexibility. In this example, the main benefit of flexibility comes from its ability to reduce the impact of variability.

---

**Want to Learn More?**

**Resource Flexibility Design**

There are many situations in which some resources cannot be made flexible, or the flexibility cost is so low that it is economically viable to make some resources flexible to perform more than two activities. In such cases, chain flexibility structure is not suitable. Operations engineers and managers then face the following questions: To achieve a certain level of process-flow measures, (i) what is the minimum number of flexible resources needed? (ii) which resource should be made flexible to perform which activity? and (iii) how should the time of a flexible resource be allocated to perform each activity?

In the online supplement of this chapter we develop some analytics that can be used to answer these flexibility design questions, focusing on improving process capacity. The reason our focus is on capacity improvement is that, improving capacity has a significant impact on improving *all* process-flow measures.

## 14.7 Bucket Brigade Lines

Making resources flexible often involves costs. However, there are situations in which the cost is not that significant. Such cases are processes in which activities are simple labor-intensive activities that require simple or no tools. For example, consider order picking operations in a warehouse in which shelves are organized in a line. Workers—order pickers—start with an empty box and move toward the line and take items from a shelf and put it in the box (according to the order content list on the box) and move to the next shelf. Considering picking an order from each shelf as one activity, it is clear that training workers to pick items from different shelves (i.e., perform different activities) does not have a significant cost. Order picking is a simple labor-intensive task that does not require any tools.

Another example of such environment is fast-food restaurants such as Subway that make sandwiches to order. The task of making a sandwich is to take a bread, move along the counter, and add whatever the customer asks to the sandwich until the sandwich is completed. Again, training a worker to add different items to a sandwich or toast the sandwich is not difficult. In this system full flexibility can be achieved at a minimum cost. However, the big problem is the control policy: how should each worker allocate their time to perform a large number of activities?

One control policy that has been widely used and has been shown to have significantly improved process performance is "Bucket Brigade" control policy. While bucket brigade

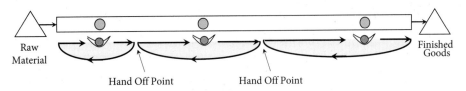

**Figure 14.9** **A bucket brigade line with three workers.**

policy can also be extended to other processes with more complex structure, it has been mainly used in processes with only one routing in which all flow units go through the same sequence of activities, for example, flow lines. The lines that follow the bucket brigade control policy are called "Bucket Brigade Lines."

### Bucket Brigade Control Policy

How does the bucket brigade policy work? Before we can answer this question, we need to describe the processes that are suitable for using bucket brigades control policy. First, the process must consist of several workstations (i.e., activities) in series and there must be less number of workers than workstations. Second, all workers must be fully trained to work in all workstations (if needed)—full resource flexibility. Third, activities should be such that they can be preempted and be done by another worker without need for setup or any other interruption. Fourth, the variability in activity times must be low and, fifth, the worker walking time between stations should be insignificant. Finally, there should always be enough flow units (i.e., raw material) in the beginning of the line, so the line is never starved due to lack of flow units.

---

**CONCEPT**

**Bucket Brigade Control Policy**

Consider a serial line with $N$ workstations and $W$ workers, where $W < N$. Workers are sequenced from Worker 1 in the beginning of the line to Worker W at the end of the line. The bucket brigade control policy consists of two rules:

- *Forward Rule:* Each Worker $i$ should process a flow unit at successive workstations and go to the next station until the worker's successor (i.e., Worker $i + 1$) takes over the unit. In that case, Worker $i$ should follow the backward rule. The last worker—Worker W—should also follow the backward rule when the worker completes the processing of the flow unit and puts it in the finished goods inventory.

- *Backward Rule:* Each Worker $i$ should walk back and take over the flow unit of the predecessor (i.e., Worker $i - 1$) and follow the forward rule. The first worker—Worker 1—must return to the first station and pick up a new flow unit and follow the forward rule.

---

The essential feature of the bucket brigade rule is that none of the workers pass their successors in forward rule and their predecessors in backward rule (see Fig. 14.9 for an example with three workers). In other words, Worker $i$, while processing units from one Station to the next, always works in a zone between Workers $i - 1$ and $i + 1$. Hence, Worker 1 is the worker that always picks the item (e.g., raw material) from the buffer of Station 1 and Worker $W$ is always the worker that completes the product and puts it in the finished goods inventory.

## Advantages of Bucket Brigade Control Policy

The bucket brigade control policy has several advantages:

- *Flexible Throughput:* By increasing (or decreasing) the number of workers in the line the throughput of the line can be increased (or decreased) without any need for changing the operations.
- *Minimum Inventory:* Lines following bucket brigade policy have the minimum inventory, which is equal to the number of workers in the line. Any inventory less than the number of workers will result in the idling of at least one worker.
- *Maximum Throughput:* If workers are properly sequenced, the bucket brigade policy can result in the maximum throughput of the line.
- *Self-Balancing:* The line achieves a balanced worker utilization without any need for redesigning activities in the workstations. When the workers are properly sequenced, the utilizations of all workers become 100 percent, and workers work at their maximum speed.

## Design Principles for Bucket Brigade Lines

The key to achieve the above advantages, as mentioned, is to sequence the workers in a particular way. Consider a six-station line as an example. Suppose that we have three workers, which we call Mr. Slow, Mr. Medium, and Mr. Fast. All three workers are fully cross-trained to work in all six stations in the line. However, they differ in their speed of how fast they process jobs in the line. If Mr. Slow is the only worker who works in the line, it takes him 15 minutes to complete a job in all six stations of the line. Thus, the throughput of the line would be $60/15 = 4$ jobs an hour. In other words, Mr. Slow has the speed of producing four jobs in an hour. Mr. Medium, on the other hand, can finish a job in the line in 12 minutes if he works alone in the line, resulting in a throughput (i.e., speed) of $60/12 = 5$ jobs an hour. Obviously, Mr. Medium seem to be more experienced and faster than Mr. Slow in doing the tasks. Mr. Fast has even a faster speed than that of Mr. Medium. If he works alone in the line, he can finish a job throughout the line in only 10 minutes, so he has a throughput (i.e., speed) of $60/10 = 6$ jobs an hour.

Now suppose we establish two additional lines—a total of three identical six-station lines—and put one of the workers in each line. Then, we get a total throughput of $4 + 5 + 6 = 15$ jobs an hour, which is the maximum throughput possible using our three workers. Furthermore, all three workers will be 100 percent utilized working at their maximum speed. However, since it is not economical to build two additional lines, we would like to use our three workers working in our existing line following bucket brigade policy. The first question of interest is whether by using one line (instead of three lines), we can still achieve the maximum throughput of 15 jobs per hour and maximum utilization of 100 percent for all three workers? The second question is how should we sequence the workers in the line? Should we put the fastest worker—Mr. Fast—as the first worker, the second worker, or the third worker in the line? What about Mr. Slow and Mr. Medium?

Well, there are six possible sequences of workers: (S, M, F), (S, F, M), (M, S, F), (M, F, S), (F, S, M), and (F, M, S)—we use S, M, F to represent slow, medium, and fast, respectively. Which sequence results in a higher throughput of the line? The answer lies in the fact that the maximum throughput of the line was achieved (in three independent lines) when all three workers were never idle and they were able to produce their maximum throughput (i.e., working at their maximum speed) without being slowed down. Thus, when we have only one line, a worker sequence that can make all three workers 100 percent utilized, that is, workers are never idle, and does not reduce workers' speeds will be able to achieve the maximum throughput.

Let's consider sequence (M, F, S), as an example, where Mr. Medium works in the beginning of the line, Mr. Slow at the end of the line, and Mr. Fast in the middle (between Mr. Medium and Mr. Slow). Under this sequence, Mr. Slow will always be 100 percent utilized, and can work at his speed of four jobs an hour. According to bucket brigade control policy, when he finishes a job at the last station he walks back and takes over the job of Mr. Fast. But, what about Mr. Fast? After his job being taken over by Mr. Slow, Mr. Fast walks back and takes over the job of Mr. Medium. Assuming short walking times, Mr. Fast and Mr. Slow (and also Mr. Medium) will start processing their new jobs at almost the same time, but in different stations in the line. Here is where the problem starts. Since Mr. Fast is faster than Mr. Slow, he processes jobs faster in subsequent stations and it is possible that he bumps to Mr. Slow before Mr. Slow finishes his job (e.g., Mr. Slow might not be done in stations 5 or 6). What should Mr. Fast do? Bucket brigade forward rule does not allow Mr. Fast to bypass Mr. Slow and continue working on stations 5 and 6 to finish his job. The best Mr. Fast can do is to wait until Mr. Slow is done in each station, or to work at the same speed as Mr. Slow so he does not pass him. This means that, while Mr. Fast is still 100 percent utilized, his speed is reduced to the speed of Mr. Slow, and thus his throughput (and the throughput of the line) does not reach the maximum of 15 jobs an hour.

As this shows, any sequence that results in a faster worker bumping to a slower successor worker before the successor starts the backward rule will reduce faster worker's speed to the speed of the successor and thus prevent the faster worker to work at the maximum speed. This, in turn, prevents the line from achieving the maximum throughput of 15 jobs an hour. The only sequence in which no worker is bumped to its successor is the one that sequences workers from slowest to fastest, that is, sequence (S, M, F). Under this sequence, all three workers are 100 percent utilized and their speeds are not reduced. In other words, under this sequence, the throughput of one line with three fully flexible workers who work under bucket brigade control policy is the same as the summation of the throughputs of the three independent lines each run by one fully flexible worker.

**Bucket Brigade Design Principle**

Consider a process in which $N$ activities are performed in $N$ workstations by $W$ workers, where $W < N$, that is, there are less workers than activities. There are always enough flow units in the beginning of the line (buffer of Workstation 1) waiting for processing. All flow units follow the same routing, and all workers are fully flexible and can perform all activities, if needed. There is no variability in process times, and setup times (e.g., walking time between workstations) are also negligible.

Consider $v_i$, the speed of Worker $i$, to be the number of flow units that the worker can finish in the line, if the worker is the only worker in the line. If bucket brigade control policy is used to process flow units, then the maximum throughput of the line is achieved if the workers are sequenced from slowest (in the beginning of the line) to fastest (at the end of the line). This arrangement of the workers results in the maximum throughput of $v_1 + v_2 + \cdots + v_w$.

In conclusion, we should note that while the above principle is for lines with no variability in process times, theoretical studies and practical implementation of the bucket brigade principle have shown that bucket brigade (or some modified version of it) significantly improves the throughput of the lines, even in systems with variability in process times and setup times. For example, after implementing the bucket brigade policy in their distribution center, Gap reported a 25-percent improvement in throughput. Ford customer service division reported a 50-percent improvement in throughput after implementing the bucket brigade policy.

## 14.8 Postponement

Mass customization processes produce high volume of customized products, and thus face large variability in demand and in process times due to the wide range of products they offer. Resource flexibility (e.g., flexible worker, flexible equipment, and flexible plants), as we discussed, can improve process capacity and reduce the negative impact of variability. Hence, using flexible resources is one effective strategy in achieving mass customization. Another strategy that is often used in mass customization is Postponement.

---

**CONCEPT**

**Postponement and Point of Differentiation**

*Postponement* or *Delayed Differentiation* is a production strategy in which same generic products (or subassemblies) are made at the earlier stages of the process. These products are then modified (or assembled) into a wide range of different final products at the later stages after the demands for products are realized or when more accurate forecasts for products become available. The point in the process when generic products are made to different final products is called *Point of Differentiation*.

---

Postponement has become a common practice in mass customization. Many manufacturing and service firms have implemented this strategy in their production and supply chains. For example, Dell uses standard and per-configured components (e.g., processors, memories, wireless boards, etc.) acquired at the earlier stages of its process and assembles its computers at the last stage of manufacturing after it receives orders from customers. McDonald restaurants have the food items (e.g., hamburger patties, toppings, etc.) ready and postpone the final stage of making sandwiches after they receive orders from customers. Many firms (e.g., Black & Decker and Bic) have postponed the packaging of the same products until they receive orders from retailers (each requesting a different packaging).

To illustrate how postponement can improve process-flow measures, we use a simple example of a process that produces two different products X and Y, each requiring two different stages of production.

**Process A (without Postponement)**

Process A, as shown in Fig. 14.10, produces products X and Y in two different and independent lines. The demand for each product follows Poisson distribution with a rate of 145

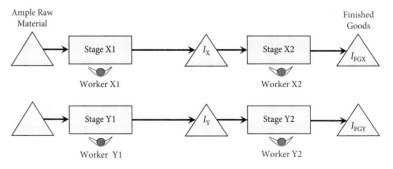

Figure 14.10    Process-flow chart of Process A—point of differentiation at the beginning of the process.

per month. The process works 24 hours a day and 30 days a month. The effective mean process time at each stage (i.e., Stages X1, X2, Y1, and Y2) is 4.8 hours; thus, all stages have an effective capacity of $(1/4.8) \times 24 \times 30 = 150$ per month. The variability of the effective process time at each stage is moderate with coefficient of variation 1. The activities at each stage are performed by one worker. To guarantee availability, products are made-to-stock at both stages of the production. The base-stock levels at both intermediate buffer and finished goods inventory (FGI) are set to 116. This base-stock level results in service level of $\alpha = 98\%$, that is, there is only 2 percent or less chance that FGI is empty and thus a customer faces shortage for a product, or intermediate buffer is empty and thus Stage 2 becomes starved. The orders are shipped from FGI to the customers when an order is received. In case of a shortage, the shortage is backlogged and the product is sent to the customers as soon as it is produced.

Using a computer simulation, we can compute the average inventory at the intermediate buffer and in FGI for Product X in Fig. 14.10 to be $I_{X1} = 29$ and $I_{FGX} = 29$, respectively—a total of $I_{X1} + I_{FGX} = 29 + 29 = 58$ units. The total average inventory in the production line for Product Y is also $I_{Y1} + I_{FGY} = 29 + 29 = 58$ units. Thus, the total average inventory of both products in Process A is $58 + 58 = 116$ products. The throughput of both processes is 145 products X and 145 products Y per hour—the same as demand, since all demand are eventually satisfied.

The goal of postponement is to push the point of differentiation to later stages in the process. In Process A in Fig. 14.10, the point of differentiation is at the beginning of the production process of products X and Y. Now consider Process B in Fig. 14.11.

### Process B (with Postponement)

Process B is the same as Process A, except that the point of differentiation in Process B is delayed to after the first stage. Specifically, Stage 1 in Process B produces a generic product (or generic component) that we call "Product Z" and keeps it in the intermediate buffer. The effective mean process time at Stage 1 for Product Z is still 4.8 hours with coefficient of variation 1, but there are two workers (i.e., workers in Stages X1 and Y1 in Fig. 14.10) who are now processing Product Z. The product is then modified at Stage 2 into products X and Y.

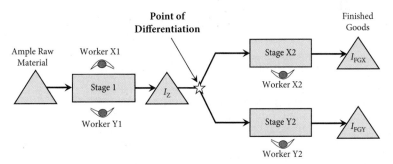

**Figure 14.11    Process-flow chart of Process B—point of differentiation after Stage 1.**

Using a computer simulation we find the base-stock level 116 for generic Product Z will achieve the same service level $\alpha = 98\%$. Also, we find a total average inventory of $I_Z = 29.5$ for Product Z, and average inventory 29 in FGI for both products X and Y, that is, $I_{FGX} = I_{FGY} = 29$. Thus, the total average inventory of Process B in Fig. 14.11 would be $29.5 + 29 + 29 = 87.5$ products. The throughput of the process is still 145 products X and Y per hour.

As these numbers show, delaying the point of differentiation reduces the average inventory in the process from 116 to 87.5—around 25-percent reduction. Since both processes A and B are using the same number of resources (i.e., workers), what can be the underlying reason for this improvement? Two reasons: capacity pooling and inventory pooling.

Note that, delaying the point of differentiation (postponing customization) beyond Stage 1 results in pooling inventories of $I_X$ and $I_Y$ into one—$I_Z$. As we described in Sec. 11.7 of Chap. 11, pooling inventories in a centralized location results in improving process-flow measures of a process. The main reason, as discussed in that section, is demand aggregation—the demand for Product Z is the aggregated demand for products X and Y, and thus has a lower variability. Also, delayed differentiation resulted in (vertical) pooling of the capacities of the two workers in stages X1 and Y1 of Process A (Fig. 14.10) into one resource pool of two workers at Stage 1 of Process B (Fig. 14.11). As we discussed in Sec. 8.5 of Chap. 8, vertical pooling of capacity also improves process-flow measures of a process.

### Postponement: A Hybrid Make-To-Stock/Make-To-Order System

As shown above, postponement in Process B reduced the inventory *before* the point of differentiation and resulted in 25-percent reduction in total average inventory of the process. But, can we also reduce inventory *after* the point of differentiation in Process B? One strategy is to restructure the process as a hybrid make-to-stock/make-to-order (MTS/MTO) process. Note that both stages in Process B are MTS systems with base-stock levels 116 at stages 1 and 2. What happens if we decide to change the MTS system at Stage 2 to a MTO system?

**Process C (Hybrid MTS/MTO process with Postponement)**
Similar to Process B, Process C produces generic Product Z and keeps them in stock with base-stock level 116; however, it does not keep any FGI of final products X and Y in stock. Process C starts turning a generic Product Z to a Product X or Y, only after it receives an order for those products. This makes the process a hybrid MTS/MTO process, as shown in Fig. 14.12.

Which Process is better? Process B or Process C?

- In Process C, the finished goods inventories of products X and Y are eliminated. This means that Process C—the hybrid MTS/MTO system—will have, on an average, $I_{FGX} + I_{FGY} = 29 + 29 = 58$ less inventory than Process B at the second stage.

- In Process B only 2 percent of orders may need to wait to get their products. In Process C, however, because there is no finished goods inventory, all orders must wait to get their products. Using a simulation model we can compute the average time an order waits in Process C before it is completed to be about 3 days.

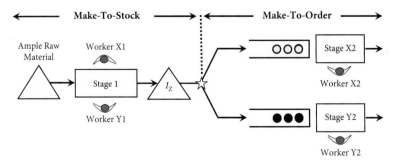

**Figure 14.12** Process-flow chart of Process C—with make-to-order after point of differentiation.

Hence, to choose between processes B and C, one should consider the trade-off between inventory and customer service. While Process C has significantly lower inventory than Process B, it offers a lower customer service—an average delay of 3 days to ship customers' orders.

The average delay of 3 days is the time that the generic Product Z of an order is waiting (after the point of differentiation) to be made into a Product X or Y. Thus, one factor that makes Process C—the hybrid MTS/MTO—a better option is when the time required to process an order after the point of differentiation is short. For example, if the effective mean process time at Stage 2 is reduced from 4.8 hours to 4 hours, then the average delay in Process C is reduced from 3 days to about 0.5 days. Therefore, one key feature that makes postponement even more effective is that the activities after the point of differentiation are short, that is, product variety (customization) can be created quickly. Consider fast-food restaurants such as McDonald or Subway. They have all the materials needed to make a sandwich (e.g., bread, cheese, vegetables) already made—a make-to-stock process. The point of differentiation is the process of making a sandwich, and the time to complete an order after this point (of differentiation) is very quick, that is, less than a minute.

### Postponement and Demand Forecasting

Demand forecasting plays a key role in managing resources, inventories, and production in manufacturing and service operations. One advantage of postponement strategy is that it results in more accurate forecasting. The reason is simple. Generic products such as Product Z in Process B (Fig. 14.11) face the aggregate demand of all products for which Product Z will be used (i.e., products X and Y). Since the aggregate demand has lower variability than each individual demand, demand forecasting is easier and more accurate.

---

**PRINCIPLE**

**Benefits of Postponement**

- Postponement improves process-flow measures (e.g., inventory, flow times) by aggregating demand at the point of differentiation. Therefore, postponement benefits from inventory and capacity pooling.
- The benefits of postponement increase as point of differentiation is pushed closer to the final stages of the process.
- If after point of differentiation products are assembled in make-to-order fashion, then the processing times after the point of differentiation should be short to reduce customer waiting time.

### Point of Differentiation in Supply Chains

Point of differentiation can be at any stage of a supply chain. As stated in the above principle, having it at the later stages provides more benefit.

- *At the Manufacturer:* Point of differentiations are often at the manufacturers, where standard (generic) products are customized into a wide range of final products within stages of production. For example, Swatch, a watch manufacturer, makes an internal mechanism that is used in almost all models. They postpone adding different straps, colors, and add-on details later in the production process.
- *At the Distribution Center:* Many firms delay the point of differentiation to their distribution center. For example, Black and Decker, Bic, and Hewlett-Packard

postpone the packaging of their products to their distribution centers, where products are differentiated and packaged for different retailers in the United States or other countries.

- *At the Retailer:* The point of differentiation can sometimes be placed at the last stage of a supply chain—the retailer. For example, hardware stores or the retailers of paint companies (e.g., Sherwin-Williams) make the paint color at their locations after a customer chooses a paint. Restaurants—the retail stores of fast-food supply chain—make the food after they receive the order from a customer.

- *At the Customer Location:* Sometimes the point of differentiation can be delayed to the customer location. For example, General Electric redesigned their circuit breaker and reduced the number of parts from 600,000 to only 300 parts. The new design allowed for assembly of the products at the customer location before it is installed.

## 14.9 Enablers of Postponement

The further the point of differentiation is pushed downstream of a supply chain, the more the benefit is provided through postponement. This is because the later the point of differentiation, the more the inventories and/or capacities are pooled. This results in lower inventory and flow times, more accurate forecasting, and thus higher customer service.

With all these benefits, postponement strategy is definitely one efficient strategy that provides significant improvement opportunities for all firms. The question, however, is whether it is always possible to implement postponement within a process or throughout a supply chain. In this section we point to three key enablers of postponement: (i) modular product, (ii) modular process, and (iii) agile supply chain network.

### 14.9.1 Modular Product

One way to enable processes to delay the point of differentiation at later stages is modular product design.

---

**CONCEPT**

**Modular Product**

A modular product consists of independent modules (e.g., parts, components, subassemblies) that can be easily and inexpensively assembled to create different versions of the product. Each version consists of different combination of modules.

---

Modular product is also referred to as *Component Commonality*, which was discussed in Sec. 11.9.4 of Chap. 11.

Modular product design allows firms to offer a wide range of products (i.e., high product variety) using small number of components (i.e., low component variety). Personal computers and laptops are good examples of modular product design. They are made of a few standard pretested components (modules) that can be assembled into a wide variety of PCs and laptops with different features. For example, for their Inspiron model laptop, Dell Computers use three screen sizes, four operating systems, seven processors, five memories, four hard drives, three laptop weights, and two choices for optical drives. This adds up to a total of 28 different components (i.e., modules) that can be used to make an Inspiron laptop computer. How many different types of Inspiron laptops can be assembled using these 28 modules? The answer is

$$3 \times 4 \times 7 \times 5 \times 4 \times 3 \times 2 = 10{,}080 \text{ Inspiron laptops}$$

Dell can offer this wide range of products with its assemble-to-order production process—a hybrid MTS/MTO system—after it receives an order from a customer.

Some restaurants also use modular components to offer a wide range of products. For example, Subway uses 18 different main ingredients in sandwiches (e.g., turkey, ham, chicken, etc.), with 5 types of bread, 3 types of cheese, 10 different veggies, 10 different sauces, and 4 different seasonings, a total of 50 ingredients (i.e., modules). This allows Subway to make

$$18 \times 5 \times 3 \times 10 \times 10 \times 4 = 108{,}000 \text{ sandwiches}$$

Considering the option of two sizes (6 inch or footlong) and the option of toasting the sandwich or not, this number increases to $108{,}000 \times 2 \times 2 = 432{,}000$ different sandwiches.

The case of Hewlett Packard (HP) LaserJet printer is another example of using modular product design to enable postponement. Before implementing postponement, the printer that was made for the markets in Europe and North America had a dedicated power supply of 110 V (for use in North America) or a 220 V (for use in Europe). This placed the point of differentiation early in the process when the printer was made. By designing a universal power supply—a module that works for both 110 V and 220 V—the point of differentiation was moved to the distribution center before the printers are loaded to be shipped to different continents.

Besides providing the opportunity for postponement, modular products have other advantages too. In a modular product, production of the modules can be done independently and in parallel. The reason is that the production of one module (e.g., a particular memory chip) can be done before, or after, or at the same time as the production of another module (e.g., a hard drive). Parallel production of the modules results in a significant reduction in time required to produce the product. Also, because modules are independent components designed to perform a specific function, it is easier to identify problems during production process and is easier to maintain or repair the product.

### 14.9.2  Modular Process

While a modular product consists of independent generic components, a modular process is a process that consists of independent stages.

---

**CONCEPT**

**Modular Process**

A modular process is a process that consists of independent subprocesses (i.e., modules) that can be moved or rearranged easily without affecting the production of the final product.

---

The flexibility that a modular process provides for manufacturing processes is analogous to what a modular product provides for a product. Being able to decouple the entire manufacturing process into independent modules—subprocesses—provides ability to resequence the process such that the modules that differentiate the product are postponed to later stages of the process.

Benetton's process resequencing is a successful example of postponement through process modularity. Before resequencing their sweater manufacturing process, Benetton was first dyeing the yarn into different colors and then knitting those yarns into finished sweaters. The point of differentiation was the dyeing process that created a variety of different colors. Because dyeing and knitting were independent modules of the production process, Benetton was able to change their orders and move (postpone) the dyeing operations after knitting, where sweaters were made. Under new sequence, the uncolored sweaters were dyed when the firm received orders or when it had more confidence about high demanding colors in the season. This saved Benetton millions of dollars in inventory.

One may consider the case of Benetton as an example of product modularity that enabled postponement. Uncolored sweaters can be considered as a module of the final product (colored sweater) if at the point of differentiation these sweaters are colored into different products. However, one should consider that the postponement could have not been done if it was not possible to move the dyeing subprocess after knitting. In other words, without process modularity, regardless of whether uncolored sweaters are considered to be a module or not, postponement was not possible in the Benetton production process.

Another example of postponement through process modularity is in paint industry. In retail stores (e.g., hardware stores) customers can choose among hundreds and thousands of colors offered in catalogs of different paint manufacturers. Before implementing postponement, paint manufacturers manufactured each color of paint and packaged them in their corresponding paint cans and shipped them to the retailers. This practice resulted in a large inventory (due to the large variety of offered colors). Also, due to the variability in demand for different colors, some colors were not sold depending on the color trend in each season.

To implement postpone strategy, paint manufacturers decoupled the process of paint production into two modules: the production of a small number of basic paint colors and the process of mixing of the pigment with those paints. By decoupling this process, they were able to postpone the mixing process outside the manufacturing plants and to the retailer. Thus, the retail stores only stock basic colors and color pigments. When a customer requests a particular color, the store identifies the right mixture of paints and pigment (determined using chromatograph) and mix those to make the exact paint color that the customer wants.

### 14.9.3 Agile Supply Chain Network

While product and process modularity provide opportunities for postponement, without an agile supply chain, the full benefits of postponement will not be captured. First, it is important that supply chain physical network be such that it moves the generic products to the point of differentiation in a cost-effective way. This includes location, number, and structure of the manufacturing and distribution centers.

Second, the supply chain network should also be able to respond rapidly to customers' orders for customized products and be able to deliver those products as quickly as possible. In addition to well-designed physical structure, this requires an efficient information technology that provides fast information sharing capability throughout the supply chain from order entry to delivering the product.

## 14.10 Summary

Mass production is a cost-efficient way to produce products in large quantities, but it is suited for producing standardized products and does not allow for product variety. Mass customization, on the other hand, is the high-volume production of customized products with high product variety. Mass customization requires a high degree of operational flexibility, which refers to the ability of an operations to adapt quickly and rapidly to changes in its environment.

While operational flexibility can be achieved through different means, this chapter illustrates how operational flexibility can be acquired by using flexible resources or postponement. Flexible resources are resources that can perform more than one activity on flow units, and postponement refers to making products in standard form in the earlier stages of a process and creating variety in later stages by adding unique components.

Flexible resources, if properly used, can increase the effective capacity of the process. To achieve process improvement, flexible resources must be used such that their utilizations are balanced. Flexible resources also improve process performance by reducing or eliminating the negative impact of variability. The effective utilization of flexible resources depends on the control policy by which each resource time is allocated to different tasks. Without a good control policy, the benefits of flexible resources are diminished or eliminated.

Finding the effective capacity of processes with flexible resources is not an easy task. This chapter shows how mathematical programming models can be developed to find the capacity of an operations with flexible resources. These models can also be used to design flexibility structures with minimum flexibility needed to achieve a certain capacity (see online supplement).

One important principle of using flexible resource is that a limited amount of flexibility—if properly designed—can achieve the process performance close to that under a full-flexibility design. One flexibility structure with a limited flexibility that performs as effective as full-flexibility structure is chain flexibility structure. In chain flexibility design, each resource performs only one additional task beyond its main task.

Postponement or delayed differentiation is a production strategy in which same generic products are made at the earlier stages of the process and are modified into a wide range of different final products at the later stages after the demands for products are realized. Postponement saves inventory and production costs since the demand for a generic product is the aggregated demand of all those final products that use the generic product. Due to larger demand for generic product and lower variability of the aggregated demand, inventory costs of those generic products are significantly reduced. Factors that enable processes to utilize postponement are: modular product, modular process, and agile supply chain. A modular product is a product that consists of independent modules (e.g., parts, components, sub-assemblies) that can be easily and inexpensively assembled to create different versions of the product. Modular process is a process that consists of independent subprocesses (i.e., modules) that can be moved or rearranged easily without affecting the production of the final product. Agile supply chain is a supply chain that is able to move generic products to the point of differentiation in a cost-effective way. This includes location, number, and structure of the manufacturer and distribution centers. Agile supply chains also have the ability to respond rapidly to changes in the market, which requires them to have efficient information technology throughout the supply chain.

## Discussion Questions

1. Describe the differences between mass production and mass customization.
2. Explain two different ways that flexible resources can improve process-flow measures of a process.
3. What is a chain flexibility structure and why does it perform so well?
4. What is a bucket brigade line and what types of operations are suited for implementing bucket brigade control policy?

5. Explain what "postponement" strategy and "point of differentiation" are in a production process.

6. What are the benefits of postponement and what are the reasons that postponement can provide those benefits?

7. Provide examples of point of differentiations at the manufacturer, at the distribution center, at the retailer, and at the customer site of a supply chain.

8. What are the three main enablers of postponement?

9. What are modular product and modular process? Provide an example for each?

## Problems

1. Mark, the inventory manager in a large warehouse, is planning to use bucket brigade line for its order picking process. Current order picking process consists of three identical lines with eight stations. Each line is assigned to a worker who works in all eight stations picking items of an order through stations 1 to 8, before starting the next order. While the orders are of almost the same size, workers have different speed. Worker 1 processes an order from Station 1 to 8 in about 5 minutes. This number is 6 and 4 minutes for workers 2 and 3, respectively. Because of the need for more storage space, Mark wants to replace two of the three lines with storage space and ask three workers to work on one line.
   a. Can the three workers work on a single line and still finish the same total number of orders as they did with three separate lines?
   b. What is the maximum throughput the single line can achieve?

2. What is the maximum capacity of Leather Office Furniture (LOF) in Sec. 14.5 if all four workers can be cross-trained to work in all four stations?

3. Office Support Equipment (OSE) is a firm that leases copiers to four major institutions in Chicago, namely, private firms, government offices, educational institutions, and small copy centers. OSE has a repair and maintenance crew with four technicians: Bill, Chad, Dan, and Paul, who provide repair and maintenance services to customers. Currently, Bill is assigned to respond to maintenance calls from private firms, Chad is assigned to government offices, Dan is assigned to educational institutions, and Paul is assigned to small copy centers. Each technician receives about the same number of calls in a month, and thus they have a balanced workload. Some customers have recently started to complain that they have to wait a long time for a technician to come and fix their broken copiers. OSE has noticed that the number of calls in the last 5 months (i.e., demand) has not increased, so they are not considering to hire a new technician. However, OSE is planning to improve its operations by making it more flexible. Specifically, OSE decided to have each technician respond to calls from two different institutions. OSE asked its repair crew to rank their preferences of which institution they would like to be assigned to. After aggregating all their responses, the following three options were those that were close to the crew preferences. In the table "Prv" stands for private firms, "Gov" stands for government offices, "Edu" stands for educational institutions, and "Cnt" stands for copy centers.

|  | Bill | Chad | Dan | Paul |
|---|---|---|---|---|
| Option 1 | Prv | Gov | Edu | Cnt |
|  | Edu | Cnt | Prv | Gov |
| Option 2 | Prv | Gov | Edu | Cnt |
|  | Cnt | Prv | Gov | Edu |
| Option 3 | Prv | Gov | edu | Cnt |
|  | Gov | Prv | Cnt | Edu |

Which one of the options do you expect to result in shorter customer waiting time?

4. About 60 percent of customers who come to Café Minoo order drink, about 10 percent of customers order sandwich, and the remaining 30 percent order sandwich and drink. Upon arrival, a customer first places the order with the cashier—Henry—and pays for it. The cashier asks the name of the customer and enters the order into the computer. The order then appears in the monitors in the drink and sandwich stations. Taking an order takes an average of 1 minute. Sam is in charge of making drinks in the drink station. The average time to prepare a drink for an order is about 1.5 minutes. Jill and Mat work in the sandwich station. The average time to prepare an order for a sandwich by Jill or Mat is around 6 minutes per order. When the order of a customer (drink or sandwich) is ready, the customer's name is called and the customer picks up the order. The Café opens at 6:00 am and closes at 8:00 pm, Monday through Friday, and is closed on weekends.

    a. Develop a linear programming model to compute the maximum number of customers that the Café can serve in an hour with its current employees.

    b. How much the total capacity of Café Minoo is increased if Henry is trained to also make drinks when needed?

5. Hampptton Phones makes two types of phones—standard and customized—on two different production lines in its assembly department. Each production line has five workstations, and each workstation has one worker, see Fig. 14.13. The assembly department works three shifts, 24 hours a day, 7 days a week. Because tasks in all five workstations are simple assembly tasks, with some training, all workers will be able to work in other stations.

The effective mean process times ($T_{\text{eff}}$) of workers in each workstation is given in the following table:

| Workstation | Standard line $T_{\text{eff}}$ (minutes) | Customized line $T_{\text{eff}}$ (minutes) |
|:-----------:|:---------------------------------------:|:-----------------------------------------:|
| 1 | 3.4 | 4.2 |
| 2 | 3.8 | 4.1 |
| 3 | 3.6 | 4.2 |
| 4 | 3.8 | 4.0 |
| 5 | 3.6 | 3.9 |

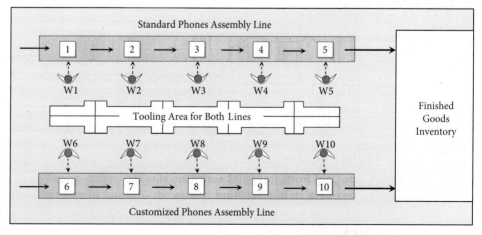

Figure 14.13    Schematic of the standard and customized production lines in Hampptton Phones in Problem 5.

The current demand is 264 and 300 per day, respectively, for standard and customized phones. However, due to introduction of new phones, the firm expects that the demand will increase to 380 and 355 per day, respectively, for standard and customized phones.

    a. Do the lines for standard and customized phones have capacity to satisfy future demands?

    b. What would be the effective capacity of each line if workers work in two stations according to a chain structure? Specifically, in the standard line, Worker 1 works in stations 1 and 2, Worker 2 works in stations 2 and 3, and so on, and Worker 5 works at stations 5 and 1. Similarly in the customized line, Worker 10 works in stations 10 and 9, Worker 9 works in stations 9 and 8, and so on, and Worker 6 works in stations 6 and 10. Can the new capacity under chain structure satisfy the new demand?

    c. Can you further increase the capacity of the lines such that they satisfy their demands without asking 10 workers to work in more than two stations?

6. Fast-&-Easy Loans (F&E Loans) is a financial institution that offers loans to home buyers. F&E works 8 hours a day (from 8:00 am to 5:00 pm with lunch break from 12:00 to 1:00), and 5 days a week. Figure 14.14 shows the process-flow chart of the loan approval process at F&E.

All applications are first reviewed by Rob. Rob first checks if the application has all the required documents. This takes about 5 minutes. If the application misses some documents, Rob sends the application to Mark, who writes a letter to the applicant and rejects the application. For applications with complete documents, Rob uploads the application into F&E information system so that others in F&E can access it. This takes about 15 minutes. Kathy then has access to the application and checks the credit record of the applicant. It takes about 40 minutes for Kathy to gather credit scores for an applicant. If the credit score is lower than a certain threshold, Kathy then directs the application to Mark, who will write a letter to the applicant and reject the application for insufficient financial credit. For those applications who have enough credit, Kathy writes a credit report, which takes her about 30 minutes. For applications with enough credit, Tim and Fred provide property survey and title search reports, respectively, and upload them to the system. This takes Tim and Fred 60 and 100 minutes, respectively. Mark makes the final decision of approving or disapproving an application after all reports are completed by Kathy, Tim, and Fred. Writing a rejection letter for incomplete applications or applications with low credit takes Mark about 5 minutes. For other applications, it takes Mark 45 minutes to make an approval/disapproval decision and write a letter to inform the applicant. Data shows that about 15 percent of applications are

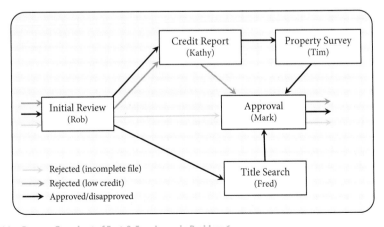

Figure 14.14   Process-flow chart of Fast-&-Easy Loans in Problem 6.

rejected because they are missing some documents and 20 percent are rejected due to lack of sufficient credit.

All five employees of F&E are experienced workers who, with some short cross-training, will be able to perform other tasks. Currently F&E receives a total of six applications per day. This number is expected to increase to nine applications per day.

   a. What is the maximum number of applications that F&E can process in a day? What is the current utilization of each employee?

   b. To increase the capacity, F&E is planning to ask Rob to also help with credit report and title search. Can this change increase the capacity of F&E beyond the future demand of nine applications per day?

7. Dell Computers is one of the pioneer companies that uses modular components to reduce its inventory costs. A personal computer (PC) is made of three main components: microprocessor (CPU), hard drive (HD), and random access memory (RAM). Now consider the case when Dell produces 30 different models of PCs, each having a different type of CPU, HD, and RAM. This implies that Dell needs to keep inventory of $3 \times 30 = 90$ different components to make the 30 models. Suppose that weekly demand for each of the 30 models is Normally distributed with mean 200 and standard deviation 40. Also, suppose that Dell orders these components from different suppliers with order lead time of 2 weeks, and pays suppliers an average of $300 for each CPU, $120 for each RAM, and $180 for each HD. Dell has an inventory carry rate of 45 percent for holding each of the 90 components.

   a. How much safety stock of each component must Dell hold in order to keep the possibility of shortage of its products to less than 3 percent? What would be the total annual safety inventory holding cost?

Now suppose that Dell decides to use modular components to offer the 30 models using only five types of HDs, three types of CPUs, and two types of RAMs. Thus, each type of HD is used in 6 different models, each type of CPU is used in 10 different models, and each type of RAM is used in 15 different models.

   b. How much does using these modular products reduce the annual safety inventory holding cost?

# Lean Operations

## 15.0 Introduction

In the 13 years of production during the 1940s and 1950s Toyota Motor Company produced 2685 automobiles, which was less than half of what Ford was producing in one day—7000 per day.[1] In 2008, however, Toyota surpassed Ford and GM and became the world's largest automaker. Toyota remained the largest auto manufacturer in 2010 with GM second and Volkswagen third. The Japanese earthquake and tsunami in 2011 put Toyota in the second place, but soon Toyota regained its position 1 year later.[2]

What happened between 1950 and 2000 that changed Toyota from a very small auto manufacturer to the largest auto manufacturer in the world? The answer is Toyota's quest for excellence and its manufacturing philosophy known as *Toyota Production System* or TPS. TPS is considered to be the next major evolution in manufacturing after Henry Ford's mass production. Toyota Production System is also known as *Lean Manufacturing* or *Lean Operations*. It evolved from being a set of manufacturing management principles to a management philosophy that can be used in the entire organization. There are several books with titles such as "Lean Enterprise" or "Lean Thinking" that describe how the basic management principles developed by Toyota extend beyond a firm's operations.

Toyota Production System enabled Toyota to achieve significant improvements in several aspects of its operations, including improving quality, reducing space requirements, reducing lead times, increasing productivity, reducing inventory, increasing equipment utilization, increasing flexibility, reducing costs, among others. But what is Toyota Production System? In its website (www.toyota.global.com) Toyota defines TPS as: *"A production system which is steeped in the philosophy of 'the complete elimination of all waste' imbuing all aspects of production in pursuit of the most efficient methods."*

As it is emphasized in the above definition, the core principle at the heart of TPS is elimination of waste. While elimination of waste was one of the main factors that helped Toyota to become the global leader of automotive industry, there are several other environmental, cultural, and geographical factors that also contributed to Toyota's success. The goal of this chapter is to review the operational and organizational principles of lean operations and to show how those principles can help improve inventory, flow times, throughput, and quality. These principles are presented as "Lean Tools" that can be implemented in both manufacturing and service processes.

Before we present those tools and to better understand lean operations, it is useful to have a brief look at Toyota's challenges and triumphs when it started manufacturing cars.

## 15.1 A Bit of History

There are several factors throughout the history that contributed to the development of Toyota Production System. One is a series of engineering innovations made by Toyoda family.

### Engineering Innovation and Genchi Genbutsu

In the 1800s, weaving was a major industry in Japan. In 1894, Sakichi Toyoda applied his skills in carpentry and built manual looms that were cheaper, but worked better than the existing looms. He then focused on improving the manual loom to a power loom that needed a power-generating mechanism. In those days, the most common way to generate power was

---

[1] Womack, J.P., D. T. Jones, and D. Roos, *The Machine that Changed the World.* Free Press, 1990.
[2] Schmitt, Bertel, Nice Try VW: Toyota Again World's Largest Automaker, Forbes, January 27, 2016.

steam engines. So, Sakichi bought a used steam engine to power his new loom. To figure out how to make his machine work with his steam engine, Sakichi experimented many different approaches with trial and error, and finally in 1924, Sakichi invented the world's first automatic loom, called the "Type-G Toyoda Automatic Loom."[3]

His approach of making his hands dirty and trying his different designs by implementing them would become part of the foundation of Toyota Production System known as *Genchi Genbutsu*. In Japanese, Genchi Genbutsu means "go and see for yourself," it is also sometimes referred to as "get your boots on." The main idea behind Genchi Genbutsu is that the best way to understand a problem is to go and see it on the ground. This practice became a part of the problem-solving culture in Toyota. When Toyota's managers are asked to solve a problem, they all go to shop floor to directly see and experience the problem.[4] This is opposite of Fredrick Taylor's Scientific Management who clearly separated the management's job and workers' jobs. According to Taylor, management should do the planning and design the job and workers should do the job.[5]

Back to Sakichi's power loom machine. Sakichi established Toyoda Automatic Loom Works in 1926, which became the parent firm of Toyota group. Sakichi never stopped improving his new machine and, with a series of new inventions, he perfected his automatic loom machine. One of his inventions was a mechanism that automatically stopped the machine when a thread broke. Sakichi's invention later evolved into a broader system called *Jidoka*, one of the two pillars of Toyota Production System.

### Jidoka

In Japanese, "jido" means automation and refers to a machine that moves on its own without needing an operator. The Toyota use of term "jido" is applied to a machine with a built-in mechanism for making judgments. Hence, in Toyota terminology, Jidoka refers to "automation with a human touch," as opposed to a standard automated machine without an intelligent mechanism that, for example, stops the machine when facing problems.

Jidoka evolved into several principles in TPS. For example, one is the practice of building in quality as production occurs. This can be achieved by having a mechanism that can stop production when a quality issue occurs. Another is to build systems that can prevent mistakes that lead to quality issues. Mistake-proofing can be achieved by designing the manufacturing process, equipment, and tools such that an operation cannot be performed incorrectly.

Since with Jidoka Sakichi's machines became mistake-proof—they automatically stopped when a problem arose—a single operator could be in charge of numerous looms, resulting in lower operator cost and higher productivity. To sell his new technology, in 1929, Sakichi sent his son Kichiro to England to negotiate with Platt Brothers to sell the patent rights of their automatic loom. The negotiated price was 100,000 English pound. Later Sakichi put his son in charge of building the car business and the 100,000 pound was used to start building Toyota Motor Corporation. Kichiro was a mechanical engineer and, like his father, he was a hands-on and learning-by-doing inventor.

### Just-In-Time

Toyota Motor Company was founded in 1937. They changed the company's name from Toyoda to Toyota, which was chosen among 27,000 suggestions made in a public contest in 1936. In building and managing of Toyota Motor Company, Kichiro followed his father's

---

[3] From Toyota's website, www.toyota.global.com.
[4] Genchi Genbutsu: More a frame of mind than a plan of action. *The Economist*, October 13, 2009.
[5] Hopp, W.J. and M.L. Spearman, *Factory Physics*. Waveland Press, Inc., 2011.

philosophy along with his own innovations. His main contribution was "Just-In-Time" or JIT, which is the other pillar of Toyota Production System (besides Jidoka). Just-in-time refers to the practice of making "only what is needed, when it is needed, and in the amount needed." Following JIT allows production systems to eliminate waste in forms of unused time, equipment, material, etc.

### Toyota Community

After World War II, the high inflation rate and the bad economy under American occupation created financial difficulties for Kichiro. To avoid bankruptcy, Toyota started cutting costs including pay cuts and asking 1600 workers to retire voluntarily. This resulted in worker strike and public demonstration. The company started negotiating with the workers and reached to a historic settlement. The settlement had a profound effect on Toyota's relationship with its workforce.

First, while the economic situation after World War II was beyond Kichiro's control, he accepted the responsibility for the failing of the company and resigned as the president. Kichiro's personal sacrifice was noticed by the workers—they all knew why he did that: for the good of the company. Kichiro was emphasizing on Toyota's philosophy of thinking beyond individual concerns and considering the long-term good of the company, as well as taking responsibility for problems.

Second, another part of the settlement was that quarter of the workforce had to be terminated according to the original plan. However, the remaining workers received two guarantees. One was the lifetime employment and the other was pay graded by seniority rather than by job function. Furthermore, the payments were tied to company's profitability through bonus payment. The implication of this settlement was that Toyota's workforce became the members of the Toyota community with high loyalty to Toyota and less incentive to leave Toyota. Workers also agreed to be flexible to do any work assigned to them. In fact, the company's official said: "If we are going to take you on for life, you have to do your part by doing the jobs that need doing."[6]

After Kichiro, Eiji the nephew of Sakichi and the younger cousin of Kichiro, who also had a mechanical engineering education was the head of Toyota. Like his cousin and uncle, Eiji was also the follower of Genchi Genbutsu and was the firm believer of learning by doing.

### Lesson's from Mass Production

Eiji Toyoda, similar to his predecessors, visited Ford and GM several times and studied Henry Ford's book *Today and Tomorrow* to adopt mass production practices in their factories. They soon realized several important facts. First, Ford's mass production is suitable for producing very large quantities with low product variety. As Henry Ford famously said: "Any customer can have a car painted any color that he wants as long as it is black." In contrast, Toyota was facing a low demand for a wide range of models. This meant that Toyota could not take advantage of economies of scale and cannot have a dedicated assembly line for each model. It needed to produce several different models on one flexible line. Second, Ford had enormous amount of cash, while Toyota did not have cash and thus needed to turn cash around quickly. This meant shortening the time from receiving an order to being paid, which requires shorter production lead time.

Third, mass production also had its own problems. Workers did not like their repetitive and fast-paced work and unions and plant managers were often in conflict. Hence, there was no sense of community and partnership. Also, higher production rate was the main goal and quality was second. The so-called "move the metal" mentality was dominant, which

---

[6] Womack, J.P., D. T. Jones, and D. Roos, *The Machine that Changed the World*. Free Press, 1990.

**Figure 15.1    Taichi Ohno, father of Toyota Production System.**[7]

encouraged pushing the product to the end of the line and avoided stopping the line for any reason (including for quality issues). Therefore, quality control (QC) workstations were located at the end of the line, with a large number of people trying to fix the quality issues. Finally, to achieve economies of scale, mass production required larger and larger specialized machines, resulting in batch production, which in turn increased the inventory throughout the plant.

Knowing that they cannot directly adopt the Ford's mass production, Eiji Toyoda gave a difficult assignment to his plant manager Taichi Ohno: Improving Toyota's manufacturing system so that it has a comparable productivity of Ford. Hence, Taichi Ohno's job was to create a manufacturing system that has short production lead time and is flexible enough to produce larger variety than what mass production offers with low cost and high quality.

To do his task, Ohno focused on implementing JIT and Jidoka in its production system. With his knowledge of manufacturing and with his motivated workforce, he gradually developed Toyota Production System over two decades of trial and error and learning by doing. By following the Genchi Genbutsu approach, he was quoted as saying:[8]

> **Taichi Ohno:** *By actually trying, various problems became known. As such problems became gradually clear, they taught me the direction of the next move. I think that we can only understand how all of these pieces fit together in hindsight.*

Toyota Production System was not noticed by North America's (auto) manufacturers until the oil crisis in 1973, which resulted in a global recession. Toyota was among few companies that had shorter financial difficulties and came back to profitability faster. Several books were published in the United States in the 1970s and 1980s that described TPS. In 1984 NUMMI, a joint venture between Toyota Motor Company and General Motors opened in California. The manufacturing world was beginning to notice the new Japanese manufacturing system and manufacturers began to adopt TPS.

In the 1990s, in the best seller book, *The Machine that Changed the World*, the authors coined the term *Lean Production* for Toyota Production System. Today the concept of lean has expanded outside manufacturing into the service industry as well as the entire organization.

In this chapter, we describe the basic components of Lean Production Systems and explain how these components reduce costs, improve quality, and enhance flexibility of operations.

---

[7] Source: https://dtc-wsuv.org/dronda17/portfolio/history.html.
[8] Suzaki, K., *The Manufacturing Challenge*. Free Press, New York, 1985, p. 250.

## 15.2 TPS—A Customer-Focused Process

Shotaro Kamiya was the leading figure of Toyota's sales as Taichi Ohno was the leading figure of Toyota's Production System. He had lots of international experience in the United States and Europe and joined Toyota as sales manager in 1935. Shotaro Kamiya established Toyota's network of dealers in Japan and also expanded Toyota's sales in the United States. After several years of working for Toyota, he became the honorary chairman of Toyota. He is known for his philosophy of "customer-first" and quoted as saying:[9]

> **Shotaro Kamiya:** *The priority in receiving benefits from automobile sales should be in the order of the customer, then the dealer, and lastly, the manufacturer. This attitude is the best approach in winning the trust of customers and dealers and ultimately bring growth to the manufacturer.*

Lean operations system is therefore customer-focused. One principle of lean operations is to identify what customers want, when they want it, and the price they are willing to pay for it, and realize that, when something goes wrong, customers want fast, accurate, and friendly help or the option to return the product. Hence, lean operations must be able to:

- accurately identify what customers value in a product or service,
- translate those values into product specifications such as functionality, price, performance, reliability, maintainability, safety, etc., and
- develop the product and production process to deliver those specifications (i.e., customer values) in a cost-effective manner.

## 15.3 Goals of Lean Operations

Lean operations has four main goals:

1. Eliminate waste
2. Increase speed and response
3. Improve quality
4. Reduce cost

In the following sections, we discuss these goals in more detail.

### 15.3.1 Eliminate Waste

Elimination of waste was originally mandated by Eiji Toyoda. He defined "Waste" or *Muda* (in Japanese) as:[10] "anything other than minimum amount of equipment, material, parts, space, and time that are absolutely needed to add value to the product."

Waste elimination is not unique to Toyota Production System. In fact, eliminating waste goes back to Fredrick Taylor's *The Principles of Scientific Management* in 1911. His approach was to decompose systems into subsystems, each specialized in one task, and optimize each subsystem in order to minimize waste. However, Toyota, and particularly Ohno and his team, saw waste and its elimination process in a different way:

- *Cost of Efficiency:* Ford's economies of scale allowed him to reduce cost per unit. To use resources (i.e., machines and workers) efficiently, they were highly utilized to minimize idle time—eliminate waste. This was considered by Ohno as overproduction resulting in large inventories with defects hidden under the piles of inventory

---

[9] Liker, J.K., *The Toyota Way: 14 Management Principles from the World's Greatest Manufacturer.* McGraw-Hill, 2004.
[10] Russell, R.B., and B. W. Taylor, *Operations Management.* Fourth Edition. Pearson Education, 2003.

and not discovered for weeks. Hence, the inventory and quality costs would be high in the way Ford was trying to achieve efficiency. Ohno's approach was different. From his perspective, it was ok to have a resource idle, if idling prevents overproduction and thus avoids large inventory.

- *Changing the Production Environment:* Ohno's approach for eliminating waste was different from traditional scientific management approach. Instead of being constrained by the factors in his production environment, he changed those factors to come up with a solution. For example, one reason that automobile manufacturers produce in large batches is the long time it takes to set up a machine for the next job. How to reduce the cost of parts in a batch? With a large market after World War II, some western manufacturers found it easier to dedicate a machine to a part and thus to avoid setups. They could produce months and months without need for a setup. Even with one machine, the scientific management approach was to find the optimal batch size to minimize the cost per unit produced. Ohno's approach was different. Instead of finding the optimal solution for his production environment, he decided to change the environment. Specifically, since he did not have money to buy additional machines, and since he believed that overproduction is a waste, he decided to eliminate setups—the environmental factor that results in these wastes.

With the above approach to eliminate waste, Ohno (and later K. Suzaki) identified seven sources of waste as follows:

1. *Overproduction:* Producing items and holding them in inventory results in production and inventory holding cost. Overproduction is concerned with the *number* of items that are produced and the *time* of production. Producing 60 items when there is a need for only 50, or producing 50 today when 50 items are needed next week are both overproduction—producing too much too early.

2. *Overprocessing:* Activities such as non-value added tasks in a process, or performing a task more than necessary (e.g., doing too much sanding of a surface), or unnecessary repetitions (e.g., checking something several times) are examples of overprocessing that must be eliminated.

3. *Wait Times:* Consider a workstation in which a worker performs a cutting operation. Two types of waiting can occur in this station, and both are considered to be waste. One is the parts waiting in the input buffer of the station to be processed, that is, in-buffer flow time. The other one occurs when the input buffer of the station is empty, resulting in the worker (the resource) to remain idle waiting for new units to arrive at the worker's input buffer. The more the worker is waiting for jobs to arrive, the lower the worker's utilization will be.

4. *Transportation:* In all processes, flow units must move among different stages, which is usually located in distance from each other. The time and effort it takes to move units among different stages of the process is considered a transportation waste. Transportation waste is often the result of poorly designed physical layout of a process. The more the layout mimics the process flow, the lower the transportation time. Specifically, two stages of the process that are next to each other in the process-flow chart should be located close to each other in the physical layout.

5. *Motion:* Performing tasks requires motion, which consumes time and energy. Any unnecessary motion that does not add value is a waste and must be eliminated. Hence, motion should be considered when designing tasks, equipment, material handling, and processes.

6. *Production Defects:* Defective products, as described in Chap. 4, result in yield loss and rework, both incur cost, waste resources, and can have negative implications

on customer perception of a firm's quality. To eliminate product defect, Toyota Production System's strategy is "Do it right the first time."

7. *Inventory:* Acquiring and holding inventory, as we discussed in Chap. 9, is costly, and hence is considered one of the main sources of waste in TPS. Inventory in a process can be the result of overproduction, high variability in the process, large batch sizes, long setup times, or poor scheduling, among other reasons. According to the just-in-time philosophy, the goal of TPS is to minimize the excess inventory—preferably to zero.

### 15.3.2 Increase Speed and Response

Another goal of lean operations is to increase the speed and response time in all aspects of a firm's functions. These include creating an efficient process to reduce the time it takes (i) to design a product, to sell the product and deliver to customers, (ii) to respond to customers' demands and inquiries about deliveries, quality, etc., and (iii) to collect payments. Lean operations achieves these goals by synchronizing and coordinating the flow in the entire supply chain. One key enablers of this is close collaboration among supply chain members as well as flexibility designed in different stages of the supply chain.

### 15.3.3 Improve Quality

Toyota's products are known for their high quality and durability. The reason is that one of the main goals of lean operations is to constantly improve quality. In fact, lean systems cannot function if raw material and work-in-process are bad. This is because bad quality material leads to different types of waste. It interrupts production schedules, results in wasted times to deal with the problem, and also in wasted times of resources to do the rework. Furthermore, it results in waiting time of other units—resources and inventory—that are waiting for quality problems to be fixed. The goal of eliminating these wastes has led lean operations to focus on improving quality as one of their main goals.

### 15.3.4 Reduce Cost

Cost reduction is another goal of lean operations—as is in any business process. For lean operations reducing cost comes naturally, since it is one direct consequence of eliminating waste and improving quality. More efficient utilization of resources is achieved by using more efficient equipment, better worker training, increasing equipment availability (e.g., by using preventive maintenance), and using flexible resources. Lean systems also use a wide range of tools developed in Toyota Production System to reduce inventory cost and quality costs.

Having discussed the four main goals of lean operations, the remaining of this chapter is devoted to present several practices that were developed by Toyota to achieve lean operations. We call these practices "*Lean Tools*" and we start with the first tool—variability reduction.

## 15.4 Lean Tool 1—Variability Reduction

We described in Chaps. 5 and 7 that variability anywhere in the process degrades process performance. Specifically, Eq. (7.2) showed how—for a produce-to-order system—both variability in the demand (i.e., coefficient of variation of demand interarrival times) and variability in process times (i.e., coefficient of variation of effective process time) have a direct impact on in-buffer flow times and hence on inventory, both of which are considered waste in lean operations. This is the same for inventory systems (i.e., produce-to-stock systems) as shown in Chap. 10. Specifically, in these systems firms need to hold additional inventory (i.e., safety inventory) to protect themselves against variability in demand or in supplier's lead time.

Since inventory and waiting times are considered wastes in lean operations, variability reduction is essential to achieve lean operations. Chapter 5 presents several strategies that can reduce variability in demand and in process times in manufacturing and service systems. Here, we describe how Toyota reduced variability in its operations.

With respect to reducing variability in its demand, Toyota took several actions:

- *Aggregating Demand:* Toyota limited the number of models produced by reducing the number of options offered for each car. Thus, the demand for the few number of offered models became the aggregated demand, which had less variability. The demand was also aggregated at the dealer level—through postponement—by dealer installing some options (e.g., radios) at the point of sales.

- *Isolating Variability:* In addition to demand aggregation, Toyota isolated the demand variability from its plants by taking two approaches:
  - It developed its production schedule at its plants months in advance and thus each plant knew exactly how many cars they needed to produce each week.
  - It also made the daily production schedule fixed through setting *takt time* (see Sec. 15.13). A fixed takt time implies a fixed inter-output time. Hence, Toyota's plants were actually facing a deterministic daily demand dictated by Toyota.

Toyota was able to reduce variability in its manufacturing facility in several ways. These included practices such as setup time reduction, preventive maintenance, work standardization, and several other approaches that we will describe later in this chapter.

Finally, Toyota was successful in reducing the variability in the deliveries made by its suppliers by establishing a close relationship with them. Since Toyota was the largest customer of its suppliers, it had enough power to influence suppliers' decisions. Toyota also helped its suppliers to reduce variability in their processes, and suppliers held enough inventory to guarantee on-time delivery to Toyota.

## 15.5 Lean Tool 2—Setup Time and Cost Reduction; the SMED

Stamping operations is an essential process in making automobiles. In stamping plants of auto manufacturers, a large sheet of material is cut to the size needed to make a part such as left or right body panel, or a door, or a hood for a car model. These cutting operations are done by blanking press. The cut sheets are then fed to a very large stamping press. Specifically, the cut sheet is placed between a lower and an upper die. These dies are made of steel and are mirror image of each other. Each part (e.g., left door, or right door) has its own die designed according to the shape of the part. With a stroke, the press puts the upper and lower dies together and the metal sheet between the dies takes the shape of the dies.

Each time a manufacturer needs to produce a different body part of a car, it needs to set up the press for the new part by replacing the current die with the new one. However, changing a die on a press is a very time-consuming task; depending on the size of the die, it may take several days to do so. The reason is that these dies weigh several tons and they require a very precise adjustment when they are installed on a stamping press. A small misalignment of dies on a press could result in a wrinkled part or in metal sheet melted on the die. This could destroy the die, which is a very expensive tool (in the order of tens of thousands of dollars).

Because of this long setup time, which also incurs costs, and to take advantage of economies of scales, most auto manufacturers in the 1950s and 1960s produced a large number of parts before they switched to producing another part. In some cases, to avoid frequent setups, some manufacturers purchased several presses and dedicated each to one part during a period of weeks or months. While this might have been a good solution for large auto manufacturers like Ford, it was not a feasible solution for Taichi Ohno, since he did not have enough budget to buy dedicated presses for each part or for each family of parts.

Taichi Ohno took a different approach. He believed that there were ways that die changing operations could be simplified and significantly shortened. A shorter setup time would allow him to change dies every 2 or 3 days instead of every 2 or 3 months. In the 1940s Ohno hired Shigeo Shingo, a consultant, to help him find ways to reduce the setup time of their stamping press. He also bought a few used presses from America and started experimenting and trying different ways—Genchi Genbutsu—to reduce die changeover times. By the 1950s Ohno and Shingo were able to reduce the die changeover operations from 1 day to only 3 minutes using a system Shingo called *Single Minute Exchange of Die* or SMED.

SMED is based on the following principles:

- *Identify all activities needed for setups and separate internal (online) activities from external (offline) activities.* To implement SMED, it is important to first list all the activities, tools, and material needed for the setup operations. Internal or online activities are those that cannot be performed until the machine is stopped. External or offline activities, however, are those that can be performed while the machine is still working.

- *Perform the external activities in advance.* Since there is no need to stop the machine to perform external setup activities, these activities must be performed in advance before stopping the machine. Applying this principle, the setup time could often be reduced by 30 to 50 percent.

- *Convert as much internal setup activities as possible to external setup activities.* This requires creative thinking such as designing new fixtures and tools to, for example, shorten the centering and alignment of dies when performing internal setup activities.

- *Standardize tasks and reduce both internal and external activities.* One can reduce the external setup activities by streamlining the workplace. This includes putting tools and dies close to where needed, and having machines and fixtures in good shape and ready to use. Internal setup activities can be shortened by using quick fasteners and pin locators that prevent misalignments.

- *Perform setup activities in parallel.* Dividing activities among multiple workers shortens the time it takes to perform external and internal setup activities.

- *Eliminate setups.* While SMED process is designed to reduce setup time, the ultimate goal of SMED is to eliminate setups. So, one should always think of finding ways (e.g., redesigning the product or the process) to eliminate the need for setups.

## 15.6 Lean Tool 3—Batch Size Reduction and One-Piece Flow

In one of their visits to U.S. auto manufacturer's plant in the 1950s, Eiji Toyoda and Taichi Ohno noticed many flaws in the way the plant was working under mass production. They saw that many equipment were making a large amount of units and keeping them in inventory, so that they could be transported later to another group of large machines making large amount of inventory. They noticed that the accounting measures rewarded managers who could produce fast and could highly utilize equipment and workers by making them work— produce—all the time. They noticed that the defective items might be hidden under the large piles of inventory and could be unnoticed for several weeks.

While in his book, *Today and Tomorrow*, Henry Ford explained the importance of having continuous flow of material, Taichi Ohno noticed that Ford plants did not always follow this practice. By moving large batches from one stage to another, Ford plants did not have a continuous flow—they had a discrete flow of large batches waiting at each stage to be moved to another.

Unlike Ford, Toyota did not have large amount of money and space to hold inventory. Its demand was small but for a wide range of products, while Ford's demand was very large for basically one type of product. Determined to use the concept of continuous flow, Taichi Ohno put effort and reduced the batch size in his plant. Smaller batch sizes have many benefits, including:

- *Smaller inventory and flow times:* As described in Sec. 8.9, reducing batch size results in reduction in wait-to-batch-time (WTBT) and wait-in-batch-time (WIBT), which in turn results in lower flow time and inventory. Hence, smaller batch size brings about all the benefits of having lower inventory and lower flow time (e.g., less space for storage, lower production lead time, etc.).

- *More flexibility and responsiveness:* Systems that use small batches have greater flexibility in changing their production schedule. This allows them to respond rapidly to customer demand and changes in market, for example, it is easy to insert a new job into the production schedule of small batches without affecting the schedule of other jobs.

- *Lower transportation time and cost:* Having smaller batch sizes allows processes to move closer together, and hence transportation time of units between stages is shortened. It is also easier to move smaller batches. For example, while moving a large batch from one stage to another stage may require a lift-truck (i.e., a capital investment), a small batch (e.g., a unit in one-piece flow) can be moved to the next stage by a worker with a simple cart (if the item is heavy).

- *Better quality product with lower quality cost:* Smaller batch sizes contribute to higher quality products in several ways. First, with smaller batch sizes, quality problems become more visible (instead of being hidden under a large pile of inventory). Second, the time between a quality problem is created until it is detected (in later stages) is reduced. The later the problem is detected, the larger the number of the defective units between the problem creation and detection stages, and thus the larger the cost of scrap and rework. Third, because of the above two, workers have less tendency to let a low-quality job pass their station, and pay more attention to quality of their work.

The best strategy is to reduce the batch size into one and design the process to move the one unit continuously through its different stages. This is called *One-Piece Flow* system. Of course, reducing batch size requires reducing setup times, which Taichi Ohno also achieved through SMED, as we described in the previous section.

## 15.7 Lean Tool 4—Flexible Resource

Recall the incident after World War II when Toyoda family had to lay off workers in their plant. It resulted in a settlement that provided lifetime employment for the workers and payment based on seniority. With lifetime employment, Taichi Ohno realized that workers are as much a fixed cost for the company as are company's machines—the cost that the company incurs for about 30 to 40 years when a worker works for the company. Hence, he decided to use his workers efficiently by enhancing their skills, knowledge, and experience. He did that by training them to do multiple tasks, maintain their machines, and check the quality of their jobs.

Focusing on eliminating waste, Ohno noticed that after loading a part on a machine, the machine operator was standing idle until the machine finished processing the part. The worker then unloaded the part. To eliminate the worker's idle time, Ohno asked his workers to operate two identical machines. Using automatic switches that turn the machine off when the processing is done, and fixtures that hold the job in place without needing an operator, a worker was able to operate more than two machines. In fact, these improvements allowed workers to operate between 5 and 10 different machines.

Ohno realized that he also needed multi-functional (i.e., flexible) machines that can perform more than one operation. So, while other manufacturers are acquiring large specialized equipment with large capacity, Ohno was purchasing small general-purpose equipment flexible enough to perform several tasks. Using flexible machines eliminates wastes. For example, using one flexible machine that can perform two tasks, instead of using two specialized machines, eliminates wastes such as operators moving between the two machines, or a machine waiting for an operator who is setting up the other machines.

Also, asking a worker to do multiple tasks was easier in Japan than in America. In his book, *Toyota Production System*, Taichi Ohno writes:

> **Taichi Ohno:** *It was possible in Japan because we lacked function-oriented unions like those in Europe and the United States. Consequently, the transition from the single- to multi-skilled operator went very smoothly, although there was initial resistance from the craftsmen. This does not mean, however, that Japanese unions are weaker than their American and European counterparts. Much of the difference lies in history and culture.*

Operational flexibility provided by multi-skilled workers and multi-functional machines gave Taichi Ohno tremendous opportunity to eliminate and reduce waste and thus produce more with less. The benefits of operational flexibility are discussed extensively in Chap. 14.

## 15.8  Lean Tool 5—Cellular Layout

With his workers operating several machines, Ohno started to reorganize the machines and put them closer to each other in what is called a *Manufacturing Cell*. While Ohno was not the first person who came up with the idea of a manufacturing cell, he was able to utilize the cells in a very efficient manner.

What exactly is a manufacturing cell? A manufacturing cell is a group of workstations (machines and/or workers) that are assigned to produce one product or a family of products that have the same processing requirements. The stations in the cell are usually arranged in a U-shaped line. Furthermore, the sequence of workstations follows the sequence of operations needed to complete the product. This minimizes the time required to move a job from one station to the next. Figure 15.2 presents an example of a manufacturing cell with nine workstations and three workers.

Manufacturing cells provide several ways to eliminate waste:

- *Flexible capacity:* The number of workers in a manufacturing cell are often less than the number of workstations, since workers are cross-trained to work in several stations. By increasing or decreasing the number of workers in a manufacturing

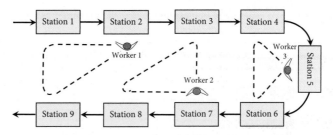

**Figure 15.2    A manufacturing cell with three workers and nine workstations.**

cell, one can increase or decrease the capacity of the cell according to changes in demand. This eliminates overproduction.

- *Flexible product mix:* Since the family of products produced in a cell require the same sequence of operations, it is easy to change the product mix within each family of product—there is no need for a significant setup or changing the process flow.

- *Small batch size:* Again, because a cell produces similar items, there is no need for setups—or setups are very short. Hence, items can be moved from one station to another in very small batch sizes. In fact, in most cases, the batch size is one, that is, workers move a workpiece to the next station after finishing its process in the current station. This brings all the benefits of smaller batch size as discussed in Sec. 15.6, that is, better quality, lower transportation time, lower inventory, and higher responsiveness.

- *Promoting teamwork:* Since workers have close interactions because they work close to each other and work on the same part, they develop a sense of teamwork which helps identifying problems and finding solutions (see Sec. 15.17). This reduces wastes such as scraps and rework.

## 15.9 Lean Tool 6—Pull Production

Assembly plant is the heart of auto manufacturers' operations. Thousands of parts and subassemblies are produced and delivered to the assembly plant from outside suppliers or from the manufacturer's stamping, engine, and transmission plants. A major problem for auto manufacturers is how to coordinate the production and delivery of those parts to the assembly plant so that assembly plant does not stop due to lack of material.

While U.S. auto manufacturers were using large inventory to protect against the lack of coordination, Taichi Ohno, for several years, was looking to find another way to achieve coordination, but with lower inventory. He finds his answer in the most unexpected place—American supermarkets. Ohno writes in his book:

**Taichi Ohno:** *In 1956, I toured U.S. production plants at General Motors, Ford and other machinery companies. But my strongest impression was the extend of supermarket's prevalence in America. The reason for this was that by late 1940s, at Toyota's machine shop that I managed, we were already studying the U.S. supermarket and applying its methods to our work.*

*Combining automobiles and supermarkets may seem odd. But for a long time, since learning about setups of supermarkets in America, we made a connection between supermarkets and just-in-time systems. A supermarket is where a customer can get (1) what is needed, (2) at the time needed, and (3) the amount needed. . . . From supermarket we got the idea of viewing the earlier process in the production line as a kind of store. The later process (customer) goes to the earlier process (supermarket) to acquire the required parts (commodities) at the time and in the quantity needed. The earlier process immediately produces the quantity just taken (restocking the shelves).*

Ohno's idea of *pulling* inventory from the earlier stages by later stages (when needed) was the opposite of what other auto manufacturers were doing, that is, *pushing* inventory to later stages according to some *work schedule*. Ohno's new idea was the birth of what we now know as a "pull system," which was in contrast with the traditional approach, the "push system."

There are many definitions for pull and push systems that capture different aspects of those systems. We define pull and push systems by contrasting their different features.

---

**Push and Pull Systems**

- *Job Release:* In pull systems, jobs are released into the process based on the *internal* status of the process; for example, based on work-in-process inventory (WIP) at different stages of the process. However, in push systems jobs are released into the process without consideration of internal status of the process; for example, jobs are released based on demand forecast, or a predetermined production schedule.

- *Control:* Pull systems control the maximum inventory in the process and observe the throughput; therefore, pull systems have *no direct control over the throughput.* Push systems, on the other hand, control the throughput of the process and observe inventory; therefore, push systems have *no direct control over the inventory* in the process.

- *Production Authorization:* In pull systems *output need*, that is, need at the downstream of a stage, authorizes production in that stage. In push systems, however, *input availability* at a stage authorizes production in that stage.

To illustrate the concept of pull and push systems, let's consider a simple process with two workstations, where all jobs are sent to Station 2 after being processed in Station 1. There is one worker at Station 1 and one worker at Station 2 processing those jobs, see Fig. 15.3.

Suppose that there is no variability in the effective process times of workers at stations 1 and 2. Specifically, it takes exactly 0.95 minutes to process a job at Station 1 or 2. Hence, both workers have an effective capacity of $60/0.95 = 63.16$ jobs per hour. The demand for the product is forecasted to have an average of 1380 per day. Considering that the line works for three shifts—24 hours a day—this translates into the demand of $1380/24 = 57.5$ per hour. Hence, managers would like to run the line such that it produces a throughput of 57.5 per hour.

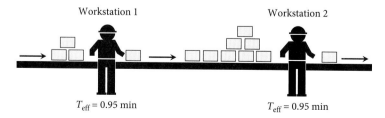

Workstation 1                    Workstation 2

$T_{\text{eff}} = 0.95$ min          $T_{\text{eff}} = 0.95$ min

**Figure 15.3    A line with two stations.**

### 15.9.1  Running the Line as a Push System

In a push system jobs are released to the process according to forecast. Hence, to have a throughput of 57.5 per hour to satisfy the demand forecast of 57.5 per hour, managers decide on the following production schedule: Releasing a job at Station 1 at exactly every $60/57.5 = 1.04$ minutes. This will result in a throughput of 57.5 jobs per hour, see Fig. 15.4. Both workers 1 and 2 are authorized to process a job as soon as a job becomes available at their input buffer. This is shown by dashed lines in Fig. 15.4 that corresponds to production authorization at each station.

What would be the inventory in the two-station line in this push system under the above job release policy? Since there is no variability in the line, it is easy to compute the inventory of jobs in each station. Consider the worker in Workstation 1. He receives a job every 1.04 minutes, and finishes the job in 0.95 minutes. Hence, the worker will be idle for $1.04 - 0.95 = 0.09$ minutes and works for 0.95 out of every 1.04 minutes—a utilization of $0.95/1.04 = 91\%$. The inventory in the buffer of Workstation 1 is always zero and Station 1 sends a job to Station 2 exactly every 1.04 minutes. The situation at Station 2 is exactly the same. The worker at that station receives a job (from Workstation 1) exactly every 1.04 minutes and finishes the job in 0.95 minutes and has utilization of 0.91 percent. Hence, the worker's throughput, and the throughput of the line, is one job every 1.04 minutes, or 57.5 jobs an hour, which is the forecasted demand.

The above line represents a perfect situation. Achieving the desired throughput with minimum in-buffer inventory—zero inventory to be exact. By running the line as a push system, we were able to achieve this perfect performance. But how does the line perform as a push system if there is variability in the process?

#### Impact of Variability on Push Systems

Let's assume that the effective process times at stations 1 and 2 have an average of 0.95 minutes with a moderate variability, that is, coefficient of variation 1. Again, to run the line under push protocol, managers release a job into the line exactly every 1.04 minutes. Since each workstation has effective capacity of $60/0.95 = 63.16$ units per hour, receiving jobs at an average rate of 57.5 units per hour results in 91-percent utilization of both stations—as in the case with no variability. Furthermore, the line would be a stable process (i.e., input rate equals output rate), and hence it would generate the throughput of 57.5 per hour. But what would be the inventory in the two-station line now?

Note that, while managers have control over how many jobs to release into the line, due to variability in the process, they do not have any control on what the inventory in the line

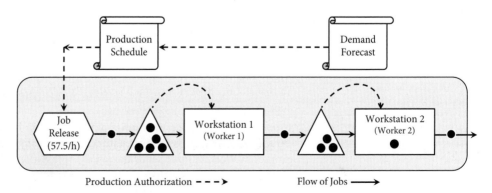

**Figure 15.4  The two-station line runs as a push system.**

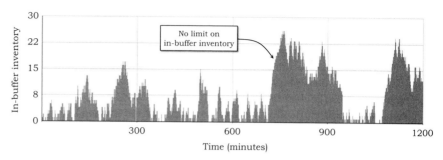

Figure 15.5   Inventory in buffer of Workstation 2 in the push system with variability.

would be. All they can do is to run the line for several days and observe the inventory at each station. Another approach to observe inventory—as we do here—is to develop a computer simulation to mimic the operations in the line and get an estimate of how inventory changes in each station of the line.

Figure 15.5 shows the inventory in the buffer of Workstation 2 during the first 1200 minutes (i.e., 20 hours) of our computer simulation of the line. As the figure shows, in the first 1200 minutes of the simulation, the inventory in the buffer of Workstation 2 varies from zero to as high as 26. In fact, if the simulation is continued for longer time, the inventory can reach as high as 68 units. Recall that to generate throughput of 57.5, worker in Workstation 2 must be 91 percent utilized. The key question is the following: Does the worker in Workstation 2 need such a large inventory to maintain his utilization at 91 percent?

Not necessarily! Recall that in our no-variability case, the worker in Station 2 was utilized 91 percent with zero inventory in the worker's buffer. Hence, to maintain worker utilization at 91 percent at Workstation 2, it is not necessary to hold a large in-buffer inventory—as large as 68 units. All that is needed is to make sure that in 91 percent of times, there is at least one unit available in the buffer of Workstation 2.

One way to achieve this, while preventing the inventory to become outrageously large, is to set a limit on the *maximum inventory* allowed in Workstation 2. This is exactly what a pull system does.

### 15.9.2  Running the Line as a Pull System

In a pull system, as mentioned before, managers do not have direct control over throughput. Specifically, they set a limit on the maximum inventory allowed in each station (or in the line) and run the line to observe the throughput. Of course, the throughput of the line depends on the maximum allowed inventory in each station. As an example, consider the following policy that sets limit on the inventory of both stations and thus runs the line as a pull system. This policy was developed by Taichi Ohno and is known as *Kanban System*:

- *Workstation 1:* The release of a job at Workstation 1 depends on the number of jobs in that station. Specifically, a job is released at Workstation 1 only if the number of jobs in Workstation 1 is less than $K_1$.

- *Workstation 2:* The release of a job (from Workstation 1) into Workstation 2 depends on the number of jobs in Workstation 2. Specifically, after a job is processed, the worker in Workstation 1 sends the job to Workstation 2 only if the number of jobs in Workstation 2 is less than $K_2$. If there already are $K_2$ jobs in Workstation 2, the worker stays idle until the in-buffer inventory at Workstation 2 reaches below $K_2$. When that occurs, the worker releases the job at Workstation 2.

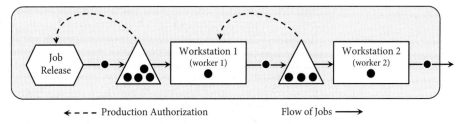

Figure 15.6    The two-station line runs as a pull (Kanban) system.

Note that the Kanban policy sets a limit on the maximum inventory in Workstations 1 and 2, respectively, as $K_1$ and $K_2$. Considering that each station has only one worker, the maximum number of in-process jobs at each station is one, and the maximum in-buffer inventory at Workstations 1 and 2 is $K_1 - 1$ and $K_2 - 1$, respectively. Also, the total (in-buffer and in-process) inventory in the line would never exceed $K_1 + K_2$ units.

Comparing Fig. 15.6 for our Kanban system with that in Fig. 15.4 we can see a fundamental difference between the push and the pull systems: The arrows representing production authorization in the pull system are in the opposite direction of that in the push system. This is because, as mentioned earlier, in push systems the availability of a job in the input buffer of a stage authorizes production in that stage, while in pull systems, the need in downstream stage (Workstation 2) authorizes production in a stage (Workstation 1).

What would be the throughput of the line if jobs are released according to the above pull policy? Again, we need to develop a computer simulation to estimate the line's throughput and its other performance measures including inventory and flow time. Let's first consider a case with $(K_1, K_2) = (2, 9)$. Note that with $K_2 = 9$, the maximum in-buffer inventory allowed in Station 2 is 8. The in-buffer inventory at Workstation 2 under this Kanban policy during 1200 minutes of operations is shown in Fig. 15.7.

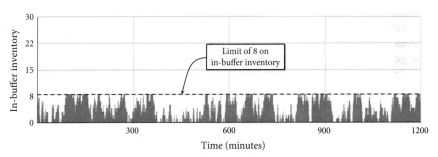

Figure 15.7    Inventory in the buffer of Workstation 2 in the pull system with $(K_1, K_2) = (2,9)$.

Compared to the inventory of the push system in Fig. 15.5, the in-buffer inventory at Workstation 2 under the pull system is drastically reduced. This is because of the maximum limit of eight units imposed by the Kanban policy. But the question is whether the reduction in in-buffer inventory at Workstation 2 would reduce the station's utilization to less than 91 percent, and hence reduce the throughput of the line.

It turns out that, as shown in Table 15.1, Workstation 2 (and Workstation 1) would have the same utilization of 91 percent as that of the push system, resulting in the same throughput of 57.5 units per hour. So, even if the in-buffer inventory at Workstation 2 is limited to a maximum of 8, Worker 2 will still have enough inventory in buffer to work 91 percent of the time and thus generating a throughput of 57.5.

Table 15.1 also shows other performance measures of the two-station line under the push and under two other pull (Kanban) systems with $(K_1, K_2) = (5, 5)$ and $(K_1, K_2) = (2, 19)$.

Table 15.1    Performance of the Two-Station Line Obtained Using Simulation

| Performance measures | Push system | Pull system with $(K_1, K_2)$ | | |
|---|---|---|---|---|
| | | (5, 5) | (2, 9) | (2, 19) |
| Throughput *(per hour)* | 57.5 | 54.3 | 57.5 | 60 |
| Utilization | 91% | 86% | 91% | 95% |
| Line's inventory | | | | |
| Average | 12.62 | 7.72 | 6.81 | 11.85 |
| Maximum | 118 | 10 | 11 | 21 |
| Line's flow time | | | | |
| Average *(in minutes)* | 13.17 | 8.53 | 7.10 | 11.85 |
| Maximum *(in minutes)* | 120.3 | 40.66 | 42.51 | 55.70 |

From Table 15.1, we make the following observations:

- *Inventory limits should be chosen carefully:* The pull system with $(K_1, K_2) = (5, 5)$ puts a maximum limit of 4 on in-buffer inventory of Workstation 2. This reduces the average and maximum inventory in the line to 7.72 and 10 units, respectively—which are lower than those in the push system. However, the throughput of the line is also reduced to 54.3 units per hour, which is less than the throughput of the push system. The reason is that limit $K_2 = 5$ is too low, resulting in more occasions that the buffer of Workstation 2 is empty. This reduces the utilization of Workstation 2 to 86 percent, compared to that of push system which is 91 percent. Hence, when setting a limit on the maximum inventory of a process, one should not set the limit too low. This will reduce the throughput of the line.

- *Pull systems are more efficient:* The pull system with $(K_1, K_2) = (2, 9)$ has average and maximum inventory 6.81 and 11, respectively, which are lower than those of the push system. This pull system, however, achieves the same throughput as that of the push system. It also has a lower average and maximum flow time of 7.10 and 42.51 minutes, respectively, which are much lower than those of the push system, which are, 13.17 and 120.3 minutes. In conclusion, under a pull (Kanban) system, if the limits on the maximum inventory are set properly, a pull system can achieve the same throughput of its push counterpart, but with lower inventory and lower flow times.

- *Producing more with less:* Compared to the system with $(K_1, K_2) = (2, 9)$, Kanban system with $(K_1, K_2) = (2, 19)$ has a larger limit on in-buffer inventory at Workstation 2. This results in average and maximum inventory of 11.85 and 21, respectively, which are higher than those under $(K_1, K_2) = (2, 9)$; however, it generates a higher throughput (i.e., 60 per hour) than the throughput when $(K_1, K_2) = (2, 9)$. The interesting observation is that the pull system under Kanban policy $(K_1, K_2) = (2, 19)$ generates more throughput than the push system and still has less inventory than that in the push system. In other words, it produces higher throughput with less inventory.

- *Increasing inventory limits increases throughput:* In general, as the limits on maximum allowed inventory increase, the throughput of the line increases. As shown in Table 15.1, increasing inventory limit at Workstation 2 from 9 in $(K_1, K_2) = (2, 9)$ to 19 in $(K_1, K_2) = (2, 19)$ increases the throughput from 57.5 to 60 per hour. There are some cases that increasing these limits may not affect throughput (e.g., increasing $K_1$ in our two-station example). But, increasing these limits never decreases the throughput.

By changing limits $K_1$ and $K_2$, managers can run the line to observe the throughput or they can develop a computer simulation to estimate the throughput of the line—as mentioned earlier, pull systems set a limit on inventory and observe throughput. What are the best limits $K_1$ and $K_2$ for a production process? The answer depends on the capacities of the stations and their variabilities. It also depends on what process-flow measures (e.g., throughput, inventory, flow time) the managers would like to achieve.

### 15.9.3 Implementing Pull—The Single-Kanban System

The pull system we chose for our two-station line example was a Kanban system. In a Kanban system, a worker is allowed to process a job and move it to the next station, only if the inventory in the next station is below a maximum level. This means that the worker in each workstation should always know the exact number of items in the buffer of the worker in the next station. This is often not easy, because a worker may not be able to see the inventory in the next station, or the inventory is piled up such that it is not easy to count it. Furthermore, even if it is possible to count inventory at the next station, it is not efficient for a worker to spend time counting inventory each time the worker finishes processing a job. Hence, some mechanism is needed to provide each worker with the information about the inventory level at the worker's next station—or even better—just tell the worker whether the worker needs to process another job or stay idle.

To achieve this, Taichi Ohno designed a system that used cards to inform workers whether they need to process another job. In Japanese, Kanban means "signal" and "card." This is where the policy gets its name. The main use of the cards in Kanban system is to authorize production. The simplest version of Kanban system that uses only one type of card is called *Single-Kanban System*. The main information on a Kanban card includes item's name, the production quantity needed, the name of the stage that needs the item, and the name of the stage that produces the item.

Figure 15.8 depicts a single-Kanban system for two stages of a process. In single-Kanban system, each stage or workstation has an output buffer, which is also the input buffer of the next stage. Each stage also has a "Kanban Post" where Kanban cards are posted. Every stage has its own number of Kanban cards used to control production in that stage. All parts (e.g., jobs, products, containers, etc.) at the output buffer of each stage have a Kanban card attached to it.

We now describe how, through Kanban cards, a demand for a part at Stage B in Fig. 15.8 triggers production at upstream Stage A.

1. The worker at Stage B, which we call Worker B, sees a Kanban card on the Kanban post in that stage. This is an authorization for producing a part. The worker then

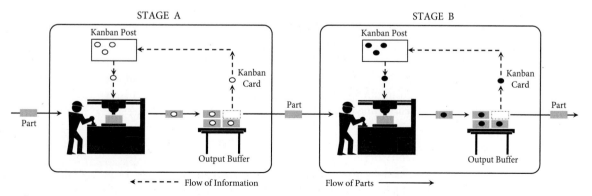

**Figure 15.8   Single-Kanban system in two stages A and B.**

takes a part from the output buffer of Stage A. The Kanban card attached to the part is separated and is posted on the Kanban post of Stage A.

2.   Each card at the Kanban post at Stage A is an indication that one part was taken from the output buffer of that stage. Hence, each card authorizes the worker at Stage A, which we call Worker A, to produce a new part.

3.   To produce a new part, Worker A takes a part from the input buffer of Stage A—which is the output buffer of its upstream stage. When Worker A finishes producing the new part, the worker takes a Kanban card from the Kanban post in Stage A, attaches it to the finished part, and puts the part in the output buffer of Stage A.

4.   Worker A repeats this process until there is no card on the Kanban post in Stage A, whereupon the worker remains idle until a Kanban card is posted on the Kanban post of Stage A. This procedure is the same for the worker at Stage B.

Three points worth mentioning here. First, each Kanban card at Kanban post of Stage A is a signal of a need at Stage B. Hence, the total number of cards at Stage A is the maximum need of Stage B, which is the maximum inventory allowed at Stage B. In our two-station line example, with $(K_1, K_2) = (2, 9)$, the maximum inventory allowed at Workstation 2 is 9. Hence, assigning nine Kanban cards at Workstation 1 will limit the maximum inventory at Workstation 2 to 9.

Second, a Kanban card can be attached to one part, or to a container that holds several parts. For example, in processes in which production is in batches, one batch of parts in a pallet or a container can have one Kanban card.

Third, single-Kanban systems can often be implemented without using any actual cards. Different mechanisms can be used to limit the maximum inventory between two stages. For example, the buffer between two stages can be designed such that it can only store up to a limit. When the worker sees that the buffer is full (no need to count), the worker stops processing more jobs. Another example is "Kanban Squares." Kanban squares are marks on the floor or on work table between two workers, each indicating the space for one unit. Workers put the finished jobs on the squares. Workers stop production when there are no empty squares remaining. In this case, the number of squares represents the maximum inventory allowed in the buffer. The more advanced form of Kanban cards are "Electronic Kanbans" that use bar codes and RF transponders to keep track of inventory at each stage of the process. When a worker takes a part from its input buffer, it scans the bar code on the part, which sends a signal to the previous stage to start producing a new part.

**Want to Learn More?**

#### Other Ways of Implementing Pull

Single-Kanban systems are not the only way to limit inventory. *Dual-Kanban systems* are another approach for implementing pull. Dual-Kanban systems are used when there are large distances between the stages of the process. The third approach to implement pull is to use a *Conwip system*. Instead of restricting inventory at each stage of the process below a certain limit (as in single-Kanban systems), Conwip systems restrict the *total inventory* in the process below a certain limit, without imposing any restriction on the inventory at each individual stage. How? You can find the answer in the online supplement of this chapter.

### 15.9.4 Benefits of Pull Systems

As our examples showed, pull systems—Kanban or Conwip, or other pull systems—can result in higher throughput, but with the same or less inventory than push systems. Why? The answer is because pull systems limit the inventory in the process. This limit prevents

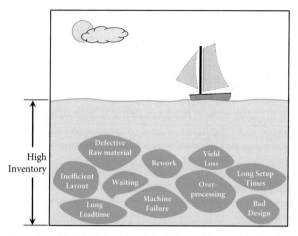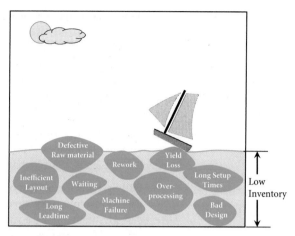

Figure 15.9    *Left:* High inventory level hides problems. *Right:* Low inventory level reveals problems.

inventory to become extremely large, which is more than enough to achieve a certain performance (e.g., a certain throughput). Besides higher throughput, the limit on inventory in pull systems provides additional benefits. Hopp and Spearman (2011) provide a list of these benefits, which we discuss below.

### Pull Systems Reduce Manufacturing Costs

Holding inventory costs money and thus adds to the manufacturing costs of products. By achieving the desired throughput with lower inventory, pull systems directly reduce the manufacturing costs.

### Pull Systems Make Problems Visible Earlier

When the inventory between two stages of a process is large, the downtime of the upstream stage—for example, due to machine failures or long setup time—does not stop the downstream stage from processing jobs. The downstream stage uses the large inventory between the stages until the upstream stage is up again. From Toyota Production System's perspective, there are two issues here. One is the main issue of having a large inventory level, which is considered a waste. The second is that, since the downstream stage is not stopped even when the upstream stage is down, the need to prevent the machine failure or reduce the setup time in the upstream stage is not felt. In other words, the large inventory hides the sources of wastes (e.g., machine downtime and long setup times). In pull systems that have lower inventory, these sources of waste become more visible. The reason is that, due to limited number of inventory between the two stages, a long setup or a long machine failure at the upstream stage results in starvation of the downstream stage much earlier and thus makes these sources of waste more pronounced.

This benefit of pull systems are often described as "river analogy," as shown in Fig. 15.9. Think of inventory as the water in a river and sources of waste as the rocks at the bottom of the river. Higher water levels hide the rocks (the left figure). This is analogous to high inventory creating operating cushion that allows the process (the boat) a smooth sail, despite the underlying problems. By lowering the water level, the rocks—the waste—are exposed and can be removed (the right figure).

### Pull Systems Result in Increasing Effective Capacity and Reducing Effective Variability

As mentioned above, inventory limits in pull systems make non-value added tasks such as machine failures, setups, and rework more visible and put pressure to reduce or eliminate them. Eliminating or reducing these supporting tasks increases the effective capacity of the process (see Chap. 4) and decreases the effective variability (see Chap. 5). Higher capacity and lower variability improve process performance.

### Pull Systems Result in Low Defective Rate

Due to the limit on inventory, pull systems cannot operate with high defective rate. Pull systems do not carry extra inventory to buffer against defective items. A high defective rate results in a large number of defective items in the buffer of each stage that cannot be used. This, in turn, leads to starvation of the stage and eventually reduces throughput. Reduction in throughput triggers action to reduce the defective rate.

### Pull Systems Result in Low-Quality Cost

Due to low inventory, quality inspection is more effective. Specifically, it is easier to spot a defective item among a small number of items than under a pile of large inventory. Hence, detecting low-quality items is easier and occurs earlier. This allows managers to fix the stage that produces low-quality items before the stage produces a large number of those items. Consequently, the cost of quality—the cost of losing the defective items or the cost of rework to fix the defective items—is lower in pull systems.

### Pull Systems Provide Flexibility and Responsiveness

Because of low inventory, pull systems are more flexible and responsive in changes in the product and in the market. To keep inventory low, pull systems keep customer orders on papers as much as possible and release the order into the process as late as possible. Hence, if for example the product design is changed, or if customers change their orders, pull systems are more flexible and responsive to incorporate those changes. This is because pull systems have less unfinished work in the process that needs to be changed; and for orders on the paper, parts will be released according to the new requirements.

### Pull Systems Result in Improving Customer Service

Because pull systems have a limit on the maximum inventory, they have less variability in the flow times—the time a customer order is released to the system until it is completed and leaves the system. Let's compare Kanban systems with push systems to illustrate this. In push systems, the inventory at each stage can vary from zero to a very large number. Hence, the flow time of a job entering each stage can vary between zero and a very large number. Consequently, the flow time at each stage of a push system has a high variability. In Kanban systems, on the other hand, there is a limit on the maximum inventory at each stage, which prevents inventory to become very large. Thus, the flow time at each stage of a Kanban system is limited and cannot be very large. Therefore, the flow time at each stage of a Kanban system has lower variability than that if the process is run as a push system. How does that improve customer service?

Well, lower variability in flow times means that it is easier to predict how long it takes for a customer order to go through the process. This helps managers to give customers more accurate due dates and delivery times, resulting in a better customer service.

## 15.10 Lean Tool 7—Quality at the Source

Lean processes, as described above, cannot operate efficiently with low-quality items passing from one stage to another. Hence, fixing quality problems at its source is one of the essential tools of lean operations. The main efforts to manage quality in lean operations can be divided into (1) prevention, (2) detection, and (3) solution. We summarize these three in this section.

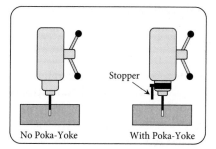 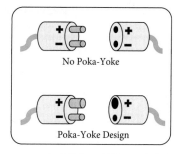

**Figure 15.10**   *Left:* Poka-Yoke in a drilling process, the stopper prevents the drill from drilling a whole longer than needed. *Right:* Poka-Yoke in product design, preventing worker from making a mistake in the assembly process.

### 15.10.1 Preventing Quality Problems

The quality issues in a product is due to either product design, or the manufacturing process that makes the product. Lean operations uses different techniques in the design and the manufacturing process to prevent errors leading to quality issues.

**Simplifying Tasks and Processes to Reduce Errors**

By breaking tasks into smaller and simpler tasks and with proper training, the chance of workers making errors in performing their tasks is reduced. A visual workplace also helps reduce worker mistakes. Standard operations sheet showing how tasks should be done, material handling routes marked clearly on the shop floor, color-coded machines, and stock point are examples of visual workplace (see Sec. 15.16 for details).

**Poka-Yoke**

Mistake-proofing, called Poka-Yoke in Japanese, is to design the product or the process such that the worker cannot make a mistake. Figure 15.10 shows two examples of Poka-Yoke in product design and manufacturing process.

Examples of Poka-Yoke in service systems are abundance. Medical laboratories ask patients to confirm their names and dates of birth before they take patients' blood. This prevents errors in mixing up blood samples. After taking a customer's order, restaurant cashiers and waiters read the order back to the customer to make sure they got it right. Some restaurants ask their waiters to put a different shaped coaster in front of diners who want decaffeinated coffee. This prevents waiters to pour regular coffee in dinners' cups by mistake. One Korean amusement part sews the pockets of its new employees' trousers closed to make sure they do not put their hands in their pockets. This helps employees to maintain a formal decorum and welcoming attitude. To make sure room cleaners correctly identify the towels that need to be replaced, some hotels put paper stripes around new towels. To make sure that passengers lock the lavatory door in airplanes, the light is not turned on until the door is locked. When customer transactions is done, ATM machines beep to signal customers to remove their cards.[11]

### 15.10.2 Early Detection of Quality Problems

Task simplification and Poka-Yoke are means to prevent quality issues. However, it is not always possible to prevent quality issues. If quality issues are to occur, lean operations principle is to detect the quality at its source as soon as it occurs. Lean operations achieves this in several ways.

---

[11]Chase, R.B. and D.M. Stewart, "Make Your Service Fail-Safe." *Sloan Management Review*, Spring (1994), 35–44.

### Workers Are Responsible for Quality

In traditional manufacturing processes, there were always quality control stations—often located at the end of the process—in which quality inspectors checked for quality problems. Taichi Ohno, however, strongly believed that workers, not inspectors, must be responsible for the product quality. Hence, in lean operations every worker in each workstation is in charge of checking the quality of the finished part in that station as well as the quality of the parts that are received from previous stations. Furthermore, in case of producing a defective item, the worker himself (instead of a rework station) is responsible to fix the quality problem. This gives workers full responsibility for product quality. The responsibilities to detect and to fix the quality problems facilitate the early detection of quality issues.

### Jidoka—Self-Stopping Machines

Recall from Sec. 15.1 that one of Sakichi Toyoda's invention was a mechanism in power loom machines that automatically stopped the machine when a thread broke. As mentioned earlier, this invention evolved into a system called *Jidoka*. The original idea of designing machines or processes that stop when they go out of tune, however, is still used in lean operations to automatically detect the quality issues and stop the process from producing more. Machines are now equipped with sensors that can detect any deviation from normal operating conditions and stop the machine to avoid producing defective items.

### Strict Rules for Compliance

To ensure that defective items do not pass to the next stage of the process, lean operations insists on strict rules of compliance. This means that every item must be checked to comply with quality standards. If a 100-percent inspection is not possible, lean operations often checks the first and the last item of a batch for quality issues. If these two items are non-defective, then it is assumed that the entire batch is non-defective.

### Statistical Quality Control

Lean operations uses statistical quality and process control methods to ensure that the process performance and its output are within acceptable range. These analytics (e.g., Statistical Process Control, or SPC) enable workers and managers to detect problems with the process performance and quality early. SPC is discussed in detail in Chap. 16.

### 15.10.3 Rapid Resolution of Problems

Fujio Cho, the president of Toyota Motor Corporation from 1999 to 2005, recalls Taichi Ohno's comment on rapid correction of quality issues as:[12]

> **Fujio Cho:** *Mr. Ohno used to say that no problem discovered when stopping the line should wait longer than tomorrow morning to be fixed. Because when making a car every minute, we know we will have the same problem again tomorrow.*

To facilitate rapid solution, Toyota Production System (TPS) emphasizes on stopping the process when a quality problem occurs.

### Jidoka—Stopping the Line: Quality before Throughput

TPS emphasizes on "quality first" even if it means stopping production (i.e., losing throughput) to fix the quality problem. In his book, *The Toyota Way*, J.K. Liker writes about Russ Scaffede who worked decades for General Motors and had an excellent reputation as a manufacturer manager who could get things done. He later worked for Toyota and his task was to

---

[12] Liker, J.K., *The Toyota Way: 14 Management Principles from the World's Greatest Manufacturer.* McGraw-Hill, 2004, p. 128.

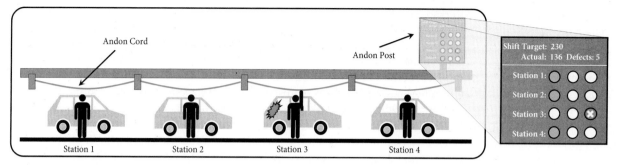

**Figure 15.11** A four-station assembly line with Andon system. The worker at Station 3 pulled the Andon cord due to a quality problem.

launch a new plant for Toyota in America. He worked very hard, with the help of his mentors (including Fujio Cho), to design and run the plant according to TPS principles. Liker writes:

> *Scaffede had learned the golden rule of automotive engine production: do not shut down the assembly plant! At General Motors, managers were judged by their ability to deliver the numbers. Get the job done no matter what—and that meant getting engines to the assembly plant to keep it running. Too many engines, that was fine. Too few, that sent you to unemployment line.*
>
> *So when Cho remarked to Scaffede that he noticed he had not shut down the assembly plant once in a whole month, Scaffede perked up: "Yes sir, we had a great month, sir. I think you will be pleased to see more months like this." Scaffede was shocked to hear from Cho:*
>
>> *Russ-san, you do not understand. If you are not shutting down the assembly plant, it means that you have no problems. All manufacturing plants have problems. So you must be hiding your problems. Please take out some inventory so the problem surface. You will shut down the assembly plant, but you will also continue to solve your problems and make even better-quality engines more efficiently.*

Later when Fuiji Cho was asked about the cultural differences in managing Toyota plants in America (i.e., Georgetown plant in Kentucky) and plants in Japan, he stated that the number one problem was to get group leaders and team members to stop the line. He explained that it took many months to "re-educate" his managers that it was necessary to stop the line, because that would result in continual improvement in the process. This goes to the core philosophy of Toyota Production System—focusing on long-term goals of improving the process, instead of focusing on myopic short-term goal of improving productivity.

Stopping the line originated from the concept of Jidoka. Recall that in Toyota terminology Jidoka refers to "automation with a human touch," and initially corresponded to machines that can stop themselves when facing problems. In his book, *Toyota Production System*, Taichi Ohno emphasizes on the importance of stopping the line. He writes:

> **Taichi Ohno:** *Stopping the machine when there is trouble forces awareness on everyone. When the problem is clearly understood, improvement is possible. Expanding this thought, we establish a rule that even in a manually operated production line the workers themselves should push the stop button to halt production if any abnormality appears.*

In a broader sense, Jidoka refers to the authority given to the workers to stop the line. But, how does a worker stop a line? In Toyota Production System, each worker has access to a mechanism that can either announce a quality problem detected in the line, or stop the line. The most commonly used mechanisms are *Andon Cord* and *Andon Switch*.

Andon cord is a cord parallel to the assembly line and close to the worker, see Fig. 15.11. By pulling the cord, a worker can stop the line if the worker detects a quality problem in the workstation. Andon switches, on the other hand, are installed in each workstation and—after being activated by the worker—stops the line. A cord, however, is more used in larger assembly lines, since it gives the worker an immediate access, wherever the worker is in the workstation. With Andon switches, the worker may need to walk to the switch, depending on where in the station the worker is.

Some lines use two Andon cords with different colors—yellow and red. The yellow color cord is used when the worker detects a quality problem and there is a possibility that the worker can fix the problem before the product reaches the end of the station. If the worker cannot fix the problem, then the worker pulls the red cord that stops the line. Also, in most cases, even if the red cord is pulled before the product reaches the station, the line does not stop until the product reaches the end of the worker's station. This has several advantages. First, the other stations in the line are still working when the worker is calling for help. Second, it does not interrupt other stations in the middle of their assembly tasks. Hence, when the line stops, all workers have already finished the work in their stations. Third, when the problem is fixed, all workers can start with the new product in their station, which is more convenient.

Andon cords are connected to *Andon Board* that displays different information about the status of stations in the line. For example, it shows which station in the line has quality problem (e.g., the worker pulled the yellow Andon cord), or which station has stopped the line (i.e., the worker pulled the red Andon cord). It also shows additional information such as actual and target production quantities, or number of defective items produced. There are several different Andon posts used in assembly lines. The simple ones use different color lights (e.g., yellow for quality problems, and red for line-stopping problems) and LED number displays. Others are large screen monitors controlled by computer programs. An example of an Andon post is also illustrated in Fig. 15.11.

### Go and See (Genchi Genbutsu)

Recall from Sec. 15.1 that according to Genchi Genbutsu, the best way to understand a problem is to go and see it on the ground. Hence, when a quality problem is detected by a worker and the worker pulls the (yellow) Andon cord to announce the problem, line supervisor rushes to the station to see the problem firsthand and help correct the problem. If the problem is not resolved quickly (before the car leaves the worker's station), the red Andon cord is pulled which brings more people to the station to see the problem and find ways to resolve it more rapidly.

### Promote Worker Involvement in Problem Solving

There is a quote from Taichi Ohno that says:[13] "Having no problems is the biggest problem of all." The underlying reason behind this statement is that TPS treats each problem as an opportunity for improvement. When a problem occurs in TPS, it is considered as a system failure and the worker is not blamed for that. Also, when a problem occurs, the best person with the most expertise to consult is the worker who performed the task. Thus, it is a natural solution to involve workers in the quality problem solving process. In addition to finding a better solution more rapidly, there are other advantages with involving the worker. It has been shown that worker involvement in decision-making processes increases worker's motivation and productivity (see Sec. 15.17).

### Find the Route Cause of the Problem—Ask "Why" Five Times

When a problem occurred, Taichi Ohno encouraged his staff to go and see the problem firsthand without preconceptions. He advised them to ask "why" five times about every matter.

---

[13] From "Ask 'why' five times about every matter." Toyota Global Site, released in March 2006.

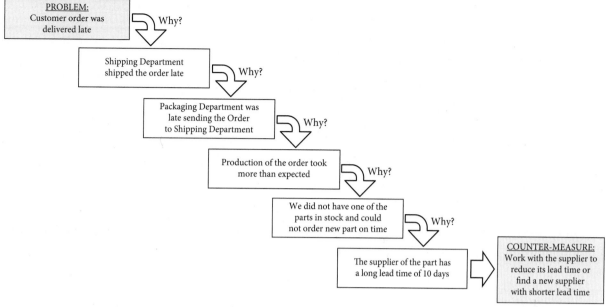

**Figure 15.12**    Asking "why" five times to find the root cause of a late delivery to a customer.

There is no magic with number "five." The idea is to keep asking why, until you get to the root cause of the problem.

Figure 15.12 shows an example of this approach to find the root cause of a customer complaint about the late delivery of the customer's order. As the figure shows, the main problem with late delivery is rooted in the long lead time of the supplier. Finding the root cause of the problem helps managers to find the correct solution for the problem—for example, finding a supplier with shorter lead time.

Note that, there might be several answers to a "why" question. For example, the reason for production taking more time than expected might be lack of supply or long repair time of a failed equipment or both. In this case, the 5-why approach should be used in each of these cases to find the root cause of (i) lack of supply, (ii) machine failure, and (iii) long repair time.

## 15.11 Lean Tool 8—Total Productive Maintenance

Machine failures interrupt operations and have serious consequences. In some cases, a machine failure stops the machine from working, resulting in loss of working capacity. In other cases, a machine failure does not stop the machine, but may (i) slow down the speed (capacity reduction), (ii) result in low-quality items that are defective or require rework, or (iii) result in an accident leading to safety issues for the worker. Since lean operations work with minimum inventory, it cannot tolerate machine failures and its consequences. Seiichi Nakajima, the "father of Total Productive Maintenance (TPM)" is quoted as saying:[14]

> **Seiichi Nakajima:** *Just-in-time manufacturing, Toyota's production system, could not exist without TPM. Trouble-free equipment leads to uninterrupted flow, improved quality, reduced waste, and lower costs.*

---

[14] 1. Nakajima, Seiichi, *Introduction to Total Productive Maintenance*, Productivity Press, 1988. 2. Nakajima, Seiichi (editor), *TPM Development Program: Implementing Total Productive Maintenance*, Productivity Press, 1989.

Hence, lean operations needs to eliminate equipment failures. In fact, the goal of lean operations is zero machine breakdown. But is that really possible? This has been often received with skepticism by managers. Seiichi Nakajima was aware of this and was quoted as saying:

> **Seiichi Nakajima:** *Americans will struggle with TPM because you expect equipment to break down and then the maintenance group will fix it. The goal of zero equipment breakdowns is being received with significant skepticism and even denial in many companies. And yet, these same companies have goals of zero defects and zero accidents.*

But is zero breakdown an idealistic and unachievable goal? Is it possible that a company, for example, has no interruptions due to equipment failures throughout a year? Well, yes. By implementing Total Productive Maintenance, Aishin Seiki, a supplier of automotive parts, has not experienced an equipment failure in more than 4 years. Before they implemented TPM, they had more than 700 equipment breakdowns a month.

---

**CONCEPT**

### Total Productive Maintenance

*Total Productive Maintenance* or TPM is a comprehensive approach to equipment maintenance with the goal of having no breakdowns, no slowdowns, no defects, and no accidents.

TPM achieves its goals through (1) Autonomous Maintenance, (2) Preventive Maintenance, and (3) Maintenance Efficiency Improvement.

### 15.11.1 Autonomous Maintenance

*Autonomous Maintenance* refers to the practice of machine operators being responsible for repairing and maintaining their own machines. This includes activities such as cleaning, lubrication, periodic inspections, and preventive repair activities. Workers often gather and interpret operating and maintenance data to identify signs of deterioration prior to a breakdown. Workers also are in charge of cleaning and organizing their tools, equipment, and workplace. This allows them to better detect unusual occurrences. For example, oil stains (leaked from a machine) are easier to spot on a clean floor than on a dirty floor.

Autonomous maintenance has several advantages: (i) It gives the operator a great sense of ownership of the equipment, and develops worker's pride in keeping the machines in top condition. (ii) It increases operator's knowledge of the equipment, which helps the operator detect possible equipment breakdowns before it occurs. (iii) It lowers the maintenance department workload to focus on larger and more critical tasks. All these increase the equipment availability, resulting in higher effective capacity.

Of course, autonomous maintenance cannot be effectively performed without worker's education and training. In fact, TPM includes systematic training of workers (to perform their maintenance tasks), maintenance personnel (to learn techniques for preventive and predictive maintenance), and managers (to learn TPM principles).

### 15.11.2 Preventive Maintenance (PM)

Suppose you have just bought a new car. How do you maintain your car? Do you use your car until it breaks down? Or, following manufacturer's maintenance instructions, you take your car for a service every 5000 to 10,000 miles? Most of us do the latter. Why? Several reasons:

- *Cost Saving:* The cost of service maintenance—for example, checking the transmission oil and replacing it, if needed—is much less than the cost of replacing a broken transmission due to lack of oil.

- *Higher Availability:* It takes less time—about 2 or 3 hours—to change transmission oil; while it takes several days to install a new transmission on your car. This means that you would not have your car available for several days.

- *Less Uncertainty:* You have control over when you would like to schedule your service maintenance. However, you do not know exactly when your transmission will fail because you did not do your maintenance.

Like you, operations systems also face the same dilemma. Their equipment are subject to aging and breakdowns. To maintain their equipment, they perform two main activities: (i) Breakdown Maintenance, and (ii) Preventive Maintenance. Breakdown maintenance is performed when an equipment fails—analogous to repairing your transmission after it fails. Preventive Maintenance (or PM) is performed before the equipment fails—analogues to changing transmission oil in your car. The goal of preventive maintenance is to prevent unplanned downtime due to breakdowns and to keep equipment in optimum working condition. Preventive maintenance includes activities such as lubrication, inspection, measuring and checking components, and replacement of various worn components. Performing a well-planned preventive maintenance program, as we explained above, can save time and cost.

Besides saving cost, preventive maintenance results in higher effective capacity and lower effective variability, both of which improve process performance. To explain this, recall that availability of a resource is

$$A = \frac{m_f}{m_f + m_r}$$

where $m_f$ is the average uptime and $m_r$ is the average downtime of a resource. Also, from Eqs. (5.4) and (5.6) we have

$$T_{\text{eff}} = \frac{T_0}{A}$$

$$\text{CV}_{\text{eff}}^2 = \text{CV}_0^2 + (1 + \text{CV}_r^2)A(1 - A)\frac{m_r}{T_0}$$

Note that under preventive maintenance the resource downtime $m_r$ is significantly lower than that under breakdown maintenance. Lower downtime $m_r$ results in higher machine availability $A$.[15] As shown in the above formulas, higher availability results in smaller effective process time $T_{\text{eff}}$, and thus in higher effective capacity $C_{\text{eff}} = 1/T_{\text{eff}}$. Also, lower downtime $m_r$, as shown above, results in lower effective variability $\text{CV}_{\text{eff}}^2$.

In conclusion, a well-planned preventive maintenance saves maintenance and repair and quality costs, increases machine availability, increases effective capacity, and reduces the effective variability of an equipment. The main question in implementing preventive maintenance programs is when should the machine be stopped to perform preventive maintenance. Here, we present two different analytics that determine the optimal time to stop an equipment to perform preventive maintenance. Each model has a different objective. The goal of the first model is to maximize effective capacity of the equipment, while the goal of the second model is to minimize the costs associated with repair and maintenance of the equipment.

### 15.11.3 A PM Model to Maximize Effective Capacity

We use the following example to illustrate the analytics and the procedure of finding the optimal preventive maintenance schedule in order to maximize effective capacity of a resource subject to breakdowns.

---

[15]Depending on the value of $m_r$ the machine availability might be lower; however, a preventive maintenance program can be developed to increase availability, see Sec. 15.11.3.

**Example 15.1**

A single punch press is the bottleneck of a process that produces exterior panels used in the production of small dehumidifiers. The press works three shifts—24 hours a day. It takes the press an average of 20 seconds to cut a panel. The press is old and has frequent breakdowns. Maintenance department has the data regarding the time to failure of the press as shown in Table 15.2. Specifically, the table shows the probability distribution of number of hours the press works until its next failure occurs, after the press is repaired and put in good operating condition. When the press fails, it takes an average of 3 hours to repair the press (i.e., breakdown maintenance).

To increase the effective capacity of the press, the maintenance department suggests that the operations on the press is stopped after 10 hours of working and a preventive maintenance (PM) is performed. The PM can put the machine back in good operating condition. The PM operations include cleaning, adjustment (if the dies are not aligned), and lubrication. It takes an average of 12 minutes to perform the PM operations. When asked by the production manager if the PM prevents press failures, the maintenance department response was: "Well, no. Even when we stop the press and perform PM for 12 minutes, there is still a chance that the press fails between two PM operations. Hence, for those failures we still need to do the repair that takes about 3 hours. However, what PM does is that it decreases the chance of a failure, compared with not doing PM."

The production manager is not convinced that the PM operation is helpful in increasing the effective capacity of the press. In fact, the production manager worries that stopping the press for preventive maintenance on top of disruptions due to breakdown maintenance—that can still occur—may decrease the effective capacity of the press. How much do you think the PM program suggested by the maintenance department increases or decreases the capacity of the press? What is the maximum capacity that can be achieved under such preventive maintenance?

Before we compute the effective capacity of the punch press under PM, we first compute the current effective capacity of the press. According to Table 15.2, the average time to failure—the press up time—is

$$m_f = \sum_{t=1}^{20} tP(t) = 1(0.004) + 2(0.008) + \cdots + 20(0.009) = 10.95 \text{ hours}$$

Table 15.2    Time to Failure Data for the Punch Press in Example 15.1

| Time to failure $t$ | Probability $p_t$ | Time to failure $t$ | Probability $p_t$ |
|---|---|---|---|
| 1 | 0.004 | 11 | 0.102 |
| 2 | 0.008 | 12 | 0.118 |
| 3 | 0.012 | 13 | 0.091 |
| 4 | 0.034 | 14 | 0.077 |
| 5 | 0.038 | 15 | 0.063 |
| 6 | 0.044 | 16 | 0.039 |
| 7 | 0.055 | 17 | 0.026 |
| 8 | 0.064 | 18 | 0.021 |
| 9 | 0.086 | 19 | 0.019 |
| 10 | 0.090 | 20 | 0.009 |

Considering the repair time with the average of $m_r = 3$ hours, the availability of the press with no PM is

$$A = \frac{m_f}{m_f + m_r} = \frac{10.95}{10.95 + 3} = 0.785 = 78.5\%$$

Since the theoretical mean process time of the press is $T_0 = 20$ seconds, the press will have the effective mean process time $T_{eff} = T_0/A = 20/0.785 = 25.48$ seconds, and thus effective capacity of $C_{eff} = 1/T_{eff} = 1/25.48 = 0.0392$ per second $\times 3600 = 141.12$ per hour.

Does the PM program suggested by the maintenance department increase the effective capacity of the process? It does, if it can increase press availability of $A = 0.785$. The following analytics shows how machine availability can be computed under a preventive maintenance program that stops operations after a predetermined number of working hours to perform PM activities.

### Computing Machine Availability Under a Preventive Maintenance (PM) Program

Consider a machine that is subject to failure. When the machine fails, repair operations are performed that put the machine back to good working condition. The repair operations take an average of $m_r$ units of time. Consider the following preventive maintenance (PM) program:

- The PM is performed each time the machine has been working for $T$ time units without failure since the last time the machine was repaired or maintained.
- If the machine fails before $T$ time units, the repair operations start and put the machine back to good working condition.
- Performing PM activities take an average of $m_p$ time units and put the machine back in good operating condition.

When the machine is put in good operating condition—after a repair or a preventive maintenance—the time it takes until the machine fails again is random and follows probability density function $f(t)$ with cumulative distribution function $F(t)$ and average $m_f$.

Note that the machine goes through repeated cycles in which the machine uptime is followed by machine downtime (during which the machine is either under repair or under preventive maintenance). Hence, the machine availability under the PM program with parameter $T$ would be

$$A(T) = \frac{\text{Average uptime in a cycle}}{\text{Average uptime in a cycle} + \text{Average downtime in a cycle}} \quad (15.1)$$

What is the average uptime in a cycle? Set the time at the beginning of a cycle to zero, and consider time $t$ when the machine would fail again. If the machine fails before $T$, that is, if $t \leq T$, the machine's uptime would be $t$. However, if the machine does not fail before $t$, that is, if $t > T$, the machine uptime would be $T$. Hence,

$$\text{Average uptime in a cycle} = \int_0^T tf(t)dt + \int_T^\infty Tf(t)dt$$

$$= \int_0^T tf(t)dt + T[1 - F(T)] \quad (15.2)$$

The average downtime in a cycle can also be obtained in the same way. If the machine fails before $T$, which occurs with probability $F(T)$, the machine will be down for repair that takes an average of $m_r$ time units. However, if the machine does not fail before $T$ time units, which occurs with probability $1 - F(T)$, the machine goes under PM which takes an average of $m_p$ time units. Hence,

$$\text{Average downtime in a cycle} = m_r F(T) + m_p[1 - F(T)] \tag{15.3}$$

Substituting Eqs. (15.3) and (15.2) in Eq. (15.1), we get the machine availability under the PM program with parameter $T$ as follows:

$$A(T) = \frac{\int_0^T tf(t)dt + T[1 - F(T)]}{\int_0^T tf(t)dt + T[1 - F(T)] + m_r F(T) + m_p[1 - F(T)]} \tag{15.4}$$

The time to failure distribution is often obtained or approximated by a discrete probability mass function. Specifically, by gathering data, one can estimate $p_t$, the probability that the machine fails after working for $t$ time units, for $t = 1, 2, 3, \ldots$ Considering $P(t)$ as the cumulative distribution of time to failure, then the above formula would turn into

$$A(T) = \frac{\sum_{t=1}^T t\,p_t + T[1 - P(T)]}{\sum_{t=1}^T t\,p_t + T[1 - P(T)] + m_r P(T) + m_p[1 - P(T)]} \tag{15.5}$$

The above discrete model assumes that failures occur at the end of a time period $t$. If $t$ is measured in minutes, for example, then the model assumes that machine failure occurs at the end of a minute (not in the middle of a minute).

Since the time to failure for the punch press is in a discrete form, we use Eq. (15.5) to compute the availability of the punch press under the preventive maintenance with parameter $T = 10$ time units. Note that, according to Table 15.2, for $T = 10$ hours we get

$$P(T = 10) = P(t \le 10) = 0.004 + 0.008 + \cdots + 0.09 = 0.435$$

$$\sum_{t=1}^{10} tp_t = 1(0.004) + 2(0.008) + \cdots + 10(0.09) = 3.217$$

Hence,

$$
\begin{aligned}
A(T = 10) &= \frac{\sum_{t=1}^{10} t\,p_t + T[1 - P(T)]}{\sum_{t=1}^{10} t\,p_t + T[1 - P(T)] + m_r P(T) + m_p[1 - P(T)]} \\
&= \frac{3.217 + 10[1 - 0.435]}{3.217 + 10[1 - 0.435] + 3(0.435) + 0.2[1 - 0.435]} \\
&= 0.8621
\end{aligned}
$$

Therefore, under the PM program suggested by the maintenance department, punch press will have effective mean process time $T_{\text{eff}} = T_0/A = 20/0.8621 = 23.20$ seconds, and thus effective capacity $C_{\text{eff}} = 1/T_{\text{eff}} = 1/23.20 = 0.0431$ per second $\times 3600 = 155.16$ per hour. Considering the capacity without PM, which was 141.12 units per hour, the PM program results in $155.16 - 141.12 = 14.04$ increase in effective capacity of the punch press per hour—a $14.04/141.12 = 10\%$ increase.

This leads us to the following question. Is there any other PM program that can result in a higher effective capacity? To answer this question, we can repeat our calculation for different values of $T = 1, 2, 3, \ldots$ This is shown in Table 15.3.

As the table shows, the PM program under which the press is stopped for preventive maintenance after $T = 3$ hours results in the maximum availability of 0.9178. This would increase the effective capacity of the punch press—the bottleneck—to 165.2 units per hour. This is $165.2 - 155.2 = 10$ units per hour more than what the effective capacity would be under the PM program recommended by the naintenance department.

**Table 15.3**  **Finding the Optimal Value of $T$ that Maximizes Press Availability**

| $T$ | $p_T$ | $P(T)$ | $\sum_{1}^{T} tp_t$ | Average uptime | Average downtime | Availability $A(T)$ |
|---|---|---|---|---|---|---|
| 1 | 0.004 | 0.004 | 0.004 | 1.000 | 0.2112 | 0.8256 |
| 2 | 0.008 | 0.012 | 0.020 | 1.996 | 0.2336 | 0.8952 |
| 3 | 0.012 | 0.024 | 0.056 | 2.984 | 0.2672 | 0.9178 |
| 4 | 0.034 | 0.058 | 0.192 | 3.960 | 0.3624 | 0.9162 |
| 5 | 0.038 | 0.096 | 0.382 | 4.902 | 0.4688 | 0.9127 |
| 6 | 0.044 | 0.14 | 0.646 | 5.806 | 0.5920 | 0.9075 |
| 7 | 0.055 | 0.195 | 1.031 | 6.666 | 0.7460 | 0.8994 |
| 8 | 0.064 | 0.259 | 1.543 | 7.471 | 0.9252 | 0.8898 |
| 9 | 0.086 | 0.345 | 2.317 | 8.212 | 1.1660 | 0.8757 |
| 10 | 0.090 | 0.435 | 3.217 | 8.867 | 1.4180 | 0.8621 |
| 11 | 0.102 | 0.537 | 4.339 | 9.432 | 1.7036 | 0.8470 |
| 12 | 0.118 | 0.655 | 5.755 | 9.895 | 2.0340 | 0.8295 |

Recall that Eq. (15.5) assumes that machine failures occur only at the end of an hour (not in the middle). Because the data presented in Table 15.3 is for values of $t$ measured in hours, a better approximation for the optimal values of $T$ can be obtained if the data is gathered for smaller time units such as minutes. Alternatively, one can fit a continuous distribution on time to failure data—find $f(t)$—and use Eq. (15.4).

### 15.11.4  A PM Model to Reduce Maintenance Cost

The cost of performing a preventive maintenance is often lower than the cost of repairing a failed machine. Nevertheless, performing PM too many times results in large PM cost. Hence, a well-planned preventive maintenance program should also reduce the total cost of repairing and maintaining the machine. While the goal of previous section was to find the optimal PM program that maximizes machine availability, the goal of this section is to find the PM program that minimizes the total repair and maintenance costs. We return to the punch press example below:

**Example 15.2**

> Consider the punch press in Example 15.1 and assume that performing a repair on the failed press costs $100. This includes labor cost and cost of losing production. On the other hand, it costs $20 each time the firm performs a preventive maintenance operations on the press. What would be the expected cost under the PM program that maximizes the effective capacity of the press? Which PM program results in the lowest expected total repair and maintenance cost?

The following analytics helps us find the expected cost under a PM program with parameter $T$.

---

**ANALYTICS**

**Computing Expected Cost Under the Preventive Maintenance Program**

Consider the preventive maintenance program that stops a machine after $T$ units of time and performs a preventive maintenance that takes an average of $m_p$ units of time as in the previous analytics. Suppose that each time the machine fails and a repair operation is performed, it costs $C_r$. Also, assume that it costs $C_p$ to perform a preventive maintenance on the machine.

Again, consider a cycle in which an uptime is followed by a downtime. The average total repair and maintenance cost per unit time is therefore obtained by

$$\text{Average cost per unit time} = \frac{\text{Average cost in a cycle}}{\text{Average length of a cycle}}$$

If machine's time to failure follows probability density function $f(t)$, then the average length of a cycle is the denominator of Eq. (15.4). The total average cost of repair and PM in a cycle can be obtained as follows. If the machine fails before $T$, which occurs with probability $F(T)$, machine will be repaired which costs $C_r$. However, if the machine does not fail before $T$ time units, which occurs with probability $1 - F(T)$, the machine goes under PM, which costs $C_p$. Hence,

$$\text{Average Cost in a Cycle} = C_r F(T) + C_p[1 - F(T)] \tag{15.6}$$

and hence

$$\text{Average Cost per Unit Time} = \frac{C_r F(T) + C_p[1 - F(T)]}{\int_0^T tf(t)dt + T[1 - F(T)] + m_r F(T) + m_p[1 - F(T)]} \tag{15.7}$$

Similarly, for the case where the time to failure is modeled with a discrete probability distribution, we have

$$\text{Average Cost per Unit Time} = \frac{C_r P(T) + C_p[1 - P(T)]}{\sum_{t=1}^{T} t\, p_t + T[1 - P(T)] + m_r P(T) + m_p[1 - P(T)]} \tag{15.8}$$

Recall that the PM program that maximizes the effective capacity of the punch press (i.e., maximizes availability) has parameter $T = 3$ hours. We can now use the above analytics to find the total expected cost under this policy. From Table 15.2, for $T = 3$ hours we get

$$P(T = 3) = P(t \leq 3) = 0.004 + 0.008 + 0.012 = 0.024$$

$$\sum_{t=1}^{3} t\, p_t = 1(0.004) + 2(0.008) + 3(0.012) = 0.056$$

On the other hand, we have $C_r = \$100$ and $C_p = \$20$. Hence,

$$\begin{aligned}
\text{Average Cost per Unit Time} &= \frac{C_r P(3) + C_p[1 - P(3)]}{\sum_{t=1}^{3} t\, p_t + 3[1 - P(3)] + m_r P(3) + m_p[1 - P(3)]} \\
&= \frac{100(0.024) + 20[1 - 0.024]}{0.056 + 3[1 - 0.024] + 3(0.024) + 0.2[1 - 0.024]} \\
&= \$6.74 \text{ per hour}
\end{aligned}$$

Hence, the PM program with $T = 3$ hours that maximizes press availability costs about $6.74 per hour. But what is the minimum cost that a PM program can achieve? To answer this, we need to find the expected cost for different values of $T$, and see which one has the lowest cost. This is done in Table 15.4.

As the table shows, the minimum expected (repair and maintenance) cost of $4.80 per hour occurs when the PM program with $T = 7$ hours is implemented. Based on Table 15.2, this policy results in availability 0.8994. The press availability under this policy is about 2 percent—about $0.9178 - 0.8994 = 0.0184$ to be exact—less than the maximum possible availability. However, the cost under this policy is $6.74 − $4.80 = $1.94 per hour less than the PM program with $T = 3$ that maximizes press capacity. Which PM program (i.e., $T = 3$

**Table 15.4  Finding the Optimal Value of $T$ that Minimizes the Expected Cost**

| $T$ | $P(t)$ | Average uptime | Average downtime | Average cycle length | Average cycle cost | Average cost per hour |
|---|---|---|---|---|---|---|
| 1 | 0.004 | 1.000 | 0.2112 | 1.211 | 20.32 | 16.78 |
| 2 | 0.012 | 1.996 | 0.2336 | 2.230 | 20.96 | 9.40 |
| 3 | 0.024 | 2.984 | 0.2672 | 3.251 | 21.92 | 6.74 |
| 4 | 0.058 | 3.960 | 0.3624 | 4.322 | 24.64 | 5.70 |
| 5 | 0.096 | 4.902 | 0.4688 | 5.371 | 27.68 | 5.15 |
| 6 | 0.14 | 5.806 | 0.5920 | 6.398 | 31.20 | 4.88 |
| 7 | 0.195 | 6.666 | 0.7460 | 7.412 | 35.60 | 4.80 |
| 8 | 0.259 | 7.471 | 0.9252 | 8.396 | 40.72 | 4.85 |
| 9 | 0.345 | 8.212 | 1.1660 | 9.378 | 47.60 | 5.08 |
| 10 | 0.435 | 8.867 | 1.4180 | 10.285 | 54.80 | 5.33 |
| 11 | 0.537 | 9.432 | 1.7036 | 11.136 | 62.96 | 5.65 |
| 12 | 0.655 | 9.895 | 2.0340 | 11.929 | 72.40 | 6.07 |

or $T = 7$) should be chosen? The answer depends on the cost of having 2 percent less effective capacity compared with the cost saving of $1.94 per hour.

### 15.11.5 Maintenance Efficiency Improvement

The third main component of Total Productive Maintenance is constant effort in improving the efficiency of activities related to repair and maintenance operations. These include, but are not limited to, the following:

- *Cost Analysis:* To develop cost-efficient maintenance programs, it is essential that all direct and indirect costs such as costs of parts, labor, downtimes, and outside contractors be well-understood and quantified.

- *Using Analytics:* One goal of TPM is to minimize long-term costs of maintenance and repair programs, taking into account its effects on production. Achieving this goal without using data and analytics is not possible. Hence, to develop maintenance and repair programs, TPM relies on data analysis and analytical models to keep the costs down and to reduce both planned and unplanned downtimes. The preventive maintenance plan should be designed for the entire life of each machine.

- *Spare Parts Management:* Availability of spare parts, when needed, is critical in reducing downtimes. Hence, it is essential to have cost-effective spare part inventory management that maintains a high service level for critical spare parts.

## 15.12 Lean Tool 9—Process Standardization

Process standardization is one essential factor in sustaining lean operations and process improvement. The concept of standardized work is much broader than just breaking activities into simple tasks and documenting the best way to implement each task. Fuiji Cho, the president of Toyota, describes process standardization as follows:[16]

> **Fuiji Cho:** *Our standardized work consists of three elements—takt time (time required to complete one job at the pace of customer demand), the sequence of doing things or sequence of processes, and how much inventory or stock on hand the individual worker needs to have*

---

[16]Liker, J.K., *The Toyota Way: 14 Management Principles from the World's Greatest Manufacturer.* McGraw-Hill, 2004.

*in order to accomplish that standardized work. Based upon these three elements, takt time, sequence, and standardized stock on hand, the standard work is set.*

We now discuss the three elements of standardization.

### 15.12.1 Takt Time—Producing at the Pace of Customer Demand

Lean operations systems such as Toyota try to minimize the finished goods inventory by planning it according to demand. This is done by computing takt time. As discussed in Chap. 13, Takt Time is a measure that translates demand of a process into a number which shows whether the process can satisfy its targeted demand. Suppose the throughput of a factory that produces mattress is 12 mattresses per hour. Thus, the average inter-throughput time of the factory is 5 minutes. The factory works 8 hours a day. The demand for mattresses is 120 per day. With the current inter-throughput of 5 minutes, can the factory satisfy its demand of 120 per day? The answer is No. The inter-throughput time of 5 minutes translates into the throughput of 12 per hour or $12 \times 8 = 96$ mattresses per day, which is less than the demand of 120 per day. What should the inter-throughput time of the factory be in order to be able to satisfy this demand? The answer is takt time.

Let's compute the takt time of the mattress factory. Since factory works 8 hours a day, to satisfy the demand of 120 per day, it should have a throughput of $120/8 = 15$ mattresses per hour, which implies an inter-throughput time of 4 minutes. Thus, to satisfy the demand of 120 mattresses per day, the factory should reduce its inter-throughput time to takt time of 4 minutes or less.

In summary,

$$\text{Takt Time} = \frac{\text{Working hours during a day}}{\text{Demand to be met during a day}} = \frac{8}{120} \text{ hours} = 4 \text{ minutes}$$

Finding takt time is the first step toward work standardization. When takt time is determined, the work sequence can be designed accordingly, so the process can have an inter-throughput time equal to (or less than) takt time. This includes the studies of all different process activities required for production of the product, which corresponds to the traditional line balancing approach. Line balancing designs the workstations in the line such that all activities within workstations would require the same amount of time (i.e., takt time) to complete (see Chap. 13 for line balancing).

### 15.12.2 Work Sequence—Finding the Best Way to Perform Tasks

The goal of this step is to find the best and the most precise work sequence in which a worker in each station can perform his or her task within takt time. The work sequence will be documented in Standard Work Sheets. Taichi Ohno emphasizes on the importance of the standard work sheets as follows:[17]

> **Taichi Ohno:** *Standard work sheets and the information contained in them are important elements of the Toyota Production System. For a production person to be able to write a standard work sheet that other workers can understand, he or she must be convinced of its importance. . . . High production efficiency has been maintained by preventing the recurrence of defective products, operational mistakes, and accidents, and by incorporating workers ideas. All of this is possible because of the inconspicuous standard work sheet.*

The standard work sheet of an activity includes content, sequence, timing of all the tasks of that activity as well as the outcome. The work sequence should be designed such that it would

---

[17] Liker, J.K., *The Toyota Way: 14 Management Principles from the World's Greatest Manufacturer.* McGraw-Hill, 2004.

be the safest and the most ergonomic way for the worker to perform the work. It should also result in a consistent outcome, regardless of which worker performs the task at which shift. Preparing a standard work sheet also allows engineers and workers to determine the tasks with high possibility of errors and design sequences or mechanisms (e.g., Poka-Yoke) to prevent errors.

In lean operation systems such as Toyota Production System, work standardization is not only about making the jobs repeatable and efficient in one plant. It is also about standardizing work throughout all production processes in the company. A Toyota worker or engineer can go to any Toyota plant in the world and see almost identical work sequences for each activity in the plant.

One criticism of standardizing work that Ford was facing when it employed this approach in its assembly line was that standardizing work creates rules that make jobs rigid and degrading. While employing work standardization, Toyota finds balance between providing workers with standard procedures and providing them with the freedom to be creative and innovative to meet target performance. This is done by involving workers in the process of developing and improving work standard. Workers know that if they design and improve a work sequence, their design will be used throughout the company. This will give the workers a sense of reward and empowerment and reduce the impression of standardized work being rigid and degrading.

Benefits of work sequencing (documented in standard work sheet) are several:

- *Consistent and High-Quality Outcomes:* By having a well-designed and precise work sequence, the performed activity will have high quality that is consistent regardless of the worker that performs the task or the plant in which the activity is performed.

- *Variability Reduction:* As discussed in Chap. 5, one approach to reduce variability in effective process time of an activity is to break the activity into its basic standard tasks and try to reduce the source of variability in those tasks. This is exactly what happens when work sequence is designed and improved, resulting in variability reduction.

- *Ease of Training:* Work sequencing documented in standard work sheets facilitates effective training of employees. These documents, as mentioned, include detailed information about how an activity should be done, from the beginning to the end.

- *Improved Safety:* Besides efficiency and quality, ergonomics of the work environment as well as worker safety are considered when work sequences are designed. When designing work sequences, for example, unsafe tasks and unsafe worker motions are identified and solutions are devised to reduce the work-related accidents.

- *Facilitating Continuous Improvement:* Work sequencing is the baseline for continuous improvement. First, without a well-defined and precise definition of an activity, it would not be clear which part can or cannot be improved. Second, when the work sequence is repeatedly performed by a worker, the process becomes more organized and opportunity for improvement surfaces and becomes easier to identify. Third, when a standard work sequence is improved, it becomes the baseline for further improvement. In fact, the process of improving work sequence is a continuous and never-ending process.

- *Adding Discipline to Company's Culture:* The process of developing work sequences and documenting them in the standard work sheets instill organized, systemic, and scientific thinking in company's employees. This, in the long run, inserts discipline in company's culture.

### 15.12.3 Standardized Stock—Minimum Inventory Needed to Operate Smoothly

After takt time is determined and work sequences are identified, the last step in work standardization is to determine the Standardized Stock—the minimum inventory needed between any two stations to achieve a certain performance. When implementing Kanban systems, for example, this step refers to finding the minimum number of Kanban cards needed to achieve a certain throughput corresponding to the takt time determined in the first step. This was discussed in detail in Sec. 15.9.

While process standardization was originally developed to improve performance in manufacturing systems, it has also been used in service systems. For example, Jefferson Pilot Financial (JPF) used takt time and work sequencing to improve their process of issuing new life-insurance policies. Insurance policies must go through several stages including initial application, underwriting, risk assessment, and policy insurance. To shorten the process of issuing a policy, JPF determined that, to satisfy demand, it needed to process 10 applications per hour. This translated to the takt time of 6 minutes. JPF then performed work sequencing and found the best way to perform tasks in each stage. Based on the takt time and the newly designed tasks, JPF was able to determine the minimum number of employees needed to satisfy the demand. Using process standardizations and other lean tools, JPF was able to reduce the time to process an application by 84 percent, the total labor cost for all applications by 28 percent, and also reduced the policy reissues (due to errors, i.e., rework) by 40 percent.

## 15.13 Lean Tool 10—Leveling Production and Schedules (Heijunka)

In Sec. 15.3.1, we presented seven sources of waste, called Muda in Japanese. Besides Muda, Toyota Production System also identifies two other sources of waste called Muri and Mura. Below, we give a brief explanation of these three sources of waste.

- *Muda—Non-Value Added:* This is what we discussed in Sec. 15.3.1. This includes tasks that do not add value to the product and result in larger lead time, inventory, and waiting time.

- *Muri—Overburdening People and Equipment:* Muri corresponds to pushing resources (workers or equipment) beyond their natural limits. While Muri increases resource utilization, it results in quality and safety issues.

- *Mura—Unevenness:* Mura refers to the unevenness of workload imposed to resources (and processes). With Mura, resources cycle through the periods with workload beyond their capacity (i.e., Muri) and periods with idleness due to lack of work (i.e., Muda). The reasons behind this unevenness in workload are irregular production schedule or variability in working environment (e.g., machine failure, rework, long lead time).

In implementing lean practices, managers' main focus is often on eliminating Muda, and they fail to eliminate Mura, since it is a more difficult task. Muda is usually easier to detect and easier to eliminate, while creating an even workload throughout the week to satisfy the variable demand is very difficult. But, the important fact is that eliminating Mura is essential in eliminating Muda and Muri. In fact, Fujio Cho, the president of Toyota Motor Corporation has said:[18]

> **Fuiji Cho:** *Once the production level is more or less the same or constant for a month, you will be able to apply pull systems and balance the assembly line. But if production levels—the*

---

[18]Liker, J.K., *The Toyota Way: 14 Management Principles from the World's Greatest Manufacturer.* McGraw-Hill, 2004.

*output—varies from day to day, there is no sense in trying to apply those other systems, because you simply cannot establish standardized work under such circumstances.*

How does Toyota eliminate Mura and create a "level" or "even" production? Through Heijunka, which refers to leveling out the work schedule.

---

**CONCEPT**

**Heijunka**

Heijunka, also known as "Uniform Plant Loading," or "Leveling Production," is the practice of leveling the quantity and type of production over a fixed period of time. Heijunka does not develop the production schedule based on actual customer demand, which is often highly variable. It uses the total demand for different products in a period and levels them out such that the same number of each product and the same mix of products are produced each day.

---

To illustrate how Heijunka levels production schedules, we use a simple example of a plant that produces three different types of refrigerators: large, medium, and small. The total monthly demand is forecasted to be around 14,000 units, which includes 2000, 8000, and 4000 units for small, medium, and large units, respectively. All three products are produced on a single assembly line that works 5 days a week and 20 days in a month. Each day consists of two 8-hour shifts.

The assembly plant works as follows. Considering the 4 weeks in a month, the weekly demand for small, medium, and large units are 500, 2000, and 1000 units. To satisfy this demand, the master production schedule for a week is to produce, 500, 2000, and 1000 units of small, medium, and large units. A *Master Production Schedule*[19] or MPS determines which product should be produced in each time period—a week in our example. The master production schedule is translated to final assembly schedule. A *Final Assembly Schedule* or FAS determines daily or hourly schedule and requirement of the assembly operations.

Figure 15.13 shows an example of a weekly FAS that is often used in scheduling final assembly. According to this weekly FAS, every Monday the plant produces the large units until the middle of Tuesday, when a changeover is preformed to set up the line for assembling medium units. The production of the medium units then starts and continues until Friday, when another changeover is performed to set up the line for small units. The production of the small units then starts and stops 4 hours before the end of the second shift on Friday, when a changeover is performed to set up the line for production of the large units on Monday morning. The changeover times for switching from large units to medium, medium to small, and small to large are about 6, 5, and 4 hours, respectively.

The schedule in Fig. 15.13 is an example of an unlevel production schedule. Assembling large units is probably more labor intensive than medium and small units, so the plant faces a higher workload on Mondays and Tuesdays—when it produces large units—than those on Wednesdays and Thursdays—when it produces medium units. The workload is the lowest on Fridays when the plant produces mostly small units.

### 15.13.1 Two Steps of Heijunka

How does Heijunka level the production? It does it in two steps:

1. Establishing a steady rate of production
2. Sequencing final assembly

---

[19]See Chap. 13 for detailed explanation of how Master Production Schedule (MPS) is developed.

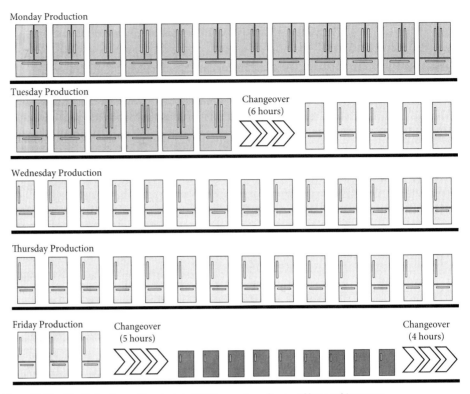

Figure 15.13  Weekly final assembly schedule (FAS) for small, medium, and large refrigerators.

### Heijunka—Step 1: Establishing a Steady Rate of Production

Lean operations require a stable daily schedule for a predictable time horizon. This reduces variability in the process and allows a steady rate of pulling materials through different stages of the process. A stable and steady production rate can be computed using takt time. Recall that the total weekly demand is $500 + 2000 + 1000 = 3500$ units. On the other hand, the assembly plant works for 5 (days) $\times$ 2 (shifts) $\times$ 8 (hours per shift) $= 80$ hours in a week, or $80 \times 60 = 4800$ minutes a week. Hence, the takt time is $4800/3500 = 1.371$ minutes $= 82.3$ seconds. The takt time of 82.3 seconds implies that the assembly line must be able to produce at the rate of one unit in about 82.3 seconds. To achieve this, process standardization (see Sec. 15.12) for assembly line operations should be performed based on the takt time of 82.3 seconds.

### Heijunka—Step 2: Sequencing Final Assembly

Once the daily production rate is set based on the takt time, the next step is to determine an even and leveled daily final assembly schedule (FAS). This can be done by first translating the weekly MPS to daily FAS. In our example, since the factory works 5 days a week, the weekly MPS of 500, 2000, and 1000 is broken into a daily FAS of 100, 400, and 200 small, medium, and large units, respectively. Second, we must determine the sequence by which different products are released to assembly line. Considering the daily demand 100, 400, and 200 (which has the ratio 1–4–2), for every small unit, four medium and two large units must be produced. The daily FAS for this sequence is shown in Fig. 15.14.

Figure 15.14 shows a FAS that releases refrigerators to the assembly line in cycles with sequence L-M-M-L-M-M-S. Specifically, after releasing one large (L) unit, two medium (M)

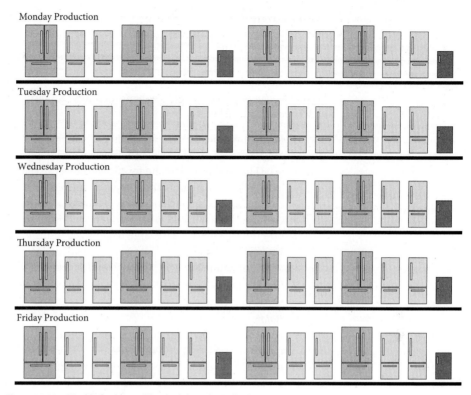

**Figure 15.14    Weekly final assembly schedule under Heijunka.**

units are released followed by another large (L) unit and two medium (M) units, and finally one small (S) unit. This cycle repeats itself throughout the day in all 5 days of a week. Units are released one every 82.3 seconds—the takt time. As opposed to the FAS in Fig. 15.13, the workload under FAS in Fig. 15.14 is the same in each day and in each hour. This is what Heijunka is all about.

Two points need to be emphasized here. First, to implement Heijunka, the changeover times must be eliminated or reduced to a very short time. This can be done through setup time reduction practices (see Sec. 15.5). Another approach is through product design. Specifically, to eliminate setups, products can be designed such that they use common parts and subassemblies. This allows for a mixed-product production line. For example, by designing different car models that use the same chassis (e.g., Honda Civic and Honda CRV), automobile manufacturers are able to assemble different car models on a single line.

Second, because of the variability in demand and in production environment it may become difficult to keep up the steady production time. How does Heijunka handle such variability and keep the production leveled? The answer depends on the magnitude and length of periods affected by high variability. Toyota handles variation in day-to-day operations (e.g., natural variability in demand) through overtime at the end of a shift or with occasional weekend shifts. To facilitate overtime, shifts are scheduled such that there is a down period between the shifts. A common approach is to schedule two shifts 8–4–8–4, which correspond to two 8-hour shifts with 4 hours down period between them. Besides overtime—if needed—down periods are used for activities such as preventive maintenance or group meetings. The larger variability that affects longer periods of time (e.g., seasonality in demand) are dealt with by adjusting the takt time—changing the production rate. Toyota develops different standardized work in advance for operations with different takt time.[20]

### 15.13.2 Benefits of Heijunka

Having a level production and steady production rate result in several advantages, some of which we introduce here:

- *Lower Inventory.* In the traditional FAS in Fig. 15.13, first a batch of 1000 large units are produced on Monday and Tuesday and kept in the finished goods inventory (FGI) to satisfy the demand for the whole week. Then a batch of 2000 medium and a batch of 500 small units are produced and kept in FGI to satisfy the weekly demand. However, the demand for these products are spread out throughout the week—customers do not buy large refrigerators only on Mondays and small ones only on Fridays. Hence, these large batch sizes result in large (cycle) inventory in FGI. Under Heijunka in Fig. 15.14 with sequence L-M-M-L-M-M-S, the batch sizes for large, medium, and small are 1, 2, and 1, respectively. These small batch sizes used to satisfy the demand throughout the day result in much lower (cycle) inventory in FGI.

- *Balanced Use of Equipment and Labor:* Different products require different labor and equipment time. For example, larger refrigerators have larger body panels and more parts to assemble, and thus need more worker time in the assembly line. Hence, under the traditional FAS in Fig. 15.13, there would be an unbalanced requirement for labor time in weekdays that produce different units. This results in Muda (e.g., labor idle time) on some days and Muri (i.e., overloaded workers) on other days. However, under Heijunka and its corresponding standardized work, the workload in each day (and each hour) is the same. Hence, the labor and machines are used in a balanced and steady way. The key is to design an efficient standardized work for the FAS job release under Heijunka.

- *Low Variability for Supplier's Demand.* Under the traditional FAS in Fig. 15.13, the supplier of parts for the large units, for example, will face large demands on Mondays and Tuesdays and zero demand in the rest of the week. Hence, the supplier of parts for the large unit faces a large variability in its demand—same for supplier of parts for medium and small units. A small change in orders placed to the supplier (due to change in manufacturer's production schedule) further intensifies the variability in supplier's demand through bullwhip effect (see Chap. 11). Under Heijunka, however, all three suppliers face a smooth daily demand that does not change by the day or by the hour. Due to low variability in their demand, a change in manufacturer's production schedule will not have a significant impact on supplier's operations—the impact of bullwhip effect is much less. Lower variability in supplier demand, results in more reliable delivery of parts to the assembly plant, lower supplier's inventory cost, and overall, lower supply chain costs.

## 15.14 Lean Tool 11—Continuous Improvement and Kaizen

While the concept of continuous improvement was advocated by W. Edward Deming (see Chap. 16), Japanese companies adopted the method, developed Kaizen, and introduced it to

---

[20] Pascal, D., *Lean Production Simplified*. CRC Press, 2015.

the West in the book "*Kaizen: The Key to Japan's Competitive Success,*" written by Masaaki Imai in 1986.[21]

### Kaizen

Kaizen comes from the Japanese words Kai (change) and zen (for the better). Kaizen is the never-ending work of continuous improvement by *everybody, everyday*, and *everywhere*. It covers all aspects of operations including material, equipment, people, processes, and methods.

The underlying idea of Kaizen is that improving performance is a continuous process; it never achieves the best, it only makes things better.

### Kaizen Focuses on Small Gradual Improvements with Minimum Cost

Kaizen is based on the premise that big improvements often come from many small changes that accumulate over time. Hence, Kaizen focuses on small and incremental improvement steps. This is an important feature since achieving large improvement often seems impossible, resulting in losing motivation and momentum to start a Kaizen effort. Also, small changes are often accomplished without a significant cost, which leads Kaizen solutions having a low cost.

### Kaizen Focuses on All Aspects of Operations

Kaizen philosophy is the improvement in all areas of an organization. Kaizen efforts expand to tasks such as minimizing cost, improving safety, meeting due dates, improving product design, improving supplier relationship, improving quality, improving equipment performance, and improving employees' skills.

### Kaizen is a Never-Ending Process

Kaizen is a continuous improvement process that never stops in its efforts. In his book, *The Toyota Way*, J.K. Liker presents an example that emphasizes on this fact. He writes that, after implementing lean, the president of Freudenberg-NOK General Partnership (FNGP) noticed something very curious: "No matter how many times his employees improved a given activity to make it leaner, they could always find more ways to remove Muda by eliminating effort, time, space, and errors." For example, in an initial Kaizen event in his Indiana facility, the team was able to increase labor productivity by 56 percent while reducing the floor space required by 13 percent. By having five additional 3-day Kaizen events in the next 3 days, the teams were able to decrease the number of people needed to perform the task from 21 to 3, increase the productivity of workers by 991 percent, and reduce the required space from 2300 square feet to 1200 square feet. The pleasant surprise was that it was believed further improvement was still possible. This and many similar cases attest that Kaizen should be a never-ending process.

### Kaizen Involves Everyone

The fundamental rule of Kaizen is to involve everyone, from CEO to the last factory worker. Everyone's knowledge—managers and employees—must be utilized to find and implement improvement ideas with minimum cost. The role of management is to create a culture of

---

[21] Imai, M., *Kaizen: The Key to Japan's Competitive Success*. McGraw-Hill, New York.

Kaizen in the organization by setting goals, providing incentives for improvement, providing resources, and enhancing employees skills. Kaizen acknowledges that employees are the main sources of improvement ideas, since they are closer to the problems. Hence, by including employees in Kaizen teams, employees take ownership and feel accountable for improvement solutions. This increases the chance of finding an improvement solution that is both implementable and sustainable.

**Kaizen Blitz**

A Kaizen Blitz (or a Kaizen Event) is an intense and short-time (2 days to 1 week) event in which a team, a department, or an organization focuses all of its resources on improving a process. Kaizen Blitz consists of a cross-functional team of all employees who are involved in the corresponding process. Kaizen Blitz starts with questioning the current method. They are known for delivering dramatic improvement to the process—improvements that do not require a large cost.

## 15.15  Lean Tool 12—Value Stream Mapping

Value Stream Mapping is a powerful tool developed as part of Toyota Production System and has now become an essential tool of lean operations. Before we show how value stream mapping is used, we first need to understand what "Value Stream" is.

**CONCEPT**

**Value Stream and Value Stream Mapping**

- *Value Stream* is the sequence of all value-added and non-value added activities required to design, produce, and deliver a specific product or service to customers.
- *Value Stream Mapping* is the process of creating a visual map of the value stream. It includes both the flow of information and flow of material and also information such as cycle time at each activity, inventory at each inventory point, and equipment downtimes.

Value Stream Mapping (VSM) is the first step toward converting a system to a lean system. Even in a lean operations, VSM is helpful to find wastes in a process and its sources. VSM can be used to study an individual workstation, a department, the entire factory, or the entire supply chain. The VSM process starts by creating a map of the current process. The visualization of the current state through VSM helps team members understand the current practice. This, in turn, helps generate improvement ideas and guides managers toward an improved future state as well as a possible implementation plan.

When mapping a value stream, it is important to define the scope of the map. This includes identifying the product (or a family of products) or the service to map and the part of the overall process that needs to be analyzed with the exact start and end of the value stream. For example, if there are customer complaints about late delivery of their orders, the entire order processing should be mapped, from the time customer places an order until the customer receives the order.

A value stream map includes more detailed information about a process than process-flow chart. Figure 15.15 shows an example of an initial value stream map of a process. The symbols used to create value stream maps are shown in Fig. 15.16.

Value stream maps contain very useful information in a compact form. Let's discuss the information provided by the value stream map in Fig. 15.15. As the figure shows, the process receives its raw material every week from its supplier and holds it in raw material inventory, which has about 3600 units of raw material. The first activity that is performed is "Cutting," done by two workers and two laser cutters. In value stream maps, C/T—which is called cycle time—corresponds to the time it takes to process a job. As the information box under the cutting activity shows, it takes about C/T = 50 minutes to process a job. As the ladder under the cutting activity shows, 40 minutes out of C/T = 50 minutes is value-added activity, the remaining time is non-value added activity. After cutting a batch of 40 jobs, the workers switch to process another type of job. This requires a changeover time of C/O = 35 minutes. The cutting station is up and working 88 percent of the time. The finished jobs are pushed to the next station, in which two workers use two milling machines to perform the next activities. After milling and painting, the products are shipped to the customers daily.

The customer places orders through electronic links (e.g., email, or linked computers). The customer (e.g., a retailer) also shares its forecast with the manufacturer. The manufacturer is also linked to the suppliers and places its orders and shares its demand through electronic links. As shown on the top part of the value stream map, using the customer order size and demand information, the manufacturer generates weekly production schedule for each of its stations, and sends them to the production supervisor. The production supervisor translates the weekly schedules into daily schedules and sends them to the workers in the cutting, milling, painting, and shipping stations.

The initial VSM in Fig. 15.15 highlights potential for improvement in several directions.

- The information box at the bottom of the map shows that a job spends a total of 23 ($= 8 + 5 + 4 + 6$) days waiting in different buffers of the process. This non-value added waiting time is called "lead time" in value stream mapping language. However, jobs are waiting that long to get a total value-added activity of only 180 ($= 40 + 70 + 60 + 10$) minutes. This is mainly because the inventories in the

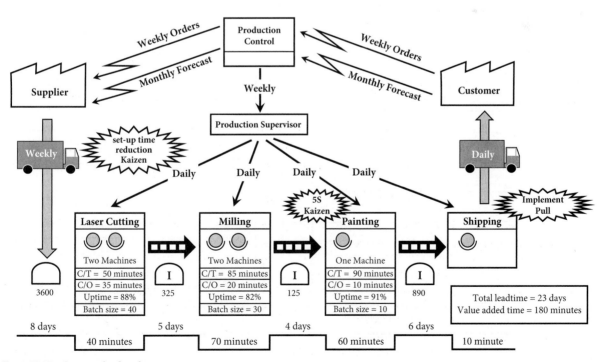

**Figure 15.15    An example of a value stream map.**

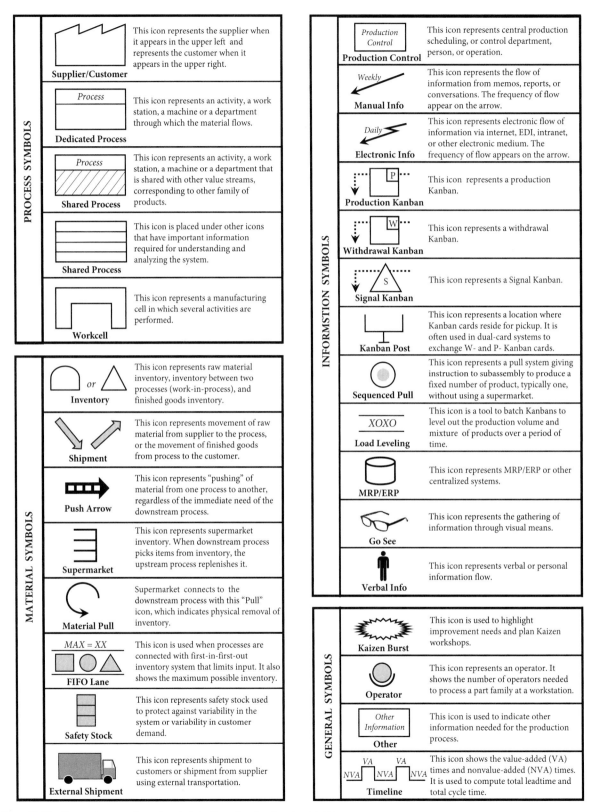

**Figure 15.16    The most commonly used symbols and icons in value stream mapping.**

buffers between stations are large. One improvement suggestion to reduce inventory, shown in the VSM, is to have a Kaizen Blitz with the goal of changing the current push system to a pull system.

- Another reason for large inventory between stations is the large batch sizes. Hence, another improvement suggestion shown in the map is to have a Kaizen event with the goal of reducing changeover times in all stations of the process that would lead to smaller batch sizes.

- In the painting station, it takes 90 minutes to process a job. However, as the map also shows, only 60 minutes of this 90 minutes is value-added time—one-third of the painting time is non-value added time. The map suggests a Kaizen event to minimize motion by implementing a 5S program (see Sec. 15.16.1).

To make the value stream mapping process work, firms should assign responsibility to one person (a manager) or a small team of people (managers) to make it happen. The value stream managers must take responsibility to develop the initial map, organize the Kaizen efforts, and implement the changes. The value stream managers should report their progress at each step to the top managers.

## 15.16 Lean Tool 13—Visual Workplace

One of the seven sources of waste (Muda) which is often not directly observed is Motion. Motion, as defined in Sec. 15.3.1 is simply moving without working. Different types of motions are Checking, Guessing, Wondering, Counting, Asking, Answering, Waiting, and Interrupting. Take asking a question as an example. A worker interrupts another worker and asks a question about tool or material. During the process of asking the question, two workers are in motion. In some cases, the second worker does not know the answer to the question. Now the two workers are asking a third person. These interruptions continue—putting more people in motion—until the answer to the question is found. This may take a long time and interrupts work. The research has shown that it takes the average person 6 to 15 minutes to get back on task after each interruption.

What triggers motion? A worker will be in motion when the worker needs to find the answer to the question of: What? When? Where? How? How many? For example, what was I supposed to do next? When I was supposed to finish it? Where is the tool I need? How do these things fit together? How many units I was supposed to produce? When answer to these questions are either not available, or are late, incomplete, or wrong, then the work environment is said to have *Information Deficit*. Hence, one way to prevent motion is to eliminate information deficit. A visual workplace eliminates information deficit using visual solutions such as signs, charts, boards, instructions, among others.

The book *Visual Workplace* defines visual workplace as follows:

### CONCEPT

**Visual Workplace**

A Visual Workplace is a self-ordering, self-explaining, self-regulating, and self-improving work environment—where what is supposed to happen does happen, on time and every time—because of visual solutions.

One way to better grasp the importance of visual workplace in lean operations is through the just-in time philosophy. Recall that JIT's goal is to produce just the right quantity of products, in just the right quality, at just the right time, in just the right place, and in the most economical way. Visual workplace does the same with respect to information. It provides the correct and right amount of information, exactly when it is needed and where it is needed. Without a visual workplace, JIT may not achieve its goals.

### Benefits of Visual Workplace

The benefits of a visual workplace are many, some of which are presented below:

- *Efficiency:* By eliminating motion, visual workplace helps increase efficiency and productivity. It has been reported that by implementing visual workplace several companies experienced up to 15-percent increase in productivity, 70-percent reduction in waiting, 96-percent improvement in quality, 60-percent reduction in floor space requirement, and 100-percent elimination of rework.[22]

- *Safety:* By providing clear, complete, and on-time information which is available at a glance, visual workplace prevents problems and mistakes, some of which can have safety consequences.

- *Faster Problem Detection:* In a visual workplace where all required information is available and all tools and material are organized and easy to access, any anomaly—which points to a possible problem—can be detected earlier. New visual solutions can be designed to prevent the problem in the future.

A visual workplace has four main components: (1) Visual Order, (2) Visual Process, (3) Visual Performance, and (4) Visual Control.

## 15.16.1  Visual Order

One advantage of lower inventory, as illustrated by the River Analogy, is that it makes waste visible. One can eliminate waste, only if one can see the waste. A cluttered, disorganized, unclean workplace hides problems. Organized and clean workplace is essential to implementing lean operations. Taichi Ohno is quoted as saying:

> **Taichi Ohno:** *Make your workplace into a showcase that can be understood by everyone at a glance.*

The book, *The Toyota Way*, has an interesting story about Donnely Mirrors (now Magna Donnely), a supplier of exterior mirrors for automobile manufacturers.[23] Before implementing lean operations, their plant was so disorganized that one day a Ford Taurus, which was being fitted it with some prototype mirrors, mysteriously disappeared. When they could not find the car, the plant filed a police report. But months later, while organizing their workplace, the car was found at the back of the plant, surrounded by inventory.

Japanese plant managers take having a clean and organized workplace very seriously. When Americans were visiting Japanese plants in the 1970s and 1980s, their first reaction was the same: "The factories were so clean you could eat off of the floor."[23] How do Japanese achieve such a clean and orderly workplace? Through their "5S Programs." 5S program consists of a number of activities aimed to make the workplace organized, clean, and visual in order to eliminate motion and errors that impact productivity and safety. Here are those 5S's (Seiri, Seiton, Seiso, Seiketsu, and Shitsuke) that translate into the following five activities (see Fig. 15.17):

---

[22]Galsworth, G.D., *Work That Makes Sense, Creating and Sustaining Visuality on the Value-Added Level*. 2011.
[23]Liker, J.K., *The Toyota Way: 14 Management Principles from the World's Greatest Manufacturer*. McGraw-Hill, 2004.

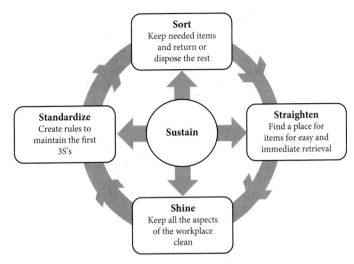

**Figure 15.17    5S cycle of creating visual order.**

- *Sort:* Sort items into three categories: retain, return, and rid. Keep only the retain items, and return or dispose the rest.
- *Straighten (Orderliness):* Find a place for everything and put everything in its place. Organize and arrange items, material, tools, equipment, and information.
- *Shine (Cleanliness):* Clean the workplace and maintain its orderliness every day. While cleaning, look for abnormalities that may cause future failures (e.g., oil leak from a machine that can result in machine failure).
- *Standardize (Create Rules):* Develop standard procedures that help maintaining the first three S's. This makes the first three S's as a part of daily work.
- *Sustain: (Self-Discipline):* Create a culture that maintains and stabilizes workplace in an ongoing process of continuous improvement. For example, use periodic management audit to sustain discipline.

Implementing a 5S program results in several benefits. It reduces waste due to motion, it makes worker training simpler and more effective, it reduces inventory, it improves worker safety, it improves quality (by reducing the possibility of errors), and it increases equipment lifetime and availability (by frequent cleaning and inspection).

Implementing 5S program can also significantly increase space utilization. One example is when Toyota formed a joint venture with GM to open Toyota's first overseas plant. Toyota took over GM's old plant in Fremont, California in 1983. Toyota wanted to add a new production line; however, it was told that there was no space available in the plant. Toyota implemented its 5S program in the plant and was able to free 30 percent of the floor space, which was more than enough for establishing the new production line. In 1984, the Fremont plant produced its first new Chevrolet Nova, and later won several awards for quality.

### 15.16.2 Visual Process

While the goal of visual order is to create a clean and organized work environment to perform the work, the goal of visual process is to use visual signs and instructions to make performing the work simple, efficient, error-free, and safe. In a visual workplace visual instructions and visual displays are located where the worker can see them. These visual solutions provide important information such as key elements of the process or activities,

Figure 15.18    Visual process solution to ordering a hamburger.

correct sequence of the stages of the process, and safety and quality requirements. Photos are used to help workers better understand the process and grasp it at a glance.

Figure 15.18 shows an example of a visual solution for the process of ordering a hamburger. Starting from top to bottom, customers can easily see their options for meat, bread, cheese, toppings, etc. This reduces motions such as searching, asking, and wondering, and therefore reduces the time to place an order. On the other hand, it also helps servers reduce the chance of making errors in the process of taking customer orders. Such a visual solution has become popular in fast-food and fast-casual restaurants that offer a large set of options in their menu.

### 15.16.3 Visual Performance

Visual performance displays and boards are used to show the actual as well as the target performance of a worker or a team of workers. The displays should be visible to workers. Figure 15.19 shows an example of a display board that shows the performance—the throughput—of three production lines. It shows the target throughput for the shift, the actual production so far (i.e., at time 14:45:30), the gap to reach the target throughput, and its corresponding efficiency.

Having both the actual and expected performance helps workers see how their work is progressing. If their actual performance is lower than what is expected, workers will be motivated to find the problem that is causing their low performance. On the other hand, by setting rewards for achieving or exceeding the target performance, workers will be motivated to work harder.

### 15.16.4 Visual Controls

Visual controls are signs or displays that (i) control the activities of a process, or (ii) provide information on whether the process is working within its standard procedure. Kanban cards

| Thursday | | | Time: **14:45:30** |
|---|---|---|---|
| | LINE 1 | LINE 2 | LINE 3 |
| Target: | **1500** | **2600** | **2100** |
| Actual: | **1233** | **2251** | **1922** |
| Gap: | **267** | **349** | **178** |
| Efficiency: | **89%** | **91%** | **93%** |

**Figure 15.19    Visual performance display board for production lines.**

are the example of the former. They are visual control signals that control production and prevent the process from producing inventory above the predetermined target (i.e., more than the number of Kanban cards). Andon displays are examples of the latter. A red Andon light signals that the quality of a job is not within the desired standard.

## 15.17 Lean Tool 14—Employee Empowerment

At the center of Toyota Production System (TPS) are the employees who work as individuals and as team members. In TPS workers have many responsibilities: they are expected to perform multiple tasks, are responsible for the quality and preventive maintenance, and are expected to engage in solving quality problems. Hence, it is essential to create a work environment in which workers are motivated, trust management, and feel secure and do not fear to say what they think.

How does Toyota create such a work environment? Traditionally, Toyota provided lifetime employment for most of their permanent employees. Today, however, only a relatively small portion of the employees have lifetime employment. Lifetime employment is not the main core principle of Toyota. Toyota firmly believes that employee involvement and empowerment is the best way that a company can reach its goals. The center of Toyota's employee involvement and empowerment is Respect for People.

### 15.17.1 Respect for People

In lean operations philosophy, people are the system's most precious resources. Lean systems treat all its people—employees and suppliers—with respect.

In his book, *Toyota Production System*, Taichi Ohno states that "respect for humanity" is as important as the concept of eliminating waste. Fujio Cho, the president of Toyota, also emphasized on the role of people:[24]

> **Fujio Cho:** *Toyota firmly believes that making up a system where the capable Japanese workers can actively participate in running and improving their workshops and be able to fully display their capabilities would be the foundation of human respect of the highest order.*

When a problem occurs, the best person with the most expertise to consult is the worker who performs the task. Furthermore, it has been shown that worker involvement in decision-making process that influences worker's jobs increases worker's motivation and productivity.

---

[24]Sugimori, Y., K. Kusunoki, F. Cho, and S. Uchikawa, "Toyota Production System and Kanban System Materialization of Just-In-Time and Respect-for-Human System." *International Journal of Production Research*, 15 (1977) 553–564.

Thus, it is natural to involve workers in the problem-solving process. Lean operations systems respect workers' talents and expertise and hence involve them in problem solving and improvement activities.

### 15.17.2 Lean Culture

Firms that adopted lean operations have realized that the most difficult obstacle in their lean transformation is their lack of "Lean Culture." The lean culture[25] emphasizes on several points. First, it should be understood that lean is part of everyone's job. It is a no-blame system that treats problems as opportunity for improvement; teamwork is the foundation of problem solving, and sharing ideas and continuous improvement are essential.[26] Hence, firms that adopt lean operations must develop lean culture and establish mechanisms that lean tools are consistently implemented to solve problems.

Leaders of many lean operations systems have used some versions of the 6-E's to develop and maintain a lean culture. The 6-E's are as follows:[27]

- *Enlist:* Seek assistance and cooperation from all of the staff
- *Enable:* Teach employees to take action to continually improve their process
- *Engage:* Keep employees involved in the process of change
- *Excite:* Encourage employees to stay active in finding and correcting problems
- *Empower:* Authorize employees to make changes happen in the areas they control
- *Encourage:* Inspire employees to tap their creative and innovative capabilities, and frequently celebrate successes

### 15.17.3 Teams in Lean Operations

Toyota Production System organizes its workers by forming teams. Teams are given training and responsibility to perform production, equipment maintenance, and housekeeping activities, to solve daily problems, and to participate in continuous improvement of the process. The team structure often consists of team members, team leaders, and group leaders.

*Team Members* perform manual and standard jobs and are responsible for problem solving and continuous improvement. *Team Leaders* take on a number of responsibilities to keep the line running smoothly and producing quality parts. This includes, for example, meeting production goals, responding to Andon calls, confirming quality, covering absenteeism, and ensuring that parts and material are supplied to the process. *Group Leaders,* on the other hand, manage several teams and have responsibilities such as workforce scheduling, monthly production planning, team morale, confirming routine quality, shift to shift coordination, team member development, and cross-training and coordination. Group leaders are also involved in process improvement and new product and process introduction. They teach short topics and, if needed, are able to get on the line and perform the task of team members.

The above view of teams at Toyota was different from those of American auto manufacturers. In early 1983, Toyota took over GM's old plant in Fremont and agreed to teach GM the principles of Toyota Production System. The plant was called New United Motor Manufacturing, Inc. or NUMMI. NUMMI plant was built by GM in 1962 and was known to be among GM's worst plants. The quality of the cars produced in the plant was known to be bad, workers had high absenteeism, they fought on the job, and sometimes sabotaged the

---

[25] Myerson, P., *Lean Supply Chain & Logistics Management.* McGraw-Hill, 2012, p. 128.

[26] Tapping, D. and T. Shucker, *Value Stream Management for the Lean Office.* CRC Press, New York, 2003.

[27] Wisner, J.D., *Operations Management and Supply Chain Process Approach.* SAGE Publications, 2017.

product. Toyota took over the plant, and 4 years later in 1984, NUMMI factory surpassed all of GM's plants in North America in productivity, quality, and inventory turn.

One reason for Toyota's success in improving GM's plant was related to Toyota's practice of Genchi Genbutsu, "go and see for yourself." With respect to management involvement in shop floor work, Taichi Ohno was quoted as saying:

> **Taichi Ohno:** *Toyota managers should be sufficiently engaged on factory floor that they have to wash their hands at least three times a day.*

This was confirmed in the a study performed by GM. Specifically, to compare how GM team leaders spend their time, GM conducted a study and compared the performance of their team leaders in their plants and in NUMMI plant. The conclusion was alarming: 52 percent of the time GM team leaders were not doing anything that could be considered as work, while NUMMI team leaders spent 90 percent of their times doing work on the shop floor. NUMMI team leaders spent 21 percent of their time filling for workers who were absent or on vacation, while GM team leaders did this only 1.5 percent of time. Furthermore, NUMMI team leaders spent 12 percent of their time observing teams' works and communicating job-related information. GM team leaders did not do that at all.[28] The conclusion was clear: GM was copying the TPS teamwork structure, but it lacked the culture and the support system.

### 15.17.4 The Role of Leadership

The TPS's approach to leadership is to groom leaders who thoroughly understand the work and the philosophy of TPS and teach it to others. Hence, Toyota's strategy is not to hire successful presidents and CEOs from outside the company. Toyota strongly believes that leaders must have experience with the day-to-day activities at the shop floor and fully comprehend the philosophy of TPS.

Toyota also expects that its leaders teach the TPS principles, create and support lean culture, serve their employees as facilitators, not "bosses," establish incentive systems to reward employees for their good performance, and create environment for continuous learning, teamwork, and continuous improvement.

One important task of the leaders is to make it clear that increase in productivity does not result in layoffs. This will eliminate employees' fear to suggest ideas for improvement. The management should work closely with the worker unions and inform them that if the company performs well, the benefits will be shared with the employees through bonuses.

### 15.17.5 Investing in Employees

Toyota makes a significant effort to hire and train its employees, looking for the right individuals to work on its teams. Toyota receives many job applications and chooses only a few after careful screening of applicants' capabilities and characteristics. After Toyota hires an individual, it constantly trains the individual several skills and teaches them the philosophy of aiming for excellence. The goal is to develop exceptional people who understand the company's culture and follow company's philosophy. This investment in hiring and training as well as the culture of respect result in committed workforce who are motivated to continually improve their operations. For example, in 1 year Toyota's Georgetown assembly plant received about 80,000 improvement suggestions from its employees and implemented 99 percent of them.[28]

---

[28]Liker, J.K., *The Toyota Way: 14 Management Principles from the World's Greatest Manufacturer.* McGraw-Hill, 2004.

## 15.18 Lean Tool 15—Supplier Partnership

For a manufacturer to operate as a lean system, it requires to receive frequent deliveries of parts and raw materials, in small batches, with short lead time delivered to the point of usage. For example, Nissan receives deliveries of seats for its vehicles every 15 minutes. It sends its orders 2 hours in advance—lead time is 2 hours. The orders include detailed information about the exact sequence for the type and color of the seats.

The other factor that is critical for lean operations is the quality of the parts delivered by the supplier. Traditionally, the manufacturer (i.e., the buyer) performed inspections of the delivered parts for quality problems and returned those with poor quality. In lean systems that operate with low inventory, poor quality items disrupt the smooth flow and have significant impact on throughput. Also, the activity of inspecting is considered waste—a non-value added activity. Hence, lean systems need to receive defect-free items from their suppliers.

How do lean systems make sure that they receive defect-free parts, in small batches, and on schedule with short lead time? The answer is through establishing a close partnership with their suppliers. This is achieved by building a long-term relationship with the supplier based on trust and mutual benefit. Toyota has been successful to do just that, better than other firms. For example, when asked about Toyota, a senior executive of a supplier to Ford, GM, Chrysler, and Toyota said:[29] *"Toyota helped us dramatically improve our production system. We started by making one component, and as we improved, Toyota rewarded us with orders for more components. Toyota is our best customer."* Several surveys also confirmed this. For example, a 2003 survey[30] about supplier relationship with the auto manufacturers in America ranked Toyota number one, followed by Honda and Nissan. Chrysler, Ford, and GM were fourth, fifth, and sixth. In the survey, suppliers said that Toyota (and Honda) were more trustworthy and were more concerned about suppliers' profitability than other manufacturers.

Building a close relationship has many benefits beyond just receiving high-quality items on time with short lead times. A well-known story about the strength of Toyota's relationship with its supplier is about p-valve. P-valve was an essential brake part that Toyota used in all its vehicles with the demand of 32,400 units per day. In February of 1997, a fire destroyed the production line of Aisin—Toyota's supplier of p-valve. Toyota had only 2 days of inventory available in its entire supply chain. If Toyota was not able to acquire more p-valves, its entire production would stop. But Toyota's suppliers came to its rescue: 200 of Toyota's suppliers self-organized and had p-valve production started within 2 days. To do that, 63 different firms took responsibility to make that happen. They used what existed of engineering documentation after the fire, and established temporary lines to produce required parts, and kept Toyota in business with almost no interruptions.

How did Toyota build such a strong relationship with its suppliers? The following practices have been identified as essential factors in building a close relationship with suppliers in a lean system:

- Low number of suppliers
- Close proximity
- Knowing how your suppliers work
- Considering supplier as a part of your firm
- Monitoring and supervising your suppliers
- Conducting joint improvement activities with suppliers

We discuss these in more details in the following sections.

---

[29] Liker, J. and T. Y. Choi, "Building Deep Supplier Relationships." *Harvard Business Review,* December 2004.
[30] OEM benchmark survey; a survey of auto suppliers by John Henke of Oakland University.

### 15.18.1 Low Number of Suppliers

Traditionally, one major source for choosing a supplier among others is price. Manufacturers often bought their parts from several suppliers to avoid being locked into one supplier. Having multiple sources enabled manufacturers to get lower prices by putting suppliers against each other. This strategy has several negative consequences. First, each supplier perceives its relationship with the manufacturer a short-term relationship, and thus feels no loyalty to the manufacturer. Second, to protect themselves, suppliers look for other manufacturers for their products, resulting in lack of focus on the manufacturer's parts.

Under lean operations, however, manufacturers reduce their number of suppliers for each part, to a few good suppliers and focus on building and maintaining a close relationship with them. In fact, a comparison between Japanese and American auto manufacturers showed that, Japanese auto manufacturers have an average of 170 suppliers while their American counterparts have about 509 suppliers. Suppliers of the Japanese manufacturers often take responsibility for an entire family of parts. This allows the supplier to utilize group technology in order to produce parts in lower costs and economic volumes. Also, suppliers will be engaged in a long-term relationship with the manufacturer, which makes suppliers make more effort to provide on-time deliveries and high-quality parts. Hence, instead of focusing on getting cheaper price from their suppliers, companies such as Toyota have smaller number of suppliers and consider quality and innovation capabilities (instead of price) as the primary determinant factors in choosing suppliers.

### 15.18.2 Close Proximity

To enable suppliers to deliver parts and materials at frequent intervals, suppliers of lean operations are often located in the proximity of the manufacturer. For example, Toyota has 12 plants in Toyota City. Suppliers of these plants are located within 50-mile radius of these plants. This, for example, allows them to deliver transmission and engines every 15 to 30 minutes. If close proximity is not possible for some suppliers, they will establish small warehouses near the manufacturer, allowing them to have frequent deliveries.

Close proximity to the manufacture also benefits suppliers. Delivering several times a day incurs high transportation cost. When suppliers are close to the manufacturer—and thus close to each other—they can coordinate their deliveries to share the transportation cost. For example, in each day, or in each hour, one of the suppliers goes to other two suppliers, load their parts and delivers them (along with its own items) to the manufacturer. By taking turns to make the deliveries, the transportation cost of each supplier becomes smaller.

### 15.18.3 Knowing How Your Suppliers Work

Following the practice of Genchi Genbutsu—go and see for yourself—Toyota insists that its managers at all levels, all the way up to the president, should know and understand how their suppliers work. By knowing how suppliers work, a lean manufacturer will have a better understanding of its suppliers' capabilities, quality, and flexibility in providing what the manufacturer wants.

Other lean manufacturers such as Honda also make sure that their engineers know how their suppliers work. For example, in 1987, before choosing Atlantic Tool and Die as one of its suppliers, Honda sent one of its engineers to spend a year there. The engineer studied the firm for 12 months, gathering data and analyzing Atlantic's process and cost structure. The engineer then suggested some improvement ideas, which were implemented in Atlantic's plant. Honda engineer's knowledge led to Honda choosing Atlantic Tool and Die as its supplier in 1988.

### 15.18.4 Considering Suppliers as a Part of Your Firm

Taichi Ohno has emphasized Toyota's suppliers being a part of Toyota's family. He is clear in the following quote, in which Ohno refer to Toyota as "parent company":

> **Taichi Ohno:** *The achievement of business performance by the parent company through bullying suppliers is totally alien to the spirit of Toyota Production System.*

How does Toyota make its suppliers a part of the firm? Here are some strategies that Toyota uses.

- *Establishing Long-Term Relationship Through Long-term Contracts:* Signing a long-term contract with suppliers have several benefits. First, it provides assurance to suppliers that they will have long-term revenue. This allows suppliers to focus on long-term efforts to improve their parts. Second, in return for offering a long-term contract, the manufacturer can negotiate the current price and future price increases—or price reduction. Third, with a long-term contract, the paper work associated with receiving and inspection can be reduced or eliminated, and instead, electronic links can be established. This leads to cost saving for both manufacturer and the supplier. Finally, a long-term contract fosters closer relationship and builds mutual trust between manufacturer and its suppliers.

- *Developing Supplier's Technical Capabilities:* Toyota understands that improving its suppliers process will result in better quality parts and lower costs. These improvements also benefit Toyota. Toyota analyzes their suppliers process and teaches them the principles of lean operations as well as techniques required to implement those principles. They also help them improve their problem-solving and innovation capabilities.

- *Sharing Information with Suppliers:* By sharing information about demand forecast and production schedule with the supplier, supplier will be able to get an estimate of the timing and the size of the order it will receive from the manufacturer. This will allow the supplier to plan its production such that it delivers the right order at the right time. Sharing information with the supplier also encourages supplier to share its information with the manufacturer, especially information about potential problems that may result in late deliveries.

- *Including Suppliers in Product Development Process:* Toyota involves the suppliers of its parts when it develops a new car model, or when it revises its current models. This provides Toyota with several benefits. First, Toyota can take advantage of suppliers' expertise in development, design, and manufacturing. Second, having suppliers during the development process prevents quality problems resulting from the manufacturing process at suppliers' facilities. Third, Toyota gets a more informed cost estimate of the parts, which helps it to design cars that cost less to produce. Finally, including suppliers in the product development process strengthen Toyota's relationship with its suppliers, and gives the suppliers the sense of being a valued partner of Toyota.[31]

### 15.18.5 Monitoring Your Suppliers' Performance

While lean manufacturers consider their suppliers as part of the firm, they closely monitor their suppliers' performance. Lean manufacturers should provide clear expectations on cost, quality, and responsiveness to their suppliers. These expectations should be obtainable and

---

[31] Wagner, S.M., and M. Hoegel, "Involving suppliers in product development: Insights from R&D directors and project managers." *Industrial Marketing Management*, 35 (2006) 936–943.

clearly understood by both the manufacturer and the suppliers. The manufacturer then gives the suppliers periodic feedback regarding their performance.

Toyota constantly monitors and evaluates the performance of its supplier. They rate their suppliers from one to five. One corresponds to a significant problem (e.g., supplier's plant burns down) and five corresponds to exemplary supplier. Two corresponds to a supplier with the possibility of shutting Toyota's plants down. If a supplier action leads to interruption in Toyota's operations, Toyota immediately sends a team of experts to the supplier's plant, studies the problem, and asks the supplier's managers to develop an action plan to address all Toyota's concerns.

One example of Toyota's approach in dealing with its suppliers' problems is that of Trim Masters Inc. (TMI), its supplier of seats that made about 25,000 seat sets a year for Avalon and Camry. The TMI's computer system went down for only 3 hours and resulted in shut down of Toyota assembly line. Toyota sent a team of experts to TMI plant immediately. Even after the computer was up, Toyota had its crew visiting TMI plant every day for 2 weeks. Furthermore, the TMI plant was put on watch and was asked by Toyota to send monthly reports and explain what improvement efforts were being made to guarantee that the problem will not occur again. The Toyota experts continued visiting TMI plant a few times a week for 6 months. Furthermore, Toyota asked TMI to examine and improve every aspect of its business, including employee hiring and training, team structures, standardized work, pull systems, problem-solving procedures, and quality control. TMI did what Toyota suggested, resulting in TMI to be ranked by J.D. Power as the top automotive seat supplier in the country in consecutive years.

### 15.18.6 Conducting Joint Improvement Activities

Toyota understands that, when its suppliers improve their quality or reduce their costs—by minimizing waste—the benefits are automatically passed on to Toyota. Therefore, Toyota spends time and effort to help its suppliers improve their processes. This includes exchanging best practices and establishing joint Kaizen events at the suppliers' facilities.

All lean companies benefited from conducting joint improvement activities with their suppliers. Honda, for example, has several of its engineers stationed in the United States. Their job is to lead Kaizen events at their suppliers and stay in touch with them long after returning to their own plants. Honda's joint improvement activities have increased suppliers' productivity by 50 percent, improved quality by 30 percent, and reduced cost by 7 percent. Suppliers shared half of the cost saving resulted from joint activities with Honda. One advantage for the suppliers is that the suppliers can use the best practices learned through their joint activities with lean manufacturers (such as Honda) and implement them on their other products that are sold to other firms, which means keeping 100 percent of the cost saving.[32]

## 15.19 Lean Operations in Service Industry

In 2000, due to economic stress and low performance, the Board of Directors of Virginia Mason Medical Center issued a mandate for change with the goal of improving organizational culture and becoming the quality leader in healthcare. Seattle's Virginia Mason Medical Center is an integrated healthcare system with 400 physicians, 5000 employees, 9 locations, and a 336-bed hospital. The leadership had the vision of Virginia Mason Production System (VMPS), which was modeled based on the lean concepts in Toyota Production System. In 2002, to have their senior leaders learn lean principles and to see how lean management works in action, Virginia Mason sent all its senior executives to Japan. Their

---

[32] Liker, J. and T. Y. Choi, "Building Deep Supplier Relationships." *Harvard Business Review,* December 2004.

executives worked on the production line at the air conditioning plant of Hitachi. They mapped the process, measured flow times and inventory and throughput. The lesson they learned, according to senior leaders, was that healthcare has many common features as manufacturing. Several similar trips were made that included managers, physicians, nurses, and staffs. The Virginia Mason Production System (VMPS) was then developed with the goal of reducing wastes and improving quality. Virginia Mason required all its 5000 employees to attend an "Introduction to Lean" course.

Virginia Mason implemented several lean tools such as value stream mapping, setup time reduction, quality at the source, visual workplace, process standardization, continuous improvement, and employee improvement. The center was able to significantly improve its operational and economical performance. Inventory was reduced by 53 percent, floor space was reduced by 41 percent, lead times were reduced by 65 percent, and setup times were reduced by 82 percent. These waste reductions resulted in significant savings in capital expenses, including $1 million for an additional hyperbaric chamber, $1 to $3 million for endoscopy suites, and $6 million for new surgery suites. None of these units were needed after implementing lean that improved the capacities of the existing units.[33]

To encourage the participation of employees in process improvement and to create job security, like Toyota, Virginia Mason adopted the "No-Layoff Policy." Kaizen events, known as "Rapid Process Improvement Workshops" at Virginia Mason, are held weekly, with the goal of achieving immediate results in eliminating wastes. Employees practice 5S to organize work areas to minimize motion and to facilitate an efficient flow of activities through workplace.

Another lean tool that helped Virginia Mason improve patient safety was "Quality at the Source." Similar to Toyota Production System, Virginia Mason gave all its employees the authority to "stop the line." Specifically, Virginia Mason established its "Patient Safety Alert System" under which each employee must stop the care process if the employee feels something is wrong. When that occurs, the employee activates a call (analogous to Andon cord) to the patient safety department. When the department receives a call, managers and other appropriate stakeholders are immediately sent to the problem area to start analyzing the problem and determine the root cause of the problem. The patient safety department received an average of three calls per month in 2002, but it increased to 17 per month by the end of 2004, including alerts regarding medication errors, equipment and facility problems, and system issues.

Virginia Mason was not the only healthcare institution that benefited from lean operations. In his book, *Lean Hospitals*, Mark Graban provides many examples of hospitals that adopted lean practices. For example, using lean principles, University of Pennsylvania Medical Center was able to reduce hospital-acquired infections, resulting in saving 57 lives and reducing costs by over $5 million over 2 years. Avera McKennan Hospital in South Dakota was able to reduce length of stay of emergency patients by 29 percent and thus avoided $1.25 million in new emergency department (ED) construction. By eliminating waste, Denver Health, Colorado, was able to achieve cost saving of $200 million over 7 years, while achieving "the lowest observed-to-expected mortality among the academic health center members of the University Health System Consortium in 2011."

It is not surprising that lean principles developed for manufacturing systems also work for service systems. From process-flow perspective, most service systems are similar to manufacturing systems. Flow units (e.g., people in theme parks, patients in a hospital, or loan applications in financial institutions) enter the process and go through several stages before they leave the process. Processing these flow units also has similar features as those in manufacturing systems: It takes time and resources to process them; they may need to wait in

---

[33] Institute for Healthcare Improvement, *Going Lean in Healthcare*, Innovation Series 2005 white paper.

the buffers of different stages (as inventory) before they leave the system; there is often setup required before processing a unit; processing may involve quality issues such as service failure and rework; and the resources may not always be available to process units (i.e., resource availability). Thus, it is not surprising that lean tools that are developed to improve process performance (i.e., minimize waste, reduce inventory, improve quality) are applicable in service systems, including food services, hotels, call centers, financial systems, among many others.

For example, to decrease the inventory in its restaurants, Taco Bell implemented just-in-time deliveries of food products, and its stores received more frequent deliveries in smaller batches. This also saved space in their restaurants. By standardizing work and building quality in its process, McDonald's can offer same quality products consistently in all its locations. By starting to make a burger after a customer places an order—a pull system—McDonald's was able to reduce inventory and improve quality (i.e., making fresh burgers). Before implementing this lean principle, McDonald's precooked its burgers and waited for the customer orders—a push system. Southwest Airlines used lean principles and eliminated non-value activities and improved customer service. Retailers such as Walmart and Target use flexible workers to do several tasks such as arranging shelves, working in cash register, and organizing inventory. HP's implementation of lean tools improved its shipment delivery services by reducing the overdue shipments to its customers. By using lean principles, West Coast Finance Company was able to significantly reduce the time for credit card approval.[34]

## 15.20 Summary

While lean operations started with developing manufacturing practices to eliminate waste, it eventually evolved into a management philosophy of continuous improvement for the entire organization, covering both manufacturing and service operations. This chapter provides a summary of lean operations tools and illustrates how those tools improve process-flow measures such as inventory, flow time, throughput, and quality.

The chapter describes how Toyota's practices such as aggregating demand and scheduling production based on takt time diminish the corrupting influence of demand variability on operations. Single minute exchange of dies (SMED) is another lean tool that reduces or eliminates setup times, allowing smaller batch sizes, which in turn reduces inventory and flow times. Using flexible resources such as cross-trained workers or multi-functional machines results in more responsive operations with less inventory. Cellular layout is another essential tool in lean operations that promotes teamwork and provides benefits such as flexible capacity, flexible product mix, and smaller batch sizes.

In pull production systems production schedule at each stage is set based on the need of downstream stages not based on the availability of units in upstream stages (like a push system). In pull systems the inventory in the process is kept below a certain limit by implementing job release policies as in Kanban system. This chapter illustrates why implementing pull systems leads to lower manufacturing costs, better problem visibility, higher quality, and higher responsiveness.

The main efforts to manage quality in lean operations focus on preventing quality problems (by simplifying tasks and Poka-Yoke), early detection of quality problems (using Jidoka), and rapid resolution of the problem (using Andon systems). The key practice of stopping the line when a quality issue is detected emphasizes on lean's philosophy of pursuing continuous improvement with the focus on long-term goals, rather than focusing on short-term goal of improving productivity. Total Productive Maintenance (TPM)—another essential part of lean operations—is a comprehensive approach to equipment maintenance

---

[34] Lee, J.Y., "JIT Works for Service Too." *CMA Magazine*, Vol. 64, No. 6, 1990, p. 20.

with the goal of having no breakdowns, no defects, and no accidents. This chapter provides analytics to design preventive maintenance schedules that maximize equipment's effective capacity or minimize total maintenance and repair costs.

This chapter also discusses process standardization, which is an essential factor in sustaining lean operations and process improvement. Process standardization is based on three elements of takt time, work sequence, and standardized stock. The main goal of process standardization is to reduce variability and to produce consistent and high-quality products. Heijunka in lean operations refers to leveling production and aims at eliminating Mura (unevenness of workload) in the process. The chapter shows how Heijunka uses the total demand for different products in a period and levels them out such that the same number of each product and the same mix of products are produced each day. Kaizen is the never-ending work of continuous improvement. It covers all aspects of operations including material, equipment, people, processes, and methods. The chapter explains why Kaizen events must focus on small gradual improvements with minimum cost and should involve everyone.

At the center of lean operations are the employees who have many responsibilities, for example, performing tasks, checking for quality issues, performing preventive maintenance, and problem solving. In lean operations philosophy, employees are the system's most precious resources. Hence, lean organizations treat all their employees with respect. They believe that employee involvement and empowerment is the best way that a company can reach its goals. Lean systems must also build trust with their suppliers and make them feel that they are a part of the firm. Lean organizations such as Toyota achieves this by using small number of suppliers, building close relationship with them, monitoring them closely, and conducting joint improvement projects with them to help them become lean.

## Discussion Questions

1.  What does Genchi Genbutsu refer to in Toyota Production System?
2.  Explain what Jidoka means and how it is evolved into several practices in lean operations.
3.  Lean operations is a customer-focused process. Describe the three capabilities that lean operations must have to become a customer-focused process.
4.  What are the main four goals of lean operations?
5.  How was Toyota's approach to eliminating waste different from traditional scientific management approach?
6.  What are the seven sources of waste in lean operations?
7.  What are the particular actions that Toyota has taken to mitigate the impact of variability in demand?
8.  How was Toyota's approach different from Ford's approach in dealing with setup time of their stamping plants?
9.  What is SMED? Describe six principles that help achieve SMED.
10. Describe four major benefits of smaller batch sizes in operations systems.
11. Explain how using flexible resources helps reduce waste in lean operations.
12. Describe four different ways that cellular layout can eliminate wastes.
13. Explain the differences between pull and push production systems from the following three perspectives: job release, control, and production authorization.
14. Describe how Kanban systems work.
15. Describe seven benefits of pull production systems.

16. What is river analogy and how is it analogous to continuous improvement in lean operations?

17. What are the two main approaches that lean operations use that can prevent quality problems from occurring?

18. Describe four different ways in lean operations that help early detection of quality problems.

19. What are the four practices in lean operations that help rapid resolution of quality problems?

20. What is Poka-Yoke? Present two examples of Poka-Yoke in manufacturing and service systems.

21. What is an Andon system in lean operations?

22. What is TPM? What are its goals and how does it achieve those goals?

23. What is the difference between autonomous maintenance and preventive maintenance, and what are the advantages of each?

24. What is takt time? and what is its role in minimizing inventory?

25. What is work sequencing in lean operations and what are its six benefits?

26. How does Toyota Production System define Muda, Muri, and Mura?

27. What is Heijunka and what are the two steps of implementing Heijunka? Also, describe the three main benefits of Heijunka.

28. What are Kaizen and Kaizen Blitz? Why is worker involvement essential in Kaizen and why does Kaizen focus on small and gradual improvements?

29. Describe what value stream and value stream mapping (VSM) are. How does VSM relate to lean operations?

30. Define visual workplace and describe three major benefits of a visual workplace.

31. What are the four main components of a visual workplace? Provide an example for each component.

32. What are 5S and 5-why practices in lean operations?

33. How do teams in lean operations function? Describe responsibilities of team members, team leaders, and group leaders in Toyota Production System.

34. What are the six principles that Toyota uses to build a close relationship with its suppliers?

35. How does Toyota make its suppliers a part of its firm?

36. Provide three examples of service firms that have used the principles of lean operations to improve performance. Which lean tool (or tools) did they use?

## Problems

1. A five-station serial push line has a throughput of 10 jobs an hour. If the line is run like a Kanban system such that it also produces 10 jobs an hour, explain why each of the following statements is true or false:
   a. The average inventory in each station of the push line is larger than that in the Kanban line.
   b. The total average inventory in the push line is larger than that in the Kanban line.
   c. Because of Little's law, the average time it takes a job to go through the push line is the same as that in the Kanban line.
   b. None of the above is true. More information is needed to make any conclusion.

2. Consider the five-station serial line in Problem 1, and suppose that the line is run as a single Kanban system. All stations have five Kanban cards, except the bottleneck station that has eight Kanban cards. Explain why each of the following statements is true or false:
   a. Reducing the number of Kanban cards in a non-bottleneck station does not affect the throughput of the line, but reduces the total inventory in the line.
   b. Reducing the number of Kanban cards in a non-bottleneck station reduces both the throughput and the total inventory in the line.
   c. Reducing the number of Kanban cards in the bottleneck station does not affect the throughput of the line, but reduces the total inventory in the line.
   d. Reducing the number of Kanban cards in the bottleneck station reduces both the throughput and the total inventory in the line.

3. Explain why each of the following statements is true or false:
   a. Pull systems generate higher throughput than push systems.
   b. For a given inventory level, pull systems generate higher throughput because they increase theoretical capacity of the process.
   c. Pull systems can generate the same throughput as their push counterparts, but with lower inventory.

4. A CNC drill is the bottleneck of a Tool and Die shop. The drill is used in the production of a variety of products. The average time it takes to drill a part is about 3 minutes. Because the drill works constantly in three working shifts (24 hours a day), it often breaks down and requires some repair. The following table shows the probability distribution of the number of hours the drill works until its next failure. When the drill is repaired, it returns to its good operating condition. The average time to repair the drill is about 2 hours.

| Time to failure | Probability | Time to failure | Probability |
|---|---|---|---|
| 1 | 0.003 | 11 | 0.113 |
| 2 | 0.004 | 12 | 0.131 |
| 3 | 0.010 | 13 | 0.110 |
| 4 | 0.034 | 14 | 0.061 |
| 5 | 0.039 | 15 | 0.067 |
| 6 | 0.041 | 16 | 0.032 |
| 7 | 0.049 | 17 | 0.029 |
| 8 | 0.055 | 18 | 0.015 |
| 9 | 0.092 | 19 | 0.009 |
| 10 | 0.100 | 20 | 0.006 |

   a. What is the effective capacity of the CNC drill?
   b. How much the effective capacity of the CNC drill increases, if the drill is stopped after 12 hours of operations and a 20-minute preventive maintenance is performed that puts the machine in its good condition?
   c. Develop a preventive maintenance schedule that maximizes the effective capacity of the CNC drill. Specifically, when should the drill be stopped for the 20-minute maintenance operation?

5. Consider the CNC drill in Problem 4 and assume that repairing the CNC drill costs about $1200. This includes the cost of labor, material, and parts and the cost of losing production. The cost of the 20-minute preventive maintenance, however, is about $350 each time it is performed. Develop a preventive maintenance schedule that results in the minimum average repair and maintenance cost.

6. A manufacturer produces three different types of microwave ovens, namely, solo microwave ovens (S ovens), grill microwave ovens (G ovens), and convection microwave ovens (C ovens). The monthly demand for these products has been steady and is about 2250, 750, and 1500 a month, respectively, for S ovens, G ovens, and C ovens. Currently these products are produced on three separate lines. The manufacturer, however, is redesigning its operations and is planning to use a mixed-model assembly line to assemble these three products. The line will work 7 hours a shift, three shifts a day, 5 days a week, and 20 days in a month. If the manufacturer wants to level the production of the assembly line using Heijunka,

   a. What should the production rate of the assembly line be?
   b. What should the daily final assembly schedule (FAS) be? How should the three types of ovens be sequenced on the assembly line?

# CHAPTER 16

# Quality Management and Control

## 16.0 Introduction

High-quality products (goods or services) result in lower rework, lower scraps, lower warranty and recall costs as well as in higher customer satisfaction, all of which lead to higher profits. Thus, in today's competitive market, quality improvement is essential for the success of a firm. But, what is quality and how is it measured? American Society for Quality (ASQ) defines quality as: "*A subjective term for which each person or sector has its own definition. In technical usage, quality can have two meanings: 1. the characteristics of a product or service that bear on its ability to satisfy stated or implied needs; 2. a product or service free of deficiencies.*" The above definition points to one important factor: quality is subjective and thus its interpretation depends on who is defining it and also depends on the product for which the quality is defined. Therefore, to better understand the quality, we must first discuss what quality means to those who use the product—customers—and those who produce the product—producers.

## 16.1 Customer's Perspective of Quality

When a customer wants to purchase a car, what are the factors that shape the customer's perception about the quality of a car? For manufactured products, eight different dimensions of quality that are important to consumers are:[1] (i) performance (the basic operating characteristics of the product, e.g., gas-mileage and acceleration of a car), (ii) special features (extra features and items added to the basic feature of the product, e.g., seat-warmer or GPS in a car), (iii) reliability (the degree with which the product functions as expected without failure), (iv) durability (expected lifetime of the product), (v) serviceability (the ease and speed of getting a product repaired, including the competence and courtesy of the repair crew), (vi) conformance (the degree with which the product meets the specified standards), (vii) aesthetics (how a product looks, feels, sounds, tastes, or smells), and (viii) safety (risk of harm and injury to the user).

While manufactured products are tangible (i.e., they can be seen and touched), service products are mainly intangible and are experienced. Hence, the dimensions of quality for service are different from those of manufactured products. Some of the quality dimensions for services are:[2] (i) completeness (is everything that the customer wanted provided?), (ii) time (how long the customer waits to receive service and is the service completed on time?), (iii) courtesy and friendliness (how customer is treated by service staff?), (iv) consistency (is the service the same each time?), (v) accessibility and convenience (how easy is it for a customer to obtain the service?), (vi) responsiveness (how does service provider handle customer requests and complaints?), (vii) competence (knowledge and skills of service provider), and (viii) tangibles (observable characteristics of the service facility, e.g., cleanliness of the facility, appearance of staff).

## 16.2 Producer's Perspective of Quality

For a product to be successful in the market, marketing department must identify the quality dimensions that are critical to the customers. Once it is determined what customers want, need, and can afford—*voice of customer*—the product development department incorporates them into the design of the product (see Fig. 16.1).

The quality dimensions identified by customers are often qualitative in nature. For example, one quality dimension important to car buyers is that the car be "Reliable." One cannot design a reliable car until the qualitative concept of "Reliability" is converted to one or more quantitative and actionable measures. One measure often used to express reliability of cars is the total number of problems that are experienced by the owner in the first 3 months or the first 3 years of ownership.[3] These measurable quantities are called "Critical To Quality."

---

**CONCEPT**

### Critical To Quality (CTQ)

*Critical To Quality* (or CTQ) are the key measurable and actionable characteristics of a product (goods or service), or a process, whose performance standards or specification limits must be met in order to satisfy the customer.

---

After identifying CTQ characteristics of a product, the product development department of a firm designs the product to meet the desired CTQ characteristics. The design process considers physical and technological limitations that affect the design. For example, customers want a laptop that is light in weight. But, there are technological constraints that limit how light a laptop can be made. After the product development department designs

---

[1] Garvin, D., "What Does Quality Really Mean." *Sloan Management Review*, 26, no. 1 (1984) 25–43.
[2] Evans, J.R. and W.M. Lindsay, *The Management and Control of Quality*. Third edition, 1996.
[3] J. D. Power & Associates studies the number of problems experienced per 100 vehicles in the first 3 years as a measure of dependability of a car. It also studies the number of problems reported by owners in the first 90 days of ownership as a measure of initial quality. Both these measures correspond to the reliability of a car.

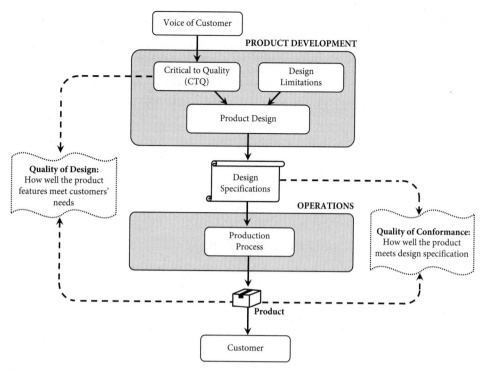

**Figure 16.1    Producer's perspective of quality: quality of design and quality of conformance.**

the product, it sends a series of design specifications to the operations department. *Design Specifications* are quantitative measures that determine every physical and performance aspect of the product (e.g., the shape and dimensions of each part, the material used to make the part, weights of the parts, etc.).

One dimension of quality that is important to the producer of the product is how well the designed product match the customers' needs and wants. This is referred to as "Quality of Design."

---

**CONCEPT**

**Quality of Design**

*Quality of Design* refers to how well a product's features and specifications match the customers' needs.

---

Having design specifications, the operations department has the task of making the product according to those design specifications. This brings us to the second dimension of quality that is important to the producer: how the produced product conforms to the design specifications.

---

**CONCEPT**

**Quality of Conformance**

*Quality of Conformance* refers to how well the features and specifications of the produced product match the designed features and specifications.

In summary, a producer's perception of the quality of its product and processes corresponds to how well its design process can develop a product that meets customers' needs (i.e., quality of design) and how well its operations can adhere to the chosen design specifications (i.e., quality of conformance). If design and production are done properly, the final product would satisfy the quality dimensions important to the customers.

## 16.3 A Bit of History

Before the introduction of mass production by Henry Ford, products were made by master craftsmen who also were responsible for quality control of their products. Under Henry Ford's mass production system (i.e., moving assembly line) several workers (instead of one expert craftsman) were responsible for producing and assembling products. This introduced variability in product quality. To identify quality problems, quality control stations were added to the production process. This made quality control an essential part of mass production.

### Walter Shewhart and Statistical Quality Control

One of the pioneers of quality improvement in mass production systems was Walter Shewhart who was working for Bell Telephone Laboratories (now known as American Telephone and Telegraph or AT&T). In 1924, he introduced the concept of Statistical Quality Control (SQC), emphasizing that products made by mass production always have variations in their features and performance. Shewhart believed that statistical methods can be used to identify, analyze, and control these variations. Shewhart used statistics to develop a graphical method that could help firms identify unusual variations in product specifications. Known as "Control Charts," these graphical techniques helped firms make the decision of whether or when their production system needs to be investigated for quality improvement. We present these charts in Sec. 16.7.

### Deming and Continuous Quality Improvement

Another person who perhaps had the most impact on quality improvement is E. Edwards Deming, who learned statistics during his PhD studies of mathematical physics. While working in the Department of Agriculture, Deming met Shewhart and learned about his approach of using statistical control charts and how they can be used to improve productivity and quality.

During World War II, Deming started teaching Shewhart's statistical process control techniques to engineers and managers of firms that were suppliers of U.S. military. More than 30,000 engineers were trained by Deming during those years. A group of those engineers later established the American Society for Quality Control (now known as American Society for Quality, or ASQ). When World War II ended, Deming was internationally known and was providing consulting services for quality improvement.

After World War II, Japan and European countries were rebuilding their cities and economies, and thus U.S. companies did not face any competition. The economy in the United States improved and the demand for products was high. Firms, therefore, focused on increasing production to meet the demand. By boosting production, American companies earned higher profits and—due to lack of competition—the quality and process improvement were not the focus of American managers. Japanese firms, however, were facing a different situation: weakened economy, crippled capacity, low quality, and low productivity. But these were about to change. The Allied sent a group of engineers who knew statistical quality control techniques to Japan to improve their communication system. This exposed Japanese to Deming's work. Deming was then invited to Japan in 1950 and taught his statistical quality control techniques to a group of Japanese managers. Deming also explained to the Japanese that the cost and effort put in improving the quality of a product will pay off later—the more the quality is built into products, the less they cost to produce and maintain.

Japanese soon put Deming's ideas and techniques to work, resulting in increase in productivity of their production processes and improvement in their product quality. Quality of Japanese (and European) products then surpassed American products. Rising prices, on the other hand, were making American consumers more careful with their purchases and they started to pay more attention to product quality. This led to a decrease in demand for American-made products and sounded an alarm for American executives.

While Deming's process and quality improvement techniques were used by Japanese engineers and manager, American companies were not much aware of Deming's approach. In fact, Deming worked with Japanese for about 30 years before he was recognized by American managers. In 1980, Americans learned about Deming when he appeared in an NBC television documentary titled "If Japan Can, Why Can't We?" The program revealed how Deming's approach to quality management had contributed to the success of Japanese companies. American companies then consulted Deming and began to implement his quality control and improvement methods. During the 1980s, American companies started to benefit from implementing Deming's quality improvement methods, and the new effort in continuous improvement of quality spread to other functions of firms such as accounting, finance, sales, and marketing. Deming continued to teach his quality improvement techniques and strategies to managers and engineers until his death. He is known as the "father of quality evolution."

Deming's contribution to quality improvement evolution is beyond just using statistical methods to identify and control variations in processes and products. Deming also promoted the idea of continuous improvement and emphasized that managers (not only the workers) are also responsible for quality control and improvement. Deming was also credited for what is known as *Deming's Wheel,* or *Plan-Do-Study-Act* or *PDSA Cycle,* see Fig. 16.2. Deming Wheel is a four-step process for continuous quality improvement. Here is how Deming Institute (founded in 1993) describes PDSA cycle:

*The cycle begins with the Plan step. This involves identifying a goal or purpose, formulating a theory, defining success metrics and putting a plan into action. These activities are followed by the Do step, in which the components of the plan are implemented, such as making a product. Next comes the Study step, where outcomes are monitored to test the validity of the plan for signs of progress and success, or problems and areas for improvement. The Act step closes the cycle, integrating the learning generated by the entire process, which can be used to adjust the goal, change methods, reformulate a theory altogether, or broaden the learning—improvement cycle from a small-scale experiment to a larger implementation Plan. These four steps can be repeated over and over as part of a never-ending cycle of continual learning and improvement.*

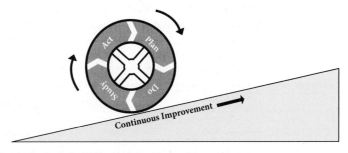

Figure 16.2   Deming's wheel: the PDSA cycle for continuous improvement.

One important point to emphasize is that, besides following Deming's quality improvement methods, the success of Japanese companies to produce high-quality products is also due to their quest for eliminating waste during the development of lean manufacturing and just-in-time in Japan. Lean practices such as identifying and eliminating quality at the source and Jidoka also contributed significantly to product quality improvement in Japan (see Chap. 15 for details).

### Juran, Focusing on Quality and Management

Like Deming, Joseph M. Juran also taught courses on quality management. His first trip to Japan was few years after he published his book titled *Quality Control Handbook*. Juran developed an influential course titled "Managing for Quality," which has been taught in Japan and 40 other countries. Juran always emphasized the role of management in quality. Like Deming, Juran is considered as a major contributor to the Japanese movement in quality improvement. In 1954 Juran was invited by the Union of Japanese Scientists and Engineers to give seminars for top and middle-level managers, to emphasize on the key role that they play in promoting quality control activities. In his book, *What Is Total Quality Control? The Japanese Way*, Kaoru Ishikawa writes:[4]

> *Juran's visit marked a transition in Japan's quality control activities from dealing primarily with technology based in factories to an overall concern for the entire management. The Juran visit created an atmosphere in which QC was to be regarded as a tool of management, thus creating an opening for the establishment of total quality control as we know it today.*

### Crosby and Cost of Poor Quality

Another leader in quality management evolution is Philip Crosby. While working at Martin Marietta in the 1960s, he developed the concept of "Zero Defect." His first book titled *Quality Is Free* sold more than 1 million copies in 1979. The book brought the issue of poor quality to top executives in the United States. Crosby's main point was that poor quality results in lower effective capacity due to yield loss and rework, as well as lost sales and customer dissatisfaction. In the long run, the total cost of poor quality outweighs the cost of investment in efforts to prevent quality problems. He emphasized that the goal of managers should be zero defects. Focusing on zero defect, better quality can be achieved with no cost.

Crosby's philosophy in responding to the quality crisis was doing it right the first time. He is also known for four Absolutes of Quality Management, which are: (i) the definition of quality is conformance to requirements (not "goodness"), (ii) the system of quality is prevention (not appraisal), (iii) the performance standard is zero defects (not "that's close enough"), and (iv) the measurement of quality is the price of nonconformance (not indices). Philip Crosby was considered as an influential contributor to quality movement who developed practical concepts to define and communicate quality and quality improvement practices. During the 1980s, his consulting firm was advising 40 percent of the Fortune 500 companies on quality management.[5]

### Feigenbaum and Total Quality Control

Armand V. Feigenbaum was a top expert at General Electrics at the age of 24. He believed that quality is not just a set of statistical techniques to control quality of the final product. He recognized that quality control applies to all activities of a firm. He developed the concept of "Total Quality Control" in an article in 1946, and later in 1951 he wrote the first edition of his

---

[4]From American Society for Quality (ASQ) website.
[5]British Library, https://www.bl.uk/people/philip-crosby.

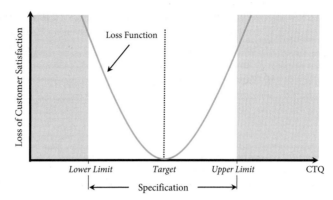

book *Total Quality Control.* He wrote: "*Total quality control is an effective system for integrating the quality development, quality maintenance, and quality improvement efforts of the various groups in an organization so as to enable production and service at the most economical levels which allow full customer satisfaction.*" The principles of Total Quality Control led to what is known as "Total Quality Management" or TQM. TQM emphasizes on the role of all people in an organization—led by top management—in continuously controlling and promoting quality of all activities of a firm. Feigenbaum also developed the concept of "Hidden Plant." According to this concept, a certain portion of capacity of every factory is wasted by not getting it right at the first time. Feigenbaum was the founding chairman of the board of the International Academy for Quality. The academy brought together leaders of the European Organization for Quality, the Union of Japanese Scientists and Engineers, and American Society for Quality (ASQ) to promote and expand quality improvement ideas and efforts.

### Taguchi and Experimental Design

While Deming was the first who recognized the benefits of moving the quality control inspection station from the end of a production line backward to earlier stages of production, Taguchi moved the quality control even further back from production stage to the design stage of a product. Genichi Taguchi was a Japanese engineer who developed a series of statistical experimental design methods—also known as Taguchi Methods—that enable manufacturers to design a high level of quality into their products at the design stage before the production begins. Taguchi Method is also referred to as "Off-Line Quality Control" or "*Robust Design*" because it is used during the design stage to develop products and processes that are robust to parameters that are outside the design engineer's control, which in turn improves performance quality. Taguchi's other contribution is *Taguchi Loss Function,* see Fig. 16.3. While the common belief among managers and engineers was that customers are satisfied if the variation in the product is within design specifications, Taguchi stated that the loss in customer satisfaction does not suddenly start when the product's actual performance violates its design specifications; any deviation from the target results in loss of customer satisfaction. The loss increases rapidly (i.e., exponentially) as the performance deviates further from the target, even if the performance is still within specifications. Also, although each individual part of a product (or component) might have a small deviation, the combined effect of variations of all parts can be very large.

### Ishikawa and Quality Circles

Kaoru Ishikawa, a professor in Tokyo University, was the quality expert who promoted the concept of "Internal Customer," that was defined as the downstream stage of a process. Like Feigenbaum, he also believed that quality is not only about the product, it also applies to the entire organization. Ishikawa contributed significantly to the success of quality circles. A *Quality Circle* is a group of employees (who perform the same tasks) meeting regularly to identify and analyze quality problems and develop solutions. Ishikawa is also known for developing Cause-and-Effect Diagram, which is a powerful tool for finding the root cause of problems (see Sec. 17.5.3).

### Global Quality Efforts and ISO 9000

The critical role of quality in global market led to an international effort to set quality standards for total quality management in organizations. This resulted in the development of International Organization for Standardization, or ISO 9000. ISO 9000 is a set of international standards on quality management and quality assurance developed to help companies effectively document the quality system elements to be implemented in order to maintain an efficient quality system (ASQ). The ISO 9000 series was first published in 1978 and has been revised and updated every 5 years. It has been recently updated and is called ISO 9000:2000. Firms can now apply for ISO certification which shows that they meet internationally accepted standard for quality management. This has become essential to suppliers, since more and more manufacturers prefer to work with an ISO-certified supplier to ensure they receive high-quality parts and material.

The ISO initiatives have gone beyond product and process quality and now covers sustainability, leading to ISO 14001. As mentioned in Chap. 1, ISO 14001 is a management system to help businesses reduce their environmental impact. To get ISO 14001 certification, firms need to show that their entire business operations is committed to reducing harmful effects on the environment, and provide evidence of continual improvement of their Environmental Management Systems.

## 16.4 Process Perspective of Quality

We have already discussed customer's and producer's perspectives of quality. To produce an acceptable product that satisfies quality standards from both the customer's and producer's perspectives, the production process must have two features. First, the process must be *capable* to produce the product that meets or exceeds design specifications. Second, if the process is capable of meeting design specifications, the process must also be stable and *in control* so its output remains within design specifications. The former corresponds to the concept of "Process Capability," and the latter corresponds to "Process Control."

As an example, consider a firm that produces food products, including peanut butter that are sold in jars containing 40 ounces of peanut butter. To make peanut butter, raw peanuts go through the process of roasting, blanching, grinding, mixing, filling, and sealing. One critical to quality (CTQ) characteristic determined by the marking department is how full the jar looks like. If the jar has a large empty space on the top, it makes the impression that the jar does not have 40 ounces of peanut butter as written on the jar. On the other hand, filling the jar more than 41 ounce can create overflow problems during the sealing process (i.e., physical limitation). Thus, the firm determines that an acceptable 40-ounce jar that meets customer quality expectation and satisfies process physical constraint is the one that contains between 39 and 41 ounces of peanut butter. In other words, the design specification for weight of the product has

$$\text{Target Value} = 40 \text{ ounces}$$
$$\text{Lower Specification Limit (LSL)} = 39 \text{ ounces}$$
$$\text{Upper Specification Limit (USL)} = 41 \text{ ounces}$$

Process Capability is concerned with whether the filling process is capable of producing jars that are within lower and upper specification limits (i.e., LSL and USL). We will discuss process capability in detail in Chap. 17 where we focus on improving quality of process output. The focus of this chapter is on process control.

Let's assume that the filling process is indeed capable of filling jars between 39 and 41 ounces. The next challenge is to make sure that the filling process is always in control and does not deviate from the desired setting. This corresponds to Process Control. Process control requires identifying and preventing factors that result in imprecise filling of the jars, for example, clogging of the filler that reduces the flow of the material in jars, or wrong setup of the machine by the worker, or low temperature of peanut butter that slows the flow.

Both process capability and process control are highly affected by the variability in the output of the process. There are two main types of variability in the output of a process that we discuss in the following section.

## 16.5 Variability in the Output of a Process

Let's return to the peanut butter filling process and focus on one of its quality characteristic, namely the amount of peanut butter in the jars. The filling process consists of a filling machine and an operator who adjusts the machine for the right amount of filling. The operator also makes sure that the jars arriving to the filling station are getting through the machine without interruption. The machine has the capacity of filling 100 jars in a minute—a total of 6000 jars in an hour. Table 16.1 shows a random sample of the weights of 100 peanut butter jars that were filled by the filling machine, when the machine was set to fill all jars with 40 ounces of peanut butter.

First observation from the table is that there is a variability in the output of the filling process. If we define

Quality Characteristic:     $X =$ Weight of peanut butter in a 40-ounce jar

then, $X$ whose 100 realizations were observed in Table 16.1, is a random variable. The empirical distribution of $X$ is shown in Fig. 16.4.

As Table 16.1 and Fig. 16.4 show, the observed values of $X$ varies between 38.975 and 41.268 ounces with mean of 40.07 ounces and standard deviation 0.47 ounces. But, since the machine is set up for 40-ounce filling, shouldn't all 100 jars contain exactly 40 ounces of peanut butter? Well, while that would be ideal, it never happens. In real world, one should expect to see variations in the weights of 40-ounce peanut butter jars—no two jars of peanut butter have the exact same weight to their last digit. In fact, no two products produced by any process are exactly the same: no two bottles of soft drinks are filled to exactly same level,

**Table 16.1    Weights of a Sample of 100 Jars of Peanut Butter (*in Ounces*)**

| | | | | | | | | | |
|---|---|---|---|---|---|---|---|---|---|
| 40.512 | 39.526 | 39.451 | 40.062 | 41.061 | 40.561 | 40.064 | 39.623 | 40.572 | 40.111 |
| 40.871 | 40.993 | 39.162 | 40.019 | 39.765 | 39.411 | 40.671 | 40.042 | 39.838 | 39.331 |
| 40.510 | 39.899 | 40.589 | 40.531 | 40.139 | 40.579 | 39.980 | 40.021 | 40.407 | 39.610 |
| 40.376 | 38.957 | 39.998 | 39.413 | 39.230 | 40.342 | 40.163 | 40.353 | 40.095 | 39.706 |
| 39.327 | 40.518 | 39.884 | 40.291 | 40.356 | 40.343 | 39.805 | 39.985 | 39.959 | 40.241 |
| 39.165 | 39.811 | 40.459 | 39.976 | 39.624 | 40.208 | 40.258 | 39.768 | 39.660 | 40.212 |
| 41.057 | 40.248 | 40.083 | 39.998 | 39.539 | 41.268 | 40.763 | 39.905 | 39.520 | 40.633 |
| 40.479 | 40.373 | 40.167 | 40.770 | 40.369 | 40.063 | 40.248 | 40.576 | 39.593 | 40.422 |
| 39.164 | 39.979 | 39.894 | 40.398 | 40.127 | 39.638 | 39.577 | 40.004 | 39.855 | 40.014 |
| 39.945 | 40.264 | 40.838 | 39.565 | 40.058 | 39.911 | 39.492 | 40.137 | 40.354 | 39.516 |

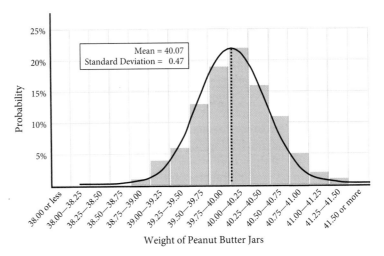

**Figure 16.4**  Empirical distribution of the weights of 100 jars of peanut butter in Table 16.1.

no two chocolate-chip cookies have the exact same size and shape, no two bags of potato chips have exactly the same amount of chips, no two cylinders of a car engine have exactly the same diameter, and no two identical twins have exactly the same height or weight.

What is causing variability in the quality characteristics of the output of processes? The answer depends on whether the variability in the output of a process is: (i) Normal variability, or (ii) Abnormal variability.

## CONCEPT

### Normal and Abnormal Variability and In-Control and Out-of-Control Processes

- *Normal Variability*, also called *Natural* or *Common-Cause* variability, is the expected inherent variation in process performance caused by unknown and random causes. It is created by a combination of many minor factors, including random variation in material, in resources (e.g., workers and machines) performance, in tools, etc. A process that is only subject to normal variability is said to be *in control*.

- *Abnormal Variability*, also called *Special-Cause* or *Assignable-Cause* variability, is a larger unusual variation for which its causes can usually be identified and eliminated. Sources of abnormal variability are often defective material, worn tools, out-of-tune machines needing adjustment, worker mistakes due to fatigue or distraction, and problems with measuring devices. A process that is subject to abnormal variability is said to be *out of control*.

To describe the difference between normal and abnormal variability, suppose that you take a bus to go to work every day. Bus is scheduled to arrive at your station at 8:00 am, and scheduled to reach at your destination (your office) at 8:20. So, your trip to office is expected to take about 20 minutes. While you expect that your trip takes 20 minutes, you know that is not always the case—you observed in the past that it takes between 16 and 24 minutes, a variation of 8 minutes. This variation is expected because of many factors, including variation in the number of red lights the bus gets, number of passengers getting in and off the bus in each station, traffic flow, etc. These are all sources of normal (i.e., common-cause) variability in your travel time.

Suppose your yesterday's trip took 21 minutes. Can you identify the exact reason of why it took 1 minute above the expected time of 20 minutes? Most probably not. Could it be because more passengers got in and off the bus, some of them being slow? Was it because the bus hit more red lights than usual? Was it because the traffic flow was slower? Or was it because of the combination of some or all of these factors? These are all common factors that can cause an increase (or decrease) in your travel time, and the exact source of 1-minute delay in your travel time cannot be clearly identified.

Now suppose you are in the bus and the bus stops due to engine failure. The driver asks passengers to wait for the next bus, which arrives shortly. You switch to the new bus and get to your office. The trip takes you 26 minutes. Can you identify the reason why your trip took 6 minutes longer than the expected time of 20 minutes? Yes, indeed. The main cause of the variation of 6 minutes above the expected travel time of 20 minutes can be assigned to the bus failure—an abnormal or special-cause variation. While your yesterday's trip was in control, your today's trip was out of control.

In our example of peanut butter filling process, the normal variability is caused by small variations in filler's valve settings and the slight changes in temperature that affect flow speed. The special-cause variability, on the other hand, is due to clogging of the filler because peanuts are not ground well, worker's error in setting up the machine, or due to machine failure.

Normal and abnormal variability affect both process control and process capability. The main task of process control is to identify whether the variability observed in the output of a process is normal or abnormal, that is, if the process is in control or out of control. If the process is out of control, then there exists a special cause for the observed variability. The sources of the abnormal variability must be identified and eliminated. Only after all sources of abnormal variability are eliminated and the process is in control, one can measure process capability to see if the process is capable of producing outputs that meet design specifications. It does not make sense to evaluate process capability if the process is out of control. No manager plans production based on an out-of-control production process; it is like planning for a two-day trip with your car, not knowing whether your car engine or brakes are working properly, that is, they may be out of control. Therefore, in the following section, we first present statistical process control techniques that can be used to determine if a process is in control. Later in Chap. 17, we present different ways in which the capability of an in-control process can be measured.

## 16.6 Statistical Process Control

The main goal of statistical process control is to monitor the stability and consistency of a process and detect whether some special causes are changing the output of the process.

---

**CONCEPT**

**Statistical Process Control (SPC)**

A process is said to be in *statistical control* (or stable) with respect to quality characteristic $X$, if the distribution of $X$ does not change over time.

---

A process that is in statistical control has significant advantage over an out-of-control process. This is shown in Fig. 16.5.[6] As the figure shows, the distribution of $X$ is not changing in an in-control process, and hence one would expect to have the same distribution for future outputs of the process. For out-of-control process, however, because the distribution of $X$

---

[6]The figure was made based on a similar figure in *Chance Encounter* by C.J. Wild and G.A.F. Seber.

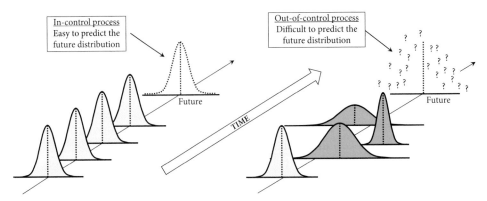

Figure 16.5    **Distribution of quality characteristics** $X$ **in in-control and out-of-control processes.**

changes over time, past performance data are not very useful in predicting the future performance of the process. In other words, we do not know how the process will perform in the future. Being able to predict the future state of a process is essential in any planning for future production.

How should a manager deal with normal (common-cause) and abnormal (special-cause) variability in the output of a process? Well, Walter Shewhart, who invented the first control charts, pointed to a very important principle. He emphasized on the fact that when a process is subject only to normal variability (i.e., when process is stable), tampering with the process and trying to adjust it with every change in process output increases variability and makes things worse. To illustrate this, we return to our peanut butter filling example. Suppose that the process is in control. What happens if we try to adjust the process every time its output deviates from the target value of 40 ounces? Let's do that. We check the next output of the filling machine and we find that it has about 38.5 ounces of peanut butter, 1.5 ounces below the target value of 40 ounces. We, therefore, adjust the machine and increase its setting by 1.5 ounce. Checking the next jar, we find that it weighs about 42.6, which is 2.6 ounces above the target value of 40 ounces. To compensate for this increase, we decrease the setting by 2.6 ounces to match the target value. If we continue in this manner—readjusting the process up and down—we will indeed move the mean and increase the variation in weights of the jars.

Hence, when a process is subject only to normal variability, one should not take any action in changing the process, until a special-cause starts to affect process performance by changing the distribution of the quality characteristic of interest. But how do we know in our peanut butter case, for example, whether a special-cause is changing the distribution of the weights of peanut butter in the jars, that is, the distribution of $X$? In other words, if the last jar is filled with 41.8 ounces, how do we know that the increase of 1.8 ounces above the target level of 40 is caused by common causes, or some special causes have put the filling process out of control? By using "Statistical Process Control Charts."

## 16.7 Statistical Process Control (SPC) Charts

Statistical Process Control charts, or SPC charts, are graphical and statistical tools that can identify whether a deviation from the target value of a quality characteristic is due to common-cause variations or if a special cause is changing the distribution of the quality characteristic. To illustrate this, let's assume we are interested to know if any special causes are affecting the *mean of the distribution* of the content of 40-ounce peanut butter jars. From

Table 16.2    Twenty Samples of Size $n = 8$ Jars Taken from the Filling Process

| Sample number | Time sample taken | Amount of peanut butter in each jar in the sample (*in ounces*) | | | | | | | | Sample mean |
|---|---|---|---|---|---|---|---|---|---|---|
| 1 | 8:00 am | 40.79 | 40.28 | 39.94 | 40.38 | 38.74 | 40.09 | 40.33 | 40.04 | 40.07 |
| 2 | 8:30 am | 40.03 | 39.10 | 39.67 | 38.89 | 39.44 | 39.51 | 38.87 | 38.64 | 39.27 |
| 3 | 9:00 am | 40.29 | 40.70 | 40.28 | 40.44 | 41.45 | 39.15 | 40.70 | 40.21 | 40.40 |
| 4 | 9:30 am | 40.31 | 40.40 | 39.92 | 39.19 | 39.77 | 41.10 | 39.95 | 40.01 | 40.08 |
| 5 | 10:00 am | 39.25 | 39.68 | 40.64 | 40.14 | 40.26 | 40.16 | 38.96 | 39.51 | 39.82 |
| 6 | 10:30 am | 40.09 | 39.71 | 40.04 | 40.54 | 39.90 | 38.93 | 40.07 | 39.41 | 39.84 |
| 7 | 11:00 am | 39.91 | 39.67 | 39.91 | 39.96 | 39.83 | 40.88 | 39.48 | 39.58 | 39.90 |
| 8 | 11:30 am | 39.32 | 40.18 | 40.10 | 39.77 | 38.78 | 38.29 | 39.10 | 40.77 | 39.54 |
| 9 | 12:00 pm | 39.94 | 41.19 | 40.23 | 40.11 | 39.92 | 40.79 | 40.44 | 39.86 | 40.31 |
| 10 | 12:30 pm | 39.49 | 39.80 | 39.93 | 41.00 | 39.71 | 39.38 | 38.79 | 39.14 | 39.65 |
| 11 | 1:00 pm | 39.35 | 40.25 | 40.50 | 40.01 | 40.06 | 40.86 | 40.86 | 39.64 | 40.19 |
| 12 | 1:30 pm | 39.93 | 40.25 | 40.13 | 40.58 | 40.22 | 39.38 | 39.22 | 39.20 | 39.86 |
| 13 | 2:00 pm | 39.70 | 39.80 | 39.85 | 39.39 | 40.34 | 40.14 | 39.89 | 39.88 | 39.87 |
| 14 | 2:30 pm | 38.82 | 41.23 | 41.85 | 39.02 | 41.89 | 39.92 | 41.89 | 41.92 | 40.82 |
| 15 | 3:00 pm | 39.27 | 40.82 | 40.43 | 40.91 | 39.37 | 39.96 | 41.28 | 40.35 | 40.30 |
| 16 | 3:30 pm | 40.36 | 40.17 | 40.78 | 39.78 | 38.90 | 40.46 | 40.41 | 40.23 | 40.14 |
| 17 | 4:00 pm | 40.21 | 39.93 | 39.85 | 40.44 | 39.67 | 40.17 | 40.22 | 40.16 | 40.08 |
| 18 | 4:30 pm | 39.85 | 39.97 | 39.08 | 41.00 | 39.38 | 40.72 | 40.62 | 39.46 | 40.01 |
| 19 | 5:00 pm | 39.14 | 41.62 | 39.85 | 40.08 | 40.43 | 39.06 | 41.08 | 40.22 | 40.18 |
| 20 | 5:30 pm | 40.08 | 40.49 | 40.92 | 40.29 | 38.70 | 40.30 | 40.98 | 40.37 | 40.27 |

the data in Fig. 16.4 we know that the amount of peanut butter in a jar is a random variable with mean $\mu_x = 40.07$ and standard deviation $\sigma_x = 0.47$.

To monitor the mean, we take a sample of eight jars every 30 minutes, weigh the amount of peanut butter in each jar, and compute the mean of eight weights. Table 16.2 shows the 20 samples taken in the last 10 hours as well as the mean of each sample. Mean of a sample is computed by simply averaging all eight weights observed in the sample.

If we compute the average of all 160 numbers in the table, we find it to be 40.03, which is very close to $\mu_x = 40.07$, our original estimate of the mean of distribution of weight of 40-ounce jars, which is our target level. Examining the sample means of the 20 samples in the last column of the table, we find that they range between 39.27 and 40.82. The average of the averages of all 20 samples, as expected, is also 40.03—the average of all 160 jars. Because the mean of our samples (i.e., 40.03) is very close to our original estimate of the mean (i.e., 40.07), can we conclude that the mean of the weights of peanut butter (i.e., $\mu_X$) has not changed and the variability in the filling machine in the last 10 hours is all due to common causes (i.e., filling process is in control)?[7]

Well, let's have a more detailed look at the means of the 20 samples by plotting them on a chart, as shown in Fig. 16.6.

As the figure shows, due to variability, the means of samples are spread around 40 ounces. Consider the mean of the second sample taken at time 8:30 am. This sample has sample mean 39.27, which is the lowest mean among the 20 samples. Can we conclude that there might have been some special causes at 8:30 am that decreased the mean of the filling process ($\mu_x$) and thus that made the machine fill less peanut butter in the jars? Or, this is just due to common-cause variation? Similarly, sample number 14, taken at 2:30 pm, has

---

[7] Note that because of Normal variability in the process, we should not expect that the mean of the data in Table 16.2 be exactly the same as the mean of the data in Table 16.1.

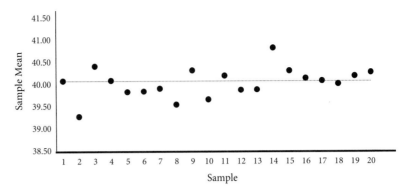

Figure 16.6    Plot of the means of the 20 samples in Table 16.2.

mean 40.82, which is the highest mean among the 20 samples. Can we conclude that the filling machine was affected by some special causes at 2:30 pm that increased the mean of the process ($\mu_x$) and made the machine fill more peanut butter in the jars? Or again, this is just due to common-cause variation?

Statistical process control charts provide answers to these questions by computing the probability of observing samples with mean 39.27 or 40.82, if only common-cause variation exists. If this probability is high, then it is highly likely that only common causes are affecting the output of the process. However, if the probability of observing such measurements is low, then chance alone cannot be the only cause; some other special causes are also affecting process performance.

How do statistical process control charts compute such probabilities? The fundamental assumption of control charts is that normal (i.e., common-cause) variations follow known probability distributions such as Normal or Poisson distributions, or can be accurately approximated by these distributions. Suppose that the distribution of quality characteristic $X$ (e.g., sample mean of eight peanut butter jars) follows a Normal distribution with mean $\mu$ and standard deviation $\sigma$.

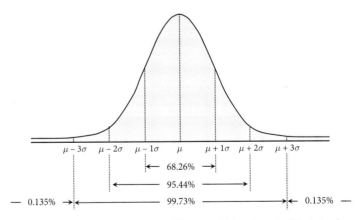

Figure 16.7    Probability of observing a sample within different multiples of standard deviation from the mean of a Normal distribution.

Figure 16.7 shows the probability that an observation of the quality characteristic $X$ falls into different segments of the Normal distribution. These segments are established by one, two, or three standard deviations from the mean. As the figure shows, 99.73 percent of the area under the Normal curve falls within minus and plus three standard deviation from the mean. This implies that if only common causes are affecting the performance of a

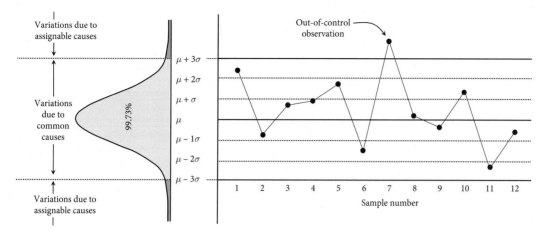

**Figure 16.8    A typical statistical process control chart.**

process (i.e., if mean and standard deviation of $X$ have not changed), we should expect to see 99.73 percent of observed measurements to be within $\mu - 3\sigma$ and $\mu + 3\sigma$. The probability that an observation that is caused by common causes to lie above $\mu + 3\sigma$ or below $\mu - 3\sigma$ is $1 - 0.9973 = 0.0027$. This implies that when only common cause variability is present, one should find only about 3 out of 1000 observations to fall beyond $\mu \pm 3\sigma$. So, if an observation lies beyond $\mu \pm 3\sigma$, it is reasonable to suspect that something other than common causes are operating and some special causes are also affecting process performance.

Figure 16.8 depicts a standard control chart side-by-side of a Normal distribution. As the figure shows, the control chart has three components:

- *Center Line* (CL) represents target value of the quality characteristic that the process must achieve.

- *Upper Control Limit* (UCL) represents the upper-limit value for variation in the quality characteristic. This limit is set three standard deviation *above* the central line. Observations above this limit signal the presence of abnormal (i.e., special-cause) variability.

- *Lower Control Limit* (LCL) represents the lower-limit value for variation in the quality characteristic. This limit is set three standard deviation *below* the central line. Observations below this limit signal the presence of abnormal (i.e., special-cause) variability.

Because LCL and UCL set three standard deviation ($\sigma$) below and above mean, respectively, they are also called *Three-Sigma Control Limits*.

In control charts, points are linked by lines, which makes it easier to see patterns that show how the points are changing over time. If a point lies above UCL or below LCL, then the process is interrupted and efforts are made to identify special causes that might have changed the distribution of $X$ by either shifting its mean or changing its standard deviation. After the special cause or causes are identified and eliminated, the process resumes its operations and new control charts are developed to monitor the process.

Let's now return to our peanut butter example and construct a control chart for the 20 sample means we observed in Table 16.2. To do that, we need to: (i) check whether the distribution of sample means is Normal, (ii) find the mean and standard deviation of the Normal distribution, and (iii) establish the three-sigma control limits.

Is the distribution of sample means Normal? The answer is yes. Statisticians have shown that when one takes independent samples of size $n$ from a population $X$ with mean $\mu_x$ and

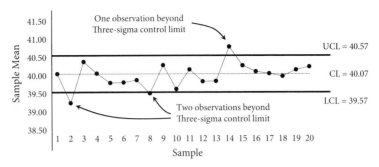

**Figure 16.9    Control chart for monitoring the mean of the amount of peanut butter in 40-ounce jars.**

standard deviation $\sigma_x$, then the means of the samples, that we denote by $\overline{X}$, follow a Normal distribution with mean $\mu_{\overline{x}} = \mu_x$, and standard deviation $\sigma_{\overline{x}} = \sigma_x/\sqrt{n}$, regardless of the distribution of the population.

Considering $\overline{X}$ as the sample mean, which is a random variable with mean $\mu_{\overline{x}}$ and standard deviation $\sigma_{\overline{x}}$, then central line (CL), upper three-sigma control limit (UCL), and lower three-sigma control limit (LCL) are established as follows:

$$\text{UCL} = \mu_{\overline{x}} - 3\sigma_{\overline{x}}, \quad \text{CL} = \mu_{\overline{x}}, \quad \text{LCL} = \mu_{\overline{x}} + 3\sigma_{\overline{x}}$$

Recall from Fig. 16.4 that weight of a jar has mean of $\mu_x = 40.07$ and standard deviation of $\sigma_x = 0.47$. We took samples of size $n = 8$ and computed their means. The mean and standard deviation of the sample means are, therefore, $\mu_{\overline{x}} = \mu_x = 40.07$, and $\sigma_{\overline{x}} = \sigma_x/\sqrt{n} = 0.47/\sqrt{8} = 0.166$. Thus, the center line and three-sigma control limits for the means of the 20 samples are as follows:

$$\text{UCL} = 40.07 - 3(0.166) = 39.57, \quad \text{CL} = 40.07, \quad \text{LCL} = 40.07 + 3(0.166) = 40.57$$

If we add these control limits to our plot in Fig. 16.6, and if we link the points in the figure by lines, we get the control chart shown in Fig. 16.9.

As the figure shows, the 2nd and the 14th samples taken at 8:30 am and 2:30 pm, respectively, are beyond the three-sigma control limits. The chart also indicates that the eighth sample taken at 11:30 am is also below the lower three-sigma control limit. Thus, it is very unlikely that the filling process was in control when producing items in those samples. Some special causes were present at 8:30 am and 11:30 am when the second and the eighth samples were taken, and reduced the average amount of peanut butter filled in jars below the target level of 40 ounces. Also, there is a very high chance that at 2:30 pm, some special causes affected the filling process and increased the amount of peanut butter filled in each jar. The firm must investigate the work condition in those times and identify and eliminate those causes.

### How Accurate Are Control Charts?

While control charts have a high probability of detecting an out-of-control process, they can result in two types of errors: (i) Type-I error, that is, a false alarm, and (ii) Type-II error, that is, failure to detect that the process is out of control. Figure 16.10 illustrates this.

If only normal variability is present and if quality characteristic $X$ follows Normal distribution, the probability that an observation lies above UCL is about 0.00135, which is the same probability that an observation lies below LCL. Hence, while it is very unlikely to have an observation beyond the control limits when only common-cause variation is present, such situations are still possible. If that occurs, control chart makes *Type-I error,* also known as "False Alarm." Specifically, when a three-sigma control limit is used and the process is in

| Control Chart Result | Actual State of the Process: Only common causes are operating | Actual State of the Process: Assignable causes are affecting the process |
|---|---|---|
| In Control | Correct Decision | Type-II Error (Failure to Detect) |
| Out of Control | Type-I Error (False Alarm) | Correct Decision |

**Figure 16.10** Type-I and type-II errors of control charts.

control, with probability $0.00135 + 0.0135 = 0.0027$ the control chart may indicate that the process is out of control—a false alarm. Hence, when the process is indeed in control, the control chart generates about 2.7 false alarms out of 1000 observations.

The probability of false alarm can be reduced if we use wider limits. For example, if we use four-sigma control limit, the probability of false alarms is reduced to 0.00006. The problem with widening the control limits, however, is that it makes the chart less sensitive to detecting an out-of-control process. Specifically, with wider control limits, most observations that are caused by special causes will lie within the widened control limits, resulting in type-II errors. *Type-II Error* occurs when special causes make the process out of control, but the control charts fail to detect that. Therefore, type-I and type-II errors go in opposite directions—widening or narrowing control limits increases one and decreases the other. This raises the following question: What control limits should we use in control charts?

### Why Three-Sigma Control Limits?

The practice of setting the upper control limit (UCL) and lower control limit (LCL) at three standard deviation above and below the mean, respectively, originated from Shewhart—the inventor of control charts. These limits are still the main limits used in developing control charts for the following reason. Normal (i.e., common-cause) variations are usually the result of a large number of small and uncontrollable factors affecting process performance, and Normal distribution is shown to well describe such situations. Experience has shown that, for Normal distribution, three-sigma control limits balance the risk of type-I and type-II errors. Even if common-cause variation does not follow a Normal distribution, Chebyshev's inequality shows that more than 89 percent of observations fall within three-sigma control limits. This number increases to 99.7 percent, as the distribution looks more like Normal distribution.

Having an observation outside the three-sigma control limits indicates a high probability that the process is out of control. It does not, however, determine what special cause is making the process out of control. Finding the special cause requires a thorough investigation of the material, equipment, labor, process, and methods used when out-of-control samples were taken.

### Signals for Out-of-Control Processes

Having an observation beyond the three-sigma control limits is only one signal that the process might be out of control. Control charts also provide other signals that point to an out-of-control process. These signals can be divided into two groups:

(i) Control limit signals
(ii) Pattern signals

Control limit signals correspond to where the observations are located with respect to the control limits, and pattern signals correspond to how observations are located with respect to each other.

### Control Limit Signals for Out-of-Control Processes

Recall that a control chart signals an out-of-control process if an observation falls beyond its three-sigma control limits. The reason is that, if only common-causes variations are present, the probability of such an event is very low (i.e., 0.0027). There are also other events that have a very low probability of occurrence in a process operating under common-cause variations. For example, in an in-control process, the probability of having one observation above or below the center line is 0.5. Hence, the probability of having eight consecutive observations on a given side of the center line would be $(0.5)^8 = 0.0039$, which is also very low. If such an event occurs, there is a high chance that some special cause is affecting process performance and the process might be out of control. Therefore, in addition to three-sigma control limits, other control limits such as two-sigma, one-sigma, and the center line can signal an out-of-control process.

The following six rules correspond to the possibility of one or more special causes affecting process performance and moving the process to an out-of-control state:

- *Rule 1:* A process is out of control if one or more points fall above $\mu + 3\sigma$ or below $\mu - 3\sigma$ control limit.

- *Rule 2:* A process is out of control if two or more consecutive points fall very close to $\mu + 3\sigma$ control limit, or if two or more consecutive points fall very close to $\mu - 3\sigma$ control limit.

- *Rule 3:* A process is out of control if two out of three consecutive points fall between $\mu + 2\sigma$ and $\mu + 3\sigma$ control limits, or if two out of three consecutive points fall between $\mu - 2\sigma$ and $\mu - 3\sigma$ control limits.

- *Rule 4:* A process is out of control if four out of five consecutive points fall above $\mu + \sigma$ control limit, or if four out of five consecutive points fall below $\mu - \sigma$ control limit.

- *Rule 5:* A process is out of control if eight or more consecutive points fall above the center line, or if eight or more consecutive points fall below the center line.

- *Rule 6:* A process is out of control if 15 or more consecutive points fall between control limits $\mu - \sigma$ and $\mu + \sigma$.

### Pattern Signals for Out-of-Control Processes

When a process is in control and only common causes are operating, the points in a control chart are randomly spread around the center line—they should not show any particular pattern. The presence of any pattern, such as increasing or decreasing patterns, signals that special causes might exist and the process might be out of statistical control. There are several known patterns that signal the presence of special causes. These patterns are shown in Fig. 16.11.

- *Trend Patterns* have an overall upward or downward movement of points as more observations are plotted (see Fig. 16.11). These patterns point to possible equipment or tool worn out, personnel fatigue, and changes in work environment. The trend pattern is characterized by six or more consecutive points increasing or decreasing in their values.

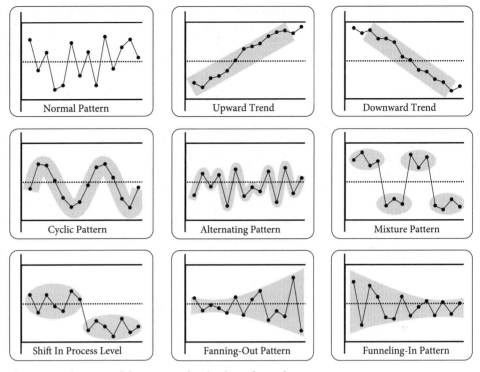

**Figure 16.11    Some control chart patterns that signal out-of-control processes.**

- *Cyclic Patterns* have an increasing trend for a period, followed by a decreasing trend in the next period. This behavior repeats in a cyclic manner, see Fig. 16.11. Special causes of cyclic patterns are factors such as periodic changes in work environment, including worker changes during different shifts, periodic maintenance of tools and equipment, and worker fatigue. A special case of cycle pattern is *Alternative Pattern* in which an upward trend is immediately followed by a downward trend (and vice versa). Alternative patterns are characterized by 15 consecutive points alternating up and down.

- *Mixture Patterns* are characterized by alternatively low and high points with a very small number of points near the center line (see Fig. 16.11). The special causes resulting in such patterns are that the high and low values are most probably generated by two different sources, for example, two different types of raw material, or two machines or workers are working in parallel—one source generates high values and other source generates low values, and the sample is taken from the mixture of the outputs of the two sources. Another underlying reason for mixture pattern can be that the operator is tampering with the process to improve the output in reaction to common-cause variation.

- *Shift in Process Level* occurs when some elements of the process change. Examples are using a new worker, a different equipment, a different procedure, or using a different source of supply. After special cause for the shift in process level is identified, if the shift degrades quality, the cause must be eliminated. If the shift improves quality, however, the special cause must be studied and made permanent.

- *Fanning-Out and Funneling-In Patterns* occur when the variability in the output of the process changes over time. Fanning-out pattern occurs when the variability increases due to worker fatigue, tool worn out, and change in raw material.

Funneling-in pattern, on the other hand, occurs when the variability in the output is decreasing. Besides changes in work environment (e.g., improvement in methods, material, worker skills), another underlying cause for funneling-in pattern might be the errors made in setting the control limits.

## 16.8 Variables and Attributes Control Charts

The first step of developing a control chart to monitor the quality of the output of a process is to determine, in detail, which quality characteristic should be measured and controlled. Developing a control chart for the quality of service in a fast-food restaurant is not possible, unless we determine which characteristic of the service we would like to control. Examples of quality characteristics in a fast-food restaurant are: number of customer complaints in a day, number of errors in taking orders, the waiting time of customers in the line to place their orders, time customers wait to get their food after placing their orders, etc. To develop a control chart, the quality characteristics must be measurable. For instance, we cannot develop a control chart for customer satisfaction, unless we quantify customer satisfaction, for example, by asking customers to rank their satisfaction between 0 (not satisfied at all) and 10 (very satisfied).

Quality characteristics of the output of a process can be divided into two groups. The first group is quality characteristics that correspond to measuring a feature of the output. Some examples are weight of peanut butter jars, diameter of cylinders of car engines, thickness of paint on cars, delivery time of products to customers, and waiting times of cars in drive-thru services. This group of quality characteristics are often continuous variables measured on continuous scales (e.g., time, weight, length). The control charts corresponding to these quality characteristics are called *Variables Control Charts*. If the measured feature, for example, the delivery time of an order, falls beyond control limits, the chart signals the presence of special causes that affect delivery times.

The second group of quality characteristics corresponds to whether the outputs of a process have a desired attribute; for example, whether or not a bag of potato chips is completely sealed, or whether or not an order of a customer is delivered before its due date. If the desired attribute exists in the output, then the output is said to be *conforming* or *non-defective*. However, if the output does not have the desired attribute, it is said to be *nonconforming* or *defective*. Thus, the bag of potato chips that is not sealed and the order that is not delivered by its due date are called defective products. Control charts that are developed to control the quality attributes of the output of a process are called Attributes Control Charts. Specifically, *Attributes Control Charts* are control charts that monitor and control the "number" of defects in one output, or the "number" of defective outputs produced by a process. These charts are based on counting the number of defective units in a sample—a discrete variable. If the number of defective units produced in an hour, for example, is beyond the control limits, the chart indicates that some special causes are affecting process output.

One advantage of variables control charts over attributes control charts is that they contain more information about the state of the process. For example, a variables control chart that monitors the delivery time of an order is more sensitive to detecting whether the delivery process is out of control than an attributes chart that only monitors the number of orders delivered after their due dates each day. However, there are many cases that variables control charts cannot be used. One such case is when there are a large number of factors affecting quality characteristics of a product. In this situation it is often not possible to quantify all those factors in order to construct variables control charts to monitor each of them. By classifying a product as either defective or non-defective, however, one can easily establish an attributes control chart to monitor the stability of the process.

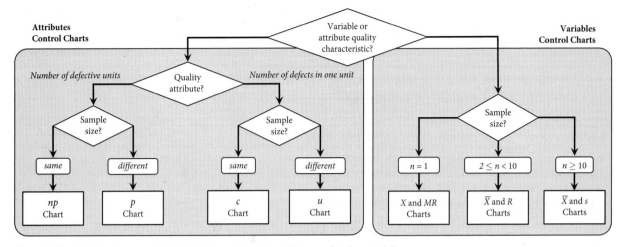

Figure 16.12    Statistical process control charts for variables and attributes quality characteristics.

Different control charts have been developed in the past decades to monitor and control variables and attributes quality characteristics of products. Figure 16.12 depicts the most commonly used variables and attributes control charts. As the figure shows, the decision of which chart to use depends on several factors such as whether the sample size is small or large, whether all samples have the same size or not, and whether we are controlling the number of defective products, or the number of defects in each product. These charts will be discussed in detail in the following sections. But before we present these charts, we need to discuss the two phases of statistical process control.

## 16.9 The Two Phases of Statistical Process Control

Developing and using Statistical Process Control (SPC) charts consist of two phases: (i) Characterization Phase, and (ii) Monitoring Phase, see Fig. 16.13.

### Characterization Phase

The goal of the characterization phase is to develop control limits that are used in the monitoring phase to detect whether the process is out of control. Suppose you would like to develop a control chart and monitor the critical to quality (CTQ) characteristics of the output of a process. The characterization phase then consists of the following steps:

**Step C1:**   Make sure that the process is in its stable state with no assignable causes affecting its output. In our peanut butter filling case, for example, we must make sure that the machine is set up correctly, the fillers are not clogged, and the machine is in its top condition. This is the state that we want the filling machine to work and fill jars (i.e., a stable and in-control state).

**Step C2:**   Take a sample of size $n$ in each period (e.g., each 15 minutes, each hour, each day) and continue until you take at least 20 samples. Use the samples and develop control limits UCL, CL, and LCL. (See Sec. 16.12 for how to choose periods and sample size $n$ for different control charts.)

**Step C3:**   Plot the samples in the control chart. If the process is in control, then go to the monitoring phase. Note that it is still possible that during the time samples are taken, the process goes out of control. Hence, if the control chart exhibits out-of-control

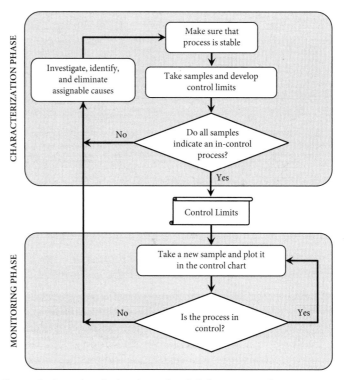

**Figure 16.13**     Characterization and monitoring phases of statistical process control.

signals, then we must investigate the process, identify the assignable causes and eliminate them, and then return to Step C1 above.

The above cycle is repeated until Step C3 indicates an in-control process. The corresponding control limits UCL, CL, and LCL will be used in the next phase—the monitoring phase.

### Monitoring Phase

The goal of the monitoring phase is to use the control limits developed in the characterization phase to monitor the process and make sure that the process remains in control. This phase consists of the following steps:

**Step M1:**   Construct a control chart with control limits obtained in the characterization phase.

**Step M2:**   Take a new sample of size $n$ and plot it in the control chart. *Do not use the new sample to revise or update the control limits.*

**Step M3:**   If the new sample shows that the process is still in control, then go to Step M2 and take the next sample. If the new sample indicates that the process is out of control, then look for assignable causes, eliminate them, return to Step C1 of the characterization phase, and develop new control limits.

In the remaining of this chapter, we present some variables and attributes control charts that are commonly used in practice. The main focus would be on developing the control limits of those charts in the characterization phase. The monitoring phase is simply taking more samples, plotting them, and looking for out-of-control signals using control limits obtained in the characterization phase. We start with variables control charts.

## 16.10 Variables Control Charts

Recall that process control charts are developed to identify whether the distribution of the quality characteristics $X$ is changing, see Fig. 16.5. When $X$ corresponds to a variable quality characteristic, for example, weight of peanut butter jars, process control charts monitor the changes in the distribution of $X$ by monitoring both the mean of quality characteristics $X$ and its variability. Therefore, variables control charts are used in pairs: one for monitoring the changes in mean and the other for monitoring changes in variability.

The mean of $X$ is controlled by a control chart called $\overline{X}$ chart (reads X-bar chart). Depending on the size of the samples taken from the output, different control charts are used for controlling the variability in $X$. This is shown in Fig. 16.11. When samples are of size 1, the $MR$ chart is used; for samples with size larger than 1 but smaller than 10, the $R$ chart is used; and for samples of size larger than or equal to 10, the $s$ chart is used.

One important point about using any pair of variables control chart is that one should always use the chart that monitors variability first. If the chart that monitors variability indicates that the process is out of control, the interpretation of the chart that monitors mean would be misleading. Thus, we first start with constructing the $R$ Chart.

### 16.10.1 The $R$ Chart

The goal of the $R$ chart is to monitor and control the variability in the quality characteristic of the output of a process when sample size is less than 10 but larger than 2. Let's return to our peanut butter example, and recall that our quality characteristic was the amount of peanut butter in the 40-ounce jars filled by the filling process, that is,

$$X = \text{Amount of peanut butter in a jar (measured in ounces)}$$

Suppose the production manager has identified and eliminated the special causes that resulted in out-of-control observations in the second, eight, and fourteenth samples in Fig. 16.9. To monitor variability of the distribution of $X$, after eliminating those special causes, we construct an $R$ chart. To do that we need to take new samples.

*Taking Samples*

The general approach to construct variable control charts, including $\overline{X}$ and $R$ charts, is to take $k$ samples—also called *subgroups*—at different points in time, where each sample takes $n$ units of the output of the process. Defining

$$x_{ij} = \text{Value of quality characteristic } X \text{ in the } j^{th} \text{ observation of the } i^{th} \text{ sample}$$

then $k$ samples of size $n$ include the following $k \times n$ observations of quality characteristic $X$:

| $n$ observations in each sample |
|---|
| Sample 1 : $x_{11}$ , $x_{12}$ , $x_{13}$ , $\cdots$ , $x_{1n}$ |
| Sample 2 : $x_{21}$ , $x_{22}$ , $x_{23}$ , $\cdots$ , $x_{2n}$ |
| Sample 3 : $x_{31}$ , $x_{32}$ , $x_{33}$ , $\cdots$ , $x_{3n}$ |
| $\cdots$  : $\cdots$ , $\cdots$ , $\cdots$ , $\cdots$ , $\cdots$ |
| Sample $k$ : $x_{k1}$ , $x_{k2}$ , $x_{k3}$ , $\cdots$ , $x_{kn}$ |

Having the observations in each sample, we can compare the variation in each sample with that of the next sample to see if the variation is changing. Table 16.3 shows 20 new samples of size $n = 8$ taken every 30 minutes of operations starting from 8:00 am to 5:30 pm.

**Table 16.3    Twenty Samples of Size $n = 8$ Taken from the Filling Process**

| Sample no. (i) | Time of sample taken | Amount of peanut butter in each jar in the sample (in ounces) | | | | | | | | $R_i$ | $\overline{X}_i$ |
|---|---|---|---|---|---|---|---|---|---|---|---|
| 1 | 8:00 am | 40.46 | 40.26 | 40.47 | 40.34 | 38.58 | 39.56 | 39.83 | 40.26 | 1.89 | 39.97 |
| 2 | 8:30 am | 40.51 | 39.60 | 40.13 | 40.09 | 38.81 | 40.01 | 40.44 | 39.75 | 1.70 | 39.92 |
| 3 | 9:00 am | 39.49 | 39.74 | 39.38 | 39.86 | 40.03 | 40.08 | 39.82 | 39.42 | 0.69 | 39.73 |
| 4 | 9:30 am | 40.69 | 39.36 | 39.49 | 39.82 | 39.67 | 40.53 | 39.66 | 39.69 | 1.34 | 39.86 |
| 5 | 10:00 am | 40.03 | 40.36 | 40.55 | 39.51 | 40.32 | 40.17 | 40.14 | 40.17 | 1.05 | 40.15 |
| 6 | 10:30 am | 39.71 | 39.93 | 39.65 | 39.20 | 40.37 | 40.16 | 40.33 | 40.28 | 1.17 | 39.95 |
| 7 | 11:00 am | 39.57 | 40.53 | 39.43 | 39.42 | 39.96 | 39.74 | 39.83 | 39.69 | 1.11 | 39.77 |
| 8 | 11:30 am | 39.58 | 40.01 | 40.11 | 40.91 | 39.52 | 39.90 | 40.41 | 40.56 | 1.39 | 40.12 |
| 9 | 12:00 pm | 39.80 | 39.96 | 39.23 | 39.44 | 39.21 | 39.90 | 39.96 | 39.60 | 0.75 | 39.64 |
| 10 | 12:30 pm | 38.50 | 39.48 | 39.70 | 40.06 | 39.82 | 40.47 | 40.76 | 39.70 | 2.26 | 39.81 |
| 11 | 1:00 pm | 39.79 | 40.37 | 40.68 | 40.50 | 38.95 | 39.31 | 40.50 | 40.69 | 1.74 | 40.10 |
| 12 | 1:30 pm | 38.92 | 40.54 | 39.97 | 40.84 | 39.76 | 40.06 | 39.59 | 40.10 | 1.92 | 39.97 |
| 13 | 2:00 pm | 39.94 | 39.26 | 40.53 | 39.94 | 38.89 | 39.53 | 39.56 | 40.74 | 1.85 | 39.80 |
| 14 | 2:30 pm | 39.56 | 39.36 | 40.56 | 39.69 | 39.90 | 39.69 | 39.53 | 39.95 | 1.20 | 39.78 |
| 15 | 3:00 pm | 39.20 | 39.80 | 39.35 | 39.68 | 39.62 | 39.42 | 39.66 | 39.17 | 0.63 | 39.49 |
| 16 | 3:30 pm | 39.69 | 39.56 | 40.19 | 39.03 | 39.31 | 38.84 | 40.47 | 38.97 | 1.63 | 39.51 |
| 17 | 4:00 pm | 40.18 | 40.01 | 38.92 | 39.48 | 39.73 | 38.40 | 40.13 | 39.58 | 1.78 | 39.55 |
| 18 | 4:30 pm | 39.29 | 39.14 | 39.92 | 38.84 | 39.96 | 40.32 | 38.80 | 39.94 | 1.53 | 39.53 |
| 19 | 5:00 pm | 39.10 | 38.74 | 39.19 | 39.54 | 39.59 | 39.80 | 39.53 | 39.57 | 1.07 | 39.38 |
| 20 | 5:30 pm | 38.99 | 39.86 | 39.31 | 38.82 | 39.87 | 39.50 | 39.44 | 39.44 | 1.04 | 39.40 |

*Developing Control Limits*

Since we would like to monitor the variation in $X$, we must measure the variation in each of the $k$ samples. There are two common measures of variation: (i) range, and (ii) standard deviation. The $R$ chart uses range to monitor the variation.[8] For values $x_{i1}, x_{i2}, \ldots, x_{in}$, in Sample $i$, the *Range* is defined as follows:

$$R_i = \text{Max}\{x_{i1}, x_{i2}, \ldots, x_{in}\} - \text{Min}\{x_{i1}, x_{i2}, \ldots, x_{in}\}$$

Range of a sample is nothing but the difference between the maximum and the minimum values observed in the sample. The larger the range, the larger the variation in a sample. As Table 16.3 shows, for example, the range for the first sample is $R_1 = 40.47 - 38.58 = 1.89$. The range for all samples are also shown in the table.

Having the range of our samples, to construct the control limits $\mu_R \pm 3\sigma_R$ for the range, we need to compute the mean $\mu_R$ and standard deviation $\sigma_R$ of the range. We can find estimates for $\mu_R$ and $\sigma_R$ using the observed values of range in the $k = 20$ samples in Table 16.3. Mean $\mu_R$ can be estimated by simply averaging all $k = 20$ values of $R_i$. If we define $\overline{R}$ as the estimate for $\mu_R$, we have

$$\overline{R} = \frac{R_1 + R_2 + \cdots + R_k}{k}$$

which would be the center line for our $R$ chart. Hence, for the peanut butter example, the $R$ chart will have the center line:

$$\text{Center Line (CL)} = \frac{1.89 + 1.70 + 0.69 + \cdots + 1.04}{20} = 1.39$$

---

[8]The $s$ chart uses standard deviation to monitor variation, see the online supplement of this chapter.

**Table 16.4**    Control Chart Factors for Different Values of Sample Size

| Sample size (n) | $d_2$ | $d_3$ | $c_4$ | Sample size (n) | $d_2$ | $d_3$ | $c_4$ |
|---|---|---|---|---|---|---|---|
| 2 | 1.128 | 0.853 | 0.7979 | 14 | 3.407 | 0.763 | 0.9810 |
| 3 | 1.693 | 0.888 | 0.8862 | 15 | 3.472 | 0.756 | 0.9823 |
| 4 | 2.059 | 0.880 | 0.9213 | 16 | 3.532 | 0.750 | 0.9835 |
| 5 | 2.326 | 0.864 | 0.9400 | 17 | 3.588 | 0.744 | 0.9845 |
| 6 | 2.534 | 0.848 | 0.9515 | 18 | 3.640 | 0.739 | 0.9854 |
| 7 | 2.704 | 0.833 | 0.9594 | 19 | 3.689 | 0.734 | 0.9862 |
| 8 | 2.847 | 0.820 | 0.9650 | 20 | 3.735 | 0.729 | 0.9869 |
| 9 | 2.970 | 0.808 | 0.9693 | 21 | 3.778 | 0.724 | 0.9876 |
| 10 | 3.078 | 0.797 | 0.9727 | 22 | 3.819 | 0.720 | 0.9882 |
| 11 | 3.173 | 0.787 | 0.9754 | 23 | 3.858 | 0.716 | 0.9887 |
| 12 | 3.258 | 0.778 | 0.9776 | 24 | 3.895 | 0.712 | 0.9892 |
| 13 | 3.336 | 0.770 | 0.9794 | 25 | 3.931 | 0.708 | 0.9896 |

To find the standard deviation of the range, it is known that if the underlying population is Normal (i.e., if $X$, the amount of peanut butter in a jar, follows Normal distribution) then there is a relationship between mean $\mu_R$ and standard deviation $\sigma_R$ of the range that depends on the sample size from which ranges are computed. Considering $s_R$ as our estimate of standard deviation $\sigma_R$, this relationship is[9]

$$s_R = \frac{d_3 \overline{R}}{d_2}$$

where values of $d_2$ and $d_3$ are called *Control Chart Factors* that depend on the sample size $n$. Montgomery (2013) provides these values for different sample sizes as shown in Table 16.4.

Using Table 16.4 for our sample of size $n = 8$ peanut butter jars, we find $d_2 = 2.847$ and $d_3 = 0.820$. Thus,

$$s_R = \frac{d_3 \overline{R}}{d_2} = \frac{0.820(1.39)}{2.847} = 0.40$$

Having estimates for mean and standard deviation of range, we can construct the center line (CL), the lower three-sigma control limit (LCL), and upper three-sigma control limit (UCL) for the $R$ chart as follows:

$$LCL = \overline{R} - 3s_R, \qquad CL = \overline{R}, \qquad UCL = \overline{R} + 3s_R$$

The two-sigma and one-sigma control limits can also be found by using 2 and 1 instead of 3 in the above formulas, respectively.

For our peanut butter example, we have $\overline{R} = 1.39$ and $s_R = 0.40$. Hence, we have center line $CL = \overline{R} = 1.39$ and

Three-sigma control limits:   $LCL = 1.39 - 3(0.40) = 0.19$,   $UCL = 1.39 + 3(0.40) = 2.59$

Two-sigma control limits:   $LCL = 1.39 - 2(0.40) = 0.59$,   $UCL = 1.39 + 2(0.40) = 2.19$

One-sigma control limits:   $LCL = 1.39 - 1(0.40) = 0.99$,   $UCL = 1.39 + 1(0.40) = 1.79$

---

[9]Factor $d_2$ captures the relationship between the average range and the standard deviation of the population for different sample size $n$, and factor $d_3$ captures the relationship between the standard deviation of the range and the standard deviation of the population.

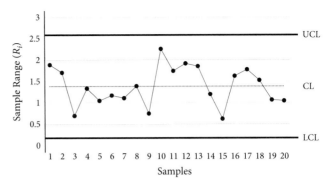

**Figure 16.14**    The $R$ chart for sample data in Table 16.3.

*Plotting and Interpreting the Chart*

The 20 observed values of range (i.e., $R_i$ in Table 16.3) are plotted in the $R$ Chart in Fig. 16.14. Interpreting the chart involves checking whether the chart exhibits any out-of-control signals—including control limit signals and out-of-control pattern signals. If any out-of-control signal is present, we conclude that the process is out of control and some special causes are changing the range and thus changing the variability in the quality characteristics of interest ($X$).

As Fig. 16.14 shows, the $R$ chart does not show any control limit signal or pattern signal for an out-of-control process—all observed values of range are within control limits and none of the non-random patterns in Fig. 16.11 are detected. Therefore, we can conclude that the variability in the amount of peanut butter in the jars is in control and there is no evidence of some special causes affecting the variability in the output of the filling process.

Because the variation in the weights of peanut butter jars is in control, we can now construct the $\overline{X}$ chart to check whether there is a change in the mean of the distribution of weights of peanut butter jars.

### 16.10.2 The $\overline{X}$ Chart

The goal of the $\overline{X}$ chart is to monitor the changes in the mean of the distribution of a quality characteristic, such as the amount of peanut butter in 40-ounce jars. As we showed in Sec. 16.6, we can do that by developing a control chart for the mean of samples taken in different time periods. Because the variability in the weights of peanut butter jars is in control, the same samples used for $R$ chart can be used to develop the $\overline{X}$ chart. While in Sec. 16.6 we have already shown how $\overline{X}$ can be constructed, we formalize it in this section.

Consider the $i^{th}$ sample that has $n$ observations $x_{i1}, x_{i2}, \ldots, x_{in}$. The mean of these observations, which we called *Sample Mean* and shown by $\overline{X}_i$, is computed as follows:

$$\overline{X}_i = \frac{x_{i1} + x_{i2} + \cdots + x_{in}}{n}$$

For example, the sample mean of the first sample is

$$\overline{X}_i = \frac{x_{11} + x_{12} + \cdots + x_{18}}{8} = \frac{40.46 + 40.26 + \cdots + 40.26}{8} = 39.97$$

Table 16.3 shows the sample means for all $k = 20$ samples taken from peanut butter filling machine.

To construct a chart to monitor the values of $\overline{X}_i$, we must compute its mean $\mu_{\overline{x}}$ and standard deviation $\sigma_{\overline{x}}$. In Sec. 16.6 we already had an estimate for these values. However, after eliminating the special causes, we need to obtain new estimates for both the mean and standard deviation of sample means. An unbiased estimate for mean $\mu_{\overline{x}}$ is to get the simple average of all $k = 20$ sample means, which we call $\overline{\overline{X}}$,

$$\overline{\overline{X}} = \frac{\overline{X}_1 + \overline{X}_2 + \cdots + \overline{X}_k}{k}$$

For our peanut butter example we have

$$\overline{\overline{X}} = \frac{39.97 + 39.92 + \cdots + 39.40}{20} = 39.77$$

Also, we use the new sample data in Table 16.3 to estimate $\sigma_{\overline{x}}$. Standard deviation $\sigma_{\overline{x}}$ can be estimated using its relationship with the average range $\overline{R}$ and sample size $n$. Specifically, considering $s_{\overline{x}}$ as an estimate for $\sigma_{\overline{x}}$, it is shown that

$$s_{\overline{X}} = \frac{\overline{R}}{d_2 \sqrt{n}}$$

where control chart factor $d_2$ depends on the sample size $n$ and can be found in Table 16.4. In our peanut butter example with sample size $n = 8$, from the table we find $d_2 = 2.847$. We have already computed $\overline{R} = 1.39$ when constructing the $R$ chart. Thus, we can estimate the standard deviation of the sample means as follows:

$$s_{\overline{X}} = \frac{\overline{R}}{d_2 \sqrt{n}} = \frac{1.39}{2.847 \sqrt{8}} = 0.17$$

Having estimates for mean and standard deviation for sample mean, we can construct the center line (CL), the three-sigma lower control limit (LCL), and the three-sigma upper control limit (UCL) for the $\overline{X}$ chart as follows:

$$\text{LCL} = \overline{\overline{X}} - 3s_{\overline{x}}, \qquad \text{CL} = \overline{\overline{X}}, \qquad \text{UCL} = \overline{\overline{X}} + 3s_{\overline{x}}$$

The two-sigma and one-sigma control limits can also be established by using 2 and 1 instead of 3 in the above formulas, respectively.

For our peanut butter example, we have $\overline{\overline{X}} = 39.77$ and $s_{\overline{x}} = 0.17$. Thus, center line CL $= \overline{\overline{X}} = 39.77$ and

Three-sigma control limits:   LCL $= 39.77 - 3(0.17) = 39.26$,  UCL $= 39.77 + 3(0.17) = 40.28$
Two-sigma control limits:   LCL $= 39.77 - 2(0.17) = 39.43$,  UCL $= 39.77 + 2(0.17) = 40.11$
One-sigma control limits:   LCL $= 39.77 - 1(0.17) = 39.60$,  UCL $= 39.77 + 1(0.17) = 39.94$

Figure 16.15 shows the $\overline{X}$ chart for the sample means of the peanut butter jars. While all observations (i.e., sample means) are within control limits, the chart clearly shows a decreasing pattern, implying an out-of-control process. The amount of peanut butter filled in each jar (and thus their mean) is decreasing, pointing to special causes such as clogging of the fillers, changes in peanut butter temperature, or out-of-tune machine.

We conclude this section by summarizing the construction of $R$ and $\overline{X}$ charts in the following analytics.

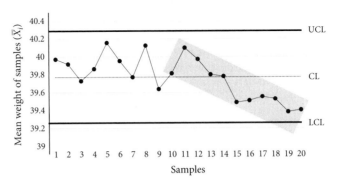

Figure 16.15    The $\overline{X}$ chart for sample means in Table 16.3.

---

**CONCEPT**

**Constructing $\overline{X}$ and $R$ Charts (*The Characterization Phase*)**

The $\overline{X}$ and $R$ charts are developed to control the mean and variability of quality characteristics that can be measured in a continuous scale (e.g., weight, time, dimension). They are used when the sample size is less than 10 (i.e., $1 < n < 10$).

- *Step 1 (Taking Samples):* To construct the $\overline{X}$ and $R$ charts for quality characteristic $X$, take $k$ samples of size $n$, where $k \geq 20$ and $n < 10$. Let $x_{i1}, x_{i2}, \ldots, x_{in}$ indicate $n$ observations in the $i^{th}$ sample, where $i = 1, 2, \ldots, k$.

- *Step 2 (Computing Range and Sample Mean):* For the $i^{th}$ sample, compute its range $R_i$, and the overall average of all ranges, $\overline{R}$, as follows:

$$R_i = \text{Max}\{x_{i1}, x_{i2}, \ldots, x_{in}\} - \text{Min}\{x_{i1}, x_{i2}, \ldots, x_{in}\} \quad \text{for all } i = 1, 2, \ldots, k$$

and

$$\overline{R} = \frac{R_1 + R_2 + \cdots + R_k}{k}$$

Also, for all $i = 1, 2, \ldots, k$, compute the sample mean $\overline{X}_i$ for the $i^{th}$ sample and the overall average of all sample means, $\overline{\overline{X}}$, as follows:

$$\overline{X}_i = \frac{x_{i1} + x_{i2} + \cdots + x_{in}}{n}, \quad \overline{\overline{X}} = \frac{\overline{X}_1 + \overline{X}_2 + \cdots + \overline{X}_k}{k}$$

- *Step 3 (Constructing the R Chart):* The three-sigma control limits and the center line for the $R$ chart are as follows:

$$\text{LCL} = \overline{R} - 3\left(\frac{d_3 \overline{R}}{d_2}\right), \quad \text{CL} = \overline{R}, \quad \text{UCL} = \overline{R} + 3\left(\frac{d_3 \overline{R}}{d_2}\right)$$

where $d_2$ and $d_3$ are control chart factors that depend on sample size $n$ and are available in Table 16.4. The one-sigma and two-sigma control limits can be computed by simply replacing 3 with 1 and 2 in the above limits, respectively. Plot values of $R_1, R_2, \ldots, R_k$ in the chart and interpret the chart. If the $R$ chart is in control, then proceed with constructing the $\overline{X}$ chart in the next step. If the $R$ chart is out of control, find and eliminate the special causes and return to Step 1.

- *Step 4 (Constructing the $\overline{X}$ Chart):* The three-sigma control limits for the $\overline{X}$ chart are as follows:

$$\text{LCL} = \overline{\overline{X}} - 3\left(\frac{\overline{R}}{d_2\sqrt{n}}\right), \quad \text{CL} = \overline{\overline{X}}, \quad \text{UCL} = \overline{\overline{X}} + 3\left(\frac{\overline{R}}{d_2\sqrt{n}}\right)$$

The one-sigma and two-sigma control limits can be computed by simply replacing 3 with 1 and 2 in the above, respectively. Plot values of $\overline{X}_1, \overline{X}_2, \ldots, \overline{X}_k$ and interpret the chart. If the $\overline{X}$ chart is out of control, find and eliminate the special causes and return to Step 1. If the $\overline{X}$ chart is in control, then proceed to the monitoring phase.

### Constructing $\overline{X}$ and $R$ Charts (*The Monitoring Phase*)

Start taking new samples of size $n$ and plot them in both $R$ chart and $\overline{X}$ chart. If you find that the process is out of control, then look for assignable causes, eliminate them, and return to Step 1 of the characterization phase.

---

### Want to Learn More?

#### Variables Control Charts with Large Sample Size

One advantage of using range as a measure of variation is the simplicity in calculating it for each sample and also its apparent relation to variation. However, as sample size $n$ increases, the range becomes less accurate in capturing the true variation of a sample. When sample size is 10 or larger, a more accurate measure of variation is the standard deviation of samples, denoted by $s$. In such cases, the *pair of $\overline{X}$ and $s$ charts* is used to control the mean and variation of the output of a process. How? Find your answer in the online supplement of this chapter. The supplement presents an example that shows how the manager of a fast-food restaurant uses $\overline{X}$ and $s$ charts to check whether the waiting time of its drive-thru customers is in control.

#### Variables Control Charts with Samples of Size One

While larger sample sizes provide more information about process output, it is not possible to take large samples in many situations due to large cost of sampling. As we described, $\overline{X}$ and $R$ charts are developed to deal with such cases when sample size $n$ is less than 10. In some processes, however, the cost of samples are so high that taking samples of size greater than one is not economical. One example is when inspecting a sample results in destructing the product, and if the product is very expensive. Another reason for taking samples of size $n = 1$ relates to the working environment. For example, in low-volume production with long inter-throughput times (e.g., project processes), it would take a long time to accumulate and analyze a sample with size larger than one. How can one monitor the output of a process by taking samples of size $n = 1$? The answer is to use the *pair of $X$ and $MR$ charts.* In the online supplement of this chapter we show how a firm uses the $X$ and $MR$ charts to monitor the time it takes to process its suppliers' checks in account payable department.

## 16.11 Attributes Control Charts

Variables control charts monitor quality characteristics that correspond to measuring a particular feature of the outputs in continuous scales such as weight, time, length, speed. Attributes control charts, however, correspond to whether the outputs of a process have a desired attribute or not. If the desired attribute does not exist in the output, then the output is called "nonconforming" or "defective." *Attribute Control Charts* are control charts that monitor and control the defective outputs of a process.

Attributes control charts, as shown in Fig. 16.12, can be divided into two groups. The first group is used when each output is classified into two types: (i) defective, or (ii) non-defective. The chart then monitors the number of defective units (i.e., the *np* chart) or the proportion of the defective units (i.e., the *p* chart) produced by a process. The *np* chart is used when samples have equal sizes, and the *p* chart is used when samples have different sizes. The second group of charts is used when each unit can have multiple defects. These charts monitor the number of defects in a sample of products produced by the process. Examples are *c* charts (used for equal sample size) and *u* charts (used for different sample sizes). We start with *np* chart.

Table 16.5    Number of Nonconforming Pizzas in 30 Samples of Size 60

| Sample number: | 1 | 2 | 3 | 4 | 5 | 6 | 7 | 8 | 9 | 10 | 11 | 12 | 13 | 14 | 15 |
|---|---|---|---|---|---|---|---|---|---|---|---|---|---|---|---|
| No. of defective pizzas: | 7 | 6 | 6 | 5 | 4 | 3 | 5 | 4 | 5 | 5 | 4 | 4 | 7 | 4 | 5 |

| Sample number: | 16 | 17 | 18 | 19 | 20 | 21 | 22 | 23 | 24 | 25 | 26 | 27 | 28 | 29 | 30 |
|---|---|---|---|---|---|---|---|---|---|---|---|---|---|---|---|
| No. of defective pizzas: | 3 | 4 | 6 | 7 | 1 | 4 | 7 | 8 | 6 | 1 | 2 | 8 | 5 | 2 | 10 |

### 16.11.1 The $np$ Chart

To illustrate how an $np$ chart can be constructed, we utilize the following example.

**Example 16.1**

A firm produces frozen pizza in three different sizes: 6-inch, 12-inch and 18-inch. Pizzas are produced in an automatic production line. The last station of the line is a vision control system that checks the diameter, size, and any abnormality in the shape of the pizza. Pizzas that do not conform with the standard are pushed out of the line by an air blasting system. Table 16.5 shows the number of nonconforming 12-inch pizzas in 30 samples of size 60. These samples are taken in the beginning of every hour for 30 hours in a three-shift working day.

Is the number of nonconforming 12-inch pizzas in control?

As the example states, the output (i.e., pizza) of the production line can be classified as nonconforming (defective) or conforming (non-defective). Obviously, due to common-cause variations, it is expected that some of the pizzas produced in the line to be nonconforming. The firm would like to monitor the number of nonconforming pizzas and identify whether some special causes are affecting its production line. Since all samples taken from the line have the same size $n = 60$, the firm can use the $np$ chart to check the stability of its production line in producing conforming pizzas.

To illustrate how $np$ chart works, we first need to define $p$ as follows:

$$p = \text{Proportion of nonconforming (defective) units produced by the process}$$

Hence, the number of nonconforming units in a sample of size $n$ is expected to be $n \times p$ or simply $np$. That is the reason the control chart that monitors the *number* of nonconforming units is called the $np$ chart.

*Taking Samples for $np$ Charts*

Considering

$$x_i = \text{Number of nonconforming units in the } i^{th} \text{ sample of size } n$$

to construct an $np$ chart, we need to take $k$ samples of size $n$, and for each sample, we should determine $x_i$, the number of nonconforming units in the sample. Hence, $x_1, x_2, x_3, \ldots, x_k$ are the number of nonconforming units in the first, second, third, ..., and $k^{th}$ samples, respectively. The number of nonconforming pizzas $(x_i)$ in each sample is shown in Table 16.5.

*Developing Control Limits for $np$ Charts*

Note that the quality characteristic that we are interested in is the number of nonconforming pizzas in $n = 60$ produced pizzas, which we denote by random variable $X$. To establish control limits, we need to estimate the mean and standard deviation of $X$ using the sample data in Table 16.5. Fortunately, we know that in a process where proportion $p$ of its output is nonconforming, the number of nonconforming units in a sample of size $n$ follows a Binomial distribution with parameters $n$ and $p$. A Binomial random variable $X$ has mean and standard deviation:

$$\mu_x = np, \qquad \sigma_x = \sqrt{np(1-p)}$$

Since we have the sample size $n = 60$, to establish the three-sigma control limits $\mu_x \pm 3\sigma_x$, all we need to do is to find $p$. If the exact value of $p$ is not known, we can estimate $p$ using the sample data. Considering $\bar{p}$ as an estimate for $p$, we can simply compute $\bar{p}$ using the $k$ samples of size $n$ as follows:

$$\bar{p} = \frac{\text{Total number of nonconforming units in all } k \text{ samples}}{\text{Total number of units in all } k \text{ samples}} = \frac{\sum_{i=1}^{k} x_i}{kn}$$

From our $k = 30$ samples in Table 16.5, we can get an estimate for the proportion of nonconforming pizzas produced by the line as follows:

$$\bar{p} = \frac{x_1 + x_2 + \cdots + x_k}{kn} = \frac{7 + 6 + 6 + \cdots + 10}{30 \times 60} = 0.082$$

Thus, an estimate for the mean is $\mu_x = n\bar{p} = 60(0.082) = 4.92$. Standard deviation $\sigma_x$ would be

$$\sigma_x = \sqrt{n\bar{p}(1-\bar{p})} = \sqrt{60(0.082)(1-0.082)} = 2.13$$

Having estimates for mean and standard deviation for the number of nonconforming units in samples of size $n = 60$, we can construct the center line (CL), the lower three-sigma control limit (LCL), and the upper three-sigma control limit (UCL) for the $np$ chart as follows:

$$LCL = n\bar{p} - 3\sqrt{n\bar{p}(1-\bar{p})}, \qquad CL = n\bar{p}, \qquad UCL = n\bar{p} + 3\sqrt{n\bar{p}(1-\bar{p})}$$

For the number of nonconforming pizzas, the center line is $CL = n\bar{p} = 60(0.0082) = 4.92$. The three-sigma control limits are as follows:

Three-sigma control limits: LCL $= 4.92 - 3(2.13) = -1.47$, UCL $= 4.92 + 3(2.13) = 11.31$

Since the number of defective pizzas cannot be negative, we set LCL $= 0$ instead of $-1.47$ above.

### Plotting and Interpreting the Chart

Figure 16.16 shows the $np$ chart for the observed number of nonconforming pizzas in Table 16.5. As the figure shows, while all the points are within three-sigma control limits, there is an obvious "Fanning-Out" pattern. This means that some special causes are increasing the variability in the number of nonconforming pizzas produced by the production line.

Before we conclude this section, we summarize the process of constructing the $np$ chart in the following analytics.

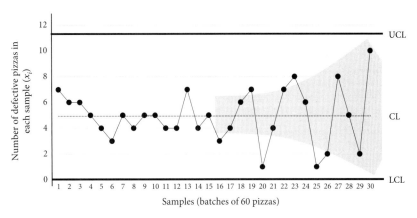

Figure 16.16    The $np$ control chart for number of nonconforming pizzas in Example 16.1.

### Constructing the $np$ Chart (*The Characterization Phase*)

The $np$ chart is used to control the number of defective units in the output of a process. The chart is used when all samples have the same size.

- *Step 1 (Taking Samples):* To construct an $np$ chart, take $k$ samples of size $n$. Let $x_1, x_2, \ldots, x_k$ indicate the number of defective units among $n$ units in Samples $1, 2, \ldots, k$, respectively.

- *Step 2 (Estimating the Proportion of Defective Units):* Use the sample data and compute an estimate for the proportion of defective units produced by the process as follows:

$$\overline{p} = \frac{x_1 + x_2 + \cdots + x_k}{kn}$$

- *Step 3 (Developing Control Limits):* The center line and the three-sigma control limits for the $np$ chart are as follows:

$$\text{LCL} = n\overline{p} - 3\sqrt{n\overline{p}(1 - \overline{p})}, \quad \text{CL} = \overline{p}, \quad \text{UCL} = n\overline{p} + 3\sqrt{n\overline{p}(1 - \overline{p})}$$

The one-sigma and two-sigma control limits can be obtained by replacing 3 with 1 and 2 in the above control limits, respectively. Plot the values $x_1, x_2, \ldots, x_k$ and interpret the chart. If the $np$ chart is out of control, find and eliminate the special causes and return to Step 1. If the $np$ chart is in control, then proceed to the monitoring phase.

  *Remark:* If an accurate estimate for the proportion of defective units, that is, $p$, is available for when the process is in control, then skip Steps 1 and 2 and use $p$ instead of $\overline{p}$ in Step 3 to establish the control limits.

### Constructing the $np$ Chart (*The Monitoring Phase*)

Start taking new samples of size $n$ and plot their number of nonconforming units in the $np$ chart. If you find that the process is out of control, then look for assignable causes, eliminate them, and return to Step 1 of the characterization phase.

### The $p$ Chart

The $np$ chart is appropriate if all $k$ samples have the same size $n$. In Example 16.1, for instance, the firm took $k = 30$ samples, where all samples had the same size $n = 60$. However, there are situations where the sample data does not have equal size. How can one control the number of defective units when each sample has a different size? By using the *p* chart. While the *np* chart monitors the *number* of defective units in each sample, the *p* chart monitors the *proportion* of defective units in each sample. The online supplement of this chapter shows how an online retailer develops a *p* chart to monitor and control the errors that its sales department makes when it processes customer orders. Check it out—*p* charts are easy to use and are very similar to *np* charts.

### 16.11.2 The *c* Charts

For a product to classify as conforming, the product must satisfy several design specifications. Defective or nonconforming units are those that do not satisfy one or more of those specifications—they have at least one nonconformity or defect. In some cases in practice, products with small number of nonconformities are still acceptable and are not considered as defective. This is because a small number of nonconformities does not affect the appearance or the functionality of the product. For example, a small number of scratches on the back of a refrigerator, or a small number of broken M&M chocolates in an M&M bag, or a small number of flaws on the surface of a glass container do not affect either the functionality or the appearance of these products. Attributes charts such as *c* chart and *u* chart monitor and control the total number of defects (nonconformities) in a single or in a collection of products.

The main difference between the *c* chart and the *np* chart is that, in *np* charts each unit in a sample is classified as conforming or nonconforming and *np* chart monitors the changes in the *number of nonconforming units* from sample to sample. However, *c* charts count the total *number of nonconformities* in each sample and monitor how that number changes from sample to sample. The sample can include only one product or a collection of several products. If the sample includes a collection of products (e.g., a box containing bottled sodas), then a *c* chart can only be used if all samples have the same size (e.g., all boxes include 24 bottles). We use Example 16.2 to illustrate how *c* charts are constructed.

 **Example 16.2**

A producer of bakery products makes a variety of cookies and candies, including alphabet cookies, which are sold in boxes that contain about 7 ounces of cookies. Because of the variability in the production process, and because cookies are transferred to the packaging station by a conveyor, it is expected that there would be some broken alphabet cookies in each box. Marketing surveys have shown that customers are not sensitive to broken cookies as long as there are just a few in each box. Hence, it is important to monitor the number of broken cookies in each box and make sure that there are no special causes affecting the performance of production and packaging stages.

Table 16.6    Number of Broken Cookies in Each 7-Ounce Box in 30 Sampled Boxes

| Sample number: | 1 | 2 | 3 | 4 | 5 | 6 | 7 | 8 | 9 | 10 | 11 | 12 | 13 | 14 | 15 |
|---|---|---|---|---|---|---|---|---|---|---|---|---|---|---|---|
| No. of broken cookies: | 4 | 6 | 5 | 2 | 7 | 1 | 4 | 10 | 3 | 7 | 2 | 6 | 6 | 5 | 3 |
| Sample number: | 16 | 17 | 18 | 19 | 20 | 21 | 22 | 23 | 24 | 25 | 26 | 27 | 28 | 29 | 30 |
| No. of broken cookies: | 7 | 8 | 5 | 4 | 0 | 4 | 7 | 6 | 5 | 9 | 6 | 6 | 6 | 0 | 5 |

To do that, quality control (QC) department randomly selects a box at the beginning of every hour and counts the number of broken cookies in the box. Table 16.6 shows the number of broken cookies in 30 samples taken by the QC department.

As the table shows, for example, the sixth box contained only one broken cookie. Does the data confirm that the production line is in control with respect to the number of broken cookies it produces?

In $c$ charts, the quality characteristic of interest is

$$X = \text{Number of defects in a sample (of one product or a collection of products)}$$

Hence, considering the unit of product to be one 7-ounce box of cookies, the quality characteristic is the number of broken cookies in the box. To develop a control chart to monitor $X$, we need to compute mean $\mu_X$ and standard deviation $\sigma_X$.

It is shown that if the number of defects occurring in a product (or in a collection of products) is completely random (caused by normal variability), then the number of defects in the product follows Poisson distribution. Specifically, if we let $c$ denote the *average* number of defects in a product, then $X$ follows a Poisson distribution as follows:

$$P(X=j) = \frac{e^{-c} \, c^j}{j!}; \quad j = 0, 1, 2, \ldots$$

which has the mean and standard deviation

$$\mu_X = c, \qquad \sigma_X = \sqrt{c}$$

It is also shown that if $c > 20$, Poisson distribution can be accurately approximated by Normal distribution.

To establish the control limits $\mu_X \pm 3\sigma_X$ on the total number of defects in a sample, we need to find an estimate for $c$. The following analytics shows how we find an estimate for $c$ and develop control limits for the $c$ chart.

---

**ANALYTICS**

**Constructing the $c$ Chart (*The Characterization Phase*)**

The $c$ chart is used to control the total number of defects (nonconformities) in a product unit. The product unit can be a single product or a collection of products.

- *Step 1 (Taking Samples):* To construct the $c$ chart, take $k \geq 20$ samples of the product unit. If each sample consists of more than a single product, then all samples must have the same number of products. Inspect each Sample $i$ and count the total number of defects in the sample and call it

$$x_i = \text{Total number of defects in the } i^{th} \text{ sample}$$

- *Step 2 (Estimating the Average Number of Defects per Product Unit):* Compute an estimate for the average number of defects in the product unit as follows:

$$\bar{c} = \frac{x_1 + x_2 + \cdots + x_k}{k}$$

- *Step 3 (Developing Control Limits):* The center line and the three-sigma control limits for the $c$ chart are as follows:

$$\text{LCL} = \bar{c} - 3\sqrt{\bar{c}}, \qquad \text{CL} = \bar{c}, \qquad \text{UCL} = \bar{c} + 3\sqrt{\bar{c}}$$

The one-sigma and two-sigma control limits can be obtained by replacing 3 with 1 and 2 in the above control limits, respectively. Plot the values of $x_1, x_2, \ldots, x_k$ in the chart and interpret the chart. If the $c$ chart is out of control, find and eliminate the special causes and return to Step 1. If the $c$ chart is in control, then proceed to the monitoring phase.

### Remark:

- If $\bar{c} < 20$, do not use one-sigma and two-sigma control limits to interpret the chart. The reason is that the signals for one-sigma and two-sigma control limits are established based on the assumption that the distribution of quality characteristic is Normal. When $\bar{c} < 20$, Normal is not a good approximation for Poisson distribution.
- If an accurate estimate for the average number of defects per product unit, that is, $c$, is available for when the process is in control, then there is no need to compute $\bar{c}$ in Step 2. Use $c$ instead of $\bar{c}$ in Step 3 to establish the control limits.

### Constructing the $c$ Chart (*The Monitoring Phase*)

Start taking new samples and plot the total number of nonconforming units in each sample in the $c$ chart. If you find that the process is out of control, then look for assignable causes, eliminate them, and return to Step 1 of the characterization phase.

Following the above analytics, we compute an estimate for the average number of broken cookies in a 7-ounce box using $k = 30$ samples as follows:

$$\bar{c} = \frac{x_1 + x_2 + \cdots + x_{30}}{30} = \frac{4 + 6 + \cdots + 5}{30} = \frac{149}{30} = 4.97$$

Thus, the center line in the $c$ control chart is $\mathrm{CL} = \bar{c} = 4.97$, and the control limits on the number of broken cookies in the 7-ounce boxes are as follows:

Three-sigma control limits: $\mathrm{LCL} = 4.97 - 3\sqrt{4.97} = -1.72$, $\mathrm{UCL} = 4.97 + 3\sqrt{4.97} = 11.66$

Because the number of broken cookies in a box cannot be negative, we set the lower control limit to $\mathrm{LCL} = 0$.

In Fig. 16.17, the observed number of broken cookies in a box are plotted in the $c$ chart for all $k = 30$ samples.

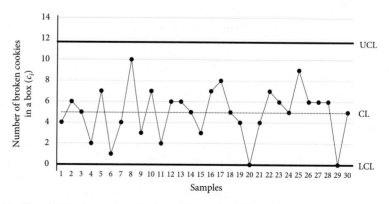

**Figure 16.17**   The $c$ chart for the number of broken cookies in a box in Example 16.2.

As Fig. 16.17 shows, the number of broken cookies in all samples are within control limits and the pattern appears random. Thus, no special causes are affecting the number of broken cookies in the boxes produced in the 30 samples.

---

### The *u* Chart

Recall that *c* charts are used to control the number of defects in a product or in a collection of products. But we can use a *c* chart only when the number of products in the collection (i.e., sample) is the same. For example, consider a major airline that would like to monitor the number of lost baggage claims it receives each day. Can we use a *c* chart to monitor the number of claims? Considering the collection of products as the total number of flights in a day, we soon realize that the number of flights is not the same each day. There might be different number of flights in each day of the week. This means that we cannot use the *c* chart—we must use the *u* chart. The online supplement of this chapter shows how a *u* chart can be developed to monitor the total number of defects when product units have different sizes. Specifically, we construct a *u* chart for monitoring the baggage processing of an airline in an airport.

## 16.12 Sample Size in Control Charts

The three critical decisions in sampling for control charts in the characterization phase are: (i) How many samples should be used to compute the control limits (i.e., what is $k$)? (ii) How often a sample should be taken (i.e., sample frequency)? and (iii) How many outputs should be in each sample (i.e., what is sample size $n$)? In the following sections we provide insights into these questions for variables and attributes control charts.

### 16.12.1 Sample Size in Variables Control Charts

#### Number of Samples ($k$)
To be able to derive accurate control limits, it is recommended that at least 20 samples are taken from the output of the process.

#### Sample Frequency
The decision of how often samples should be taken depends on the cost of sampling and the benefit of adjusting the process, if it becomes unstable. It is recommended to take more frequent samples and then reduce the frequency as it becomes more clear that the process is in control. For example, we can start by taking a sample of 10 units every 30 minutes, and then reduce it to samples of size 10 units every 3 hours, if we find that the process is in control. Irrespective of whether samples of 10 units are taken every 30 minutes or every 3 hours, one should make sure that the 10 units are produced as closely together in time as possible—preferably, 10 consecutive units. Furthermore, the 10 units must be taken from the same working condition. For example, one should not take five units from the end of the first shift and the remaining five units from the beginning of the second shift. Finally, one should increase the frequency of sampling if one suspects the presence of special causes.

#### Sample Size ($n$)
With respect to the size of a sample, smaller sample sizes are preferred. One reason is that small sample sizes have higher chance of revealing changes in a process than larger sample sizes. For example, an increase in the mean of a process may not be captured in a sample of size 10, when a shift occurs after the first five units in the sample are produced. Two samples of size 5 have a higher chance of detecting the shift, since the mean of the second five units is more likely to be higher (or lower) than the mean of the first five units. Another reason

that favors smaller sample sizes is that smaller sample sizes also have lower costs, especially in cases that the inspection of a unit destructs the unit. For variable control charts, samples of size 5 or 6 are commonly used.

When the variation due to special causes is small, it might be difficult to detect it with small sample sizes. In such case, sample sizes larger than 5 or 6 are used, since they give tighter control limits, making the control charts more sensitive to shifts in the process. For larger sample sizes, however, different variables control charts may be used. For example, as Fig. 16.11 shows, when sample size $n$ is larger than 10, $s$ chart must be used instead of $R$ chart (see the online supplement of this chapter).

### 16.12.2 Sample Size in Attributes Control Charts

To construct $np$ charts and $p$ charts, the common practice is to take at least 20 samples, that is, $k \geq 20$. Sample size $n$ must be large enough in order to have an average of at least two (some prefer at least five) nonconforming units in each sample, that is, $np \geq 2$ (or $np \geq 5$). Since the proportion of nonconforming units in production processes is often less than 0.05 or 0.01, to have an average of at least two (or five) nonconforming units in a sample, the sample size becomes large. For example, if $p = 0.01$ of the outputs of a process are nonconforming, then to have an average of at least two nonconforming units in a sample (i.e., to have $np \geq 2$), samples size $n$ should be at least 200.

Similar to $np$ and $p$ charts, the practical recommendation to construct $c$ and $u$ charts is to take at least $k \geq 20$ samples. Similarly, sample size $n$ must be large enough to observe an average of two nonconforming units (or two nonconformities in $u$ chart) in each sample.

## 16.13 Summary

Customers of a product and producer of the product have two different perspectives of product quality. Customers' perception of quality is shaped by product's features such as performance, special features, reliability, durability, serviceability, conformance, aesthetics, and safety. For intangible products such as services, some of the quality dimensions that are important to customers are completeness of service, speed of service, courtesy and friendliness, consistency, accessibility, responsiveness, competence of service provider, and tangibles. The producer's perception of the quality corresponds to how well its design process can develop a product that meets customers' needs (i.e., quality of design) and how well its operations can adhere to the chosen design specifications (i.e., quality of conformance). If design and production are done properly, the final product would satisfy the quality dimensions important to the customers.

First step in controlling quality is to identify critical to quality (or CTQ) of a product. CTQ are the key measurable and actionable characteristics of a product (goods or service) whose performance standards or specification limits must be met in order to satisfy the customer. Because of the variability in production process, the measure of CTQ varies in the output of the process. Some of the variations in the output are due to normal variability and some are due to abnormal variability. Abnormal variability is the result of assignable causes that change the underlying distribution of CTQ. Statistical process control (SPC) charts are visual statistical tools that are designed to detect whether the changes in the output of a process are due to normal variability (process is in control) or due to abnormal variability (process is out of control). When an SPC chart signals an out-of-control process, the process must be stopped and investigation should start to find the assignable cause or causes that put the process out of control.

There are many different types of SPC charts. The two main types are variables control charts and attributes control charts. Variables control charts are used to monitor CTQ characteristics that correspond to measuring a feature of the outputs that are often variables measured in continuous scale (e.g., weight, time, length). Attributes control charts are used to monitor whether outputs of a process have a desired attribute (e.g., being defective or non-defective).

One pair of variables control charts used to monitor the mean and variation (range) in the output of a process are $\overline{X}$ and $R$ charts. These charts are used when the sample size is less than 10. When the sample size is 10 or larger, the pair of $\overline{X}$ and $s$ chart is more effective for monitoring mean and variation (standard deviation). In situations in which samples are of size one—because samples of larger size take a long time to gather, or they do not provide additional information—the pair of $X$ chart and $MR$ chart is used to monitor the CTQ measure in each individual output. For $(\overline{X}, s)$ chart and $(X, MR)$ chart see the online supplement of this chapter.

Most commonly used attributes control charts are $np$ charts, $p$ charts, $c$ charts, and $u$ charts. The $np$ chart is used when the goal is to monitor the number of defective units in samples of equal size. If the sample sizes are different, then $p$ chart must be used to monitor the fraction of defective units in each sample.

In some cases in practice, products are considered defective if they have at least one nonconformity (defect). In these cases, $p$ and $np$ charts are used. In some other cases in practice, products with small number of nonconformities (i.e., defects) are still acceptable and are not considered as defective, unless the number of defects are above a certain limit. This is because a small number of defects does not affect the appearance or the functionality of the product. Attributes control charts such as $c$ chart and $u$ chart monitor the total (or the average) number of defects in a single or a collection of products. The $c$ chart is used when the sample sizes are the same, and the $u$ chart is used when the samples are of different sizes. For $p$ chart and $u$ chart see the online supplement of this chapter.

## Discussion Questions

1. Describe eight dimensions of quality that are important to customers of physical goods. How do they differ for customers of services?

2. What is critical to quality and how does it relate to design specifications of a product?

3. Explain two main dimensions of quality that are important to producer of a product.

4. What is Deming's wheel and what are its four main steps?

5. Briefly describe the contribution of the following people to quality control and management:
   a. Walter Shewhart
   b. E. Edward Deming
   c. Joseph M. Juran
   d. Phillip Crosby
   e. Armand V. Feigenbaum
   f. Genichi Taguchi
   g. Kaoru Ishikawa

6. What are lower specification limit (LSL) and upper specification limit (USL) and how are they determined?

7. What are the differences in normal and abnormal variability in process performance? How do they correspond to the stability of a process?

8. What does it mean to have a process in statistical control, and what does an SPC chart do?

9. What are the control limits in SPC charts and why the three-sigma control limits are commonly used in practice?

10. What do type-I and type-II errors mean in SPC charts?

11. Describe the six control limit signals of SPC charts that can signal an out-of-control process.

12. Describe eight pattern signals of SPC charts that can signal an out-of-control process.

13. What are the differences between variables control charts and attributes control charts?

14. What is the difference between $np$ chart and $c$ chart, and when should each one be used?

15. Describe the two main phases of developing and using statistical process control charts.

16. When taking samples for variables control charts, how many samples should be used to compute the control limits? How often a sample should be taken? What is the appropriate sample size? What about attributes control charts?

17. What is the appropriate control chart for monitoring each of the following:
    a. Time it takes to deliver express mail.
    b. Acidity of the batch of vinegar produced in each run.
    c. Fraction of packages delivered after their delivery guarantee dates.
    d. Number of packages delivered to wrong addresses in a week.
    e. Average time it takes to fix customer complaints.
    f. Total revenue of a hospital each month.
    g. Number of mistakes in bills an insurance company issues each month.
    h. Number of delayed flights of an airline each day.

## Problems

1. Explain why each of the following statements is true or false:
    a. A process has just made a defective output, the process is therefore out of control.
    b. A process has been producing outputs that are all within design specifications, the process is therefore in control.

2. Jack wants to use two-sigma control limits for his control chart, but Jill suggests using three-sigma control limits. Jill makes the following arguments to convince Jack to use three-sigma control limits. Explain why each of the arguments is true or false:
    a. Three-sigma control limits can detect an out-of-control process faster than two-sigma control limits.
    b. Three-sigma control limits result in a lower chance of type-1 error than two-sigma control limits.
    c. Three-sigma control limits result in a lower chance of type-2 error than two-sigma control limits.
    d. Three-sigma control limits generate less false alarms than two-sigma control limits.
    e. Using three-sigma control limits, when a process is in control, 99.73 percent of its output will be within design specifications.

3. Currently Sanjay takes samples of size 6 to monitor the average weight of cereal boxes filled by his filling process using the $\overline{X}$ chart. If Sanjay increases his sample size to 8, explain why each of the following statements is true or false:
    a. The control limits of the chart will become closer to its mean.
    b. On an average, less number of samples is needed to detect if the process is out of control.
    c. The chance of type-1 error increases.
    d. The chance of type-2 error increases.

4. Hospital discharge process is a complex process and thus it often takes several hours. Caroline, the manager of NMMEH hospital, has received complaints from some of her patients about the long discharge time. She decided to monitor the average discharge time each day

to make sure things are in control. She randomly chose four patients in each of the last 15 days, and obtained their discharge times, as shown in the following table:

| Day | Discharge times (*in days*) | | | |
|---|---|---|---|---|
| 1 | 162 | 184 | 103 | 173 |
| 2 | 86 | 82 | 133 | 116 |
| 3 | 90 | 141 | 130 | 146 |
| 4 | 100 | 71 | 84 | 102 |
| 5 | 70 | 92 | 107 | 104 |
| 6 | 95 | 123 | 116 | 151 |
| 7 | 166 | 116 | 126 | 70 |
| 8 | 98 | 147 | 123 | 157 |
| 9 | 172 | 99 | 119 | 114 |
| 10 | 160 | 169 | 106 | 125 |
| 11 | 107 | 119 | 95 | 171 |
| 12 | 100 | 135 | 162 | 119 |
| 13 | 141 | 185 | 167 | 170 |
| 14 | 130 | 147 | 159 | 135 |
| 15 | 194 | 122 | 148 | 114 |

Develop a control chart that helps Caroline monitor the discharge time of patients in her hospital. Has the patient discharge process been in control in the last 15 days?

5. The weight of a professional tennis racket must be around 300 grams. When the production process that produces these rackets is set up properly, the weight of produced rackets is Normally distributed with mean of 300 grams and standard deviation of 6 grams. The production manager takes a sample of eight rackets and measures their weights. They are: 292.8, 306.7, 303.1, 305.3, 294.9, 306.1, 299.5, and 312.1 grams. Do you think that the production process is out of control? Specifically, are the average and standard deviation of the rackets produced by the process in control?

6. Healthy-&-Organic (H&O) produces a large number of breakfast cereal and granola bars. One of its popular product is blueberry granola cereal, sold in boxes that contain 28 ounces of cereal. Every 3 hours, a sample of five boxes is taken from the filling process and their weights are measured and used to monitor the performance of the filling process. The following table shows the weights (in ounces) of the last 14 samples taken:

| Sample | Weight of cereal boxes (*in ounces*) | | | | |
|---|---|---|---|---|---|
| 1 | 28.07 | 23.62 | 29.84 | 26.27 | 26.84 |
| 2 | 26.15 | 30.14 | 30.37 | 26.37 | 28.25 |
| 3 | 26.34 | 26.59 | 31.58 | 29.11 | 27.63 |
| 4 | 27.63 | 26.66 | 28.10 | 25.50 | 29.55 |
| 5 | 26.33 | 28.29 | 29.46 | 30.79 | 26.65 |
| 6 | 28.91 | 28.13 | 28.10 | 27.47 | 27.67 |
| 7 | 26.10 | 26.43 | 29.16 | 24.72 | 23.69 |
| 8 | 27.80 | 29.52 | 28.21 | 28.58 | 26.21 |
| 9 | 26.37 | 26.82 | 29.29 | 27.77 | 27.70 |
| 10 | 27.83 | 27.48 | 29.40 | 30.52 | 25.81 |
| 11 | 25.66 | 30.70 | 27.83 | 28.22 | 26.83 |
| 12 | 29.13 | 28.67 | 30.26 | 27.88 | 30.45 |
| 13 | 28.47 | 28.54 | 28.07 | 28.93 | 26.93 |
| 14 | 25.38 | 25.74 | 25.94 | 30.16 | 29.61 |

Develop a control chart to monitor the filling process? Has the process been in control?

7.  The blueberry granola cereal of H&O in Problem 6 must have at least 2 ounces of dried blueberry in each 28-ounce box. To monitor this, H&O takes samples of 40 boxes every 2 hours and measures the amount of blueberry in the box. The following table shows the number of boxes with less than 2 ounces of blueberry (i.e., number of defectives) in the last 20 samples:

| Sample | Number of defectives | Sample | Number of defectives |
|--------|----------------------|--------|----------------------|
| 1 | 4 | 11 | 2 |
| 2 | 2 | 12 | 0 |
| 3 | 2 | 13 | 0 |
| 4 | 0 | 14 | 2 |
| 5 | 2 | 15 | 1 |
| 6 | 5 | 16 | 3 |
| 7 | 6 | 17 | 3 |
| 8 | 3 | 18 | 2 |
| 9 | 3 | 19 | 2 |
| 10 | 5 | 20 | 1 |

Develop a control chart to monitor the amount of blueberry in the cereal box. Does the control chart show an in-control process?

8.  Steve knows that when its injection molding machine is working properly, only 3 percent of plastic bottles produced by the machine are defective. He receives a phone call from one of its customers who complains that 12 of the 200 bottles that the customer received yesterday were defective. This surprises Steve and he wonders if something has gone wrong with the injection machine or the material used to make the bottles yesterday. What do you think?

9.  Cashmere R Us is a firm that sells cashmere clothing and yarns. One of its product is a 2-ply yarn of cashmere and silk, which is sold in 67-yarn skein in a variety of colors. The quality control department takes random samples of 60 skeins each day and looks for nonconformities such as discolorations, non-uniformity of cashmere and silk plies, and the way the skeins are formed. Skeins that do not conform with the quality standards are considered defective and are recycled.

The following table shows the number of defective skeins in each sample in the last 20 days:

| Day | Number of defective skeins | Day | Number of defectives skeins |
|-----|----------------------------|-----|-----------------------------|
| 1 | 3 | 11 | 6 |
| 2 | 0 | 12 | 0 |
| 3 | 2 | 13 | 1 |
| 4 | 1 | 14 | 0 |
| 5 | 4 | 15 | 3 |
| 6 | 3 | 16 | 3 |
| 7 | 3 | 17 | 0 |
| 8 | 5 | 18 | 4 |
| 9 | 1 | 19 | 6 |
| 10 | 5 | 20 | 0 |

Develop a control chart to monitor the number of defective skeins and investigate whether the process is in control.

10. A manufacturer of plastic products produces a large plastic cart used for carrying heavy items. The main body of the cart is made using thermoforming equipment and is checked for quality issues by a quality control (QC) worker. The QC worker checks each produced body for nonconformities such as scratches, chipped corners, surface defects, and so on. If the total number of nonconformities is 10 or above, the worker rejects the part. The following table shows the number of nonconformities (i.e., #NC) in the last 18 carts checked by the worker:

| Sample | #NC | Sample | #NC | Sample | #NC |
|--------|-----|--------|-----|--------|-----|
| 1 | 4 | 7 | 4 | 13 | 1 |
| 2 | 1 | 8 | 2 | 14 | 2 |
| 3 | 4 | 9 | 6 | 15 | 0 |
| 4 | 0 | 10 | 1 | 16 | 0 |
| 5 | 0 | 11 | 2 | 17 | 1 |
| 6 | 1 | 12 | 2 | 18 | 2 |

Develop a control chart to monitor the quality of the carts produced by the thermoforming process. Is the process out of control?

11. An airline has the same number of flights every day departing its hub airport in a large city. The average number of flights that are delayed due to mechanical failure in a normal day is about 5. Yesterday, the number of delayed flights was 10—twice the average. The head of maintenance team is wondering if she should investigate whether there was a particular reason for the large number of delayed flights yesterday. What do you think?

12. The Police department is under pressure from the Mayor's office to take actions to reduce the number of bicycles stolen throughout the city. As the first step, the deputy police chief has asked you to set a system to monitor the number of bicycles stolen each day. Asking for data, you receive the following table that shows the number of stolen bicycles (i.e., # Stolen) in the last 3 weeks, that is, last 21 days:

| Day | # Stolen | Day | # Stolen |
|-----|----------|-----|----------|
| 1 | 16 | 12 | 8 |
| 2 | 17 | 13 | 19 |
| 3 | 7 | 14 | 10 |
| 4 | 15 | 15 | 14 |
| 5 | 13 | 16 | 13 |
| 6 | 21 | 17 | 9 |
| 7 | 8 | 18 | 12 |
| 8 | 7 | 19 | 13 |
| 9 | 12 | 20 | 18 |
| 10 | 10 | 21 | 11 |
| 11 | 15 | — | — |

Develop a control chart to monitor the number of stolen bicycles each day. Does your chart provide any useful information for the police department?

# CHAPTER 17

# Quality Improvement and Six Sigma

## 17.0 Introduction

During the 1970s and 1980s American manufacturers were facing fierce competition from Japanese and European companies that were producing products with lower price and higher quality. American managers realized the critical role of quality in a firm's success and thus quality became a competitive advantage. To be able to maintain their market shares, firms initiated several quality improvement programs to produce products that met or exceeded customer expectations.

Although the main focus was on improving quality of the final product, in the 1980s and 1990s the quality improvement programs were extended to all functional areas and business processes of the manufacturing firms. Later those programs were further extended from manufacturing processes to service processes: government, healthcare, and education.

The focus of Chap. 16 was to present different methods to monitor and *control* the outputs of a process and stop the process when an assignable cause interferes with the process and puts the process in an out-of-control state. An investigation is then done to identify and eliminate the assignable cause and put the process back into in-control state. This chapter takes one step beyond just keeping the process in an in-control state. It illustrates the principles that can help operations engineers and managers to further *improve* the quality of an in-control process. These principles can also be used to identify and eliminate assignable causes in an out-of-control process.

The chapter begins by providing general principles for a successful quality improvement program that are applicable to improving quality of both products and processes. It then focuses on Six Sigma approach to quality improvement and describes its connection to lean operations, leading to the new practice of Lean Six Sigma.

## 17.1 Quality Improvement—Lessons Learned

While many companies have invested in quality improvement programs, studies have shown that only about one-third of those programs led to significant results. Several reasons have been identified that hinder the success of quality improvement efforts, ranging from improvement process to organizational structure. This section presents some lessons learned from successful and failed quality improvement programs, as pointed out in Juran's influential *Quality Handbook*.[1]

### Lesson #1—Break the Overall Quality Improvement Efforts to Several Projects

Often times managers set a general and extensive plan for quality improvement. Such plans often fail because the plans aim to do too many things at the same time. An important observation is that improvement occurs project by project. Thus, such large quality improvement plans must be divided into several smaller projects. This approach also has the advantage that several projects can be worked on concurrently.

Consider an effort to improve patient satisfaction in the emergency department (ED) of a hospital. Studies have shown that patient satisfaction in ED is affected by many factors such as perceived wait time, actual wait time before seen by a doctor, empathy/attitude of staff (e.g., doctors and nurses), technical competence of staff, pain management, and lack of information about the process and expected waiting time. Thus, the effort to improve patient satisfaction should be broken into several projects, each focusing on one of these factors (or a set of related factors).[2]

### Lesson #2—Choose the Right Project to do First

When choosing which project to do first, one should consider the following:

- *The first project must be among "Vital Few"*: It has been observed that majority of the benefits of quality improvement comes from a "vital few" number of projects. This is known as Pareto Phenomenon (see Sec. 11.1 of Chap. 11). In our ED example, the patient waiting time before seen by a doctor has shown to have the most impact on patient satisfaction.[3] Thus, the project that focuses on reducing patient waiting time in ED is among the vital few that should be considered first.

  Notice that in addition to vital few, there are "useful many" projects that also contribute to quality improvement, but their contribution is not as significant as those of the vital few. In our ED example, providing wait time information is among "useful many" projects that can improve patient satisfaction, but not as much as reducing patient's actual waiting time. Hence, if there are not enough resources to work on several projects, then it is more beneficial to do the vital few projects first.

- *The first project must be a winner*: The first project should be the one that has been waiting for a long time for a solution. It should also have a high chance of a successful completion in a few months. Finally, the results should be significant

---

[1] Juran, J.M., and A.B. Godfrey. *Juran's Quality Handbook*. Fifth edition, McGraw-Hill Professional, 1998.

[2] Shari Jule Welch, S.J. "Twenty Years of Patient Satisfaction Research Applied to the Emergency Department: A Qualitative Review." *American Journal of Medical Quality OnlineFirst*, December 4, 2009.

[3] In this study patient satisfaction was measured by likelihood of patients recommending the hospital ED to others.

enough to be recognized as a successful effort. These will increase the confidence of quality improvement teams and will increase the excitement to do the next project, promoting the culture of continuous improvement.

### Lesson #3—Set Clear Goals for Quality Improvement Projects

Each quality improvement project must have a clear goal. Specifically, the goal

- should *clearly describe the mission* of the quality improvement team,
- should *quantify* the intended amount of improvement, and
- should include a *time table* for achieving the desired goal.

In our ED example, one goal can be "Decreasing the average waiting time of patients from 1.5 hours to 1 hour in 2 months." Other examples with clear goals are: "Decreasing the number of defective items produced in an assembly line from 8 percent to 3 percent in the next 6 weeks," or "increasing on-time deliveries of packages from 80 percent to 95 percent in the next 10 months." Goals such as "decreasing customer waiting time in ED," or "decreasing the number of defective items produced in the assembly line," or "improving the quality of delivery operations," are vague and are not useful.

### Lesson #4—Quality Improvement Is a Team Effort

Most quality improvement projects are multi-functional in nature. Consider the project with the goal of reducing patient waiting times in emergency department. Staff, nurses, and doctors are all involved in the process of providing care for a patient. Thus, a multi-functional team consisting of doctors, nurses, staff, and operations manager is more effective in identifying the sources of long waiting times in ED and finding ways to improve it.

A team must have a leader who organizes the activities of the team and is responsible to make sure that the project is completed before the deadline. The leader is either chosen by upper management or elected by team members. The team must also make close connection with workers who are involved with the problem (if they are not part of the team). This, as we will describe later, reduces the worker resistance when the quality improvement results are implemented.

### Lesson #5—Diagnostics First, Then Improvement

Obviously one cannot improve quality unless one identifies the sources of low quality. The diagnostic process includes analyzing the symptoms of the problem, looking for potential causes that may have resulted in those symptoms, examining each cause, and finding the actual cause (or causes) of the quality problems. Having identified the actual causes, the team can then look for potential solution to eliminate the causes.

### Lesson #6—Quality Improvement Costs in the Short Term, But Pays Off Much More in the Long Term

Quality improvement projects require firms to invest money and time. The implementation of the quality improvement practices may also cost firms. However, the return on these investments is much higher in the long term. This is often not clear to some managers due to the following reasons. First, the cost of poor quality is not clear. Accounting practices capture only a small fraction of internal cost of poor quality. A large fraction of internal cost of poor quality is hidden within different overhead costs. External cost of poor quality (e.g., cost of losing a customer's next purchase due to late delivery of the customer's last purchase) is even more difficult to capture. Second, managers are motivated by short-term financial measures of success, which do not capture the long-term benefits of quality improvement. Third, some managers have the wrong belief that "higher quality costs more." Lean operations practices are one example of achieving higher quality with lower cost.

### Lesson #7—Successful Quality Improvement Projects Must be Rewarded and Cloned

One factor that encourages more quality improvement efforts and instills it into the culture of a firm is the public recognition of the successful quality improvement projects. This can be done by giving awards, certificates, or plaques to project teams, or even to people who complete training courses on quality. Recognition can also be in the form of bonus, salary increase, and promotions.

Another aspects of recognition is when the results of a successful project are applied in other units of the firm. As mentioned in Chap. 15, Toyota workers take pride in the fact that their ideas that improved their process will be used in all Toyota units that have the same process. This is called "Cloning the Project," and is practiced in firms such as retailers, hospitals, auto industry that have similar processes in their branches.

### Lesson #8—Quality Improvement Must be Institutionalized

Quality improvement efforts in an organization should not be left to occasional initiatives made by a small group of people. To achieve continuous quality improvement, the quality improvement culture must be built into all levels of an organization. Some of the ways to do that are as follows:

- Quality improvement goals must be included in a firm's Business Plan,
- Quality improvement must be a part of job description at all levels,
- Upper management should be assigned to support and lead quality improvement initiatives,
- Periodic reviews must be established to examine the improvement achieved in each period, and
- Reward and recognition mechanisms should be established to recognize good performance on quality improvement.

### Lesson #9—Implementation of Quality Improvement Practices Often Faces Resistance

Some quality improvement projects that were successful in finding good solutions to improve quality have failed during implementation. One common reason for such a failure is resistance from people who are supposed to implement those results. The root of this problem is in human nature because humans are resistant to changes in their beliefs, habits, and cultural patterns. Many researches on human behavior have identified practices to deal with cultural resistance in organizations, some of them are as follows:

- *Allow Participation:* One very effective approach to eliminate or reduce resistance to implementing quality improvement projects is to involve the people who will be affected by changes recommended in the projects. That is another reason for having workers who perform activities in a process as a part of the team that looks for improving the quality of the process.
- *Give Enough Time:* It is critical to give time to people to understand the reason for changes and to evaluate the impact of those changes. In addition to the overall benefits of the changes for the firm, people need time to evaluate and understand their own benefits (or costs) coming from implementing the changes.
- *Start Slow:* Before full implementation of the changes, first implement the changes in a small-scale tryout and evaluate the risks for people who are affected by the changes

as well as for advocates of changes. This will provide useful information about how to proceed with the full implementation of the changes.

- *Avoid Surprises:* One reason that people resist to changes is that they like the predictability of their current situations. A surprise (such as a significant change) will bring uncertainty and disturb the peace. Therefore, significant changes must be gradually presented and implemented to avoid surprises.

- *Choose the Right Time:* There is always a right and a wrong time to implement a change. Choose the time that the resistance is expected to be lower.

- *Focus on the Main Change:* Outcomes of quality improvement processes often consist of several changes, some minor and some major. Focus on the major changes, so that the negotiations are not distracted by long discussions and disagreements over some minor changes.

- *Involve Informal Leaders:* While every firm has a formal organizational chart that shows different levels of leadership (e.g., management) according to their job description, it also has an informal organizational chart that is based on social ties of the firm's employees. Informal organizational charts and informal leaders are not known. Social scientists have been using social network tools to identify informal leaders of organizations. An informal leader in an assembly department, for example, might be an assembly line worker who has more influence on the people in that department than middle or upper management. Thus, to analyze and implement a quality improvement project in the assembly department, one should include both the formal leader (manager of the department) and the informal leader of the assembly department. Convincing the informal leader is a huge step toward a smooth implementation of changes.

- *Do Not Ignore Resistance and Be Flexible:* Ignoring resistance and simply enforcing the changes will backfire. Deal with resistance directly and be flexible. You may need to revise some of the proposed changes to accommodate some of the objections—compromise something to get something.

*Lesson #10—Quality Improvement Is a Never-Ending Journey for Perfection*

Because of the fierce competition in global market, firms that stop their quality improvement efforts will gradually find themselves further and further away from operations frontiers of quality. Customer expectation for quality is constantly increasing. Thus, to keep their market shares, firms must improve their quality year by year. This requires continual quality improvement efforts. What helps quality improvement efforts to become a never-ending journey is to pursue the ideal goal of achieving perfection—having all products and processes free of errors and defects.

The above 10 lessons are essential in the success of quality improvement in any organization. One of the most successful approach to quality improvement that follows these lessons is Six Sigma. Six Sigma focuses on identifying and eliminating the sources of defects and errors with the goal of minimizing (common cause) variations anywhere in the process. Six Sigma is established based on the concept of process capability. Hence, before we present the Six Sigma approach, we first need to introduce Process Capability.

## 17.2 Process Capability—An Essential Metric for Quality Improvement

Recall Lesson #4 for a successful quality improvement that emphasizes on having a clear *quantifiable* measure as the goal for quality improvement projects. Suppose you are in charge of leading a team to improve the quality of the peanut butter filling process in our example in the previous chapter. What would be a quantifiable metric that you can set as a goal to achieve? How can you measure the quality of the output of the filling machine, so you can improve that measure?

The answer is "Process Capability." Process capability is a metric that measures the quality of a process by its capability in producing non-defective (error-free) outputs. Therefore, to set a goal for the project of improving the quality of the filling machine, you first need to compute process capability of the machine.

To illustrate what exactly process capability is and how it is computed, we need to review design specification limits discussed in Chap. 16. Recall in Sec. 16.2 of Chap. 16 that by understanding what customers want, need, and can afford, firms identify critical to quality (CTQ) factors that are then translated into design specifications. Hence, to have an acceptable product to satisfy customers (i.e., a non-defective and error-free product), the quality characteristics of the output of the production process must be between lower specification limit (LSL) and upper specification limit (USL). For instance, in our peanut butter example, the lower and upper design specification limits were LSL = 39 and USL = 41 ounces.

Statistical process control charts, on the other hand, are used to detect the presence of special causes that affect the output of the process. Specifically, statistical process control charts monitor the quality characteristics of the outputs, and if they are within lower control limit (LCL) and upper control limit (UCL), then the process is in statistical control. But, here is the question: Do the outputs of an in-control process meet design specifications (i.e., are acceptable to customers)? Not necessarily!

To illustrate this, suppose that the special cause that resulted in out-of-control filling of peanut butter jars that was detected by the $\overline{X}$ chart in Fig. 16.15 was identified and eliminated. The firm takes another $k = 20$ samples of size $n = 5$ every half an hour as shown in Table 17.1.

Figure 17.1 shows the $R$ and $\overline{X}$ charts for these samples. As the figure shows, both variability and mean of the filling process are in control. Hence, the amount of peanut butter filled in jars by the filling process is only affected by common cause (normal) variations. Does this mean that all filled jars are within design specifications LSL = 39 and USL = 41?

To check that, we develop the histogram of all 100 jars we examined in Table 17.1. This is shown in Fig. 17.2. The figure also shows the lower and upper design specification limits LSL = 39 and USL = 41 ounces.

As Fig. 17.2 shows, the filling process has generated four jars that are not acceptable: Two jars with weights less than lower specification limit of 39 ounces, and two jars with weights more than upper specification limit of 41 ounces. As this example shows, in-control

Table 17.1   Twenty Samples of Size $n = 5$ of Jars of Peanut Butter

| Sample 1 | Sample 2 | Sample 3 | Sample 4 | Sample 5 | Sample 6 | Sample 7 | Sample 8 | Sample 9 | Sample 10 |
|---|---|---|---|---|---|---|---|---|---|
| 40.18 | 40.10 | 39.63 | 38.63 | 40.33 | 40.50 | 39.84 | 39.56 | 39.44 | 40.34 |
| 39.40 | 40.18 | 39.51 | 39.67 | 40.17 | 39.59 | 40.36 | 40.27 | 40.82 | 39.57 |
| 39.95 | 39.79 | 40.08 | 40.87 | 39.88 | 40.16 | 39.76 | 40.42 | 40.02 | 38.93 |
| 39.86 | 41.34 | 40.67 | 39.93 | 39.96 | 40.39 | 39.68 | 39.88 | 40.11 | 39.50 |
| 40.32 | 40.24 | 39.35 | 40.06 | 40.37 | 40.64 | 40.28 | 40.07 | 40.02 | 39.27 |

| Sample 11 | Sample 12 | Sample 13 | Sample 14 | Sample 15 | Sample 16 | Sample 17 | Sample 18 | Sample 19 | Sample 20 |
|---|---|---|---|---|---|---|---|---|---|
| 39.69 | 40.42 | 40.84 | 41.19 | 39.55 | 40.69 | 40.44 | 39.67 | 39.97 | 40.10 |
| 39.92 | 39.22 | 39.69 | 39.06 | 40.52 | 39.32 | 39.33 | 40.12 | 39.93 | 39.97 |
| 39.51 | 39.54 | 40.10 | 40.02 | 39.54 | 39.99 | 40.48 | 39.92 | 40.67 | 40.46 |
| 39.29 | 40.53 | 40.40 | 39.38 | 39.57 | 39.46 | 40.32 | 40.78 | 39.75 | 40.21 |
| 39.82 | 40.21 | 40.99 | 40.07 | 40.19 | 40.09 | 39.94 | 40.11 | 39.32 | 40.02 |

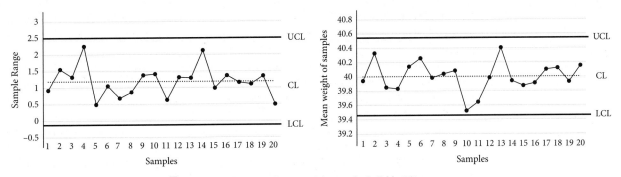

**Figure 17.1**　The $R$ chart (left) and the $\overline{X}$ chart (right) for peanut butter weight samples in Table 17.1.

processes can still produce outputs that are unacceptable (i.e., defective). The reason is that while the outputs of an in-control process are not affected by any special causes, they still are affected by common causes. If there are many common causes, and if these common causes have significant impact on the output of a process, the outputs may not satisfy design specification limits.

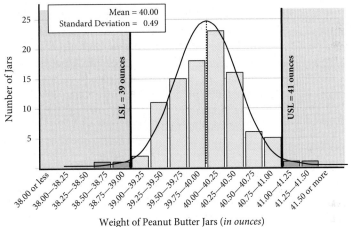

**Figure 17.2**　Histogram of the weights of 100 jars of peanut butter in Table 16.1.

This example also points to the important difference between process control limits (UCL and LCL) and design specification limits (LSL and USL). As discussed before, design specification limits are established based on the performance variation that is acceptable to customers. Thus, they actually represent the *voice of the customer*—an external factor. Process control limits, however, are limits that are established based on the past performance of a stable process (i.e., from observed samples) with the goal of maintaining the stability of the output of the process—an internal factor. Hence, in some sense, they represent the *voice of the process*. Therefore, processes can be in control, but not capable of meeting product specifications (see Fig. 17.3). Common cause variations in in-control processes can result in these processes producing products that do not meet product specifications.

---

**CONCEPT**

### Process Capability

*Process Capability* is the ability of an *in-control* process to produce outputs that meet or exceed product design specifications.

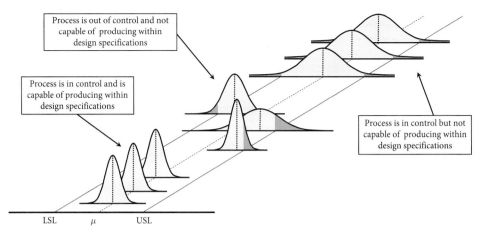

Figure 17.3   In-control and out-of-control processes and their capabilities.

How is the capability of a process determined? There are several approaches that are commonly used to measure the capability of a process to meet its design specifications. We discuss three of these measures here, namely (i) Fraction of Output within Specification, (ii) Process Capability Ratios, and (iii) Sigma Capability Measure. The important fact to remember is that process capability is measured only after the special causes that degrade the outputs of a process are eliminated—only when process is in control. If we want to measure if the filling process is capable of filling 40 ounces of peanut butter in jars, it does not make sense to measure it when some of the filling pipes are already clogged (a special cause). We fix the machine and measure its capability after this and other special causes are eliminated.

### 17.2.1 Fraction of Output Within Specification and DPMO

One way to measure process capability is to compute the fraction of the outputs of the process that meets or exceeds design specification.

---

**CONCEPT**

**Measuring Process Capability—Fraction of Output within Specification and DPMO**

To measure process capability of a (stable) process, take a large sample of the output of the process. Inspect each unit and determine whether it meets design specifications.

- *Fraction of Output Within Specification* is

$$\text{Fraction of output within specification} = \frac{\text{Number of inspected units that are within specification}}{\text{Total number of units inspected}}$$

The larger the fraction of output within specification, the higher the process capability.

- *Defects Per Million Opportunities* or DPMO is the number of defective units (or errors) among 1 million opportunities to make defective units (or errors). DPMO can be computed as follows:

$$\text{DPMO} = (1 - \text{Fraction of output within specification}) \times 1{,}000{,}000$$

The smaller the DPMO, the higher the process capability.

In our peanut butter example in Fig. 17.2, we have a total of 4 jars among 100 jars that were not within specification limits LSL = 39 and USL = 41 ounces; other 96 jars were within specification limits. Thus, for the filling process

$$\text{Fraction of output within specification} = \frac{\text{Number of units within specification}}{\text{Total number of units}} = \frac{96}{100} = 0.96$$

Hence, our filling process is 96 percent capable of meeting product design specification. In other words, 96 percent of the output of filling process weigh within LSL = 39 and USL = 41 ounces. The DPMO for the filling process is, therefore,

$$\text{DPMO} = (1 - 0.96) \times 1{,}000{,}000 = 40{,}000$$

which implies that, about 40,000 out of every 1,000,000 jars filled by the filling process do not meet the specifications and are defective.

Often times probability distributions such as Normal distribution can be used to approximate fraction of outputs within specification. To illustrate this, suppose that we would like to use Normal distribution to compute the fraction of outputs within specification. Considering $X$ as the amount of peanut butter (in ounces) in a 40-ounce jar, as Fig. 17.2 shows, the 100 samples in the figure have mean of $\mu_x = 40$ and standard deviation $\sigma_X = 0.49$. The figure also shows that Normal distribution is a reasonable approximation for the observed samples. The fraction of outputs within specification can be approximated by the area under Normal distribution between LSL = 39 and USL = 41. Specifically,

$$\begin{aligned}
\text{Fraction of Output within Specification} &= P(39 \leq X \leq 41) \\
&= P(X \leq 41) - P(X \leq 39) \\
&= \Phi\left(\frac{41 - 40}{0.49}\right) - \Phi\left(\frac{39 - 40}{0.49}\right) \\
&= 0.979 - 0.021 = 0.958
\end{aligned}$$

which is close to 0.96—the capability we computed using the observed number of non-defective units (i.e., 96) among the 100 samples.

### 17.2.2 Process Capability Ratios

Consider a process that produces outputs where quality characteristic of interest is $X$. Suppose that, when process is in control, $X$ has mean $\mu$ and standard deviation $\sigma$. Recall that in Normal distribution, the probability that an observation falls within $3\sigma$ above or below the mean is very large, that is, about 99.73 percent. Hence, if a process has a mean $\mu$ which is $3\sigma$ above its lower specification limit, that is,

$$\mu - \text{LSL} \geq 3\sigma \quad \Longrightarrow \quad \frac{\mu - \text{LSL}}{3\sigma} \geq 1$$

and if the mean is $3\sigma$ below its upper specification limit,

$$\text{USL} - \mu \geq 3\sigma \quad \Longrightarrow \quad \frac{\text{USL} - \mu}{3\sigma} \geq 1$$

then at least 99.73 percent of the output of the process will be within specification limit. As ratios $(\mu - \text{LSL})/3\sigma$ and $(\text{LSL} - \mu)/3\sigma$ become larger than one, the process becomes more capable of producing outputs with specification limits. Hence, one way to measure process capability is with these two ratios. These ratios are called "process capability ratios."

## Measuring Process Capability—Process Capability Ratios

To measure process capability of a (stable) process, take a large sample of the output of the process and compute its mean $\mu$ and standard deviation $\sigma$ of the quality characteristics of interest. *Process Capability Ratio $C_{pk}$* is defined as follows:

$$C_{pk} = \text{Min} \left\{ \frac{\text{USL} - \mu}{3\sigma} , \ \frac{\mu - \text{LSL}}{3\sigma} \right\}$$

In a *Centered Process*, which is a process with its mean being at the center of the specification limits, process capability ratio $C_{pk}$ simplifies to $C_p$, where

$$\text{For Centered Processes} \quad \Longrightarrow \quad C_p = \frac{\text{USL} - \text{LSL}}{6\sigma}$$

A process with capability ratio $C_{pk} = 1$ (or $C_p = 1$) is considered a capable process with a very large fraction of its output within design specification. Obviously, the higher the $C_{pk}$, the more capable the process.

Some points worth mentioning here. First, process capability ratio $C_{pk}$ is a conservative measure since it chooses the smaller (i.e., minimum) of the two ratios $(\mu - \text{LSL})/3\sigma$ and $(\text{LSL} - \mu)/3\sigma$ as a measure of process capability. Second, for a process that is centered at its target specification, capability ratio $C_p$ can be interpreted as "Specification Width" (i.e., USL − LSL) over "Process Width" (i.e., $6\sigma$). The reason $6\sigma$ presents process width is that most of process outputs (i.e., about 99.73%) vary within six standard deviation around mean—three standard deviation above and three standard deviation below mean. For instance, the peanut butter filling process in Fig. 17.2, which is a centered process with standard deviation $\sigma = 0.49$, has a capability ratio

$$C_p = \frac{\text{USL} - \text{LSL}}{6\sigma} = \frac{41 - 39}{6(0.49)} = 0.68$$

Third, while the conceptual foundation of process capability ratios is based on Normal distribution, these ratios are used to measure process capability even when the quality characteristics do not follow Normal distribution.

 **Example 17.1**

> Fasten-Tight, also known as FT, is a supplier of fasteners to more than 80 manufacturers. FT's order lead time is 72 hours (i.e., 3 days). FT's manufacturers practice lean operations and thus would like to receive their orders not earlier and not later than 3 days after they place an order. Orders delivered much earlier than 3 days increase the inventory at the manufacturers' and orders delivered much later than 3 days can cause shortage and stop manufacturers' production processes. FT has identified that orders delivered between 60 and 84 hours are still considered acceptable by its manufacturers. Data regarding past deliveries show that the delivery times to manufacturers have an average of 80 hours and standard deviation 3 hours. What is the capability of FT's delivery process?

The lower and upper specification limits that are acceptable to customers (i.e., the manufacturers) are LSL = 60 and USL = 84 hours. The delivery process has mean $\mu = 80$ hours and standard deviation $\sigma = 3$ hours. Note that the center of specification limits—the target value—is 72 hours. Thus, FT's delivery process is not a centered process. Hence, we use process capability ratio $C_{pk}$ to measure the capability of FT's delivery process as follows:

$$C_{pk} = \text{Min} \left\{ \frac{\text{USL} - \mu}{3\sigma} , \frac{\mu - \text{LSL}}{3\sigma} \right\} = \text{Min} \left\{ \frac{84 - 80}{3(3)} , \frac{80 - 60}{3(3)} \right\} = \text{Min} \{0.44 , 2.22\} = 0.44$$

Therefore, the delivery process has capability ratio $C_{pk} = 0.44$. Does this mean that 44 percent of the time FT is capable of delivering orders between 60 and 84 hours? Absolutely not! To elaborate, let's assume that delivery time follows Normal distribution. The fraction of orders that are delivered within 60 to 84 hours is then

$$\text{Fraction of orders within specification} = P(60 \leq X \leq 84)$$
$$= \Phi \left( \frac{84 - 80}{3} \right) - \Phi \left( \frac{60 - 80}{3} \right)$$
$$= 0.909$$

Hence, FT delivery process is capable of delivering 90.9 percent of its orders within the time acceptable to its manufacturers. This corresponds to DPMO $= (1 - 0.909) \times 1,000,000 = 91,000$, which implies that about 91,000 deliveries among 1 million deliveries would not meet manufacturers' requirement.

### 17.2.3 Sigma Capability Measure

The third process capability metric that is often used is sigma capability measure. Similar to capability ratios $C_{pk}$ and $C_p$, sigma capability measure is based on the distance between the mean of the process and upper and lower specification limits (i.e., USL $- \mu$ and $\mu -$ LSL). However, this measure is scaled based on one standard deviation of the process (not three standard deviation).

---

**CONCEPT**

**Measuring Process Capability—Sigma Capability Measure**

To measure process capability of a (stable) process, take a large sample of the output of the process and compute its mean $\mu$ and standard deviation $\sigma$ of the quality characteristics of interest. *Sigma Capability Measure*, or simply *Sigma Measure S*, of the process is defined as follows:

$$S = \text{Min} \left\{ \frac{\text{USL} - \mu}{\sigma} , \frac{\mu - \text{LSL}}{\sigma} \right\}$$

If the process is centered—the mean of the process is at the center of specification limits—sigma measure $S$ simplifies to

$$S = \frac{\text{USL} - \text{LSL}}{2\sigma}$$

A process with sigma measure $S$ is called a *S-sigma process*.

---

Hence, sigma capability is how many standard deviation (sigma) are between the closest design specification and the mean of the process. To see the relationship between sigma measure and process capability ratios $C_{pk}$ and $C_p$, consider a centered 3-sigma process. In this process, the lower and the upper specification limits are three standard deviation below and above the mean, respectively. Hence, 3-sigma process will have $C_p = 1$. Similarly, a 6-sigma process will have $C_p = 2$.

The delivery process of Fasten-Tight (FT) in Example 17.1 with $C_{pk} = 0.44$ will have sigma capability measure:

$$S = \text{Min}\left\{ \frac{\text{USL} - \mu}{\sigma} , \frac{\mu - \text{LSL}}{\sigma} \right\} = \text{Min}\left\{ \frac{84 - 80}{3} , \frac{80 - 60}{3} \right\} = \text{Min}\{1.33 , 6.67\} = 1.33$$

In other words, the delivery process of Fasten-Tight is a 1.33-sigma process.

Often times design specification limits are one-sided, that is, there is either an upper specification limit or a lower specification limit. The following example shows how process capability for such processes can be computed.

### Example 17.2

> Consider the alphabet cookie producer in Example 16.2 of Chap. 16. The marketing department has indicated that having more than eight broken cookies results in customer dissatisfaction and can have long-term impact on future sales. Considering this new information, what is sigma capability measure of the firm's production process?

The information about the acceptable number of broken cookies in a box corresponds to upper specification limit USL = 8 cookies. Obviously, there is no lower specification limit on the number of broken cookies, below which customers will be dissatisfied. Hence, the capability of the production process should be measured against upper specification limit. Considering the mean and standard deviation of the number of broken cookies in a box that we computed in Example 16.2 of Chap. 16, that is, $\mu = \bar{c} = 4.97$ and $\sigma = \sqrt{\bar{c}} = \sqrt{4.97} = 2.23$, we have

$$S = \frac{\text{USL} - \mu}{\sigma} = \frac{8 - 4.97}{2.23} = 1.36$$

Therefore, the production process is a 1.36-sigma process.

To have a better understanding of the $S$-sigma capability measure, Fig. 17.4 shows examples of different (centered) processes with different sigma capability measures. The design specification has target value 100 and lower specification USL = 70 and upper specification limit USL = 130. As the figure shows, Process E, which is a 6-sigma process has such a small

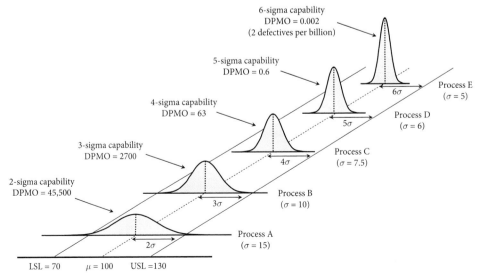

**Figure 17.4  Examples of processes with different sigma capability measures.**

standard deviation (i.e., $\sigma = 5$) that we can fit 6 standard deviation between mean (i.e., 100) and the upper specification limit (i.e., 130) and also 6 standard deviation between mean (i.e., 30) and the lower design specifications (i.e., 70).

## 17.3  Quality Improvement—The Six Sigma Approach

The concept of Six Sigma quality was developed by Motorola in 1986 when Motorola was trying to achieve 6-sigma capability and reduce its number of defects and mistakes to less than 3.4 per million opportunities. It all started in mid and late 1980s when Bill Smith, an engineer in Motorola, noticed the correlation between product performance in the market and amount of modifications required at the manufacturing points. He and his colleagues realized that minimizing the defects at each stage of production resulted in better performance in the marketplace. The CEO of Motorola, Bob Galvin, recognized the importance of Bill's findings and launched a long-term program called "The Six Sigma Quality Program." The program set 6 sigma as the required process capability level to achieve the standard of 3.4 defects per million opportunities.

Implementing Six Sigma methodologies, Motorola became the first company that won the prestigious Malcolm Baldrige National Quality Award in 1988. In addition, Motorola has estimated that, as a result of implementing Six Sigma, it has saved more than $17 billion. In 1996, Jack Welch, the CEO of General Electric, launched Six Sigma at his company and became its global promoter. The GE Annual Report in 1997 stated that Six Sigma added $300 million to GE's bottom line. Other pioneer firms that used Six Sigma approach and expanded it from a shop floor quality improvement to reducing variability and improving quality in all activities of organizations were Allied Signal (now part of Honeywell) and Texas Instruments.[4]

After Six Sigma was proven to be an effective problem-solving and improvement methodology, many companies started their own Six Sigma initiatives. By implementing Six Sigma, Allied Signal reported cost saving of $800 million in less than 2 years; Bank of America improved customer satisfaction and also saved more than $2 billion in less than 3 years;[5] Air Canada gained $450 million in 3 years; American Express benefited $200 million in 2 years; and Sun Microsystems gained 1170 million in 3 years. Since 1987, 53 percent of Fortune 500 companies and 82 percent of Fortune 100 companies have used Six Sigma to some degree. It has saved Fortune 500 companies an estimated $427 billion since 1987.

While initial activities to achieve Six Sigma capability were around statistical methods, it was gradually evolved into a comprehensive quality management and problem-solving system known as "Six Sigma." According to a 2011 survey by ASQ, 79 percent of quality improvement professionals said that Six Sigma is a very effective system in improving productivity and efficiency in their firms. The participants also found Six Sigma to be effective in raising quality levels in their firms (74 percent); reducing costs (73 percent); helping individuals in their organization to be competitive in the marketplace or to pursue the organization's core mission (68 percent); having a positive impact on employee safety (56 percent); and improving innovation (46 percent).

Six Sigma is now a *business improvement approach* focusing on finding and eliminating errors and defects in every aspect of manufacturing and service organizations, including product development, marketing, finance, accounting, and operations. It is used in finance and accounting departments, for example, to eliminate errors or to reduce the costing errors as well as to reduce the time to close books.

---

[4]Meredith, J., and S. Shafer, *Operations Management for MBAs*. Fifth Edition, John Wiley and Sons, 2013.
[5]Jones, M.H., Jr., "Six Sigma ... at a Bank?" *Six Sigma Forum Magazine*, (February 2004) 13–17.

Six Sigma focuses on understanding customer needs and uses data-driven approaches and rigorous statistical tools (e.g., process control charts) to improve process performance. Six Sigma methodologies have been able to increase firms' profitability by reducing costs, increasing productivity, improving quality, increasing customer satisfaction, and improving employees' morale. The term "Six Sigma," with uppercase S's, is a registered trademark of Motorola.

What exactly is Six Sigma and how it generates such a significant financial gain for firms? Six Sigma has three main pillars:

- Striving for 6-sigma capability,
- Six Sigma improvement cycle, and
- Organizational involvement.

We devote the remaining of this chapter to describe these three pillars.

## 17.4 Striving for Six Sigma Capability

The core of Six Sigma is to achieve 6-sigma capability in all functional areas of a firm. This is done, mainly, by reducing variability in those processes, resulting in a very small number of defects per million opportunities. In fact, some interpret Six Sigma as a system of achieving 3.4 defects per million opportunities, that is, DPMO = 3.4. You may have noticed, however, that as shown in Fig. 17.4, a process with 6-sigma capability will have about two defects per billion opportunities, not 3.4 defects per million opportunities. Where does this difference come from?

The answer is in the way Motorola originally computed DPMO for a 6-sigma process. Motorola assumed that, as a centered process continues producing its outputs, its mean can shift as much as 1.5 standard deviation to the left or to the right (see Fig. 17.5). This is a practical assumption, since there is always a possibility that the mean of a process shifts to the left or to the right, and the managers are not aware of that.

Consider Case I in the figure where the process mean shifted by 1.5 standard deviation to the right. In this case, USL is only 4.5 standard deviation away from the mean of the 6-sigma process. For Normal distribution, the chance of observing an output beyond 4.5 standard deviation is 0.0000034. On the other hand, LSL is 7.5 standard deviation away from the mean of the process. The probability observing an output 7.5 standard deviation below the mean is 0.000000000003, which is almost zero. Thus, in Case I, where the mean of the 6-sigma process shifts to the right by 1.5 standard deviation, fraction 0.0000034 of the outputs—3.4 out of 1 million outputs—are defective. Similarly, in Case II where the

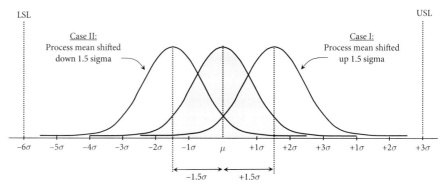

**Figure 17.5**    Motorola's approach to compute DPMO assumes a 1.5 $\sigma$ shift in the process mean.

mean of the 6-sigma process shifts 1.5 standard deviation to the left, the process will have DPMO = 3.4. Thus, being conservative and considering the possibility of a shift in mean, Motorola computes DPMO of a 6-sigma process as 3.4 defects per million opportunities.[6]

### Why Six Sigma Capability?

In a 6-sigma process fraction 0.999999998 of its output are within specification. Why should firms try to achieve such a high standard? Considering the potential costs of achieving such a high standard, why not achieve fraction of output within specification of 0.999, which is large enough? Well, there are many reasons for achieving 6-sigma capability, five of which are: (i) high cost of a defective output, (ii) accumulation of errors, (iii) robustness to changes in the process, (iv) early detection of quality problems, and (v) striving for perfection.

### 17.4.1 High Cost of Defective Outputs

Achieving a 6-sigma capability is critical in processes that have a very large number of outputs and the cost of one defective output is very high. To illustrate this, suppose processes work with fraction of output within specification of 0.999, which is about 3-sigma capability. In such situations, we would have about 5000 surgical procedures done incorrectly every week, 200,000 prescriptions filled incorrectly each year, and 20,000 pieces of lost mail every hour. In a 6-sigma process with fraction of output within specification of 0.999999998, however, we would have about one incorrect surgical procedure every 2 years, one prescription incorrectly filled every 2.5 years, and one piece of mail lost every day. Therefore, when the process works in high volume (e.g., mail delivery) or when the cost of a defective output is high (e.g., surgical procedure), then achieving a 6-sigma capability becomes critical. If you are the manager of a firm that produces airplanes, will you settle with less than 6-sigma capability for your welding process that produces airplanes wings?

### 17.4.2 Accumulation of Errors

The second reason for having a 6-sigma process is the compounding effect of errors in products that are made of several parts, or products that are produced in several stages. As an example, suppose a product consists of 80 parts and each part is produced in a different production process. If each process is a 3-sigma process, then the probability that each part is within specification is 0.997. Assuming that the product is non-defective if all of its 80 parts are within specifications, then the probability that the product is not defective is $(0.997)^{80} = 0.786$. In other words, while only 3 out of 1000 parts produced in each process are defective, about one out of every four final product will be defective. If each part was produced in a 6-sigma process, however, only 16 products out of 100 million products will be defective. Similar analogy can be made if one product (e.g., a part) is manufactured through 80 consecutive stages, with each being a 3-sigma process. In this system, about one out of every four parts exiting the last stage of the production is defective, while in a 6-sigma process, this number is 16 out of 100 million.

Ford experienced this firsthand when it was simultaneously manufacturing the same car model, some with U.S.-made and some with Japanese-made transmissions. After selling those cars, Ford noticed that customers were requesting the model with Japanese-made transmission over those with U.S.-made transmission. This was a puzzle for Ford, since both transmissions were made to the same design specifications. After taking apart several transmissions of each type, Ford engineers found that, while all parts of both transmissions were

---

[6]Venkataraman, R.R., and J. K. Pinto, *Operations Management, Managing Global Supply Chains*. SAGE Publications, 2018.

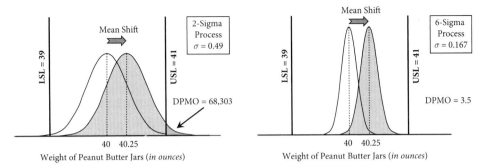

Figure 17.6  Defect per million opportunities (DPMO) for the filling machine when mean shifts from 40 to 40.25 ounces. *Left:* The current 2-sigma process. *Right:* The improved 6-sigma process.

within specified limits, the variation in the parts of Japanese-made transmission was much less (i.e., a higher sigma capability) than that in U.S.-made transmissions. The low variation in all parts resulted in better performance of the final product—the transmission—as well as in fewer problems.

### 17.4.3 Robustness to Changes in the Process

The third reason for achieving 6-sigma capability is that the outputs of 6-sigma processes are not significantly affected by the changes in the process. To illustrate this, recall that in our example, the amount of peanut butter filled by the filling machine was well-approximated by a Normal distribution with mean $\mu_x = 40$ and standard deviation $\sigma_x = 0.49$. Considering lower and upper specifications LSL $= 39$ and USL $= 41$, respectively, the process has sigma capability

$$S = \frac{\text{USL} - \text{LSL}}{2\sigma} = \frac{41 - 39}{2(0.49)} \simeq 2$$

Thus, the filling machine has DPMO $= 45,500$. Now suppose that the machine goes out of control and the mean shifts from 40 to 40.25, but standard deviation remains the same. However, the managers are not aware of this shift (see the left panel in Fig. 17.6). What is the impact of this shift on the machine's output, namely DPMO? To compute DPMO, we first need to compute the fraction of output within specification as follows:

$$\begin{aligned}
\text{Fraction of Output within Specification} &= P(39 \leq X \leq 41) \\
&= \Phi\left(\frac{41 - 40.25}{0.49}\right) - \Phi\left(\frac{39 - 40.25}{0.49}\right) \\
&= 0.9317
\end{aligned}$$

Thus

$$\text{DPMO} = (1 - 0.917) \times 1{,}000{,}000 = 68{,}303$$

Thus, if the mean shift is not detected and the machine is not centered, we will have an increase of about $68{,}303 - 45{,}500 = 22{,}803$ more defective jars produced per million.

Now suppose that by performing preventive maintenance and replacing old parts, we are able to reduce the standard deviation of the filling machine from 0.49 to 0.167. This makes the machine a 6-sigma process, which produces about DPMO $= 0.002$ defective items (jars) per million. Again, suppose that the mean of this new improved machine shifts from 40 to 40.25 ounces. If the managers do not notice this shift, how many defective jars the machine will produce? Repeating a similar calculation we find that, after the mean shift, the machine produces DPMO $= 3.5$ defective jars per million.

Notice that, if the mean shift is not detected, the 6-sigma machine will produce only about 3.5 defective jars per million. In other words, the quality of the output of the 6-sigma machine is very robust with respect to changes in its environment. If, for example, an assignable cause affects mean or variability of the process, the impact on the quality of the outputs is very small. This is another advantage of 6-sigma processes.

### 17.4.4 Early Detection of Quality Problems

While a 2-sigma process (like the current filling machine) may have a significant increase in its number of defective outputs if a mean shift occurs, one may argue that—using appropriate control charts—the shift can be immediately detected before a large number of defective outputs are produced. Well, let's investigate this. Suppose that we use an $\overline{X}$ chart to monitor the changes in the mean of the filling machine by taking samples of $n = 12$ jars every 30 minutes. Considering that the mean and standard deviation of the output (when it is stable) is known to be $\mu = 40$ and $\sigma = 0.49$, the control limits for the $\overline{X}$ chart will be CL $= \mu = 40$, and[7]

$$\text{LCL} = \mu - 3(\sigma/\sqrt{n}) = 40 - 3(0.49/\sqrt{12}) = 39.58$$
$$\text{UCL} = \mu + 3(\sigma/\sqrt{n}) = 40 + 3(0.49/\sqrt{12}) = 40.42$$

Here is the main question: After the mean shift occurs, how long it takes, on an average, for the $\overline{X}$ chart to detect that?

Suppose that the mean has just shifted to 40.25, but we do not know that, and we are using the above control limits to detect that. What is the probability that the next sample (i.e., mean of $n = 12$ jars) lies beyond our control limits? Recall that $\overline{X}$ is our sample mean, which is a random variable, changing from sample to sample. Also, recall that for samples of size $n = 12$, random variable $\overline{X}$ follows a Normal distribution with mean $\mu_{\overline{x}} = \mu = 40.25$ (i.e., the shifted mean) and standard deviation $\sigma_{\overline{X}} = \sigma/\sqrt{n} = 0.49/\sqrt{12} = 0.14$. Thus, the probability of the next sample of size $n = 12$ has a mean that violates the control limits LCL $= 39.58$ or UCL $= 40.42$ is

$$1 - P(39.58 \leq \overline{X} \leq 40.42) = 1 - \left[ \Phi\left( \frac{40.42 - 40.25}{0.14} \right) - \Phi\left( \frac{39.58 - 40.25}{0.14} \right) \right]$$
$$= 0.11$$

Hence, although the mean has shifted to 40.25, the chance that the next sample in our $\overline{X}$ chart violates the control limits is only 11 percent. How long it takes to get a sample that violates control limits? Well, it has been found that, if the probability that a sample violates control limits is $q$, then the average number of samples until we get the *first* sample that violates control limits is $1/p$. Since we have $p = 0.11$, the average number of samples that should be expected until we observe a sample that violates control limit is $1/p = 1/0.11 = 9.1$.

Recall that a sample is taken every 30 minutes—two samples an hour. Thus, after the mean shift, the time it takes, on an average, to observe the first sample that indicates an out-of-control machine is $9.1/2 = 4.55$ hours, or 4 hours and 33 minutes. During these four and a half hours, the machine would have produced many defective jars, some of which may have already been transferred to finished goods inventory, or been shipped to a customer.

---

[7] Since we already know the values of mean $\mu$ and standard deviation $\sigma$ of the outputs of the machine, we can set the three-sigma control limits using those values, that is, no need to take samples to estimate $\mu$ and $\sigma$.

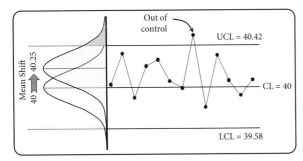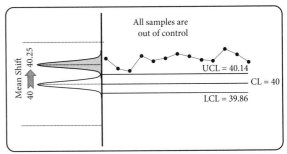

Let's now focus on the improved 6-sigma machine that has $\mu = 40$ and $\sigma = 0.167$. The control limits for the $\overline{X}$ chart for this machine are, $CL = 40$, and

$$LCL = \mu - 3(\sigma/\sqrt{n}) = 40 - 3(0.167/\sqrt{12}) = 39.86$$
$$UCL = \mu + 3(\sigma/\sqrt{n}) = 40 + 3(0.167/\sqrt{12}) = 40.14$$

which are a much tighter control limits than those of the current machine.

Now suppose that the mean shifts from 40 to 40.25. For sample size $n = 12$, random variable $\overline{X}$ follows a Normal distribution with mean $\mu_{\overline{X}} = \mu = 40.25$ (i.e., the shifted mean) and standard deviation $\sigma_{\overline{X}} = \sigma/\sqrt{n} = 0.167/\sqrt{12} = 0.048$. Thus, the probability that the next sample of size $n = 12$ has a mean that violate the control limits $LCL = 39.86$ or $UCL = 40.14$ is

$$1 - P(39.86 \leq \overline{X} \leq 40.14) = 1 - \left[ \Phi\left( \frac{40.14 - 40.25}{0.048} \right) - \Phi\left( \frac{39.86 - 40.25}{0.048} \right) \right]$$
$$= 0.99$$

In other words, the first sample after the mean shift will violate the control limits with probability 0.99. In other words, the $\overline{X}$ charts will immediately detect that the machine is out of control after taking the first sample.

As this example shows, 6-sigma processes also have the advantage that any change in their outputs can be very quickly detected—in other words, control charts become more effective in detecting assignable causes in 6-sigma processes. Why? The answer is again in the low variability in the output of 6-sigma processes. This is shown in Fig. 17.7. As the figure shows, 6-sigma processes have very tight control limits. Thus, any small change (e.g., a mean shift) moves all the samples beyond the control limits and signals an out-of-control process.

### 17.4.5  Striving for Perfection

Recall Lesson #10 that says quality improvement efforts are never-ending journeys with the ultimate goal of striving for perfection. This is the third pillar of Six Sigma. In fact, there is a philosophical reason why Six Sigma's goal is to achieve 6-sigma capability. By setting such a high standard for quality, Six Sigma has the goal of developing the organizational culture of a never-ending pursuit of perfection. This culture motivates efforts for continuous improvement. The variability reduction through increasing sigma capability brings many benefits, one of which is the improvement in quality.

To achieve a 6-sigma process, firms need to reduce the variability in their outputs and, if necessary, center the mean of the process between lower and upper specification limits.

**Example 17.3**

> Consider the alphabet cookie producer in Example 17.2. The production manager has decided to increase the capability of its production process from a 1.36-sigma process to a 6-sigma process. How much the manager should reduce variability in its production line in order to achieve the goal?

Recall that when only common-cause variations are present, the number of defects in a production unit is well-approximated by a Poisson distribution, which has a mean of $\mu = c$ and standard deviation $\sigma = \sqrt{c}$, where $c$ is the average number of defects in a production unit. Therefore, to have the improved production process with $S = 6$-sigma capability, we must reduce the average number of broken cookies in each box to $c$ such that

$$S = \frac{USL - \mu}{\sigma} = \frac{8 - c}{\sqrt{c}} = 6$$

After some search, we find that $c = 1.26$ satisfies the above. Hence, to achieve 6-sigma capability, the production managers must find ways to reduce the number of broken cookies in a box, such that the average number of broken cookies in the boxes coming out of the production process be at most 1.26. But how? This can be done through DMIAC improvement process of Six Sigma that we discuss next.

## 17.5 Six Sigma Improvement Cycle

As we mentioned, Six Sigma is beyond just reaching 6-sigma capability. It is a problem-solving methodology focusing on continuous improvement with strong emphasis on data analysis using quantitative methods and statistical analysis. The basic framework of Six Sigma to improve quality is a structured problem-solving procedure called DMAIC (improvement cycle). DMAIC stands for Define, Measure, Analyze, Improve, and Control (see Fig. 17.8). Using DMAIC improvement cycle, organizations can systematically monitor and improve processes in virtually any aspects of their business.

### 17.5.1 Define

In the first step of DMAIC improvement cycle—Define—critical to quality (CTQ) characteristic of the output of a process that is important to customers must be identified. Customers could be market customers who buy the product (external customers), or they could be the

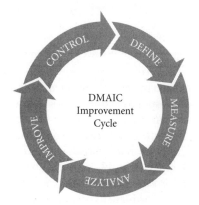

**Figure 17.8    DMAIC improvement cycle of Six Sigma.**

Table 17.2    Number of Guests' Complaints During a 6-Week Period

| Complaints | Week 1 | Week 2 | Week 3 | Week 4 | Week 5 | Week 6 | Total | Percent |
|---|---|---|---|---|---|---|---|---|
| Errors in reservation | — | — | 1 | — | 2 | — | 3 | 4% |
| Room not cleaned well | 3 | 8 | 5 | 2 | 10 | 8 | 36 | 55% |
| Discourteous staff | — | — | — | — | — | 1 | 1 | 1% |
| Long wait for check-in | 2 | — | 4 | 3 | 1 | 2 | 12 | 18% |
| Issues with TV or Internet | 1 | — | 2 | 1 | — | 3 | 7 | 11% |
| Poor room service | — | — | — | — | 2 | — | 2 | 3% |
| Errors in check-out charges | 2 | 1 | — | 2 | — | — | 5 | 8% |
| | | | | | | | 66 | 100% |

next stage of the operations within an organization (internal customers). Having identified CTQ, the gap between the output of the process and CTQ measures must be computed. One tool that helps to better understand the process is process mapping tools such as *Process-Flow Chart* (see Chap. 1) and *Value Stream Mapping* (see Chap. 15). Value stream mapping, for example, helps identify potential problems such as long wait times, excess inventories, and capacity problems. Also, it is important to determine all *key input variables* as well as *key output variables*.

### 17.5.2 Measure

The goal of the second step of DMAIC improvement cycle—Measure—is to evaluate the current performance of the process. To achieve this, data about several performance measures and key input and key output variables (identified in the previous step) must be gathered. The data is then used to measure these variables and determine the current performance of the process in terms of performance measures such as quality, cost, flow time, inventory, and throughput. Finally, the capability of the process is computed in terms of sigma capability measure.

One commonly used tool in data collection is Pareto Chart. *Pareto Chart* is simply a rearranged frequency distribution (i.e., histogram) in which the quality problems in the histogram are arranged from the most frequently occurring to the least frequently occurring. This allows the user to visually identify the top one or two most frequently occurring problems that often account for most of the quality problem occurrences. Obviously, firms should fix those vital few problems first (Lesson #2).

To illustrate how a Pareto chart is constructed, consider the data in Table 17.2 that depicts the number of different types of complaints made by guests of a hotel during a 6-week period.

To construct a Pareto chart, we first need to determine the total number of complaints made during the 6-week period, which is 66. Among those 66 complaints, 3 of them were about errors in reservation. This constitutes $3/66 = 0.04 = 4\%$ of the total complaints and is shown in the last column of Table 17.2. The table also shows what percent of total complaints corresponds to other types of complaints. Figure 17.9 shows the Pareto chart for the quality problem (i.e., complaints) in Table 17.2. Note that the Pareto chart presents quality issues from the most frequent (on the left) to least frequent (on the right).

As Pareto chart in Fig. 17.9 shows, "Room not cleaned well" is the single quality issue that accounts for 55 percent of the complaints made in the 6 weeks. "Long wait for check-in" is the second most occurring complaint that accounts for 18 percent of the complaints. These two quality issues together constitute 73 percent of the complaints—this is shown in the cumulative percent line in the Pareto chart. Hence, hotel managers can devote their

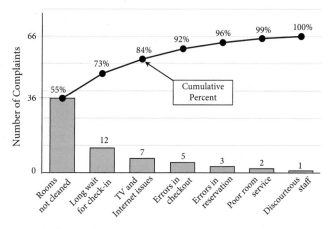

Figure 17.9   Pareto chart for the complaints in Table 17.2.

resources to solve these two problems, starting with looking into why rooms are not as clean as customers expect.

### 17.5.3  Analyze

After identifying a quality problem, the goal of the "Analyze" step of DMAIC is to use the data obtained in the "Measure" step to understand why the quality problem is occurring and to identify the root causes of variation that result in the quality problem.

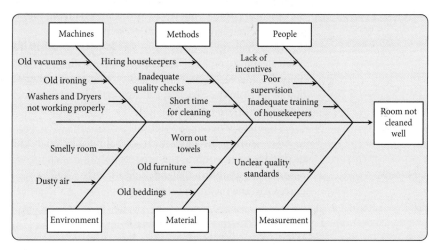

Figure 17.10   Cause-and-effect diagram for hotel "Rooms Not Cleaned" quality problem.

One very useful tool that is commonly used in this step is *Cause-and-Effect Diagram* (also known as *Fishbone* or *Ishikawa* diagram). Figure 17.10 shows an example of a cause-and-effect diagram for finding the root cause of hotel rooms not cleaned properly. As the figure shows, cause-and-effect diagram has six diagonals representing six major groups of causes for quality problems in a process: machines, methods, people, environment, material, and measurements. Each branch of these six diagonals represents one potential cause for the quality problem of "Room Not Cleaned."

The cause-and-effect diagram is proven to be a very powerful tool in improving quality through Six Sigma initiatives. Usually, cause-and-effect diagram is developed by Six Sigma team members during a brainstorming session. One advantage of this team effort—besides

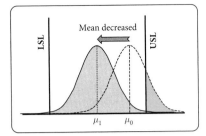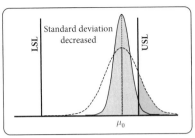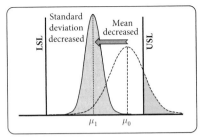

**Figure 17.11**    *Left:* Mean of process output is decreased (mean moved away from the closest specification limit). *Middle:* Standard deviation of process output is decreased. *Right:* Both mean and standard deviation of process output are decreased.

better results coming from teamwork—is that it makes people focus on finding the root cause of the problem rather than finding someone to blame.

### 17.5.4 Improve

The objective of the "Improve" step is threefold. First, the Six Sigma team must develop alternative solutions to eliminate the quality problem. The team must evaluate those solutions and choose the best solution among them. The solution may require the Six Sigma team to modify, redesign, or re-engineer some parts or the entire process. The team must also determine the maximum acceptable ranges of the key variables in the solution and develop a system for measuring deviations of those variables. Second, Six Sigma team must perform a pilot test of the solution to confirm that the solution indeed improves quality. Often during the pilot test the original solution is revised and improved several times according to the outcomes of the pilot test. Third, after finalizing the solution, the team must get the final approval to implement the solution permanently.

One challenging question in this step is to confirm if the alternative solution can indeed result in improvement. The improvement usually corresponds to increasing process capability. To increase process capability, one must either (i) move the mean of the process output far from the closest specification limit, or (ii) reduce the standard deviation of process output, or (iii) do both, see Fig. 17.11.

To check whether a solution indeed improved capability of a process, samples must be taken during the pilot study and should be analyzed to determine whether the mean or standard deviation of the process output has changed. We use the following example to describe the analytics that can help identify whether the improvement effort increased (or decreased) the mean and standard deviation of the process output.

  **Example 17.4**

A fast-food restaurant chain identifies that the upper specification limit for waiting time of drive-thru customers is USL = 100 seconds. The marketing department has identified this limit through customer surveys and study of the waiting times in other competing fast-food chains. The chain announces to its restaurants that they should improve their drive-thru operations. One restaurant close to a highway has average drive-thru waiting time of 90 seconds with standard deviation 20 seconds. The restaurant manager has streamlined its drive-thru operations by standardizing the ordering process and training staff, and thus the manager believes that the improvement efforts have increased sigma capability of its process by reducing both average and standard deviation of customer waiting times. Figure 17.12 presents a sample of waiting times of 50 drive-thru customers.

Based on the above data, do you think that the manager's effort has indeed improved the capability of the drive-thru process?

| | A | B | C | D | E |
|---|---|---|---|---|---|
| 1 | 99 | 102 | 91 | 86 | 108 |
| 2 | 115 | 85 | 88 | 104 | 92 |
| 3 | 115 | 75 | 85 | 91 | 99 |
| 4 | 80 | 77 | 93 | 81 | 90 |
| 5 | 69 | 101 | 95 | 89 | 75 |
| 6 | 69 | 73 | 72 | 84 | 93 |
| 7 | 59 | 78 | 90 | 97 | 106 |
| 8 | 85 | 71 | 121 | 66 | 95 |
| 9 | 108 | 60 | 93 | 61 | 82 |
| 10 | 89 | 90 | 101 | 77 | 79 |

Figure 17.12    Waiting times for $n = 50$ drive-thru customers in Example 17.4.

We can find the average and standard deviation of the $n = 50$ samples in Fig. 17.12, using the following Excel functions:

$$\text{Sample Mean} = \overline{X} = \text{AVERAGE(A1 : E10)} = 87.68$$

$$\text{Sample Standard Deviation} = s = \text{STDEV(A1 : E10)} = 14.49$$

While sample mean 87.68 seconds is lower than the average waiting time before improvement efforts, that is, $\mu_x = 90$ seconds, we cannot conclude with certainty that the improvement efforts have been successful in reducing the average waiting time. Due to variability in waiting times, it might be that—just by pure chance—the waiting times in this particular sample of 50 customers ended up to be low. So, how do we find out that the average waiting times were indeed reduced after implementing improvement solutions?

We can take statisticians' approach and investigate the answer to the following question: If the improvement effort did not reduce customer's waiting time, and the customer waiting time still has mean $\mu_x = 90$ seconds, what is the probability of observing such a sample in Fig. 17.12 that has an average waiting time of 87.68 (or lower)? If the probability of such an event is very low, then we can conclude with high level of confidence that manager's improvement efforts were effective and have reduced the average waiting times of drive-thru customers.

Statisticians call this probability "*p-value*," and if it is less than 0.05, they conclude that the sample provides statistically significant evidence that the average waiting time has been reduced.

$$p\text{-}value = \text{P(sample mean} \leq 87.68 \mid \text{average waiting time not changed)}$$
$$= \text{P}(\overline{X} \leq 87.68 \mid \mu_x = 90)$$

Statisticians have shown that, when the real standard deviation of the waiting times is not known—which is the case here since we do not know if the improvement effort also changed the standard deviation of waiting times—then the above probability can be found using a $t$ distribution with $n - 1$ degrees of freedom. Excel provides this probability for $t$ distribution as follows:

$$p\text{-}value = \text{T.DIST}\left(\frac{\overline{X} - \mu_x}{s/\sqrt{n}}, n - 1, \text{True}\right) = \text{T.DIST}\left(\frac{87.68 - 90}{14.49/\sqrt{50}}, 50 - 1, \text{True}\right) = 0.131$$

Hence, if the improvement efforts have not been effective, and the average waiting time is still $\mu_x = 90$ seconds, then the probability of observing such a sample is 0.131, which is larger than 5 percent. Hence, the sample does not provide enough evidence for us to conclude that the average waiting time of drive-thru customers has been reduced. But what about variability reduction? Can we—based on the sample of 50 waiting times in Fig. 17.12—conclude that improvement efforts have reduced the variability (i.e., standard deviation) of the waiting times?

Note that the sample has a standard deviation $s = 14.49$ seconds, which is less than standard deviation of waiting times before improvement, that is, $\sigma_x = 20$ seconds. So, the question is the following: If the improvement efforts have not reduced the standard deviation of the waiting times and the standard deviation is still $\sigma_x = 20$, what is the probability of observing the sample in Fig. 17.12 with standard deviation $S = 14.49$ or lower? Mathematically speaking, we are looking for the following probability value (i.e., p-value):

$$p\text{-}value = P(\text{sample standard deviation} \leq 14.49 \mid \text{standard deviation of waiting time}$$
$$\text{not changed})$$
$$= P(S \leq 14.49 \mid \sigma_x = 20)$$

where $S$ is the random variable representing the standard deviations of all samples of size $n$. Fortunately, Excel also has the following function to compute the above probability. The function is based on Chi-Square probability distribution that links the variance of samples taken from a population and the variance of population.

$$p\text{-}value = \text{CHISQ.DIST}\left(\frac{(n-1)S^2}{\sigma_x^2}, n-1, \text{True}\right)$$
$$= \text{CHISQ.DIST}\left(\frac{(50-1)(14.49)^2}{(20)^2}, 50-1, \text{True}\right)$$
$$= 0.002$$

which is very close to zero and much less than 5 percent. Thus, it is very unlikely that the 50 samples in Fig. 17.12 are taken from a population with standard deviation of $\sigma_x = 20$ seconds. It must have been taken from a population with standard deviation lower than $\sigma_x = 20$. This means that the standard deviation of the customer waiting time in the drive-thru has been reduced.

In conclusion, while the manager's effort to reduce the average waiting time in the drive-thru has not been successful, those efforts did reduce the variability in the waiting time. As discussed earlier, reducing variability improves process capability.

Before we return to DMAIC improvement cycle, we summarize the process of checking whether process output (i.e., mean and standard deviation) is improved in the following analytics.

---

**ANALYTICS**

### Checking if Process Capability (i.e., Mean or Standard Deviation of Process Output) Is Improved

Consider a process where its Critical to Quality $X$, has mean $\mu_x$ and standard deviation $\sigma_x$. The process has gone through improvement initiatives that are believed to change standard deviation of the process and/or mean of the process. To check this, a sample of size $n$ of process output is taken and its critical to quality are measured as: $x_1, x_2, x_3, \ldots, x_n$. How can we use the data in the sample and check whether the process output is improved?

- *Step 1:* Compute sample mean ($\overline{X}$) and sample standard deviation ($s$) as follows:

$$\overline{X} = \frac{x_1 + x_2 + x_3 + \cdots + x_n}{n}, \qquad s = \sqrt{\frac{\sum_{i=1}^{n}(x_i - \overline{X})^2}{n-1}}$$

or use Excel functions AVERAGE( ) and STDEV( ) to compute the above.

- **Step 2:** Also compute $\mathcal{T}$ and $\mathcal{X}$ as follows:

$$\mathcal{T} = \frac{\overline{X} - \mu_X}{s/\sqrt{n}}, \qquad \mathcal{X} = \frac{(n-1)s^2}{\sigma_X^2}$$

- **Step 2: (Did mean change?)**
  - *Mean decreased?* To check if the mean was *decreased,* compute the following probability value using Excel:
    $$p\text{-}value = \text{T.DIST}(\mathcal{T},\ n-1,\ \text{True})$$
    If p-value is less than 0.05, we have convincing evidence that the mean of the output is decreased.
  - *Mean increased?* To check if the mean was *increased,* compute the following probability value using Excel:
    $$p\text{-}value = 1 - \text{T.DIST}(\mathcal{T},\ n-1,\ \text{True})$$
    If p-value is less than 0.05, we have convincing evidence that the mean of the output is increased.
- **Step 3: (Did standard deviation change?)**
  - *Standard deviation decreased?* To check if standard deviation was *decreased,* compute the following probability value using Excel:
    $$p\text{-}value = \text{CHISQ.DIST}(\mathcal{X},\ n-1,\ \text{True})$$
    If p-value is less than 0.05, we have convincing evidence that the standard deviation of the output was decreased.
  - *Standard deviation increased?* To check if the standard deviation was *increased,* compute the following probability value using Excel:
    $$p\text{-}value = 1 - \text{CHISQ.DIST}(\mathcal{X},\ n-1,\ \text{True})$$
    If p-value is less than 0.05, then we have convincing evidence that standard deviation of the output was increased.

One important quality measure is the proportion of defective units produced by a process. Defective units are either discarded or send back for rework, both of which result in waste of material and resource time. Due to large costs of defective units, many quality improvement efforts focus on reducing the proportion of defective units produced by a process. After the process improvement is completed, the critical question becomes whether the improvement effort was indeed effective in reducing defective units. We use the following example to illustrate the analytics required to answer such questions.

 **Example 17.5**

Fifth-Ninth Bank has a call center that responds to a variety of customers' inquiries. To provide high quality of service, Jean, the manager of the call center, has been monitoring customer complaints. The customer complaint is realized when a caller asks an agent to connect the caller to the agent's manager. Data have shown that the proportion of callers who made complaints has been steady and is about 8 percent. Receiving a mandate from the upper management to improve quality, Jean followed DMAIC improvement steps and implemented improvement strategies to reduce the number of complaints. Jean is wondering if her improvement initiatives decreased the proportion of customers' complaint below 8 percent. Checking the data, she finds that 54 out of 765 customers (i.e., about $54/765 = 7.06\%$) who called the call center in the last 10 days have complained. This is smaller than the proportion of complaints before improvement, that is, smaller than 8 percent. Because 765 is a large sample, Jean feels confident that her improvement initiatives worked. What do you think?

To check if Jean's improvement ideas really worked, we need the following analytics.

---

**ANALYTICS**

**Checking if the Proportion of Defective Units Decreased**

Consider a process with proportion $p$ of its output being defective. The process has gone through improvement initiatives that are believed to have decreased the defective rate of the process. How can we check if the proportion of defective units has decreased?

- *Step 1:* Take a sample of $n$ units from the output of the process and find $x$, the number of defective units among the $n$ units. Then compute an estimate for proportion of defective units as follows:

$$\bar{p} = \frac{x}{n}$$

- *Step 2:* Also compute $\mathcal{Z}$ as follows:

$$\mathcal{Z} = \frac{\bar{p} - p}{\sqrt{p(1-p)/n}}$$

- *Step 3:* To check if the proportion of defective unit is decreased, compute the following probability value using Excel:

$$p\text{-}value = \Phi(\mathcal{Z}) = \text{NORM.DIST}(\mathcal{Z}, 0, 1, \text{True})$$

If p-value is less than 0.05, we have convincing evidence that the proportion of defective units in the output of the process is decreased.

In Example 17.5, we have sample of size $n = 765$ callers with $x = 54$ complaints (i.e., defective units). Thus, following Step 2 we have $\bar{p} = 54/765 = 0.0706$. According to Step 3, on the other hand,

$$\mathcal{Z} = \frac{\bar{p} - p}{\sqrt{p(1-p)/n}} = \frac{0.0706 - 0.08}{\sqrt{0.08(1-0.08)/765}} = -0.958$$

Since $p\text{-}value = \text{NORM.DIST}(-0.958, 0, 1, \text{True}) = 0.168 > 0.05$, we do not have convincing evidence that Jean's improvement efforts have reduced the fraction of complaints. Sorry Jean!

Now, back to DMAIC improvement cycle and its last step—Control.

### 17.5.5 Control

After the approval is gained and the solution is implemented in the improvement step, the "Control" step of the DMAIC requires the Six Sigma team to hand off the project to the process manager. This includes providing the process manager with the data on key process measures before and after implementing the solution as well as information such as new process maps, new procedures, new equipment, training and evaluation procedures. The main objective of the final step of DMAIC improvement cycle is to establish a system that ensures that the improvement achieved by the solution is sustained. This is done by monitoring and controlling the critical variables of the process (e.g., using control charts), mistake-proofing the process, and developing action plans if the key variables of the process go beyond their acceptable ranges—go out of control.

All five steps of the DMAIC improvement cycle encourage creative thinking for identifying the problem and its causes and generating solutions in the context of the product and/or the process. The first three steps focus mainly on the current process and the last two steps involve the new redesigned and improved process. The completion of the last step is not the end of DMIAC improvement cycle—it is in fact the beginning of the next improvement cycle, since Six Sigma philosophy is to continuously search for quality problems and eliminate or improve them.

## 17.6 Six Sigma and Organizations' Involvement

The third pillar of Six Sigma is the commitment from top managers to Six Sigma as well as the expertise of people who implement Six Sigma programs. Top management must promote Six Sigma, provide resources, and eliminate obstacles for implementation of Six Sigma initiatives. Of course, the rigorous quantitative steps of Six Sigma cannot be implemented unless all employees have proper training in using technical tools (e.g., statistical process control) to identify and eliminate the quality problems.

Six Sigma institutionalizes quality improvement (i.e., Lesson #9) by instilling an informal hierarchy within the formal organizational structure. Specifically, the informal organizational structure of Six Sigma defines the following roles, which are given martial arts titles that reflect the hierarchy and the level of skills of the employees:

- *Champions*: Champions are the key individuals who are from top management and are leaders with the responsibility for Six Sigma implementation across the organization.

- *Master Black Belts (MBB)*: Master Black Belts, unlike Champions, spend 100 percent of their time to Six Sigma. They are identified by Champions and serve as Six Sigma experts for the organization. MBBs have rigorous training in statistics and strong management skills as well as good experience in waste and variability reduction. They train and coach Black Belts and Green Belts.

- *Black Belts (BB)*: Like Master Black Belts, Six Sigma Black Belts devote 100 percent of their time to Six Sigma programs. They operate under Master Black Belts and mentor Green Belts. Their main focus is on Six Sigma project execution, as opposed to Champions and Master Black Belts who mainly focus on finding opportunities for Six Sigma projects. Black Belts have complete knowledge of philosophies and principles of Six Sigma, including DMIAC and lean concepts as well as statistical tools and team management.

- *Green Belts*: Green Belts are Six Sigma team members with part-time commitment (typically 25-percent time commitment) to Six Sigma programs. They are trained and are allowed to only work on carefully defined small Six Sigma projects in which their main task is to support Black Belts with gathering data and executing the project.

- *Yellow Belts*: Six Sigma Yellow belt is the entry level in the Six Sigma employee hierarchy. They have some basic knowledge and understanding of Six Sigma tools and problem-solving process. They can gather data, perform simple data analysis, and develop process maps.

The above organizational roles bring several advantages to Six Sigma. First, it creates full-time positions with the responsibility of advocating and implementing business process improvement. Second, the titles and visibility of "belts" motivate employees to go through rigorous training, which, in turn, enhances employees' career from both intellectual and financial perspectives. This also provides companies with significant financial gains. A study has shown that a Black Belt Project can save a firm an average of $175,000. The study also observed that Black Belts who spend 100 percent of their time on quality improvement projects can execute five or six projects a year.[8] Thus, there is high demand for Master Black Belts and Black Belts in various industries. In fact, a survey in 2012 has shown that, in the United States, workers with any level of Six Sigma training earned about $16,826 more, on an average, than the workers without any Six Sigma training.[9]

## 17.7 Lean Six Sigma

Lean operations has its own approach to improve quality, which we highlight below (see Chap. 15):

- *Variability Reduction:* Like Six Sigma, lean operations also focuses on variability reduction in all aspects of the operations.

- *Preventing Quality Problems:* Lean operations tries to prevent quality problems to occur by simplifying tasks and procedures, and also by making tasks and processes mistake-proof (Poka-Yoke).

- *Early Detection of Quality Problems:* Lean operations achieve this by making workers responsible for quality, using self-stopping machines and processes when quality problems occur (Jidoka), following strict rules for compliance to quality metrics, and by utilizing statistical quality control tools.

- *Rapid Resolution of Quality Problems:* This element of quality improvement of lean operations includes the practice of Go-and-See (Genchi Genbutsu) that requires managers and workers to go and observe the quality problem at its source. It also promotes worker involvement in solving the quality problem, and makes significant effort to find the root cause of the problem.

- *Kaizen:* Kaizen in lean operations is a never-ending process for continuous improvement that involves everyone. It focuses on small gradual improvements with minimum cost and focuses on all aspects of operations.

Reviewing the above list of lean operations practices for quality improvement, it is easy to see that there are many similarities between Six Sigma and lean operations. Like lean operations, Six Sigma focuses on improving quality through practices such as variability reduction, team work, and finding the root cause of the problem; and like Six Sigma, lean operations uses approaches similar to DMAIC to solve problems. The center of both Six Sigma and lean philosophies is striving for perfection through continuous and gradual improvement. So, it is a natural next step for organizations to merge their Six Sigma program and lean initiatives, and that is what has happened in recent years. The tools and methods used in lean and Six Sigma complement each other—with lean increasing efficiency and eliminating wastes and Six Sigma reducing variability and improving quality. In today's competitive and continually changing global market, full potential for improvement and competitiveness cannot be achieved if lean and Six Sigma are implemented in isolation.

---

[8]"Six Sigma: A Breakthrough Strategy for Profitability." *Quality Progress*, 31 (5) (1998) 60–64.
[9]"Rewards for Master Black Belts." *Quality Progress*, 46 (12) (2013) 52–53.

The merged system has brought the benefits of both approaches. For example, by using lean and Six Sigma tools, Honeywell was able to double its capacity at its European chemical plants and reduce its cost by 50 percent, resulting in turning its annual loss of $1 million per year to annual profit of $3.4 million per year.[10]

## 17.8 Summary

This chapter presents principles for improving quality of a product or a process. The first step before starting a quality improvement project is to have a clear quantifiable measure as the goal for quality improvement. Process capability is a metric that measures the quality of a process by its ability to produce non-defective (error-free) outputs. Non-defective outputs are defined as those that satisfy design specifications. The chapter introduces three commonly used metrics of process capability, namely (i) fraction of output within specification, (ii) process capability ratios, and (iii) sigma capability measure.

This chapter also describes Six Sigma quality improvement approach. Six Sigma quality was developed by Motorola and later evolved into an effective problem-solving and quality improvement methodology. Six Sigma has three main pillars: (i) striving for 6-sigma capability, (ii) Six Sigma improvement cycle, and (iii) organizational involvement. Striving for 6-sigma capability is more crucial in systems in which a defective output can result in a very high cost, or when a product (a process) consists of several parts (several stages) and thus errors accumulate and increase the chance of a product failure. When a process is a 6-sigma process, it becomes very robust to abnormal variability affecting its performance, and if an assignable cause puts the process out of control, it would be quickly detected by control charts.

Implementing Six Sigma in an organization requires the commitment from top managers as well as the expertise of people who implement Six Sigma programs. Six Sigma cannot be implemented unless all employees have proper training in using technical tools needed to identify and eliminate quality problems. Six Sigma institutionalizes quality improvement by instilling an informal hierarchy within the formal organization. It defines different roles and full-time positions within the organization, which are given martial arts titles that reflect the hierarchy and the level of skills of the employees. These titles motivate employees to go through rigorous training, which in turn, enhances employees' careers, and ultimately improves the quality at all levels of an organization.

## Discussion Questions

1. What are the two important considerations when choosing which quality improvement project to do first?

2. What are the three main characteristics of clear goals for quality improvement projects?

3. Describe ways that quality improvement culture can be built into all levels of an organization.

4. Describe several ways that one can reduce the resistance to initiating and implementing quality improvement projects.

5. What is process capability? Describe four different metrics used in measuring process capability.

6. What is Six Sigma?

---

[10]Meredith, J., and S. Shafer, *Operations Management for MBAs.* Fifth Edition, John Wiley and Sons, 2013.

7. In a 6-sigma process, 99.9999998% of process output is within design specifications. Explain five reasons why firms try to achieve such a high standard.

8. Describe the five steps of DMAIC improvement cycle.

9. What are the similarities between lean operations approach to quality control and improvement and that of Six Sigma approach?

## Problems

1. Explain why each of the following statements is true or false:
    a. An in-control process produces outputs that are within design specifications.
    b. A 6-sigma process is always in control.

2. The time to discharge a patient in NMMEH hospital varies. Some patients are discharged in less than an hour, and discharge times for some patients take more than 3 hours. Patient survey data shows that, if the discharge time takes more than 3 hours, patient satisfaction is significantly reduced. Hence, hospital manager would like to have all the patients discharged in less than 3 hours. The data shows that, in a normal day with no abnormal variability, patient discharge time follows a Normal distribution with mean 140 minutes and standard deviation 50 minutes.
    a. Compute DPMO of the discharge process.
    b. Compute capability ratio of the discharge process.
    c. Compute sigma capability of the discharge process.
    d. The hospital manager believes that by implementing some improvement idea, the manager can reduce the average discharge time to 2 hours. How much does the manager need to reduce the variability in the discharge time in order to make the process a 3-sigma process?

3. The manager of the hospital in Problem 2 organizes several Kaizen events, and after 2 months, is able to make the discharge process a 3-sigma process. Before improvement, the manager was using a control chart with three-sigma control limits to monitor the discharge times. We call that Chart C1 and the discharge system before improvement System S1. The manager makes a new control chart with three-sigma control limit, which we call Chart C2 to monitor the discharge time in the improved system, which we call System S2. The sample size and sample frequency in both charts are the same. Explain why each of the following statements is true or false?
    a. The probability of observing a discharge time beyond the upper control limits in Chart C2 is lower than that in Chart C1.
    b. Chart C2 has a lower possibility of having a false alarm (i.e., suggesting that the process is out of control, while it is not).
    c. Suppose both systems S1 and S2 are out of control. It takes Chart C2 less time (i.e., less number of samples) to detect that S2 is out-of-control than the time it takes C1 to detect that S1 is out of control.

4. Healthy-&-Organic (H&O) produces blueberry granola cereal, sold in boxes that contain 28 ounces of cereal. The marketing department has identified that due to federal regulations and market consideration, each cereal box must have between 27 and 29 ounces of cereal. The data shows that the weight of the boxes filled in H&O is well-approximated by a Normal distribution with mean 28 ounce and standard deviation 0.5 ounce. How much do DPMO, capability ratio, and sigma capability change if the mean of the filling process shifts from 28 to 27.5 (with standard deviation unchanged) and the shift is not detected?

5. Suppose H&O in Problem 4 detects the mean shift and fixes the problem, so the filling process fills boxes with an average of 28 ounces and standard deviation of 0.5. H&O then

develops a control chart to monitor if the mean of filling processes moves again. The sample of size 8 is taken every hour and the result is plotted on the chart.

   a. What would be the three-sigma control limits of the control chart?
   b. If the mean of the filling process shifts from 28 to 27.9, what is the probability that the shift is detected in the next sample?
   c. On an average, how many samples are needed to detect the mean shift?
   c. Answer Parts (a) to (c) if the sample size was 12.
   d. Answer Parts (a) to (c), if the filling process was a 6-sigma centered process.

6. The headquarters of a fast-casual restaurant chain started a quality improvement project to improve the quality of service in its restaurants. When customers pay for their food, they are given a coupon for free drink next time, if they fill out the online customer satisfaction survey. One of its restaurants in downtown is receiving more complaints than others. The following table shows the summary of the complaints of customers who visited the restaurant on Saturday of the last 6 weeks, that is, the busiest day of the week:

| Quality issues | Weeks | | | | | |
|---|---|---|---|---|---|---|
| | 1 | 2 | 3 | 4 | 5 | 6 |
| Low-quality food | 2 | — | 3 | 2 | — | — |
| Unfriendly servers | 2 | 3 | 2 | 2 | 5 | 3 |
| Error in check | — | — | 1 | — | — | — |
| Restaurant ambience | 1 | — | 2 | — | — | — |
| Price too high | 2 | 1 | 2 | 1 | — | — |
| Long wait for service | 3 | 3 | 4 | 5 | 6 | 4 |
| Cleanliness | — | 1 | — | — | — | 1 |

Develop a Pareto chart. Which complaints would you recommend that the restaurant should work on to improve customer service?

7. NMMEH hospital in Problem 2 hired a consulting firm to help them further improve the process of discharging patients. Before the consulting firm was hired, the time it took to discharge patients had an average of 140 minutes and standard deviation 35 minutes. The consulting firm has finished its study and made several suggestions that were implemented in the hospital. The hospital manager, however, is not completely convinced if the discharge process has actually improved. He looks at the discharge time of the last eight patients, which are 132, 101, 135, 120, 152, 128, 149, and 138 minutes, and is wondering if implementing the consulting firm's suggestions actually improved the discharge process. What do you think?

8. Steve knows that when its injection molding machine is working properly, only 3 percent of plastic bottles produced by the machine are defective. The bottles are sold to the producers of soft drinks and fruit juices. Steve got a phone call from his supplier of plastic resin informing him about a new type of plastic resin that melts in a different temperature, resulting in a better formation of bottles. This, supplier claimed, would reduce the number of defective bottles made by the injection molding machines. The resin is more expensive than the current resin that Steve uses to make bottles. Steve decided to try the new resin to make bottles. Looking for the number of defective bottles produced using the new resin, Steve finds that, among 1200 bottles produced in the first batch, only 30 bottles were defective. How confident can Steve be that the new resin has reduced its percent of defective bottles to less than 3 percent?

9. Basil Car Rental (BCR) has a location in Chicago airport with a large number of customers each hour arriving to pick up their cars, and a large number of customers returning their rental cars to take their flights out of Chicago. BCR has simplified the process of receiving

a car from a customer, so those customers do not wait to return their cars. Returned cars are processed (i.e., inspected for mechanical problems, vacuumed, washed, and filled with gas) before they become available for rent. Often times arriving customers who want to pick up their cars must wait for their cars because, either there are no cars available for rent, or their cars are being processed. Data shows that currently 15 percent of arriving customers must wait for their cars, and the waiting time has an average of 20 minutes with standard deviation of 10 minutes.

Recently, BCR's marketing department pointed out that customers are complaining on social media about the long waiting times to get their already reserved cars. BCR has taken actions to improve its car processing operations and believes that it has been able to reduce both the mean and the variability of the time it takes to process returned cars. The question is whether these improvements had any impact on the waiting time of customers for their cars. To check that, BCR studied the waiting time of all customers in 1 day. The data shows that BCR had 285 customers who arrived to pick up their cars, and only 36 of them had to wait to get their cars. The following table shows the waiting times of those 36 customers:

| 26 | 18 | 36 | 21 | 19 | 22 |
|----|----|----|----|----|----|
| 36 | 15 | 22 | 24 | 7  | 8  |
| 16 | 6  | 11 | 15 | 24 | 23 |
| 11 | 10 | 22 | 43 | 14 | 18 |
| 4  | 31 | 27 | 1  | 21 | 20 |
| 7  | 23 | 36 | 41 | 25 | 23 |

Based on this information, do you believe that BCR's improvement efforts had any impact on customers waiting to get their cars?

# Appendix: Standard Normal Table

Normal distribution has the following density function:

$$f(x) = \frac{1}{\sigma_D \sqrt{2\pi}} \, e^{-\frac{1}{2}[(x-\overline{D})/\sigma_D]^2}$$

where $\overline{D}$ is the average demand and $\sigma_D$ is the standard deviation of the demand.

Normal distribution with mean 0 and standard deviation 1 is called *standard Normal distribution*, and random variable $Z$ is used to show it. Standard Normal distribution has the density and cumulative distribution functions as follows:

$$\phi(z) = \frac{1}{\sqrt{2\pi}} \, e^{-\frac{1}{2}(z)^2}$$

$$\Phi(z) = P(Z \le z) = \int_{-\infty}^{z} \phi(u) du$$

The relationship between a Normal random variable $X$—with mean $\overline{D}$ and standard deviation $\sigma_D$—and standard Normal random variable $Z$ is as follows:

$$Z = \frac{X - \overline{D}}{\sigma_D}$$

Thus, standard Normal random variable $Z$ will have mean of zero and standard deviation 1. The above relationship between $Z$ and $X$ results in the following relationships between Normal and standard Normal distributions and their loss functions:

$$f(x) = \left(\frac{1}{\sigma_D}\right) \phi(z)$$

$$F(x) = P(D \le x) = \Phi(z) \quad \text{where} \quad z = \frac{x - \overline{D}}{\sigma_D}$$

$$\mathcal{L}(x) = \sigma \mathcal{L}(z)$$

$$\mathcal{L}^2(x) = \sigma^2 \mathcal{L}^2(z)$$

$\mathcal{L}(z)$ and $\mathcal{L}^2(z)$ are the first- and second-order loss functions for standard Normal distribution.

This appendix provides the above values for standard Normal distribution. The values are computed using their corresponding Excel functions.

| $z$ | $\phi(z)$ | $\Phi(z)$ | $\mathcal{L}(z)$ | $\mathcal{L}^2(z)$ | $z$ | $\phi(z)$ | $\Phi(z)$ | $\mathcal{L}(z)$ | $\mathcal{L}^2(z)$ |
|---|---|---|---|---|---|---|---|---|---|
| −2.99 | 0.0046 | 0.0014 | 2.9904 | 4.9699 | −2.39 | 0.0229 | 0.0084 | 2.3928 | 3.3552 |
| −2.98 | 0.0047 | 0.0014 | 2.9804 | 4.9401 | −2.38 | 0.0235 | 0.0087 | 2.3829 | 3.3313 |
| −2.97 | 0.0048 | 0.0015 | 2.9704 | 4.9103 | −2.37 | 0.0241 | 0.0089 | 2.3730 | 3.3075 |
| −2.96 | 0.0050 | 0.0015 | 2.9604 | 4.8807 | −2.36 | 0.0246 | 0.0091 | 2.3631 | 3.2839 |
| −2.95 | 0.0051 | 0.0016 | 2.9505 | 4.8511 | −2.35 | 0.0252 | 0.0094 | 2.3532 | 3.2603 |
| −2.94 | 0.0053 | 0.0016 | 2.9405 | 4.8217 | −2.34 | 0.0258 | 0.0096 | 2.3433 | 3.2368 |
| −2.93 | 0.0055 | 0.0017 | 2.9305 | 4.7923 | −2.33 | 0.0264 | 0.0099 | 2.3334 | 3.2134 |
| −2.92 | 0.0056 | 0.0018 | 2.9205 | 4.7631 | −2.32 | 0.0270 | 0.0102 | 2.3235 | 3.1901 |
| −2.91 | 0.0058 | 0.0018 | 2.9105 | 4.7339 | −2.31 | 0.0277 | 0.0104 | 2.3136 | 3.1669 |
| −2.90 | 0.0060 | 0.0019 | 2.9005 | 4.7049 | −2.30 | 0.0283 | 0.0107 | 2.3037 | 3.1438 |
| −2.89 | 0.0061 | 0.0019 | 2.8906 | 4.6759 | −2.29 | 0.0290 | 0.0110 | 2.2938 | 3.1209 |
| −2.88 | 0.0063 | 0.0020 | 2.8806 | 4.6470 | −2.28 | 0.0297 | 0.0113 | 2.2839 | 3.0980 |
| −2.87 | 0.0065 | 0.0021 | 2.8706 | 4.6183 | −2.27 | 0.0303 | 0.0116 | 2.2740 | 3.0752 |
| −2.86 | 0.0067 | 0.0021 | 2.8606 | 4.5896 | −2.26 | 0.0310 | 0.0119 | 2.2641 | 3.0525 |
| −2.85 | 0.0069 | 0.0022 | 2.8506 | 4.5611 | −2.25 | 0.0317 | 0.0122 | 2.2542 | 3.0299 |
| −2.84 | 0.0071 | 0.0023 | 2.8407 | 4.5326 | −2.24 | 0.0325 | 0.0125 | 2.2444 | 3.0074 |
| −2.83 | 0.0073 | 0.0023 | 2.8307 | 4.5043 | −2.23 | 0.0332 | 0.0129 | 2.2345 | 2.9850 |
| −2.82 | 0.0075 | 0.0024 | 2.8207 | 4.4760 | −2.22 | 0.0339 | 0.0132 | 2.2246 | 2.9627 |
| −2.81 | 0.0077 | 0.0025 | 2.8107 | 4.4478 | −2.21 | 0.0347 | 0.0136 | 2.2147 | 2.9405 |
| −2.80 | 0.0079 | 0.0026 | 2.8008 | 4.4198 | −2.20 | 0.0355 | 0.0139 | 2.2049 | 2.9184 |
| −2.79 | 0.0081 | 0.0026 | 2.7908 | 4.3918 | −2.19 | 0.0363 | 0.0143 | 2.1950 | 2.8964 |
| −2.78 | 0.0084 | 0.0027 | 2.7808 | 4.3640 | −2.18 | 0.0371 | 0.0146 | 2.1852 | 2.8745 |
| −2.77 | 0.0086 | 0.0028 | 2.7708 | 4.3362 | −2.17 | 0.0379 | 0.0150 | 2.1753 | 2.8527 |
| −2.76 | 0.0088 | 0.0029 | 2.7609 | 4.3086 | −2.16 | 0.0387 | 0.0154 | 2.1655 | 2.8310 |
| −2.75 | 0.0091 | 0.0030 | 2.7509 | 4.2810 | −2.15 | 0.0396 | 0.0158 | 2.1556 | 2.8094 |
| −2.74 | 0.0093 | 0.0031 | 2.7409 | 4.2535 | −2.14 | 0.0404 | 0.0162 | 2.1458 | 2.7879 |
| −2.73 | 0.0096 | 0.0032 | 2.7310 | 4.2262 | −2.13 | 0.0413 | 0.0166 | 2.1360 | 2.7665 |
| −2.72 | 0.0099 | 0.0033 | 2.7210 | 4.1989 | −2.12 | 0.0422 | 0.0170 | 2.1261 | 2.7452 |
| −2.71 | 0.0101 | 0.0034 | 2.7110 | 4.1718 | −2.11 | 0.0431 | 0.0174 | 2.1163 | 2.7240 |
| −2.70 | 0.0104 | 0.0035 | 2.7011 | 4.1447 | −2.10 | 0.0440 | 0.0179 | 2.1065 | 2.7029 |
| −2.69 | 0.0107 | 0.0036 | 2.6911 | 4.1177 | −2.09 | 0.0449 | 0.0183 | 2.0966 | 2.6818 |
| −2.68 | 0.0110 | 0.0037 | 2.6811 | 4.0909 | −2.08 | 0.0459 | 0.0188 | 2.0868 | 2.6609 |
| −2.67 | 0.0113 | 0.0038 | 2.6712 | 4.0641 | −2.07 | 0.0468 | 0.0192 | 2.0770 | 2.6401 |
| −2.66 | 0.0116 | 0.0039 | 2.6612 | 4.0375 | −2.06 | 0.0478 | 0.0197 | 2.0672 | 2.6194 |
| −2.65 | 0.0119 | 0.0040 | 2.6512 | 4.0109 | −2.05 | 0.0488 | 0.0202 | 2.0574 | 2.5988 |
| −2.64 | 0.0122 | 0.0041 | 2.6413 | 3.9844 | −2.04 | 0.0498 | 0.0207 | 2.0476 | 2.5782 |
| −2.63 | 0.0126 | 0.0043 | 2.6313 | 3.9581 | −2.03 | 0.0508 | 0.0212 | 2.0378 | 2.5578 |
| −2.62 | 0.0129 | 0.0044 | 2.6214 | 3.9318 | −2.02 | 0.0519 | 0.0217 | 2.0280 | 2.5375 |
| −2.61 | 0.0132 | 0.0045 | 2.6114 | 3.9056 | −2.01 | 0.0529 | 0.0222 | 2.0183 | 2.5172 |
| −2.60 | 0.0136 | 0.0047 | 2.6015 | 3.8796 | −2.00 | 0.0540 | 0.0228 | 2.0085 | 2.4971 |
| −2.59 | 0.0139 | 0.0048 | 2.5915 | 3.8536 | −1.99 | 0.0551 | 0.0233 | 1.9987 | 2.4771 |
| −2.58 | 0.0143 | 0.0049 | 2.5816 | 3.8277 | −1.98 | 0.0562 | 0.0239 | 1.9890 | 2.4571 |
| −2.57 | 0.0147 | 0.0051 | 2.5716 | 3.8020 | −1.97 | 0.0573 | 0.0244 | 1.9792 | 2.4373 |
| −2.56 | 0.0151 | 0.0052 | 2.5617 | 3.7763 | −1.96 | 0.0584 | 0.0250 | 1.9694 | 2.4176 |
| −2.55 | 0.0154 | 0.0054 | 2.5517 | 3.7507 | −1.95 | 0.0596 | 0.0256 | 1.9597 | 2.3979 |
| −2.54 | 0.0158 | 0.0055 | 2.5418 | 3.7253 | −1.94 | 0.0608 | 0.0262 | 1.9500 | 2.3784 |
| −2.53 | 0.0163 | 0.0057 | 2.5318 | 3.6999 | −1.93 | 0.0620 | 0.0268 | 1.9402 | 2.3589 |
| −2.52 | 0.0167 | 0.0059 | 2.5219 | 3.6746 | −1.92 | 0.0632 | 0.0274 | 1.9305 | 2.3396 |
| −2.51 | 0.0171 | 0.0060 | 2.5119 | 3.6495 | −1.91 | 0.0644 | 0.0281 | 1.9208 | 2.3203 |
| −2.50 | 0.0175 | 0.0062 | 2.5020 | 3.6244 | −1.90 | 0.0656 | 0.0287 | 1.9111 | 2.3011 |
| −2.49 | 0.0180 | 0.0064 | 2.4921 | 3.5994 | −1.89 | 0.0669 | 0.0294 | 1.9013 | 2.2821 |
| −2.48 | 0.0184 | 0.0066 | 2.4821 | 3.5746 | −1.88 | 0.0681 | 0.0301 | 1.8916 | 2.2631 |
| −2.47 | 0.0189 | 0.0068 | 2.4722 | 3.5498 | −1.87 | 0.0694 | 0.0307 | 1.8819 | 2.2442 |
| −2.46 | 0.0194 | 0.0069 | 2.4623 | 3.5251 | −1.86 | 0.0707 | 0.0314 | 1.8723 | 2.2255 |
| −2.45 | 0.0198 | 0.0071 | 2.4523 | 3.5005 | −1.85 | 0.0721 | 0.0322 | 1.8626 | 2.2068 |
| −2.44 | 0.0203 | 0.0073 | 2.4424 | 3.4761 | −1.84 | 0.0734 | 0.0329 | 1.8529 | 2.1882 |
| −2.43 | 0.0208 | 0.0075 | 2.4325 | 3.4517 | −1.83 | 0.0748 | 0.0336 | 1.8432 | 2.1697 |
| −2.42 | 0.0213 | 0.0078 | 2.4226 | 3.4274 | −1.82 | 0.0761 | 0.0344 | 1.8336 | 2.1514 |
| −2.41 | 0.0219 | 0.0080 | 2.4126 | 3.4032 | −1.81 | 0.0775 | 0.0351 | 1.8239 | 2.1331 |
| −2.40 | 0.0224 | 0.0082 | 2.4027 | 3.3792 | −1.80 | 0.0790 | 0.0359 | 1.8143 | 2.1149 |

| $z$ | $\phi(z)$ | $\Phi(z)$ | $\mathcal{L}(z)$ | $\mathcal{L}^2(z)$ | $z$ | $\phi(z)$ | $\Phi(z)$ | $\mathcal{L}(z)$ | $\mathcal{L}^2(z)$ |
|---|---|---|---|---|---|---|---|---|---|
| $-1.79$ | 0.0804 | 0.0367 | 1.8046 | 2.0968 | $-1.19$ | 0.1965 | 0.1170 | 1.2473 | 1.1836 |
| $-1.78$ | 0.0818 | 0.0375 | 1.7950 | 2.0788 | $-1.18$ | 0.1989 | 0.1190 | 1.2384 | 1.1712 |
| $-1.77$ | 0.0833 | 0.0384 | 1.7854 | 2.0609 | $-1.17$ | 0.2012 | 0.1210 | 1.2296 | 1.1588 |
| $-1.76$ | 0.0848 | 0.0392 | 1.7758 | 2.0431 | $-1.16$ | 0.2036 | 0.1230 | 1.2209 | 1.1466 |
| $-1.75$ | 0.0863 | 0.0401 | 1.7662 | 2.0254 | $-1.15$ | 0.2059 | 0.1251 | 1.2121 | 1.1344 |
| $-1.74$ | 0.0878 | 0.0409 | 1.7566 | 2.0078 | $-1.14$ | 0.2083 | 0.1271 | 1.2034 | 1.1223 |
| $-1.73$ | 0.0893 | 0.0418 | 1.7470 | 1.9902 | $-1.13$ | 0.2107 | 0.1292 | 1.1946 | 1.1104 |
| $-1.72$ | 0.0909 | 0.0427 | 1.7374 | 1.9728 | $-1.12$ | 0.2131 | 0.1314 | 1.1859 | 1.0985 |
| $-1.71$ | 0.0925 | 0.0436 | 1.7278 | 1.9555 | $-1.11$ | 0.2155 | 0.1335 | 1.1773 | 1.0866 |
| $-1.70$ | 0.0940 | 0.0446 | 1.7183 | 1.9383 | $-1.10$ | 0.2179 | 0.1357 | 1.1686 | 1.0749 |
| $-1.69$ | 0.0957 | 0.0455 | 1.7087 | 1.9211 | $-1.09$ | 0.2203 | 0.1379 | 1.1600 | 1.0633 |
| $-1.68$ | 0.0973 | 0.0465 | 1.6992 | 1.9041 | $-1.08$ | 0.2227 | 0.1401 | 1.1514 | 1.0517 |
| $-1.67$ | 0.0989 | 0.0475 | 1.6897 | 1.8871 | $-1.07$ | 0.2251 | 0.1423 | 1.1428 | 1.0402 |
| $-1.66$ | 0.1006 | 0.0485 | 1.6801 | 1.8703 | $-1.06$ | 0.2275 | 0.1446 | 1.1342 | 1.0289 |
| $-1.65$ | 0.1023 | 0.0495 | 1.6706 | 1.8535 | $-1.05$ | 0.2299 | 0.1469 | 1.1257 | 1.0176 |
| $-1.64$ | 0.1040 | 0.0505 | 1.6611 | 1.8369 | $-1.04$ | 0.2323 | 0.1492 | 1.1172 | 1.0063 |
| $-1.63$ | 0.1057 | 0.0516 | 1.6516 | 1.8203 | $-1.03$ | 0.2347 | 0.1515 | 1.1087 | 0.9952 |
| $-1.62$ | 0.1074 | 0.0526 | 1.6422 | 1.8038 | $-1.02$ | 0.2371 | 0.1539 | 1.1002 | 0.9842 |
| $-1.61$ | 0.1092 | 0.0537 | 1.6327 | 1.7875 | $-1.01$ | 0.2396 | 0.1562 | 1.0917 | 0.9732 |
| $-1.60$ | 0.1109 | 0.0548 | 1.6232 | 1.7712 | $-1.00$ | 0.2420 | 0.1587 | 1.0833 | 0.9623 |
| $-1.59$ | 0.1127 | 0.0559 | 1.6138 | 1.7550 | $-0.99$ | 0.2444 | 0.1611 | 1.0749 | 0.9515 |
| $-1.58$ | 0.1145 | 0.0571 | 1.6044 | 1.7389 | $-0.98$ | 0.2468 | 0.1635 | 1.0665 | 0.9408 |
| $-1.57$ | 0.1163 | 0.0582 | 1.5949 | 1.7229 | $-0.97$ | 0.2492 | 0.1660 | 1.0582 | 0.9302 |
| $-1.56$ | 0.1182 | 0.0594 | 1.5855 | 1.7070 | $-0.96$ | 0.2516 | 0.1685 | 1.0499 | 0.9197 |
| $-1.55$ | 0.1200 | 0.0606 | 1.5761 | 1.6912 | $-0.95$ | 0.2541 | 0.1711 | 1.0416 | 0.9092 |
| $-1.54$ | 0.1219 | 0.0618 | 1.5667 | 1.6755 | $-0.94$ | 0.2565 | 0.1736 | 1.0333 | 0.8988 |
| $-1.53$ | 0.1238 | 0.0630 | 1.5574 | 1.6599 | $-0.93$ | 0.2589 | 0.1762 | 1.0250 | 0.8885 |
| $-1.52$ | 0.1257 | 0.0643 | 1.5480 | 1.6443 | $-0.92$ | 0.2613 | 0.1788 | 1.0168 | 0.8783 |
| $-1.51$ | 0.1276 | 0.0655 | 1.5386 | 1.6289 | $-0.91$ | 0.2637 | 0.1814 | 1.0086 | 0.8682 |
| $-1.50$ | 0.1295 | 0.0668 | 1.5293 | 1.6136 | $-0.90$ | 0.2661 | 0.1841 | 1.0004 | 0.8582 |
| $-1.49$ | 0.1315 | 0.0681 | 1.5200 | 1.5983 | $-0.89$ | 0.2685 | 0.1867 | 0.9923 | 0.8482 |
| $-1.48$ | 0.1334 | 0.0694 | 1.5107 | 1.5832 | $-0.88$ | 0.2709 | 0.1894 | 0.9842 | 0.8383 |
| $-1.47$ | 0.1354 | 0.0708 | 1.5014 | 1.5681 | $-0.87$ | 0.2732 | 0.1922 | 0.9761 | 0.8285 |
| $-1.46$ | 0.1374 | 0.0721 | 1.4921 | 1.5531 | $-0.86$ | 0.2756 | 0.1949 | 0.9680 | 0.8188 |
| $-1.45$ | 0.1394 | 0.0735 | 1.4828 | 1.5383 | $-0.85$ | 0.2780 | 0.1977 | 0.9600 | 0.8092 |
| $-1.44$ | 0.1415 | 0.0749 | 1.4736 | 1.5235 | $-0.84$ | 0.2803 | 0.2005 | 0.9520 | 0.7996 |
| $-1.43$ | 0.1435 | 0.0764 | 1.4643 | 1.5088 | $-0.83$ | 0.2827 | 0.2033 | 0.9440 | 0.7901 |
| $-1.42$ | 0.1456 | 0.0778 | 1.4551 | 1.4942 | $-0.82$ | 0.2850 | 0.2061 | 0.9360 | 0.7807 |
| $-1.41$ | 0.1476 | 0.0793 | 1.4459 | 1.4797 | $-0.81$ | 0.2874 | 0.2090 | 0.9281 | 0.7714 |
| $-1.40$ | 0.1497 | 0.0808 | 1.4367 | 1.4653 | $-0.80$ | 0.2897 | 0.2119 | 0.9202 | 0.7622 |
| $-1.39$ | 0.1518 | 0.0823 | 1.4275 | 1.4510 | $-0.79$ | 0.2920 | 0.2148 | 0.9123 | 0.7530 |
| $-1.38$ | 0.1539 | 0.0838 | 1.4183 | 1.4367 | $-0.78$ | 0.2943 | 0.2177 | 0.9045 | 0.7439 |
| $-1.37$ | 0.1561 | 0.0853 | 1.4092 | 1.4226 | $-0.77$ | 0.2966 | 0.2206 | 0.8967 | 0.7349 |
| $-1.36$ | 0.1582 | 0.0869 | 1.4000 | 1.4086 | $-0.76$ | 0.2989 | 0.2236 | 0.8889 | 0.7260 |
| $-1.35$ | 0.1604 | 0.0885 | 1.3909 | 1.3946 | $-0.75$ | 0.3011 | 0.2266 | 0.8812 | 0.7171 |
| $-1.34$ | 0.1626 | 0.0901 | 1.3818 | 1.3807 | $-0.74$ | 0.3034 | 0.2296 | 0.8734 | 0.7084 |
| $-1.33$ | 0.1647 | 0.0918 | 1.3727 | 1.3670 | $-0.73$ | 0.3056 | 0.2327 | 0.8658 | 0.6997 |
| $-1.32$ | 0.1669 | 0.0934 | 1.3636 | 1.3533 | $-0.72$ | 0.3079 | 0.2358 | 0.8581 | 0.6910 |
| $-1.31$ | 0.1691 | 0.0951 | 1.3546 | 1.3397 | $-0.71$ | 0.3101 | 0.2389 | 0.8505 | 0.6825 |
| $-1.30$ | 0.1714 | 0.0968 | 1.3455 | 1.3262 | $-0.70$ | 0.3123 | 0.2420 | 0.8429 | 0.6740 |
| $-1.29$ | 0.1736 | 0.0985 | 1.3365 | 1.3128 | $-0.69$ | 0.3144 | 0.2451 | 0.8353 | 0.6656 |
| $-1.28$ | 0.1758 | 0.1003 | 1.3275 | 1.2995 | $-0.68$ | 0.3166 | 0.2483 | 0.8278 | 0.6573 |
| $-1.27$ | 0.1781 | 0.1020 | 1.3185 | 1.2862 | $-0.67$ | 0.3187 | 0.2514 | 0.8203 | 0.6491 |
| $-1.26$ | 0.1804 | 0.1038 | 1.3095 | 1.2731 | $-0.66$ | 0.3209 | 0.2546 | 0.8128 | 0.6409 |
| $-1.25$ | 0.1826 | 0.1056 | 1.3006 | 1.2600 | $-0.65$ | 0.3230 | 0.2578 | 0.8054 | 0.6328 |
| $-1.24$ | 0.1849 | 0.1075 | 1.2917 | 1.2471 | $-0.64$ | 0.3251 | 0.2611 | 0.7980 | 0.6248 |
| $-1.23$ | 0.1872 | 0.1093 | 1.2827 | 1.2342 | $-0.63$ | 0.3271 | 0.2643 | 0.7906 | 0.6169 |
| $-1.22$ | 0.1895 | 0.1112 | 1.2738 | 1.2214 | $-0.62$ | 0.3292 | 0.2676 | 0.7833 | 0.6090 |
| $-1.21$ | 0.1919 | 0.1131 | 1.2650 | 1.2087 | $-0.61$ | 0.3312 | 0.2709 | 0.7759 | 0.6012 |
| $-1.20$ | 0.1942 | 0.1151 | 1.2561 | 1.1961 | $-0.60$ | 0.3332 | 0.2743 | 0.7687 | 0.5935 |

| $z$ | $\phi(z)$ | $\Phi(z)$ | $\mathcal{L}(z)$ | $\mathcal{L}^2(z)$ | $z$ | $\phi(z)$ | $\Phi(z)$ | $\mathcal{L}(z)$ | $\mathcal{L}^2(z)$ |
|---|---|---|---|---|---|---|---|---|---|
| −0.59 | 0.3352 | 0.2776 | 0.7614 | 0.5858 | 0.01 | 0.3989 | 0.5040 | 0.3940 | 0.2460 |
| −0.58 | 0.3372 | 0.2810 | 0.7542 | 0.5782 | 0.02 | 0.3989 | 0.5080 | 0.3890 | 0.2421 |
| −0.57 | 0.3391 | 0.2843 | 0.7471 | 0.5707 | 0.03 | 0.3988 | 0.5120 | 0.3841 | 0.2383 |
| −0.56 | 0.3410 | 0.2877 | 0.7399 | 0.5633 | 0.04 | 0.3986 | 0.5160 | 0.3793 | 0.2344 |
| −0.55 | 0.3429 | 0.2912 | 0.7328 | 0.5559 | 0.05 | 0.3984 | 0.5199 | 0.3744 | 0.2307 |
| −0.54 | 0.3448 | 0.2946 | 0.7257 | 0.5486 | 0.06 | 0.3982 | 0.5239 | 0.3697 | 0.2269 |
| −0.53 | 0.3467 | 0.2981 | 0.7187 | 0.5414 | 0.07 | 0.3980 | 0.5279 | 0.3649 | 0.2233 |
| −0.52 | 0.3485 | 0.3015 | 0.7117 | 0.5343 | 0.08 | 0.3977 | 0.5319 | 0.3602 | 0.2197 |
| −0.51 | 0.3503 | 0.3050 | 0.7047 | 0.5272 | 0.09 | 0.3973 | 0.5359 | 0.3556 | 0.2161 |
| −0.50 | 0.3521 | 0.3085 | 0.6978 | 0.5202 | 0.10 | 0.3970 | 0.5398 | 0.3509 | 0.2125 |
| −0.49 | 0.3538 | 0.3121 | 0.6909 | 0.5132 | 0.11 | 0.3965 | 0.5438 | 0.3464 | 0.2091 |
| −0.48 | 0.3555 | 0.3156 | 0.6840 | 0.5064 | 0.12 | 0.3961 | 0.5478 | 0.3418 | 0.2056 |
| −0.47 | 0.3572 | 0.3192 | 0.6772 | 0.4996 | 0.13 | 0.3956 | 0.5517 | 0.3373 | 0.2022 |
| −0.46 | 0.3589 | 0.3228 | 0.6704 | 0.4928 | 0.14 | 0.3951 | 0.5557 | 0.3328 | 0.1989 |
| −0.45 | 0.3605 | 0.3264 | 0.6637 | 0.4861 | 0.15 | 0.3945 | 0.5596 | 0.3284 | 0.1956 |
| −0.44 | 0.3621 | 0.3300 | 0.6569 | 0.4795 | 0.16 | 0.3939 | 0.5636 | 0.3240 | 0.1923 |
| −0.43 | 0.3637 | 0.3336 | 0.6503 | 0.4730 | 0.17 | 0.3932 | 0.5675 | 0.3197 | 0.1891 |
| −0.42 | 0.3653 | 0.3372 | 0.6436 | 0.4665 | 0.18 | 0.3925 | 0.5714 | 0.3154 | 0.1859 |
| −0.41 | 0.3668 | 0.3409 | 0.6370 | 0.4601 | 0.19 | 0.3918 | 0.5753 | 0.3111 | 0.1828 |
| −0.40 | 0.3683 | 0.3446 | 0.6304 | 0.4538 | 0.20 | 0.3910 | 0.5793 | 0.3069 | 0.1797 |
| −0.39 | 0.3697 | 0.3483 | 0.6239 | 0.4475 | 0.21 | 0.3902 | 0.5832 | 0.3027 | 0.1766 |
| −0.38 | 0.3712 | 0.3520 | 0.6174 | 0.4413 | 0.22 | 0.3894 | 0.5871 | 0.2986 | 0.1736 |
| −0.37 | 0.3725 | 0.3557 | 0.6109 | 0.4352 | 0.23 | 0.3885 | 0.5910 | 0.2944 | 0.1707 |
| −0.36 | 0.3739 | 0.3594 | 0.6045 | 0.4291 | 0.24 | 0.3876 | 0.5948 | 0.2904 | 0.1677 |
| −0.35 | 0.3752 | 0.3632 | 0.5981 | 0.4231 | 0.25 | 0.3867 | 0.5987 | 0.2863 | 0.1649 |
| −0.34 | 0.3765 | 0.3669 | 0.5918 | 0.4171 | 0.26 | 0.3857 | 0.6026 | 0.2824 | 0.1620 |
| −0.33 | 0.3778 | 0.3707 | 0.5855 | 0.4113 | 0.27 | 0.3847 | 0.6064 | 0.2784 | 0.1592 |
| −0.32 | 0.3790 | 0.3745 | 0.5792 | 0.4054 | 0.28 | 0.3836 | 0.6103 | 0.2745 | 0.1564 |
| −0.31 | 0.3802 | 0.3783 | 0.5730 | 0.3997 | 0.29 | 0.3825 | 0.6141 | 0.2706 | 0.1537 |
| −0.30 | 0.3814 | 0.3821 | 0.5668 | 0.3940 | 0.30 | 0.3814 | 0.6179 | 0.2668 | 0.1510 |
| −0.29 | 0.3825 | 0.3859 | 0.5606 | 0.3883 | 0.31 | 0.3802 | 0.6217 | 0.2630 | 0.1484 |
| −0.28 | 0.3836 | 0.3897 | 0.5545 | 0.3828 | 0.32 | 0.3790 | 0.6255 | 0.2592 | 0.1458 |
| −0.27 | 0.3847 | 0.3936 | 0.5484 | 0.3772 | 0.33 | 0.3778 | 0.6293 | 0.2555 | 0.1432 |
| −0.26 | 0.3857 | 0.3974 | 0.5424 | 0.3718 | 0.34 | 0.3765 | 0.6331 | 0.2518 | 0.1407 |
| −0.25 | 0.3867 | 0.4013 | 0.5363 | 0.3664 | 0.35 | 0.3752 | 0.6368 | 0.2481 | 0.1382 |
| −0.24 | 0.3876 | 0.4052 | 0.5304 | 0.3611 | 0.36 | 0.3739 | 0.6406 | 0.2445 | 0.1357 |
| −0.23 | 0.3885 | 0.4090 | 0.5244 | 0.3558 | 0.37 | 0.3725 | 0.6443 | 0.2409 | 0.1333 |
| −0.22 | 0.3894 | 0.4129 | 0.5186 | 0.3506 | 0.38 | 0.3712 | 0.6480 | 0.2374 | 0.1309 |
| −0.21 | 0.3902 | 0.4168 | 0.5127 | 0.3454 | 0.39 | 0.3697 | 0.6517 | 0.2339 | 0.1285 |
| −0.20 | 0.3910 | 0.4207 | 0.5069 | 0.3403 | 0.40 | 0.3683 | 0.6554 | 0.2304 | 0.1262 |
| −0.19 | 0.3918 | 0.4247 | 0.5011 | 0.3353 | 0.41 | 0.3668 | 0.6591 | 0.2270 | 0.1239 |
| −0.18 | 0.3925 | 0.4286 | 0.4954 | 0.3303 | 0.42 | 0.3653 | 0.6628 | 0.2236 | 0.1217 |
| −0.17 | 0.3932 | 0.4325 | 0.4897 | 0.3254 | 0.43 | 0.3637 | 0.6664 | 0.2203 | 0.1194 |
| −0.16 | 0.3939 | 0.4364 | 0.4840 | 0.3205 | 0.44 | 0.3621 | 0.6700 | 0.2169 | 0.1173 |
| −0.15 | 0.3945 | 0.4404 | 0.4784 | 0.3157 | 0.45 | 0.3605 | 0.6736 | 0.2137 | 0.1151 |
| −0.14 | 0.3951 | 0.4443 | 0.4728 | 0.3109 | 0.46 | 0.3589 | 0.6772 | 0.2104 | 0.1130 |
| −0.13 | 0.3956 | 0.4483 | 0.4673 | 0.3062 | 0.47 | 0.3572 | 0.6808 | 0.2072 | 0.1109 |
| −0.12 | 0.3961 | 0.4522 | 0.4618 | 0.3016 | 0.48 | 0.3555 | 0.6844 | 0.2040 | 0.1088 |
| −0.11 | 0.3965 | 0.4562 | 0.4564 | 0.2970 | 0.49 | 0.3538 | 0.6879 | 0.2009 | 0.1068 |
| −0.10 | 0.3970 | 0.4602 | 0.4509 | 0.2925 | 0.50 | 0.3521 | 0.6915 | 0.1978 | 0.1048 |
| −0.09 | 0.3973 | 0.4641 | 0.4456 | 0.2880 | 0.51 | 0.3503 | 0.6950 | 0.1947 | 0.1029 |
| −0.08 | 0.3977 | 0.4681 | 0.4402 | 0.2835 | 0.52 | 0.3485 | 0.6985 | 0.1917 | 0.1009 |
| −0.07 | 0.3980 | 0.4721 | 0.4349 | 0.2792 | 0.53 | 0.3467 | 0.7019 | 0.1887 | 0.0990 |
| −0.06 | 0.3982 | 0.4761 | 0.4297 | 0.2749 | 0.54 | 0.3448 | 0.7054 | 0.1857 | 0.0972 |
| −0.05 | 0.3984 | 0.4801 | 0.4244 | 0.2706 | 0.55 | 0.3429 | 0.7088 | 0.1828 | 0.0953 |
| −0.04 | 0.3986 | 0.4840 | 0.4193 | 0.2664 | 0.56 | 0.3410 | 0.7123 | 0.1799 | 0.0935 |
| −0.03 | 0.3988 | 0.4880 | 0.4141 | 0.2622 | 0.57 | 0.3391 | 0.7157 | 0.1771 | 0.0917 |
| −0.02 | 0.3989 | 0.4920 | 0.4090 | 0.2581 | 0.58 | 0.3372 | 0.7190 | 0.1742 | 0.0900 |
| −0.01 | 0.3989 | 0.4960 | 0.4040 | 0.2540 | 0.59 | 0.3352 | 0.7224 | 0.1714 | 0.0882 |
| 0.00 | 0.3989 | 0.5000 | 0.3989 | 0.2500 | 0.60 | 0.3332 | 0.7257 | 0.1687 | 0.0865 |

| $z$ | $\phi(z)$ | $\Phi(z)$ | $\mathcal{L}(z)$ | $\mathcal{L}^2(z)$ | $z$ | $\phi(z)$ | $\Phi(z)$ | $\mathcal{L}(z)$ | $\mathcal{L}^2(z)$ |
|---|---|---|---|---|---|---|---|---|---|
| 0.61 | 0.3312 | 0.7291 | 0.1659 | 0.0849 | 1.21 | 0.1919 | 0.8869 | 0.0550 | 0.0233 |
| 0.62 | 0.3292 | 0.7324 | 0.1633 | 0.0832 | 1.22 | 0.1895 | 0.8888 | 0.0538 | 0.0228 |
| 0.63 | 0.3271 | 0.7357 | 0.1606 | 0.0816 | 1.23 | 0.1872 | 0.8907 | 0.0527 | 0.0222 |
| 0.64 | 0.3251 | 0.7389 | 0.1580 | 0.0800 | 1.24 | 0.1849 | 0.8925 | 0.0517 | 0.0217 |
| 0.65 | 0.3230 | 0.7422 | 0.1554 | 0.0784 | 1.25 | 0.1826 | 0.8944 | 0.0506 | 0.0212 |
| 0.66 | 0.3209 | 0.7454 | 0.1528 | 0.0769 | 1.26 | 0.1804 | 0.8962 | 0.0495 | 0.0207 |
| 0.67 | 0.3187 | 0.7486 | 0.1503 | 0.0754 | 1.27 | 0.1781 | 0.8980 | 0.0485 | 0.0202 |
| 0.68 | 0.3166 | 0.7517 | 0.1478 | 0.0739 | 1.28 | 0.1758 | 0.8997 | 0.0475 | 0.0197 |
| 0.69 | 0.3144 | 0.7549 | 0.1453 | 0.0724 | 1.29 | 0.1736 | 0.9015 | 0.0465 | 0.0193 |
| 0.70 | 0.3123 | 0.7580 | 0.1429 | 0.0710 | 1.30 | 0.1714 | 0.9032 | 0.0455 | 0.0188 |
| 0.71 | 0.3101 | 0.7611 | 0.1405 | 0.0696 | 1.31 | 0.1691 | 0.9049 | 0.0446 | 0.0184 |
| 0.72 | 0.3079 | 0.7642 | 0.1381 | 0.0682 | 1.32 | 0.1669 | 0.9066 | 0.0436 | 0.0179 |
| 0.73 | 0.3056 | 0.7673 | 0.1358 | 0.0668 | 1.33 | 0.1647 | 0.9082 | 0.0427 | 0.0175 |
| 0.74 | 0.3034 | 0.7704 | 0.1334 | 0.0654 | 1.34 | 0.1626 | 0.9099 | 0.0418 | 0.0171 |
| 0.75 | 0.3011 | 0.7734 | 0.1312 | 0.0641 | 1.35 | 0.1604 | 0.9115 | 0.0409 | 0.0166 |
| 0.76 | 0.2989 | 0.7764 | 0.1289 | 0.0628 | 1.36 | 0.1582 | 0.9131 | 0.0400 | 0.0162 |
| 0.77 | 0.2966 | 0.7794 | 0.1267 | 0.0615 | 1.37 | 0.1561 | 0.9147 | 0.0392 | 0.0158 |
| 0.78 | 0.2943 | 0.7823 | 0.1245 | 0.0603 | 1.38 | 0.1539 | 0.9162 | 0.0383 | 0.0155 |
| 0.79 | 0.2920 | 0.7852 | 0.1223 | 0.0591 | 1.39 | 0.1518 | 0.9177 | 0.0375 | 0.0151 |
| 0.80 | 0.2897 | 0.7881 | 0.1202 | 0.0578 | 1.40 | 0.1497 | 0.9192 | 0.0367 | 0.0147 |
| 0.81 | 0.2874 | 0.7910 | 0.1181 | 0.0567 | 1.41 | 0.1476 | 0.9207 | 0.0359 | 0.0143 |
| 0.82 | 0.2850 | 0.7939 | 0.1160 | 0.0555 | 1.42 | 0.1456 | 0.9222 | 0.0351 | 0.0140 |
| 0.83 | 0.2827 | 0.7967 | 0.1140 | 0.0543 | 1.43 | 0.1435 | 0.9236 | 0.0343 | 0.0136 |
| 0.84 | 0.2803 | 0.7995 | 0.1120 | 0.0532 | 1.44 | 0.1415 | 0.9251 | 0.0336 | 0.0133 |
| 0.85 | 0.2780 | 0.8023 | 0.1100 | 0.0521 | 1.45 | 0.1394 | 0.9265 | 0.0328 | 0.0130 |
| 0.86 | 0.2756 | 0.8051 | 0.1080 | 0.0510 | 1.46 | 0.1374 | 0.9279 | 0.0321 | 0.0127 |
| 0.87 | 0.2732 | 0.8078 | 0.1061 | 0.0499 | 1.47 | 0.1354 | 0.9292 | 0.0314 | 0.0123 |
| 0.88 | 0.2709 | 0.8106 | 0.1042 | 0.0489 | 1.48 | 0.1334 | 0.9306 | 0.0307 | 0.0120 |
| 0.89 | 0.2685 | 0.8133 | 0.1023 | 0.0478 | 1.49 | 0.1315 | 0.9319 | 0.0300 | 0.0117 |
| 0.90 | 0.2661 | 0.8159 | 0.1004 | 0.0468 | 1.50 | 0.1295 | 0.9332 | 0.0293 | 0.0114 |
| 0.91 | 0.2637 | 0.8186 | 0.0986 | 0.0458 | 1.51 | 0.1276 | 0.9345 | 0.0286 | 0.0111 |
| 0.92 | 0.2613 | 0.8212 | 0.0968 | 0.0449 | 1.52 | 0.1257 | 0.9357 | 0.0280 | 0.0109 |
| 0.93 | 0.2589 | 0.8238 | 0.0950 | 0.0439 | 1.53 | 0.1238 | 0.9370 | 0.0274 | 0.0106 |
| 0.94 | 0.2565 | 0.8264 | 0.0933 | 0.0430 | 1.54 | 0.1219 | 0.9382 | 0.0267 | 0.0103 |
| 0.95 | 0.2541 | 0.8289 | 0.0916 | 0.0420 | 1.55 | 0.1200 | 0.9394 | 0.0261 | 0.0100 |
| 0.96 | 0.2516 | 0.8315 | 0.0899 | 0.0411 | 1.56 | 0.1182 | 0.9406 | 0.0255 | 0.0098 |
| 0.97 | 0.2492 | 0.8340 | 0.0882 | 0.0402 | 1.57 | 0.1163 | 0.9418 | 0.0249 | 0.0095 |
| 0.98 | 0.2468 | 0.8365 | 0.0865 | 0.0394 | 1.58 | 0.1145 | 0.9429 | 0.0244 | 0.0093 |
| 0.99 | 0.2444 | 0.8389 | 0.0849 | 0.0385 | 1.59 | 0.1127 | 0.9441 | 0.0238 | 0.0090 |
| 1.00 | 0.2420 | 0.8413 | 0.0833 | 0.0377 | 1.60 | 0.1109 | 0.9452 | 0.0232 | 0.0088 |
| 1.01 | 0.2396 | 0.8438 | 0.0817 | 0.0368 | 1.61 | 0.1092 | 0.9463 | 0.0227 | 0.0086 |
| 1.02 | 0.2371 | 0.8461 | 0.0802 | 0.0360 | 1.62 | 0.1074 | 0.9474 | 0.0222 | 0.0084 |
| 1.03 | 0.2347 | 0.8485 | 0.0787 | 0.0352 | 1.63 | 0.1057 | 0.9484 | 0.0216 | 0.0081 |
| 1.04 | 0.2323 | 0.8508 | 0.0772 | 0.0345 | 1.64 | 0.1040 | 0.9495 | 0.0211 | 0.0079 |
| 1.05 | 0.2299 | 0.8531 | 0.0757 | 0.0337 | 1.65 | 0.1023 | 0.9505 | 0.0206 | 0.0077 |
| 1.06 | 0.2275 | 0.8554 | 0.0742 | 7 0.0329 | 1.66 | 0.1006 | 0.9515 | 0.0201 | 0.0075 |
| 1.07 | 0.2251 | 0.8577 | 0.0728 | 0.0322 | 1.67 | 0.0989 | 0.9525 | 0.0197 | 0.0073 |
| 1.08 | 0.2227 | 0.8599 | 0.0714 | 0.0315 | 1.68 | 0.0973 | 0.9535 | 0.0192 | 0.0071 |
| 1.09 | 0.2203 | 0.8621 | 0.0700 | 0.0308 | 1.69 | 0.0957 | 0.9545 | 0.0187 | 0.0069 |
| 1.10 | 0.2179 | 0.8643 | 0.0686 | 0.0301 | 1.70 | 0.0940 | 0.9554 | 0.0183 | 0.0067 |
| 1.11 | 0.2155 | 0.8665 | 0.0673 | 0.0294 | 1.71 | 0.0925 | 0.9564 | 0.0178 | 0.0066 |
| 1.12 | 0.2131 | 0.8686 | 0.0659 | 0.0287 | 1.72 | 0.0909 | 0.9573 | 0.0174 | 0.0064 |
| 1.13 | 0.2107 | 0.8708 | 0.0646 | 0.0281 | 1.73 | 0.0893 | 0.9582 | 0.0170 | 0.0062 |
| 1.14 | 0.2083 | 0.8729 | 0.0634 | 0.0275 | 1.74 | 0.0878 | 0.9591 | 0.0166 | 0.0060 |
| 1.15 | 0.2059 | 0.8749 | 0.0621 | 0.0268 | 1.75 | 0.0863 | 0.9599 | 0.0162 | 0.0059 |
| 1.16 | 0.2036 | 0.8770 | 0.0609 | 0.0262 | 1.76 | 0.0848 | 0.9608 | 0.0158 | 0.0057 |
| 1.17 | 0.2012 | 0.8790 | 0.0596 | 0.0256 | 1.77 | 0.0833 | 0.9616 | 0.0154 | 0.0056 |
| 1.18 | 0.1989 | 0.8810 | 0.0584 | 0.0250 | 1.78 | 0.0818 | 0.9625 | 0.0150 | 0.0054 |
| 1.19 | 0.1965 | 0.8830 | 0.0573 | 0.0244 | 1.79 | 0.0804 | 0.9633 | 0.0146 | 0.0053 |
| 1.20 | 0.1942 | 0.8849 | 0.0561 | 0.0239 | 1.80 | 0.0790 | 0.9641 | 0.0143 | 0.0051 |

| $z$ | $\phi(z)$ | $\Phi(z)$ | $\mathcal{L}(z)$ | $\mathcal{L}^2(z)$ | $z$ | $\phi(z)$ | $\Phi(z)$ | $\mathcal{L}(z)$ | $\mathcal{L}^2(z)$ |
|---|---|---|---|---|---|---|---|---|---|
| 1.81 | 0.0775 | 0.9649 | 0.0139 | 0.0050 | 2.41 | 0.0219 | 0.9920 | 0.0026 | 0.0008 |
| 1.82 | 0.0761 | 0.9656 | 0.0136 | 0.0048 | 2.42 | 0.0213 | 0.9922 | 0.0026 | 0.0008 |
| 1.83 | 0.0748 | 0.9664 | 0.0132 | 0.0047 | 2.43 | 0.0208 | 0.9925 | 0.0025 | 0.0008 |
| 1.84 | 0.0734 | 0.9671 | 0.0129 | 0.0046 | 2.44 | 0.0203 | 0.9927 | 0.0024 | 0.0007 |
| 1.85 | 0.0721 | 0.9678 | 0.0126 | 0.0044 | 2.45 | 0.0198 | 0.9929 | 0.0023 | 0.0007 |
| 1.86 | 0.0707 | 0.9686 | 0.0123 | 0.0043 | 2.46 | 0.0194 | 0.9931 | 0.0023 | 0.0007 |
| 1.87 | 0.0694 | 0.9693 | 0.0119 | 0.0042 | 2.47 | 0.0189 | 0.9932 | 0.0022 | 0.0007 |
| 1.88 | 0.0681 | 0.9699 | 0.0116 | 0.0041 | 2.48 | 0.0184 | 0.9934 | 0.0021 | 0.0006 |
| 1.89 | 0.0669 | 0.9706 | 0.0113 | 0.0040 | 2.49 | 0.0180 | 0.9936 | 0.0021 | 0.0006 |
| 1.90 | 0.0656 | 0.9713 | 0.0111 | 0.0039 | 2.50 | 0.0175 | 0.9938 | 0.0020 | 0.0006 |
| 1.91 | 0.0644 | 0.9719 | 0.0108 | 0.0037 | 2.51 | 0.0171 | 0.9940 | 0.0019 | 0.0006 |
| 1.92 | 0.0632 | 0.9726 | 0.0105 | 0.0036 | 2.52 | 0.0167 | 0.9941 | 0.0019 | 0.0006 |
| 1.93 | 0.0620 | 0.9732 | 0.0102 | 0.0035 | 2.53 | 0.0163 | 0.9943 | 0.0018 | 0.0005 |
| 1.94 | 0.0608 | 0.9738 | 0.0100 | 0.0034 | 2.54 | 0.0158 | 0.9945 | 0.0018 | 0.0005 |
| 1.95 | 0.0596 | 0.9744 | 0.0097 | 0.0033 | 2.55 | 0.0154 | 0.9946 | 0.0017 | 0.0005 |
| 1.96 | 0.0584 | 0.9750 | 0.0094 | 0.0032 | 2.56 | 0.0151 | 0.9948 | 0.0017 | 0.0005 |
| 1.97 | 0.0573 | 0.9756 | 0.0092 | 0.0031 | 2.57 | 0.0147 | 0.9949 | 0.0016 | 0.0005 |
| 1.98 | 0.0562 | 0.9761 | 0.0090 | 0.0031 | 2.58 | 0.0143 | 0.9951 | 0.0016 | 0.0005 |
| 1.99 | 0.0551 | 0.9767 | 0.0087 | 0.0030 | 2.59 | 0.0139 | 0.9952 | 0.0015 | 0.0004 |
| 2.00 | 0.0540 | 0.9772 | 0.0085 | 0.0029 | 2.60 | 0.0136 | 0.9953 | 0.0015 | 0.0004 |
| 2.01 | 0.0529 | 0.9778 | 0.0083 | 0.0028 | 2.61 | 0.0132 | 0.9955 | 0.0014 | 0.0004 |
| 2.02 | 0.0519 | 0.9783 | 0.0080 | 0.0027 | 2.62 | 0.0129 | 0.9956 | 0.0014 | 0.0004 |
| 2.03 | 0.0508 | 0.9788 | 0.0078 | 0.0026 | 2.63 | 0.0126 | 0.9957 | 0.0013 | 0.0004 |
| 2.04 | 0.0498 | 0.9793 | 0.0076 | 0.0026 | 2.64 | 0.0122 | 0.9959 | 0.0013 | 0.0004 |
| 2.05 | 0.0488 | 0.9798 | 0.0074 | 0.0025 | 2.65 | 0.0119 | 0.9960 | 0.0012 | 0.0004 |
| 2.06 | 0.0478 | 0.9803 | 0.0072 | 0.0024 | 2.66 | 0.0116 | 0.9961 | 0.0012 | 0.0003 |
| 2.07 | 0.0468 | 0.9808 | 0.0070 | 0.0023 | 2.67 | 0.0113 | 0.9962 | 0.0012 | 0.0003 |
| 2.08 | 0.0459 | 0.9812 | 0.0068 | 0.0023 | 2.68 | 0.0110 | 0.9963 | 0.0011 | 0.0003 |
| 2.09 | 0.0449 | 0.9817 | 0.0066 | 0.0022 | 2.69 | 0.0107 | 0.9964 | 0.0011 | 0.0003 |
| 2.10 | 0.0440 | 0.9821 | 0.0065 | 0.0021 | 2.70 | 0.0104 | 0.9965 | 0.0011 | 0.0003 |
| 2.11 | 0.0431 | 0.9826 | 0.0063 | 0.0021 | 2.71 | 0.0101 | 0.9966 | 0.0010 | 0.0003 |
| 2.12 | 0.0422 | 0.9830 | 0.0061 | 0.0020 | 2.72 | 0.0099 | 0.9967 | 0.0010 | 0.0003 |
| 2.13 | 0.0413 | 0.9834 | 0.0060 | 0.0020 | 2.73 | 0.0096 | 0.9968 | 0.0010 | 0.0003 |
| 2.14 | 0.0404 | 0.9838 | 0.0058 | 0.0019 | 2.74 | 0.0093 | 0.9969 | 0.0009 | 0.0003 |
| 2.15 | 0.0396 | 0.9842 | 0.0056 | 0.0018 | 2.75 | 0.0091 | 0.9970 | 0.0009 | 0.0003 |
| 2.16 | 0.0387 | 0.9846 | 0.0055 | 0.0018 | 2.76 | 0.0088 | 0.9971 | 0.0009 | 0.0002 |
| 2.17 | 0.0379 | 0.9850 | 0.0053 | 0.0017 | 2.77 | 0.0086 | 0.9972 | 0.0008 | 0.0002 |
| 2.18 | 0.0371 | 0.9854 | 0.0052 | 0.0017 | 2.78 | 0.0084 | 0.9973 | 0.0008 | 0.0002 |
| 2.19 | 0.0363 | 0.9857 | 0.0050 | 0.0016 | 2.79 | 0.0081 | 0.9974 | 0.0008 | 0.0002 |
| 2.20 | 0.0355 | 0.9861 | 0.0049 | 0.0016 | 2.80 | 0.0079 | 0.9974 | 0.0008 | 0.0002 |
| 2.21 | 0.0347 | 0.9864 | 0.0047 | 0.0015 | 2.81 | 0.0077 | 0.9975 | 0.0007 | 0.0002 |
| 2.22 | 0.0339 | 0.9868 | 0.0046 | 0.0015 | 2.82 | 0.0075 | 0.9976 | 0.0007 | 0.0002 |
| 2.23 | 0.0332 | 0.9871 | 0.0045 | 0.0014 | 2.83 | 0.0073 | 0.9977 | 0.0007 | 0.0002 |
| 2.24 | 0.0325 | 0.9875 | 0.0044 | 0.0014 | 2.84 | 0.0071 | 0.9977 | 0.0007 | 0.0002 |
| 2.25 | 0.0317 | 0.9878 | 0.0042 | 0.0013 | 2.85 | 0.0069 | 0.9978 | 0.0006 | 0.0002 |
| 2.26 | 0.0310 | 0.9881 | 0.0041 | 0.0013 | 2.86 | 0.0067 | 0.9979 | 0.0006 | 0.0002 |
| 2.27 | 0.0303 | 0.9884 | 0.0040 | 0.0013 | 2.87 | 0.0065 | 0.9979 | 0.0006 | 0.0002 |
| 2.28 | 0.0297 | 0.9887 | 0.0039 | 0.0012 | 2.88 | 0.0063 | 0.9980 | 0.0006 | 0.0002 |
| 2.29 | 0.0290 | 0.9890 | 0.0038 | 0.0012 | 2.89 | 0.0061 | 0.9981 | 0.0006 | 0.0002 |
| 2.30 | 0.0283 | 0.9893 | 0.0037 | 0.0012 | 2.90 | 0.0060 | 0.9981 | 0.0005 | 0.0001 |
| 2.31 | 0.0277 | 0.9896 | 0.0036 | 0.0011 | 2.91 | 0.0058 | 0.9982 | 0.0005 | 0.0001 |
| 2.32 | 0.0270 | 0.9898 | 0.0035 | 0.0011 | 2.92 | 0.0056 | 0.9982 | 0.0005 | 0.0001 |
| 2.33 | 0.0264 | 0.9901 | 0.0034 | 0.0010 | 2.93 | 0.0055 | 0.9983 | 0.0005 | 0.0001 |
| 2.34 | 0.0258 | 0.9904 | 0.0033 | 0.0010 | 2.94 | 0.0053 | 0.9984 | 0.0005 | 0.0001 |
| 2.35 | 0.0252 | 0.9906 | 0.0032 | 0.0010 | 2.95 | 0.0051 | 0.9984 | 0.0005 | 0.0001 |
| 2.36 | 0.0246 | 0.9909 | 0.0031 | 0.0009 | 2.96 | 0.0050 | 0.9985 | 0.0004 | 0.0001 |
| 2.37 | 0.0241 | 0.9911 | 0.0030 | 0.0009 | 2.97 | 0.0048 | 0.9985 | 0.0004 | 0.0001 |
| 2.38 | 0.0235 | 0.9913 | 0.0029 | 0.0009 | 2.98 | 0.0047 | 0.9986 | 0.0004 | 0.0001 |
| 2.39 | 0.0229 | 0.9916 | 0.0028 | 0.0009 | 2.99 | 0.0046 | 0.9986 | 0.0004 | 0.0001 |
| 2.40 | 0.0224 | 0.9918 | 0.0027 | 0.0008 | 3.00 | 0.0044 | 0.9987 | 0.0004 | 0.0001 |

# References

Ackoff, R.L., *The Art of Problem Solving*. John Wiley & Sons, New York, NY, 1978.

Anupindi, R., S. Chopra, S.D. Deshmukh, J.A. Van Mieghem, and E. Zemel. *Managing Business Process Flows: Principles of Operations Management*. Third edition, Prentice Hall, 2012.

Anonymous, "Rewards for Master Black Belts." *Quality Progress*, 46 (2013) 52–53.

Arcus, L.A., "COMSOAL: A Computer Method of Sequencing Operations for Assembly Lines." *International Journal of Production Research*, 4 (1966).

Arthur, J., *Lean Six Sigma for Hospitals*. McGraw-Hill, 2011.

Askin, R.G. and J.B. Goldberg, *Design and Analysis of Lean Production Systems*. John Wiley & Sons, 2002.

Baker, K.R. and A.G. Merten, "Scheduling with Parallel Processors and Linear Delay Costs." *Naval Research Logistics*, 20 (1973) 793–804.

Barron, M. and D. Targett, *The Manager's Guide to Business Forecasting*. Basil Blackwell, Ltd., Oxford, UK, 1985.

Belouadah, H. and C.N. Potts, "Scheduling Identical Parallel Machines to Minimize Total Weighted Completion Time." *Discrete Applied Mathematics*, 48 (1994) 201–218.

Bitran, G.R., J. Ferrer, and P.R. Oliveira, "Managing Customer Experiences: Perspectives on the Temporal Aspects of Service Encounters." *MSOM* 10 (2008) 61–83.

Blockley, D., *Engineering: A Very Short Introduction*. Oxford University Press, 2012.

Boyer, K.K. and R. Verma, *Operations & Supply Chain Management for the 21st Century*. South-Western, 2010.

Bozarth, C.C. and R.B. Handfield, *Introduction to Operations and Supply Chain Management*. Second edition, Pearson Education, Inc., 2008.

Brown, M.G., D.E. Hitchcock, and M.L. Willard, *Why TQM Fails and What to Do About It*. Irwin, Burr Ridge, IL, 1994.

Brue, G., *Six Sigma for Managers*. McGraw-Hill, 2002.

Buzacott, J.A. and M. Mandelbaum, "Flexibility in Manufacturing and Services: Achievements, Insights and Challenges." *Flexible Service and Manufacturing Journal*, 20 (2008) 13–58.

Buzacott, J.A. and J.G. Shanthikumar, *Stochastic Models of Manufacturing Systems*. Prentice Hall, Inc., Englewood Cliffs, NJ, 1993.

Cachon, G.P. and M.A. Lariviere, "Supply Chain Coordination with Revenue-Sharing Contracts: Strengths and Limitations." *Management Science*, 51 (2005) 30–44.

Cachon, G.P. and C. Terwiesch, *Matching Supply with Demand: An Introduction to Operations Management*. Third edition, McGraw-Hill/Irwin, 2013.

Canel, C., D. Rosen, and E.A. Anderson, "Just-in-Time Is Not Just for Manufacturing: A Service Perspective." *Industrial Management and Data Systems*, 100/2 (2000) 51–60.

Carmon, Z. and D. Kahneman, *The Experienced Utility of Queueing: Real-Time Affect and Retrospective Evaluations of Simulated Queues*. Duke University and Princeton University.

Catalina, S., "Six Sigma as a Strategic Tool for Companies." *Young Economists Journal*, 7/19 (2012) 94–102.

Chase, R.B., N.T. Aquilano, and F.R. Jacobs, *Production and Operations Management*. Eighth edition, Irwin/McGraw-Hill, 1995.

Chase, R.B. and D.M. Stewart, "Make Your Service Fail-Safe." *Sloan Management Review*, Spring (1994) 35–44.

Chopra, S. and P. Meindl, *Supply Chain Management, Strategy, Planning and Operations*. Prentice Hall, NJ, 2001.

CNN.com, "Queuing Psychology: Can Waiting in Line be Fun?"

Collier, D.A. and J.R. Evans. *OM²*. South-Western, Cengage Learning, 2010.

Crockett, R. and J. McGregor, "Six Sigma Still Pays Off at Motorola." *Business Week*, December 4 (2006) 7/19 50.

Davenport, T.H. and J.G. Harris, *Competing on Analytics: The New Science of Winning*. Harvard Business School Publishing Corporation, 2007.

Dennis P., *Lean Production Simplified*. Third edition, CRC Press, 2015.

Eckes, G., *The Six Sigma Revolution: How GE and Others Turned Process into Profits.* Wiley, 2006.

Feitzinger, E. and H.L. Lee, "Mass Customization at Hewlett-Packard: The Power of Postponement." *Harvard Business Review*, May (2009).

Fitzsimmons, J.A., K.J. Fitzsimmons, and S. Berdoloi, *Service Management, Operations, Strategy, Informtation Technology.* Eighth edition, McGraw-Hill/Irwin, 2011.

Galsworth, G.D., *Visual Workplace, Visual Thinking.* Visual-Lean Press, 2005.

Galsworth, G.D., *Work That Makes Sense, Creating and Sustaining Visuality on the Value-Added Level.* Visual-Lean Press, 2011.

Garvin, D., "What Does Quality Really Mean." *Sloan Management Review,* 26 (1984) 25–43.

George, M.L., *Lean Six Sigma for Service.* McGraw-Hill, 2003.

Georgoff, D.M. and R.G. Murdick, "Manager's Guide to Forecasting." *Harvard Business Review*, January (1986).

Goldratt, E.M. and J. Cox, *The Goal: A Process of Ongoing Improvement.* Third edition, North River Press, 2004.

Graban, M., *Lean Hospitals, Improving Quality, Patient Safety, and Employee Engagement.* Third edition, CRC Press, 2016.

Hayes, R.H. and S.C. Wheelwright, "Link Manufacturing Process and Product Life Cycles." *Harvard Business Review*, 57 (1979) 133–140.

Heizer, J. and B. Render, *Operations Management.* Fifth edition, Prentice Hall, 1999.

Hillier, F.S., M.S. Hillier, and G.J. Lieberman, *Introduction to Management Science: A Modeling and Case Studies Approach with Spreadsheets.* Irwin/McGraw-Hill, Burr Ridge, IL, 2000.

Hindo, B. and B. Grow, "Six Sigma so Yesterday: In an Innovative Economy, It's No Longer a Cure-All." *Business Week*, June 11 (2007) 11.

Hopp, W.J. and M.L. Spearman, "To Pull or Not to Pull: What Is the Question?" *MSOM,* 6 (2004) 133–148.

Hopp, W.J. and M.L. Spearman, *Factory Physics.* Waveland Press, Inc., 2011.

Hopp, W.J. and M.P. Van Oyen, "Agile Workforce Evaluation: A Framework for Cross-Training and Coordination." *IIE Transactions,* 36 (2004) 919–940.

How US Manufacturers are Thriving (or Not) in a World of Ongoing Volatility and Uncertainty. The US Manufacturing Flexibility Index. *Accenture Strategy*, 2014.

Imai, M., *Kaizen: The Key to Japan's Competitive Success.* McGraw-Hill, New York, NY, 1986.

Institute for Healthcare improvement, *Going Lean in Healthcare*, Innovation Series 2005 white paper.

International Federation of Robotics, *History of Industrial Robots: From the First Installation Until Today.* 2012.

Iravani, S.M.R., "Design and Control Principles of Flexible Workplace in Manufacturing Systems." *Encyclopedia of Operations Research and Management Science.* Wiley, 2011.

Iravani, S.M.R., B. Kolfal, and M.P. Van Oyen, "Process Flexibility and Inventory Flexibility via Product Substitution." *Flexible Services and Manufacturing,* 26 (2014) 320–343.

Jones, M.H., "Six Sigma … At a Bank?" *Six Sigma Forum Magazine*, February (2004) 13–17.

Jordan, W.C. and S. Graves, "Principles on the Benefits of Manufacturing Process Flexibility." *Management Science,* 41 (1995) 577–594.

Jordan, W.C., R.R. Inman, and D.E. Blumenfeld, "Chained Cross-Training of Workers for Robust Performance." *IIE Transactions,* 36 (2004) 953–967.

Juran, M.J., *Managerial Breakthrough.* McGraw-Hill, New York, NY, 1964, revised edition, 1995.

Juran, M.J. and B. Godfery, *Juran's Quality Handbook.* McGraw-Hill, 1998.

Kahneman, D. and A. Tversky, "Prospect Theory: An Analysis of Decision under Risk." *Econometrica,* 47 (2) (1979) 263.

Kleiner, A., "What Does It Mean to be Green?" *Harvard Business Review,* 69 (1991) 38–47.

Lee, J.Y., "JIT Works for Service Too." *CMA Magazine,* 64/6 (1990).

Lee, H.L., E. Feitzinger, and C. Billington, "Getting Ahead of Your Competition through Design for Mass Customization." *Target,* 13 (1997) 8–17.

Lee, H.L., V. Padmanabhan, and S. Whang, "The Bullwhip Effect in Supply Chains." *Sloan Management Review,* Spring (1997).

Liker, J.K., *The Toyota Way, 14 Management Principles from the World's Greatest Manufacturer.* McGraw-Hill, 2004.

Liker, J. and T.Y. Choi, "Building Deep Supplier Relationships." *Harvard Business Review,* December (2004).

Little, J.D.C., "Little's Law as Viewed on Its 50th Anniversary." *Operations Research,* 59 (2011) 536–549.

Maister, D.H., *The Psychology of Waiting in Lines.* Harvard Business School, Boston, MA, Note 9-684-064, Rev May 1984.

Mandelbaum, A. and M.I. Reiman, "On Pooling in Queueing Networks." *Management Science,* 44 (1998) 1971–1981.

Mead, M., *Cultural Patterns and Technical Change.* UNESCO, Paris; also published by Mentor Books, New American Library of World Literature, New York, NY, 1955.

Meredith, J. and S. Shafer, *Operations Management for MBAs.* Fifth edition, John Wiley & Sons, Inc., 2013.

M'Hallah, R. and R.I. Bulfin, "Minimizing the Weighted Number of Tardy Jobs on a Single Machine." *European Journal of Operational Research*, 145 (2003) 45–56.

Monden, Y. "Adaptable Kanban System Helps Toyota Maintain Just-in-Time Production." *Industrial Engineering*, 13 (1981) 28–40.

Montgomery, D.C., *Introduction to Statistical Quality Control.* Seventh edition, John Wiley & Sons, Inc., 2013.

Moss, S., C. Dale, and G. Brame, "Sequence-Dependent Scheduling at Baxter International." *Interfaces*, 30 (2000) 70–80.

Myerson, P., *Lean Supply Chain & Logistics Management.* McGraw-Hill, 2012.

Nahmias, S. and T.L. Olsen, *Production and Operations Analysis.* Waveland Press Inc., 2015.

Nakajima, S., *Introduction to Total Productive Maintenance.* Productivity Press, 1988.

Nakajima, S. (editor), *TPM Development Program: Implementing Total Productive Maintenance.* Productivity Press, 1989.

Nakane, J. and R.W. Hall, "Ohno's Method, Creating a Survival Work Culture." *Target*, 18 (2002) 6.

Norman D., "Designing Waits That Work." *MIT Sloan Management Review*, (2009).

Norton, M.I. and R.W. Buell, "Think Customers Hate waiting? Not So fast ...." *Harvard Business Review*, May (2011).

Ohno, T., *Toyota Production System, Beyond Large-Scale Production.* Productivity Press, 1988.

Ord, K. and R. Fildes, *Principles of Business Forecasting.* South-Western, Cengage Learning, 2013.

Pascal, D., *Lean Production Simplified.* CRC Press, 2015.

Pentico, D.W., "Assignment Problems: A Golden Anniversary Survey." *European Journal of Operational Research*, 176 (2007) 774–793.

Pinedo, M., *Scheduling, Theory, Algorithms and Systems.* Prentice Hall, 1995.

Plossi, G.W. and O.W. Wight, *Production and Inventory Control.* Prentice Hall, 1967.

Rarindrass, A.R. and D.P. Warsing, *Supply Chain Engineering Models and Applications.* CRC Press, 2013.

Reid, R.D. and N.R. Sanders, *Operations Management: An Integrated Approach.* Fourth edition, John Wiley & Sons, Inc., 2010.

Richtel, M., "The Long-Distance Journey of a Fast-Food Order." *Wall Street Journal,* April 11 (2006).

Rosenfeld, D., "Optimal Management of Tax-Sheltered Employee Reimbursement Programs." *Interfaces*, 16 (1986) 68–72.

Russell, R.B. and B.W. Taylor, *Operations Management.* Fourth edition, Pearson Education, 2003.

Schein, E.H., "How Can Organizations Learn Faster? The Challenge of Entering the Green Room." *Sloan Management Review*, (1993) 85–92.

Schmitt, B., "Nice Try VW: Toyota Again World's Largest Automaker." *Forbes*, January 27, 2016.

Schroeder, R.G., S.M. Goldstein, and M.J. Rugtusanatham, *Operations Management in the Supply Chain.* Sixth edition, McGraw-Hill, 2013.

Schwarz, L., "Chapter 8: The Economic Order-Quantity (EOQ) Model." *Building Intuition: Insights From Basic Operations Management Models and Principles.* Springer, 2008, pp. 135–154.

Sellers, P., "The Dumbest Marketing Ploy." *Fortune*, 126, 5, October (1992), 88–93.

Shah, R. and P.T. Ward, "Defining and Developing Measures of Lean Production." *Journal of Operations Management*, 25 (2007) 785–805.

Shaw, K.A., *Operations Methods, Managing Waiting Line Applications.* Second edition, Business Expert Press, 2011.

Shunko, M., J. Niederhoff, and Y. Rosokha, "Humans Are Not Machines: The Behavioral Impact of Queueing Design on Service Time." *Management Science*, 64 (2018) 453–473.

Silver, E.A., D.E. Pyke, and P. Peterson, *Inventory Management and Production Planning and Scheduling.* John Wiley & Sons, Inc., 1998.

Silveria, G.D., D. Borenstein, and F. S. Fogliatto, "Mass Customization: Literature Review and Research Directions." *International Journal of Production Economics*, 72 (2001) 1–13.

Simchi-Levi, D., P. Kaminsky, and E. Simchi-Levi, *Designing & Managing the Supply Chain: Concepts, Strategies & Case Studies.* Second edition, McGraw-Hill/Irwin, 2003.

Sipper, D. and R.L. Bilfin, *Production Planning, Control, and Integration.* McGraw-Hill, 1997.

Slack, N. and M. Lewis. *Operations Strategy.* Pearson Education Limited, Harlow, UK, 2003.

Smart, T., "Jack Welsh's Cyber-Czar." 5/August(1996) 82–83.

Snyder, L.V. and Z.M. Shen, *Fundamentals of Supply Chain Theory.* John Wiley & Sons, Inc., 2011.

Spear, S. and H.K. Bowen, "Decoding the DNA of Toyota Production System." *Harvard Business Review,* September/October (1999) 97–106.

Srinivasan, A. and B. Kurey, "Creating a Culture of Quality." *Harvard Business Review,* April (2014) 23.

Staats, B.R. and D.M. Upton, "Lean Knowledge Work." *Harvard Business Review,* October (2011) 100.

Stevenson, W.J., *Production Operations Management.* Sixth edition, Irwin/McGraw-Hill, 1999.

Stewart, T.A., "Rate Your Readiness to Change." *Fortune,* February 7 (1994) 106–110.

Stone, A., "Why Waiting Is Torture?" *New York Times,* August 18 (2012).

Sugimori, Y., K. Kusunoki, F. Cho, and S. Uchikawa, "Toyota Production System and Kanban System Materialization of Just-In-Time and Respect-for-Human System." *International Journal of Production Research,* 15 (1997) 553–564.

Sule, D.R., *Production Planning and Industrial Scheduling.* Second edition, CRC Press, 2008.

Suzaki, K., *The Manufacturing Challenge.* New York Free Press, 1985.

Swank, C.K., "The Lean Service Machine." *Harvard Business Review,* (2003) 123–129.

Tapping, D. and T. Shucker, *Value Stream Management for the Lean Office.* CRC Press, 2003.

Tapping, D., T. Luyster, and T. Shuker, *Value Stream Management.* Productivity Press, 2002.

*The Economist,* "Genchi Genbutsu: More a Frame of Mind Than a Plan of Action." October 13, 2009, online extra.

Venkataraman, R.R. and J.K. Pinto, *Operations Management, Managing Global Supply Chains.* SAGE Publication Inc., 2018.

Venkateswatru, P., "Costs and Saving of Six Sigma Programs: An Empirical Study." *The Quality Management Journal,* 19 (2012) 39.

Vijaya, S.M., "Synergies of Lean Six Sigma." *IUP Journal of Operations Management,* 12 (2013) 21–31.

Virts, J. and R. Garrett, "Weighting Risk in Capacity Management." *Harvard Business Review,* 48 (1970).

Wagner, S.M. and M. Hoegel, "Involving Suppliers in Product Development: Insights from R&D Directors and Project Managers." *Industrial Marketing Management,* 35 (2006) 936–943.

Webster, S., *Principles of Supply Chain Management.* Dynamic Ideas, 2009.

Welch, S.J., "Twenty Years of Patient Satisfaction Research Applied to the Emergency Department: A Qualitative Review." *American Journal of Medical Quality,* 25 (2010) 64–72.

Wilson, J.H. and D. Alison-Koerber, "Combining Subjective and Objective Forecasts Improve Results." *The Journal of Business Forecasting Methods & Systems,* 11/3 (1992) 15.

Wilson, J.H. and B. Keating, *Business Forecasting.* Fifth edition, McGraw-Hill, 2007.

Winston, W.L., *Operations Research, Applications and Algorithms.* Second edition, PWS-KENT, Boston, MA, 1991.

Winston, L.W. and S.C. Albright, *Practical Management Science.* Third edition, South-Western, Cengage Learning, 2009.

Wisner, J.D., *Operations Management and Supply Chain Process Approach.* SAGE Publications, 2017.

Womack, J.P. and D.T. Jones, "Lean Consumption." *Harvard Business Review,* (2005) 58–68.

Womack, J.P. and D.T. Jones, *Lean Thinking, Banish Waste and Create Wealth in Your Corporation.* Free Press, 2003.

Womack, J.P., D.T. Jones, and D. Roos, *The Machine that Changed the World.* Free Press, 1990.

Zipkin, P.H., *Foundations of Inventory Management.* McGraw-Hill, 2000.

Zipkin, P.H., "The Limits of Mass Customization." *MIT Sloan Management Review,* Spring (2001) 81.

# Index

# Index